Business and Management

1.19, 2.32, 2.61, 2.96, 3.89, 3.105, 3.114, 4.76, 4.78, 4.90, 4.112, 4.131, 4.171, 5.9, 5.38, 5.93, 5.97, 5.98, 5.127, 5.132, 5.139, 5.142, 5.157, 6.95, 6.110, 7.77, 8.11, 8.50, 8.51, 8.75, 8.109, 8.110, 8.113, 8.133, 8.134, 8.145, 8.165, 9.29, 9.72, 9.146, 9.151, 9.222, 9.226, 9.239, 9.251, 9.256, 10.15, 11.28, 11.76, 12.23, 12.58, 12.108, 12.113, 12.142, 13.19, 13.26, 13.29, 13.58, 14.42, 14.59, 14.81

Demographics and Population Statistics

1.28, 1.43, 2.25, 3.80, 4.56, 4.64, 4.129, 4.149, 4.176, 5.43, 5.94, 5.95, 5.128, 7.14, 7.83, 7.90, 9.153, 9.195, 9.197, 11.50, 11.57, 12.83, 13.66

Economics and Finance

1.11, 2.14, 2.103, 3.110, 3.117, 4.20, 4.33, 4.91, 4.120, 4.141, 4.152, 4.157, 4.160, 4.178, 4.188, 5.9, 5.42, 5.87, 5.126, 5.133, 5.140, 6.20, 6.42, 6.117, 7.9, 7.21, 7.107, 8.45, 8.78, 8.107, 8.116, 8.160, 9.30, 9.50, 9.77, 9.148, 9.232, 9.257, 10.28, 10.60, 10.72, 10.102, 10.165, 10.167, 11.103, 12.61, 12.141, 12.150, 12.155, 12.161, 12.164, 13.13, 13.63, 14.35, 14.37, 14.100, 14.125

Education and Child Development

2.6, 2.21, 2.26, 3.17, 3.22, 3.47, 3.87, 3.108, 3.126, 4.94, 4.141, 5.14, 5.65, 6.76, 7.17, 7.76, 8.115, 9.12, 9.22, 9.46, 9.181, 9.192, 10.72, 13.46, 13.62, 14.102

Fuel Consumption and Cars

0.11, 1.5, 1.27, 1.31, 1.37, 2.5, 2.23, 2.87, 2.95, 2.97, 2.104, 2.106, 3.16, 3.29, 3.46, 4.22, 4.83, 4.85, 4.111, 4.142, 4.169, 4.190, 5.10, 5.33, 5.88, 6.94, 6.111, 6.118, 7.18, 7.57, 8.71, 8.139, 8.144, 9.89, 9.96, 9.114, 9.155, 10.45, 10.74, 10.91, 10.150, 11.23, 11.30, 12.54, 12.77, 12.140, 14.14, 14.62, 14.122, 14.140

Manufacturing and Product Development

1.6, 1.14, 1.16, 1.17, 1.29, 1.32, 1.33, 1.38, 1.39, 1.42, 1.52, 1.53, 2.12, 2.13, 2.62, 2.90, 2.102, 3.18, 3.26, 3.30, 3.83, 3.91, 3.93, 3.119, 3.121, 3.123, 3.128, 3.132, 3.133, 4.24, 4.52, 4.65, 4.84, 4.96, 4.112, 4.167, 4.172, 4.183, 5.12, 5.32, 5.37, 5.56, 5.62, 5.101, 5.103, 5.125, 5.141, 5.157, 6.12, 6.14, 6.45, 6.56, 6.59, 6.61, 6.83, 6.112, 6.116, 6.123, 6.124, 7.10, 7.11, 7.13, 7.41, 7.43, 7.54, 7.55, 7.59, 7.81, 7.82, 7.84, 7.86, 7.98, 7.100,

7.104, 8.15, 8.67, 9.08, 9.138, 8.145 8.16, 8.175, 9.49, 9.82, 9.84, 9.90, 9.91, 9.97, 9.116, 9.118, 9.119, 9.120, 9.217, 9.219, 9.221, 9.225, 9.227, 9.228, 9.230, 9.231, 9.233, 9.234, 9.237, 9.245, 9.255, 10.14, 10.16, 10.19, 10.21, 10.23, 10.46, 10.48, 10.50, 10.52, 10.53, 10.55, 10.56, 10.64, 10.73, 10.87, 10.115, 10.120, 10.144, 10.145, 10.147, 10.158, 10.164, 11.16, 11.24, 11.29, 11.51, 11.63, 11.90, 11.91, 11.93, 12.21, 12.31, 12.104, 12.106, 12.162, 12.167, 12.170, 14.15, 14.16, 14.57, 14.96, 14.101, 14.132, 14.135, 14.138, 14.147, 14.148

Marketing and Consumer Behavior

1.5, 1.6, 1.7, 1.12, 1.18, 1.40, 1.41, 1.43, 1.45, 2.6, 2.7, 2.8, 2.11, 2.29, 2.31, 2.33, 2.38, 2.88, 3.51, 3.56, 3.90, 3.95, 4.35, 4.37, 4.55, 4.58, 4.60, 4.63, 4.67, 4.79, 4.81, 4.92, 4.112, 4.126, 4.127, 4.141, 4.142, 4.173, 4.179, 4.185, 4.194, 5.13, 5.36, 5.64, 5.89, 5.100, 5.102, 5.130, 5.148, 5.149, 5.154, 6.18, 6.19, 6.22, 6.43, 6.46, 6.54, 6.81, 6.100, 6.108, 6.126, 7.9, 7.11, 7.72, 7.74, 7.88, 7.89, 7.92, 8.34, 8.77, 8.102, 8.105, 8.108, 8.111, 8.153, 8.158, 9.13, 9.16, 9.53, 9.74, 9.83, 9.93, 9.94, 9.147, 9.152, 9.156, 9.164, 9.186, 9.253, 10.18, 10.27, 10.61, 10.102, 10.109, 10.119, 11.19, 11.99, 12.163, 13.12, 13.15, 13.16, 13.17, 13.22, 13.23, 13.43, 13.44, 13.56, 13.60, 13.65, 13.69, 13.71, 14.97, 14.137, 14.149

Medicine and Clinical Studies

0.4, 0.10, 0.13, 1.5, 1.6, 1.7, 1.15, 1.42, 1.46, 2.5, 2.105, 3.28, 3.124, 4.27, 4.66, 4.86, 4.111, 4.134, 4.155, 4.162, 4.165, 4.180, 4.184, 5.9, 5.12, 5.18, 5.153, 5.155, 5.158, 6.21, 6.57, 6.99, 7.9, 7.10, 7.96, 8.17, 8.81, 8.117, 8.148, 8.152, 8.156, 8.162, 8.169, 8.174, 9.24, 9.47, 9.52, 9.56, 9.113, 9.149, 9.150, 9.183, 9.194, 9.238, 9.244, 9.246, 10.20, 10.24, 10.49, 10.58, 10.73, 10.74, 10.90, 10.113, 10.122, 10.138, 11.22, 11.49, 11.85, 11.87, 12.26, 12.27, 12.29, 12.33, 12.87, 12.105, 12.110, 12.111, 12.154, 12.160, 12.168, 13.50, 14.13, 14.22, 14.23, 14.32, 14.33, 14.55, 14.103, 14.146

Physical Sciences

1.5, 1.49, 2.57, 2.58, 2.99, 3.45, 3.50, 3.52, 3.85, 3.127, 3.129, 3.130, 3.131, 4.19, 4.23, 4.153, 4.174, 5.121, 5.123, 5.131, 6.16, 6.60, 6.79, 6.101, 6.119, 7.10, 7.11, 7.50, 7.87, 7.93, 7.102, 8.31, 8.35, 8.42, 8.70, 8.136,

8.137, 8.143, 8.151, 8.167, 9.17, 9.45, 9.55, 9.75, 9.95, 9.160, 9.161, 9.188, 9.193, 9.215, 9.216, 9.235, 9.248, 9.250, 10.26, 10.59, 10.62, 10.63, 10.79, 10.83, 10.148, 10.151, 10.156, 10.170, 11.27, 11.31, 11.89, 11.96, 12.19, 12.30, 12.50, 12.51, 12.55, 12.62, 12.63, 12.76, 12.86, 12.88, 12.103, 12.109, 12.114, 12.139, 12.143, 12.144, 12.145, 12.148, 12.173, 13.68, 14.41, 14.75, 14.82, 14.120, 14.127, 14.136

Psychology and Human Behavior

0.12, 1.9, 1.12, 1.13, 2.19, 2.20, 2.30, 2.43, 2.55, 2.98, 4.28, 4.62, 4.87, 4.95, 4.97, 4.103, 4.128, 4.132, 4.156, 4.161, 4.181, 4.189, 5.17, 5.61, 5.66, 5.104, 5.119, 5.135, 5.147, 6.17, 6.23, 6.50, 6.96, 6.103, 7.10, 7.22, 7.44, 7.79, 8.47, 8.79, 8.114, 8.147, 8.163, 8.166, 9.51, 9.165, 9.180, 9.185, 10.85, 10.88, 10.102, 10.112, 10.117, 10.121, 10.162, 11.84, 12.25, 12.75, 12.82, 13.11, 13.24, 13.47, 13.53, 13.55, 13.59, 13.64, 13.67, 14.77, 14.99, 14.121, 14.150

Public Health and Nutrition

1.5, 1.6, 1.12, 1.43, 1.44, 1.54, 2.27, 2.56, 2.59, 2.89, 2.108, 3.24, 3.53, 3.61, 3.109, 3.115, 3.118, 4.29, 4.31, 4.54, 4.57, 4.133, 4.168, 4.186, 5.12, 5.59, 5.60, 5.91, 5.96, 5.134, 5.138, 5.156, 5.158, 6.48, 6.49, 6.75, 6.98, 6.102, 6.106, 6.107, 6.115, 6.122, 7.9, 7.15, 7.39, 7.47, 7.51, 7.70, 7.75, 7.101, 8.18, 8.32, 8.49, 8.83, 8.87, 8.100, 8.150, 8.168, 8.170, 8.172, 8.176, 9.31, 9.78, 9.85, 9.87, 9.111, 9.162, 9.163, 9.179, 9.218, 9.224, 9.242, 9.252, 10.25, 10.74, 10.81, 10.84, 10.86, 10.89, 10.106, 10.110, 10.114, 10.149, 10.160, 10.171, 11.15, 11.20, 11.56, 11.58, 11.61, 11.75, 11.86, 11.94, 11.101, 11.102, 12.32, 12.52, 12.59, 12.79, 12.151, 12.157, 12.172, 13.18, 13.20, 13.51, 13.57, 14.39, 14.61, 14.78, 14.117, 14.129, 14.131, 14.141, 14.142, 14.144

Public Policy and Political Science

1.5, 1.10, 1.12, 1.41, 1.47, 2.5, 2.24, 2.34, 2.42, 3.49, 3.107, 3.111, 3.113, 3.116, 4.82, 4.98, 4.105, 4.125, 4.135, 4.150, 4.154, 5.10, 5.34, 5.35, 5.58, 5.92, 5.150, 6.44, 6.78, 6.93, 6.97, 7.11, 7.40, 7.107, 8.41, 8.80, 8.101, 8.104, 8.106, 8.118, 9.21, 9.25, 9.26, 9.27, 9.28, 9.44, 9.48, 9.81, 9.86, 9.122, 9.145, 9.166, 9.182,

9.184, 9.187, 9.189, 9.190, 9.196, 9.240, 9.241, 9.247, 10.17, 10.107, 10.153, 10,163, 10.166, 11.33, 11.79, 12.85, 13.28, 13.45, 13.52, 13.54, 13.61, 14.34, 14.79, 14.95, 14.98

Sports and Leisure

0.5, 0.6, 0.8, 1.6, 1.34, 2.5, 2.10, 2.11, 2.12, 2.13, 2.28, 2.36, 2.41, 2.64, 2.91, 2.107, 3.20, 3.23, 3.27, 3.44, 3.82, 3.92, 3.120, 3.122, 3.125, 3.135, 4.7, 4.9, 4.25, 4.26, 4.59, 4.77, 4.88, 4.89, 4.99, 4.102, 4.111, 4.112, 4.119, 4.121, 4.124, 4.158, 4.159, 4.170, 4.175, 4.193, 5.9, 5.19, 5.40, 5.41, 5.57, 5.63, 5.90, 5.99, 5.105, 5.124, 5.129, 5.144, 5.146, 6.13, 6.25, 6.51, 6.52, 6.62, 6.63, 6.77, 6.80, 6.114, 7.12, 7.20, 7.24, 7.49, 7.52, 7.53, 7.56, 7.73, 7.78, 7.103, 7.106, 8.33, 8.46, 8.48, 8.52, 8.72, 8.76, 8.82, 8.84, 8.85, 8.140, 8.141, 8.154, 8.159, 8.171, 9.19, 9.54, 9.76, 9.112, 9.115, 9.154, 9.158, 9.220, 9.223, 10.54, 10.74, 10.80, 10.116, 10.140, 10.141, 10.142, 10.146, 10.157, 11.18, 11.32, 11.54, 11.59, 11.78, 11.82, 11.88, 11.97, 12.20, 12.24, 12.48, 12.53, 12.64, 12.112, 12.116, 12.169, 13.21, 13.42, 13.48, 13.49, 14.17, 14.36, 14.58, 14.80, 14.119, 14.126, 14.134, 14.151

Technology and the Internet

1.50, 2.100, 2.101, 3.15, 3.32, 3.62, 4.61, 4.112, 4.142, 5.120, 5.145, 6.105, 6.113, 7.9, 7.71, 7.94, 8.38, 8.43, 8.73, 8.119, 8.135, 8.164, 9.18, 9.80, 9.123, 9.191, 10.78, 10.108, 10.118, 11.92, 12.84, 12.158, 14.56, 14.60, 14.133

Travel and Transportation

1.12, 1.35, 1.41, 1.48, 1.51, 2.6, 2.8, 2.14, 2.35, 2.37, 2.39, 2.66, 2.93, 3.13, 3.25, 3.48, 3.57, 3.58, 3.81, 3.88, 3.94, 3.106, 3.136, 4.21, 4.30, 4.32, 4.36, 4.53, 4.93, 4.104, 4.111, 4.121, 4.122, 4.163, 4.166, 4.182, 4.187, 4.191, 4.192, 4.193, 4.194, 5.9, 5.10, 5.12, 5.122, 5.143, 6.15, 6.120, 6.121, 7.10, 7.19, 7.99, 8.16, 8.36, 8.40, 8.74, 8.103, 8.112, 8.142, 9.20, 9.23, 9.43, 9.117, 9.143, 9.243, 10.22, 10.51, 10.72, 10.73, 10.93, 10.111, 10.139, 10.143, 10.152, 11.17, 11.80, 11.83, 11.95, 12.80, 12.152, 12.166, 12.171, 13.27, 13.72, 14.18, 14.20, 14.40, 14.74, 14.113, 14.130, 14.139, 14.143, 14.152

INTRODUCTORY
STATISTICS

Omar Harran/Moment/Getty Images

SECOND EDITION

INTRODUCTORY STATISTICS

A Problem-Solving Approach

Stephen Kokoska

Bloomsburg University

W. H. FREEMAN
& COMPANY

A **Macmillan** Education Imprint

Senior Publisher: Terri Ward
Senior Acquisitions Editor: Karen Carson
Marketing Manager: Cara LeClair
Development Editor: Leslie Lahr
Associate Editor: Marie Dripchak
Senior Media Editor: Laura Judge
Media Editor: Catriona Kaplan
Associate Media Editor: Liam Ferguson
Editorial Assistant: Victoria Garvey
Marketing Assistant: Bailey James
Photo Editor: Robin Fadool
Cover Designer: Vicki Tomaselli
Text Designer: Jerry Wilke
Managing Editor: Lisa Kinne
Senior Project Manager: Dennis Free, Aptara®, Inc.
Illustrations and Composition: Aptara®, Inc.
Production Coordinator: Julia DeRosa
Printing and Binding: QuadGraphics
Cover credit: Omar Harran/Moment/Getty Images

Library of Preassigned Control Number: 2014950583

Student Edition Hardcover (packaged with EESEE/CrunchIt! access card):
ISBN-13: 978-1-4641-1169-3
ISBN-10: 1-4641-1169-3

Student Edition Loose-leaf (packaged with EESEE/CrunchIt! access card):
ISBN-13: 978-1-4641-5752-3
ISBN-10: 1-4641-5752-9

Instructor Complimentary Copy:
ISBN-13: 978-1-4641-7986-0
ISBN-10: 1-4641-7986-7

Printed in the United States of America

First printing

W. H. Freeman and Company
41 Madison Avenue
New York, NY 10010
Houndmills, Basingstoke RG21 6XS, England
www.whfreeman.com

BRIEF CONTENTS

Optional Sections

(available online at **www.whfreeman.com/introstats2e** and on **LaunchPad**):

CONTENTS

Optional Sections
(available online at **www.whfreeman.com/introstats2e** and on **LaunchPad**):

Students frequently ask me why they need to take an introductory statistics course. My answer is simple. In almost every occupation and in ordinary daily life, you will have to make data-driven decisions, inferences, as well as assess risk. In addition, you must be able to translate complex problems into manageable pieces, recognize patterns, and most important, solve problems. This text helps students develop the fundamental lifelong tool of solving problems and interpreting solutions in real-world terms.

One of my goals was to make this problem-solving approach accessible and easy to apply in many situations. I certainly want students to appreciate the beauty of statistics and the connections to so many other disciplines. However, it is even more important for students to be able to apply problem-solving skills to a wide range of academic and career pursuits, including business, science and technology, and education.

Introductory Statistics: A Problem-Solving Approach, Second Edition, presents long-term, universal skills for students taking a one- or two-semester introductory-level statistics course. Examples include guided, explanatory Solution Trails that emphasize problem-solving techniques. Example solutions are presented in a numbered, step-by-step format. The generous collection and variety of exercises provide ample opportunity for practice and review. Concepts, examples, and exercises are presented from a practical, realistic perspective. Real and realistic data sets are current and relevant. The text uses mathematically correct notation and symbols and precise definitions to illustrate statistical procedures and proper communication.

This text is designed to help students fully understand the steps in basic statistical arguments, emphasizing the importance of assumptions in order to follow valid arguments or identify inaccurate conclusions. Most important, students will understand the process of statistical inference. A four-step process (Claim, Experiment, Likelihood, Conclusion) is used throughout the text to present the smaller pieces of introductory statistics on which the larger, essential statistical inference puzzle is built.

NEW TO THIS EDITION

In this thoroughly updated new edition, Steve Kokoska again combines a classic approach to teaching statistics with contemporary examples, pedagogical features, and use of technology. He blends solid mathematics with lucid, often humorous, writing and a distinctive stepped "Solution Trail" problem-solving approach, which helps students understand the processes behind basic statistical arguments, statistical inference, and data-based decision making.

LaunchPad

Introductory Statistics is accompanied by its own dedicated version of W. H. Freeman's breakthrough online course space, which offers the following:

- Pre-built Units for each chapter, curated by experienced educators, with media for each chapter organized and ready to assign or customize to suit the course.
- All online resources for the text in one location, including an interactive e-Book, LearningCurve adaptive quizzing, Try It Now exercises, StatTutors, video technology manuals, statistical applets, CrunchIt! and JMP statistical software, EESEE case studies, and statistical videos.
- Intuitive and useful analytics, with a Gradebook that allows instructors to see how the class is progressing, for individual students and as a whole.
- A streamlined and intuitive interface that lets instructors build an entire course in minutes.

New Solution Trail Exercises

Kokoska's unique "Solution Trail" framework appears in the text margins alongside selected examples. This feature, highly praised by reviewers, serves as a unique guide for approaching and solving the problems before moving to the solution steps within the example. To allow students to put this guidance to use, exercise sets now feature questions that ask students to create their own solution trails.

New Concept Check Exercises

Strengthening the book's conceptual coverage, these exercises open each exercise set with true/false, fill-in-the-blank, and short-answer questions that help students solidify their understanding of the reading and the essential statistical ideas.

New Chapter 0

This introductory chapter eases students into the course and Kokoska's approach. It includes about a dozen exercises that instructors can assign for the first day of class, helping students settle into the course more easily.

Revised Chapter Openers that include "Looking Forward/ Looking Back"

"Looking Back" recaps key concepts learned in prior chapters. "Looking Forward" lists the key concepts to be covered within the chapter.

New "Last Step" Exercises Based on Opening Scenarios

The chapter-opening question is presented again as an exercise at the end of the chapter, to close the concept and application loop, as a last step. In addition, this gives instructors the option of making the scenarios assignable and assessable.

Try It Now References

Most examples include a reference to a specific related exercise in the end-of-chapter set. With this, students can test their understanding of the example's concepts and techniques immediately.

Approximately 40% New and Updated Exercises and Examples

Approximately 100 new examples and almost 800 new exercises are included in this new edition.

More Statistical Technology Integration

In addition to presenting Excel, Minitab, and TI output and instruction, the new edition incorporates sample output screens and guidance for both CrunchIt!, W. H. Freeman's web-based statistical software, and JMP. (CrunchIt! and JMP packages are available free of charge in LaunchPad.)

FEATURES

Focus on Statistical Inference   The main theme of this text is statistical inference and decision making through interpretation of numerical results. The process of statistical inference is introduced in a variety of contexts, all using a similar, carefully delineated, four-step approach: Claim, Experiment, Likelihood, and Conclusion.

Can the Florida Everglades be saved?

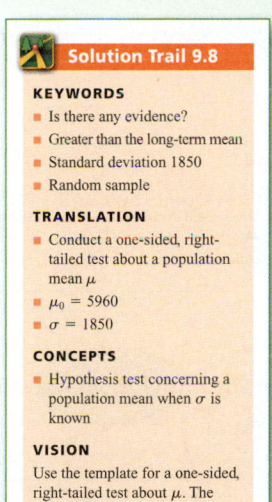

Burmese pythons have invaded the Florida Everglades and now threaten the wildlife indigenous to the area. It is likely that people were keeping pythons as pets and somehow a few animals slithered into Everglades National Park. The first python was found in the Everglades in 1979, and these snakes became an officially established species in 2000.[1] The Everglades has an ideal climate for the pythons, and the large areas of grass allow the snakes plenty of places to hide.

In January 2013, the Florida Fish and Wildlife Conservation Commission started the Python Challenge. The purpose of the contest was to thin the python population, which could be tens of thousands, and help save the natural wildlife in the Everglades. There were 800 participants, with prizes for the most pythons captured and for the longest. At the end of the competition, 68 Burmese pythons had been harvested.

Suppose a random sample of pythons captured during the Challenge was obtained. The length (in feet) of each python is given in the following table.

9.3	3.5	5.2	8.3	4.6	11.1	10.5	3.7	2.8	5.9
7.4	14.2	13.6	8.3	7.5	5.2	6.4	12.0	10.7	4.0
11.1	3.7	7.0	12.2	5.2	8.1	4.2	6.1	6.3	13.2
3.9	6.7	3.3	8.3	10.9	9.5	9.4	4.3	4.6	5.8
4.1	5.2	4.7	5.8	6.4	3.8	7.1	4.6	7.5	6.0

The tabular and graphical techniques presented in this chapter will be used to describe the shape, center, and spread of this distribution of python lengths and to identify any outliers.

Chapter Opener Each chapter begins with a unique, real-world question, providing an interesting introduction to new concepts and an application to begin discussion. The chapter question is presented again as an exercise at the end of the chapter, to close the concept and application loop, as a last step.

◄ **Looking Back**
- Recall that $\bar{x}$, $\hat{p}$, and s^2 are the point estimates for the parameters μ, p, and σ^2.
- Remember how to construct and interpret confidence intervals.
- Think about the concept of a sampling distribution for a statistic and the process of standardization.

► **Looking Forward**
- Use the available information in a sample to make a specific decision about a population parameter.
- Understand the formal decision process and learn the four-part hypothesis test procedure.
- Conduct formal hypothesis tests concerning the population parameters μ, p, and σ^2.

Looking Back and Looking Forward At the beginning of almost every chapter, "Looking Back" includes reminders of specific concepts from earlier chapters that will be used to develop new skills. "Looking Forward" offers the learning objectives for the chapter.

Solution Trail The Solution Trail is a structured technique and visual aid for solving problems that appears in the text margins alongside selected examples. Solution Trails serve as guides for approaching and solving the problems before moving to the solution steps within the example. The four steps of the Solution Trail are

1. Find the *keywords*.
2. Correctly *translate* these words into statistics.
3. Determine the applicable *concepts*.
4. Develop a *vision* for the solution.

The *keywords* lead to a *translation* into statistics. Then, the statistics question is solved with the use of specific *concepts*. Finally, the keywords, translation, and concepts are all used to develop a *vision* for the solution. This method encourages students to think conceptually before making calculations. Selected exercises ask students to write a formal Solution Trail.

Solution Trail 9.8

KEYWORDS
- Is there any evidence?
- Greater than the long-term mean
- Standard deviation 1850
- Random sample

TRANSLATION
- Conduct a one-sided, right-tailed test about a population mean μ
- $\mu_0 = 5960$
- $\sigma = 1850$

CONCEPTS
- Hypothesis test concerning a population mean when σ is known

VISION
Use the template for a one-sided, right-tailed test about μ. The underlying population distribution is unknown, but n is large and σ is known. Determine the appropriate alternative hypothesis and the corresponding rejection region, find the value of the test statistic, and draw a conclusion.

Step-by-Step Solutions The solutions to selected examples are presented in logical, systematic steps. Each line in a calculation is explained so that the reader can clearly follow each step in a solution.

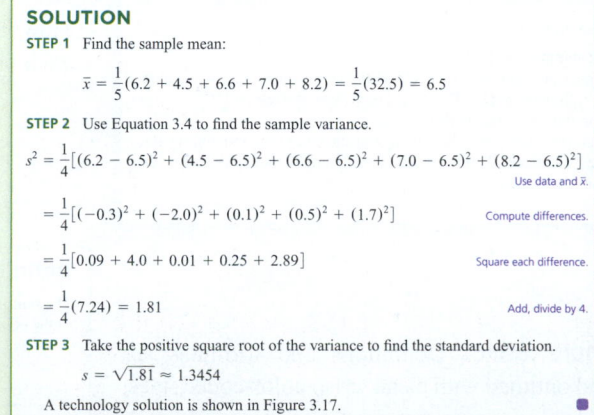

SOLUTION

STEP 1 Find the sample mean:

$$\bar{x} = \frac{1}{5}(6.2 + 4.5 + 6.6 + 7.0 + 8.2) = \frac{1}{5}(32.5) = 6.5$$

STEP 2 Use Equation 3.4 to find the sample variance.

$$s^2 = \frac{1}{4}[(6.2 - 6.5)^2 + (4.5 - 6.5)^2 + (6.6 - 6.5)^2 + (7.0 - 6.5)^2 + (8.2 - 6.5)^2]$$
Use data and $\bar{x}$.

$$= \frac{1}{4}[(-0.3)^2 + (-2.0)^2 + (0.1)^2 + (0.5)^2 + (1.7)^2]$$
Compute differences.

$$= \frac{1}{4}[0.09 + 4.0 + 0.01 + 0.25 + 2.89]$$
Square each difference.

$$= \frac{1}{4}(7.24) = 1.81$$
Add, divide by 4.

STEP 3 Take the positive square root of the variance to find the standard deviation.

$$s = \sqrt{1.81} \approx 1.3454$$

A technology solution is shown in Figure 3.17.

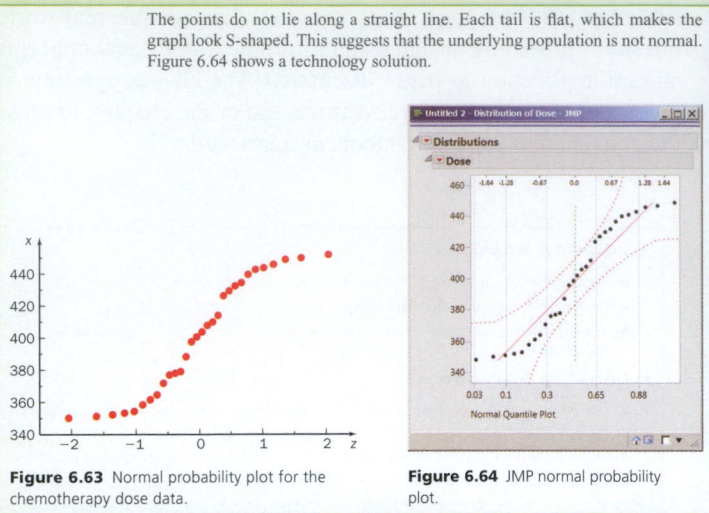

The points do not lie along a straight line. Each tail is flat, which makes the graph look S-shaped. This suggests that the underlying population is not normal. Figure 6.64 shows a technology solution.

Figure 6.63 Normal probability plot for the chemotherapy dose data.

Figure 6.64 JMP normal probability plot.

Technology Solutions Wherever possible, a technology solution using CrunchIt!, JMP, the TI-84, Minitab, or Excel is presented at the end of each text example. This allows students to focus on concepts and interpretation.

A Closer Look The details provided in these sections offer straightforward explanations of various definitions and concepts. The itemized specifics, including hints, tips, and reminders, make it easier for the reader to comprehend and learn important statistical ideas.

Theory Symbols More advanced material, which may be found in "A Closer Look" and regular exposition as appropriate, is offset with a blue triangle. This material can be skipped by the typical reader, but provides more complete explanations to various topics.

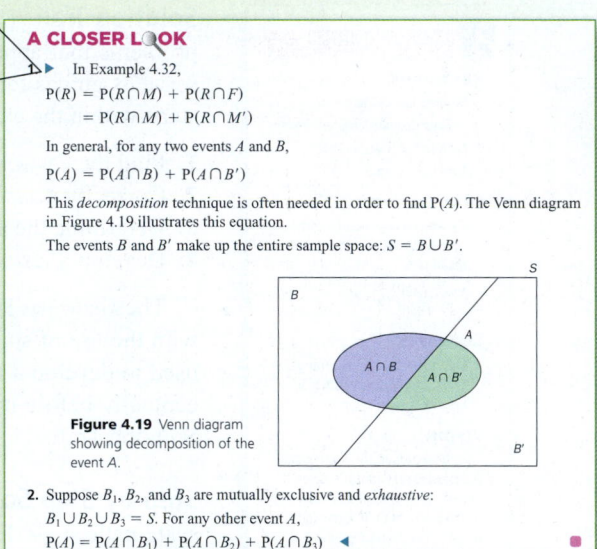

A CLOSER LOOK

1. ▶ In Example 4.32,
$$P(R) = P(R \cap M) + P(R \cap F)$$
$$= P(R \cap M) + P(R \cap M')$$

In general, for any two events A and B,
$$P(A) = P(A \cap B) + P(A \cap B')$$

This *decomposition* technique is often needed in order to find $P(A)$. The Venn diagram in Figure 4.19 illustrates this equation.

The events B and B' make up the entire sample space: $S = B \cup B'$.

Figure 4.19 Venn diagram showing decomposition of the event A.

2. Suppose B_1, B_2, and B_3 are mutually exclusive and *exhaustive*: $B_1 \cup B_2 \cup B_3 = S$. For any other event A,
$$P(A) = P(A \cap B_1) + P(A \cap B_2) + P(A \cap B_3)$$ ◀

How to Construct a Standard Box Plot

Given a set of n observations $x_1, x_2, \ldots, x_n$:

1. Find the five-number summary $x_{min}, Q_1, \tilde{x}, Q_3, x_{max}$.
2. Draw a (horizontal) measurement axis. Carefully sketch a box with edges at the quartiles: left edge at Q_1, right edge at Q_3. (The height of the box is irrelevant.)
3. Draw a vertical line in the box at the median.
4. Draw a horizontal line (whisker) from the left edge of the box to the minimum value (from Q_1 to x_{min}). Draw a horizontal line (whisker) from the right edge of the box to the maximum value (from Q_3 to x_{max}).

How To Boxes This feature provides clear steps for constructing basic graphs or performing essential calculations. How To boxes are color-coded and easy to locate within each chapter.

Definition/Formula Boxes Definitions and formulas are clearly marked and outlined with clean, crisp color-coded lines.

Definition

The **sample (arithmetic) mean**, denoted $\bar{x}$, of the n observations $x_1, x_2, \ldots, x_n$ is the sum of the observations divided by n. Written mathematically.

$$\bar{x} = \frac{1}{n}\Sigma x_i = \frac{x_1 + x_2 + \cdots + x_n}{n} \qquad (3.1)$$

Technology Corner This feature, at the end of most sections, presents step-by-step instructions for using CrunchIt!, the TI-84, Minitab, and Excel to solve the examples presented in that section. Keystrokes, menu items, specific functions, and screen illustrations are presented.

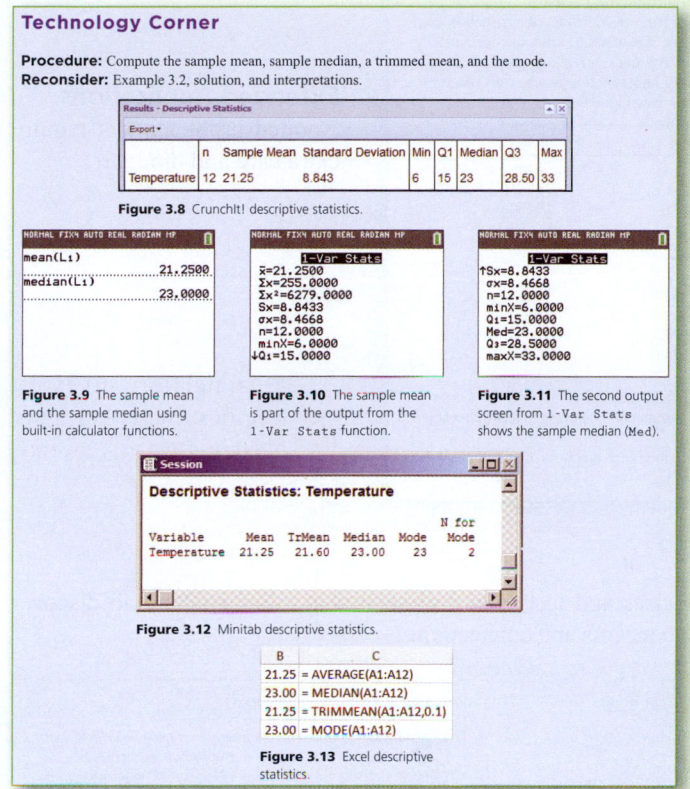

Helpful Icons

Data Set icons indicate when a data set is available online, and also the name of the data set.

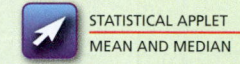
Statistical Applet icons indicate statistical applets that are available in LaunchPad.

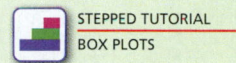
Stepped Tutorial icons indicate detailed tutorials for specific calculations.

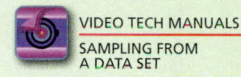
Video Tech Manual icons indicate video instructions for solving certain kinds of problems using statistical software.

Solution Trail icons within the exercise sets indicate the opportunity for students to create their own Solution Trails.

Grouped Exercises Kokoska offers a wide variety of interesting, engaging exercises on relevant topics, based on current data, at the end of each section and chapter. These problems provide plenty of opportunity for practice, review, and application of concepts. Answers to odd-numbered section and chapter exercises are given at the back of the book. Exercises are grouped according to:

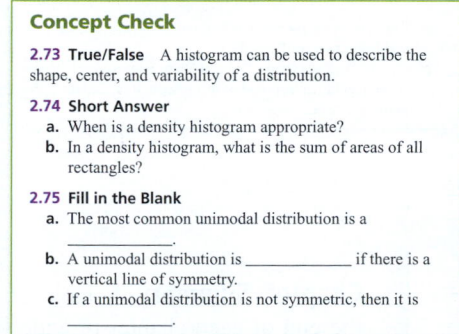

Concept Check
True/False, Fill-in-the-Blank, and Short-Answer exercises designed to reinforce the basic concepts presented in the section.

Practice
Basic, introductory problems to familiarize students with the concepts and solution methods.

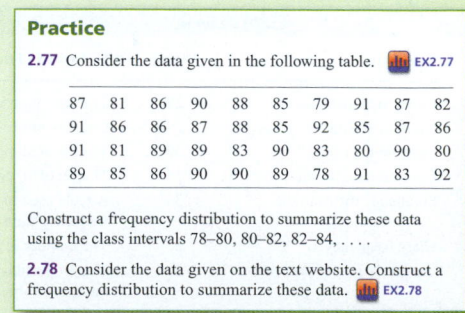

Applications

2.86 Biology and Environmental Science A weather station located along the Maine coast in Kennebunkport collects data on temperature, wind speed, wind chill, and rain. The maximum wind speed (in miles per hour) for 50 randomly selected times in February 2013 are given on the text website.[27] ▥ MAXWIND

a. Construct a frequency distribution to summarize these data, and draw the corresponding histogram.

b. Describe the shape of the distribution. Are there any outliers?

Applications

Realistic, appealing exercises to build confidence and promote routine understanding. Many exercises are based on interesting and carefully researched data.

Extended Applications

2.92 Biology and Environmental Science Fruits such as cherries and grapes are harvested and placed in a shallow box or crate called a lug. The size of a lug varies, but one typically holds between 16 and 28 pounds. A random sample of the weight (in pounds) of full lugs holding peaches was obtained, and the data are summarized in the following table.

Class	Frequency
20.0–20.5	6
20.5–21.0	12
21.0–21.5	17
21.5–22.0	21
22.0–22.5	28
22.5–23.0	25
23.0–23.5	19
23.5–24.0	15
24.0–24.5	11
24.5–25.0	10

a. Complete the frequency distribution.

b. Construct a histogram corresponding to this frequency distribution.

c. Estimate the weight w such that 90% of all full peach lugs weigh more than w.

Extended Applications

Applied problems that require extra care and thought.

CHALLENGE

2.107 Sports and Leisure An *ogive*, or *cumulative relative frequency polygon*, is another type of visual representation of a frequency distribution. To construct an ogive:

■ Plot each point (upper endpoint of class interval, cumulative relative frequency).

■ Connect the points with line segments.

Figures 2.52 and 2.53 show a frequency distribution and the corresponding ogive. The observations are ages. The values to be used in the plot are shown in bold in the table.

Class	Frequency	Relative frequency	Cumulative relative frequency
12–16	8	0.08	**0.08**
16–20	10	0.10	**0.18**
20–24	20	0.20	**0.38**
24–28	30	0.30	**0.68**
28–32	15	0.15	**0.83**
32–36	10	0.10	**0.93**
32–40	7	0.07	**1.00**
Total	100	1.00	

Figure 2.52 Frequency distribution.

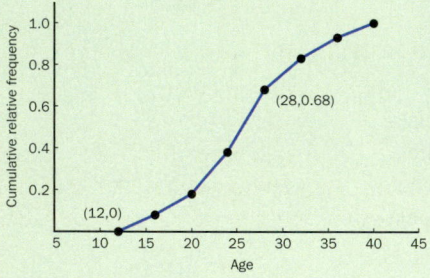

Figure 2.53 Resulting ogive.

A random sample of game scores from Abby Sciuto's evening bowling league with Sister Rosita was obtained, and the data are given on the text website. ▥ BOWLING

a. Construct a frequency distribution for these data.

b. Draw the resulting ogive for these data.

Challenge

Additional exercises and technology projects that allow students to discover more advanced concepts and connections.

Last Step

Each set of chapter exercises concludes with the "Last Step." This exercise is connected to the chapter-opening question and the solution involves the skills and concepts presented in the chapter.

LAST STEP

2.109 Can the Florida Everglades be saved? In January 2013, the Florida Fish and Wildlife Conservation Commission started the Python Challenge. The purpose of the contest was to thin the python population, which could be tens of thousands, and help save the natural wildlife in the Everglades. At the end of the competition, 68 Burmese pythons had been harvested. Suppose a random sample of pythons captured during the Challenge was obtained and the length (in feet) of each is given in the following table: ▥ PYTHON

9.3	3.5	5.2	8.3	4.6	11.1	10.5	3.7	2.8	5.9
7.4	14.2	13.6	8.3	7.5	5.2	6.4	12.0	10.7	4.0
11.1	3.7	7.0	12.2	5.2	8.1	4.2	6.1	6.3	13.2
3.9	6.7	3.3	8.3	10.9	9.5	9.4	4.3	4.6	5.8
4.1	5.2	4.7	5.8	6.4	3.8	7.1	4.6	7.5	6.0

a. Construct a frequency distribution, stem-and-leaf plot, and histogram for these data.

b. Use these tabular and graphical techniques to describe the shape, center, and spread of this distribution, and to identify any outlying values.

CHAPTER 2 SUMMARY

Concept	Page	Notation / Formula / Description
Categorical data set	29	Consists of observations that may be placed into categories.
Numerical data set	29	Consists of observations that are numbers.
Discrete data set	30	The set of all possible values is finite, or countably infinite.
Continuous data set	30	The set of all possible values is an interval of numbers.
Frequency distribution	33	A table used to describe a data set. It includes the class, frequency, and relative frequency (and cumulative relative frequency, if the data set is numerical).
Class frequency	33	The number of observations within a class.
Class relative frequency	33	The proportion of observations within a class: class frequency divided by total number of observations.

Chapter Summary A table at the end of each chapter provides a list of the main concepts with brief descriptions, proper notation, and applicable formulas, along with page numbers for quick reference.

Macmillan Education LaunchPad

W. H. Freeman's new online homework system, **LaunchPad,** offers our quality content curated and organized for easy assignability in a simple but powerful interface. We've taken what we've learned from thousands of instructors and hundreds of thousands of students to create a new generation of W. H. Freeman/Macmillan technology.

Curated Units. Combining a curated collection of videos, homework sets, tutorials, applets, and e-Book content, LaunchPad's interactive units give instructors building blocks to use as is or as a starting point for their own learning units. Thousands of exercises from the text can be assigned as online homework, including many algorithmic exercises. An entire unit's worth of work can be assigned in seconds, drastically reducing the amount of time it takes to have a course up and running.

Easily customizable. Instructors can customize the LaunchPad Units by adding quizzes and other activities from our vast collection of resources. They can also add a discussion board, a dropbox, and RSS feed, with a few clicks. LaunchPad allows instructors to customize their students' experience as much or as little as they like.

Useful analytics. The Gradebook quickly and easily allows instructors to look up performance metrics for classes, individual students, and individual assignments.

Intuitive interface and design. The student experience is simplified. Students' navigation options and expectations are clearly laid out at all times, ensuring that they can never get lost in the system.

Assets integrated into LaunchPad include:

Interactive e-Book. Every LaunchPad e-Book comes with powerful study tools for students, video and multimedia content, and easy customization for instructors. Students can search, highlight, and bookmark, making it easier to study and access key content. And teachers can ensure that their classes get just the book they want to deliver: customize and rearrange chapters, add and share notes and discussions, and link to quizzes, activities, and other resources.

Macmillan Education LearningCurve provides students and instructors with powerful adaptive quizzing, a gamelike format, direct links to the e-Book, and instant feedback. The quizzing system features questions tailored specifically to the text and adapts to students' responses, providing material at different difficulty levels and topics based on student performance.

FREEMAN SolutionMaster offers an easy-to-use web-based version of the instructor's solutions, allowing instructors to generate a solution file for any set of homework exercises.

New Stepped Tutorials are centered on algorithmically generated quizzing with step-by-step feedback to help students work their way toward the correct solution. These new exercise tutorials (two to three per chapter) are easily assignable and assessable. Icons in the textbook indicate when a Stepped Tutorial is available for the material being covered.

Statistical Video Series consists of StatClips, StatClips Examples, and Statistically Speaking "Snapshots." View animated lecture videos, whiteboard lessons, and documentary-style footage that illustrate key statistical concepts and help students visualize statistics in real-world scenarios.

New Video Technology Manuals available for TI-83/84 calculators, Minitab, Excel, JMP, SPSS, R, Rcmdr, and CrunchIT! provide brief instructions for using specific statistical software.

Updated StatTutor Tutorials offer multimedia tutorials that explore important concepts and procedures in a presentation that combines video, audio, and interactive features. The newly revised format includes built-in, assignable assessments and a bright new interface.

Updated Statistical Applets give students hands-on opportunities to familiarize themselves with important statistical concepts and procedures, in an interactive setting that allows them to manipulate variables and see the results graphically. These new applets now include a "Quiz Me" function that allows them to be both assignable and assessable. Icons in the textbook indicate when an applet is available for the material being covered.

CrunchIt! is a web-based statistical program that allows users to perform all the statistical operations and graphing needed for an introductory statistics course and more. It saves users time by automatically loading data from the text, and it provides the flexibility to edit and import additional data.

JMP **JMP Student Edition** (developed by SAS) is easy to learn and contains all the capabilities required for introductory statistics, including pre-loaded data sets from *Introductory Statistics: A Problem-Solving Approach*. JMP is the commercial data analysis software of choice for scientists, engineers, and analysts at companies around the globe (for Windows and Mac).

Stats@Work Simulations put students in the role of the statistical consultant, helping them better understand statistics interactively within the context of real-life scenarios.

EESEE Case Studies (Electronic Encyclopedia of Statistical Examples and Exercises), developed by The Ohio State University Statistics Department, teach students to apply their statistical skills by exploring actual case studies using real data.

Data files are available in ASCII, Excel, TI, Minitab, SPSS (an IBM Company),* and JMP formats.

Student Solutions Manual provides solutions to the odd-numbered exercises in the text. Available electronically within LaunchPad, as well as in print form.

Interactive Table Reader allows students to use statistical tables interactively to seek the information they need.

Instructor's Solutions Manual contains full solutions to all exercises from *Introductory Statistics: A Problem-Solving Approach*. Available electronically within LaunchPad.

Test Bank offers hundreds of multiple-choice questions. Also available on CD-ROM (for Windows and Mac), where questions can be downloaded, edited, and resequenced to suit each instructor's needs.

---

*SPSS was acquired by IBM in October 2009.

Lecture PowerPoint Slides offer a detailed lecture presentation of statistical concepts covered in each chapter of *Introductory Statistics: A Problem-Solving Approach*.

Additional Resources Available with *Introductory Statistics: A Problem-Solving Approach*

Companion Website www.whfreeman.com/introstats2e This open-access website includes statistical applets, data files, and self-quizzes. The website also offers three optional sections covering the normal approximation to the binomial distribution (Section 6.5), polynomial and qualitative predictor models (Section 12.6), and model selection procedures (Section 12.7). Instructor access to the Companion Website requires user registration as an instructor and features all of the open-access student web materials, plus:

- Instructor version of **EESEE** with solutions to the exercises in the student version.
- **PowerPoint Slides** containing all textbook figures and tables.
- **Lecture PowerPoint Slides**
- **Tables and Formulas cards** offer tables, key concepts, and formulas for use as a study tool or during exams (as allowed by the instructor); available as downloadable PDFs.

Special Software Packages Student versions of JMP and Minitab are available for packaging with the text. JMP is available inside LaunchPad at no additional cost. Contact your W. H. Freeman representative for information or visit www.whfreeman.com.

i-clicker is a two-way radio-frequency classroom response solution developed by educators for educators. Each step of i-clicker's development has been informed by teaching and learning. To learn more about packaging i-clicker with this textbook, please contact your local sales rep or visit www.iclicker.com.

ACKNOWLEDGMENTS

I would like to thank the following colleagues who offered specific comments and suggestions on the second-edition manuscript throughout various stages of development:

Jonathan Baker, *Ohio State University*
Andrea Boito, *Penn State Altoona*
Alexandra Challiou, *Notre Dame of Maryland University*
Carolyn K. Cuff, *Westminster College*
Greg Davis, *University of Wisconsin Green Bay*
Richard Gonzalez, *University of Michigan*
Justin Grieves, *Murray State University*
Christian Hansen, *Eastern Washington University*
Christopher Hay-Jahans, *University of Alaska Southeast*
Susan Herring, *Sonoma State University*
Chester Ismay, *Arizona State University*
Ananda Jayawardhana, *Pitt State University*
Phillip Kendall, *Michigan Technological University*
Bashir Khan, *St. Mary's University*
Barbara Kisilevsky, *Queens University*
Tammi Kostos, *McHenry County College*
Adam Lazowski, *Sacred Heart University*
Jiexiang Li, *College of Charleston*
Edgard Maboudou, *University of Central Florida*
Tina Mancuso, *Sage College*
Scott McClintock, *West Chester University*
Jackie Miller, *University of Michigan*
Daniel Ostrov, *Santa Clara University*
William Radulovich, *Florida State College at Jacksonville*
Enayetur Raheem, *University of Wisconsin - Green Bay*
Daniel Rothe, *Alpena Community College*
James Stamey, *Baylor University*
Sunny Wang, *St. Francis Xavier University*
Derek Webb, *Bemidji State University*
Daniel Weiner, *Boston University*
Mark Werner, *University of Georgia*
Nancy Wyshinski, *Trinity College*

A special thanks to Ruth Baruth, Terri Ward, Karen Carson, Cara LeClair, Lisa Kinne, Tracey Kuehn, Julia DeRosa, Vicki Tomaselli, Robin Fadool, Marie Dripchak, Liam Ferguson, Catriona Kaplan, Laura Judge, and Victoria Garvey of W. H. Freeman and Company.

Designer Jerry Wilke and illustrator Cambraia Fernandez, led by Vicki Tomaselli, offered the creativity, expertise, and hard work that went into the design of this new edition.

I am very grateful to Jackie Miller for her insights, suggestions, and editorial talent throughout the production of the second edition. She is doggedly accurate in her accuracy reviews and page proof examination. Thanks to Dennis Free of Aptara for his patience and typesetting expertise. Much appreciated are the copy editing skills brought to the project by Lynne Lackenbach; her time and perseverance helped to add cohesion and continuity to the flow of the material. Many thanks to Aaron Bogan for bringing his attention to detail and knowledge of statistics to the accuracy review of the solutions manuals. And I could not have completed this project without Karen Carson and Leslie Lahr. Both have superb editing skills, a keen eye for style, a knack for eliciting the best from an author, and unwavering support.

My sincere thanks go to the authors and reviewers of the supplementary materials available with *Introductory Statistics: A Problem-Solving Approach*, Second Edition; their hard work, expertise, and creativity have culminated in a top-notch package of resources:

Test Bank written by Julie Clark, Hollins University

Test Bank and iClicker slides accuracy reviewed by John Samons, Florida State College at Jacksonville

Practice Quizzes written by James Stamey, Baylor University

Practice Quizzes accuracy reviewed by Laurel Chiappetta, University of Pittsburgh

iClicker slides created by Paul Baker, Catawba College

Lecture PowerPoints created by Susan Herring, Sonoma State University

I am very grateful to the entire Antoniewicz family for providing the foundation for a wide variety of problems, including those that involve nephelometric turbidity units, floor slip testers, and crazy crawler fishing lures.

I continue to learn a great deal with every day of writing. I believe this kind of exposition has made me a better teacher.

To Joan, thank you for your patience, understanding, inspiration, and tasty treats.

ABOUT THE AUTHOR

Credit: Eric Foster

Steve received his undergraduate degree from Boston College and his M.S. and Ph.D. from the University of New Hampshire. His initial research interests included the statistical analysis of cancer chemoprevention experiments. He has published a number of research papers in mathematics journals, including *Biometrics, Anticancer Research,* and *Computer Methods and Programs in Biomedicine;* presented results at national conferences; and written several books. He has been awarded grants from the National Science Foundation, the Center for Rural Pennsylvania, and the Ben Franklin Program.

Steve is a long-time consultant for the College Board and conducted workshops in Brazil, the Dominican Republic, and China. He was the AP Calculus Chief Reader for four years and has been involved with calculus reform and the use of technology in the classroom. He has been teaching at Bloomsburg University for 25 years and recently served as Director of the Honors Program.

Steve has been teaching introductory statistics classes throughout his academic career, and there is no doubt that this is his favorite course. This class (and text) provides students with basic, lifelong, quantitative skills that they will use in almost any job and teaches them how to think and reason logically. Steve believes very strongly in data-driven decisions and conceptual understanding through problem solving.

Steve's uncle, Fr. Stanley Bezuszka, a Jesuit and professor at Boston College, was one of the original architects of the so-called new math in the 1950s and 1960s. He had a huge influence on Steve's career. Steve helped Fr. B. with test accuracy checks, as a teaching assistant, and even writing projects through high school and college. Steve learned about the precision, order, and elegance of mathematics and developed an unbounded enthusiasm to teach.

INTRODUCTORY
STATISTICS

0 Why Study Statistics

The Science of Intuition

In the movie *Erin Brokovich,* actress Julia Roberts plays a feisty, unemployed, single mother of three children. After losing a lawsuit because of her bad behavior in the courtroom, Erin pressures her lawyer Ed Masry, for a job and he conceded. Despite having no legal background, Erin begins working on a real estate case involving Pacific Gas and Electric (PG&E) and the purchase of a home in Hinkley, California.

Erin visits the seller, Donna Jensen, and learns that her husband has Hodgkin's disease and that many Hinkley residents have concerns about the environment. After further investigation, Erin discovers that several residents of Hinkley have suffered from autoimmune disorders and various forms of cancer.

In fact, so many people in Hinkley suffer from similar rare diseases that Erin concludes it could not be a coincidence. This is a very natural, intuitive conclusion, and it is the essence of statistical inference. Erin observed an occurrence that was so rare and extraordinary that she instinctively concluded it could not be due to pure chance or luck. There had to be another reason. Her logic was correct: The unusually high incidence of cancer in Hinkley suggested that something abnormal was happening.

Indeed, PG&E had dumped water contaminated with the chemical chromium 6 into unlined storage pools. The polluted water seeped into the groundwater and eventually into local wells, and many people became ill with various medical problems.

We all have this same natural instinctive reaction when we see something that is extraordinary. Sometimes we think, "Wow, that's incredibly lucky." More often we question the observed outcomes, "There must be some other explanation."

This natural reaction is the foundation of statistical inference. We make these kinds of decisions every single day. We gather evidence, we make an observation, and we conclude that the outcome is either reasonable or extraordinary. The purpose of statistics is simply to quantify this typical, everyday, deductive process. We need to learn about probability so that we know for sure when an outcome is really rare. And we need to study the concepts of randomness and uncertainty.

The most important point here is that this process is not unusual or exceptional. The purpose of this text is to translate this common practice into statistical terms and models. This will make you better prepared to interpret outcomes, draw appropriate conclusions, and assess risk.

Here is another example of an extraordinary event involving a daily lottery number. The 1980 Pennsylvania Lottery scandal, or the Triple Six Fix, involved a three-digit daily lottery number. Nick Perry was the announcer for the Daily Number and the plan's architect. With the help of partners, Nick was able to weight all of the balls except for the ones numbered 4 and 6. This meant that the winning three-digit lottery number would be a combination of 4s and 6s. There were thus only eight possible winning lottery numbers, 444, 446, 464, 466, 644, 646, 664, and 666, and the conspirators were certain that the plan would work.

The winning number on the day of the fix was 666. Ignoring the connection to the *Book of Revelations*, lottery officials discovered that there were very unusual betting patterns that day, all on the eight possible lottery numbers involving 4 and 6. This extraordinary occurrence suggested that the unusual bets were not due to pure chance. This conclusion, along with an anonymous tip, helped in a grand jury investigation leading to convictions and jail time for several men.

The Statistical Inference Procedure

The crucial prevailing theme in this text is statistical inference and decision making through problem solving. Computation is important and is shown throughout the text. However, calculators and computers remove the drudgery of hand calculations and allow us to concentrate more on interpretation and drawing conclusions. Most problems in this text contain a part asking the reader to interpret the numerical result or to draw a conclusion.

The process of questioning a rare occurrence or claim can be described in four steps.

Claim: This is the status quo, the ordinary, typical, and reasonable course of events—what we assume to be true.

Experiment: To check a claim, we conduct a relevant experiment or make an appropriate observation.

Likelihood: Here we consider the likelihood of occurrence of the observed experimental outcome, assuming the claim is true. We will use many techniques to determine whether the experimental outcome is a reasonable observation (subject to some variability), or whether it is an exceptionally rare occurrence. We need to consider carefully and quantify our natural reaction to the relevant experiment. Using probability rules and concepts, we will convert our natural reaction to an experimental outcome into a precise measurement.

Conclusion: There are always only two possible conclusions.

1. If the outcome is reasonable, then we cannot doubt the original claim. The natural conclusion is that nothing out of the ordinary is occurring. More formally, there is no evidence to suggest that the claim is false.

2. If the experimental outcome is rare or extraordinary, we usually disregard the lucky alternative, and we think something is wrong. A rare outcome is a contradiction. Strange occurrences naturally make us question a claim. In this case we believe there is evidence to suggest that the claim is false.

Let's try to apply these four steps to the PG&E case in *Erin Brokovich*. The claim or status quo is that the cancer incidence rate in Hinkley is equivalent to the national incidence rate. Recent figures from the American Cancer Society suggest that the cancer incidence rate is approximately 551 in 100,000 for men and 419 in 100,000 for women.[1]

The experiment or observed outcome is the cancer incidence rate for the population living in Hinkley. In the movie, it is implied that Erin counts the number of people in Hinkley who have developed cancer.

Erin determines that the likelihood, or probability, of observing that many people in Hinkley who have developed cancer is extremely low. Subject to reasonable variability, we should not see that many people with cancer in this location.

The conclusion is that this rare event is not due to pure chance or luck. There is some other reason for this rare observation. The implication in the movie is that there is evidence to suggest that something else is affecting the health of the people in Hinkley.

Problem Solving

Solution Trail

KEYWORDS
- Normally distributed
- Mean
- Standard deviation

TRANSLATION
- Normal random variable
- $\mu = 34$
- $\sigma = 0.5$

CONCEPTS
- Normal probability distribution
- Standardization

VISION
Define a normal random variable and translate each question into a probability statement. Standardize and use cumulative probability associated with Z if necessary.

Perhaps one of the most difficult concepts to teach is problem solving. We all struggle to solve problems: thinking about where to begin, what assumptions we can make, and which rules and techniques to use. One reason many students consider statistics a difficult course is because almost every problem is a word problem. These word problems have to be translated into mathematics.

The Solution Trail in this text is a prescriptive technique and visual aid for problem solving. To decipher a word problem, start by identifying the keywords and phrases. Here are the four steps identified in each Solution Trail for solving many of problems in this text.

1. Find the *keywords*.
2. Correctly *translate* these words in statistics.
3. Determine the applicable *concepts*.
4. Develop a *vision*, or strategy, for the solution.

Many of the examples presented in this text have a corresponding Solution Trail in the margin to aid in problem solving. An example of a Solution Trail appears in the margin. Note that many of these terms and symbols may be unfamiliar to you at this point. Right now, just focus on the idea that the Solution Trial involves keywords, a translation, concepts, and a vision.

The keywords in the problem lead to a translation into statistics. The statistics question is then solved by using the appropriate, specific concepts. The keywords, translation, and concepts are used to develop a grand vision for solving the problem.

This solution technique is not applicable to every problem, but it is most appropriate for finding probabilities through hypothesis testing, which is the foundation of most introductory statistics courses. Some exercises in this text ask you to write each step in the Solution Trail formally. As you become accustomed to using this solution style, it will become routine, natural, and helpful.

With a Little Help from Technology

Although it is important to know and understand underlying formulas, their derivations, and how to apply them, we will use and present several different technology tools to supplement problem solving. Your focus should be on the interpretation of results, not the actual numerical calculations.

Four common technology tools are presented in this text.

1. **CrunchIt!** is available in LaunchPad, the publisher's online homework system, and is accessed under the Resources tab. The opening screen (Figure 0.1) looks like a spreadsheet with pull-down menus at the top. You can enter data in columns, Var1, Var2, etc., import data from a file, and export and save data.

   Most Statistics, Graphics, and Distribution Calculator functions start with input screens. Output is displayed in a new screen. Figure 0.2 shows the input screen for a bar chart with summarized data, and Figure 0.3 shows the resulting graph.[2]

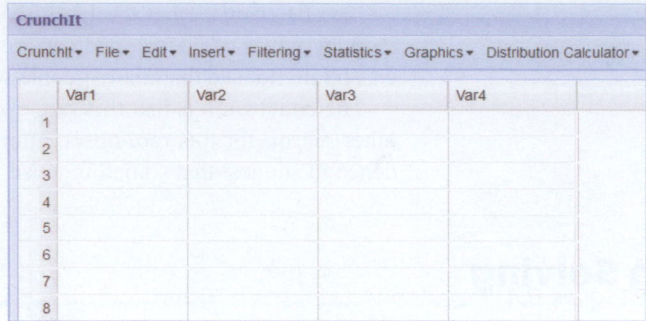

Figure 0.1 CrunchIt! opening screen.

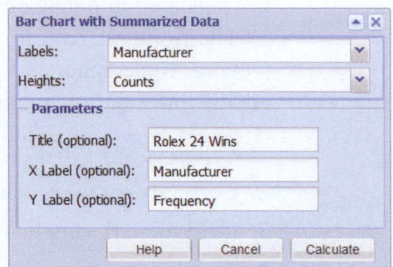

Figure 0.2 Bar chart input screen.

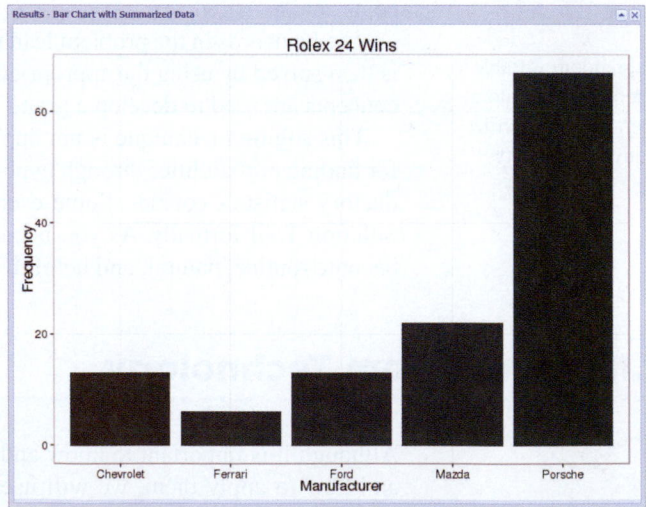

Figure 0.3 CrunchIt! bar chart.

2. The **Texas Instruments TI-84 Plus C** graphing calculator includes many common statistical features such as confidence intervals, hypothesis tests, and probability distribution functions. Data are entered and edited in the stat list editor as shown in Figure 0.4. Figure 0.5 shows the results from a one-sample t test, and Figure 0.6 shows a visualization of this hypothesis test.

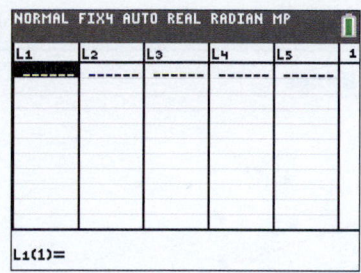

Figure 0.4 The stat list editor.

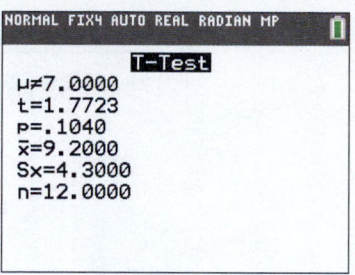

Figure 0.5 One-sample *t*-test output.

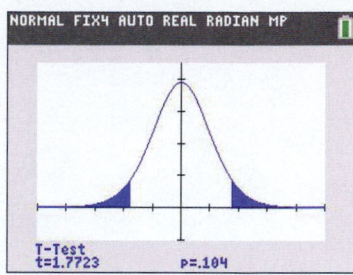

Figure 0.6 One-sample *t*-test visualization.

3. **Minitab** is a powerful software tool for analyzing data. It has a logical interface, including a worksheet screen similar to a common spreadsheet. Data, graph, and statistics tools can be accessed through pull-down menus, and most commands can also be entered in a session window. Figure 0.7 shows a bar chart of the number of Rolex 24 sports car race wins by automobile manufacturer.

4. **Excel 2013** includes many common chart features accessible under the Insert tab. There are also probability distribution functions that allow the user to build templates for confidence intervals, hypothesis tests, and other statistical procedures. The Data Analysis tool pack provides additional statistical functions. Figure 0.8 shows some descriptive statistics associated with the ages of 100 stock brokers at a New York City firm.

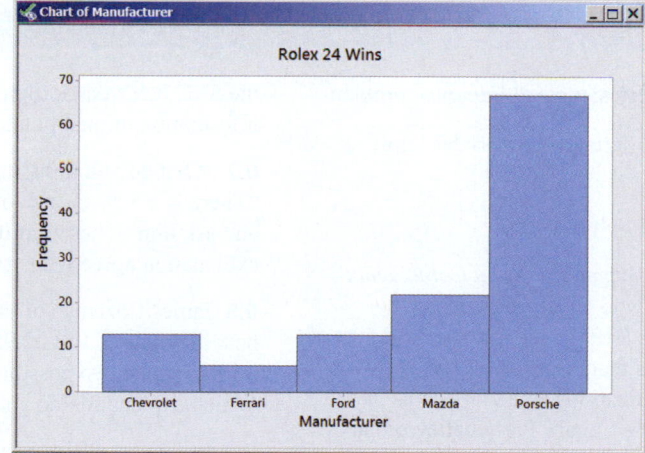

Figure 0.7 Minitab bar chart.

	D	E
	Age	
Mean		42.7700
Standard Error		0.6543
Median		43
Mode		45
Standard Deviation		6.5426
Sample Variance		42.8052
Kurtosis		-0.5767
Skewness		0.1419
Range		28
Minimum		29
Maximum		57
Sum		4277
Count		100

Figure 0.8 Excel descriptive statistics.

In addition to these tools, JMP statistical software is used by scientists, engineers, and others who want to explore or mine data. Various statistical tools and dynamic graphics are available, and this software features a friendly interactive interface. Figure 0.9 shows a scatter plot of the price of used Honda Accords versus the age of the vehicle, the least-squares regression line, and confidence bands for the true mean price for each age.

Many other technology tools and statistical software packages are also available. For example, R is free statistical software, SPSS is used primarily in the social sciences, and SAS incorporates a proprietary programming language. Regardless of your technology choice, remember that careful and thorough interpretation of the results is an essential part of using software properly.

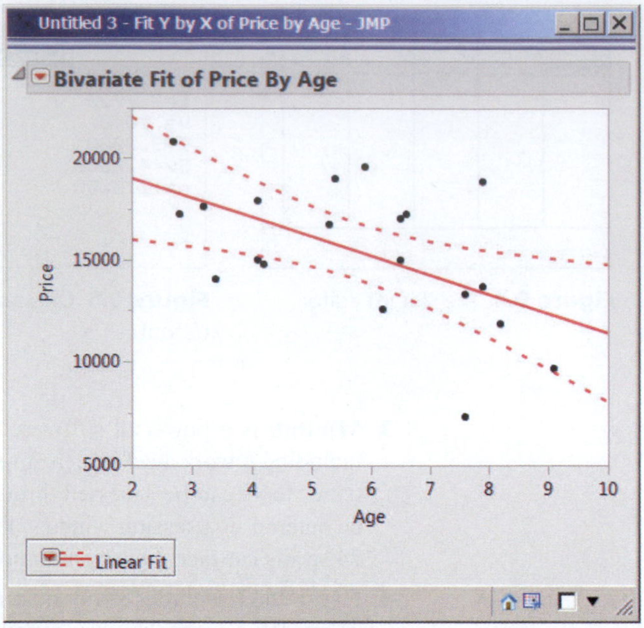

Figure 0.9 JMP scatter plot, regression line, and confidence bands.

CHAPTER 0 EXERCISES

0.1 Name the four parts of every statistical inference problem.

0.2 Apply the four statistical inference steps to the Triple Six Fix.

0.3 Name the four parts of the Solution Trail.

0.4 The Canary Party recently began the *Not a Coincidence* campaign to highlight women who have been affected by Merck's human papillomavirus (HPV) vaccine, Gardasil.[3] As of November 2013, a report states that there have been 31,741 adverse events, 10,849 hospitalizations, and 144 deaths due to HPV vaccines. Explain why The Canary Party believes that there must be something wrong with the vaccine.

0.5 It had been very rare for an NBA player to suffer a major knee injury while on the court. Derrick Rose tore an anterior cruciate ligament (ACL) in his left knee in 2012. Rose was the first player to suffer an ACL tear since Danny Manning in 1995 and Bernard King in 1985. Since the injury to Derrick Rose, at least six NBA players have experienced similar injuries—torn ACLs. State two possible explanations for this rare rash of injuries. Which explanation do you think is more plausible? Why?

0.6 In the movie, *Wall Street*, corporate raider Gordon Gekko and his partner Bud Fox made a lot of money trading stocks. However, several of the trades attracted the attention of the Securities and Exchange Commission (SEC). Why do you think the SEC believed Gordon and Bud may have had inside information or manipulated the price of certain stocks?

0.7 What do you think it means when a weatherperson says, "There is a 50% chance of rain today." Contact a weatherperson and ask him or her what this statement means. Does this explanation agree with yours?

0.8 James Bozeman of Orlando won the Florida lotto twice. He beat the odds of 1 in 22,957,480 twice to win a total of $13 million. State two possible explanations for this occurrence. Which explanation do you think is more reasonable? Why?

0.9 In January 2014, 33 whales died off the coast of Florida. Twenty-five were found on Kice Island in Collier County. Blair Mase, a marine mammal scientist with the National Oceanic and Atmospheric Administration (NOAA) indicated that NOAA was carefully investigating these deaths.[4] Explain why NOAA believes the whales did not die as a result of natural causes and is investigating the deaths.

0.10 In January 2014, 62 people became sick after dining at one of two restaurants that share a kitchen in Muskegon County, Michigan.[5] The illnesses occurred over a four-day period, and county health officials began an immediate investigation.
  a. Explain why officials investigated the source of these illnesses.
  b. Apply the four statistical inference steps to this situation.

0.11 In 2009 and 2010, Toyota issued a costly recall of over 9 million vehicles because of possibly out-of-control gas pedals. There had been at least 60 reported cases of runaway vehicles, some of which resulted in at least one death.[6]

    a. State two possible reasons for this observed high number of runaway vehicles.

    b. Why do you think Toyota issued this recall?

0.12 Suppose there were 15 home burglaries in a small town during the entire year. None occurred on a Thursday. Do you think there is evidence to suggest that something very unusual is happening on Thursdays in this town to prevent burglaries on this day of the week? Why or why not?

0.13 Recently, the Sedgwick County Health Department reported at least 27 cases of whooping cough in one month. This observed count was more than in any month in the previous five years. Do you think health officials should be concerned about this outbreak of whooping cough? Why or why not?

0.14 To understand the definitions and formulas in this text, you will need to feel comfortable with mathematical notation. To review and prepare for the notation we will use, make sure you are familiar with the following:

    a. Subscript notation—for example, $x_1, x_2, \ldots$

    b. Summation notation—for example, $\sum_{i=1}^{n} x_i$

    c. The definition of a function.

1 An Introduction to Statistics and Statistical Inference

Is it safe to eat rice?

Arsenic is a naturally occurring element that is found mainly in the Earth's crust. Some people are exposed to high levels of arsenic in their jobs, or near hazardous waste sites, or in some areas of the country in which there are high levels of arsenic in the surrounding soil, rocks, or even water. Exposure to small amounts of arsenic can cause skin discoloration, and long-term exposure has been associated with higher rates of some forms of cancer. Excessive exposure can cause death.

In 2012, the U.S. Food and Drug Administration (FDA) and *Consumer Reports* announced test results that revealed many brands of rice contain more arsenic in a single serving than is allowed by the Environmental Protection Agency (EPA) in a quart of drinking water.[1] Trace amounts of arsenic may also be found in flour, juices, and even beer. Earlier in that year, a study conducted at Dartmouth College detected arsenic in cereal bars and infant formula.

The FDA has established a safe level of arsenic in drinking water, 10 parts per billion (ppb). However, there is no equivalent safe maximum level for food. Suppose the FDA is conducting an extensive study to determine whether to issue any warnings about rice consumption. One hundred random samples of rice are obtained, and each is carefully measured for arsenic.

The methods presented in this chapter will enable us to identify the population of interest and the sample, and to understand the definition and importance of a random sample. Most important, we will characterize the deductive process used when an an extraordinary event is observed and cannot be attributed to luck.

CONTENTS

1.1   Statistics Today

Statistics data are everywhere: in newspapers, magazines, the Internet, the evening weather forecast, medical studies, and even sports reports. They are used to describe typical values and variability, and to make decisions that affect every one of us. It is important to be able to read and understand statistical summaries and arguments with a critical eye. This chapter presents the basic elements of every statistics problem—a population and a sample and their connection to probability and statistics. Two common methods for data collection, observational sampling and experimentation, are also introduced.

Statistics data are used by professionals in many different disciplines. Actuaries are probably the biggest users of statistics. They conduct statistical analyses, assess risk, and estimate financial outcomes. An actuary helped compute your last automobile insurance bill.

Statistical analyses are used in a variety of settings. The National Agricultural Statistics Service publishes statistics on food production and supply, prices, farm labor, and even the price of land. Pollsters use statistical methods to predict a candidate's chances of winning an election. Using complex statistical analyses, companies make decisions about new products.

Traditional statistical techniques and new sophisticated methods are used every day in making decisions that affect our lives directly. Pharmaceutical companies use a battery of standard statistical tests to determine a new drug's efficacy and possible side effects. Data mining, a combination of computer science and statistics, is a new technique used for constructing theoretical models and detecting patterns. This technique is used by many companies to understand customers better and to respond quickly to their needs. Predictive microbiology is used to ensure that our food is not contaminated and is safe to consume. Given certain food properties and environmental parameters, a mathematical model is used to predict safety and shelf life.

Statistics is the science of collecting and interpreting data, and drawing logical conclusions from available information to solve real-world problems. This text presents several numerical and graphical procedures for organizing and summarizing data. The constant theme throughout the course, however, is statistical inference using a four-step approach: claim, experiment, likelihood, and conclusion.

Here are some examples of statistics in the news.

1. **Statistical inference:** As reported in the *Archives of Internal Medicine*,[2] researchers discovered a decline in the incidence of heart attacks in one Minnesota county following the implementation of new smoke-free workplace laws. The incidence of heart attacks decreased by 33%, from 150.8 to 100.7 per 100,000 people. The authors concluded that second-hand smoke affects the cardiovascular system nearly as much as active smoking.

2. **Summary statistics:** In July 2012, *Time Newsfeed* reported that the *average* Canadian is now richer that the *average* American. The story summarized an article published in the Toronto-based *Globe and Mail* and concluded that the *average* net worth of a Canadian household was approximately $40,000 more that a typical American household. The *average* net worths in this report are summary statistics that suggest the *middle*, or *central tendency*, of a data set.

3. **Probability and odds:** Every year, approximately 1 in 5 Americans gets the flu. WebMD suggests that the best way to prevent the flu is to get a flu shot, or influenza vaccine. Some people have a higher risk of getting the flu. For example, children and infants, pregnant women, and seniors have a greater chance of getting the flu. Individuals with disabilities are at higher risk because of their lack of mobility. People with certain health conditions have a greater chance of getting the flu, and people traveling to certain areas of the world may be subject to a higher probability of getting the flu. A solid background in probability is necessary to understand statistical inference.

4. **Likelihood and inference:** Recent research suggests that there is a link between the incidence of lung cancer and cancer-causing chemical pollutants near large industrial facilities, especially oil refineries.[3] The incidence of lung cancer in oil refinery counties was higher than in non–oil refinery counties. The chance of this happening was so small that the researchers concluded that oil refinery products significantly affect lung carcinogenesis.

5. **Relative frequency and probability:** According to Oddee, some vending machines mysteriously fall over and crush 13 people per year.[4] If all 312 million Americans are equally likely to be killed by a vending machine, then the probability that a randomly selected individual will be crushed by a vending machine during a particular year is $0.00000004167 = 13/312,000,000$. The relative frequency of occurrence is a good estimate of probability and is often used to develop statistical models and make predictions.

There has been an explosion of numerical information, in stories like those above, in business, in consumer reports, and even in casual conversation. Interpretation of graphs and evaluation of statistical arguments are no longer reserved for academics and researchers. It is essential for all of us to be able to understand arguments based on acquired data. This numerical, or quantitative, literacy is a vital life-long tool.

No matter how you are employed or where you live, you will have to make decisions based on available information or data. Here are some questions you may have to consider.

1. Do you have enough information (data) to make a confident decision? How were the data obtained? If more information is necessary, how will these data be gathered?

2. How are the data summarized? Are the graphical and/or numerical techniques appropriate? Does the summary represent the data accurately?

3. What is the appropriate statistical technique for analyzing the data? Are the conclusions reasonable and reliable?

WARNING

DO NOT TIP OR ROCK THIS VENDING MACHINE

TIPPING OR ROCKING MAY CAUSE SERIOUS INJURY OR DEATH

ANTI-THEFT DEVICE PREVENTS OBTAINING FREE PRODUCT

Stephen Finn/Shutterstock

1.2 Populations, Samples, Probability, and Statistics

There are two very general applications of statistics: *descriptive statistics* and *inferential statistics*. Descriptive statistics involve summarizing and organizing the given information, graphically and/or numerically. The focus of this text is statistical inference. The procedures of inferential statistics allow us to use the given data to draw conclusions and assess risk.

Definition

Descriptive statistics: Graphical and numerical methods used to describe, organize, and summarize data.

Here's a dictionary definition for inference: a deduction or logical conclusion.

Inferential statistics: Techniques and methods used to analyze a *small*, specific set of data so as to draw a conclusion about a large, more general collection of data.

Example 1.1 Mishandled Baggage

The U.S. Department of Transportation publishes information concerning automobiles, public transportation, railroads, and waterways. Much of this information can be summarized or organized—in tables or charts, with a variety of graphs, and numerically—to describe typical values and variability. These summary *descriptive* procedures might be used to indicate preference for a certain airline or to promote safety records. Figure 1.1 shows a bar graph of the total mishandled baggage reports for certain airlines in July 2012.

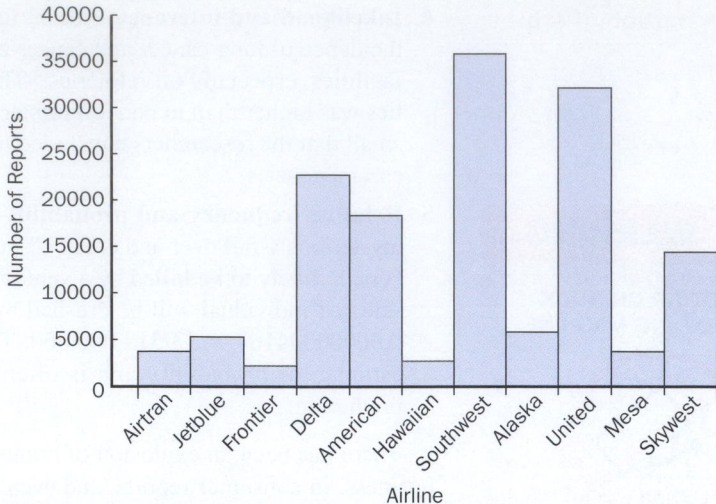

Figure 1.1 Mishandled baggage reports.   (*Source:* Office of Aviation Enforcement and Proceedings, U.S. Department of Transportation.)

Example 1.2 Summary Statistics

The U.S. Census Bureau maintains a huge database of people and households, business and industry, geography, and other special topics. This information may be neatly organized using tables, bar charts, pie charts, histograms, or stem-and-leaf plots, or summarized numerically using the mean, median, quartiles, percentiles, variance, or standard deviation. These simple *descriptive* statistics reveal characteristics of the entire data set. Part of a table is shown in Table 1.1.   ■

Example 1.3 Hand Sanitizers that Kill Germs

A department manager for Walgreens claims that Germ-X HandiSani hand sanitizer kills 99.99% of common harmful germs and bacteria in as little as 15 seconds. For a hand sanitizer to be effective, the alcohol concentration must be at least 60%. To check the germ-killing claim, several randomly selected bottles of the hand sanitizer are obtained and the alcohol concentration in each is carefully measured. These concentrations are used to determine whether there is any evidence to suggest sanitizer is ineffective. The collected data are used in *inferential* statistics to draw a conclusion regarding a claim.   ■

Example 1.4 Magnetic Fuel Savers

Many companies sell magnetic fuel savers for stoves, which are designed to condition liquefied propane gas (LPG) prior to combustion to increase power output, reduce emissions, and save gas. An independent agency tests these devices by recording the amount of gas necessary to boil a specific volume of water. Each test boil is classified by stove brand, shape and thickness of the pot, and burner size. The data collected are used to determine whether there is a difference in efficiency. If there is a difference in the amount of LPG gas used, further *inferential* statistical techniques will be used to isolate this difference. It may be due to the stove brand, type of pot, or burner size.   ■

TRY IT NOW   GO TO EXERCISE 1.5

Whether we are summarizing data or making an inference, every statistics problem involves a *population* and a *sample*. Consider the definitions on the page that follows.

Table 1.1 A portion of table summarizing health and safety characteristics in owner-occupied units (national)

Characteristics	Total	Region			
		Northeast	Midwest	South	West
Total	76,091	13,480	18,032	29,119	15,460
Safety equipment					
Smoke detectors:					
Working smoke detector	70,801	12,893	17,015	26,476	14,417
Powered by:					
Electricity	4,506	853	954	1,791	908
Batteries	40,094	7,553	10,372	14,221	7,949
Both	25,763	4,409	5,622	10,253	5,479
Not reported	438	78	68	211	81
Carbon monoxide detectors:					
Working carbon monoxide detector	35,215	9,311	10,832	9,129	5,944
Powered by:					
Electricity	7,248	1,764	2,811	1,647	1,026
Batteries	15,895	4,467	4,533	4,205	2,689
Both	11,810	2,998	3,416	3,205	2,191
Not reported	262	80	72	72	38
Mold					
Housing units with mold in last 12 months	2,015	527	524	611	353
Kitchen	213	28	49	88	49
Bathroom(s)	683	99	180	265	139
Bedroom(s)	378	78	69	137	94
Living room	219	45	52	86	35
Basement	611	276	216	94	26
Other room	277	75	65	76	60
Mold not present	72,817	12,762	17,231	27,953	14,870
Not reported	1,259	191	277	554	238
Musty smells					
Housing units with musty smells in last 12 months	11,238	2,377	2,999	4,000	1,861
Daily	772	233	216	202	121
Weekly	5,235	808	1,083	2,316	1,028
Monthly	354	94	112	85	63
A few times	4,877	1,242	1,589	1,397	649
Musty smells not present	63,563	10,912	14,760	24,544	13,346
Not reported	1,291	191	273	574	253

All numbers are in thousands.
Source: U.S. Census Bureau.

Definition

A **population** is the entire collection of individuals or objects to be considered or studied.

A **sample** is a subset of the entire population, a small selection of individuals or objects taken from the entire collection.

A **variable** is a characteristic of an individual or object in a population of interest.

A CLOSER LOOK

1. A population consists of all objects of a particular type. There are usually infinitely many objects in a population, or at least so many that we cannot look at all of them.

2. A sample is simply a (usually) small part of a population.

3. A variable may be a *qualitative* (categorical) or a *quantitative* (numerical) attribute of each individual in a population.

The **Solution Trail** is a technique and visual aid for problem solving (illustrated in the next example). It is a guide to help us plan how to solve a problem. Look at the Solution Trail before you read the steps of the solution. Start this hike by identifying keywords and phrases. The four steps to solving each problem are

1. Find the **keywords**.

2. Correctly **translate** these words into statistics.

3. Determine the applicable **concepts**.

4. Develop a **vision** for the solution.

The keywords lead to a translation into statistics. The statistics question is solved using specific concepts. The keywords, translation, and concepts are all used to develop a vision for the solution. This technique is not applicable or necessary in every problem. It is most appropriate for probability through hypothesis testing, the foundation of most introductory statistics courses.

The following examples illustrate the relationships among populations, samples, and variables.

Solution Trail 1.5

KEYWORDS
- Magnesium level
- Slice of whole-grain bread
- One hundred slices

TRANSLATION
- Characteristic of each slice
- All slices of whole-grain bread
- Subset of all slices

CONCEPTS
- Variable
- Population
- Sample

VISION
Determine the set of all objects of interest, the subset, and the attribute to be measured.

Example 1.5 High Anxiety

Various research studies suggest that whole-grain foods may be a natural help for those people who suffer from anxiety, or maybe even statistics anxiety! Whole grains generally contain high levels of magnesium, and magnesium deficiency can lead to anxiety.[5] A new study is concerned with the magnesium level in a slice of whole-grain bread. One hundred slices of whole grain bread are selected at random from various markets, and the magnesium in each is carefully measured and recorded. Describe the population, sample, and variable in this problem.

SOLUTION

STEP 1 The population consists of all slices of whole-grain bread in the entire world. Although this population is not infinite, we certainly could not examine every single slice.

STEP 2 The sample is the 100 slices selected at random. This is a subset of or selection from the population.

STEP 3 The variable in this problem is the magnesium level. This characteristic will be carefully measured for each slice, and the data will be summarized or used to draw a conclusion.

Example 1.6 Asleep at the Wheel

The Centers for Disease Control and Prevention released results from a study that indicated approximately 4% of all adults in the United States said they had fallen asleep at least once while driving in the last month. Nodding off while driving seems to be more common in men, and some officials claim the percentage of all U.S. adults who have fallen asleep while driving is greater than 4%. To check this claim, 10 adult drivers were selected from across the country. Each person was asked if he or she had fallen asleep while driving, and the results were recorded. Describe the population, sample, and variable in this problem.

SOLUTION

STEP 1 The population consists of all adult drivers in the United States. This population is not infinite, but it is so large that it would be impossible to contact every adult driver.

STEP 2 The sample consists of the 10 adult drivers selected.

STEP 3 The variable in the problem is whether the driver has fallen asleep at the wheel, a yes/no response.

TRY IT NOW   GO TO EXERCISE 1.7

A CLOSER LOOK

Example 1.6 raises some important issues regarding the sample of 10 adult drivers.

1. How large a sample is necessary for us to be confident in our conclusion? Ten adult drivers may not seem like enough. But how many do we need? 100? 1000? We will consider the problem of sample size in Chapter 7 and beyond.

2. This problem does not say how the sample was obtained. Perhaps the first 10 drivers who recently renewed their licenses were selected. Or maybe only those from one state were included. To draw a valid conclusion, we need to be certain the sample is *representative* of the entire population. The formal definition of a representative sample is presented in Section 1.3.

Statistical inference is based on, and follows from, basic probability concepts. *Probability* and inferential *statistics* are both related to a population and a sample, but from different perspectives. For the rest of this chapter, *statistics* really means *inferential statistics*.

Definition

To solve a **probability** problem, certain characteristics of a population are assumed to be known. We then answer questions concerning a sample from that population.

In a **statistics** problem, we assume very little about a population. We use the information about a sample to answer questions concerning the population.

Figure 1.2 illustrates this definition. Picture an entire population of individuals or objects. Suppose we know *everything* about the population and we select a sample from this population. A probability problem would involve answering a question concerning the sample. In a typical statistics problem, we assume very little about the population. We select a sample and analyze it completely. We use this information to draw a conclusion about the population.

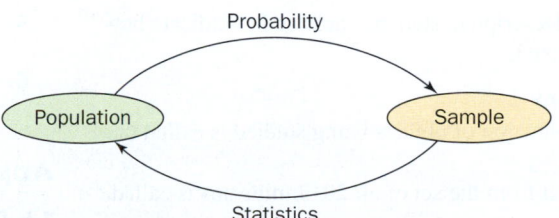

Figure 1.2 Relationships among probability, statistics, population, and sample.

In Figure 1.2, it may seem like we can start our study anywhere in this circular diagram. However, we need to understand probability before we can learn statistics. A solid background in probability is necessary before we can actually do statistical inference.

Example 1.7 Most-Watched Television Finales

According to Koldcast Entertainment Media, the final episode of M*A*S*H, with almost 106 million viewers, is still the most-watched television finale ever.[6] Other top television finales include *Cheers*, *Seinfeld*, *Breaking Bad*, and *Friends*. Consider the population consisting of all television viewers and a sample of 20 from this population.

Population: All television viewers at the time M*A*S*H was aired

Sample: The 20 television viewers from this population

Here is a probability question. The final episode of M*A*S*H was watched by 47% of all television viewers. What is the probability that 10 or more (of the 20) selected viewers watched this episode? We know something about the population, and try to answer a question about the sample.

Here is a statistics question. Suppose we interview the 20 viewers in the sample and find that 9 of the 20 watched the final episode of M*A*S*H. What can we conclude about the percentage of *all* television viewers who watched the final episode of M*A*S*H? We know about the sample and try to answer a question about the (whole or general) population.   ●

Example 1.8 Trouble Falling Asleep

A recent study of sleep habits by Statistics Canada indicated that 35% of those women surveyed had difficulty falling asleep and staying asleep, whereas only 25% of men experienced the same troubles. The study also showed that Canadian women tend to sleep longer than Canadian men.[7] Consider the population consisting of all Canadian women and a sample of 100 from this population.

A probability question: Suppose 35% of all Canadian women have difficulty falling asleep. What is the probability that at most 30 (of the 100) women in the sample have difficulty falling asleep?

A statistics question: All 100 women selected are asked to complete an extensive questionnaire. The information indicates that 45 of the 100 have difficulty falling asleep. What does this suggest about the proportion of all Canadian women who have difficulty falling asleep?   ●

TRY IT NOW   GO TO EXERCISE 1.12

SECTION 1.2 EXERCISES

Concept Check

1.1 True/False   Inferential statistics are used to draw a conclusion about a population.

1.2 True/False   Descriptive statistics are used to indicate how the data were collected.

1.3 Fill in the Blank
  a. The entire collection of objects being studied is called the _____.
  b. A small subset from the set of all 2013 minivans is called a _____.
  c. Consider the amount of sugar in breakfast cereals. This characteristic of breakfast cereal (objects) is called a _____.

Practice

1.4 Probability/Statistics   In each of the following problems, write a probability and a statistics question associated with the given information.

  a. It has been reported that 62% of all people use a social media.
  b. Only 37% of all people in the United States are eligible to donate blood.
  c. Fifty-two percent of all old washing machines are front-loading.
  d. Forty percent of the people in the Washington, DC, area travel over the holiday season.

Applications

1.5 Descriptive or Inferential Statistics   Determine whether each of the following is a descriptive or an inferential statistics problem.

  a. The Nebraska Department of Transportation maintains records concerning all trucks stopped for inspection. A report of these inspections lists the proportion of all trucks stopped, by cargo.
  b. Eric Knudsen, a researcher at Stanford University Medical Center, obtains a random sample of wild owls and measures how far each can turn its neck. The data are used

to conclude that an owl can turn its neck more than 120 degrees from the forward position.

c. A Navy research facility runs several tests to check the structural integrity of a new submarine. A laboratory report states the vessel can withstand pressure at depths of at most 800 ft.

d. A safety inspector in Atlanta selects a sample of apartment buildings and checks the fire ladders on each. The proportion of broken ladders in the sample is used to estimate the proportion of broken fire ladders in the entire city.

e. Like most states, a large portion of the New York State budget is spent on health care, pensions, and education. The pie chart in Figure 1.3 shows the percentage of the budget spent on each item for Fiscal Year 2013.[8]

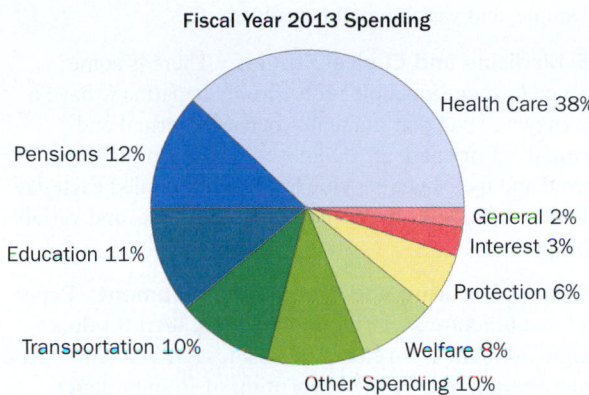

Fiscal Year 2013 Spending

Figure 1.3 Pie chart illustrating New York State Fiscal Year 2013 spending.

f. A report from the Louisiana Department of Agriculture and Forestry lists the prices paid for raw forest products at the first point of sale.

1.6 Descriptive or Inferential Statistics Determine whether each of the following is a descriptive or an inferential statistics problem.

a. The bar chart in Figure 1.4 shows the number of bankruptcies for certain industries in Quebec during a recent year.[9]

b. The resting heart rate was measured for adult males from two separate groups: those who exercise at least three days per week, and those who do not exercise regularly. The resulting data are used to suggest that regular exercise decreases the resting heart rate in adult males.

c. Interior Exterior Remodeling, in Northridge, California, maintains a comprehensive list of each home constructed by type, size, exterior color, etc.

d. Researchers at the Center for Food Safety selected a sample of frozen toaster apple strudel sold in grocery stores. Measurements indicated the producer was baking each piece of strudel with less apple than advertised on the box.

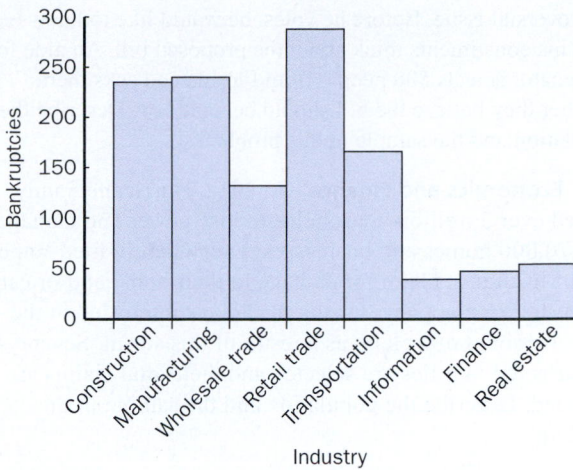

Figure 1.4 Total bankruptcies in Quebec.

e. A report issued by the athletic department at Brigham Young University listed each item in a trainer's bag and the number of times each was used.

f. The manager at the Bear Pause Theater in Hackensack, Minnesota, surveyed patrons and summarized opinions associated with seating comfort, movie sound, and snacks.

1.7 Medicine and Clinical Studies Managers at Cedarcrest Hospital in Newington, Connecticut, are interested in the length of stay (in days) of patients admitted for open-heart surgery. Hospital managers have decided to limit their investigation to open-heart patients who were operated on within the last year. Thirty open-heart surgery patients admitted to the hospital within the last year are selected. What is the population of interest, the sample, and the variable in this problem? Write a Solution Trail for this problem.

1.8 Marketing and Consumer Behavior T-shirt labels irritate many people's skin, so Calico Graphics of Wolfeboro, New Hampshire, would like to produce shirts without a label. The company wants to know whether there is an advantage to producing this type of T-shirt. Fifty people are surveyed about whether they cut the tags off their T-shirts. What is the population of interest, the sample, and the variable in this problem? Write a Solution Trail for this problem.

1.9 Psychology and Human Behavior Managers at Citigroup, Inc., in New York, are concerned about the number of employees who eat and/or drink at their desks while working. Some managers believe this is an unnecessary distraction, and spills can cause computer failures and ruin documents. Thirty-five employees are selected, and each is questioned about eating/drinking while working. Describe the population and the sample in this problem.

1.10 Public Policy and Political Science Senator Marco Rubio of Florida is unsure of his vote on an emotional and controversial issue. Before he votes, he would like to know what his constituents think about the proposed bill. An aide for the senator selects 500 people from Florida and asks them whether they believe the bill should become law. Describe the population and the sample in this problem.

1.11 Economics and Finance   In 2012, Hurricane Sandy caused over 2 million households to lose power and damaged over 70,000 homes and businesses. Every family filed some sort of insurance claim for damage to their home and/or car. An insurance company serving the area is interested in the typical amount of a claim as a result of this storm. Seventy-five affected families are selected and their total claims are recorded. Describe the population and the sample in this problem.

1.12 Probability or Statistics   In each of the following problems, identify the population and the sample, and determine whether the question involves probability or statistics.

a. Seventy-five percent of all people who buy a dining room table purchase matching chairs. Five people who purchased a dining room table within the last month are selected at random. What is the probability that all five purchased matching chairs?

b. Twenty-five people entering a rest area and food court on Highway 59 near Houston are selected at random. Of these 25 people, 20 purchased food from at least one of the eateries. Estimate the true proportion of people stopping at this rest area who purchase food.

c. Historical records indicate 1 out of every 500 people using a particularly steep water slide suffer some kind of injury. Fifty people using the slide are selected at random. How many do you expect to be injured?

d. A building inspector in Henderson, Nevada, is checking public buildings with doors that open automatically. One hundred doors are randomly selected. Careful inspection reveals that 12 doors are broken. Use this information to estimate the percentage of automatic doors in Henderson that are broken.

e. One thousand people entering Los Angeles International Airport (LAX) are selected at random. Each person is asked to complete a short survey regarding travel. The survey results show 637 carry a frequent-flier card. Is there evidence to suggest the true proportion of travelers entering LAX who carry frequent flier cards is greater than 0.60?

f. The Risdall Advertising Agency reports 65% of all women have purchased perfume within the last three months. Thirty-four women are selected at random. Is it likely more than 20 of these women purchased perfume within the last three months?

g. Representatives from the Occupational Safety and Health Administration inspected several for-profit and Medicare nursing homes for any violations. The resulting data will be used to determine whether there is any evidence to suggest the quality of treatment is different in the two types of nursing homes.

1.13 Psychology and Human Behavior   During each summer, many families spend part of their vacation time at a beach along the East or West Coast. Due to the popularity of movies like *Jaws*, and recent shark attacks on surfers, swimmers, snorkelers, and spearfishermen, Americans have become increasingly concerned about water activities. Research suggests that 46% of all shark attacks are on divers.[10] One thousand records of shark attacks are selected, and each is categorized by victim group.

a. What is the population of interest?
b. What is the sample?
c. Describe the variable of interest.

1.14 Manufacturing and Product Development   Spray-on tans, or fake tans, contain several chemicals that have been linked to allergies, diabetes, and obesity.[11] Twenty fake tan products are selected and the amount of dihydroxyacetone (the active ingredient) in each is measured. Describe the population, the sample, and variable in this problem.

1.15 Medicine and Clinical Studies   There is some evidence to suggest people with chronic hepatitis C have a liver enzyme level that fluctuates between normal and abnormal.[12] Fifty patients diagnosed with hepatitis C are selected and their liver enzyme levels are recorded each day for one month. Describe the population, sample, and variable in this problem.

1.16 Manufacturing and Product Development   Paper towel manufacturers constantly advertise their products' strength, amount of stretch, and softness. A consumer group is interested in testing the absorption of Bounty paper towels. Thirty-five rolls are selected, and the amount of absorption for a single paper towel from each roll is recorded.

a. What is the population of interest?
b. What is the sample?
c. Describe the variable of interest.

Extended Applications

1.17 Manufacturing and Product Development   While much of the cheddar cheese consumed around the world is processed, some is still produced in the traditional manner: made in small batches, wrapped in cloth to breathe, and allowed to age. Most traditional cheddar is aged one to two years; like fine wines, older cheddars assume their own character and flavor. Suppose 75% of all cheddars are aged less than two years, and a sample of 20 cheddar cheeses from around the world is obtained.

a. Describe the population and the sample in this problem.
b. Write a probability question and a statistics question involving this population and sample.

1.18 Marketing and Consumer Behavior   Magazines, newspapers, and books have become more readily available in digital format. In addition, the quality of readers, for example, the Kindle, Nook, and iPad, has increased. A recent study suggests that 21% of adults read an ebook within the

past year.[13] Suppose a sample of 500 adults in the United States is obtained.
  a. Describe the population and the sample in this problem.
  b. Write a probability question and a statistics question involving this population and sample.

1.19 Business and Management One of the main reasons that U.S. companies shift jobs overseas is labor costs. Although the compensation gap between the United States and China has decreased recently, the tax code still rewards companies for making certain investments overseas. A recent study suggests that 4% of large companies have plans to relocate jobs back to the United States.[14] Seventy-five large companies are selected, and each is surveyed to determine if it plans to move jobs back to the United States.
  a. What is the population of interest?
  b. What is the sample?
  c. Describe the variable of interest.
  d. Write a probability question and a statistics question involving this population and sample.

1.3 Experiments and Random Samples

Statisticians analyze data from two types of experiments: observational studies and experimental studies. The definitions are given below.

Definition

In an **observational study**, we observe the response for a specific variable for each individual or object.

In an **experimental study**, we investigate the effects of certain conditions on individuals or objects in the sample.

The collected data in an observational study may be summarized in a variety of ways, or used to draw a conclusion about the entire population. The following is an example of an observational study.

Example 1.9 Is There Time for Breakfast?

Lorraine GaNun, a guidance counselor at Rice School in Marlton, New Jersey, is interested in the amount of time each student spends in the morning eating breakfast. Some students wake up an hour before the bus arrives, have a leisurely breakfast, read the newspaper comics, and complete last-minute homework. Others roll out of bed and onto the school bus. Mrs. GaNun decides to measure the amount of time from wake-up to school bus arrival. A random sample of students is selected, and each is asked for the amount of school-day preparation time. The data are summarized graphically and numerically in this observational study.

In almost all statistical applications, it is important for the data to be *representative* of the relevant population. A representative sample has characteristics similar to those of the entire population, and therefore can be used to draw a conclusion about the (general) population. The following definition describes a method for obtaining data in an *observational study* to ensure the resulting sample is representative of the corresponding population.

Definition

A (**simple**) **random sample** (SRS) of size n is a sample selected in such a way that every possible sample of size n has the same chance of being selected.

How can we be absolutely certain *every* possible sample of size *n* is equally likely?

A CLOSER LOOK

1. In practice, a random sample may be very difficult to achieve. Statisticians employ various techniques, including random number tables and random number generators, to select a random sample.

2. If a sample is not random, then it is *biased*. There are many different kinds of bias and factors that contribute to a biased sample.

3. *Nonresponse bias* is very common when data are collected using surveys. The majority of people who receive a survey in the mail simply discard it. The original collection of people receiving the survey may be random, but the final sample of completed surveys is not. Because the sample is biased, it is impossible to draw a valid conclusion.

4. *Self-selection bias* occurs when the individuals (or objects) choose to be included in the sample, as opposed to being selected. For example, a television news program may ask viewers to respond to a yes/no question by dialing one of two phone numbers to cast their vote. Viewers *choose* to participate, and usually those with strong opinions (either way) vote. There are many more who did not have the opportunity to respond— every single sample is *not* equally likely. Certainly this sample is biased, and hence no valid conclusion is possible.

5. If the population is infinite, then the number of simple random samples is also infinite. For finite populations, the formula for the number of possible random samples is presented in Chapter 7.

6. A simple random sample is vital for sound statistical practice. Before doing any analysis, you should always ask how the data were obtained. If there is any evidence of a pattern in selection, if the observations are associated or linked in some way, or if there is some connection among the observations, then the sample is not random. There is simply no way to transform *bad* data into good statistics.

Example 1.10 Town Facility Master Plan

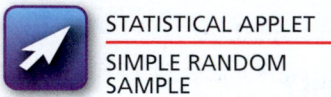

STATISTICAL APPLET

SIMPLE RANDOM SAMPLE

Mark Traeger, a member of the Sandown, New Hampshire, planning board, is conducting a survey of residents concerning proposed changes in facilities over the next 10 years. He plans to choose 100 residents from the total town population of 5143 and will ask each selected person to complete a short questionnaire. The results will be summarized and presented at the next town meeting.

Traeger would like a simple random sample of size 100, a representative sample of the entire town population. Here is one basic selection procedure. Write each person's name on a piece of paper and place all of them in a hat. Thoroughly mix the papers and then select 100 names.

Although this procedure is clear-cut and uncomplicated, it can be very tedious if the number of individuals or objects in the population is large. In addition, it is hard to guarantee a thorough mixing of the slips of paper.

More practical methods for selecting a simple random sample include the use of a random number table or a random number generator (available in most statistical software packages). In this example, we might assign each resident a number, from 1 to 5143, and use a random number generator to produce a list of 100 numbers in this range. The residents associated with these 100 numbers would comprise the random sample.

TRY IT NOW   GO TO EXERCISE 1.29

Researchers often investigate the effects of certain conditions on individuals or objects. The data obtained are from an *experimental study*. Individuals are randomly assigned to specific groups, and certain factors are systematically controlled, or imposed, in order to investigate and isolate specific effects. The following example is of an experimental study.

Example 1.11 To Fertilize or Not to Fertilize

The manager of Gardener's Supply Company claims that a new organic fertilizer, in comparison with the leading brand, increases the yield and size of tomatoes. To test this claim, tomato plants are randomly assigned to one of two groups. One group is grown using the leading fertilizer and the other is cultivated using the new product. At harvest time the size and weight of each tomato is recorded, along with the total yield per plant. The collected data from this experiment are used to compare the two fertilizers.

TRY IT NOW   GO TO EXERCISE 1.28

In an experimental study, researchers must be careful to ensure that significant effects are indeed due to an imposed *treatment*, or controlled factor. *Confounding* occurs when several factors together contribute to an effect, but no single cause can be isolated. Suppose the tomato plants in one of the groups in Example 1.11 are watered more and/or exposed to more sunlight and warmer temperatures. If the tomato plants that received the new fertilizer were subject to these different (favorable) growing conditions, a difference in yield cannot be attributed to the new product.

The focus of this text is statistical inference, most of which is based on determining the likelihood of an observed experimental outcome. This strategy will be used informally in the early chapters of this book. Formal procedures will be presented beginning in Chapter 9. For now, we will follow the four-step process presented below.

Statistical Inference Procedure

The process of checking a claim can be divided into four parts.

Claim: This is a statement of what we assume to be true.

Experiment: To check the claim, we conduct a relevant experiment.

Likelihood: This considers the likelihood of occurrence of the observed experimental outcome assuming the claim is true. We will use many techniques to determine whether the experimental outcome is a reasonable observation (subject to reasonable variability), or whether it is a rare occurrence.

Conclusion: There are only two possible conclusions. (1) If the outcome is reasonable, then we cannot doubt the claim. We usually write, "There is no evidence to suggest the claim is false." (2) If the outcome is rare, we disregard the lucky alternative, and question the claim. A rare outcome is a contradiction. It shouldn't happen (often) if the claim is true. In this case we write, "There is evidence to suggest the claim is false."

Example 1.12 Cell Phone Chargers

The Wireless Emporium ships a box containing 1000 cell phone chargers and claims 999 are in perfect condition and only 1 is defective. Upon receipt of the shipment, a quality control inspector reaches into the box, mixes the chargers around a bit, selects one at random, and it's defective!

Claim: There were 999 good cell phone chargers and 1 defective charger in the box.

Experiment: The quality control inspector selected one cell phone charger from the box, tested it, and found it to be defective.

Likelihood: One of two things has happened.

1. The quality control inspector could be incredibly lucky. Intuitively, the chance of selecting the one defective charger from among the 1000 total chargers is very small. It is *possible* to select the one defective charger, but it is very *unlikely*.

2. The claim (999 perfect chargers, 1 defective) is false. Because the chance of selecting the single defective charger is so small, it is more likely the manufacturer (Wireless Emporium) lied about the number of defective chargers in the shipment. (Perhaps there are really 999 *defective* chargers and only one good charger in the box.)

We have found evidence the claim is false by showing that the observed experimental outcome is unreasonable, an outcome so rare that it should almost never happen if the claim is really true.

Conclusion: Typically, statistical inference discounts the lucky alternative. Selecting the single defective charger is an extremely rare occurrence. Therefore, there is evidence to suggest the manufacturer's claim is false, because this outcome is very rare. ■

We will use this four-step process to check a claim in many different contexts. The method for determining likelihood is the key to this valuable tool for logical reasoning.

SECTION 1.3 EXERCISES

Concept Check

1.20 True/False   In an observational study, we record the response for a specific variable for each individual or object.

1.21 True/False   In an experimental study, we investigate the effects of certain conditions on at least three different groups.

1.22 True/False   The number of simple random samples is always infinite.

1.23 True/False   A simple random sample is representative of the entire population of interest.

1.24 Fill in the Blank
   a. If a sample is not random, then it is _____.
   b. It is very common to experience _____ when data is collected using surveys.
   c. _____ occurs when individuals ask to be included in an survey.

1.25 Statistical Inference   Name the four parts of every statistical inference problem.

1.26 Liar, Liar   Suppose an experimental outcome is very rare. What two things could have happened?

Applications

1.27 Fuel Consumption and Cars   The administration at the University of Nebraska in Lincoln is interested in student reaction to a planned parking garage on campus. A dormitory near the proposed site is selected and several Student Senate members volunteer to solicit responses. One Thursday evening, the volunteers each take a specific dorm wing, knock on doors, and record student answers to several prepared questions.
   a. Is this an observational or an experimental study?
   b. Describe the sample in this problem.
   c. Is this a random sample? Justify your answer.

1.28 Demographics and Population Statistics   State Farm Insurance Company would like to estimate the proportion of volunteer firefighters across the country who are full-time teachers. The 25 largest volunteer fire companies in the United States are identified. Each is contacted and asked to complete a short survey regarding the number of volunteers and the occupation of each volunteer.
   a. Is this an observational or an experimental study?
   b. Describe the sample in this problem.
   c. Is this a random sample? Justify your answer.

1.29 Manufacturing and Product Development   The Visniak Bottling Plant in Cheektowaga, New York, has been accused of systematically underfilling 12-oz bottles of soda. An inspection team enters the plant one afternoon and selects bottled soda ready for shipment from various locations within the plant. The contents of each selected bottle are carefully measured.
   a. Describe the population and the sample in this problem.
   b. Is this a random sample? Justify your answer.

1.30 Biology and Environment Science   Oregon Scientific has come under suspicion of purposely shipping defective wireless weather stations. The Attorney General's office in Delaware would like to estimate the proportion of defective products being shipped by this company. Describe a method for obtaining a simple random sample of shipped wireless weather stations.

1.31 Fuel Consumption and Cars The Massachusetts State Police union is interested in the number of miles driven by each officer during an 8-hour shift. Twelve officers are selected from the 11:00 P.M. to 7:00 A.M. shift, and the number of miles traveled by each officer is recorded.
   a. Is this an observational or an experimental study?
   b. Describe the population and the sample in this problem.
   c. Is this a random sample? Justify your answer.

1.32 Manufacturing and Product Development Gillette claims a new disposable razor provides a *closer* shave than any other brand currently on the market. One hundred men who are observed buying a disposable razor are selected and asked to participate in a shaving study.
   a. Describe the population and the sample in this problem.
   b. Is this a random sample? Justify your answer.

1.33 Manufacturing and Product Development Midwest Pet Supplies claims its K9 Chain Link Dog Kennel can be set up in less than 30 minutes. An investigative reporter would like to check this claim. Describe a method for obtaining a simple random sample of customers who set up this kennel.

1.34 Sports and Leisure A National Football League coach is permitted to initiate two challenges to referee calls per game (outside of the final 2 minutes in each half). If both challenges are successful, then the coach is given a third. During a challenge, the referee reviews the play in question on a replay monitor on the field, and the call is either confirmed or the challenge is upheld. The NFL reports the time required to resolve a coach's challenge is less than 5 minutes. A sports statistician would like to check this claim. Describe a method for obtaining a simple random sample of challenges during NFL games.

1.35 Travel and Transportation The Department of Public Works in Bismarck, North Dakota, would like to estimate the number of potholes per mile (after a long, snowy winter). Each selected mile-long stretch will be thoroughly examined for potholes, and the number in each section will be recorded.
   a. Describe a method for obtaining a simple random sample of mile-long road segments.
   b. Is this an observational or experimental study?

1.36 Biology and Environmental Science The Faber Floral Company in Kankakee, Illinois, claims to have developed a special spray for roses that causes the blossom to last longer than an untreated flower. Fifty long-stemmed roses are obtained and randomly assigned to one of two groups: treated versus untreated. The treated roses are sprayed, and the lifetime of each blossom is carefully recorded.
   a. Is this an observational or an experimental study?
   b. What is the variable of interest?
   c. Describe a technique to randomly assign each rose to a group.

1.37 Fuel Consumption and Cars Electric and plug-in electric cars are designed to save gasoline and help the environment. In addition, there are certain tax credits for these types of hybrid automobiles.[15] Although there are certainly benefits to owning a hybrid car, many people complain about the slow acceleration, repair expense, and overall comfort. Thirty-five passengers are randomly selected. Each is blind-folded and taken for a ride in a traditional combustion-engine automobile and in a comparably sized hybrid car (over the same route). The passenger is then asked to select the car with the most comfortable ride.
   a. Is this an observational or an experimental study?
   b. What is the variable of interest?
   c. Describe possible sources of bias in these results.

1.38 Manufacturing and Product Development The ceramic tile used to construct the floors in a mall must be sturdy, easy to clean, and long-lasting. Before installing a specific tile, a construction firm orders a box of 25 tiles and uses a standard strength test on each. The results are used to determine whether the tiles will be used throughout the new mall.
   a. Describe the population and the sample in this problem.
   b. Is this a random sample? If so, justify your answer. If not, describe a technique for obtaining a random sample.

1.39 Manufacturing and Product Development Many comforters contain both white feathers and down in order to provide a warm, soft cover. A bed-and-bath company would like to expand its line of products and sell comforters for queen- and king-size beds. Before manufacturing begins, a random sample of comforters is obtained from other companies and the proportions of white feathers, down, and other components are measured and recorded. These data will be used to determine the exact mixture of feathers and down for the new line of comforters.
   a. Is this an observational or an experimental study?
   b. What are the variables of interest?
   c. Describe a method for obtaining a random sample of comforters from current manufacturers.

1.40 Marketing and Consumer Behavior Disney World is going to initiate the use of wireless tracking wristbands for visitors to the Orlando, Florida, theme park.[16] The new *MagicBand* has several functions: it serves as a hotel room key and park entry pass, and is linked to customer credit card information. Visitors wearing these wristbands will have immediate access to certain rides. Suppose a random sample of Disney World visitors wearing wristbands is obtained and the wait time for Big Thunder Mountain is recorded for each.
   a. Is this an observational or an experimental study?
   b. What are the variables of interest?
   c. Describe a method for obtaining a random sample of visitors wearing wristbands.

CHAPTER 1 SUMMARY

Concept	Page	Notation / Formula / Description
Descriptive statistics	11	Graphical and numerical methods used to describe, organize, and summarize data.
Inferential statistics	11	Techniques and methods used to draw a conclusion or make an inference.
Population	13	The entire collection of individuals or objects to be considered or studied.
Sample	13	A subset of the entire population.
Variable	13	A characteristic of an individual or object in a population of interest.
Probability problem	15	Certain properties of a population are assumed known. Questions involve a sample taken from this population.
Statistics problem	15	Information about a sample is used to answer questions concerning a population.
Observational study	19	We observe the response for a specific variable for each individual or object in the sample.
Experimental study	19	We investigate the effects of certain conditions on individuals or objects in the sample.
Simple random sample of size n	19	A sample selected in such a way that every possible sample of size n has the same chance of being selected.
Statistical inference procedure	21	Four-step process: Claim, Experiment, Likelihood, and Conclusion.

CHAPTER 1 EXERCISES

APPLICATIONS

1.41 Descriptive or Inferential Statistics Determine whether each of the following is a descriptive or an inferential statistics problem.
  a. The Society of Government Economists conducted a salary and working conditions survey of top bank executives in the United States. A report issued by this group included a table that listed the number of bank executives in each state with salaries above $1 million.
  b. The Flowers Canada Growers obtained a sample of people who sent roses for Valentine's Day and recorded the color of the roses purchased. This information was used to construct a table listing the proportion of each color of rose purchased on Valentine's Day.
  c. The Intergovernmental Panel of Climate Change collected data associated with global warming and predicted the extinction of up to 30% of plant and animal species in the world.
  d. American Express conducted a survey of travelers at Los Angeles International Airport. The information was used to estimate the proportion of all travelers who make a purchase in an airport duty-free shop.

1.42 Descriptive or Inferential Statistics Determine whether each of the following is a descriptive or an inferential statistics problem.
  a. A report by NASA listed each weather satellite orbiting the Earth and the number of years each has been in service.

  b. The Agricultural Research Service obtained samples of natural cocoa from a variety of sources and measured the total antioxidant capacity in each sample. The resulting data were used to suggest that eating a moderate amount of chocolate may help prevent cancer, heart disease, and stroke.
  c. The U.S. Patent Office issued a report listing every company that was granted a patent in 2013 and the number of patents awarded to each company.
  d. A researcher at Emory University used brain scans to conclude that zen meditation may help treat disorders characterized by distracting thoughts.

1.43 Descriptive or Inferential Statistics Determine whether each of the following is a descriptive or an inferential statistics problem.
  a. The Food Channel conducted a blind taste test to determine the best chocolate for baking. A random sample of adults was obtained, and each was asked to select the best chocolate from among 10 varieties. The final report listed each chocolate along with the number of people who rated it the best.
  b. After an extensive survey, the Association of Realtors in Chicago concluded that the mean price of a single-family home was less the $500,000.
  c. After conducting several measurements, the Beijing Municipal Environmental Monitoring Center issued a warning that indicated the density of PM2.5 (fine particulate matter, a measure of air pollution) was over the safe limit.
  d. International Living issued a report listing percentages of Americans retired and living in each foreign country.

1.44 Public Health and Nutrition   Parents Association would like to determine the proportion of teenagers who have the ability to prepare an entire meal. A sample of teenagers was obtained and all were asked if they can cook. Describe the population of interest, the sample, and the variable of interest in this problem.

1.45 Marketing and Consumer Behavior   Hallmark is interested in the proportion of adults who sent a greeting card on Mother-in-Law Day. A sample of 400 adults was obtained and all were asked whether they sent a greeting card on this holiday, which started in 2002. Describe the population and the sample in this problem.

1.46 Medicine and Clinical Studies   A recent study by the American Academy of Neurology suggests that soft drinks, iced tea, and even fruit drinks may lead to depression.[17] One thousand individuals who regularly consume soft drinks were selected and each was evaluated for signs of depression. Describe the population, the sample, and the variable in this problem.

1.47 Public Policy and Political Science   The Office of the Privacy Commissioner of Canada's (OPC) Contributions Program is interested in reaction to a proposal to allow police to obtain cell phone records without a subpoena. One thousand people in British Columbia were called and each was asked to respond to several questions.
   a. Is this an observational or an experimental study?
   b. Describe the sample in this problem.
   c. Is this a random sample? Justify your answer.

1.48 Travel and Transportation   Amtrak would like to estimate the proportion of travelers on the Sunset Limited, from New Orleans to Los Angeles, who utilize the Sightseer Lounge en route. At the end of the trip on March 15, an Amtrak representative stopped every third person getting off the train and asked them if they used the Sightseer Lounge to buy food, a drink, or souvenirs.
   a. Is this an observational or an experimental study?
   b. Describe the sample in this problem.
   c. Is this a random sample? If not, suggest a method for obtaining a random sample.

1.49 Physical Sciences   The Air Liquide Company has developed a new deicing chemical for airplanes, consisting of glycol and several proprietary additives. The new chemical was designed to keep aircraft wings ice-free for a longer period of time. Ten typical Dehaviland commuter airplanes were obtained and randomly assigned to one of two groups: new chemical versus old chemical. Each plane was subject to constant icing conditions in a controlled environment and treated with one of the chemicals. The length of time until ice formed on the wings was recorded for each plane.
   a. Is this an observational or an experimental study?
   b. What is the variable of interest?
   c. Describe a technique to randomly assign each plane to a chemical group.

EXTENDED APPLICATIONS

1.50 Technology and Internet   NationMaster.com reported that the most recent software piracy rate in the United States was 20%. They define the piracy rate as the number of units of pirated software deployed divided by the total number of units of software installed. One thousand installed software titles are selected. Each is carefully examined to determine if the software was pirated.
   a. What is the population of interest?
   b. What is the sample?
   c. Describe the variable of interest.
   d. Write a probability question and a statistic question involving this population and sample.

1.51 Travel and Transportation   The Channel Tunnel, or Chunnel, is a 31.4-mile railroad tunnel beneath the English Channel between Folkstone, Kent, in England and Coquelles in France. To ensure passenger safety, engineers selected the 35 deepest areas in the tunnel and measured the pressure on each section.
   a. Is this an observational or an experimental study?
   b. What is the variable of interest?
   c. Is this sample random? If so, justify your answer. If not, describe a technique to obtain a random sample.

1.52 Manufacturing and Product Development   In January 2013, flaws were discovered in two Boeing 787 Dreamliners aircraft. Japan Airlines found cracks in the cockpit window in one jet and a minor oil leak in another.[18] To assure the public that the jet is safe, the FAA selected 20 Dreamliners currently operated by American Airlines and carefully inspected each for any flaws.
   a. Is this an observational or an experimental study?
   b. What is the variable of interest?
   c. Is this sample random? If so, justify your answer. If not, describe a technique to obtain a random sample.

1.53 Public Policy and Political Science   The Thirty Bench Wine Makers in Beamsville, Ontario, would like to determine if the alcohol content of its wine is determined by the grape variety. Samples from two Riesling wines, made from different grape varieties, were obtained and the alcohol content in each bottle was carefully measured.
   a. Is this an observational or an experimental study?
   b. What is the variable of interest?

LAST STEP

1.54 Is it safe to eat rice?   A rice manufacturer claims that a single serving contains at most 10 ppb of arsenic. Suppose a random sample of 100 rice servings is obtained and each is measured for arsenic.
   a. Identify the population of interest and the sample.
   b. Apply the statistical inference procedure to draw a conclusion when an extraordinary, rare event is observed.

2 Tables and Graphs for Summarizing Data

◄ **Looking Back**

- Realize the difference between a sample and the population.
- Recognize the importance of a simple random sample in the statistical inference procedure.
- Understand the difference between descriptive and inferential statistics.

▶ **Looking Forward**

- Be able to classify a data set as categorical or numerical, discrete or continuous.
- Learn several graphical summary techniques.
- Construct bar charts, pie charts, stem-and-leaf plots, and histograms.

 ## Can the Florida Everglades be saved?

Burmese pythons have invaded the Florida Everglades and now threaten the wildlife indigenous to the area. It is likely that people were keeping pythons as pets and somehow a few animals slithered into Everglades National Park. The first python was found in the Everglades in 1979, and these snakes became an officially established species in 2000.[1] The Everglades has an ideal climate for the pythons, and the large areas of grass allow the snakes plenty of places to hide.

In January 2013, the Florida Fish and Wildlife Conservation Commission started the Python Challenge. The purpose of the contest was to thin the python population, which could be tens of thousands, and help save the natural wildlife in the Everglades. There were 800 participants, with prizes for the most pythons captured and for the longest. At the end of the competition, 68 Burmese pythons had been harvested.

Suppose a random sample of pythons captured during the Challenge was obtained. The length (in feet) of each python is given in the following table.

9.3	3.5	5.2	8.3	4.6	11.1	10.5	3.7	2.8	5.9
7.4	14.2	13.6	8.3	7.5	5.2	6.4	12.0	10.7	4.0
11.1	3.7	7.0	12.2	5.2	8.1	4.2	6.1	6.3	13.2
3.9	6.7	3.3	8.3	10.9	9.5	9.4	4.3	4.6	5.8
4.1	5.2	4.7	5.8	6.4	3.8	7.1	4.6	7.5	6.0

The tabular and graphical techniques presented in this chapter will be used to describe the shape, center, and spread of this distribution of python lengths and to identify any outliers.

CONTENTS

Michael R. Rochford/University of Florida/AP

2.1 Types of Data

As members of an information society, we have access to all kinds of descriptive statistics: in newspapers, in research journals, and even via the Internet. Whether the information is obtained from a carefully designed experiment or an observational study, the first step is to organize and summarize the data. Tables, charts, and graphs reveal characteristics about the shape, center, and variability of a data set, or distribution. For example, Figure 2.1 shows a stacked bar chart of the number of automobile crashes related to hand-held cell phone use in certain New Jersey counties over the past several years.

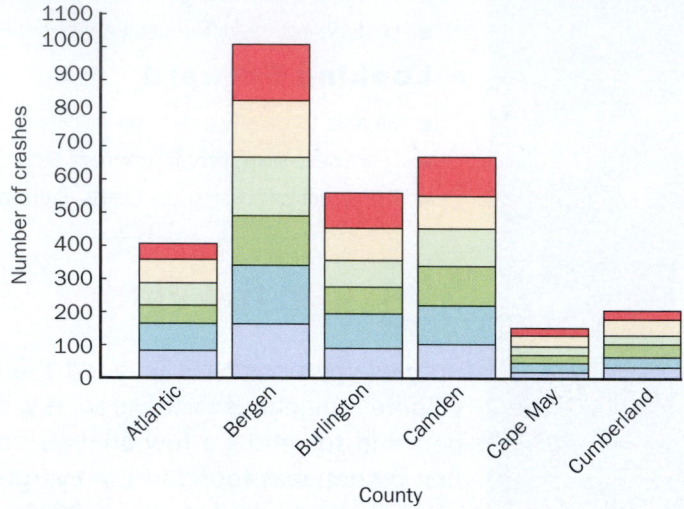

Figure 2.1 Automobile accidents in six New Jersey counties over several years.

The *shape* of a distribution may be symmetric or skewed. The *center* of a distribution refers to the position of the majority of the data, and measures of variability indicate the *spread* of the data. The *variability* (or *dispersion*) of a distribution describes how much the measurements vary, and how compact or how spread out the data are. Although they are not suitable for making inferences, the tabular and graphical techniques introduced in this chapter help to describe the distribution of data and identify unusual characteristics.

The summary table or graph to be used, and later the statistical analysis to be performed, depends on the type of data. Consider golfers arriving at a public country club on a Saturday morning. Here are several characteristics we could record: brand of golf clubs, handicap, whether the patron wears a golf hat, even the number of days since the golfer last played at this course.

Definition

A data set consisting of observations on only a single characteristic, or attribute, is a **univariate** data set.

We'll do more with bivariate data in Chapter 12.

If we measure, or record, two observations on each individual, the data set is **bivariate**.

If there are more than two observations on the same person, the data set is **multivariate**.

Suppose we record only the make of car driven by each person who arrives at the country club—a univariate data set. The observations, for example, Ford, Honda, or Lexus, are *categorical*. There is no natural ordering of the data, and each observation falls into only one category or class. We might instead ask each person who arrives how long it took to reach the country club. This time the responses, for example, 10, 15, or 45 minutes, are *numerical*.

Definition

A **categorical**, or **qualitative**, univariate data set consists of non-numerical observations that may be placed in categories.

A **numerical**, or **quantitative**, univariate data set consists of observations that are numbers.

The following examples illustrate the two basic types of data sets.

Example 2.1 Oscar Night

A random sample of actresses attending the Oscars was obtained, and the designer of each gown was recorded. The responses are given in the following table.

Gucci	Versace	Dior	Vera Wang	Dior	Versace
Vera Wang	Ralph Lauren	Valentino	Gucci	Dior	

Each response is non-numerical, because there is no natural ordering. This is a (univariate) *categorical* data set.

Example 2.2 Priority Mail

The U.S. Postal Service offers Priority Mail Flat Rate boxes with which customers can expect delivery within two days of packages weighing up to 70 pounds. A random sample of Small Boxes shipped from Post Offices in Oklahoma was obtained and each was weighed. The resulting weights (in pounds) are given in the following table.

DATA SET

MAILWTS

The number of lightning strikes is discrete. For instance, the number of possible lightning strikes can be 1, or 2, or 3, and so on; but not, for example, 2.5. However, if we had an instrument that could measure barometric pressure accurately enough, any number between 960 and 1070 millibars is possible, for example, 995.466347789.

2.0	6.1	7.3	6.4	8.0	8.2	9.9	5.2	7.7	6.7
6.9	4.8	10.8	9.2	3.2	7.9	8.5	8.9	6.6	8.1

Because each observation is numerical, this is a (univariate) *numerical* data set.

We can classify numerical data even further. Consider the following examples.

On a hot summer day in the Southeast, suppose we record the number of lightning strikes within a specified county during the next 24 hours. The possible values are 0, 1, 2, 3, up to, say, 10. There are only a finite number of possible numerical values, and these values are **discrete**, isolated points on a number line (Figure 2.2). Instead, suppose we record the barometric pressure, in millibars, at 4:00 P.M. The possible values are not discrete and isolated. Rather, the barometric pressure can (theoretically) be *any* number in the **continuous** interval 960 to 1070 (Figure 2.3).

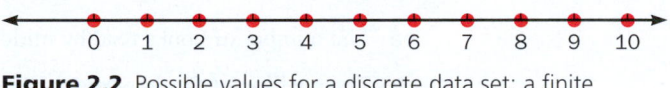

Figure 2.2 Possible values for a discrete data set: a finite number of values, isolated on a number line.

Figure 2.3 Possible values for a continuous data set: numerical values on some interval.

Definition

A numerical data set is **discrete** if the set of all possible values is finite, or countably infinite. Discrete data sets are usually associated with *counting*.

A numerical data set is **continuous** if the set of all possible values is an interval of numbers. Continuous data sets are usually associated with *measuring*.

A CLOSER L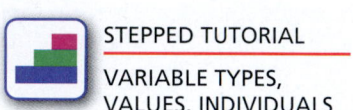OK

1. To decide whether a data set is discrete or continuous, consider *all* the *possible* values. Finite or countably infinite means *discrete*. An interval of possible values means *continuous*.

2. Countably infinite means there are infinitely many possible values, but they are countable. You may not ever be able to finish counting all of the possible values, but there exists a method for actually counting them.

3. The interval for a continuous data set can be *any* interval, of any length, open or closed. The exact interval may not be known, only that there is *some* interval of possible values.

4. In practice, we have no measurement device that is precise enough to return *any* number in some interval. We may only be able to achieve up to 10 digits of accuracy. So a continuous data set may contain *any* number in some interval in *theory*, but not in reality.

Mathematically, a set is countably infinite if it can be put into one-to-one correspondence with the counting numbers (1, 2, 3, 4, . . .). The three dots, . . . , mean the list continues in the same manner.

STEPPED TUTORIAL

VARIABLE TYPES, VALUES, INDIVIDUALS

The classifications of univariate data are shown in Figure 2.4. Here is an example to illustrate these classifications.

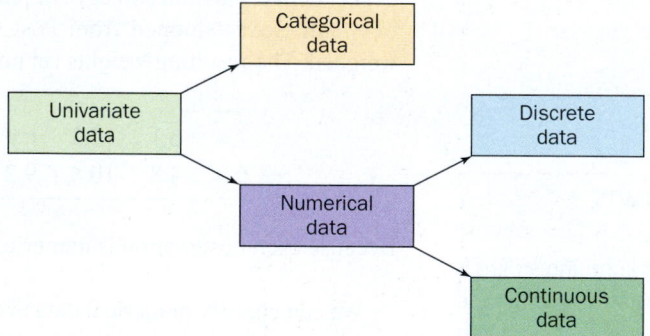

Figure 2.4 Classifications of univariate data.

Example 2.3 Univariate Data Classifications

A researcher obtained the following observations. Classify each resulting data set as categorical or numerical. If the data set is numerical, determine whether it is discrete or continuous.

a. The number of books read by middle-school students during an academic year.

b. The position of the drawbridge in Belmar, New Jersey, at noon on days in July. Assume the drawbridge is not moving, and is either open or closed to boat traffic.

c. The length of time (in minutes) it takes to get a haircut.

d. The number of garage sales advertised in a local newspaper.

e. The types of candy received at houses on Halloween.

f. The air pressure in footballs at the beginning of college games.

g. The type of plumbing problem reported by the next person who contacts the Plumbing Pros.

SOLUTION

a. The observations are numbers, so the data set is numerical. The set of possible values is finite. We don't know the maximum number of books read, but the possible numbers in the data set represent counts. The data set is discrete.

b. The observations are categorical: Up (open) or down (closed). There is no natural ordering; the possible responses fall into groups or classes. This data set is categorical.

c. The observations are numbers and the set of possible values is some interval, perhaps 5 to 45 minutes. This is a numerical continuous data set.

d. The observations are numbers and the set of possible values is finite. We can count the number of advertised garage sales. The minimum number may be 0 and the maximum may be 25. This is a numerical discrete data set.

e. The observations may be Milky Way, Snickers, Nestle Crunch, etc. Although there may be some personal preference and an individual ranking, this is a categorical data set.

f. The observations are numbers and the set of possible values is some interval, say 12.5 to 13.5 psi. This is a numerical continuous data set.

g. The observations are dripping faucets, leaking pipes, plugged sinks, etc. There may be some preference for the Plumbing Pros, but this is a categorical data set. ∎

TRY IT NOW   GO TO EXERCISE 2.5

Ways of summarizing and displaying categorical data are discussed in Section 2.2, and tables and graphs for numerical data are presented in Sections 2.3 and 2.4.

SECTION 2.1 EXERCISES

Concept Check

2.1 True/False   A data set obtained by recording the height and weight of every person entering a doctor's office is univariate.

2.2 True/False   Every data set is multivariate.

2.3 True/False   A data set consisting of 37 times, in seconds, for pedestrians to cross a certain city street is univariate.

2.4 Fill in the Blank
   a. A _____ univariate data set consists of observations that are numbers.
   b. A _____ univariate data set consists of non-numerical observations.
   c. If the set of all possible values for a numerical data set is finite, then the data set is _____.
   d. If the set of all possible values for a numerical data set is some interval of numbers, then the data set is _____.

Practice

2.5 Univariate Data Classifications   A set of observations is obtained as indicated below. In each case, classify the resulting data set as categorical or numerical. If the data set is numerical, determine whether it is discrete or continuous.
   a. The weights of several reams of paper.
   b. The number of cars towed from the Pennsylvania Turnpike during given 24-hour periods.
   c. The first ingredient in the product listing of boxes of cereal.
   d. The number of games the Red Sox win during several seasons.
   e. The amount of sand used on roads during winters in a small town.
   f. The diagnoses of patients in an emergency ward.

2.6 Univariate Data Classifications   A set of observations is obtained as indicated below. In each case, classify the resulting data set as categorical or numerical. If the data set is numerical, determine whether it is discrete or continuous.
   a. The lengths of the spans of bridges in New York State.
   b. The number of people hired by a company during certain weeks.
   c. The cloud ceiling at airports around the country.
   d. The temperature of the coffee purchased at several fast-food restaurants.
   e. The type of notebook used by students in a statistics class.

f. The classifications of Forward Operating Air Force bases (Main Air Base, Air Facility, Air Site, or Air Point).

2.7 Univariate Data Classifications   A set of observations is obtained as indicated below. In each case, classify the resulting data set as categorical or numerical. If the data set is numerical, determine whether it is discrete or continuous.

a. The number of steps on apartment fire escapes.
b. The number of leaves on maple trees.
c. The reason several automobiles fail inspection.
d. The weight of fully loaded tractor trailers.
e. The area of several Nebraska farms.
f. The cellular calling plan selected by customers.

2.8 Univariate Data Classifications   A set of observations is obtained as indicated below. In each case, classify the resulting data set as categorical or numerical. If the data set is numerical, determine whether it is discrete or continuous.

a. The number of engine revolutions per minute in automobiles.
b. The thickness of the polar ice cap in several locations.
c. The state in which families vacationed last summer.
d. The type of Internet connection in county households.
e. The make of watch worn by people entering a certain department store.
f. The number of raisins in 24-ounce boxes.

2.9 Numerical Observations   A set of numerical observations is obtained as described below. Classify each resulting data set as discrete or continuous.

a. The widths of posters at an art gallery.
b. The time it takes to compile computer programs.
c. The number of radioactive particles that escape from special containers during a one-hour period.
d. The time it takes to bake batches of banana muffins.
e. The concentration of carbon monoxide in homes during the winter.
f. The number of pages in best-selling murder-mystery novels.

2.10 Numerical Observations   A set of numerical observations is obtained as described below. Classify each resulting data set as discrete or continuous.

a. The weight of baseball bats.
b. The area of selected dorm rooms.
c. The number of bees in hives.
d. The height of a storm surge during hurricanes.
e. The amount of ink used in office printers during a week.
f. The number of fish in office aquariums.

2.11 Numerical Observations   A set of numerical observations is obtained as described below. Classify each resulting data set as discrete or continuous.

a. The time it takes giant slalom skiers to cover a race course.
b. The number of magazines available for sale at newsstands.
c. The number of black squares in crossword puzzles.
d. The length of time spent waiting in line at grocery-store checkout lanes.
e. The number of French fries in a small order from fast-food restaurants.
f. The length in words of email messages received.

2.12 Univariate Data Classifications   Classify each data set as categorical, discrete, or continuous.

a. A random sample of mature Eastern tent caterpillars is obtained from a tree branch in a neighborhood yard. The length of each caterpillar is recorded.
b. Randomly selected prime-time television shows are selected and the number of violent acts is recorded for each show.
c. A representative sample of employees from a large company is obtained, and the overtime hours for the past month are recorded for each employee.
d. An HMO selects a random sample of subscribers and records the number of office visits over the past year for each patient.
e. Thirty-six apples are randomly selected from an orchard. Each is graded for quality of appearance: excellent, good, fair, or poor.
f. A random sample of mattresses is obtained and the firmness (medium, medium firm, firm, or extra firm) of each is recorded.

2.13 Univariate Data Classifications   Classify each data set as categorical, discrete, or continuous.

a. A random sample of cheeses is obtained and the number of months each is allowed to age before sale is recorded.
b. Sixteen universities are selected and each computer network system is carefully analyzed. The computer virus threat is assessed for each campus: low, medium, or high.
c. A random sample of Waterford Normandy dinner plates is selected and the weight of each plate is recorded.
d. Thirty-five new customers at a health club are selected and the body-fat percentage of each member is computed and recorded.
e. A random sample of CDs is obtained from a local music store. The company that produced each CD is noted.
f. A collection of pens is obtained from employees at a large company. For each pen, the outside diameter of the barrel at its widest point is measured and recorded.

2.14 Univariate Data Classifications   Classify each data set as categorical, discrete, or continuous.

a. A random sample of Hudson River ferry trips is obtained and the number of riders on each trip is recorded.
b. A random sample of military helicopters is obtained and the weight of each is recorded.
c. A random sample of communities in Canada is selected and the number of full-time police officers employed is recorded.
d. A random sample of locations in the United States is selected. Temperature data are used to determine whether or not a new record high temperature was set during the past year.
e. A random sample of stock analysts is obtained and each is asked to rate a specific stock as buy, sell, or hold.
f. A random sample of cross-country flights is obtained and the number of controllers each pilot talks to during the flight is recorded.

2.2 Bar Charts and Pie Charts

The natural summary measures for a categorical data set are the number of times each category occurred and the proportion of times each category occurred. These values are usually displayed in a table as in Table 2.1.

Table 2.1 A frequency distribution summarizing the results of a survey on computer security threats

Class	Frequency	Relative Frequency
Physical damage	130	0.26
Natural events	50	0.10
Loss of essential services	75	0.15
Compromise of information	35	0.07
Technical failures	95	0.19
Compromise of functions	115	0.23
Total	500	1.00

Definition

A **frequency distribution** for categorical data is a summary table that presents categories, counts, and proportions.

1. Each unique value in a categorical data set is a label, or **class**. In Table 2.1, the classes are physical damage, natural events, loss of essential services, etc.

2. The **frequency** is the count for each class. In Table 2.1, the frequency for the compromise of information class is 35 (i.e., 35 computer security threats were due to compromise of information).

3. The **relative frequency,** or sample proportion, for each class is the frequency of the class divided by the total number of observations. In Table 2.1, the relative frequency for the technical failure class is 95/500 = 0.19.

A frequency distribution for a categorical data set is illustrated in the next example.

Example 2.4 Cruise Ship Destinations

A random sample of cruise ships leaving from the Port of New York showed the following destinations.

Bermuda	Southampton	Mediterranean	Southampton	Caribbean
Southampton	Bermuda	Southampton	Caribbean	Caribbean
Caribbean	Bermuda	Mediterranean	Caribbean	Southampton
Caribbean	Southampton	Mediterranean	Southampton	Southampton
Bahamas	Bermuda	Bahamas	Southampton	Southampton

Jimmy Lopes/FeaturePics

Construct a frequency distribution to describe these data. What proportion of cruise ships did not go to Southampton?

SOLUTION

STEP 1   Each unique destination is a label, or **class.** This is a categorical data set. There are five unique classes and 25 observations in total.

STEP 2   Draw a table and list each unique class in the left-hand column. Find the **frequency** and **relative frequency** for each class. For example, because Bermuda appears four times in the sample, the frequency for this class is 4. The relative frequency for Bermuda is 4/25 = 0.16.

Class	Tally	Frequency	Relative frequency	
Bahamas	||	2	0.08	(= 2/25)
Bermuda	||||	4	0.16	(= 4/25)
Caribbean	~~||||~~|	6	0.24	(= 6/25)
Mediterranean	|||	3	0.12	(= 3/25)
Southampton	~~||||~~ ~~||||~~	10	0.40	(= 10/25)
Total		25	1.00	

The proportion of cruise ships that did go to Southampton is 10/25 = 0.40. The total proportion is always 1.00. Therefore, the proportion of cruise ships that did not go to Southampton is 1.00 − 0.40 = 0.60.   ●

TRY IT NOW   GO TO EXERCISE 2.19

A CLOSER LOOK

A tally mark is a short line drawn for each count up to four. On number five, draw a diagonal line across the other four. Count in sets of five.

1. If you have to construct a frequency distribution by hand, an additional *tally column* is helpful. Insert this after the class column, and use a *tally mark* or *tick mark* to count observations as you read them from the table.

2. The last (*total*) row is optional, but it is a good check of your calculations. The frequencies should sum to the total number of observations, and the relative frequencies should sum to 1.00 (subject to round-off error).

3. There is no rule for ordering the classes. In Example 2.4, the classes happen to be presented in alphabetical order.   ●

A **bar chart** is a graphical representation of a frequency distribution for categorical data. An example of a bar chart is shown in Figure 2.5.[2]

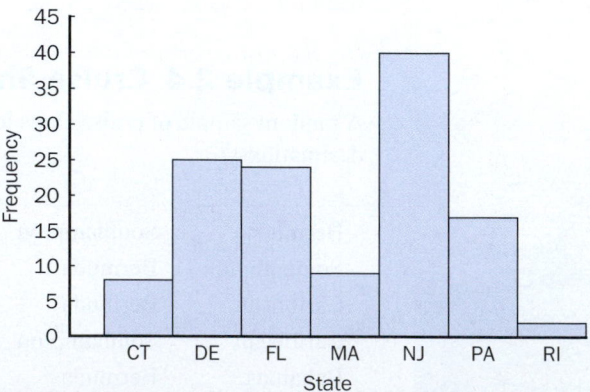

Figure 2.5 Bar chart showing the number of Nathan's World Famous Beef Hot Dogs franchises in certain states as of March 25, 2012.

How to Construct a Bar Chart

1. Draw a horizontal axis with equally spaced tick marks, one for each class.
2. Draw a vertical axis for the frequency (or relative frequency) and use appropriate tick marks. Label each axis.
3. Draw a rectangle centered at each tick mark (class) with height equal to, or proportional to, the frequency of each class (also called the class frequency). The bars should be of equal width, but do not necessarily have to abut one another; there can be spaces between them.

Example 2.5 Cruise Ship Destinations, Continued

Construct a bar chart for the cruise ship data in Example 2.4.

SOLUTION

STEP 1   Use the frequency distribution for the cruise ship data. There are five classes, and the frequencies range from 2 to 10.

STEP 2   Draw a horizontal and a vertical axis. On the horizontal axis, draw five ticks for the five classes and label them with the class names. Because the greatest frequency is 10, draw and label tick marks from 0 to at least 10 on the vertical axis.

STEP 3   The height of each vertical bar is determined by the frequency of the class. For example, the frequency of trips to Bermuda is 4, so the height of the bar representing Bermuda is 4. The resulting bar chart is shown in Figure 2.6. A technology solution is shown in Figure 2.7.

Either frequency or relative frequency may be used on the vertical axis. Both are acceptable because the resulting graphical representations of the distribution are identical. The only difference between the two graphs is the labels on the vertical axis. Unless it is stated otherwise, frequencies are used.

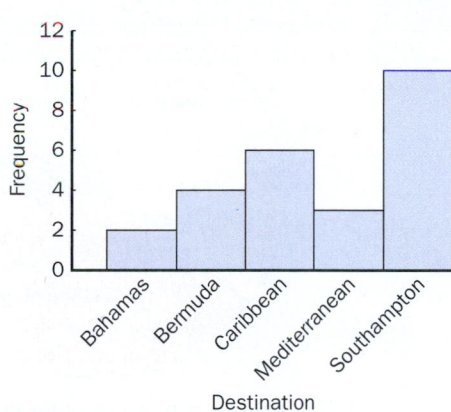

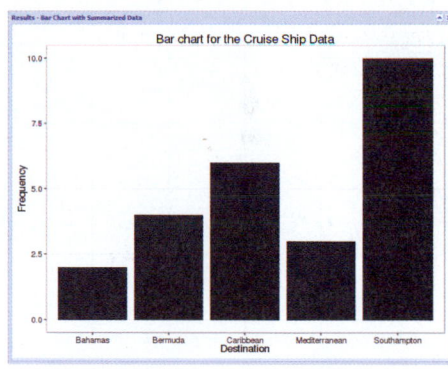

Figure 2.6 Bar chart for the cruise ship data.   **Figure 2.7** CrunchIt! bar chart.

TRY IT NOW   GO TO EXERCISE 2.21

A **pie chart** is another graphical representation of a frequency distribution for categorical data. An example of a pie chart is shown in Figure 2.8.[3]

How to Construct a Pie Chart

1. Divide a circle (or pie) into slices or wedges so that each slice corresponds to a class.
2. The size of each slice is measured by the angle of the slice. To compute the angle of each slice, multiply the relative frequency by 360° (the number of degrees in a whole or complete circle).

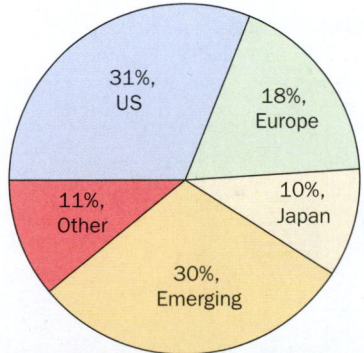

Figure 2.8 Projected global spending on medicine in 2016.

3. The first slice of a pie chart is usually drawn with an edge horizontal and to the right (0°). The angle is measured counterclockwise. Each successive slice is added counterclockwise with the appropriate angle.

Example 2.6 Cruise Ship Destinations: Another Stop

Construct a pie chart for the cruise ship data in Example 2.4.

SOLUTION

STEP 1 Add a column to the frequency distribution for slice angle. Use the relative frequency of each class to find the slice angle.

Class	Relative frequency	Angle
Bahamas	0.08	$28.8° = (0.08 \times 360°)$
Bermuda	0.16	$57.6° = (0.16 \times 360°)$
Caribbean	0.24	$86.4° = (0.24 \times 360°)$
Mediterranean	0.12	$43.2° = (0.12 \times 360°)$
Southampton	0.40	$144.0° = (0.40 \times 360°)$
Total	1.00	360.0°

STEP 2 Draw a circle and mark slices using the angles in the frequency distribution. Draw the first slice with an edge extending from the center of the circle to the right. The remaining slices are drawn moving around the pie counterclockwise. It may be helpful to use a protractor and compass to draw the circle and measure the angles. See Figure 2.9. A technology solution is shown in Figure 2.10.

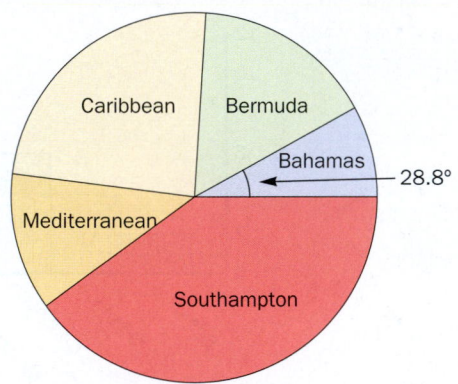

Figure 2.9 Pie chart for the cruise ship data.

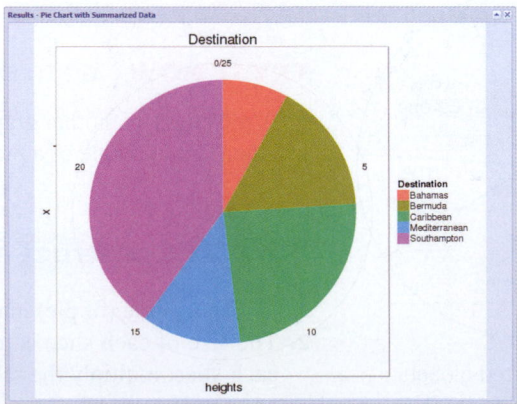

Figure 2.10 CrunchIt! pie chart.

Note: Because Southampton corresponds to the biggest slice of the pie (chart), this class has the greatest frequency (and relative frequency); it is the destination that occurred most often in the sample.

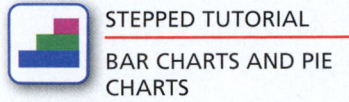

STEPPED TUTORIAL

BAR CHARTS AND PIE CHARTS

TRY IT NOW   GO TO EXERCISE 2.23

A CLOSER LOOK

1. A pie chart is hard to draw accurately by hand, even with a protractor and compass. A graphing calculator or computer is quicker and more efficient for constructing this graph.

2. There are lots of pie-chart variations, for example, exploding pie charts and 3D pie charts. Each is simply a visual representation of a frequency distribution for categorical data.

Technology Corner

Procedure: Construct a bar chart.
Reconsider: Example 2.5, solution, and interpretations.

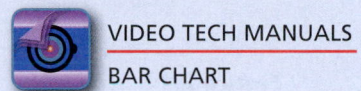

VIDEO TECH MANUALS

BAR CHART

CrunchIt!

CrunchIt! has a built-in function to construct a bar chart from original or summarized data.

1. Enter the classes in column Var1 and the frequencies in column Var2. Rename each column if desired.
2. Select Graphics; Bar Chart. Choose Var1 for Labels and Var2 for Heights. Optionally enter a Title, X Label, and Y Label. Click the Calculate button. Refer to Figure 2.7.

TI-84 Plus C

The TI-84 Plus C does not accept categorical data. Therefore, there is no built-in function to construct a bar chart. However, you may assign a number to each class, and use the Histogram statistical plot to construct a bar chart.

1. Enter integers corresponding to each class in list L1 and the frequency for each class in the corresponding row in list L2 (Figure 2.11).
2. Press STATPLOT and select Plot1 from the STAT PLOTS menu.
3. Turn the plot On and select Type histogram. For Xlist, enter the name of the list containing the categories. For Freq, enter the name of the list containing the frequencies. Select a Color (Figure 2.12).
4. Consider each *class* to have width 1. Enter appropriate WINDOW settings. Press GRAPH to display the bar chart (Figure 2.13).

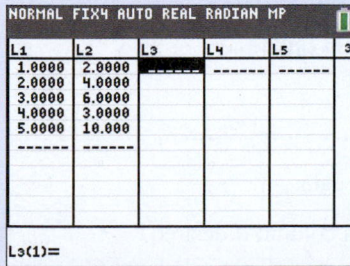

Figure 2.11 The *categories* and frequencies.

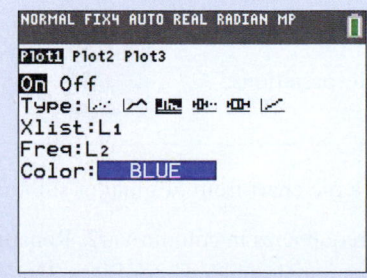

Figure 2.12 The Plot1 setup screen.

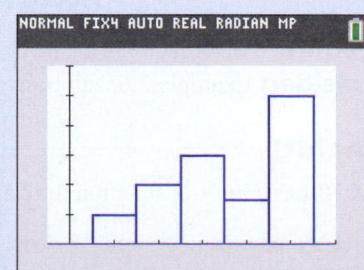

Figure 2.13 TI-84 Plus C bar chart.

Minitab

The input can be either the entire data set in a single column or a summary table of categories and frequencies in two columns.

1. Enter the data into column `C1`.
2. Select Graph; Bar chart. Choose Counts of unique values and Simple.
3. Enter `C1` under Categorical variables.
4. Edit graph attributes as necessary, for example, the axes labels, plot title, and gaps between clusters. See Figure 2.14.

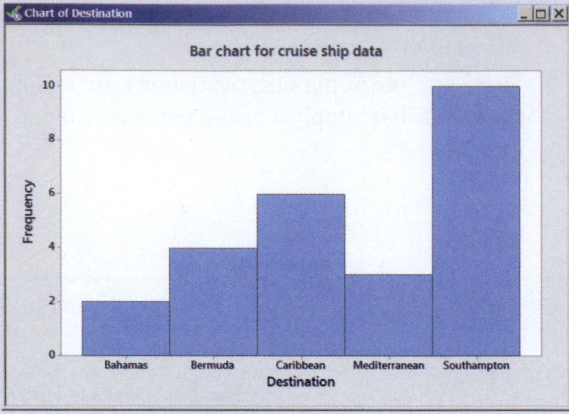

Figure 2.14 Minitab bar chart.

Excel

The built-in functions Frequency or Sumif may be used to construct a frequency distribution. Assume this summary information is available.

1. Enter the categories into column A and the corresponding frequencies into column B.
2. Select the range of cells `A1:B5`. Under the Insert tab, select Column; 2-D Column; Clustered Column.
3. Use Chart Tools to format the bar chart as necessary. See Figure 2.15.

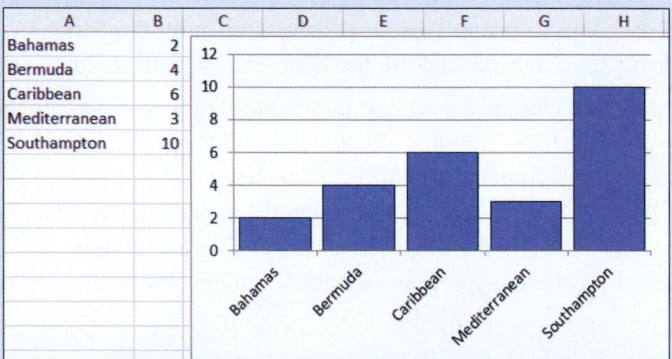

Figure 2.15 Excel bar chart.

Procedure: Construct a pie chart.
Reconsider: Example 2.6, solution, and interpretations.

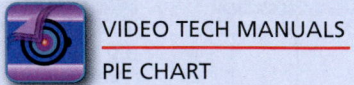
VIDEO TECH MANUALS
PIE CHART

CrunchIt!

CrunchIt! has a built-in function to construct a pie chart from original or summarized data.

1. Enter the classes in column Var1 and the frequencies in column Var2. Rename each column if desired.
2. Select Graphics; Pie Chart. Choose Var1 for Labels and Var2 for Sizes. Optionally enter a Title. Click the Calculate button. Refer to Figure 2.7.

TI-84 Plus C

A pie chart can be constructed using the CellSheet App for the TI-84 Plus C.

1. Open a new spreadsheet in the CellSheet app.
2. Enter the categories in row 1 and the corresponding frequencies in row 2. See Figure 2.16.
3. Select MENU; `Charts`; `Pie`. Enter the range for the categories, the range for the frequencies, select `Number` or `Percent` for display, and enter a `Title` if desired. See Figure 2.17.
4. Highlight `Draw` and press ENTER to display the pie chart. See Figure 2.18.

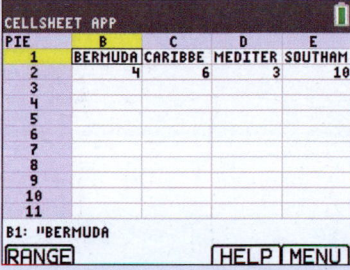

Figure 2.16 The categories and frequencies in a CellSheet spreadsheet.

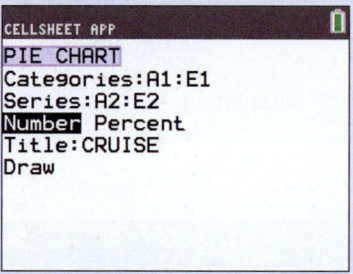

Figure 2.17 The PIE CHART setup screen.

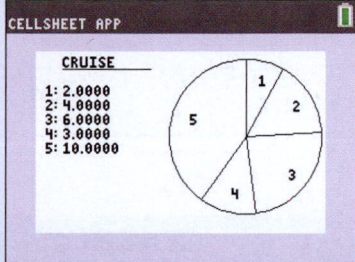

Figure 2.18 TI-84 Plus C pie chart.

Minitab

The input can be either the entire data set in a single column or a summary table of categories and frequencies in two columns.

1. Enter the data into column C1.
2. Select Graph; Bar chart. Choose Counts of unique values and Simple.
3. Enter C1 under Categorical variables.
4. Edit graph attributes as necessary, for example, the axes labels, plot title, and gaps between clusters. See Figure 2.19.

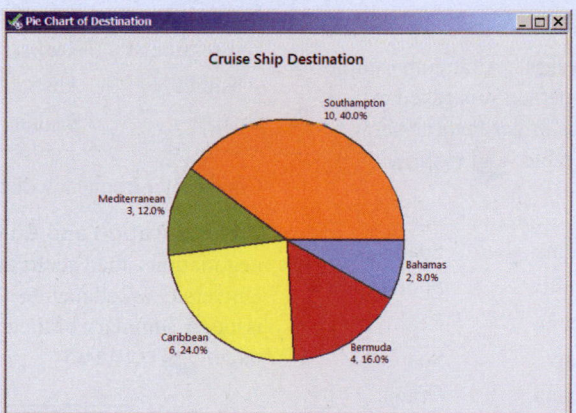

Figure 2.19 Minitab pie chart.

Excel

The built-in functions Frequency or Sumif may be used to construct a frequency distribution. Assume this summary information is available.

1. Enter the categories into column A and the corresponding frequencies into column B.
2. Select the range of cells A1 : B5. Under the Insert tab, select Column; 2-D Column; Clustered Column.
3. Use Chart Tools to format the bar chart as necessary. See Figure 2.20.

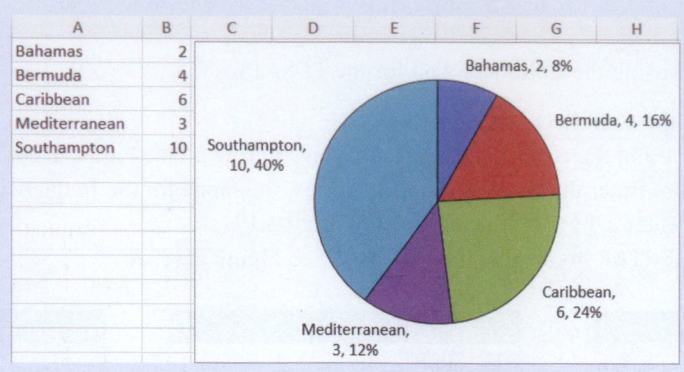

Figure 2.20 Excel pie chart.

SECTION 2.2 EXERCISES

Concept Check

2.15 True/False   A frequency distribution is a summary table for categorical data.

2.16 True/False   The relative frequency for each class in a frequency distribution is a sample proportion.

2.17 True/False   A bar chart is constructed using the frequency for each class.

2.18 True/False   All the slices in a pie chart should have approximately the same angle.

Applications

2.19 Psychology and Human Behavior   A random sample of TV viewers was obtained and each person was asked to select the entertainment category of his or her favorite show. The results are given in the following table. ▪▪▪ **TVSHOW**

Comedy	Comedy	Drama	Soap
Reality	Sports	Reality	Soap
Sports	Comedy	Drama	Sports
Drama	Soap	Soap	Reality
Educational	Reality	Drama	Soap
Sports	Educational	Drama	Drama
Comedy	Soap	Reality	Comedy
Sports	Drama	Drama	Soap
Soap	Reality	Reality	Soap
Soap	Drama	Educational	Drama
Comedy	Comedy		

Construct a frequency distribution for these data.

2.20 Psychology and Human Behavior   A random sample of patrons visiting the Rena Branston Gallery in San Francisco

was obtained, and each was asked the type of art the patron most enjoy viewing. The results are given in the following table. ▪▪▪ **ARTSTYLE**

Abstract	Abstract	Surrealist	Expressionist
Realist	Realist	Realist	Realist
Surrealist	Abstract	Abstract	Abstract
Realist	Realist	Abstract	Abstract
Surrealist	Surrealist	Abstract	Realist
Expressionist	Expressionist	Abstract	Surrealist
Surrealist	Surrealist	Abstract	Abstract
Surrealist	Surrealist	Abstract	Realist
Expressionist	Realist	Realist	Expressionist
Abstract	Abstract	Abstract	Expressionist
Realist	Realist		

Construct a frequency distribution for these data.

2.21 Education and Child Development   To prepare for negotiations, the Faculty Association at Eastern Michigan University asked members to name the most important contract issue. A summary of their responses is given in the following table. ▪▪▪ **CONTRACT**

Issue	Frequency
Salary	50
Health insurance	100
Retirement benefits	75
Class size	60
Temporary faculty	90
Parking	25

a. Find the relative frequency for each issue.

b. Construct a bar chart for these data using frequency on the vertical axis.

2.22 Biology and Environmental Science   The following table lists the number of dairy farms for various counties in Vermont.[4]   **DAIRY**

County	Frequency
Addison	145
Bennington	18
Caledonia	80
Chittenden	42
Essex	12
Franklin	210
Grand Isle	17
Lamoille	38
Orange	93
Orleans	141
Rutland	71
Washington	39
Windham	25
Windsor	39

a. Find the relative frequency for each county.
b. Construct a bar chart for these data using relative frequency on the vertical axis.

2.23 Fuel Consumption and Cars   The business manager of a Chrysler automobile dealership sent a survey to randomly selected owners in order to gauge customer satisfaction. One question was, "How likely are you to buy another car of the same make and model?" Survey participants could answer Very Likely (VL), Likely (L), Neutral (N), Unlikely (U), or Very Unlikely (VU). The results are given on the text website.   **CARSATIS**
a. Construct a frequency distribution for these data.
b. Use the table in part (a) to construct a pie chart for these data.

2.24 Public Policy and Political Science   According to the *Atlanta Journal*, U.S. Senator Saxby Chambliss will not seek reelection in 2014.[5] A random sample of Georgia voters was obtained and each was asked to consider certain potential successors. The political affiliation of each voter is given on the website for this book: Democrat (D), Republican (R), Independent (I).   **VOTING**
a. Construct a frequency distribution for these data.
b. Use the table in part (a) to construct a pie chart for these data.

2.25 Demographics and Population Statistics   The following table lists the number of Nobel Prize laureates in each category.[6]   **PRIZECT**

Nobel Prize	Frequency
Physics	194
Chemistry	163
Medicine	201
Literature	109
Peace	125
Economic Sciences	71

a. Find the relative frequency for each prize.
b. Construct a pie chart for these data.

2.26 Education and Child Development   The grade distribution for a large psychology class at Louisiana State University is given in the following frequency distribution: **PSYCHGRD**

Grade	Frequency	Relative frequency
A	10	
B	43	
C	54	
D	26	
F	15	

a. Find the relative frequency for each grade.
b. Construct a bar chart using frequency on the vertical axis and a pie chart from the frequency distribution.
c. How many students were in this psychology class? What proportion of students passed (i.e., received a D or better)?

2.27 Public Health and Nutrition   A random survey of 200 customers who purchased ice cream at Brigham's showed the following proportions: **ICECREAM**

Ice cream	Relative frequency
The Big Dig	0.100
Cashew Turtle	0.185
Chocolate Chip	0.260
Pistachio	0.150
Strawberry	0.080
Vanilla with Oreos	0.225

a. Find the frequency of each ice cream (class).
b. Construct a bar chart using frequency on the vertical axis and a pie chart for these ice cream data.

2.28 Sports and Leisure   A random sample of long-time subscribers to *Popular Woodworking* was obtained, and each person was asked to name the brand of table saw he or she uses. The results are given in the following table: **TABLESAW**

DeWalt	DeWalt	Craftsman
DeWalt	Craftsman	Delta
Craftsman	Craftsman	Delta
DeWalt	Black & Decker	Makita
Black & Decker	DeWalt	Delta
Makita	Delta	Makita
Makita	DeWalt	Black & Decker
Delta	Delta	Makita
Black & Decker	Makita	Craftsman
DeWalt		

a. Construct a frequency distribution for these data.
b. Carefully sketch a bar chart using frequency on the vertical axis and a pie chart for these data.

c. What proportion of people in this sample use a Craftsman or Black & Decker table saw?

d. What proportion of people in this sample do not use a Delta table saw?

2.29 Marketing and Consumer Behavior Suppose there were 253 exhibitors at the NFPA World Fire Safety Conference in Chicago, Illinois, in June 2013. Each exhibitor was classified according to the type of product or service offered for sale. The proportions are given in the following table: **FIREXHIB**

Product	Proportion
Alarms	0.2964
Training	0.0632
Extinguishers	0.0514
Pumps	0.0237
Sprinklers	0.0632
Building materials	0.0751
Electrical equipment	0.1265
Hazmat storage	0.0870
Security products	0.1621
Signaling systems	0.0514

a. Find the number of exhibitors in each classification.

b. Carefully sketch a bar chart and a pie chart using the proportions for each class.

2.30 Psychology and Human Behavior Using the Library of Congress classification scheme, the Brookings Public Library in South Dakota recorded the type of book borrowed by 30 randomly selected patrons. The data are given in the following table: **BOOKS**

Medicine	Science	Medicine	Medicine
Science	Education	Medicine	Science
Education	Law	Medicine	Technology
Education	Technology	Literature	Education
Technology	Medicine	Technology	Science
Science	Literature	Medicine	Literature
Law	Literature	Law	Technology
Technology	Education		

a. Construct a frequency distribution for these data.

b. Carefully sketch a bar chart using relative frequency on the vertical axis and pie chart for these data.

c. Do you think the public library should try to purchase more books in one particular subject area? Why or why not?

2.31 Marketing and Consumer Behavior Cardinal Glass Industries produces several products for residential buildings, for vehicles, and for ordinary consumer use. The proportion of each type of manufactured product is given in the following table: **GLASS**

Building window	Vehicle window	Containers	Tableware	Lamps
0.35	0.15	0.10	0.25	0.15

Construct a bar chart and a pie chart for these data using the proportions in the table.

2.32 Business and Management In the 2012 Canadian Lawyer Corporate Counsel Survey, each company/organization was classified by sector. The sectors and corresponding proportions are given in the following table.[7] **CASECTOR**

Sector	Proportion
Government	0.246
Professional services	0.062
Technology	0.136
Industry, manufacturing	0.154
Service	0.104
Resource-based	0.098
Financial	0.142
Nonprofit	0.058

a. Construct a bar chart and a pie chart for these data using the proportions in the table.

b. Suppose 225 companies participated in this survey. Find the frequency, the number of companies, for each sector.

2.33 Marketing and Consumer Behavior A survey of new homes built in the Sleepy Creek Mountains of West Virginia produced the following results for the type of siding: **SIDING**

Siding	Frequency
Aluminum	20
Brick	15
Stucco	12
Vinyl	45
Wood	24

a. Find the relative frequency for each siding classification.

b. Construct a bar chart using frequency on the vertical axis and a pie chart for these data.

2.34 Public Policy and Political Science There are many *think tanks* in the world, consisting of groups of independent scholars with academic, government, and/or private experience. These think-tank scholars publish articles in appropriate journals and offer advice on politics, economics, and governmental policy matters. The following table lists the number of think tanks (TTs) in the world in 2012 by region.[8] **THINK**

Region	No. of TTs
Africa	554
Asia	1194
Europe	1836
Latin America and the Caribbean	721
Middle East and North Africa	339
North America	1919
Oceania	40

a. Find the relative frequency associated with each region.

b. Construct a bar chart using frequency on the vertical axis and a pie chart for these data.

2.35 Travel and Transportation In late October 2012, Hurricane Sandy had a devastating effect on the coast of the northeastern United States. This *superstorm* caused an estimated $65.6 billion in damages and was the largest Atlantic hurricane by diameter ever recorded. A survey was conducted to determine the primary source of local transportation information for people affected by the storm. The following table lists each source and its frequency.[9] **SANDY**

Source	Frequency
Official websites and alerts	265
Social media	198
News websites	152
News TV/radio	147
Friends/family	115
Community groups	45
Smartphone apps	45
Other	30

a. Find the relative frequency associated with each source.

b. Construct a pie chart for these data.

Extended Applications

2.36 Sports and Leisure Complete the following frequency distribution from a random sample of people visiting Atlantic City casinos.

Class	Frequency	Relative frequency
Bally's	40	
Caesars	25	0.125
Harrah's	32	
Resorts		0.110
Sands	25	
Trump Plaza		0.280
Total		1.000

a. What is the size of the random sample?

b. Which casino is most preferred by people in this survey? Justify your answer.

2.37 Travel and Transportation Families traveling to Walt Disney World in Florida often rent a car rather than use airport and hotel shuttle buses. A recent survey asked families to indicate the rental car agency used. The results are presented in the following bar chart.

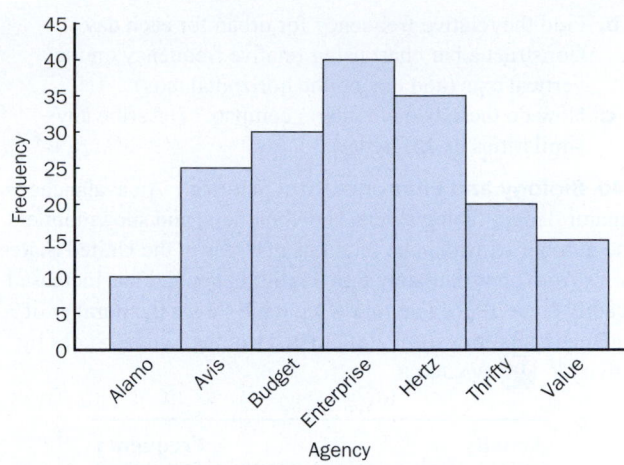

a. Construct a frequency distribution for these survey results.

b. How many observations were in this data set?

c. What proportion of people did not use Hertz or Enterprise?

d. Construct a pie chart for these data.

2.38 Marketing and Consumer Behavior One thousand customers entering the Mall of America in Bloomington, Minnesota, were randomly selected and asked to rank the variety of stores. The results are given in the following table.

Response	Frequency	Relative frequency
Excellent	50	
Very good	152	
Good	255	
Fair		0.4250
Poor		0.1180

a. Complete the frequency distribution.

b. Construct a bar chart using frequency on the vertical axis and a pie chart for these data.

c. What proportion of customers did not rank the store variety as very good or excellent?

2.39 Travel and Transportation The following table shows the number of fatal vehicle crashes in 2011 by day of the week for rural and urban roads or streets.[10] **CRASH**

Day of week	Rural	Urban
Monday	557	448
Tuesday	383	359
Wednesday	348	328
Thursday	351	330
Friday	415	386
Saturday	464	404
Sunday	676	514

a. Find the relative frequency for rural for each day. Construct a bar chart using relative frequency on the vertical axis (and day on the horizontal axis).

b. Find the relative frequency for urban for each day. Construct a bar chart using relative frequency on the vertical axis (and day on the horizontal axis).

c. How do these two bar charts compare? Describe any similarities or differences.

2.40 Biology and Environmental Science An avalanche is a major danger facing skiers, snowboarders, and snowmobilers. The number of avalanche fatalities per year in the United States varies from approximately 5 to 35, but in general has increased steadily since 1956. The following table shows the number of avalanche fatalities from 2002 to 2011 in the United States by activity.[11] **AVALANCH**

Activity	Frequency
Climber	30
Hiker	2
Rec snowplayer	1
Resident	8
Ski inbounds	9
Ski out of bounds	21
Ski tour	53
Snowboard inbounds	1
Snowboard out of bounds	10
Snowboard tour	23
Snowmobiler	107
Snowshoer	12
Work other	1
Work patrol	3
Ski helicoptor	2
Snowmobiler other	1

a. Find the relative frequency of fatalities for each activity.

b. Construct a bar chart using frequency on the vertical axis, and a bar chart using relative frequency on the vertical axis. Which of these two graphs do you think is a better graphical description of avalanche fatalities by activity? Why?

2.41 Sports and Leisure The following table shows the intended game involved in accidental hunting accidents in Texas in the years 2009–2011.[12] **HUNTING**

Animal hunted	Frequency
Dove	14
White-tailed deer	11
Rabbit/hare	5
Hog	21
Quail/pheasant	6
Turkey	3
Duck/goose	4
Coyote	2
Squirrel/prairie dog	4
Nongame bird/snake	5
Raccoon	2

a. Construct a bar chart for these data.

b. Construct a pie chart for these data.

c. Is it reasonable to conclude that hunters in Texas are more likely to be injured hunting hogs than any other animal? Why or why not?

2.42 Public Policy and Political Science A side-by-side or a stacked bar chart may be used to compare categorical data obtained from two (or more) different sources or groups. Figures 2.21 and 2.22 show an example of each—a comparison of test grades in two different sections of an introductory statistics course. The blue rectangles represent students from Section 01; the green rectangles represent students from Section 02. **RATINGS**

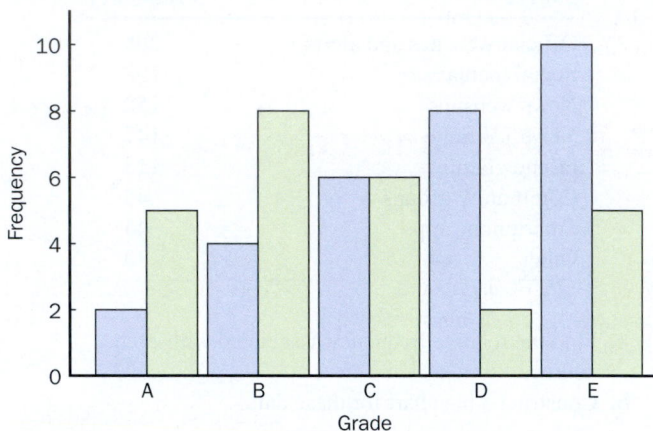

Figure 2.21 Side-by-side bar chart. Bars corresponding to the same category are placed side by side for easy comparison.

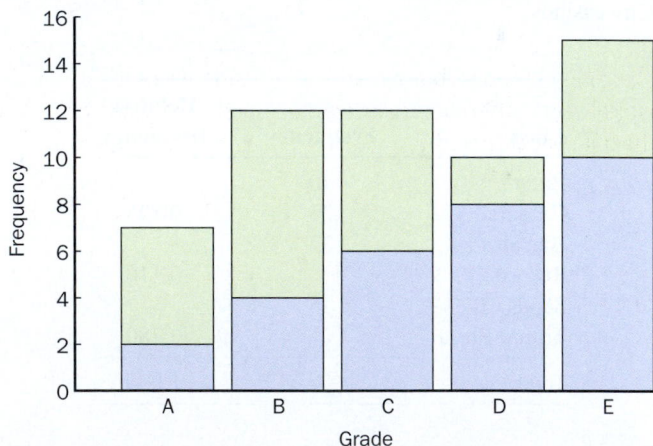

Figure 2.22 Stacked bar chart. Within each category, bars are stacked for comparison.

In January 2013, a poll by ABC News indicated that President Obama's popularity had reached a three-year high. Suppose the frequency of occurrence of each response, grouped by sex, is given in the following table.

Rating	Men frequency	Women frequency
Excellent	368	350
Very good	550	375
Good	426	165
Fair	450	360
Poor	206	250

a. Compute the relative frequency for each rating, for both groups.

b. Construct a side-by-side bar chart using the relative frequency of each class.

c. Why should relative frequency be used for comparison in the side-by-side bar chart rather than frequency?

2.43 Psychology and Human Behavior   The following table shows the number of property-crime violations in Manitoba and Saskatchewan in 2011.[13] 🏠 PROPERTY

Violation	Manitoba	Saskatchewan
Breaking and entering	9,305	9,079
Theft of motor vehicles	3,919	4,967
Theft over $5,000 (non-motor vehicle)	427	510
Theft under $5,000 (non-motor vehicle)	17,933	19,756
Mischief	26,361	31,741

a. Find the relative frequency for Manitoba for each violation.

b. Find the relative frequency for Saskatchewan for each violation.

c. Construct a side-by-side bar chart using the relative frequency of each class.

2.3 Stem-and-Leaf Plots

We will eventually need a quantitative measure of very far away.

This section introduces the stem-and-leaf plot, a graphical technique for describing numerical data. In Section 2.4, you will learn about some other tables and graphs for summarizing numerical data. The goal of all of these techniques is the same: to get a quick idea of the distribution of the data in terms of *shape*, *center*, and *variability*. In addition, we are always watching for *outliers*, values that are *very far away* from the rest.

A *stem-and-leaf plot* is a relatively new graphical procedure used to describe numerical data. It is fairly easy to construct, even by hand, and most statistical software packages have options for drawing this graph. A stem-and-leaf plot is a combination of sorting and graphing. One advantage of this plot is that the actual data are used to create the graph; we do not lose the original data values as we do when using tally marks to count them.

The center of a distribution, or typical value, often occurs where the data are clustered.

A stem-and-leaf plot can be used to describe the *shape*, *center*, and *variability* of the distribution. In Section 2.4, some specific terms and expressions used to describe shape are defined and illustrated. To estimate the center of a distribution, or to find a typical value, first arrange the observations in increasing order. Simply approximate a middle value, or range of values, in this list. More precise definitions and computations are presented in Chapter 3. The variability refers to the spread or compactness of the data. In addition, we always check for outliers.

How to Create a Stem-and-Leaf Plot

There are, of course, exceptions to this two-digit rule.

To create a stem-and-leaf plot, each observation in the data set must have at least two digits. Think of each observation as consisting of two pieces (a stem and a leaf). For example, suppose we consider the number of people watching a movie, and in one theater there are 372 people. The number 372 could be split into the pieces 37 (the first two digits) and 2 (the last digit).

1. Split each observation into a

   Stem: one or more of the leading, or left-hand, digits; and a

   Leaf: the trailing, or remaining, digit(s) to the right.

   Each observation in the data set must be split at the same place, for example between the tens place and the ones place.

2. Write a sequence of stems in a column, from the smallest occurring stem to the largest. Include all stems between the smallest and largest, even if there are no corresponding leaves.

3. List all the digits of each leaf next to its corresponding stem. It is not necessary to put the leaves in increasing order, but make sure the leaves line up vertically.

4. Indicate the units for the stems and leaves.

STEPPED TUTORIAL

STEMPLOTS

DATA SET

WTRFALL

Yoshio Tomii/SuperStock

Example 2.7 Waterfall Heights

Kerepakupai Meru, or Angel Falls, is the highest waterfall in the world.[14] Because the falls are so high, 979 meters, by the time water reaches the canyon below, it has vaporized into a giant mist cloud. Suppose the following table lists the total height, in meters, of several waterfalls in the world.

693	745	631	635	625	629	739	738	732	725
720	719	715	715	707	707	706	705	700	680
674	671	671	665	660	660	650	646	645	640
640	638	620	620	612	610	610	610	610	610
610	610	610	610	610	600	600	600	651	727

Construct a stem-and-leaf plot for these data.

SOLUTION

STEP 1   There are only two options for splitting each observation:

   a. split between the hundreds place and the tens place (e.g., split 693 as 6 and 93); or

   b. split between the tens place and the ones place (e.g., split 693 as 69 and 3).

   If we split between the hundreds and tens place, there will be only two stems, because the only numbers in the hundreds place are 6 and 7. The resulting plot will not reveal much about the distribution of the data. The better split is between the tens place and the ones place.

STEP 2   Scan the data to find the smallest and largest stems, and list all of the stems in a vertical column. Write each leaf next to its corresponding stem. For example,

$$693 \ \Rightarrow \ 69 \ | \ 3 \qquad \text{A 3 is placed in the 69 stem row.}$$
$$\qquad\qquad \uparrow \quad \uparrow$$
$$\qquad\quad \text{stem} \ \ \text{leaf}$$

For 745, a 5 is placed in the 74 stem row.
For 631, a 1 is placed in the 63 stem row.
For 635, a 5 is placed in the 63 stem row.

STEP 3   Continue in this manner, to produce the following stem-and-leaf plot:

60	0 0 0
61	2 0 0 0 0 0 0 0 0 0 0
62	5 9 0 0
63	1 5 8
64	6 5 0 0
65	0 1
66	5 0 0
67	4 1 1
68	0
69	3
70	7 7 6 5 0
71	9 5 5
72	5 0 7
73	9 8 2
74	5

Stem = 10
Leaf = 1

A technology solution:

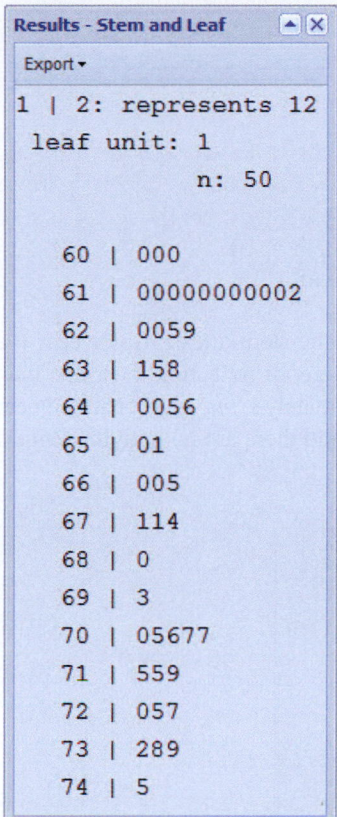

Results - Stem and Leaf   ▲ ✕

Export ▾

1 | 2: represents 12

 leaf unit: 1

            n: 50

   60 | 000
   61 | 00000000002
   62 | 0059
   63 | 158
   64 | 0056
   65 | 01
   66 | 005
   67 | 114
   68 | 0
   69 | 3
   70 | 05677
   71 | 559
   72 | 057
   73 | 289
   74 | 5

Figure 2.23 CrunchIt! stem-and-leaf plot.

Note that Stem = 10 means the rightmost digit in the stem is in the tens place and Leaf = 1 means the leftmost digit in each leaf is in the ones place. Reading from the graph, the smallest waterfall height is 600 meters and the largest is 745 meters. The *center* of a data set is a typical value or values near the middle of the observations when they are arranged in (increasing) order. For these data, the center appears to be in the 64 or 65 stem row. There are no outlying values. Figure 2.23 shows a technology solution.

TRY IT NOW   GO TO EXERCISE 2.51

A CLOSER LOOK

1. As a general rule of thumb, try to construct the plot with 5 to 20 stems. With fewer than 5, the graph is too compact; with more than 20, the observations are too spread out. Neither extreme reveals much about the distribution.

2. Sometimes, to help us find the center of the data, we put the leaves in increasing order, to make an **ordered** stem-and-leaf plot.

3. Some advantages of a stem-and-leaf plot: each observation is a visible part of the graph and (in an ordered stem-and-leaf plot, as when using a computer) the data are sorted. However, a stem-and-leaf plot can get very big, very fast.

If a stem-and-leaf plot is made for a very large data set, the stems may be divided, usually in half or fifths. Consider the following example.

Example 2.8 Hotel Room Rates

The I-95 Exit Guide allows travelers to easily find hotels at exits along I-95 within $\frac{1}{2}$ mile of an I-95 exit.[15] Suppose a random sample of room rates (in dollars) for hotels along I-95 was obtained, and the data are given in the following table.

75	84	78	79	72	73	50	90	85	69
76	77	77	78	61	58	80	81	75	75
89	89	74	73	79	86	85	94	64	72
77	83	78	81	91	70	93	75	78	54
55	60	63	69	65	73	93	81	79	79
77	75	61	71	68	72	77			

Construct an ordered stem-and-leaf plot for these data.

SOLUTION

STEP 1 If we split each observation between the tens place and the ones place, there will be five stems. However, the leaves will extend far to the right and the shape, center, and spread of the distribution will be unclear.

STEP 2 Divide each stem in half. The first 5-stem row holds numbers 50–54, the second 5-stem row holds numbers 55–59, the first 6-stem row holds numbers 60–64, etc.

STEP 3 The resulting stem-and-leaf plot, with divided stems, offers a better graphical description of the distribution. Note that the leaves have been ordered.

Stem-and-leaf plot for the hotel rate data

5	0 4
5	5 8
6	0 1 1 3 4
6	5 8 9 9
7	0 1 2 2 2 3 3 3 3 4
7	5 5 5 5 5 6 7 7 7 7 7 8 8 8 8 9 9 9 9
8	0 1 1 1 3 4
8	5 5 6 9 9
9	0 1 3 3 4

Stem = 10
Leaf = 1

STEP 4   Notice that we can draw a straight line across the stem-and-leaf plot near the 75–79 row and the graph is almost a mirror image, or reflection, over this line. Therefore, the distribution of the data is approximately *symmetric*, centered near 75–79. In addition, the distribution is compact and there are no outlying values.

Figure 2.24 shows a technology solution.

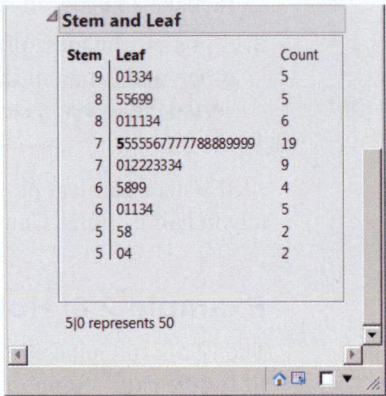

Figure 2.24 JMP stem-and-leaf plot.

TRY IT NOW   GO TO EXERCISE 2.55

Two sets of data can be compared graphically using a back-to-back stem-and-leaf plot. Two plots are constructed using the same stem column. List the leaves for one data set to the left, and those for the other to the right.

Example 2.9 Cholesterol Levels

Your total cholesterol level is the sum of low-density lipoproteins (LDLs) and high-density lipoproteins (HDLs). A total cholesterol level of less than 200 mg/dL (milligrams per deciliter) is desirable, whereas 240 mg/dL or higher is considered high risk.[16] According to the Centers for Disease Control and Prevention, the average total cholesterol for adult Americans is about 200 mg/dL. Suppose a random sample of total cholesterol levels was obtained for men and women. The data are given in the following table.

	Men					Women			
110	124	132	147	157	183	190	201	211	154
164	172	180	193	201	186	212	213	169	173
210	224	112	158	165	177	195	203	203	189
173	181	193	205	216	207	158	218	213	205
194	194	179	185	185	205	204	189	179	177

Construct a back-to-back ordered stem-and-leaf plot for these data.

SOLUTION

STEP 1   The following graph is a back-to-back stem-and-leaf plot for these data.

Men		Women
2 0	11	
4	12	
2	13	
7	14	
8 7	15	4 8
5 4	16	9
9 3 2	17	3 7 7 9
5 5 1 0	18	3 6 9 9
4 4 3 3	19	0 5
5 1	20	1 3 3 4 5 5 7
6 0	21	1 2 3 3 8
4	22	

Stem = 10
Leaf = 1

STEP 2   The center column of numbers (11, 12, 13, . . .) represents the stems for both groups. The stem-and-leaf plot for the men's data is constructed to the left, while the plot for the women's data is constructed to the right. Note that the leaves have been placed in increasing order, starting from the stem and proceeding outward.

STEP 3   The distribution for the men seems more spread out, or has more variability, while the distribution of women's HDL-cholesterol levels is more compact and seems centered at a slightly greater value. ●

TRY IT NOW   GO TO EXERCISE 2.57

In mathematics, *truncate* means discard the digits to the right of a specific place. In order to round a number to a certain position, consider the digit to the right of the rounding position. If this digit is a 5 or greater, then round up. Otherwise, leave the rounding digit unchanged (and replace all digits to the right with 0).

When constructing a stem-and-leaf plot, if there are two or more digits in each leaf, the trailing digits may be *truncated* or the entire leaf may be *rounded*. Suppose a data set includes the total yardage for randomly selected golf courses. Consider three observations, 6518, 6523, and 6576, and suppose each observation is split between the hundreds place and the tens place.

The following diagram shows the 65 stem row for three stem-and-leaf plots. The first is constructed with two-digit leaves. The second is constructed by simply truncating the ones, or last, digit. The third plot is constructed by rounding each leaf to the nearest ten.

Two-digit leaf	Truncate each leaf	Round each leaf
65 | 18 leaf = 18	65 | 18 leaf = 1	65 | 18 rounds to 20, leaf = 2
65 | 23 leaf = 23	65 | 23 leaf = 2	65 | 23 rounds to 20, leaf = 2
65 | 76 leaf = 76	65 | 76 leaf = 7	65 | 76 rounds to 80, leaf = 8

Stem	Leaves		Stem	Leaves		Stem	Leaves
⋮	⋮		⋮	⋮		⋮	⋮
65	18 23 76		65	1 2 7		65	2 2 8
⋮	⋮		⋮	⋮		⋮	⋮

Stem = 100   Stem = 100   Stem = 100
Leaf = 10   Leaf = 10   Leaf = 10

Technology Corner

Procedure: Construct a stem-and-leaf plot.
Reconsider: Example 2.7, solution, and interpretations.

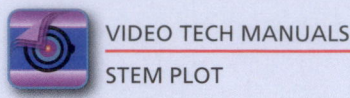

VIDEO TECH MANUALS

STEM PLOT

There is no built-in command on the TI-84 Plus C or in Excel to construct a stem-and-leaf plot. Some calculator programs are available, and there are several add-ins for Excel for drawing a stem-and-leaf plot.

CrunchIt!

1. Enter the data in column Var1. Rename the column if desired.
2. Select Graphics; Stem and Leaf. Choose Var1 for the Sample and optionally enter a Title. Click the Calculate button. Refer to Figure 2.23 (page 47).

Minitab

1. Enter the data into column C1.
2. Select <u>G</u>raph; Stem-and-Lea<u>f</u>.
3. Enter C1 under Graph variables. Click <u>O</u>K.
4. The increment (distance between stems) is automatically selected and the leaves are placed in order. The numbers in the left column represent the cumulative counts from each end. The stem row containing the middle value is marked by only a count, in parentheses, of the number of observations in that row. N is the total number of observations. See Figure 2.25.

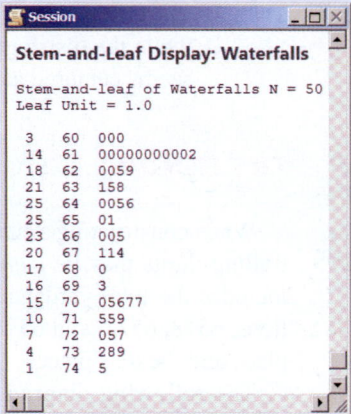

Figure 2.25 Minitab stem-and-leaf plot.

SECTION 2.3 EXERCISES

Concept Check

2.44 True/False　The stem in a stem-and-leaf plot must be only one digit.

2.45 True/False　When constructing a stem-and-leaf plot, one can omit stem rows that contain no leaves.

2.46 True/False　There may be more than one way to split each observation into a stem and a leaf.

2.47 True/False　When constructing a stem-and-leaf plot, every observation must be split into a stem and a leaf in the same way.

2.48 Short Answer　We try to construct a stem-and-leaf plot with 5–20 stems. What happens if we use fewer than 5 or more than 20 stems?

Practice

2.49　Construct a stem-and-leaf plot for the following data.

4.7	5.1	6.6	3.9	5.0	2.9	3.6	5.5	4.2	5.1
4.9	5.4	6.1	4.1	3.6	6.4	4.7	4.1	5.7	3.6
6.8	3.5	6.4	6.4	7.1	2.7	5.8	5.2	5.9	5.7

Determine a range of numbers to indicate the center of the data. Within this range, select one number that is a *typical* value for this data set. EX2.49

2.50 Construct a stem-and-leaf plot for the data given on the text website. EX2.50

2.51 Construct a stem-and-leaf plot for the data given on the text website. Split each observation between the tens place and the ones place, and divide each stem in half. Determine a range of numbers to indicate the center of the data. Within this range, select one number that is a *typical* value for this data set. EX2.51

2.52 Construct a stem-and-leaf plot for the data given on the website for this book. Use the stem-and-leaf plot to identify any outliers in this distribution. EX2.52

2.53 Consider the following stem-and-leaf plot:

50	3
51	5
52	3 7
53	4 6
54	3 3 9
55	0 0 3 3 7
56	1 1 1 3 3 4 6 7 7
57	0 0 1 1 3 4 4 4 4 5 7 8 8
58	0 1 2 2 3 4 4 6 6 6 7 7
59	3 3 3 5 5 6 9
60	1 1 2

Stem = 10
Leaf = 1

a. List the actual observations in the 54 stem row.
b. What is a typical value for this data set?
c. Do the data seem to be evenly distributed, or does one end tail off more slowly than the other?
d. Does the stem-and-leaf plot suggest there are any outliers in this data set? If so, what are they?

2.54 Consider the data given in the table below: EX2.54

1717	1719	1645	3739	3024	3664	3830
2991	2430	2730	3469	5086	2119	3021
3292	2844	3426	2067	3215	2767	3124
2573	2840	2449	2584	1505	1390	1645
2497	3466	3228	3192			

a. Construct a stem-and-leaf plot by splitting each observation between the thousands place and the hundreds place.
b. Construct a stem-and-leaf plot by splitting each observation between the hundreds place and the tens place (using two-digit leafs).
c. Which plot presents a better picture of the distribution? Why?

Applications

2.55 Psychology and Human Behavior A random sample of patients involved in a psychology experiment was

selected and the reaction time (in seconds) for each was recorded. REACTIME

a. Construct a stem-and-leaf plot by splitting each observation between the ones place and the tenths place. Truncate the hundredths digit so that each leaf has a single digit.
b. Construct a stem-and-leaf plot by splitting each observation between the ones place and the tenths place. Round each leaf to the nearest tenth so that each leaf has a single digit.
c. Describe any differences between the two plots. What is a typical value?

2.56 Public Health and Nutrition The owner of Copperfield Racquet and Health Club randomly selected 50 people and recorded the number of calories burned after 20 minutes on a treadmill. CALBURN

a. Construct a stem-and-leaf plot by splitting each observation between the tens place and the ones place.
b. Construct a stem-and-leaf plot by splitting each observation between the tens place and the ones place, and by dividing each stem in half.
c. Which stem-and-leaf plot is *better*? Why?

2.57 Physical Sciences A random sample of hot water temperatures (°F) on lower floors and upper floors in the Renaissance Dallas Hotel was obtained. WATEMP

a. Construct a back-to-back stem-and-leaf plot to compare these two distributions.
b. Using the plot in part (a), describe any similarities and/or differences between the distributions.

2.58 Physical Sciences The intensity of light is measured in foot-candles or in lux. In full daylight, the light intensity is approximately 10,700 lux, and at twilight the light intensity is about 11 lux. The recommended level of light in offices is 500 lux.[17] A random sample of 50 offices was obtained and the lux measurement at a typical work area was recorded for each. The data are given in the following table: LUXMEAS

468	526	463	520	481	521	536	492	509	520
497	487	506	464	474	516	503	481	562	514
503	482	531	486	488	508	495	536	504	514
529	518	495	497	471	458	494	519	511	490
435	520	499	492	519	466	450	482	514	475

a. Construct a stem-and-leaf plot for these light-intensity data.
b. What is a typical light intensity? Are there any outliers? If so, what are they?

2.59 Public Health and Nutrition Every patient who visits a hospital emergency room is classified by the immediacy with which the patient should be seen. The American College of Emergency Physicians suggests that approximately 92% of all patients who visit emergency rooms can be classified as urgent, that is, need attention within 1 minute to 2 hours.[18] Suppose a random sample of hospital emergency rooms was obtained and yearly records were examined. The percentage of patient visits classified as urgent is given on the text website. **URGENT**

a. Construct a stem-and-leaf plot for these data.

b. What is a typical value? Are there any outliers? If so, what are they?

2.60 Biology and Environmental Science The Port of Tacoma (Washington) handled approximately 1.7 million TEUs (20-foot equivalent units) in 2012.[19] A random sample of domestic containers was obtained and the volume of each (in TEUs) was recorded. **DOMVOL**

a. Construct a stem-and-leaf plot for these data. Split each observation between the tenths place and the hundredths place.

b. Describe the shape of the container volume distribution in terms of shape, center, and spread. Are there any outliers? If so, what are they?

2.61 Business and Management A random sample of gasoline stations in Philadelphia was obtained. The number of years each station has been in operation is given on the text website. **GASYEAR**

a. Construct a stem-and-leaf plot for these data.

b. What is a typical number of years a station has been in operation? Are there any outliers? If so, what are they?

2.62 Manufacturing and Product Development Home Depot conducted a survey on the lifetime of dishwashers. Forty random users were contacted and asked to report the number of years their dishwasher lasted before needing replacement. **DISHLIFE**

a. Construct a stem-and-leaf plot for these data. Split each observation between the ones place and the tenths place.

b. Describe the distribution of dishwasher lifetimes in terms of shape, center, and spread.

c. What is a typical lifetime? Are there any outliers? If so, what are they?

2.63 Biology and Environmental Science The Great Pumpkin Commonwealth promotes the hobby of growing giant pumpkins. This group establishes standards and regulations so that the each pumpkin is of high quality and to ensure fairness in the competition for the largest pumpkin. A random sample of the largest pumpkins from 2012 was obtained, and the data are given on the text website.[20] **PUMPKIN**

a. Construct a stem-and-leaf plot for these data.

b. What is a typical weight for these giant pumpkins? Are there any outliers in the data set? If so, what are they?

Extended Applications

2.64 Sports and Leisure A greyhound race handicapper uses several factors to predict the winner, such as past performance, track condition, early speed, form, and competition. Races on a 5/16-mile track were randomly selected at the Naples-Fort Myers track, and the winning time (in seconds) was recorded for each. The data are given in the following table.[21] **GREYRACE**

30.78	30.00	30.47	30.81	30.02	30.76
30.47	30.70	30.17	30.58	30.56	30.44
30.35	30.35	30.41	31.37	30.57	30.52
30.06	30.59	30.56	30.38	30.82	31.05
30.67	30.31	29.95	30.21	29.98	30.59

a. Construct a stem-and-leaf plot for these data. Split each observation between the tenths place and the hundredths place.

b. What is a typical winning time? If a dog has never run better than 31.20 seconds in a 5/16-mile race, do you think it has a chance of winning? Justify your answer.

c. Could a stem-and-leaf plot be constructed with the split between the ones place and the tenths place? How about between the tens place and the ones place? Explain.

2.65 Biology and Environmental Science Many piano sellers recommend a special humidifier, especially for more expensive pianos. This device is installed inside the piano and works to keep the instrument in tune by maintaining a stable humidity. To test whether a humidifier really helps, several pianos with and without humidifiers were tuned and then checked six months later. Middle C was used as a measure of how well each piano stayed in tune. In a perfectly tuned piano, middle C has a frequency of 256 cycles per second. The frequency (in cycles per second) of middle C for each group, after six months, is given on the text website. **MIDDLEC**

a. Construct a back-to-back stem-and-leaf plot for this data.

b. Use the plot in part (a) to describe any differences between the groups. Based on this plot, do you think a humidifier helps a piano stay in tune? Justify your answer.

2.66 Travel and Transportation There are national standards for every road sign, pavement marking, and traffic signal. However, there are no formal state policies regarding the duration of an amber light. According to the Center for Sustainable Mobility at the Virginia Tech Transportation Institute, the amber light time is set to 4.2 seconds on a 45-mph road.[22] Longer times would be used on roads with higher speed limits. Suppose a random sample of traffic signals for 45-mph roads in Norman, Oklahoma, was selected, and the duration of the amber light was recorded for each. **SIGNAL**

a. Construct a stem-and-leaf plot for these data. Divide each stem into five parts.

b. Based on the plot in part (a), do you believe this city has set the amber light duration to meet federal recommendations? Justify your answer.

2.67 Biology and Environmental Science The 2011 Maine Sea Scallop Survey was conducted in November 2011 between West Quoddy Head and Matinicus Island.[23] Each sea scallop catch was divided into three size categories: seed, sublegal, and harvestable (≥ 101.6 mm). Based on information in this report, a random sample of shell heights (in millimeters) is given on the text website. **SCALLOP**

a. Construct a stem-and-leaf plot for these data.

b. Describe the distribution of shell heights in terms of shape, center, and spread.

c. Estimate the proportion of sea scallops that are harvestable.

2.4 Frequency Distributions and Histograms

Stem-and-leaf plots can be used to describe the shape, center, and variability of a numerical data set, but they can become huge and complex if the number of observations is large. A summary table like a frequency distribution for categorical data would be helpful. However, when the data set is numerical, there are no natural categories, as for qualitative data. The solution is to use intervals as categories, or classes. We can then construct a frequency distribution for continuous data (similar to the categorical case), and a histogram (analogous to a bar chart for categorical data). For a random sample of days in 2011 and 2012, Figure 2.26 shows a histogram of silver prices (in dollars per ounce).[24]

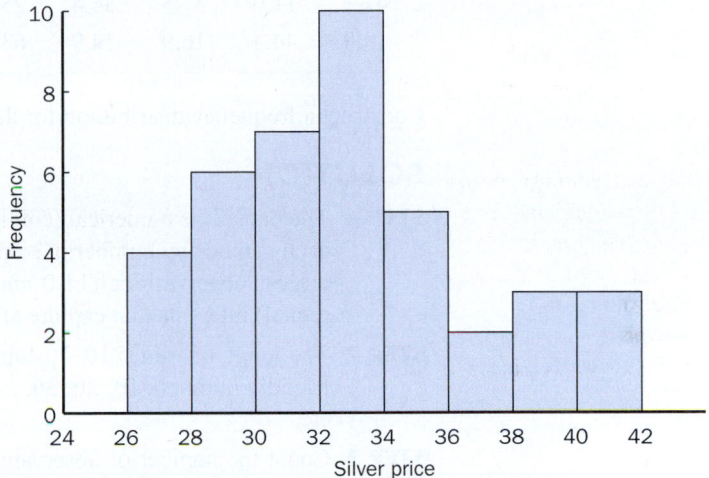

Figure 2.26 An example of a histogram.

Definition

A **frequency distribution** for numerical data is a summary table that displays classes, frequencies, relative frequencies, and cumulative relative frequencies.

Here is a procedure for constructing a frequency distribution, along with the necessary definitions.

How to Construct a Frequency Distribution for Numerical Data

In other words, partition the measurement axis into 5–20 subintervals.

1. Choose a range of values that captures all of the data. Divide it into nonoverlapping (usually equal) intervals. Each interval is called a **class**, or **class interval**. The endpoints of each class are the *class boundaries*.
2. We use the left-endpoint convention. An observation equal to an endpoint is allocated to the class with that value as its lower endpoint. Hence, the lower class boundary is always included in the interval, and the upper class boundary is never included. This ensures that each observation falls into exactly one interval.
3. As a rule of thumb, there should be 5–20 intervals. Use *friendly* numbers, for example, 10–20, 20–30, etc., not 15.376–18.457, 18.457–21.538, etc.
4. Count the number of observations in each class interval. This count is the **class frequency** or simply the **frequency**.
5. Compute the proportion of observations in each class. This ratio, the class frequency divided by the total number of observations, is the **relative frequency**.

6. Find the **cumulative relative frequency** (CRF) for each class: the sum of all the relative frequencies of classes up to and including that class. This column is a *running total* or *accumulation* of relative frequency, by row.

DATA SET

TORQUE

Example 2.10 Nuts and Bolts

Torque is a measure of the force needed to cause an object to rotate. It is usually measured in foot-pounds (ft-lb). As part of a quality-control program, Whirlpool inspectors measure the initial torque needed to loosen the balancing bolts on each leg of a clothes washer. A random sample of these measurements is given in the following table.

20.4	24.1	28.4	53.4	62.1	31.7	57.2	45.7	38.1	51.1
41.3	11.0	37.5	36.4	25.6	43.5	23.1	24.2	35.5	26.4
13.0	44.4	16.9	14.9	63.7					

Construct a frequency distribution for these data.

SOLUTION

STEP 1 The data set is numerical (continuous). The observations are measurements, and each can be any number in some interval. Scan the data to find the smallest and largest observations (11.0 and 63.7). Choose between 5 and 20 reasonable (equal) intervals that capture all of the data.

STEP 2 The range of values 10–70 captures all of the data. Divide this range using the friendly numbers 10, 20, 30, . . . into the class intervals 10–20, 20–30, 30–40, etc.

STEP 3 Count the number of observations in each interval. For example, in the interval 10–20, there are four observations (16.9, 14.9, 11.0, and 13.0), so the frequency is 4.

STEP 4 Compute the proportion of observations in each class. For example, in the interval 10–20, the relative frequency is 4 (observations) divided by 25 (total number of observations).

STEP 5 Find the CRF for each class. For example, for the class 30–40, the cumulative relative frequency is the sum of the relative frequencies of this class and of all those listed above it: 0.16 + 0.28 + 0.20 = 0.64.

Class	Frequency	Relative frequency		Cumulative relative frequency	
10–20	4	0.16	(=4/25)	0.16	(=0.16)
20–30	7	0.28	(=7/25)	0.44	(=0.16 + 0.28)
30–40	5	0.20	(=5/25)	0.64	(=0.44 + 0.20)
40–50	4	0.16	(=4/25)	0.80	(=0.64 + 0.16)
50–60	3	0.12	(=3/25)	0.92	(=0.80 + 0.12)
60–70	2	0.08	(=2/25)	1.00	(=0.92 + 0.08)
Total	25	1.00			

STEP 6 As for categorical data, if you must construct a frequency distribution by hand, an additional *tally column* is helpful (as introduced in Section 2.2). Insert this after the class column, and use a *tally mark* or *tick mark* to count observations as you read them from the table. ●

TRY IT NOW   GO TO EXERCISE 2.77

The last (*total*) row in a frequency distribution is optional, but it is a good check of your calculations. The frequencies should sum to the total number of observations

(25 in Example 2.10), and the relative frequencies should sum to 1.00 (subject to round-off error).

The CRF of the first class row is equal to the relative frequency of the first class. There are no other observations before the first class. The CRF of the last class should be 1.00 (subject to round-off error). You must accumulate all of the data by the last class.

CRF gives the proportion of observations in that class and all previous classes. In Example 2.10, the CRF of the class 40–50 is 0.80. Interpretation: the proportion of torque measurements less than 50 is 0.80.

A CLOSER LOOK

This idea of working *backward* from cumulative relative frequency to obtain relative frequency is a handy technique for answering many probability questions.

1. Suppose you were given just the CRF for each class. To find the *relative frequency* for a class, take the class CRF and subtract the previous class CRF. In Example 2.10, to find the relative frequency for the class 50–60: 0.92 (CRF for the class 50–60) − 0.80 (CRF for the previous class 40–50) = 0.12.

2. If the data set is numerical and discrete, use the same procedure outlined above for constructing a frequency distribution. If the number of discrete observations is small, then each value may be a class, or category. In addition, certain liberties are sometimes acceptable in listing the classes. For example, suppose a discrete data set consists of integers from 1 to 30. One might use the classes 1–5, 6–10, 11–15, 16–20, 21–25, and 26–30. This is not a strict partition of the interval 1–30 even though these classes are disjoint, or do not overlap. They do not allow for all numbers between 1 and 30. For example, the value 5.5 is between 1 and 30 but does not fall into any of these classes. However, these classes work fine in this case, because each observation is an integer. The resulting frequency distribution is perfectly valid.

A *histogram* is a graphical representation of a frequency distribution, a plot of frequency versus class interval. Given a frequency distribution, here is a procedure for constructing a histogram.

How to Construct a Histogram

1. Draw a horizontal (measurement) axis and place tick marks corresponding to the class boundaries.
2. Draw a vertical axis and place tick marks corresponding to frequency. Label each axis.
3. Draw a rectangle above each class with height equal to frequency.

STEPPED TUTORIAL

HISTOGRAMS

Example 2.11 Nuts and Bolts, Continued

Construct a frequency histogram for the torque data presented in Example 2.10. For reference, here is the frequency distribution from Example 2.10:

Class	Frequency	Relative frequency	Cumulative relative frequency
10–20	4	0.16	0.16
20–30	7	0.28	0.44
30–40	5	0.20	0.64
40–50	4	0.16	0.80
50–60	3	0.12	0.92
60–70	2	0.08	1.00

SOLUTION

STEP 1 Draw a horizontal axis and place tick marks corresponding to the class boundaries, or endpoints: 10 through 70 by tens.

STEP 2 Draw a vertical axis for frequency and place appropriate tick marks by checking the frequency distribution. The frequencies range from 0 to 7, so draw tick marks at 0 to at least 7 on the vertical axis.

STEP 3 Draw a rectangle above each class with height equal to frequency. The resulting histogram is shown in Figure 2.27. Figure 2.28 shows a technology solution.

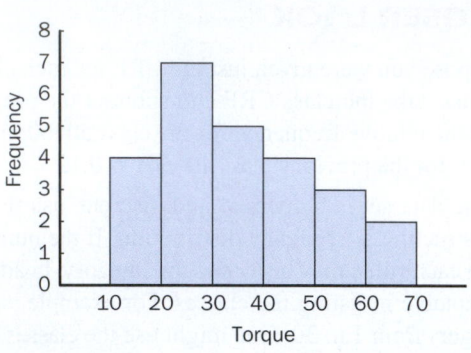

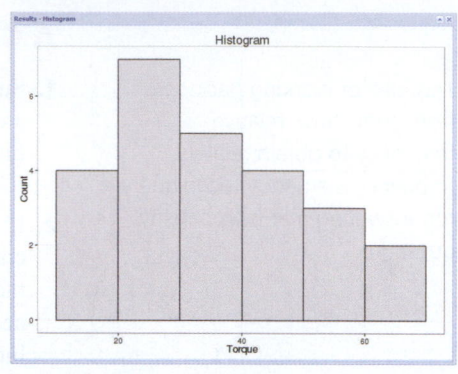

Figure 2.27 Frequency histogram for torque.   **Figure 2.28** CrunchIt! histogram.

TRY IT NOW   GO TO EXERCISE 2.86

A CLOSER LOOK

1. A histogram tells us about the shape, center, and variability of the distribution. In addition, we can quickly identify any outliers.

2. If you must draw a histogram by hand, then you need to construct the frequency distribution first. However, calculators and computers construct histograms directly from the data. The frequency distribution is in the background and is usually not displayed.

3. To construct a *relative* frequency histogram, plot relative frequency versus class interval. The only difference between a frequency histogram and a relative frequency histogram is the scale on the vertical axis. The two graphs are identical in appearance. In Example 2.12 both a frequency histogram and a relative frequency histogram are shown.

4. Histograms should not be used for inference. They provide a quick look at the distribution of data and only *suggest* certain characteristics.

Histogram usually means *frequency histogram.*

Example 2.12 Highway Tunnels

A random sample of highway tunnel lengths (in feet) was obtained, and the resulting frequency distribution is shown in the following table:

Class	Frequency	Relative frequency	Cumulative relative frequency
0–500	16	0.08	0.08
500–1000	28	0.14	0.22
1000–1500	54	0.27	0.49
1500–2000	48	0.24	0.73
2000–2500	36	0.18	0.91
2500–3000	18	0.09	1.00
Total	200	1.00	

© Carol/Alamy

Use this table to construct a frequency histogram and a relative frequency histogram for these data.

SOLUTION

STEP 1 For each graph, draw a horizontal axis and place tick marks at the class boundaries: 0, 500, 1000, . . . , 3000.

STEP 2 For the frequency histogram:

    a. Draw a vertical axis for frequency. Since the largest frequency is 54, use the tick marks at 0, 10, 20, . . . , 60.

    b. Draw a rectangle above each class with height equal to frequency. The resulting frequency histogram is shown in Figure 2.29.

STEP 3 For the relative frequency histogram:

    a. Draw a vertical axis for relative frequency. Because the largest relative frequency is 0.27, use the tick marks at 0, 0.05, 0.10, . . . , 0.30.

    b. Draw a rectangle above each class with height equal to relative frequency. The resulting relative frequency histogram is shown in Figure 2.30.

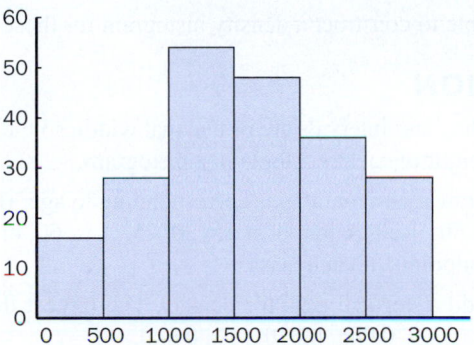
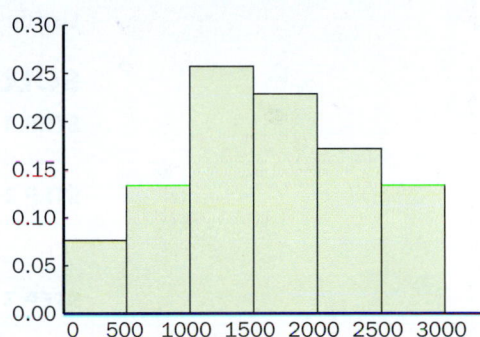

Figure 2.29 Frequency histogram for the tunnel-length data.

Figure 2.30 Relative frequency histogram for the tunnel-length data.

TRY IT NOW GO TO EXERCISE 2.89

If the class widths are *unequal* in a frequency distribution, then neither the frequency nor the relative frequency should be used on the vertical axis of the corresponding histogram. To account for the unequal class widths, set the area of each rectangle equal to the relative frequency. In this case, the height of each rectangle is called the *density*, and it is equal to the relative frequency divided by the class width.

How to Find the Density

To find the density for each class:

1. Set the *area* of each rectangle equal to relative frequency.
The *area* of each rectangle is *height* times class *width*.
Area of rectangle = Relative frequency
$$= (\text{Height}) \times (\text{Class width})$$

2. Solve for the height.
Density = Height = (Relative frequency)/(Class width)

If two classes have the same frequency, but one class has double the width, then the corresponding rectangle in a traditional histogram would have double the area. This misrepresents the distribution.

The following example shows an extended frequency distribution with the density of each class included, and the corresponding density histogram.

Example 2.13 Accident Demographics

Younger drivers tend to be involved in more automobile crashes than older drivers. This may be attributed to risk and inexperience. Suppose the following table shows the number of automobile accidents in Michigan in 2013 for each driver age group.[25] The width of each class and the density calculations are also shown.

Class	Frequency	Relative frequency	Width	Density
16–18	18,157	0.0494	2	0.0247 (=0.0494/2)
18–21	40,122	0.1091	3	0.0364 (=0.1091/3)
21–25	45,247	0.1231	4	0.0308 (=0.1231/4)
25–30	43,106	0.1172	5	0.0234 (=0.1172/5)
30–40	73,846	0.2008	10	0.0201 (=0.2008/10)
40–50	78,442	0.2133	10	0.0213 (=0.2133/10)
50–60	68,781	0.1871	10	0.0187 (=0.1871/10)
Total	367,701	1.0000		

Use this table to construct a density histogram for these data.

SOLUTION

STEP 1   The class intervals are of unequal width, so the class density must be used as the height of each rectangle in a histogram.

STEP 2   Draw a horizontal axis corresponding to age. Because the classes range from 16 to 60, use tick marks at 15, 20, 25, . . . , 60, or tick marks corresponding to the endpoints of each class.

STEP 3   Add a vertical axis for density. The largest density is 0.0364, so use the tick marks 0, 0.005, 0.010, . . . , 0.040.

STEP 4   Draw a rectangle above each class with height equal to density. The resulting density histogram is shown in Figure 2.31. A technology solution is shown in Figure 2.32.

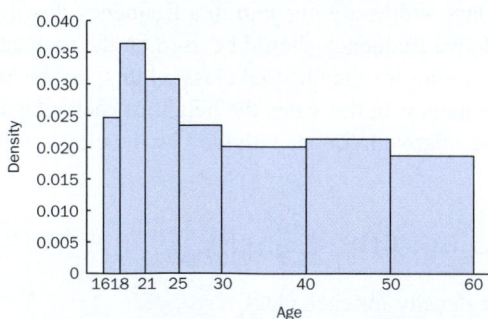

Figure 2.31 Histogram for unequal class widths: density histogram for the age data.

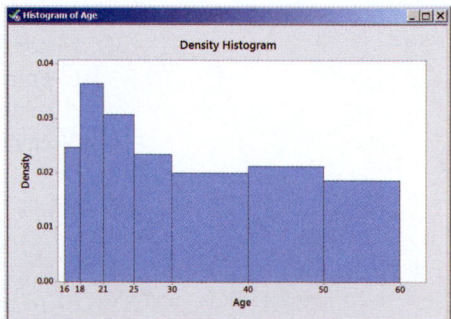

Figure 2.32 A technology solution: Minitab density histogram.

TRY IT NOW   GO TO EXERCISE 2.94

Shape of a Distribution

Because the relative frequency is equal to the area of each rectangle in a *density* histogram, the sum of the *areas* of all the rectangles is 1. This is an important concept as we begin to associate area with probability.

The *shape* of a distribution, represented in a histogram, is an important characteristic. To help describe the various shapes, we draw a smooth curve along the tops of the rectangles that captures the general nature of the distribution (as shown in Figure 2.33). To help identify and describe distributions quickly, a *smoothed histogram* is often drawn on a graph without a vertical axis, without any tick marks on the measurement axis, and without any rectangles (as shown in Figure 2.34).

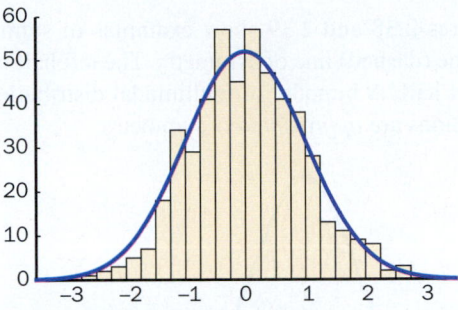

Figure 2.33 Smooth curve that captures the general shape of the distribution.

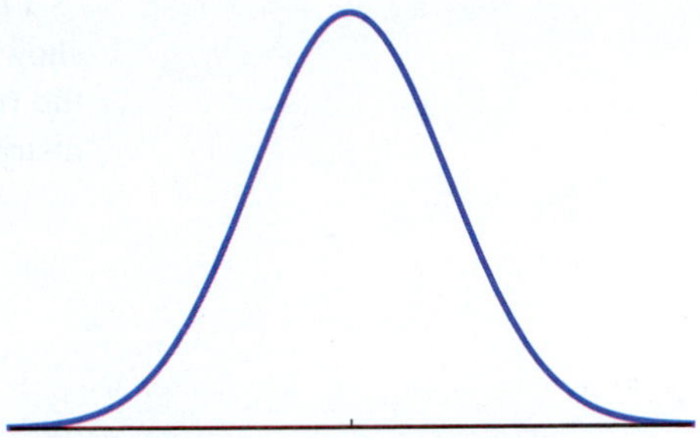

Figure 2.34 Typical smoothed histogram.

The first important characteristic of a distribution is the number of peaks.

Definition

1. A **unimodal** distribution has one peak. This is very common; almost all distributions have a single peak.
2. A **bimodal** distribution has two peaks. This shape is not very common and may occur if data from two different populations are accidentally mixed.
3. A **multimodal** distribution has more than one peak. A distribution with more than two distinct peaks is very rare.

Examples of these three types of distributions are shown in Figures 2.35–2.37.

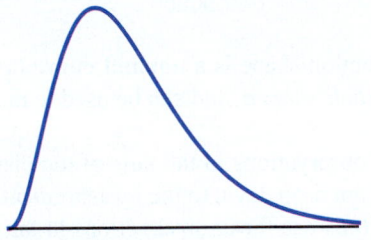

Figure 2.35 Unimodal histogram.

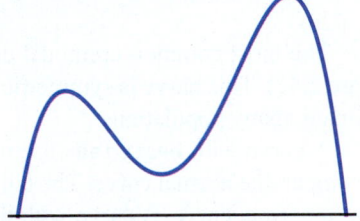

Figure 2.36 Bimodal histogram.

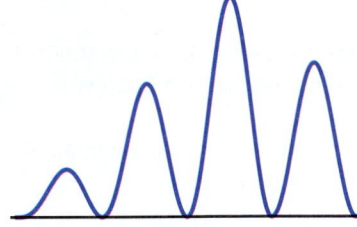

Figure 2.37 Multimodal distribution with four peaks.

The following characteristics are used to further classify and identify unimodal distributions.

Definition

1. A unimodal distribution is **symmetric** if there is a vertical line of symmetry in the distribution.
2. The **lower tail** of a distribution is the leftmost portion of the distribution, and the **upper tail** is the rightmost portion of the distribution.

3. If a unimodal distribution is not symmetric, then it is **skewed**.

   (a) In a **positively skewed** distribution or a distribution that is **skewed to the right**, the upper tail extends farther than the lower tail.

   (b) In a **negatively skewed** distribution, or a distribution that is **skewed to the left**, the lower tail extends farther than the upper tail.

Figures 2.38 and 2.39 show examples of symmetric, unimodal distributions. Each shows the (dashed) line of symmetry. The left half of the distribution is a mirror image of the right half. A bimodal or multimodal distribution may also be symmetric, and many distributions are *approximately* symmetric.

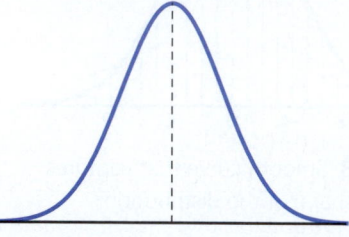

Figure 2.38 Symmetric distribution.

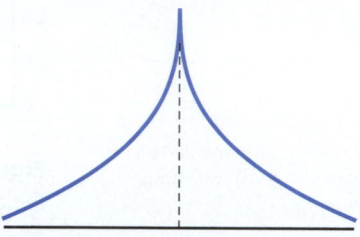

Figure 2.39 Symmetric distribution.

Examples of skewed distributions are shown in Figures 2.40 and 2.41. Positively skewed distributions are more common. The distribution of the lifetime of an electronics part might be positively skewed.

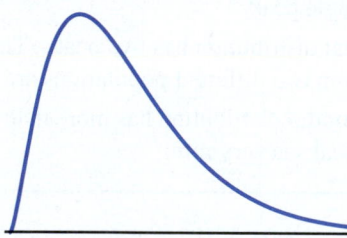

Figure 2.40 Positively skewed distribution.

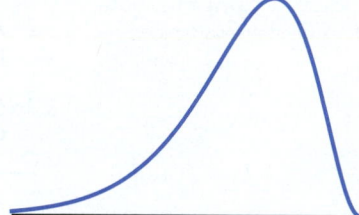

Figure 2.41 Negatively skewed distribution.

We will learn much more about the normal curve in Chapter 6.

The vertical cross section of a bell is a normal curve.

The most common unimodal distribution shape is a **normal curve** (as shown in Figure 2.42). This curve is symmetric and *bell-shaped*, and can be used to model, or approximate, many populations.

A curve with **heavy tails** has more observations in the tails of the distribution than a comparable normal curve. The tails do not drop down to the measurement axis as quickly as a normal curve. A curve with **light tails** has fewer observations in the tails of the distribution than a comparable normal curve. The tails drop to the measurement axis quickly. Examples of curves with heavy and light tails are shown in Figures 2.43 and 2.44. Both of these characteristics are subtle and tricky to spot.

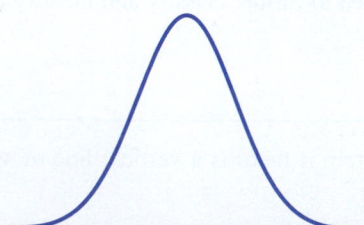

Figure 2.42 Normal curve.

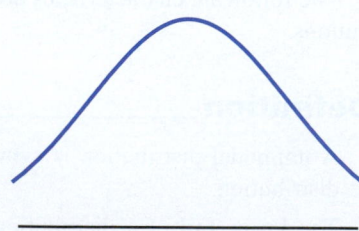

Figure 2.43 A distribution with heavy tails.

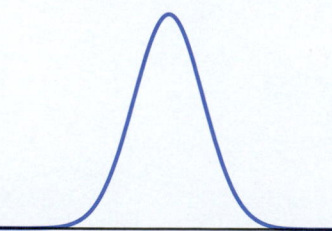

Figure 2.44 A distribution with light tails.

Example 2.14 Radiation Exposure

The U.S. Nuclear Regulatory Commission was established to regulate commercial, industrial, academic, and medical uses of nuclear materials.[26] The NRC is charged with monitoring our health and safety and protecting the environment. Part of this responsibility involves monitoring the radiation exposure at nuclear power reactors and other facilities. The individual radiation dose per year is measured in rem (roentgen equivalent man). Suppose a sample of 50 individual radiation measurements was obtained from employees and the data are given in the following table:

0.62	0.29	0.06	0.09	0.10	0.24	0.06	0.38	0.32	0.46
0.71	0.53	0.28	0.19	0.16	0.40	0.08	0.24	0.57	0.11
0.30	0.32	0.18	0.29	0.15	0.13	0.42	0.18	0.28	0.39
0.14	0.18	0.27	0.20	0.37	0.22	0.26	0.31	0.11	0.29
0.19	0.12	0.22	0.21	0.12	0.05	0.22	0.26	0.49	0.43

a. Construct a frequency distribution and a histogram for these data using the class intervals 0–0.10, 0.10–0.20, etc.

b. Describe the shape, center, and spread of the distribution.

c. What proportion of observations are less than 0.40 rem?

d. What proportion of observations are at least 0.50 rem?

SOLUTION

STEP 1 The class intervals are given. Create a frequency distribution and compute the frequency, relative frequency, and cumulative relative frequency for each class.

Class	Frequency	Relative frequency		Cumulative relative frequency	
0.00–0.10	5	0.10	(= 5/50)	0.10	(= 0.10)
0.10–0.20	14	0.28	(= 14/50)	0.38	(= 0.10 + 0.28)
0.20–0.30	15	0.30	(= 15/50)	0.68	(= 0.38 + 0.30)
0.30–0.40	7	0.14	(= 7/50)	0.82	(= 0.68 + 0.14)
0.40–0.50	5	0.10	(= 5/50)	0.92	(= 0.82 + 0.10)
0.50–0.60	2	0.04	(= 2/50)	0.96	(= 0.92 + 0.04)
0.60–0.70	1	0.02	(= 1/50)	0.98	(= 0.96 + 0.02)
0.70–0.80	1	0.02	(= 1/50)	1.00	(= 0.98 + 0.02)
Total	50	1.00			

Use the frequency distribution to sketch the histogram (Figure 2.45). A technology solution is shown in Figure 2.46.

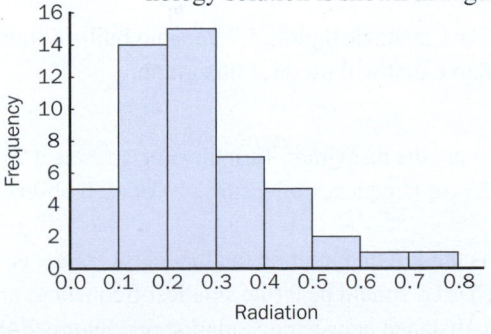

Figure 2.45 Histogram for the radiation data.

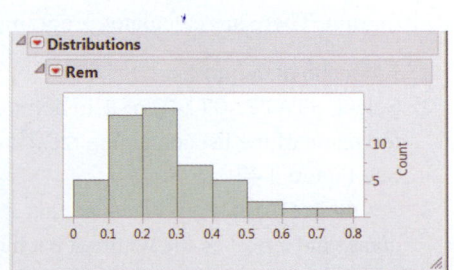

Figure 2.46 JMP histogram for the radiation data.

Other estimates of the center are also valid here. 0.24 is reasonable. So is 0.27. More precise measurements of the center of a data set are presented in Chapter 3.

The distribution is positively skewed. There are more observations in the lower tail than in the upper tail. The upper tail extends farther than the lower tail.

To estimate the center of the distribution, use the histogram to identify a value such that approximately half of the observations are below that number and half are above that number. A number between 0.2 and 0.3 appears to divide the ordered data in half. Typical values for this data set are in this range, and an estimate of the center is 0.25.

The variability is typically described as either compact (data that are compressed or squeezed together) or spread out (observations that extend over a wide range). Although this is somewhat subjective for now, this data set is fairly compact. All of the observations lie between 0.05 and 0.71 (even though the smallest class boundary is 0.00 and the largest class boundary is 0.80).

STEP 2 Using the cumulative relative frequency column of the frequency distribution, the proportion of observations less than 0.40 is 0.82.

STEP 3 There are two ways to find the proportion of observations that are at least 0.50.

    a. Add the relative frequencies that correspond to the classes that are at least 0.50.

$$\underbrace{0.04}_{0.50-0.60} + \underbrace{0.02}_{0.60-0.70} + \underbrace{0.02}_{0.70-0.80} = 0.08$$

    b. Find the cumulative relative frequency up to 0.50 and subtract this value from 1.

    Proportion of observations ≥ 0.50

$$= 1 - (\text{proportion of observations} < 0.50)$$

$$= 1 - 0.92 = 0.08$$

TRY IT NOW   GO TO EXERCISE 2.95

Technology Corner

Procedure: Construct a histogram.
Reconsider: Example 2.11, solution, and interpretations.

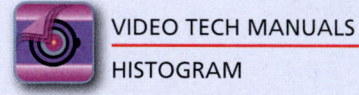

VIDEO TECH MANUALS

HISTOGRAM

CrunchIt!

CrunchIt! has a built-in function to construct a histogram.

1. Enter the data into column Var1.
2. Select Graphics; Histogram. Choose Sample (column) Var1. Optionally enter the Bin Width and Start Bins At. Optionally enter a Title and X Label. The Y Label is Count by default. Click the Calculate button. Refer to Figure 2.28 (page 56).

TI-84 Plus C

A histogram is one of the six types of TI-84 Plus C statistical plots. There is no built-in function to construct a density histogram. There are calculator programs available that will produce this graph.

1. Enter the data into list L1.
2. Select STATPLOT; Plot1 to define, or set up, the histogram. Turn the plot On, select Type histogram, set Xlist to the name of the list containing the data, set Freq (frequency of occurrence of each observation) to 1, and select a Color. See Figure 2.47.
3. Set the WINDOW parameters so that Xmin is the left endpoint on the first class, Xmax is the right endpoint on the last class, and Xscl is the width of each class. Ymin should be 0 (the smallest frequency) and Ymax should be at least the largest frequency. Set Yscl to a reasonable distance between tick marks. See Figure 2.48.

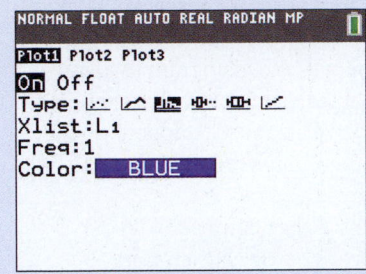

Figure 2.47 TI-84 Plus C Plot1 setup screen.

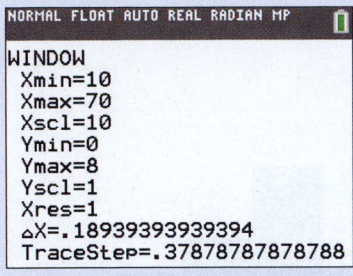

Figure 2.48 The WINDOW settings.

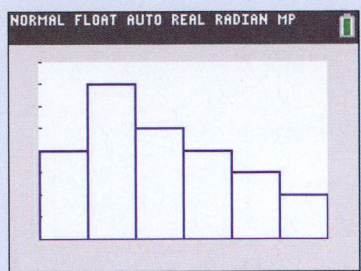

Figure 2.49 TI-84 Plus C histogram.

4. Press GRAPH to display the histogram. See Figure 2.49. Note: The TRACE key is used to move on the graph between rectangles (classes). The corresponding class boundaries and frequency are displayed.

Minitab

The input may be either a single column containing the data or summary information in two columns: observations and frequencies.

1. Enter the data into column C1.
2. Select Graph; Histogram and highlight Simple histogram. Click OK.
3. Enter C1 in the Graph variables window. Click OK to view the histogram.
4. Edit the horizontal axis scale to use the correct class intervals. Under the Binning tab, select Interval type: Cutpoint and enter the Midpoint/Cutpoint positions (class boundaries). See Figure 2.50.

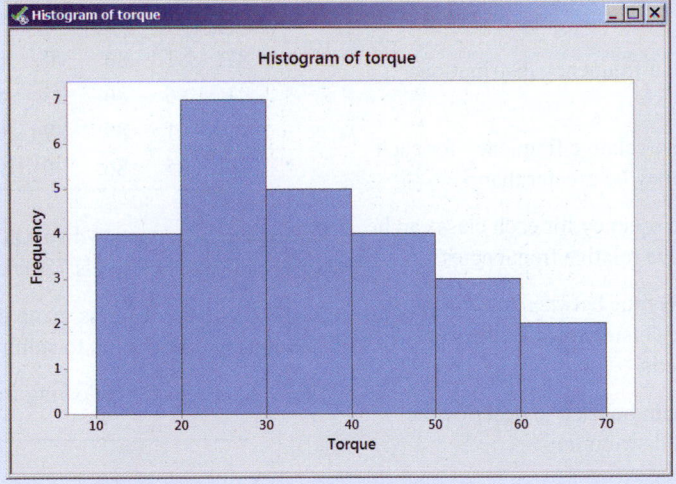

Figure 2.50 Minitab histogram.

Excel

The input may be either a single column containing the data or summary information in two columns: right endpoint of each class (bin limits) and corresponding frequencies. There are several methods to construct a frequency distribution in Excel using FREQUENCY, SUMIF, or COUNTIFS, for example.

1. Enter the data into column A and the right endpoint of each class into column B.
2. Under the Data tab, select Data Analysis and choose Histogram. Enter the Input Range, the Bin Range, the Output Range, and select Chart Output.

3. Each class is labeled with its right endpoint. In addition, Excel places observations on a boundary in the smaller class. See Figure 2.51.

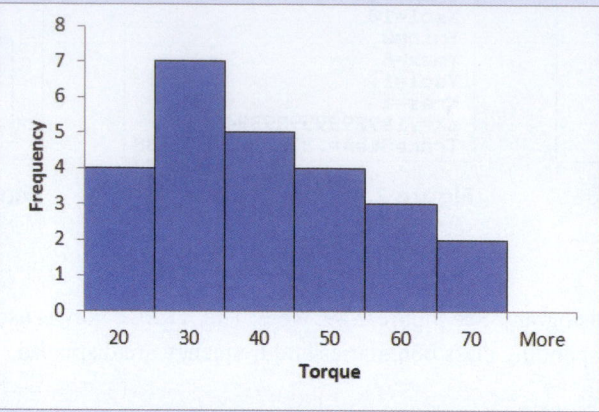

Figure 2.51 Excel histogram.

SECTION 2.4 EXERCISES

Concept Check

2.68 True/False   The classes in a frequency distribution may overlap.

2.69 True/False   The classes in a frequency distribution should have the same width.

2.70 True/False   The cumulative relative frequency for each class in a frequency distribution may be greater than 1.

2.71 True/False   The relative frequency for each class can be determined by using the cumulative relative frequencies.

2.72 True/False   The only difference between a frequency histogram and a relative frequency histogram (for the same data) is the scale on the vertical axis.

2.73 True/False   A histogram can be used to describe the shape, center, and variability of a distribution.

2.74 Short Answer
a. When is a density histogram appropriate?
b. In a density histogram, what is the sum of areas of all rectangles?

2.75 Fill in the Blank
a. The most common unimodal distribution is a _____.
b. A unimodal distribution is _____ if there is a vertical line of symmetry.
c. If a unimodal distribution is not symmetric, then it is _____.

2.76 True/False   A bimodal distribution cannot be symmetric.

Practice

2.77 Consider the data given in the following table.   📊 EX2.77

87	81	86	90	88	85	79	91	87	82
91	86	86	87	88	85	92	85	87	86
91	81	89	89	83	90	83	80	90	80
89	85	86	90	90	89	78	91	83	92

Construct a frequency distribution to summarize these data using the class intervals 78–80, 80–82, 82–84, ...

2.78 Consider the data given on the text website. Construct a frequency distribution to summarize these data.   📊 EX2.78

2.79 Consider the following frequency distribution.

Class	Frequency	Relative frequency	Cumulative relative frequency
400–410	5	0.0758	0.0758
410–420	8	0.1212	0.1970
420–430	10	0.1515	0.3485
430–440	12	0.1818	0.5303
440–450	9	0.1364	0.6667
450–460	8	0.1212	0.7879
460–470	5	0.0758	0.8637
470–480	4	0.0606	0.9243
480–490	3	0.0455	0.9698
490–500	2	0.0303	1.0001

Draw the corresponding *frequency* histogram. (Notice the last entry in the Cumulative Relative Frequency column is not exactly 1. This is due to round-off error.)

2.80 Consider the following frequency distribution.

Class	Frequency	Relative frequency	Cumulative relative frequency
0.5–1.0	6	0.03	0.03
1.0–1.5	8	0.04	0.07
1.5–2.0	10	0.05	0.12
2.0–2.5	16	0.08	0.20
2.5–3.0	24	0.12	0.32
3.0–3.5	34	0.17	0.49
3.5–4.0	36	0.18	0.67
4.0–4.5	22	0.11	0.78
4.5–5.0	18	0.09	0.87
5.0–5.5	12	0.06	0.93
5.5–6.0	8	0.04	0.97
6.0–6.5	4	0.02	0.99
6.5–7.0	2	0.01	1.00

Draw the corresponding *relative frequency* histogram.

2.81 Complete the following frequency distribution.

Class	Frequency	Relative frequency	Cumulative relative frequency
100–150	155		
150–200	120		
200–250	130		
250–300	145		
300–350	150		
350–400	100		
Total			

2.82 Complete the following frequency distribution.

Class	Frequency	Relative frequency	Cumulative relative frequency
1.0–1.1		0.05	
1.1–1.2	20		
1.2–1.3		0.15	
1.3–1.4	65		
1.4–1.5		0.25	
1.5–1.6	35		
1.6–1.7	25		
1.7–1.8			
Total	300		

2.83 Complete the following frequency distribution and draw the corresponding histogram.

Class	Frequency	Relative frequency	Cumulative relative frequency
0–25			0.150
25–50			0.350
50–75			0.525
75–100			0.675
100–125			0.800
125–150			0.900
150–175			0.975
175–200			1.000
Total	1000		

2.84 Consider the data given on the text website. EX2.84
 a. Construct a frequency distribution to summarize these data using the class intervals 0–1, 1–2, 2–3, etc., and draw the corresponding histogram.
 b. Use the histogram to describe the shape of the distribution. Are there any outliers?

2.85 Consider the data given on the text website. EX2.85
 a. Construct a frequency distribution to summarize these data and draw the corresponding histogram.
 b. Use the histogram to describe the shape of the distribution.
 c. Use the frequency distribution to estimate the *middle* of the data: a number M such that 50% of the observations are below M and 50% are above M.
 d. Use the frequency distribution to estimate a number Q_1 such that 25% of the observations are below Q_1 and 75% are above Q_1.
 e. Use the frequency distribution to estimate a number Q_3 such that 75% of the observations are below Q_3 and 25% are above Q_3.

Applications

2.86 Biology and Environmental Science   A weather station located along the Maine coast in Kennebunkport collects data on temperature, wind speed, wind chill, and rain. The maximum wind speed (in miles per hour) for 50 randomly selected times in February 2013 are given on the text website.[27] MAXWIND
 a. Construct a frequency distribution to summarize these data, and draw the corresponding histogram.
 b. Describe the shape of the distribution. Are there any outliers?

2.87 Fuel Consumption and Cars   The quality of an automobile battery is often measured by cold cranking amps (CCA), a measure of the current supplied at 0°F. Thirty automobile batteries were randomly selected and subjected to subfreezing temperatures. The resulting CCA data are given in the following table. BATTERY

63	87	302	4	259	106	198	55	99	134
122	514	91	117	325	39	30	164	75	16
340	199	77	217	64	320	145	84	47	232

a. Construct a frequency distribution to summarize these data, and draw the corresponding histogram.

b. Describe the shape of the distribution.

c. Estimate the middle of the distribution, a number M such that 50% of the data are below M and 50% are above M.

2.88 Marketing and Consumer Behavior   The weights of a diamond and other precious stones are usually measured in carats. One carat is traditionally equal to 200 milligrams. A random sample of the weights (in carats) of conflict-free loose diamonds is given in the following table.[28] **DIAMOND**

0.23	0.27	0.30	0.25	0.27	0.26	0.40	0.40
0.51	0.58	0.61	0.80	0.90	1.02	0.92	1.01
0.76	0.90	1.14	1.11	1.38	1.52	0.96	1.16
1.52	1.05	1.51	1.36	0.91	1.22	1.54	1.35
1.76	2.00	1.01	1.69	1.51	2.00	1.45	1.38

a. Construct a frequency distribution and a histogram for these data.

b. Multiply each observation in the table by 200, to convert the weights into milligrams. Construct a frequency distribution and a histogram for these new, transformed data.

c. Compare the two histograms. Are the shapes similar? Describe any differences.

2.89 Public Health and Nutrition   Vitamin B_3 (niacin) helps to detoxify the body, aids digestion, can ease the pain of migraine headaches, and helps promote healthy skin. A random sample of adults in the United States and in Europe was obtained and the daily intake of niacin (in milligrams) was recorded. The data are summarized in the following table.

Class	United States frequency	Europe frequency
0–3	15	4
3–6	23	6
6–9	21	12
9–12	14	17
12–15	12	32
15–18	9	25
18–21	3	20
21–24	2	10

a. Construct two *relative frequency* histograms, one for the United States and one for Europe.

b. Describe the shape of each histogram. Does a comparison of the two histograms suggest any differences in niacin intake between the two samples? Explain.

2.90 Manufacturing and Product Development   In the United States, yarn is often sold in hanks. For woolen yarn, one hank is approximately 1463 meters. A quality control inspector uses a special machine to quickly measure each hank. A random sample was obtained during the manufacturing process, and the length (in meters) of each hank is given on the text website. **YARN**

a. Construct a histogram for these data.

b. Describe the distribution in terms of shape, center, and variability.

2.91 Sports and Leisure   The National Hockey League is concerned about the number of penalty minutes assessed to each player. While some people in attendance hope to see a lot of fighting (and penalty minutes), the League Office believes most fans are interested in good, clean hockey. A sample of total penalty minutes per player during the 2012–2013 regular season was obtained, and the data are given in the following table.[29] **PENALTY**

39	38	14	22	26	15	65	39	24	44
21	29	16	18	19	17	37	40	56	17
39	21	19	19	15	25	46	14	32	30
25	14	22	17	15	71	14	13	23	24

a. Construct a histogram for these data. Describe the distribution in terms of shape, center, and variability. Write a Solution Trail for this problem.

b. Find a value m for the number of minutes such that 90% of all players have fewer than m penalty minutes.

Extended Applications

2.92 Biology and Environmental Science   Fruits such as cherries and grapes are harvested and placed in a shallow box or crate called a lug. The size of a lug varies, but one typically holds between 16 and 28 pounds. A random sample of the weight (in pounds) of full lugs holding peaches was obtained, and the data are summarized in the following table.

Class	Frequency
20.0–20.5	6
20.5–21.0	12
21.0–21.5	17
21.5–22.0	21
22.0–22.5	28
22.5–23.0	25
23.0–23.5	19
23.5–24.0	15
24.0–24.5	11
24.5–25.0	10

a. Complete the frequency distribution.

b. Construct a histogram corresponding to this frequency distribution.

c. Estimate the weight w such that 90% of all full peach lugs weigh more than w.

2.93 Travel and Transportation   Maglev trains operate in Germany and Japan at speeds of up to 300 miles per hour. Magnets create a frictionless system in which the train operates at a distance of 100–150 millimeters from the rail. The size of this air gap is monitored constantly to ensure a safe ride. A random sample of the size of air gaps (in millimeters) at one specific location in the track was obtained. The frequency distribution for this data is shown in the following table.

Class	Frequency	Relative frequency	Cumulative relative frequency
100–105			0.050
105–110			0.425
110–115			0.625
115–120			0.750
120–125			0.850
125–130			0.925
130–135			0.975
135–140			1.000
Total	200		

a. Complete the frequency distribution.
b. Draw a histogram corresponding to this frequency distribution.
c. What proportion of air gaps were between 110 and 125 millimeters?

2.94 Biology and Environmental Science Many scientists have warned that global warming is causing the polar ice caps to melt and, therefore, sea levels around the world to rise. A random sample of the sea level (in millimeters) at Rockport, Massachusetts, from 1987 to 2011 was obtained and is summarized in the following table.[30]

Class	Frequency	Relative frequency	Width	Density
1400–1500	1			
1500–1600	2			
1600–1800	41			
1800–2000	192			
2000–2100	79			
2100–2200	60			
2200–2300	15			
2300–2800	10			
Total	400			

a. Complete the frequency distribution.
b. A traditional frequency histogram or relative frequency histogram is not appropriate in this case. Why not?
c. Construct a density histogram corresponding to this frequency distribution.

2.95 Fuel Consumption and Cars The total cost of owning an automobile includes the amount spent on repairs. Before purchasing a new car, many consumers research the past quality of specific makes and models. The data in the following table lists the number of problems per 100 vehicles over the three years 2010–2012. **CARCOST**

Lexus	71	Porsche	94	Lincoln	112
Toyota	112	Mercedes	115	Buick	118
Honda	119	Acura	120	Ram	122
Suzuki	122	Mazda	124	Chevrolet	125
Ford	127	Cadillac	128	Subaru	132
BMW	133	GMC	134	Scion	135
Nissan	137	Infiniti	138	Kia	140
Hyundai	141	Audi	147	Volvo	149
Mini	150	Chrysler	153	Jaguar	164
Volkswagen	174	Jeep	178	Mitsubishi	178
Dodge	190	Land Rover	220		

Source: J. D. Powers 2013 Dependability Survey

a. Construct a histogram for these data.
b. Describe the distribution in terms of shape, center, and variability.
c. Find a number Q_1 such that 25% of the problem data are less than Q_1. Find a number Q_3 such that 25% of the problem data are greater than Q_3.
d. How many values should be between Q_1 and Q_3? Find the actual number of values between Q_1 and Q_3. Explain any difference between these two values.

CHAPTER 2 SUMMARY

Concept	Page	Notation / Formula / Description
Categorical data set	29	Consists of observations that may be placed into categories.
Numerical data set	29	Consists of observations that are numbers.
Discrete data set	30	The set of all possible values is finite, or countably infinite.
Continuous data set	30	The set of all possible values is an interval of numbers.
Frequency distribution	33	A table used to describe a data set. It includes the class, frequency, and relative frequency (and cumulative relative frequency, if the data set is numerical).
Class frequency	33	The number of observations within a class.
Class relative frequency	33	The proportion of observations within a class: class frequency divided by total number of observations.

Class cumulative relative frequency	34	The proportion of observations within a class and every class before it: the sum of all the relative frequencies up to and including the class.
Bar chart	34	A graphical representation of a frequency distribution for categorical data with a vertical bar for each class.
Pie chart	35	A graphical representation of a frequency distribution for categorical data with a slice, or wedge, for each class.
Stem-and-leaf plot	45	A graph used to describe numerical data. Each observation is split into a stem and a leaf.
Histogram	55	A graphical representation of a frequency distribution for numerical data.
Density histogram	58	A graphical representation of a frequency distribution for numerical data containing class intervals of unequal width.
Unimodal distribution	59	A distribution with one peak.
Bimodal distribution	59	A distribution with two peaks.
Multimodal distribution	59	A distribution with more than one peak.
Symmetric distribution	59	A distribution with a vertical line of symmetry.
Positively skewed distribution	60	A distribution in which the upper tail extends farther than the lower tail.
Negatively skewed distribution	60	A distribution in which the lower tail extends farther than the upper tail.
Normal curve	60	The most common distribution, a bell-shaped curve.

CHAPTER 2 EXERCISES

2 APPLICATIONS

2.96 Business and Management   A laborshed is a region from which an employment center draws its workforce. In order to understand the potential workforce in a laborshed, the Walker County Development Authority in Alabama sampled residents and reported the data in the following table. **WORKERS**

Employment status	Frequency
Employed (white collar)	125
Employed (blue collar)	200
Unemployed	30
Homemaker	50
Retired	95

 a. Add a relative frequency column to the table.
 b. Construct a bar chart and a pie chart for these data.

2.97 Fuel Consumption and Cars   The coefficient of drag (C_d) is a measure of a car's aerodynamics. This unitless number is related directly to the speed of the car, overall performance, and miles per gallon. A low coefficient of drag indicates good performance. A random sample of new automobiles was examined, and the coefficient of drag was computed. The results are given on the text website. **DRAGCOEF**
 a. Construct a stem-and-leaf plot for these data.
 b. Use the plot in part (a) to describe the distribution in terms of shape, center, and variability.

2.98 Psychology and Human Behavior   In January 2013, Harris Interactive released results of a survey in which adults were asked to name their favorite TV personality.[31] Ellen DeGeneres captured the top spot, with Mark Harmon second. Jon Stewart, Jay Leno, and Jim Parsons round out the top five, and Bill O'Reilly, Anderson Cooper, and Oprah Winfrey also received strong support. Suppose the results from this survey are given in the following table. **TVHOST**

TV personality	Frequency
Ellen DeGeneres	638
Mark Harmon	532
Jon Stewart	402
Jay Leno	376
Jim Parsons	350
Others	320

 a. Find the relative frequency for each category.
 b. Construct a pie chart for these data.
 c. What proportion of adults selected Jay Leno or Jim Parsons?
 d. What proportion of adults did not select Jon Stewart?

2.99 Physical Science   Construction equipment used to build homes, businesses, and roads (for example, cranes, backhoes, and front loaders) can be exceptionally loud. The noise level in dBA (A-weighted decibels) measured 50 feet away from several construction-related machines are given on the text website.[32] **NOISE**

a. Construct a frequency distribution for these data.
b. Draw the corresponding histogram.
c. What proportion of construction equipment had a peak noise level below 80 dBA?
d. What proportion of construction equipment have peak noise levels of at least 90 dBA?

2.100 Technology and the Internet Many computer sellers and most software vendors maintain help lines for customers. A random sample of the duration (in minutes) of customer support calls to Amazon.com was obtained, and the resulting stem-and-leaf plot is given below.

Stem	Leaf
0	11223344555566678888999
1	0001222222335668999
2	012334556678
3	000123478
4	334468
5	125
6	15
7	7

Stem = 10
Leaf = 1

a. Describe the shape of this distribution of the duration of technical support calls.
b. Use the plot to construct a frequency distribution using the class intervals 0–5, 5–10, 10–15, etc.
c. What proportion of support calls last less than 15 minutes?
d. If a call lasts at least 25 minutes, a supervisor monitors the conversation. What proportion of calls were monitored?

2.101 Technology and the Internet Many police departments have been experimenting with and implementing state-of-the-art emergency 9-1-1 equipment. This equipment is designed to allow a faster response time without voice contact. Caller information is displayed on a monitor, printed, and then processed. To compare the two procedures (old and new), a random sample of police response times (in minutes) was obtained. The data are given on the text website. **POLICE**

a. Construct a back-to-back stem-and-leaf plot for these data.
b. Use the plot in part (a) to describe any similarities and/or differences between the distributions.
c. Based on the plot in part (a), which procedure is better? Justify your answer.

2.102 Manufacturing and Product Development Microwave ovens are often rated by their output power, for example, 900 watts. However, the actual output of a microwave oven tends to decrease with age. If the *actual* output is more than 400 watts below the *rated* output, then service is recommended. A random sample of five-year-old, 1000-watt-rated microwave ovens was obtained and tested for output. **MICRO**

a. Construct a frequency distribution for these data and draw the corresponding histogram.
b. Based on this random sample, what proportion of five-year-old, 1000-watt microwave ovens need service?

c. Suppose the performance of these microwave ovens is *graded* by actual output power, according to the following chart.

Power	Grade
900–1000	Excellent
800–900	Very good
700–800	Good
600–700	Fair
500–600	Poor
0–500	Not serviceable

Classify each power output, construct a frequency distribution by grade, and draw the resulting pie chart.

EXTENDED APPLICATIONS

2.103 Economics and Finance PwC, World Bank, and IFC released a study about paying taxes around the world. The report includes measures of the world's tax systems associated with a standardized business. One of the measures, Time to Comply (in hours), for each of the 185 countries in the study, is given on the text website.[33] **TAXPAY**

a. Use the class intervals 0–200, 200–400, etc., to construct a frequency distribution and draw the corresponding histogram.
b. Describe the distribution in terms of shape, center, and variability.
c. What is a typical Time to Comply? Are there any outliers?
d. What proportion of countries had a time to comply of at least 800 hours?

2.104 Fuel Consumption and Cars Remanufactured parts are common in the automotive industry. To ensure quality, Hite Parts Exchange routinely checks the maximum output of rebuilt alternators. Each day a random sample is obtained and the output delivered (in amps) at 2500 rpm is recorded. The results from a recent day are presented in the following table.

Class	Frequency
30.0–32.0	8
32.0–33.0	7
33.0–34.0	10
34.0–34.5	25
34.5–35.0	30
35.0–35.5	40
35.5–36.0	45
36.0–50.0	5
Total	170

a. Find the width and the density for each class.
b. Construct a density histogram for these data.

2.105 Medicine and Clinical Studies A common cold usually lasts from 3 to 14 days. Some studies suggest echinacea, zinc, or vitamin C can prevent colds and/or shorten their duration. In a new study of the effect of vitamin C, patients with colds were randomly assigned to a placebo group or a vitamin C group. The duration of each cold (in days) was recorded, and the data are summarized in the following table.

Duration	Placebo frequency	Vitamin C frequency
3	0	3
4	0	6
5	8	7
6	7	10
7	21	18
8	10	15
9	26	17
10	15	10
11	8	9
12	3	2
13	1	3
14	1	0

a. Use appropriate graphical procedures to compare the placebo and vitamin C data sets.

b. Do the graphs suggest any differences in shape, center, or variability?

c. Is there any graphical evidence to suggest vitamin C reduced the duration of a cold?

2.106 Fuel Consumption and Cars   The performance of a gas furnace can be measured by the annual fuel utilization efficiency (AFUE). This number depends on many furnace properties, and is an indication of the proportion of fuel energy delivered as heat energy during an entire heating season. The U.S. Department of Energy (DOE) requires all new gas furnaces to operate at an AFUE of at least 78%.[34] A gas company selected a random sample of customers, carefully tested each furnace, and recorded the AFUE number. The data are given on the text website.   ▌▌ **FURNACE**

a. Construct a stem-and-leaf plot for these data.

b. Construct a frequency distribution for these data and draw the corresponding histogram.

c. Describe the distribution in terms of shape, center, and variability. Are there any outliers? If so, what are they?

d. Using the frequency distribution in part (a), approximately what proportion of furnaces do not meet the DOE's minimum AFUE requirement?

e. The gas company classifies each AFUE reading according to the following scheme: 90 or above, excellent; at least 80 but below 90, good; at least 70 but below 80, fair; and less than 70, poor. Classify each reading in the table above, and construct a bar chart for these classification data.

CHALLENGE

2.107 Sports and Leisure   An *ogive*, or *cumulative relative frequency polygon*, is another type of visual representation of a frequency distribution. To construct an ogive:

■ Plot each point (upper endpoint of class interval, cumulative relative frequency).

■ Connect the points with line segments.

Figures 2.52 and 2.53 show a frequency distribution and the corresponding ogive. The observations are ages. The values to be used in the plot are shown in bold in the table.

Class	Frequency	Relative frequency	Cumulative relative frequency
12–16	8	0.08	**0.08**
16–20	10	0.10	**0.18**
20–24	20	0.20	**0.38**
24–28	30	0.30	**0.68**
28–32	15	0.15	**0.83**
32–36	10	0.10	**0.93**
32–40	7	0.07	**1.00**
Total	100	1.00	

Figure 2.52 Frequency distribution.

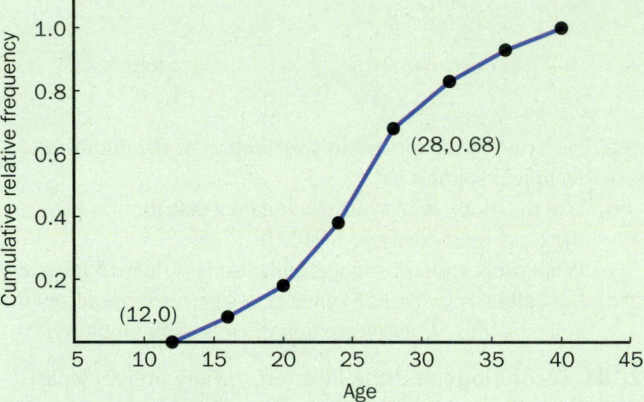

Figure 2.53 Resulting ogive.

A random sample of game scores from Abby Sciuto's evening bowling league with Sister Rosita was obtained, and the data are given on the text website.   ▌▌ **BOWLING**

a. Construct a frequency distribution for these data.

b. Draw the resulting ogive for these data.

2.108 Public Health and Nutrition   A *doughnut graph* is another graphical representation of a frequency distribution for categorical data. To construct a doughnut graph:

1. Divide a (flat) doughnut (or washer) into pieces, so that each piece (bite of the doughnut) corresponds to a class.

2. The size of each piece is measured by the angle made at the center of the doughnut. To compute the angle of each piece, multiply the relative frequency times 360° (the number of degrees in a whole, or complete, circle).

The manager at a Whole Foods Market obtained a random sample of customers who purchased at least one popular herb (for cooking or medicinal purposes). Figure 2.54 and 2.55 show a frequency distribution and the corresponding doughnut graph.

Herb	Frequency	Relative frequency
Echinacea	25	0.125
Ephedra	15	0.075
Feverfew	20	0.100
Garlic	35	0.175
Ginkgo	40	0.200
Kava	30	0.150
Saw palmetto	20	0.100
St. John's wort	15	0.075
Total	200	1.000

Figure 2.54 Frequency distribution.

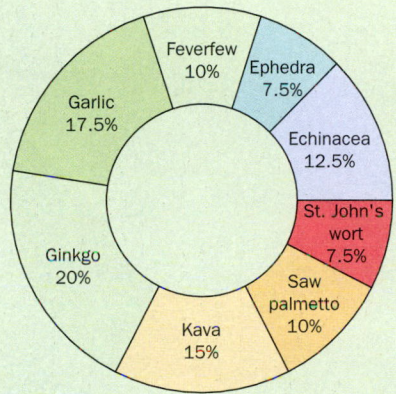

Figure 2.55 Resulting doughnut graph.

A random sample of house fires in Bismarck, North Dakota, was selected and the cause of each was recorded. The resulting data are shown in the following table.

Class	Frequency
Smoking or smoking materials	70
Heating equipment	85
Cooking and cooking equipment	205
Children playing with matches	105
Arson / suspicious	35

a. Find the relative frequency for each class.
b. Draw a doughnut graph for these data.

LAST STEP

2.109 Can the Florida Everglades be saved? In January 2013, the Florida Fish and Wildlife Conservation Commission started the Python Challenge. The purpose of the contest was to thin the python population, which could be tens of thousands, and help save the natural wildlife in the Everglades. At the end of the competition, 68 Burmese pythons had been harvested. Suppose a random sample of pythons captured during the Challenge was obtained and the length (in feet) of each is given in the following table: **PYTHON**

9.3	3.5	5.2	8.3	4.6	11.1	10.5	3.7	2.8	5.9
7.4	14.2	13.6	8.3	7.5	5.2	6.4	12.0	10.7	4.0
11.1	3.7	7.0	12.2	5.2	8.1	4.2	6.1	6.3	13.2
3.9	6.7	3.3	8.3	10.9	9.5	9.4	4.3	4.6	5.8
4.1	5.2	4.7	5.8	6.4	3.8	7.1	4.6	7.5	6.0

a. Construct a frequency distribution, stem-and-leaf plot, and histogram for these data.
b. Use these tabular and graphical techniques to describe the shape, center, and spread of this distribution, and to identify any outlying values.

3 Numerical Summary Measures

◀ **Looking Back**

- Be familiar with several common tabular and graphical summary procedures.
- Be able to construct a bar chart, pie chart, frequency distribution, stem-and-leaf plot, and histogram.

▶ **Looking Forward**

- Learn how to compute and interpret common numerical summary measures that describe central tendency, variability, or relative standing.
- Learn how to measure *distance* in statistics.
- Find a five-number summary and construct box plots.

How efficient is the Canadian Pacific Railway?

The Canadian Pacific Railway (CPR) was incorporated in 1881 and played an important role in the development of western Canada. It is primarily a freight railway with over 14,000 miles of track.

To increase efficiency, officials at CPR monitor several variables, including train speed, cars on each train, and terminal dwell time. In addition, the type and amount of freight is carefully recorded for each train. The following table shows the number of carloads of grain mill products for 30 randomly selected weeks in 2011 and 2012.[1]

572	711	582	663	612	577	650	550	590	659
610	611	557	685	683	629	626	637	634	723
718	707	673	697	808	755	438	569	684	637

The procedures presented in this chapter will be used to describe the *center* and *variability* of these data, and to search for any unusual observations.

CONTENTS

3.1 Measures of Central Tendency

As we learned in Chapter 2, tabular and graphical procedures provide some very useful summaries of data. However, these techniques are not sufficient for statistical inference. For example, because there are no definite rules for constructing a histogram, two people may construct very different looking displays for the same data, which could lead to different conclusions. The numerical summary measures presented in this chapter are more precise, combine information from the data into a single number, and allow us to draw a conclusion about an entire population. The two most common types of numerical summary measures describe the *center* and the *variability* of the data.

A numerical summary measure is a single number computed from a sample that conveys a specific characteristic of the entire sample. Measures of *central tendency* indicate where the majority of the data is centered, bunched, or clustered. There are many different measures of central tendency. They all combine information from a sample into a single number, and each has advantages and disadvantages.

To properly define and understand numerical summary measures, the following notation will be used.

Note: A *capital*, or *uppercase*, *X* has a very different meaning (introduced in Chapter 5).

x: This stands for a specific, fixed observation on a variable. In general, lowercase letters are used to represent observations on a variable; y and z are also commonly used.

n: This is usually used to denote the number of observations in a data set, or the sample size. If there are two relevant data sets, then m and n may be used to denote their sample sizes. Or, if there are two (or more) relevant data sets, then $n_1, n_2, n_3, \ldots$ may be used to denote their sample sizes.

The three dots, . . . , mean the list continues in the same manner.

$x_1, x_2, x_3, \ldots, x_n$: This refers to a set of fixed observations on a variable. The subscripts indicate the order in which the observations were selected, not magnitude. For example, x_5 is the fifth observation drawn from a population, not the fifth largest.

$\sum\limits_{i=1}^{n} x_i = x_1 + x_2 + \cdots + x_n$: This is an example of summation notation, often used to write long mathematical expressions more concisely. Here, the sum of n observations can be written more compactly by using the notation on the left side. Σ is the Greek capital letter sigma; i is the *index of summation*; 1 is the *lower bound*; and n is the *upper bound*. To make the notation more compact and less threatening, we will usually omit the subscript $i = 1$ and superscript n. Unless specifically indicated, each summation applies to all values of the variable. For example, the following notation is used to represent the sum of each squared observation: $\sum x_i^2 = x_1^2 + x_2^2 + \cdots + x_n^2$.

The following example illustrates the use of this notation and some of the computations used throughout this text.

Example 3.1 Sum Practice

Suppose $x_1 = 5, x_2 = 9, x_3 = 12, x_4 = -6, x_5 = 17,$ and $x_6 = -2$. Compute the following sums:

a. $(\sum x_i)^2$ **b.** $\sum x_i^2$ **c.** $\sum (x_i - 7)^2$

SOLUTION

In each case, i is the index of summation, 1 is the lower bound, and 6 is the upper bound. Apply the definition of summation notation to each expression.

a. In words, expression (a) says add all of the observations, and square the result.

$$
\begin{aligned}
(\sum x_i)^2 &= (x_1 + x_2 + x_3 + x_4 + x_5 + x_6)^2 \qquad &&\text{Expand summation notation.}\\
&= [5 + 9 + 12 + (-6) + 17 + (-2)]^2 \qquad &&\text{Use given data.}\\
&= (35)^2 = 1225 \qquad &&\text{Add, and square the sum.}
\end{aligned}
$$

b. In words, expression (b) says square each observation, and add the resulting values.

$$\Sigma x_i^2 = x_1^2 + x_2^2 + x_3^2 + x_4^2 + x_5^2 + x_6^2 \qquad \text{Expand summation notation.}$$
$$= (5)^2 + (9)^2 + (12)^2 + (-6)^2 + (17)^2 + (-2)^2 \qquad \text{Use given data.}$$
$$= 25 + 81 + 144 + 36 + 289 + 4 \qquad \text{Square each observation.}$$
$$= 579 \qquad \text{Add.}$$

c. In words, expression (c) says subtract 7 from each observation, square each difference, and add the resulting values.

$$\Sigma(x_i - 7)^2 = (x_1 - 7)^2 + (x_2 - 7)^2 + (x_3 - 7)^2 + (x_4 - 7)^2 + (x_5 - 7)^2 + (x_6 - 7)^2$$

Expand summation notation.

$$= (5 - 7)^2 + (9 - 7)^2 + (12 - 7)^2 + (-6 - 7)^2 + (17 - 7)^2 + (-2 - 7)^2$$

Use given data.

$$= (-2)^2 + (2)^2 + (5)^2 + (-13)^2 + (10)^2 + (-9)^2 \qquad \text{Compute each difference.}$$
$$= 4 + 4 + 25 + 169 + 100 + 81 = 383 \qquad \text{Square each difference, and add.}$$

TRY IT NOW GO TO EXERCISE 3.2

The most common measure of central tendency is the *sample*, or *arithmetic, mean*.

Definition

$\bar{x}$ is read as "x bar."

The **sample (arithmetic) mean**, denoted $\bar{x}$, of the n observations $x_1, x_2, \ldots, x_n$ is the sum of the observations divided by n. Written mathematically.

$$\bar{x} = \frac{1}{n}\Sigma x_i = \frac{x_1 + x_2 + \cdots + x_n}{n} \qquad (3.1)$$

A CLOSER LOOK

1. The notation $\bar{x}$ is used to represent the sample mean for a set of observations denoted by $x_1, x_2, \ldots, x_n$. Similarly, $\bar{y}$ would represent the sample mean for a set of observations denoted by $y_1, y_2, \ldots, y_n$.

2. The **population mean** is denoted by μ, the Greek letter mu.

Example 3.2 Base Camp Temperature

DATA SET
DENALI

Denali National Park and Preserve in Alaska covers over 6 million acres and includes the tallest mountain in North America, Mount McKinley. Over 1200 climbers reached the peak of Mount McKinley in 2012. The temperature (in degrees Fahrenheit) at the 7200-foot base camp for 12 randomly selected days is given in the following table.[2]

6	11	20	19	23	28	30	8	23	25	29	33

Find the sample mean temperature at the base camp.

SOLUTION

Use Equation 3.1 to find the sample mean.

$$\bar{x} = \frac{1}{12}\Sigma x_i = \frac{1}{12}(x_1 + x_2 + \cdots + x_{12}) \qquad \text{Add all the numbers, and divide by } n = 12.$$

$$= \frac{1}{12}(6 + 11 + 20 + 19 + 23 + 28 + 30 + 8 + 23 + 25 + 29 + 33)$$

$$= \frac{1}{12}(255) = 21.25°F$$

Galyna Andrushko/Shutterstock

Figure 3.1 shows the sample mean using CrunchIt!.

Results - Descriptive Statistics		
Export ▾		
	n	Sample Mean
Temperature	12	21.25

Figure 3.1 The sample mean using CrunchIt!.

A CLOSER LOOK

1. $\bar{x}$ is a *sample characteristic*. It describes the center of a fixed collection of data. There is no set rule to determine the number of included decimal places. Often, at least one extra decimal place to the right is used to write the result; then the sample mean has one more decimal place than the original data values.

2. The sample mean is *an* average. There are many other averages, for example, the geometric mean, the harmonic mean, a weighted mean, the median, and the mode. People usually associate *the* average with the sample mean.

3. μ is a *population characteristic*. It describes the center of an entire population. If the population happens to be of finite size N, then μ is the sum of all the values divided by N. Most populations of interest are infinite, or at least very large, and therefore μ is an unknown constant that cannot be measured. It seems reasonable to use $\bar{x}$ to estimate and draw conclusions about μ.

4. The population mean μ is a fixed constant. $\bar{x}$ varies from sample to sample. It is reasonable to think that two sample means computed using samples from the same population should be close, but different.

If a data set contains outliers—observations *very far away from the rest*—then the sample mean may not be a very good measure of central tendency. An outlier has lots of influence on the sample mean, and tends to pull the mean in its direction. Example 3.3 shows how an outlier can affect the sample mean.

DATA SET

DENALI2

Example 3.3 Base Camp Temperature (Modified)

Modify the data in Example 3.2: Suppose one temperature at the base camp was 72, not 33. So the data set is now

6	11	20	19	23	28	30	8	23	25	29	72

The observation 72 is an obvious outlier. The new sample mean is

$$\bar{y} = \frac{1}{12}(6 + 11 + 20 + 19 + 23 + 28 + 30 + 8 + 23 + 25 + 29 + 72)$$

$$= \frac{1}{12}(294) = 24.5°F$$

Because $\bar{x} = 21.25$, $\bar{y} > \bar{x}$. The sample mean is pulled in the direction of the outlier, and is therefore not necessarily an adequate measure of central tendency. The *sample median* is another measure of central tendency that is not as sensitive to outlying values.

Definition

$\tilde{x}$ is read as "x tilde."

The **sample median**, denoted $\tilde{x}$, of the n observations $x_1, x_2, \ldots, x_n$ is the *middle number* when the observations are arranged in order from smallest to largest.

1. If n is odd, the sample median is the single middle value.

2. If n is even, the sample median is the mean of the two middle values.

A CLOSER LOOK

1. The median divides the data set into two parts, so that half of the observations lie below and half lie above the median.

2. Only one calculation is necessary to find the median (no calculations are needed if n is odd). Put the observations in ascending order of magnitude (*not* the order in which the observations were selected), and find the middle value.

3. Similarly, $\tilde{y}$ represents the sample median for a set of observations denoted by $y_1, y_2, \ldots, y_n$.

4. The **population median** is denoted by $\tilde{\mu}$.

Example 3.4 Median Calculations

The following three examples show how to find the median under various circumstances, and the effect of an outlying value. The observations are already arranged in order from smallest to largest.

DATA SET

PUBLISH

Observations	Median
a. 10 11 14 16 17	There are $n = 5$ observations. The middle number is in the third position. $\tilde{x} = 14$.
b. 10 11 14 16 57	There are still $n = 5$ observations. The middle number is in the third position, and $\tilde{x} = 14$. The outlier 57 does not affect the median.
c. 10 11 14 16 17 20	There are $n = 6$ observations. There is no single middle value. The median is the mean of the observations in the third and fourth positions. $\tilde{x} = \frac{1}{2}(14 + 16) = 15$.

Example 3.5 Nonfarm Employment

The number of people employed in the publishing industry in Oregon over the last 12 years is given in the following table. Find the median number of people employed.

16,100	16,900	15,200	13,900	13,500	14,300
15,200	15,900	15,700	14,500	14,100	14,000

Source: Oregon Employment Departments.

SOLUTION

STEP 1 Arrange the observations in order.

13,500 13,900 14,000 14,100 14,300 14,500 15,200 15,200 15,700 15,900 16,100 16,900

STEP 2 There are $n = 12$ observations. The median is the mean of the two middle values (in the sixth and seventh positions).

$$\tilde{x} = \frac{1}{2}(14{,}500 + 15{,}200) = 14{,}850$$

Figure 3.2 shows a technology solution.

TRY IT NOW GO TO EXERCISE 3.13

A CLOSER LOOK

1. In general, the sample mean is not equal to the sample median, $\bar{x} \neq \tilde{x}$. If the distribution of the sample is symmetric, then $\bar{x} = \tilde{x}$. If the sample distribution is approximately symmetric, then $\bar{x} \approx \tilde{x}$.

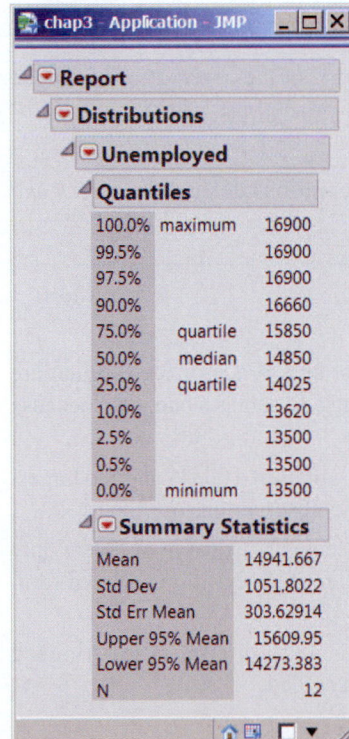

Figure 3.2 The sample median and other summary statistics found using JMP.

2. In general, the population mean is not equal to the population median, $\mu \neq \tilde{\mu}$. If the distribution of the population is symmetric, then $\mu = \tilde{\mu}$.

3. The relative positions of $\bar{x}$ and $\tilde{x}$ suggest the shape of a distribution. The *smoothed histograms* in Figures 3.3–3.5 illustrate three possibilities:

   a. If $\bar{x} > \tilde{x}$, the distribution of the sample is positively skewed, or skewed to the right (Figure 3.3).

   b. If $\bar{x} \approx \tilde{x}$, the distribution of the sample is approximately symmetric (Figure 3.4).

   c. If $\bar{x} < \tilde{x}$, the distribution of the sample is negatively skewed, or skewed to the left (Figure 3.5).

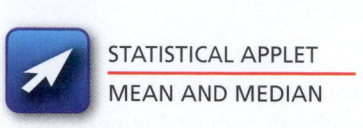

STATISTICAL APPLET

MEAN AND MEDIAN

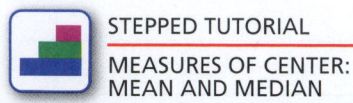

STEPPED TUTORIAL

MEASURES OF CENTER:
MEAN AND MEDIAN

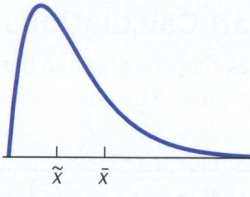

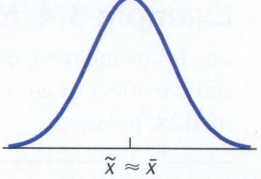

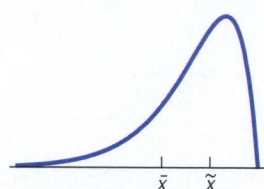

Figure 3.3 Positively skewed distribution.

Figure 3.4 Approximately symmetric distribution.

Figure 3.5 Negatively skewed distribution.

Recall: A histogram consists of rectangles drawn above each class with height proportional to frequency or relative frequency. We draw a curve along the tops of the rectangles to *smooth out* the histogram and display an enhanced graphical representation of the distribution.

Because the sample mean is extremely sensitive to outliers, and the sample median is very insensitive to outliers, it seems reasonable to search for a compromise measure of central tendency. A *trimmed mean* is moderately sensitive to outliers.

Definition

A **100p% trimmed mean**, denoted $\bar{x}_{\text{tr}(p)}$, of the n observations $x_1, x_2, \ldots, x_n$ is the sample mean of the *trimmed* data set.

1. Order the observations from smallest to largest.

2. Delete, or trim, the smallest 100p% and the largest 100p% of the observations from the data set.

3. Compute the sample mean for the remaining data.

100p is the **trimming percentage**, the percentage of observations deleted from *each end* of the ordered list.

A CLOSER LOOK

1. We compute a trimmed mean by deleting the smallest and largest values, which are possible outliers. Some statisticians believe that deleting any data is a bad idea, because every observation contributes to the big picture.

2. A **100p% trimmed mean** is computed by deleting the smallest 100p% and the largest 100p% of the observations. Therefore, 2(100p)% of the observations are removed.

3. There is no set rule for determining the value of p. It seems reasonable to delete only a few observations, and to select p so that np (the number of observations deleted from each end of the ordered data) is an integer.

4. Here is a specific example using the notation: $\bar{x}_{\text{tr}(0.05)}$ is a $(100)(0.05) = 5\%$ trimmed mean. In this example, 10% of the observations are discarded.

Example 3.6 Overtime and Stress

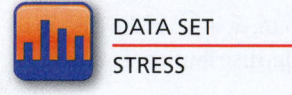

DATA SET

STRESS

According to an article in *The Guardian*,[3] Americans spend more time at their jobs than workers in Germany do. Dr. Paul Landsbergis, an epidemiologist at Mt. Sinai Medical Center, studies job stress, and he warns that too many overtime hours may increase the chance of heart disease. Suppose the following December overtime hours for tellers at the Kaw Valley State Bank and Trust Company in Topeka, Kansas, were obtained. Find a 10% trimmed mean.

0.2 0.8 1.5 1.5 1.6 1.7 1.7 1.8 2.0 2.0 2.2 2.5 2.7 2.7 3.0 3.0 3.2 3.5 4.0 5.0

SOLUTION

STEP 1   The trimming percentage is 10%. $p = 10/100 = 0.10$. Find the number of observations to delete from each end of the ordered list.

There are $n = 20$ observations.

$$np = (20)(0.10) = 2$$ Trim 2 observations from each end.

Note that np may not be an integer. Computer software packages have algorithms for dealing with this problem.

STEP 2   The resulting data set is

0.2 0.8 1.5 1.5 1.6 1.7 1.7 1.8 2.0 2.0 2.2 2.5 2.7 2.7 3.0 3.0 3.2 3.5 4.0 5.0

STEP 3   Find the sample mean for the remaining data.

$$\bar{x}_{tr(0.10)} = \frac{1}{16}(1.5 + 1.5 + 1.6 + \cdots + 3.0 + 3.2 + 3.5) = \frac{1}{16}(36.6) = 2.29$$

2.29 hours is the 10% trimmed mean.

TRY IT NOW   GO TO EXERCISE 3.15

Another commonly used measure of central tendency is the *mode*.

Definition

The **mode**, denoted M, of the n observations $x_1, x_2, \ldots, x_n$ is the value that occurs most often, or with the greatest frequency.

If all the observations occur with the same frequency, then the mode does not exist.

If two or more observations occur with the same greatest frequency, then the mode is not unique. If there are two modes, the distribution is bimodal, three modes, trimodal, etc.

The mode is easy to compute and, intuitively, it does return a reasonable measure of central tendency. For example, consider a bell-shaped distribution. A random sample from this distribution should contain lots of (identical) values near the center. Therefore, the mode should suggest the middle of the distribution (Figure 3.7). For symmetric distributions, the mean, the median, and the mode will be about the same.

A CLOSER LOOK

1. Sometimes only a data summary table, or *grouped data*, is available. Let $x_1, x_2, \ldots, x_k$ be a set of (representative) observations with corresponding frequencies $f_1, f_2, \ldots, f_k$. For example, x_7 occurs f_7 times. The total number of observations is $n = \sum f_i$. If the data are grouped, there are corresponding formulas for the (approximate) measures of central tendency defined above.
2. Remember, there are many other *averages*, for example, the weighted mean, the geometric mean, and the harmonic mean.

The natural summary measures for observations on a qualitative variable are simply the frequency and relative frequency of occurrence for each category. We have already

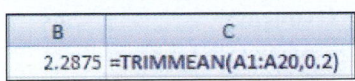

B	C
2.2875	=TRIMMEAN(A1:A20,0.2)

Figure 3.6 Calculation of a trimmed mean using Excel.

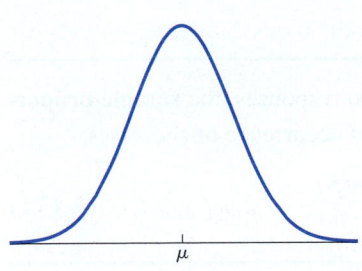

Figure 3.7 We expect the mode M of a sample from this distribution to be near the population mean μ.

The remainder of this section describes summary measures for qualitative data.

done this! Recall Example 2.4, in which 25 cruise ships were randomly selected and the destination of each ship was recorded. Each response was categorical (destination), and the data were summarized in a table listing only category, frequency of occurrence for each category, and relative frequency of occurrence for each category.

Suppose now that the commuter students at a small college are asked to complete a survey to identify the make of car they use to drive to school. Numerical summary measures for this categorical variable should include frequencies and relative frequencies, or proportions, as shown in the following table. (The cumulative relative frequency is used only for numerical data sets, and doesn't really make sense here because there is no natural ordering.)

Category	Frequency	Relative frequency
Buick	137	0.0938
Chevrolet	288	0.1973
Ford	202	0.1384
Honda	336	0.2301
Mazda	175	0.1199
Saturn	322	0.2205
Total	1460	1.0000

A *dichotomous* or *Bernoulli* variable is a special categorical variable that has only two possible responses. One response is often associated with, or called, a *success*, denoted *S*, and the other response is called a *failure*, denoted *F*. The two possible actual responses are ignored. For example, suppose a medical researcher selects children at random and asks them all whether they have had an ear infection within the past year. The response *had an ear infection* might be a *success*, and *had no ear infection* would be a *failure*. The same numerical measures are used to summarize observations on this kind of categorical variable: frequency and relative frequency of occurrence for each response. The relative frequency of successes has a special name.

Definition

$\hat{p}$ is read as "p hat."

For observations on a categorical variable with only two responses, the **sample proportion of successes**, denoted $\hat{p}$, is the relative frequency of occurrence of successes:

$$\hat{p} = \frac{\text{number of } S\text{'s in the sample}}{\text{total number of responses}} = \frac{n(S)}{n} \tag{3.2}$$

A CLOSER LOOK

The symbol *p* is used in notation to represent several quantities: the population proportion of successes, the sample proportion of successes, and in the definition of the trimmed mean. The context in which the notation is used implies the appropriate concept.

1. The **population proportion of successes** is denoted by *p*.

2. The success response is *not* necessarily associated with a good thing. For example, a researcher may be interested in the proportion of laboratory animals that die when they are exposed to a certain toxic chemical. A *success* may be associated with the death of an animal.

3. The sample proportion of successes $\hat{p}$ can be thought of as a sample mean in disguise. Suppose every *S* is changed to a 1, and every *F* to a 0. The sample mean for this new numerical data is

$$\bar{x} = \frac{1}{n}(\text{a sum of 0's and 1's}) = \frac{n(S)}{n} = \hat{p}$$

DATA SET
SEATBELT

Example 3.7 Seatbelt Checkpoint

In many states it is against the law to drive without a fastened seatbelt. The State Police recently established a checkpoint along a heavily traveled road. A success was recorded for a driver wearing a seatbelt, and a failure recorded otherwise. The observations from this checkpoint are given in the following table.

S	S	F	F	S	S	F	S	F	S	F	S	S	S
S	F	S	S	S	S	F	S	S	F	S	F	S	F

The sample contains 28 observations and 18 successes. The sample proportion of successes is

$$\hat{p} = \frac{n(S)}{n} = \frac{18}{28} = 0.6429$$

Approximately 64% of the drivers stopped at the checkpoint were wearing their seatbelts. It is reasonable to assume the value of $\hat{p}$ is *close* to the population proportion of successes—in this example, the true proportion of drivers who wear a seatbelt. ●

TRY IT NOW   GO TO EXERCISE 3.10

Technology Corner

Procedure: Compute the sample mean, sample median, a trimmed mean, and the mode.
Reconsider: Example 3.2, solution, and interpretations.

CrunchIt!

CrunchIt! has a built-in function to find certain descriptive statistics, including the sample mean and the sample median. There is no built-in function to compute a trimmed mean nor a sample mode.

1. Enter the data into a column.
2. Select Statistics; Descriptive Statistics. Choose the appropriate column and click the Calculate button. See Figure 3.8.

Results - Descriptive Statistics
Export ▾

	n	Sample Mean	Standard Deviation	Min	Q1	Median	Q3	Max
Temperature	12	21.25	8.843	6	15	23	28.50	33

Figure 3.8 CrunchIt! descriptive statistics.

TI-84 Plus C

There are several ways to find the sample mean and the sample median using the graphing calculator. There is no built-in function to compute a trimmed mean nor a sample mode.

1. Enter the data into list L1.
2. Use the command LIST ; MATH; mean to compute the sample mean. Use the command LIST ; MATH; median to compute the sample median. See Figure 3.9.
3. The function STAT ; CALC; 1-Var Stats returns several summary statistics. The sample mean is on the first output screen and the sample median is on the second, denoted by Med. See Figures 3.10 and 3.11.

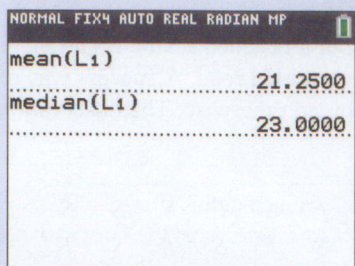

Figure 3.9 The sample mean and the sample median using built-in calculator functions.

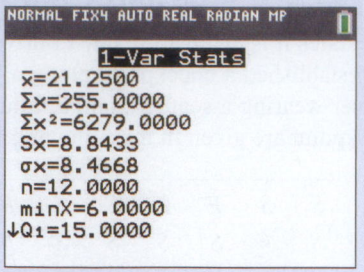

Figure 3.10 The sample mean is part of the output from the 1-Var Stats function.

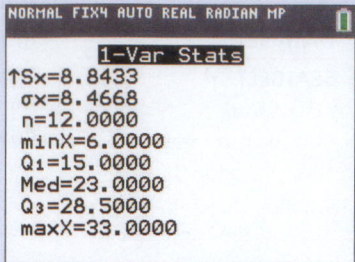

Figure 3.11 The second output screen from 1-Var Stats shows the sample median (Med).

Minitab

There are several ways to find the summary statistics using Minitab. In addition to the general Describe command, there are Calc; Column statistics functions, Calc; Calculator functions, and various macros.

1. Enter the data into column C1.
2. Select Stat; Basic Statistics; Display Descriptive Statistics. Enter C1 in the Variables window.
3. Choose the Statistics option button and check the summary statistics Mean, Median, Mode, and Trimmed mean. Note: Minitab computes only a 5% trimmed mean. Other macros allow any percentage. See Figure 3.12.

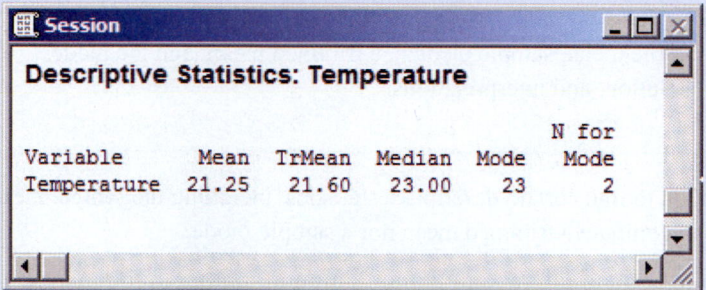

Figure 3.12 Minitab descriptive statistics.

Excel

Excel has built-in functions for these four descriptive statistics. Under the Data tab, Data Analysis; Descriptive Statistics can also be used to compute several summary statistics simultaneously.

1. Enter the data into column A.
2. Use the appropriate Excel function to compute the sample mean, sample median, trimmed mean, and mode. Note: the second argument in TRIMMEAN is the total proportion of data trimmed. Excel rounds the number of trimmed observations down to the nearest multiple of 2. See Figure 3.13.

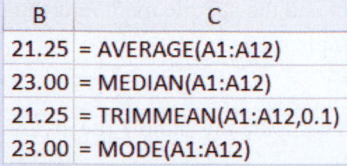

Figure 3.13 Excel descriptive statistics.

SECTION 3.1 EXERCISES

Concept Check

3.1 Fill in the Blank
  a. The two most common types of numerical summary measures describe the _____ and the _____ of the data.
  b. Measures of central tendency suggest where the data is _____.

3.2 True/False
  a. The sample mean and the population mean are always the same value.
  b. The sample mean and the sample median can be the same value.
  c. The sample mean is sensitive to outliers.
  d. When computing a trimmed mean, we discard the same number of observations from each end of the ordered list.
  e. The mode may not exist for a specific data set.
  f. It is reasonable to assume that the sample proportion of successes is close to the population proportion of successes.

Practice

3.3 Compute each summation using the following random sample. EX3.3

$$x_1 = -15 \quad x_2 = 6 \quad x_3 = 40 \quad x_4 = 13 \quad x_5 = 38$$

  a. $\sum x_i$
  b. $\sum x_i^2$
  c. $\sum (x_i - 10)$
  d. $\sum (x_i - 5)^2$
  e. $\sum (2x_i)$
  f. $2\sum x_i$

3.4 Suppose the following random sample is obtained.

43.3	52.7	67.7	52.1	54.7
54.1	46.7	47.2	48.5	45.8

Compute the following sums. EX3.4
  a. $\sum x_i^2$
  b. $\sum x_i^3$
  c. $\sum (x_i - 50)^2$
  d. $(\sum x_i)^2$
  e. $\sum (x_i - 51.28)$
  f. $\sum \frac{x_i}{7}$

3.5 Compute the mean for each sample with known sum.
  a. $\sum x_i = 1057, n = 10$
  b. $\sum x_i = 356, n = 27$
  c. $\sum x_i = 250.5, n = 36$
  d. $\sum x_i = 1.355, n = 11$
  e. $\sum x_i = -37.4, n = 15$
  f. $\sum x_i = 496.81, n = 28$

3.6 Find the position, or location, of the sample median in an ordered data set of size n.
  a. $n = 22$
  b. $n = 37$
  c. $n = 117$
  d. $n = 64$

3.7 Find the sample mean and the sample median for each data set. EX3.7
  a. 5, 3, 7, 9, 11, 5, 6, 7, 7
  b. −7, 10, 25, 22, 36, −24, 0, 1, 12, 9, −11
  c. 5.4, 3.3, 6.0, 10.1, 13.6, 7.7, 16.6, 28.9, 4.6
  d. −103.7, −110.4, −109.1, −99.7, −115.6

3.8 Consider the data given in the following table. EX3.8

5	7	8	27	3	15	7	6	4	5	5	1

Find the sample median. Note that this summary statistic is a better measure of central tendency than the sample mean for this data set. Why?

3.9 Use the values of the sample mean and the sample median to determine whether the distribution is symmetric, skewed to the left, or skewed to the right.
  a. $\bar{x} = 37, \ \tilde{x} = 49$
  b. $\bar{x} = 63.5, \ \tilde{x} = 62.75$
  c. $\bar{x} = -37, \ \tilde{x} = -16$
  d. $\bar{x} = -12.56, \ \tilde{x} = 12.56$

3.10 Compute the indicated trimmed mean for each data set. EX3.10
  a. {24, 36, 26, 30, 28, 35, 33, 33, 34, 27} $\bar{x}_{tr(0.10)}$
  b. {72, 76, 76, 77, 85, 76, 80, 86, 62, 70} $\bar{x}_{tr(0.20)}$
  c. {182, 169, 180, 166, 173, 101, 188, 124, 182, 137, 100, 137, 118, 111, 137, 181, 189, 130, 168, 133} $\bar{x}_{tr(0.20)}$
  d. {5.5, 7.5, 7.3, 6.4, 5.3, 9.5, 7.2, 5.8, 7.0, 6.7, 9.0, 8.1, 8.4, 5.8, 5.4, 7.2, 7.4, 7.5, 5.9, 7.5} $\bar{x}_{tr(0.15)}$

3.11 Find the mode for each data set, if it exists. EX3.11
  a. 3, 5, 6, 7, 3, 4, 6, 6, 8, 11, 13, 2, 1
  b. −17, −10, 0, 3, −5, 4.3, 12, 0, 5, −2.1, 1.7, −7
  c. 6.6, 7.3, 5.2, 6.2, 8.3, 9.8, 4.1, 3.7

3.12 Find the sample proportion of successes for each data set. EX3.12
  a. S, F, S, F, F, F, F, F, S, S, S, F, F, S
  b. F, S, S, F, S, F, S, S, S, S, S, S, S, S, S, F, S, S, S, S, S
  c. S, F, S, F, F, F, F, F, S, S, S, F, S, F, F, S, F, S, S, S, S, S, F, F, F, F, F, F, F, S, S, F, S, F, F

Applications

3.13 Travel and Transportation   Tractor trailers tend to exceed the speed limit (65 mph) on one downhill stretch of Route 80 in Pennsylvania. Using a radar gun, the following tractor trailer speeds (in mph) were observed. RADAR

81	66	67	69	79	62	70	73	67	60	61
67	74	65	77	74	64	71	64	67	61	

  a. Find the sample mean, $\bar{x}$.
  b. Find the sample median, $\tilde{x}$.
  c. What do your answers to parts (a) and (b) suggest about the shape of the distribution of speeds?

3.14 Biology and Environmental Science   The 2012 estimated wheat production (in 1000 metric tons) for several countries is given on the text website.[4] WHEAT
  a. Find the sample mean and the sample median for these data.

b. What do the summary statistics in part (a) suggest about the shape of the distribution of wheat yield?

3.15 Technology and the Internet   Steam is a software platform used to distribute and manage multiplayer online games. A day was selected at random and the number of simultaneous peak users for certain games is given in the following table.[5] **GAMES**

930	858	827	849	744	849	763	753	781
948	678	742	769	754	782	862	894	861

a. Find the sample mean and the sample median.
b. Suppose the last observation had been 2861 instead of 861. Find the sample mean and sample median for this revised data set. Explain how this change in the data affects the mean and median found in part (a).

3.16 Fuel Consumption and Cars   The text website contains a table that lists the atmospheric CO_2 concentration (in ppm) for 36 months ending in October 2012.[6] **ATMOCO2**
a. Find the sample mean and sample median for this data.
b. A certain group considers any monthly concentration less than 390 a success (not harmful to the environment). Find the sample proportion of successes.

3.17 Education and Child Development   The Math SAT scores for all students in an introductory statistics class at Edinboro University are given on the text website. **EDINSAT**
a. Find the sample mean and the sample median.
b. Find a 5% trimmed mean.
c. Using these three numerical summary measures, describe the shape of the distribution.

3.18 Manufacturing and Product Development   A random sample of 12-ounce cans of Dr Pepper soda was obtained from E. M. Heaths supermarket. The exact amount of soda (in ounces) in each can was measured, and the data are given on the text website. **SODACAN**
a. Find the sample mean and the sample median.
b. What do the summary statistics in part (a) suggest about the shape of the distribution of the amount of soda in each can?
c. Suppose any amount of 12 ounces or greater is considered a *success*. Find the sample proportion of successes.

3.19 Biology and Environmental Science   There are approximately 10,000 commercial fishing harvesters in Maine, and these businesses contribute one-third of the value of all New England fisheries. Some of the species caught off the Maine coast include cod, salmon, and flounder. The number of pounds (in millions) caught by Maine fisheries for several recent years is given on the text website.[7] **CATCH**
a. Find the sample mean and the sample median for this data set.
b. Which statistic is a better measure of central tendency for these data? Justify your answer.

3.20 Sports and Leisure   Some critics of Major League Baseball believe the ball is *juiced* (livelier) because it is

manufactured to give hitters an advantage. To investigate this claim, a sample of the earned run average (ERA) for American League starting pitchers for the 2012 season was obtained.[8] **PITCH**
a. Find the sample mean and the sample median.
b. Suppose the pitcher with the highest ERA plays in Colorado, where the air is thin, and the home runs are many. To eliminate such outliers, find a 4% trimmed mean.
c. Find the mode for the original data set, if it exists.

3.21 Biology and Environmental Science   The water temperature (in degrees Fahrenheit) during the summer of 2012 at several locations off the coast of Florida is given in the following table.[9] **TEMP2012**

79	80	80	80	80	81	79	82	84	86
81	81	83	84	84	84	80	81	83	84
84	85	86	86	79	84	79	80	79	79

a. Find the sample mean and the sample median.
b. Find a 10% trimmed mean for these data.
c. Find the mode for the original data, if it exists.

3.22 Education and Child Development   An educational study was designed to compare cooperative learning versus traditional lecture style. Two sections of an introductory statistics class were used. Seven students were randomly selected from each section. The scores on the second test (a 30-item exam) are given in the following table. **EDSTYLE**

Traditional	21	28	25	25	21	19	23
Cooperative	25	30	28	25	24	24	29

Which group of students did better *on average*? Justify your answer.[10]

Extended Applications

3.23 Sports and Leisure   The gold medal in the women's 10-meter platform diving at the 2012 Summer Olympics in London was won by Roulin Chen from China. Many of the participants in this competition included a 407C, an inward $3\frac{1}{2}$ somersault, as one of their dives. The text website contains the scores for some of these dives for various participants.[11] **DIVING**
a. Find the sample mean and the sample median for this data set.
b. Find the mode, if it exists.
c. Multiply each score by the degree of difficulty, 3.2. Find the sample mean for this new data set. How does this sample mean compare with the sample mean found in part (a)?

3.24 Public Health and Nutrition   Residents in the greater Toronto area have complained that there has been an increase in aircraft noise.[12] This increased noise may lead to sleep disturbances and other health problems. Suppose the noise level (in dBA) of several aircraft was measured using one flight path. The data are given in the following table. **AIRNOISE**

75	72	71	65	63	72	70	68

a. Find the sample mean for these data.
b. Some researchers believe that the wind velocity gradient can add as much as 4 dB to each reading. Add 4 dB to each observation in the data set. Compute the new sample mean. How does this compare with the sample mean found in part (a)?

3.25 Travel and Transportation The following table contains the estimated unlinked light rail transit passenger trips (in thousands) for various cities during June 2012. **TRANSIT**

995.1	5132.7	190.9	1018.0	2593.8	4055.9
849.6	1691.2	22.5	597.5	6415.3	752.4
933.4	1426.4	409.9	1867.0	517.0	245.8
3605.5	2381.4	655.3	155.0	1822.0	896.2

Source: American Public Transportation Association.

a. Find the sample mean number of unlinked light rail transit trips.
b. Use each June observation to estimate the yearly number of trips. That is, multiply each observation by 12. Find the sample mean for this new data set. How does this sample mean compare with the sample mean found in part (a)?

3.26 Manufacturing and Product Development A new quality control program was recently started at a Hyundai manufacturing facility. Several times each day, randomly selected panels from a stamping press are inspected for defects. A nondefective panel is a success (S). A defective panel is a failure (F) and must be restamped at an additional cost. During a recent inspection, the following 32 observations were recorded:

S	S	S	S	F	S	F	F	S	S	S	S	S	S
S	S	F	S	S	S	S	S	S	S	S	F	S	S
S	S	S	S										

a. Find the sample proportion of successes.
b. Change each S to a 1, and each F to a 0. Find the sample mean for these new data. How does the mean compare with the sample proportion of successes found in part (a)?
c. Suppose 8 additional panels were selected and inspected (for a total of 40 panels). Is it possible for the sample proportion of successes to be 0.9? Why or why not?

3.27 Sports and Leisure The playing time for rookies in the National Basketball Association (NBA) depends on many factors, including position and performance. A random sample of playing times per game (in minutes) for rookies in the NBA during the 2011–2012 season was obtained. The data are given in the following table. **ROOKIES**

34.2	30.5	29.4	29.4	25.5	23.1	20.2	19.5	10.5
18.9	18.6	16.7	15.2	15.0	14.6	13.5	13.2	12.8

Source: National Basketball Association.

a. Find the sample mean and the sample median.
b. What do the summary statistics in part (a) suggest about the shape of the distribution of playing time for rookies?
c. Can you change the maximum observation (34.2) so that the sample mean is equal to the sample median? Why or why not?

3.28 Medicine and Clinical Studies In a random sample of 13 patients with calcaneus bone fractures, the sample mean number of days until fracture healing was $\bar{x} = 37.85$ and the sample median was $\tilde{x} = 40$. Suppose an additional patient is added to the sample so that $x_{14} = 44.5$.
a. Find the sample mean for all 14 patients.
b. Is there any way to determine the sample median for all 14 patients? Explain.

3.29 Fuel Consumption and Cars The estimated oil reserves (in millions of barrels) of four wells are given by

$$x_1 = 1078 \quad x_2 = 5833 \quad x_3 = 10,772 \quad x_4 = 7320$$

a. Find x_5 so that the mean for all five observations is 6883.4.
b. Find x_5 so that the sample mean is equal to the sample median.

3.30 Manufacturing and Product Development A consumer group has tested the drying time for 15 samples of exterior latex paint. The sample mean drying time is 83.8 minutes. What must the 16th drying time be if the 16th observation decreases the mean drying time by 30 seconds? By 1 minute?

3.31 Biology and Environmental Science The beaches along the coast of New Hampshire are famous for chilly waters, even during the hottest summer days. A recent sample of the water temperature on 24 randomly selected summer days was obtained. The following temperatures are in degrees Fahrenheit (°F). **BEACH**

58	58	53	53	59	57	54	61	56	60
57	61	56	55	59	60	55	53	55	58
59	53	59	63						

a. Find the sample mean and the sample median.
b. Convert each temperature to degrees Celsius (°C). Use the formula $C = \frac{F - 32}{1.8}$. Find the mean for all the water temperatures in degrees Celsius.
c. What is the relationship between the sample means in parts (a) and (b)?

3.32 Technology and the Internet A recent survey of students at Minneapolis North High School included a question about the number of computers at home. The (grouped) data are summarized below. **HOMECOMP**

Number of computers	Frequency of occurrence
0	3
1	27
2	23
3	7
4	3
5	1

Find the sample mean and the sample median number of computers at home.

3.2 Measures of Variability

Measures of central tendency are only one characteristic of a data set. These numerical summary measures alone are not sufficient to describe a sample completely. It is possible to have two very *different* data sets with (approximately) the same mean (and median). Figures 3.14 and 3.15 show two smoothed histograms to illustrate the problem.

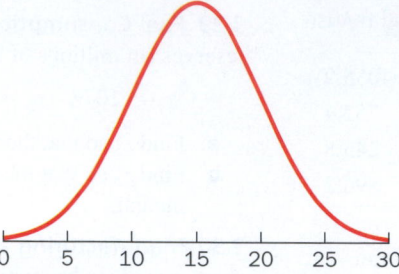

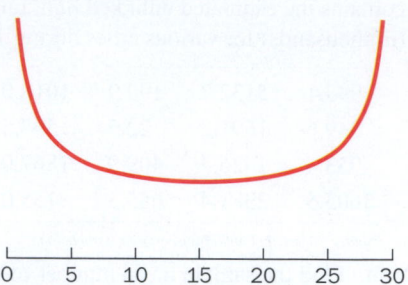

Figure 3.14 Sample 1: $x_1, x_2, \ldots, x_n$. The smoothed histogram suggests a compact distribution.

Figure 3.15 Sample 2: $y_1, y_2, \ldots, y_m$. The smoothed histogram suggests the data are more disperse, or spread out.

The measures of central tendency (sample mean and sample median) are approximately the same ($\bar{x} \approx \bar{y} \approx 15$ and $\tilde{x} \approx \tilde{y} \approx 15$), but the data in Sample 1 are more compact because more of the data are clustered about the mean $\bar{x} = 15$. To describe the difference between the data sets, we need to consider variability.

Definition

The (**sample**) **range**, denoted R, of the n observations $x_1, x_2, \ldots, x_n$ is the largest observation minus the smallest observation. Written mathematically,

$$R = x_{\max} - x_{\min} \tag{3.3}$$

where $x_{\max}$ denotes the maximum, or largest, observation, and $x_{\min}$ stands for the minimum, or smallest, observation.

A CLOSER LOOK

1. In theory, the sample range does measure, or describe, variability. A data set with a small range has little variability and is compact. A data set with a large range has lots of variability and is spread out.

2. The sample range is used in many quality control applications. For example, a production supervisor may want to maintain small variability in a manufacturing process. The sample range may be used to determine whether the process is still well controlled, or whether there is abnormal variation.

Despite being very easy to compute and a logical measure, the sample range is not adequate for describing variability. The sample range may not accurately represent the variability of a distribution if the maximum and minimum values are outliers.

The sample range for each data set summarized by the smoothed histograms in Figures 3.14 and 3.15 is approximately the same: $R \approx 30 - 0 = 30$. In fact, the two data sets have approximately the same mean. Therefore, it is necessary to use a better, more sensitive measure of variability. To derive a more precise measure of variability, consider how far each observation lies from the mean.

A graph may be used to visualize the spread of data and to suggest another measurement. A **dot plot** is a graph that simply displays a dot corresponding to each observation

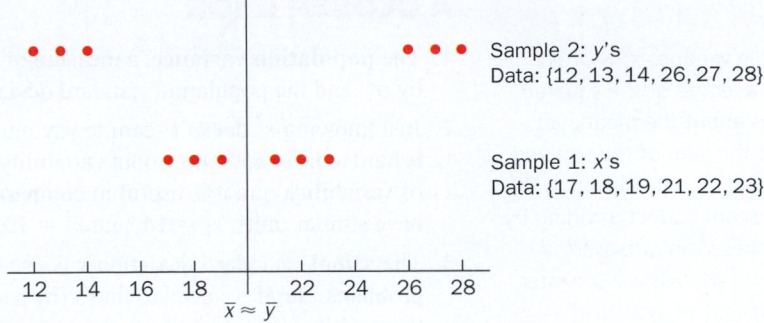

Sample 2: y's
Data: {12, 13, 14, 26, 27, 28}

Sample 1: x's
Data: {17, 18, 19, 21, 22, 23}

Figure 3.16 Stacked dot plot.

along a number line. The *stacked* dot plot in Figure 3.16 may be used to compare the variability in Sample 1 (x's) versus Sample 2 (y's).

In Sample 1, the data set is compact; each observation is *very close to the mean*. In Sample 2, the data set is more spread out; each observation is *far away from the mean*. This analysis of Figure 3.16 suggests that a better measure of variability might include the distances from the mean.

Definition

Given a set of n observations $x_1, x_2, \ldots, x_n$, the **ith deviation about the mean** is $x_i - \bar{x}$.

A CLOSER LOOK

1. Given a data set, to calculate the ith deviation about the mean, find $\bar{x}$, then compute the difference $x_i - \bar{x}$. For example, the seventh deviation about the mean is the value $x_7 - \bar{x}$.

2. We usually do not need any one deviation about the mean; all of the deviations about the mean together will be used to find a suitable measure of variability.

3. If the ith deviation about the mean is positive, then the observation is to the right of the mean: If $x_i - \bar{x} > 0$, then $x_i > \bar{x}$.

   If the ith deviation about the mean is negative, then the observation is to the left of the mean: If $x_i - \bar{x} < 0$, then $x_i < \bar{x}$.

A data set with little variability should have small deviations about the mean, and the squares of the deviations should be small. A data set with lots of variability should have large deviations about the mean, and the squares of the deviations should be large. This idea is used to define the **sample variance**.

Definition

The **sample variance**, denoted s^2, of the n observations $x_1, x_2, \ldots, x_n$ is the sum of the squared deviations about the mean divided by $n - 1$. Written mathematically,

$$s^2 = \frac{1}{n-1}\sum(x_i - \bar{x})^2$$

$$= \frac{1}{n-1}[(x_1 - \bar{x})^2 + (x_2 - \bar{x})^2 + \cdots + (x_n - \bar{x})^2] \tag{3.4}$$

The **sample standard deviation**, denoted s, is the positive square root of the sample variance. Written mathematically,

$$s = \sqrt{s^2} \tag{3.5}$$

The sample variance s^2 is often called an *average* of the squared deviations about the mean, yet we divide the sum of the squared deviations by $n - 1$. Although this does not seem correct, dividing by $n - 1$ makes s^2 an *unbiased* estimator of σ^2. We will see later in the text that an unbiased statistic is, in some sense, a good thing. There are $n - 1$ *degrees of freedom*, a kind of dimension of variability, associated with the sample variance s^2.

A CLOSER LOOK

1. The **population variance**, a measure of variability for an entire population, is denoted by σ^2, and the population standard deviation is denoted by σ, the Greek letter sigma.

2. Just knowing s^2 doesn't seem to say much about variability. If $s^2 = 6$, for example, it is hard to infer anything about variability. However, the sample variance s^2 *is* a measure of variability, and it is useful in comparisons. For example, if Sample 1 and Sample 2 have similar units, $s_1^2 = 14$, and $s_2^2 = 10$, then the data in Sample 2 are more compact.

3. The sample standard deviation s is used (rather than s^2) in many statistical inference problems. So, if we need to find s (by hand), we need to compute s^2 first, and then take the positive square root to find s.

4. The units for the sample standard deviation are the same as for the original data. And a value of $s = 0$ means there is no variability in the data set.

5. The notation s_x^2 is used to represent the sample variance for a set of observations denoted by $x_1, x_2, \ldots, x_n$. Similarly, s_y^2 represents the sample variance for a set of observations $y_1, y_2, \ldots, y_n$. ●

DATA SET

ZUCCHINI

Example 3.8 Zucchini Weight

Welliver Farms in Bloomsburg, Pennsylvania, sells a wide variety of fruits and vegetables and frequently donates crates of zucchini to the local food cupboard. Five of the donated zucchini were randomly selected, and each was carefully weighed. The weights, in ounces, were 6.2, 4.5, 6.6, 7.0, and 8.2. Find the sample variance and the sample standard deviation for these data.

SOLUTION

STEP 1 Find the sample mean:

$$\bar{x} = \frac{1}{5}(6.2 + 4.5 + 6.6 + 7.0 + 8.2) = \frac{1}{5}(32.5) = 6.5$$

STEP 2 Use Equation 3.4 to find the sample variance.

$$s^2 = \frac{1}{4}\left[(6.2 - 6.5)^2 + (4.5 - 6.5)^2 + (6.6 - 6.5)^2 + (7.0 - 6.5)^2 + (8.2 - 6.5)^2\right]$$

Use data and $\bar{x}$.

$$= \frac{1}{4}\left[(-0.3)^2 + (-2.0)^2 + (0.1)^2 + (0.5)^2 + (1.7)^2\right]$$

Compute differences.

$$= \frac{1}{4}\left[0.09 + 4.0 + 0.01 + 0.25 + 2.89\right]$$

Square each difference.

$$= \frac{1}{4}(7.24) = 1.81$$

Add, divide by 4.

Figure 3.17 Sample variance and sample standard deviation.

STEP 3 Take the positive square root of the variance to find the standard deviation.

$$s = \sqrt{1.81} \approx 1.3454$$

A technology solution is shown in Figure 3.17. ●

Equation 3.4 is the definition of the sample variance and may be used to find s^2, but there is actually a more efficient technique for computing s^2.

Definition

The computational formula for the sample variance is

$$s^2 = \frac{1}{n-1}\left[\sum x_i^2 - \frac{1}{n}(\sum x_i)^2\right] \tag{3.6}$$

This is a convenient shortcut method for calculating s^2 without having to find all the deviations about the mean. Suppose $x_1, x_2, \ldots, x_n$ is a set of observations. To find s^2, Equation 3.6 says:

1. Find the sum of the *squared* observations, $\sum x_i^2$.
2. Find the sum of the observations, $\sum x_i$.
3. Square the sum of the observations, $(\sum x_i)^2$.
4. Multiply the square of the sum of the observations by $1/n$, $\frac{1}{n}(\sum x_i)^2$.
5. Subtract the two quantities, and multiply the difference by $1/(n-1)$,

$$s^2 = \frac{1}{n-1}\left[\sum x_i^2 - \frac{1}{n}(\sum x_i)^2\right]$$

Example 3.9 Zucchini Weight (Continued)

Use the computational formula for s^2 to find the sample variance for the data in Example 3.8. The zucchini weights are 6.2, 4.5, 6.6, 7.0, and 8.2.

SOLUTION

STEP 1 Find the sum of the squared observations:

$$\sum x_i^2 = 6.2^2 + 4.5^2 + 6.6^2 + 7.0^2 + 8.2^2$$
$$= 38.44 + 20.25 + 43.56 + 49.0 + 67.24 = 218.49$$

STEP 2 Find the sum of the observations:

$$\sum x_i = 6.2 + 4.5 + 6.6 + 7.0 + 8.2 = 32.5$$

STEP 3 Square this sum and multiply by $1/n$:

$$\frac{1}{5}(\sum x_i)^2 = \frac{1}{5}(32.5)^2 = 211.25$$

STEP 4 Subtract the two quantities, and multiply by $1/(n-1)$:

$$s^2 = \frac{1}{4}(218.49 - 211.25) = \frac{1}{4}(7.24) = 1.81$$

(the same answer as above).

TRY IT NOW   GO TO EXERCISE 3.31

It can be shown that Equation 3.4 and Equation 3.6 are equivalent. Exercise 3.49 at the end of this section asks for a proof. If you must find a sample variance by hand, then use the computational formula. It has fewer calculations (is more efficient) and is usually more accurate (has less round-off error). In fact, most calculator and computer programs that find the sample variance use the computational formula.

The sample variance is always greater than or equal to zero: $s^2 \geq 0$. This is easy to see by looking at the definition in Equation 3.4. We sum *squared* deviations about the mean (always greater than or equal to zero) and divide by a positive number $(n-1)$. There are two *special cases*.

1. $s^2 = 0$: This occurs if all the observations are the same. If all the observations are equal to some constant c, the mean is c, and all the deviations about the mean are zero. Hence, $s^2 = 0$. This makes sense intuitively also: If all the observations are the same, there is no variability.

2. $n = 1$: This is a strange case, but it can occur. If $n = 1$, there is no variability—or, another way to think of this—we cannot measure variability. The denominator in Equation 3.4 is zero, and anything divided by zero is undefined.

The sample variance (and the sample standard deviation) can be greatly influenced by outliers. An observation very far away from the rest has a large deviation about the mean, a large squared deviation about the mean, and therefore contributes a lot to the sum (in the definition of the sample variance). The *interquartile range* is another measure of variability, and it is resistant to outliers.

Definition

Note that the definition for Q_1 and Q_3 involves the median, not the mean.

Let $x_1, x_2, \ldots, x_n$ be a set of observations. The **quartiles** divide the data into four parts.

1. The **first (lower) quartile**, denoted Q_1 (Q_L), is the median of the lower half of the observations when they are arranged in ascending order.
2. The **second quartile** is the median $\tilde{x} = Q_2$.
3. The **third (upper) quartile**, denoted Q_3 (Q_U), is the median of the upper half of the observations when they are arranged in ascending order.
4. The **interquartile range**, denoted IQR, is the difference IQR $= Q_3 - Q_1$.

In smoothed histograms, the area under the curve between two points corresponds to the proportion of observations between those points. Interpreting Figure 3.18: 25% of the observations are between Q_1 and $\tilde{x}$.

The quartiles are illustrated in Figure 3.18.

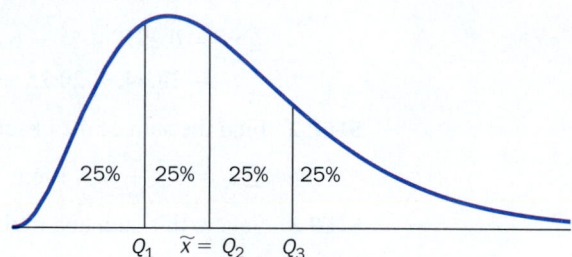

Figure 3.18 Smoothed histogram and quartiles.

A CLOSER LOOK

There is a very intuitive method for finding the quartiles. Arrange the data in order from smallest to largest. The median, $\tilde{x} = Q_2$, is the middle value. The first quartile, Q_1, is the median of the lower half, and the third quartile, Q_3, is the median of the upper half. In practice, a more general method is used for locating the *position*, or *depth*, of the first and third quartiles (in the ordered data set). ●

How to Compute Quartiles

Suppose $x_1, x_2, \ldots, x_n$ is a set of n observations.

1. Arrange the observations in ascending order, from smallest to largest.
2. To find Q_1, compute $d_1 = n/4$.
   a. If d_1 is a whole number, then the depth of Q_1 (position in the ordered list) is $d_1 + 0.5$. Q_1 is the mean of the observations in positions d_1 and $d_1 + 1$ in the ordered list.
   b. If d_1 is not a whole number, round up to the next whole number for the depth of Q_1.
3. To find Q_3, compute $d_3 = 3n/4$.
   a. If d_3 is a whole number, then the depth of Q_3 is $d_3 + 0.5$. Q_3 is the mean of the observations in positions d_3 and $d_3 + 1$ in the ordered list.
   b. If d_3 is not a whole number, round up to the next whole number for the depth of Q_3.

DATA SET

PULSE

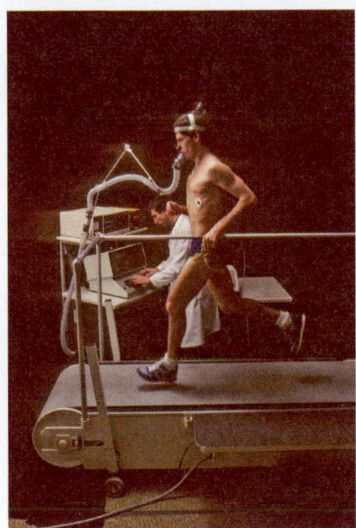

© Tom Tracy Photography/Alamy

Example 3.10 Pulse Rates

The following 10 observations represent the resting pulse rate for patients involved in an exercise study:

68	71	64	58	61	76	73	62	72	66

a. Find the first quartile, the third quartile, and the interquartile range.

b. Suppose there are 12 patients in the study, with $x_{11} = 78$ and $x_{12} = 81$. Find the first quartile, the third quartile, and the interquartile range for this modified data set.

SOLUTION

STEP 1   Arrange the observations in order from smallest to largest.

Observation	58	61	62	64	66	68	71	72	73	76
Position	1	2	3	4	5	6	7	8	9	10

STEP 2   Find the depth of the first quartile.

$$d_1 = \frac{n}{4} = \frac{10}{4} = 2.5$$

Because d_1 is not a whole number, round up. The depth of the first quartile is 3.

Q_1 is in the third position in the ordered list.
Using the table above, $Q_1 = 62$.

STEP 3   Find the depth of the third quartile.

$$d_3 = \frac{3n}{4} = \frac{(3)(10)}{4} = 7.5$$

Because d_3 is not a whole number, round up. The depth of the third quartile is 8.

Q_3 is in the eighth position in the ordered list.
Using the table above, $Q_3 = 72$.

STEP 4   Find the interquartile range IQR $= Q_3 - Q_1$.

IQR $= 72 - 62 = 10$

A technology solution is shown in Figures 3.19 and 3.20.

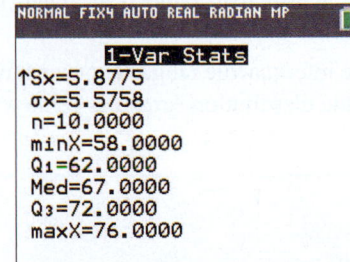

Figure 3.19 `1-Var Stats` is used to compute the quartiles.

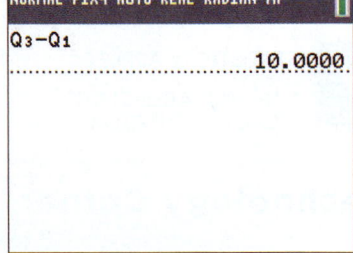

Figure 3.20 Compute IQR on the Home screen.

STEP 5   If there are 12 patients in the study, arrange the observations in order from smallest to largest in the modified data set.

Observation	58	61	62	64	66	68	71	72	73	76	78	81
Position	1	2	3	4	5	6	7	8	9	10	11	12

STEP 6   Find the depth of the first quartile.

$$d_1 = \frac{n}{4} = \frac{12}{4} = 3$$

Because d_1 is a whole number, add 0.5. The depth of the first quartile is 3.5.

Q_1 is the mean of the observations in the third and fourth positions in the ordered list.

$$Q_1 = \frac{1}{2}(62 + 64) = 63$$

STEP 7 Find the depth of the third quartile.

$$d_3 = \frac{3n}{4} = \frac{(3)(12)}{4} = 9$$

Because d_3 is a whole number, add 0.5.
The depth of the third quartile is 9.5.

Q_3 is the mean of the observations in the ninth and tenth positions in the ordered list.

$$Q_3 = \frac{1}{2}(73 + 76) = 74.5$$

STEP 8 Find the interquartile range.

$$IQR = 74.5 - 63 = 11.5$$

A technology solution is shown in Figures 3.21 and 3.22.

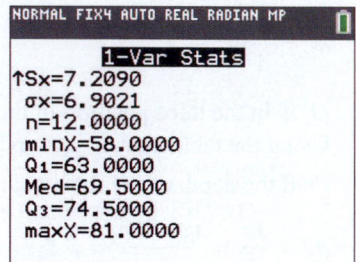

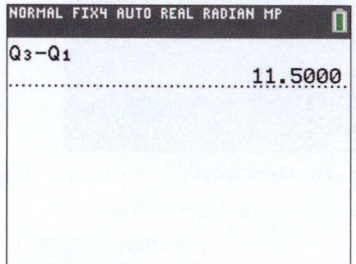

Figure 3.21 `1-Var Stats` is used to compute the quartiles.

Figure 3.22 Compute IQR on the Home screen.

TRY IT NOW GO TO EXERCISE 3.39

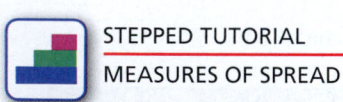

STEPPED TUTORIAL
MEASURES OF SPREAD

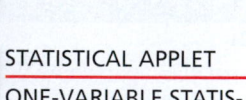

STATISTICAL APPLET
ONE-VARIABLE STATIS-TICAL CALCULATOR

A CLOSER LOOK

1. The interquartile range is the length of an interval that includes the middle half (middle 50%) of the data.

2. The interquartile range is not sensitive to outlying values. The lower and/or upper 25% of the distribution can be extreme without affecting Q_1 and/or Q_3.

Technology Corner

Procedure: Compute the sample variance, sample standard deviation, first quartile, third quartile, and interquartile range.
Reconsider: Example 3.10(b), solution, and interpretations.

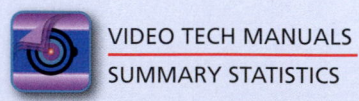

VIDEO TECH MANUALS
SUMMARY STATISTICS

CrunchIt!

CrunchIt! has a built-in function to find certain descriptive statistics, including the sample standard deviation, first quartile, and third quartile. There is no built-in function to compute the sample variance nor the interquartile range.

1. Enter the data into a column.
2. Select Statistics; Descriptive Statistics. Choose the appropriate column and click the Calculate button. See Figure 3.23.

Results - Descriptive Statistics

Export ▾

	n	Sample Mean	Standard Deviation	Min	Q1	Median	Q3	Max
Pulse	12	69.17	7.209	58	63	69.50	74.50	81

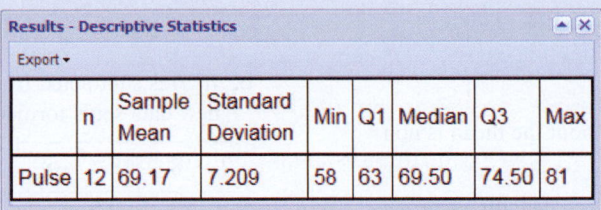

Figure 3.23 CrunchIt! descriptive statistics.

TI-84 Plus C

1. Enter the data into list `L1`.
2. Select `LIST`; `MATH`; `variance`. Take the square root of the variance to find the standard deviation. See Figure 3.17.
3. Select `STAT`; `CALC`; `1-Var Stats`.
4. The quartiles are displayed on the second output screen. Refer to Figure 3.21. Note: The sample standard deviation is displayed on the first output screen and the value is stored in the statistic variable `Sx`.
5. Compute the interquartile range on the Home screen. Use the TI-84 Plus statistics variables that represent the quartiles. Refer to Figure 3.22.

Minitab

1. Enter the data into column `C1`.
2. Select Stat; Basic Statistics; Display Descriptive Statistics. Enter `C1` in the Variables window.
3. Choose the Statistics option button and check the summary statistics Standard deviation, variance, First quartile, Third quartile, and Interquartile range. Note: Minitab computes quartiles using a slightly different algorithm. See Figure 3.24.

Session

Descriptive Statistics: Pulse

Variable	StDev	Variance	Q1	Q3	IQR
Pulse	7.21	51.97	62.50	75.25	12.75

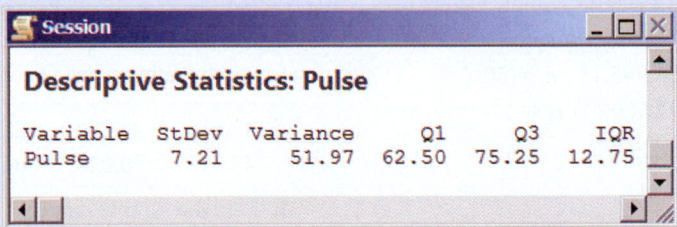

Figure 3.24 Measures of variability computed using Minitab.

Excel

Use built-in functions to compute the sample standard deviation, sample variance, quartiles, and interquartile range.

1. Enter the data into column `A`.
2. Use the function `STDEV.S` to compute the sample standard deviation; `VAR.S` to compute the sample variance; `QUARTILE` to compute the first and third quartile; and compute the interquartile range using the results. See Figure 3.25. Note: Excel computes quartiles using another different algorithm.

B	C
7.2090	= STDEV.S(A1:A12)
51.9697	= VAR.S(A1:A12)
62.5000	= QUARTILE.EXC(A1:A12,1)
75.2500	= QUARTILE.EXC(A1:A12,3)
12.7500	= B4-B3

Figure 3.25 Measures of variability computed using Excel.

Concept Check

3.33 True/False Every deviation about the mean is non-negative.

3.34 True/False The sample standard deviation is greater than or equal to 0.

3.35 True/False The sample standard deviation and the population standard deviation are always the same value.

3.36 True/False The computational formula for the sample variance is used only for large data sets.

3.37 True/False Quartiles divide the data into four parts.

Practice

3.38 Find the sample range, sample variance, and sample standard deviation for each data set. **EX3.38**
  a. {2.7, 6.0, 5.7, 5.4, 4.0, 3.1, 6.6, 5.7, 6.1, 3.0}
  b. {18.5, 23.5, 15.7, 15.7, 36.3, 20.8, 21.1, 20.2, 26.8, 19.9, 17.6, 17.5, 21.5, 22.4, 25.7}
  c. {23.94, −31.04, 37.09, 22.64, −61.23, 1.59, 23.09, 1.14}
  d. {0.13, 0.96, −0.50, 0.10, −1.65, −0.14, 1.43, −2.57, −1.28, −0.24, −0.90, −1.27, 1.53, 3.00, −1.28, 1.04, −0.90, 2.44, 1.70, 3.13}

3.39 Compute the sample variance and the sample standard deviation for each sample with known sum(s).

  a. $\sum x_i = 1219.29$ $\sum x_i^2 = 58{,}945.1$ $n = 30$
  b. $\sum x_i = 35.2918$ $\sum x_i^2 = 7748.98$ $n = 17$
  c. $\sum x_i = 218.291$ $\sum x_i^2 = 3615.96$ $n = 15$
  d. $\sum (x_i - \bar{x})^2 = 49.784$ $n = 21$

3.40 Find the depth of the first quartile and the third quartile in an ordered data set of size n.
  a. $n = 60$
  b. $n = 37$
  c. $n = 100$
  d. $n = 48$

3.41 Find the first quartile, the third quartile, and the interquartile range for each data set. **EX3.41**
  a. {20, 17, 37, 33, 29, 50, 20, 33}
  b. {13.1, 7.8, 11.9, 2.3, 6.7, 2.3, 7.4, 2.7, 8.9, 6.6, 6.8, 5.1, 2.2, 5.6, 5.5, 2.1, 7.7, 13.9, 1.6, 1.7}
  c. {−15, −13, −7, −15, −22, −12, −21, −21, −26, −17}
  d. {43.6, 44.1, 59.5, 52.3, 50.9, 39.7, 42.4, 58.5, 40.9, 38.5, 44.2, 60.3, 72.2, 34.8, 46.0, 54.7, 51.0, 54.3, 49.7, 62.9, 44.6, 61.3, 52.4, 43.9, 68.8, 59.2, 57.1, 70.5, 52.3, 49.5}

3.42 Consider the following data set: **EX3.42**

21	28	38	12	33	47	51	11	81	36

  a. Find the sample variance and the sample standard deviation.

  b. If 20 is subtracted from each observation in part (a), a new data set is formed:

1	8	18	−8	13	27	31	−9	61	16

Find the sample variance and the sample standard deviation for this new data set. How are these values related to the sample variance and the sample standard deviation found in part (a)?

  c. If each observation in part (a) is multiplied by 20, the following data set is formed:

420	560	760	240	660
940	1020	220	1620	720

Find the sample variance and the sample standard deviation for this new data set. How are these values related to the sample variance and the sample standard deviation found in part (a)?

3.43 How does an outlier affect each of the following?
  a. The sample variance
  b. The sample standard deviation
  c. The first quartile and the third quartile
  d. The interquartile range

Applications

3.44 Sports and Leisure The following times (in seconds) are from the ISU World Cup 2012/2013, Montreal, women's 500-meter speed skating event, October 26–28, 2012.[13] **SPDSKT**

47.611	46.206	45.299	45.405	47.611
47.149	59.028	46.188	54.118	45.416

  a. Find the sample range, R.
  b. Find the sample variance, s^2, and the sample standard deviation, s.
  c. Find the first and third quartiles, Q_1 and Q_3, and the interquartile range, IQR.

3.45 Physical Sciences The following table includes some of the data from an experiment performed by H. S. Lew for the Center for Building Technology at the U.S. National Institute of Standards and Technology (NIST). These data are used to certify computational results and evaluate statistical software. Each observation represents the deflection of a steel-concrete beam while subjected to periodic pressure. **BEAM**

−213	−564	−35	−15	141	115	−420
−360	203	−338	−431	194	−220	−513
154	−125	−559	92	−21	−579	

Source: National Institute of Standards and Technology.

  a. Find the sample standard deviation.
  b. Find the interquartile range.
  c. Which statistic, s or IQR, is a better measure of variability for this data set? Why?

3.46 Fuel Consumption and Cars The gross vehicle weight rating (in pounds) for several 2013 automobiles is given in the following table:[14] **AUTOWT**

| 5369 | 5612 | 6305 | 6355 | 6891 | 5137 | 6371 | 6327 |
| 6472 | 3925 | 4201 | 4680 | 4734 | 4178 | 5730 | 5899 |

a. Find the sample variance and the sample standard deviation.
b. Find the first and third quartiles.
c. Find the interquartile range and the *quartile deviation* (another measure of variability), QD = $(Q_3 - Q_1)/2$.

3.47 Education and Child Development Many educators believe that success in school is related directly to the amount of time spent completing homework assignments. A research study compared the academic ability of 17-year-olds who spend less than one hour on homework every day and those who spend more than two hours on homework every day. The National Assessment of Educational Progress (NAEP) scores for each student in each group are given in the following table.[19] **HOMEWORK**

Less than one hour

| 290 | 289 | 291 | 289 | 289 | 294 | 288 | 291 |
| 293 | 290 | 290 | 291 | 290 | 290 | 296 | 292 |

More than two hours

| 303 | 305 | 302 | 297 | 294 | 303 | 299 | 297 | 303 | 299 |
| 300 | 295 | 297 | 297 | 297 | 293 | 296 | 297 | 302 | 294 |

Source: National Center for Education Statistics.

a. Find the sample variance, sample standard deviation, and interquartile range of the progress scores for students who spend less than one hour on homework.
b. Find the sample variance, sample standard deviation, and interquartile range of the progress scores for students who spend more than two hours on homework.
c. Use your answers to parts (a) and (b) to determine which data set has more variability.

3.48 Travel and Transportation Air Canada recently discontinued a regularly scheduled flight from Montreal to Iqaluit. The route was not profitable because of rising fuel costs. Before the flight was canceled, seven days were randomly selected and the number of passengers recorded. The data are given in the following table. **PASSENGER**

| 51 | 76 | 47 | 61 | 53 | 68 | 79 |

a. Compute s^2 using the definition in Equation 3.4.
b. Compute s^2 using the computational formula in Equation 3.6.
c. How do your answers to parts (a) and (b) compare?

3.49 Public Policy and Political Science The president of the United States has the authority to grant clemencies, pardons, and commutations of sentences to convicted criminals. A sample of U.S. presidents was obtained, and the number of presidential clemency actions for each was recorded. The data are given in the following table.[15] **CLEMENCY**

President	Clemency actions
Calvin Coolidge	1545
Jimmy Carter	566
Woodrow Wilson	2480
John F. Kennedy	575
Thomas Jefferson	119
Millard Fillmore	170
Rutherford B. Hayes	893
Richard Nixon	926
Andrew Jackson	386
James Madison	196
Zachary Taylor	38
Martin Van Buren	168
Ulysses S. Grant	1332
Lyndon B. Johnson	1187
George W. Bush	176

a. Find Q_1, Q_3, and IQR for the clemency actions data.
b. Find s^2 and s.
c. Franklin D. Roosevelt had the highest number of clemency actions of any president, 3687. Add this value to the data set. Find IQR and s^2 for this expanded data set.
d. How do IQR and s^2 compare in these two data sets? Explain why these values are the same/different.

3.50 Physical Sciences The following operating temperatures (°F) for a certain steam turbine were measured on 10 randomly selected days. **TURBINE**

| 298 | 313 | 305 | 292 | 283 | 348 | 291 | 286 | 346 | 304 |

a. Find Q_1, Q_3, and IQR.
b. Find s^2 and s.
c. Suppose the smallest observation (283) is changed to 226. Find IQR and s^2 for this modified data set.
d. How do IQR and s^2 compare in these two data sets? Which measurement is more sensitive to outliers?

3.51 Marketing and Consumer Behavior Two measures designed to give a relative measure of variability are the **coefficient of variation**, denoted CV, and the **coefficient of quartile variation**, denoted CQV. These measures are defined by

$$CV = 100 \cdot \frac{s}{\bar{x}} \qquad CQV = 100 \cdot \frac{Q_3 - Q_1}{Q_3 + Q_1}$$

The areas (in square feet) for homes constructed in two new residential developments in San Antonio (one in North Central and one on the city's West Side) were recorded and are given in the following table. **HOMES**

East-side development

| 2038 | 1939 | 2024 | 1990 | 2109 | 2102 | 1918 | 2022 |

West-side development

| 2061 | 2383 | 2638 | 2142 | 2382 | 1489 | 2070 | 2340 |
| 1725 | 2368 | 1674 | 1877 | | | | |

a. Compute CV and CQV for each development.
b. Compare the coefficient of variation and the coefficient of quartile variation for each development. Which data set has more variability?

3.52 Physical Sciences Solar wind released from the Sun can affect power grids on Earth, the northern and southern lights, and even the tails of comets. The following table lists the proton density in protons per cubic centimeter (p/cc) for several times in November 2012.[16] **SUNWIND**

| 2.8 | 15.7 | 0.7 | 0.5 | 0.6 | 2.7 | 2.2 | 2.7 | 3.1 | 0.5 |
| 5.0 | 5.5 | 1.3 | 10.9 | 3.2 | 3.4 | 2.7 | 1.2 | 1.7 | 0.8 |

a. Find the sample variance and the sample standard deviation.
b. Find the Q_1, Q_3, and IQR for these data.
c. Remove the two largest proton densities from the data set. Answer parts (a) and (b) for this reduced data set. Compare the sample standard deviation and IQR in these two data sets and explain how these values have changed.

3.53 Public Health and Nutrition The Center for Science in the Public Interest (CSPI), a consumer group concerned about nutrition labeling, has defined a new measure of breakfast cereal called the nutritional index (NI), which is based on calories, vitamins, minerals, and sugar content per serving. A larger NI indicates greater nutritional value. The NI was measured for randomly selected cereals sold by Kellogg's and General Mills. The results are given in the following table. **CEREALNI**

Kellogg's

| 86 | 70 | 77 | 79 | 71 | 80 | 88 | 62 | 81 | 82 |
| 75 | 83 | 70 | 67 | 72 | 68 | 74 | 80 | 62 | 74 |

General Mills

| 54 | 49 | 50 | 31 | 46 | 29 | 81 | 63 | 41 | 60 |
| 66 | 68 | 39 | 59 | 47 | 80 | 41 | 91 | 41 | 33 |

a. Find s^2, s, and IQR for Kellogg's.
b. Find s^2, s, and IQR for General Mills.
c. Use the results in parts (a) and (b) to compare the variability in NI for the two companies.

3.54 Biology and Environmental Science Stage data (in feet-NGVD) for the Mississippi River at Baton Rouge at various times in 2012 are given on the text website.[17] **MSRIVER**

a. Find Q_1, Q_3, and IQR for these stream velocity data.
b. How large could the minimum stage be without changing the IQR?
c. Find the coefficient of quartile variation, CQV (defined in Exercise 3.51).

Extended Applications

3.55 Biology and Environmental Science The number of wildland fires has increased dramatically over the last decade,

and the number of acres burned has approximately doubled. The following table lists the number of wildland fires in 2012 as of November 2 in selected states.[18] **WILDFIRE**

State	AK	CA	CO	CT	GA	FL
Fires	398	7737	1447	180	2878	2217

State	IN	KY	LA	MA	MD	ME
Fires	67	896	828	1446	143	551

State	NC	NH	NJ	OH
Fires	2575	312	994	218

a. Find the sample variance and the sample standard deviation.
b. Find Q_1, Q_3, and IQR for these data.
c. Verify that the sum of the deviations about the mean is 0 (subject to round-off error).

3.56 Marketing and Consumer Behavior An Internet search for the best deal on a 12-megapixel digital camera revealed the following prices (in U.S. dollars).[19] **BESTDEAL**

| 160 | 169 | 783 | 90 | 129 | 188 |
| 300 | 295 | 600 | 579 | 356 | 553 |

a. Find Q_1, Q_3, and IQR for these price data.
b. Suppose the highest price (783) is changed to 699. Find Q_1, Q_3, and IQR for this modified data set.
c. How large could the maximum price be without changing IQR?
d. How much could the minimum price be raised before Q_1 changes?

3.57 Travel and Transportation A typical road bridge is constructed to last 50 years. The mean age of all bridges in the United States is approximately 43 years. The number of structurally deficient bridges in each state and the District of Columbia as of 2009 is given on the text website.[20] **BRIDGES**

a. Find the sample variance and the sample standard deviation of the number of structurally deficient bridges.
b. Suppose each state is able to repair 10% of all structurally deficient bridges. Find the sample variance and the sample standard deviation for this new data set.
c. How do your answers to parts (a) and (b) compare?

3.58 Travel and Transportation In its annual study, the International Telework Association & Council asks survey participants how many miles they must drive to work each day. Six study participants were selected at random and their mileage was recorded: **MILEAGE**

| 25 | 39 | 16 | 35 | 18 | 45 |

a. Find each deviation about the mean.
b. Verify that the sum of the deviations about the mean is 0 (subject to round-off error).
c. Prove that, in general, $\sum(x_i - \bar{x}) = 0$. (Hint: Write as two separate sums, and use the definition of the sample mean.)

3.59 Biology and Environmental Science The Virginia Estuarine & Coastal Observing System monitors the Chesapeake Bay and records values of several variables, including salinity, temperature, and turbidity. The wind speed (in miles per hour) at selected locations on November 6, 2012, is given in the following table.[21] **CHESABAY**

2	5	11	17	9	7	11	10
6	9	24	27	8	8	10	28
30	25	14	24	10	18	18	25

a. Find the sample variance and the sample standard deviation for these wind speed data.
b. Convert each observation to meters per second (multiply each observation by 0.44704). Find the sample variance and the sample standard deviation for the wind-speed data in meters per second.
c. How do your answers to parts (a) and (b) compare?

3.60 Proof Prove that Equation 3.4 (the definition of the sample variance) can be written as Equation 3.6. That is, show that

$$\frac{1}{n-1}\sum(x_i - \bar{x})^2 = \frac{1}{n-1}\left[\sum x_i^2 - \frac{1}{n}(\sum x_i)^2\right]$$

3.61 Public Health and Nutrition A nutritional study recently found the following number of calories in one slice of plain pizza at 10 different national chains.[27] **PIZZA**

228	281	274	408	364	259	317	299	302	231

Source: Food Science and Human Nutrition, Colorado State University.

a. Find the sample variance and the sample standard deviation.
b. Add 15 (calories) to each observation. Find the sample variance and the sample standard deviation for this modified data set.
c. How do your answers to parts (a) and (b) compare?
d. Suppose a data set (x's) has variance s_x^2 and standard deviation s_x. A new (transformed) data set is created using the equation $y_i = x_i + b$, where b is a constant. How are the variance and standard deviation of the new data set (s_y^2 and s_y) related to s_x^2 and s_x?

3.62 Technology and the Internet A benchmark computer program was executed on eight different machines, and the following times to completion (in seconds) were recorded: **PROGRAMS**

12.592	14.152	12.396	6.801
13.646	12.075	15.377	7.602

a. Find the sample variance and the sample standard deviation.
b. Multiply each observation by 7. Find the sample variance and the sample standard deviation for this modified data set.
c. How do your answers to parts (a) and (b) compare?
d. Suppose a data set (x's) has variance s_x^2 and standard deviation s_x. A new (transformed) data set is created using the equation $y_i = ax$, where a is a constant. How are the variance and standard deviation of the new data set (s_y^2 and s_y) related to s_x^2 and s_x?

3.63 Transformed Data Combine the results obtained in the previous two exercises. Suppose a data set (x's) has variance s_x^2 and standard deviation s_x. A new (transformed) data set is created using the equation $y_i = ax + b$, where a and b are constants. How are the variance and standard deviation of the new data set (s_y^2 and s_y) related to s_x^2 and s_x?

3.64 Is This Possible? Consider the set of observations
$$\{5, 7, 3, 2, 4, 6, 9, 11, 13\}$$
Can you find a subset of size $n = 7$ with $\bar{x} = 5$ and $s^2 = 6$? If not, why not?

Challenge

3.65 Biology and Environmental Science A whale-watching tour off the coast of Maine is considered a success if at least one whale is sighted. Thirty-two randomly selected summer tours are classified in the following table:

S	S	S	S	F	S	S	S	S	S	S	S
S	S	S	S	S	S	S	F	S	S	S	S
S	F	S	S	S	S	S	S				

a. Find the sample proportion of successes.
b. Change each S to a 1 and each F to a 0. Find the sample variance for these new data. Write the sample variance in terms of the sample proportion.
c. If a population happens to be of finite size N, then the population mean and population variance are defined by

$$\mu = \frac{1}{N}\sum x_i \qquad \sigma^2 = \frac{1}{N}\sum(x_i - \mu)^2$$

Suppose the table represents an entire population. Find the population variance for the data (consisting of 0's and 1's). Write the population variance in terms of the sample proportion.

3.66 Other Summary Statistics Many other summary statistics can also be used to describe various characteristics of a numerical data set. Suppose $x_1, x_2, \ldots, x_n$ is a set of observations. For $r = 1, 2, 3, \ldots$, the **rth moment about the mean** $\bar{x}$ is defined as

$$m_r = \frac{1}{n}\sum(x_i - \bar{x})^r$$

For example, the second moment about the mean is

$$m_2 = \frac{1}{n}\sum(x_i - \bar{x})^2$$

Certain moments about the mean are used to define the **coefficient of skewness** (g_1) and the **coefficient of kurtosis** (g_2):

$$g_1 = \frac{m_3}{m_2^{3/2}} \qquad g_2 = \frac{m_4}{m_2^2}$$

The statistic g_1 is a measure of the lack of symmetry, and g_2 is a measure of the extent of the peak in a distribution.

Use technology to compute the values g_1 and g_2 for various distributions: skewed, symmetric, unimodal, uniform. Use your results to determine the values of g_1 that suggest more skewness in the distribution, and the values of g_2 that indicate a flatter, more uniform distribution.

3.3 The Empirical Rule and Measures of Relative Standing

Measures of central tendency and measures of variability are used to describe the general nature of a data set. These two types of measures may be combined to describe the distribution of a data set more precisely. In addition, these values may be used to define measures of relative standing, quantities used to compare observations from different data sets (with different units), or even to draw a conclusion or make an inference.

The first result combines the mean and the standard deviation to describe a distribution.

Chebyshev's Rule

What happens if $k = 1$?

Let $k > 1$. For *any* set of observations, the proportion of observations within k standard deviations of the mean [lying in the interval $(\bar{x} - ks, \bar{x} + ks)$, where s is the standard deviation] is at least $1 - (1/k^2)$.

Recall interval notation: (a, b) denotes an open interval, with the endpoints not included, from a to b. Therefore, $(\bar{x} - ks, \bar{x} + ks)$ means the set of all x's such that $\bar{x} - ks < x < \bar{x} + ks$.

The diagram in Figure 3.26, and the accompanying table, illustrate this idea. For any set of observations, the smoothed histogram shows that the proportion of observations captured in the interval $(\bar{x} - ks, \bar{x} + ks)$ is at least $1 - (1/k^2)$. For example, the proportion of observations within 1.5 standard deviations of the mean is at least 0.56 (or 56%). The proportion of observations within 3 standard deviations of the mean is at least 0.89 (or 89%).

Recall: In smoothed histograms, the area under the curve between two points a and b corresponds to the proportion of observations between a and b.

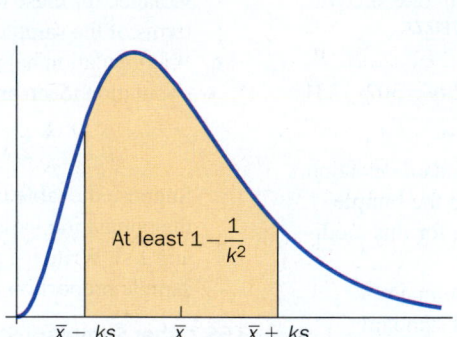

k	$1 - \dfrac{1}{k^2}$
1.5	$1 - \dfrac{1}{1.5^2} \approx 0.56$
2.0	$1 - \dfrac{1}{2.0^2} \approx 0.75$
2.3	$1 - \dfrac{1}{2.3^2} \approx 0.81$
3.0	$1 - \dfrac{1}{3.0^2} \approx 0.89$

Figure 3.26 Illustration of Chebyshev's rule.

A CLOSER LOOK

1. **Chebyshev's rule** simply helps to describe a set of observations using symmetric intervals about the mean. If we move k standard deviations from the mean in both directions, then the proportion of observations captured is at least $1 - (1/k^2)$.

2. The total area under the curve (the sum of all the proportions) is 1. Hence, Chebyshev's rule also implies that the proportion of observations in the tails of the distribution, outside the interval $(\bar{x} - ks, \bar{x} + ks)$, is at most $1/k^2$.

A symmetric interval about the mean is centered at the mean and has endpoints that are the same distance from the mean.

3. As indicated in the statement of Chebyshev's rule and as suggested in the table, you may use any value of k greater than 1, including decimals. The two most common values for k are $k = 2$ and $k = 3$. The actual proportions of observations within 2 and within 3 standard deviations can be compared to the values predicted by Chebyshev's rule and the empirical rule (page 100). In addition, $k = 2$ and $k = 3$ provide the fundamental background to statistical inference.

4. Chebyshev's rule is very conservative because it applies to any set of observations. Usually, the proportion of observations within k standard deviations of the mean is bigger than $1 - (1/k^2)$.

5. Chebyshev's rule may also be used to describe a population. If the mean and standard deviation are known, then μ and σ may be used in place of $\bar{x}$ and s. For any population, the proportion of observations that lie in the interval $(\mu - k\sigma, \mu + k\sigma)$ is at least $1 - (1/k^2)$. ●

Example 3.11 Automobile Battery Lifetime

In a random sample of the lifetime (in months) of a Honda Odyssey automobile battery, $\bar{x} = 54$ and $s = 5.3$. Use Chebyshev's rule with $k = 2$ and $k = 3$ to describe this distribution of battery lifetimes.

SOLUTION

STEP 1 For $k = 2$: $1 - \dfrac{1}{k^2} = 1 - \dfrac{1}{2^2} = 1 - \dfrac{1}{4} = \dfrac{3}{4} = 0.75$

At least 3/4 (or 75%) of the observations lie in the interval $(\bar{x} - 2s, \bar{x} + 2s) =$ $(54 - 2(5.3), 54 + 2(5.3)) = (43.4, 64.6)$.

STEP 2 For $k = 3$: $1 - \dfrac{1}{k^2} = 1 - \dfrac{1}{3^2} = 1 - \dfrac{1}{9} = \dfrac{8}{9} \approx 0.89$

At least 8/9 (or 89%) of the observations lie in the interval $(\bar{x} - 3s, \bar{x} + 3s) =$ $(54 - 3(5.3), 54 + 3(5.3)) = (38.1, 69.9)$.

STEP 3 Note also:

At most 1/4 (or 25%) of the observations lie *outside* the interval (43.4, 64.6).

At most 1/9 (or 11%) of the observations lie *outside* the interval (38.1, 69.9). ●

Example 3.12 How Long Was "In-A-Gadda-Da-Vida"?

In 1968, the psychedelic rock band Iron Butterfly recorded the 17-minute song "In-A-Gadda-Da-Vida." Most popular songs are much shorter, for example, "Viva la Vida" by Coldplay is approximately 4 minutes long. Suppose that in a random sample of the length (in minutes) of songs produced by hard rock bands, $\bar{x} = 3.35$ and $s = 0.5$.

a. Find the approximate proportion of observations between 2.35 and 4.35 minutes.

b. Find the approximate proportion of observations less than 1.85 or greater than 4.85 minutes.

c. Approximately what proportion of songs lasts more than 5 minutes?

SOLUTION

No values of k are specified, so use $k = 2$ and $k = 3$.

a. $(\bar{x} - 2s, \bar{x} + 2s) = (3.35 - 2(0.5), 3.35 + 2(0.5)) = (2.35, 4.35)$ $k = 2$
At least $1 - (1/4) = 3/4$ (or 75%) of the observations lie between 2.35 and 4.35 minutes.

b. $(\bar{x} - 3s, \bar{x} + 3s) = (3.35 - 3(0.5), 3.35 + 3(0.5)) = (1.85, 4.85)$ $k = 3$
At least $1 - (1/9) = 8/9$ (or 89%) of the observations lie between 1.85 and 4.85 minutes.
At most 1/9 (or 11%) of the observations are less than 1.85 or greater than 4.85 minutes.

c. Since Chebyshev's rule measures intervals in terms of the number of standard deviations from $\bar{x}$, find out how far 5 is from $\bar{x}$ in standard deviations.

$\bar{x} + ks = 3.35 + k(0.5) = 5 \Rightarrow k = 3.3$

We cannot assume anything about the shape of the distribution.

$1 - \dfrac{1}{k^2} = 1 - \dfrac{1}{3.3^2} \approx 0.91$

At least 0.91 (or 91%) of the observations lie in the interval
$(\bar{x} - 3.3s, \bar{x} + 3.3s) = (3.35 - 3.3(0.5), 3.35 + 3.3(0.5)) = (1.7, 5.0)$

Solution Trail 3.12

KEYWORDS

■ Approximate proportion of observations between.

TRANSLATION

■ What proportion of observations is captured by the interval?

CONCEPTS

■ Chebyshev's rule.

VISION

We don't know anything about the shape of the distribution of the length of songs. However, Chebyshev's rule applies to any distribution, tells us about the proportion of observations captured by certain intervals, and may be used here if the questions involve symmetric intervals about the mean.

Therefore, at most $1 - 0.91 = 0.09$ (or 9%) of the observations are outside this interval, either less than 1.7 or greater than 5.0 minutes. We cannot assume that the distribution is symmetric, so we do not know what part of the 9% is less than 1.7 and what part is more than 5.0 minutes. To be conservative, the best we can say is that at most 9% of the observations are more than 5 minutes long. ●

TRY IT NOW   GO TO EXERCISE 3.65

A normal curve is bell-shaped and symmetric, centered at the mean.

If a set of observations can be reasonably modeled by a normal curve, then we can describe this distribution more precisely. The *empirical rule* involves the mean and standard deviation also, and the results apply to three specific symmetric intervals about the mean.

The Empirical Rule

If the shape of the distribution of a set of observations is approximately normal, then:

1. The proportion of observations within **one standard deviation** of the mean is approximately 0.68.

2. The proportion of observations within **two standard deviations** of the mean is approximately 0.95.

3. The proportion of observations within **three standard deviations** of the mean is approximately 0.997.

Figure 3.27 illustrates the empirical rule, the symmetric intervals about the mean, and the proportions. The empirical rule conclusions are more accurate than Chebyshev's rule because we know (assume) more about the shape of the distribution (normality).

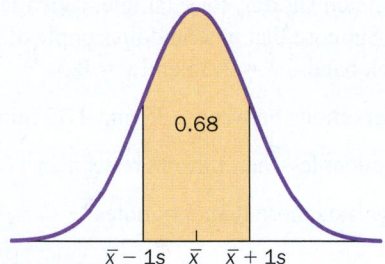 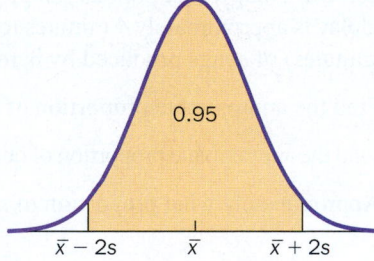 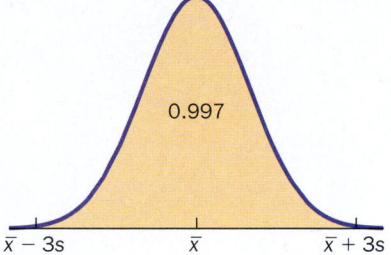

Figure 3.27 Symmetric intervals and proportions associated with the empirical rule.

A CLOSER LOOK

For now, the reasons for the proportions 0.68, 0.95, and 0.997 remain a mystery. We will discover where these numbers come from in Chapter 6.

1. Given a set of observations, the empirical rule may be used to *check* normality. To test for normality numerically, find the mean, standard deviation, and the three symmetric intervals about the mean $(\bar{x} - ks, \bar{x} + ks)$, $k = 1, 2, 3$. Compute the *actual* proportion of observations in each interval. If the actual proportions are close to 0.68, 0.95, and 0.997, then normality seems reasonable. Otherwise, there is evidence to suggest the shape of the distribution is not normal. This process is sort of a *backward* empirical rule.

2. The empirical rule may also be used to describe a population. If the distribution of the population is approximately normal, and the mean and standard deviation are known, then μ and σ may be used in place of $\bar{x}$ and s.

3. The proportion of observations beyond three standard deviations from the mean is $1 - 0.997 = 0.003$ (pretty small). Therefore, if the shape of a (population) distribution is approximately normal, it would be unusual to have an observation more than three standard deviations from the mean. What if there is one? (See Example 3.14.) ●

Solution Trail 3.13

KEYWORDS

- Approximately normal;
- Approximate proportion of observations between

TRANSLATION

- What proportion of observations is captured by the interval?

CONCEPTS

- The empirical rule.

VISION

Since the shape of the distribution is approximately normal, the empirical rule may be used to determine the proportion of observations captured by certain intervals, related in some way to three special symmetric intervals about the mean.

Example 3.13 Expensive Speeding Tickets

Some of the world's most expensive speeding tickets are issued in Finland and Canada. Over a long weekend in August 2012, there were 3556 speeding tickets issued in Alberta, Canada.[22] The cost of each ticket depends on the speed of the car and the posted limit. In a random sample of these ticket fines (in Canadian dollars), suppose the shape of the distribution is approximately normal, with $\bar{x} = 130$ and $s = 25$.

a. Approximately what proportion of observations is between 80 and 180?

b. Approximately what proportion of observations is greater than 205 or less than 55?

c. Approximately what proportion of observations is greater than 205?

d. Approximately what proportion of observations is between 105 and 180?

SOLUTION

a. Find the values one, two, and three standard deviations about the mean in each direction. See Figure 3.28. Notice that

$$(130 - 2(25), 130 + 2(25)) = (80, 180)$$

So, 80 to 180 is a symmetric interval about the mean, two standard deviations in each direction. The empirical rule states that approximately 0.95 (or 95%) of the observations lie in this interval. See Figure 3.28.

b. Notice that

$$(130 - 3(25), 130 + 3(25)) = (55, 205)$$

So, 55 to 205 is a symmetric interval about the mean, three standard deviations in each direction. The empirical rule states that approximately 0.997 (or 99.7%) of the observations lie in this interval. The remaining proportion, $1 - 0.997 = 0.003$ (or 0.3%), of observations lie outside this interval, greater than 205 or less than 55. See Figure 3.29.

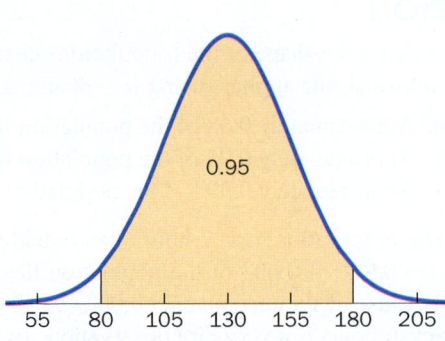

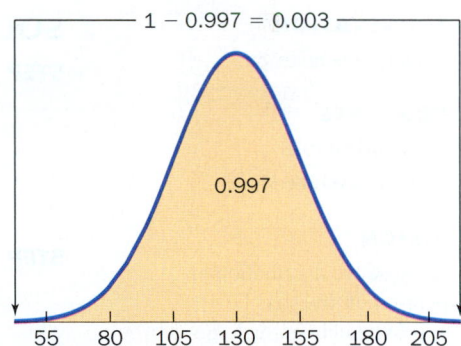

Figure 3.28 Approximately 0.95 (or 95%) of the observations lie within two standard deviations of the mean, in the interval (80, 180).

Figure 3.29 Approximately $1 - 0.997 = 0.003$ (or 0.3%) of the observations lie outside the interval (55, 205).

c. Because a normal distribution is symmetric about the mean, the remaining proportion outside three standard deviations from the mean ($1 - 0.997 = 0.003$) is divided evenly between the two tails. Therefore, approximately $0.003/2 = 0.0015$ (or 0.15%) of the observations are greater than 205. See Figure 3.30.

d. (105, 180) is not a symmetric interval about the mean. However, approximately 0.68 of the observations lie in the interval (105, 155) (one standard deviation from the mean). Approximately 0.95 of the observations lie in the interval (80, 180) (two standard deviations from the mean). This means that $0.95 - 0.68 = 0.27$ of the observations lie in the intervals (80, 105) and (155, 180). Because a normal distribution is symmetric, $0.27/2 = 0.135$ of the observations lie between 155 and 180. Therefore, a total of approximately $0.68 + 0.135 = 0.815$ (or 81.5%) of the observations lie between 105 and 180. See Figure 3.31.

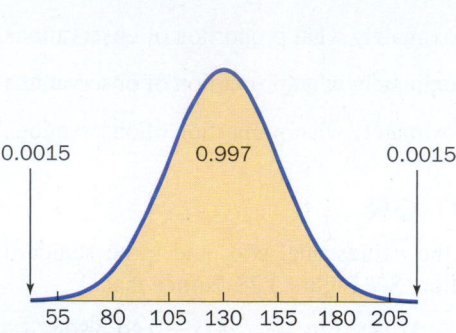

Figure 3.30 Approximately 0.0015 (or 0.15%) of the observations are greater than 205.

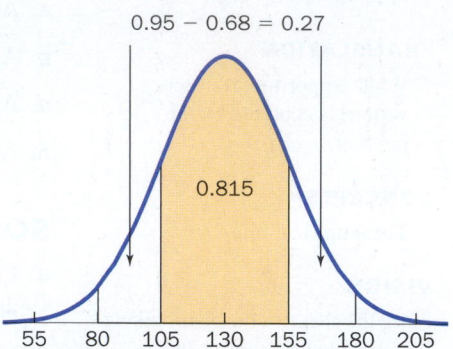

Figure 3.31 Approximately 0.27 (or 27%) of the observations lie in the intervals (80, 105) and (155,180).

TRY IT NOW   GO TO EXERCISE 3.70

Example 3.14 When Will the Pain Stop?

First Horizon Pharmaceutical has just developed a new medicine for treatment of routine aches and pains. The company claims the distribution of pain-relief times (in hours) is approximately normal, with mean $\mu = 8$ and standard deviation $\sigma = 0.2$. A patient with a typical muscle ache is randomly selected and the medicine is administered. The patient reports pain relief for only 7 hours. Is there any evidence to refute the manufacturer's claim?

SOLUTION

STEP 1 Because the shape of the (population) distribution is approximately normal, the empirical rule applies (using $\mu = 8$ and $\sigma = 0.2$).

    a. Approximately 0.68 of the population lies in the interval (7.8, 8.2).
    b. Approximately 0.95 of the population lies in the interval (7.6, 8.4).
    c. Approximately 0.997 of the population lies in the interval (7.4, 8.6).

STEP 2 The observation $x = 7$ hours lies outside the largest interval (7.4, 8.6). Only $1 - 0.997 = 0.003$ of the population lies outside this interval. More precisely (because of symmetry), only $0.003/2 = 0.0015$ of the population lies below 7.4. Seven hours is a very rare observation. Two things may have occurred.

    a. Seven hours is an incredibly lucky observation. Even though the proportion of observations below 7.4 is small, it is still possible for the manufacturer's claim to be true and for the pain reliever to last only 7 hours in this patient.
    b. The manufacturer's claim is false. Because an observation of 7 hours is so rare, it is more likely that one of the assumptions is wrong. The shape of the distribution may not be normal, the mean may be different from 8, and/or the standard deviation might be different from 0.2.

STEP 3 Typically, statistical inference discounts the *lucky* alternative. Therefore, because 7 hours is such an unlikely observation, there is evidence to suggest the manufacturer's claim is false. Something is awry. We would rarely see pain relief of only 7 hours if the claim is true.

Solution Trail 3.14

KEYWORDS

- Approximately normal
- Evidence to refute the claim?

TRANSLATION

- Draw a conclusion.

CONCEPTS

- Empirical rule
- Inference procedure.

VISION

Because the distribution of pain-relief times is approximately normal, the empirical rule may be used to determine how often observed times in certain intervals occur. If the observed pain-relief time is rare, then we should question the manufacturer's claim.

Note: We may be too quick to make an inference based on only a single observation. We will learn how to use more observations (information) to reach a more confident conclusion. ■

One method for comparing observations from different samples (with different units) is to use a *standardized score*. For a given observation, this relative measure is used to determine the distance from the mean in standard deviations.

Definition

Suppose $x_1, x_2, \ldots, x_n$ is a set of n observations with mean $\bar{x}$ and standard deviation s. The **z-score** corresponding to the ith observation, x_i, is given by

$$z_i = \frac{x_i - \bar{x}}{s} \tag{3.7}$$

z_i is a measure associated with x_i that indicates the distance from $\bar{x}$ in standard deviations.

A CLOSER LOOK

1. z_i may be positive or negative (or zero). A positive z-score indicates the observation is to the right of the mean. A negative z-score indicates the observation is to the left of the mean.

2. A z-score is a measure of relative standing; it indicates where an observation lies in relation to the rest of the values in the data set. There are other methods of standardization, but this is the most common.

3. Given a set of n observations, the sum of all the z-scores is 0; $\sum z_i = 0$. Can you prove this? ■

Statisticians tend to measure distances *in standard deviations, not miles, feet, inches, or meters. We often ask, "How many standard deviations from the mean is a given observation?"*

Example 3.15 Starting Salary

Most college career counselors agree that starting salary is associated with academic major. Even if a person's first job is not related directly to his course of study, his salary may still be related to his academic major. A recent survey of academic major and starting salary of graduates showed the following information:

Major	Mean	Standard deviation
English	$38,100	$3,600
Computer science	$57,690	$5,370

A computer science major who responded to the survey received a starting salary of $64,000, and an English major received an offer of $47,000. Which salary is *better*, in terms of statistics?

SOLUTION

STEP 1 The higher starting salary is probably better (subject to working conditions, benefits, location, etc.), but to answer this question in terms of statistics, consider the z-scores.

STEP 2 Computer science major: $z = \dfrac{64,000 - 57,690}{5,370} \approx 1.18$

Solution Trail 3.15

KEYWORDS
- Which salary is better?

TRANSLATION
- Which salary is farther away from the mean (to the right) in standard deviations?

CONCEPTS
- z-score.

VISION
Compute and compare the z-scores for each salary. This will allow us to determine how many *statistical steps* each observation is from the mean. The higher the z-score, the *better* the salary.

$64,000 is approximately 1.18 standard deviations to the right of the mean.

English major: $z = \dfrac{47{,}000 - 38{,}100}{3{,}600} \approx 2.47$

$47,000 is approximately 2.47 standard deviations to the right of the mean.

STEP 3   The English major's starting salary is actually better, because the salary is much higher than those of most English majors. ●

TRY IT NOW   GO TO EXERCISE 3.73

Example 3.16 Pet Return Policy

The owner of the Jungle Pet Store is trying to establish a policy for the return of animals. In a random sample of the lifetime (in months) of pet guinea pigs, $\bar{x} = 72$ and $s = 12$. One of the guinea pigs in this sample lived 62 months. Is this a reasonable lifetime, or should the store provide some sort of refund (or a new guinea pig)?[23]

SOLUTION

STEP 1   To determine whether 62 months is a *reasonable* lifetime, consider the z-score corresponding to this observation.

STEP 2 $z = \dfrac{62 - 72}{12} = -0.83$

The observation (62 months) is only 0.83 standard deviations to the left of the mean. Because 62 months is within 1 standard deviation of the mean (regardless of the shape of the distribution), this is a very conservative, reasonable observation.

STEP 3   The guinea pig lived a very *normal* life. No refund is necessary. ●

TRY IT NOW   GO TO EXERCISE 3.74

Another indication of relative standing is a percentile. Do you remember all of those *standardized* tests in grade school? The results were usually reported in terms of percentiles. The 90th percentile was a good score and the 25th percentile meant more homework in your future.

Definition

Let $x_1, x_2, \ldots, x_n$ be a set of observations. The **percentiles** divide the data set into 100 parts. For any integer r $(0 < r < 100)$, the **rth percentile**, denoted p_r, is a value such that r percent of the observations lie at or below p_r (and $100 - r$ percent lie above p_r).

The rth percentile has the same units as the observations, not a percent. Figure 3.32 shows a smoothed histogram and illustrates the location of the 75th percentile on the measurement axis.

A CLOSER LOOK

1. The 50th percentile is the median, $p_{50} = \tilde{x}$.
2. The 25th percentile is the first quartile and the 75th percentile is the third quartile: $p_{25} = Q_1$, $p_{75} = Q_3$. ●

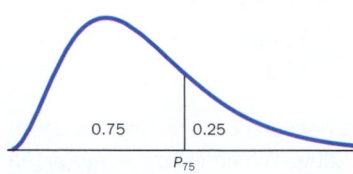

Figure 3.32 The 75th percentile is illustrated using a smoothed histogram.

Remember, the area under the curve between *a* and *b* corresponds to the proportion of observations between *a* and *b*. So the total area under the curve is 1.

How to Compute Percentiles

Suppose $x_1, x_2, \ldots, x_n$ is a set of n observations.

1. Arrange the observations in ascending order, from smallest to largest.

2. To find p_r, compute $d_r = \dfrac{n \cdot r}{100}$.

    a. If d_r is a whole number, then the depth of p_r (position in the ordered list) is $d_r + 0.5$. p_r is the mean of the observations in positions d_r and $d_r + 1$ in the ordered list.

    b. If d_r is not a whole number, round up to the next whole number for the depth of p_r.

© Andy Selinger/age fotostock

DATA SET

BRIDGE

Example 3.17 Camping Out

There are 4524 campsites managed by the Minnesota Department of Natural Resources Division of Parks and Trails.[24] The number of campsites utilized each day is carefully monitored, and on a randomly selected day there were 3000 campsites in use, a number that lies at the 75th percentile. Interpret these results.

SOLUTION

Here, 3000 is a single observation from the population of number of campsites used per day and percentiles divide these observations into 100 parts. Because 3000 lies at the 75th percentile, 75% of the days had 3000 or fewer campsites used, and on $100 - 75 = 25\%$ of the days, more than 3000 campsites were utilized.

Note: We do not know anything about the shape of the distribution, nor do we know the mean or standard deviation. There is no way of telling how *far* 3000 is from the mean in standard deviations.

Example 3.18 Scenic Stroll Across the Brooklyn Bridge

The Brooklyn Bridge in New York City is a popular tourist attraction, and many people enjoy a walk along the pedestrian walkway. A walk across the bridge takes approximately 25–60 minutes.[25] A random sample of people walking across the bridge was obtained, and their times are given in the following table:

44	51	43	31	50	53	59	49	55	25
28	30	60	42	36	54	31	33	48	39
37	44	48	51	34	38	58	59	53	59

Find the time at which it took 20% of the walkers to make it across the bridge.

SOLUTION

STEP 1 Order the data from smallest to largest. A portion of this ordered list is given in the following table:

Observation	25	28	30	31	31	33	34	36	37	38
Position	1	2	3	4	5	6	7	8	9	10

STEP 2 Compute
$$d_{20} = \frac{n \cdot r}{100} = \frac{30 \cdot 20}{100} = 6$$

STEP 3 Because d_{20} is a whole number, add 0.5. The depth of p_{20} is $d_{20} + 0.5 = 6 + 0.5 = 6.5$.

Solution Trail 3.18

KEYWORDS

- Find the time at which it took 20% of the walkers to make it across the bridge.

TRANSLATION

- Find the time, t, such that 20% of all times are less than or equal to t.

CONCEPTS

- Percentiles.

VISION

Find p_{20} (in minutes) so that 20% of the observations lie below and 80% lie above. Follow the steps for computing percentiles.

STEP 4 The 20th percentile, p_{20}, is the mean of the sixth and seventh observations.

$$p_{20} = \frac{1}{2}(33 + 34) = 33.5$$

Figure 3.33 shows a technology solution.

Figure 3.33 The 20th percentile computed using CrunchIt! Note: CrunchIt! computes percentiles using a slightly different algorithm.

Results - Descriptive Statistics   Export ▾

	n	Sample Mean	Standard Deviation	Min	Q1	Median	Q3	Max	20.0%
Times	30	44.73	10.56	25	36	46	53	60	33.20

STEP 5 Twenty percent of the walkers made it across the bridge before 33.5 minutes, and 80% made it after 33.5 minutes.

Technology Corner

Procedure: Compute the rth percentile.
Reconsider: Example 3.18, solution, and interpretation. The TI-84 does not have a built-in function to compute the rth percentile.

CrunchIt!

CrunchIt! has a built-in function to find certain descriptive statistics, including the rth percentile.

1. Enter the data into a column.
2. Select Statistics; Descriptive Statistics. Choose the appropriate column and enter the desired value of r. Click the Calculate button. Refer to Figure 3.33.

Minitab

The Minitab calculator function `Percentile` can be used to compute the rth percentile in a Session Window.

1. Enter the data into column `C1`.
2. In the Session Window, compute and print the appropriate percentile (Figure 3.34).

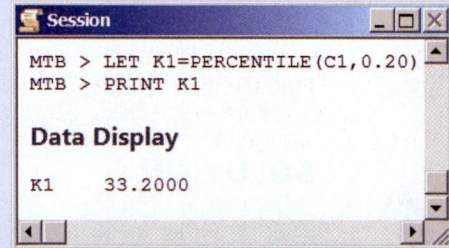

Figure 3.34 The 20th percentile computed using Minitab.

```
MTB > LET K1=PERCENTILE(C1,0.20)
MTB > PRINT K1

Data Display

K1      33.2000
```

Excel

1. Enter the data into column A.
2. Use the function `PERCENTILE.EXC` to compute the 20th percentile. See Figure 3.35.

Figure 3.35 The Excel function `PERCENTILE`.

B	C
33.2	= PERCENTILE.EXC(A1:A30,0.2)

SECTION 3.3 EXERCISES

Concept Check

3.67 True/False Chebyshev's rule applies to any set of data.

3.68 True/False The conclusion in Chebyshev's rule applies to a symmetric interval about the mean.

3.69 True/False In a smoothed histogram, the area under the curve between two points a and b corresponds to the proportion of observations between a and b.

3.70 Short Answer Why is Chebyshev's rule conservative?

3.71 True/False The empirical rule applies to any set of observations.

3.72 Short Answer If the shape of a distribution is approximately normal, what does it mean if an observation is more than three standard deviations from the mean?

3.73 True/False A z-score is a measure of relative standing.

3.74 True/False All quartiles are percentiles.

Practice

3.75 For each data set with $\bar{x}$ and s given, find a symmetric interval k standard deviations about the mean, and use Chebyshev's rule to compute the approximate proportion of observations within this interval.
  a. $\bar{x} = 50$, $\quad s = 5$, $\quad\quad k = 2$
  b. $\bar{x} = 352$, $\quad s = 10.5$, $\quad k = 3$
  c. $\bar{x} = 17$, $\quad s = 3.5$, $\quad\quad k = 1.6$
  d. $\bar{x} = 36.5$, $\quad s = 10.45$, $\quad k = 1.75$
  e. $\bar{x} = 158$, $\quad s = 25$, $\quad\quad k = 2.5$
  f. $\bar{x} = -55$, $\quad s = 0.125$, $\quad k = 2.8$
  g. $\bar{x} = 1.7$, $\quad s = 25.8$, $\quad\quad k = 2.25$

3.76 Assume the distribution of each data set is approximately normal, with $\bar{x}$ and s given. Find the intervals (referred to by the empirical rule) that are one, two, and three standard deviations about the mean. Carefully sketch the corresponding normal curve for each data set, indicating the endpoints of each interval.
  a. $\bar{x} = 20$, $\quad\quad s = 5$
  b. $\bar{x} = 37$, $\quad\quad s = 0.2$
  c. $\bar{x} = 675$, $\quad\quad s = 250$
  d. $\bar{x} = -5.5$, $\quad s = 12$
  e. $\bar{x} = 98.6$, $\quad s = 1.7$
  f. $\bar{x} = 5280$, $\quad s = 150$

3.77 For each data set with $\bar{x}$ and s given, find the z-score corresponding to the given observation x.
  a. $\bar{x} = 8$, $\quad\quad s = 3$, $\quad\quad x = 17$
  b. $\bar{x} = 100$, $\quad s = 16$, $\quad\quad x = 80$
  c. $\bar{x} = 15$, $\quad\quad s = 3$, $\quad\quad x = 17.5$
  d. $\bar{x} = 27$, $\quad\quad s = 4.5$, $\quad\quad x = 22$
  e. $\bar{x} = 122$, $\quad s = 32$, $\quad\quad x = 175$
  f. $\bar{x} = -105$, $\quad s = 33$, $\quad\quad x = -90$
  g. $\bar{x} = 6.55$, $\quad s = 0.25$, $\quad x = 6$

  h. $\bar{x} = 64$, $\quad\quad s = 8.75$, $\quad\quad x = 100$
  i. $\bar{x} = 0.025$, $\quad s = 0.0018$, $\quad x = 0.027$
  j. $\bar{x} = 407$, $\quad\quad s = 16$, $\quad\quad x = 500$

3.78 For each data set with $\bar{x}$ and s given, find an observation corresponding to the z-score given.
  a. $\bar{x} = 25$, $\quad\quad s = 5$, $\quad\quad z = 2.3$
  b. $\bar{x} = 9.8$, $\quad\quad s = 1.2$, $\quad z = -0.7$
  c. $\bar{x} = -456$, $\quad s = 37$, $\quad z = 1.25$
  d. $\bar{x} = 37.6$, $\quad s = 5.9$, $\quad z = -1.96$
  e. $\bar{x} = 55$, $\quad\quad s = 0.05$, $\quad z = 3.5$
  f. $\bar{x} = 3.14$, $\quad s = 0.5$, $\quad z = 1.28$
  g. $\bar{x} = 2.35$, $\quad s = 0.94$, $\quad z = -2.5$
  h. $\bar{x} = 0.529$, $\quad s = 1.9$, $\quad z = 0.55$

3.79 Find the position, or depth, of the indicated percentile in an ordered data set of size n.
  a. $n = 150$, $\quad p_{80}$
  b. $n = 257$, $\quad p_{35}$
  c. $n = 36$, $\quad p_{60}$
  d. $n = 75$, $\quad p_{40}$
  e. $n = 100$, $\quad p_{20}$
  f. $n = 5035$, $\quad p_{70}$

Applications

3.80 Demographics and Population Statistics The FBI uses public assistance in tracking criminals by maintaining the "Ten Most Wanted Fugitives" list. A fugitive is removed from this list if she is captured, the charges are dropped, or she no longer fits a certain profile. In a random sample of fugitives, the mean time on the list was 26.5 months, with a standard deviation of 4.3 months.
  a. What values are one standard deviation away from the mean? What values are two standard deviations away from the mean?
  b. Without assuming anything about the shape of the distribution of times, approximately what proportion of times is between 17.9 months and 35.1 months? Write a Solution Trail for this problem.

3.81 Travel and Transportation Royal Caribbean Cruises recently ordered a cruise ship similar to *Oasis of the Seas*, the world's largest cruise ship. This ship has over 12 restaurants, four pools, a parklike area with trees, and even zip lines. A random sample of large passenger liners was obtained, and the cruising speed of each was recorded. The sample mean was 25.6 knots and the standard deviation was 3.4 knots. Assume the shape of the speed distribution is approximately normal.
  a. What values are two standard deviation away from the mean? What values are three standard deviations away from the mean?
  b. Approximately what proportion of speeds is between 22.2 and 29.0 knots?

3.82 Sports and Leisure During the Hawaiian International Billfish Tournament, teams tag and release Pacific blue marlin.

During the 2012 tournament, the team headed by Sue Vermillion caught a 638-pound fish, a weight in the 85th percentile of all blue marlin caught. Interpret this value.

3.83 Manufacturing and Product Development   There are approximately 3 million parts in a Boeing 777, and suppliers are all over the world. It takes considerable coordination and organization to assemble this aircraft. From the time the first part is moved from the factory to delivery of an aircraft, the mean time to assemble a Boeing 777 is 83 days.[26] Suppose the standard deviation is 6 days.

   a. What values are one standard deviation away from the mean? What values are two standard deviations away from the mean?
   b. Without assuming anything about the shape of the distribution of times, approximately what proportion of assembly times is between 71 and 95 days?
   c. Without assuming anything about the shape of the distribution of times, approximately what proportion of assembly times is either less than 65 or greater than 101 days?
   d. Assuming the distribution of times is normal, what proportion of assembly times is between 71 and 95? Either less than 65 or greater than 101?

3.84 Biology and Environmental Science   The Commonwealth of Pennsylvania is concerned about the dwindling number of family-owned farms and the number of smaller, less efficient farms. For a random sample, the total acreage of each farm was recorded. The mean was 1125 acres, with a standard deviation of 250. The shape of the distribution of areas is not normal.

   a. Approximately what proportion of areas is between 625 and 1625 acres?
   b. Approximately what proportion of areas is between 375 and 1875 acres?
   c. Approximately what proportion of areas is less than 375 acres?
   d. Approximately what proportion of areas is between 750 and 1500 acres?

3.85 Physical Sciences   During the spring, many rivers are monitored very carefully so as to be able to warn residents of an impending flood. The depth (in feet) of the Susquehanna River at the Bloomsburg bridge is measured and reported daily. In a random sample of depths, $\bar{x} = 16.7$, $s = 2.1$, and the shape of the distribution is approximately normal.

   a. Approximately what proportion of depths is between 14.6 and 18.8 feet?
   b. Approximately what proportion of depths is less than 14.6 feet?
   c. Approximately what proportion of depths is between 14.6 and 23 feet?

3.86 Biology and Environmental Science   Many farmers use the height of their corn on July 4th as an indication of the entire crop. In a random sample of corn-stalk heights on July 4th in Columbia County, $\bar{x} = 25.6$, $s = 0.9$ inches, and a histogram of the observations is bell-shaped.

   a. Approximately what proportion of observations is between 23.8 and 27.4 inches?
   b. Approximately what proportion of observations is between 22.9 and 26.5 inches?
   c. Approximately what proportion of observations is less than 27.4 inches?

3.87 Education and Child Development   The Iowa Test of Basic Skills (ITBS) is a multiple-choice exam given to students in various grades in each state each year. The purpose is to test fundamental skills in reading, mathematics, language, social studies, and science. Scores are reported as state and/or national percentile points. Results from the 2010–2011 academic year indicate that students in grades 7–9 at Carlisle High School in Arkansas scored at the 53rd and 69th (national) percentiles in mathematics and science, respectively.[27]

   a. Interpret these values.
   b. The 50th percentile (in any subject area) is the national average. Explain the meaning of *average* in this context.
   c. Suppose a seventh grader scored at the 99th percentile (nationally) in mathematics. Interpret this result.

3.88 Travel and Transportation   Bicycle delivery services are utilized in many metropolitan areas because they can provide rush deliveries and are not subject to traffic jams or parking restrictions. Suppose an architectural firm would like to evaluate two bicycle delivery services in New York City. The first service has a mean and standard deviation for delivery (in minutes) of 37 and 5. The second service has a mean of 42 with a standard deviation of 7. The company sent two test packages to the same location, one with each delivery service. The times to delivery were 33 and 35 minutes, respectively. Use z-scores to determine which service performed better.

3.89 Business and Management   The Green Mill Restaurant and Bar in Wausau, Wisconsin, is advertising quick lunches with a mean waiting time of 11 minutes and a standard deviation of 2.5 minutes. The general manager (Rob Meyer, a former statistician) also claims that the distribution of waiting times is approximately normal.

   a. Suppose your waiting time is 13 minutes. Is there any reason to believe the general manager's claim is false? (Use a z-score.) Write a Solution Trail for this problem.
   b. Suppose your waiting time is 20 minutes. Now, is there any evidence to refute the general manager's claim?

3.90 Marketing and Consumer Behavior   The time spent in a grocery store is an important issue for shoppers and for companies trying to market new products. Men tend to spend less time in a grocery store than women, and people spend more time in the store on weekends. A random sample of shoppers at a local grocery store was obtained and the shopping time (in minutes) for each was recorded. **GROCERY**

   a. Construct a histogram for these data.
   b. Use your histogram in part (a) to approximate the following percentiles: (i) 45th, (ii) 80th, (iii) 10th.
   c. Compute the exact percentiles in part (b) and compare your results.

Extended Applications

3.91 Manufacturing and Product Development The engine in a tractor trailer is designed to last 1,000,000 miles before a rebuild or overhaul. The engines are also designed to run nonstop and have between 400 and 600 horsepower.[28] A random sample of tractor trailers was obtained and the horsepower was measured for each engine. **ENGINEHP**

a. Find the mean and the standard deviation of these horsepower measurements.
b. Find the actual proportion of observations within one standard deviation of the mean, within two standard deviations of the mean, and within three standard deviations of the mean.
c. Using the results in part (b), do you think the shape of the distribution of horsepower measurements is normal? Why or why not?

3.92 Sports and Leisure In 1974, Erno Rubik created an imaginative and best-selling puzzle—the Rubik's cube. Many countries hold competitions in which participants try to solve this puzzle as quickly as possible. The text website provides a sample from the list of record times (in seconds) in official world competitions.[29] **CUBETIME**

a. Find the actual proportion of observations within one standard deviation of the mean, within two standard deviations of the mean, and within three standard deviations of the mean.
b. Using the results in part (a), do you think the shape of the distribution of national record times is normal? Why or why not?
c. Construct a histogram for these data. Describe the shape of the distribution.

3.93 Manufacturing and Product Development Paint viscosity is a measure of thickness that determines whether the paint will cover in a single coat. A random sample of latex paint viscosities (in KU, or Krebs units) was obtained, and the data are given in the following table: **VISCOS**

113	124	141	115	115	129	113	129	112	112

a. Find the mean and the standard deviation for this data.
b. Find the z-score for each observation.
c. Find the mean and the standard deviation for all of the z-scores.
d. For *any* set of observations, can you predict the mean and standard deviation of the corresponding z-scores? Try to prove this result.

Challenge

3.94 Travel and Transportation According to the Massachusetts Bay Transportation Authority, the ride from Chestnut Hill to Boston's Logan Airport on the MBTA takes less than 45 minutes. A random sample of travel times (in minutes) was obtained, and the results are given in the following table: **MBTA**

46.5	38.3	39.1	41.1	42.0	37.6	41.6	45.5
39.0	34.8	36.5	38.6	38.4	44.4	42.4	

a. Find the mean ($\bar{x}$) and the standard deviation (s) for this data set.
b. Compute each z-score and find $\sum z_i^2$.
c. Find a general formula for $\sum z_i^2$ for any data set.

3.95 Reconsider Example 3.16. Find a good *minimum* guaranteed life. That is, if a guinea pig fails to reach such an age, then the store would provide a refund.

3.4 Five-Number Summary and Box Plots

A box plot, or box-and-whisker plot, is a compact graphical summary that conveys information about central tendency, symmetry, skewness, variability, and outliers. A standard box plot is constructed using the minimum and maximum values in the data set, the first and third quartiles, and the median. This collection of values is called the *five-number summary*.

Definition

The **five-number summary** for a set of n observations $x_1, x_2, \ldots, x_n$ consists of the minimum value, the maximum value, the first and third quartiles, and the median.

Recall: The range of a data set is the largest observation (maximum value) minus the smallest observation (minimum value). This descriptive statistic was our first attempt at measuring variability in a data set.

These five numbers do provide a glimpse of symmetry, central tendency, and variability in a data set. For example, minimum and maximum values that are very far apart suggest lots of variability. If the median is approximately halfway between the minimum and maximum values and approximately halfway between the first and third quartiles, that suggests the distribution is symmetric. A box plot is constructed as described below.

How to Construct a Standard Box Plot

Given a set of n observations $x_1, x_2, \ldots, x_n$:

1. Find the five-number summary $x_{min}, Q_1, \tilde{x}, Q_3, x_{max}$.
2. Draw a (horizontal) measurement axis. Carefully sketch a box with edges at the quartiles: left edge at Q_1, right edge at Q_3. (The height of the box is irrelevant.)
3. Draw a vertical line in the box at the median.
4. Draw a horizontal line (whisker) from the left edge of the box to the minimum value (from Q_1 to x_{min}). Draw a horizontal line (whisker) from the right edge of the box to the maximum value (from Q_3 to x_{max}).

Recall: x_{min} denotes the minimum value and x_{max} denotes the maximum value.

Figure 3.36 illustrates this step-by-step procedure and shows a standard box plot with the five numbers indicated on a measurement axis. Note that the length of the box is the interquartile range. The box contains the middle half of the values.

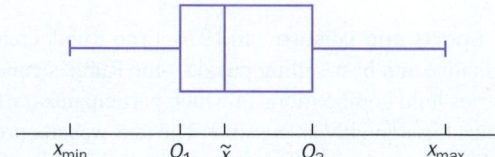

Figure 3.36 Standard box plot.

The position of the vertical line in the box (median) and the lengths of the horizontal lines (whiskers) indicate symmetry or skewness, and variability. Figure 3.37 shows a standard box plot for a distribution of data that is skewed to the right. The lower half of the data is in the interval from 3 to 4.5, while the upper half of the data is much more spread out, from 4.5 to 11. Figure 3.38 shows a standard box plot for a fairly symmetric distribution with lots of variability. The lower and upper half of the data are evenly distributed, but the whiskers extend far from each edge of the box. That is, 25% of the data are between 0 and 4, and 25% are between 7 and 11.

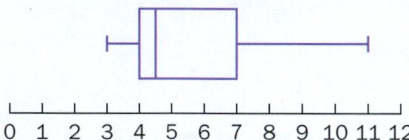

Figure 3.37 Standard box plot for data skewed to the right.

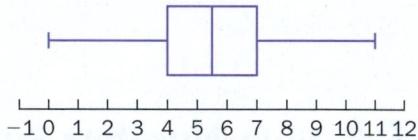

Figure 3.38 Standard box plot for a symmetric distribution.

Example 3.19 Blood Pressure

DATA SET

SYSTOLIC

There is some evidence to suggest that consumption of nonalcoholic red wine may decrease systolic and diastolic blood pressure.[30] Suppose the systolic blood pressure for 30 randomly selected subjects involved in this research study is given in the following table. Construct a standard box plot for these data.

177	122	128	191	180	142	197	196	67	160
167	138	107	188	102	116	138	114	188	176
148	175	169	203	135	142	168	181	168	150

SOLUTION

STEP 1 Find the five-number summary:

$$x_{min} = 67 \quad Q_1 = 135 \quad \tilde{x} = 163.5 \quad Q_3 = 180 \quad x_{max} = 203$$

STEP 2 Draw a measurement axis and sketch a box with edges at $Q_1 = 135$ and $Q_3 = 180$.

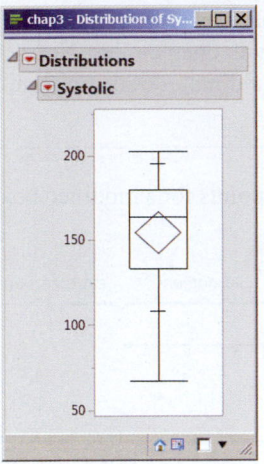

Figure 3.40 JMP quantile box plot.

STEP 3   Draw a vertical line at the median, $\tilde{x} = 163.5$.

STEP 4   Draw a horizontal line from $Q_1 = 135$ to $x_{\min} = 67$, and another horizontal line from $Q_3 = 180$ to $x_{\max} = 203$. The resulting box plot is shown in Figure 3.39.

Figure 3.39 Standard box plot for the systolic blood pressure data. Tick marks for Q_1, $\tilde{x}$, and Q_3 are added to this graph for clarity.

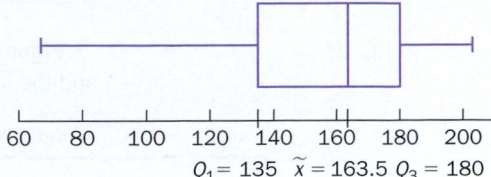

STEP 5   The box plot suggests the data are negatively skewed, or skewed to the left. The lower half of the data is much more spread out than the upper half. A technology solution is shown in Figure 3.40.

A CLOSER LOOK

1. A box plot has only one measurement axis, and it may be horizontal or vertical. Many software packages, including CrunchIt! and Minitab, draw box plots with a vertical measurement axis by default. The construction and interpretations are the same. The Transpose option in Minitab produces a box plot with a horizontal measurement axis.

2. The software does *not* usually include tick marks on the measurement axis for the five-number summary. The tick marks and scale are selected simply for convenience.

There are some disadvantages to using a standard box plot based on the five-number summary to describe a data set. By examining the graph, there is no way of knowing how many observations are between each quartile and the extreme. Each whisker is drawn from the quartile to the extreme, regardless of the number of observations in between. In addition, there are no provisions for identifying outliers. A standard (graphical) technique for distinguishing outliers is important because these values play an important role in statistical inference. Therefore, many statisticians prefer to use a modified box plot to describe a data set graphically. This type of graph still conveys information about center, variability, symmetry, and skewness, but it is also more precise and plots outliers.

How to Construct a Modified Box Plot

Given a set of n observations $x_1, x_2, \ldots, x_n$:

1. Find the quartiles, the median, and the interquartile range:

   $Q_1, \tilde{x}, Q_3, \text{IQR} = Q_3 - Q_1.$

2. Compute the two inner *fences* (low and high) and the two outer (low and high) *fences* using the following formulas:

   $\text{IF}_L = Q_1 - 1.5(\text{IQR})$ $\text{IF}_H = Q_3 + 1.5(\text{IQR})$
   $\text{OF}_L = Q_1 - 3(\text{IQR})$ $\text{OF}_H = Q_3 + 3(\text{IQR})$

   Think of the interquartile range as a *step*. The inner fences are 1.5 steps away from the quartiles, and the outer fences are 3 steps away from the quartiles.

3. Draw a (horizontal) measurement axis. Carefully sketch a box with edges at the quartiles: left edge at Q_1, right edge at Q_3. Draw a vertical line in the box at the median.

4. Draw a horizontal line (whisker) from the left edge of the box to the most extreme observation within the low inner fence. This line will extend from Q_1 to at most IF_L. Draw a horizontal line (whisker) from the right edge of the box to the most extreme observation within the high inner fence. This line will extend from Q_3 to at most IF_H.

5. Any observations between the inner and outer fences (between IF_L and OF_L, or between IF_H and OF_H) are classified as *mild outliers* and are plotted separately with shaded circles.

Any observations outside the outer fences (less than OF_L, or greater than OF_H) are classified as *extreme outliers* and are plotted separately with open circles. Note: Some statistical packages will use other symbols for outliers and may not distinguish mild and extreme outliers.

Figure 3.41 shows the relationship between construction points for a modified box plot and the location of any outliers.

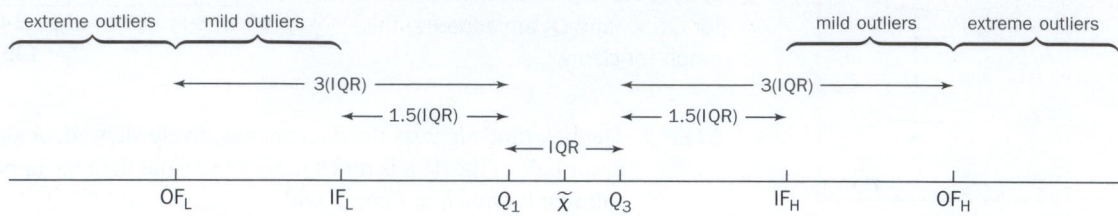

Figure 3.41 Construction points for a modified box plot.

Example 3.20 Sled Dog Trips

DATA SET

SLEDDOGS

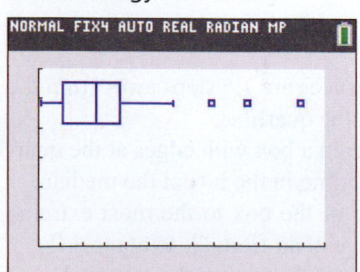

Winterdance Dogsled Tours

The Ignace, Ontario, fishing and hunting resort Agimac River Outfitters offers guided sled dog trips on wooded trails and along beautiful lakes.[31] Most trips last approximately $2\frac{1}{2}$ hours, but there are all-day trips and the weather conditions may affect the length of a scheduled trip. A random sample of sled dog trips was obtained and the length (in hours) of each was recorded. The data are given in the following table. Construct a modified box plot for these data.

0.7	1.7	0.8	2.7	1.4	0.8	2.6	3.5	1.3	0.6
9.5	2.6	2.7	11.9	3.1	7.9	0.6	6.1	5.2	1.9
2.2	3.7	0.5	1.1	1.4	3.1	0.1	2.6	4.3	4.5

SOLUTION

STEP 1 Find the quartiles, the median, and the interquartile range.

$$Q_1 = 1.1 \quad \tilde{x} = 2.6 \quad Q_3 = 3.7 \quad IQR = 3.7 - 1.1 = 2.6$$

STEP 2 Find the inner and outer fences.

$$IF_L = 1.1 - (1.5)(2.6) = 1.1 - 3.9 = -2.8$$
$$IF_H = 3.7 + (1.5)(2.6) = 3.7 + 3.9 = 7.6$$
$$OF_L = 1.1 - (3)(2.6) = 1.3 - 7.8 = -6.5$$
$$OF_H = 3.7 + (3)(2.6) = 3.7 + 7.8 = 11.5$$

STEP 3 Draw a (horizontal) measurement axis. Carefully sketch a box with edges at the quartiles: left edge at Q_1, right edge at Q_3. Draw a vertical line in the box at the median.

A technology solution:

NORMAL FIX4 AUTO REAL RADIAN MP

Figure 3.42 TI-84 Plus C modified box plot.

STEP 4 Draw a horizontal line (whisker) from the left edge of the box to the most extreme observation within the inner fence IF_L (0.1).

Draw a horizontal line (whisker) from the right edge of the box to the most extreme observation within the inner fence IF_H (6.1).

STEP 5 Plot any mild outliers, observations between -6.5 and -2.8, or between 7.6 and 11.5. There are two mild outliers, 7.9 and 9.5.

Plot any extreme outliers, observations less than -6.5 or greater than 11.5. There is one extreme outlier, 11.9.

STEP 6 The resulting modified box plot is shown in Figure 3.43. The box plot suggests the data are positively skewed, or skewed to the right. The upper half of the data is much more spread out than the lower half. There are two mild outliers and one extreme outlier.

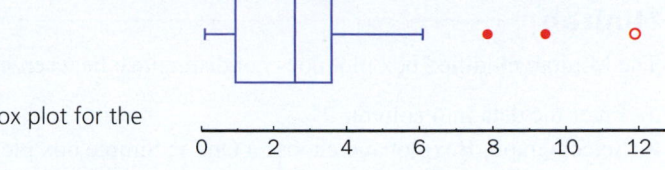

Figure 3.43 Modified box plot for the sled dog trip data.

Note: IF_L and OF_L are negative even though an observed trip time cannot be less than 0 hours. That's OK. This is a correct statistical calculation, not a contradiction, even though it seems odd. Figure 3.42 shows a technology solution.

TRY IT NOW   GO TO EXERCISE 3.91

A CLOSER LOOK

When we compare two (or more) data sets graphically, the corresponding box plots may be placed on the same measurement axis (one above the other using a horizontal axis, or side-by-side with a vertical axis). Figure 3.44 shows three box plots on the same measurement axis, representing the number of gallons of gasoline pumped in randomly selected vehicles at three different stations.

TRY IT NOW   GO TO EXERCISE 3.94

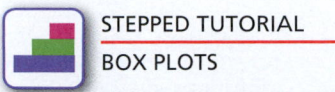

STEPPED TUTORIAL
BOX PLOTS

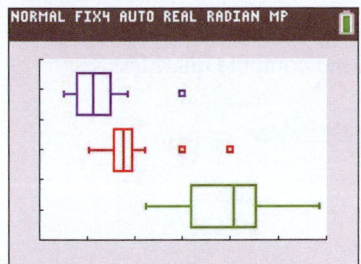

Figure 3.44 TI-84 Plus C box plots for gasoline data.

Technology Corner

VIDEO TECH MANUALS
BOX PLOTS

Procedure: Construct a box plot.
Reconsider: Example 3.20, solution, and interpretations.

CrunchIt!

CrunchIt! has a built-in function to construct a box plot.

1. Enter the data into a column.
2. Select Graphics; Box Plot. Choose the appropriate column, optionally enter a Title, X Label, and Y Label, and click the calculate button. The resulting box plot is shown in Figure 3.45.

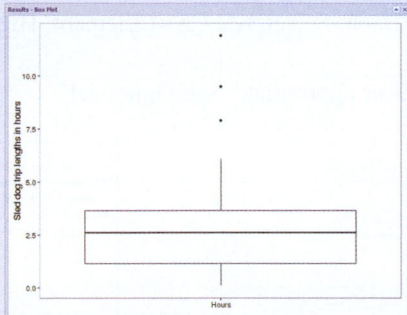

Figure 3.45 CrunchIt! box plot of the sled dog trip data.

TI-84 Plus C

The TI-84 Plus has two built-in statistical plots, a standard and a modified box plot. The modified box plot does not distinguish between mild and extreme outliers.

1. Enter the data into list `L1` .
2. Press `STATPLOT` and select `Plot1` from the `STATPLOTS` menu.
3. Turn the plot `On` and select `Type` box plot (modified or standard). Set `Xlist` to the name of the list containing the data and `Freq` to 1. Choose a `Mark` for outliers (if constructing a modified box plot) and select a `Color`.
4. Enter appropriate `WINDOW` settings. Press `GRAPH` to display the box plot. A modified box plot is shown in Figure 3.42.

Minitab

The Minitab modified box plot does not distinguish between mild and extreme outliers.

1. Enter the data into column C1.
2. Select Graph; Boxplot and choose a One Y; Simple box plot.
3. Enter C1 in the Graph variables window. Select the Scale options button and check Transpose value and category scales to construct a box plot with a horizontal measurement axis.
4. Select the Data view options button and check Interquartile range box, and Outlier symbols for a modified box plot. Note there are only two outliers in this box plot due to the numerical method Minitab uses to compute quartiles. See Figure 3.46.

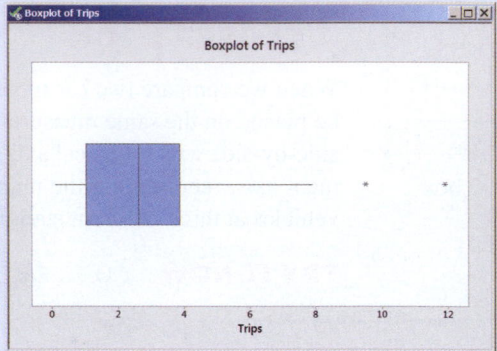

Figure 3.46 Minitab modified box plot.

Excel

The following steps may be used to construct a standard box plot. Additional calculations and options are necessary to construct a modified box plot.

1. Enter the data into column A.
2. Find the five-number summary in the order shown. Highlight these five cells.
3. Under the Insert tab, select Insert Line Chart; 2-D Line; Line with Markers.
4. Under Chart tools; Design, select Switch Row/Column.
5. Right-click on the point representing the minimum value. Select Format Data Series; Line and choose No line. Repeat this process for each point.
6. Select any point on the graph. Select Add Chart Element; High-Low Lines and Add Chart Element; Up/Down Bars; Up/Down Bars.
7. Format other graph items as appropriate. See Figure 3.47.

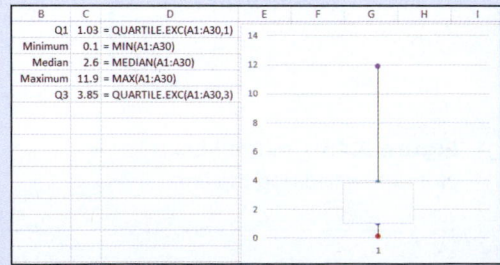

Figure 3.47 Excel standard box plot.

SECTION 3.4 EXERCISES

Concept Check

3.96 True/False The five-number summary for a set of observations is determined by finding the most extreme observations.

3.97 True/False In a standard box plot, a line is drawn in the box at the sample mean.

3.98 True/False A box plot may reveal whether a distribution is symmetric.

3.99 True/False A modified box plot always has markers for mild and extreme outliers.

Practice

3.100 Find the five-number summary for each data set. **EX3.100**

a. {34, 40, 34, 32, 32, 40, 35, 35, 28, 35}
b. {57, 65, 70, 71, 67, 56, 52, 66, 74, 57, 67, 78}
c. {94, 80, 91, 94, 83, 92, 83, 93, 96, 80, 87, 98, 81, 93}
d. {2.3, 1.8, 2.1, 1.0, 2.4, 2.3, 0.4, 9.8, 0.6, 1.4, 3.1, 10.9, 3.8, 0.5, 0.9, 2.2, 1.3, 1.3}
e. {166.8, 103.1, 119.9, 141.9, 110.6, 189.8, 121.6, 141.6, 133.6, 178.2, 158.9, 145.9, 139.1, 148.6, 135.0, 174.0, 152.4, 119.7, 196.9, 118.7, 159.7, 150.3, 113.8, 108.9, 163.2}
f. {−33.8, −9.8, −18.5, −11.5, −36.3, −33.1, −21.1, −26.2, −25.4, −32.1, −35.9, −28.0, −38.2, −12.0, −29.2, −40.1, −13.1}

3.101 Construct a standard box plot for each five-number summary.

a. $x_{min} = 15.3$, $Q_1 = 21.8$, $\tilde{x} = 25.3$, $Q_3 = 28.2$, $x_{max} = 34.2$
b. $x_{min} = 70.9$, $Q_1 = 167.8$, $\tilde{x} = 187.1$, $Q_3 = 225.3$, $x_{max} = 329.3$
c. $x_{min} = 0.06$, $Q_1 = 5.3$, $\tilde{x} = 13.7$, $Q_3 = 30.8$, $x_{max} = 122.3$
d. $x_{min} = 10.1$, $Q_1 = 10.7$, $\tilde{x} = 11.3$, $Q_3 = 12.5$, $x_{max} = 26.7$

3.102 For each data set with Q_1 and Q_3 given, find the interquartile range and the inner and outer fences.

a. $Q_1 = 22$, $\quad Q_3 = 46$
b. $Q_1 = 1255$, $\quad Q_3 = 1306$
c. $Q_1 = 65.75$, $\quad Q_3 = 75.21$
d. $Q_1 = 914.9$, $\quad Q_3 = 1140.5$
e. $Q_1 = 1.275$, $\quad Q_3 = 4.07$
f. $Q_1 = 0.265$, $\quad Q_3 = 2.51$
g. $Q_1 = -33.67$, $\quad Q_3 = -23.90$
h. $Q_1 = 98.43$, $\quad Q_3 = 98.81$

3.103 For each data set with Q_1 and Q_3 given, determine whether the observation x is a mild outlier, an extreme outlier, or neither.

a. $Q_1 = 20$, $\quad Q_3 = 29$, $\quad x = 35$
b. $Q_1 = 486.1$, $\quad Q_3 = 510.9$, $\quad x = 440$
c. $Q_1 = 5.18$, $\quad Q_3 = 6.32$, $\quad x = 4.2$
d. $Q_1 = 96.3$, $\quad Q_3 = 101.1$, $\quad x = 116.5$
e. $Q_1 = 68.92$, $\quad Q_3 = 69.07$, $\quad x = 68.4$
f. $Q_1 = 101.26$, $\quad Q_3 = 144.59$, $\quad x = 132.6$

3.104 For each box plot, find the five-number summary. (Estimate these numbers as best you can by using the tick marks on each graph.)

a.

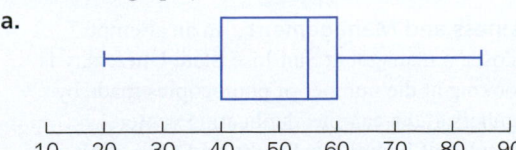

b.

c.

d.

e.

f.

g.

Applications

3.105 Business and Management A Roth's supermarket in Salem, Oregon, is using a new statistical tool to help in ordering bottles of raspberry iced tea. A random sample of the number of bottles sold per day is given in the following table. Construct a modified box plot for these data. Describe the distribution of the number of bottles sold. **BOTTLES**

48	52	46	58	50	46	59	51	46	48
45	47	50	48	49	49	49	48	48	45

3.106 Travel and Transportation Major airlines compete for customers by advertising on-time arrival. A random sample of flights arriving at the Philadelphia International Airport was obtained and the actual arrival time was compared to the scheduled arrival time. The differences (in minutes) are given in the following table (negative numbers indicate the flight arrived before the scheduled arrival time).[32] Construct a modified box plot for these data. Describe the distribution in terms of symmetry, skewness, and variability. Are there any outliers? If so, are they mild or extreme? **ARRIVAL**

14	64	−5	−5	74	215	19	202	113	90
135	121	6	−18	−17	−5	82	116	18	82
210	98	90	175	6	42	35	54	10	12

3.107 Public Policy and Political Science A recent study reported the school tax bills (in dollars) for randomly selected families in Hillsboro, New Hampshire, a small rural community. Use the following modified box plot to describe the distribution of the data.

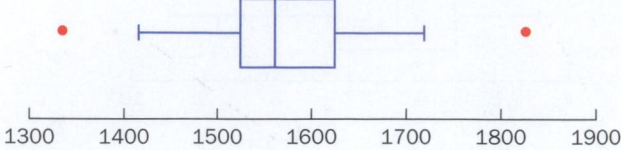

3.108 Education and Child Development Dr. Jan Remer, a psychologist in Hermitage, Tennessee, randomly selected six-year-olds and recorded the time (in minutes) each child needed to complete a 20-piece picture puzzle. The data were used to predict *readiness* for first grade. The following standard box plot for the data was drawn. Describe the distribution of completion times.

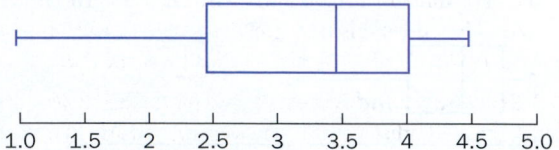

3.109 Public Health and Nutrition As part of a new physical fitness program, the Crooked Oak Middle School in Oklahoma City, Oklahoma, records the number of sit-ups each sixth-grade student can do in one minute. A random sample for males and females was obtained and the following modified box plots were drawn. Describe the male and female data separately. What similarities and/or differences do the box plots suggest?

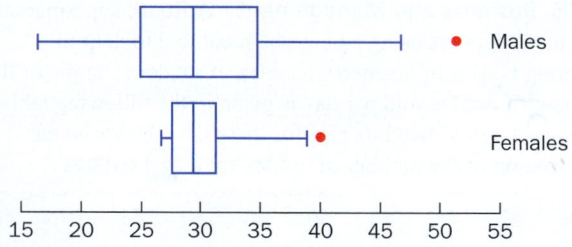

3.110 Economics and Finance Many government program budgets are determined by the annual inflation rate. The inflation rate (as a percent change in the consumer price index since 1913) for the United States for the years 1965–2011 are given on the text website.[33] Construct a modified box plot for these data. Describe the distribution of inflation rate over the past 40 years in terms of symmetry, skewness, variability, and outliers. **INFRATE**

3.111 Public Policy and Political Science According to US Vending Company, the national average commission per machine per month is approximately $35.[34] Suppose the Culver City, California, City Council recently entered into an agreement with Coca-Cola Bottling Company. As part of this contract, two US Vending snack machines were installed, one at Veteran's Park and the other at the

Teen Center. The 2012 monthly commission checks (in dollars) associated with these machines are given in the following table: **VENDING**

51.24	27.60	37.80	44.76	24.36	14.04
119.52	123.48	38.48	35.28	16.92	47.25
25.65	31.95	40.50	22.50	33.75	59.85
36.90	18.45	31.95	18.45	19.50	22.65

a. Construct a modified box plot for these data. Describe the distribution in terms of symmetry, skewness, variability, and outliers.
b. Construct a standard box plot for these data. Which plot do you think is more descriptive? Why?

3.112 Biology and Environmental Science Several weather centers across the country carefully track and record data for tropical disturbances during the hurricane season. The following table from the 2012 Atlantic hurricane season contains the minimum central pressure (in millibars) for each named tropical storm and hurricane.[35] **STORMS**

998	992	987	990	980	1000	965
1004	968	1006	970	968	964	978
997	1005	969	940	1000		

Construct a modified box plot for the hurricane data. Describe the distribution in terms of symmetry, skewness, variability, and outliers. How would the graph change if a standard box plot were used?

Extended Applications

3.113 Public Policy and Political Science The amber-light time at an intersection usually varies according to the speed limit.[36] A random sample of amber-light times (in seconds) at intersections with different speed limits is given below. Construct a modified box plot for each set of data on the same measurement axis. Describe and compare the distributions. Do the box plots suggest that amber-light times are longer at intersections with a higher speed limit? **AMBERLT**

30 mph

3.7	3.7	3.4	2.2	3.4	3.6	3.5	3.8	2.6	2.1
3.1	3.9	4.3	1.8	3.0	1.6	4.5	2.7	2.2	2.4

50 mph

4.1	4.7	4.3	4.1	4.8	5.4	2.9	3.9	2.1	6.9
4.6	3.3	6.6	5.0	5.5	3.6	3.7	5.9		

3.114 Business and Management In an attempt to control costs, a manager at San José State University is carefully looking at the number of photocopies made by faculty members at the campus duplicating center. Random samples of liberal arts faculty and of natural science faculty were obtained. The number of copies made by each faculty member is given on the text website. 📊 COPIES

a. Construct a modified box plot for each data set on the same measurement axis.
b. Compare the distributions in terms of symmetry, skewness, variability, and outliers.
c. Is there any graphical evidence to suggest that one group of faculty uses the copy center more than the other?

3.115 Public Health and Nutrition The U.S. Food and Drug Administration has become concerned about claims made by companies selling vitamin and mineral tablets. A random sample of 400-mg vitamin C tablets was obtained and analyzed by an independent laboratory for exact vitamin C content. The results (in milligrams) are given on the text website. 📊 VITC

a. Construct a modified box plot for the vitamin C data.
b. Describe the distribution in terms of symmetry, skewness, variability, and outliers.
c. Does the box plot suggest any graphical evidence that the claim of a 400-mg content is wrong?

3.116 Public Policy and Political Science If you receive a jury summons, you are obligated to appear in court. There are, however, several general (honest) instant, temporary, and hardship excuses to avoid jury duty. For example, firefighters and physicians may be automatically excused from serving. The compensation for serving on a jury varies by state; some states reimburse child-care expenses and/or transportation costs; some states do not compensate jurors for the first few days of service. A sample of states was obtained, and the juror pay per day as of January 1, 2012, for each was recorded.[37] The data are given in the following table. 📊 JURY

State	Pay	State	Pay
Alabama	10.00	Alaska	25.00
Arkansas	50.00	California	15.00
Delaware	20.00	Hawaii	30.00
Idaho	10.00	Illinois	4.00
Kentucky	12.50	Maine	10.00
Maryland	15.00	Massachusetts	50.00
Missouri	6.00	Nebraska	35.00
New Mexico	40.00	North Carolina	12.00
Pennsylvania	9.00	Texas	6.00
Virginia	30.00	Wyoming	40.00

a. Construct a modified box plot for the jury pay per day data.
b. Describe the distribution in terms of symmetry, skewness, variability, and outliers.
c. Suppose that, in January 2013, the $9.00 rate in Pennsylvania was raised to $60.00 and all other rates given remained the same. Change the value for Pennsylvania to $60.00 and construct a new modified box plot. How does this new box plot compare with the one in part (a)? Describe any similarities and/or differences.

3.117 Economics and Finance In January 2006, Wisconsin joined at least 18 other states by posting on websites the names of people and businesses that owe back taxes. The law in Wisconsin requires the Department of Revenue to list those who owe at least $25,000. These "websites of shame" are designed to help states collect additional tax money during tight budget times. A random sample of names from the Wisconsin list as of October 4, 2012, was obtained, and the taxes owed by each individual or business are given on the text website.[38] Construct a modified box plot for these data and describe the distribution in terms of symmetry, skewness, variability, and outliers. 📊 TAXLIST

CHAPTER 3 SUMMARY

Concept	Page	Notation / Formula / Description
Sample mean	75	$\bar{x} = \frac{1}{n}\sum x_i$: the sum of the observations divided by n.
Population mean	75	μ: the mean of an entire population.
Sample median	76	$\tilde{x}$: the middle value of the ordered data.
Population median	77	$\tilde{\mu}$: the middle value of an entire population.
Trimmed mean	78	$\bar{x}_{tr(p)}$: the sample mean of a *trimmed* data set.
Mode	79	The value that occurs most often.
Sample proportion of successes	80	$\hat{p} = \frac{n(S)}{n}$: the relative frequency of occurrence of successes.
Population proportion of successes	80	p: the true proportion of successes in an entire population.
Sample range	86	R: the largest observation (x_{max}) minus the smallest observation (x_{min}).
Deviation about the mean	87	$x_i - \bar{x}$.
Sample variance	87	$s^2 = \frac{1}{n-1}\sum(x_i - \bar{x})^2 = \frac{1}{n-1}\left[\sum x_i^2 - \frac{1}{n}(\sum x_i)^2\right]$

Sample standard deviation	87	$s = \sqrt{s^2}$: the positive square root of the sample variance.
Population variance	88	σ^2: the variance for an entire population.
Population standard deviation	88	$\sigma = \sqrt{\sigma^2}$: the positive square root of the population variance.
Quartiles	90	The quartiles divide the data into four parts. Q_1 is the first quartile and Q_3 is the third quartile. Q_2 is the median.
Interquartile range	90	$IQR = Q_3 - Q_1$
Chebyshev's rule	98	For any set of observations, the proportion of observations within k standard deviations of the mean is at least $1 - \dfrac{1}{k^2}$.
Empirical rule	100	If a distribution is approximately normal, the proportion of observations within one, two, and three standard deviations about the mean is approximately 0.68, 0.95, and 0.997, respectively.
z-score	103	$z_i = \dfrac{x_i - \bar{x}}{s}$, how far an observation is from the mean in standard deviations.
Percentiles	104	The percentiles divide a data set into 100 parts.
Five-number summary	109	$x_{min}, Q_1, \tilde{x}, Q_3, x_{max}$
Box plot	110	A graphical description of a data set, constructed using the five-number summary. The graph conveys information about central tendency, symmetry, skewness, and variability.
Modified box plot	111	A graphical description of a data set, constructed using $\tilde{x}, Q_1, Q_3$, IQR, and the inner and outer fences. This box plot also indicates any outliers.

CHAPTER 3 EXERCISES

3 APPLICATIONS

3.118 Public Health and Nutrition   Most multivitamins contain calcium for strong bones and to lower the risk of heart disease. A random sample of multivitamins was obtained, and the calcium content for each (in milligrams) is given in the following table: **CALCIUM**

156	151	173	201	182	166	173	180	174	185
160	178	173	169	203	190	187	202	173	171

a. Find the mean, the variance, and the standard deviation.
b. Find the proportion of observations within one, two, and three standard deviations about the mean.
c. Using the proportions obtained in part (b), do you think the distribution of observations is normal? Why or why not?

3.119 Manufacturing and Product Development   Many boxed cake mixes include special high-altitude baking instructions. To determine any difference between baking times at low and high altitudes, the consumer group Public Citizen made several similar cakes in nine-inch round pans in Miami and Denver, and carefully recorded the time to bake (in minutes). The data are given in the following table. **CAKEMIX**

Low-altitude times (Miami)

25.1	25.6	24.9	23.7	25.5	22.4	24.7	24.2	25.6
24.8	23.9	24.4	24.7	24.4	26.4	24.7	24.7	26.8
24.9	24.3							

High-altitude times (Denver)

22.8	30.0	27.3	30.3	28.3	31.1	27.0	26.8	26.3
29.1	23.5	26.2	29.2	23.0				

a. Construct a modified box plot for each data set on the same measurement axis.
b. Describe each box plot in terms of center, shape, spread, and outliers.
c. Describe the similarities and differences between the two distributions.

3.120 Sports and Leisure   The longest running U.S. produced, fictional-content television show by number of episodes is *WWE Raw*.[39] Other long-running shows of this type include *Gunsmoke, Lassie, Ozzie and Harriett*, and *Bonanza*. *Law and Order* is currently number 6, with approximately 500 shows. The number of episodes for some of the top 115 shows are given in the following table: **TVSHOW**

633	588	505	500	456	452	435	430	369	361
357	344	336	331	291	296	286	284	278	271
260	254	243	227	223	216	213	212	180	160

a. Find the median, the first and third quartiles, and the interquartile range.

b. Find the 30th and the 95th percentiles.

c. Suppose *Dallas* currently has 357 episodes (as of May 2014). Using the data in the table, in what percentile does this episode count lie?

3.121 Manufacturing and Product Development The most popular wind turbine sold by General Electric has a rated mean electrical generating capacity of 1.5 megawatts (MW)[40] with a standard deviation of 0.07 MW. A quality control engineer is trying to develop a plan for routine maintenance based on *z*-scores.

a. Suppose a randomly selected wind turbine is inspected and found to have a generating capacity of 1.54 MW. Is there any reason to believe this generating capacity is unusual? Why or why not?

b. Suppose another randomly inspected wind turbine has a generating capacity of 1.3 MW. Is there any reason to believe this generating capacity is unusual? Why or why not?

3.122 Sports and Leisure String tension in tennis rackets is usually measured in pounds. Recommended string tensions are usually in the mid-60s (pounds) for oversize rackets, and high 50s to low 60s for mid-overs. Higher tensions tend to decrease the size of the "sweet spot" and reduce power, but increase control. This book's website presents the results from a random sample of string tension from tennis rackets of players on the professional tour. 🔊 RACKETS

a. Find the range, sample variance, interquartile range, coefficient of variation, and coefficient of quartile variation for each type of racket. (CV and CQV were defined in Exercise 3.51.)

b. Using the results from part (a), compare the variability in string tension for the two types of rackets.

c. Construct a modified box plot for each type of racket on the same measurement axis. Does this graphical comparison support your numerical comparison in part (b)?

3.123 Manufacturing and Product Development Many homes that use forced hot air for heat have air ducts installed in every room. A system using galvanized pipe is constructed to distribute heat throughout the house. A random sample of six-inch diameter, five-foot long, 28-gauge galvanized pipe was obtained from various manufacturers and the weights (in pounds) are given on the text website. 🔊 AIRDUCT

a. Find the sample mean and the sample median.

b. Use your results from part (a) to describe the symmetry of the distribution.

c. Find a 10% trimmed mean. Is the use of a trimmed mean to measure central tendency justified (or necessary) in this case? Why or why not?

3.124 Medicine and Clinical Studies Although caffeine is believed to be safe in moderate amounts, some health experts suggest that 300 mg of caffeine (the amount in about three

cups of coffee) is a moderate intake.[41] The amount of caffeine in a cup of coffee varies according to coffee bean, brewing technique, filter, etc. A random sample of eight-ounce cups of coffee was obtained and the caffeine content (in milligrams) was measured. The data are given in the following table: 🔊 CAFFEIN

89	75	90	115	88	96	107	106	93
97	95	101	115	112	100	71	109	89

a. Find the mean, median, variance, and standard deviation.

b. Construct a modified box plot for these data.

c. Use your results from parts (a) and (b) to describe the data.

d. Based on your results in parts (a) and (b), do you believe a person who drinks three cups of coffee ingests a moderate amount of caffeine? Justify your answer.

3.125 Sports and Leisure *World of Warcraft* is one of the most popular multiplayer video games. The number of subscribers to *World of Warcraft* from 1st quarter 2005 to 3rd quarter 2012 (in millions) is given on the text website.[42]

🔊 GAMING

a. Find the mean, variance, and standard deviation.

b. Find the proportion of observations within one, two, and three standard deviations about the mean. Use these proportions to determine whether the distribution of subscribers is approximately normal.

c. Blizzard Entertainment has decided to advertise more if the number of subscribers per quarter drops below a certain threshold. Using the data above, find the number of subscribers, *c*, so that 90% of all values are at or below *c*.

3.126 Education and Child Development The time (in minutes) it takes to read a certain passage is part of an elementary school assessment test. Two different groups were given the same passage to read. One group received a standard reading curriculum, and the other was given reading instruction based on the "whole language" paradigm. The results are given on the text website. 🔊 READING

a. Find the mean, variance, and standard deviation for each group.

b. Construct a modified box plot for each group and display the graphs on the same measurement axis.

c. Based on your results in parts (a) and (b), describe any differences in reading speed distributions.

3.127 Physical Sciences A standard often used for measuring brightness is lux. For example, bright moonlight has 0.1 lux and bright sunshine has 100,000 lux. The light required for general office work is approximately 400 lux. A random sample of the brightness in office cubicles was obtained, and the data are given on the text website. 🔊 CUBELUX

a. Find the mean, variance, and standard deviation.

b. Construct a modified box plot for these data. Classify any outliers as mild or extreme.

c. Using the data in the table, in what percentile does 400 lux lie?

d. Use Chebyshev's rule to describe this data set ($k = 2, 3$).

3.128 Manufacturing and Product Development The density of tires is an important selling point for serious mountain bike riders. The tire industry uses a type A durometer to measure the indentation hardness for mountain bike tires. Suppose the distribution of tire hardness is approximately normal, with mean 45 and standard deviation 7.

a. Carefully sketch the normal curve for tire hardness.
b. Is a tire hardness of 30 unusually soft? Justify your answer.
c. A certain bicycle shop claims the hardness of all its tires is in the 84th percentile. If this is true, what is the minimum hardness of any tire in the store?

3.129 Physical Sciences There were approximately 17,000 earthquakes around the world in 2012.[43] A random sample of the magnitudes (on the Richter scale) of these earthquakes during November is given on the text website.
ılıı QUAKES

a. Find the mean, median, variance, and standard deviation of the magnitudes.
b. Find the 40th and the 80th percentiles.
c. How likely is a magnitude of 4.8? Justify your answer.

EXTENDED APPLICATIONS

3.130 Physical Sciences Hydraulic fracturing, or fracking, is a method used to extract natural gas from deep shale deposits. This process involves over 500 chemicals and millions of gallons of water. In a random sample of fracking wells, the mean depth was 8000 feet.[44] Assume the standard deviation is 450 feet and the distribution of depths is approximately normal.

a. What proportion of wells have depths between 7100 and 8900 feet?
b. What proportion of wells have depths less than 6650 feet?
c. What proportion of wells have depths between 7550 and 9350 feet?
d. Suppose a new fracking well was drilled in 2012 to a depth of 8255. Is there any evidence to suggest that the mean depth of wells has changed?

3.131 Physical Sciences A building code officer inspected random home fire extinguishers for pressure (in psi), and the data are given on the text website. **ılıı** FIREX

a. Construct a modified box plot for these data.
b. Use the empirical rule to decide whether this distribution of pressures is approximately normal.
c. Create a new set of observations, $y_i = \ln(x_i)$, where ln is the natural logarithm function. Construct a modified box plot for this new set of data. Use this graph and the empirical rule to decide whether the distribution of the transformed data is approximately normal.

3.132 Manufacturing and Product Development America's favorite candy is M&Ms, with over $670 million in annual sales. M&Ms were originally sold in tubes and are now available in several versions and can even be personalized.[45] Suppose the manufacturer (Mars) claims the mean weight of a single M&M is 0.91 gram with standard deviation 0.04 gram.

a. Without assuming anything about the shape of the distribution of M&M weights, what proportion of M&Ms have weights between 0.83 and 0.99 gram?
b. Suppose a random M&M has weight 0.74 gram. Do you believe the manufacturer's claim about the mean weight? Justify your answer.

3.133 Manufacturing and Product Development The actual width of a 2×4 piece of lumber is approximately $1\frac{3}{4}$ inches but can vary considerably. The Lumber Yard in Martinsburg, West Virginia, advertises consistent dimensions for better building, and claims all 2×4s sold have a mean width of $1\frac{3}{4}$ inches with a standard deviation of 0.02 inch.

a. Assume the distribution of widths is approximately normal. Find a symmetric interval about the mean that contains almost all of the 2×4 widths.
b. Suppose a random 2×4 has width 1.79 inches. Is there any evidence to suggest The Lumber Yard's claim is wrong? Justify your answer.
c. Suppose a random 2×4 has width 1.68 inches. Is there any evidence to suggest The Lumber Yard's claim is wrong? Justify your answer.

3.134 Biology and Environmental Science Some fish have been found to have mercury levels greater than 1 ppm (parts per million), a level considered safe by the U.S. Food and Drug Administration. Suppose the mean mercury level for smallmouth bass in the Susquehanna River is 0.7 ppm with standard deviation 0.1 ppm, and the distribution of mercury level is approximately normal.

a. Is it likely a fisherman will catch a smallmouth bass with mercury level greater than 1 ppm? Justify your answer.
b. Suppose the standard deviation is 0.05 ppm. Now, is it likely a fisherman will catch a smallmouth bass with mercury level greater than 1 ppm? Justify your answer.
c. Carefully sketch the normal curves for parts (a) and (b) on the same measurement axis.

3.135 Sports and Leisure The longest-running Broadway show, with over 10,000 performances, is *Phantom of the Opera*, surpassing *Cats* in 2006. A sample of Broadway shows was obtained, and the number of performances of each was recorded.[46] **ılıı** SHOWS

a. Find the sample mean and the sample median number of performances. What do these values suggest about the shape of the distribution?
b. Find the sample variance and the sample standard deviation. Find the proportion of observations within one standard deviation of the mean, within two standard deviations of the mean, and within three standard deviations of the mean. What do these proportions suggest about the shape of the distribution?
c. Find the first quartile, the third quartile, and the interquartile range. Construct a modified box plot for the performance data. Use this graph to describe the distribution in terms of symmetry, skewness, variability, and outliers. Does your description based on the box plot agree with your answers to parts (a) and (b)? Why or why not?

d. Find out how many performances there have been for *Phantom of the Opera* and add this value to the data set. How will this value affect the sample mean, sample median, sample variance, and quartiles? Find these values and verify your predictions.

LAST STEP

3.136 How efficient is the Canadian Pacific Railway?   To increase efficiency, officials at the Canadian Pacific Railway monitor several variables including train speed, cars on each train, and terminal dwell time. In addition, the type and amount of freight is carefully recorded for each train. The following table shows the number of

carloads of grain mill products for 30 randomly selected weeks in 2011 and 2012. **RAILWAY**

572	711	582	663	612	577	650	550	590	659
610	611	557	685	683	629	626	637	634	723
718	707	673	697	808	755	438	569	684	637

a. Compute the summary statistics for these data, including the mean, median, variance, standard deviation and quartiles.
b. Construct a box plot for these data.
c. Describe the distribution. Identify any outliers.
d. Find the proportion of observations within one, two, and three standard deviations of the mean. Compare the results to the empirical rule.
e. Find the 90th percentile.

4 Probability

What are the chances of winning a prize in Monopoly Sweepstakes?

In 2012, McDonald's once again offered customers an opportunity to play the Monopoly Game. The objective of this contest was to collect McDonald's Monopoly game pieces corresponding to the properties from the original Monopoly board game. Customers could collect the game pieces with the purchase of certain McDonald's products, and some game pieces were instant winners. However, special collections of properties were worth big prizes, including cash, cars, and vacations.

Each set of properties consists of two to four game pieces, corresponding to squares on the Monopoly board. In the McDonald's game, one game piece in each set of properties is rare. Here is a list of a few property collections, the rare piece, and the probability of finding the rare piece.

St. James Place, Tennessee Avenue, and New York Avenue: $10,000 cash. Probability of finding Tennessee Avenue: 0.00000000193

Park Place, Boardwalk: $1,000,000 annuity. Probability of finding Boardwalk: 0.00000000326

Reading Railroad, Pennsylvania Railroad, B&O Railroad, Short Line: EA Sports fan trip. Probability of finding Short Line: 0.00000000185

The techniques presented in this chapter will be used to determine the probability of winning at least one of the property prizes. Hint: It's going take a few Big Macs![1]

CONTENTS

Jordan Siemens/Getty Images

4.1 Experiments, Sample Spaces, and Events

To understand probability concepts, we need to think carefully about **experiments**. Consider the activity, or act, of tossing a coin, selecting a card from a standard poker deck, counting the number of contaminants in 1 cm^3 of drinking water, or even testing a cell phone for defects before shipment. In every one of these activities, the outcome is uncertain. For example, when we test a new cell phone, we do not know (for sure) whether it will be defect-free. This idea of uncertainty leads to the definition of an experiment.

Definition

An **experiment** is an activity in which there are at least two possible outcomes and the result of the activity cannot be predicted with absolute certainty.

Here are some examples of experiments.

1. Roll a six-sided die and record the number that lands face up.

   We cannot say with certainty that the number face up will be a 1, or a 2, etc., so this activity is an experiment.

2. Using a radar gun, record the speed of a pitch at a Red Sox baseball game.

   We're not sure whether the pitch will be a fastball, curveball, slider, etc. And even if we steal the signal from the catcher, we cannot predict the speed of the pitch with certainty.

3. Count the number of patients who arrive at the emergency room of a city hospital during a 24-hour period.

   Although past records might help us form an estimate, there is no way of predicting the exact number of emergency room patients during a 24-hour period.

4. Select two Keurig Home Brewers and inspect each for flaws in materials and workmanship.

   Even though a strict quality control process might be in place, there is no way of knowing whether both Keurigs will be flawless, one will contain a flaw, or both will have flaws.

Because we don't know for sure what will happen when we conduct an experiment, we need to consider *all* possible outcomes. This sounds easy (just think about all the things that can happen), but it can be tricky. Sometimes it involves a lot of counting, but often outcomes can be visualized using a *tree diagram*. Consider the following examples.

Example 4.1 Social Security Numbers

Suppose a U.S. citizen is selected and the last digit of her Social Security number is recorded. How many possible outcomes are there, and what are they?

> This is an experiment, because we cannot predict the last digit with certainty.

SOLUTION

STEP 1 The last digit of a person's Social Security number can be any integer from 0 to 9.

STEP 2 There are 10 possible outcomes.

The outcomes are 0, 1, 2, 3, 4, 5, 6, 7, 8, and 9. ■

Example 4.2 Buckle Up

Two drivers on the Pennsylvania Turnpike are selected at random and checked for compliance with the seatbelt law. How many possible outcomes are there, and what are they?

SOLUTION

STEP 1 If a driver is wearing a seatbelt, denote this observation by R (for restrained), and if he is not wearing a seatbelt, use U (for unrestrained).

There are lots of other ways to denote these four outcomes. There is no single *correct* notation. Write the outcomes so that others can understand and interpret your list.

STEP 2 Each outcome is a pair of observations, one on each driver. There are four possible outcomes, and here is a way to write them: RR, RU, UR, UU.

The first letter indicates the observation on the first driver, and the second letter indicates the observation on the second driver.

RU is a different outcome from UR. RU means the first driver was wearing a seatbelt and the second driver was not. UR means the first driver was not wearing a seatbelt and the second driver was. ∎

Tree diagrams will also be extremely useful for determining probabilities in problems involving Bayes's rule. Problems of this type are presented in Section 4.5.

All of the outcomes from the experiment in Example 4.2 can be determined by constructing a **tree diagram**, a visual road map of possible outcomes. Figure 4.1 is a tree diagram associated with this experiment.

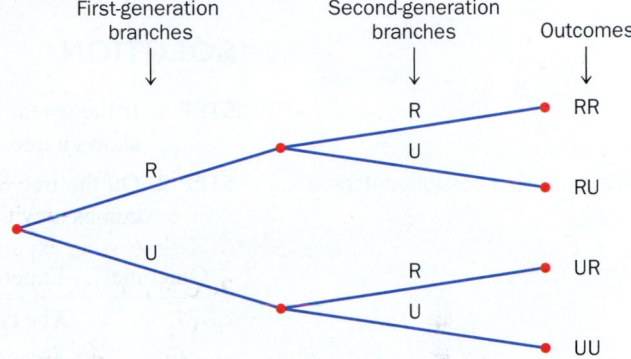

Figure 4.1 Tree diagram for Example 4.2.

The *first-generation branches* indicate the possible choices associated with the first driver, and the *second-generation branches* represent the choices for the second driver. A path from left to right represents a possible experimental outcome.

Example 4.3 Buckle Up (Continued)

Extend the previous example. How many outcomes are there if we stop *three* drivers and record their seatbelt status?

SOLUTION

Now there are eight possible outcomes: RRR, RRU, RUR, RUU, URR, URU, UUR, UUU.

Figure 4.2 is a tree diagram for this extended experiment. Again, every path from left to right represents a possible outcome.

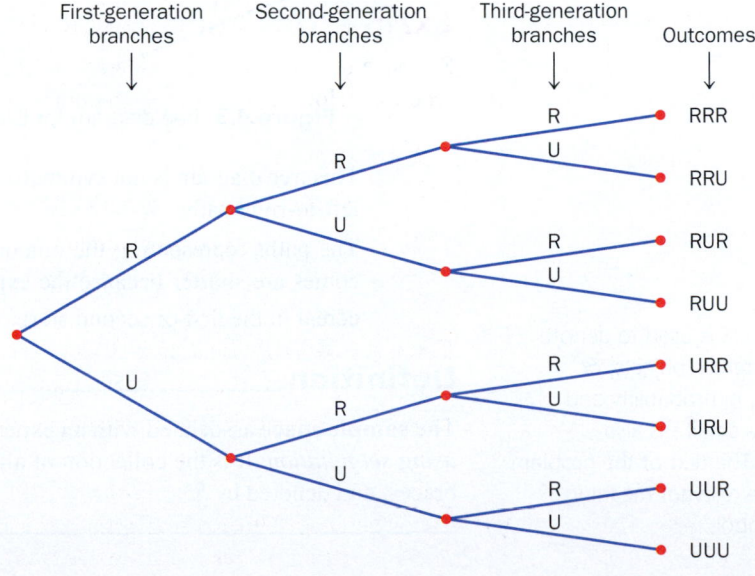

Figure 4.2 Tree diagram for Example 4.3.

Tree diagrams are also used to prove the *multiplication rule* (Section 4.3), an arithmetic technique used to count the number of possible outcomes in certain experiments.

1. Tree diagrams are a fine technique for finding all the possible outcomes for an experiment. However, they can get very big, very fast.

2. A tree diagram does not have to be symmetric, as they are in Figures 4.1 and 4.2. The branches and paths depend on the experiment. Consider the next example. ●

Example 4.4 Breakfast of Champions

A consumer in Clarkdale, Arizona, is searching for a box of his favorite breakfast cereal. He will check all three grocery stores in town if necessary, but will stop if the cereal is found. The experiment consists of searching for the cereal. How many possible outcomes are there, and what are they?

SOLUTION

STEP 1 If the cereal is in stock, use the letter I; if it is out of stock, use O. Figure 4.3 shows a tree diagram for this experiment.

Why isn't IO a possible outcome?

STEP 2 On the tree diagram, there are four possible paths from left to right. The outcomes are

Outcome	Experiment result
I	The cereal is in stock in store 1.
OI	The cereal is not in stock in store 1, but it is in stock in store 2.
OOI	The cereal is not is stock in stores 1 and 2, but it is in stock in store 3.
OOO	The cereal is not in stock in any store.

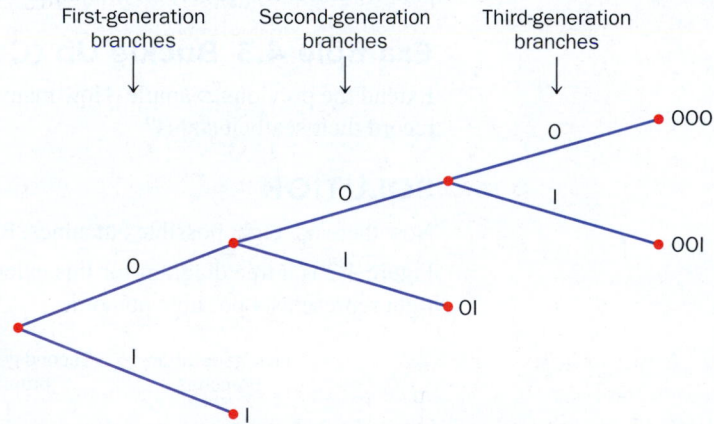

Figure 4.3 Tree diagram for Example 4.4.

This tree diagram is not symmetric, but all possible outcomes are represented by left-to-right paths.

The paths representing the outcomes have different lengths. Some of the outcomes are shorter because the experiment ends early if the consumer finds the cereal in the first or second store. ●

The symbol S is used to denote several different objects, or quantities, in probability and statistics; a small s is also common. The text of the problem reveals the relevant meaning of the symbol.

Definition

The **sample space** associated with an experiment is a listing of all the possible outcomes *using set notation*. It is the collection of all outcomes written mathematically, with curly braces, and denoted by S.

Example 4.5 Sample Spaces

Find the sample space for each of the four experiments above.

SOLUTION

We determined the outcomes for each experiment. Write the sample space using set notation.

STEP 1 Last digit of Social Security number: $S = \{0, 1, 2, 3, 4, 5, 6, 7, 8, 9\}$.

STEP 2 Seatbelt experiment: $S = \{RR, RU, UR, UU\}$.

STEP 3 Extended seatbelt experiment:
$$S = \{RRR, RRU, RUR, RUU, URR, URU, UUR, UUU\}.$$

STEP 4 Cereal experiment: $S = \{I, OI, OOI, OOO\}$.

TRY IT NOW   GO TO EXERCISE 4.19

Given an experiment and the sample space, we usually study and find the probability of specific collections of outcomes, called **events**.

Definition

1. An **event** is any collection (or set) of outcomes from an experiment (any subset of the sample space).

2. A **simple event** is an event consisting of exactly one outcome.

3. An event has **occurred** if the resulting outcome is contained in the event.

A CLOSER LOOK

1. An event may be given in standard set notation, or it may be defined in words. If a written definition is given, we need to translate the words into mathematics in order to identify the event outcomes.

2. Notation:

   (a) Events are denoted with capital letters, for example, $A, B, C, \ldots$.

   (b) Simple events are often denoted by $E_1, E_2, E_3, \ldots$.

3. It is possible for an event to be empty. An event containing no outcomes is denoted by $\{\}$ or $\emptyset$ (the empty set).

Example 4.6 College Dining

Translate *at most* and *at least* carefully. These expressions appear frequently in probability and statistics questions.

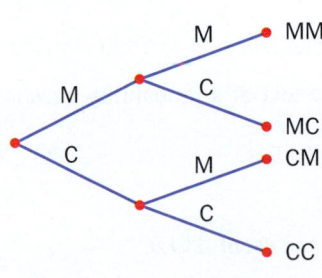

Two resident students at Bucknell University are selected and asked if they purchased a meal plan (M) or cook for themselves (C). The experiment consists of recording the response from both students.

There are four possible outcomes. A tree diagram is shown in the margin. The sample space is $S = \{MM, MC, CM, CC\}$.

There are four relevant simple events:

$E_1 = \{MM\}, E_2 = \{MC\}, E_3 = \{CM\}, E_4 = \{CC\}$.

Here are some other events, in words and in set notation.

Let A be the event that both students made the same choice.
$A = \{MM, CC\}$.

Let B be the event that at most one student purchased a meal plan.
$B = \{CC, MC, CM\}$ contains observations with at most one M.

Let D be the event that at least one student cooks for himself.
$D = \{CM, MC, CC\}$ contains observations with one or more Cs.

UPS delivery routes include as many right turns as possible.

© B Christopher/Alamy

Example 4.7 On-Time Delivery

A UPS driver may deliver to floors 2 through 6 in an office building and use one of three elevators (labeled A, B, and C). The experiment consists of recording the floor and elevator used.

There are 15 possible outcomes because there are three elevators for each of the five floors. A tree diagram works again. The sample space is

$S = \{2A, 3A, 4A, 5A, 6A, 2B, 3B, 4B, 5B, 6B, 2C, 3C, 4C, 5C, 6C\}.$

The number in each outcome represents the floor, and the letter represents the elevator.

Let E be the event that the delivery is made on an odd floor using elevator B.

$E = \{3B, 5B\}.$

Let F be the event that the delivery is made on an even floor.

This definition says nothing about the elevator used. There are no restrictions on the elevator in this event.

$F = \{2A, 4A, 6A, 2B, 4B, 6B, 2C, 4C, 6C\}.$

Let G be the event that the delivery is made using elevator C.

$G = \{2C, 3C, 4C, 5C, 6C\}.$

TRY IT NOW   GO TO EXERCISE 4.22

When an experiment is conducted, only one outcome can occur. For example, if the UPS driver used elevator B to deliver to the third floor, the experimental outcome is 3B. The observed outcome may be included in several relevant events. In the delivery example above, if the outcome 4C is observed, then the events F and G have occurred. The event E did not occur.

Given an experiment, the sample space, and some relevant events, we often combine events in various ways to create and study new events. Events are really sets, so the methods of combining events are set operations.

Definition

Let A and B denote two events associated with a sample space S.

A′ is read as "A prime" or "A complement."

1. The event **A complement**, denoted A', consists of all outcomes in the sample space S that are *not* in A.

2. The event **A union B**, denoted $A \cup B$, consists of all outcomes that are in A or B or both.

3. The event **A intersection B**, denoted $A \cap B$, consists of all outcomes that are in both A and B.

4. If A and B have no elements in common, they are **disjoint** or **mutually exclusive**, written $A \cap B = \{\}.$

A CLOSER LOOK

1. The event A' is also called **not A**. The word *not* in the text of a probability question usually means you need to find the complement of an event.

2. *Or* usually means **union**; A or B means $A \cup B$.

3. *And* usually means **intersection**; A and B means $A \cap B$.

4. Any outcome in *both* A and B is included only once in the event $A \cup B$.

5. The three events defined above could be denoted using any new symbols. A', $A \cup B$, and $A \cap B$ are traditional mathematical symbols to denote complement, union, and intersection.

6. It is possible for one of these new events to contain all the outcomes in the sample space.

Example 4.8 One-Coat Coverage

Home Depot sells Behr Premium Plus Ultra interior paint in one of three finishes: flat (F), satin (T), or gloss (G). The manager is interested in customer preferences and conducts an experiment by recording the interior paint finish for the next two customers who buy paint.

The sample space for this experiment has nine outcomes. See Figure 4.4.

$S = \{FF, FT, FG, TF, TT, TG, GF, GT, GG\}$.

Consider the following events.

$A = \{FF, TT, GG\}$ Both buy the same finish.
$B = \{FF, FT, TF, TT\}$ Neither buys gloss.
$C = \{FF, FT, FG, TF, GF\}$ At least one buys flat.
$D = \{FT, TF, TG, GT\}$ Exactly one buys satin.

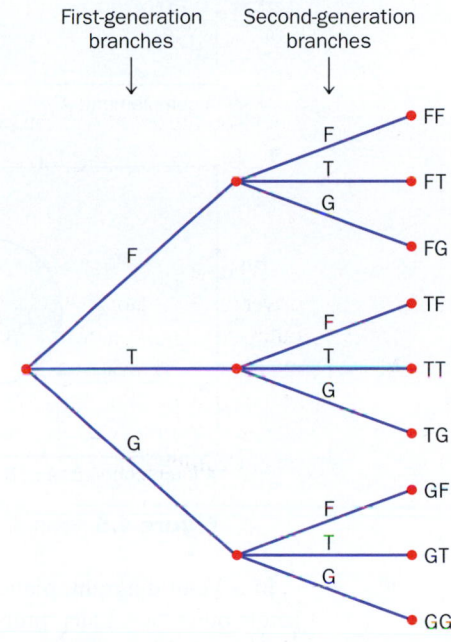

Figure 4.4 Tree diagram for Example 4.8.

Here are some new events created from the four given events:

$D' = \{FF, FG, TT, GF, GG\}$
     = D complement, neither or both buy satin.
     = All outcomes in S *not* in D.

$A \cup C = \{FF, TT, GG, FT, FG, TF, GF\}$
     = Both buy the same finish or at least one buys flat.
     = All outcomes in A *or* C (or both).

$A \cap D = \{\ \}$
     = Both buy the same finish and exactly one buys satin.
     = All outcomes in A *and* D. A and D are disjoint.

$(A \cup C)' = \{TG, GT\}$
     = A union C, complement.
     = All outcomes in S *not* in $A \cup C$.

$(A \cap D)' = S$
     = A intersection D, complement.
     = All outcomes in S *not* in $A \cap D$.

TRY IT NOW   GO TO EXERCISE 4.12

A **Venn diagram** may be used to visualize a sample space and events, to determine outcomes in combinations of events, and to answer probability questions in later sections. To construct a Venn diagram draw a rectangle to represent the sample space. Various figures (often circles) are drawn inside the rectangle to represent events. The Venn diagrams in Figure 4.5 illustrate various combinations of events.

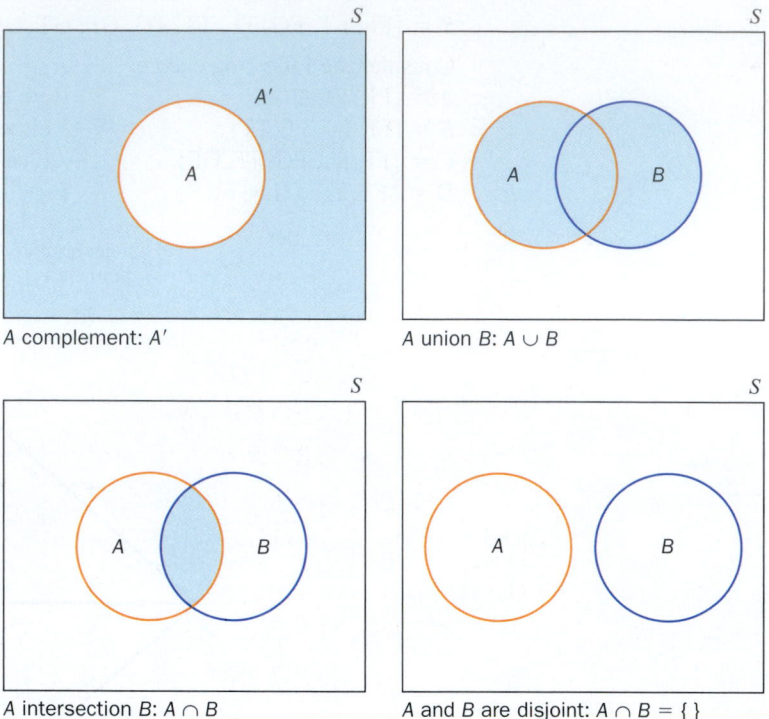

A complement: A' A union B: $A \cup B$

A intersection B: $A \cap B$ A and B are disjoint: $A \cap B = \{\ \}$

Figure 4.5 Venn diagrams.

In a Venn diagram, plane regions represent events. We often add labeled points to denote outcomes. Later, probabilities assigned to events will be added to the diagrams.

The definitions of union, intersection, and disjoint events can be extended to a collection consisting of more than two events.

Definition

Let $A_1, A_2, A_3, \ldots, A_k$ be a collection of k events.

1. The event $A_1 \cup A_2 \cup \cdots \cup A_k$ is a **generalized union** and consists of all outcomes in at least one of the events $A_1, A_2, A_3, \ldots, A_k$.

2. The event $A_1 \cap A_2 \cap \cdots \cap A_k$ is a **generalized intersection** and consists of all outcomes in every one of the events $A_1, A_2, A_3, \ldots, A_k$.

3. The k events $A_1, A_2, A_3, \ldots, A_k$ are **disjoint** if no two have any element in common.

Example 4.9 Priority Request

A university computer technician attaches a priority code to each help request. The range is 0 to 9, with 0 as the lowest priority and 9 as the highest priority. Consider an experiment in which a random request is selected and the priority is recorded. The sample space is

$S = \{0, 1, 2, 3, 4, 5, 6, 7, 8, 9\}$

and consider the events

$A = \{0, 1, 2, 3, 4\}$ $B = \{3, 4, 5, 6\}$
$C = \{7, 8\}$ $D = \{2, 4, 6, 9\}$

a. List the outcomes in the event $A \cup B \cup C$ and illustrate these three events using a Venn diagram.

b. List the outcomes in the event $A \cup B \cup D$ and illustrate these three events using a Venn diagram.

c. List the outcomes in each of the following events:

   i. $A \cap B \cap C$ **ii.** $A \cap B \cap D$

   iii. $(A \cup B)'$ **iv.** $(A \cup B \cup D)'$

SOLUTION

STEP 1 The event $A \cup B \cup C$ includes all the outcomes in at least one of the events A, B, or C. $A \cup B \cup C = \{0, 1, 2, 3, 4, 5, 6, 7, 8\}$.

Figure 4.6 shows the relationships among the events A, B, and C, and the sample space S.

STEP 2 The event $A \cup B \cup D$ includes all the outcomes in at least one of the events A, B, or D. $A \cup B \cup D = \{0, 1, 2, 3, 4, 5, 6, 9\}$.

Figure 4.7 shows the relationships among the events A, B, and D, and the sample space S.

STEP 3 $A \cap B \cap C = \{ \}$ There are no outcomes in all three events.

$A \cap B \cap D = \{4\}$ 4 is the only outcome in all three events.

$(A \cup B)' = \{7, 8, 9\}$ All outcomes in S not in $A \cup B$.

$(A \cup B \cup D)' = \{7, 8\} = C$ All outcomes in S not in $A \cup B \cup D$.

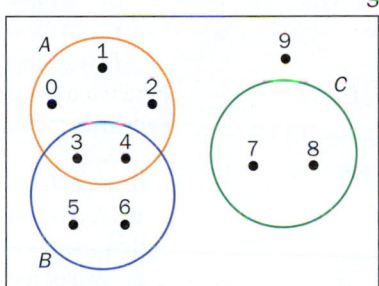

Figure 4.6 The events A, B, and C in Example 4.9.

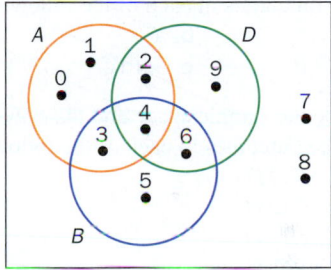

Figure 4.7 The events A, B, and D in Example 4.9.

TRY IT NOW   GO TO EXERCISE 4.13

SECTION 4.1 EXERCISES

Concept Check

4.1 True/False   In an experiment, the result of the activity is always pretty certain.

4.2 True/False   A tree diagram is always symmetric.

4.3 True/False   A sample space consists of all possible outcomes.

4.4 True/False   A simple event occurs very rarely.

4.5 Short Answer

   a. The word _____ is usually associated with the complement of an event.

   b. The word _____ is usually associated with union.

   c. The word _____ is usually associated with intersection.

Practice

4.6 An experiment consists of rolling a six-sided die, recording the number face up, and then tossing a coin and recording head or tail. Carefully sketch a tree diagram and find the sample space for this experiment.

4.7 A basketball player is going to select a sneaker with red, blue, green, or black stripes, and in either low- or high-top style. An experiment consists of recording the color and style. Carefully sketch a tree diagram and find the sample space for this experiment.

4.8 An experiment consists of selecting one letter from B, I, N, G, O, and one of five rows. How many possible outcomes are there in this experiment? Carefully sketch the corresponding tree diagram.

4.9 One playing card is selected from a regular 52-card deck. An experiment consists of recording the denomination (ace, 2, 3, 4, 5, 6, 7, 8, 9, 10, jack, queen, king) and suit (club, diamond, heart, or spade). How many possible outcomes are there in this experiment?

4.10 Consider an experiment with sample space
$S = \{0, 1, 2, 3, 4, 5, 6, 7, 8, 9\}$
and the events
$A = \{0, 2, 4, 6, 8\}$
$B = \{1, 3, 5, 7, 9\}$
$C = \{0, 1, 2, 3, 4\}$
$D = \{5, 6, 7, 8, 9\}$
Find the outcomes in each of the following events.
  a. A' **b.** C' **c.** D'
  d. $A \cup B$ **e.** $A \cup C$ **f.** $A \cup D$

4.11 Use the sample space and the events in Exercise 4.10 to find the outcomes in each of the following events.
  a. $B \cap C$ **b.** $B \cap D$ **c.** $A \cap B$
  d. $A \cap C$ **e.** $(B \cap C)'$ **f.** $B' \cup C'$

4.12 Consider an experiment with sample space
$S = \{a, b, c, d, e, f, g, h, i, j, k\}$
and the events
$A = \{a, c, e, g\}$ $B = \{b, c, f, j, k\}$
$C = \{c, f, g, h, i\}$ $D = \{a, b, d, e, g, h, j, k\}$
Find the outcomes in each of the following events.
  a. A' **b.** C' **c.** D'
  d. $A \cap B$ **e.** $A \cap C$ **f.** $C \cap D$

4.13 Use the sample space and the events in Exercise 4.12 to find the outcomes in each of the following events.
  a. $A \cup B \cup D$
  b. $B \cup C \cup D$
  c. $B \cap C \cap D$
  d. $A \cap B \cap C$

4.14 Use the sample space and the events in Exercise 4.12 to find the outcomes in each of the following events.
  a. $(A \cap B \cap C)'$
  b. $A \cup B \cup C \cup D$
  c. $(B \cup C \cup D)'$
  d. $B' \cap C' \cap D'$

4.15 The Venn diagram below shows the relationship between two events.

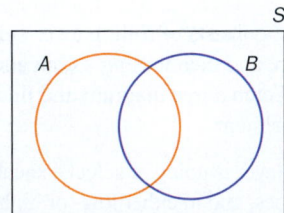

Redraw the Venn diagram for each part of this problem and carefully shade in the region corresponding to each new event.
  a. $(A \cup B)'$ **b.** $(A \cap B)'$ **c.** $A' \cap B$
  d. $A \cap B'$ **e.** $A' \cap B'$ **f.** $A' \cup B'$

4.16 The Venn diagram below shows the relationship among three events.

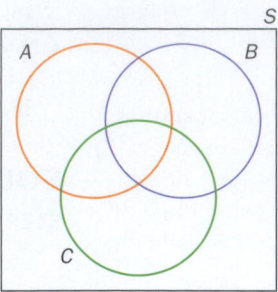

Redraw the Venn diagram for each part of this problem and carefully shade in the region corresponding to each new event.
  a. $A \cup B \cup C$ **b.** $A \cap B \cap C$ **c.** $A \cup C$
  d. $B \cap C$ **e.** $B \cap C'$ **f.** $(A \cup B)' \cap C$
  g. $(A \cup B \cup C)'$ **h.** $A' \cap B' \cap C'$ **i.** $B \cap C \cap A'$

4.17 Consider an experiment with sample space
$S = \{YYY, YYN, YNY, YNN, NYY, NYN, NNY, NNN\}$
  a. Find the outcomes in each of the following events.
    A = Exactly one Y.
    B = Exactly two Ns.
    C = At least one Y.
    D = At most one N.
Find the outcomes in each event described in words and write each as a combination of the events A, B, C, and D.
  b. Exactly one Y or at most one N
  c. Two or more Ns
  d. Exactly two Ns and at least one Y
  e. Two or more Ys

4.18 Consider an experiment with sample space
$S = \{0, 1, 2, 3, 4, 5, 6, 7, 8, 9\}$
and the events
$A = \{0, 1, 2, 7, 8, 9\}$
$B = \{0, 1, 2, 4, 8\}$
$C = \{0, 1, 3, 9\}$
$D = \{1, 4, 9\}$
Draw a separate Venn diagram to illustrate the relationships among each collection of events and the sample space S.
  a. B and C
  b. A and D
  c. A, B, and C
  d. A, C, and D

Applications

4.19 Physical Sciences An experiment consists of recording the time zone (E, C, M, P) and strength (L, M, H) for the next earthquake in the 48 contiguous states United States.
  a. Carefully sketch a tree diagram to illustrate the possible outcomes for this experiment.
  b. Find the sample space S for this experiment.

4.20 Economics and Finance   Three taxpayers are selected at random and asked whether they itemized their tax deductions last year or used the standard deduction. An experiment consists of recording each response. Construct a tree diagram to represent this experiment and find the outcomes in the sample space.

4.21 Travel and Transportation   Two people who work in New York City are selected at random and asked how they get to work: drive, take a train, or take a bus. An experiment consists of recording each response. Construct a tree diagram to represent this experiment and find the outcomes in the sample space.

4.22 Fuel Consumption and Cars   In early 2013 it was revealed that the Winnipeg Police Service had worked 627 overtime shifts over the last six months specifically for traffic enforcement.[2] The cost to Canadian taxpayers was approximately \$900,000, and some people believe the time and money should be spent on crime prevention instead. An experiment consists of selecting a ticketed car at random and recording:
   a. Whether the driver has a valid registration.
   b. Whether the automobile is properly insured.
   c. The time of day during the traffic stop: morning, afternoon or evening.
Construct a tree diagram to represent this experiment and find the outcomes in the sample space.

4.23 Physical Sciences   A construction crew excavating a site for a building foundation must remove the rock and prepare a trench for concrete footers. An experiment consists of recording the type of rock present (I, igneous; S, sedimentary; M, metamorphic) and the number of days needed to prepare the site (1 to 5).
   a. Carefully sketch a tree diagram to illustrate the possible outcomes for this experiment.
   b. Find the sample space S for this experiment.

4.24 Manufacturing and Product Development   One of four calculator batteries is bad. An experiment consists of testing each battery until the dead one is found.
   a. How many possible outcomes are there for this experiment?
   b. Is the outcome GBGG (Good, Bad, Good, Good) possible? Why or why not?

4.25 Sports and Leisure   An experiment consists of recording the number of pins knocked down on each roll during a frame of a bowling game. A bowler may take a maximum of two rolls per frame. How many outcomes are in the sample space for this experiment? Hint: If the first roll is a 10 (a strike), the experiment is over.

4.26 Sports and Leisure   A sports statistician must carefully chart opposition football plays in preparation for the next game. An experiment consists of recording the type of play (pass or rush) and the yards gained ($-99, -98, -97, \ldots, -2, -1, 0, 1, 2, \ldots, 97, 98, 99$) on a randomly selected first down. How many outcomes are in the sample space for this experiment?

4.27 Medicine and Clinical Studies   The Emergency Room in a rural hospital is staffed in four six-hour shifts (1, 2, 3, 4). During any shift, an Emergency Room patient is attended to by either a general physician (G), a surgeon (R), or an intern (I). An experiment consists of coding the next Emergency Room patient by shift and attending doctor. Consider the following events.
A = The attending doctor is the general physician.
B = The patient is admitted during the second shift.
C = The patient is admitted during shift 3 or is seen by the intern.
D = The patient is admitted during shift 4 and is seen by the general physician.
   a. Find the sample space S for this experiment.
   b. List the outcomes in each of the events A, B, C, and D.
   c. List the outcomes in the events $A \cup B$ and $A \cap B$.

4.28 Psychology and Human Behavior   Drivers entering the Quaker Bride Mall parking lot at the main entrance may turn left, right, or go straight. An experiment consists of recording the direction of the next car entering the mall and the vehicle style (sedan, SUV, van, or pickup). Consider the following events.
A = The next vehicle is a van.
B = The next vehicle is a sedan or pickup.
C = The next vehicle turns left.
D = The next vehicle goes straight or turns right.
   a. Find the sample space S for this experiment.
   b. List the outcomes in *each* of the events A, B, C, and D.
   c. List the outcomes in the events $C \cup D$ and $C \cap D$.

4.29 Public Health and Nutrition   Each patient with a regular appointment at Dr. Kenneth Heise's dental office is classified by the number of cavities found (assume 4 is the maximum) and as late (L) or on time (T) for the appointment.
   a. Find the sample space S for this experiment.
   b. Describe the following events in words.
   $A = \{0L, 1L, 2L, 3L, 4L\}$
   $B = \{3L, 4L, 3T, 4T\}$
   $C = \{1L, 3L, 1T, 3T\}$
   $D = \{0L, 0T\}$
   $E = \{0L, 0T, 1L, 2L, 3L, 4L\}$
   $F = \{4T\}$

4.30 Travel and Transportation   Every passenger arriving at the Las Vegas McCarran International Airport is classified as American (A) or Foreign (F), and by the number of checked bags (assume 5 is the maximum).
   a. Find the sample space S for this experiment.
   b. Describe the following events in words.
   $A = \{A0, F0\}$
   $B = \{F0, F1, F2, F3, F4, F5\}$
   $C = \{A1, F1, A2, F2\}$
   $D = \{F0, F5\}$
   $E = \{A1, F1, A3, F3, A5, F5\}$

4.31 Public Health and Nutrition   A researcher working for a Five Guys fast-food restaurant in Savannah, Georgia, selects random customers and classifies each according to sex [male (M) or female (F)], fresh-cut French fries or not [(C) or (N)], and age group [young (Y), middle aged (D), or senior (R)].

  a. Find the sample space S for this experiment.

  b. Describe the following events in words.

$A = \{MCY, MCD, MCR, MNY, MND, MNR\}$

$B = \{MCR, FCR\}$

$C = \{MCY, MNY, FCY, FNY\}$

$D = \{MNY, MND, MNR, FNY, FND, FNR\}$

Extended Applications

4.32 Sports and Leisure   A single six-sided die is rolled. If the number face up is even, then the experiment is over. If the number face up is odd, then the die is rolled again. The experiment continues until the number face up is even.

  a. Carefully sketch (part of) a tree diagram to illustrate the possible outcomes for this experiment.

  b. Find the sample space for this experiment.

4.33 Economics and Finance   A taxpayer in need of advice will call the IRS repeatedly until she can get through (no busy signal). If she receives a busy signal, she will hang up and try again later, and will stop calling as soon as she reaches an agent. An experiment consists of recording the calling pattern. A possible outcome is BBH: a busy signal (B) on the first two calls, and (finally) help (H) on the third call.

  a. How many possible outcomes are there in this experiment?

  b. List some of the outcomes for this experiment.

4.34 Marketing and Consumer Behavior   Musicnotes.com sells sheet music in the following genres: rock, jazz, new age, and country. An experiment consists of recording the preferred genre for the next customer, and the number of songs purchased (assume 5 is the maximum). Consider the following events.

A = The next customer prefers rock.

B = The next customer prefers jazz and buys at least three songs.

C = The next customer buys at most two songs.

D = The next customer prefers country and buys one song.

  a. Find the sample space S for this experiment.

  b. Find the outcomes in each of the following events.

    i. A' **ii.** $A \cup C$ **iii.** $A \cap D$

    iv. $C \cap D$ **v.** $A \cap C \cap D$ **vi.** $(A \cap B)$

4.35 Marketing and Consumer Behavior   Verizon Wireless offers smartphone and tablet customers a variety of data plans.

Each user must select the number of gigabytes depending on the planned volume of emails, music, and videos.[3] An experiment consists of selecting a Verizon Wireless smartphone customer and recording the data plan (1, 2, 3, 4, 5, or 6 GB) and whether the customer used more data than planned last month. Consider the following events.

A = The customer used more data than planned.

B = The customer has a 1-, 2-, or 3-GB plan.

C = The customer has a 5- or 6-GB plan, and used more data than planned.

D = The customer has a 2-, 4-, or 6-GB plan.

  a. Find the sample space S for this experiment.

  b. Find the outcomes in each of the following events.

    i. B' **ii.** $A \cup B$ **iii.** $A \cap B$

    iv. $C \cap D$ **v.** $A \cap B \cap D$ **vi.** $(A \cap D)'$

4.36 Travel and Transportation   An experiment consists of selecting a random passenger on a train from Washington, DC, Union Station to Trenton, NJ, and recording the purpose of travel (business or pleasure) and the number of pieces of luggage (0 to 4). Consider the following events:

A = The passenger is traveling on business.

B = The passenger has no luggage.

C = The passenger has at most one piece of luggage.

D = The passenger has three pieces of luggage or is traveling for pleasure.

  a. Find the sample space S for this experiment.

  b. Find the outcomes in each of the following events.

    i. $A \cup B$ **ii.** $A \cap B$ **iii.** $B \cup C$

    iv. $B \cap C$ **v.** $A \cap D$ **vi.** $A \cap B \cap C \cap D$

4.37 Marketing and Consumer Behavior   Raman's Coffee and Chai offers Chai tea in the following variations: hot or cold; with whipped cream or without; and in small, medium, or large size. An experiment consists of recording these three options for the next customer.

  a. Carefully sketch a tree diagram to illustrate the possible outcomes for this experiment.

  b. Find the sample space S for this experiment.

  c. Consider the following events.

A = The next customer order is small.

B = The next customer order is cold.

C = The next customer order is small or hot.

Find the outcomes in each of the following events.

    i. $A \cup B$ **ii.** $B \cup C$ **iii.** $B \cap C$ **iv.** C'

4.2 An Introduction to Probability

P(A) works like a function. The inputs are events; the outputs are probabilities.

Given an experiment, some events are more likely to occur than others. For any event A, we need to assign a number to A that corresponds to this intuitive *likelihood of occurrence*. The likelihood that A will occur is simply the probability of the event A. For example, the probability that an asteroid 100 meters in diameter will strike the Earth in any given year is 0.001 (a pretty unlikely event). The probability of wind gusts over 40 miles per hour at the Mount Washington Observatory on any given winter day is 0.07 (a more likely event). The notation P(A) is used to denote this likelihood, the probability of an event A. To begin our discussion of probability, consider the following *working definition*.

Definition

The probability of an event A is a number between 0 and 1 (including those endpoints) that measures the likelihood A will occur.

1. If the probability of an event is close to 1, then the event is likely to occur.

2. If the probability of an event is close to 0, then the event is not likely to occur.

Would you enroll in a class where the probability of receiving an A is 1?

If the probability of an event A is 1, then the event is a certainty: It will occur. If the probability of an event B is 0, then B is definitely not going to occur. What about events with probabilities in between? How do we decide to assign a probability of 0.3, for example, to an event C? We need a reasonable, all-purpose rule for linking an event to its likelihood of occurrence. The natural (theoretical) definition for assigning a probability to an event is very intuitive.

Definition

The **relative frequency of occurrence of an event** is the number of times the event occurs divided by the total number of times the experiment is conducted.

Example 4.10 Pick a Card, Any Card

It seems like the answer should be 1/4. Why?

In a regular 52-card deck there are 13 clubs, 13 diamonds, 13 hearts, and 13 spades. Suppose an experiment consists of selecting one card from the deck and recording the suit. What is the probability of selecting a club?

SOLUTION

STEP 1 Let C be the event that a club is selected. We want the probability of the event C, which is denoted by P(C).

STEP 2 To estimate the probability of C, it seems reasonable to conduct the experiment several times and see how often a club is selected. If C occurs often (we get a club a lot of the time), then the likelihood (probability) should be high. If C occurs rarely, then the probability should be close to 0.

Relative frequency was defined in Chapter 2 in the context of frequency distributions.

STEP 3 To estimate the likelihood of selecting a club, we use the *relative frequency* of occurrence of a club, which is the frequency divided by total trials, or

$$\text{Relative frequency} = \frac{\text{number of times a club is selected}}{\text{total number of selections}}$$

After every selection, the observed card is placed back in the deck. The deck is shuffled, and another selection is made.

STEP 4 Suppose that after 10 tries, a club was selected only twice. The relative frequency is $2/10 = 0.2$. This is an estimate of P(C). It's quick and easy, but it doesn't seem too accurate.

STEP 5 Suppose we try the experiment a few more times. With more observations we should be able to make a better guess at P(C). The table below shows values for N, the number of trials, and $\hat{p}$, the relative frequency of occurrence of a club.

N	10	50	100	200	300	400	500	600	700
$\hat{p}$	0.2	0.3	0.29	0.23	0.223	0.205	0.228	0.252	0.267
N	800	900	1000	1100	1200	1300	1400	1500	
$\hat{p}$	0.245	0.243	0.227	0.254	0.260	0.256	0.261	0.249	

For example, after 300 draws, the relative frequency of occurrence of the event C was 0.223.

STEP 6 Figure 4.8 shows a plot of relative frequency versus number of trials. The graph shows a remarkable pattern. As N increases, the points are noticeably closer to the dashed line. The relative frequencies seem to be homing in on one number (around 0.25); this relative frequency, whatever it is, should be the probability of the event C.

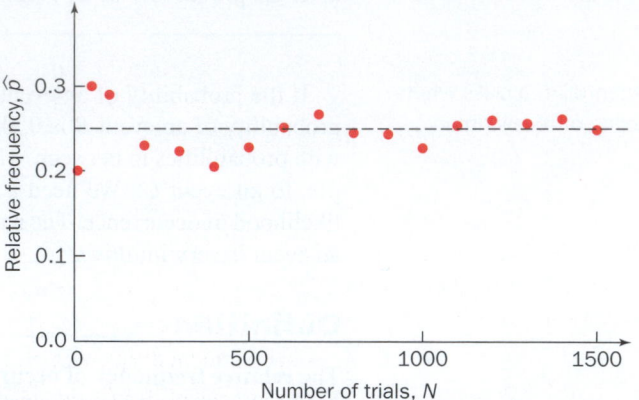

Figure 4.8 Scatter plot of relative frequency versus number of trials.

STEP 7 In the long run, the relative frequencies tend to stabilize, or even out, and become almost constant. They close in on one number, the *limiting relative frequency*. The probability of the event C is the limiting relative frequency.

If an experiment is conducted N times and an event occurs n times, then the probability of the event is *approximately n/N* (the relative frequency of occurrence). The **probability of an event A**, $P(A)$, is the *limiting* relative frequency, the proportion of time the event A will occur in the long run. This is a basic and sensible definition, a rule for assigning probability to an event. Given an event, all we need to do is find the limiting relative frequency.

Although this definition makes sense, and Example 4.10 and Figure 4.8 support and illustrate our intuition, there is a real practical problem. We cannot conduct experiments over and over, compute relative frequencies, and only then estimate the true probability. How will we ever know the true limiting relative frequency? How large should N be? When are we close enough? Will we ever hit the limiting relative frequency exactly? The definition is nice, but there seems little hope of ever finding the true probability of an event. Fortunately, there is another way to determine the exact probability in some cases. Consider the next two examples.

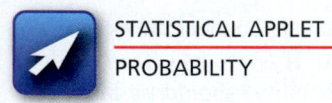

STATISTICAL APPLET

PROBABILITY

Example 4.11 Call It in the Air

Suppose an experiment consists of tossing a fair coin and recording the side that lands face up. The event H is the coin landing with heads face up. Find $P(H)$.

SOLUTION

There are only two possible outcomes on each flip of the coin, and they are both *equally likely* to occur. In the long run, we expect heads to occur half of the time.

Therefore, $P(H) = 1/2$.

If we were to conduct this experiment over and over, the relative frequency of occurrence of H would close in on 1/2.

Without flipping the coin thousands of times, making estimates, or guessing at the limiting relative frequency, we are certain the probability is 1/2.

Example 4.12 Roll the Die

An experiment consists of tossing a fair six-sided die and recording the number that lands face up. Consider the event $E = \{1\}$, rolling a one. Find $P(E)$.

SOLUTION

There are six possible outcomes on each roll of the die, and they are all equally likely to occur. In the long run, we expect 1 to occur one-sixth of the time.

The relative frequency of occurrence of a 1 would get closer and closer to 1/6 as the number of rolls gets larger and larger.

Therefore, $P(E) = 1/6$.

We *can* identify the exact limiting relative frequency.   ●

These two examples suggest it is indeed possible to find the limiting relative frequency! They are special cases, however, because in each experiment, all of the outcomes are *equally likely*.

Properties of Probability

The word *chance* is also used to express likelihood. A 10% chance means the probability is 0.10.

1. For any event A, $0 \leq P(A) \leq 1$.

   The probability of any event is a limiting *relative frequency*, and a relative frequency is a number between 0 and 1. An event with probability close to 0 is very unlikely to occur, and an event with probability close to 1 is very likely to occur.

2. For any event A, $P(A)$ is the sum of the probabilities of all of the outcomes in A.

   To compute $P(A)$, just add up the probability of each outcome or simple event in A.

3. The sum of the probabilities of all possible outcomes in a sample space is 1: $P(S) = 1$.

   The sample space S is an event. If an experiment is conducted, S is guaranteed to occur.

4. The probability of the empty set is 0: $P(\{\ \}) = P(\emptyset) = 0$. This event contains no outcomes.

In the next example, the probability (limiting relative frequency) of each simple event is assumed to be known. We will use the properties above and some earlier definitions to develop some common tools and strategies for solving similar probability questions.

Example 4.13 Try the Easy Button

There are five sales associates (indicated by their employee number) on duty in a Staples office supply store: three women (3, 4, and 5) and two men (1 and 2). An experiment consists of classifying the next customer's action. He or she will make a purchase from one of the sales associates (indicated by number) or buy nothing (99). The probability of each simple event is given in the table below.

Simple event	1	2	3	4	5	99
Probability	0.08	0.12	0.10	0.25	0.15	0.30

Consider the following events.

$A = \{1, 2\}$
   = The next customer buys something from a male sales associate.
$B = \{3, 4, 5\}$
   = The next customer buys something from a female sales associate.
$C = \{99\}$
   = The next customer buys nothing.
$D = \{1, 4\}$
   = The next customer buys from one of these two sales associates.
Find $P(A)$, $P(C)$, $P(B \cup D)$, $P(A \cap D)$, and $P(A \cap B)$.

Bloomberg/Getty Images

SOLUTION

STEP 1 $P(A) = P(1) + P(2)$ Add the probabilities of each simple event in A.
                      $= 0.08 + 0.12 = 0.20$

STEP 2 $P(C) = P(99) = 0.30$ There is only one outcome in C.

STEP 3 $P(B \cup D) = P(1, 3, 4, 5)$ Find the outcomes in the event $B \cup D$.

$\qquad\qquad\qquad = P(1) + P(3) + P(4) + P(5)$ Add up the probabilities of each simple event.

$\qquad\qquad\qquad = 0.08 + 0.10 + 0.25 + 0.15 = 0.58$

STEP 4 $P(A \cap D) = P(1) = 0.08$ The intersection is one simple event.

$\qquad\qquad\qquad\qquad\qquad\qquad\qquad\qquad\qquad$ Check the probability in the table above.

STEP 5 $P(A \cap B) = P(\{\ \}) = 0$ The intersection is empty, so the probability is 0.

TRY IT NOW   GO TO EXERCISE 4.44

> Think about tossing a fair coin, or rolling a fair die, or randomly selecting a student in a class to answer a question.

To find probabilities in the previous example, we looked at each event piece by piece. We broke down each event into simple events. Let's apply the same properties in an **equally likely outcome experiment.**

Suppose an experiment has n equally likely outcomes, $S = \{e_1, e_2, e_3, \ldots, e_n\}$. Each simple event has the same chance of occurring, so the probability of each is $1/n$; $P(e_i) = 1/n$. The limiting relative frequency of e_i is $1/n$. This is exactly what we found in Examples 4.11 and 4.12. Consider an event $A = \{e_1, e_2, e_3, e_4, e_5\}$. To find $P(A)$, add up the probabilities of each simple event in A.

$$P(A) = P(e_1) + P(e_2) + P(e_3) + P(e_4) + P(e_5)$$

$$= \frac{1}{n} + \frac{1}{n} + \frac{1}{n} + \frac{1}{n} + \frac{1}{n} = \frac{5}{n}$$

$$= \frac{\text{number of outcomes in } A}{\text{number of outcomes in the sample space } S} = \frac{N(A)}{N(S)}$$

Finding Probabilities in an Equally Likely Outcome Experiment

> You will not always see the phrase *equally likely outcomes* in these probability questions. We will identify some keywords and work with familiar experiments that imply equally likely outcomes.

In an equally likely outcome experiment, the probability of *any* event A is the number of outcomes in A divided by the total number of outcomes in the sample space S. Finding the probability of any event, in this case, means counting the number of outcomes in A, counting the number of outcomes in the sample space S, and dividing.

$$P(A) = \frac{N(A)}{N(S)}$$

Section 4.3 presents some special counting rules to help compute probabilities associated with common experiments and events. However, we can solve some of these problems already, and even use our results to make a statistical inference.

Example 4.14 Bank Teller Jobs

The Beneficial Savings Bank in Tabernacle, New Jersey, has five tellers: 1 and 2 are trainees; 3, 4, and 5 are veterans. Tellers 2, 3, and 4 are female, and tellers 1 and 5 are male. At the end of the day, two tellers will be randomly selected and all of their transactions for the day will be audited.

a. What is the probability that both trainees will be selected for the audit?

b. What is the probability that one male and one female will be selected for the audit?

c. What is the probability that two females will be selected for the audit?

SOLUTION

The experiment consists of selecting two tellers at random. The outcomes consist of two tellers who can be represented by their numbers. Therefore, 12 represents the outcome that tellers 1 and 2 were selected. The order of selection does not matter. For example, 12 and 21 both represent the event tellers 1 and 2 were selected. We can (a) list all possible outcomes systematically, (b) sketch a tree diagram, or (c) use combinations (to be presented in Section 4.3). There are 10 outcomes in the sample space.

$$S = \{12, 13, 14, 15, 23, 24, 25, 34, 35, 45\}.$$

a. Let A = both trainees are selected for the audit. Because the trainees are tellers 1 and 2, there is only one outcome in the event A: $A = \{12\}$.

$$P(A) = \frac{\text{number of outcomes in } A}{\text{number of outcomes in } S} = \frac{N(A)}{N(S)} = \frac{1}{10} = 0.10$$

b. Let B = one male and one female teller are selected. Tellers 2, 3, and 4 are female, and tellers 1 and 5 are male. Check the sample space carefully to list the outcomes in B.

$$B = \{12, 13, 14, 25, 35, 45\} \Rightarrow P(B) = \frac{N(B)}{N(S)} = \frac{6}{10} = 0.60$$

c. Let C = two females are selected. Tellers 2, 3, and 4 are female. Check the sample space again, and pick out the matching outcomes.

$$C = \{23, 24, 34\} \Rightarrow P(C) = \frac{N(C)}{N(S)} = \frac{3}{10} = 0.30$$

The next example involves an equally likely outcome experiment and an inference question. We'll need to compute the likelihood of the *observed* event to help us draw a conclusion.

Example 4.15 Buttered Bagels

Suppose Bloomin Bagels sells only two different varieties of bagels: plain (P) and cinnamon raisin (C). The owner believes the demand for each kind is the same and the shop should continue to bake these varieties in equal numbers. Five customers are selected at random. Each customer buys only one bagel and the bagel purchased is noted.

a. Find the probability that exactly one person buys a plain bagel.

b. Suppose all five customers purchase a plain bagel. Is there any evidence to suggest that demand is weighted more toward one variety?

SOLUTION

The experiment consists of selecting five customers at random and recording their bagel purchase. Each outcome is a sequence of five letters: Cs and/or Ps. For example, the outcome CCPCP stands for: the first customer buys a cinnamon raisin bagel, the second customer buys a cinnamon raisin bagel, the third customer buys a plain bagel, the fourth buys a cinnamon raisin bagel, and the fifth buys a plain bagel. There are 32 possible outcomes; a systematic listing helps, and a tree diagram works (but is big). (The multiplication rule also works here. This very useful counting technique is presented in Section 4.3.) Here is the sample space:

$S = \{$PPPPP, PPPPC, PPPCP, PPPCC, PPCPP, PPCPC, PPCCP, PPCCC, PCPPP, PCPPC, PCPCP, PCPCC, PCCPP, PCCPC, PCCCP, PCCCC, CPPPP, CPPPC, CPPCP, CPPCC, CPCPP, CPCPC, CPCCP, CPCCC, CCPPP, CCPPC, CCPCP, CCPCC, CCCPP, CCCPC, CCCCP, CCCCC$\}$.

a. Let A = exactly one person buys a plain bagel. Check the sample space and carefully list all the outcomes in A.

$$A = \{\text{PCCCC, CPCCC, CCPCC, CCCPC, CCCCP}\}$$

$$P(A) = \frac{N(A)}{N(S)} = \frac{5}{32} = 0.15625$$

Equally likely outcomes.

b. The claim is that the demand for each type of bagel is equal. If this is true, then all of the outcomes in the sample space S are equally likely.

The experiment consists of observing the bagel purchase for the next five customers. Let B = the observed outcome, everyone buys a plain bagel.

Find the likelihood of the event B occurring. There is only one outcome in B, so the probability of the event B is

$$P(B) = \frac{N(B)}{N(S)} = \frac{1}{32} = 0.03125$$

Count and divide.

The conclusion: Because this probability is so small, all five people buying a plain bagel is a rare event. But it happened! This suggests the assumption is wrong—there is evidence to suggest that the demand for each type of bagel is *not* equal.

Note: There is really evidence to suggest that *some* assumption is wrong. It could be, for example, that the five customers were not selected at random. To draw a conclusion about the demand for these two types of bagels, we must accept all other assumptions are true.

TRY IT NOW   GO TO EXERCISE 4.52

Consider an experiment, two events A and B, and known probabilities $P(A)$ and $P(B)$. Suppose we use A and B to create a new event using complement, union, or intersection. Sometimes we can use the *known probabilities* $P(A)$ and $P(B)$ to calculate the probability of the *new event* quickly. We may not have to break down the new event into simple events, or even count all the outcomes in the new event (if it is an equally likely outcome experiment). The **complement rule** and the **addition rule for two events** are two rules that help with probability calculations.

The Complement Rule

For any event A, $P(A) = 1 - P(A')$.

A CLOSER LOOK

1. The complement rule is easy to visualize and justify by looking at a Venn diagram. Figure 4.9 shows an event A and its complement A'.

   Remember, the area of a region represents probability. $P(A) + P(A') = P(S) = 1$, which can be written as $P(A) = 1 - P(A')$ or $P(A') = 1 - P(A)$.

2. The complement rule is incredibly handy; it is used in various contexts throughout probability and statistics. The problem is, how do you know when to use it? Look for keywords such as *not*, *at least*, and *at most*. A rule of thumb: If you are faced with a very long probability calculation involving many simple events, or one that may require lots of counting, try looking at the complement.

Example 4.16 Law and Order

Three public defenders are assigned to cases randomly. An experiment consists of recording the lawyer (by number) assigned to the next three cases. The outcome 132 means lawyer 1 was assigned case 1, lawyer 3 was assigned case 2, and lawyer 2 was assigned case 3.

a. Find the probability that all three cases are assigned to different lawyers.

b. Find the probability that lawyer 2 is not assigned to any of the three cases.

c. Find the probability that lawyer 2 is assigned to at least one case.

Solution Trail 4.15b

KEYWORDS

■ All five purchase a plain bagel
■ Is there any evidence?

TRANSLATION

■ Experimental outcome
■ Draw a conclusion

CONCEPTS

■ Inference procedure

VISION

Find the probability of the experimental outcome that all five customers buy a plain bagel. Compute how likely this outcome is, and draw a conclusion about the claim.

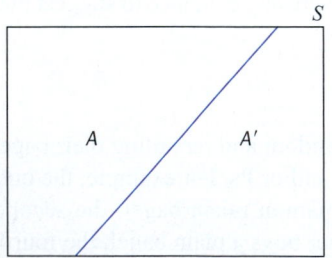

Figure 4.9 Venn diagram for visualizing the complement rule.

$(A')'$ is read as "A complement, complement." What is $(A')'$? All outcomes not in A', which is A!

SOLUTION

There are 27 possible outcomes; a tree diagram works.
Note: Each case can be assigned to one of three lawyers:

Number of possible assignments for each case.

$$\underset{\text{Case 1}}{3} \times \underset{\text{Case 2}}{3} \times \underset{\text{Case 3}}{3} = 27$$

Here's the sample space:

$S = \{111, 112, 113, 121, 122, 123, 131, 132, 133,$
$\quad\ 211, 212, 213, 221, 222, 223, 231, 232, 233,$
$\quad\ 311, 312, 313, 321, 322, 323, 331, 332, 333\}.$

a. Let A = all three cases are assigned to different lawyers. Find all the outcomes in S with a 1, a 2, and a 3.

$A = \{123, 132, 213, 231, 312, 321\}$

$P(A) = \dfrac{N(A)}{N(S)} = \dfrac{6}{27} = 0.2222$ Equally likely outcomes.

b. Let B = lawyer 2 is not assigned to any of the three cases. Find all the outcomes without a 2.

$B = \{111, 113, 131, 133, 311, 313, 331, 333\}$

$P(B) = \dfrac{N(B)}{N(S)} = \dfrac{8}{27} = 0.2963$

> 8/27 is really *approximately* equal to 0.2963. Many answers in this text are rounded (here, to four decimal places) and an equal sign is used for simplicity and convenience.

c. Let C be the event lawyer 2 has at least one case. The outcomes in C include those with one 2, two 2s, and three 2s. That seems like a lot of counting. This is a good opportunity to use the complement rule.

$P(C) = 1 - P(C')$ Complement rule.
$\quad\quad = 1 - P(\text{lawyer 2 is assigned 0 cases})$ Interpretation of C'.
$\quad\quad = 1 - P(B)$ $C' = B$ in this example.
$\quad\quad = 1 - \dfrac{8}{27} = 1 - 0.2963 = 0.7037$ ◼

TRY IT NOW   GO TO EXERCISE 4.54

Solution Trail 4.16c

KEYWORDS
- At least one case

TRANSLATION
- Let C be the event that lawyer 2 has at least one case

CONCEPTS
- Complement rule

VISION
Consider the complement, C', the event that lawyer 2 has no cases. Count the outcomes that have no 2s and use the complement rule: $P(C) = 1 - P(C')$.

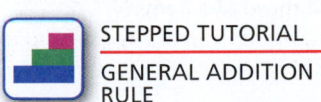

STEPPED TUTORIAL
GENERAL ADDITION RULE

The Addition Rule for Two Events

1. For any two events A and B, $P(A \cup B) = P(A) + P(B) - P(A \cap B)$.
2. For any two *disjoint* events A and B, $P(A \cup B) = P(A) + P(B)$.

A CLOSER LOOK

1. Figure 4.10 helps illustrate and justify this rule.

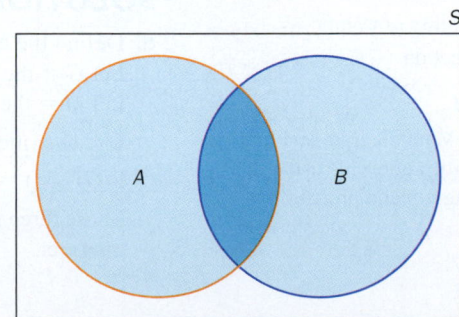

Figure 4.10 Venn diagram for illustrating the addition rule.

To find the probability of the union, start by adding $P(A) + P(B)$. This sum includes the region of intersection, $P(A \cap B)$, twice. Adjust this total by subtracting the intersection area, once.

2. $A \cup B = B \cup A$; order doesn't matter here. So, $P(A \cup B) = P(B \cup A)$.

3. The first, more general, formula *always* works. If A and B are disjoint, then $P(A \cap B) = 0$.

4. ▶ The addition rule can be extended.
   For any three events A, B, and C:

$$P(A \cup B \cup C) = P(A) + P(B) + P(C)$$
$$- P(A \cap B) - P(A \cap C) - P(B \cap C)$$
$$+ P(A \cap B \cap C)$$

You can also visualize and derive this by using a Venn diagram. In this case, the sum $P(A) + P(B) + P(C)$ includes the double intersections twice and the triple intersection three times. We therefore need to adjust the total accordingly. ◀

5. Let $A_1, A_2, A_3, \ldots, A_k$ be a collection of k *disjoint* events.

$$P(A_1 \cup A_2 \cup \cdots \cup A_k) = P(A_1) + P(A_2) + \cdots + P(A_k).$$

If the events are disjoint, to find the probability of a union, just add up the corresponding probabilities. This is especially useful in questions that ask about the number of individuals or objects with a specific attribute. For example, suppose 10 people are asked whether they received a flu shot this winter. The probability of at least 3 is the probability of 0, plus the probability of 1, plus the probability of 2, plus the probability of 3.

> Beware of these common errors: $P(A + B)$—you can't add two events; $P(A) \cup P(B)$—you can't union two numbers.

6. Complement, union, and intersection are operations applied to *events*. It doesn't make sense to take the union of probabilities (which are numbers). Similarly, addition and subtraction are operations on real numbers. You shouldn't try to add or subtract events. ●

> Probabilities are given as percentages in Example 4.17. Divide each by 100 to convert to a probability.

Solution Trail 4.17

KEYWORDS
- 70% use sugar
- 35% use milk
- 25% use both

TRANSLATION
- $P(G) = 0.70$
- $P(M) = 0.35$
- $P(G \cap M) = 0.25$

CONCEPTS
- Probability of events
- Intersection

VISION
Create a Venn diagram and find probabilities of events using the Venn diagram and probability rules.

Example 4.17 Milk or Sugar?

Marketing research by The Coffee Beanery in Detroit, Michigan, indicates that 70% of all customers put sugar in their coffee, 35% add milk, and 25% use both. Suppose a Coffee Beanery customer is selected at random.

a. Draw a Venn diagram to illustrate the events in this problem.

b. What is the probability that the customer uses at least one of these two items?

c. What is the probability that the customer uses neither?

d. What is the probability that the customer uses just sugar?

e. What is the probability that the customer uses just one of these two items?

SOLUTION

a. Define the events given in the problem.
   Let G = the customer adds sugar; $P(G) = 0.70$.
   Let M = the customer adds milk; $P(M) = 0.35$.

   Use both means uses sugar *and* milk, which means intersection. Therefore,

   $P(G \cap M) = 0.25$.

   These three probabilities add up to more than 1. That's OK because the events M and G intersect.

Remember, area of a region corresponds to probability. To complete the picture, start at the inside and work your way out.

  i. The shaded area represents the probability that the customer uses both sugar *and* milk, that is, $P(G \cap M)$. We know that $P(G \cap M) = 0.25$. Because $P(G) = 0.70$, the remaining area representing G corresponds to $0.70 - 0.25 = 0.45$.

  ii. Similarly, because $P(G \cap M) = 0.25$ and $P(M) = 0.35$, the remaining area representing M corresponds to $0.35 - 0.25 = 0.10$.

  iii. The total probability in the entire sample space must sum to 1; the remaining probability is $1 - (0.45 + 0.25 + 0.10) = 0.20$.

Figure 4.11 is the Venn diagram that corresponds to this problem.

Figure 4.11 Venn diagram for Example 4.17.

b. The probability of using *at least one* item means using sugar, or milk, or both. That's a union of two events.

$$P(G \cup M) = P(G) + P(M) - P(G \cap M) \qquad \text{Addition rule for two events.}$$
$$= 0.35 + 0.70 - 0.25 = 0.80 \qquad \text{Use the known probabilities.}$$

The Venn diagram supports this answer. Look at the region that represents $P(G \cup M)$, and add up the corresponding probabilities.

c. *Uses neither* means does *not* use sugar *or* milk. Because $G \cup M$ means sugar or milk, neither suggests the complement of $G \cup M$.

$$P[(G \cup M)'] = 1 - P(G \cup M) \qquad \text{Complement rule applied to the event } G \cup M.$$
$$= 1 - 0.80 = 0.20 \qquad \text{Use the previous answer.}$$

In the Venn diagram, this is the region outside $G \cup M$.

d. *Uses just* sugar means uses sugar *but not both* sugar and milk. This is not simply $P(G)$, because this probability includes more than just sugar. Start with the probability of using sugar, and subtract the probability of using both.

$$P(\text{just sugar}) = P(G) - P(G \cap M) \qquad \text{Use the Venn diagram.}$$
$$= 0.70 - 0.25 = 0.45 \qquad \text{Use the known probabilities.}$$

e. *Uses just one* of these items means uses sugar or milk, but not both. Start with the union, subtract off the intersection.

$$P(\text{exactly one}) = P(G \cup M) - P(G \cap M) \qquad \text{Use the Venn diagram.}$$
$$= 0.80 - 0.25 = 0.55 \qquad \text{Use the known probabilities.}$$

TRY IT NOW GO TO EXERCISE 4.61

Example 4.18 Movie Receipts

The Cheswick Theatre in Pittsburgh, Pennsylvania, has six screens, each showing a different movie. Receipts from a recent weekend were used to compile the following table, showing the probability of watching each movie.

Movie	M_1	M_2	M_3	M_4	M_5	M_6
Probability	0.10	0.25	0.20	0.30	0.10	0.05

Consider the following events.

$A = \{M_1, M_2\}$ (movies rated PG)
$B = \{M_2, M_3, M_6\}$ (action adventures)
$C = \{M_4, M_5\}$ (dramas)
$D = \{M_6\}$ (foreign film)

Suppose a patron is randomly selected. Find the probability he

a. watched a movie rated PG or an action adventure.

b. watched a movie rated PG or a drama.

c. watched a movie rated PG, or a drama, or a foreign film.

SOLUTION

Find the probability of A, B, C, and D. Break down each event and look at the individual outcomes.

$P(A) = P(M_1) + P(M_2) = 0.10 + 0.25 = 0.35$

$P(B) = P(M_2) + P(M_3) + P(M_6) = 0.25 + 0.20 + 0.05 = 0.50$

$P(C) = P(M_4) + P(M_5) = 0.30 + 0.10 = 0.40$

$P(D) = P(M_6) = 0.05$

a. *Or* means union. Find the corresponding events, and translate everything into mathematics.

$$P(A \cup B) = P(A) + P(B) - P(A \cap B) \qquad \text{(General) addition rule.}$$
$$= P(A) + P(B) - P(M_2) \qquad \text{Find the events in } A \cap B.$$
$$= 0.35 + 0.50 - 0.25 = 0.60 \qquad \text{Use known probabilities.}$$

b. Part (b) is the same kind of question; *or* means union.

$$P(A \cup C) = P(A) + P(C) \qquad \text{A and C are disjoint.}$$
$$= 0.35 + 0.40 = 0.75 \qquad \text{Use known probabilities.}$$

c. *Or* means union again in part (c), but with three events.

$$P(A \cup C \cup D) = P(A) + P(C) + P(D) \qquad \text{Three } \textit{disjoint} \text{ events.}$$
$$= 0.35 + 0.40 + 0.05 = 0.80 \qquad \text{Use known probabilities.}$$

●

SECTION 4.2 EXERCISES

Concept Check

4.38 True/False The probability of any event is always a number between 0 and 1.

4.39 Fill in the Blank
  a. If the probability of an event is close to 1, then the event is _____.
  b. If the probability of an event is close to 0, then the event is _____.
  c. The probability of an event is _____.

4.40 True/False There is no way of knowing the sum of all possible outcomes in a sample space.

4.41 True/False In an equally likely outcome experiment, the probability of any event A is the number of outcomes in A.

4.42 True/False For any event A, $P(A) + P(A') = 1$.

4.43 Fill in the Blank For any two events A and B, $P(A \cup B) = P(A) + P(B) -$ _____.

Practice

4.44 Consider an experiment with the probability of each simple event given in the table below.

Simple event	e_1	e_2	e_3	e_4
Probability	0.07	0.09	0.13	0.18

Simple event	e_5	e_6	e_7
Probability	0.22	0.15	0.16

The events A, B, C, and D are defined by

$A = \{e_1, e_2, e_3\}$ $\qquad$ $B = \{e_2, e_4, e_6, e_7\}$
$C = \{e_1, e_5, e_7\}$ $\qquad$ $D = \{e_3, e_4, e_5, e_6, e_7\}$

Find the following probabilities.
  a. P(A)   **b.** P(C)   **c.** P(D)
  d. P(A∪B)   **e.** P(A∩C)   **f.** P(B∩D)
  g. P(A')   **h.** P(A∩C')   **i.** P(A'∩D)
  j. P(C')   **k.** P(B∩C∩D)   **l.** P[(B∪C)']
How do you know there is no other possible simple event in this experiment?

4.45 An experiment consists of rolling a special 18-sided die. All of the numbers, 1 through 18, are equally likely. Find the probability of each event.
  a. A = rolling an even number.
  b. B = rolling a number divisible by 3.
  c. C = rolling a number less than 7.
  d. D = rolling at least a 10.

4.46 An experiment consists of rolling a special 22-sided die. All of the numbers, 1 through 22, are equally likely. Find the probability of each event.
  a. A = rolling a number greater than 10 and even.
  b. B = rolling a prime number or a number divisible by 5.
  c. C = rolling at most an 11.
  d. D = rolling a number divisible by 2 and 3.

4.47 Consider an experiment, the events A and B, and probabilities P(A) = 0.55, P(B) = 0.45, and P(A∩B) = 0.15. Find the probability of:
  a. A or B occurring.
  b. A and B occurring.
  c. Just A occurring.
  d. Just A or just B occurring.

4.48 Consider an experiment, the events A and B, and probabilities P(A) = 0.26, P(B) = 0.68, and P(A∪B) = 0.80. Find each probability.
  a. P(A∩B)   **b.** P(A')
  c. P[(A∩B)']   **d.** P[(A∪B)']

4.49 Consider an experiment, the events A and B, and probabilities P(A) = 0.355, P(B) = 0.406, and P(A∩B) = 0.229. Find each probability.
  a. P(A∪B)   **b.** P[(A∪B)']
  c. P(B')   **d.** P[(A∩B)']

4.50 Carefully sketch a Venn diagram showing the relationship between two events. Add probabilities to the appropriate regions so that the following statements are true: P(A∩B) = 0.31, P(A) = 0.57, and P(B) = 0.48.

4.51 Carefully sketch a Venn diagram showing the relationships among three events. Add probabilities to the appropriate regions so that the following statements are true:
P(A) = 0.46   P(B) = 0.35   P(C) = 0.44
P(A∩B) = 0.05   P(A∩C) = 0.18
P(B∩C) = 0.14   P(A∩B∩C) = 0.03

Applications

4.52 Manufacturing and Product Development   Valassis, a marketing services company, offers a cafeteria-style benefit program; an employee may select three benefits from five. The five possible benefits are health insurance, life insurance, a prescription plan, dental insurance, and vision insurance.
  a. How many different benefit packages can an employee select? List them.
  b. If all benefit packages are equally likely, what is the probability that an employee selects a package that includes health insurance?
  c. If all benefit packages are equally likely, what is the probability that an employee selects a package that includes life insurance and a prescription?

4.53 Travel and Transportation   Suppose a bridge has 10 toll booths in the east-bound lane: four are only for E-Z Pass holders, two are only for exact change, one takes only tokens, and the remainder are manned by toll collectors who accept only cash. During heavy-traffic hours it is difficult to see the signs indicating the type of toll booth. Suppose a driver selects a toll booth randomly.
  a. What is the probability that an exact-change toll booth is selected?
  b. What is the probability that a manual-collection toll booth or the token toll booth is selected?
  c. What is the probability that an E-Z Pass toll booth is not selected?
  d. Suppose the driver has only tokens. What is the probability of selecting the appropriate toll booth?

4.54 Public Health and Nutrition   As of February 16, 2013, the risk of contracting the flu in the United States was still elevated. However, the number of new cases was decreasing. The following table lists the proportion of reported cases of influenza A by region since September 30, 2012.[4]

Region	Proportion
1	0.076
2	0.074
3	0.215
4	0.080
5	0.154
6	0.065
7	0.062
8	0.089
9	0.104
10	0.081

Suppose a reported case is selected at random.
  a. What is the probability that the case is from Region 1, 2, 3, or 4?
  b. What is the probability that the case is from Region 9 or 10?
  c. What is the probability that the case is not from Region 5?

4.55 Marketing and Consumer Behavior   Delorenzo's Pizza offers five different toppings on its pizzas: pepperoni, sausage, olives, mushrooms, and anchovies. A large pizza comes with any two different toppings.
  a. How many different two-topping pizzas are possible?
  b. Suppose that all of the pizzas are equally likely. What is the probability that the next pizza ordered has at least one meat topping?

c. What is the probability that the next pizza ordered does not have anchovies?

d. Suppose one more large pizza choice is added: plain cheese with no toppings. Answer parts (b) and (c) with this added assumption.

4.56 Demographics and Population Statistics The following table lists the proportion of employed people by each major industry in Japan as of December 2012.[5]

Major industry	Proportion
Agriculture and forestry	0.0328
Construction	0.0845
Manufacturing	0.1722
Information and communications	0.0330
Transport and postal activities	0.0585
Wholesale and retail trade	0.1786
Scientific, professional, technical	0.0367
Accommodations	0.0666
Personal services	0.0412
Education	0.0511
Medical	0.1247
Services	0.0797
Government	0.0404

Suppose an employed person in Japan is selected at random.

a. What is the probability that the person works in construction or manufacturing?

b. What is the probability that the person does not work in wholesale or retail trade?

c. What is the probability that the person does not work in education nor medical?

4.57 Public Health and Nutrition The number of reported cases of vaccine-preventable diseases in Canada in 2011 is given in the following table.[6]

Disease	Frequency
Hib meningitis	38
Measles	759
Mumps	282
Pertussis	676
Other	8

Suppose a reported case is selected at random.

a. What is the probability that the case is measles?

b. What is the probability that the case is mumps or pertussis?

c. What is the probability that the case is not Hib meningitis?

4.58 Marketing and Consumer Behavior A marketing firm can place an advertisement using several media. The table below shows the probability that a randomly selected person in a targeted region will see the advertisement in the given medium.

Medium	Newspaper	Radio	Magazine	TV
Probability	0.15	0.10	0.08	0.30

Medium	Internet	Billboard	Not seen
Probability	0.12	0.05	0.20

Consider the following events.
A = {Magazine, Newspaper}
B = {TV, Radio, Internet}
C = {Magazine, Newspaper, Internet, Billboard}

Find the following probabilities.

a. $P(A)$, $P(B)$, $P(C)$

b. $P(A \cup B)$, $P(A \cap B)$, $P(B \cap C)$

c. $P(A')$, $P(A' \cap C)$, $P(A \cap B \cap C)$

d. $P(B' \cap C')$, $P[(B \cup C)']$

4.59 Sports and Leisure The Florida Cash 3 daily midday lottery number consists of three digits, each 0–9.

a. How many possible midday numbers are there?

b. If all of the midday numbers are equally likely, find the probability that all three digits are the same.

c. If all of the midday numbers are equally likely, find the probability that all three digits are 8s or 9s.

d. There is an evening number, also consisting of three digits, 0–9. If all of the midday and evening numbers are equally likely, what is the probability that the two numbers are the same?

4.60 Marketing and Consumer Behavior The following table lists the most popular U.S. convention centers.[7] Suppose the probability given represents the likelihood that a randomly selected U.S. convention will be held at that site.

Site	Probability
Orlando	0.310
Chicago	0.225
Las Vegas	0.098
Washington, DC	0.075
Dallas	0.064
Atlanta	0.055
Phoenix	0.033
Other	0.140

Suppose a convention is randomly selected. Consider the events:
A = {Convention in Orlando or Chicago}
B = {Convention not in Washington, DC}
C = {Convention in Las Vegas}

Find the following probabilities.

a. $P(A)$, $P(B)$, $P(C)$

b. $P(A \cap B)$, $P(A \cup C)$, $P(A \cap C)$

c. $P(A' \cup C)$, $P(A \cup B \cup C')$

4.61 Technology and the Internet Tablet computers have become very popular and fill a gap between smartphones and PCs. A recent survey indicated that of those people who own tablets, 70% use the device to play games and 44% use the device to access bank accounts.[8] Suppose 30% do both—play games and access bank accounts—and suppose a tablet user is selected at random.

a. What is the probability that the tablet user plays games or accesses bank accounts?

b. What is the probability that the tablet user does not play games nor access bank accounts?

c. What is the probability that the tablet user only plays games? Only accesses bank accounts?

Extended Applications

4.62 Psychology and Human Behavior According to the 2011–2012 APPA National Pet Owners Survey, approximately 33% of all U.S. households own a cat and 39% of all U.S. households own a dog.[9] Suppose 10% of all U.S. households own both a cat and a dog.

   a. Carefully sketch a Venn diagram with probabilities to illustrate the relationship between the two events $C =$ household owns a cat, and $D =$ household owns a dog.

   b. What is the probability that a randomly selected U.S. household owns a cat or a dog?

   c. What is the probability that a randomly selected household owns neither a cat nor a dog?

   d. What is the probability that the U.S. household owns only a cat?

4.63 Marketing and Consumer Behavior Of all those people who enter Uncle's Stereo, a discount electronics store in New York City, 28% purchase a digital camera, 5% buy a home theater receiver, and 4% buy both. Suppose a customer is selected at random.

   a. What is the probability that the customer buys a digital camera or a home theater receiver?

   b. What is the probability that the customer buys either a digital camera or a home theater receiver, but not both?

   c. What is the probability that the customer buys only a digital camera?

   d. What is the probability that the customer does not buy a home theater receiver?

4.64 Demographics and Population Statistics The following table shows the ABO and Rh blood-type probabilities for people in the United States.[10] (This table is called a *joint probability table*. Each number in the table can be thought of as the probability of an intersection; for example, the probability of blood type A *and* negative Rh is 0.06.)

		ABO type			
		O	A	B	AB
Rh type	Positive	0.38	0.34	0.09	0.03
	Negative	0.07	0.06	0.02	0.01

Suppose a U.S. resident is selected at random. Find the following probabilities.

   a. The person has Rh-positive blood.

   b. The person has type B blood.

   c. The person does not have type O blood.

   d. The person has type AB or Rh-negative blood.

4.65 Manufacturing and Product Development A tire manufacturer has started a program to monitor production. In every batch of eight tires, two will be randomly selected and tested for defects electronically. An experiment consists of recording the condition of these two tires: defect-free (G) or reject (B). Suppose two of the eight tires in a batch actually have serious defects.

   a. List the outcomes in this experiment.

   b. What is the probability that both tires selected will be defect-free?

   c. What is the probability that at least one of the tires selected will have a defect?

   d. What is the probability that both tires selected will have a defect?

4.66 Medicine and Clinical Studies The number of Emergency Room visits has increased over the past several years in the United States. One reason for this increase may be that in difficult economic times people tend to postpone routine health care. This results in more visits to the ER. Of all patients who visit an Emergency Room, suppose 22% are seen in less than 15 minutes, 13% are admitted to the hospital, and 5% are seen in less than 15 minutes and admitted to the hospital.[11] Suppose a patient who made an Emergency Room visit is selected at random.

   a. Carefully sketch a Venn diagram showing the relationship between the events seen in less than 15 minutes and admitted to the hospital, and add probabilities to the appropriate regions.

   b. What is the probability that the patient was seen in less than 15 minutes or admitted to the hospital?

   c. What is the probability that the patient was seen in less than 15 minutes but not admitted to the hospital?

   d. What is the probability that the patient was neither seen in less than 15 minutes nor admitted to the hospital?

Challenge

4.67 Reconsider Example 4.15. Suppose the owner records the type of bagel purchased for the next *ten* customers. Find the probability that everyone buys a plain bagel. What do you think about the assumption of equal demand now?

4.3 Counting Techniques

In an equally likely outcome experiment, computing probabilities means counting. To find the probability of an event A, count the number of outcomes in the event A and divide by the number of outcomes in the entire sample space S: $P(A) = N(A)/N(S)$. If $N(S)$ is large, drawing a tree diagram or listing all of the possible outcomes is impractical. For certain experiments, the following rules may be used instead to count outcomes in an event and/or a sample space.

The Multiplication Rule

Suppose an outcome in an experiment consists of an ordered list of k items selected using the following procedure:

1. There are n_1 choices for the first item.
2. There are n_2 choices for the second item, no matter which first item was selected.
3. The process continues until there are n_k choices for the kth item, regardless of the previous items selected.

There are $N(S) = n_1 \cdot n_2 \cdot n_3 \cdots n_k$ outcomes in the sample space S.

A CLOSER L👁OK

1. You can picture (and even prove) this rule by drawing a tree diagram and counting the number of paths from left to right.
2. To use this rule, think of each choice as a slot, or a position, to fill.

Number of choices for each slot.

$\downarrow$ $\downarrow$ $\downarrow$

$$\underbrace{n_1}_{\text{Item 1}} \times \underbrace{n_2}_{\text{Item 2}} \times \cdots \times \underbrace{n_k}_{\text{Item } k} = n_1 \cdot n_2 \cdots n_k$$

3. This counting technique can also be used for events, not just for sample spaces. ●

Example 4.19 Surround Sound

A home theater system consists of a receiver, surround-sound speakers, and a Blu-ray player. Vann's Inc. store sells 7 different receivers, 12 types of speakers, and 9 different Blu-ray players. How many possible systems can be constructed?

SOLUTION

STEP 1 This is a counting problem, and there are three slots to fill: receiver, speakers, and Blu-ray player. We'll assume that all components are compatible, and that the choice of any one item does not depend on any other item.

STEP 2 Here's how to apply the multiplication rule:

$$\underbrace{7}_{\text{Receiver}} \times \underbrace{12}_{\text{Speakers}} \times \underbrace{9}_{\text{Blu-ray player}} = 756$$

There are 756 possible systems. ●

Example 4.20 License Plates

A Connecticut license plate consists of three letters followed by three numbers.

a. How many different license plates are possible?

b. How many license plates end in 555?

SOLUTION

a. This is a counting problem. There are six slots to fill: three letters followed by three numbers. There are 26 possible letters for each of the first three positions, and 10 possible numbers for each of the last three positions. Use the multiplication rule.

$$\underbrace{26}_{\text{Letter}} \times \underbrace{26}_{\text{Letter}} \times \underbrace{26}_{\text{Letter}} \times \underbrace{10}_{\text{Number}} \times \underbrace{10}_{\text{Number}} \times \underbrace{10}_{\text{Number}} = 17{,}576{,}000$$

There are 17,576,000 possible different license plates.

The actual number of possible license plates is smaller, because some three-letter words aren't allowed.

b. If the license plate ends in 555, then each of the number positions is fixed; there is only one choice. We are still free to choose any letter in each of the first three positions. The multiplication rule still works.

$$\underset{\text{Letter}}{26} \times \underset{\text{Letter}}{26} \times \underset{\text{Letter}}{26} \times \underset{\text{Number}}{1} \times \underset{\text{Number}}{1} \times \underset{\text{Number}}{1} = 17{,}576$$

There are 17,576 license plates that end in 555.

Example 4.21 Five-of-a-Kind

In the game of Yahtzee, five fair dice are rolled and the numbers that land face up are recorded.

a. How many different rolls are possible?

b. What is the probability of rolling a Yahtzee (all five dice show the same number)?

SOLUTION

a. There are five slots to fill, one for each die. Use the multiplication rule.

$$\underset{\text{Die 1}}{6} \times \underset{\text{Die 2}}{6} \times \underset{\text{Die 3}}{6} \times \underset{\text{Die 4}}{6} \times \underset{\text{Die 5}}{6} = 7776$$

There are 7776 possible rolls, or outcomes, in the sample space.

b. There are only six possible Yahtzees: 11111, 22222, 33333, 44444, 55555, and 66666. Because all the outcomes are equally likely (fair dice), the probability of rolling a Yahtzee is

$$P(\text{Yahtzee}) = \frac{\text{number of Yahtzees}}{\text{number of different rolls}} = \frac{6}{7776} = 0.0007716$$

Keeweeboy/Dreamstime.com

Example 4.22 Win, Place, or Show

Suppose there are 12 entries in the Preakness Stakes horse race. An experiment consists of recording the *finish*: the first-, second-, and third-place horse. For example, the outcome (7, 9, 2) means horse 7 came in first, horse 9 came in second, and horse 2 came in third.

a. How many different finishes are possible?

b. What is the probability of a finish with horse 4 or 5 in first place?

c. What is the probability that horse 7 will not finish first, second, or third?

SOLUTION

a. There are three positions to fill, but the number of choices in the second slot depends on the first choice and the number of choices in the third slot depends on the first two choices. Even though we are drawing from the same, reduced, collection, we can still use the multiplication rule.

There are 12 horses that could finish first. Once a first-place horse is selected, there are only 11 left that could come in second. After a first and second-place horse are selected, there are only 10 possible for third place. The multiplication rule is used here to count the number of *permutations*.

$$\underset{\text{First}}{12} \times \underset{\text{Second}}{11} \times \underset{\text{Third}}{10} = 1320$$

There are 1320 possible different finishes. N(*S*) = 1320.

b. Let A be the event horse 4 or 5 wins the race. We'll assume that all of the outcomes are equally likely so that $P(A) = N(A)/N(S) = N(A)/1320$.

There are two choices for first place (horse 4 or 5). There are now 11 choices for second place (the horse not selected for first, plus the remaining 10), and 10 choices for third place.

The multiplication rule is used to find the number of outcomes in A:

$$\underset{\text{First}}{\underline{2}} \times \underset{\text{Second}}{\underline{11}} \times \underset{\text{Third}}{\underline{10}} = 220$$

Finally, $P(A) = 220/1320 = 0.1667$.

c. The word *not* suggests the use of a complement, but a direct approach may be easier. Let the event B = horse 7 does not finish first, second, or third. $P(B) = N(B)/N(S) = N(B)/1320$.

Use the multiplication rule again to count the number of outcomes in the event B. We do not want horse 7 in the top three. That leaves 11 possible horses for first place, 10 for second, and 9 for third.

$$\underset{\text{First}}{\underline{11}} \times \underset{\text{Second}}{\underline{10}} \times \underset{\text{Third}}{\underline{9}} = 990$$

$P(B) = 990/1320 = 0.75$.

TRY IT NOW   GO TO EXERCISE 4.76

The following notation is often used to write large numbers associated with counting problems more concisely.

Definition

For any positive whole number n, the symbol $n!$ (read "**n factorial**") is defined by

$$n! = n(n-1)(n-2)\cdots(3)(2)(1)$$

In addition, $0! = 1$ (0 factorial is 1).

A CLOSER LOOK

1. To find $n!$, just start with n, multiply by $(n-1)$, then $(n-2)$, . . . , down to 1. For example,

$7! = (7)(6)(5)(4)(3)(2)(1) = 5040$

$10! = (10)(9)(8)(7)(6)(5)(4)(3)(2)(1) = 3,628,800$

2. Factorials get really big, really fast. Try finding 50!. If you absolutely have to find a large factorial, then you should probably use a good calculator or computer.

Consider a generalization of the horse-racing problem. Suppose there are n items to choose from, r positions to fill, and the order of selection matters. There are n choices for the first position, $n-1$ choices for the second position, and $n-2$ choices for the third position. This process continues until there are $n-(r-1)$ choices for the rth position. The product of these numbers is the total number of **permutations**.

Definition

Given a collection of n different items, an ordered arrangement, or subset, of these items is called a **permutation**. The number of permutations of n items, taken r at a time, is given by

$$_nP_r = n(n-1)(n-2)\cdots[n-(r-1)]$$

$_nP_r$ is also referred to as *n* items permuted *r* at a time.

Using the definition of factorial,

$$_nP_r = \frac{n!}{(n-r)!}$$

In the denominator, do the subtraction first, then the factorial.

A CLOSER LOOK

1. All *n* items must be *different* in order for this formula to be used.

2. A distinguishing characteristic of a permutation is that order matters. For example, if the outcome AB is *different* from the outcome BA, that suggests a permutation. Suppose an experiment consists of selecting two students from a class of 35. The first one selected will be the president and the second will be the vice president. Order certainly matters here; we will be counting permutations. If the two students selected will form a committee, however, then the order of selection does not matter. Counting in this case involves a *combination*, which will be introduced a little later.

3. Here is an example to justify this formula.

$$_{12}P_3 = (12)(11)(10)$$

Definition of $_{12}P_3$, *n* = 12, *r* = 3.

$$= (12)(11)(10) \times \frac{(9)(8)(7)(6)(5)(4)(3)(2)(1)}{(9)(8)(7)(6)(5)(4)(3)(2)(1)}$$

Multiply by 1 in a nice form.

$$= \frac{(12)(11)(10)(9)(8)(7)(6)(5)(4)(3)(2)(1)}{(9)(8)(7)(6)(5)(4)(3)(2)(1)}$$

Rewrite as one fraction.

$$= \frac{12!}{9!} = \frac{12!}{(12-3)!} = \frac{n!}{(n-r)!}$$

Definition of factorial.

Example 4.23 Vending Machine Selection

A vending machine has room for six types of soda. The soda can be arranged in any order to correspond with the selection buttons on the front of the machine. If the operator has 10 different types of soda to choose from, how many machine selection arrangements are possible?

SOLUTION

If you compute $_nP_r$ by hand, there is always a lot of canceling. $_nP_r$ is a count, so the answer has to be an integer.

STEP 1 There are *n* = 10 items, we need to choose *r* = 6, and the order in which the soda is arranged matters. For example, if capital letters represent soda types, then the arrangement ABCDEF is *different* from ABCEDF. We must count the number of permutations of 10 items, taken 6 at a time.

STEP 2 $_{10}P_6 = \dfrac{10!}{(10-6)!} = \dfrac{10!}{4!}$

Definition of $_nP_r$, using factorials.

$$= \frac{(10)(9)(8)(7)(6)(5)(4)(3)(2)(1)}{(4)(3)(2)(1)}$$

Definition of factorial.

$$= (10)(9)(8)(7)(6)(5) = 151{,}200$$

Cancel; multiply.

There are 151,200 ordered arrangements of soda types in the vending machine. Figure 4.12 shows a technology solution.

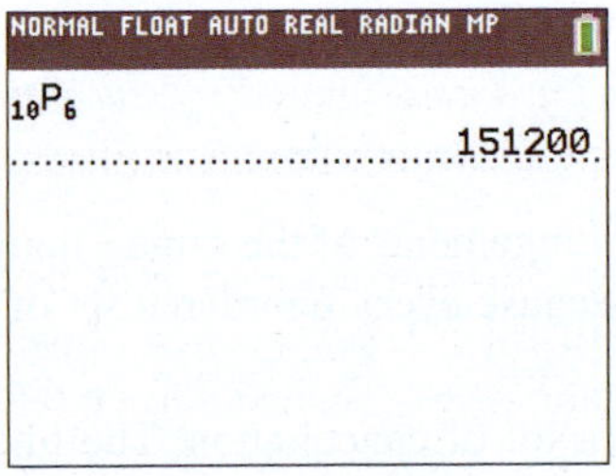

```
NORMAL FLOAT AUTO REAL RADIAN MP

10P6
                        151200
```

Figure 4.12 TI-84 Plus C permutation function.

Example 4.24 Sheldon and Leonard

A fan of *The Big Bang Theory* has taped nine episodes from the most recent season of this show. However, he only has time to watch four episodes. Suppose he selects four shows at random.

a. How many different ordered arrangements of episodes are possible?

b. If the season finale is recorded, what is the probability that he will select and watch this episode last?

Solution Trail 4.24b

KEYWORDS
- Selects at random

TRANSLATION
- Equally likely outcomes

CONCEPTS
- $P(A) = N(A)/N(S)$

VISION
Count the number of arrangements in which the final recording is selected last, and divide this count by the total number of ordered arrangements

SOLUTION

a. There are $n = 9$ episodes to choose from. We need to count the number of ordered arrangements of $r = 4$ recordings.

$$_9P_4 = \frac{9!}{(9-4)!} = \frac{9!}{5!}$$ Definition of $_nP_r$, using factorials.

$$= \frac{(9)(8)(7)(6)(5)(4)(3)(2)(1)}{(5)(4)(3)(2)(1)}$$ Definition of factorial.

$$= (9)(8)(7)(6) = 3024$$ Cancel; multiply.

There are 3024 different ordered arrangements of four episodes.

b. Let A = the last recording selected is the season finale. There are four positions to fill, but the last slot is fixed (with the season finale). The first three positions can be filled by any of the remaining eight recordings, in any order.

$$\overbrace{\underset{\text{Rec 1}}{8} \times \underset{\text{Rec 2}}{7} \times \underset{\text{Rec 3}}{6}}^{_8P_3} \times \underset{\text{Rec 4}}{1} = 336$$

$$P(A) = \frac{336}{3024} = 0.1111$$

Figure 4.13 shows a technology solution.

TRY IT NOW   GO TO EXERCISE 4.91

In many experiments, the order in which the items are selected does *not* matter, for example, selecting five manufactured items from a batch of 50 for inspection, choosing nine people from 35 for a search committee, or picking three tax returns from 100 for a federal audit. In each case, the order of selection is not important; the collection, or group selected, is a single outcome. These *unordered* arrangements are called **combinations**.

Definition

Given a collection of n different items, an unordered arrangement, or subset, of these items is called a **combination**. The number of combinations of n items, taken r at a time, is given by

$$_nC_r = \binom{n}{r} = \frac{n!}{r!(n-r)!} = \frac{_nP_r}{r!}$$

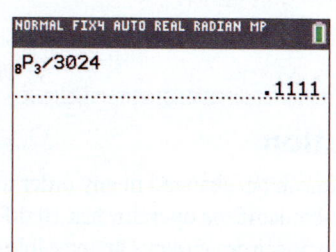

Figure 4.13 Find the number of permutations and divide by the total number of outcomes.

A CLOSER LOOK

1. $\binom{n}{r}$ is read as "n choose r."

2. To find $_nC_r$ from $_nP_r$ we need to *collapse* all ordered arrangements of the same r items into one possible outcome. Dividing by $r!$ does this because every unordered set of r distinct items can be arranged in $r!$ ways.

3. If you have to calculate $_nC_r$ by hand, there is always a lot of cancellation. The final answer must be an integer because it is a count.

Example 4.25 Jury Duty

How many different ways are there to select a jury of 12 people from a pool of 20?

SOLUTION

STEP 1 There are $n = 20$ prospective jurors, and we need to choose $r = 12$, without regard to order. A jury is an *unordered* arrangement of 12 people. We need to count the number of combinations of 20 items, taken 12 at a time.

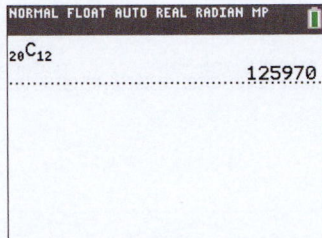

Figure 4.14 The TI-84 Plus C combination function.

Solution Trail 4.26

KEYWORDS
- Randomly selects

TRANSLATION
- Equally likely outcomes

CONCEPTS
- Probability of an event means counting
- Combinations

VISION

The order in which the cartons are selected does not matter. To find the number of outcomes in the sample space, count combinations. To count the number of outcomes in each event, use the multiplication rule and the formula for $_nC_r$.

STEP 2 $_{20}C_{12} = \dbinom{20}{12}$

$$= \frac{20!}{12!(20-12)!} = \frac{20!}{12!18!}$$ Definition of $\dbinom{n}{r}$.

$$= \frac{(20)(19)(18)(17)(16)(15)(14)(13)}{8!} = 125{,}970$$ Cancellation; computation.

There are 125,970 ways to select a jury of 12 from a pool of 20 candidates. Figure 4.14 shows a technology solution.

Example 4.26 Hardwood Floors

Lumber Liquidators ships $2\frac{1}{4}$-inch solid oak wood flooring in cartons containing 20 square feet. Suppose that two cartons in a shipment of 11 contain defective pieces. An installer randomly selects five cartons.

a. What is the probability that there are no defective pieces in any of the five cartons selected?

b. What is the probability that the installer picks exactly one carton that contains defective pieces?

SOLUTION

There are $n = 11$ cartons, and we need to choose $r = 5$, without regard to order.

$$_{11}C_5 = \dbinom{11}{5} = \frac{11!}{5!(11-5)!} = \frac{11!}{5!6!} = \frac{(11)(10)(9)(8)(7)}{5!} = 462$$

There are 462 outcomes in the sample space, all equally likely.

a. Let A = select no cartons that contain defective pieces. Count the number of ways to select no cartons that contain defective pieces. This is the same as choosing five good cartons. There are nine good cartons, so we count the number of ways to select five cartons from the nine *good* cartons, without regard to order.

$$N(A) = \dbinom{9}{5} = \frac{9!}{5!(9-5)!} = \frac{9!}{5!4!} = \frac{(9)(8)(7)(6)}{4!} = 126$$

Because this is an equally likely outcome experiment,

$$P(A) = \frac{N(A)}{N(S)} = \frac{126}{462} = 0.2727$$

b. Let B = select one carton that contains defective pieces (and, therefore, four good cartons). To find the number of outcomes in B, there are two *cases* to consider: the number of ways to select one bad carton, and the number of ways to select four good cartons.

$$\dbinom{2}{1} \times \dbinom{9}{4} = 2 \times 126 = 252$$

↑ The number of ways to select one bad carton from two, without regard to order

↑ The number of ways to select four good cartons from nine, without regard to order

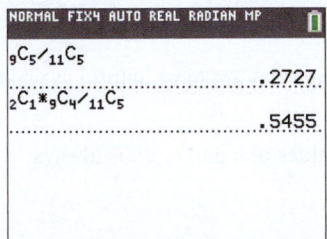

Figure 4.15 Probability calculations.

$$P(B) = \frac{N(B)}{N(S)} = \frac{252}{462} = 0.5455$$

Figure 4.15 shows a technology solution.

TRY IT NOW   GO TO EXERCISE 4.85

Technology Corner

Procedure: Compute permutations and combinations.
Reconsider: Examples 4.23 and 4.25, solutions, and interpretations.

TI-84 Plus C

There are built-in functions to compute permutations, combinations, and even factorials.

1. On the Home screen, enter the value of n.
2. Select MATH ; PROB ; nPr or MATH ; PROB ; nCr.
3. Enter the value of r. See Figures 4.13 and 4.14.

Minitab

Use the functions Permutations and Combinations in either a Session window or in Calc; Calculator. See Figure 4.16.

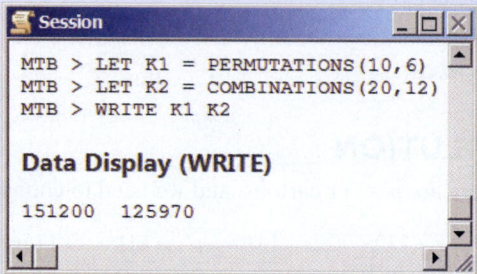

```
MTB > LET K1 = PERMUTATIONS(10,6)
MTB > LET K2 = COMBINATIONS(20,12)
MTB > WRITE K1 K2

Data Display (WRITE)

151200   125970
```

Figure 4.16 Minitab functions for permutations and combinations.

Excel

Use the function PERMUT to compute permutations and the function COMBIN to compute combinations. Use intermediate results to compute probabilities. See Figure 4.17.

A	B
151200	= PERMUT(10,6)
125970	= COMBIN(20,12)

Figure 4.17 Excel functions for permutations and combinations.

SECTION 4.3 EXERCISES

Concept Check

4.68 Fill in the Blank A _____ is a visualization of the multiplication rule.

4.69 Fill in the Blank An ordered arrangement is called a _____ .

4.70 Fill in the Blank An unordered arrangement is called a _____ .

4.71 Short Answer Counting rules are most helpful in what kind of an experiment?

4.72 True/False For fixed values of n and r, $_nP_r$ is always greater than or equal to $_nC_r$.

Practice

4.73 Find the number of permutations indicated.

a. $_8P_4$ b. $_{11}P_7$ c. $_{12}P_4$
d. $_{10}P_{10}$ e. $_{10}P_1$ f. $_{10}P_0$
g. $_9P_2$ h. $_{20}P_2$ i. $_{100}P_2$

4.74 Find the number of combinations indicated.

a. $\binom{9}{5}$ b. $\binom{9}{4}$ c. $\binom{14}{7}$

d. $\binom{10}{10}$ e. $\binom{10}{1}$ f. $\binom{10}{0}$

g. $\binom{12}{3}$ h. $\binom{16}{7}$ i. $\binom{20}{18}$

4.75 How many permutations of the letters in the word HISTOGRAM are possible?

4.76 A businessman's outfit consists of a pair of pants, a shirt, and a tie. Suppose he can choose from among 5 pairs of pants, 8 shirts, and 15 ties.
a. How many different outfits are possible?
b. Suppose a winter outfit includes a sweater and he can select one of 7 sweaters. Now how many different winter outfits are possible?

4.77 A disc jockey has 20 songs to choose from but can play only 7 in the next half-hour. How many different playlists are possible?

4.78 A grocery store has 6 cashiers on duty, 10 baggers, and 4 people who will help customers load their groceries into a car. How many different checkout crews are possible?

4.79 A television station is developing a new identifying three-note theme. How many different three-note themes are possible if there are 20 notes to choose from and no note can be repeated?

4.80 A small basket contains 17 good apples and 3 rotten apples.
a. How many different handfuls of six apples are possible?
b. How many different handfuls of five good apples and one rotten apple are possible?
c. How many different handfuls of three good apples and three rotten apples are possible?

Applications

4.81 Manufacturing and Product Development   Suppose Target sells a *combination* lock, which is really a *permutation* lock, with 40 numbers, 0 to 39. The combination for each lock is set at the factory and consists of three numbers.
a. How many lock combinations are possible if numbers can be repeated?
b. If all lock combinations are equally likely, what is the probability of selecting a lock with only single-digit numbers in the combination?
c. Answer parts (a) and (b) if the lock combination must be three different numbers.

 4.82 Public Policy and Political Science   Suppose 14 carpenters report to the union hall hoping for a chance to

work. Three of the 14 do not have their union card, and 6 carpenters will be selected at random for construction jobs.
a. What is the probability that all six carpenters selected will have their union card? Write a Solution Trail for this problem.
b. What is the probability that exactly one carpenter selected will not have a union card?
c. What is the probability that at least one carpenter selected will not have a union card?

4.83 Fuel Consumption and Cars   Suppose Geico offers automobile insurance with specific levels of coverage according to the table below.

Coverage levels
Medical: $10,000; $20,000; $50,000; $100,000
Bodily injury liability: $50,000; $100,000
Property damage liability: $25,000; $50,000; $100,000
Uninsured motorists: $50,000; $100,000; $200,000
Comprehensive: $250,000; $500,000; $1,000,000

Suppose an automobile policy must have all five coverages.
a. How many different automobile policies are possible?
b. How many policies have comprehensive coverage of at least $500,000?
c. How many policies have bodily injury liability and property damage liability of $100,000?

4.84 Manufacturing and Product Development   eBags offers backpacks in 5 styles, 3 sizes, and 10 colors.
a. How many different backpacks does the company offer?
b. Midnight blue and dark green are the two most popular colors. How many different backpacks in these colors does the company offer?
c. The urban-style backpack is the least popular. If the company eliminates this style, how many different backpacks will it offer?

4.85 Fuel Consumption and Cars   A small tool-and-die shop manufactures kneuter valves. A shipment of 15 valves to a Swedish automobile assembly plant contains three defective values. Suppose the assembly plant randomly selects four valves from the shipment.
a. What is the probability that all four valves will be defect-free?
b. What is the probability that the plant will select all three defectives?
c. What is the probability that the plant will select at least one defective?

4.86 Medicine and Clinical Studies   A physician routinely visits a local nursing home on Thursday mornings to examine patients. Suppose the facility has 20 residents, but the physician only has time to check 8. The supervisor places 8 random patients on an ordered list and presents the schedule to the physician.
a. How many different schedules are possible?
b. If there are 15 women and 5 men in the facility, what is the probability that all appointments will be with women?

4.87 Psychology and Human Behavior   A telemarketer has 12 people on his contact list. Suppose he will randomly select 8 people to call during the next shift.
   a. How many different calling schedules are possible?
   b. Suppose only 2 of the 12 will definitely purchase the product when contacted. What is the probability that these 2 people will be the first 2 called?
   c. Suppose another 2 of the 12 will ask to be placed on the do-not-call list when contacted. What is the probability that these 2 people will not be called?

4.88 Sports and Leisure   In preparation for the coming season, a bass fisherman decides to buy 5 random lures out of the 10 new ones in the local tackle shop.
   a. How many different collections of 5 new lures are possible?
   b. Suppose 1 of the 10 lures is a Crazy Crawler. What is the probability that the fisherman will not select this lure? Write a Solution Trail for this problem.
   c. Suppose 3 of the 10 are Excalibur lures. What is the probability that at least 1 of the 5 selected will be an Excalibur lure?

4.89 Classic Vinyl   A music collector has 15 unopened classic rock albums in her collection. Suppose she decides to select three to sell at an upcoming auction.
   a. How many different ways are there to select three albums from her collection?
   b. Suppose the albums are selected at random and five are by the Beatles. What is the probability that all three selected will be by the Beatles?
   c. What is the probability that none of the three will be by the Beatles?
   d. Suppose that one album is by the Doors. What is the probability that this album is not selected?

4.90 Business and Management   The Gagosian art gallery in New York City has 20 stored paintings but has just made room to display several of them. Seven paintings will be randomly selected and offered to the public for sale.
   a. How many different collections of 7 paintings are possible?
   b. Suppose 10 of the 20 stored works are by the same local artist. What is the probability that all 7 of the selected paintings will be by this artist?
   c. The featured room in the gallery receives the most attention, and the order in which the paintings are displayed in this room is related to buyer interest. Suppose the 7 selected paintings will be placed in this featured room. How many different arrangements are possible?

4.91 Economics and Finance   The purchasing agent for a state office building placed a call for bids on replacing the entry doors. Suppose that eight sealed bids are received by the deadline. The bids will be opened in random order.
   a. In how many different ways can the bids be opened?
   b. What is the probability that the lowest bid will be opened last?

4.92 Marketing and Consumer Behavior   In remodeling a kitchen, a builder decides to place a splashguard behind the sink consisting of 8 six-inch-square ceramic tiles decorated with different botanical herbs. The tiles will be installed in a custom-made wooden panel. The tile supplier has 12 different herb designs to choose from, and the builder selects 8 of these 12 at random. Suppose the order in which the tiles are arranged on the splashguard does not matter.
   a. Two of the 12 herb tiles contain a blue tint that matches the kitchen color scheme. What is the probability that these 2 tiles will be included in the splashguard?
   b. The family actually grows 5 of the 12 herbs in a backyard garden. What is the probability that all 5 of these will be included on the splashguard?

4.93 Travel and Transportation   A PennDOT road line-painting crew consists of a foreman, a driver, and a painter. Suppose a supervisor is preparing the schedule to paint lines on roads in Johnstown and 10 foremen, 15 drivers, and 17 painters are available.
   a. How many different crews are possible?
   b. Suppose the crews are selected at random, and there is one foreman who has a severe personality conflict with one driver. What is the probability that neither of these individuals will be on the road painting crew?
   c. Eight of the painters have been cited by a supervisor for improper painting. What is the probability that the crew will include one of these painters?

4.94 Education and Child Development   A university library is preparing a display case of books written by faculty members. There are 25 new faculty books, but there is room for only 10 in the display case. Suppose 10 books are selected at random.
   a. How many different faculty book collections can be displayed?
   b. If 15 of the new books are written by faculty members from the College of Science and Technology, what is the probability that all 10 displayed books are written by faculty members from this college?
   c. If none of the 10 displayed books is written by faculty members from the College of Science and Technology, is there any evidence to suggest the selection process was not random? Justify your answer.

4.95 Psychology and Human Behavior   In a family with five children, two of the five are selected at random each evening to do the dishes. The first one selected washes, and the second one dries.
   a. How many different wash–dry crews are possible?
   b. Suppose there are two girls and three boys in the family. If the two girls are selected to wash and dry, is there any evidence to suggest the selection process was not random? Justify your answer.

Extended Applications

4.96 Manufacturing and Product Development   A remote-control garage door opener has a series of 10 two-position (0 or 1) switches used to set the access code. The code is initially set at the factory, and the switch sequence on

the remote control and the opener must match in order to use the system.
  a. How many different access codes are possible?
  b. If all access codes are equally likely, what is the probability that a randomly selected system will have a code with exactly one 0?
  c. To increase security and ensure that customers will have different access codes, new systems have 10 three-position switches (0, 1, or 2). Answer parts (a) and (b) using the new system.

4.97 Psychology and Human Behavior   An annual family picture following Thanksgiving dinner is arranged with all 10 family members in a row in front of a fireplace.
  a. How many different arrangements of family members are possible?
  b. Suppose the family includes one set of twins, and all arrangements are equally likely. What is the probability that the twins will be in the middle two places (positions 5 and 6)?
  c. What is the probability that the twins will be side by side in the picture?
  d. Suppose the family includes five males and five females. What is the probability that the picture arrangement will alternate male, female, male, female, etc., or female, male, female, male, etc.?

4.98 Public Policy and Political Science   A special committee on community development has four members from the town council. The full town council has 14 members, six Democrats and eight Republicans.
  a. How many different committees on community development are possible?
  b. Suppose the committee members are selected at random. What is the probability of a committee consisting of all Republicans?
  c. Suppose every member of the committee selected is a Democrat. Do you believe the selection process was random? Justify your answer.

4.99 Sports and Leisure   Texas hold 'em poker has become very popular in gambling casinos and is seen on ESPN and the Travel Channel. In November 2012, Greg Merson won the World Series of Poker in Las Vegas and a cool $8.53 million. The game is played with a standard 52-card deck, and starts with each player being dealt two (random) cards face down (hole cards). There is a round of betting, the dealer then flips three cards face up (the flop), betting, one card is flipped (the turn), betting, a fifth card is flipped (the river), and more betting. Let's focus on the two hole cards, called a (pre-flop) *hand*, in this problem.
  a. How many (two-card, pre-flop) hands are possible in Texas hold 'em?
  b. What is the probability that a pre-flop hand consists of two aces?
  c. What is the probability that a pre-flop hand consists of a pair, that is, two cards of the same rank?
  d. What is the probability that a pre-flop hand consists of two cards of the same suit?

Challenge

4.100 The Complement Rule   Reconsider Example 4.22. Verify the probability in part (c) using the complement rule.

4.101 Combination Patterns   Find the following sums.
  a. $\binom{2}{0} + \binom{2}{1} + \binom{2}{2}$
  b. $\binom{3}{0} + \binom{3}{1} + \binom{3}{2} + \binom{3}{3}$
  c. $\binom{4}{0} + \binom{4}{1} + \binom{4}{2} + \binom{4}{3} + \binom{4}{4}$
  d. $\binom{n}{0} + \binom{n}{1} + \binom{n}{2} + \cdots + \binom{n}{n}$

4.102 Sports and Leisure   Consider a regular deck of 52 playing cards. For a five-card poker hand, find the probability of:
  a. One pair.
  b. Two pairs.
  c. Three of a kind: three cards of the same rank and two others of different ranks, for example, JJJ74.
  d. A straight: five cards in sequence; the ace can be either high or low.
  e. A flush: five cards of the same suit.

4.103 Psychology and Human Behavior   How many different ways are there to arrange n people at a round table? (*Hint*: A simple rotation of a seating plan, shifting each person around the table but keeping the order the same, is not a different arrangement.)

4.104 Travel and Transportation   Suppose there are n items of which n_1 are of one type, n_2 are of a second type, . . . , and n_k are of the kth type, and $n_1 + n_2 + \cdots + n_k = n$. The number of unordered arrangements of the n items is a *generalized combination* given by

$$\binom{n}{n_1\, n_2 \cdots n_k} = \frac{n!}{n_1!\, n_2! \cdots n_k}$$

(Think about arranging a string of colored Christmas tree lights.) Suppose the Amtrak Auto Train from Washington, DC, to Florida has 10 sleeper cars, 2 diner cars, and 14 car carriers. Discounting the engine and caboose, how many *different* arrangements of cars in the train are there?

4.105 Public Policy and Political Science   The U.S. Senate Committee on Homeland Security and Governmental Affairs has 16 members. The full Senate has 53 Democrats, 45 Republicans, and 2 Independents.
  a. How many different 16-member Senate committees are possible?
  b. If the committee members are selected at random, what is the probability of a committee consisting of all Democrats?
  c. What is the probability that the committee consists of 14 Democrats and 2 Independents?

4.4 Conditional Probability

The probability questions we have considered so far have all been examples of *unconditional* probability. No special conditions were imposed, nor was any extra information given. However, sometimes two events are related so that the probability of one *depends* on whether the other has occurred. In this case, knowing something extra may affect the probability assignment. This type of situation usually involves two events. The *extra* information may be expressed as an event separate from the event whose probability is desired.

Example 4.27 Morning Commute

Consider a banker who commutes 30 miles to work every day. Because of several factors (weather, road construction, family obligations, etc.), the probability that she makes it to work on time on any random day is 0.5. If the event T is

T = the banker makes it to work on time,

then $P(T) = 0.5$. This is an *unconditional* probability statement: No extra information related to the event T is known or given.

Suppose a random day is selected, and the road conditions are terrible because of a snowstorm. The probability that the banker arrives at work on time is surely lower, perhaps around 0.1. Knowing the extra information (a snowstorm) changes the probability assignment for T.

The statement, "What is the probability that the banker arrives at work on time if it is snowing?" is a *conditional* probability question. The extra information is that it's snowing outside. If the event F is defined as

F = a snowstorm,

The vertical bar, |, in the probability statement is read as "given."

then this conditional probability is written as $P(T \mid F) = 0.1$; the probability that the banker arrives at work on time, *given* that it is snowing, is 0.1.

Suppose another random day is selected, but this time the banker wakes up before the alarm goes off and leaves the house early. The probability that she makes it to work on time is certainly higher, say, close to 0.95. Once again, knowing some extra information changes the probability assignment for T. If the event E is

E = the banker leaves her house early,

then $P(T \mid E) = 0.95$.   ●

Knowing something extra *may* change the probability assignment. How do we use any added information to compute the (possibly) new probability? Consider the next example.

Example 4.28 Roll the Die

Consider an experiment in which a fair, six-sided die is rolled and the number landing face up is recorded. The sample space is $S = \{1, 2, 3, 4, 5, 6\}$. Consider the following events:

$$A = \{1\} = \text{roll a 1} \quad \text{and} \quad B = \{1, 3, 5\} = \text{roll an odd number.}$$

Finding $P(A)$ is an unconditional probability question because no extra information is known. Because all of the outcomes in the experiment are equally likely, and there is one outcome in A and there are six outcomes in the sample space, $P(A) = 1/6$.

Suppose someone rolls the die, covers it with her hands, peeks at the number, and reports, "I rolled an odd number." With this added information, the probability of a 1 is now $P(A \mid B) = \frac{1}{3}$. This conditional probability is reasonable because now we only have to consider three possibilities—that is, we have reduced the sample space from six outcomes to three, and the number of outcomes in A is 1. ■

The idea of reducing, or shrinking, the sample space is key to calculating conditional probabilities. The definition of **conditional probability**, and some justification for it, are given below.

Definition

What goes wrong with this definition if $P(B) = 0$?

Suppose A and B are events with $P(B) > 0$. The **conditional probability of the event A given that the event B has occurred**, $P(A \mid B)$, is

$$P(A \mid B) = \frac{P(A \cap B)}{P(B)}$$

A CLOSER LOOK

1. The unconditional probability of an event A can be written as

$$P(A) = \frac{P(A)}{1} = \frac{P(A)}{P(S)} = \frac{\text{probability of the event } A}{\text{probability of the relevant sample space}}$$

We use this same reasoning to find $P(A \mid B)$.

2. Given that B has occurred, the relevant sample space has changed. It is *reduced* from S to B. (See Figure 4.18.)

3. Given that B has occurred, the only way A can occur is if $A \cap B$ has occurred, because the sample space has been reduced to B.

4. $P(A \mid B)$ is the probability that A has occurred, $P(A \cap B)$, divided by the probability of the relevant sample space $P(B)$.

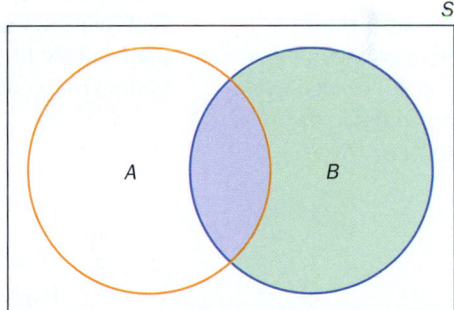

Figure 4.18 An illustration for calculating conditional probability.

In the following example, the formula for conditional probability is used to confirm our intuitive answer to the die problem above.

Example 4.29 Roll the Die (Continued)

The experiment consists of rolling a fair, six-sided die and recording the number that lands face up. $S = \{1, 2, 3, 4, 5, 6\}$. Consider the following events:

$A = \{1\} = $ roll a 1   and $B = \{1, 3, 5\} = $ roll an odd number.

Find $P(A \mid B)$, the probability of rolling a one given that an odd number was rolled.

SOLUTION

STEP 1 We will need the following probabilities:

$$P(B) = 3/6 \text{ and } P(A \cap B) = P(1) = 1/6.$$

STEP 2 $P(A \mid B) = \dfrac{P(A \cap B)}{P(B)} = \dfrac{1/6}{3/6} = \dfrac{1}{6} \cdot \dfrac{6}{3} = \dfrac{1}{3}$

This answer agrees with the intuitive answer above (thank goodness).

A CLOSER LOOK

Here are some facts about union, intersection, and conditional probability to help translate and solve many of the problems that follow.

1. $P(A \cup B) = P(B \cup A)$.

   This is always true, because $A \cup B = B \cup A$ (all the outcomes in A or B or both).

2. $P(A \cap B) = P(B \cap A)$.

   This is also always true, because $A \cap B = B \cap A$.

When are these two conditional probabilities equal?

3. $P(A \mid B) \neq P(B \mid A)$.

   These two probabilities *could* be equal, but in general they are different.

   It's all right to switch A and B with union and intersection, but *not* with conditional probability.

4. The keywords *given* and *suppose* often signal partial information and, therefore, indicate a conditional probability question.

Example 4.30 Do You Have a Reservation?

You can also think of this table as representing all of the simple events in an equally likely outcome experiment. For example, let the outcome HA mean a person rated the prices high and the food 1 star. The probability of HA is the number of outcomes in HA divided by the number of outcomes in the sample space: N(HA)/N(S) = 25/510.

The Zagat Survey, started in 1979 by two Yale-educated lawyers, invites diners to rate and review restaurants. The first survey included only New York City restaurants, but the company now offers dining guides to thousands of restaurants worldwide.[12] Suppose Zagat asked 510 people selected at random to rate Charlie Trotter's Restaurant in Chicago according to price (low, medium, or high) and food (1, 2, 3, or 4 stars). The results of this survey are presented in the **two-way**, or **contingency**, **table** below. The numbers in this table represent frequencies. For example, in the third row and fourth column, 30 people rated the prices high and the food 4 stars. The last column contains the sum for each row, and similarly, the bottom row contains the sum for each column. These sums are often called *marginal totals*.

		Food rating				
		1 star (A)	2 stars (B)	3 stars (C)	4 stars (D)	
Price	Low (L)	20	35	90	15	160
	Medium (M)	50	80	95	25	250
	High (H)	25	5	40	30	100
		95	120	225	70	510

Assume that these results are representative of the entire population of Chicago, so the relative frequency of occurrence is the true probability of the event. A person from Chicago is randomly selected.

a. Find the probability that the person rates the prices medium.

b. Find the probability that the person rates the food 2 stars.

c. Suppose the person selected rates the prices high. What is the probability that he rates the restaurants 1 star?

d. Suppose the person selected does not rate the food 4 stars. What is the probability that she rates the prices high?

SOLUTION

STEP 1 This is an unconditional probability question, asking only about the event M. Compute the relative frequency of occurrence of M, that is, the proportion of responses that rated the restaurant medium priced.

$$P(M) = \frac{50 + 80 + 95 + 25}{510} = \frac{250}{510} = 0.4902$$

STEP 2 This is just another unconditional probability question. Find the relative frequency of occurrence of 2 stars.

$$P(B) = \frac{35 + 80 + 5}{510} = \frac{120}{510} = 0.2353$$

STEP 3 This is a conditional probability question; the key word is *suppose*. The *given* information is *rates prices high*. We need the probability of the event A given that the event H has occurred. Using the formula for conditional probability,

$$P(A \mid H) = \frac{P(A \cap H)}{P(H)} = \frac{25/510}{100/510} = \frac{25}{510} \cdot \frac{510}{100} = \frac{25}{100} = 0.2500$$

This probability can also be obtained directly by *reducing* the sample space in the two-way table. The shaded row is the reduced sample space.

	1 star (A)	2 stars (B)	3 stars (C)	4 stars (D)	
Low (L)	20	35	90	15	160
Medium (M)	50	80	95	25	250
High (H)	25	5	40	30	100
	95	120	225	70	510

In the reduced sample space, 25 outcomes are in the event A. Therefore,

$$P(A \mid H) = \frac{\text{number of outcomes in } A \text{ and in the reduced sample space}}{\text{number of outcomes in the reduced sample space}}$$

$$= \frac{25}{100} = 0.2500$$

STEP 4 Solve this conditional probability question by reducing the sample space via the two-way table.

	1 star (A)	2 stars (B)	3 stars (C)	4 stars (D)	
Low (L)	20	35	90	15	160
Medium (M)	50	80	95	25	250
High (H)	25	5	40	30	100
	95	120	225	70	510

There are 440 ($= 95 + 120 + 225$) outcomes in the reduced sample space, and 70 ($= 25 + 5 + 40$) people rated the prices high.

$$P(H \mid D') = \frac{70}{440} = 0.1591$$

TRY IT NOW GO TO EXERCISE 4.124

Example 4.31 The Changing Labor Force

Over the past several decades, the nature of the U.S. labor force has changed dramatically. More women are searching for jobs, more men are staying home with children, and senior citizens are remaining in their jobs longer. According to the U.S. Census Bureau, for married-couple family groups, 84.9% of all fathers are employed; and in 57.5% of these households, both parents are employed.[13] Suppose a married couple family group is selected at random. If the father is employed, what is the probability that the mother is also employed?

SOLUTION

STEP 1 Consider the events:

F = the father is employed.

M = the mother is employed.

The statement of the problem includes two probabilities involving these two events.

$$P(F) = 0.849$$ ⟶ Percentage converted to unconditional probability.

$$P(F \cap M) = 0.575$$ ⟶ The word *both* means intersection.

STEP 2 $P(M \mid F) = \dfrac{P(M \cap F)}{P(F)}$ ⟶ Translated conditional probability; definition.

$$= \dfrac{0.575}{0.849} = 0.6773$$ ⟶ Use known probabilities.

If the father is employed in a married-couple family, the probability that the mother is also employed is 0.6773. ●

TRY IT NOW GO TO EXERCISE 4.122

Solution Trail 4.31

KEYWORDS

- *If* the father is employed
- Probability that the mother is also employed

TRANSLATION

- Given the event *the father is employed*, find the probability of the event *the mother is employed*

CONCEPTS

- Conditional probability

VISION

Use the formula for conditional probability to find $P(M \mid F)$.

The solution here requires careful translation of the words into mathematics.

Steps for Calculating a Conditional Probability

To find the conditional probability of the event A given that the event B has occurred:

a. Calculate $P(B)$ and $P(A \cap B)$.

b. Find $P(A \mid B) = \dfrac{P(A \cap B)}{P(B)}$

Example 4.32 Sex, Marital Status, and the Census

The U.S. Constitution directs the government to conduct a census of the population every 10 years. Population totals are used to allocate congressional seats, electoral votes, and funding for many government programs. The U.S. Census Bureau also compiles information related to income and poverty, living arrangements for children, and marital status. The following *joint probability table* lists the probabilities corresponding to marital status and sex of persons 18 years and over.[14]

		Marital status				
		Married (R)	Never married (N)	Widowed (W)	Divorced or separated (D)	
Sex	Male (M)	0.282	0.147	0.013	0.043	0.485
	Female (F)	0.284	0.121	0.050	0.060	0.515
		0.566	0.268	0.063	0.103	1.000

Suppose a U.S. resident 18 years or older is selected at random.

a. Find the probability that the person is female and widowed.

b. Suppose the person is male. What is the probability that he was never married?

c. Suppose the person is married. What is the probability that the person is female?

The body of the table contains *intersection* probabilities: the probability of a row event *and* a column event. For example, the probability that a person is male and divorced is 0.043, the intersection of the first row and the fourth column. The probabilities obtained by summing across rows or down columns are called *marginal* probabilities. The total probability in the table is 1.000.

SOLUTION

STEP 1 The keyword is *and*, which means intersection. The probability of female (F) and widowed (W) is found by reading the appropriate cell.

	Married (R)	Never married (N)	Widowed (W)	Divorced (D)	
Male (M)	0.282	0.147	0.013	0.043	0.485
Female (F)	0.284	0.121	0.050	0.060	0.515
	0.566	0.268	0.063	0.103	1.000

$$P(F \cap W) = 0.050$$

STEP 2 The keyword is *suppose*. That suggests conditional probability. The extra information is *male*.

$$P(N \mid M) = \frac{P(N \cap M)}{P(M)}$$ Translated conditional probability; definition.

$$= \frac{0.147}{0.485} = 0.303$$ Use known probabilities.

STEP 3 This is another conditional probability. This time, the event R is given.

$$P(F \mid R) = \frac{P(F \cap R)}{P(R)} = \frac{0.284}{0.566} = 0.502$$

TRY IT NOW   GO TO EXERCISE 4.125

A CLOSER LOOK

1. ▶ In Example 4.32,

$$P(R) = P(R \cap M) + P(R \cap F)$$
$$= P(R \cap M) + P(R \cap M')$$

In general, for any two events A and B,

$$P(A) = P(A \cap B) + P(A \cap B')$$

This *decomposition* technique is often needed in order to find $P(A)$. The Venn diagram in Figure 4.19 illustrates this equation.

The events B and B' make up the entire sample space: $S = B \cup B'$.

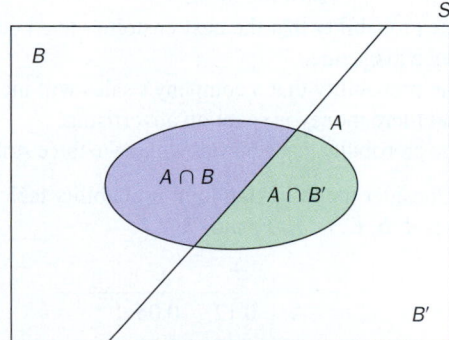

Try to draw the Venn diagram to illustrate this equality.

Figure 4.19 Venn diagram showing decomposition of the event A.

2. Suppose B_1, B_2, and B_3 are mutually exclusive and *exhaustive*: $B_1 \cup B_2 \cup B_3 = S$. For any other event A,

$$P(A) = P(A \cap B_1) + P(A \cap B_2) + P(A \cap B_3)$$ ◀

SECTION 4.5 EXERCISES

Concept Check

4.106 True/False In an unconditional probability statement, no extra relevant information related to the event in question is given.

4.107 True/False Extra information always changes a probability assignment.

4.108 Fill in the Blank In the conditional probability statement $P(A \mid B)$, the relevant sample space is _____.

4.109 True/False
   a. $P(A \cup B) = P(B \cup A)$
   b. $P(A \cap B) = P(B \cap A)$
   c. $P(A \mid B) = P(B \mid A)$

4.110 Short Answer Suppose B_1, B_2, B_3, and B_4 are mutually exclusive and exhaustive events. For *any other* event A, write $P(A)$ as a sum of probabilities involving the events B_1, B_2, B_3, and B_4.

Practice

4.111 Identify each of the following statements as a conditional or unconditional probability question.
   a. The probability that a randomly selected car will start in the morning.
   b. The probability that a person will remember to bring home a loaf of bread after work if he leaves a Post-It note reminder on the steering wheel.
   c. The probability that the next batter will get a hit in a baseball game.
   d. The probability that a randomly selected heart transplant operation will be successful.
   e. Of all one-way streets in a large city, the probability that the street has more than two lanes.

4.112 Identify each of the following statements as a conditional or unconditional probability question.
   a. The probability that a randomly selected circuit board will be defective, given that it was manufactured during the third shift.
   b. The probability that a waitress receives a tip of more than 18% of the cost of the meal.
   c. The probability that the next customer in a bookstore will buy a magazine.
   d. The probability that a company's sales will increase, given that more money is spent on advertising.
   e. The probability that a bowler will make three strikes in a row.

4.113 Consider the following joint probability table describing the events A, B, C, D, E, F, and G.

	F	G
A	0.12	0.05
B	0.15	0.07
C	0.17	0.04
D	0.19	0.02
E	0.11	0.08

a. Verify that this is a valid joint probability table; that is, each probability must be greater than or equal to 0, and the sum of all probabilities must equal 1.
b. Compute the marginal probabilities.
c. Find $P(A \cap F)$, $P(B \cap G)$, and $P(D \cap G)$.
d. Find $P(A \mid G)$, $P(F \mid D)$, and $P(E \mid C)$.
e. Verify that $P(C) = P(C \cap F) + P(C \cap G)$.

4.114 Consider the following joint probability table.

	B_1	B_2	B_3
A_1	0.095	0.016	0.007
A_2	0.205	0.188	0.003
A_3	0.155	0.238	0.093

a. Find $P(A_1)$, $P(A_2)$, and $P(A_3)$.
b. Find $P(B_1)$, $P(B_2)$, and $P(B_3)$.
c. Find $P(A_1 \cap B_1)$, $P(A_2 \cap B_2)$, and $P(A_3 \cap B_3)$.
d. Find $P(A_1 \mid B_1)$, $P(B_1 \mid A_1)$, and $P(A_1' \mid B_1')$.
e. Find $P(B_2 \mid A_2)$, and $P(B_3 \mid A_3)$.

4.115 Consider the following joint probability table.

	C_1	C_2	C_3	
A	0.135	0.125	0.206	0.466
B	0.145	0.174	0.215	0.534
	0.280	0.299	0.421	1.000

a. Find $P(A)$ and $P(C_2)$.
b. Find $P(A \cap C_1)$ and $P(B \cap C_3)$.
c. Find $P(C_2 \mid B)$, $P(A \mid C_3)$, and $P(A \mid C_3')$.
d. Verify that $P(B) = P(B \cap C_1) + P(B \cap C_2) + P(B \cap C_3)$. Carefully sketch a Venn diagram to illustrate this equality.

4.116 A recent survey classified each person according to the following two-way table.

	B_1	B_2	B_3	
A_1	178			815
A_2		150	244	
A_3	165	202		
	466	583	985	

a. Complete the two-way table.
b. How many people participated in this survey?

Assume that the results from this survey are representative of the entire population, and one person from this population is randomly selected.
c. Find $P(A_1)$, $P(A_2)$, and $P(A_3)$.
d. Find $P(B_1 \cap A_1)$, $P(B_2 \cap A_2)$, and $P(B_3 \cap A_3)$.
e. Find $P(A_3 \mid B_1)$, $P(B_2 \mid A_2)$, and $P(A_3 \mid B_1')$.

4.117 Consider an experiment and three events A, B, and C defined in the Venn diagram below.

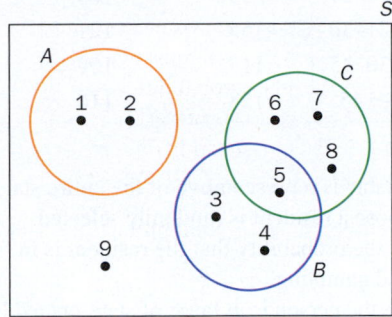

The following table gives the probability of each outcome.

Outcome	1	2	3	4	5
Probability	0.01	0.12	0.11	0.10	0.15

Outcome	6	7	8	9
Probability	0.25	0.14	0.08	0.04

Find the following probabilities.
a. P(A), P(B), and P(C).
   Why don't these three probabilities sum to 1?
b. P($A \cap B$) and P($B \cap C$).
c. P($B \mid C$) and P($C \mid B$).
d. P($A \mid B'$), P($C \mid A'$), and P[$1 \mid (A \cup B)'$].
e. P($3 \mid B$), P($4 \mid B$), and P($5 \mid B$).
   Why do these three probabilities sum to 1?

4.118 Consider an experiment and three events A, B, and C defined in the Venn diagram below.

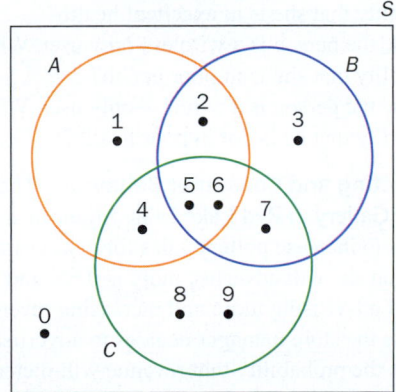

The following table gives the probability of each outcome.

Outcome	0	1	2	3	4
Probability	0.135	0.130	0.142	0.128	0.147

Outcome	5	6	7	8	9
Probability	0.083	0.072	0.063	0.055	0.045

Find the following probabilities.
a. P(A), P(B), and P(C).
b. P($A \cap B$) and P($B \cap C$).
c. P($A \mid B$), P($B \mid C$), and P[$(A \cap B) \mid C$].
d. P($0 \mid C'$), P($7 \mid C$), and P[$(A \cup B) \mid C'$].
e. P($2 \mid B$), P($3 \mid B$), and P($7 \mid B$).

Applications

4.119 Sports and Leisure Consider a regular 52-card deck of playing cards. Suppose two cards are drawn at random from the deck without replacement.
a. What is the probability that the second card is an ace, given that the first card is a king?
b. What is the probability that the second card is an ace, given that the first card is an ace?
c. What is the probability that the second card is a heart, given that the first card is a heart?
d. Suppose two cards are drawn at random from the deck with replacement. What is the probability that the second card is a heart, given that the first card is a heart?

4.120 Economics and Finance In the United States, there is some evidence to suggest a link between people who participate in an office football pool and those who cheat on their income taxes. Suppose 25% of all people participate in an office football pool. The IRS estimates that 15% of all people participate in an office football pool and cheat on their income tax return. Suppose a person is randomly selected. If the person is known to participate in an office football pool, what is the probability that she cheats on her income tax return? Write a Solution Trail for this problem.

4.121 Trail Users According to Trail Count, the annual survey of San Jose's off-street bicycle and pedestrian trail users, approximately 24% use trails daily.[15] Suppose 12% use trails daily and exercise, and 8% use trails daily for commuting. Suppose a trail user is randomly selected.
a. Given that the person uses trails daily, what is the probability that he uses the trails for exercise?
b. Given that the person uses trails daily, what is the probability that she uses the trails for commuting?
c. If the person uses trails daily, what is the probability that he does not use the trial for commuting?

4.122 Travel and Transportation In a particularly rural area in upstate New York, 80% of all people use chains on their car tires (for winter driving), 60% carry a snow shovel in their car and use chains, and 15% carry a shovel but do not use chains. Suppose a person from this area is selected at random.
a. If the person uses chains, what is the probability that he carries a shovel? Write a Solution Trail for this problem.
b. Given that the person does not use chains, what is the probability that she carries a shovel?

4.123 Biology and Environmental Science The Florida Sea Grants Agents, Florida Fish and Wildlife Commission, and volunteer divers work together in the Great Goliath Grouper Count. The table below lists the probability of each size grouper in two regions.[16]

Region		Length < 3 ft	3–5 ft	> 5 ft
	Sarasota	0.059	0.347	0.079
	Monroe	0.030	0.277	0.208

Suppose a grouper from one of these regions is selected at random.

a. Suppose the grouper is 3–5 feet long. What is the probability that it is from Monroe?

b. Suppose the grouper is from Sarasota. What is the probability that the grouper is longer than 5 feet?

c. Suppose the grouper is longer than 3 feet. What is the probability that it is not from Sarasota?

4.124 Sports and Leisure An assistant football coach at a Division II school helps his team prepare for the next opponent by charting plays. He looks at game films and records the down distance (first, second, or third down, and a categorical measure of the number of yards needed for a first down) and type of play (rush or pass), to look for tendencies. The two-way table below shows the number of plays that fall into each category. (Fourth-down plays are not charted, because they usually involve a punt or a field-goal attempt.)

		1st	2nd short	2nd long	3rd short	3rd long
Play	Rush	126	35	46	65	12
	Pass	87	16	67	23	59

Suppose this table represents the true tendencies of the next opponent.

a. What is the probability that the opponent rushes the ball?

b. Suppose the opponent has a first down. What is the probability of a pass?

c. Suppose it is a first or second down. What is the probability that the opponent rushes the ball?

d. Suppose the opponent passes the ball. What is the probability that it is a third down?

4.125 Public Policy and Political Science As a result of decreasing revenue, economic conditions, and deep budget deficits, many states have tried to legalize gambling or, in states where gambling is already legal, expand casino operations. For example, Washington state and Pennsylvania have joined multistate lotteries, several states have raised taxes on receipts from riverboat gambling, and at least four states have had ballot questions about starting new types of games.[17] To measure public opinion in Kansas, a random sample of residents was selected and each response was categorized according to revenue preference and age. The results are given in the following two-way table.

		Gambling	Liquor stores	Other
Age	18–21	33	68	12
	21–30	55	121	50
	30–45	117	109	132
	≥ 45	158	110	90

Assume this table is representative of the entire state's population and suppose a resident is randomly selected.

a. What is the probability that the resident is in favor of legalized gambling?

b. Suppose the person is in favor of state-owned liquor stores. What is the probability that the person is 30–45 years old?

c. Suppose the person selected is under 21. What is the probability that the person is in favor of some other option?

d. Suppose the resident is not in favor of legalized gambling. What is the probability that the respondent is 21–30 years old?

e. If the person selected is under 21 or at least 45, what is the probability that this resident is in favor of state-owned liquor stores?

4.126 Marketing and Consumer Behavior Because wireless technology has become reliable and prevalent, many people are terminating their landline service. According to a survey by the Centers for Disease Control and Prevention, 35.8% of households have wireless phones only. In addition, 52.5% have a landline with wireless, 9.4% have a landline without wireless, 0.2% have a landline with unknown wireless, and 2.1% are phoneless.[18] Suppose the survey indicated that 22.2% were wireless-only users and in excellent health, 9.3% were wireless-only users and in adequate health, and 4.3% were wireless-only users and in poor health. Suppose a person who completed the survey is selected at random.

a. Suppose the person is a wireless-only user. What is the probability that she is in excellent health?

b. Suppose the person is a wireless-only user. What is the probability that she is in poor health?

c. Suppose the person is a wireless-only user. What is the probability that he is not in poor health?

4.127 Marketing and Consumer Behavior The manager at Of The Land Gallery in Red Lake Falls, Minnesota, is looking at many ways to increase pottery sales for next year. The probability that she will advertise more is 0.65, and the probability of advertising more and increasing revenue is 0.35.

a. Suppose the store manager decides to advertise more. What is the probability that revenue will increase?

b. If the store manager does advertise more, what is the probability that revenue will not increase?

c. What is the probability of not advertising more and revenue increasing?

4.128 Psychology and Human Behavior A random sample of adult drivers was obtained, 1000 men and 900 women. A survey showed that 640 men rely on GPS systems and 450 women rely on them.[19] Suppose a person included in this survey is randomly selected.

a. What is the probability that the person selected is a woman and relies on GPS systems?

b. Suppose the person selected is a man. What is the probability that he relies on a GPS system?

c. Suppose the person selected relies on a GPS system. What is the probability that the person is a woman?

Extended Applications

4.129 Demographics and Population Statistics According to the U.S. Census Bureau, 87.1% of people living in the United States are native-born and 12.9% are foreign-born. In addition, 10.16% are native-born and had no health insurance during the last year, and 3.78% are foreign-born and had no health insurance during the last year.[20] Suppose a person living in the United States is selected at random.

a. Suppose the person is native-born. What is the probability that he had no health insurance during the last year?

b. Suppose the person is foreign-born. What is the probability that she had no health insurance during the last year?

c. Suppose the person is foreign-born. What is the probability that he had health insurance during the last year?

4.130 Biology and Environmental Science Homeowners who cultivate small backyard gardens are often worried about pests (for example, rabbits and groundhogs) ruining plants. Some gardeners protect their gardens with a fence, others spread chemicals around the perimeter of the garden to keep animals away, and some do nothing. The joint probability table below shows the relationships among these garden protection methods and success.

		Garden defense		
		Fence	Chemicals	Nothing
Result	Pests	0.05	0.08	0.34
	No pests	0.30	0.20	0.03

Suppose a backyard gardener is selected at random.

a. Suppose the garden had pests. What is the probability the gardener used nothing?

b. Suppose the gardener used chemicals. What is the probability there were pests?

c. Given that the garden had no pests, which method of defense did the gardener most likely use? Justify your answer.

4.131 Business and Management Bank of America, Time Warner Cable, and Delta Airlines are some of the companies ranked worst in customer service according to the American Consumer Satisfaction Index.[21] Suppose the table below represents the results of a survey of customer satisfaction for Comcast, another company on the Worst Customer Service list.

		Customer service			
		Excellent	Good	Fair	Poor
Region	North	0.102	0.059	0.062	0.004
	Midwest	0.105	0.105	0.144	0.007
	South	0.075	0.084	0.213	0.040

Suppose this table is representative of all Comcast user customer satisfaction and a customer is selected at random.

a. What is the probability that the customer is from the Midwest and customer service is fair?

b. If the customer is from the North, what is the probability that customer service is excellent?

c. Suppose customer service is poor. What is the probability that the customer is from the South?

d. If the customer service is good or fair, which region is the customer most likely from? Justify your answer.

4.132 Psychology and Human Behavior The following partial two-way table lists the number of adult criminal cases in Canada by case type and sentence.[22]

		Type of sentence			
		Custody	Conditional sentence	Probation	
Criminal code	Crimes against the person	16,067	2,465		55,006
	Property crimes	22,178		34,353	60,456
	Administration of justice	28,186	1,460	20,378	50,024
	Other criminal code offenses		456	5,708	10,288
	Criminal code offenses (traffic)	7,387	751	7,141	15,279
	Other federal statute	7,980	2,732	10,534	21,246
		85,922		114,588	

a. Complete the table.

b. Suppose the case was a crime against the person. What is the probability that it resulted in custody?

c. Suppose the case results in probation. What is the probability that it was a conviction for a criminal code offense (traffic)?

d. Suppose the case was a crime against the person. What is the probability that it resulted in custody?

4.133 Public Health and Nutrition The McPherson Middle School in Clyde, Ohio, is set to examine its school lunch program. A survey of 2200 students asked students about their lunch type and how they got to school in the morning. The following (partial) two-way table is assumed to represent the entire student body.

		Arrival mode			
		Bus	Car	Walk	
Lunch	Carries		466	142	
	Buys	345		500	967
		970			

a. Complete the two-way table.

b. Suppose a student at the school is randomly selected. What is the probability that the student carries a lunch and gets to school by car?

c. Suppose the student takes the bus to school. What is the probability that the student buys lunch?

d. Suppose the student does not walk to school. What is the probability that the student carries a lunch?

e. If the student buys lunch, how did he or she most likely get to school?

4.134 Medicine and Clinical Studies   In the movie *A Christmas Story*, Ralph "Ralphie" Parker wanted an official Red Ryder carbine-action 200-shot range model BB gun (with a compass in the stock). Everyone in the movie (including Santa) tells Ralphie he will shoot his eye out with this present. According to the Centers for Disease Control and Prevention, approximately 30,000 people visited an Emergency Room last year for BB gun accidents. The most common injuries were to the face, head, neck, and eye.[23] In a survey of BB– and pellet gun–related injuries treated at hospitals, each injury was classified by primary body part injured and victim–shooter relationship. The results are given in the table below.

		Victim-shooter relationship				
	Self	Friend/ acquaintance	Relative	Stranger	Other/ shooter not seen	Not stated
Extremity	200	128	89	17	25	189
Trunk	57	40	21	5	10	58
Face	51	35	20	4	14	49
Head/neck	40	34	19	6	12	44
Eye	23	20	9	3	7	26
Other	2	3	3	2	1	5

(Body part injured)

Suppose this table is representative of the entire population of BB– and pellet gun–related injuries, and a person seen in an Emergency Room suffering from this type of injury is selected at random.

a. What is the probability that the injury is to the eye and the shooter is a relative?

b. Suppose the injury was caused by a stranger. What is the probability that the body part injured is an extremity?

c. Suppose the injury is to the head/neck. What is the probability that the shooter is a friend/acquaintance?

d. Suppose the injury is not to the eye. What is the probability that the shooter is a relative?

Challenge

4.135 Public Policy and Political Science   A survey of voters in a certain district asked if they favored a return to stronger isolationism. The following *three-way* table classifies each response by sex, political party, and response.

	Male			Female		
	Dem	Rep	Ind	Dem	Rep	Ind
Yes	202	126	105	234	101	95
No	124	288	85	312	66	150

Suppose a random voter is selected from this district.

a. What is the probability that the voter is in favor of isolationism, a female, and a Republican?

b. What is the probability that the voter is not in favor of isolationism?

c. Suppose the voter is female. What is the probability that she is a Democrat?

d. Suppose the voter is not in favor of isolationism. What is the probability that the voter is a Republican and male?

e. Suppose the voter is not an Independent. What is the probability that he or she is in favor of isolationism?

4.5 Independence

If extra information is given, sometimes we simply say, "So what?"

In the last section we learned about conditional probability, that is, how knowing extra information *may* change a probability assignment. Often, however, additional information has no effect on the probability assignment. Consider the following examples.

EXAMPLE 4.33 The Common Cold

Hundreds of different viruses can cause the common cold. Many people are able to develop a resistance to some of these viruses, but they may still contract a cold from a different virus. Catching a cold is *not* related to cold temperatures or bad weather, exercise, diet, or enlarged tonsils or adenoids.[24]

Let the event C = catching a cold. Suppose the (unconditional) probability that a certain person contracts a cold this winter is 0.45: P(C) = 0.45.

If this person decides to exercise more this winter, the *cold* facts above mean this extra exercise has no effect on contracting a cold. What is the probability that this person contracts a cold this winter given the event E = they exercise more?

$$P(C \mid E) = P(C) = 0.45.$$

lightwavemedia/Shutterslock

Knowing extra information here does not change the conditional probability assignment. Intuitively, the events C (contracting a cold) and E (exercising more) are unrelated, or *independent*.

Example 4.34 No Purchase Necessary

There are lots of sweepstakes in which a consumer is automatically entered by making a purchase. However, almost all sweepstakes entry rules explain that there is "No purchase necessary to enter." A person may make a purchase to enter the sweepstakes, or instead enter by mailing a postcard or completing an online form. If this statement in the rules is true, then making a purchase *cannot* change the probability of winning the sweepstakes.

Suppose the event $A =$ winning the sweepstakes, and the event $B =$ making a purchase.

$$P(A \mid B) = P(A) \quad \text{and} \quad P(A \mid B') = P(A).$$

Whether or not you make a purchase has no effect on the chance of your entry being the winner. The events winning the sweepstakes and making a purchase are *independent*.

In these two examples, the occurrence or nonoccurrence of one event has no effect on the occurrence of the other. In this case, the two events are **independent**.

Definition

Two events A and B are **independent** if and only if

$$P(A \mid B) = P(A)$$

If A and B are *not* independent, they are said to be **dependent** events.

A CLOSER LOOK

One way to verify independent events: Is $P(A \mid B) = P(A)$? If so, then A and B are independent; if not, they are dependent.

1. If we know the events A and B are independent, then

   $$P(A \mid B) = P(A) \quad \text{and} \quad P(B \mid A) = P(B).$$

   Similarly, if either one of these equations is true, then the other is also true, and the events are independent.

2. If A and B are independent events, then so are all combinations of these two events and their complements.

   Mathematical translation: If $P(A \mid B) = P(A)$, then $P(A \mid B') = P(A)$, $P(A' \mid B) = P(A')$, and $P(A' \mid B') = P(A')$.

3. Unfortunately, independent events *cannot* be shown on a Venn diagram. In problems that involve independent events, we'll have to translate the words into a probability question and then use an appropriate formula.

4. It is reasonable to think of independent events as unrelated. One might conclude that they are therefore disjoint. This is not true!

   Suppose A and B are mutually exclusive and $P(A) \neq 0$ (there is some positive probability associated with the event A). Then

   $$P(A \mid B) = 0 \neq P(A)$$

   The probability of A given B has to be 0, because A and B are disjoint. Once B occurs, A cannot occur. Hence, disjoint events are **dependent**.

In Section 4.4, we learned the formula for finding conditional probability:

$$P(A \mid B) = \frac{P(A \cap B)}{P(B)} \quad \text{or} \quad P(B \mid A) = \frac{P(A \cap B)}{P(A)}$$

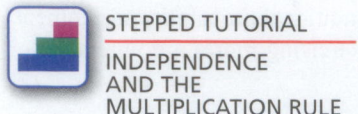

We can solve both of these equations for $P(A \cap B)$ to obtain the following **probability multiplication rule.**

The Probability Multiplication Rule

For any two events A and B,

$$P(A \cap B) = P(B) \cdot P(A \mid B)$$
$$= P(A) \cdot P(B \mid A) \quad \text{Always true.}$$
$$= P(A) \cdot P(B) \qquad \textbf{Only true if } A \textbf{ and } B \textbf{ are independent.}$$

A CLOSER LOOK

1. The real skill in applying this rule is knowing which equality to use. The first two equalities are *always* true. Use one of these only if A and B are dependent and you need to find $P(A \cap B)$. Read the problem carefully to determine which conditional and unconditional probabilities are given.

   If A and B are independent, use the third equality to compute the probability of intersection. The word *independent* will not always appear in the problem. It may be implied or can be inferred from the type of experiment described.

2. If events are dependent, a *modified tree diagram* can be used to apply the probability multiplication rule. In Figure 4.20 the probability of *traveling* along any branch is written along the appropriate leg. Second-generation branch probabilities are conditional. For example, $P(C \mid A)$ (the probability of C given A) is the probability of taking path C, given path A.

On this road map, to determine a final probability, we *multiply* probabilities along the way.

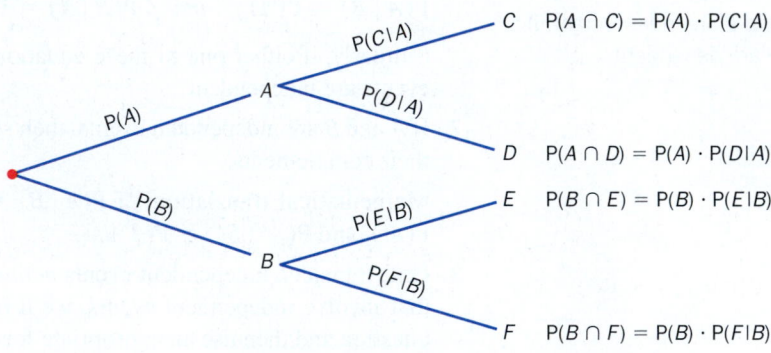

Figure 4.20 The probability multiplication rule on a tree diagram.

All probabilities coming from a single node must sum to 1. To find the probability of traveling along a complete path from left to right (equivalent to the probability of an intersection), we *multiply* probabilities along the path.

A modified tree diagram is useful here also. Try drawing one to illustrate this extended rule.

3. The probability multiplication rule can be extended. For any three events A, B, and C:
$$P(A \cap B \cap C) = P(A) \cdot P(B \mid A) \cdot P(C \mid A \cap B).$$

4. If the events $A_1, A_2, \ldots, A_k$ are mutually independent, then $P(A_1 \cap A_2 \cap \cdots \cap A_k) = P(A_1) \cdot P(A_2) \cdots P(A_k).$

   In words, if the events are mutually independent, the probability of an intersection is the product of the corresponding probabilities.

There are lots of probability rules, formulas, and diagrams in this section. Here are some examples (along with Solution Trails to help you translate the words into mathematics) to illustrate these concepts.

Example 4.35 Mobile Shoppers

Cyber Monday is a huge online shopping day. More people are now using their mobile devices, smart phones or tablets, to shop and make purchases on this Monday after Thanksgiving. Approximately 18% of Cyber Monday shoppers used their mobile device to look for deals.[25] If a person used a mobile device to shop, the probability of making a purchase was 0.45. Suppose a Cyber Monday shopper is selected at random. What is the probability that the shopper used a mobile device and made a purchase?

SOLUTION

STEP 1 Define the following events:

V = used a mobile device;

M = made a purchase.

We are given that the probability the person used a mobile device is 0.18, that is, $P(V) = 0.18$. In addition, we are told that the probability the shopper made a purchase if he used a mobile device is 0.45. This is a conditional probability statement that can be written $P(M \mid V) = 0.45$.

STEP 2 $P(V \cap M) = P(V) \cdot P(M \mid V)$ Probability Multiplication Rule.

$\qquad = (0.18)(0.45) = 0.081$ Use the given probabilities.

Note: Using the probability multiplication rule, we can also write:

$P(V \cap M) = P(M) \cdot P(V \mid M)$

This is a correct application of the rule, but it doesn't help in this problem, because the probabilities on the right-hand side are not given. An accurate but inappropriate use of the probability multiplication rule is evident because the probabilities given in the problem and in the equality are mismatched. Simply try the other equality.

TRY IT NOW GO TO EXERCISE 4.154

Example 4.36 It's Made for Sleep

Better Bedding in East Hartford, Connecticut, claims that 99.4% of all its mattress deliveries are on time. Suppose two mattress deliveries are selected at random.

a. What is the probability that both mattresses will be delivered on time?

b. What is the probability that both mattresses will be delivered late?

c. What is the probability that exactly one mattress will be delivered on time?

SOLUTION

a. Let M_i = mattress i is delivered on time; $P(M_i) = 0.994$ (given).

Both mattresses delivered on time means mattress 1 is on time *and* mattress 2 is on time.

$P(M_1 \cap M_2) = P(M_1) \cdot P(M_2)$ Both mattresses on time; independent events.

$\qquad = (0.994)(0.994)$ Probability of each mattress delivered on time.

$\qquad = 0.988036$

The probability that both mattresses will be delivered on time is 0.9880.

Solution Trail 4.35

KEYWORDS

- Used their mobile device *and* made a purchase

TRANSLATION

- What is the probability of the intersection of the event *used a mobile device* and the event *made a purchase?*

CONCEPTS

- Probability multiplication rule

VISION

Determine whether the events are independent, determine which probabilities are given, and use the appropriate form of the probability multiplication rule.

Solution Trail 4.36

KEYWORDS

- Both mattresses will be delivered on time

TRANSLATION

- What is the probability of the intersection of the event *mattress 1 is delivered on time* and the event *mattress 2 is delivered on time?*

CONCEPTS

- Probability multiplication rule

VISION

Determine whether the events are independent, determine which probabilities are given, and use the appropriate form of the probability multiplication rule.

b. Both mattresses delivered late means mattress 1 is delivered late *and* mattress 2 is delivered late. Delivered late is the complement of delivered on time.

$$P(M_1' \cap M_2') = P(M_1') \cdot P(M_2') \qquad \text{Independent events.}$$
$$= [1 - P(M_1)] \cdot [1 - P(M_2)] \qquad \text{Complement rule.}$$
$$= (0.006)(0.006)$$
$$= 0.000036$$

The probability that both mattresses will be delivered late is 0.000036.

c. Exactly one mattress on time means

Mattress 1 is on time *and* mattress 2 is late, *or*

Mattress 1 is late *and* mattress 2 is on time.

$$P(\text{Exactly one mattress is on time}) = P[(M_1 \cap M_2') \cup (M_1' \cup M_2)]$$

We don't usually see this step, but there really is a union of two events in the background. [Notice the *or* separating (a) and (b) above.] These two events are disjoint, and the probability of the union of disjoint events is the sum of the corresponding probabilities.

$$= P(M_1 \cap M_2') + P(M_1' \cap M_2)$$
$$= P(M_1) \cdot P(M_2') + P(M_1') \cdot P(M_2) \qquad \text{Independent events.}$$
$$= (0.994)(0.006) + (0.006)(0.994) \qquad \text{Use known probabilities.}$$
$$= 0.011928$$

The probability that exactly one mattress will be on time is 0.0119.

Note: There is another way to solve part (c). With two mattresses to deliver, one of three things must happen: 0 are on time, 1 is on time, or 2 are on time; and the probabilities of these three events must sum to 1. (Why?) From part (a), P(2 on time) = 0.9880; from part (b), P(0 on time) = 0.000036.

$$P(1 \text{ on time}) = 1 - [P(0 \text{ on time}) + P(2 \text{ on time})] \qquad \text{Complement rule.}$$
$$= 1 - (0.000036 + 0.988036) \qquad \text{Use known probabilities.}$$
$$= 1 - 0.988072$$
$$= 0.011928 \qquad \blacksquare$$

> In Chapter 5, we will convert all (symbolic) outcomes into real numbers, and use the probabilities of experimental outcomes to find the probabilities associated with real numbers.

TRY IT NOW   GO TO EXERCISE 4.151

Example 4.37 Immunizations

Federal health officials have reported that the proportion of children (ages 19 to 35 months) who received a full series of inoculations against vaccine-preventable diseases, including diphtheria, tetanus, measles, and mumps, increased up until 2006, but has stalled since. The CDC reports that 14 states have achieved a vaccination coverage rate of at least 80% for the 4:3:1:3:3:1 series.[26] The probability that a randomly selected toddler in Alabama has received a full set of inoculations is 0.792, for a toddler in Georgia, 0.839, and for a toddler in Utah, 0.711.[27] Suppose a toddler from each state is randomly selected.

a. Find the probability that all three toddlers have received these inoculations.

b. Find the probability that none of the three has received these inoculations.

c. Find the probability that exactly one of the three has received these inoculations.

SOLUTION

a. Define the following three events:

A = toddler A from Alabama has received these inoculations;

G = toddler G from Georgia has received these inoculations; and

U = toddler U from Utah has received these inoculations.

Assume these three events are independent.

$P(A \cap G \cap U)$ *All three* means intersection.

$= P(A) \cdot P(G) \cdot P(U)$ Independent events.

$= (0.792)(0.839)(0.711) = 0.4725$ Use given probabilities.

The probability that all three toddlers have received these inoculations is 0.4725.

b. *None of the three has received inoculations* means toddler A has *not* received the inoculations *and* toddler G has *not* received the inoculations *and* toddler U has *not* received the inoculations. Translate this sentence into mathematics using intersection and complement.

$P(A' \cap G' \cap U')$ Math translation: intersection.

$= P(A') \cdot P(G') \cdot P(U')$ Independent events.

$= [1 - P(A)] \cdot [1 - P(G)] \cdot [1 - P(U)]$ Complement rule.

$= (1 - 0.792)(1 - 0.839)(1 - 0.711)$ Use given probabilities.

$= (0.208)(0.161)(0.289)$ Simplify.

$= 0.0097$

The probability that none of the three has received the inoculations is 0.0097.

c. To write a probability statement for *exactly one has received the inoculations,* ask "How can that happen?" Toddler A has received the inoculations and toddlers G and U have not, or toddler G has received the inoculations and toddlers A and U have not, or toddler U has received the inoculations and toddlers A and G have not. Translate this sentence into probability using intersection and complement.

$P(A \cap G' \cap U') + P(A' \cap G \cap U') + P(A' \cap G' \cap U)$
    Three ways exactly one toddler has received the inoculations.

$= P(A) \cdot P(G') \cdot P(U') + P(A') \cdot P(G) \cdot P(U') + P(A') \cdot P(G') \cdot P(U)$
    Independent events.

$= (0.792)(0.161)(0.289) + (0.208)(0.839)(0.289) + (0.208)(0.161)(0.711)$
    Known probabilities; complement rule.

$= 0.0369 + 0.0504 + 0.0238$

$= 0.1111$

The probability that exactly one toddler of the three has received the inoculations is 0.1111.   ■

TRY IT NOW   GO TO EXERCISE 4.152

Example 4.38 Winter Tires

Winter tires are designed to reduce automobile crashes and improve driver safety in a wide range of winter weather conditions. Despite the advantages of using winter tires, there is increased cost, decreased fuel economy, and the aggravation of mounting and installation. According to a recent survey, 40% of drivers in Manitoba use winter tires (event M), 43% in Ontario (event O) do, and 98% in Quebec (event Q) do.[28] Suppose one driver from each jurisdiction is randomly selected. Find the probability that at least one driver uses winter tires.

Solution Trail 4.38

KEYWORDS

■ At least one driver uses winter tires

TRANSLATION

■ The words *at least one* means the event *one, two, or three drivers use winter tires*. This is the same as the complement of the event *none of the drivers uses winter tires*

CONCEPTS

■ Complement rule

VISION

Define the event *none of the drivers uses winter tires* and use the complement rule to find the probability that at least one does.

SOLUTION

P(at least one driver uses winter tires)

$$= 1 - \text{P}(0 \text{ drivers use winter tires}) \qquad \text{Complement rule.}$$
$$= 1 - \text{P}(M' \cap O' \cap Q') \qquad \text{All three do } not \text{ use winter tires.}$$
$$= 1 - \text{P}(M') \cdot \text{P}(O') \cdot \text{P}(Q') \qquad \text{Independent events.}$$
$$= 1 - [1 - \text{P}(M)] \cdot [1 - \text{P}(O)] \cdot [1 - \text{P}(Q)] \qquad \text{Complement rule.}$$
$$= 1 - (1 - 0.40)(1 - 0.43)(1 - 0.98) \qquad \text{Use given probabilities.}$$
$$= 1 - (0.60)(0.57)(0.02)$$
$$= 1 - 0.0068 = 0.9932$$

The probability that at least one driver uses winter tires is 0.9932.

Challenge: Find this probability using a direct approach, without using the complement rule.

TRY IT NOW   GO TO EXERCISE 4.153

Example 4.39 A Traveling Salesperson

During frequent trips to a certain city, a traveling salesperson stays at hotel A 50% of the time, at hotel B 30% of the time, and at hotel C 20% of the time. When checking in, there is some problem with the reservation 3% of the time at hotel A, 6% of the time at hotel B, and 10% of the time at hotel C. Suppose the salesperson travels to this city.

a. Find the probability that the salesperson stays at hotel A and has a problem with the reservation.

b. Find the probability that the salesperson has a problem with the reservation.

c. Suppose the salesperson has a problem with the reservation; what is the probability that the salesperson is staying at hotel A?

Solution Trail 4.39a

KEYWORDS

■ Stays at hotel A, problem with the reservation

TRANSLATION

■ What is the probability of the intersection of the event A and the event R?

CONCEPTS

■ Probability multiplication rule

VISION

Determine whether the events are independent, determine which probabilities are given, and use the appropriate form of the probability multiplication rule.

SOLUTION

Define the following events:

A = stays at hotel A; B = stays at hotel B;

C = stays at hotel C; and R = problem with the reservation.

Convert all the given percentages into probabilities.

The phrase *of the time* indicates conditional probability.

$$\text{P}(A) = 0.50 \qquad \text{P}(B) = 0.30 \qquad \text{P}(C) = 0.20$$
$$\text{P}(R \mid A) = 0.03 \qquad \text{P}(R \mid B) = 0.06 \qquad \text{P}(R \mid C) = 0.10$$

This experiment can be represented with a modified tree diagram (Figure 4.21). Remember, the probabilities along all paths coming from a node must sum to 1, and second-generation branch probabilities are conditional.

To find P($R' \mid A$), apply the complement rule to a conditional probability statement:

P($R' \mid A$) = 1 − P($R \mid A$).

a. The events A and R are dependent. The likelihood of a problem with a reservation depends on the hotel.

$$\text{P}(A \cap R) = \text{P}(A) \cdot \text{P}(R \mid A) \qquad \text{Probability multiplication rule.}$$
$$= (0.50)(0.03) \qquad \text{Use known probabilities.}$$
$$= 0.0150$$

The probability of staying at hotel A and having a problem with the reservation is 0.0150.

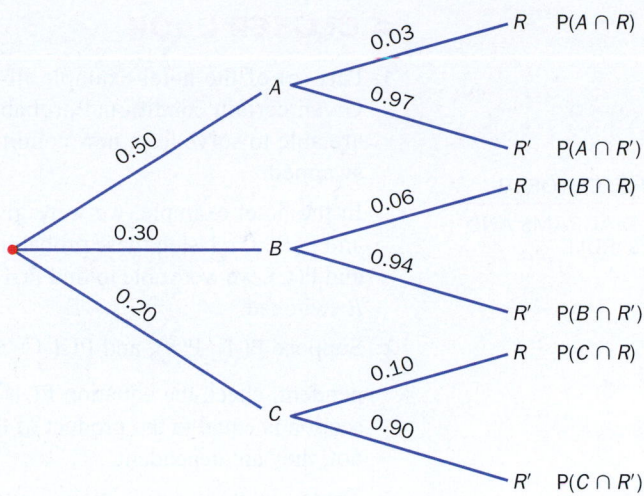

Figure 4.21 Tree diagram for Example 4.39.

Solution Trail 4.39b

KEYWORDS
- Problem with the reservation

TRANSLATION
- The event R

CONCEPTS
- Unconditional probability

VISION

To find $P(R)$, ask, "How can that happen?" Which paths from left to right involve the event R? The tree diagram suggests that three compound events (paths) involve a problem with the reservation.

b. $P(R) = P(A \cap R) + P(B \cap R) + P(C \cap R)$ Decomposition of R.

$\quad = P(A) \cdot P(R \mid A) + P(B) \cdot P(R \mid B) + P(C) \cdot P(R \mid C)$ Probability multiplication rule.

$\quad = (0.50)(0.03) + (0.30)(0.06) + (0.20)(0.10)$ Use known probabilities.

$\quad = 0.0150 + 0.0180 + 0.0200 = 0.0530$

The probability of a problem with the reservation (regardless of the hotel) is 0.0530. Figure 4.22 shows this decomposition of R using a Venn diagram.

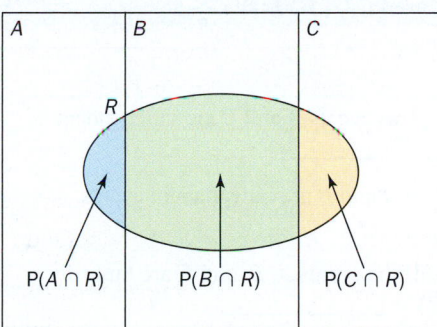

Figure 4.22 Venn diagram showing decomposition of R.

Solution Trail 4.39c

KEYWORDS
- Problem with the reservation
- Staying at hotel A

TRANSLATION
- Given the event R, find the probability of the event A

CONCEPTS
- Conditional probability with the event R given

VISION

Find $P(A \mid R)$.

c. $P(A \mid R) = \dfrac{P(A \cap R)}{P(R)}$ Formula for conditional probability.

$\quad = \dfrac{0.0150}{0.0530}$ Use answers to (a) and (b).

$\quad = 0.2830$

The probability that the salesperson stayed at hotel A, given a problem with the reservation, is 0.2830.

TRY IT NOW   GO TO EXERCISE 4.163

A CLOSER LOOK

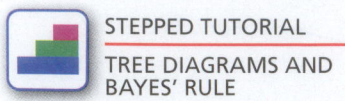

STEPPED TUTORIAL

TREE DIAGRAMS AND BAYES' RULE

1. Part (c) of the hotel example illustrates **Bayes' rule**. This theorem loosely states: Given certain conditional probabilities (and other unconditional probabilities), we are able to solve for a new conditional probability where the events are inverted, or swapped.

   In the hotel example, we were given the conditional probabilities $P(R \mid A)$, $P(R \mid B)$, and $P(R \mid C)$. Using these probabilities and the unconditional probabilities $P(A)$, $P(B)$, and $P(C)$, we were able to find $P(A \mid R)$, a conditional probability with the events A and R switched.

2. Suppose $P(A)$, $P(B)$, and $P(A \cap B)$ are known. To decide whether A and B are independent, check the equation $P(A \cap B) \overset{?}{=} P(A) \cdot P(B)$. If the probability of the intersection is equal to the product of the probabilities, then the events are independent. If not, they are dependent.

3. There are many applications in probability and statistics that involve repeated sampling from a population *with replacement*. In this case, each draw is independent of any other draw.

Other applications involve sampling *without replacement*, for example, exit polls and telephone surveys. Consider each individual response as an event. These events are definitely dependent. However, if the population is *large enough* and the sample is small relative to the size of the population, then the events are *almost independent*. Calculating probabilities assuming independence results in little loss of accuracy. Exercise 4.134 illustrates this idea.

SECTION 4.5 EXERCISES

Concept Check

4.136 Fill in the Blank Two events A and B are independent if and only if $P(A \mid B) = $ _____.

4.137 Fill in the Blank If A and B are independent events, $P(A \cap B) = $ _____.

4.138 Fill in the Blank If the events A, B, and C are mutually independent, $P(A \cap B \cap C) = $ _____.

4.139 True/False For any two events A and B, $P(A \cap B) = P(A) \cdot P(B \mid A)$.

4.140 True/False When sampling from a population with replacement, each draw is independent of any other draw.

Practice

4.141 Decide whether each pair of events is independent or dependent.
   a. A = make an error on your income tax return, and B = file Form 1040 long.
   b. C = put together a swing set correctly, and D = read the directions.
   c. E = run out of milk, and F = the refrigerator breaks down.
   d. G = break your pencil lead while writing, and H = feel overly stressed.

4.142 Decide whether each pair of events is independent or dependent.
   a. A = a randomly selected CD has a scratch, and B = a random email message is spam.
   b. C = one paper towel is enough to completely clean a spill, and D = you use a generic paper towel.
   c. E = no accidents are reported in 24 hours in a county, and F = there are no storms in the area.
   d. G = your automobile insurance bill increases, and H = you had one speeding ticket within the last year.

4.143 Suppose the following probabilities are known: $P(A) = 0.25$, $P(B \mid A) = 0.34$, and $P(C \mid A \cap B) = 0.62$.
   a. Find $P(A \cap B)$, $P(B' \mid A)$, and $P(A \cap B')$.
   b. Find $P(A \cap B \cap C)$, $P(C' \mid A \cap B)$, and $P(A \cap B \cap C')$.
   c. Are the events A and B independent? Justify your answer.

4.144 Suppose the events A, B, and C are independent and $P(A) = 0.55$, $P(B) = 0.45$, and $P(C) = 0.35$. Find the following probabilities.
   a. $P(A \cap B)$, $P(A \cap C)$, and $P(B \cap C)$.
   b. $P(A \cap B \cap C)$ and $P(A' \cap B' \cap C')$.
   c. $P(A \cap B' \cap C')$ and $P(A' \cap B \cap C)$.

4.145 Suppose the following probabilities are known: $P(A) = 0.40$, $P(B \mid A) = 0.25$, $P(C \mid A) = 0.45$, and $P(D \mid A) = 0.30$.

a. Find P($A \cap B$), P($A \cap C$), and P($A \cap D$).

b. Are the events A and B independent? Justify your answer.

c. Find P($B' \mid A$). If the event A occurs, are there any other events in addition to B, C, and D that can occur? Justify your answer.

4.146 Suppose the probability that an individual has blue eyes is 0.41. Four people are randomly selected.

a. Find the probability that all four have blue eyes.

b. Find the probability that none has blue eyes.

c. Find the probability that exactly two have blue eyes.

4.147 Consider the modified tree diagram below.

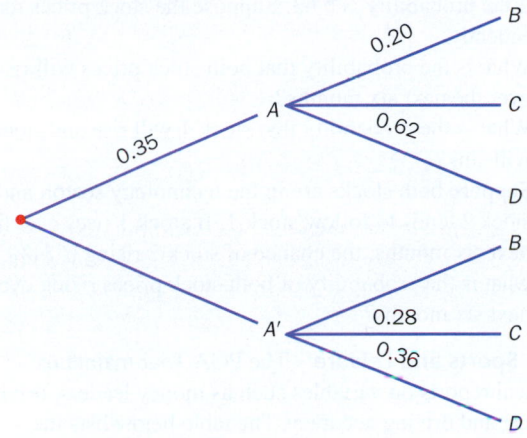

a. Identify and determine each missing path probability.

b. Find P($A \cap C$) and P($A' \cap B$).

c. Find P(D).

4.148 Consider the modified tree diagram below.

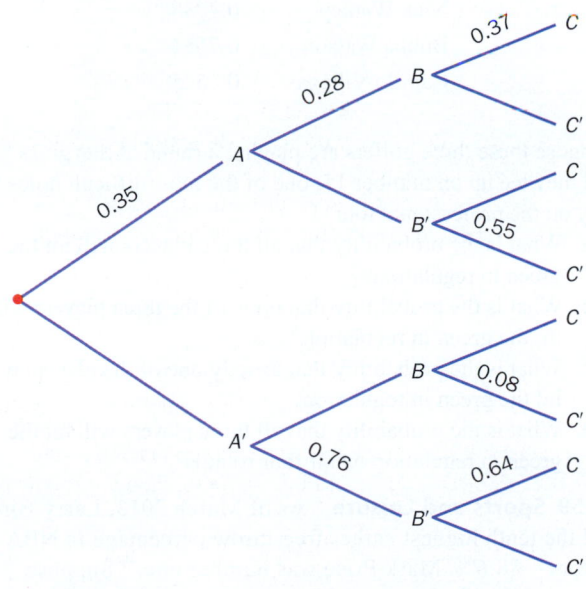

a. Identify and determine each missing path probability.

b. Find P($A \cap B \cap C$) and P($A' \cap B \cap C'$).

c. Find P(C). Are the events B and C independent? Justify your answer.

Applications

4.149 Demographics and Population Statistics Fishing is often considered a quiet, serene pastime. However, the job of fisherman is actually very dangerous—*The Deadliest Catch* on The Discovery Channel chronicles the risky lives of fishermen on the Bering Sea. According to recent data, the fatality rate for fishermen is 0.0012.[29] Suppose two fishermen are selected at random.

a. What is the probability that both fishermen will be fatally injured during the year?

b. What is the probability that neither will be fatally injured during the year?

c. What is the probability that exactly one will be fatally injured during the year?

4.150 Public Policy and Political Science The port of South Louisiana is the largest-tonnage port in the United States. Inspectors randomly select ships at one of the facilities and check for safety violations. Past records indicate that 90% of all ships inspected have no safety violations. Suppose two ships are selected at random.

a. What is the probability that both ships have safety violations? Write a Solution Trail for this problem.

b. What is the probability that neither ship has a safety violation?

c. What is the probability that exactly one ship has a safety violation?

4.151 Snakes on a Plane India has reported the greatest number of venomous snake bites and fatalities in the world. The four snakes that reportedly do the most biting are the King Cobra, Indian Krait, Russell's Viper, and the Saw Scaled Viper. The snake bite fatality rate in India is 0.20.[30] Suppose two people in India bitten by a Krait are selected at random.

a. What is the probability that both people will die?

b. What is the probability that exactly one person will die?

c. What is the probability that at most one person will die?

4.152 Economics and Finance In a 2013 survey conducted by the Bank of Montreal, Canadians indicated that they were feeling more optimistic about the economy. It was found that 38% of the respondents believed that their employer would hire additional people during the year.[31] Suppose three Canadian workers are selected at random.

a. What is the probability that all three believe their employer will hire additional people in the coming year?

b. What is the probability that at least one believes that her employer will hire additional people in the coming year?

c. What is the probability that none of the three believes that their employer will hire additional people in the coming year?

4.153 Physical Sciences The San Francisco Bay Area is near several geological fault lines and is therefore vulnerable to the constant threat of earthquakes. According to a study by the U.S. Geological Survey, the probability of a magnitude 6.7 or greater earthquake in the Greater Bay Area in the next 30 years is 0.63. The probability of a large earthquake within the next 30 years along four major fault lines is given in the following table.[32]

Fault line	Probability
North San Andreas	0.21
Hayward	0.31
San Gregorio	0.06
Concord–Green Valley	0.03

Suppose earthquakes occur in this area independently of one another.

a. What is the probability that there will be a major earthquake within the next 30 years in all four fault regions? Write a Solution Trail for this problem.

b. What is the probability that there will be no major earthquake within the next 30 years in any of the four regions?

c. What is the probability that there will be a major earthquake within the next 30 years in at least one of the four regions?

d. Suppose each probability is doubled if we consider the next 40 years. What is the probability that there will be a major earthquake within the next 40 years in at least one of the four regions?

4.154 Public Policy and Political Science Recent national elections suggest that the political ideology of adults in the United States is very evenly divided. In a USA Today/Gallup survey, 29% were Republicans, 30% were Democrats, and 41% were Independents.[33] In addition, 51% of all Republicans, 16% of all Democrats, and 28% of all Independents describe their political views as conservative. Suppose an adult in the United States is selected at random.

a. What is the probability that the adult is a Republican and describes her political views as conservative?

b. What is the probability that the adult is a Democrat and describes his political views as conservative?

c. Suppose the adult described his views as conservative. What is the probability that he is an Independent?

4.155 Medicine and Clinical Studies According to the Alzheimer's Association, approximately 13% of older Americans have Alzheimer's disease.[34] This is the sixth leading cause of death in the United States, and there is no treatment to prevent, cure, or even slow the disease. Suppose four older Americans are selected at random.

a. What is the probability that all four have Alzheimer's disease?

b. What is the probability that exactly one has Alzheimer's disease?

c. What is the probability that at least two have Alzheimer's disease?

4.156 Psychology and Human Behavior There are almost 36 million homes in the United States that have four TVs. In homes where there are TVs, 88% have a TV in the living room, 68% have a TV in the bedroom, and 17% have a TV in the kitchen.[35] Suppose a home with TVs is selected at random.

a. What is the probability that the home has a TV in all three rooms?

b. What is the probability that the home has a TV in only the living room?

c. What is the probability that the home has a TV in exactly two of the three rooms?

4.157 Economics and Finance Detailed analysis of two technology stocks indicates over the next six months the probability that the price of stock 1 will rise is 0.42 and for stock 2 the probability is 0.63. Suppose the stock prices react independently.

a. What is the probability that both stock prices will rise over the next six months?

b. What is the probability that stock 1 will rise and stock 2 will sink?

c. Suppose both stocks are in the technology sector, and stock 2 tends to follow stock 1. If stock 1 rises over the next six months, the chance of stock 2 rising is 81%. Now what is the probability of both stock prices rising over the next six months?

4.158 Sports and Leisure The PGA Tour maintains statistical reports on variables such as money leaders, driving distance, and driving accuracy. The table below lists the probability that selected players were able to hit the green "in regulation."[36] According to the PGA Tour, a green is considered hit in regulation if any portion of the ball is touching the putting surface after the green in regulation stroke has been taken.

Golfer	Probability
Nick Watney	0.7738
Bubba Watson	0.7654
Camilo Villegas	0.7525

Suppose these three golfers are playing a round at Sawgrass and they tee up on number 11, one of the most difficult holes to play on the professional tour.

a. What is the probability that all three players will hit the green in regulation?

b. What is the probability that none of the three players will hit the green in regulation?

c. What is the probability that exactly one of the players will hit the green in regulation?

d. What is the probability that all three players will hit the green in regulation on all four rounds?

4.159 Sports and Leisure As of March 2013, Larry Bird had the tenth highest career free-throw percentage in NBA history—88.6%. Mark Price was number one.[37] Suppose Larry were still playing and he steps up to the free-throw line for two shots. It is unlikely that the two shots are independent. If he misses the first shot, the probability that he makes the second is 0.95, and if he makes the first shot, the probability that he makes the second is 0.85.

a. What is the probability that he makes both shots?

b. What is the probability that he misses both shots?

c. What is the probability that he makes only one shot?

4.160 Economics and Finance   As Baby Boomers reach retirement age, many are beginning to carefully examine their savings plans and government benefits. This generation is worried about remaining financially independent as many companies cut or even eliminate pension plans. According to a report commissioned by the Canadian bank CIBC, approximately 25% of retiring Baby Boomers expect to carry some debt into their retirement.[38] Suppose three Baby Boomers are selected at random.

a. What is the probability that exactly two of the three expect to carry some debt into retirement?

b. What is the probability that all three expect to carry some debt into retirement?

c. Suppose another random sample of five Baby Boomers is obtained. What is the probability that exactly one adult from each sample expects to carry some debt into retirement?

4.161 Psychology and Human Behavior   A recent survey revealed that more people in the United Kingdom believe in space aliens than in God.[39] One in 10 people has reported seeing a UFO, and 20% of the respondents believe that UFOs have landed on Earth. Suppose three people from the United Kingdom are selected at random.

a. What is the probability that all three believe UFOs have landed on Earth?

b. What is the probability that none of the three believes UFOs have landed on Earth?

c. What is the probability that exactly one of the three believes UFOs have landed on Earth?

Extended Applications

4.162 Medicine and Clinical Studies   When a person has a certain type of leukemia, a physician may perform a bone marrow transplant in order to restore a healthy blood supply. Among the general population, the chances of an acceptable bone marrow match are 1 in 20,000.[40] Suppose a person needs a bone marrow transplant and four people from the general population are selected at random.

a. What is the probability that none of the four will match?

b. What is the probability that at least one will match?

c. How many people would have to be tested in order for the probability of at least one match to be 0.50?

4.163 Travel and Transportation   A family trying to arrange a vacation is using the Internet to name their own price for a rental car. The software reports that 50% of all people name a price of $30 per day, 40% bid $25 per day, and 10% bid $20 per day. The Internet company also reports that 90% of all $30 bids are accepted, 60% of all $25 bids are accepted, and only 5% of all $20 bids are accepted.

a. What is the probability that the family will submit a bid of $25 and have it accepted?

b. What is the probability that their bid will be accepted?

c. Suppose their bid is accepted. What is the probability that it is for $20?

4.164 Biology and Environmental Science   Opponents of the U.S. Navy SURTASS LFA Sonar System argue that it constitutes a substantial risk to marine life, causing extraordinary numbers of stranded, or beached, whales. Consider the following statements concerning the use of this system near a remote island in the South Pacific.

- On any given day, the probability of a mass stranding of whales in this area is 0.01.
- The probability of a military exercise on any given day is 0.001.
- If there is a military exercise, the probability of a mass stranding is 0.17.

a. Define events and write a probability statement for each fact above.

b. On a randomly selected day, what is the probability of a mass stranding of whales and a military exercise?

c. Are the events mass stranding and military exercise independent? Justify your answer.

4.165 Medicine and Clinical Studies   A tine test is a common method used to determine whether a person has been exposed to tuberculosis. Approximately 5% of people in the United States have been exposed to tuberculosis.[41] Using the tine test, 95% of all people who have been exposed test positive, and 98% of those not exposed test negative. Suppose a person is randomly selected and given the tine test.

a. What is the probability that the person tests positive and has been exposed to tuberculosis?

b. What is the probability that the person tests positive?

c. Suppose the test is positive; what is the probability that the person actually has been exposed?

4.166 Travel and Transportation   There are four major air carriers with flights from Boston to Los Angeles; 32% of all passengers take American Airlines, 25% take Jet Blue, 17% take United, and 26% take Virgin America. Data from 2012 indicate that 20% of all American Airlines flights from Boston to Los Angeles are late, 23% of Jet Blue flights from Boston to Los Angeles are late, 19% of United flights from Boston to Los Angeles are late, and 11% of Virgin America flights from Boston to Los Angeles are late.[42] Suppose a passenger taking a flight from Boston to Los Angeles is randomly selected.

a. What is the probability that the passenger takes American Airlines and is late?

b. What is the probability that the passenger is late? On time?

c. Suppose the passenger arrives late. Which airline did the passenger most likely fly?

4.167 Manufacturing and Product Development   The Italian *Aspide* missile, a licensed version of the U.S. *Sparrow*, has a sophisticated homing guidance system and single-shot hit probability of 0.80.[43] Suppose an enemy plane is within range of three missile firing stations, all three fire an *Aspide* surface-to-air missile, and the missiles operate independently.

   a. What is the probability that the plane is hit?

   b. What is the probability that all three missiles miss?

   c. How many missiles would have to be fired at the plane in order to be 99.99% sure it would be hit?

4.168 Public Health and Nutrition Many more adults in the United States have celiac disease than a decade ago. A new study suggests that approximately 1% of all U.S. adults have celiac disease and should avoid eating foods with gluten.[44] Suppose five adults in the United States are selected at random.

   a. What is the probability that exactly one of the five has celiac disease?

   b. What is the probability that only the first adult and the fifth adult selected have celiac disease?

   c. Suppose all five adults have celiac disease. Do you believe the claim concerning the percentage of adults with celiac disease? Justify your answer.

4.169 Fuel Consumption and Cars After a minor collision, a driver must take his car to one of two body shops in the area. Consider the following events.

D = driver takes his car to shop D.
L = driver takes his car to shop L.
T = the work is completed on time.
B = the cost is less than or equal to the estimate (under budget).

The following modified tree diagram describes the relationships among these events.

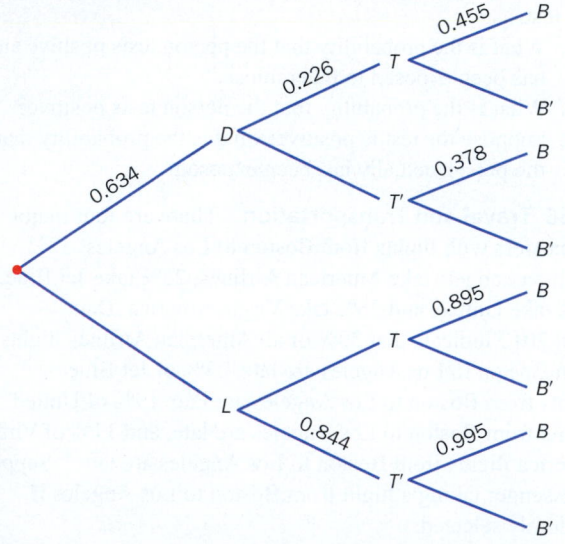

a. Complete the tree diagram by filling in the missing path probabilities.

b. What is the probability that the car is repaired under budget, on time, and with company D?

c. What is the probability that the cost of the repair is over the estimate?

d. What is the probability that the car is repaired under budget, given that it is ready on time?

4.170 Sports and Leisure Suppose two cards are drawn without replacement from a regular deck of 52 playing cards. Consider the events

A_1 = an ace is selected on the first draw;
A_2 = an ace is selected on the second draw.

a. Find $P(A_2 \mid A_1)$ and $P(A_2)$. Are the events A_1 and A_2 independent? Justify your answer.

b. Suppose the two cards are drawn without replacement from *six* regular 52-card decks shuffled together. Find $P(A_2 \mid A_1)$ and $P(A_2)$ for this experiment. Are the events A_1 and A_2 independent? Justify your answer.

c. In part (b), the events are *almost independent*. For six decks, find $P(A_1 \cap A_2)$ exactly, and then find the same probability assuming the two events are independent (with the probability of an ace on any draw being 24/312).

Challenge

4.171 The Traveling Salesperson Reconsider Example 4.39. Suppose the salesperson has a problem with the reservation. In which hotel did the salesperson most likely stay?

4.172 The Grapes of China The Chinese wine industry has become very large. However, there are many quality control problems.[45] Wine batches are systematically examined for alcohol content, total sugar, volatile acid, and other food additives. Suppose the Yantai Best Cellar Consulting Company claims that only 3% of all bottles are unqualified, or defective. Suppose six bottles of Yantai wine are selected at random.

a. What is the probability that none of the six bottles will be unqualified?

b. What is the probability that at least one of the bottles will be unqualified?

c. Suppose all six bottles are unqualified. Do you believe the company's claim? Justify your answer.

CHAPTER 4 SUMMARY

Concept	Page	Notation / Formula / Description
Experiment	124	An activity in which there are at least two possible outcomes and the result cannot be predicted with certainty.
Sample space	126	S: a listing of all possible outcomes, using set notation.

CHAPTER 4 EXERCISES

APPLICATIONS

4.173 Marketing and Consumer Behavior   A home decorating store received a shipment of 20 different Tiffany-style lamps and the store manager selects three lamps at random for display.

  a. How many different displays are possible?
  b. Suppose three of the lamps were damaged during shipping. What is the probability that two of the lamps selected will be broken?
  c. What is the probability that at least one of the lamps selected will be broken?

4.174 Physical Sciences   At a state middle-school science fair, students launch bottle rockets designed and built from

plastic two-liter beverage containers. An experiment consists of recording the general appearance of a rocket [bad (B), good (G), or excellent (E)] and the maximum altitude [low (L), medium (M), or high (H)]. Consider the following events.

A = The rocket is rated as excellent.
B = The rocket flies to a high altitude.
C = The rocket is rated as bad or flies low.
D = The rocket is good and flies to a medium altitude.

  a. Find the sample space S for this experiment.
  b. List the outcomes in each of the events A, B, C, and D.
  c. List the outcomes in $A \cup B$, $B \cup C$, and D'.
  d. List the outcomes in $A \cap B$, $C \cap D$, and $(B \cup D)'$.

4.175 Sports and Leisure   The Boston Bruins play in the Northeast Division of the Eastern Conference in the National

Hockey League. For each game played against another team in the Eastern Conference, the division [Atlantic (A), Northeast (N), or Southeast (S)] and the outcome [win (W), loss (L), tie (T), or overtime loss (O)] are recorded. Consider the following events.

E = The opponent is in the Southeast Division.
F = The Bruins win the game.
G = The opponent is in the Northeast or the game is an overtime loss.
H = The Bruins lose and the opponent is from the Atlantic Division.

   a. Carefully sketch a tree diagram to illustrate the possible outcomes for this experiment.
   b. Find the sample space for this experiment.
   c. Find the outcomes in each of the events E, F, G, and H.
   d. List the outcomes in $E \cup F$, $F \cap G$, and H'.
   e. List the outcomes in $E \cup H'$, $E \cup F \cup G'$, and $F \cup G'$.

4.176 Demographics and Population Statistics   In a recent population survey, the U.S. Census Bureau reported the following classifications and corresponding probabilities.[46]

Educational attainment	Probability
No degree	0.1317
High school graduate	0.3001
Some college, no degree	0.1946
Associate degree	0.0916
Bachelor's degree	0.1844
Master's degree	0.0708
Professional degree	0.0132
Doctorate degree	0.0136

Consider the events:
A = has a bachelor's, master's or doctorate degree
B = does not have an associate degree
C = does not have a degree
Find the following probabilities.
   a. $P(A)$, $P(B)$, and $P(C)$.
   b. $P(A \cap B)$, $P(A \cup C)$, and $P(B \cup C)$.
   c. $P(C')$, $P(A' \cup B)$, and $P(B' \cap C')$.

4.177 Biology and Environmental Science   The germination rate for pumpkin seeds is directly related to the prevailing weather conditions. The Autumn Gold is a popular medium-sized pumpkin and ripens to a deep orange. If conditions are seasonable, the probability of germination is 0.85.[47] If it is dry, suppose the probability that a random seed will germinate is 0.75. Recent weather history suggests there is a 40% chance of a dry start to the growing season. Suppose an Autumn Gold pumpkin seed is randomly selected.
   a. What is the probability that the growing season will be dry and the seed will germinate?
   b. What is the probability that the seed will germinate?
   c. Suppose the seed does not germinate. What is the probability that the growing season had a dry start?

4.178 Economics and Finance   Many Americans use savings bonds to supplement retirement funds or to pay for qualified higher-education expenses. The U.S. Treasury even sells savings bonds online. Approximately one in every six Americans owns savings bonds.[48] Suppose four Americans are randomly selected.
   a. What is the probability that all own savings bonds?
   b. What is the probability that none of the four owns savings bonds?
   c. What is the probability that exactly two of the four own savings bonds?

4.179 Marketing and Consumer Behavior   At Elmo's, an old-fashioned barber shop in Melbourne, Florida, 70% of all customers get a haircut, 40% get a shave, and 15% get both.
   a. What is the probability that a randomly selected customer gets a shave or a haircut?
   b. What is the probability that a randomly selected customer gets neither?
   c. What is the probability that a randomly selected customer gets only a shave?
   d. What is the probability that a randomly selected customer gets a shave, given that he gets a haircut?
   e. Suppose two customers are selected at random. What is the probability that both get only a haircut?

4.180 Medicine and Clinical Studies   More and more people are trying herbal remedies, including gooseberry juice, eucalyptus oil, and crushed ajwain, for relief from the common cold. The following joint probability table shows the relationship between having tried an herbal remedy and highest degree earned.

	Highest degree earned			
	Vocational	High school	College degree	Graduate degree
Tried	0.23	0.17	0.06	0.05
Not tried	0.04	0.12	0.15	0.18

Suppose one person is randomly selected.
   a. What is the probability that the person has tried an herbal remedy, given that the highest degree earned is from college?
   b. If the person has not tried an herbal remedy, what is the probability that the highest degree earned is from high school?
   c. Suppose the person has not earned a graduate degree. What is the probability that the person has tried an herbal remedy?
   d. Suppose two people are selected at random. What is the probability that exactly one has tried an herbal remedy?

4.181 Psychology and Human Behavior   Do you believe in ghosts? According to a recent survey, 21% of people in Sweden believe in ghosts.[49] Suppose that of those who believe in ghosts, 20% said they have had a spiritual encounter with a ghost. Suppose a Swede is selected at random.

a. If the person believes in ghosts, what is the probability that she has never had an encounter with a ghost?

b. What is the probability that the person believes in ghosts and has had an encounter with a ghost?

c. What is the probability that the person believes in ghosts and has not had an encounter with one?

4.182 Travel and Transportation   A super-commuter is a person who commutes to work from one large metro area to another by car, rail, bus, or even air. Super-commuters are not necessarily elite business travelers, but rather middle-income individuals who are willing to commute long distances in order to secure affordable housing or better schools. According to a recent study, 13% of workers in Houston are super-commuters, 8.6% in Phoenix, and 7.5% in Atlanta.[50] Suppose three workers are selected at random, one from each city.

a. What is the probability that all three are super-commuters?

b. What is the probability that none of the three are super-commuters?

c. What is the probability that only the worker from Houston is a super-commuter?

d. What is the probability that exactly two of the workers are super-commuters?

4.183 Manufacturing and Product Development   At a glass manufacturing facility, crystal stemware is carefully inspected for correct dimensions, quality, and production trends. After lengthy studies, the factory is known to produce 15% defectives. Most of these pieces are discovered through inspection and are reworked or discarded. Suppose two pieces are randomly selected for inspection.

a. What is the probability that both pieces are defect-free?

b. What is the probability that neither piece is defect-free?

c. Suppose at least one of the pieces has a flaw; what is the probability that both are defective?

4.184 Medicine and Clinical Studies   There is a constant shortage of organ donors in the United States. Fewer people are donating organs and ever more people are on waiting lists. One solution to this problem involves compensating organ donors. A recent poll suggests that 60% of Americans support some form of compensation in terms of future health care for people who make organ donations while alive, for example, kidneys, bone marrow, or liver.[51] Suppose four Americans are selected at random.

a. What is the probability that all four support compensation for organ donors?

b. What is the probability that none of the four supports compensation for organ donors?

c. What is the probability that exactly two of the four support compensation for organ donors?

4.185 Marketing and Consumer Behavior   Americans drink a lot of coffee, and they put all sorts of extras into their coffee to enhance the drink, including flavor shots and flavored creams. Research data indicates that 62% of all coffee drinkers put creamer in their coffee.[52] Of those people who use creamer, 40% say they would drink more coffee if

their preferred flavors were offered. Suppose a coffee drinker is selected at random.

a. Suppose the coffee drinker uses creamer. What is the probability that he would not drink more even if his preferred flavor were offered?

b. What is the probability that the coffee drinker uses creamer and would drink more if his preferred flavor were offered?

c. Suppose three coffee drinkers are selected at random. What is the probability that exactly one uses creamer?

EXTENDED APPLICATIONS

4.186 Public Health and Nutrition   Tobacco smoke contains more than 7000 chemicals, many that are toxic and several that are known to cause cancer. As a result, many smokers try various methods to quit. Consider a group of smokers who want to quit. In this group, 2.7% have tried an electronic cigarette, or e-cigarette. Of those who have tried e-cigarettes, 31% quit smoking after six months.[53] Of those people who tried some other method, suppose 16% quit smoking after six months. Suppose a smoker who would like to quit is selected at random.

a. What is the probability that the smoker tried e-cigarettes and quit smoking after six months?

b. What is the probability that the smoker quit after six months?

c. Suppose the smoker quit smoking after six months. What is the probability that she tried e-cigarettes?

4.187 Travel and Transportation   In a study of the worldwide commercial jet fleet through 2011, 37% of all fatal accidents occurred when the plane was on final approach or landing.[54] Suppose four fatal jet accidents are selected at random.

a. What is the probability that all four occurred during final approach or landing?

b. What is the probability that none of the four occurred during final approach or landing?

c. What is the probability that exactly one of the four occurred during final approach or landing?

4.188 Economics and Finance   Customers at a Publix grocery store in Charleston, South Carolina, can pay for purchases with cash, a debit card, or a credit card. Fifty-five percent of all customers use cash and 38% use a debit card. Careful research has shown of those paying with cash, 75% use coupons; of those using a debit card, 35% use coupons; and of those using a credit card, only 10% use coupons. Suppose a customer is randomly selected.

a. What is the probability that the customer pays with a credit card and does not use coupons?

b. What is the probability that the customer does not use coupons?

c. If the customer does not use coupons, what is the probability that he paid with a debit card?

4.189 Psychology and Human Behavior   According to a recent survey, 11% of men ages 50 to 64 now color their hair.[55]

Many men feel this is necessary in order to remain competitive in the workplace. Suppose four men ages 50 to 64 are selected at random.

a. What is the probability that all four color their hair?
b. What is the probability that exactly one of the four colors his hair?
c. Suppose none of the four colors his hair. Is there any evidence to suggest that the study's claim is not correct? Justify your answer.

4.190 Fuel Consumption and Cars Auto Parts Warehouse offers a wide variety of parts and accessories for cars. Consider the following events:

A = a randomly selected customer purchases a manual
B = a randomly selected customer purchases trim accessories
C = a randomly selected customer purchases a car-care product

Suppose the following probabilities are known.
$P(A) = 0.44$, $P(B) = 0.52$, $P(C) = 0.39$,
$P(A \cap B) = 0.19$, $P(A \cap C) = 0.10$, $P(B \cap C) = 0.23$,
$P(A \cap B \cap C) = 0.08$.

a. Carefully sketch a Venn diagram illustrating the relationship among these three events and label each region with the corresponding probability.
b. Find the probability of just event A occurring.
c. Find the probability of none of the events (A, B, or C) occurring.
d. Find $P(A \mid C)$, $P(B \mid A \cap C)$, and $P(A \cap B \cap C \mid A)$.

4.191 Travel and Transportation Pasco County in Florida has special evacuation plans in the event of a hurricane. Suppose residents can take one of five different major highways out of the county. Department of Transportation officials have produced the following table indicating the probability that a resident will use a selected road.

Road	A	B	C	D	E
Probability	0.20	0.18	0.26	0.32	0.04

Suppose three Pasco County residents are selected at random and a hurricane strikes.

a. What is the probability that all three will take the same escape route?
b. What is the probability that exactly one will take escape route E?
c. What is the probability that two will take escape route C?
d. Suppose all three Pasco County residents hear a traffic report indicating that route A is flooded and impassable. What is the probability that all three will take route B?

4.192 Travel and Transportation Some researchers believe that the severity of an automobile accident is related to the type of vehicle. For example, pickup trucks tend to be involved in more fatal crashes than other types of vehicles. The following table presents the number of vehicles involved in each type of crash.[56] Suppose this table is representative of all crashes in the United States and one crash is selected at random.

		Crash		
		Fatal	Injury	Property
Vehicle type	Passenger car	18,350	1,506,595	3,686,062
	Pickup	8,452	376,156	1,001,893
	Utility	6,924	447,946	1,187,911
	Van	2,494	174,299	453,197
	Other light truck	32	67,828	222,940
	Large truck	3,215	53,411	239,298
	Bus	221	9,968	47,387
	Other	1,152	6,282	12,429

a. What is the probability that the crash results in injury only?
b. Suppose the crash involves a utility vehicle. What is the probability that it is fatal?
c. Suppose the crash involves property damage only. What is the probability that the vehicle is a large truck?
d. Suppose the crash is not fatal. What is the probability that it involves a bus?
e. Suppose the crash does not involve a passenger car. What is the probability that it results in injury only?
f. Are the events fatal crash and van independent? Justify your answer.
g. Suppose two crashes are selected at random. What is the probability that both were fatal and involved an unknown type of vehicle?

CHALLENGE

4.193 Free Nights During the month of August, one guest at the Golden Nugget in Las Vegas will be selected at random to participate in a contest to win free lodging. A fair quarter will be tossed until the first head is recorded. If the first head occurs on toss x, the contestant will win x free nights' stay at the Golden Nugget.

So, if a head is obtained on the first coin toss, the contest is over, and the guest wins one free night. If the first head appears on the fourteenth toss (13 tails and then a head), the guest wins 14 free nights. Theoretically, a guest could win any number of free nights, 1, 2, 3, 4, . . . , although it seems unlikely someone could win, for example, 100 free nights.

a. Use technology to model this contest. Try your simulation 10 times and record the number of free nights awarded each time. Did anyone win five or more free nights' stay?
b. Consider the event A = the guest wins five or more free nights at the Golden Nugget. Simulate the contest $n = 50$ times and compute the relative frequency of occurrence of the event A. Repeat this process for $n = 100, 150, 200, \ldots, 2000$.
c. Construct a plot of the relative frequency versus the number of simulations. Describe any patterns.
d. Use your results in (b) and (c) to estimate the probability of winning five or more free nights at the Golden Nugget.

e. Find the exact probability of winning five or more free nights at the Golden Nugget. *Hint:* Consider the complement of the event A.

4.194 What are the chances of winning a prize in Monopoly Sweepstakes? The object in Monopoly Sweepstakes was to collect game pieces with the purchase of certain products. Some game pieces were instant winners. However, special collections of properties were worth big prizes. There were nine rare property prizes involving the following game pieces: Mediterranean Avenue, Vermont Avenue, Virginia Avenue, Tennessee Avenue, Kentucky Avenue, Ventnor Avenue, Pennsylvania Avenue, Boardwalk, and Short Line. The probability of finding the winning combination involving each of these rare properties is given in the following table.

Rare property	Probability of winning
Mediterranean Avenue	1/402,602
Vermont Avenue	1/578,695,060
Virginia Avenue	1/12,953,122
Tennessee Avenue	1/518,330,833
Kentucky Avenue	1/161,914,024
Ventnor Avenue	1/499,516,192
Pennsylvania Avenue	1/158,948,243
Boardwalk	1/306,939,484
Short Line	1/539,566,072

Find the probability of winning at least one of these rare property prizes.

5 Random Variables and Discrete Probability Distributions

Is a flu shot really effective?

Each year, the Centers for Disease Control and Prevention (CDC) recommend a flu shot for certain groups of people who are classified as *at risk* for serious complications from the most common strains of influenza virus. Adults aged 50 or older, residents of nursing homes, people with chronic heart or lung conditions, and even people who simply hate the flu, are all advised to get a shot when the shots become available, usually during the fall.

Approximately 135 million doses of flu vaccine were distributed during the 2012–2013 flu season. The vaccine was designed to protect individuals against the three most common types of flu predicted to occur. The CDC reported that the flu vaccine was 56% effective. That is, 56% of all people receiving the flu vaccine who were exposed to a flu virus did not contract the flu.

To check this claim (56% effective), a random sample of 50 at-risk people who received a flu shot was selected. During the flu season, all 50 were exposed to the flu and 29 actually contracted the disease. The techniques presented in this chapter will allow us to compute the likelihood of at least 29 people (out of 50) contracting the flu. This result will be used to determine whether there is any evidence that the claim is false.

CONTENTS

© Jose Luis Pelaez/Corbis

5.1 Random Variables

The idea of *assignment* suggests the need for a *function*.

A *function f* is a rule that takes an input value and returns an output value (according to the rule). Suppose the function *f* is defined by $f(x) = x^2 + 4$. This rule indicates that *f* takes an input *x* and *assigns*, or *maps*, *x* to the value $x^2 + 4$. For example, the function *f* assigns the input 1 to the output 5 because $f(1) = 1^2 + 4 = 5$. A random variable is just a special kind of function.

Definition

A **random variable** is a function that assigns a unique numerical value to each outcome in a sample space.

A CLOSER LOOK

1. Such functions are called random variables because their values cannot be predicted with certainty before the experiment is performed.

2. Capital letters, such as X and Y, are used to represent random variables.

3. A random variable is a rule for assigning each outcome in a sample space to a unique real number. If *e* is an experimental outcome and *x* is a real number, here is a formal way to picture this assignment: $X(e) = x$. The random variable X takes an outcome *e* and maps, or assigns, it to the number *x*. The number *x* is associated with the outcome *e*, and is a value the random variable can take on, or assume.

$X: S \rightarrow R$
A random variable maps elements of a sample space to the real numbers.

4. Figures 5.1 and 5.2 help us understand how a random variable works. These figures illustrate the random variable X as the link between experimental outcomes and numerical values. The rule for a random variable may be given by a formula, as a table, or even in words. Note that several outcomes may be assigned to the same number, but each outcome is assigned to only one number.

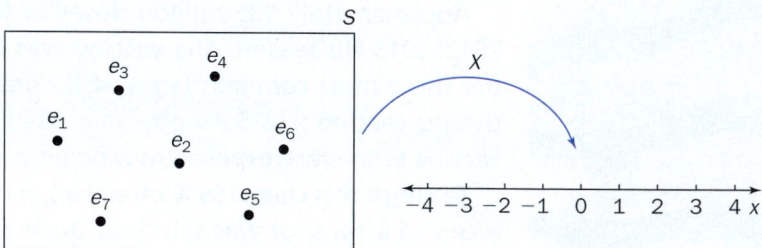

Figure 5.1 A random variable assigns a numerical value to each outcome.

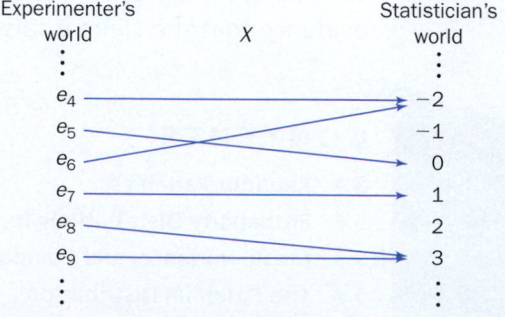

Figure 5.2 Another visualization of the definition of a random variable.

The next example shows how a specific random variable maps outcomes to numbers. The notation will get shorter and more concise as the concept of assignment becomes clearer.

Example 5.1 That Sinking Feeling

St. Petersburg Times/ZUMAPRESS/Newscom

In February 2013, a sinkhole suddenly opened up in the bedroom of a home in a Tampa, Florida, suburb. The homeowner died in this incident and the house was razed because officials feared it could collapse at any time. The Florida Department of Environmental Protection maintains a database of subsidence incident reports and has recorded over 3000 sinkholes since 1970.[1] The most common type of sinkhole in Florida is a collapse sinkhole. Suppose three sinkholes in Florida are selected at random and each is classified as a collapse sinkhole (C) or some other type (O). Let the random variable X be defined to be the number of collapse sinkholes out of the three selected.

SOLUTION

STEP 1 The experiment consists of recording the type of each sinkhole. Each outcome consists of a sequence of three letters, with each letter a C or an O. There are eight possible outcomes (from the multiplication rule). Here is the sample space:

$$S = \{OOO, OOC, OCO, OCC, COO, COC, CCO, CCC\}$$

STEP 2 The random variable X takes each outcome and returns the number of collapse sinkholes (Cs). Here is a table that illustrates this mapping and the values the random variable X can assume:

Outcome	Value of X
OOO	0
OOC	1
OCO	1
OCC	2
COO	1
COC	2
CCO	2
CCC	3

More formally, one can write:

$$X(OOO) = 0, \ X(OOC) = 1, X(OCO) = 1, \ \ldots$$

A number is assigned to each outcome. Note that the outcomes OOC, OCO, and COO are mapped to the same number (1), and the outcomes OCC, COC, and CCO are mapped to 2.

STEP 3 Here's the key: We are no longer interested in the sequence of letters, or outcomes, but rather focus on the *numbers* associated with the outcomes. We need to consider the number of possible values X can assume and the probability that X assumes each value.

The statement $X = 1$ is an event defined in terms of a random variable.

STEP 4 To find, say, the probability that X takes on the value 1, think about which outcomes are assigned to 1 and sum the probabilities of those outcomes. The probability that the random variable X equals 1 is

$$P(X = 1) = P(OOC) + P(OCO) + P(COO)$$

because these three outcomes are mapped to 1. As shown in Figure 5.3, the random variable X links these three outcomes and their associated probabilities to the number 1.

x is a possible value of the
random variable *X*.

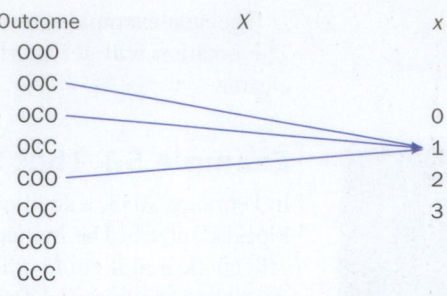

Outcome	X	x
OOO		
OOC		
OCO		0
OCC		1
COO		2
COC		3
CCO		
CCC		

Figure 5.3 The random variable *X* maps
three outcomes to the number 1.

There are two types of random variables. The type depends on the number of possible values the random variable can assume.

Definition

A random variable is **discrete** if the set of all possible values is finite, or countably infinite. A random variable is **continuous** if the set of all possible values is an interval of numbers.

These definitions are analogous to those for discrete and continuous data sets. The following remarks are also similar.

A CLOSER LOOK

1. Discrete random variables are usually associated with *counting*, and continuous random variables are usually associated with *measuring*.

2. To decide whether a random variable is discrete or continuous, consider *all* the *possible* values the random variable could assume. Finite or countably infinite means discrete. An interval of possible values means continuous.

3. Recall: Countably infinite means there are infinitely many possible values, but they are countable. You may not ever be able to finish counting all of the possible values, but there exists a *method* for actually counting them.

4. The interval of possible values for a continuous random variable can be *any* interval, of any length, open or closed. The exact interval may not be known, only that there is *some* interval of possible values.

5. In practice, no measurement device is precise enough to return *any* number in some interval. In *theory*, a continuous random variable may assume any value in some interval (but not in reality).

Remember, an experiment may result in a numerical value right away, not a symbol or a token. In this case we do not need any extra link or connection to the real numbers. The description of the experiment is the same as the definition of the random variable. The values the random variable can assume are the possible distinct experimental outcomes. In the following example, several experiments are described and each associated random variable is identified.

Example 5.2 Discrete or Continuous

Consider each experiment below and determine whether the associated random variable is discrete or continuous.

a. A Kohl's department store has 65 cash registers. At the end of the day, the receipts are carefully audited to determine whether each cash register balances. Let the random variable X be the number of cash registers that balance on a randomly selected day.

b. Patients undergoing a tonsillectomy are administered a general anesthetic. Let the random variable Y be the length of time from injection of the anesthetic until a patient is rendered unconscious.

c. Schlage manufactures and sells a maximum-security double-cylinder deadbolt lock for homes. At the facility where the locks are made and assembled, finished locks are randomly selected and carefully checked for defects. If a defective lock is found, the assembly line is shut down. An experiment consists of recording whether the selected lock is good (G) or defective (B). The sample space is $S = \{B, GB, GGB, GGGB, GGGGB, \ldots\}$. Let the random variable X be the number of locks inspected until a defect is found.

d. Let the random variable Y be the length of the largest fish caught on the next party boat arriving back to the dock in Belmar, New Jersey.

SOLUTION

a. There is no need to use a collection of symbols to represent experimental outcomes for the cash registers. The possible values for X (and the distinct experimental outcomes) are finite: $0, 1, 2, 3, \ldots, 65$. These values are distinct, disconnected points on a number line. The random variable X is discrete.

b. Y is a measurement, the time elapsed until a patient is unconscious. The possible values for Y are any number in some *interval*, say, 0 to 60 minutes. The random variable Y is continuous.

c. The values X can assume are $1, 2, 3, 4, \ldots$. The number of possible values is countably infinite; the values are disconnected on a number line. The random variable X is discrete.

d. Y is a measurement, and can (theoretically) take on any value in some interval. The possible values for Y are any number is some interval, say 5 to 25 inches. The random variable Y is continuous.

SECTION 5.1 EXERCISES

Concept Check

5.1 True/False The set of all possible values for a random variable can be infinite.

5.2 True/False A random variable may assign more than one numerical value to an outcome.

5.3 True/False A random variable can be both discrete and continuous.

5.4 Fill in the Blank A random variable is a special kind of _____.

5.5 Fill in the Blank A random variable maps elements of the _____ to the _____.

5.6 Fill in the Blank Discrete random variables are usually associated with _____.

5.7 Fill in the Blank Continuous random variables are usually associated with _____.

5.8 Short Answer If X is a discrete random variable, explain how to find $P(X = 2)$.

Practice

5.9 Classify each random variable as discrete or continuous.
  a. The number of boll weevils in one acre of a Louisiana cotton farm.
  b. The volume of ice cream in one scoop.
  c. The area of a randomly selected baseball field including foul territory.
  d. The number of late deliveries in one month by a package delivery service.

e. The number of girls born in a rural hospital during the next year.
f. The interest rate on a savings account at a randomly selected bank in Philadelphia.
g. The number of tickets sold in the next Powerball lottery.
h. The number of oil tankers registered to a certain country at a given time.

5.10 Classify each random variable as discrete or continuous.
a. The number of visitors to the Museum of Science in Boston on a randomly selected day.
b. The camber-angle adjustment necessary for a front-end alignment.
c. The total number of pixels in a photograph produced by a digital camera.
d. The number of days until a rose begins to wilt after purchase from a flower shop.
e. The running time for the latest James Bond movie.
f. The blood alcohol level of the next person arrested for DUI in a particular county.

5.11 Classify each random variable as discrete or continuous.
a. The number of people requesting vegetarian meals on a flight from New York to London.
b. The exact thickness (in millimeters) of a paper towel.
c. The time it takes a driver to react after the car in front stops suddenly.
d. The number of escapees in the next prison breakout.
e. The length of time a deep-space probe remains in contact with Earth.
f. The number of points on a randomly selected buck. The definition of a point is an antler projection at least one inch in length from the base to tip. The brow tine and main beam tip shall be counted as points regardless of length.

5.12 Classify each random variable as discrete or continuous.
a. The number of votes necessary to elect a new Pope.
b. The amount of sugar in a 16-ounce sweetened bottled drink purchased in a New York City cafe.
c. The total number of riders on all forms of public transportation in the United States during the year.
d. The number of residents in an assisted-living center who suffer from hardening of the arteries.
e. The amount of lead measured in the soil of a children's playground.
f. The time it takes an automobile to pass through the George Massey Tunnel in Vancouver, British Columbia.

Applications

5.13 Marketing and Consumer Behavior   T. J. Maxx sells home fashions and men's, women's, boys', and girls' apparel. An experiment consists of classifying the next two items purchased, each as men's, women's, boys', or girls' apparel. Let the random variable X be the number of sales of women's or girls' apparel.
a. List the outcomes in the sample space.
b. What are the possible values for X? Is X discrete or continuous? Justify your answer.

5.14 Education and Child Development   An experiment consists of showing a four-year-old child an interactive instructional video and then asking the child to tie his shoe-laces. The random variable Y is the length of time the child takes to tie the first shoelace. Is Y discrete or continuous? Justify your answer.

5.15 Biology and Environmental Science   The Waynes-burg Lions Club receives a shipment of 300 Christmas trees from Wending Creek Farms in Coudersport, Pennsylvania, to sell as a fundraiser. Classify each of the following random variables as discrete or continuous.
a. The number of trees over six feet tall.
b. The moisture content (expressed as a percentage) of a randomly selected tree.
c. The number of Douglas fir trees in the shipment.
d. The diameter of the trunk at the bottom of a randomly selected tree.

5.16 Biology and Environmental Science   To map the current of bottom water in a certain part of the Atlantic Ocean, a dye is released and used to trace the water flow. Let the random variable X be the maximum distance (in meters) from release at which the dye is detected after one day. Is X discrete or continuous? Justify your answer.

5.17 Psychology and Human Behavior   An experiment consists of recording the behavior of a randomly selected Duluth cab driver as a traffic signal changes from red to green. Let the random variable X be the acceleration (in ft/s^2) of the cab one second after the light changes. Is X discrete or continuous? Justify your answer.

5.18 Medicine and Clinical Studies   A report in the journal *Cancer* suggests that women who take aspirin on a regular basis lower their risk of a certain kind of melanoma.[2] Every woman who visits the Turtle Creek Medical Center in Dallas, Texas, during the next business day will be asked whether she regularly takes an aspirin. Let X be the number of women who take an aspirin daily. Is X discrete or continuous? Justify your answer.

5.19 Sports and Leisure   The Boston Red Sox play their spring training games in JetBlue Park, Fort Myers, Florida. Suppose a game against the Tampa Bay Rays is selected. Classify each of the following random variables as discrete or continuous.
a. The price of a randomly selected ticket.
b. The time it takes to complete the game.
c. Whether the game is postponed due to rain or is completed.
d. The speed of the first pitch in the bottom of the third inning.
e. The number of fans in attendance.
f. The number of hot dogs sold during the entire game.
g. The total number of errors in the game.
h. The weight in ounces of the bat used by the third hitter in the fifth inning.

5.2 Probability Distributions for Discrete Random Variables

A random variable is a rule that assigns each experimental outcome to a real number. To complete the description of a discrete random variable so that we can understand and answer questions involving the random variable, we need to know all the possible values the random variable can assume and all the associated probabilities. This collection of values and probabilities is called a *probability distribution*. Because random variables are used to model populations, a probability distribution is a theoretical description of a population.

A random variable provides the link between experimental outcomes and real numbers. An experimental outcome and the probability assigned to that outcome are both associated with exactly the same value of the random variable. This connection determines probability assignments for a random variable.

Definition

The **probability distribution for a discrete random variable** X is a method for specifying *all* of the possible values of X and the probability associated with each value.

A CLOSER LOOK

1. A probability distribution for a discrete random variable may be presented in the form of an itemized listing, a table, a graph, or a function.

2. A probability mass function (pmf) is denoted with a small p, and is the probability that a discrete random variable is equal to some specific value. In symbols, it is defined by

$$p(x) = \underbrace{P(X = x)}_{\text{Rule}}.$$

In words, the *rule* for the function p evaluated at an input x is the probability of an event, the probability that the random variable X takes on the specific value x. The function p and its probability rule are used interchangeably.

Suppose X is a discrete random variable. Then $p(7)$ means find the probability that the random variable X equals 7, or $P(X = 7)$.

A probability distribution is constructed using the definition of a random variable and the links between experimental outcomes and real numbers. The next example illustrates this concept.

Example 5.3 Construct a Probability Distribution

DATA SET

EG5.3

Suppose an experiment has eight possible outcomes, each denoted by a sequence of three letters, each an N or a D. The probability of each outcome is given in the following table.

Outcome	NNN	NND	NDN	DNN	NDD	DND	DDN	DDD
Probability	0.336	0.224	0.144	0.084	0.096	0.056	0.036	0.024

The random variable X is defined to be the number of Ds in an outcome. Find the probability distribution for X.

SOLUTION

STEP 1 The probability distribution for X consists of all the possible values X can assume along with the associated probabilities. The table below shows the random variable assignment and a technique for calculating the probability of each value.

Experiment			Probability distribution	
Probability	Outcome	X	Value, x	Probability
0.336	NNN ⟶		0	P(X = 0) = 0.336
0.224	NND ⟍			
0.144	NDN ⟶		1	P(X = 1) = 0.224 + 0.144 + 0.084 = 0.452
0.084	DNN ⟋			
0.096	NDD ⟍			
0.056	DND ⟶		2	P(X = 2) = 0.096 + 0.056 + 0.036 = 0.188
0.036	DDN ⟋			
0.024	DDD ⟶		3	P(X = 3) = 0.024

To find the probability that X takes on a specific value x, find all the outcomes that are mapped to x, and add the probabilities of these outcomes.

STEP 2 The random variable X takes on the values 0, 1, 2, and 3. The probability distribution can be presented in a table as shown below:

Looking at just this (probability distribution) table, how do you know X cannot assume any other value?

x	0	1	2	3
$p(x)$	0.336	0.452	0.188	0.024

TRY IT NOW   GO TO EXERCISE 5.29

A CLOSER LOOK

1. Think about this process of constructing a probability distribution. To find the probability that X takes on the value x, look back at the experiment and find all the outcomes that are mapped to x. *Drag along* these probabilities and sum them.

2. The probability distribution for a random variable X is a reference for use in answering probability questions about the random variable. For example, we'll need to answer probability questions such as "Find P(X = 3)." Think of X = 3 as an *event* stated in terms of a random variable. The details needed for answering this question are in the probability distribution.

The next example illustrates various methods for presenting a probability distribution.

Example 5.4 Public Defender

Don't worry about where these actual probabilities came from here. In this example, focus only on the *methods* for conveying all the values and probabilities.

Suppose the random variable Y represents the number of arraignments in a day before a certain judge in which the accused uses a public defender. The probabilities of Y taking on various values are as follows: 5/15 for no public defender; 4/15 for one; 3/15 for two; 2/15 for three; and 1/15 for four. Here are several ways to represent the probability distribution for Y.

 DATA SET

DEFENDER

SOLUTION

STEP 1 A complete *listing* of all possible values and associated probabilities (use either the probability mass function, p, or the assignment rule):

P(Y = 0) = 5/15

P(Y = 1) = 4/15

P(Y = 2) = 3/15

P(Y = 3) = 2/15

P(Y = 4) = 1/15

The random variable Y can take on the values 0, 1, 2, 3, or 4, and the probability of each value is given. There can be no other value of Y, because the probabilities sum to 1.

STEP 2 A table of values and probabilities:

y	0	1	2	3	4
$p(y)$	5/15	4/15	3/15	2/15	1/15

This kind of table is the most common way to present a probability distribution for a discrete random variable. It concisely lists all the values Y can assume and the associated probabilities.

STEP 3 A probability histogram:

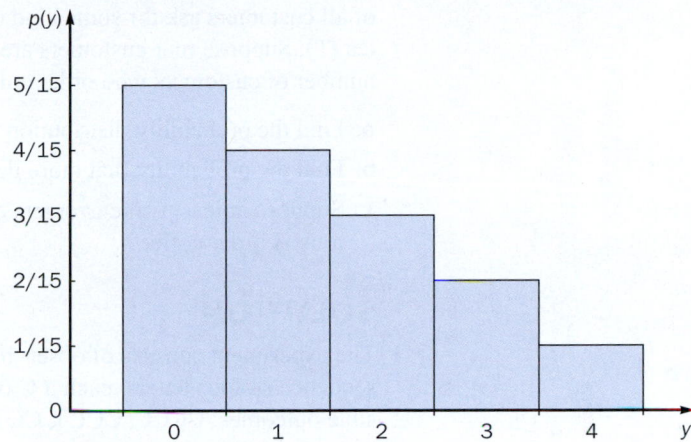

The distribution of Y is represented graphically. A rectangle is drawn for each value y, centered at y, with height equal to $p(y)$.

STEP 4 A point representation:

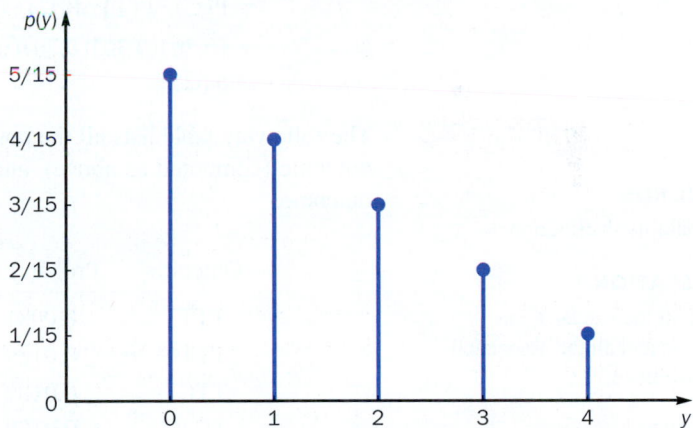

Plot the points $(y, p(y))$ and draw a line from $(y, 0)$ to $(y, p(y))$.

STEP 5 A formula:

$$p(y) = \frac{5 - y}{15} \quad y = 0, 1, 2, 3, 4$$

This shows the rule for the probability mass *function*. For example, to find $p(2)$, the probability $Y = 2$, let $y = 2$ in the formula to find:

$$p(2) = \frac{5 - 2}{15} = \frac{3}{15}$$

TRY IT NOW   GO TO EXERCISE 5.30

All of the techniques presented in the previous example are valid methods for presenting a probability distribution. Use the style that is most convenient or appropriate, or what is called for in the question. Often, a graphical representation of the distribution will be helpful. Sometimes, having a formula for the probability distribution is more useful. In the next example we'll construct another probability distribution and consider some probability questions involving a random variable.

Example 5.5 Who Wants Coffee?

The Hard Rock Cafe in Dallas carefully monitors customer orders and has found that 70% of all customers ask for some kind of coffee (C), while the remainder order a specialized tea (T). Suppose four customers are selected at random. Let the random variable X be the number of customers who order coffee.

a. Find the probability distribution for X.

b. Find the probability that more than two customers order coffee.

c. Suppose at least two customers order coffee. What is the probability that all four customers order coffee?

SOLUTION

The experiment consists of observing four customer choices. Each outcome consists of a sequence of four letters, each a C or a T. From the multiplication rule, there are 16 possible outcomes: CCCC, CCCT, CCTC, etc.

Because the customers are selected at random, each choice is independent, and the probability of each outcome is obtained by multiplying the corresponding probabilities. For example,

$$P(CTCT) = P(C \cap T \cap C \cap T)$$ First customer buys coffee *and* second customer buys tea *and* . . .

$$= P(C) \cdot P(T) \cdot P(C) \cdot P(T)$$ Events are independent. Multiply corresponding probabilities.

$$= (0.70)(0.30)(0.70)(0.30)$$ $P(T) = 1 - P(C)$

$$= 0.0441$$

The following table lists all the possible experimental outcomes, the probability of each outcome (computed as above), and the value of the random variable assigned to each outcome.

Outcome	Probability	x	Outcome	Probability	x
TTTT	0.0081	0	CTTC	0.0441	2
TTTC	0.0189	1	CTCT	0.0441	2
TTCT	0.0189	1	CCTT	0.0441	2
TCTT	0.0189	1	TCCC	0.1029	3
CTTT	0.0189	1	CTCC	0.1029	3
TTCC	0.0441	2	CCTC	0.1029	3
TCTC	0.0441	2	CCCT	0.1029	3
TCCT	0.0441	2	CCCC	0.2401	4

This table shows that the values of X are 0, 1, 2, 3, and 4.

a. Use the links in the table to construct the probability distribution for X.

$$p(0) = P(X = 0) = P(TTTT) = 0.0081$$

There is only one outcome assigned to a 0, and the probability of that outcome is 0.0081.

Solution Trail 5.5a

KEYWORDS

■ Probability distribution

TRANSLATION

■ Find all the values X can assume and all the associated probabilities.

CONCEPTS

■ Connection between experimental outcomes and real numbers

VISION

First, think about the experiment. Use the definition of the random variable to link experimental outcomes with values of the random variable, and drag along all of the probabilities. Construct a table listing all the values of X and the associated probabilities.

Solution Trail 5.5b

KEYWORDS

■ More than two

TRANSLATION

■ > 2

CONCEPTS

■ Find the probability that the random variable X takes on a value greater than 2.

VISION

Use the probability distribution to determine which values are greater than 2, and add the associated probabilities.

$p(1) = P(X = 1)$ *Definition of a probability mass function.*

$\quad = P(\text{TTTC or TTCT or TCTT or CTTT})$ *These outcomes are mapped to 1.*

$\quad = P(\text{TTTC}) + P(\text{TTCT}) + P(\text{TCTT}) + P(\text{CTTT})$

Or means union; the outcomes are disjoint.

$\quad = 0.0189 + 0.0189 + 0.0189 + 0.0189 = 0.0756$

Continue in this manner to obtain the probability distribution for X.

x	0	1	2	3	4
$p(x)$	0.0081	0.0756	0.2646	0.4116	0.2401

b. $P(X > 2) = P(X = 3) + P(X = 4)$ *Only values greater than 2.*

$\quad\quad\quad\quad\; = 0.4116 + 0.2401 = 0.6517$ *Use the probability distribution table.*

The probability of more than two customers ordering coffee is 0.6517.

Note: How would this probability change if the question asked for the probability that *two or more* customers order coffee?

c. Given that X is at least 2, find the probability that X is exactly 4.

Solution Trail 5.5c

KEYWORDS

■ Suppose at least two
■ All four

TRANSLATION

■ *Given* at least two
■ Find the probability that $X = 4$.

CONCEPTS

■ Conditional probability

VISION

Given that at least two customers order coffee, the number of values X can assume is reduced. Use the definition of conditional probability with events involving the random variable.

$P(X = 4 \mid X \geq 2) = \dfrac{P(X = 4 \cap X \geq 2)}{P(X \geq 2)}$ *Definition of conditional probability.*

$\quad\quad\quad\quad\quad\;\; = \dfrac{P(X = 4)}{P(X \geq 2)}$ *Intersection of $X = 4$ and $X \geq 2$.*

$\quad\quad\quad\quad\quad\;\; = \dfrac{0.2401}{0.2646 + 0.4116 + 0.2401}$ *Use the probability distribution.*

$\quad\quad\quad\quad\quad\;\; = \dfrac{0.2401}{0.9163} = 0.2620$

Given that at least two people order coffee, the probability that exactly four order coffee is 0.2620.

Here is a way to picture this conditional probability using the probability distribution table:

x	0	1	2	3	4
$p(x)$	0.0081	0.0756	0.2646	0.4116	0.2401

Given that X is either 2, 3, or 4, the *reduced*, or relevant, probability is 0.9163. The proportion of time that X is equal to 4, given X is 2, 3, or 4, is 0.2401/0.9163.   ●

TRY IT NOW   GO TO EXERCISE 5.36

The probability distribution of a random variable reveals which values of the random variable are most likely to occur. This information is extremely helpful in making a statistical inference. Consider the following example.

Example 5.6 In Case There Is a Power Outage

The Carson City Hospital has three emergency generators for use in case of a power failure. Each generator operates independently, and the manufacturer claims the probability that each generator will function properly during a power failure is 0.95. Suppose a power failure occurs and all three generators fail. Do you have reason to doubt the manufacturer's claim? Justify your answer.

Solution Trail 5.6

KEYWORDS

- Reason to doubt the manufacturer's claim?

TRANSLATION

- Use the available evidence to draw a reasonable conclusion.

CONCEPTS

- Inference procedure

VISION

Consider the claim, the experiment, and the likelihood of the experimental outcome. Then draw a conclusion. Define a random variable, and use the probability distribution to determine the probability that all three generators fail.

SOLUTION

STEP 1 In the event of a power failure, let F stand for a generator that fails, and let S represent a generator that functions properly, that is, starts. There are eight possible experimental outcomes. Let X be the number of failures.

STEP 2 The table below lists each outcome, the probability of each outcome, and the value of the random variable associated with each outcome.

Outcome	Probability	x
SSS	0.8574	0
SSF	0.0451	1
SFS	0.0451	1
SFF	0.0024	2
FSS	0.0451	1
FSF	0.0024	2
FFS	0.0024	2
FFF	0.0001	3

Note: The probabilities are rounded to four places to the right of the decimal.

Because each generator operates independently, the probability of each outcome is the product of the corresponding probabilities. For example,

$$P(SFS) = P(S \cap F \cap S)$$
$$= P(S) \cdot P(F) \cdot P(S)$$
$$= (0.95)(0.05)(0.95) = 0.0451$$

STEP 3 Use the links in the table to construct the probability distribution for X.

x	0	1	2	3
$p(x)$	0.8574	0.1353	0.0072	0.0001

STEP 4 Use the four-step inference procedure.

Claim: The probability that each generator will function properly during a power failure is 0.95.

Experiment: The value of the random variable observed is $x = 3$.

Likelihood: The likelihood of the observed outcome is $P(X = 3) = 0.0001$.

Conclusion: Because this probability is so small, the outcome of observing three failures is very rare. But it happened! This small probability suggests the assumption is wrong. There is evidence to suggest that the claim of 0.95 (start probability) is wrong. ●

TRY IT NOW   GO TO EXERCISE 5.42

As the previous examples suggest, the following properties must be true for every probability distribution for a discrete random variable X.

Properties of a Valid Probability Distribution for a Discrete Random Variable

1. $0 \leq p(x) \leq 1$

The probability that X takes on any value, $p(x) = P(X = x)$, must be between 0 and 1.

2. $\sum_{\text{all } x} p(x) = 1$

The sum of all the probabilities in a probability distribution for a discrete random variable must equal 1.

The following example involves a probability distribution for a discrete random variable and illustrates these two properties.

Example 5.7 Rewards for Donating

The Central Blood Bank in Pittsburgh, Pennsylvania, offers bonus points to donors, which can be redeemed at an online store.[3] The number of points a donor has earned is a random variable Y. Suppose Y has the following probability distribution:

y	0	100	150	200	250	300
$p(y)$	0.01	0.05	?	0.25	0.35	0.30

a. Find $p(150)$.

b. Find $P(100 \le Y \le 250)$ and $P(100 < Y < 250)$.

c. Construct the corresponding probability histogram.

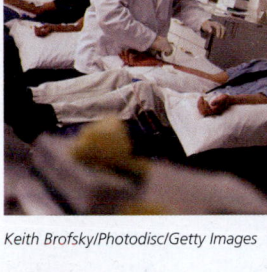

Keith Brofsky/Photodisc/Getty Images

DATA SET

DONORS

SOLUTION

a. The sum of all the probabilities must equal 1.

$$p(150) = 1 - [p(0) + p(100) + p(200) + p(250) + p(300)]$$
$$= 1 - (0.01 + 0.05 + 0.25 + 0.35 + 0.30)$$
$$= 1 - 0.96 = 0.04$$

b. The values Y takes on between 100 and 250 inclusive are 100, 150, 200, and 250.

$$P(100 \le Y \le 250) = p(100) + p(150) + p(200) + p(250)$$
$$= 0.05 + 0.04 + 0.25 + 0.35 = 0.69$$

The values Y takes on *strictly* between 100 and 250 are 150 and 200.

$$P(100 < Y < 250) = p(150) + p(200)$$
$$= 0.04 + 0.25 = 0.29$$

In this example, including (or excluding) an endpoint (a single value) changes the probability assignment. It is important to remember that a single value *may* make a difference in a probability assignment for a discrete random variable.

c. To construct the probability histogram, draw a rectangle for each value y, centered at y, with height equal to $p(y)$.

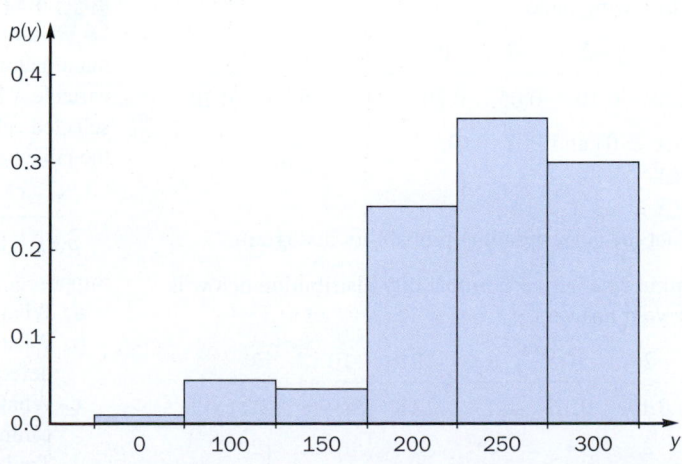

TRY IT NOW   GO TO EXERCISE 5.26

SECTION 5.2 EXERCISES

Concept Check

5.20 True/False A probability distribution is a theoretical model of a population.

5.21 True/False The sum of all the probabilities in a probability distribution for a discrete random variable must equal 1.

5.22 True/False For a discrete random variable, under certain circumstances $p(x)$ could be less than 0.

5.23 Fill in the Blank The probability distribution for a discrete random variable is a method for specifying _____ and _____.

5.24 Short Answer Briefly describe several methods to represent a probability distribution.

Practice

5.25 The probability distribution for the random variable X is given in the following table:

x	1	2	3	4	5	6	7
$p(x)$	0.35	0.20	0.15	0.12	?	0.08	0.03

a. Find $p(5)$.
b. Find $P(2 \le X \le 6)$ and $P(2 < X \le 6)$.
c. Find $P(X < 4)$.
d. Find the probability that X takes on the value 1 or 7.

5.26 The probability distribution for the random variable Y is given in the following table:

y	10	20	25	30	45	50
$p(y)$	0.155	0.237	0.184	0.122	?	0.258

a. Find $p(45)$.
b. Find $P(Y \ge 25)$ and $P(Y > 25)$.
c. Find the probability Y is divisible by 10.
d. Construct the corresponding probability histogram.

5.27 The probability distribution for the random variable X is given in the following table:

x	−3	−2	−1	0	1	2	3
$p(x)$	0.20	0.10	0.05	0.30	0.05	0.10	0.20

a. Find $P(X \ge 0)$ and $P(X > 0)$.
b. Find $P(X^2 > 1)$.
c. Find $P(X \ge 2 \mid X \ge 0)$.
d. Construct the corresponding probability histogram.

5.28 Determine whether each probability distribution below is valid. Justify your answers.

a.

x	2	4	6	8	10	12
$p(x)$	0.15	0.16	0.17	0.18	0.19	0.20

b.

x	2	4	6	8	10	12
$p(x)$	0.25	0.25	0.25	−0.25	0.25	0.25

c.

x	2	4	6	8	10	12
$p(x)$	0.05	0.20	0.25	0.25	0.20	0.05

5.29 The table below lists all of the possible outcomes for an experiment, the probability of each outcome, and the value of a random variable assigned to each outcome. Use this table to construct the probability distribution for X. Construct the corresponding probability histogram.

Outcome	Probability	x	Outcome	Probability	x
AA	0.01	1	CA	0.03	3
AB	0.02	2	CB	0.06	3
AC	0.03	3	CC	0.09	3
AD	0.04	4	CD	0.12	4
BA	0.02	2	DA	0.04	4
BB	0.04	2	DB	0.08	4
BC	0.06	3	DC	0.12	4
BD	0.08	4	DD	0.16	4

5.30 The table on the text website lists all of the possible outcomes for an experiment, the probability of each outcome, and the value of a random variable assigned to each outcome. Use this table to construct the probability distribution for Y. Construct the corresponding probability histogram. **EX5.30**

5.31 The probability distribution for a discrete random variable X is given by the formula.

$$p(x) = \frac{x(x + 1)}{112} \quad x = 1, 2, \ldots, 6$$

a. Verify that this is a valid probability distribution.
b. Find $P(X = 4)$.
c. Find $P(X > 2)$.
d. Find the probability that X takes on the value 3 or 4.
e. Construct the corresponding probability histogram.

Applications

5.32 Manufacturing and Product Development A wooden kitchen cabinet is carefully inspected at the manufacturing facility before it is sent to a retailer. The random variable X is the number of defects found in a randomly selected cabinet. The probability distribution for X is given in the table below.

x	0	1	2	3	4	5
$p(x)$	0.900	0.050	0.025	0.020	0.004	0.001

Suppose a cabinet is selected at random.
a. What is the probability that the cabinet is defect free?
b. What is the probability that the cabinet has at most two defects?
c. What is the probability that two randomly selected cabinets both have at least three defects? Write a Solution Trail for this problem.
d. Find $P(2 \le X \le 4)$ and $P(2 < X < 4)$.

5.33 Fuel Consumption and Cars An automobile insurance policy depends on many factors, including the vehicle type, year, make, and model, type of coverage, your driving history, your insurance score based on claims history, payment history, and credit score, and your state of residence.[4] Suppose that for some driver category, the probability distribution for the random variable Y, the amount (in dollars) paid to policyholders for claims in one year, is given in the table below.

y	0	500	1000	5000	10000
$p(y)$	0.65	0.20	0.10	0.04	0.01

a. Find $P(Y > 0)$.
b. Find $P(Y \leq 1000)$.
c. What is the probability that a randomly selected driver is paid $5000?
d. Suppose two drivers are selected at random. What is the probability that both are paid $1000?
e. Suppose two drivers are selected at random. What is the probability that at least one is paid $500 or more?

5.34 Public Policy and Political Science Camden, New Jersey, has one of the highest crime rates in the United States. As a result, a new Camden County Police Department will include 400 officers and be responsible for an area covering nine square miles.[5] Suppose the number of times a police cruiser from this new department drives through the Whitman Park neighborhood during a one-hour period is a random variable X, with probability distribution as given in the table below.

x	0	1	2	3
$p(x)$	0.3679	0.3679	0.1839	0.0613

x	4	5	6	7
$p(x)$	0.0153	0.0031	0.0005	0.0001

Suppose a one-hour period is randomly selected.
a. What is the probability that no police cruiser will drive through the neighborhood?
b. What is the probability that at least one police cruiser will drive through the neighborhood?
c. What is the probability that at most two police cruisers will drive through the neighborhood?
d. What is the probability that more than seven police cruisers will drive through the neighborhood?
e. Suppose at least two police cruisers were sighted in the neighborhood during the one-hour period. What is the probability that there were at least four during this time?

5.35 Public Policy and Political Science According to a January 2013 survey by the Pew Research Center, the percentage of Americans who trust the government in Washington has decreased steadily since Bill Clinton left office.[6] Today, approximately 30% of Americans say they trust the government (always or most of the time). Suppose three Americans are selected at random. Let X be the number who trust the government.

a. Construct the probability distribution for X. Construct the corresponding probability histogram.
b. What is the probability that all three Americans trust the government?
c. What is the probability that at least one of the three trusts the government?

5.36 Marketing and Consumer Behavior Twinkies are an American tradition, sold for over 83 years, originating in a Chicago bakery, and forever linked to legal lingo. (You may have read about the "Twinkie defense," associated with a 1979 murder trial in San Francisco.) Approximately 12% of households in the United States buy Twinkies.[7] Suppose four households are selected at random and let Y be the total number of households that buy Twinkies.
a. Construct the probability distribution for Y. Write a Solution Trail for this problem.
b. What is the probability that at least one household buys Twinkies?
c. Suppose at least two households buy Twinkies. What is the probability that all four households buy Twinkies?

5.37 Manufacturing and Product Development Staples has six special drafting pencils for sale, two of which are defective. A student buys two of these six drafting pencils, selected at random. Let the random variable X be the number of defective pencils purchased. Construct the probability distribution for X.

5.38 Business and Management Two packages are independently shipped from Fort Collins, Colorado, to the Convention Center in Kansas City, Missouri, and each is guaranteed to arrive within four days. The probability that a package arrives within one day is 0.10, within two days is 0.15, within three days is 0.25, and on the fourth day is 0.50. Let the random variable X be the total number of days for both packages to arrive. Construct the probability distribution for X.

Extended Applications

5.39 Biology and Environmental Science Agway, a farm and garden supply store, sells winter fertilizer in 50-pound bags. For customers who purchase this product, the probability distribution for the random variable X, the number of bags sold, is given in the table below.

x	1	2	3	4	5
$p(x)$	0.55	0.35	0.07	0.02	0.01

Suppose a person buying winter fertilizer is randomly selected.
a. What is the probability that the customer buys more than two bags?
b. What is the probability that the customer does not buy two bags $[P(X \neq 2)]$?
c. Find the probability that two randomly selected customers each buy one bag.
d. Suppose two customers are randomly selected. What is the probability that the total number of bags purchased will be at least eight?

e. Let the random variable Y be the number of *pounds* sold to a randomly selected customer buying winter fertilizer. Find the probability distribution for Y.

5.40 Sports and Leisure Suppose the probability that a person *says* he or she was at the Woodstock Festival and Concert is 0.20. An experiment consists of randomly selecting people in Green Bay and asking them whether they were at Woodstock. The experiment stops as soon as one person says he or she was there. The random variable X is the number of people stopped and questioned (until one person says he or she was there). Let Y and N stand for a Yes and No response, respectively.

a. List the first several outcomes in the sample space.
b. Find the probability of each outcome in part (a).
c. Find the value of the random variable associated with each outcome in (a).
d. Find a formula for the probability distribution of X.

5.41 Sports and Leisure A game show contestant on *Let's Make a Deal* selects two envelopes with prize money enclosed. Two of the envelopes contain $100, one envelope contains $250, two envelopes contain $500, and the last envelope contains $1000. Let the random variable M be the maximum of the two prizes.

a. Find the probability distribution for M.
b. Suppose two contestants independently select prize envelopes on two different days. What is the probability that both win the top prize?

5.42 Economics and Finance A Subway restaurant in Scotsbluff, Nebraska, recently installed a new computer system that allows customers to pay for meals with a bank debit card. The manager of the restaurant claims this new system will decrease the waiting time, and that the probability of getting a meal in under two minutes (with this system in place) is 0.75. Suppose four customers are selected at random. Let the random variable X be the number of customers who receive their meal in under two minutes.

a. Construct the probability distribution for X.
b. Suppose none of the four customers receives their meal in under two minutes. Is there any evidence to suggest the manager's claim is false? Justify your answer.

5.43 Demographics and Population Statistics Most physicians in Canada graduated from a medical school in Canada. However, some attended medical school in the United States or another foreign country. The probability that a physician in Alberta attended a Canadian medical school is 0.684; in British Columbia, 0.705; in Quebec, 0.891; and in Ontario, 0.736.[8] Suppose one physician is independently selected from each province, and let the random variable Y be the number of physicians who attended a Canadian medical school.

a. Construct a probability distribution for Y.
b. What is the probability that at least one physician attended a Canadian medical school?
c. Suppose another group of physicians is selected, one from each province. What is the probability that at least three physicians from each group attended a Canadian medical school?

5.3 Mean, Variance, and Standard Deviation for a Discrete Random Variable

David Cannon/Getty Images

Just as there are descriptive measures of a sample (for example, $\bar{x}$, s^2, and s), there are corresponding descriptive measures of a population (μ, σ^2, and σ). As we said in Chapter 3, these population *parameters* describe the center and variability of the *entire* population. They are usually unknown values we would like to estimate.

However, because a random variable may be used to model a population, these (population) descriptive measures are inherent in and determined by the probability distribution. This section presents the methods used to compute the mean, variance, and standard deviation of a random variable (or population). The next example suggests a definition of *expected value*.

Example 5.8 The Caddy Pool

A teenager is a member of the caddy program at the Montebello Country Club. Each morning he arrives at the golf course and enters his name into the caddy pool. The probability of being selected on any day is 4/5, and if selected he will earn $50. On days he is not selected, he earns nothing. How much money does this caddy earn per day *on average*? Or, in the long run, how much does the caddy earn each day?

SOLUTION

STEP 1 This question is concerned with the amount earned each day *on average*, not on any one particular day. Consider the probabilities given and consider five typical days.

On four of five days, the caddy earns $50. On the fifth day, he earns $0. The total earned for the five *typical* days is $200. To find the *average* amount earned each day, divide by five: $200/5 = $40.

STEP 2 Consider a random variable X that takes on only two values, 0 and 50, with probabilities 0.20 and 0.80, respectively. Another way to compute the *average* earned each day is to use this probability distribution.

$$40 \;=\; 0 \;\times\; 0.20 \;+\; 50 \;\times\; 0.80$$

$$\quad\uparrow\qquad\uparrow\qquad\uparrow\qquad\uparrow$$

Value   Probability   Value   Probability

The long-run average earnings per day can be found by using a probability distribution. Multiply each value by its corresponding probability, and sum these products. ■

Definition

Let X be a discrete random variable with probability mass function $p(x)$. The **mean**, or **expected value**, of X is

$$\underbrace{E(X) = \mu = \mu_X}_{\text{Notation}} = \underbrace{\sum_{\text{all } x}[x \cdot p(x)]}_{\text{Calculation}}. \qquad (5.1)$$

A CLOSER LOOK

1. The capital E stands for *expected value* and is a function. The function E takes a random variable as an input and returns the expected value. More generally, E accepts as an input any *function* of a random variable. For example, suppose $f(X)$ is a function of a discrete random variable X. The expected value of $f(X)$ is

$$E[f(X)] = \sum_{\text{all } x}[f(x) \cdot p(x)] \qquad (5.2)$$

2. μ is the mean, or expected value, of a random variable (which may model a population). If necessary, the associated random variable is used as a subscript for identification, for example, μ_X or μ_Y.

3. The mean is easy to compute. Multiply each value of the random variable by its corresponding probability, and add the products.

4. The mean of a random variable is a *weighted average* and is only what happens on average. The mean may not be any of the possible values of the random variable. ●

Example 5.9 Road Construction Next 30 Miles

DATA SET

PAVING

Thirty miles of Interstate Highway I-16 near Macon, Georgia, were recently repaved. At the end of each work day, the project supervisor estimated the number of hours behind or ahead of schedule. Suppose this estimate is a discrete random variable, X, with probability distribution as given in the following table:

x	-20	-10	30	60
$p(x)$	0.2	0.3	0.4	0.1

Find the mean of X.

SOLUTION

STEP 1 X is a discrete random variable. To find the mean, use Equation 5.1.

STEP 2 $\mu = \sum\limits_{\text{all } x}[x \cdot p(x)]$ Equation 5.1.

$= (-20)(0.2) + (-10)(0.3) + (30)(0.4) + (60)(0.1)$

Multiply each value by its probability, and sum.

$= (-4) + (-3) + (12) + (6) = 11$

STEP 3 The mean, or long-run value, of X is 11. On average, the project supervisor estimated the project was 11 hours ahead of schedule. In this example the mean is *not* a possible value of X.

As the sample size increases, the sample mean, $\bar{x}$, will tend to, or approach, the population mean $\mu = 11$.

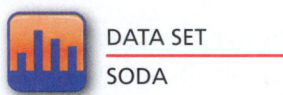
DATA SET

SODA

Example 5.10 The Health Risks of Soda

According to a Gallup poll, almost half of all Americans drink at least one glass of soda per day.[9] There is some evidence that this habit leads to increased health risks, especially diabetes, heart disease, obesity, and high cholesterol. Suppose X is a random variable that represents the number of glasses of soda that a randomly selected American drinks each day. The probability distribution for X is given in the following table.

x	0	1	2	3	4	5
$p(x)$	0.55	0.28	0.09	0.04	0.03	0.01

Find the expected number of glasses of soda that an American drinks each day.

SOLUTION

STEP 1 X is a discrete random variable (it takes on only a finite number of values). Find the mean using Equation 5.1.

$\mu = \sum\limits_{\text{all } x}[x \cdot p(x)]$ Equation 5.1.

$= (0)(0.55) + (1)(0.28) + (2)(0.09) + (3)(0.04) + (4)(0.03) + (5)(0.01)$

Sum of each value times probability.

$= 0 + 0.28 + 0.18 + 0.12 + 0.12 + 0.05 = 0.75$

STEP 2 The mean number of glasses of soda per day for Americans is 0.75.

The **variance** and **standard deviation** of a random variable measure the spread of the distribution. The variance is computed using the expected value function, and the standard deviation together with the mean can be used to determine the most likely values of the random variable.

Definition

Let X be a discrete random variable with probability mass function $p(x)$. The **variance** of X is

$$\underbrace{\text{Var}(X) = \sigma^2 = \sigma_X^2}_{\text{Notation}} = \underbrace{\sum_{\text{all } x}[(x-\mu)^2 \cdot p(x)]}_{\text{Calculation}} = \underbrace{E[(X-\mu)^2]}_{\substack{\text{Definition in terms} \\ \text{of expected value}}} \qquad (5.3)$$

The **standard deviation** of X is the positive square root of the variance:

$$\underbrace{\sigma = \sigma_X}_{\text{Notation}} = \underbrace{\sqrt{\sigma^2}}_{\text{Calculation}} \tag{5.4}$$

A CLOSER LOOK

1. In words, the variance is the expected value of the *squared deviations about the mean.*

2. The symbol Var stands for variance and is a function. The function Var takes a random variable as an input and returns the variance.

3. To compute the variance using Equation 5.3:

   a. Find the mean, μ, of X using Equation 5.1.

   b. Find each difference: $(x - \mu)$.

   c. Square each difference: $(x - \mu)^2$.

   d. Multiply each squared difference by the associated probability.

   e. Sum the products.

4. There is a *computational formula* for the variance of a random variable.

Computational Formula for σ^2

$$\sigma^2 = E(X^2) - E(X)^2 = E(X^2) - \mu^2 \tag{5.5}$$

In words, the variance is the expected value of X squared minus the expected value of X, squared.

Example 5.11 Children in Day Care

DATA SET

DAYCARE

Suppose the discrete random variable X, the age of a randomly selected child at the Looney Toons Child Care Center in Cleveland, Ohio, has the probability distribution given in the following table.

x	1	2	3	4	5	6	7
$p(x)$	0.05	0.10	0.15	0.25	0.20	0.15	0.10

a. Find the expected value, variance, and standard deviation of X.

b. Find the probability that the random variable X takes on a value within one standard deviation of the mean.

SOLUTION

a. *Expected value*: Find the expected value of X using Equation 5.1.

$$\begin{aligned}
E(X) &= \sum_{\text{all } x}[x \cdot p(x)] \\
&= (1)(0.05) + (2)(0.10) + (3)(0.15) + (4)(0.25) + (5)(0.20) \\
&\quad + (6)(0.15) + (7)(0.10) \\
&= 0.05 + 0.20 + 0.45 + 1.00 + 1.00 + 0.90 + 0.70 \\
&= 4.30 = \mu
\end{aligned}$$

Variance: Find the variance of X using Equation 5.3.

$$\text{Var}(X) = \sum_{\text{all } x} (x - \mu)^2 \cdot p(x) \qquad \text{Equation 5.3.}$$

$$= (1 - 4.30)^2(0.05) + (2 - 4.30)^2(0.10) + (3 - 4.30)^2(0.15)$$
$$+ (4 - 4.30)^2(0.25) + (5 - 4.30)^2(0.20) + (6 - 4.30)^2(0.15)$$
$$+ (7 - 4.30)^2(0.10) \qquad \text{Sum over all values of } x.$$

$$= (10.89)(0.05) + (5.29)(0.10) + (1.69)(0.15)$$
$$+ (0.09)(0.25) + (0.49)(0.20) + (2.89)(0.15)$$
$$+ (7.29)(0.10) \qquad \text{Square each difference.}$$

$$= 0.5445 + 0.5290 + 0.2535 + 0.0225$$
$$+ 0.0980 + 0.4335 + 0.7290 \qquad \text{Compute each product.}$$

$$= 2.6100 = \sigma^2 \qquad \text{Sum the products.}$$

Here is a tabular method for computing the variance using the definition. Sum the last column to obtain σ^2.

x	$x - \mu$	$(x - \mu)^2$	$p(x)$	$(x - \mu)^2 \cdot p(x)$	
1	−3.30	10.89	0.05	0.5445	
2	−2.30	5.29	0.10	0.5290	
3	−1.30	1.69	0.15	0.2535	
4	−0.30	0.09	0.25	0.0225	Sum this column.
5	0.70	0.49	0.20	0.0980	
6	1.70	2.89	0.15	0.4335	
7	2.70	7.29	0.10	0.7290	
				2.6100	$\leftarrow \sigma^2$

Variance: Using the *computational formula*.

Find $E(X^2)$, the expected value of X^2.

$$E(X^2) = \sum_{\text{all } x} \left[x^2 \cdot p(x) \right] \qquad \text{Equation 5.2.}$$

$$= 1^2(0.05) + 2^2(0.10) + 3^2(0.15) + 4^2(0.25) + 5^2(0.20)$$
$$+ 6^2(0.15) + 7^2(0.10) \qquad \text{Sum over all values of } x.$$

$$= 1(0.05) + 4(0.10) + 9(0.15) + 16(0.25) + 25(0.20)$$
$$+ 36(0.15) + 49(0.10) \qquad \text{Square each } x.$$

$$= 0.05 + 0.40 + 1.35 + 4.00 + 5.00 + 5.40 + 4.90 \qquad \text{Compute each product.}$$

$$= 21.10 \qquad \text{Sum the products.}$$

Use this result to find the variance.

$$\sigma^2 = E(X^2) - \mu^2 \qquad \text{Equation 5.5.}$$

$$= 21.10 - (4.30)^2 \qquad \text{Use previous results.}$$

$$= 21.10 - 18.49 \qquad \text{Find } \mu^2.$$

$$= 2.61 \qquad \text{Find the difference.}$$

Variance: Using a tabular method.

Start at the middle of each row and work toward the ends. Sum the outer columns to obtain μ and $E(X^2)$.

Solution Trail 5.11b

KEYWORDS

- Within one standard deviation of the mean

TRANSLATION

- In the interval $(\mu - \sigma, \mu + \sigma)$, within one *step* in each direction from the mean

CONCEPTS

- Probability distribution
- Probability statement

VISION

The probability distribution is given. Use the *Translation* to write a mathematical probability statement. Find the value(s) of X that lie in the interval, and add the corresponding probabilities.

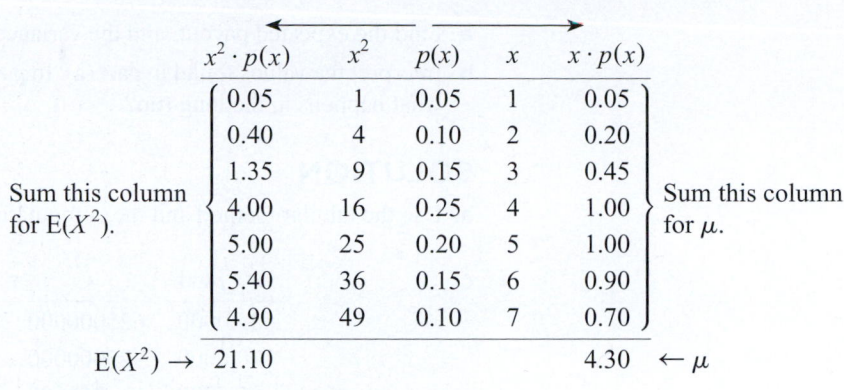

$x^2 \cdot p(x)$	x^2	$p(x)$	x	$x \cdot p(x)$
0.05	1	0.05	1	0.05
0.40	4	0.10	2	0.20
1.35	9	0.15	3	0.45
4.00	16	0.25	4	1.00
5.00	25	0.20	5	1.00
5.40	36	0.15	6	0.90
4.90	49	0.10	7	0.70

Sum this column for $E(X^2)$.

Sum this column for μ.

$$E(X^2) \rightarrow \quad 21.10 \qquad\qquad\qquad 4.30 \quad \leftarrow \mu$$

$$\sigma^2 = E(X^2) - \mu^2 = 21.10 - (4.30)^2 = 2.61$$

Standard deviation: The positive square root of the variance.

$$\sigma = \sqrt{\sigma^2} = \sqrt{2.61} \approx 1.6155$$

b. $P(\mu - \sigma \le X \le \mu + \sigma)$ Translation to a probability statement.

$= P(4.30 - 1.6155 \le X \le 4.30 + 1.6155)$ Use values for μ and σ.

$= P(2.6845 \le X \le 5.9155)$ Compute the difference and sum.

$= P(X = 3) + P(X = 4) + P(X = 5)$ Find values of X in the interval.

$= 0.15 + 0.25 + 0.20$ Use corresponding probabilities.

$= 0.60$ Compute the sum.

The probability that X takes on a value within one standard deviation of the mean is 0.60.

A CLOSER LOOK

1. The computational formula for the variance is quicker and produces less round-off error. Use Equation 5.5 to find the variance of a discrete random variable.

2. In Example 5.11, the random variable X is not (approximately) normal; the empirical rule does *not* apply. In addition, even though Chebyshev's rule applies to any distribution, it should not be used if the probability distribution is known. Chebyshev's rule provides only an *estimate*, a lower bound for the probability that X is within k standard deviations of the mean. The exact probability can be determined using the known probability distribution. (Actually, Chebyshev's rule can't help at all here, because k must be greater than 1.)

3. Neither the TI-84 Plus C nor Minitab has a built-in menu function to find the mean, variance, and standard deviation of a discrete random variable. However, it's easy to perform list operations on the calculator and column operations in Minitab or Excel in order to produce these summary statistics. See the technology manuals for details.

Example 5.12 Rockin' 5's

DATA SET

ROCKIN5

One of the newest instant scratch-off games in the North Carolina Education Lottery is Rockin' 5's. The payout (in dollars) is a discrete random variable with probability distribution as given in the following table.[10]

x	25000	5000	500	100	50	25
$p(x)$	0.000001	0.000004	0.000022	0.000800	0.002667	0.006667

x	20	10	5	4	2	0
$p(x)$	0.006667	0.013333	0.040000	0.046667	0.100000	0.783172

a. Find the expected payout, and the variance and standard deviation of the payout.

b. Interpret the values found in part (a). In particular, if it costs $2.00 to purchase a ticket, what happens in the long run?

SOLUTION

a. Use the tabular method and the computation formula for the variance.

$x^2 \cdot p(x)$	x^2	$p(x)$	x	$x \cdot p(x)$
625.0000	625000000	0.000001	25000	0.0250
100.0000	25000000	0.000004	5000	0.0200
5.5000	250000	0.000022	500	0.0110
8.0000	10000	0.000800	100	0.0800
6.6675	2500	0.002667	50	0.1334
4.1669	625	0.006667	25	0.1667
2.6668	400	0.006667	20	0.1333
1.3333	100	0.013333	10	0.1333
1.0000	25	0.040000	5	0.2000
0.7467	16	0.046667	4	0.1867
0.4000	4	0.100000	2	0.2000
0.0000	0	0.783172	0	0.0000

$E(X^2) \rightarrow$ 755.4811   1.2894 $\leftarrow \mu$

$$\sigma^2 = E(X^2) - \mu^2 = 755.4811 - (1.2894)^2 = 753.8187$$
$$\sigma = \sqrt{\sigma^2} = \sqrt{753.8187} = 27.4558$$

b. The mean payout from this instant scratch-off game is approximately $1.29, with variance $753.82 and standard deviation $27.46. If it costs $2.00 to play (purchase a ticket), in the long run the player loses $2.00 - 1.29 = 0.71$, or 71 cents, on average. This doesn't seem like much. However, from the State's point of view, every time someone buys a ticket, North Carolina makes 71 cents (on average). ●

SECTION 5.3 EXERCISES

Concept Check

5.44 True/False   The mean of a discrete random variable must be a possible value of the random variable.

5.45 True/False   The expected value of a discrete random variable can be negative.

5.46 True/False   The standard deviation of a discrete random variable is always nonnegative.

5.47 True/False   The standard deviation of a discrete random variable could be 0.

5.48 True/False   The computational formula for σ^2 should only be used when the number of possible values for the random variable is small.

5.49 Fill in the Blank   Suppose $f(X)$ is a function of a discrete random variable. $E[f(X)] = $ _____.

5.50 Fill in the Blank   The variance of a discrete random variable is the expected value of _____.

Practice

5.51 Suppose X is a discrete random variable. Complete the table below and find the mean, variance, and standard deviation of X.

$x^2 \cdot p(x)$	x^2	$p(x)$	x	$x \cdot p(x)$
		0.10	2	
		0.16	4	
		0.20	6	
		0.24	8	
		0.18	10	
		0.12	12	

5.52 The probability distribution for a random variable X is given in the table below.

x	5	10	15	20
$p(x)$	0.10	0.15	0.70	0.05

a. Find the mean, variance, and standard deviation of X.
b. Find the probability X takes on a value smaller than the mean.
c. Using the probability distribution, explain why the value of the mean of X makes sense.

5.53 Suppose Y is a discrete random variable with probability distribution given in the table below.

y	-20	-10	0	10	20
$p(y)$	0.30	0.15	0.10	0.15	0.30

a. Find μ, σ^2, and σ.
b. Find $P(\mu - 2\sigma \leq Y \leq \mu + 2\sigma)$.
c. Find $P(Y \geq \mu)$ and $P(Y > \mu)$.

5.54 Suppose the random variable X has the probability distribution given in the table below.

x	2	3	5	7	11	13
$p(x)$	0.15	0.25	0.15	0.10	0.30	0.05

a. Find the mean, variance, and standard deviation of X.
b. Suppose the random variable Y is defined by $Y = 2X + 1$. Find the mean, variance, and standard deviation of Y.
c. Suppose the random variable W is defined by $W = X^2 + 1$. Find the mean, variance, and standard deviation of W.

5.55 Suppose X is a discrete random variable with probability distribution as given in the table below.

x	1	2	3	5	8	13	21
$p(x)$	0.05	0.10	0.15	0.20	0.25	0.20	0.05

a. Find the mean, variance, and standard deviation of X.
b. Find the probability X is more than one standard deviation from the mean.
c. Find $P(X \leq \mu + 2\sigma)$.

Applications

5.56 Manufacturing and Product Development   In a quarter-pound bag of red pistachio nuts, some shells are too difficult to pry open by hand. Suppose the random variable X, the number of pistachios in a randomly selected bag that cannot be opened by hand, has the probability distribution given in the table below.

x	0	1	2	3	4	5
$p(x)$	0.500	0.250	0.100	0.050	0.075	0.025

a. Is this a valid probability distribution? Justify your answer.
b. Find the mean, variance, and standard deviation of X.
c. Find the probability X takes on a value less than the mean.
d. Suppose two bags of pistachios are selected at random. What is the probability that both bags have four or more pistachios too difficult to open by hand?

5.57 Sports and Leisure   Suppose the number of rides that a visitor enjoyed at Disney World during a day is a random variable with probability distribution given in the table below.[11]

x	5	6	7	8	9	10	11	12
$p(x)$	0.04	0.07	0.09	0.12	0.20	0.30	0.13	0.05

a. Find the mean number of rides for a Disney World visitor.
b. Find the variance and standard deviation of the number of rides for a Disney World visitor.
c. Find the probability that the number of rides for a randomly selected Disney World visitor is within one standard deviation of the mean. Write a Solution Trail for this problem.
d. Find the probability that the number of rides for a randomly selected Disney World visitor is less than one standard deviation above the mean.

5.58 Public Policy and Political Science   Approximately 100 children's products are recalled every year.[12] In particular, children's clothing is recalled for a variety of reasons, for example, drawstrings that are too long and pose a hazard, small buttons that may break off and cause choking, and material that fails to meet federal flammability standards. Suppose the number of recalls of children's clothing during a given month is a random variable with probability distribution given in the table below.

x	0	1	2	3	4	5	6
$p(x)$	0.005	0.185	0.275	0.305	0.200	0.020	0.010

a. Find the mean, variance, and standard deviation of the number of recalls of children's clothing during a given month.
b. Suppose the number of recalls in a given month is at least three. What is the probability that the number of recalls that month will be at least five?
c. If the number of recalls in a given month is above more than one standard deviation from the mean, the federal government issues a special warning directed toward parents. What is the probability that a special warning will be issued during a given month?

5.59 Public Health and Nutrition   A bill introduced into the Virginia State Senate stipulated that the owner of a tanning facility must identify and document the skin type of every customer, and must advise every customer of the maximum time of recommended exposure in the tanning device.[13] Past records indicate that most sessions range from 10 to 30 minutes. Suppose the duration of a tanning session (in minutes) at the Solar Planet tanning facility in Herendon, Virginia, is a discrete random variable with probability distribution given in the table below.

x	10	12	15	20	25	30
p(x)	0.30	0.25	0.15	0.12	0.10	0.08

a. Find the mean, variance, and standard deviation of the duration of a tanning session time.
b. Find the probability that a randomly selected session has a duration within one standard deviation of the mean.
c. Find the probability that a randomly selected session has a duration within two standard deviations of the mean.
d. Suppose a sunlamp lasts for 100 hours. After approximately how many tanning sessions will the sunlamp have to be replaced?

5.60 Public Health and Nutrition   For nurses, additional patients contribute heavily to increased stress and job burnout. Suppose the number of patients assigned to each nurse at the Banner Thunderbird Medical Center in Glendale, Arizona, is a random variable with probability distribution given in the table below.[14]

x	3	4	5	6	7	8
p(x)	0.07	0.12	0.18	0.37	0.17	0.09

a. Find the mean, variance, and standard deviation of the number of patients assigned to each nurse.
b. Find the probability that the number of assigned patients is greater than one standard deviation to the right of the mean.
c. Suppose three nurses are selected at random. What is the probability that exactly two of the three have five assigned patients?

5.61 Psychology and Human Behavior   A certain elevator in Tampa's tallest office building, 100 North Tampa, is used heavily between 8:00 A.M. and 9:00 A.M. as employees arrive for work. Suppose the number of people who board the elevator on the ground floor going up is a random variable with probability distribution given in the table below.

x	1	2	3	4	5	6
p(x)	0.002	0.010	0.050	0.060	0.080	0.090

x	7	8	9	10	11	12
p(x)	0.100	0.120	0.140	0.150	0.150	0.048

a. Find μ, σ^2, and σ.
b. For a randomly selected elevator ride from the ground floor going up, what is the probability that the number of riders is within one standard deviation of the mean?
c. For two randomly selected elevator rides from the ground floor going up, what is the probability that both trips have a number of riders more than two standard deviations from the mean?

Extended Applications

5.62 Manufacturing and Product Development   A cordless drill has several torque settings for driving different screws into different materials. A manufacturer models the torque setting required for a randomly selected task with a probability distribution given by

$$p(x) = \frac{(x - 12)^2}{247} \qquad x = 1, 5, 10, 15, 20$$

a. Verify that this is a valid probability distribution.
b. Find the mean, variance, and standard deviation of X.
c. A torque setting is classified as *rare* if it is more than one standard deviation from the mean. Find the probability of a task requiring a rare torque setting.

5.63 Six Degrees of Kevin Bacon   The actor Kevin Bacon has been in so many movies that *almost everyone* in Hollywood can be connected to him within six degrees. Let $n = 1,326,359$, the number of actors linked to Kevin Bacon. The probability distribution for the Bacon Number of a randomly selected Hollywood personality is given in the following table.[15]

x	p(x)	x	p(x)	x	p(x)
0	1/n	1	2511/n	2	262,544/n
3	839,562/n	4	204,764/n	5	15,344/n
6	1397/n	7	204/n	8	32/n

a. Find the mean, variance, and standard deviation for the Bacon Number.
b. Find $P(X \geq \mu - \sigma)$.
c. The number of movies, Y, in which a Hollywood personality has appeared is related to the Bacon Number by the formula $Y = 2X + 5$. Find the mean, variance, and standard deviation of the number of movies in which a Hollywood personality has appeared.

5.64 Marketing and Consumer Behavior   While the temperature range of most household ovens is approximately 200–600°F, most consumers only use four or five common settings. Suppose the probability distribution for the oven-temperature setting for a randomly selected use is given in the table below.

x	300	325	350	375	400	500
p(x)	0.040	0.205	0.400	0.075	0.200	0.080

a. Find the mean, variance, and standard deviation of the oven temperature settings.
b. Suppose three different uses are randomly selected. Find the probability that the temperature settings for all three are at least 400°F.
c. Suppose three different uses are randomly selected. Find the probability that exactly one use is for 350°F.

5.65 Education and Child Development   An elementary class rarely remains the same size from the beginning of the school year until the end. Families move in and out of the district, some students are reassigned, and scheduling conflicts necessitate changes. Suppose the change in the number of students in a class is a random variable with probability distribution given by

$$p(x) = \frac{|x| + 1}{19} \qquad x = -3, -2, -1, 0, 1, 2, 3$$

a. Verify that this is a valid probability distribution.
b. Find the mean, variance, and standard deviation of the change in class size.
c. Suppose two classes are selected at random. Find the probability that both classes remain the same size for the entire year.

5.66 Psychology and Human Behavior   Many organizations publish wedding guidelines with specific suggestions regarding the reception, wedding cake, flowers, and even themes and styles. However, the number of bridesmaids and groomsmen is usually a very personal decision made by the bride and groom. For semiformal weddings, one to six bridesmaids are typical, with a possible flower girl and/or ring bearer.[16] Suppose the probability distribution for the number of bridesmaids at a semiformal wedding is given in the table below.

x	1	2	3	4	5	6
$p(x)$	0.05	0.21	0.34	0.27	0.08	0.05

a. Find the mean, variance, and standard deviation for the number of bridesmaids at a semiformal wedding.
b. Suppose a randomly selected semiformal wedding has at least four bridesmaids. What is the probability that it has exactly six bridesmaids?
c. Suppose four semiformal weddings are randomly selected. What is the probability that all four have at least three bridesmaids?

Challenge

5.67 Dichotomous Random Variable   Suppose the random variable X takes on only two values, according to the probability distribution given below.

x	0	1
$p(x)$	0.4	0.6

a. Find the mean, variance, and standard deviation of X.
b. Suppose $P(X = 1) = 0.7$, and therefore $P(X = 0) = 1 - 0.7 = 0.3$. Find the mean, variance, and standard deviation of X.
c. Suppose $P(X = 1) = 0.8$. Find the mean, variance, and standard deviation of X.
d. Suppose $P(X = 1) = p$ and $P(X = 0) = 1 - p = q$. Find the mean, variance, and standard deviation of X in terms of p and q.
e. For what values of p (and q) is the variance of X greatest?

5.68 Linear Function   Suppose X is a discrete random variable with mean μ_X and variance σ_X^2. Let Y be a linear function of X, such that $Y = aX + b$, where a and b are constants. Find the mean and variance of Y in terms of μ_X and σ_X^2.

5.69 Variance Computation Formula   Suppose X is a discrete random variable that takes on a finite number of values. Prove the variance *computation formula*. That is, show that

$$E[(X - \mu)^2] = E(X^2) - \mu^2$$

Hint: Write $E[(X - \mu)^2]$ as a sum using the probability mass function $p(x)$. Expand and simplify.

5.70 Standardization   Suppose X is a discrete random variable with mean μ_X and variance σ_X^2. Let Y be defined in terms of X by

$$Y = \frac{X - \mu_X}{\sigma_X}$$

Find the mean, variance, and standard deviation of Y.

5.4 The Binomial Distribution

In the previous sections, the general definition and probability distribution for a discrete random variable were introduced. This section presents a specific discrete random variable that is common and very important. The **binomial random variable** can be used to model many real-world populations and to do more formal inference.

As with any random variable, there is a related experiment in the background. Consider the following experiments (and look for similarities).

1. Simply toss a coin 50 times and record the sequence of heads and tails.

2. For a random sample of 100 voters, ask each one whether he or she is going to vote for a particular candidate. Record the sequence of yes and no responses.

3. Select a random sample of 25 customers at a fast-food restaurant and record whether or not each pays with exact change.

4. Drill a series of randomly selected test oil wells. Each well will either yield oil worth drilling or be classified as dry. Record the result for each well.

There are four common properties in all of these experiments. These properties are used to describe a **binomial experiment**, and they are necessary to define a **binomial random variable**.

Properties of a Binomial Experiment

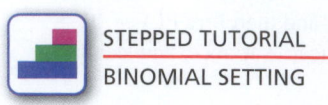
STEPPED TUTORIAL
BINOMIAL SETTING

1. The experiment consists of *n* identical trials.

2. Each trial can result in only one of two possible (mutually exclusive) outcomes. One outcome is usually designated a success (S) and the other a failure (F).

3. The outcomes of the trials are independent.

4. The probability of a success, *p*, is constant from trial to trial.

A CLOSER LOOK

1. A *trial* is a small part of the larger experiment. A trial results in a single occurrence of either a success or a failure. For example, flipping a coin once, or drilling one test oil well, is a single trial. A typical binomial experiment might consist of $n = 50$ trials.

2. A *success* does not have to be a *good* thing. For example, the experiment may consist of injecting animals with a potential carcinogen and checking for the development of tumors. A *success* might be an animal that develops at least one tumor. *Success* and *failure* could stand for *heads* and *tails*, *acceptable* and *not acceptable*, or even *dead* and *alive*.

3. Trials are *independent* if whatever happens on one trial has no effect on any other trial. For example, any one voter response has no effect on any other voter response.

4. The probability of a success on every trial is exactly the same. For example, the probability of the tossed (fair) coin landing with head face up is always 1/2.

In a binomial experiment, outcomes consist of sequences of Ss and Fs. For example, SSFSFSFS is a possible outcome in a binomial experiment with $n = 8$ trials.

The Binomial Random Variable

The **binomial random variable** maps each outcome in a binomial experiment to a real number, and is defined to be the *number of successes* in *n* trials.

Notation

Why is P(F) = 1 − *p*?

1. The probability of a success is denoted by *p*. Therefore, $P(S) = p$ and $P(F) = 1 - p = q$.

2. A binomial random variable X is completely determined by the number of trials *n* and the probability of a success *p*. If we know those two values, then we will be able to answer any probability question involving X.

   The shorthand notation $X \sim B(n, p)$ means X is (distributed as) a binomial random variable with *n* trials and probability of a success *p*.

   For example, $X \sim B(25, 0.4)$ means X is a binomial random variable with 25 trials and probability of success 0.4.

Our goal now is to find the probability distribution for a binomial random variable. Given *n* and *p*, we want to find the probability of obtaining *x* successes in *n* trials, $P(X = x) = p(x)$. We will solve this problem by first considering a simple case with $n = 5$.

For example, suppose five people are selected at random. Let p be the probability that a randomly selected person snores.

Example 5.13 Binomial Experiment with $n = 5$

Consider a binomial experiment with $n = 5$ trials and probability of success p.

a. A typical outcome with two successes is SFFSF. Find the probability of this outcome.

b. Another possible outcome with two successes is FSFSF. Find the probability of this outcome.

c. Compare your results from (a) and (b).

d. Find the probability that $X = 2$ successes.

SOLUTION

a. The probability of this outcome is

$$P(SFFSF) = P(S \cap F \cap F \cap S \cap F)$$

Probability of a success on the first trial *and* a failure on the second trial *and* . . . .

$$= P(S) \cdot P(F) \cdot P(F) \cdot P(S) \cdot P(F)$$

Trials are independent (property of a binomial experiment).

$$= p \cdot (1 - p) \cdot (1 - p) \cdot p \cdot (1 - p)$$

$P(S) = p$, $P(F) = 1 - p$.

$$= p^2(1 - p)^3$$

Multiplication is commutative.

b. The probability of this outcome is

$$P(FSFSF) = P(F) \cdot P(S) \cdot P(F) \cdot P(S) \cdot P(F)$$

Trials are independent.

$$= (1 - p) \cdot p \cdot (1 - p) \cdot p \cdot (1 - p)$$

$P(S) = p$, $P(F) = 1 - p$.

$$= p^2(1 - p)^3$$

Multiplication is commutative.

c. The results are identical. Every other outcome with two successes, and therefore three failures, has exactly the sample probability, $p^2(1 - p)^3$. Therefore, the probability of an outcome depends on the *number of successes* (and failures), *not* on the order in which they appear.

d. To compute the probability that $X = 2$ successes, find all the outcomes mapped to a 2, and add the corresponding probabilities. However, every outcome that is mapped to a 2 has the *same* probability, so all we need to know is *how many* outcomes are mapped to a 2.

$$P(X = 2) = (\text{number of outcomes with 2 successes})p^2(1 - p)^3 \qquad \blacksquare$$

The number of successes and the number of failures must sum to n, the total number of trials.

▶ Generalizing, suppose $X \sim B(n, p)$. We want to find the probability of obtaining x successes in n trials. The probability of any *single outcome* with x successes, and therefore $n - x$ failures, is $p^x(1 - p)^{n-x}$. The probability of obtaining x successes is

$$P(X = x) = (\text{number of outcomes with } x \text{ successes})p^x(1 - p)^{n-x}$$

We need a method for quickly counting the number of outcomes with x successes. Recall: For any positive whole number n, the symbol $n!$ (read "n factorial") is defined by

$$n! = n(n - 1)(n - 2) \cdots (3)(2)(1)$$

In addition, $0! = 1$ (0 factorial is 1). Given a collection of n items, the number of combinations of size x is given by

$$_nC_x = \binom{n}{x} = \frac{n!}{x! \, (n - x)!}$$

The number of outcomes with x successes is determined using combinations. Suppose $X \sim B(n, p)$. The number of outcomes with x successes is $\binom{n}{x}$. We can now write an expression for the probability of obtaining x successes in n trials. ◀

Can you figure out why this is true?

Solution Trail 5.14a

KEYWORDS

- 75%
- 10 bottles
- Selected at random
- Exactly six

TRANSLATION

- Binomial experiment

CONCEPTS

- Binomial probability distribution

VISION

Assume this problem describes a binomial experiment: $n = 10$ trials (bottles), P(ice wine) $= p = 0.75$, p is the same on each trial, and the trials are independent. Define the binomial random variable and write a probability statement.

The Binomial Probability Distribution

Suppose X is a binomial random variable with n trials and probability of a success p: $X \sim \mathrm{B}(n, p)$. Then

$$p(x) = \mathrm{P}(X = x) = \underbrace{\binom{n}{x}}_{\substack{\text{Number of outcomes} \\ \text{with } x \text{ successes}}} \underbrace{p^x(1 - p)^{n-x}}_{\substack{\text{Probability of } x \text{ successes and } n - x \\ \text{failures in any single outcome}}} \quad x = 0, 1, 2, 3, \ldots, n \qquad (5.6)$$

Example 5.14 Dessert Wine

Ice wine is made from grapes that are allowed to freeze naturally while still on the vine. Seventy-five percent of all ice wine in Canada is made in Ontario.[17] The remaining 25% is made in other provinces. Suppose 10 bottles of Canadian ice wine are selected at random.

a. Find the probability that exactly six bottles are from Ontario.

b. Find the probability that at least seven bottles are from Ontario.

SOLUTION

Let X be the number of bottles of ice wine (out of the 10 selected) that come from Ontario. The experiment exhibits the properties of a binomial experiment, so $X \sim \mathrm{B}(10, 0.75)$ ($n = 10, p = 0.75$).

a. Translate the words into a probability statement involving the random variable X. *Exactly six* means $X = 6$. Use Equation 5.6 to find the relevant probability.

$$P(X = 6) = \binom{10}{6}(0.75)^6(1 - 0.75)^{10-6} \qquad \text{\small Equation 5.6.}$$

$$= (210)(0.75)^6(0.25)^4 \qquad \text{\small Compute } {}_{10}C_6.$$

$$= 0.1460$$

The probability that exactly 6 of the 10 randomly selected bottles of ice wine come from Ontario is 0.1460.

b. The probability that X is greater than or equal to 7 means the probability that X is 7 or 8 or 9 or 10.

$$P(X \geq 7) = P(X = 7 \text{ or } X = 8 \text{ or } X = 9 \text{ or } X = 10)$$

$$= P(X = 7) + P(X = 8) + P(X = 9) + P(X = 10)$$

Or means union; the outcomes are disjoint.

$$= \binom{10}{7}(0.75)^7(0.25)^3 + \binom{10}{8}(0.75)^8(0.25)^2$$

$$+ \binom{10}{9}(0.75)^9(0.25)^1 + \binom{10}{10}(0.75)^{10}(0.25)^0 \qquad \text{\small Use Equation 5.6 four times.}$$

$$= 0.2503 + 0.2816 + 0.1877 + 0.0563$$

Compute combinations and powers, and multiply.

$$= 0.7759$$

The probability that at least seven bottles of ice wine are from Ontario is 0.7759. ●

Solution Trail 5.14b

KEYWORDS

- At least seven

TRANSLATION

- Seven or more, X is 7 or greater.

CONCEPTS

- Binomial probability distribution

VISION

We already have the distribution for X. We need a probability statement. Use Equation 5.6 to calculate the necessary probabilities.

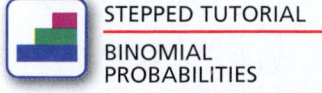

STEPPED TUTORIAL

BINOMIAL PROBABILITIES

Note:

1. The two most important elements for solving this problem are (1) the probability distribution and (2) the probability statement.

2. Often, the properties of a binomial experiment will not be stated explicitly in the problem. Usually we must read into the problem to *see* the *n* trials, to *identify* a success, to *recognize* independence, and to *presume* that the probability of a success remains constant from trial to trial.

A CLOSER LOOK

1. Even for small values of *n*, many of the probabilities associated with a binomial random variable are a little tedious to calculate, and are subject to lots of round-off error. Technology helps, and Table 1 in Appendix A presents **cumulative probabilities** for a binomial random variable, for various values of *n* and *p*.

2. Cumulative probability is an important concept. If $X \sim B(n, p)$, the probability that X takes on a value *less than or equal to x* is cumulative probability. Accumulate all the probability associated with values up to and including *x*. Symbolically, cumulative probability is

$$P(X \le x) = \sum_{k=0}^{x} P(X = k) \tag{5.7}$$

$$= P(X = 0) + P(X = 1) + P(X = 2) + \cdots + P(X = x)$$

3. Graphically, cumulative probability is like standing on a special staircase, looking down (or back), and measuring the height. The steps are labeled 0, 1, 2, . . . , *n*, the height of step *x* is $P(X = x)$, and the total height of the staircase is 1. Figure 5.4 illustrates $P(X \le 3)$. The number of steps is $n + 1$, and the height of each step depends on *n* and *p*. In this example, $n = 10$ and $p = 0.25$. The largest steps (the highest probabilities) are associated with $X = 1, 2$, and 3. Steps 7, 8, 9, and 10 are hard to see because $P(X = 7)$, $P(X = 8)$, $P(X = 9)$, and $P(X = 10)$ are so small.

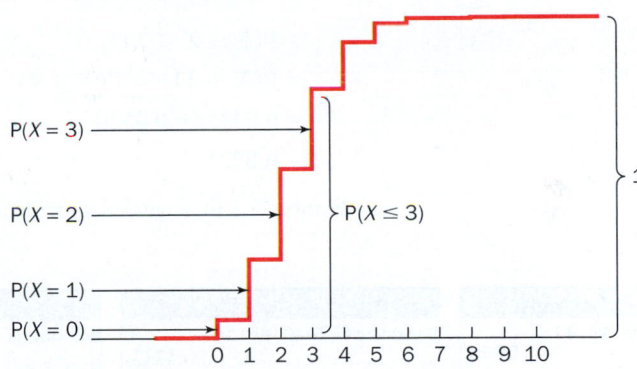

Figure 5.4 Staircase analogy to cumulative probability.

Every probability question about a binomial random variable can be answered using *cumulative* probability. There may also be other, faster methods, but cumulative probability always works. The following example illustrates some of the techniques for converting to and using cumulative probability.

Example 5.15 A Slice Above

Approximately 40% of all pizza orders are carry-out.[18] Suppose 20 pizza orders are randomly selected.

a. Find the probability that at most 8 are carry-out orders.

b. Find the probability that exactly 10 are carry-out orders.

c. Find the probability that at least 7 are carry-out orders.

d. Find the probability that between 5 and 11 (inclusive) are carry-out orders.

SOLUTION

Let X be the number of pizza orders (out of the 20 selected) that are carry-out. X is a binomial random variable with $n = 20$ and $p = 0.40$: $X \sim B(20, 0.40)$.

a. The probability that at most 8 are carry-out orders

$$= P(X \le 8) \qquad \text{Translate the words into mathematics.}$$
$$= 0.5956 \qquad \text{Cumulative probability; use Table 1 in Appendix A.}$$

b. The probability that exactly 10 are carry-out orders

$$= P(X = 10) \qquad \text{Translate the words into mathematics.}$$
$$= P(X \le 10) - P(X \le 9) \qquad \text{Convert to cumulative probability.}$$
$$= 0.8725 - 0.7553 \qquad \text{Use Table 1 in Appendix A.}$$
$$= 0.1172$$

This solution may also be found by using the probability mass function for a binomial random variable (Equation 5.6). In addition, most statistical software can compute binomial probabilities for single values.

c. The probability that at least 7 are carry-out orders

$$= P(X \ge 7) \qquad \text{Translate the words into mathematics.}$$
$$= 1 - P(X < 7) \qquad \text{The complement rule.}$$
$$= 1 - P(X \le 6) \qquad \text{The first value } X \text{ takes on that is } less \ than \ 7 \text{ is } 6.$$
$$= 1 - 0.2500 \qquad \text{Use Table 1 in Appendix A.}$$
$$= 0.7500$$

d. The probability that between 5 and 11 (inclusive) are carry-out orders

$$= P(5 \le X \le 11) \qquad \text{Translate the words into mathematics.}$$
$$= P(X \le 11) - P(X \le 4) \qquad \text{Convert to cumulative probability.}$$
$$= 0.9435 - 0.0510 \qquad \text{Use Table 1 in Appendix A.}$$
$$= 0.8925$$

Figures 5.5 through 5.8 show technology solutions.

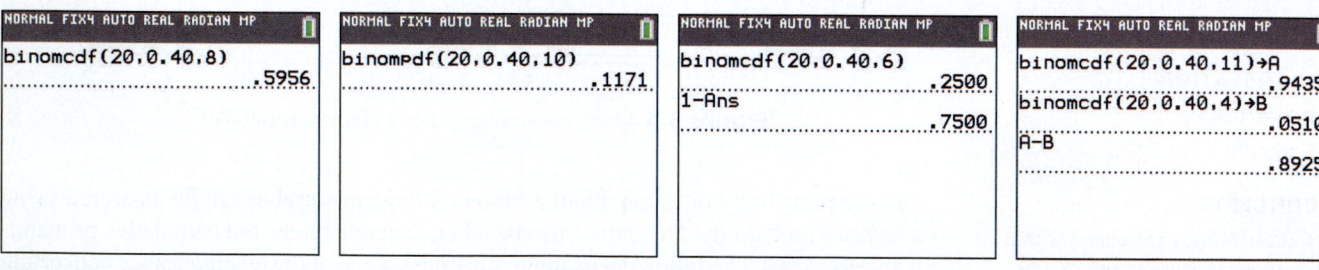

Figure 5.5 $P(X \le 8)$; cumulative probability.

Figure 5.6 $P(X = 10)$; using the probability mass function.

Figure 5.7 $P(X \ge 7)$; using cumulative probability.

Figure 5.8 $P(5 \le X \le 11)$; using cumulative probability.

The next example shows how the binomial distribution can be used to make an inference.

Example 5.16 Lower Your Cholesterol

The drug Lipitor, made by Pfizer, is used to lower cholesterol levels. It was first sold in 1997 and is the best-selling drug of all time. Based on clinical trials, Pfizer claims that approximately 10% of patients using Lipitor in a 40-mg dose will experience arthralgia, or joint pain, which is considered an adverse reaction.[19] Suppose 25 people who need Lipitor are selected at random. Each is given a 40-mg dose (per day), and the number of people who experience joint pain is recorded.

a. Find the probability that at most one person will experience joint pain.

b. Suppose seven people experience joint pain. Is there any evidence to suggest that Pfizer's claim is wrong? Justify your answer.

SOLUTION

Let X be the number of people (out of the 25 selected) who experience joint pain after taking Lipitor. X is a binomial random variable with $n = 25$ and $p = 0.10$: $X \sim B(25, 0.10)$.

Translate the words into a mathematical probability statement, convert to cumulative probability if necessary, and use Table I in the Appendix.

a. The probability that at most one will experience joint pain

$$= P(X \leq 1) \qquad \text{Translate the words into mathematics.}$$

$$= 0.2712 \qquad \text{Already cumulative probability; use Table 1 in Appendix A.}$$

The probability of at most one person experiencing joint pain is 0.2712.

b. Pfizer claims $p = 0.10$. This implies that the random variable X has a binomial distribution with $n = 25$ and $p = 0.10$.

Claim: $p = 0.10 \quad \rightarrow \quad X \sim B(25, 0.10)$.

The experimental outcome is that seven people experience joint pain.

Experiment: $x = 7$.

It seems reasonable to consider $P(X = 7)$ and draw a conclusion based on this probability. However, to be conservative (to give the person making the claim the benefit of the doubt), we always consider a *tail probability*. We accumulate the probability in a tail of the distribution, and if it is small, then there is evidence to suggest the claim is false.

So, which tail? It depends on the mean of the distribution (and later on, the alternative hypothesis). Formulas for the mean, variance, and standard deviation of a binomial random variable are given below. Intuitively, however, the mean of a binomial random variable is $\mu = np$. If $n = 25$ and $p = 0.10$, we expect to see $\mu = (25)(0.10) = 2.5$ people experience joint pain. Because $x = 7$ is to the right of the mean, we'll consider a right-tail probability. See Figure 5.9.

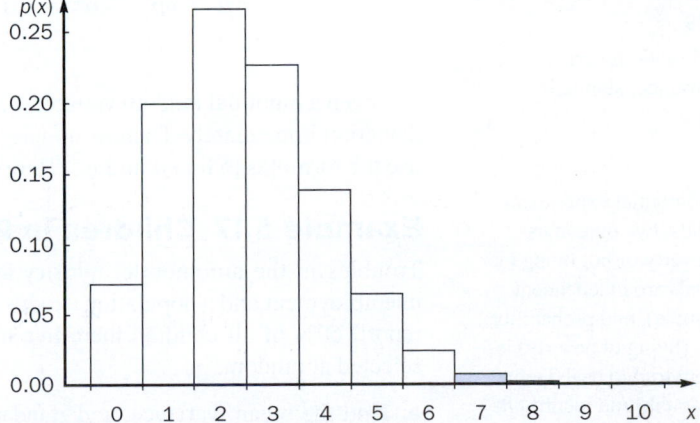

Figure 5.9 A portion of the probability histogram for the random variable X in Example 5.16. The right-tail probability $P(X \geq 7)$ is the sum of the heights of the rectangles above 7, 8, 9, . . . , 25.

Likelihood:

$$P(X \geq 7) = 1 - P(X < 7)$$ The complement rule.

$$= 1 - P(X \leq 6)$$ The first value X takes on that is *less than* 7 is 6.

$$= 1 - 0.9905$$ Cumulative probability; use Table 1 in Appendix A.

$$= 0.0095$$

Conclusion: Because this tail probability is so small (less than 0.05), it is very unusual to observe seven or more people with joint pain. But it happened! This is either an incredibly lucky occurrence, or someone is lying. We usually discount the lucky possibility, and conclude that there is evidence to suggest Pfizer's claim is false.

Figure 5.10 shows a technology solution.

Probability	JMP Formula
0.2712	= Binomial Distribution(0.10,25,1)
0.0095	= 1 - Binomial Distribution(0.10,25,6)

Figure 5.10 Probability calculations using JMP.

A random variable is often described, or characterized, by its mean and variance (or standard deviation): μ and σ^2 (or σ). If we know μ and σ, we can use Chebyshev's rule to determine the most likely values of the random variable. For *any* population (random variable), most (at least 89%) of the values are within three standard deviations of the mean. This fact provides another approach to statistical inference, for determining the likelihood of an experimental outcome.

To find the mean and variance of a binomial random variable, we could use the mathematical definitions (Equations 5.1 and 5.3). These formulas are used to produce the general results below. However, the mean is intuitive. Consider a binomial random variable $X \sim B(10, 0.5)$. We *expect* to see $(10)(0.5) = 5 = np$ successes in 10 trials. (Think about tossing a fair coin 10 times.) Similarly, if $X \sim B(100, 0.75)$, we expect to see $100(0.75) = 75 = np$ successes. The mean of a binomial random variable with n trials and probability of a success p is $\mu = np$.

Mean, Variance, and Standard Deviation of a Binomial Random Variable

If X is a binomial random variable with n trials and probability of a success p, $X \sim B(n, p)$, then

$$\mu = np \qquad \sigma^2 = np(1 - p) \qquad \sigma = \sqrt{np(1 - p)} \tag{5.8}$$

Given a binomial random variable, n, and p, we know the mean, variance, and standard deviation immediately. There is no need to create a table of values and probabilities, and use the formulas to find μ and σ^2. Here is an example to illustrate the use of this concept.

Example 5.17 Children in Poverty

Troubles in the automobile industry and the global economic downturn caused high unemployment and a population exodus in Detroit. According to the 2013 State of Detroit report, 60% of all children there live in poverty.[20] Suppose 100 children in Detroit are selected at random.

a. Find the mean, variance, and standard deviation of the number of children living in poverty.

b. Suppose 55 of the 100 children are living in poverty. Is there any evidence to suggest the report's claim is false? Justify your answer.

Solution Trail 5.17a

KEYWORDS

- 60%
- 100 children

TRANSLATION

- $p = 0.60$
- $n = 100$

CONCEPTS

- Binomial random variable, mean, variance, standard deviation

VISION

Consider a binomial experiment: $n = 100$ trials, two outcomes (living in poverty or not living in poverty), trials are independent (random sample), and probability of a success (living in poverty) is constant from trial to trial. Define a random variable and identify its probability distribution.

Solution Trail 5.17b

KEYWORDS

- Is there any evidence?

TRANSLATION

- Use the experimental outcome to draw a conclusion concerning the report's claim.

CONCEPTS

- Inference procedure

- Most likely values of a binomial random variable

VISION

Follow the four-step inference procedure. Use the mean and standard deviation to determine the most likely values of the random variable.

SOLUTION

a. Let X be the number of children (out of the 100 selected) who are living in poverty. X is a binomial random variable with $n = 100$ and $p = 0.60$: $X \sim B(100, 0.60)$.

Use Equation 5.8 to find the mean, variance, and standard deviation.

$$\mu = np = (100)(0.60) = 60$$
$$\sigma^2 = np(1 - p) = (100)(0.60)(1 - 0.60) = (100)(0.60)(0.40) = 24$$
$$\sigma = \sqrt{\sigma^2} = \sqrt{24} \approx 4.9$$

The expected number of children living in poverty is 60, with a variance of 24 and a standard deviation of approximately 4.9.

b. The report claims $p = 0.60$. This implies that the random variable X has a binomial distribution with $n = 100$ and $p = 0.60$.

Claim: $p = 0.60 \rightarrow X \sim B(100, 0.60)$.

The experimental outcome is that 55 children are living in poverty.

Experiment: $x = 55$.

Likelihood: From part (a), $\mu = 60$ and $\sigma = 4.9$. Most observations are within three standard deviations of the mean. Therefore, most values of X are in the interval

$$(\mu - 3\sigma, \mu + 3\sigma) = (60 - 3(4.9), 60 + 3(4.9))$$
$$= (60 - 14.7, 60 + 14.7)$$
$$= (45.3, 74.7)$$

Conclusion: Because 55 lies in this interval, it is a reasonable observation. There is no evidence to lead us to doubt the claim of $p = 0.60$. ◼

A CLOSER LOOK

1. Recall: In statistics, we usually measure *distance* in standard deviations, not miles, feet, inches, meters, or other units. We often want to know how many standard deviations from the mean is a given observation.

2. The inference problem in part (b) of Example 5.17 can also be answered using the *tail probability* approach. (Try it!) This method leads to the same conclusion and is generally more precise than constructing an interval about the mean. A more formal process for checking claims (hypothesis tests) is introduced in Chapter 9.

3. Whenever we test a claim, there are only two possible conclusions:

    a. There *is* evidence to suggest the claim is false.

    b. There *is no* evidence to suggest the claim is false.

    Note that, in either case, we never state with absolute certainty that the claim is true or the claim is false. This is because we never look at the *entire* population, only at a sample. With a large, random (representative) sample, we can be pretty confident in our conclusion, but never absolutely sure. ●

Technology Corner

Procedure: Compute probabilities associated with a binomial random variable.
Reconsider: Example 5.15, solution, and interpretations.

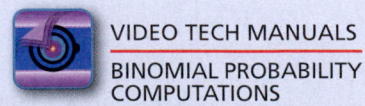

VIDEO TECH MANUALS

BINOMIAL PROBABILITY COMPUTATIONS

CrunchIt!

CrunchIt! has a built-in function to compute probabilities associated with a binomial random variable.

1. Select Distribution Calculator; Binomial. Enter the values for n and p, and select an appropriate inequality symbol or the equals sign. Enter the value for x and click Calculate. See Figure 5.11.
2. To find the probability that X takes on a single value, use the same menu choices and enter values for n and p. Select the equals sign and enter the value for x. Click Calculate. See Figure 5.12.

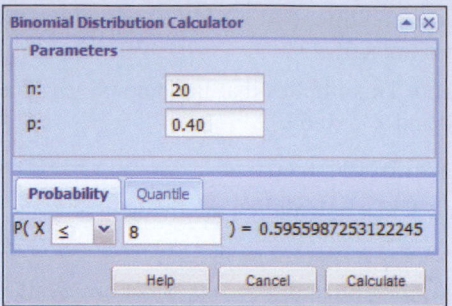

Figure 5.11 Cumulative probability.

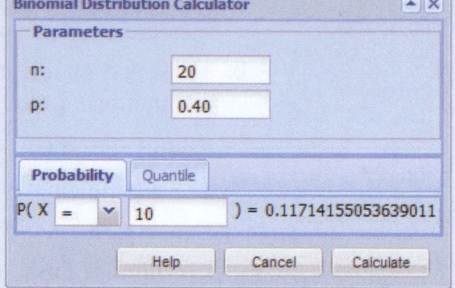

Figure 5.12 The probability mass function.

TI-84 Plus C

Suppose $X \sim B(n, p)$. There are built-in functions to compute a value of the probability mass function (the probability X takes on a single value) and cumulative probability.

1. For cumulative probability, use `DISTR`; DISTR; `binomcdf`. Enter values for the number of `trials` (n), p (p), and `x value` (x). Position the cursor on `Paste` and tap `ENTER`. The appropriate calculator command is copied to the Home screen. Tap `ENTER` again to compute the resulting probability. Refer to Figure 5.5, page 216.
2. To find the probability that X takes on a single value, use `DISTR`; DISTR; `binompdf`. Enter values for the number of `trials` (n), p (p), and `x value` (x). Position the cursor on `Paste` and tap `ENTER`. The appropriate calculator command is copied to the Home screen. Tap `ENTER` again to compute the resulting probability. Refer to Figure 5.6, page 216.

Minitab

There are built-in functions accessed via input windows or the command language to compute a value of the probability mass function (the probability that X takes on a single value) and cumulative probability. The command language may be necessary to perform additional calculations involving probabilities.

1. Select Calc; Probability Distributions; Binomial. Choose Cumulative probability. Enter the Number of trials (n), the Event probability (p), and the Input constant (x). See Figure 5.13.
2. To find the probability that X takes on a single value, select Calc; Probability Distributions; Binomial. Choose Probability. Enter the Number of trials (n), the Event probability (p), and the Input constant (x). See Figure 5.14.

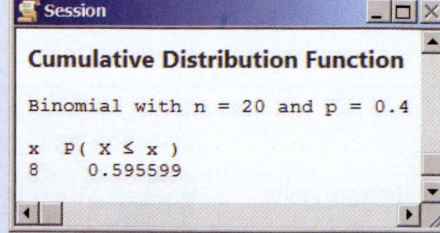

Figure 5.13 Cumulative probability.

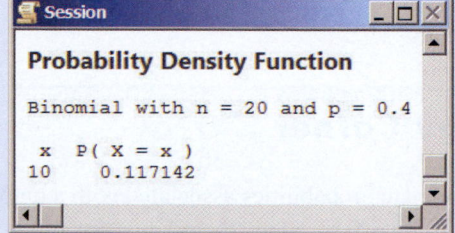

Figure 5.14 The probability mass function.

Excel

There is a single built-in function to find the probability that X takes on a single value or cumulative probability. The last argument of the function is either `True` for cumulative probability or `False` for the probability mass function. Additional spreadsheet calculations may be necessary to find the final answer.

1. To find cumulative probability, use the function `BINOMDIST`. Enter x, n, p, and `True`. See Figure 5.15.
2. To find the probability that X takes on a single value, use `BINOMDIST`. Enter x, n, p, and `False`. See Figure 5.15.

Figure 5.15 Excel function for finding cumulative probability and for evaluating the probability mass function.

A	B
0.5956	= BINOMDIST(8,20,0.4,TRUE)
0.1171	= BINOMDIST(10,20,0.4,FALSE)

SECTION 5.4 EXERCISES

Concept Check

5.71 True/False A binomial random variable is completely described by the number of trials, n.

5.72 True/False There can be three or more outcomes in each trial of a binomial experiment.

5.73 True/False For a binomial random variable, $P(F) = 1 - P(S)$.

5.74 True/False Every probability question about a binomial random variable can be answered using cumulative probability.

5.75 True/False Suppose $X \sim B(n, p)$. Then $E(X) = np$.

5.76 True/False The most common values for a binomial random variable are less than the mean.

5.77 Fill in the Blank The binomial random variable is a count of _____.

5.78 Fill in the Blank Suppose $X \sim B(n, p)$. The number of outcomes with x successes is _____.

5.79 Short Answer Name the four properties of a binomial experiment.

5.80 Short Answer Write a probability expression involving a binomial random variable X that represents cumulative probability.

Practice

5.81 Suppose $X \sim B(15, 0.25)$. Find the following probabilities.
a. $P(X \le 2)$
b. $P(X < 2)$
c. $P(X = 7)$
d. $P(X > 6)$
e. $P(3 \le X \le 10)$

5.82 Suppose $X \sim B(20, 0.40)$. Find the following probabilities.
a. $P(X \ge 12)$
b. $P(X \ne 10)$
c. $P(X \le 15)$
d. $P(2 < X \le 8)$

5.83 Suppose $X \sim B(25, 0.70)$. Find the following probabilities.
a. $P(X \ge 1)$
b. $P(X \ge 10)$
c. $P(X \ge 17.5)$
d. $P(10.1 \le X \le 19)$

5.84 Suppose X is a binomial random variable with $n = 25$ and $p = 0.80$.
a. Find the mean, variance, and standard deviation of X.
b. Find the probability X is within one standard deviation of the mean.
c. Find the probability X is more than two standard deviations from the mean.

5.85 Suppose X is a binomial random variable with $n = 30$ and $p = 0.40$.
a. Find the mean, variance, and standard deviation of X.
b. Find the intervals $\mu \pm \sigma$, $\mu \pm 2\sigma$, and $\mu \pm 3\sigma$.
c. Find $P(X > \mu + 3\sigma)$.
d. Find $P(X \le \mu - 2\sigma)$.

5.86 Suppose X is a binomial random variable with $n = 10$ and $p = 0.50$.
a. Create a table of values of X and associated probabilities. (*Hint*: This is quick and easy using technology.)
b. Use the table in part (a) and the definitions of expected value and variance (Equations 5.1 and 5.3) to find μ, σ^2, and σ.
c. Use Equation 5.8 to find μ, σ^2, and σ. Check these answers with those in part (b).

Applications

5.87 Economics and Finance Approximately 90% of freshmen at Marquette University receive financial aid.[21] Suppose 20 Marquette freshmen are randomly selected.
a. Find the probability that at most 15 freshmen receive financial aid.
b. Find the probability that at least 12 freshmen receive financial aid.

c. Find the expected number of freshmen who receive financial aid.

d. Suppose at least 15 freshmen receive financial aid. What is the probability that all 20 freshmen receive financial aid?

5.88 Fuel Consumption and Cars   The battery manufacturer Varta sells a car battery with 800 cold-cranking amps and advertises great performance even in bitingly cold weather. Varta claims that after sitting on a frozen Minnesota lake for 10 days at temperatures below 32°F, this battery will still have enough power to start a car. Suppose the actual probability of starting a car following this experiment is 0.75, and 15 randomly selected cars (equipped with this battery) are subjected to these grueling conditions.

a. Find the probability that fewer than 10 cars will start. Write a Solution Trail for this problem.

b. Find the probability that more than 12 cars will start.

c. Suppose 9 cars actually start. Is there any evidence to suggest that the probability of starting a car is different from 0.75? Justify your answer.

5.89 Marketing and Consumer Behavior   Levain Bakery, on West 74th Street in New York City, is trying to determine the number of loaves of raisin bread to make each day. Over the past few months the store has baked 50 loaves each day and has sold out with probability 0.80. Suppose the owner continues this practice and 30 days are selected at random.

a. What is the expected number of days on which all 50 loaves will be sold?

b. Find the probability of selling all 50 loaves on at least 20 days.

c. Find the probability of selling all 50 loaves on at most 18 days.

5.90 Sports and Leisure   A Six Flags Great Adventure Theme Park now offers a "wild safari" drive-thru with more than 1000 animals roaming freely on over 400 acres. The park claims that the probability of some car damage by an animal during a safari drive-thru is 0.60. Suppose 20 cars are selected at random.

a. Find the probability that exactly 10 cars will be damaged.

b. Find the probability that at least 15 cars will be damaged.

c. Find the probability that no more than 12 cars will be damaged.

d. Suppose 19 cars are damaged. Is there any evidence to suggest the claim of 0.60 is false? Justify your answer. Write a Solution Trail for this problem.

5.91 Public Health and Nutrition   Parents tend to be very good at diagnosing their children's routine medical problems, such as an ear infection, sinus infection, or strep throat. If an ailment is identified correctly, a trip to the doctor's office may be avoided. A physician may confer with a parent by telephone, and simply call a pharmacy with a prescription for an antibiotic. Suppose parents are correct 90% of the time, and 50 families with a child suffering from some minor illness are selected at random.

a. Find the mean, variance, and standard deviation of the number of parents who identify their child's illness correctly.

b. Find the probability that at least 42 parents are correct.

c. Find the probability that between 42 and 47 (inclusive) parents are correct.

d. Suppose 41 parents are actually correct. Is there any evidence to suggest that fewer than 90% of parents are correct? Justify your answer.

5.92 Public Policy and Political Science   A building inspector enforces building, electrical, mechanical, plumbing, and energy code requirements for the safety and health of people in a certain city, county, or state. In Santa Cruz County, the probability that a building inspector will find at least one code violation at a commercial building is 0.25. Suppose 30 commercial buildings are selected at random.

a. Find the mean, variance, and standard deviation of the number of commercial buildings with at least one violation.

b. Find the probability that the number of commercial buildings with at least one violation will be within one standard deviation of the mean.

c. Find the probability that the number of commercial buildings with at least one violation will be more than two standard deviations from the mean.

d. Suppose the actual number of commercial buildings with at least one violation is 10. Is there any evidence to suggest that code violations are found in more than 25% of commercial buildings? Justify your answer.

5.93 Business and Management   The Sundance Film Festival is held every January in Park City, Utah. Individuals from the film community determine various awards for independent film makers, and audience awards are also presented. In 2012, approximately 17% of the feature films in the documentary category were directed by women.[22] Suppose a sample of 35 documentary films are selected at random.

a. What is the probability that exactly five films were directed by women?

b. What is the probability that at least eight films were directed by women?

c. Suppose three films were directed by women. Is there any evidence to suggest that the proportion of films directed by women is different from 0.17? Justify your answer.

5.94 Demographics and Population Statistics   In Illinois, a typical DUI (Driving Under the Influence) offender is a 34-year-old male, arrested between 11 P.M. and 4 A.M. on a weekend, and has a BAC (blood alcohol content) of 0.16. Eighty-five percent of all drivers arrested in Illinois are first-time offenders.[23] Suppose 40 people arrested for DUI in Illinois are selected at random.

a. What is the probability that at least 12 are first-time offenders?

b. What is the probability that between 7 and 10 (inclusive) are first-time offenders?

c. Suppose 4 of those arrested are first-time offenders. Is there any evidence to suggest that the proportion of first-time offenders arrested for DUI in Illinois has changed? Justify your answer.

5.95 Demographics and Population Statistics There is some evidence to suggest that businesses are moving out of states where unions are prevalent. In California, 18.4% of all workers belong to a union, and in Arkansas, 3.7%.[24] Suppose 20 workers from California and 20 workers from Arkansas are selected at random.
  a. Find the probability that at most two California workers belong to a union.
  b. Find the probability that none of the Arkansas workers belongs to a union.
  c. Find the probability that at least one worker from each state belongs to a union.

5.96 Public Health and Nutrition A recent report suggests that one-third of all U.S. children are overweight or obese. One possible cause is the availability of junk foods and sugary snacks in schools. Approximately 40% of all students buy and eat one or more snacks at school.[25] Suppose 40 school children are selected at random.
  a. Find the mean, variance, and standard deviation of the number of school children (out of the 40 selected) who buy and eat a snack at school.
  b. Construct intervals one, two, and three standard deviations from the mean.
  c. Suppose 27 students (out of the 40 selected) buy and eat a snack at school. How many standard deviations from the mean is this observation? What does this *distance* measure indicate about the likelihood of observing 27 students who buy and eat a snack at school?

5.97 Business and Management According to the CICA Business Monitor, 11% of Canada's executive chartered accountants are pessimistic about the economy.[26] To check this claim, 50 chartered accountants were randomly selected and asked whether they feel pessimistic about the economy in 2013.
  a. If the claim is true, find the probability that exactly seven chartered accountants are pessimistic about the economy.
  b. If the claim is true, find the probability that at most four chartered accountants are pessimistic about the economy.
  c. Suppose 12 chartered accountants are pessimistic about the economy. Is there any evidence to suggest that the claim is false? Justify your answer.

Extended Applications

5.98 Business and Management In the movie *Lethal Weapon*, the character played by Joe Pesci is concerned about drive-thru windows at fast-food restaurants. Suppose the probability that an order at a drive-thru window at a fast-food restaurant will be filled correctly is 0.75. Twenty orders are selected at random.
  a. What is the probability that exactly 15 orders will be filled correctly?
  b. What is the probability that at most 12 orders will be filled correctly?
  c. What is the probability that between 10 and 14 (inclusive) orders will be filled correctly?

  d. Suppose *two* groups of 20 random orders are independently selected. What is the probability that at least 16 orders will be filled correctly in both groups?

5.99 Sports and Leisure The reality television series *Splash!* features celebrities attempting to learn how to dive. The first episode aired in January 2013 and earned a 23.6% audience share. That is, 23.6% of all TVs in use during the show time period were tuned to a station airing *Splash!*.[27] Twenty people who watched TV during that time period were selected at random.
  a. Find the probability that at least six watched *Splash!*.
  b. Find the expected number of people who watched *Splash!*. Find the probability that the number of people who watched *Splash!* is less than the mean.
  c. Suppose that at most four people watched *Splash!*. What is the probability that no one watched *Splash!*?

5.100 Marketing and Consumer Behavior More children are being rushed to the hospital because they were able to push down and twist the cap on a medication bottle and were poisoned by a common drug. A recent research study suggested that 25% of all preschool children can open a medication bottle and 10 preschool children are selected at random. Let the random variable X be the number of children who can open the bottle.
  a. Construct a probability histogram for the random variable X.
  b. Find the mean, variance, and standard deviation of X. Indicate the mean on the graph from part (a).
  c. Find $P(\mu - \sigma \le X \le \mu + \sigma)$ and indicate this probability on the graph from part (a).
  d. Suppose these 10 children try to open a medicine bottle with a new design for the cap and one child is able to open the bottle. Is there any evidence to suggest that the new cap is more effective in stopping children from opening the bottle? Justify your answer.

5.101 Manufacturing and Product Development A company has developed a very inexpensive explosive-detection machine for use at airports. However, if an explosive is actually in a suitcase, the probability of it being detected by this machine is only 0.60. Therefore, several of these machines will be used simultaneously to screen each piece of luggage independently. Suppose a piece of luggage actually contains an explosive.
  a. If three machines screen this luggage, what is the probability that exactly one will detect the explosive? What is the probability that none of the three will detect the explosive?
  b. If four machines screen this luggage, what is the probability that at least one device will detect the explosive?
  c. If five machines screen this luggage, what is the probability that at least one device will detect the explosive?
  d. How many machines are necessary for screening in order to be certain that at least one device will detect the explosive with probability 0.999 or greater?

5.102 Marketing and Consumer Behavior   Forever 21 sells women's flip-flops in (oddly enough) 21 different colors. Despite this vast available color selection, 50% of all flip-flop purchases are in white. Suppose 30 buyers are selected at random.

   a. Find the mean, variance, and standard deviation of the number of buyers who purchase white flip-flops.

   b. Find the probability that the number of white flip-flops purchased will be within two standard deviations of the mean. Compare this with the predicted result from Chebyshev's rule.

   c. Suppose two groups of 30 customers are independently selected. What is the probability of at least one group having exactly 15 people who buy white flip-flops?

5.103 Manufacturing and Product Development   In early 2013 there was a worldwide glut of solar panels. This forced prices down and made this form of energy production affordable to many more people. This oversupply of solar panels may have caused some manufacturers to cut corners. Sainty Solar is a leading producer and claims the proportion of its solar panels that are defective is 0.02. Solar Solutions, a leading installer, receives a shipment of 50,000 solar panels. Before accepting the entire lot, Solar Solutions selects a random sample of 25 panels and thoroughly tests each one. If four or more panels are found to be defective, the entire shipment will be sent back. Otherwise the shipment will be accepted.

   a. Suppose the claim is true: The actual proportion of defectives is $p = 0.02$. What is the probability that the shipment will be rejected? (This is one type of *error probability*. The company would be making a mistake if this event occurred. It would be rejecting the shipment when the proportion of defectives is as claimed.)

   b. Suppose the actual proportion of defectives is $p = 0.05$. What is the probability that the shipment will be accepted? (This is another type of error probability. In this case, the company would also be making a mistake. It would be accepting the shipment when the proportion of defectives is too high.)

   c. Suppose the actual proportion of defectives is $p = 0.07$. What is the probability that the shipment will be accepted?

5.104 Psychology and Human Behavior   As a result of stricter training requirements, fewer big fires, higher-paying jobs in cities, and changes in society, the number of volunteer firefighters is declining. Approximately three-fourths of all firefighters in the United States are volunteers, and the total number of volunteer firefighters has decreased steadily over the last two decades. Suppose 30 U.S. firefighters are selected at random.

   a. Find the probability that exactly 22 of the firefighters are volunteers.

   b. Find the probability that more than 25 of the firefighters are volunteers.

   c. Suppose 17 of the firefighters are volunteers. Is there any evidence to suggest that the proportion of volunteer firefighters has decreased? Justify your answer.

   d. Suppose 50 firefighters are selected at random from the West and 50 firefighters are selected from the Northeast. What is the probability that at least 40 of the firefighters will be volunteers in both groups?

5.105 Sports and Leisure   Most ski resorts operate beginner, intermediate, and advanced terrain in order to appeal to people with varying abilities. The table below lists several ski areas in Canada and the proportion of skiers who attempt the advanced terrain during their visit.

Ski area	Probability
Big White	0.28
Kicking Horse	0.60
Norquay	0.44

Suppose 20 skiers are randomly selected from each area.

   a. Find the probability that exactly five skiers at Big White will attempt the advanced terrain.

   b. Find the probability that more than eight skiers at Kicking Horse will attempt the advanced terrain.

   c. Find the probability that between 12 and 16 (inclusive) skiers at Norquay will attempt the advanced terrain.

   d. Find the probability that at most five skiers at all three locations will attempt the advanced terrain.

5.5 Other Discrete Distributions

There are many other common discrete probability distributions. This section presents three of these distributions along with brief background, properties, and examples. Remember that many of the problems involving these distributions are solved using the same general technique:

1. Define a random variable and identify its probability distribution (distribution statement).

2. Translate the words into a probability question where the event is stated in terms of the random variable (probability statement).

3. If necessary, try to convert the probability statement into an equivalent expression involving *cumulative* probability. Use tables and technology wherever possible.

The **geometric distribution** is closely related to the binomial distribution. In a binomial experiment, n (the number of trials) is fixed and the number of successes varies. The binomial random variable is the number of successes in n trials. In a geometric experiment, *the number of successes is fixed* at 1, and the number of trials varies.

Properties of a Geometric Experiment

1. The experiment consists of identical trials.

2. Each trial can result in only one of two possible outcomes: a success (S) or a failure (F).

3. The trials are independent.

4. The probability of a success, p, is constant from trial to trial.

The experiment ends when the first success is obtained.

The Geometric Random Variable

The **geometric random variable** is the number of trials necessary to realize the first success.

Think of an experiment in which you continue to phone a friend until you get through. The number of calls necessary until the first success (reaching your friend) is the value of a geometric random variable.

The derivation of the probability distribution involves the properties given above. Let X be a geometric random variable, the number of trials until the first success (including the trial on which the success is obtained). Given p, the probability of a success, find the probability of needing x trials, $P(X = x) = p(x)$.

$$P(X = 1) = P(S) = p$$

$X = 1$ means the first trial results in a success, and the experiment is over. The probability of a success is simply p.

$$P(X = 2) = P(F \cap S) = P(F) \cdot P(S) = (1 - p)p$$

$X = 2$ means the first trial is a failure *and* the second trial is a success. Because trials are independent, we multiply the corresponding probabilities.

$$P(X = 3) = P(F \cap F \cap S) = P(F) \cdot P(F) \cdot P(S)$$
$$= (1 - p)(1 - p)p = (1 - p)^2 p$$

Why isn't FSF a possible outcome in a geometric experiment?

$X = 3$ means the first two trials are failures and the third trial is a success. We use independence again, and multiply the corresponding probabilities.

$$P(X = 4) = P(F \cap F \cap F \cap S) = P(F) \cdot P(F) \cdot P(F) \cdot P(S)$$
$$= (1 - p)(1 - p)(1 - p)p = (1 - p)^3 p$$

$X = 4$ means the first three trials are failures and the fourth trial is a success. We use independence again, and multiply the corresponding probabilities.

In general,

$$P(X = x) = \underbrace{P(F) \cdot P(F) \cdots P(F)}_{x - 1 \text{ failures}} \cdot P(S) = \underbrace{(1 - p)(1 - p) \cdots (1 - p)}_{x - 1 \text{ terms}} \cdot p$$

$$= (1 - p)^{x-1} p$$

$X = x$ means the first $x - 1$ trials are failures and the xth trial is the first success. This generalization is the formula for the probability distribution.

The Geometric Probability Distribution

Suppose X is a geometric random variable with probability of a success p. Then

$$p(x) = P(X = x) = (1 - p)^{x-1}p \qquad x = 1, 2, 3, \ldots \qquad (5.9)$$

$$\mu = \frac{1}{p} \quad \text{and} \quad \sigma^2 = \frac{1 - p}{p^2} \qquad (5.10)$$

A CLOSER LOOK

1. The geometric random variable is discrete. The number of possible values is countably infinite: 1, 2, 3, ...

2. The geometric distribution is completely characterized, or defined, by one parameter, p.

3. We do not need a table to find cumulative probabilities associated with a geometric random variable because there is an easy formula for computing these values. If X is a geometric random variable with probability of success p, then

$$P(X \le x) = 1 - (1 - p)^x \qquad (5.11)$$

4. Equation 5.9 is a valid probability distribution.

▶ Each probability is between 0 and 1, and the sum of all the probabilities is an *infinite series*. The sum

$$\sum_{x=1}^{\infty} P(X = x) = \sum_{x=1}^{\infty} (1 - p)^{x-1}p$$

is called a *geometric series* and it does sum to 1! ◀

There is a formula for the sum of a geometric series. Can you use it to show that this sum is 1?

Several years ago the Bekins Moving Company used a clever jingle in many of their advertisements. The song started, "Bekins men are careful, quick, and kind, Bekins takes a load off of your mind. . . ."

Solution Trail 5.18

KEYWORDS
- First person to have moved

TRANSLATION
- First success

CONCEPTS
- Geometric probability distribution

VISION

Consider a geometric experiment. Each trial is a call to ask if the person has moved in the past year, $P(S) = P(\text{moved}) = 0.12$, $P(S)$ is the same on each trial, and the trials are independent. Define the geometric random variable and write a probability statement for each part.

Example 5.18 Bekins Men are Careful, Quick, and Kind

The number of people who changed residences in the United States has declined steadily since 1985. However, perhaps due to the sluggish economy, the U.S. Census Bureau estimates that approximately 12% of all people changed residences in 2012.[28] Suppose researchers at Bekins randomly call people in the United States and ask if they have moved in the last year.

a. What is the probability that the fourth person called will be the first to have moved in the past year?

b. What is the probability that it will take at least six calls before speaking to someone who has moved in the past year?

SOLUTION

a. Let X be the number of calls necessary until the first mover is found. X is a geometric random variable with $P(S) = 0.12 = p$.

The probability that the first mover (success) is found on the fourth call:

$$P(X = 4) = (1 - p)^{4-1}p \qquad \text{Equation 5.9.}$$

$$= (1 - 0.12)^3(0.12) = (0.88)^3(0.11) = 0.0818 \qquad \text{Use } p = 0.12.$$

The probability that the first mover is found on the fourth call is 0.0818.

b. *At least six calls before speaking to someone who has moved in the past year* means the first success will occur on the sixth call or later.

The probability that at least six calls will be needed is

$P(X \geq 6)$

$= 1 - P(X < 6)$ The complement rule.

$= 1 - P(X \leq 5)$ The first value *X* takes on that is *less than 6* is 5.

$= 1 - [1 - (1 - p)^5]$ Use Equation 5.11.

$= 1 - [1 - (0.88)^5]$ Use $p = 0.12$.

$= 1 - 0.4723 = 0.5277$ Expand and simplify.

The probability that it will take six or more calls to find the first mover is 0.5277.

Figures 5.16 and 5.17 show technology solutions.

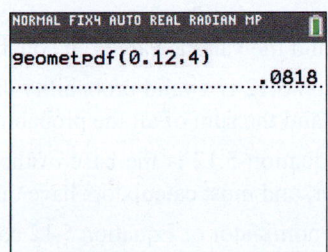

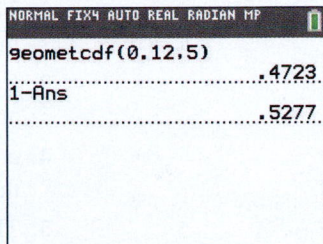

Figure 5.16 $P(X = 4)$ using the probability mass function.

Figure 5.17 $P(X \geq 6)$ using cumulative probability.

TRY IT NOW   GO TO EXERCISE 5.118

The distribution is named after the French mathematician Simeon Denis Poisson (1781–1840).

The **Poisson probability distribution** has many practical applications and is often associated with *rare* events. A **Poisson random variable** is a count of the number of occurrences of a certain event in a given unit of time, space, volume, distance, etc., for example, the number of arrivals to a hospital Emergency Room in a certain 30-minute period, the number of asteroids that pass through Earth's orbit during a given year, or the number of bacteria in a milliliter of drinking water.

Properties of a Poisson Experiment

1. The probability that a single event occurs in a given interval (of time, length, volume, etc.) is the same for all intervals.

2. The number of events that occur in any interval is independent of the number that occur in any other interval.

These properties are often referred to as a *Poisson process* and can be difficult to verify.

The Poisson Random Variable

The **Poisson random variable** is a count of the number of times the specific event occurs during a given interval.

The Poisson distribution is completely determined by the mean, denoted by the Greek letter lambda, λ. Because the Poisson distribution is often used to count rare events, the mean number of events per interval is usually small. The probability distribution is given below.

The Poisson Probability Distribution

Suppose X is a Poisson random variable with mean λ. Then

$$p(x) = P(X = x) = \frac{e^{-\lambda}\lambda^x}{x!} \qquad x = 0, 1, 2, 3, \ldots \qquad (5.12)$$

$$\mu = \lambda \qquad \sigma^2 = \lambda \qquad (5.13)$$

A CLOSER LOOK

1. The Poisson random variable is discrete. The number of possible values is countably infinite: $0, 1, 2, 3, \ldots$.

2. The Poisson distribution is completely characterized by only one parameter, λ. The mean and the variance are both equal to the same value, λ.

3. Equation 5.12 is a valid probability distribution. All of the probabilities are between 0 and 1, and the sum of all the probabilities is 1 (another infinite series).

4. e in Equation 5.12 is the base of the natural logarithm. $e \approx 2.71828$ is an irrational number, and most calculators have this special constant built in.

5. The denominator of Equation 5.12 contains $x!$ (x factorial).
   Recall: $x! = x(x-1)(x-2)\cdots(3)(2)(1)$ and $0! = 1$.

6. Table 2 in Appendix A contains values for $P(X \leq x)$ (cumulative probability) for various values of λ.

Solution Trail 5.19

KEYWORDS

- Mean number of fatalities per month
- Two each month

TRANSLATION

- Fixed time
- $\lambda = 2$

CONCEPTS

- Poisson probability distribution

VISION

Consider a Poisson distribution. The probability of a single fatality is the same every month, and the number of fatalities in any month is independent of the number of fatalities in any other month. The mean of the Poisson distribution is given.

Example 5.19 Look Out Below!

Skydiving is the ultimate thrill for some, and there were over 3.1 million jumps in 2012. Despite an improved safety record, there are approximately two skydiving fatalities each month.[29] Suppose this is the mean number of fatalities per month and a random month is selected.

a. Find the probability that exactly three fatalities will occur.

b. Find the probability that at least five fatalities will occur.

c. Find the probability that the number of fatalities will be within one standard deviation of the mean.

SOLUTION

Let X be the number of skydiving fatalities per month. X has a Poisson distribution with $\lambda = 2$.

a. The probability of *exactly three* means $P(X = 3)$.

$$
\begin{aligned}
P(X = 3) &= \frac{e^{-2}2^3}{3!} = 0.1804 && \text{Or, use Equation 5.12.}\\
&= P(X \leq 3) - P(X \leq 2) && \text{Convert to cumulative probability.}\\
&= 0.8571 - 0.6767 = 0.1804 && \text{Use Table II in the Appendix.}
\end{aligned}
$$

b. *At least five* means five or more: $X \geq 5$.

$$
\begin{aligned}
P(X \geq 5) &= 1 - P(X < 5) && \text{The complement rule.}\\
&= 1 - P(X \leq 4) && \text{The first value } X \text{ takes on that is } \textit{less than 5} \text{ is 4.}\\
&= 1 - 0.9473 && \text{Use Table II in the Appendix.}\\
&= 0.0527
\end{aligned}
$$

c. *Within one standard deviation of the mean is the interval* $(\mu - \sigma, \mu + \sigma)$.
$\mu = 2 = \sigma^2 \;\rightarrow\; \sigma = \sqrt{2} = 1.4142$.

$$P(\mu - \sigma \leq X \leq \mu + \sigma)$$

$\quad = P(2 - 1.4142 \leq X \leq 2 + 1.4142)$ Use values for μ and σ.

$\quad = P(0.5858 \leq X \leq 3.4142)$ Compute the difference and sum.

$\quad = P(1 \leq X \leq 3)$ Use properties of the Poisson distribution.

$\quad = P(X \leq 3) - P(X \leq 0)$ Convert to cumulative probability.

$\quad = 0.8571 - 0.1353$ Use Table II in the Appendix.

$\quad = 0.7218$ Compute the difference.

Figures 5.18 through 5.20 show technology solutions.

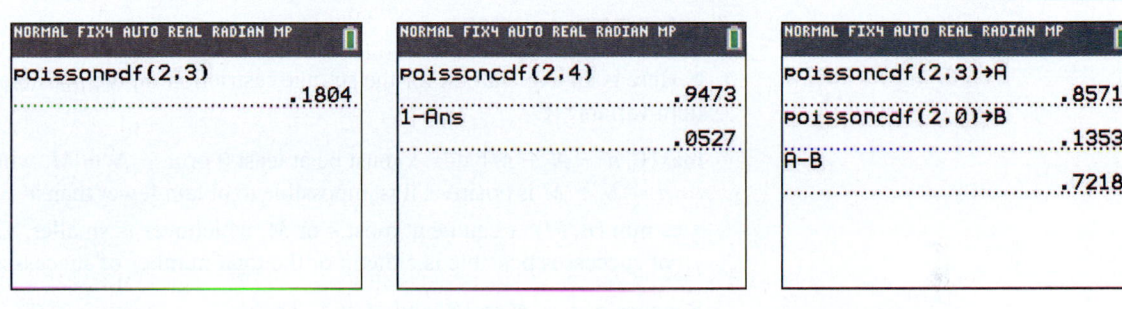

Figure 5.18 $P(X = 3)$.   **Figure 5.19** $P(X \geq 5)$.   **Figure 5.20** $P(1 \leq X \leq 3)$. ●

TRY IT NOW   GO TO EXERCISE 5.119

The **hypergeometric probability distribution** arises from an experiment in which there is sampling without replacement from a finite population. Each element in the population is labeled a success or failure. The **hypergeometric random variable** is a count of the number of successes in the sample. For example, consider a shipment of 12 automobile tires, of which two are defective, and a random sample of four tires. A hypergeometric random variable may be defined as a count of the number of *good* tires selected.

Properties of a Hypergeometric Experiment

1. The population consists of N objects, of which M are successes and $N - M$ are failures.

2. A sample of n objects is selected *without* replacement.

3. Each sample of size n is equally likely.

The Hypergeometric Random Variable

The **hypergeometric random variable** is a count of the number of successes in a random sample of size n.

The hypergeometric probability distribution is completely determined by n, N, and M. The probability of obtaining x successes is derived using many concepts introduced earlier: independence, the multiplication rule, equally likely outcomes, and combinations.

The Hypergeometric Probability Distribution

Suppose X is a hypergeometric random variable characterized by sample size n, population size N, and number of successes M. Then

$$p(x) = P(X = x) = \frac{\binom{M}{x}\binom{N-M}{n-x}}{\binom{N}{n}} \tag{5.14}$$

$$\max(0, n - N + M) \leq x \leq \min(n, M)$$

$$\mu = n\frac{M}{N}, \quad \sigma^2 = \left(\frac{N-n}{N-1}\right)n\frac{M}{N}\left(1 - \frac{M}{N}\right) \tag{5.15}$$

A CLOSER LOOK

1. ▶ Here is an explanation for the strange restriction on the possible values for the random variable X.

   $\max(0, n - N + M) \leq x$: x must be at least 0 or $n - N + M$, whichever is bigger. If $n - N + M$ is positive, it is impossible to obtain fewer than $n - N + M$ successes.

   $x \leq \min(n, M)$: x can be at most n or M, whichever is smaller. The greatest number of successes possible is either n or the total number of successes in the population.

   Suppose $n = 5$, $N = 10$, and $M = 6$. Then:

   $\max(0, n - N + M) = \max(0, 5 - 10 + 6) = \max(0, 1) = 1$ and
   $\min(n, N) = \min(5, 10) = 5 \Rightarrow 1 \leq x \leq 5$

   It is impossible to obtain less than 1 success. Also, the greatest number of successes possible is 5. ◀

2. The hypergeometric random variable is discrete. All of the probabilities are between 0 and 1, and the probabilities do sum to 1.

3. Recall: $\binom{n}{r}$ is a *combination*. The number of combinations of size r is given by

$$_nC_r = \binom{n}{r} = \frac{n!}{r!(n-r)!}$$ ●

Solution Trail 5.20

KEYWORDS
- 10 twin-size comforters
- 2 stitched incorrectly
- 4 selected at random

TRANSLATION
- $N = 10$ (finite population)
- 2 failures, therefore $M = 8$ successes
- Sample size $n = 4$

CONCEPTS
- Hypergeometric probability distribution

VISION
All four comforters selected at random *without* replacement and there are 10 comforters to choose from. Each comforter selected is either a success (stitched correctly) or a failure. Consider a hypergeometric distribution.

Example 5.20 Hello Kitty

A Target store has 10 Hello Kitty twin-size comforters for sale. Two of the 10 comforters have been stitched incorrectly at the factory and will split open when used. Suppose four of the comforters are randomly selected.

a. What is the probability that exactly two comforters will be stitched correctly?

b. What is the probability that at least three comforters will be stitched correctly?

SOLUTION

Let X be the number of successes in the sample. X has a hypergeometric distribution with $n = 4$, $N = 10$, and $M = 8$. Translate each question into a probability statement, convert to cumulative probability if necessary, and use Equation 5.14 and/or technology.

a. *Exactly two* means $X = 2$.

$$P(X = 2) = \frac{\binom{M}{x}\binom{N-M}{n-x}}{\binom{N}{n}} = \frac{\binom{8}{2}\binom{10-8}{4-2}}{\binom{10}{4}}$$

Use Equation 5.14.

$$= \frac{\binom{8}{2}\binom{2}{2}}{\binom{10}{4}}$$

In the numerator, from the 8 good comforters, choose 2, from the 2 bad comforters, choose 2. In the denominator, $\binom{10}{4}$ is the total number of ways to choose 4 comforters from 10.

$$= \frac{(28)(1)}{210} = 0.1333$$

Use the formula for a combination.

The probability of selecting exactly two correctly stitched comforters is 0.1333.

b. *At least three* means three or more. The maximum number of successes is $\min(n, M) = \min(4, 8) = 4$. In this case, three or more means 3 or 4.

$$P(X \geq 3) = P(X = 3) + P(X = 4)$$

Consider the values X can assume that are greater than or equal to 3.

$$= \frac{\binom{8}{3}\binom{2}{1}}{\binom{10}{4}} + \frac{\binom{8}{4}\binom{2}{0}}{\binom{10}{4}}$$

Use Equation 5.14.

$$= \frac{(56)(2)}{210} + \frac{(70)(1)}{210}$$

Use the formula for a combination.

$$= 0.5333 + 0.3333 = 0.8666$$

Note: This problem can also be solved using cumulative probability:

$$P(X \geq 3) = 1 - P(X \leq 2)$$

Figures 5.21 and 5.22 show technology solutions.

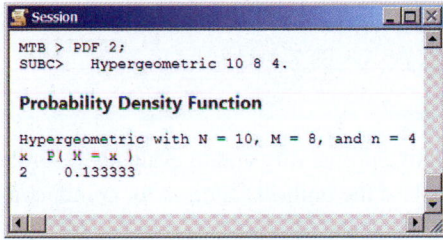

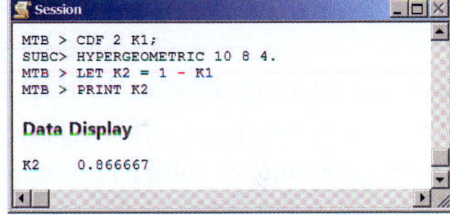

Figure 5.21 Minitab session window output using the Hypergeometric Distribution input window.

Figure 5.22 $P(X \geq 3)$ using the Minitab command language.

TRY IT NOW   GO TO EXERCISE 5.123

Technology Corner

Procedure: Compute probabilities associated with a geometric, Poisson, or hypergeometric distribution.
Reconsider: Examples 5.18, 5.19, and 5.20, solutions, and interpretations.

CrunchIt!

CrunchIt! has a built-in function to compute probabilities associated with a geometric and a Poisson variable. The CrunchIt! geometric random variable is defined to be the number of failures until the first success is achieved.

1. Suppose X is a geometric random variable with $P(X) = p$. Select Distribution Calculator; Geometric. Enter the value for p and select an appropriate inequality symbol or the equals sign. Enter the value for $x - 1$ and click Calculate. See Figures 5.23 and 5.24.

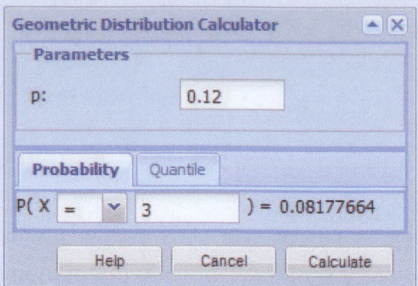

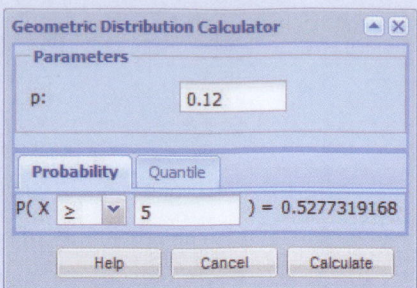

Figure 5.23 The probability mass function; use $x - 1$.

Figure 5.24 Right-tail probability.

2. Suppose X is a Poisson random variable with mean λ. Select Distribution Calculator; Poisson. Enter the value for lambda and select an appropriate inequality symbol or the equals sign. Enter the value for x and click Calculate. See Figures 5.25 and 5.26.

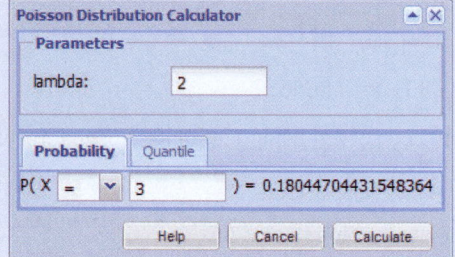

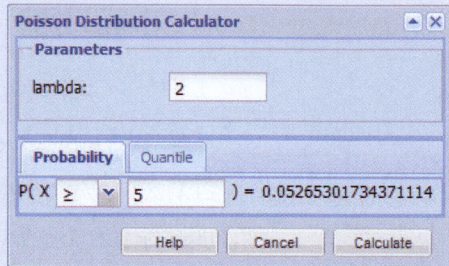

Figure 5.25 The probability mass function.

Figure 5.26 Cumulative probability.

TI-84 Plus C

There are built-in functions to compute cumulative probability and to evaluate the probability mass function associated with a geometric and Poisson random variable. Use the built-in function for combinations to compute probabilities associated with a hypergeometric random variable.

1. Suppose X is a geometric random variable with P(S) $= p$. Use the functions in the ⌷DISTR⌷; DISTR menu. Use the function `geometpdf` to find the probability that X takes on a single value and `geometcdf` to find cumulative probability. In each case, enter a value of p (p) and x value (x). Position the cursor on `Paste` and tap ⌷ENTER⌷. The appropriate calculator command is copied to the Home screen. Tap ⌷ENTER⌷ again to compute the desired probability. Refer to Figures 5.16 and 5.17, page 227.

2. Suppose X is a Poisson random variable with mean λ. Use the functions in the ⌷DISTR⌷; DISTR menu. Use the function `poissonpdf` to find the probability that X takes on a single value and `poissoncdf` to find cumulative probability. In each case, enter a value of p (p) and x value (x). Position the cursor on `Paste` and tap ⌷ENTER⌷. The appropriate calculator command is copied to the Home screen. Tap ⌷ENTER⌷ again to compute the desired probability. Refer to Figures 5.18 through 5.20, page 229.

Minitab

There are built-in functions to compute cumulative probability and to evaluate the probability mass function associated with a geometric, Poisson, or hypergeometric random variable. These functions may be accessed through a graphical input window: Calc; Probability Distributions, or by using the command language.

1. Suppose X is a geometric random variable with P(S) $= p$. In a session window, use the commands PDF or CDF to evaluate the probability mass function or to compute cumulative probability. See Figures 5.27 and 5.28.

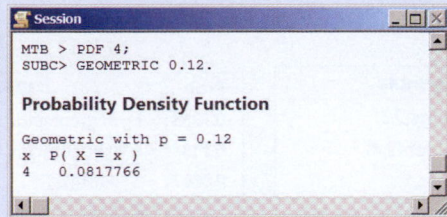

Figure 5.27 P(X = 4) in Example 5.18. **Figure 5.28** P(X ≥ 6) in Example 5.18.

2. Suppose X is a Poisson random variable with mean λ. In a session window, use the command PDF or CDF to evaluate the probability mass function or to compute cumulative probability. See Figures 5.29 and 5.30.

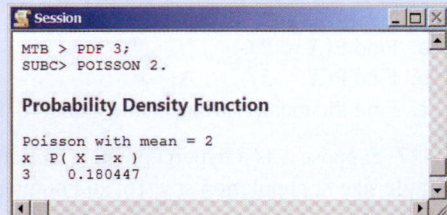

Figure 5.29 P(X = 3) in Example 5.19. **Figure 5.30** P(X ≥ 5) in Example 5.19.

3. Suppose X is a hypergeometric random variable with parameters n, N, and M. In a session or <u>C</u>alc; Probability <u>D</u>istributions input window, evaluate the probability mass function or compute cumulative probability. Refer to Figures 5.21 and 5.22, page 231.

Excel

There are built-in functions to compute cumulative probability and to evaluate the probability mass function associated with a Poisson and a hypergeometric random variable. Use the built-in function for a binomial distribution, BINOM.DIST, to find probabilities associated with the geometric distribution.

1. Suppose X is a geometric random variable with $P(S) = p$. Use the following formulas:

$P(X = x) = $ BINOM.DIST $(1, x, p, \text{FALSE})/x$

$P(X \le x) = 1 - $ BINOM.DIST $(0, x, p, \text{FALSE})$

See Figure 5.31.

2. Suppose X is a Poisson random variable with $\lambda = L$. To evaluate the probability mass function, use the function POISSON.DIST with the last argument set to False. To compute cumulative probability, use the function POISSON.DIST with the last argument set to True. See Figure 5.32.

3. Suppose X is a hypergeometric random variable with parameters n, N, and M. To evaluate the probability mass function, use the function HYPGEOM.DIST with the last argument set to False. To compute cumulative probability, use the function HYPGEOM.DIST with the last argument set to True. See Figure 5.33.

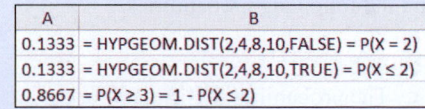

A	B
0.0818	= BINOM.DIST(1,4,0.12,FALSE)/4 = P(X = 4)
0.4723	= 1-BINOM.DIST(0,5,0.12,FALSE) = P(X ≤ 5)
0.5277	= P(X ≥ 6) = 1 - P(X ≤ 5)

A	B
0.1804	= POISSON.DIST(3,2,FALSE) = P(X = 3)
0.9473	= POISSON.DIST(4,2,TRUE) = P(X ≤ 4)
0.0527	= P(X ≥ 5) = 1 - P(X ≤ 4)

A	B
0.1333	= HYPGEOM.DIST(2,4,8,10,FALSE) = P(X = 2)
0.1333	= HYPGEOM.DIST(2,4,8,10,TRUE) = P(X ≤ 2)
0.8667	= P(X ≥ 3) = 1 - P(X ≤ 2)

Figure 5.31 Probabilities associated with a geometric random variable. **Figure 5.32** Probabilities associated with a Poisson random variable. **Figure 5.33** Probabilities associated with a hypergeometric random variable.

Note: Similar calculations are available using JMP. See Figures 5.34–5.36.

Prob	JMP Formula
0.0818	= Binomial Probability(0.12,4,1)/4
0.4723	= 1 - Binomial Probability(0.12,5,0)
0.5277	= 1 - Prob[[2]]

Figure 5.34 Probabilities associated with a geometric random variable.

Prob	JMP Formula
0.1804	= Poisson Probability(2,3)
0.9473	= Poisson Distribution(2,4)
0.0527	= 1 - Prob[[2]]

Figure 5.35 Probabilities associated with a Poisson random variable.

Prob	JMP Formula
0.1333	= Hypergeometric Probability(10,8,4,2)
0.1333	= Hypergeometric Distribution(10,8,4,2)
0.8667	= 1 - Prob[[2]]

Figure 5.36 Probabilities associated with a hypergeometric random variable.

SECTION 5.5 EXERCISES

Concept Check

5.106 True/False   In a geometric experiment, the probability of a success varies from trial to trial.

5.107 True/False   A geometric experiment ends when the first success is observed.

5.108 True/False   The number of possible values for a geometric random variable is infinite.

5.109 True/False   For a Poisson random variable, the mean is equal to the variance.

5.110 Fill in the Blank   A Poisson random variable is often used to count _____.

5.111 Short Answer   Explain the difference between a hypergeometric experiment and a binomial experiment.

Practice

5.112 Suppose X is a geometric random variable with probability of success 0.35. Find the following probabilities.
 a. $P(X = 4)$ **b.** $P(X \geq 3)$
 c. $P(X \leq 2)$ **d.** $P(X \geq \mu)$

5.113 Suppose X is a geometric random variable with mean $\mu = 4$. Find the following probabilities.
 a. $P(X = 1)$ **b.** $P(3 \leq X \leq 7)$
 c. $P(X > \mu + 2\sigma)$

5.114 Suppose X is a Poisson random variable with $\lambda = 2$. Find the following probabilities.
 a. $P(X = 0)$ **b.** $P(2 \leq X \leq 8)$
 c. $P(X > 5)$ **d.** $P(X \leq 6)$

5.115 Suppose X is a Poisson random variable with $\lambda = 4.5$. Find the following probabilities.
 a. $P(X > \mu)$
 b. $P(X = 2)$
 c. The probability X is either 4 or 5.
 d. $P(X \leq \mu + 2\sigma)$

5.116 Suppose X is a hypergeometric random variable with $n = 5$, $N = 12$, and $M = 6$.

 a. Find $P(X = 2)$.
 b. Find $P(X = 5)$.
 c. Find the mean, variance, and standard deviation of X.

5.117 Suppose X is a hypergeometric random variable with sample size 8, population size 16, and number of successes in the population 12.
 a. List the possible values for X.
 b. Find the mean, variance, and standard deviation of X.
 c. Find $P(X = 5)$.
 d. Find $P(X = 8)$.

Applications

5.118 According to the Anxiety and Depression Association of America (ADAA), approximately 4% of all adults have attention-deficit/hyperactive disorder (ADHD).[30] An experiment consists of selecting adults at random and asking them if they have ADHD.
 a. What is the probability that the fifth adult selected will be the first with ADHD?
 b. What is the probability that at least eight adults will be selected before identifying a person with ADHD?
 c. What is the mean number of adults that must be selected before identifying a person with ADHD?
 d. Suppose the 35th adult is the first with ADHD. Is there any evidence to suggest the ADAA claim is false? Justify your answer.

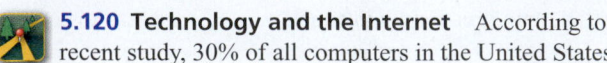 **5.119 Psychology and Human Behavior**   According to recent FBI statistics, the mean number of bank robberies per day in the Southern Region of the United States is 4.32.[31] Suppose a day is selected at random.
 a. What is the probability of exactly two bank robberies in the Southern Region? Write a Solution Trail for this problem.
 b. What is the probability that there will be more than eight bank robberies on that day in the Southern Region?
 c. Suppose two days are selected at random. What is the probability that there will be no robberies in the Southern Region on both days?

5.120 Technology and the Internet   According to a recent study, 30% of all computers in the United States are infected with some form of malware.[32] Suppose a computer repair specialist carefully checks every machine left at his store.

a. What is the probability that the second computer examined will be the first to have malware? Write a Solution Trail for this problem.

b. What is the probability that the tenth machine examined will be the first to have malware?

c. What is the mean number of computers examined before one will be infected with malware?

d. What is the probability that at least five computers will be examined before one will have malware?

5.121 Physical Sciences Of all cities in the United States, Amherst, New York, has the fewest number of days per year clear of clouds, 4.4.[33] Other cities with very few clear days include Buffalo, New York, Lakewood, Washington, and Seattle, Washington. Suppose a random year is selected.

a. What is the probability that Amherst will have exactly three days clear of clouds?

b. What is the probability of fewer than six days clear of clouds?

c. What is the probability of at least nine days clear of clouds?

d. Suppose that between 2 and 10 (inclusive) days are clear of clouds. What is the probability of more than five days clear of clouds?

5.122 Travel and Transportation Bad weather is to blame for some of the worst highway crashes in Canada. In December 2012 there was a 27-vehicle pile-up on Highway 40, and in February 2013 a 50-car pile-up shut down Highway 401 near Woodstock, Ontario. Highway 63 in Alberta has a notorious reputation; there are approximately four accidents every week on this road.[34] Suppose a week is randomly selected.

a. Find the probability that there are no more than four crashes.

b. Find the probability that the number of crashes is more than $\mu + 2\sigma$.

c. To obtain government funding for safety improvements, there must be five weeks in a row with six or more crashes. What is the probability of this happening?

5.123 Physical Sciences On a Friday night in late March 2013, there were hundreds of reports of meteor sightings along the East Coast of the United States. People from North Carolina to Canada contacted the American Meteor Society to report what they saw and heard. Suppose that in a group of 25 people at the National Mall, 15 actually saw the meteor. A patrolman randomly selects five people from this group.

a. What is the probability that none of the five people saw the meteor?

b. What is the probability that at least four people saw the meteor?

c. What is the probability that at most two people saw the meteor?

5.124 Sports and Leisure The NCAA Men's and Women's Swimming and Diving Committee recently recommended a *no recall false-start* rule.[35] This proposal means that unless a false start is blatant, the race will continue. The student-athlete committing the false start will be disqualified following the

event. Suppose the probability of a false start in any swimming event is 0.07, and swimming events are selected at random.

a. What is the probability that the first false start will occur in the fourth event selected?

b. What is the probability that the first false start will occur after the 15th event selected?

c. What is the mean number of events selected before a false start occurs?

d. Suppose there are 26 events at a swimming and diving meet. What is the probability that the first false start will occur at this meet?

5.125 Manufacturing and Product Development Flat-panel TV displays in televisions and computer monitors often develop *dead pixels*, pixels that become locked in one state—for example, red at all times. Manufacturers maintain that dead pixels are a natural defect and there are various return policies after the discovery of a dead pixel. Suppose the mean number of dead pixels in a new Samsung 64-inch plasma TV is 2.5. One of these TVs is randomly selected and inspected for dead pixels.

a. What is the probability that there will be no dead pixels?

b. If the number of dead pixels is more than $\mu + 3\sigma$, the assembly line is automatically stopped and examined. What is the probability that the assembly line will be stopped?

c. What is the probability that the number of dead pixels will be within two standard deviations of the mean?

5.126 Economics and Finance Most banks charge a monthly fee for a checking account, in addition to ATM, overdraft, and other fees. However, 72% of credit unions offer checking accounts with no monthly fees.[36] Suppose a bank auditor selects credit unions at random.

a. What is the probability that the second credit union selected will be the first to offer free checking?

b. What is the probability that the first credit union to offer free checking will be one of the first three?

c. Suppose the first credit union to offer free checking is the 10th selected. Is there any evidence to suggest that the claim (72%) is false? Justify your answer.

5.127 Business and Management Fifteen lobstermen have their boats anchored at a small pier along the New Hampshire coast. Five of these lobstermen have been fined within the past year for commercial lobster-size violations. Suppose four lobstermen are selected at random.

a. What is the probability that exactly two have been fined for violations within the past year? Write a Solution Trail for this problem.

b. What is the probability that all four have been fined for violations within the past year?

c. What is the probability that at least one has been fined for violations within the past year?

5.128 Demographics and Population Statistics Buchtal, a manufacturer of ceramic tiles, reports 3.9 job-related accidents per year. Accident categories include trip, fall, struck by equipment, transportation, and handling. Suppose a year is selected at random.

a. What is the probability that there will be no job-related accidents?

b. What is the probability that the number of accidents that year will be between two and five (inclusive)?

c. If the number of accidents is more than three standard deviations above the mean, the company insurance carrier will raise the rates. What is the probability of an increase in the company's insurance bill?

5.129 Sports and Leisure Amusement park rides are great family fun, but over 4400 children are injured on rides every year. According to a recent study, on average, one child is treated in a hospital Emergency Room every two hours as a result of an injury from an amusement park ride.[37] Suppose a two-hour period is selected at random.

a. What is the probability that no children will be treated in an Emergency Room as a result of an injury from an amusement park ride?

b. What is the probability that at most three children will be treated in an Emergency Room as a result of an injury from an amusement park ride?

c. Suppose that six children are treated in an Emergency Room as a result of an injury on an amusement park ride. Is there any evidence to suggest that the claim (of one every two hours) is wrong? Justify your answer.

Extended Applications

5.130 Marketing and Consumer Behavior The Sweet Leaf Iced Teas Company is sponsoring a conventional bottle-cap sweepstakes game. Under each bottle cap there is a note saying either "You are not a winner," or the prize awarded. Suppose there are 20 of the game bottles on a shelf in the supermarket, and two of them are winners. A customer randomly selects six bottles from the shelf.

a. What is the probability of selecting no winning bottles?

b. What is the probability of selecting both winning bottles?

c. What is the mean number of winning bottles selected?

d. How many bottles would the customer have to purchase in order to expect one winning bottle?

5.131 Physical Sciences There are over 1 million earthquakes worldwide of magnitude 2–2.9 each year. However, the mean number of earthquakes of magnitude 8 or higher is approximately one per year.[38] Suppose a random year is selected.

a. What is the probability of exactly two earthquakes of magnitude 8 or higher?

b. What is the probability of at most four earthquakes of magnitude 8 or higher?

c. Suppose there are three earthquakes of magnitude 8 or higher. Is there any evidence to suggest that the mean is different from one? Justify your answer.

5.132 Business and Management Managers at CafePress acknowledge that a variety of errors may occur in customer orders received via telephone. A recent audit revealed that the probability of some type of error in a telephone order is 0.20. In an attempt to correct these errors, a supervisor randomly selects telephone orders and carefully inspects each one.

a. What is the probability that the third telephone order selected will be the first to contain an error?

b. What is the probability that the supervisor will inspect between two and six (inclusive) telephone orders before finding an error?

c. What is the probability that the inspector will examine at least seven orders before finding an error?

d. What is the probability that the first error will be on the fourth telephone order or later?

e. Suppose the first four telephone orders contain no errors. What is the probability that the first error will be on the eighth order or later?

5.133 Economics and Finance The manager of Capitol Park Plaza, an apartment complex in Washington, DC, collects the rent from each tenant on the first day of every month. Past records indicate that the mean number of tenants who do not pay the rent on time in any given month is 4.7. Consider the rent collection for the next month.

a. Find the probability that every tenant will pay the rent on time.

b. Find the probability that at least seven tenants will be late with their rent.

c. Suppose the number of delinquent rent payments in a month is independent of the number in every other month. What is the probability that at most three tenants will be late with their rent in two consecutive months?

5.134 Public Health and Nutrition The mean number of commercially prepared meals per week for a typical American is 4.[39] This might seem a little high. But consider young professionals who eat lunch out several times each week, and families that order out for pizza or stop at a fast-food restaurant routinely.

a. What is the probability that a randomly selected American does not eat out during a week? Eats out once during the week? Twice during the week?

b. Suppose two Americans are selected at random. What is the probability that the total number of meals for the two Americans during a week is 0? 1? 2?

c. Suppose the mean number of times an American eats out during a two-week period is 8. What is the probability that a randomly selected American does not eat out during a two-week period? Once during the two-week period? Twice during the two-week period?

d. How do your answers in parts (b) and (c) compare? What property does this suggest about a Poisson random variable?

5.135 Psychology and Human Behavior The percentage of Americans who claim to have no religious affiliation is the highest since 1930, approximately 20%.[40] In a group of 30 police officers, 6 have no religious affiliation. Suppose 4 officers from this group are selected at random.

a. What is the probability that exactly 1 officer will have no religious affiliation?

b. What is the probability that at most 2 officers will have no religious affiliation?

c. Suppose the group consists of 50 police officers, 10 with no religious affiliation. Find the probabilities in parts (a) and (b) given this new, larger group.

d. Suppose 4 officers are selected at random from across the country. What is the probability that exactly 1 will have no religious affiliation? At most 2 will have no religious affiliation?

e. Compare all of these probabilities. Explain how the hypergeometric distribution is related to the binomial distribution.

Challenge

5.136 Approaching Poisson Suppose X is a Poisson random variable with $\lambda = 2$. Let the random variable Y have a probability distribution as given in the following table.

y	$P(Y = y)$
0	$P(X = 0) = 0.1353$
1	$P(X = 1) = 0.2707$
2	$P(X = 2) = 0.2707$
3	$P(X \geq 3) = 0.3233$

Find the expected value of Y.

Suppose the distribution of Y is changed slightly, at the right tail, as given in the following table.

y	$P(Y = y)$
0	$P(X = 0) = 0.1353$
1	$P(X = 1) = 0.2707$
2	$P(X = 2) = 0.2707$
3	$P(X = 3) = 0.1804$
4	$P(X \geq 4) = 0.1429$

Find the expected value of Y.

Suppose the distribution of Y is changed again, once more at the right tail.

y	$P(Y = y)$
0	$P(X = 0) = 0.1353$
1	$P(X = 1) = 0.2707$
2	$P(X = 2) = 0.2707$
3	$P(X = 3) = 0.1804$
4	$P(X = 4) = 0.0902$
5	$P(X \geq 5) = 0.0527$

Find the expected value of Y.

Continue in this manner. To what number is E(Y) converging, and why does this make sense?

5.137 A Committed Relationship Suppose X is a geometric random variable with probability of success $p = 0.40$ and Y is a binomial random variable with the same probability of success $p = 0.40$. For $a = 1, 2, 3, \ldots, 10$, construct a table with the following probabilities.

a. $P(X = a)$
b. $P(Y = 1)/a$, where $Y \sim B(a, 0.40)$
c. $P(X \leq a)$
d. $1 - P(Y = 0)$, where $Y \sim B(a, 0.40)$

Carefully examine the table and write a general formula to explain each equality. Can you prove these results?

CHAPTER 5 SUMMARY

Concept	Page	Notation / Formula / Description
Random variable	188	A function that assigns a unique numerical value to each outcome in a sample space.
Discrete random variable	190	The set of all possible values is finite, or countably infinite.
Continuous random variable	190	The set of all possible values is an interval of numbers.
Probability distribution for a discrete random variable	193	A method for conveying all the possible values of the random variable and the probability associated with each value.
Mean, or expected value, of a discrete random variable X	203	$\mu = E(X) = \sum_{\text{all } x}[x \cdot p(x)]$
Variance of a discrete random variable X	204	$\sigma^2 = \text{Var}(X) = \sum_{\text{all } x}[(x - \mu)^2 \cdot p(x)]$
Properties of a binomial experiment	212	1. n identical trials. 2. Each trial can result in only a success (S) or a failure (F). 3. Trials are independent. 4. Probability of a success is constant from trial to trial.
Binomial random variable	212	The number of successes in n trials.

Binomial probability distribution	214	If $X \sim B(n, p)$ then $$p(x) = \binom{n}{x}p^x(1 - p)^{n-x}, x = 0, 1, 2, 3, \ldots, n,$$ where $\mu = np$, $\sigma^2 = np(1 - p)$, and $\sigma = \sqrt{np(1 - p)}$.
Cumulative probability	215	$P(X \leq x)$
Geometric random variable	225	The number of trials necessary to realize the first success.
Geometric probability distribution	226	$$p(x) = (1 - p)^{x-1}p, \quad x = 1, 2, 3, \ldots$$ where $\mu = \dfrac{1}{p}$, $\sigma^2 = \dfrac{1 - p}{p^2}$, and $\sigma = \sqrt{\dfrac{1 - p}{p^2}}$.
Poisson random variable	227	A count of the number of times a specific event occurs during a given interval.
Poisson probability distribution	228	$$p(x) = \dfrac{e^{-\lambda}\lambda^x}{x!}, \quad x = 0, 1, 2, 3, \ldots,$$ where $\mu = \lambda$, $\sigma^2 = \lambda$, and $\sigma = \sqrt{\lambda}$.
Hypergeometric random variable	229	A count of the number of successes in a random sample of size n from a population of size N.
Hypergeometric probability distribution	230	$$p(x) = \dfrac{\binom{M}{x}\binom{N - M}{n - x}}{\binom{N}{n}}, \quad \max(0, n - N + M) \leq x \leq \min(n, M),$$ where $\mu = n\dfrac{M}{N}$, $\sigma^2 = \left(\dfrac{N - n}{N - 1}\right)n\dfrac{M}{N}\left(1 - \dfrac{M}{N}\right)$, and $$\sigma = \sqrt{\left(\dfrac{N - n}{N - 1}\right)n\dfrac{M}{N}\left(1 - \dfrac{M}{N}\right)}.$$

CHAPTER 5 EXERCISES

5 APPLICATIONS

5.138 Public Health and Nutrition   Emergency defibrillators are now located in many public buildings. However, the U.S. Food and Drug Administration (FDA) is concerned about the reliability of these devices. Approximately 45,000 devices failed during the past seven years.[41] Suppose the FDA claims that the probability of an emergency defibrillator working correctly is 0.90, and 30 of these devices are selected at random and tested.

  a. Find the probability that exactly 28 devices will work correctly.
  b. Find the probability that at least 25 devices will work correctly.
  c. Suppose only 20 of the devices work correctly. Is there any evidence to suggest that the proportion of emergency defibrillators that work correctly has changed? Justify your answer.

5.139 Business and Management   IKEA is a Swedish company that sells ready-to-assemble furniture. Shoppers contact customer service with regard to finding a nearby store, online shipping questions, and even for help with assembly. IKEA classifies all telephone calls to its customer support staff by the amount of time the customer is on hold. If the customer is on hold for no more than 60 seconds, then the call is classified as successful (actually, this sounds like a miracle). The supervisor in technical support claims 80% of all calls are successful. Suppose 25 calls to technical support are selected at random.

  a. Find the mean, variance, and standard deviation of the number of successful calls.
  b. Find the probability that at least 18 calls will be successful.
  c. Suppose 21 calls are successful. Is there any evidence to suggest the supervisor's claim is false? Justify your answer.

5.140 Economics and Finance   Bank overdraft fees range from $10 to $38, consumers believe they are annoying and excessive, and these fees are a huge revenue source for banks.[42] The Overdraft Protection Act of 2013 is designed to limit overdraft fees in a variety of ways and to require fees to be *reasonable and proportional* to the amount of the overdraft. Let X be the amount of an overdraft fee for a randomly selected bank. The probability distribution for X is given in the table below.

x	10	12	15	20	25
$p(x)$	0.02	0.06	0.08	0.10	0.16

x	27	30	35	38
$p(x)$	0.28	0.15	0.07	0.08

a. Find the mean, variance, and standard deviation of the overdraft amount.

b. Find the probability that a randomly selected bank has an overdraft fee greater than $25.

c. Find the probability that a randomly selected bank has an overdraft fee less than $\mu - \sigma$.

d. Suppose three banks are selected at random. What is the probability that at least one bank has an overdraft fee less than $20?

5.141 Manufacturing and Product Development Thales Alenia Space is a European company that manufactures communications satellites. Researchers at the company have determined that the most common reason for a satellite to fail once it is in orbit is a problem related to opening and initiating the solar panels. Suppose the probability of a failure related to the solar panels is 0.08.

a. What is the probability that the fifth satellite launched will be the first to fail due to a solar-panel problem?

b. What is the expected number of satellites launched until the first one fails due to a solar-panel problem?

c. Thales Alenia Space is preparing an advertising campaign in which they claim to have had 20 successful launches in a row. What is the probability that the first failure due to a solar-panel problem will occur after the 20th launch?

5.142 Business and Management An easy-assembly, no-tools-required, gas grill comes with detailed step-by-step instructions. Even though each grill is carefully packaged, there are often missing pieces. This can aggravate the customer and increase the cost to the producer, who must provide phone support and ship the missing parts. Suppose the mean number of missing pieces per packaged grill is 0.7, and one grill is randomly selected from the stockroom.

a. What is the probability that there will be no missing pieces in the package?

b. If there are more than five missing pieces, the producer identifies the packager and issues a warning. What is the probability of a warning being issued at the packaging plant?

c. Suppose three grills are randomly selected. What is the probability that each will have no more than one missing piece?

5.143 Travel and Transportation Because of forced spending cuts in Spring 2013, the Federal Aviation Administration identified 149 air traffic control towers that would be closed.[43] In a group of 20 air traffic control towers in the Midwest, five will be closed. Suppose six of the 20 air traffic control towers are selected at random.

a. What is the probability that none of the six air traffic control towers will be closed?

b. What is the probability that at most two of the air traffic control towers will be closed?

c. What is the probability that at least four will be closed?

5.144 Sports and Leisure Over a period of 12 years, there were approximately 2.42 shark attacks per year at the beaches in North Carolina.[44] Consider the number of shark attacks during the thirteenth year.

a. What is the probability that there will be no shark attacks?

b. What is the probability that between two and five shark attacks inclusive will occur?

c. If there is evidence to suggest that the mean number of shark attacks per year has increased, the Coast Guard will begin more patrols to adequately protect the public. Suppose there are eight shark attacks in the thirteenth year. Is there evidence to suggest the need for more patrols? Justify your answer.

d. Find out exactly how many shark attacks there were in a recent year in North Carolina. Determine the probability of this occurrence.

e. Florida has the highest number of shark attacks per year, approximately 22.5, followed by Hawaii (4.33 per year), and California (3.17 per year). What is the probability that there will be no attacks in all four states in a given year?

5.145 Technology and the Internet In February 2013, the Chinese Ministry of Industry and Information Technology announced plans to expand broadband connections (or faster) to 70% of Chinese households.[45] Suppose the plan is successful and 40 Chinese households are selected at random.

a. What is the probability that exactly 25 households have broadband coverage?

b. What is the probability that at most 30 households have broadband coverage?

c. What is the probability that more than 33 households have broadband coverage?

d. Suppose the number of households that have broadband coverage is within two standard deviations of the mean. What is the probability that the actual number that have broadband coverage is within one standard deviation of the mean?

5.146 Sports and Leisure In 2012–2013, Carnival Cruise Lines experienced onboard problems with four ships: *Triumph*, *Elation*, *Dream*, and *Legend*. The technical issues included loss of power, steering problems, and even a fire in an engine room. Despite these setbacks, the cruise industry contributes almost $38 billion to the U.S. economy, and approximately 20% of all people in the United States have taken a cruise.[46] Suppose 25 people from the United States are selected at random.

a. Find the probability that exactly three people have been on a cruise.

b. Find the probability that at most two people have been on a cruise.

c. Suppose seven people have taken a cruise. Is there any evidence to suggest that the percentage of people who have taken a cruise has increased? Justify your answer.

5.147 Psychology and Human Behavior The army emphasizes cleanliness and neatness in a military barracks. Each cadet is responsible for maintaining his or her area in top condition. Periodic inspections are held, and those receiving top scores are rewarded. Suppose the mean number of violations discovered per cadet during a barracks inspection is 2.7.

a. What is the probability that a randomly selected cadet will have exactly three violations during an inspection?

b. If a cadet has six or more violations, he or she is assigned to KP duty for one week. What is the probability that a randomly selected cadet will be assigned to KP duty following a barracks inspection?

c. If every member of a 10-cadet unit has no violations, then each will receive a weekend pass. What is the probability of this happening following a barracks inspection?

5.148 Marketing and Consumer Behavior In the country's 2012 budget, Canada decided to stop production of the penny.[47] The government indicated that it costs 1.6 cents to produce each penny, some see the penny as a burden to the economy, and approximately 10% of all Canadians believe the penny is a nuisance. Suppose 50 Canadians are selected at random and asked if they believe the penny is a nuisance.

a. Find the mean, variance, and standard deviation of the number of Canadians who believe the penny is a nuisance.

b. What is the probability that at most three Canadians believe the penny is a nuisance?

c. Find the probability that the number of Canadians who believe the penny is a nuisance will be within two standard deviations of the mean.

5.149 Marketing and Consumer Behavior The movie *Les Misérables*, an adaptation of Victor Hugo's novel, starred Hugh Jackman, Russell Crowe, Anne Hathaway, and Amanda Seyfried, and won many awards. The Flixster movie site, Rotten Tomatoes, rated the movie at 74% on the Tomatometer. However, 81% of all people who saw the movie liked it.[48] Suppose 30 people who saw the movie are selected at random.

a. What is the probability that at most 20 people liked the movie?

b. What is the probability that at least 25 people liked the movie?

c. Suppose 18 people liked the movie. Is there any evidence to suggest that the claim (81%) is wrong? Justify your answer.

5.150 Public Policy and Political Science In a recent nationwide study, it was reported that U.S. adults continue to believe that big companies and lobbyists have too much power. In particular, 85% of those polled indicated that PACs (Political Action Committees) have too much influence in Washington. Suppose 50 U.S. adults are selected at random.

a. What is the probability that at least 45 U.S. adults think PACs have too much power?

b. What is the probability that between 38 and 42 (inclusive) U.S. adults think PACs have too much power?

c. Suppose 35 U.S. adults think PACs have too much power. Is there any evidence to suggest the poll results are wrong? Justify your answer.

5.151 The CBS show *NCIS* stars Mark Harmon as Special Agent Leroy Jethro Gibbs in which his team investigates military-related criminal cases. The character Abigail Sciuto is frequently seen drinking the high-energy caffeine-laden drink Caf-Pow (she keeps a spare in the lab refrigerator). The mean number of Caf-Pows Abby consumes per show is 3. Suppose a 2013 *NCIS* episode is selected at random.

a. What is the probability that Abby has no Caf-Pows on the show?

b. What is the probability that she is shown having at most 3 Caf-Pows on the show?

c. Suppose Abby has 6 Caf-Pows. Is there any evidence that the number of Caf-Pows per show has changed? Justify your answer?

EXTENDED APPLICATIONS

5.152 Discrete Uniform Random Variable Suppose X is a random variable with probability distribution given by

$$p(x) = \frac{1}{5} \quad x = 1, 2, 3, 4, 5$$

a. Find the mean, variance, and standard deviation of X.

b. Suppose $p(x) = 1/6$, $x = 1, 2, 3, 4, 5, 6$. Find the mean, variance, and standard deviation of X.

c. Suppose $p(x) = 1/n$, $x = 1, 2, 3, \ldots, n$. Find the mean, variance, and standard deviation of X in terms of n.

5.153 Medicine and Clinical Studies According to a Pew Research Center survey, approximately 35% of Americans attempt to diagnose a medical condition online.[49] Highmark insurance company is concerned about the rising number of online diagnosers and the resulting failure to consult a physician. They have decided to select 25 policyholders at random. If the number of online diagnosers is 11 or fewer, then no action will be taken. Otherwise, they will begin a campaign to remind policyholders that they should always consult a physician to confirm a medical condition.

a. Suppose the true proportion of online diagnosers is 0.35. What is the probability that Highmark will begin a new reminder campaign?

b. Suppose the true proportion of online diagnosers is 0.40. What is the probability that no action will be taken? What if the true proportion is 0.50?

c. Suppose the decision rule is changed such that if the number of online diagnosers is 12 or fewer, then no action will be taken. Answer parts (a) and (b) using this rule.

5.154 Marketing and Consumer Behavior Kohl's is running a sale in which customers may save as much as 40% on any purchase. Once a customer decides to make a purchase, he selects two sales prize tickets at random from a large bin at the front of the store. Each ticket has a percentage marked on it, and the probability of selecting each ticket is given in the table below.

Percentage	10%	20%	30%	40%
Probability	0.50	0.35	0.10	0.05

The larger of the two percentages selected is used for the purchase.

a. Let X be the maximum of the two prize ticket percentages. Find the probability distribution for X.

b. Find the mean, variance, and standard deviation of X.

c. What is the probability that a customer will receive at least 20% off on his or her purchase?

5.155 Medicine and Clinical Studies   A recent study suggests that total knee joint replacement surgery may be related to weight gain. Researchers who studied records from the Mayo Clinic Health system report that 30% of these patients gained 5% or more of their body weight following surgery.[50] Suppose this percentage is the same at all hospitals in the United States and 40 people who had total knee joint replacement surgery are selected at random.

   a. What is the probability that exactly 14 will experience a weight gain?

   b. Find the largest value w such that the probability of w or fewer patients who experience a weight gain is at most 0.20.

   c. Suppose 16 patients experience a weight gain. Is there any evidence to suggest that the proportion of patients with knee joint replacement surgery who experience a weight gain has changed? Justify your answer.

5.156 Public Health and Nutrition   A recent study suggested that 75% of all meals marketed specifically to babies and toddlers and sold in grocery stores had high sodium content.[51] This is of deep concern because increased salt in the diet can cause hypertension, which may lead to cardiovascular disease. Suppose 30 toddler meals are randomly selected and the sodium content in each is carefully measured.

   a. What is the probability that exactly 23 meals will have high sodium content?

   b. What is the probability that at least 25 will have high sodium content?

   c. Suppose at most 20 meals have high sodium content. What is the probability that at most 15 will have high sodium content?

CHALLENGE

5.157 A Day on the Dock   Two crews work on a receiving dock at a fabric manufacturing plant. The first crew unloads four shipments every day and the second crew unloads seven shipments every day. A supervisor records whether each shipment is complete (a success) or missing items (a failure).

Suppose X_1 is a binomial random variable, representing the number of complete shipments for crew 1, with parameters $n_1 = 4$ and $p = 0.6$. Similarly, let X_2 be a binomial random variable, representing the number of complete shipments for crew 2, with parameters $n_2 = 7$ and $p = 0.6$. Assume X_1 and X_2 are independent.

   a. Use technology to generate a random observation for X_1 (the number of complete shipments for crew 1) and a random observation for X_2 (the number of complete

shipments for crew 2). Add these two values to compute a random total number of complete shipments for crews 1 and 2.

Repeat this process to generate 1000 random total number of complete shipments for crews 1 and 2. Compute the relative frequency of occurrence of each observation.

Suppose Y is a binomial random variable with $n = 11$ and $p = 0.6$. Use technology to construct a table of probabilities for $Y = 0, 1, 2, 3, \dots, 11$. Compare these probabilities with the relative frequencies obtained above.

   b. Suppose a new receiving crew is added and it unloads five shipments each day. Let X_3 be a binomial random variable, representing the number of complete shipments for crew 3, with parameters $n_3 = 5$ and $p = 0.6$.

Use technology to generate random observations for X_1, X_2, and X_3. Add these three values to compute a random total number of complete shipments for crews 1, 2, and 3.

Repeat this process to generate 1000 random total number of complete shipments for crews 1, 2, and 3. Compute the relative frequency of occurrence of each observation.

Suppose Y is a binomial random variable with $n = 16$ and $p = 0.6$. Use technology to construct a table of probabilities for $Y = 0, 1, 2, 3, \dots, 16$. Compare these probabilities with the relative frequencies obtained above.

   c. Suppose another receiving crew is added and it unloads nine shipments each day. Let X_4 be a binomial random variable, representing the number of complete shipments for crew 4, with parameters $n_2 = 9$ and $p = 0.6$. Let Y represent the total number of complete shipments for all four crews.

      i. Find $P(Y = 15)$ (the probability of exactly 15 total complete shipments).

      ii. Find $P(Y \le 12)$.

      iii. Find $P(Y > 16)$.

      iv. How many total complete shipments can be expected?

 **LAST STEP**

5.158 Is a flu shot really effective?   In 2012–2013 the CDC reported that the flu vaccine was 56% effective. That is, 56% of all people receiving the flu vaccine who were exposed to a flu virus did not contract the flu. To check this claim (56% effective), a random sample of 50 at-risk people who received a flu shot was selected. During the flu season, all 50 were exposed to the flu and 29 actually contracted the disease. Is there any evidence to suggest the claim is false, that the chance of contracting the flu is greater than 44%?

6 Continuous Probability Distributions

◀ **Looking Back**

- Remember how to completely describe and compute probabilities associated with a discrete random variable.
- Recall the characteristics of and probability computations associated with the binomial, geometric, Poisson, and hypergeometric random variables.

▶ **Looking Forward**

- Learn how to completely describe a continuous random variable.
- Compute probabilities associated with a continuous random variable.
- Understand the characteristics of the normal distribution and compute probabilities involving a normal random variable.

What's better, faster or slower?

LTE, or long-term evolution, is a wireless communications standard for mobile phones and data terminals. The first LTE service became available in 2009 in Oslo and Stockholm. Since then, countries all over the world have adopted this technology. Verizon was the first wireless company to offer LTE service in the United States, in 2010.

For mobile phone users, data download speed is perhaps the most important characteristic of an LTE network. Faster transmission rates mean web pages, songs, and videos download more quickly. OpenSignal uses a mobile application to measure wireless download speed, and in 2013 AT&T was the fastest at 13 Mbps (megabits per second), Verizon Wireless's mean download speed was 10 Mbps, Sprint Nextel's was 7.7 Mbps, and MetroPCS was in last place with a speed of 1.2 Mbps.[1]

According to OpenSignal, the mean download speed in the United States in 2013 was 9.6 Mbps. Suppose the standard deviation is 2.3 Mbps and the distribution of download speeds is approximately normal. The concepts presented in this chapter will allow us to determine the most reasonable download speeds for customers and to decide when a customer has a legitimate complaint about download speed.

CONTENTS

David Vernon/E+/Getty Images

6.1 Probability Distributions for a Continuous Random Variable

Suppose X is a continuous random variable; X takes on any value in some interval of numbers. A **continuous probability distribution** completely describes the random variable and is used to compute probabilities associated with the random variable.

Definition

A **probability distribution for a continuous random variable** X is given by a smooth curve called a **density curve**, or **probability density function** (pdf). The curve is defined so that the probability that X takes on a value between a and b $(a < b)$ is the area under the curve between a and b.

A CLOSER LOOK

1. Probability in a continuous world is *area under a curve*. Figures 6.1–6.3 illustrate the correspondence between the probability of an event (defined in terms of a continuous random variable) and the area under the density curve.

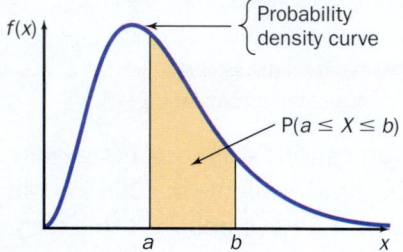

Figure 6.1 The probability is the area of the shaded region.

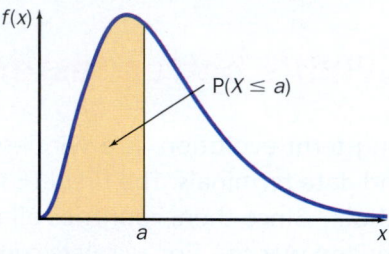

Figure 6.2 The shaded area is $P(X \le a)$.

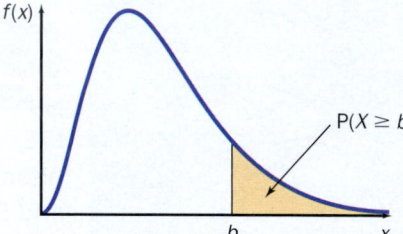

Figure 6.3 The shaded area is $P(X \ge b)$.

2. The density curve, or **probability density function**, is usually denoted by f. It is a *function*, defined for *all* real numbers. $f(x)$ is *not* the probability that the random variable X equals the specific value x. Rather, the function f leads to, or conveys, probability through area.

Remember that $f(x)$ is *not* a probability. The density function *leads to* probability.

3. The shape of the graph of a density function can vary considerably. However, a density function must satisfy the following two properties:

   a. f must be defined so that the total area under the curve is 1. The total probability associated with any random variable must be 1. $f(x)$, a specific value of the density function, *may* be greater than 1 (while the total area under the curve is still exactly 1).

   b. $f(x) \ge 0$ for all x. Therefore, the entire graph lies on or above the x axis. See Figure 6.4.

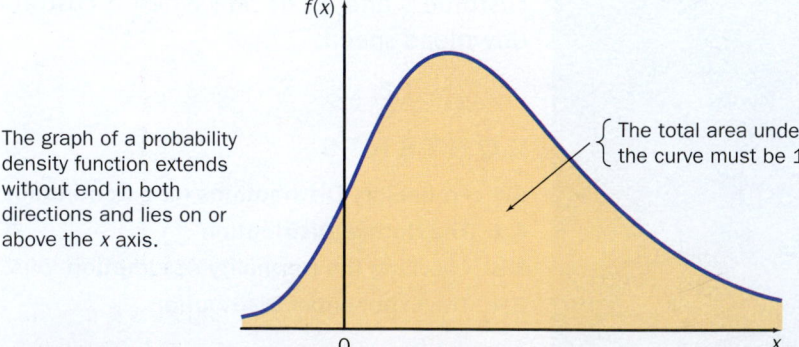

The graph of a probability density function extends without end in both directions and lies on or above the x axis.

The total area under the curve must be 1.

Figure 6.4 A valid probability density function.

P(X = a) translated: Find the area under the curve between a and a. This is asking for the area of a line segment. There is no second dimension. Hence the area is 0.

4. If X is a continuous random variable with density function f, the probability that X equals *any* one specific value is 0. That is, $P(X = a) = 0$ for any a. The reason: There is no area under a single point.

▶ This seems like a contradiction. Certainly we can observe specific values of X, yet the probability of observing any single value is 0. Recall: Probability is a *limiting relative frequency*. There are an (uncountably) infinite number of values for any continuous random variable. Therefore, the limiting relative frequency of occurrence of any single value is 0.

This is not necessarily true for a discrete random variable.

Because no probability is associated with a single point, the following four probabilities are all the same:

$$P(a \leq X \leq b) = P(a < X \leq b) = P(a \leq X < b) = P(a < X < b) \qquad (6.1)$$

In fact, we can remove as many single points as we want from any interval, and the probability will stay the same. The only reasonable probability questions concerning continuous random variables involve intervals. And we can almost always sketch a graph to visualize these probabilities, or regions. ◀

So, how do we find *area under a curve*, and therefore probability? In general, this is a calculus question. Don't panic. We'll use a little geometry, tables, and technology to find the necessary area (probability).

The (continuous) **uniform distribution** provides a good opportunity to illustrate the connection between area under the curve and probability. For this random variable, the total probability, 1, is distributed evenly, or uniformly, between two points. Computing probabilities associated with this random variable reduces to finding the area of a rectangle.

Definition

The random variable X has a **uniform distribution** on the interval $[a, b]$ if

$$f(x) = \begin{cases} \dfrac{1}{b-a} & \text{if } a \leq x \leq b \\ 0 & \text{otherwise} \end{cases} \qquad -\infty < a < b < \infty \qquad (6.2)$$

$$\mu = \frac{a+b}{2} \qquad \sigma^2 = \frac{(b-a)^2}{12} \qquad (6.3)$$

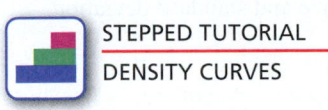
STEPPED TUTORIAL

DENSITY CURVES

A CLOSER LOOK

1. a and b can be any real numbers, as long as a is less than b ($a < b$).

2. All of the probability (action) is between a and b. The probability density function is the constant $1/(b-a)$ between a and b, and zero outside of this interval. Hence, there is no area and no probability outside the interval $[a, b]$. Figure 6.5 shows a graph of the uniform probability density function.

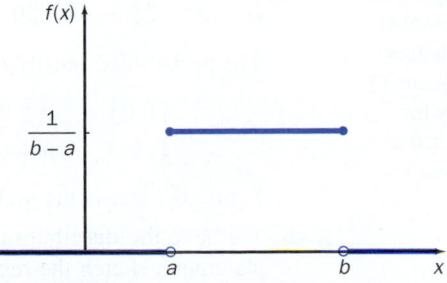

Figure 6.5 The graph of the probability density function for a uniform random variable.

3. Equation 6.2 is a valid probability density function because $f(x) \geq 0$ for all x, and the total area under the curve is 1. The area under the curve for $x < a$ is zero, and the area under the curve for $x > b$ is zero. Between a and b, the area under the curve is the area of a rectangle (area = width × height). See Figure 6.6.

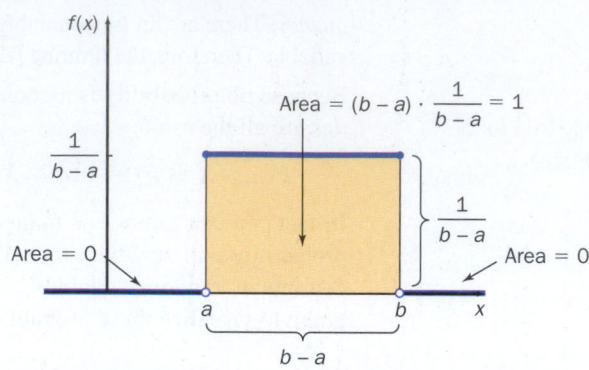

Figure 6.6 The total area under the curve is 1. Here, the density *curve* consists of three line segments.

The following example involves a uniform distribution and illustrates visualizing and calculating probabilities associated with a continuous random variable.

Example 6.1 Reef Dives

Bonaire, an Island in the Dutch Caribbean, is considered one of the top 10 diving destinations in the world. There is generally 60–100 feet of visibility, the current is mild, and at least 58 dive sites can be reached from shore. Guides take tourist groups on commercial boats to scuba dive and snorkel at selected locations. A careful examination of boat records has shown that the time it takes to reach a randomly selected dive site has a uniform distribution between 5 and 25 minutes. Suppose a dive site is selected at random.

a. Carefully sketch a graph of the probability density function.

b. Find the probability that it takes at most 10 minutes to reach the dive site.

c. Find the probability that it takes between 10 and 20 minutes to reach the dive site.

d. Find the mean time it takes to a dive site, and the variance and standard deviation.

SOLUTION

a. Let X be the time it takes to reach a dive site. X is uniform between the times $a = 5$ and $b = 25$. Use Equation 6.2 to find

$$\frac{1}{b-a} = \frac{1}{25-5} = \frac{1}{20} = 0.05$$

The probability density function is

$$f(x) = \begin{cases} 0.05 & \text{if } 5 \leq x \leq 25 \\ 0 & \text{otherwise} \end{cases}$$

Figure 6.7 shows the graph of the probability density function.

b. We have the distribution of X. Translate the question in part (b) into a probability statement, sketch the region corresponding to the probability statement, and find the area of that region.

At most means up to and including 10. We need the probability that X is less than or equal to 10: P($X \leq 10$). See Figure 6.8.

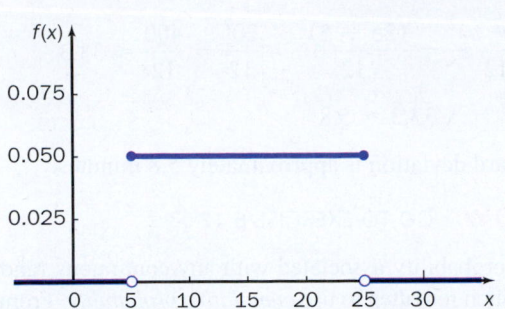 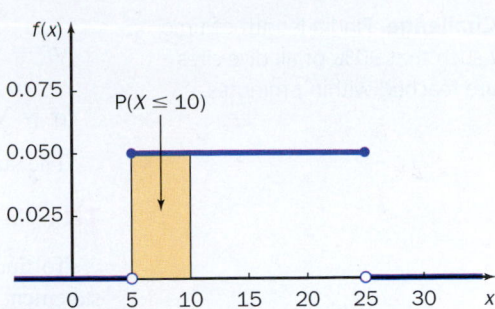

Figure 6.7 The graph of the probability density function for a uniform random variable on the interval $a = 5$ to $b = 25$.

Figure 6.8 The area of the shaded region is $P(X \le 10)$.

The probability statement $P(X \le 10)$ simplifies to $P(5 \le X \le 10)$ in this case because there is no probability (area) for x less than 5.

$P(X \le 10)$ = area under the density curve between 5 and 10

      = area of a rectangle

      = width × height

      = $(5)(0.05) = 0.25$

The probability that it takes at most 10 minutes is 0.25.

$P(10 \le X \le 20)$

   $= P(10 < X \le 20)$

   $= P(10 \le X < 20)$

   $= P(10 < X < 20)$

c. The probability that it takes between 10 and 20 minutes to reach a dive site in terms of the random variable X is $P(10 \le X \le 20)$. Even though the word *inclusive* is not used in the question, we chose to write the interval including the endpoints. It doesn't really matter! Remember: In a continuous world, single values contribute no probability and do not change the probability calculation.

$P(10 \le X \le 20)$ = area under the density curve between 10 and 20

           = area of a rectangle

           = width × height

           = $(10)(0.05) = 0.50$

The probability that it takes between 10 and 20 minutes is 0.50. See Figure 6.9.

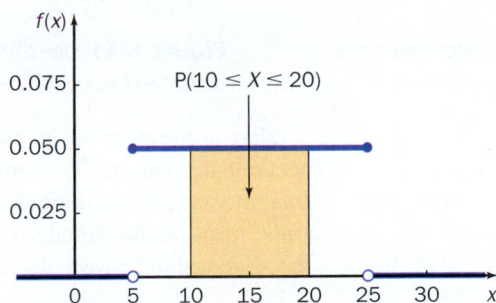

Figure 6.9 The area of the shaded region is $P(10 \le X \le 20)$.

d. Use Equation 6.3 to find the mean and variance.

$$\mu = \frac{a + b}{2} = \frac{5 + 25}{2} = \frac{30}{2} = 15$$

The mean time it takes to reach a dive site is 15 minutes. Because the uniform distribution is *symmetric*, the mean is the middle of the distribution, and the mean is equal to the median.

Challenge: Find a length of time t such that 90% of all dive sites are reached within t minutes.

$$\sigma^2 = \frac{(b-a)^2}{12} = \frac{(25-5)^2}{12} = \frac{20^2}{12} = \frac{400}{12} \approx 33.3$$

$$\sigma = \sqrt{\sigma^2} = \sqrt{33.3} \approx 5.8$$

The standard deviation is approximately 5.8 minutes.

TRY IT NOW   GO TO EXERCISE 6.12

To find a probability associated with any continuous random variable, the probability statement is often rewritten to use *cumulative probability*. From Chapter 5, cumulative probability means *accumulate* probability up to and including a fixed value. Find all the area under the density curve to the *left* of the fixed value. For a continuous random variable X, the cumulative probability up to x is $P(X \leq x)$. Figure 6.10 illustrates this cumulative probability.

Suppose X is a continuous random variable, and a and b are constants. Here are some typical probability statements involving X, and equivalent expressions using cumulative probability.

It doesn't matter whether we use $\leq$ or $<$, because one point contributes no probability. However, for consistency and accuracy throughout this text, cumulative probability will mean *up to and including x*; we use $\leq$ (not $<$).

$$P(X \geq b) = 1 - P(X < b) \qquad \text{(Figure 6.11)} \qquad \text{The complement rule.}$$
$$= 1 - \underbrace{P(X \leq b)}_{\text{Cumulative probability}} \qquad \text{A single value contributes no probability.}$$

$$P(a \leq X \leq b) = P(X \leq b) - P(X < a) \qquad \text{(Figure 6.12)}$$
$$= P(X \leq b) - P(X \leq a) \qquad \text{A single value contributes no probability.}$$

Find all the probability up to b, find all the probability up to a, and subtract. The difference is the probability that X lies in the interval from a to b.

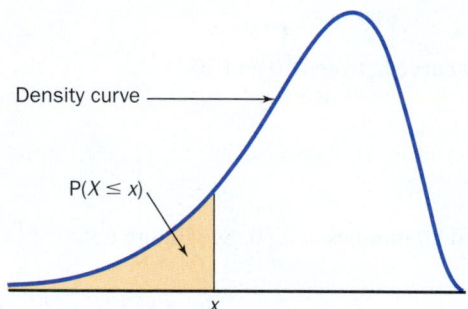

Figure 6.10 The shaded area is the cumulative probability $P(X \leq x)$.

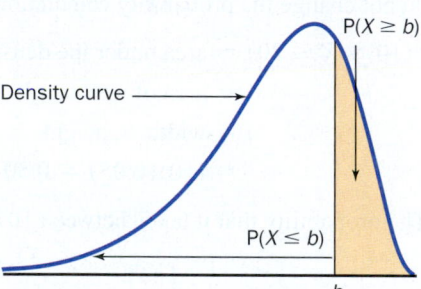

Figure 6.11 Use the complement rule to convert to cumulative probability.

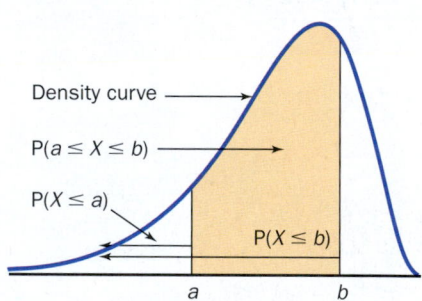

Figure 6.12 The shaded area is $P(X \geq b)$.

Here is one more way to picture cumulative probability. As x moves from left to right, we accumulate more and more probability. As x increases, cumulative probability also increases. Imagine starting at an altitude of zero and walking up a (smooth) hill. At any point along the walk, measure the altitude. This distance is the cumulative probability. Figure 6.13 shows the relationship between the area under the density curve and the altitude.

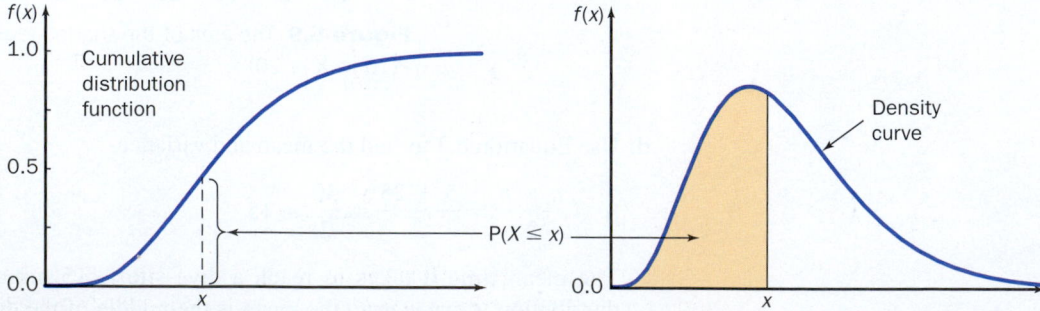

Figure 6.13 Picturing cumulative probability: The altitude is equal to the shaded area.

A CLOSER L🔍OK

Why is 1 the maximum value of the cumulative distribution function?

1. The drawing on the left in Figure 6.13 is a graph of a **cumulative distribution function**. This function starts at 0 and is always increasing, until it reaches a maximum value of 1.

2. The mean, μ, and the variance, σ^2, for a continuous random variable are computed using calculus. Although we will not consider any of these calculations, we will interpret and use these values as usual. μ is a measure of the *center* of the distribution, and σ^2 (or σ) is a measure of the spread, or *variability*, of the distribution.

Figure 6.14 shows density functions for the random variables X and Y.

a. The mean of X is less than the mean of Y, $\mu_X < \mu_Y$, because the *center* of the distribution of X is to the left of the *center* of the distribution of Y.

b. The standard deviation of X is greater than the standard deviation of Y, $\sigma_X > \sigma_Y$, because the distribution of X is more spread out, and thus has more variability, than the distribution of Y.

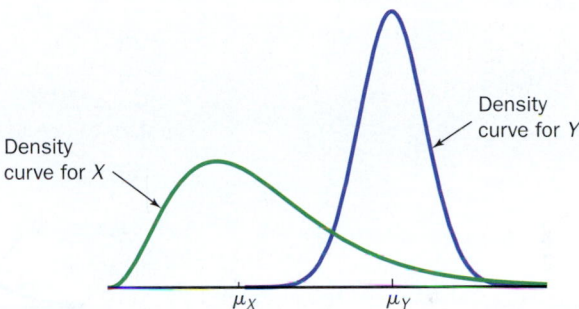

Density curve for X

Density curve for Y

μ_X μ_Y

Figure 6.14 Density functions for X and Y. The mean and the variance (and standard deviation) convey the same information as before about the center and variability.

The following example illustrates the use of cumulative probability to compute probability associated with a continuous random variable.

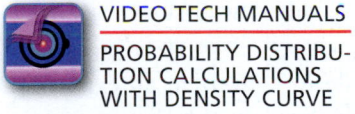

VIDEO TECH MANUALS

PROBABILITY DISTRIBU-
TION CALCULATIONS
WITH DENSITY CURVE

Example 6.2 Keeping Good Time

Each Citizen Eco-Drive wristwatch is carefully tested for accuracy before being packaged and shipped. If the watch gains or loses time during the 24-hour testing period, it is sent to a technician for adjustment. The time inconsistency (in seconds) is a random variable, X, with the probability density function shown in Figure 6.15. A negative value of X indicates that the watch lost time, and a positive value indicates that the watch gained time. Cumulative probability for X is illustrated in Figure 6.16 and can be computed using

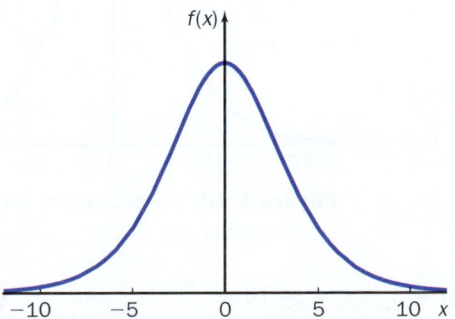

Figure 6.15 The probability density function for the time inconsistency of a wristwatch.

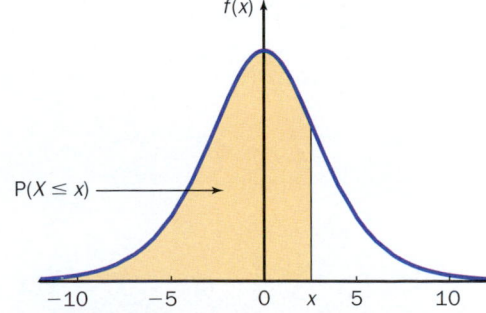

Figure 6.16 The cumulative probability for the time inconsistency of a wristwatch.

Recall: e is the base of the natural logarithm; $e \approx 2.71828$. Most calculators have a specific key for e.

the equation that follows. The cumulative probability (the area of the shaded region in Figure 6.16) is

$$P(X \le x) = \frac{1}{1 + e^{-x/2}} \qquad \text{for all } x \qquad (6.4)$$

Suppose a watch is randomly selected.

a. What is the probability that the watch is 5 seconds slow or slower?

b. What is the probability that the watch is more than 10 seconds fast?

c. What is the probability that the watch is between 3 seconds slow and 3 seconds fast?

SOLUTION

a. If the watch is 5 seconds slow or slower, this means $X \le -5$. The probability $P(X \le -5)$ is cumulative probability already. The calculation and the graphical interpretation (Figure 6.17) follow.

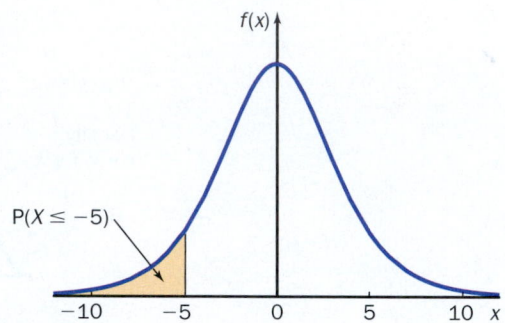

Figure 6.17 The cumulative probability for watch inconsistency, $P(X \le 5)$.

$$P(X \le -5) = \frac{1}{1 + e^{-(-5)/2}} = 0.0759 \qquad \text{Use Equation 6.4.}$$

The probability that a randomly selected watch is 5 seconds slow or slower is 0.0759.

b. If the watch is more than 10 seconds fast, this means $X > 10$. To compute the corresponding probability, use the complement rule to convert to an expression involving cumulative probability, and use Equation 6.4. See Figure 6.18.

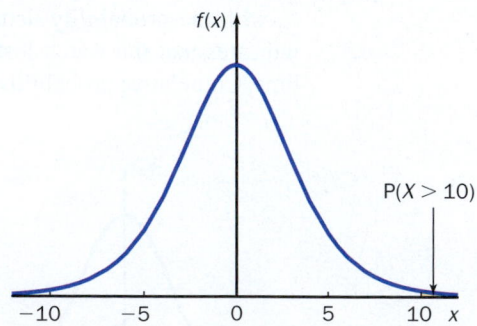

Figure 6.18 The probability for watch inconsistency, $P(X > 10)$.

$$P(X > 10) = 1 - P(X \le 10) \qquad \text{The complement rule.}$$

$$= 1 - \frac{1}{1 + e^{-10/2}} \qquad \text{Use Equation 6.4.}$$

$$= 1 - 0.9933 = 0.0067 \qquad \text{Simplify.}$$

The probability that a randomly selected watch is at least 10 seconds fast is 0.0067.

c. If the watch is between 3 seconds slow and 3 seconds fast, this means $-3 \leq X \leq 3$. To compute the corresponding probability, find the difference between two cumulative probabilities. See Figure 6.19.

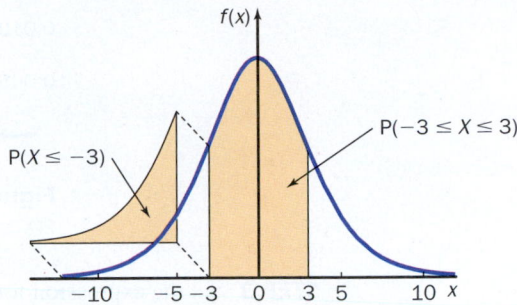

Figure 6.19 The probability for watch inconsistency, $P(-3 \leq X \leq 3)$.

$$P(-3 \leq X \leq 3) = P(X \leq 3) - P(X < -3)$$

$$= P(X \leq 3) - P(X \leq -3) \qquad \text{A single point contributes no probability.}$$

$$= \left(\frac{1}{1 + e^{-3/2}}\right) - \left(\frac{1}{1 + e^{-(-3)/2}}\right) \qquad \text{Use Equation 6.4.}$$

$$= 0.8176 - 0.1824 = 0.6352 \qquad \text{Simplify.}$$

The probability that a randomly selected watch is between 3 seconds slow and 3 seconds fast is 0.6352.

TRY IT NOW   GO TO EXERCISE 6.22

In some problems a probability is given and we need to work *backward* to find a solution. Consider the following example.

Example 6.3 Voting Time

The time it takes to vote may affect the outcome of an election. For example, some people may decide not to vote in districts with long lines and delays. The mean time to vote in the 2012 election was approximately 14 minutes, and Florida voters had the longest wait.[2] Suppose the time to vote for a randomly selected person in Florida, X, has a uniform distribution between 10 and 60 minutes. Find the time t such that 75% of all people have to wait at most t minutes to vote.

This is a *backward* problem, because we know the probability (0.75) and need to find a starting point (*t*), a value of X that produces this probability.

SOLUTION

STEP 1 Because X has a uniform distribution with $a = 10$ and $b = 60$, the probability density function is

$$f(x) = \begin{cases} \dfrac{1}{50} = 0.02 & 10 \leq x \leq 60 \\ 0 & \text{otherwise} \end{cases}$$

We need to find the value of t such that $P(X \leq t) = 0.75$.

STEP 2 From Figure 6.20, the probability that it takes at most t minutes to vote is

$$P(X \leq t) = \text{area under the density curve from 10 to } t$$

$$= \text{area of a rectangle}$$

$$= \text{width} \times \text{height}$$

$$= (t - 10)(0.02)$$

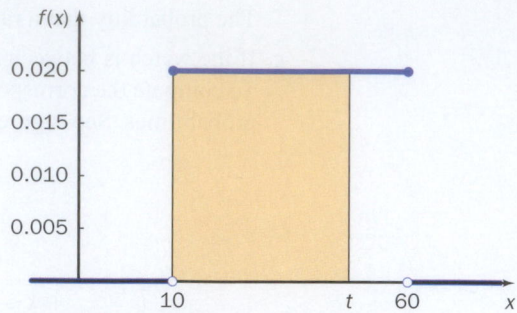

Figure 6.20 The area of the shaded region is $P(X \leq t)$.

STEP 3 Set the expression for probability equal to 0.75 and solve for t.

$$(t - 10)(0.02) = 0.75$$

$$t - 10 = \frac{0.75}{0.02} = 37.5 \qquad \text{Divide both sides by 0.02.}$$

$$t = 37.5 + 10 = 47.5 \qquad \text{Add 10 to both sides.}$$

Seventy-five percent of all voters vote within 47.5 minutes.

TRY IT NOW   GO TO EXERCISE 6.15

SECTION 6.1 **EXERCISES**

Concept Check

6.1 True/False   The graph of a probability density function may extend below the x axis.

6.2 True/False   For a continuous random variable X with probability density function f, $P(X = x) = f(x)$.

6.3 True/False   For a continuous random variable there is no probability associated with a single value.

6.4 True/False   The mean, μ, and the variance, σ^2, of a continuous random variable describe the center and spread of the distribution.

6.5 Short Answer   Explain how to compute probabilities for a continuous random variable.

6.6 Short Answer   Explain why a cumulative distribution function can never have a value greater than 1.

Practice

6.7 Suppose X is a uniform random variable with $a = 0$ and $b = 16$.
  a. Carefully sketch a graph of the probability density function for X.
  b. Find the mean, variance, and standard deviation of X.
  c. Find $P(X \geq 4)$.
  d. Find $P(2 \leq X < 12)$.
  e. Find $P(X \leq 7)$.

6.8 Suppose X is a uniform random variable with $a = -5$ and $b = 25$.
  a. Carefully sketch a graph of the probability density function for X.
  b. Find the mean, variance, and standard deviation of X.
  c. Find $P(-10 < X < -1)$.
  d. Find $P(X > 0)$ and $P(X \geq 0)$.
  e. Find $P(X \geq 20 \,|\, X \geq 10)$.

6.9 Suppose X is a uniform random variable with $a = 50$ and $b = 100$.
  a. Find the mean, variance, and standard deviation of X.
  b. Find $P(\mu - \sigma \leq X \leq \mu + \sigma)$.
  c. Find $P(X \geq \mu + 2\sigma)$.
  d. Find a value c such that $P(X \leq c) = 0.20$.

6.10 Suppose X is a uniform random variable with $a = 25$ and $b = 65$.
  a. Find the mean, variance, and standard deviation of X.
  b. Find the probability that X is more than two standard deviations from the mean.
  c. Find a value c such that $P(X \geq c) = 0.40$.
  d. Suppose two values of X are selected at random. What is the probability that both values are between 30 and 40?

6.11 Suppose X is a continuous random variable with probability density function given by

$$f(x) = \begin{cases} \dfrac{x}{8} & \text{if } 0 \leq x \leq 4 \\ 0 & \text{otherwise} \end{cases}$$

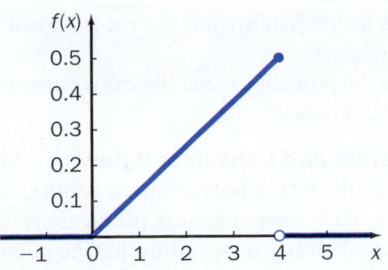

a. Find $P(X \leq 1)$.
b. Find $P(X > 3)$.
c. Find $P(X > 4)$.
d. Find $P(2 \leq X \leq 3)$.
e. Find $P(X \leq 2 \,|\, X \leq 3)$.
f. Find a value c such that $P(X \leq c) = 0.5$. Explain why c is *not* equal to 2.

Applications

6.12 Manufacturing and Product Development A Gold Canyon candle is designed to last nine hours. However, depending on the wind, air bubbles in the wax, the quality of the wax, and the number of times the candle is re-lit, the actual burning time (in hours) is a uniform random variable with $a = 6.5$ and $b = 10.5$. Suppose one of these candles is randomly selected.

a. Find the probability that the candle burns at least seven hours.
b. Find the probability that the candle burns at most eight hours.
c. Find the mean burning time and the probability that the burning time of a randomly selected candle will be within one standard deviation of the mean.
d. Find a time t such that 25% of all candles burn longer than t hours.

6.13 Sport and Leisure According to Major League Baseball rules, a baseball should weigh between 5 and 5.25 ounces and have a circumference of between 9 and 9.25 inches. Suppose the weight of a baseball (in ounces) has a uniform distribution with $a = 5.085$ and $b = 5.155$, and the circumference (in inches) has a uniform distribution with $a = 9.0$ and $b = 9.1$.

a. Find the probability that a randomly selected baseball has a weight greater than 5.14 ounces. Write a Solution Trail for this problem.
b. Find the probability that a randomly selected baseball has a circumference less than 9.03 inches.
c. Suppose the weight and the circumference are independent. Find the probability that a randomly selected baseball will have a weight between 5.11 and 5.13 ounces and a circumference between 9.04 and 9.06 inches.

6.14 Manufacturing and Product Development Pre-manufactured wooden roof trusses allow builders to complete projects faster and with lower on-site labor costs. The connector plates for trusses are made from Grade A steel and are hot-dip galvanized. The thickness of a truss connector (in inches) varies slightly and has a uniform distribution with $a = 0.036$ and $b = 0.050$.

a. If the manufacturer will only use connectors with a minimum thickness of 0.04 inch, what proportion of connectors is rejected?
b. Suppose a truss connector is selected at random. Find the probability that the truss connector has a thickness between 0.042 and 0.045 inch.
c. Find the mean, variance, and standard deviation of the thickness of a truss connector.

6.15 Travel and Transportation When the Department of Transportation (DOT) repaints the center lines, edge lines, or no-passing-zone lines on a highway, epoxy paint is sometimes applied. This paint is more expensive than latex but lasts longer. If this paint splashes onto a vehicle, it has to be completely sanded off, and that area of the vehicle has to be repainted. The DOT has warned motorists that the drying time for this epoxy paint (in minutes) has a uniform distribution with $a = 30$ and $b = 60$. Suppose epoxy paint is applied to a small section of center line.

a. What is the probability that the paint will be dry within 45 minutes?
b. What is the probability that the paint will be dry in between 40 and 50 minutes?
c. Find a value t such that the probability of the paint taking at least t minutes to dry is 0.75.
d. If the DOT road crew removes all of the cones on the center line 55 minutes after painting, what is the probability that the paint will still be wet?

6.16 Physical Sciences In Grafton, a rural area in Vermont, the distance (in meters) between telephone poles has a uniform distribution with $a = 40$ and $b = 65$. Suppose two consecutive telephone poles are selected at random.

a. What is the probability that the distance between the poles is less than 60 meters?
b. What is the probability that the distance between the poles is between 45 and 55 meters? Write a Solution Trail for this problem.
c. Any distance between poles greater than 50 meters is considered to be *environment friendly*. What is the probability that the distance is environment friendly?

Extended Applications

6.17 Psychology and Human Behavior Some of the common medications to help people with insomnia fall asleep include Ambien, Lunesta, and Sonata. Ambien CR is an extended-release variation and is formulated so that an individual will fall asleep within 30 minutes.[3] The probability density function for X, the time (in minutes) it takes to fall asleep after taking an Ambien CR tablet, is given below.

$$f(x) = \begin{cases} 0.05 & \text{if } 0 \leq x \leq 10 \\ -0.0025(x - 30) & \text{if } 10 < x \leq 30 \\ 0 & \text{otherwise} \end{cases}$$

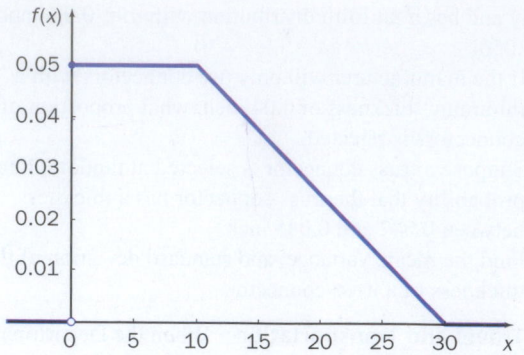

a. Verify that this is a valid probability density function.
b. If a randomly selected person takes an Ambien CR tablet at bedtime, what is the probability that he will fall asleep within 5 minutes?
c. What is the probability that the person will fall asleep between 20 and 30 minutes after taking a tablet?
d. Find a value t such that the probability of falling asleep within t minutes after taking a tablet is 0.75.
e. If it takes less than 15 minutes to fall asleep after taking a tablet, people consider the medication a success. Suppose 20 people are selected at random. What is the probability that exactly 14 people fall asleep successfully? What is the probability that at least 16 people fall asleep successfully? What is the probability that at most 10 people fall asleep successfully?

6.18 Marketing and Consumer Behavior The city of Kingston, Ontario, has approximately 3700 parking spaces. Metered parking is available on some streets and in certain parking garages, and the maximum length of stay at a metered spot is between one and three hours.[4] The probability density function for the length of time a car is parked (in hours) at a metered spot in a certain lot is given below. Suppose a car parked at a metered spot in this lot is selected at random.

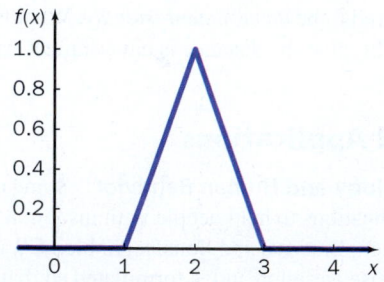

a. What is the probability that the car is parked for less than 2 hours?
b. What is the probability that the car is parked for less than 1.4 hours?

c. What is the probability that the car is parked for more than 2.6 hours?
d. What is the probability that the car is parked for between 1.4 and 2.6 hours?

6.19 Marketing and Consumer Behavior Marini's candy store on the beach boardwalk in Santa Cruz sells candy in bulk. Customers can mix products from over 100 barrels. The probability distribution for the number of pounds of candy purchased by a randomly selected customer is shown below.

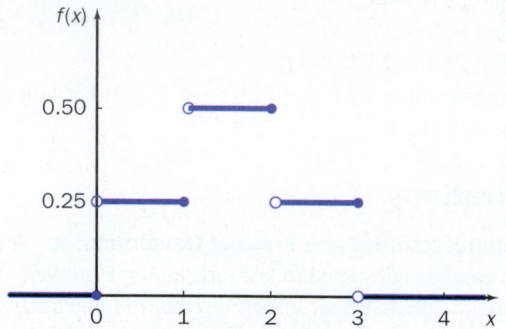

a. Verify that this is a valid probability density function.
b. Find the probability that the next customer buys at most 2 pounds of candy.
c. Find the probability that the next customer buys more than 1 pound of candy.
d. Suppose the next customer buys at most 1.5 pounds of candy. What is the probability that she buys at most 0.5 pound of candy?

6.20 Economics and Finance On any given trading day, the fluctuation, or change, in the price (in dollars) of JP Morgan Chase stock, listed on the New York Stock Exchange, is between −2.00 and 2.00. Suppose the change in price is a random variable with the probability density function shown below.

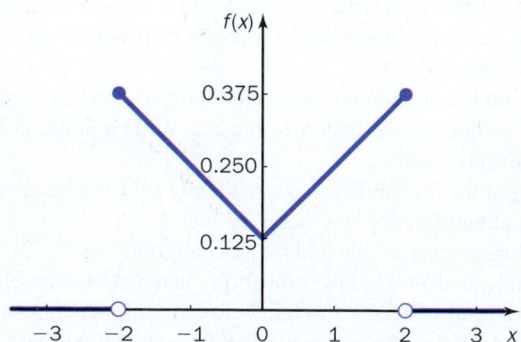

a. Verify that this is a valid probability density function.
b. What is the probability that the stock price increases by at least $1.00 on a randomly selected day?

c. What is the probability that the change in stock price is between -1.00 and 1.00?

d. Find a value c such that $P(-c \leq X \leq c) = 0.90$.

6.21 Medicine and Clinical Studies Although we all experience inflammation associated with a bruise or sprain, inflammation can also affect the cells in the body. C-reactive protein (CRP) is a measure of inflammation and can be part of a routine blood test. For healthy adults, the CRP level is less than 5 milligrams per liter of blood. Suppose the probability distribution for X, the CRP level in healthy adults, is given as follows:

$$f(x) = \begin{cases} -0.08(x - 5) & \text{if } 0 \leq x \leq 5 \\ 0 & \text{otherwise} \end{cases}$$

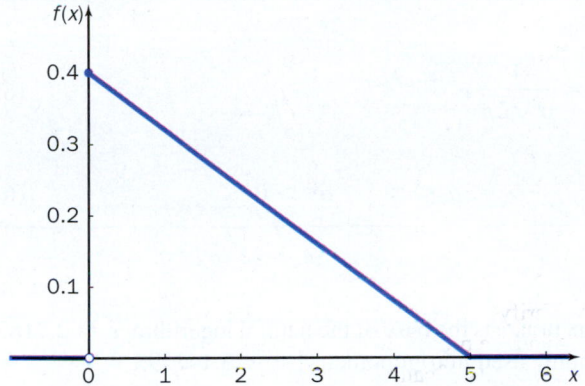

a. Verify that this is a valid probability distribution.

b. What is the probability that a randomly selected healthy adult will have a CRP level less than 2.5?

c. What is the probability that a randomly selected healthy adult will have a CRP level between 2 and 3?

d. Find a value c such that 5% of healthy adults have a CRP level of at least c.

e. If a patient has a CRP level of at least 4, then additional testing is done. What is the probability that a healthy adult will need additional testing?

f. What is the probability that the fifth healthy adult will be the first to need additional testing?

Challenge

6.22 Marketing and Consumer Behavior Dinner customers at the Primanti Brothers restaurant in Pittsburgh, Pennsylvania, often experience a long wait for a table. For a randomly selected customer who arrives at the restaurant between 6:00 P.M. and 7:00 P.M., the waiting time (in minutes) is a continuous random variable such that

$$P(X \leq x) = \begin{cases} 1 - e^{-0.05x} & \text{if } x \geq 0 \\ 0 & \text{otherwise} \end{cases}$$

Suppose a dinner customer is randomly selected.

a. What is the probability that the person must wait for a table for at most 20 minutes?

b. What is the probability that the person must wait for a table for over one half-hour?

c. What is the probability that the person must wait for a table for between 15 and 30 minutes?

6.23 Psychology and Human Behavior Parents with children under age 16 often spend a lot of time during the day driving their kids to various places, for example, to/from after-school activities, music practice, sports practices and games, the library, and a friend's home. Suppose a family has k child(ren) under 16 ($k = 1, 2, 3, 4, 5$), and let the random variable X_k be the time (in hours) spent *taxiing* during the day. X_k has a uniform distribution with $a = 0$ and $b = k$. For example, for a family with two children, X_2 has a uniform distribution with $a = 0$ and $b = 2$.

a. For a family with three children, what is the probability that parents will spend less than one hour driving kids on a randomly selected day?

b. For a family of four children, what is the mean number of hours spent driving kids? What is the probability that the driving time will be greater than two standard deviations from the mean?

c. For a family with five children under 16, find a time t such that the probability of driving kids more than t hours is 0.25.

d. Suppose five families are selected at random, the first with one child under 16, the second with two children under 16, etc. What is the probability that all five families drive less than 30 minutes on a randomly selected day? What is the probability that all five families drive more than 90 minutes on a randomly selected day?

6.24 Suppose X is a continuous random variable such that

$$P(X \leq x) = \begin{cases} 1 - e^{-x^2/8} & \text{if } x \geq 0 \\ 0 & \text{otherwise} \end{cases}$$

a. Find $P(X \leq 4)$.

b. Find $P(X > 2)$.

c. Find $P(1 \leq X \leq 3)$.

d. Find $P(X \leq 2 \mid X \leq 4)$.

6.25 Sports and Leisure A figure skating routine is designed to last six minutes. The amount of time (in minutes) less than or greater than six minutes is a random variable, X, with a probability density function given by

$$f(x) = \begin{cases} \sqrt{\dfrac{2}{\pi} - x^2} & \text{if } -\sqrt{\dfrac{2}{\pi}} \leq x \leq \sqrt{\dfrac{2}{\pi}} \\ 0 & \text{otherwise} \end{cases}$$

If the value of X is negative, then the routine was shorter than six minutes; if the value of X is positive, the routine went too long.

a. Carefully sketch a graph of the density function.

b. Find the probability that a randomly selected performance is within $1/\sqrt{\pi}$ minutes of 6. That is, find

$$P\left(-\frac{1}{\sqrt{\pi}} \leq X \leq \frac{1}{\sqrt{\pi}}\right)$$

6.2 The Normal Distribution

The **normal probability distribution** is very common and is the most important distribution in all of statistics. This *bell-shaped* density curve can be used to model many natural phenomena, and the normal distribution is used extensively in statistical inference. Recall that a random variable is completely described by certain *parameters*—for example, a binomial random variable by n and p, and a Poisson random variable by λ. A normal distribution is completely characterized, or determined, by its mean μ and variance σ^2 (or by its mean μ and standard deviation σ).

The Normal Probability Distribution

Suppose X is a normal random variable with mean μ and variance σ^2. The probability density function is given by

$$f(x) = \frac{1}{\sigma\sqrt{2\pi}}\, e^{-(x-\mu)^2/2\sigma^2} \tag{6.5}$$

and

$$-\infty < x < \infty \qquad -\infty < \mu < \infty \qquad \sigma^2 > 0 \tag{6.6}$$

A CLOSER LOOK

We've seen e before, in the Poisson distribution.

1. In this probability density function, e is the base of the natural logarithm; $e \approx 2.71828$. π is another constant, commonly used in trigonometry; $\pi \approx 3.14159$.

2. We use the shorthand notation $X \sim \mathrm{N}(\mu, \sigma^2)$ to indicate that X is (distributed as) a normal random variable with mean μ and variance σ^2. For example, $X \sim \mathrm{N}(5, 36)$ means that X is a normal random variable with mean $\mu = 5$ and variance $\sigma^2 = 36$ (and $\sigma = 6$).

3. Equation 6.6 means that x can be any real number (the density curve continues forever in both directions), the mean μ can be any real number (positive or negative), and the variance can be any positive real number.

Bell-shaped means: Place a bell on a table and pass a plane (a piece of paper) *through* the bell perpendicular to the table. The intersection of the plane and the bell is a *bell-shaped curve*.

4. For *any* mean μ and variance σ^2, the density curve is symmetric about the mean μ, unimodal, and bell-shaped as shown in Figure 6.21.

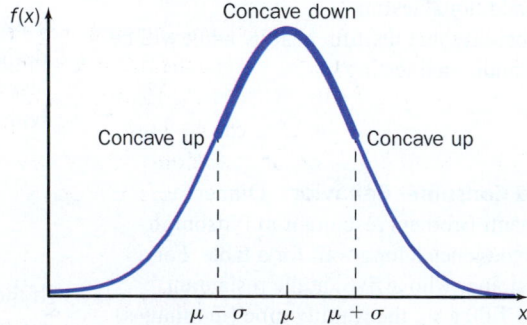

Figure 6.21 Graph of the probability density function for a normal random variable with mean μ and variance σ^2.

▶ The graph of the probability density function changes *concavity* at $x = \mu - \sigma$ and again at $x = \mu + \sigma$. ◀

The mean is equal to the median because the normal distribution is symmetric.

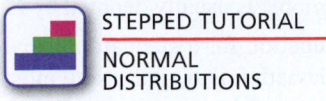

▶ It can be shown (using calculus) that the total area under this density curve is 1 (even though it extends forever in both directions, getting closer and closer to the x axis but never touching it). ◀

5. The mean μ is a *location* parameter, and the variance σ^2 determines the spread of the distribution. As the variance increases, the total area under the probability density function (1) is rearranged. The graph is compressed down and pushed out (on the tails). Figures 6.22 and 6.23 show the effects of μ and σ^2 on the location (center) and spread of the density curve.

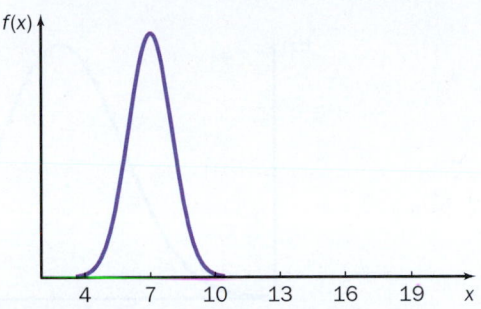

Figure 6.22 Normal probability density function with $\mu = 7$ and small σ^2.

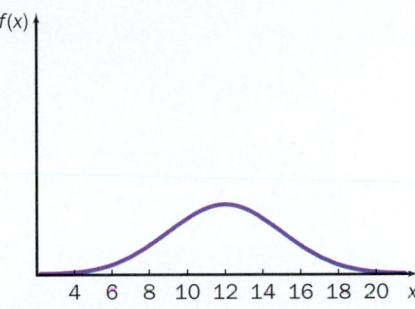

Figure 6.23 Normal probability density function with $\mu = 12$ and large σ^2. ●

Suppose X is a normal random variable with mean μ and variance σ^2: $X \sim \mathrm{N}(\mu, \sigma^2)$. The probability X lies in some interval, for example $[a, b]$, is the area under the density curve between a and b (Figure 6.24).

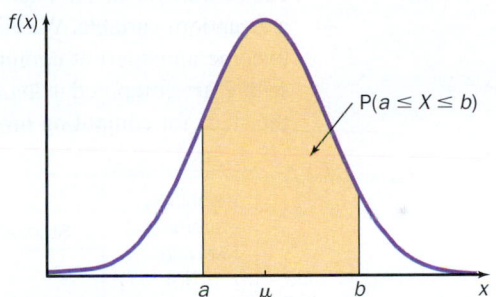

Figure 6.24 The shaded region corresponds to $P(a \leq X \leq b)$.

The shaded region in Figure 6.24 is not a simple geometric figure; it's bounded by a curve! Consequently, there is no *nice* formula for the area of this region, corresponding to $P(a \leq X \leq b)$. However, a probability statement associated with *any* normal random variable can be *transformed* into an equivalent expression involving a **standard normal random variable** (defined below). Cumulative probabilities associated with this distribution are provided in Appendix A, Table III.

The Standard Normal Random Variable

The normal distribution with $\mu = 0$ and $\sigma^2 = 1$ (and $\sigma = 1$) is called the **standard normal distribution**. A random variable that has a standard normal distribution is called a **standard normal random variable**, usually denoted Z. The probability density function for Z is given by

Let $\mu = 0$ and $\sigma = 1$ in Equation 6.5.

$$f(z) = \frac{1}{\sqrt{2\pi}} e^{-z^2/2} \qquad -\infty < z < \infty \qquad (6.7)$$

A CLOSER LOOK

1. In Equation 6.7 the independent variable z is used to define the probability density function simply because the standard normal random variable is usually denoted by Z.

2. Figure 6.25 shows a graph of the probability density function for a standard normal random variable. The mean is $\mu = 0$ and the standard deviation is $\sigma = 1$. Note most of the probability (area) is within three standard deviations of the mean, between -3 and 3. The shorthand notation $Z \sim N(0, 1)$ means Z is a normal random variable with mean 0 and variance 1.

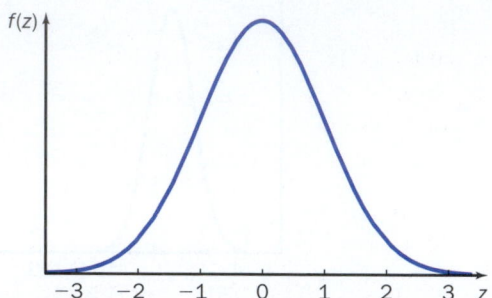

Figure 6.25 Graph of the probability density function for a standard normal random variable.

We will often refer to a standard normal distribution as a *Z world*.

3. The standard normal distribution is not common, but it is used extensively as a *reference* distribution. Any probability statement involving any normal random variable can be transformed into an equivalent expression (with the same probability) involving a Z random variable. We will learn how to *standardize* shortly. Therefore, you need to become an expert at computing probabilities in the Z world. Probabilities associated with Z are computed using cumulative probability, as shown below. Figure 6.26 shows the steps for computing probabilities associated with a normal random variable.

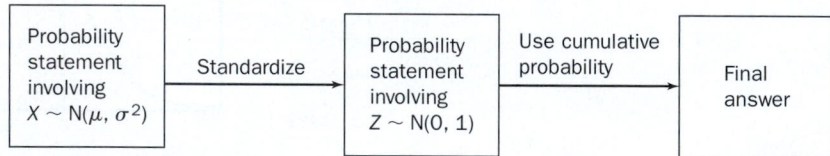

Figure 6.26 Strategy for computing a probability associated with any normal random variable.

Probabilities associated with a standard normal random variable, Z, are computed using cumulative probability. Table III in Appendix A contains values for $P(Z \leq z)$ for selected values of z. Figure 6.27 shows the geometric region corresponding to $P(Z \leq z)$, and Figure 6.28 illustrates the use of Table III in Appendix A. Locate the units and tenths digits in z along the left side of the table. Find the hundredths digit in z across the top row. The intersection of this row and column, in the body of the table, contains the cumulative probability.

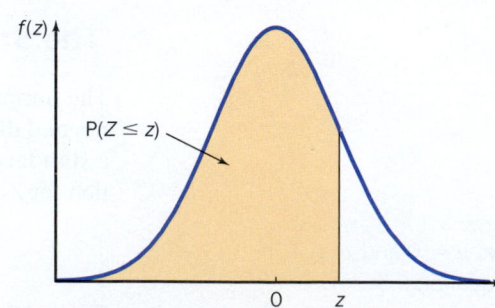

Figure 6.27 The shaded area under the standard normal density curve corresponds to $P(Z \leq z)$.

z	0.00	0.01	0.02	0.03	0.04	0.05	0.06	0.07	0.08	0.09
⋮	⋮	⋮	⋮	⋮	⋮	⋮	⋮	⋮	⋮	⋮
1.0	0.8413	0.8438	0.8461	0.8485	0.8508	0.8531	0.8554	0.8577	0.8599	0.8621
1.1	0.8643	0.8665	0.8686	0.8708	0.8729	0.8749	0.8770	0.8790	0.8810	0.8830
1.2	0.8849	0.8869	0.8888	0.8907	0.8925	0.8944	0.8962	0.8980	0.8997	0.9015
1.3	0.9032	0.9049	0.9066	0.9082	0.9099	0.9115	0.9131	0.9147	0.9162	0.9177
1.4	0.9192	0.9207	0.9222	0.9236	0.9251	0.9265	0.9279	0.9292	0.9306	0.9319
⋮	⋮	⋮	⋮	⋮	⋮	⋮	⋮	⋮	⋮	⋮

Figure 6.28 $P(Z \le 1.23) = 0.8907$ in Table III, Appendix A.

The following example illustrates the use of Table III in Appendix A to find probabilities associated with Z.

Example 6.4 Probability Calculations Associated with the Standard Normal Distribution

Use Table III in Appendix A to find each probability associated with the standard normal distribution.

a. $P(Z \le 1.45)$

b. $P(Z \ge -0.6)$

c. $P(-1.25 \le Z \le 2.13)$

d. Find the value b such that $P(Z \le b) = 0.90$.

SOLUTION

a. This expression is already *cumulative probability*. Go directly to Table III in the Appendix, and find the intersection of row 1.4 and column 0.05. See Figure 6.29.

$P(Z \le 1.45) = 0.9265$ Cumulative probability; use Table III in Appendix A.

Figure 6.30 shows a technology solution.

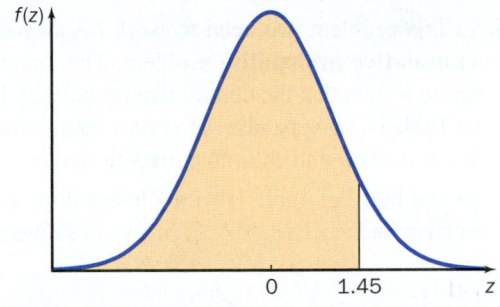

Figure 6.29 The area of the shaded region is $P(Z \le 1.45)$.

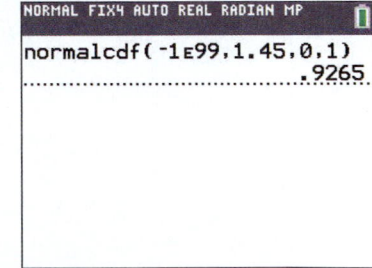

Figure 6.30 $P(Z \le 1.45)$.

b. This is a right-tail probability. Convert to cumulative probability and use Table III in the Appendix. See Figure 6.31.

$$P(Z \ge -0.6) = 1 - P(Z < -0.6) \qquad \text{The Complement Rule.}$$
$$= 1 - P(Z \le -0.6) \qquad \text{One value doesn't matter.}$$
$$= 1 - 0.2743 = 0.7257 \qquad \text{Use Table III in the Appendix.}$$

Figure 6.32 shows a technology solution.

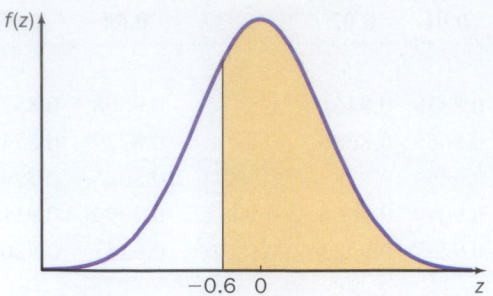

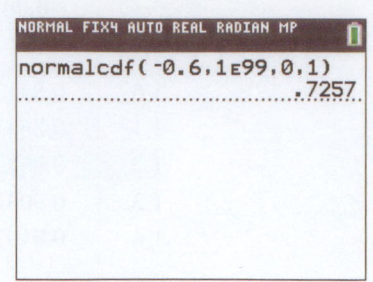

Figure 6.31 The area of the shaded region is $P(Z \geq -0.6)$.

Figure 6.32 $P(Z \geq -0.6)$.

c. Find all the probability up to 2.13, find all the probability up to -1.25, and subtract. The difference is the probability that Z lies in this interval. See Figure 6.33.

$$P(-1.25 \leq Z \leq 2.13)$$

$$= P(Z \leq 2.13) - P(Z < -1.25) \qquad \text{Use cumulative probability.}$$

$$= P(Z \leq 2.13) - P(Z \leq -1.25) \qquad \text{One value doesn't matter.}$$

$$= 0.9834 - 0.1056 = 0.8778 \qquad \text{Use Table III in the Appendix.}$$

Figure 6.34 shows a technology solution.

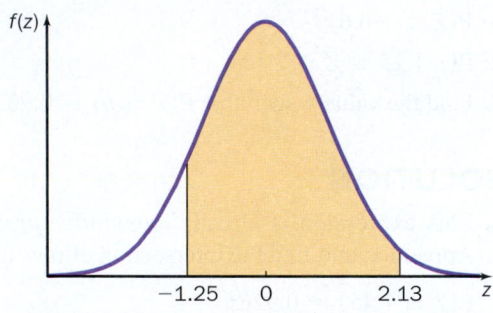

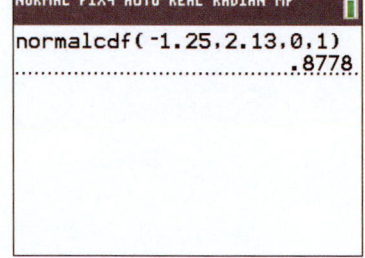

Figure 6.33 The area of the shaded region is $P(-1.25 \leq x \leq 2.13)$.

Figure 6.34 $P(-1.25 \leq Z \leq 2.13)$.

d. In this problem, we need to work *backward* to find the solution. This is an **inverse cumulative probability** problem. The cumulative probability is given. We need the value b such that the cumulative probability is 0.90. See Figure 6.35. Search the body of Table III in Appendix A to find a cumulative probability as close to 0.90 as possible. Read the row and column entries to find b.

In the body of Table III, the closest cumulative probability to 0.90 is 0.8997. This corresponds to $1.28 \approx b$. Figure 6.36 shows a technology solution.

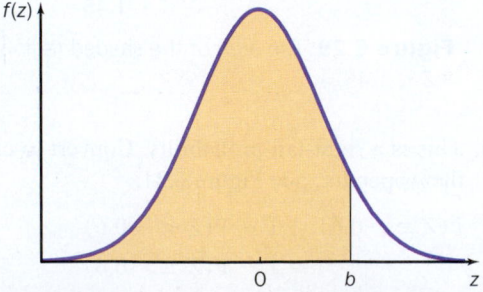
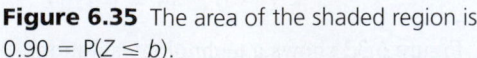

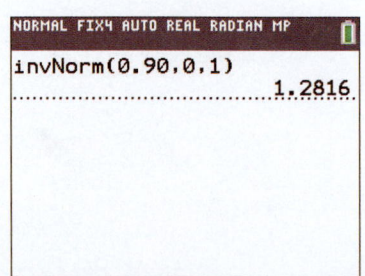

Figure 6.35 The area of the shaded region is $0.90 = P(Z \leq b)$.

Figure 6.36 Inverse cumulative probability.

Note: Linear interpolation can be used to find a more exact answer. The technology solution presented in Figure 6.36 uses a special inverse cumulative probability functions.

TRY IT NOW GO TO EXERCISES 6.32 AND 6.34

Interpolation

Recall: interpolation is a method of approximation. It is often used to estimate a value at a position between two given values in a table. Linear interpolation assumes that the two known values lie on a straight line.

In Example 6.4(d), 0.90 is between the Table III known cumulative probabilities 0.8997 and 0.9015. Suppose the two points (1.28, 0.8997) and (1.29, 0.9015) lie on a straight line. The approximate z value corresponding to the cumulative probability 0.90 is

$$1.28 + (0.01)(0.90 - 0.8997)/(0.9015 - 0.8997) = 1.2817$$

The following rule provides the connection between *any* normal random variable and the standard normal random variable.

Standardization Rule

If X is a normal random variable with mean μ and variance σ^2, then a standard normal random variable is given by

$$Z = \frac{X - \mu}{\sigma} \qquad (6.8)$$

A CLOSER LOOK

There are other types of standardization. $Z = (X - \mu)/\sigma$ is the most common.

1. The process of converting from X to Z is called **standardization**. Z is a *standardized* random variable.

2. Using this rule, any probability involving a normal random variable can be transformed into an equivalent expression involving a Z random variable. We can then convert to cumulative probability if necessary, and use Table III in the Appendix.

3. The rule above is illustrated in Figure 6.37, using cumulative probability.

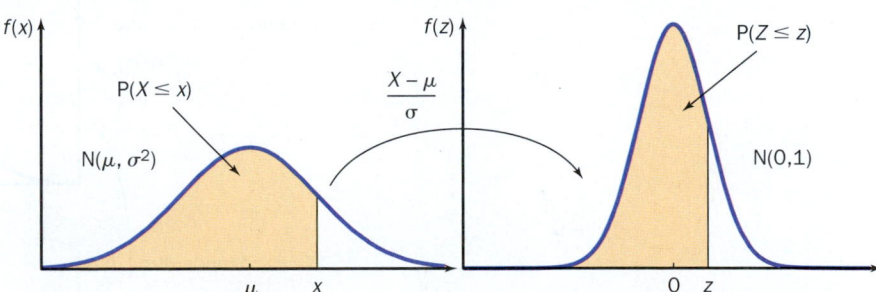

Figure 6.37 An illustration of standardization. The areas of the shaded regions are equal.

The following calculation shows why the two shaded regions in Figure 6.37 have the same area, and how to use the rule to compute probabilities involving any normal random variable.

Assume: $X \sim N(\mu, \sigma^2)$.

Remember the phrase: Whatever you do to one side of the inequality, you have to do to the other side.

$P(X \leq x)$ The original (cumulative) probability statement.

$$= P\left(\frac{X - \mu}{\sigma} \leq \frac{x - \mu}{\sigma}\right)$$ Work *within* the probability statement. Subtract the mean of X and divide by the standard deviation of X, on both sides of the inequality (standardize).

$$= P(Z \leq z)$$ Apply the standardization rule *within* the probability statement. The expression with X is transformed into Z. The expression with x becomes some fixed value z. Use Table III in the Appendix to find this probability.

The examples below involve normal random variables and standardization. The hardest part of these types of problems is (as before) (1) to define and identify the probability distribution, and (2) to write a probability statement. Given a probability statement involving a normal random variable, all we have to do is standardize and use cumulative probability. Even for backward problems (with a known probability), we still standardize and still use cumulative probability. Note that the technology solutions presented do *not* require standardization.

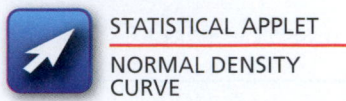

STATISTICAL APPLET

NORMAL DENSITY CURVE

Example 6.5 Probability Calculations Associated with a Normal Random Variable

Suppose X is a normal random variable with mean 10 and variance 4: $X \sim N(10, 4)$, and $\sigma = \sqrt{4} = 2$.

a. Find $P(X > 12.5)$.

b. Find $P(9 \leq X \leq 10)$.

c. Find the value b such that $P(X \leq b) = 0.75$.

SOLUTION

a. X is normal. We know the mean and standard deviation. Standardize and use cumulative probability associated with Z.

$$P(X > 12.5) = P\left(\frac{X - 10}{2} > \frac{12.5 - 10}{2}\right)$$ Standardize.

$$= P(Z > 1.25)$$ Equation 6.8; simplify.

$$= 1 - P(Z \leq 1.25)$$ The complement rule.

$$= 1 - 0.8944 = 0.1056$$ Use Table III in the Appendix.

Figure 6.38 illustrates this solution.

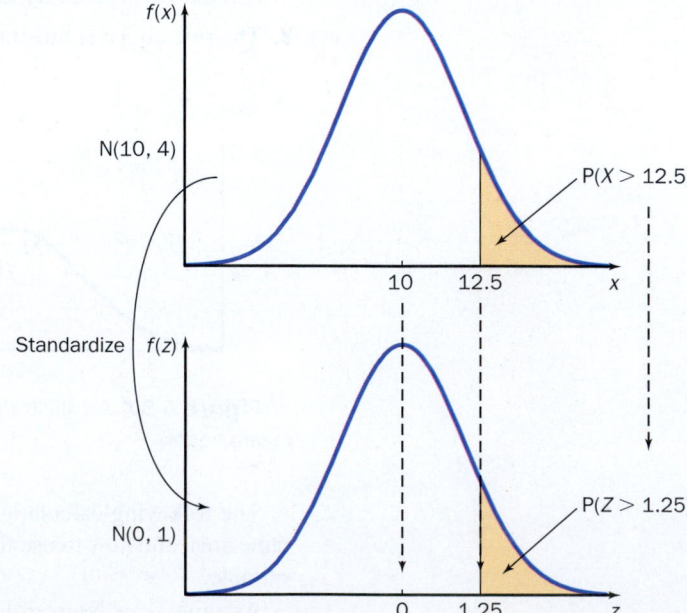

Figure 6.38 Example 6.5 part (a) standardization illustrated: 10 is transformed to 0. 12.5 is transformed to 1.25. The areas of the shaded regions are the same.

Standardization illustrated part (b):

$P(9 \leq X \leq 10)$

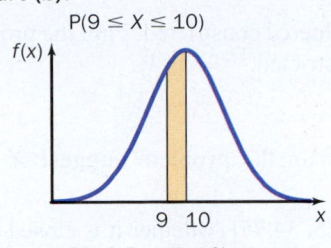

$P(-0.5 \leq Z \leq 0)$

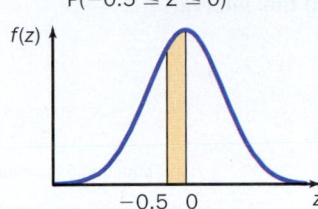

Standardization illustrated part (c):

$P(X \leq b) = 0.75$

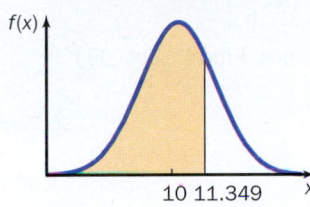

$P\left(Z \leq \dfrac{b-10}{2}\right)$

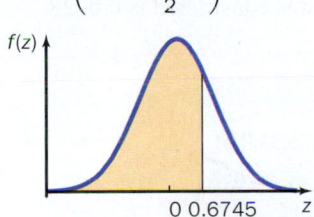

b. Standardize again. Work within the probability statement to write an equivalent expression involving Z.

$$P(9 \leq X \leq 10) = P\left(\frac{9-10}{2} \leq \frac{X-10}{2} \leq \frac{10-10}{2}\right) \qquad \text{Standardize.}$$

$$= P(-0.5 \leq Z \leq 0) \qquad \text{Use Equation 6.8; simplify.}$$

$$= P(Z \leq 0) - P(Z < -0.5) \qquad \text{Use cumulative probability.}$$

$$= P(Z \leq 0) - P(Z \leq -0.5) \qquad \text{One value doesn't matter.}$$

$$= 0.5000 - 0.3085 = 0.1915 \qquad \text{Use Table III in the Appendix.}$$

c. Convert the expression into cumulative probability involving Z. Because the probability is already given, this is an inverse cumulative probability problem. Work backward in Appendix Table III.

$$P(X \leq b) = P\left(\frac{X-10}{2} \leq \frac{b-10}{2}\right) \qquad \text{Standardize.}$$

$$= P\left(Z \leq \frac{b-10}{2}\right) = 0.75 \qquad \text{Equation 6.8.}$$

There is no other simplification within the probability statement. However, the resulting probability statement involves Z, and is cumulative probability. Find a value in the body of Table III, Appendix A, as close to 0.75 as possible. Set the corresponding z equal to $\left(\dfrac{b-10}{2}\right)$, and solve for b.

$$\frac{b-10}{2} = 0.6745 \qquad \text{Table III; interpolation.}$$

$$b - 10 = 1.349 \qquad \text{Multiply both sides by 2.}$$

$$b = 11.349 \qquad \text{Add 10 to both sides.}$$

Therefore, $P(X \leq 11.349) = 0.75$ and hence $b = 11.349$.

Figures 6.39–6.41 show technology solutions:

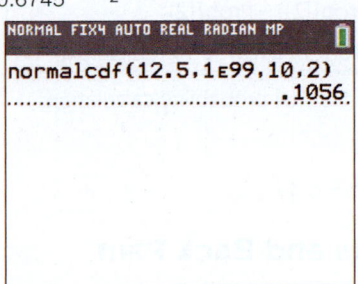

Figure 6.39 $P(X \geq 12.5)$.

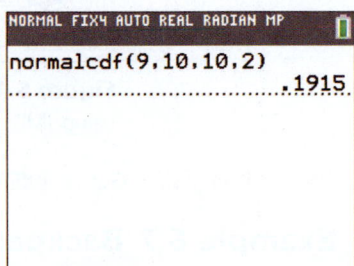

Figure 6.40 $P(9 \leq X \leq 10)$.

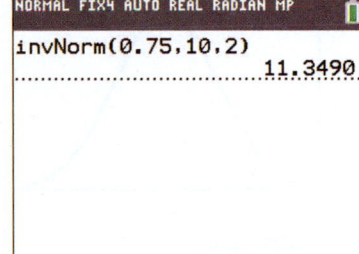

Figure 6.41 Inverse cumulative probability.

TRY IT NOW   GO TO EXERCISES 6.39 AND 6.40

Example 6.6 Seat Pitch

Seat pitch on a passenger airline is the distance from the back of one seat to the front of the one directly behind it. The greater the seat pitch, the more comfortable the seat and the less likely you are to travel with your knees against your chest. Some seats—for example, bulkhead seats—have a larger seat pitch.[5] However, the seat pitch for all economy seats is normally distributed with mean 34 inches and standard deviation 0.5 inch.

Solution Trail 6.6

KEYWORDS
- Normally distributed
- Mean
- Standard deviation

TRANSLATION
- Normal random variable
- $\mu = 34$
- $\sigma = 0.5$

CONCEPTS
- Normal probability distribution
- Standardization

VISION
Define a normal random variable and translate each question into a probability statement. Standardize and use cumulative probability associated with Z if necessary.

Recall that [33.25, 34.75] means $33.25 \leq X \leq 34.75$.

Standardization illustrated part (b):

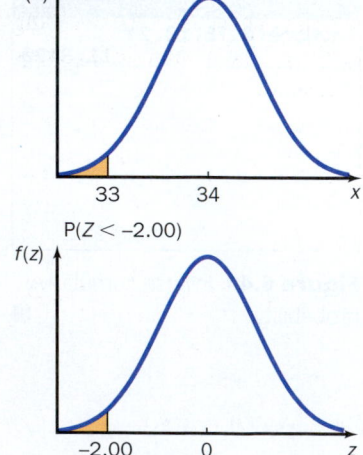

a. For a randomly selected economy seat, find the probability that the seat pitch is between 33.25 and 34.75 inches (considered comfortable).

b. Any seat with a seat pitch less than 33 inches is considered constricted. Find the probability that a randomly selected economy seat is constricted.

SOLUTION

a. Let X be the seat pitch in inches. The keywords in the problem suggest $X \sim$ N(34, 0.25), $\sigma = 0.5$.

Between 33.25 and 34.75 means in the interval [33.25, 34.75] (whether it is closed or open doesn't matter). Find the probability that X lies in this interval.

$$P(33.25 \leq X \leq 34.75)$$

$$= P\left(\frac{33.25 - 34}{0.5} \leq \frac{X - 34}{0.5} \leq \frac{34.75 - 34}{0.5}\right) \qquad \text{Standardize.}$$

$$= P(-1.50 \leq Z \leq 1.50) \qquad \text{Equation 6.8; simplify.}$$

$$= P(Z \leq 1.50) - P(Z \leq -1.50) \qquad \text{Use cumulative probability.}$$

$$= 0.9332 - 0.0668 = 0.8664 \qquad \text{Use Table III in the Appendix.}$$

The probability that a randomly selected economy seat has seat pitch between 33.25 and 34.75 inches is 0.8664.

b. A seat is constricted if the value of X is less than 33 inches. Find $P(X < 33)$.

$$P(X < 33) = P\left(\frac{X - 34}{0.5} < \frac{33 - 34}{0.5}\right) \qquad \text{Standardize.}$$

$$= P(Z < -2.00) \qquad \text{Equation 6.8; simplify.}$$

$$= 0.0228 \qquad \text{Cumulative probability; use Table III in the Appendix.}$$

The probability that a randomly selected economy seat is constricted is 0.0228. Figure 6.42 shows a technology solution.

Prob	JMP Formula
0.9332	= Normal Distribution(34.75,34,0.5)
0.0668	= Normal Distribution(33.25, 34,0.5)
0.8664	= Prob[[1]] - Prob[[2]]
0.0228	= Normal Distribution(33,34,0.5)

Figure 6.42 Normal probability calculations using JMP.

TRY IT NOW   GO TO EXERCISE 6.43

Example 6.7 Backpacks and Back Pain

Chronic back pain has become common in children because so many carry overfilled and overweight backpacks. Heavy school books, notebooks, calculators, and computer equipment, all crammed into a backpack and lugged around all day, increase the chance of neck and shoulder muscle spasms and lower-back pain. Research has shown that the total weight carried is directly related to the volume of a backpack. The volume of a randomly selected backpack sold commercially is normally distributed with mean 600 cubic inches and standard deviation 100 cubic inches. Find a symmetric interval about the mean volume, $[\mu - b, \mu + b]$, such that 95% of all backpack volumes lie in this interval.

SOLUTION

STEP 1   Let X be the volume (in cubic inches, in^3) of a randomly selected backpack. The information given indicates that X is a normal random variable with mean $\mu = 600$ and standard deviation $\sigma = 100$: $X \sim$ N(600, 10,000).

Solution Trail 6.7

KEYWORDS

- Normally distributed
- Mean
- Standard deviation
- 95%

TRANSLATION

- Normal random variable
- $\mu = 600$
- $\sigma = 100$
- Probability 0.95

CONCEPTS

- Normal probability distribution
- Standardization

VISION

Define a random variable and translate the question into a probability statement. A probability is given (0.95), suggesting an inverse cumulative probability question.

Standardization illustrated:

$P(X \le 600 - b) = 0.025$

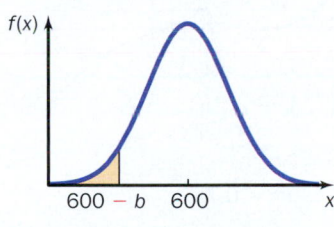

$P\left(Z \le \dfrac{-b}{100}\right) = 0.025$

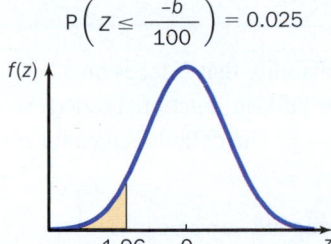

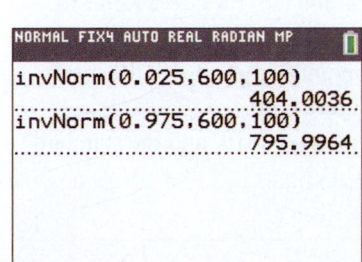

Figure 6.44 A technology solution: Use inverse cumulative probability to find each endpoint.

Find a symmetric interval about the mean such that 95% of all backpack volumes lie in this interval translates as: Find a value of b such that $P(600 - b \le X \le 600 + b) = 0.95$. Figure 6.43 illustrates this probability statement.

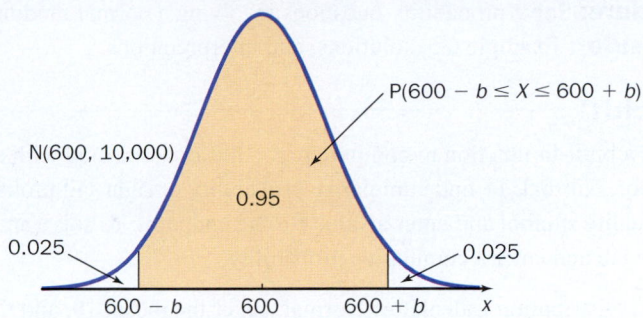

Figure 6.43 A graphical representation of the probability statement.

STEP 2 This problem reduces to finding the value for b. This question involves a normal random variable, so we will certainly have to standardize. And, because the probability is given, this is a backward problem. To use Table III in the Appendix, we need a cumulative probability statement. We need another interpretation of Figure 6.43 involving cumulative probability *and* b. Here are two possibilities.

a. $P(X \le 600 - b) = 0.025$.

The area (or probability) in the tails of the distribution is $1 - 0.95 = 0.05$ (the complement rule). The distribution is symmetric, so the probability to the left of $(600 - b)$ is $0.05/2 = 0.025$.

b. $P(X \le 600 + b) = 0.975$.

The probability to the left of $(600 + b)$ is $0.95 + 0.025 = 0.975$.

STEP 3 We'll use the expression in (a).

$$P(X \le 600 - b) = P\left(\frac{X - 600}{100} \le \frac{(600 - b) - 600}{100}\right) = 0.025 \quad \text{Standardize.}$$

$$= P\left(Z \le \frac{-b}{100}\right) \quad \text{Use Equation 6.8; simplify.}$$

There is no further simplification within the probability statement. The resulting expression involves Z and is a cumulative probability. Find a value in the body of Table III in the Appendix as close to 0.025 as possible.

Set the corresponding z equal to $\dfrac{-b}{100}$, and solve for b.

$$\frac{-b}{100} = -1.96 \qquad \text{Table III in the Appendix.}$$

$$-b = -196.00 \qquad \text{Multiply both sides by 100.}$$

$$b = 196.00 \qquad \text{Multiply both sides by } -1.$$

The value of b is 196 and the symmetric interval about the mean is

$$P(600 - b \le X \le 600 + b) = P(600 - 196 \le X \le 600 + 196)$$

$$= P(404 \le X \le 796) = 0.95$$

95% of all backpacks have a volume between 404 and 796 in^3.

Technology can be used to find the endpoints of the interval without solving for b. See Figure 6.44.

TRY IT NOW GO TO EXERCISE 6.46

Technology Corner

Procedure: Solve probability questions involving a normal random variable.
Reconsider: Example 6.5, solutions, and interpretations.

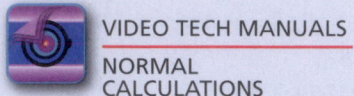

VIDEO TECH MANUALS
NORMAL
CALCULATIONS

CrunchIt!

There is a built-in function to compute probabilities associated with a normal random variable. Select Distribution Calculator; Normal. To find cumulative probability or right-tail probability, select the Probability tab. Choose an appropriate inequality symbol and enter a value for the endpoint. To solve an inverse cumulative probability problem, select the Quantile tab and enter a cumulative probability.

1. Select Distribution Calculator; Normal. Enter the mean, 10, and the standard deviation, 2. Under the Probability tab, select > and enter 12.5 for the endpoint. See Figure 6.45.
2. To find the probability that X takes on a value in an interval, use cumulative probability.
$P(9 \leq X \leq 10) = P(X \leq 10) - P(X \leq 9)$.
3. Select Distribution Calculator; Normal. Enter the mean, 10, and the standard deviation, 2. Under the Quantile tab, enter the cumulative probability and click Calculate. See Figure 6.46.

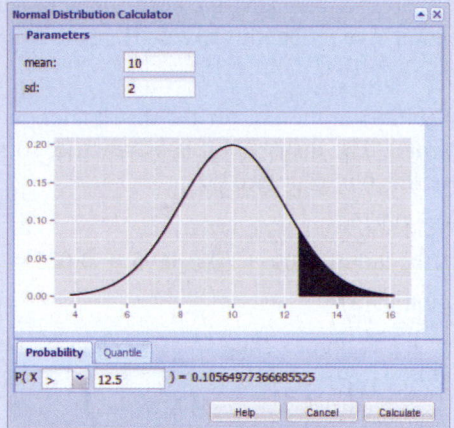
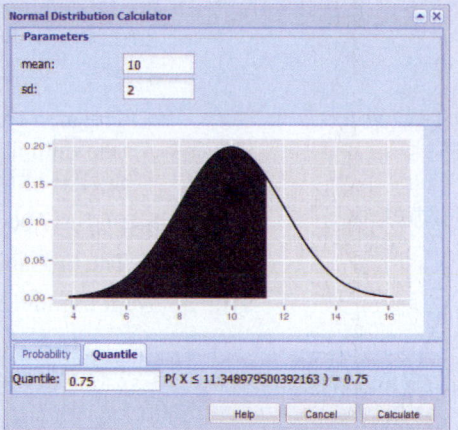

Figure 6.45 $P(X \geq 12.5)$.

Figure 6.46 A solution to $P(X \leq b) = 0.75$.

TI-84 Plus C

The built-in function `normalcdf` is used to find (calculator) cumulative probability: the probability that X takes on a value between a and b. This function takes four arguments: a (lower), b (upper), μ, and σ. The built-in function `invNorm` takes three arguments: p (area), μ, and σ. This function returns a value x such that $P(X \leq x) = p$. The default values for μ and σ are 0 and 1, respectively.

1. Select DISTR; DISTR; `normalcdf`. Enter the left endpoint (`lower`), 12.5, the right endpoint (`upper`), 1E99 (calculator infinity), the mean (μ), 10, and the standard deviation (σ), 2. Highlight Paste and tap ENTER. Refer to Figure 6.39.
2. Select DISTR; DISTR; `normalcdf`. Enter the left endpoint (`lower`), 9, the right endpoint (`upper`), 10, the mean (μ), 10, and the standard deviation (σ), 2. Highlight Paste and tap ENTER. Refer to Figure 6.40.
3. Select DISTR; DISTR; `invNorm`. Enter the cumulative probability (`area`), 0.75, the mean (μ), 10, and the standard deviation (σ), 2. Highlight Paste and tap ENTER. Refer to Figure 6.41.

Minitab

There are several built-in functions to compute cumulative probability, tail probability, and inverse cumulative probability. These functions may be accessed through a graphical input window or by using the command language.

1. In a session window, use the function CDF and the complement rule to find $P(X > 12.5)$. See Figure 6.47.
2. Select Graph; Probability Distribution Plot; View Probability. In the Distribution menu, select Normal. Enter the Mean and Standard deviation. Under the Shaded Area tab, choose X Value and Middle. Enter the X value 1 (9) and X value 2 (10). Click OK. Minitab displays a distribution plot with the shaded area corresponding to probability. See Figure 6.48.
3. Select Calc; Probability Distributions; Normal. Choose Inverse cumulative probability, enter the Mean, 10, and the Standard deviation, 2. Select Input constant (p), and enter 0.75. Click OK. The value of x is displayed in the session window. See Figure 6.49.

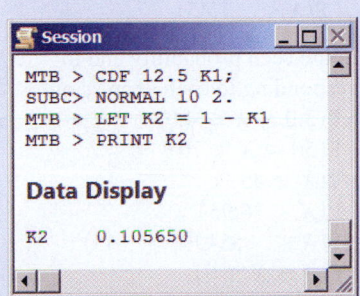

Figure 6.47 $P(X > 12.5)$ using the command language.

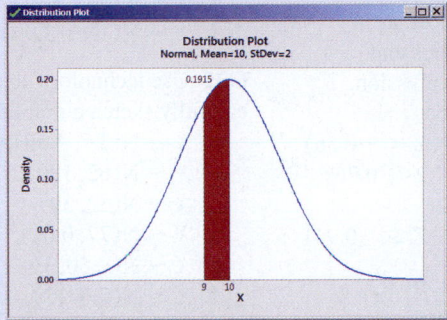

Figure 6.48 $P(9 \leq X \leq 10)$ using Probability Distribution Plot.

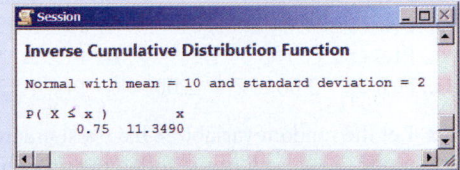

Figure 6.49 Inverse cumulative distribution function output.

Excel

There are built-in functions to compute cumulative probability associated with a standard normal random variable (NORM.S.DIST) and a normal random variable with arbitrary mean and standard deviation (NORM.DIST). The functions NORM.S.INV and NORM.INV are the corresponding inverse cumulative probability functions.

1. Use the function NORM.DIST to find $P(X \leq 12.5)$. Use the complement rule to find $P(X > 12.5)$. See Figure 6.50.
2. Use the function NORM.DIST to find $P(X \leq 9)$ and $P(X \leq 10)$. Compute the difference to find $P(9 \leq X \leq 10)$. See Figure 6.51.
3. Use the function NORM.INV. Enter the cumulative probability, 0.75, the mean, 10, and the standard deviation, 2. See Figure 6.52.

A	B
0.8944	= NORM.DIST(12.5,10,2,TRUE)
0.1056	= 1 - A1

Figure 6.50 $P(X > 12.5)$.

A	B
0.5000	= NORM.DIST(10,10,2,TRUE)
0.3085	= NORM.DIST(9,10,2,TRUE)
0.1915	= A1 - A2

Figure 6.51 $P(9 \leq X \leq 10)$.

A	B
11.3490	= NORM.INV(0.75,10,2)

Figure 6.52 Inverse cumulative probability.

SECTION 6.2 EXERCISES

Concept Check

6.26 True/False The probability density function for any normal random variable is bell-shaped.

6.27 True/False The mean and variance of a normal random variable determine the location and spread of the distribution.

6.28 Fill in the Blank The standard normal random variable has mean _____ and variance _____.

6.29 Fill in the Blank Any probability statement involving a normal random variable can be converted to an equivalent statement involving a standard normal random variable through the process of _____.

6.30 Multiple Choice For any normal random variable X, the statement $P(X \leq x)$ is (a) cumulative probability; (b) inverse cumulative probability; (c) standardized.

Practice

6.31 Let the random variable Z have a standard normal distribution. Find each of the following probabilities and carefully sketch a graph corresponding to each expression.
- **a.** $P(Z \le 2.16)$
- **b.** $P(Z < 2.16)$
- **c.** $P(Z \le -0.47)$
- **d.** $P(0.73 > Z)$
- **e.** $P(-1.75 \ge Z)$
- **f.** $P(-0.35 \le Z \le 0.65)$
- **g.** $P(Z < 5)$
- **h.** $P(Z \le -4)$
- **i.** $P(Z \le 4)$

6.32 Let the random variable Z have a standard normal distribution. Find each of the following probabilities and carefully sketch a graph corresponding to each expression.
- **a.** $P(-1.33 > Z)$
- **b.** $P(Z < 2.35)$
- **c.** $P(Z > 2.59)$
- **d.** $P(-1.56 < Z < -0.56)$
- **e.** $P(0.13 < Z < 2.44)$
- **f.** $P(-0.05 < Z < 0.76)$
- **g.** $P(Z \ge 2.67)$
- **h.** $P(Z \le 1.42)$
- **i.** $P(Z \le -2.00 \cup Z \ge 2.00)$
- **j.** $P(-1.82 < Z \le -0.94)$

6.33 Let the random variable Z have a standard normal distribution. Find each of the following probabilities.
- **a.** $P(-1.00 \le Z \le 1.00)$
- **b.** $P(-2.00 \le Z \le 2.00)$
- **c.** $P(-3.00 \le Z \le 3.00)$

Do you recognize these three probabilities? What rule are they associated with?

6.34 Let the random variable Z have a standard normal distribution. Solve each expression for b. Carefully sketch a graph corresponding to each probability statement.
- **a.** $P(Z \le b) = 0.8686$
- **b.** $P(Z < b) = 0.1867$
- **c.** $P(Z < b) = 0.0016$
- **d.** $P(Z \ge b) = 0.2643$
- **e.** $P(Z > b) = 0.9382$
- **f.** $P(Z \ge b) = 0.5000$
- **g.** $P(b < Z) = 0.0192$
- **h.** $P(b > Z) = 0.9938$
- **i.** $P(-b < Z < b) = 0.7995$
- **j.** $P(-b \le Z \le b) = 0.5527$

6.35 Let the random variable Z have a standard normal distribution. Solve each expression for b. Carefully sketch a graph corresponding to each probability statement.
- **a.** $P(Z \le b) = 0.5100$
- **b.** $P(Z > b) = 0.1080$
- **c.** $P(Z \ge b) = 0.0500$
- **d.** $P(Z \le b) = 0.0100$
- **e.** $P(-b \le Z \le b) = 0.8000$
- **f.** $P(-b < Z < b) = 0.6535$

6.36 Let the random variable Z have a standard normal distribution. Recall the definition for percentiles. $P(Z \le 1.0364) = 0.85$, so 1.0364 is the 85th percentile. Find each of the following percentiles for a standard normal distribution.
- **a.** 10th
- **b.** 27th
- **c.** 85th
- **d.** 40th
- **e.** 49th
- **f.** 61st

6.37 Let Z be a standard normal random variable and recall the calculations necessary to construct a box plot.
- **a.** Find the first and third quartiles for a standard normal distribution.
- **b.** Find the inner fences for a standard normal distribution.
- **c.** Find the probability that Z is beyond the inner fences.
- **d.** Find the outer fences for a standard normal distribution.
- **e.** Find the probability that Z is beyond the outer fences.

6.38 Compute each probability and carefully sketch a graph corresponding to each expression.
- **a.** $X \sim N(3, 0.0225)$, $\quad P(X \le 3.25)$
- **b.** $X \sim N(52, 49)$, $\quad P(X > 60)$
- **c.** $X \sim N(-7, 1)$, $\quad P(X \le -4.5)$
- **d.** $X \sim N(235, 121)$, $\quad P(X > 200)$
- **e.** $X \sim N(242, 132)$, $\quad P(X \ge 350)$
- **f.** $X \sim N(1.17, 3.94)$, $\quad P(X < -1.45)$

6.39 Use technology to compute each probability and to carefully sketch a graph corresponding to each expression.
- **a.** $X \sim N(3.7, 4.55)$, $\quad P(3.0 \le X \le 4.0)$
- **b.** $X \sim N(62, 100)$, $\quad P(50 < X < 70)$
- **c.** $X \sim N(32, 30)$, $\quad P(X \ge 45)$
- **d.** $X \sim N(77, 0.01)$, $\quad P(X < 76.95)$
- **e.** $X \sim N(-50, 16)$, $\quad P(X < -55 \cup X > -45)$
- **f.** $X \sim N(7.6, 12)$, $\quad P(8 \le X \le 9)$

6.40 Use technology to solve each expression for b.
- **a.** $X \sim N(17, 28)$, $\quad P(X < b) = 0.75$
- **b.** $X \sim N(303, 70)$, $\quad P(X \le b) = 0.05$
- **c.** $X \sim N(0, 25)$, $\quad P(-b \le X \le b) = 0.90$
- **d.** $X \sim N(-12, 2)$, $\quad P(X > b) = 0.35$
- **e.** $X \sim N(37, 2.25)$, $\quad P(\mu - b \le X \le \mu + b) = 0.68$
- **f.** $X \sim N(26.35, 7.21)$, $\quad P(X < b) = 0.11$

6.41 Suppose X is a normal random variable with mean 25 and standard deviation 6: $X \sim N(25, 36)$.
- **a.** Find the first and third quartiles for X.
- **b.** Find the inner fences for X.
- **c.** Find the probability that X is beyond the inner fences.
- **d.** Find the outer fences for X.
- **e.** Find the probability that X is beyond the outer fences.

Applications

6.42 Economics and Finance San Francisco is one of the most expensive cities in which to live in the United States. As of February 2013, the mean rent for a one-bedroom apartment in the Mission District was $2600.[6] Assume that the distribution of rents is approximately normal and the standard deviation is $200. A one-bedroom apartment in the Mission District is selected at random.
- **a.** Find the probability that the rent is less than $2450.
- **b.** Find the probability that the rent is between $2500 and $2650.
- **c.** Find a rent r such that 90% of all rents are less than r dollars per month. Write a Solution Trail for this problem.

6.43 Marketing and Consumer Behavior According to an annual survey conducted by TheKnot.com, the mean cost of a wedding in 2012 was $28,427.[7] This is still less than the high in 2008, but it reflects increasing confidence in the economy. Suppose the cost for a wedding is normally distributed, with a standard deviation of $1500, and a wedding is selected at random.

a. Find the probability that the wedding costs more than $31,000.

b. Find the probability that the wedding costs between $26,000 and $30,000.

c. Find the probability that the wedding costs less than $25,000.

6.44 Public Policy and Political Science   The President of the United States gives a State of the Union Address every year in late January or early February. Since Lyndon Johnson's address in 1966, Richard Nixon gave some of the shortest messages, and Bill Clinton presented a 1-hour and 28-minute address in 2000. The mean length for these addresses is 51.75 minutes and the standard deviation is 14.37 minutes. Assume the length of a State of the Union Address is normally distributed.

a. What is the probability that the next State of the Union Address will be between 45 and 55 minutes long?

b. What is the probability that the next State of the Union Address will be more than 90 minutes long?

c. What is the probability that the next two State of the Union Addresses will be less than 30 minutes long?

6.45 Manufacturing and Product Development   A standard Versa-lok block used in residential and commercial retaining wall systems has mean weight 37.19 kg.[8] Assume the standard deviation is 0.8 kg and the distribution is approximately normal. A standard block unit is selected at random.

a. What is the probability that the block weighs more than 38 kg? Write a Solution Trail for this problem.

b. What is the probability that the block weighs between 36 and 37 kg?

c. If the block weighs less than 35.5 kg, it cannot be used in certain commercial construction projects. What is the probability that the block cannot be used?

6.46 Marketing and Consumer Behavior   Movie trailers are designed to entice audiences by showing scenes from coming attractions. Several trailers are usually shown in a theater before the start of the main feature, and most are available via the Internet. The duration of a movie trailer is approximately normal, with mean 150 seconds and standard deviation 30 seconds.

a. What is the probability that a randomly selected trailer lasts less than 1 minute?

b. Find the probability that a randomly selected trailer lasts between 2 minutes and 3 minutes 15 seconds.

c. Any movie trailer that lasts beyond 4 minutes and 30 seconds is considered too long. What proportion of movie trailers is too long?

d. Find a symmetric interval about the mean such that 99% of all movie trailer durations lie in this interval.

6.47 Biology and Environmental Science   The salinity, or salt content, in the ocean is expressed in parts per thousand (ppt). The number varies with depth, rainfall, evaporation, river runoff, and ice formation. The mean salinity of the oceans is 35 ppt.[9] Suppose the distribution of salinity is normal and the standard deviation is 0.52 ppt, and suppose a random sample of ocean water from a region in the tropical Pacific Ocean is obtained.

a. What is the probability that the salinity is more than 36 ppt?

b. What is the probability that the salinity is less than 33.5 ppt?

c. A certain species of fish can only survive if the salinity is between 33 and 35 ppt. What is the probability that this species can survive in a randomly selected area?

d. Find a symmetric interval about the mean salinity such that 50% of all salinity levels lie in this interval. What are the endpoints of this interval called?

6.48 Public Health and Nutrition   Many people grab a granola bar for breakfast or for a snack to make it through the afternoon slump at work. A Kashi GoLean Crisp Chocolate Caramel bar is 45 grams, and the mean amount of protein in each bar is 8 grams.[10] Suppose the distribution of protein in a bar is normally distributed and the standard deviation is 0.15 gram, and a random Kashi bar is selected.

a. What is the probability that the amount of protein is less than 7.75 grams?

b. What is the probability that the amount of protein is between 7.8 and 8.2 grams?

c. Suppose the amount of protein is at least 8.1 grams. What is the probability that it is more than 8.3 grams?

d. Suppose three bars are selected at random. What is the probability that all three will be between 7.7 and 8.3 grams?

6.49 Public Health and Nutrition   Many typical household cleaners contain toxic chemicals. 2-Butoxyethanol is found in multipurpose cleaners and is a very powerful solvent. The EPA has a safety standard for this chemical when used in the workplace, but cleaning at home in a confined area can cause levels to rise well above this standard. The mean percent of 2-butoxyethanol in Rain-X Glass Cleaner is 3.[11] Suppose the distribution is approximately normal and the standard deviation is 1%, and a random bottle of Rain-X Glass Cleaner is selected.

a. What is the probability that the percentage of 2-butoxyethanol is less than 2.5?

b. What is the probability that the percentage of 2-butoxyethanol is between 2.2 and 3.5?

c. Suppose the EPA has established a limit of 5% 2-butoxyethanol in all consumer products. What is the probability that a bottle exceeds this limit?

6.50 Psychology and Human Behavior   In many U.S. families, both parents work outside the home, while children spend time at daycare centers or are cared for by other relatives. The mean amount of time fathers spend with their child(ren) is 7.3 hours per week.[12] Suppose this time distribution is approximately normal with standard deviation 0.75 hour, and suppose a father is randomly selected.

a. What is the probability that the father spends at least 8 hours with his child in a given week?

b. What is the probability that the father spends between 6 and 7 hours with his child in a given week?

c. If the child sees his or her father for less than 6 hours per week, the parental bond is weakened. What is the probability that this special bond will be weakened in a given week?

d. What is the probability that the father will spend at least 9 hours with his child in each of five randomly selected weeks?

Extended Applications

6.51 Sports and Leisure   People who ride in hot-air balloons usually fly just above the treetops at 200–500 feet. In populated areas, however, they usually stay at an altitude of at least 1000 feet. The amount of flying time possible in a hot-air balloon depends on many factors, including the number of propane burners, the number of people in the basket, and the weather. Assume the time spent aloft is normally distributed with mean 1.5 hours and standard deviation 0.45 hour. Suppose a hot-air balloon flight is selected at random.

  a. What is the probability that the flight time is between 1 and 2 hours?
  b. What is the probability that the flight time is more than 1 hour and 15 minutes?
  c. Find a value t such that 10% of all flights last less than t hours.
  d. Suppose a person offering hot-air balloon rides charges $50 for each ride of at least 1 hour, and $1.00 for every minute after 1 hour. What proportion of rides costs more than $100?

6.52 Sports and Leisure   The Daytona 500, often referred to as The Great American Race, is a spectacular sporting event, complete with a pre-race show. Jimmie Johnson won this race in 2013, when the mean speed per lap for all racers was 159.25 mph.[13] Assume the speed is normally distributed with a standard deviation of 16 mph, and a driver and lap are selected at random.

  a. What is the probability that the speed on this lap is less than 155 mph?
  b. What is the probability that the speed on this lap is between 140 and 150 mph?
  c. The fastest recorded speed is 212 mph at Talladega in 1986. What is the probability that the speed on this lap will set a new record?
  d. What is the probability that the four leaders will all have a speed of at least 165 on this lap?

6.53 Biology and Environmental Science   The amount of timber harvested and sold is associated with the housing market and the general economy. In 2012, the total amount of timber harvested in the United States was 2,500,321 mbf (thousand board feet).[14] It takes approximately 11 mbf to construct a typical 1900-square-foot-home. Assume the volume of timber harvested per acre is normally distributed with mean 30 mbf and standard deviation 6.25 mbf. Suppose an acre of timber is selected at random.

  a. What is the probability that the volume of timber harvested is between 25 and 40 mbf?
  b. What is the probability that the volume of timber harvested is less than 20 mbf?
  c. Suppose the acre has already produced 35 mbf. What is the probability that the volume harvested will be more than 40 mbf?
  d. A logging company selects three random acres to harvest during a week. The company will make a profit if all three acres produce more than 32 mbf. What is the probability that the logging company makes a profit?

6.54 Marketing and Consumer Behavior   Kraft Foods recently announced that the Kool-Aid mascot, that big red pitcher of the powdered drink mix with arms and legs, will receive a makeover as they unveil a new liquid mix. Kool-Aid bursts are distributed in various flavors including tropical punch, berry blue, and grape. Kraft Foods claims that the mean amount of Kool-Aid in each burst bottle is 200 milliliters (ml).[15] Assume the amount of drink in each bottle is normally distributed with standard deviation 1.25 ml. Suppose a bottle of berry blue is selected at random.

  a. What is the probability that the amount of drink will be between 199 and 201 ml?
  b. If the amount of drink is more than 202 ml, when the bottle is opened there will be a spill. What is the probability of a spill?
  c. Suppose there are 196 ml in the bottle of berry blue. Is there any evidence to suggest the claim made by Kraft Foods is false? Justify your answer.

6.55 Biology and Environmental Science   Many backyard gardeners prefer Silver Queen Hybrid corn. This late-season variety is very sweet and has tender, white kernels. In some locations in the Northeast, gardeners have trouble harvesting this variety because of its longer growing time. The temperature of the soil should be at least 65°F before planting, and the growing time is approximately normal with mean 92 days and standard deviation 5 days.

  a. What is the probability that a randomly selected seed will mature in less than 90 days?
  b. What is the probability that a randomly selected seed will mature in between 95 and 100 days?
  c. Suppose a row in a backyard garden contains 12 plants. What is the probability that four will be ready for dinner by the 95th day?
  d. Find a value h such that 99% of all plants are ready to be harvested within h days.

6.56 Manufacturing and Product Development   Violin bows are made from various woods to accommodate musicians' preferences and demands. Some commonly used woods include snakewood, ironwood, hakia, and pernambuco. While the bows are carefully handcrafted, they vary slightly in weight. Suppose a bowmaker claims the weight of his bows is normally distributed with mean 60 grams and standard deviation 3.2 grams.

  a. What is the probability that the weight of a randomly selected ironwood bow is between 58 and 62 grams?
  b. Good musicians can detect an *unacceptable* bow weight, i.e., a weight that differs from the mean by more than two standard deviations. What is the probability that a bow weight is unacceptable?
  c. Any manufactured bow that weighs more than 66 grams is reworked in order to decrease the weight. What is the probability that a randomly selected ironwood bow will need rework?
  d. Suppose the weight of a randomly selected bow is 55 grams. Is there any evidence to suggest the mean weight is less than 60 grams? Justify your answer.

6.57 Medicine and Clinical Studies   Repeated industrial tasks often cause work-related muscle disorders. Measurements of joint angles (of the shoulder and elbow, for example) required to complete a certain task can be used to predict future injuries. The shoulder joint angle required to fasten an aluminum door frame on an assembly line varies according to the worker's height, arm length, and location. The shoulder joint angle for this task is normally distributed with mean 23.7 degrees and standard deviation 1.9 degrees. Suppose an employee is randomly selected.

a. What is the probability that the shoulder joint angle will be between 20 and 25 degrees?

b. What is the probability that the joint angle will be less than 18 degrees?

c. If the joint angle is more than 28 degrees, there is a good chance the employee will suffer from a muscle disorder. What is the probability that the employee will suffer from a muscle disorder?

d. If the joint angle is between 21.7 and 25.7 degrees, then management believes the ergonomics of the task are adequate. If five employees are randomly selected, what is the probability that four of the five have adequate ergonomics?

6.58 Biology and Environmental Science   Many lakes are carefully monitored for pH concentration, total phosphorus, chlorophyll, nitrogen, and total suspended solids. These data are used to characterize the condition of the lake and to chart year-to-year variability. Based on information from the Lake Partner Program, Ontario Ministry of the Environment, Aberdeen Lake has a mean total phosphorus concentration of 14.6 mg/liter and standard deviation 5.8 mg/liter. Suppose a day is selected at random, and a total phosphorus measure from Aberdeen Lake is obtained.

a. What is the probability that the total phosphorus is less than 13 mg/liter?

b. What is the probability that the total phosphorus differs from the mean by more than 5 mg/liter?

c. Suppose the total phosphorus is less than 20 mg/liter. What is the probability that it is less than 14 mg/liter?

d. If the total phosphorus measurement is 27 mg/liter, is there any evidence to suggest the mean has increased?

6.59 Manufacturing and Product Development   High-pressure washers have become popular for cleaning siding, decks, and windows. This equipment is available in various engine types and horsepower. Suppose the power rating (in horsepower, hp) for a residential pressure washer is normally distributed with mean 20 hp and standard deviation σ.

a. The probability that a randomly selected power rating is within 2.5 hp of the mean is 0.7229. Find the value of σ.

b. A leading consumer magazine advised its readers to purchase pressure washers with a power rating of 15 hp or more. What proportion of pressure washers have this rating?

c. If the power rating is more than 26.5 hp, the pressure washer will crack, or even break, certain windows. What is the probability that a pressure washer could break a window?

6.60 Physical Sciences   Hydroelectric projects are carefully monitored, and their energy capability is predicted for several years into the future. Suppose the Klamath Hydro Project, located on the upper Klamath River in south-central Oregon, generates electricity according to a normal distribution. The Pacific Northwest Utilities Conference Committee claims the mean electricity generated per year is 35 megawatts (MW).

a. The probability that the Klamath Hydro Project generates less than 34 MW during any randomly selected year is 0.3540. Find the standard deviation.

b. Suppose the years are independent, and the hydro project will record a profit in a given year if it is able to generate at least 37.8 MW that year. What is the probability that the project will record a profit for four consecutive years?

c. Suppose the electricity generated during a certain year is 33.5 MW. Is there any evidence to suggest that the claim by the Pacific Northwest Utilities Conference Committee is false? Justify your answer.

6.61 Manufacturing and Product Development   Dining-room chairs come in many different woods, styles, and shapes. The height of the seat of a randomly selected oak dining-room chair is approximately normal with mean 85 centimeters (cm) and standard deviation 1.88 cm.

a. Find a value h such that 99% of all dining-room chairs have height less than h.

b. Consumer testing indicates that any chair seat higher than 90 cm is uncomfortable to use when eating. What is the probability that a randomly selected dining-room chair is uncomfortable?

c. Find the first and third quartiles of the dining-room chair height distribution.

d. There is some evidence to suggest that, after five years of use, the mean height of these chairs has decreased, due to wear, erosion, and humidity. Suppose that after five years, the probability the height is more than 86 cm is 0.0718. Find the mean height after five years.

6.62 Sports and Leisure   Tianlang Guan was the youngest person ever to participate in the Masters Golf Tournament. The 14-year-old from China played the Augusta, Georgia, course with the confidence of a professional, but very slowly. He was warned about his slow play and was assessed a one-stroke penalty on the par-4 17th hole on the second day of the tournament.[16] The PGA tour maintains a 40-second time limit to play a stroke, but also has several exceptions to this rule that allow for an additional 20 seconds. The mean time for all golf shots is 38 seconds.[17] Assume the time for all golf shots is normally distributed with standard deviation 9 seconds, and suppose a golf shot is selected at random.

a. What is the probability that a randomly selected shot takes between 25 and 35 seconds?

b. If the shot takes more than 60 seconds, the golfer is assessed a penalty stroke. What is the probability that the golfer will be assessed a penalty stroke?

c. Suppose a golfer takes 72 strokes to complete the round. What is the probability that at least 60 of these shots take less than 45 seconds?

Challenge

6.63 Sports and Leisure The International Tennis Federation (ITF) establishes the specifications for tennis balls. The diameter of a tennis ball used in any tournament must be between 2.5 and 2.625 inches. Suppose the diameter of a tennis ball is approximately normal with mean 2.5625 inches and standard deviation 0.04 inch.

a. What is the probability that a randomly selected tennis ball will meet ITF diameter specifications?

b. Suppose six tennis balls will be used in a tournament game. What is the probability that exactly one will not meet ITF diameter specifications? Assume independence.

6.3 Checking the Normality Assumption

Almost every inferential statistics procedure requires certain assumptions, for example, that observations are selected independently or that variances are equal (for analysis of variance). And many statistical techniques are valid *only* if the observations are from a normal distribution. If an inference procedure requires normality, and the population distribution is not normal, then the conclusions are worthless. Therefore, it seems reasonable to be able to perform some kind of check for normality, to make sure there is no evidence to refute this assumption.

Until now we have been using the normal distribution as a model for describing the variability of a random variable X, and we have been assuming that we know the values of the population mean μ and the population variance σ^2. If those values are not known, the sample mean $\bar{x}$ and the sample standard deviation s can be used as estimates of the unknown parameters μ and σ. However, we still cannot be sure that the normal distribution is an appropriate model to describe a particular set of observations. We need a way to check whether a set of observations does seem to come from a population with a normal distribution. There are four different methods we can use to look for any evidence of non-normality. Three of them use techniques that we have seen before; the fourth one is a new method.

Given a set of observations, $x_1, x_2, x_3, \ldots, x_n$, the following four methods may be used to check for any evidence of non-normality, for example, a distribution that is not bell-shaped, a skewed distribution, or a distribution with heavy tails.

1. Graphs

Construct a histogram, a stem-and-leaf plot, and/or a dot plot. Examine the shape of the distribution for any indications that the distribution is not bell-shaped and symmetric. In a random sample, the distribution of the sample should be similar to the distribution of the population.

2. Backward Empirical Rule

To use the empirical rule to test for normality, find the mean, the standard deviation, and the three symmetric intervals about the mean $(\bar{x} - ks, \bar{x} + ks)$, $k = 1, 2, 3$. Compute the *actual* proportion of observations in each interval. If the actual proportions are close to 0.68, 0.95, and 0.997, then normality seems reasonable. Otherwise, there is evidence to suggest that the shape of the distribution is not normal.

3. IQR/s

Find the interquartile range, IQR, and standard deviation, s, for the sample, and compute the ratio IQR/s. If the data are approximately normal, then IQR/$s \approx 1.3$.

$P(Z \le -0.6745) = 0.25$, and
$P(Z \le 0.6745) = 0.75$.

Here is some justification for this ratio. Consider a standard normal random variable, Z ($\mu = 0$, $\sigma = 1$). The first quartile for Z is -0.6745 and the third quartile is 0.6745. The interquartile range divided by the standard deviation is $[0.6745 - (-0.6745)]/1 = 1.349$. In a random sample, the interquartile range should be close to the population interquartile range, and the standard deviation should be close to the population standard de-viation. Any normal distribution can be standardized, or compared to Z, so IQR/$s \approx 1.3$.

4. Normal Probability Plot

A normal probability plot is a scatter plot of each observation versus its corresponding standardized normal score. For a normal distribution, the points will fall along a straight line.

The *standardized normal scores* are expected values. For example, in repeated samples of size n from the Z distribution, on average the smallest value is z_1, on average the next largest value is z_2 , etc., on average the largest value is z_n.

How to Construct a Normal Probability Plot

Suppose $x_1, x_2, \ldots, x_n$ is a set of observations.

1. Order the observations from smallest to largest and let $x_{(1)}, x_{(2)}, \ldots, x_{(n)}$ represent the set of ordered observations.
2. Find the standardized normal scores for a sample of size n in Table IV in the Appendix: $z_1, z_2, \ldots, z_n$.
3. Plot the ordered pairs $(z_i, x_{(i)})$.

Most of the standardized normal scores are always between -2.0 and $+2.0$, because approximately 95% of all observations lie within two standard deviations of the mean.

If the scatter plot is nonlinear, there is evidence to suggest the data did not come from a normal distribution. Most statistical software (the TI-84 Plus C and Minitab included) automatically computes the expected Z values. Table IV in the Appendix provides standardized normal scores for some values of n.

Figures 6.53–6.56 are examples of normal probability plots.

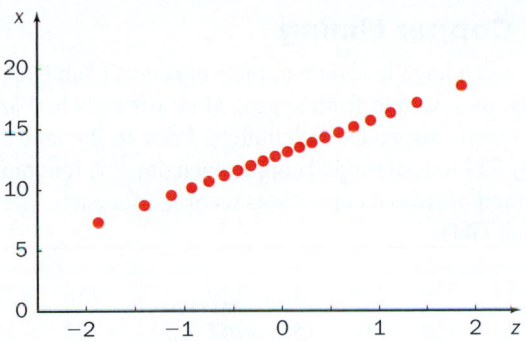

Figure 6.53 A normal probability plot. The points lie along an approximate straight line. There is no evidence of non-normality.

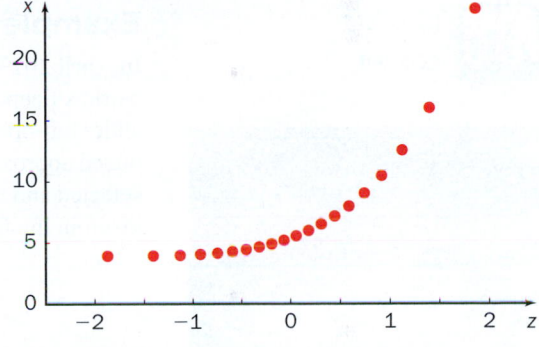

Figure 6.54 A normal probability plot. The curved graph suggests that the distribution is not normal and is skewed.

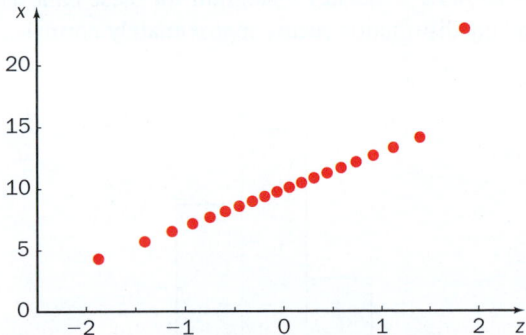

Figure 6.55 A normal probability plot. The plot suggests that the distribution is not normal and that the data set contains an outlier.

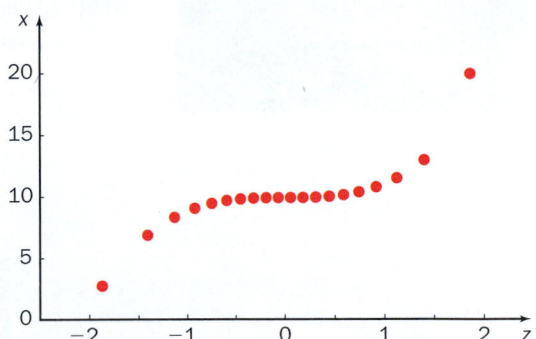

Figure 6.56 A normal probability plot. The plot suggests that the distribution is not normal and has *heavy tails*.

The data axis can be horizontal or vertical. To use a horizontal data axis, plot the points $(x_{(i)}, z_i)$. Figure 6.57 shows a normal probability plot with the data plotted on the vertical axis and Figure 6.58 shows a normal probability plot (using the same data and standardized normal scores) with the data plotted on the horizontal axis.

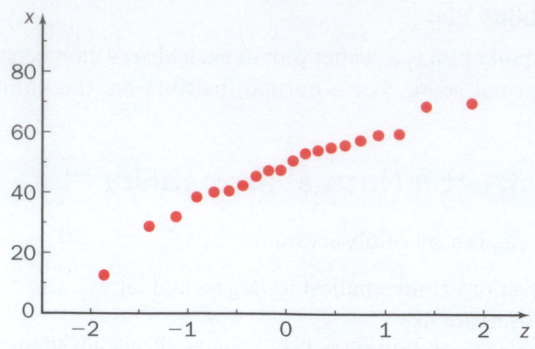

Figure 6.57 A normal probability plot with the data plotted on the vertical axis.

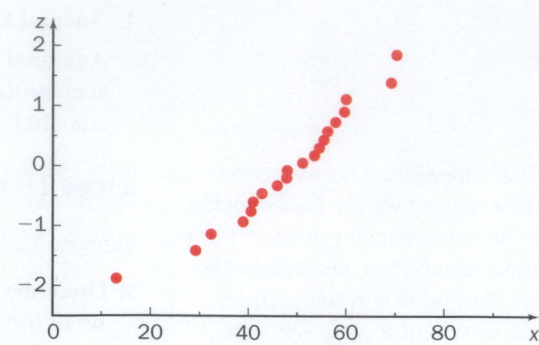

Figure 6.58 A normal probability plot with the data plotted on the horizontal axis.

Interpretation of a normal probability plot is very subjective, and even if the axes are reversed, we are still looking for the points to lie along a straight line.

All four methods can be used to check the normality assumption, and any one (or several) may suggest the data did *not* come from a normal distribution. Because we are searching for evidence of non-normality, even if we fail to reject the normality assumption in each test, we still cannot say with *absolute certainty* that the data came from a normal distribution.

DATA SET

COPPER

George Frey /Landov

Example 6.8 Copper Mining

In April 2013 there was a huge landslide at the Kennecott Utah Copper mine, one of the world's deepest open pits, visible from space. Mine officials had anticipated this landslide, but operations were suspended indefinitely. Prior to the landslide, Kennecott produced approximately 753 tons of refined copper each day.[18] A random sample of days was selected and the amount of refined copper was recorded for each. The 20 observations are given in the following table:

757	751	749	753	745	749	738	746	732	750
741	743	760	741	758	762	745	752	735	767

Is there any evidence to suggest that this distribution is not normally distributed?

SOLUTION

STEP 1 Figure 6.59 shows a frequency histogram for these data. There are no obvious outliers, and the distribution seems approximately normal.

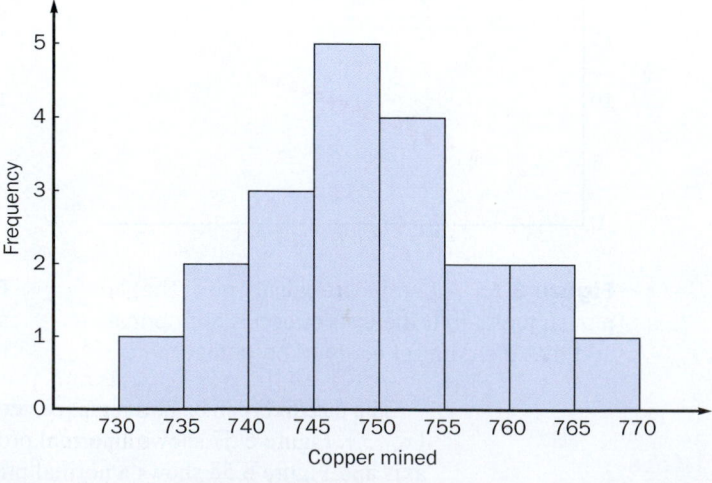

Figure 6.59 Frequency histogram for the copper mine data.

STEP 2 The sample mean and the sample standard deviation are $\bar{x} = 748.70$ and $s = 9.17$. The following table lists three symmetric intervals about the mean, the number of observations in each interval, and the proportion of observations in each interval (recall that $n = 20$).

Interval	Frequency	Proportion
$(\bar{x} - s, \bar{x} + s) = (739.53, 757.87)$	13	0.65
$(\bar{x} - 2s, \bar{x} + 2s) = (730.36, 767.04)$	20	1.00
$(\bar{x} - 3s, \bar{x} + 3s) = (721.19, 776.21)$	20	1.00

The actual proportions are *close* to those given by the empirical rule (0.68, 0.95, and 0.997).

STEP 3 The quartiles are $Q_1 = 742$, $Q_3 = 755$.

$$IQR/s = (755 - 742)/9.17 = 1.4177$$

This ratio is *close* to 1.3.

STEP 4 The table below lists each observation along with the corresponding normal score from Table IV in the Appendix.

Observation	Normal score	Observation	Normal score
732	−1.87	749	0.06
735	−1.40	750	0.19
738	−1.13	751	0.31
741	−0.92	752	0.45
741	−0.74	753	0.59
743	−0.59	757	0.74
745	−0.45	758	0.92
745	−0.31	760	1.13
746	−0.19	762	1.40
749	−0.06	767	1.87

Plot these points to obtain the normal probability plot, as shown in Figure 6.60. Figure 6.61 shows a technology solution. The points lie along an approximately straight line.

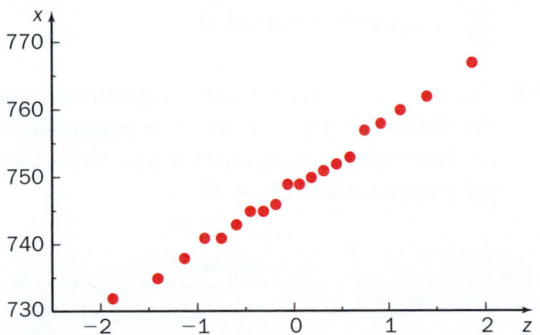

Figure 6.60 Normal probability plot for the copper mine data.

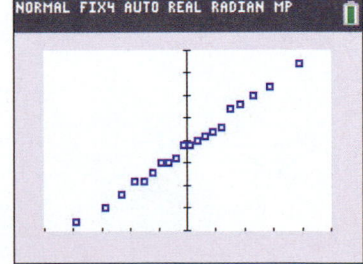

Figure 6.61 Normal probability plot.

The histogram, backward empirical rule, IQR/s, and normal probability plot show no significant evidence of non-normality. Remember, however, that this decision is very subjective.

TRY IT NOW GO TO EXERCISE 6.76

DATA SET

DOSAGE

Example 6.9 Chemotherapy Protocol

A certain protocol for chemotherapy states that the total dose for patients under the age of 12 is no greater than 450 mg/m^2 within six months. A random sample of 30 patients undergoing this form of chemotherapy was obtained, and their medical records were examined to determine the total dose of the drug over the past six months. The data are given in the following table.

350	351	352	353	354	358	361	364	371	376
377	378	387	396	399	402	406	408	412	424
427	430	432	437	440	441	443	446	447	449

Is there any evidence to suggest the distribution of six-month total dosage is not normally distributed?

SOLUTION

STEP 1 Figure 6.62 shows a frequency histogram for these data. Although the graph seems symmetric, it is not bell-shaped. Most of the data are concentrated in the tails of the distribution. This suggests the data are not from a normal distribution.

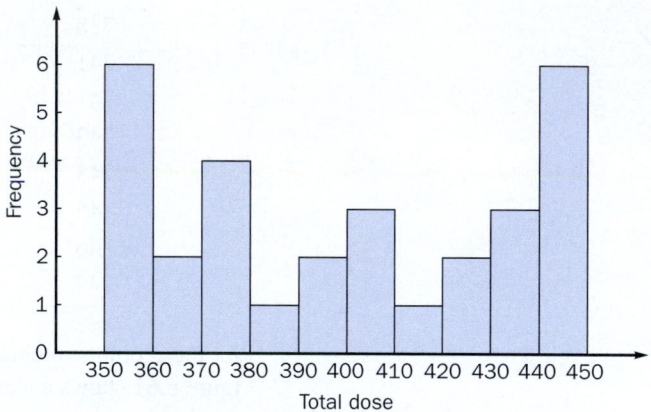

Figure 6.62 Frequency histogram for the cumulative chemotherapy dose data.

STEP 2 The sample mean is $\bar{x} = 399.03$ and the sample standard deviation is $s = 34.94$. The following table lists three symmetric intervals about the mean, the number of observations in each interval, and the proportion of observations in each interval (computed using $n = 30$).

Interval	Frequency	Proportion
$(\bar{x} - s, \bar{x} + s) = (364.09, 433.97)$	15	0.50
$(\bar{x} - 2s, \bar{x} + 2s) = (329.15, 468.91)$	30	1.00
$(\bar{x} - 3s, \bar{x} + 3s) = (294.21, 503.85)$	30	1.00

The first two proportions (0.50 and 1.00) are significantly different from those given by the empirical rule (0.68 and 0.95). This suggests the population of total chemotherapy doses is not normal.

STEP 3 The quartiles are $Q_1 = 364.00$ and $Q_3 = 432.00$.

$$\text{IQR}/s = (432.00 - 364.00)/34.94 = 1.9462$$

This ratio is significantly different from 1.3, so there is more evidence to suggest the underlying population is not normal.

STEP 4 The following table lists each observation along with the corresponding normal score.

Observation	Normal score	Observation	Normal score	Observation	Normal score
350	−2.04	377	−0.38	427	0.47
351	−1.61	378	−0.29	430	0.57
352	−1.36	387	−0.21	432	0.67
353	−1.18	396	−0.12	437	0.78
354	−1.02	399	−0.04	440	0.89
358	−0.89	402	0.04	441	1.02
361	−0.78	406	0.12	443	1.18
364	−0.67	408	0.21	446	1.36
371	−0.57	412	0.29	447	1.61
376	−0.47	424	0.38	449	2.04

The normal probability plot is shown in Figure 6.63.

The points do not lie along a straight line. Each tail is flat, which makes the graph look S-shaped. This suggests that the underlying population is not normal. Figure 6.64 shows a technology solution.

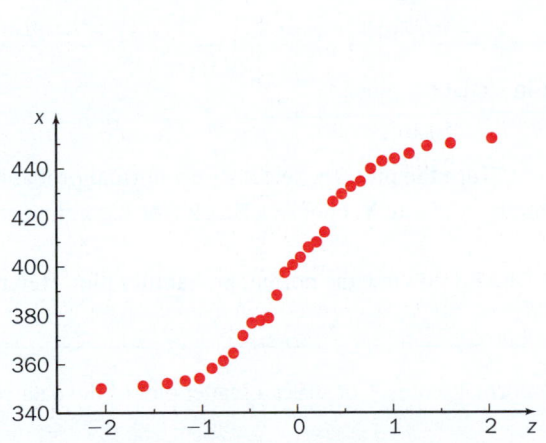

Figure 6.63 Normal probability plot for the chemotherapy dose data.

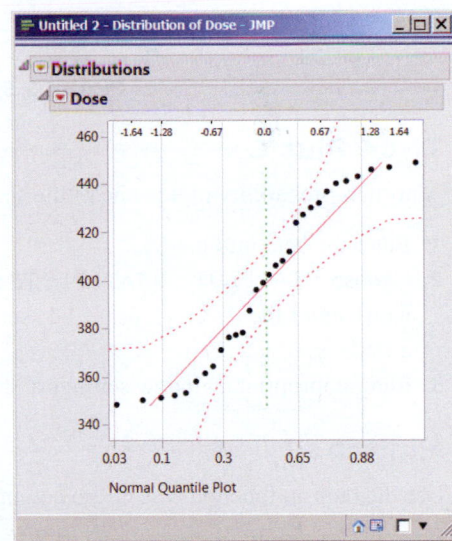

Figure 6.64 JMP normal probability plot.

The histogram, backward empirical rule, IQR/s, and the normal probability plot all indicate that this sample did not come from a normal population.

TRY IT NOW GO TO EXERCISE 6.79

TECHNOLOGY CORNER

Procedure: Construct a normal probability plot.
Reconsider: Example 6.8, solution, and interpretations.

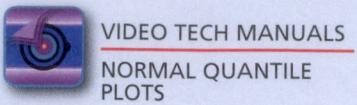

VIDEO TECH MANUALS

NORMAL QUANTILE PLOTS

CrunchIt!

A QQ Plot (quantile plot) is used to construct a normal probability plot.

1. Enter the data into a column. Rename the column if desired.
2. Select Graphics; QQ Plot. Select the Sample (name of the column) from the drop-down menu. Click Calculate to view the graph. See Figure 6.65. CrunchIt! adds a straight line to this plot for visual reference.

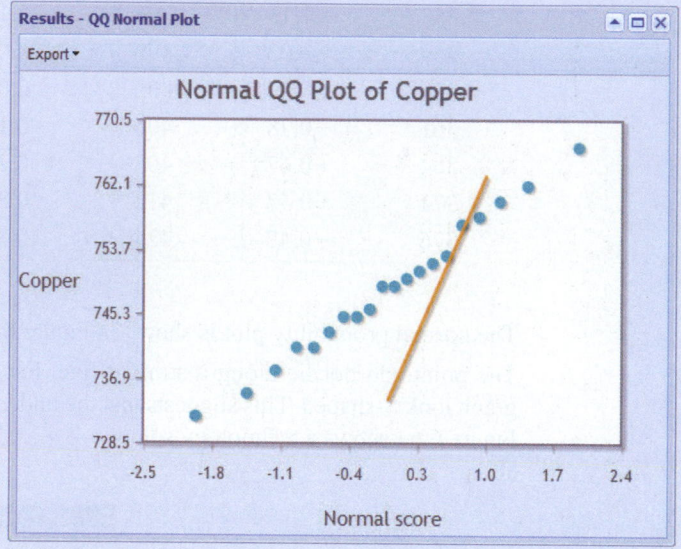

Figure 6.65 A QQ Plot constructed using CrunchIt!.

TI-84 Plus C

A normal probability plot is one of the six built-in statistical plots.

1. Enter the data into list L1.
2. Choose `STATPLOT`; STAT PLOTS; Plot1. Turn the plot On, select Type normal probability plot (the last graph icon), enter the Data List, L1, set the Data Axis to Y, choose a Mark (for the points on the graph), and select a Color.
3. Enter appropriate window settings and press `GRAPH` to view the normal probability plot. Refer to Figure 6.61.

Minitab

Use the built-in function NSCOR to compute the normal scores. Construct a scatter plot of the data versus the normal scores.

1. Enter the data into column C1.
2. Compute the normal scores in a session window (or by using the Minitab Calculator) and store the results in column C2: LET C2 = NSCOR(C1).
3. Construct a scatter plot.
   a. In a session window: PLOT C1*C2.
   b. In a graphical input window: Graph; Scatterplot. Select Simple and let the Y variable be the data column (C1) and the X variable be the normal scores column (C2). See Figure 6.66.

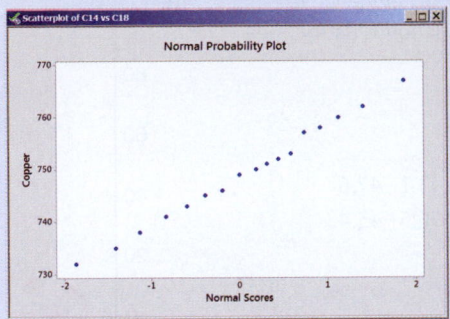

Figure 6.66 A normal probability plot constructed using Minitab.

Excel

To construct a normal probability plot, compute the normal scores using the formula below. Construct a scatter plot of the data versus the normal scores.

1. Enter the data into column C, in increasing order, and the numbers 1 to $n = 20$ in column A.
2. Set the cell B1 equal to `NORM.S.INV((A1 - 3/8)/(20 + 1/4))`. Copy this result and paste into the cells B2-B20. These are (approximately) the normal scores.
3. Highlight the data range B1:C20. Under the Insert tab, select Scatter; Scatter. See Figure 6.67.

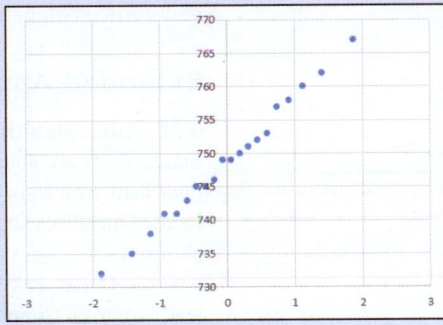

Figure 6.67 A normal probability plot constructed using Excel.

SECTION 6.3 EXERCISES

Concept Check

6.64 Short Answer Name four methods to search for evidence of nonnormality.

6.65 True/False In a normal probability plot, the data axis can be horizontal or vertical.

6.66 True/False For a normal distribution, the points in a normal probability plot will fall along a straight line or a bell-shaped curve.

6.67 True/False Most standardized normal scores are between -2.0 and $+2.0$.

6.68 Fill in the Blank For a normal distribution, $IQR/s \approx$ _____.

6.69 True/False In a random sample, the distribution of the sample should be similar to the distribution of the population.

Practice

6.70 Consider the following 20 observations:

15.4	13.9	14.9	16.2	16.6	15.4	17.2	18.5
19.3	13.0	16.5	20.2	16.4	15.3	18.5	17.9
15.5	17.4	16.3	14.3				

Construct a normal probability plot. Is there any evidence to suggest the data are from a non-normal population? Justify your answer.

6.71 Consider the following 20 observations:

52.0	52.1	58.8	88.0	49.9	18.7	43.1	47.6
90.0	49.8	54.8	35.1	56.1	53.2	76.5	45.4
34.1	19.5	58.7	25.7				

Construct a normal probability plot. Is there any evidence to suggest the data are from a non-normal population? Justify your answer.

6.72 Examine each normal probability plot below. Is there any evidence to suggest the data are from a non-normal population? Justify your answer.

a.

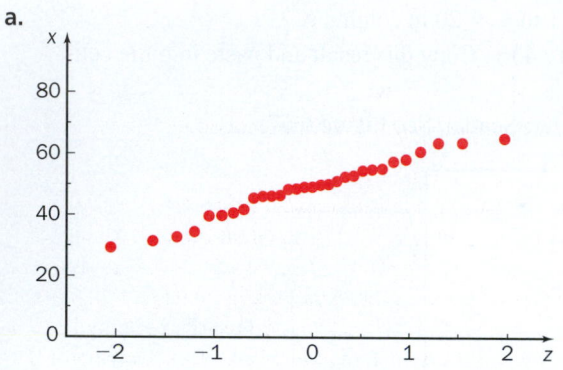

b.

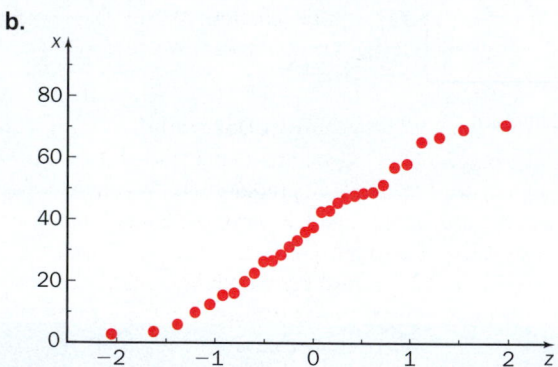

c.

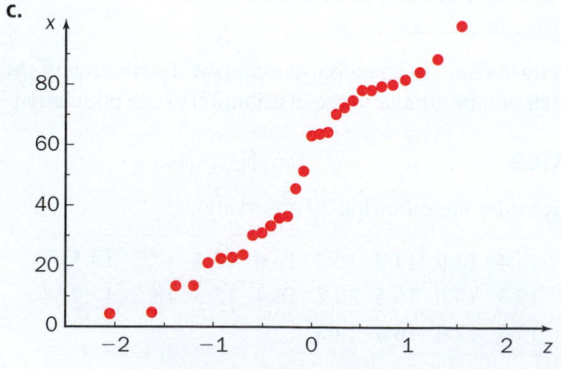

d.

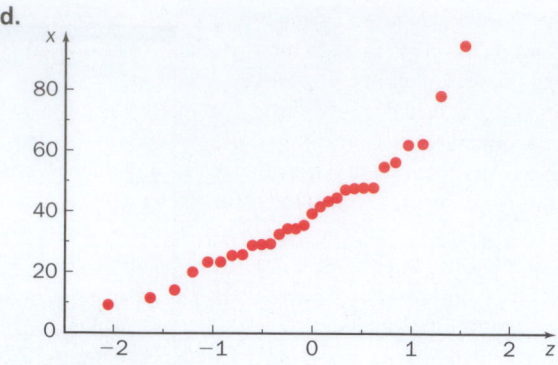

6.73 Consider the given data. Use the four methods presented in this section to determine whether there is any evidence to suggest the data are from a non-normal population. **EX6.73**

6.74 Consider the following data:

5.32	9.87	11.25	10.94	5.58
6.29	7.47	10.75	6.22	8.00

Use the four methods presented in this section to determine whether there is any evidence to suggest the data are from a non-normal population.

Extended Applications

6.75 Public Health and Nutrition All across the United States, there are abandoned factories that were once used to melt lead. Soil tests in neighborhoods near some of these factory locations suggest the dirt is contaminated and that children should not play in or near the area.[19] Hedley Street near the Delaware River in Philadelphia is the site of two old lead factories. Suppose samples of soil around homes in the area were obtained and the lead level in each was carefully measured (in parts per million, ppm). The data are given in the following table: **SOILTEST**

2223	641	1275	1796	802	691	1716	1157
1340	1504	1006	778	1476	914	1108	2110
1704	2106	1069	1398	1680	1705	2376	1577
1601	2201	1809	1847	1721	1069		

a. Construct a normal probability plot for these data. Is there any evidence to suggest non-normality?
b. Use the backward empirical rule to check for evidence of non-normality.

6.76 Education and Child Development Some people claim children who practice yoga are more physically fit, self-confident, and self-aware. A random sample of pre-teens (ages 10–12) practicing yoga was obtained and their meditation (or quiet breathing) times (in minutes) per day were recorded. Use the four methods presented in this section to determine whether there is any evidence to suggest the data are from a non-normal population. **YOGA**

6.77 Sports and Leisure Many NBA players express themselves on the court through their sneakers. High-school basketball players often insist on wearing the same expensive sneakers worn by their favorite player. A random sample of sneakers worn by Notre Dame Prep (in Massachusetts) basketball players was obtained, and the retail price (in dollars) for each pair of sneakers is given in the table below: **SNEAKERS**

96.70	112.05	120.70	106.40	86.60
126.40	134.75	76.75	142.20	116.70

Use the four methods presented in this section to determine whether there is any evidence to suggest the data are from a non-normal population.

6.78 Public Policy and Political Science Bicycle paths are usually planned and constructed according to certain guidelines (for example, the American Association of State Highway and Transportation Officials guidelines for construction). There are construction standards for width, offset from the road, maximum grade, and horizontal and vertical clearances. A random sample of bicycle-path widths (in feet) was obtained. **BIKEPATH**
   a. Find the sample mean and the sample standard deviation for these data.
   b. Compute the intervals $(\bar{x} - s, \bar{x} + s)$, $(\bar{x} - 2s, \bar{x} + 2s)$, and $(\bar{x} - 3s, \bar{x} + 3s)$.
   c. Find the proportion of observations in each interval in part (b). Is there any evidence to suggest the data are from a non-normal population?

6.79 Physical Sciences Near-Earth objects (NEOs) are comets and asteroids that have entered the Earth's *neighborhood*. Between March 4 and March 10, 2013, four asteroids were in our neighborhood; the largest was approximately 600,000 miles away. The National Aeronautics and Space Administration maintains a list of NEOs deemed to have the potential to collide with Earth. The estimated absolute magnitude (H, related to diameter) of several of these objects is given on the text website.[20] Use the methods presented in this section to determine whether there is any evidence to suggest the data are from a non-normal population. **NEOS**

6.80 Sports and Leisure A typical round of golf takes approximately 3.5 hours (walking, without an electric cart). Many weekend golfers take more time as a result of lost golf balls, thinking about certain shots, and talking to other players. The manager at the Wawona Golf Course in Wawona, California, obtained a random sample of round times (in hours), and the data are reported in the table below. **GOLF**

3.86	4.92	4.15	3.83	4.34	4.56	4.24	4.33
4.36	4.09	4.30	4.23	4.28	4.63	4.34	3.73
4.66	4.40	4.45	4.42				

   a. Construct a stem-and-leaf plot for these data.
   b. Compute the ratio IQR/s.
   c. Construct a normal probability plot for these data.
   d. Is there any evidence to suggest these data are from a non-normal population? Justify your answer.

6.81 Marketing and Consumer Behavior The predominant acid in frozen concentrated orange juice (FCOJ) is citric acid, and the amount is usually given as a percentage. Degrees Brix is a measure of the total soluble solids in FCOJ and is also a percentage. The Brix/acid ratio is computed by simple division, and 12 is considered an ideal ratio. A random sample of FCOJ was obtained from different sellers, and the Brix/acid ratio was measured for each. These measurements were used to construct the following normal probability plot:

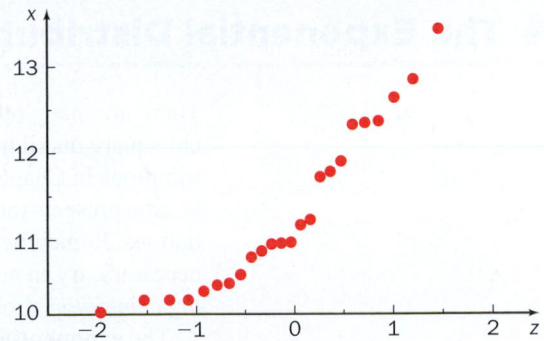

Does this plot suggest the data are from a non-normal distribution? Justify your answer.

6.82 Biology and Environmental Science There are approximately 1500 black bears in Great Smoky Mountains National Park. Park visitors are warned that these animals are dangerous and unpredictable, and that it is illegal to disturb or displace a black bear. To track and protect these animals, suppose a random sample of male black bears was obtained and the weight of each (in pounds) was recorded.[21] Use the four methods presented in this section to determine whether there is any evidence to suggest the data are from a non-normal population. **BEARS**

6.83 Manufacturing and Product Development Pig iron is a mixture of iron ore, charcoal from coal, and limestone, melted together under very high pressure. This type of iron is usually refined and used to produce wrought iron, cast iron, or steel. A random sample of the production and shipment of pig iron (in 100 metric tonnes) per month in Canada was recorded.[22] Use the four methods presented in this section to determine whether there is any evidence to suggest the data are from a non-normal population. **PIGIRON**

Challenge

6.84 Normal Scores Generate 500 random samples of size 10 from a standard normal distribution. Order each sample from smallest to largest. Find $\bar{x}_{(1)}$, the sample mean of the 500 smallest values from each sample. Consider the next largest value in each sample. Find $\bar{x}_{(2)}$, the sample mean of these 500 values. Continue in this manner to find $\bar{x}_{(3)}, \bar{x}_{(4)}, \dots,$ and $\bar{x}_{(10)}$, the mean of the 500 largest values from each sample. Compare these 10 sample means, $\bar{x}_{(1)}, \bar{x}_{(2)}, \dots, \bar{x}_{(10)}$ with the standardized normal scores for $n = 10$ in Table IV in the Appendix.

Try a similar procedure for $n = 20$. Generate 500 random samples of size 20 from a standard normal distribution. Order each sample from smallest to largest. Find $\bar{x}_{(1)}, \bar{x}_{(2)}, \ldots, \bar{x}_{(20)}$ and compare these values with the standardized normal scores for $n = 20$ in Table IV in the Appendix.

Explain why these sample means should be good estimates of the standardized normal scores.

Generate 500 random observations from a normal distribution with mean 50 and standard deviation 10. Arrange the observations in order from smallest to largest and denote this ordered list $x_{(1)}, x_{(2)}, \ldots, x_{(500)}$.

Form the ordered pairs $(x_{(1)}, 1/500), (x_{(2)}, 2/500), \ldots,$ $(x_{(i)}, i/500), \ldots, (x_{(500)}, 1)$. Plot these points in a rectangular coordinate system and describe the shape of the graph. Which curve does this graph approximate? Why?

6.4 The Exponential Distribution

There are many other common continuous distributions, for example, the t distribution, chi-square distribution, and F distribution. We will learn a little about each of these distributions in Chapters 8 and 9 as we study confidence intervals and hypothesis tests. This section presents the exponential distribution, which is related to several continuous distributions. Remember that probability in a continuous world is area under the curve, and if necessary, try to convert any probability statement into an equivalent expression involving *cumulative* probability.

The **exponential probability distribution** is often used to model the time to failure of an electronic part, or the waiting time between events. This distribution is completely characterized by one parameter, λ.

The Exponential Probability Distribution

Suppose X is an exponential random variable with parameter λ (with $\lambda > 0$). The probability density function is given by

$$f(x) = \begin{cases} \lambda e^{-\lambda x} & \text{if } x \geq 0 \\ 0 & \text{otherwise} \end{cases} \tag{6.9}$$

A CLOSER LOOK

1. The symbol e in Equation 6.9 is the base of the natural logarithm ($e \approx 2.71828$). The constant e is also used in the Poisson distribution and the normal distribution.

2. If the exponential distribution is used to model the lifetime of a light bulb, machine, or even a human being, then λ represents the failure rate.

Notice that $f(x) = \lambda$ when $x = 0$ (because $e^0 = 1$).

3. The exponential distribution has positive probability only for $x \geq 0$. Figures 6.68 and 6.69 show the graphs of a general probability density function for an exponential random variable and a probability density function with $\lambda = 2$.

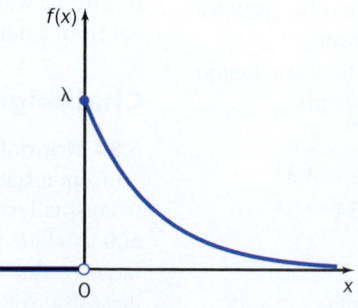

Figure 6.68 A graph of the probability density function for an exponential random variable with parameter λ.

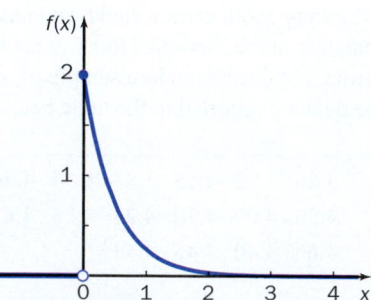

Figure 6.69 A graph of the probability density function for an exponential random variable with $\lambda = 2$.

4. The mean and variance for an exponential random variable, X, with parameter λ are

$$E(X) = \mu = \frac{1}{\lambda} \qquad \sigma^2 = \frac{1}{\lambda^2} \tag{6.10}$$

Probabilities associated with an exponential random variable with parameter λ are computed using cumulative probability. We do not need a table for these calculations! Figure 6.70 illustrates cumulative probability associated with an exponential random variable. Remember, there is no area (or probability) for $x < 0$.

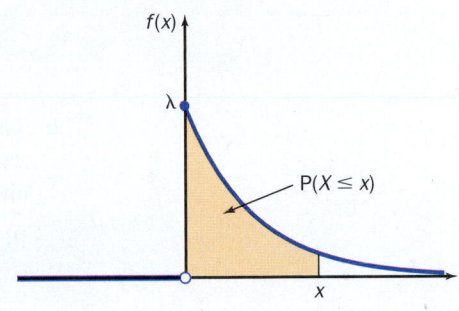

Figure 6.70 A graph of the probability density function for an exponential random variable. The shaded area corresponds to $P(X \leq x)$.

The formula for cumulative probability is given by

$$P(X \leq x) = \begin{cases} 0 & \text{if } x < 0 \\ 1 - e^{-\lambda x} & \text{if } x \geq 0 \end{cases} \tag{6.11}$$

If $x \geq 0$, the probability that X assumes a value greater than x is a right-tail probability, and is given by

$$P(X > x) = 1 - P(X \leq x) = 1 - (1 - e^{-\lambda x}) = 1 - 1 + e^{-\lambda x} = e^{-\lambda x} \tag{6.12}$$

The following example illustrates the use of cumulative probability and the formula for right-tail probability to find probabilities associated with an exponential random variable.

Example 6.10 Relief from the Common Cold

Some pharmacists recommend Zicam to relieve many symptoms due to the common cold. After the prescribed dose is taken, suppose the length of time (in hours) until symptoms return is a random variable, X, that has an exponential distribution with parameter $\lambda = 0.1$.

a. Carefully sketch a graph of the probability density function for X. Find the mean, variance, and standard deviation of X.

b. What is the probability that the length of time until symptoms return is less than the mean?

c. What is the probability that the length of time until symptoms return is at least 12 hours?

d. What is the probability that the length of time until symptoms return is between 8 and 16 hours?

SOLUTION

a. Use Equation 6.9 to sketch the graph (Figure 6.71) and Equation 6.10 to compute the mean, variance, and standard deviation.

$$\mu = \frac{1}{\lambda} = \frac{1}{0.1} = 10$$

$$\sigma^2 = \frac{1}{\lambda^2} = \frac{1}{0.1^2} = 100 \qquad \sigma = \sqrt{\sigma^2} = \sqrt{100} = 10$$

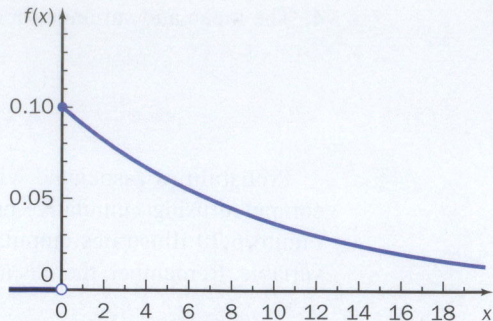

Figure 6.71 A graph of the probability density function for an exponential random variable with $\lambda = 0.1$.

b. The length of time until symptoms return is modeled by an exponential random variable ($\lambda = 0.1$). The mean is 10. Translate the question into a probability statement, and use cumulative probability where appropriate.

$P(X < 10)$ Translation to a probability statement.

$= 1 - e^{-0.1(10)}$ Use the formula for cumulative probability.

$= 1 - e^{-1}$ Simplify.

$= 1 - 0.3679 = 0.6321$

The probability that the length of time until symptoms return is less than the mean is 0.6321. Figure 6.72 illustrates this probability.

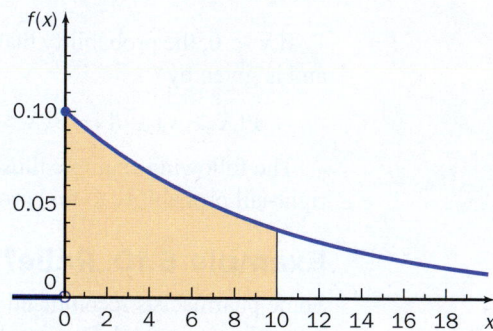

Figure 6.72 The shaded region represents the probability that the symptoms return in less than 10 hours.

c. Translate the question into a probability statement, and use cumulative probability where appropriate. *At least 12 hours* means 12 hours or more.

$P(X \geq 12)$ Translation to a probability statement.

$= e^{-0.1(12)}$ Use the formula for a right-tail probability.

$= e^{-1.2} = 0.3012$ Simplify.

The probability the symptoms will return in 12 hours or more is 0.3012.

d. Translate the question into a probability statement, and use cumulative probability where appropriate.

$P(8 \leq X \leq 16)$ Translation to a probability statement.

$= P(X \leq 16) - P(X < 8)$ Use cumulative probability.

$= [1 - e^{-0.1(16)}] - [1 - e^{-0.1(8)}]$ Formula for cumulative probability.

$= 0.7981 - 0.5507 = 0.2474$

The probability that the symptoms will return in between 8 and 16 hours is 0.2474. Figures 6.73 through 6.75 show technology solutions.

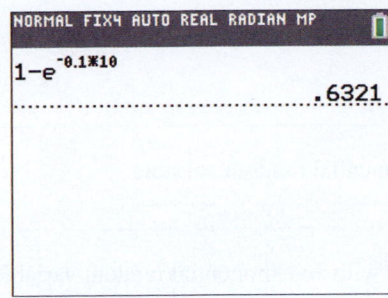

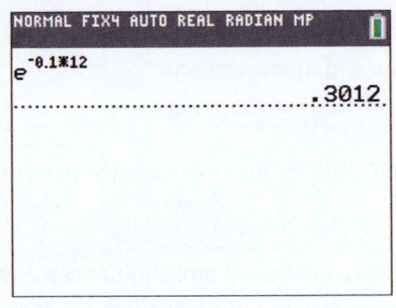

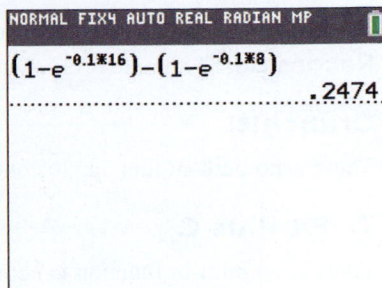

Figure 6.73 P(X < 10): Use the formula for cumulative probability.

Figure 6.74 P(X ≥ 12): Use the formula for right-tail probability.

Figure 6.75 P(8 ≤ X ≤ 16): Compute each cumulative probability and subtract.

TRY IT NOW   GO TO EXERCISE 6.93

Example 6.11 Heat-Pump Lifetime

Christian Delbert/Shutterstock

A Carrier XH heat pump, designed to heat a home in the winter and cool the home in the summer, lasts, on average, 16 years. The lifetime (in years) of this system can be modeled by an exponential random variable, X, with $\lambda = 0.0625$. Suppose a heat pump is selected at random.

a. What is the probability that the heat pump will last for at least five years?

b. Suppose the heat pump lasts for five years. What is the probability that it will last for at least another five years?

SOLUTION

a. Translate the question into a probability statement. Use cumulative probability or right-tail probability where appropriate. *At least five years* means five years or longer.

$$P(X \geq 5) \qquad \text{Translation to a probability statement.}$$

$$= e^{-0.0625(5)} = 0.7316 \qquad \text{Right-tail probability.}$$

The probability that the heat pump lasts for at least five years is 0.7316.

b. Use the definition of conditional probability, and cumulative or right-tail probability where appropriate.

$$P(X \geq 5 + 5 \mid X \geq 5) \qquad \text{Translation to a probability statement.}$$

$$= \frac{P[(X \geq 10) \cap (X \geq 5)]}{P(X \geq 5)} \qquad \text{Definition of conditional probability.}$$

$$= \frac{P(X \geq 10)}{P(X \geq 5)} \qquad (X \geq 10) \cap (X \geq 5) = X \geq 10.$$

$$= \frac{e^{-0.0625(10)}}{e^{-0.0625(5)}} = \frac{0.5353}{0.7316} \qquad \text{Right-tail probabilities.}$$

$$= 0.7316$$

This *conditional* probability of lasting an additional five years is the same as the *unconditional* probability of lasting an initial five years. According to this model, the fact that it has lasted five years does not affect the probability of it lasting an additional five years. This unrealistic result is called the *memoryless property* of an exponential random variable. At any point in time, the exponential random variable *forgets* or *ignores* what happened earlier: "The future is independent of the past."

TRY IT NOW   GO TO EXERCISE 6.100

Solution Trail 6.11b

KEYWORDS
- Suppose the heat pump lasts for five years

TRANSLATION
- Given that the heat pump lasts for five years

CONCEPTS
- Conditional probability

VISION
In formulating the conditional probability statement, remember that X is a continuous random variable. The events, stated in terms of X, must describe an interval. Given that the heat pump lasts for five years means it lasts for *at least* five years.

Technology Corner

Procedure: Compute probabilities associated with an exponential random variable.
Reconsider: Example 6.11, solution, and interpretations.

CrunchIt!

There is no built-in function to compute probabilities associated with an exponential random variable.

TI-84 Plus C

There is no built-in function to compute (cumulative) probabilities associated with an exponential random variable. Use the formulas for cumulative probability and right-tail probability.

1. $P(X < 10)$: Use the formula for cumulative probability. Refer to Figure 6.73.
2. $P(X > 12)$: Use the formula for right-tail probability. Refer to Figure 6.74.
3. $P(8 \leq X \leq 16)$: Compute each cumulative probability and subtract. Refer to Figure 6.75.

Minitab

There are built-in functions to compute (strict) cumulative probability and inverse cumulative probability. These functions may be accessed through a graphical input window or by using the command language.

1. In a session window, use the function CDF to find $P(X < 10)$, or select Calc; Probability Distributions; Exponential. Choose Cumulative probability and enter the Scale = μ = 10 and Threshold = 0. Select Input constant (x) and enter 10. See Figure 6.76. Click OK. The cumulative probability is displayed in the session window. See Figure 6.77.

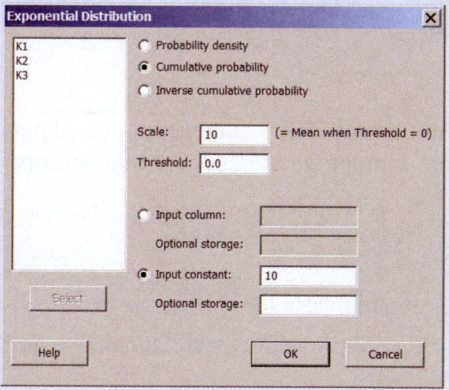

Figure 6.76 Graphical input window.

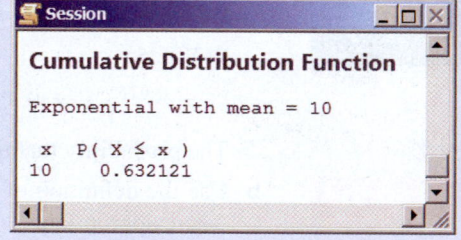

Figure 6.77 Cumulative distribution function output.

2. In a session window, use the function CDF to find $P(X \leq 12)$. Use the complement rule to find $P(X \geq 12)$. See Figure 6.78.

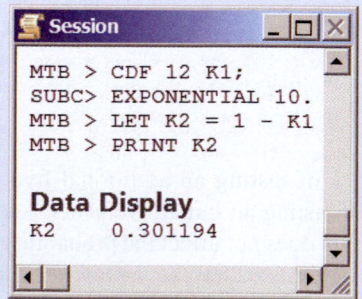

Figure 6.78 $P(X \geq 12)$.

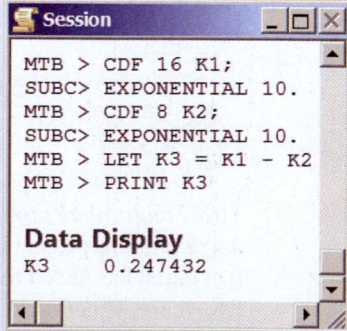

Figure 6.79 $P(8 \leq X \leq 16)$.

3. Use the function CDF to find P($X \leq 16$). Note that Minitab uses the mean to completely describe an exponential random variable. Store the result in K1. Use the function CDF to find P($X \leq 8$). Store the result in K2. Find the difference, and print the result. See Figure 6.79.

Excel

The built-in function EXPON.DIST may be used to compute (strict) cumulative probability.

1. Use the function EXPON.DIST to find P($X < 10$). The arguments for this function are x, λ, and the logical value True to compute cumulative probability. See Figure 6.80.
2. Use the function EXPON.DIST to find P($X \leq 12$). Use the complement rule to find P($X \geq 12$). See Figure 6.80.
3. Use the function EXPON.DIST to find P($X \leq 16$). Use the function EXPON.DIST to find P($X \leq 8$). Find the difference. See Figure 6.82.

A	B
0.6321	= EXPON.DIST(10,0.1,TRUE)

Figure 6.80 P($X < 10$): Use the function for cumulative probability.

A	B
0.6988	= EXPON.DIST(12,0.1,TRUE)
0.3012	= 1 - A1

Figure 6.81 P($X \geq 12$): Use the function for cumulative probability and the complement rule.

A	B
0.7981	= EXPON.DIST(16,0.1,TRUE)
0.5507	= EXPON.DIST(8,0.1,TRUE)
0.2474	= A1 - A2

Figure 6.82 P($8 \leq X \leq 16$).

SECTION 6.4 EXERCISES

Concept Check

6.85 True/False For any exponential random variable, the mean is equal to the standard deviation.

6.86 True/False An exponential random variable can only take on values greater than or equal to 0.

6.87 Short Answer Suppose X is an exponential random variable with parameter λ.
   a. P($X \leq x$) =
   b. P($X > x$) =

Practice

6.88 Suppose X is an exponential random variable with $\lambda = 0.2$.
   a. Carefully sketch a graph of the density curve for X.
   b. Find the mean, variance, and standard deviation of X.
   c. Find P($X < 3$).
   d. Find P($1 < X < 9$).

6.89 Suppose X is an exponential random variable with $\lambda = 0.25$.
   a. Carefully sketch a graph of the density curve for X.
   b. Find P($0.01 \leq X \leq 0.05$).
   c. Find P($X > 0.06$).

6.90 Suppose X is an exponential random variable with $\lambda = 0.025$.
   a. Find P($X > 30$) and illustrate this probability using an appropriate density curve.

   b. Find P($X > 20$).
   c. Find P($X > 50 \mid X \geq 30$)

6.91 Suppose X is an exponential random variable and P($X \leq 20$) = 0.7981. Find the value of λ.

6.92 Suppose X_i ($i = 1, 2, 3, 4$) is an exponential random variable with $\lambda = i/10$. If the X_i's are independent, find the probability that the values of all four random variables are less than 1.

Applications

6.93 Public Policy and Political Science *The Weekly Standard* is a politically conservative magazine (and website), described in ads as humorous and witty. This magazine is read by some of the most powerful men and women in government, politics, and the media. According to a survey conducted by magazine staff, the mean amount spent during the past 12 months for personal or vacation travel is $6159. Suppose the distribution of the amount spent by a subscriber during the past 12 months for personal or vacation travel can be modeled by an exponential random variable.
   a. Find the value of λ.
   b. Find the probability that a randomly selected subscriber spent more than $8000 for personal or vacation travel during the past 12 months.
   c. Find the probability that a randomly selected subscriber spent between $5000 and $10,000 for personal or vacation travel during the past 12 months.

d. If a subscriber spent less than $500 for personal or vacation travel during the past 12 months, the circulation department will no longer hand-deliver each issue. What is the probability that a randomly selected subscriber will no longer receive hand-delivered issues?

6.94 Fuel Consumption and Cars The purpose of an automobile's timing belt is to provide a connection between the camshaft and the crankshaft. This allows the valves to open and close in sync with the pistons. Suppose the duration of a timing belt (in miles) can be modeled by an exponential random variable with mean 100,000 miles.
  a. Find the value of λ.
  b. Find the probability that a randomly selected timing belt lasts for more than 120,000 miles.
  c. Find the probability that a randomly selected timing belt lasts for between 75,000 and 125,000 miles.
  d. If the timing belt on a new car breaks within 70,000 miles, the dealer will install a new belt free of charge. What is the probability the dealer will be forced to install a new timing belt free of charge on a randomly selected car?

6.95 Business and Management Suppose the dentists at the Eastin Center for Modern Dentistry in Coeur D'Alene, Idaho, never see a patient on time. The amount of time (in minutes) past the appointment time a random patient must wait to see a dentist has an exponential distribution with $\lambda = 0.01$. Suppose a patient is selected at random, and the appointment time has passed.
  a. What is the probability that the dentist will see the patient within five minutes after the appointment time? Write a Solution Trail for this problem.
  b. Find the probability that the dentist will see the patient between 10 and 20 minutes after the appointment time.
  c. Find the probability that the patient will have to wait at least 30 minutes past the appointment time.

6.96 Psychology and Human Behavior In Columbia, South Carolina, on a randomly selected Friday night between 9:00 P.M. and 2:00 A.M., the time (in minutes) between calls to a 911 dispatcher has an exponential distribution with $\lambda = 0.125$. Suppose a Friday evening in Columbia is selected at random.
  a. If a 911 call is taken at 10:07 P.M., what is the probability that the next call will occur before 10:30 P.M.?
  b. If a 911 call is taken at 11:30 P.M., what is the probability that the next call will occur after 11:45 P.M.?
  c. Find the mean, variance, and standard deviation for the time between 911 calls.
  d. Find a value t such that if a 911 call is received between 9:00 P.M. and 2:00 A.M., then the probability that the next 911 call will occur more than t minutes later is 0.75.

6.97 Public Policy and Political Science During the Cold War there were frequent radio tests of alert warnings. The regular program was interrupted and replaced by a long high-pitched signal. An announcer would break in with a message similar to, "This has been a test of the emergency broadcasting system." Suppose the time (in hours) between these tests had an exponential distribution with mean 24 hours.

  a. If a test occurred at 6:00 A.M., what is the probability that another one would occur before 6:00 P.M.?
  b. If a test occurred at 10:00 P.M., what is the probability that another one would not occur until after 6:00 A.M.?
  c. If a test occurred at 9:00 A.M., what is the probability that the next test would occur between 12:00 P.M. and 1:00 P.M.?

Extended Applications

6.98 Public Health and Nutrition The U.S. Public Health Service's Advisory Committee on Immunization Practices recommends a tetanus booster every 10 years. Suppose the lifetime (in years) of a tetanus booster shot has an exponential distribution with $\lambda = 0.05$.
  a. Find the probability that a randomly selected tetanus booster shot is still protecting against tetanus after more than 10 years.
  b. Suppose a physician recommends a booster shot after a period when only 10% of all tetanus shots still have an effect. How long should you wait before getting another booster shot?
  c. Suppose two people independently receive tetanus shots. What is the probability that both shots still have an effect after more than five years?

6.99 Medicine and Clinical Studies A patient's blood cholesterol level is often checked during a routine physical examination. A blood sample is taken (from the finger or arm) and tested for total cholesterol and HDL (high-density lipoprotein, the "good cholesterol") cholesterol levels in milligrams per deciliter (mg/dL). If total cholesterol is less than 200 mg/dL, then no action is taken. If a patient's total cholesterol is approximately 300 mg/dL, a new drug together with a strict diet is prescribed to reduce this number to a safe level. The amount of time (in days) it takes to reduce total cholesterol to a safe level has an exponential distribution with $\lambda = 1/15$. A patient with total cholesterol of 300 mg/dL is randomly selected and placed on this drug-and-diet regimen.
  a. Find the mean number of days until the total cholesterol level is safe (less than 200 mg/dL).
  b. Find the probability that the total cholesterol level will be safe in a number of days within two standard deviations of the mean.
  c. Find a value d such that 75% of all such patients have a safe cholesterol level with d days.
  d. Suppose two patients are selected at random. Find the probability that this drug-and-diet regimen works for at least one of the patients within 10 days.

6.100 Marketing and Consumer Behavior A toy manufacturer routinely tests new toys in a controlled environment before deciding whether to actually market a toy. Research has shown that the amount of time (in minutes) a randomly selected child plays with a new toy has an exponential distribution with $\lambda = 0.05$. Suppose a new toy is presented for study.
  a. What is the probability that a randomly selected child will play with the toy for at most 10 minutes?

b. What is the probability that a randomly selected child will play with the toy for between 5 and 20 minutes?

c. Suppose a child plays with the toy for at least 15 minutes. What is the probability that the child will play with the toy for another 20 minutes?

d. If four different children each play independently with a new toy for at least 25 minutes, then the toy is immediately brought to market. What is the probability of this happening for a newly designed toy?

6.101 Physical Sciences Four large water pumps supply Bellingham, Washington, with water. If water pump i ($i = 1, 2, 3, 4$) breaks down, then the time to repair it has an exponential distribution with $\lambda = 1/(2i)$ hours. Suppose the water pumps operate independently, and when breakdowns occur, the repair times are also independent.

a. Suppose water pump 1 breaks down. What is the probability that it will take more than 30 minutes to repair?

b. Answer part (a) for water pumps 2, 3, and 4.

c. Suppose all four water pumps break simultaneously. What is the probability that at least one of the four water pumps will not be repaired within one hour?

6.102 Public Health and Nutrition From the instant a fresh-baked chocolate-chip cookie is taken out of the oven, the time (in minutes) the wonderful aroma lasts has an exponential distribution with $\lambda = 1/30$. Suppose a chocolate-chip cookie is done baking and is taken out of the oven.

a. Find the mean, variance, and standard deviation for the time the aroma lasts.

b. What is the probability that the aroma lasts for at least 40 minutes?

c. What is the probability that the aroma lasts for between 30 and 50 minutes?

d. Find a value t such that the probability that the aroma lasts for at most t minutes is 0.90.

e. Suppose a second batch of cookies is taken out of the oven 10 minutes after the first. What is the probability that there will be no aroma from either batch 35 minutes after the first batch was done?

6.103 Psychology and Human Behavior Sensory memory in humans is very short-term and lasts approximately 200–500 milliseconds after an individual perceives an object.[23] Suppose the length of time (in milliseconds) an object remains in sensory memory for a randomly selected adult has an exponential distribution with $\lambda = 0.003$. A flashing grid of letters is displayed to a randomly selected adult and the sensory memory is measured.

a. What is the probability that the sensory memory will last for at most 200 milliseconds?

b. Find the median time the sensory memory will last.

c. Find a time t such that only 5% of all adults' sensory memory lasts at least t milliseconds.

Challenge

6.104 Memoryless Property Suppose X is an exponential random variable with parameter λ. If a and b are constants (>0), confirm the *memoryless property* by showing that $P(X \geq a) = P(X \geq a + b \mid X \geq b)$.

CHAPTER 6 SUMMARY

Concept	Page	Notation / Formula / Description
Probability distribution for a continuous random variable	244	A smooth curve (density curve) defined such that the probability X takes on a value between a and b is the area under the curve between a and b.
Uniform distribution	245	If X has a uniform distribution on the interval $[a, b]$, then $$f(x) = \begin{cases} \dfrac{1}{b-a} & \text{if } a \leq x \leq b \\ 0 & \text{otherwise} \end{cases} \qquad \mu = \frac{a+b}{2}, \sigma^2 = \frac{(b-a)^2}{12}$$
Normal distribution	256	If X is a normal random variable with mean μ and variance σ^2, then the probability density function is given by $$f(x) = \frac{1}{\sigma\sqrt{2\pi}} e^{-(x-\mu)^2/2\sigma^2}$$
Standard normal distribution	257	A normal distribution with $\mu = 0$ and $\sigma^2 = 1$ is the standard normal distribution. The standard normal random variable is usually denoted by Z: $Z \sim N(0, 1)$.
Standardization	261	If X is a normal random variable with mean μ and variance σ^2, then $$Z = \frac{X - \mu}{\sigma}$$ is a standard normal random variable.

| Normal probability plot | 272 | A scatterplot of each observation versus its corresponding expected value from a Z distribution. For a normal distribution, the points will fall along a straight line. |
| Exponential distribution | 282 | If X is an exponential random variable with parameter λ, then the probability density function is given by |

$$f(x) = \begin{cases} \lambda e^{-\lambda x} & \text{if } x \geq 0 \\ 0 & \text{otherwise} \end{cases} \qquad \mu = \frac{1}{\lambda}, \sigma^2 = \frac{1}{\lambda^2}$$

CHAPTER 6 EXERCISES

6 APPLICATIONS

6.105 Technology and the Internet In the advertising department of SBC Communications, Inc., all of the computers are part of a local area network, connected to one main printer. When an advertising employee prints a document, the job is placed in a queue. It may take several minutes before the document begins to print, due to the number of other print jobs and the complexity of the document. The time (in minutes) until the document starts to print has an exponential distribution with $\lambda = 0.40$. Suppose a randomly selected document is sent to the main printer.

a. What is the mean time until the document begins to print?

b. What is the probability that the document will begin to print within 30 seconds?

c. What is the probability that the document will need more than 5 minutes before starting to print?

d. Find a value t such that only 2% of all documents take at least t minutes before starting to print.

6.106 Public Health and Nutrition The amount of sodium in a randomly selected eight-ounce serving of chicken noodle soup has a normal distribution with mean $\mu = 1737$ mg and standard deviation $\sigma = 150$ mg.[24] Suppose an eight-ounce serving is randomly selected.

a. What is the probability that the amount of sodium is more than 1500 mg?

b. What is the probability that the amount of sodium is between 1700 and 2000 mg?

c. Find a symmetric interval about the mean such that 90% of all eight-ounce servings have amounts of sodium in that interval.

d. Suppose a randomly selected eight-ounce serving has a sodium level of 2100 mg. Is there any evidence to suggest the mean sodium level reported above is false?

6.107 Public Health and Nutrition The Food and Drug Administration (FDA) reviews all advertisements for drugs to check for omissions regarding a drug's risk; inadequate, incorrect, or inconsistent labeling; misleading claims; unsupported comparative claims; and unapproved purposes. Lengthy legal reviews of advertisements have increased the total review time. Suppose the length of time between a request for a review and final approval has a normal distribution with mean $\mu = 21$

days and $\sigma = 4$ days. Consider a randomly selected advertisement submitted to the FDA for review.

a. What is the probability that the advertisement will be reviewed in less than 14 days?

b. What is the probability that the advertisement will be reviewed in between 15 and 19 days?

c. Suppose the advertisement takes at least 20 days for review. What is the probability that it will take less than 30 days for review?

d. Suppose two independent advertisements are submitted for review simultaneously. What is the probability that both will take more than 30 days for review?

6.108 Marketing and Consumer Behavior Canister vacuums are often rated according to their ease of use, noise level, emissions, cleaning ability, and length of the power cord. The length (in feet) of the electric cord on a canister vacuum is a random variable, X, with a uniform distribution on the interval $[20, 30]$. Suppose a canister vacuum is selected at random.

a. Carefully sketch a graph of the probability density curve for the random variable X.

b. What is the probability that the power cord has a length less than 22 feet?

c. What is the probability that the power cord has a length greater than 26 feet?

d. Find a value f such that 75% of all power cords have lengths greater than f feet.

6.109 Biology and Environmental Science A random sample of the rice production (in 1000 tons) per year in various provinces in China is given.[25] Is there any evidence to suggest that the data are from a non-normal population? **RICE**

6.110 Business and Management The manager for the Wingate Inn claims the time it takes to make a room reservation over the phone is approximately normally distributed with mean four minutes and standard deviation 45 seconds. Suppose a call placed to the inn to make a room reservation is selected at random.

a. What is the probability that it will take less than three minutes to make the reservation?

b. What is the probability that it will take between $3\frac{1}{2}$ and $4\frac{1}{2}$ minutes to make the reservation?

c. Find the first and the third quartile times.

d. Suppose it takes seven minutes to make the reservation. Is there any evidence to suggest the inn's claim ($\mu = 4$ minutes) is false? Justify your answer.

6.111 Fuel Consumption and Cars   The Fiat 500E is an electric car that can travel approximately 80–100 miles on a full charge.[26] The time it takes to fully recharge depends on the percent depleted, but is approximately normal with mean 2.2 hours and standard deviation 0.6 hour (on a 240-volt line). Suppose a random Fiat 500E is recharged.

a. What is the probability that the amount of time to fully recharge is between 1 and 2 hours?

b. What is the probability that the amount of time to fully recharge is more than 3 hours?

c. If the amount of time to fully recharge is less than 30 minutes, the car could still travel another 50 miles. What is the probability that the car could travel another 50 miles?

d. Suppose three Fiat 500E's are selected at random. What is the probability that all three will have a time to fully recharge greater than 2.5 hours?

6.112 Manufacturing and Product Development   Oriental rugs are made from various wools and woven in several different countries. The pile height (in millimeters) of an Oriental rug varies slightly and can be modeled by a uniform random variable on the interval [6, 10].

a. Carefully sketch a graph of the density function for pile height.

b. What is the probability that a randomly selected Oriental rug will have pile height less than 7 mm?

c. What is the probability that a randomly selected Oriental rug will have pile height between 8.5 and 9.5 mm?

d. Find a value h such that 90% of all Oriental rugs have pile height less than h.

6.113 Technology and the Internet   The life expectancy of a laser-printer toner cartridge varies considerably according to the toner and drum type, and how the cartridge is used. Printed pages containing lots of graphics require more toner, while pages with mostly text require considerably less toner. Page coverage is usually measured as a proportion (or percentage). For example, a typical text page has approximately 0.05 coverage. A random sample of printed pages from an office printer was obtained, and the page coverage was carefully measured. Is there any evidence to suggest the data are from a non-normal population?   📊 **TONER**

6.114 Sports and Leisure   Jockeys in the United States and England work very hard to keep their weight down. Many participate in weight-loss programs, carefully monitor their diet, and exercise regularly. The weight of a male jockey is approximately normal with mean 52 kg and standard deviation 1.2 kg. Suppose a male jockey is randomly selected.

a. What is the probability that the jockey weighs more than 53 kg?

b. What is the probability that the jockey weighs between 50 and 54 kg?

c. Find a value w such that 80% of all male jockeys weigh more than w.

d. Suppose the jockey selected weighs 57 kg. Is there any evidence to suggest the claimed mean (52 kg) is wrong? Justify your answer.

6.115 Public Health and Nutrition   The amount of caffeine in a cup of coffee varies considerably, even if it is brewed by the same person, using the same brewing method and ingredients. Suppose the amount of caffeine in an eight-ounce cup of Maxwell House coffee is approximately normally distributed with mean 120 mg and standard deviation 7.75 mg.[27]

a. What is the probability that a randomly selected cup of Maxwell House coffee will have less than 110 mg of caffeine?

b. What is the probability that a randomly selected cup of Maxwell House coffee will have between 115 and 130 mg of caffeine?

c. A recent article in a medical journal suggests that an eight-ounce cup of coffee with more than 140 mg of caffeine could cause a person's heart to race. What is the probability that a randomly selected cup will have more than 140 mg of caffeine?

6.116 Manufacturing and Product Development   The quality of a kitchen knife is often measured by the sharpness and total lifetime of the blade. One test for sharpness involves mounting the knife with the blade vertical and lowering a specially designed pack of paper onto the blade. The sharpness is measured by the depth of the cut. A greater depth indicates a sharper knife. Suppose the depth of the cut for a randomly selected knife has a normal distribution with mean 92 mm and standard deviation 21 mm.

a. What is the probability that a randomly selected kitchen knife has a sharpness measure less than 75 mm?

b. A kitchen knife with a sharpness measure of at least 100 mm qualifies as a steak knife. What proportion of kitchen knives are steak knives?

c. The Kitchen Gadgets Association would like to set a maximum sharpness for butter knives. Find a value c such that 15% of all knives have sharpness less than c.

d. Suppose a randomly selected kitchen knife has sharpness greater than 90 mm. What is the probability that it has sharpness greater than 100 mm?

6.117 Economics and Finance   Some of the variables that affect the monthly payment of a new-car loan are the total amount borrowed, the interest rate, and the length of the loan. During the fourth quarter of 2012, the mean length of a new-car loan was 65 months, a record high according to Experian.[28] Suppose the length of a new-car loan is approximately normal with standard deviation nine months.

a. What is the probability that a new-car loan is for at most 48 months?

b. What is the probability that a new-car loan length is between 50 and 70 months?

c. Find a symmetrical interval about the mean such that 95% of all new-car loan lengths fall in this interval.

d. Suppose the amount borrowed on a new-car loan is also approximately normal with mean $20,000 and standard deviation $5000. If the length of the loan and the amount borrowed are independent, what is the probability that the loan will be for more than $27,000 and for less than 60 months?

6.118 Fuel Consumption and Cars   Even though automobiles are becoming more fuel-efficient, the cost of gasoline is still taking a huge chunk of our personal income. According to the U.S. Energy Information Administration, the mean amount spent on gasoline in 2012 was $2912.[29] A random sample of U.S. households was obtained, and each was asked for the amount (in dollars) spent on gasoline during the last year. Is there any evidence to suggest that the data are from a non-normal distribution? 🔢 GASCOST

6.119 Physical Sciences   Large reservoirs of oil found underground are under very high pressure, which allows the oil to be pumped to the surface. All oil fields contain some water, and as water is pumped back into a well to maintain high pressure, the water content increases. Suppose the proportion of water in a randomly selected barrel of oil pumped to the surface has a normal distribution with mean 0.12 and standard deviation 0.025.

a. What is the probability that a randomly selected barrel of oil has a proportion of water less than 0.12?
b. What is the probability that a randomly selected barrel of oil has a proportion of water between 0.15 and 0.17?
c. The higher the proportion of water in oil, the more expensive it is to separate the oil from the water. If the proportion of water is greater than 0.20, then the well is too expensive to operate and maintain. What is the probability that a randomly selected well is too expensive?

6.120 Travel and Transportation   The Washington, DC, Metro has some long and deep escalators. The longest continuous escalator is at the Wheaton (Red Line) stop. It is 230 feet long and approximately 140 feet deep.[30] For those who simply ride the escalator (without extra steps), it takes just over 2.5 minutes to complete the journey. However, because many people walk while on the escalator, the amount of time to travel the 230 feet is approximately normal with mean 1.9 minutes and standard deviation 0.3 minute. Suppose an escalator rider is selected at random.

a. What is the probability that it takes less than 2 minutes to ride the escalator?
b. What is the probability that it takes between 1.5 and 2.5 minutes to ride the escalator?
c. Suppose that five escalator riders are selected at random. What is the probability that three of the five take at least 2 minutes and 20 seconds to ride the escalator?

EXTENDED APPLICATIONS

6.121 Travel and Transportation   The U.S. Customs and Border Protection Agency maintains a table of the estimated wait times for reaching the primary inspection booth when crossing the Canada/U.S. border. The processing goal for

passenger vehicles is 15 minutes.[31] Suppose the time to cross the border at Alexandria Bay is a random variable, X, with probability density function given by

$$f(x) = \begin{cases} 0.02\,x & \text{if } 0 \le x \le 10 \\ 0 & \text{otherwise} \end{cases}$$

a. Carefully sketch a graph of the density function.
b. Find the probability that it takes less than five minutes for a randomly selected passenger vehicle to cross the border at Alexandria Bay.
c. Find the probability that it takes more than eight minutes for a randomly selected passenger vehicle to cross the border at Alexandria Bay.
d. Find the probability that it takes between two and six minutes for a randomly selected passenger vehicle to cross the border at Alexandria Bay.
e. Suppose it takes less than two minutes for a randomly selected passenger vehicle to cross the border at Alexandria Bay. What is the probability that it takes less than one minute?

6.122 Public Health and Nutrition   Meat or poultry classified as *lean* has less than four grams of saturated fat. Suppose the amount of saturated fat in a randomly selected piece of lean meat or poultry is a random variable, X, with probability density function given by

$$f(x) = \begin{cases} 0.1 & \text{if } 0 \le x < 1 \\ 0.2 & \text{if } 1 \le x < 2 \\ 0.3 & \text{if } 2 \le x < 3 \\ 0.4 & \text{if } 3 \le x < 4 \\ 0 & \text{otherwise} \end{cases}$$

a. Carefully sketch a graph of the density function.
b. What is the probability that a randomly selected piece of lean meat or poultry has less than 1.5 grams of saturated fat?
c. What is the probability that a randomly selected piece of lean meat or poultry has more than three grams of saturated fat?
d. What is the probability that a randomly selected piece of lean meat or poultry has between two and four grams of saturated fat?
e. Suppose a randomly selected piece of lean meat or poultry has at most three grams of saturated fat. What is the probability that it has at most one gram of saturated fat?

6.123 Manufacturing and Product Development
Four people working independently on an assembly line all perform the same task. The time (in minutes) to complete this task for person i ($i = 1, 2, 3, 4$) has a uniform distribution on the interval $[0, i]$. Suppose each person begins the task at the same time.

a. What is the probability that person 2 takes less than 90 seconds to complete the task?
b. What is the mean completion time for each person?
c. What is the probability that all four people complete the task in less than 30 seconds?
d. What is the probability that exactly one person completes the task in less than one minute?

6.124 Probability Calculations Using a Density Function
The probability density function for a random variable X is given by

$$f(x) = \begin{cases} -\dfrac{1}{4}x + \dfrac{1}{2} & \text{if } 0 \le x < 2 \\ -\dfrac{1}{4}x + 1 & \text{if } 2 \le x < 4 \\ 0 & \text{otherwise} \end{cases}$$

a. Carefully sketch a graph of the density function.
b. Find $P(X < 1)$.
c. Find $P(X \ge 3)$.
d. Find $P(1 < X < 3)$.

6.125 Greek Yogurt Chobani yogurt is made using a special straining technique to remove excess liquid. This process yields a thicker, creamier yogurt with more protein per serving than a regular yogurt.[32] Suppose the amount of protein in a six-ounce cup of Chobani yogurt is normal with mean 15.5 grams and standard deviation 0.8 gram. A six-ounce cup of Chobani yogurt is selected at random from the assembly line.

a. What is the probability that there are more than 16.5 grams of protein in the cup of yogurt?

b. If the amount of protein in a cup of yogurt is between 14 and 17 grams, the manufacturing process is consider to be in control. What is the probability that the process is considered in control? Out of control?
c. Suppose the cup of yogurt has less than 16 grams of protein. What is the probability that it has more than 15 grams?
d. Suppose that six cups of yogurt are selected at random. What is the probability that at least three cups have at least 16.5 grams of protein?

LAST STEP

6.126 What's better, faster or slower? For mobile phone users, data download speed is perhaps the most important characteristic of an LTE network. Faster transmission rates mean web pages, songs, and videos download faster. According to OpenSignal, the mean download speed in the United States was 9.6 Mbps. Suppose the standard deviation is 2.3 Mbps and the distribution of download speeds is approximately normal. Wireless customers will begin to complain and even switch providers if they notice a significant decrease in download speeds. Find the industry-wide complaint download speed, s, such that 2% of all download speeds are less than s.

7 Sampling Distributions

◄ **Looking Back**

- Recall that parameters such as μ, σ^2, and p completely characterize, or describe, a population.
- These parameters are constant and usually unknown, but we would like to estimate these values, or draw a conclusion about these parameters.

► **Looking Forward**

- Understand and utilize methods to find a sampling distribution (for $\overline{X}$ and for $\hat{P}$) and use it in statistical inference.
- Discover the central limit theorem, and solve probability and inference problems using this result.

How much was the typical foreclosure check?

In May 2013, millions of Americans received checks from the federal government as part of a program to resolve the housing foreclosure crisis.[1] Check recipients had contested foreclosure for a variety of reasons. For example, some claimed that the mortgage company did not follow the proper procedure and others asserted that a mistake had been made in the foreclosure process. Thirteen mortgage servicing companies had been cited for errors, misrepresentations, and "robo-signing" of documents.

The checks were in amounts from $300 to $125,000. Suppose a spokesperson from the U.S. Treasury claimed that the mean check amount was $800 with standard deviation $125.

A random sample of 40 Americans compensated under this program was obtained. The amount of each check was recorded, and the sample mean check amount for these Americans was $755. The concepts presented in this chapter will allow us to find the distribution of the sample mean, to answer probability questions about this random variable, and to make inferences about a population mean. These ideas all rely on the *central limit theorem*, the most important result in probability and statistics!

CONTENTS

7.1 Statistics, Parameters, and Sampling Distributions

The terms *parameter* and *statistic* have been used in previous chapters in intuitive contexts. The following definitions distinguish between measures represented by symbols such as μ, σ^2, and p, and the quantities used to estimate these values.

Definition

A **parameter** is a numerical descriptive measure of a population.

A **statistic** is any quantity computed from values in a sample.

A CLOSER LOOK

1. A parameter is a population quantity. It is used to describe some characteristic of a population. Usually we cannot measure a parameter; it is an unknown constant that we would like to estimate.

2. A statistic is *any* sample quantity. There are infinitely many quantities one could compute using the data in a sample. For example, $\bar{x}$ and $\tilde{x}$ are statistics, as is the sum of the smallest and the largest values divided by 2.

The difference between a parameter and a statistic is key: Parameters describe populations, and statistics describe samples. We use statistics to make inferences about parameters. Therefore, the properties of a statistic are important.

Example 7.1 Parameter Versus Statistic

In each of the following statements, identify the **boldface** number as the value of a population parameter or a sample statistic.

a. In a recent survey of Americans, **52**% of Republicans say global warming is happening.

b. A spokesman for a large insurance agency reported that the proportion of all women with some form of life insurance is **0.32**.

c. The U.S. Department of Transportation recently reported that the mean age of all highway bridges in the United States is **42** years.

d. The manager of a large hotel located near Disney World indicated that 20 selected guests had a mean length of stay equal to **5.6** days.

e. In a recent survey about the Canadian economy, **7.2**% of the individuals who responded were unemployed.

SOLUTION

a. 52% is a statistic. This number describes a characteristic of a *sample* of Republicans.

b. 0.32 is a parameter. This number describes a characteristic of the *entire* population of women.

c. 42 is a parameter. This number is a characteristic of *all* the highway bridges in the United States.

d. 5.6 is a statistic. This number describes a characteristic of the *sample* of 20 guests.

e. 7.2 is a statistic. This number describes a characteristic of the *sample* of individuals who responded to the survey.

TRY IT NOW   GO TO EXERCISE 7.9

Suppose the mean is computed for a sample of size n from a population of interest. Denote this value $\bar{x}_1$. In a second sample of size n from the same population, let $\bar{x}_2$ be the mean. It is reasonable to expect $\bar{x}_1$ and $\bar{x}_2$ to be *close to each other* but not equal. The important realization is that $\bar{x}_1$ will be *different* from $\bar{x}_2$. In fact, the sample mean will differ from sample to sample. This statistic is subject to sampling variability. Therefore, the sample mean, $\overline{X}$, is a random variable, and $\overline{X}$ has a mean, a variance, a standard deviation, and a probability distribution. This distribution is called a **sampling distribution**.

Any statistic is a random variable, because the value of the statistic differs from sample to sample. One cannot predict the value of a statistic with absolute certainty. To make a reliable inference based on a specific statistic, we need to know the properties of the distribution of the statistic.

> **Statistics are random variables!**

Definition

The **sampling distribution** of a statistic is the probability distribution of the statistic.

A CLOSER LOOK

1. A sampling distribution (like any random variable) describes the long-run behavior of the statistic.

2. We can use either of two techniques to obtain, or find, a sampling distribution.

   a. Recall that to approximate the distribution of a population, we construct a histogram (or stem-and-leaf plot) using values from the population. If the sample is representative, then the histogram should be similar in shape, center, and spread to the population distribution.

   Similarly, to approximate the distribution of a statistic, we obtain (many) values of the statistic and construct a histogram. The resulting graph approximates the sampling distribution of the statistic.

   For example, to approximate the sampling distribution of the mean of a sample of size $n = 10$ from a population: (i) obtain several samples of size 10 from the population; (ii) compute the sample mean for each sample; (iii) construct a histogram using all the sample means. The histogram approximates the sampling distribution of the sample mean.

   b. In some cases, the exact sampling distribution of a statistic can be obtained. If the statistic is a discrete random variable, the sampling distribution includes all the values the statistic assumes and the associated probabilities. If the statistic is a continuous random variable, the sampling distribution consists of a probability density curve.

Example 7.2 Stuffed Animals

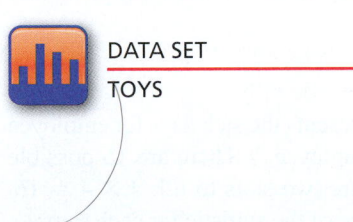

DATA SET

TOYS

A young girl has started a collection of stuffed animals and currently has five koala bears on her dresser. The plush toys vary in color, design, and size. These factors, along with different stuffing materials, contribute to the various weights of the five koalas: 10.9, 14.5, 17.1, 18.1, and 17.6 ounces.

Suppose a sample of three koalas is selected and the sample median weight is computed. Find the sampling distribution for the sample median.

SOLUTION

> The order of selection does not matter here. Therefore, this is a combination.

STEP 1 There are $\binom{5}{3} = 10$ ways to select three koalas from the five in the population.

List all the possible samples, the computed value of the statistic for each sample, and the probability of selecting each sample. Use this table to construct the sampling distribution.

STEP 2 Here is the resulting table.

Sample	$\tilde{x}$	Probability	Sample	$\tilde{x}$	Probability
10.9, 14.5, 17.1	14.5	0.1	10.9, 18.1, 17.6	17.6	0.1
10.9, 14.5, 18.1	14.5	0.1	14.5, 17.1, 18.1	17.1	0.1
10.9, 14.5, 17.6	14.5	0.1	14.5, 17.1, 17.6	17.1	0.1
10.9, 17.1, 18.1	17.1	0.1	14.5, 18.1, 17.6	17.6	0.1
10.9, 17.1, 17.6	17.1	0.1	17.1, 18.1, 17.6	17.6	0.1

The median for each sample is the middle value. The probability of each sample is $1/10 = 0.1$, because we assume that each sample is equally likely.

STEP 3 Of all the possible samples, only three values are possible for the sample median in this case. Sum the probabilities associated with each value. The probability distribution for the random variable $\tilde{X}$ lists the values (of $\tilde{X}$) and the associated probabilities.

$\tilde{x}$	14.5	17.1	17.6
$p(\tilde{x})$	0.3	0.4	0.3

TRY IT NOW   GO TO EXERCISE 7.12

A CLOSER LOOK

1. In Example 7.2, $\tilde{X}$ is a discrete random variable. Using Equations 5.1, 5.3, and 5.4, the mean, variance, and standard deviation for $\tilde{X}$ are $\mu = 16.47$, $\sigma^2 = 1.71$, and $\sigma = 1.31$.

2. The koalas were selected *without* replacement. Suppose three bears are selected *with* replacement. That is, select a stuffed animal, record its weight, and place it back on the dresser. The same koala could be selected two or three times. This sampling scheme changes the probability distribution for $\tilde{X}$.

Challenge: Find the sampling distribution for $\tilde{X}$ if sampling is done with replacement. Hint: There are $5 \times 5 \times 5 = 125$ possible samples.

DATA SET

SICKDAY

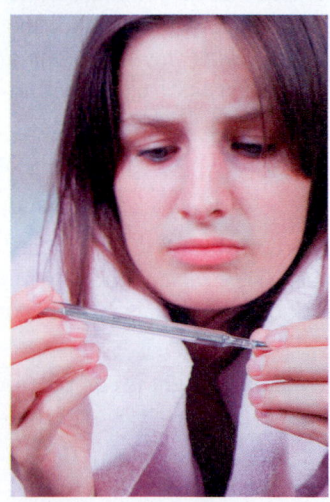

Anyra/Dreamstime.com

Each *event* is given in terms of a random variable.

Example 7.3 Sick Days

Employees at the Hilmar Cheese Company are allowed up to three sick days during a calendar year.[2] Suppose the probability distribution for X, the number of sick days used by an employee during a year, is given in the following table.

x	0	1	2	3
$p(x)$	0.40	0.35	0.20	0.05

Two employees are independently selected at random, and the number of sick days is recorded for each. Consider the statistic M, the maximum number of sick days taken by either employee. Find the probability distribution for M.

SOLUTION

STEP 1 A sample consists of two numbers: The first represents the sick days for employee 1, and the second denotes the sick days for employee 2. There are 16 possible samples. Using the multiplication rule, there are two slots to fill: $4 \times 4 = 16$. List all the possible samples, the computed value of the statistic for each sample, and the probability of selecting each sample.

STEP 2 The probability associated with each sample is computed using the independence assumption. For example, the probability the first employee used one sick day *and* the second employee used two sick days is given by

$$P(X = 1 \cap X = 2)$$ Independent events.

$$= P(X = 1) \cdot P(X = 2)$$ Use the probability distribution for X.

$$= (0.35)(0.20) = 0.07$$

STEP 3   Use the table below to construct the sampling distribution.

Sample	m	Probability	Sample	m	Probability
0, 0	0	0.1600	2, 0	2	0.0800
0, 1	1	0.1400	2, 1	2	0.0700
0, 2	2	0.0800	2, 2	2	0.0400
0, 3	3	0.0200	2, 3	3	0.0100
1, 0	1	0.1400	3, 0	3	0.0200
1, 1	1	0.1225	3, 1	3	0.0175
1, 2	2	0.0700	3, 2	3	0.0100
1, 3	3	0.0175	3, 3	3	0.0025

The maximum for each sample is the largest of the two values.

> Remember, a capital letter, such as M, represents a random variable. The corresponding lowercase letter m denotes a specific value the random variable M can assume.

STEP 4   There are four possible values for the discrete random variable M. Sum the probabilities associated with each value. The probability distribution is given in the table below.

m	0	1	2	3
$p(m)$	0.1600	0.4025	0.3400	0.0975

> **Challenge**: Find the mean, variance, and standard deviation for the random variable M.

TRY IT NOW   GO TO EXERCISE 7.14

In almost all observational studies it is assumed that the data are obtained from a **simple random sample**. Usually the sampling is done *without* replacement. Consider an exit poll or a study of the time spent each week on lawn care by homeowners. An individual is selected from the population and an observation is recorded. The individual is *not* placed back into the population. There is no chance the individual will be selected again.

If sampling is done without replacement, individual responses are dependent. However, if the population is *large enough* and the sample is *small relative to the size of the* population, then the responses are *almost independent*. Calculating probabilities assuming independence results in little loss of accuracy. As a rule of thumb, if the sample size is at most 5% of the total population, then successive observations can be considered independent. Even though sampling is done without replacement, the data are assumed to be part of a simple random sample.

Recall the following definition from Chapter 1.

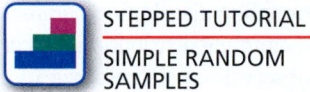

STEPPED TUTORIAL

SIMPLE RANDOM SAMPLES

Definition

A (**simple**) **random sample** (SRS) of size n is a sample selected in such a way that every possible sample of size n has the same chance of being selected.

A CLOSER LOOK

1. Suppose the population is finite of size N and the sample is of size n. The number of possible simple random samples is $\binom{N}{n}$.

2. A random sample consists of individuals or objects, and a variable is a characteristic of an individual or object. A value of the variable is obtained for each member of the random sample. We often refer to the *values* of the variable as the random sample rather than the *individuals*. For example, consider a study in which the amount of the trace element chromium is measured in coal from around the world. The random

sample consists of pieces of coal (objects) and the values are the chromium measurements. However, it is common practice to say, "Consider the random sample of chromium measurements."

3. Unless stated otherwise, all data presented in this text are obtained from a simple random sample.

DATA SET

FEED

Note: The population mean is the sum of all the observations divided by $n = 20$: $\mu = 15.8$.

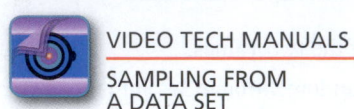

VIDEO TECH MANUALS

SAMPLING FROM A DATA SET

Example 7.4 Cattle Feeds

The following table lists the percentage of protein for the entire population of 20 different cattle feeds.[3]

18	19	34	26	11	3	20	9	12	11
13	24	4	10	24	10	25	15	12	16

Find an approximate sampling distribution for the sample mean of five observations from this population.

SOLUTION

STEP 1 There are $\binom{20}{5} = 15{,}504$ possible samples of size 5. Instead of considering every one of these samples, select some (say, 100) samples of size 5, compute the mean for each sample, and construct a histogram of the sample means.

STEP 2 The table below lists the first few samples of size 5 and the mean for each sample.

Sample					$\bar{x}$	Sample					$\bar{x}$
25	13	18	10	12	15.6	24	3	13	20	10	14.0
3	10	13	12	11	9.8	4	24	10	9	19	13.2
19	9	10	11	16	13.0	20	16	26	24	15	20.2
					$\vdots$						$\vdots$

STEP 3 Figure 7.1 shows a histogram of the sample means for 100 samples of size 5.

STEP 4 There are some very interesting, curious results here. Even though the population is not normally distributed (the population is finite; and consider a graph of the 20 cattle feeds in Figure 7.2), the shape of the sampling distribution appears to be approximately normal! In addition, the *center* of the sampling distribution of the mean is approximately the *population* mean ($\mu = 15.8$). Although the relationship between the original population parameters and the sampling distribution parameters is not clear, there is certainly *less* variability in the sampling distribution than in the population distribution ($\sigma = 8.11$). These three observations suggest the exact distribution of the sample mean, as discussed in the next section.

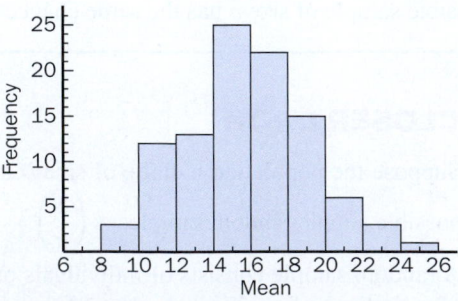

Figure 7.1 Histogram of the sample means.

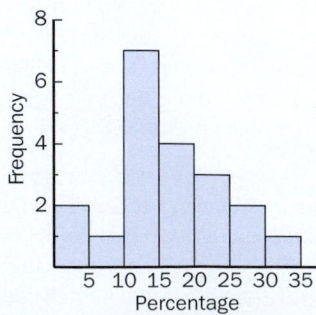

Figure 7.2 Histogram of the original population.

SECTION 7.1 EXERCISES

Concept Check

7.1 Fill in the Blank A parameter describes a _____. A statistic describes a _____.

7.2 True/False Statistics are used to estimate parameters.

7.3 True/False Any statistic is a random variable.

7.4 True/False An exact sampling distribution can never be obtained.

7.5 Short Answer Describe a method to approximate a sampling distribution.

7.6 Fill in the Blank If sampling is done without replacement, individual responses are _____.

7.7 Short Answer Suppose a population is finite of size N and the sample is of size n. How many possible simple random samples are there?

7.8 Short Answer Describe three possible relationships between the distribution of the sample mean and the distribution of the original population.

Applications

7.9 Parameter Versus Statistic In each of the following statements, identify the **boldface** number as the value of a population parameter or a sample statistic.

a. A researcher in Boston conducted a study to investigate the prevalence of Alzheimer's disease in people over the age of 85. The results indicated that **47.2%** of the patients studied experienced symptoms consistent with the disease.

b. A brokerage firm reported that the mean dividend paid by all Fortune 500 companies for the year 2003 was **$0.62**.

c. A consumer magazine tested a random sample of 19 green teas for taste and health benefits. The mean cost per cup of tea was reported to be **14** cents.

d. A computer manufacturer claims that the mean lifetime of laptop batteries is **6.7** hours.

e. A new law requires that all propane tanks (used, for example, on barbecue grills) must have new, safer valves before they can be refilled. A gas company that refills tanks took a random sample of customers on a recent Saturday morning and reported the mean age of the propane tanks to be **3.45** years.

7.10 Parameter Versus Statistic In each of the following statements, identify the **boldface** number as the value of a population parameter or a sample statistic.

a. A half-ton pickup truck is designed to safely carry a maximum of 1000 pounds. The Department of Transportation is concerned that drivers are hauling much heavier loads. The State Police randomly stopped 50 pickup trucks on an interstate highway and carefully weighed the contents in the truck bed. The mean weight was **1037** pounds.

b. A toy manufacturer issued a recall on a small wooden toy car because the wheels could break off and pose a choking hazard for small children. The proportion of buyers who took advantage of the recall was reported to be **0.45**.

c. A recent study indicated that more people today are rising before 6:00 A.M. each weekday to prepare for work and to help children get ready for school. In a random sample of 500 adults, **42%** said they get up each weekday before 6:00 A.M.

d. In a random sample of dentists, **80%** recommended a certain product to help whiten teeth.

e. During a recent winter, the month of January was particularly cold. A power company in Pennsylvania reported that the mean number of kilowatts used by each customer during January was **1346**.

7.11. Parameter Versus Statistic In each of the following statements, identify the **boldface** number as the value of a population parameter or a sample statistic.

a. The maker of a new, enhanced water product conducted a survey to determine if people are drinking enough liquids each day. In a random sample of Americans, **90%** were chronically dehydrated.

b. Last year, the proportion of all robberies in Washington, D.C., that involved a cellphone was **0.42**.

c. NASA reported that the mean time for all spacewalks at the International Space Station was **45** minutes.

d. The International Energy Agency reported that **93%** of Canada's renewable energy comes from hydroelectric sources.

e. In a random sample of publishing companies, **23%** of sales were ebooks.

7.12 Sports and Leisure The Sports Authority sells five different medicine balls with weights (in kilograms) given in the table below. 📊 MEDBALL

10	12	15	18	25

a. Find the population mean and the population median.

b. Suppose a random sample of size 3 is selected from this population without replacement. Find the sampling distribution of the sample mean. Find the mean, variance, and standard deviation for the sample mean.

c. Suppose a random sample of size 3 is selected from this population without replacement. Find the sampling distribution of the sample median. Find the mean, variance, and standard deviation for the sample median.

d. Compare the mean of the sample mean with the population mean. Compare the mean of the sample median with the population median.

7.13 Manufacturing and Product Development The coverage of a gallon of paint depends on the surface, the type and quality of the paint, and the applicator. A local hardware store stocks 30 different paints. 📊 **PAINT**

  a. Use a computer or calculator to draw 50 random samples of size 5 without replacement, and compute the mean for each sample.

  b. Construct a histogram of the sample means.

  c. Use the histogram to approximate the sampling distribution of the mean. What is the approximate shape of the distribution? What is the approximate value of the mean of the sample mean?

  d. Find the population mean. How does this compare with the approximate mean of the sampling distribution?

7.14 Demographics and Population Statistics   For planning purposes, U.S. Bancorp has determined the probability distribution for the retirement age, X, of employees in the mortgage work group. The probability distribution for X is given in the table below. 📊 **RETIRE**

x	64	65	66
$p(x)$	0.1	0.7	0.2

  a. Find the mean of X.

  b. Suppose two employees from this work group are selected at random. Find the exact probability distribution for the sample mean, $\overline{X}$.

  c. Find the mean of $\overline{X}$. How does this compare with your answer in part (a)?

7.15 Public Health and Nutrition   The U.S. Food & Drug Administration regulates the use of certain terms used on food labels—for example, reduced fat, low fat, and light. Lean meat must contain less than 4.5 grams of saturated fat per serving.[4] Suppose the probability distribution for the amount of saturated fat in a randomly selected serving of lean meat is given in the table below. 📊 **USFDA**

x	0	1	2	3	4
$p(x)$	0.50	0.25	0.10	0.10	0.05

  a. Suppose two servings of lean meat are selected at random. Find the exact probability distribution for the sample median amount of saturated fat, $\tilde{X}$.

  b. Find the mean, variance, and standard deviation of $\tilde{X}$.

7.16 Biology and Environmental Science   In April 2013, Eddie ("Flathead Ed") Wilcoxson caught a flathead catfish that weighed 76.52 pounds on Bartlett Lake in Arizona.[5] The catfish was 53.5 inches long and its girth measured 34.75 inches. Suppose five catfish were caught on the same lake with weights (in pounds) given in the following table. 📊 **CATFISH**

11	15	32	21	27

A random sample of three of these catfish is selected without replacement.

  a. Find the sampling distribution of the sample mean $\overline{X}$.

  b. Find the sampling distribution of the total length for all three catfish, T.

7.17 Education and Child Development   The number of copies sold (in millions) for the all-time best-selling children's books are given in the table below. 📊 **BOOKS**

9.9	9.7	7.1	7.0	6.8

Suppose two of these books are selected at random without replacement.

  a. Find the sampling distribution of the sample mean.

  b. Find the sampling distribution of the total number of books sold.

Extended Applications

7.18 Fuel Consumption and Cars   An automobile manufacturer lists several specifications for every one of its cars. For example, the length, width, wheelbase, turning circle, curb weight, and interior room measurements are readily available to customers. The acceleration time from 0 to 60 miles per hour (in seconds) for 10 cars is given in the table below.[6] 📊 **CARS**

2.5	2.8	3.0	3.5	3.4	3.7	3.9	4.1	4.7	5.1

  a. Use a computer or calculator to draw 30 random samples of size 3 without replacement, and compute the standard deviation for each sample.

  b. Construct a histogram of the sample standard deviations.

  c. Use the histogram to describe the shape of the sampling distribution.

  d. Compute the *population* standard deviation. Find an approximate mean of the sampling distribution. How do these two numbers compare?

7.19 Travel and Transportation   American Airlines offers limited first-class seating to passengers on a flight from Newark to Los Angeles. The probability distribution for the number of passengers in first class on a randomly selected flight is a random variable, X. The probability distribution for X is given in the table below. 📊 **FLIGHT**

x	5	6	7	8
$p(x)$	0.50	0.30	0.15	0.05

  a. Find the variance of X.

  b. Suppose two American Airlines cross-country flights are randomly selected. Find the exact probability distribution for the sample variance, S^2.

  c. Find the mean of S^2. How does this compare with your answer in part (a)?

7.20 Sports and Leisure   The times (in minutes) for three men in the 3M Half Marathon are given in the following table.[7] 📊 **RUNTIME**

66.08	68.58	77.10

Suppose a random sample of two of these times is selected with replacement.

a. Find the sampling distribution of the minimum time.

b. Find the sampling distribution of the total time.

7.21 Economics and Finance   The Kentucky Derby is run every May at Churchill Downs. This extravagant affair features mint juleps and unusual woman's hats. The jockeys with the most Kentucky Derby mounts are given in the following table.[8] **DERBY**

Jockey	Mounts
Bill Shoemaker	26
Pat Day	22
Eddie Arcaro	21
Laffit Pincay Jr.	21

Suppose a random sample of two of these jockeys is selected at random without replacement.

a. Find the sampling distribution for the maximum number of mounts.

b. Find the sampling distribution for the total number of mounts.

7.22 Psychology and Human Behavior   Five people own and operate the Victrola Coffee Shop in Seattle. Each person is married, and the number of years each has been married is given below. **COFFEE**

5	3	7	2	12

Suppose a random sample of two of the owners is selected with replacement. Let D be a statistic defined to be the absolute value of the difference in the number of years each has been married. For example, if 2 and 7 were selected, the value of D would be $|2 - 7| = |-5| = 5$. Find the sampling distribution of D.

7.23 Simple Random Sample   Consider the population consisting of the numbers 1 through 50.

a. Find the population mean μ.

b. Use technology to select 50 random samples of size 10 from this population. Find the sample mean for each sample.

c. Construct a histogram of the 50 sample means. Where is the histogram centered?

7.24 Sports and Leisure   The Toronto Rock are in the National Lacrosse League. The total number of shots on goal for each player during the 2013 season is given in the following table.[9] **LACROSSE**

215	129	138	133	114	68	70	33
45	6	8	15	9	18	4	4
9	15	0	12	5	4	0	

a. Find the population mean number of shots on goal, μ.

b. Find the population standard deviation of shots on goal, σ.

c. Use technology to select 100 random samples of size 10 from this population. Find the sample mean for each sample.

d. Construct a histogram of the 100 sample means. Where is the histogram centered? How does this value compare to μ?

e. Find the standard deviation for the 100 sample means. How does this value compare to σ?

Challenge

7.25 Which Estimator Is Better?   Each of the following graphs shows the probability distribution for two statistics that could be used to estimate a parameter θ (describing an underlying population). Select the statistic that would be a *better* estimator of θ and justify your answer.

a.

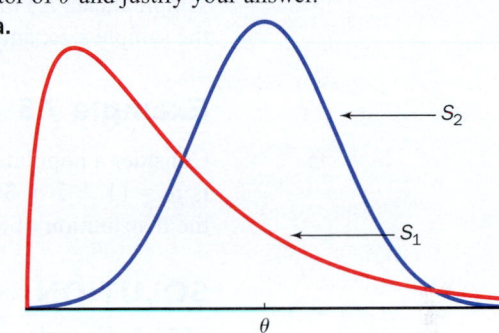

b.

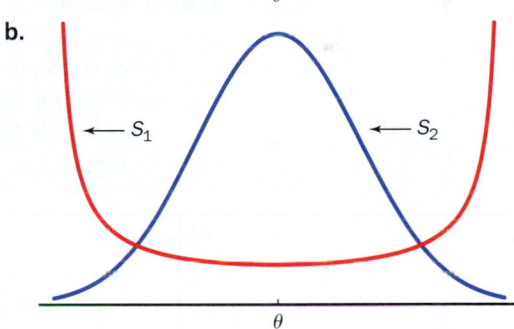

c.

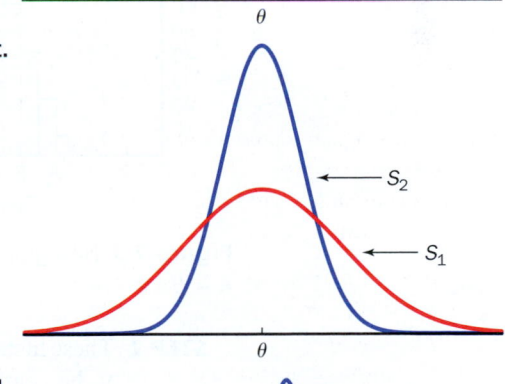

d.

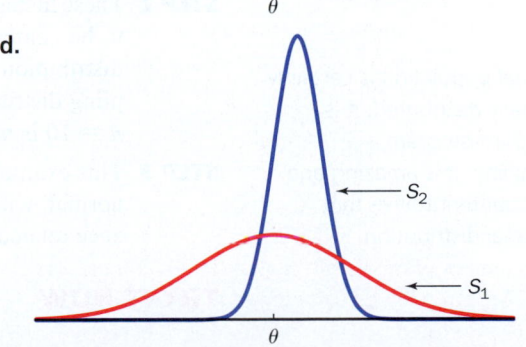

7.2 The Sampling Distribution of the Sample Mean and the Central Limit Theorem

It seems reasonable to use a value of the sample mean, $\bar{x}$, to make an inference about the population mean μ. However, as discussed in the previous section, the sample mean varies from sample to sample. This **sampling variability** makes it difficult to know how far a specific $\bar{x}$ is from μ, or even whether the sample mean is an overestimate or underestimate. In this section, the sampling variability of the sample mean is completely characterized. The probability distribution of the sample mean, $\overline{X}$, can be used to make a sensible guess about the true value of the population mean.

To make a reliable estimate, we need to know the exact probability distribution of the sample mean. As the next few examples will show, the distribution of $\overline{X}$ is related to n, the sample size, and to the parameters of the original, or underlying, population.

Example 7.5 Approximate Distribution of $\overline{X}$

Consider a population consisting of the numbers 1, 2, 3, ..., 20. The population mean is $\mu = (1 + 2 + 3 + \cdots + 20)/20 = 10.5$. Use frequency histograms to approximate the distribution of the mean, $\overline{X}$.

SOLUTION

STEP 1   Consider a random sample of n observations selected with replacement. For $n = 5$, five numbers are selected at random from the population, and the sample mean is computed. This procedure is repeated 500 times. A histogram of the resulting 500 sample means is shown in Figure 7.3. For $n = 10$, a similar procedure is followed. The resulting histogram of the sample means is shown in Figure 7.4.

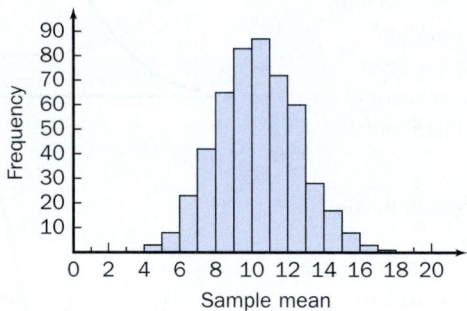

Figure 7.3 Histogram of sample means for $n = 5$.

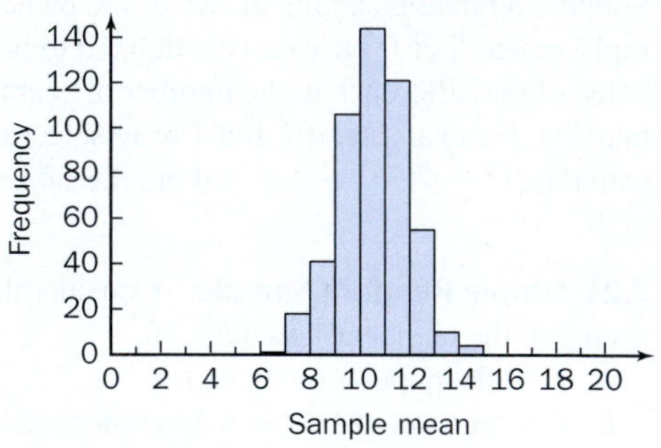

Figure 7.4 Histogram of sample means for $n = 10$.

STEP 2   These histograms suggest some very surprising results. Each distribution appears to be centered near the population mean, 10.5. In addition, the shape of each distribution is approximately normal! Also notice that the variability of the sampling distribution decreases as n increases. The sampling distribution for $\overline{X}$ for $n = 10$ is more compact than that for $n = 5$.

The original population is certainly not normally distributed; it is finite, and a histogram is a horizontal line. It is amazing and perhaps counterintuitive that $\overline{X}$ has a normal distribution.

STEP 3   This example implies that the distribution of the sample mean is approximately normal, with mean equal to the underlying, original, population mean and variance related to the sample size, n.   ●

TRY IT NOW   GO TO EXERCISE 7.59

In the next example, the exact sampling distribution of the mean is obtained.

DATA SET

ESSAY

Example 7.6 Essay Sources

Mary Dunn, a teacher at The Hill School, has assigned a short research paper in which each student must write about an endangered species. Ms. Dunn requires each student to correctly reference at least two sources. Past experience indicates that the number of sources cited is a random variable, X, with probability distribution given in the table below.

x	2	3	4
$p(x)$	0.5	0.3	0.2

Suppose a random sample of three student papers is selected.

a. Find the sampling distribution of $\overline{X}$, the sample mean number of sources cited.

b. Find the mean and the variance of the random variable X.

c. Find the mean and the variance of the random variable $\overline{X}$.

d. How do the results from (b) compare with the results from (c)?

SOLUTION

a. A sample consists of three numbers: The first represents the number of sources for student 1, the second for student 2, and the third for student 3. By the multiplication rule, there are $3 \times 3 \times 3 = 27$ possible samples. For each sample, list the sample mean and the probability of selecting that sample.

The probability associated with each sample is computed using independence. For example, the probability of observing the sample 2, 2, 4 is

$$P(X = 2 \cap X = 2 \cap X = 4)$$ Intersection of three events.

$$= P(X = 2) \cdot P(X = 2) \cdot P(X = 4)$$ Independence.

$$= (0.5)(0.5)(0.2) = 0.05$$ Given probabilities.

Use the table below to construct the sampling distribution.

Sample	$\overline{x}$	Probability	Sample	$\overline{x}$	Probability
2, 2, 2	2	0.125	3, 3, 4	10/3	0.018
2, 2, 3	7/3	0.075	3, 4, 2	3	0.030
2, 2, 4	8/3	0.050	3, 4, 3	10/3	0.018
2, 3, 2	7/3	0.075	3, 4, 4	11/3	0.012
2, 3, 3	8/3	0.045	4, 2, 2	8/3	0.050
2, 3, 4	3	0.030	4, 2, 3	3	0.030
2, 4, 2	8/3	0.050	4, 2, 4	10/3	0.020
2, 4, 3	3	0.030	4, 3, 2	3	0.030
2, 4, 4	10/3	0.020	4, 3, 3	10/3	0.018
3, 2, 2	7/3	0.075	4, 3, 4	11/3	0.012
3, 2, 3	8/3	0.045	4, 4, 2	10/3	0.020
3, 2, 4	3	0.030	4, 4, 3	11/3	0.012
3, 3, 2	8/3	0.045	4, 4, 4	4	0.008
3, 3, 3	3	0.027			

The probability distribution for $\overline{X}$ is given in the table below. It lists all the values $\overline{X}$ can assume and the corresponding probabilities.

$\overline{x}$	2	7/3	8/3	3	10/3	11/3	4
$p(\overline{x})$	0.125	0.225	0.285	0.207	0.114	0.036	0.008

b. The mean and variance for X are

$$E(X) = (2)(0.5) + (3)(0.3) + (4)(0.2) = 2.7$$

$$\text{Var}(X) = (2 - 2.7)^2(0.5) + (3 - 2.7)^2(0.3) + (4 - 2.7)^2(0.2) = 0.61$$

c. The mean and variance for $\overline{X}$ are

$$E(\overline{X}) = (2)(0.125) + (7/3)(0.225) + \cdots + (4)(0.008) = 2.7$$

$$\text{Var}(\overline{X}) = (2 - 2.7)^2(0.125) + \cdots + (4 - 2.7)^2(0.008) = 0.2033$$

d. Notice the following extraordinary relationship between the means and the variances.

$$E(\overline{X}) = 2.7 = E(X)$$

The mean of the sample mean is equal to the original population mean!

$$\text{Var}(\overline{X}) = 0.2033 = \frac{0.61}{3} = \frac{\text{Var}(X)}{n}$$

The variance of $\overline{X}$ is equal to the original variance divided by the sample size.   ●

Here is one more approach to illustrate the very important connections between the distribution of the sample mean and the distribution of the original population.

Consider three original, or underlying, distributions: (1) the standard normal distribution, a normal distribution with mean 0 and standard deviation 1; (2) a uniform distribution with parameters $a = 0$ and $b = 1$; and (3) an exponential distribution with parameter $\lambda = 0.5$. The probability density function and the mean for each distribution are shown in the first row of Figure 7.5.

Consider the following process for each distribution. Select 500 samples of size $n = 2$ and compute the mean for each sample. Construct a histogram of the sample means. The resulting *smoothed* histogram is shown in the second row of Figure 7.5. Repeat this procedure for $n = 5$, 10, and 20, for each underlying distribution. These smoothed histograms are also given in Figure 7.5.

Notice the following incredible patterns.

1. If the underlying population is normal, the distribution of the sample mean appears to be normal, regardless of the sample size. See Figure 7.5(a).

2. Even if the underlying population is *not* normal, the distribution of the sample mean becomes *more normal* as n increases. See Figures 7.5(b) and 7.5(c).

3. The sampling distribution of the mean is centered at the mean of the underlying population. See Figures 7.5(a), 7.5(b), and 7.5(c).

4. As the sample size, n, increases, the variance of the distribution of the sample mean decreases. See Figures 7.5(a), 7.5(b), and 7.5(c).

The previous examples and observations lead to the following properties concerning the sample mean, $\overline{X}$.

Properties of the Sample Mean

Let $\overline{X}$ be the mean of observations in a random sample of size n drawn from a population with mean μ and variance σ^2.

1. The mean of $\overline{X}$ is equal to the mean of the underlying population.

   In symbols: $\mu_{\overline{X}} = \mu$.

2. The variance of $\overline{X}$ is equal to the variance of the underlying population divided by the sample size.

   In symbols: $\sigma_{\overline{X}}^2 = \dfrac{\sigma^2}{n}$.

The standard deviation of $\overline{X}$ is $\sigma_{\overline{X}} = \sqrt{\dfrac{\sigma^2}{n}} = \dfrac{\sigma}{\sqrt{n}}$.

3. If the underlying population is distributed normally, then the distribution of $\overline{X}$ is also *exactly* normal for any sample size.
    In symbols: $\overline{X} \sim N(\mu, \sigma^2/n)$

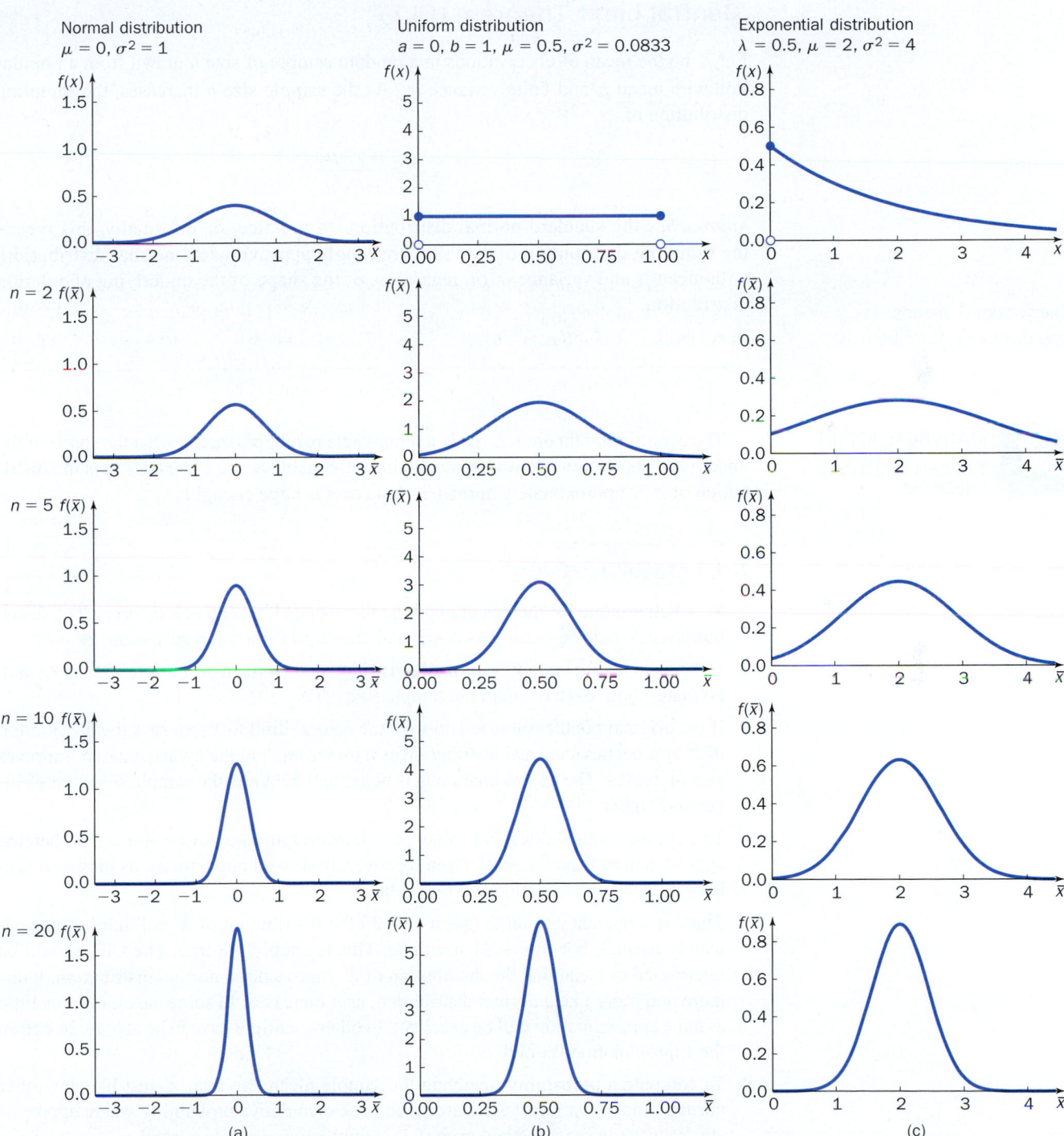

Figure 7.5 Three different original, or underlying, populations, and approximations to the distribution of the sample mean for various sample sizes *n*.

The mean of the sampling distribution of $\overline{X}$ is the same as the mean of the underlying distribution that we want to estimate, so the sample mean is an **unbiased estimator** of the population mean μ.

Even if the underlying population is *not* normal, the previous examples suggest the distribution of $\overline{X}$ *becomes more normal* as the sample size increases. This amazing result is the most important idea in all of statistics, the **central limit theorem**.

Central Limit Theorem (CLT)

Let $\overline{X}$ be the mean of observations in a random sample of size n drawn from a population with mean μ and finite variance σ^2. As the sample size n increases, the sampling distribution of

$$Z = \frac{\overline{X} - \mu}{\sigma/\sqrt{n}}$$

approaches the standard normal distribution. In practice, or informally, this means the sampling distribution of $\overline{X}$ will increasingly approximate a normal distribution, with mean μ and variance σ^2/n, regardless of the shape of the underlying population distribution.

The symbol $\overset{\bullet}{\sim}$ means "is approximately distributed as."

In symbols: $\overline{X} \overset{\bullet}{\sim} \mathrm{N}(\mu, \sigma^2/n)$.

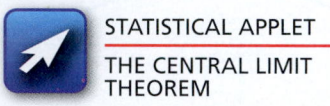

STATISTICAL APPLET

THE CENTRAL LIMIT THEOREM

The central limit theorem is really a remarkable result. No matter what the shape of the underlying population (skewed, bimodal, lots of variability, etc.), the CLT says the distribution of $\overline{X}$ is approximately normal, as long as n is large enough!

A CLOSER LOOK

1. ▶ A better name for this result might be the *normal convergence theorem*. The distribution of $\overline{X}$ *converges*, or gets closer and closer, to a normal distribution. ◀

2. If the original population is normally distributed, then the distribution of $\overline{X}$ is normal, no matter how large or small the sample size (n) is.

3. If the original population is *not* normal, the central limit theorem says the distribution of $\overline{X}$ approaches a normal distribution as n increases, and the approximation improves as n increases: The approximation gets better and better as the sample size n gets bigger and bigger.

   There is no magical threshold value for n. However, in most cases, if $n \geq 30$, then the approximation is pretty good. Even for severely skewed populations, as long as n is at least 30, the approximation is reasonable.

   There is a tendency to think that if $n = 29$ the distribution of $\overline{X}$ will not be approximately normal, but if $n = 31$ it will be. This is simply not true. The CLT should be interpreted to mean that the distribution of $\overline{X}$ approaches a normal distribution, looks more and more like a normal distribution, as n increases. In some cases, for n as little as 5 the approximation will be excellent. In others, n might have to be at least 26 before the approximation is good.

4. To compute a probability involving the sample mean, we treat $\overline{X}$ just like any other normal random variable: Standardize and use cumulative probability where appropriate. We use the same method even if $\overline{X}$ is only *approximately* normal.

5. The expression for the variance of $\overline{X}$ mathematically confirms our observations regarding the variability of the sampling distribution. As n increases, the distribution of the

Solution Trail 7.7

KEYWORDS
- Normally distributed
- Mean 50, standard deviation 4, 25 trips
- Sample mean

TRANSLATION
- Normal
- $\mu = 50, \sigma^2 = 16, n = 25$
- $\overline{X}$

CONCEPTS
- Properties of $\overline{X}$

VISION

The underlying distribution is normal, so the distribution of $\overline{X}$ is *exactly* normal. Find the mean, variance, and standard deviation of $\overline{X}$, translate each question into a probability statement, standardize, and use cumulative probability where appropriate.

sample mean becomes more compact. $\sigma_{\overline{X}}^2 = \sigma^2/n$ and σ^2 is constant. This fraction becomes smaller as n (the denominator) increases.

6. A more general version of the CLT includes a statement about the sum of independent observations, T. If n is sufficiently large, the distribution of T approaches a normal distribution with mean $n\mu$ and variance $n\sigma^2$.

   In symbols: $T \overset{\bullet}{\sim} \mathrm{N}(n\mu, n\sigma^2)$.

An even more general version of the CLT concludes, essentially, that any statistic that is a sum or a mean tends toward a normal distribution as n increases. This is useful in many inference problems because the appropriate statistic is often a sum or a mean. It is easy to compute probabilities associated with a normal statistic (random variable). Therefore, the likelihood, or tail probability, associated with an observed value of the statistic is a straightforward calculation.

In addition, the central limit theorem helps to explain why so many real-world distributions are approximately normal. Almost any statistic can be decomposed into a sum of other variables. For example, the height of a tomato plant after six weeks might be directly related to, or is the *sum* of, the effects of a large number of independent variables including the amount of water, fertilizer, and sunlight it has received. Therefore, the distribution of the height should be approximately normal. This single theorem explains empirical evidence that suggests almost every *measurement* distribution is approximately normal.

The following examples illustrate the properties of $\overline{X}$ and the central limit theorem.

Standardization illustrated:

$P(\overline{X} < 48)$

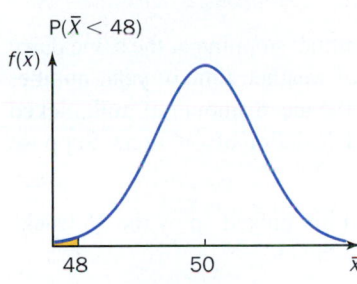

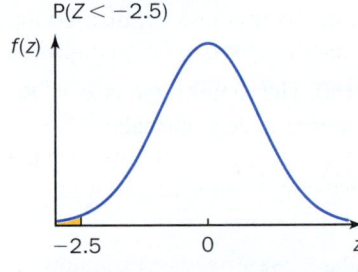

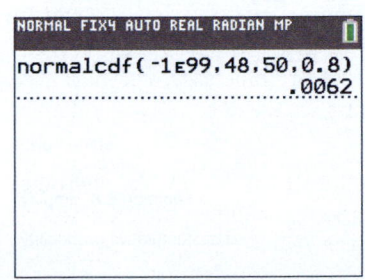

Figure 7.6 $P(\overline{X} < 48)$.

Example 7.7 Green Line Time

The Massachusetts Bay Transportation Authority (MBTA) Green Line from Eliot to Lechmere in Boston has trolleys leaving regularly throughout the day beginning at 5:00 A.M. Although the length of the trip and the number of stops are constant, the time taken for each trip varies due to weather, traffic, and time of day. According to the MBTA, the mean time taken for the trip is 50 minutes. Suppose the travel time is normally distributed with standard deviation $\sigma = 4$ minutes. A random sample of 25 trips is obtained, and the time for each is recorded.

a. Find the probability that the sample mean time will be less than 48 minutes.

b. Find the probability that the sample mean will be within one minute of the population mean (50 minutes).

SOLUTION

a. The underlying distribution is normal with $\mu = 50$ and $\sigma^2 = 16$. The sample size is $n = 25$. The sample mean is (exactly) normally distributed:

$$\overline{X} \sim \mathrm{N}(\mu, \sigma^2/n) = \mathrm{N}(50, 16/25); \qquad \sigma_{\overline{X}} = \sqrt{16/25} = 4/5 = 0.80.$$

To solve part (a), begin with a probability statement involving the random variable $\overline{X}$. We need the probability that $\overline{X}$ will be less than 48 minutes.

$$P(\overline{X} < 48) = P\left(\frac{\overline{X} - 50}{0.80} < \frac{48 - 50}{0.80}\right) \qquad \text{Standardize.}$$

$$= P(Z < -2.5) \qquad \text{Equation 6.8; simplify.}$$

$$= 0.0062 \qquad \text{Use Table III in the Appendix.}$$

The probability that the sample mean time for the 25 trips will be less than 48 minutes is 0.0062. Figure 7.6 shows a technology solution.

Remember, one point in a continuous world (Z) contributes no probability:

$$P(Z < -1.25) = P(Z \leq -1.25).$$

b. *Within 1 minute of the population mean* means $49 \leq \overline{X} \leq 51$. Write the probability statement, and standardize again.

$$P(49 \leq \overline{X} \leq 51)$$

$$= P\left(\frac{49 - 50}{0.8} \leq \frac{\overline{X} - 50}{0.8} \leq \frac{51 - 50}{0.8}\right) \quad \text{Standardize.}$$

$$= P(-1.25 \leq Z \leq 1.25) \quad \text{Equation 6.8; simplify.}$$

$$= P(Z \leq 1.25) - P(Z < -1.25) \quad \text{Use cumulative probability.}$$

$$= 0.8944 - 0.1056 \quad \text{Use Table III in the Appendix.}$$

$$= 0.7888$$

Standardization illustrated:

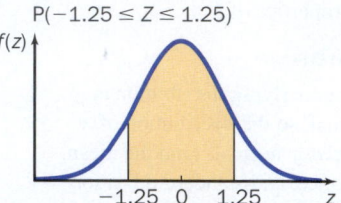

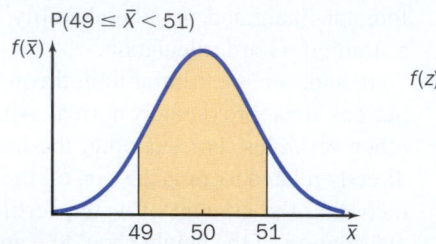

Figure 7.7 $P(49 \leq \overline{X} \leq 51)$.

The probability that the sample mean time for the 25 trips will be between 49 and 51 minutes is 0.7888. Figure 7.7 shows a technology solution.

TRY IT NOW GO TO EXERCISE 7.39

Example 7.8 Milk Deliveries

In upstate New York, milk tanker trucks follow a daily routine, stopping at the same dairy farms every day. Farm output, however, varies because of weather, time of year, number of cows, and other factors. From years of recorded data, the mean amount of milk picked up by a truck for processing is 7750 liters, with a standard deviation of 150 liters. Suppose 36 trucks are randomly selected.

a. Find the probability that the sample mean amount of milk picked up by the 36 trucks is more than 7800 liters.

b. Find a value m such that the probability that the sample mean is less than m is 0.1.

SOLUTION

a. The underlying distribution has $\mu = 7750$ and $\sigma = 150$. The sample size is $n = 36$. By the central limit theorem, the distribution of $\overline{X}$ is approximately normal:

$$\overline{X} \overset{\bullet}{\sim} N(\mu, \sigma^2/n) = N(7750, 150^2/36) = N(7750, 625)$$

The standard deviation of $\overline{X}$ is

$$\sigma_{\overline{X}} = \sqrt{\sigma^2/n} = \sqrt{625} = 25 \ (\text{or } \sigma_{\overline{X}} = \sigma/\sqrt{n} = 150/6 = 25)$$

To solve part (a), start with a probability statement involving $\overline{X}$. We need the probability that the sample mean will be *more than* 7800 liters.

$$P(\overline{X} > 7800) = P\left(\frac{\overline{X} - 7750}{25} > \frac{7800 - 7750}{25}\right) \quad \text{Standardize.}$$

$$= P(Z > 2) \quad \text{Equation 6.8; simplify.}$$

$$= 1 - P(Z \leq 2) \quad \text{Use cumulative probability.}$$

$$= 1 - 0.9772 \quad \text{Use Table III in the Appendix.}$$

$$= 0.0228$$

The probability that the sample mean amount of milk picked up by the 36 trucks will be greater than 7800 is 0.0228.

Standardization illustrated:

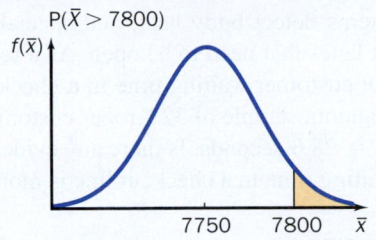
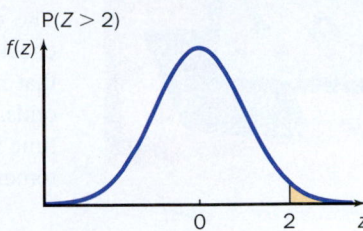

b. To solve part (b), translate the question into a probability statement. Convert the expression into cumulative probability involving Z. Because the probability is already given, this is an inverse cumulative probability problem. Work backward in Table III.

$$P(\overline{X} < m) = P\left(\frac{\overline{X} - 7750}{25} < \frac{m - 7750}{25}\right)$$ Standardize.

$$= P\left(Z < \frac{m - 7750}{25}\right) = 0.1$$ Equation 6.8.

There is no further simplification within the probability statement. However, the resulting probability statement involves Z and is cumulative probability. Find a value in the body of Table III, as close to 0.1 as possible. Set the corresponding z value equal to $\left(\dfrac{m - 7750}{25}\right)$, and solve for m.

$$\frac{m - 7750}{25} = -1.28$$ Table III in the Appendix.

$$m - 7750 = -32$$ Multiply both sides by 25.

$$m = 7718$$ Add 7750 to both sides.

The probability that the sample mean will be less than $m = 7718$ is (approximately) 0.1.

Standardization illustrated:

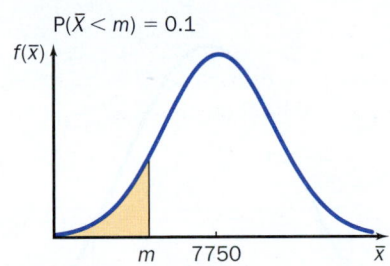
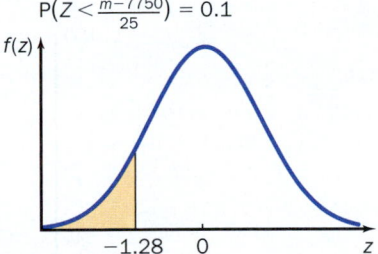

Figure 7.8 shows a technology solution.

Prob	JMP Formula
0.9772	= Normal Distribution(7800,7750,25)
0.0228	= 1 - Prob[[1]]
7717.9612	= Normal Quantile(0.1,7750,25)

Figure 7.8 JMP calculations for $P(\overline{X} < 7800)$ and $P(\overline{X} < m) = 0.1$.

TRY IT NOW   GO TO EXERCISE 7.56

The last example in this section involves an inference using the random variable $\overline{X}$.

Cheryl A. Guerrero/Los Angeles Times/MCT

Solution Trail 7.9

KEYWORDS

- Is there any evidence?

TRANSLATION

- Statistical inference

CONCEPTS

- Central limit theorem

VISION

This inference question concerns a population mean. Use the CLT and the distribution of $\overline{X}$ to determine the likelihood of an observed sample mean of $\overline{x} = 28.0$ or greater.

Example 7.9 Supermarket Checkout Lines

Kroger Supermarkets are now using infrared cameras similar to those used by the military and law enforcement personnel to determine customer waiting time in checkout lines.[10] The cameras detect body heat, and special software is used to control the number of checkout lanes that need to be open. As a result of this new system, the company claims that mean customer waiting time in a checkout line is 26 seconds. Suppose $\sigma = 9$ seconds, a random sample of 32 Kroger customers is selected, and the sample mean waiting time is $\overline{x} = 28.0$ seconds. Is there any evidence to suggest that the population mean customer waiting time in a checkout line is more than 26 seconds?

SOLUTION

STEP 1 Assume the underlying distribution has $\mu = 26$ and $\sigma = 9$. We do not know the shape of the underlying distribution of customer waiting times, but the sample size is large: $n = 32$ (≥ 30). By the central limit theorem, the distribution of $\overline{X}$ is approximately normal.

STEP 2 **Claim:** $\mu = 26 \to \overline{X} \overset{\bullet}{\sim} N(26, 9^2/32)$, $\qquad \sigma_{\overline{X}} = 9/\sqrt{32}$.

Experiment: $\overline{x} = 28.0$.

Likelihood: We are looking for evidence that the mean customer waiting time is *greater* than 26 seconds, so find the right-tail probability (to be conservative).

$$P(\overline{X} \geq 28.0) = P\left(\frac{\overline{X} - 15}{5/\sqrt{40}} \geq \frac{28.0 - 26.0}{9/\sqrt{32}}\right) \qquad \text{Standardize.}$$

$$= P(Z \geq 1.26) \qquad \text{Equation 6.8; simplify.}$$

$$= 1 - P(Z < 1.26) \qquad \text{Use cumulative probability.}$$

$$= 1 - 0.8962 \qquad \text{Table III in the Appendix.}$$

$$= 0.1038$$

Conclusion: This probability is large (>0.05), so there is no evidence to suggest the mean customer waiting time in a checkout line is greater than 26. Even if the mean waiting time is 26, an observation of $\overline{x} = 28.0$ is not very unlikely, and so we cannot doubt that the mean is really 26. Figure 7.9 illustrates the distribution of $\overline{X}$, and the right-tail probability; Figure 7.10 shows a technology solution.

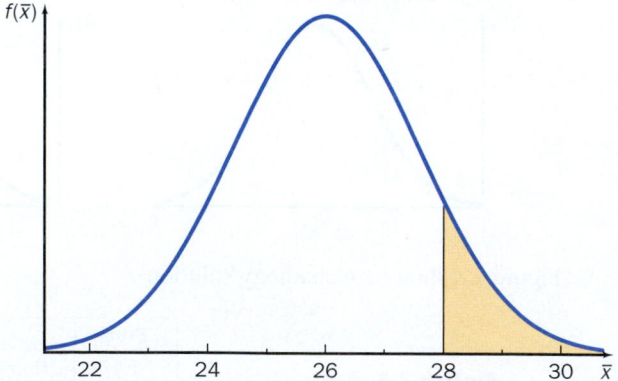

Figure 7.9 The distribution of $\overline{X}$. The shaded region represents the right-tail probability $P(\overline{X} \geq 28)$.

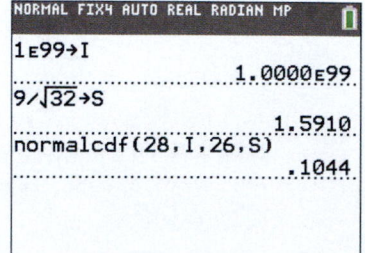

Figure 7.10 Use calculator variables to make the input easier.

TRY IT NOW GO TO EXERCISE 7.42

Technology Corner

Procedure: Solve probability questions involving the sample mean $\overline{X}$.
Reconsider: Example 7.7, solution, and interpretations.

Note: In these probability questions $\overline{X}$ is either normally distributed or approximately normally distributed. In either case, the technology procedures are similar to those described in Section 6.2.

CrunchIt!

CrunchIt! has a built-in function to compute probabilities associated with a normal random variable. Select Distribution Calculator; Normal. To find cumulative probability or right-tail probability, select the Probability tab. Choose an appropriate inequality symbol and enter a value for the endpoint. To solve an inverse cumulative probability problem, select the Quantile tab and enter a cumulative probability.

1. Select Distribution Calculator; Normal. Enter the mean, 50, and the standard deviation, 0.8. Under the Probability tab, select ≤ and enter 48 for the endpoint. Click the Calculate Tab. See Figure 7.11.
2. To find the probability that $\overline{X}$ takes on a value in an interval, use cumulative probability.
   $$P(49 \leq \overline{X} \leq 51) = P(\overline{X} \leq 51) - P(\overline{X} \leq 49).$$

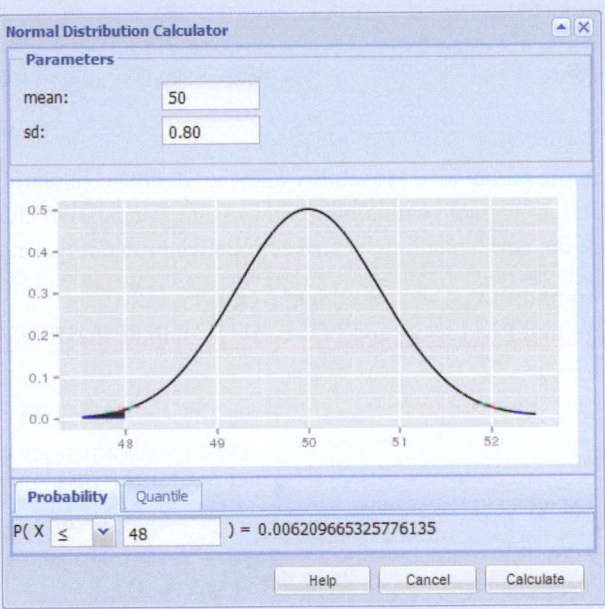

Figure 7.11 CrunchIt! returns a graph of the corresponding probability density function. The shaded area corresponds to $P(\overline{X} < 48)$.

TI-84 Plus C

The built-in function `normalcdf` is used to find (calculate) cumulative probability: the probability that X takes on a value between a and b. This function takes four arguments: a (lower), b (upper), μ, and σ. The built-in function `invNorm` takes three arguments: p (area), μ, and σ. This function returns a value x such that $P(X \leq x) = p$. The default values for μ and σ are 0 and 1, respectively.

1. Select DISTR; DISTR; `normalcdf`. Enter the left endpoint (lower), −1E99 (negative calculator infinity), the right endpoint (upper), 48, the mean (μ), 50, and the standard deviation (σ), 0.8. Highlight Paste and tap ENTER. Refer to Figure 7.6.
2. Select DISTR; DISTR; `normalcdf`. Enter the left endpoint (lower), 49, the right endpoint (upper), 51, the mean (μ), 50, and the standard deviation (σ), 0.8. Highlight Paste and tap ENTER. Refer to Figure 7.7.

Minitab

There are several built-in functions to compute cumulative probability, tail probability, and inverse cumulative probability. These functions may be accessed through a graphical input window or by using the command language.

1. In a session window, use the function CDF to find $P(\overline{X} < 48)$. See Figure 7.12.
2. Select Graph; Probability Distribution Plot; View Probability. In the Distribution menu, select Normal. Enter the Mean and Standard deviation. Under the Shaded Area tab, choose X Value and Middle. Enter the X value 1 (49) and X value 2 (51). Click OK. Minitab displays a distribution plot with the shaded area corresponding to probability. See Figure 7.13.

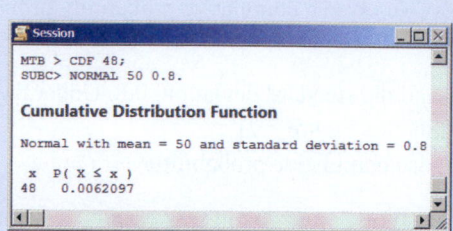

Figure 7.12 $P(\overline{X} < 48)$ using the command language.

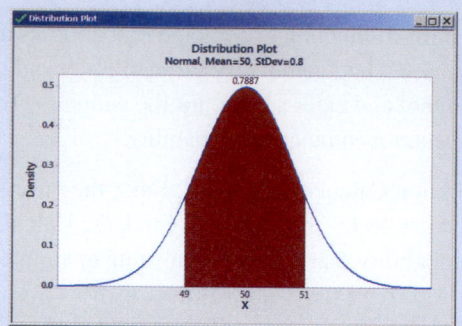

Figure 7.13 $P(49 \le \overline{X} \le 51)$ using Probability Distribution Plot.

Excel

There are built-in functions to compute cumulative probability associated with a standard normal random variable (NORM.S.DIST) and a normal random variable with arbitrary mean and standard deviation (NORM.DIST). The functions NORM.S.INV and NORM.INV are the corresponding inverse cumulative probability functions.

1. Use the function NORM.DIST to find $P(\overline{X} < 48)$. See Figure 7.14.
2. Use the function NORM.DIST to find $P(\overline{X} \le 49)$ and $P(\overline{X} \le 51)$. Compute the difference to find $P(49 \le X \le 51)$. See Figure 7.15.

A	B
0.0062	= NORM.DIST(48,50,0.8,TRUE)

Figure 7.14 $P(\overline{X} < 48)$: cumulative probability.

A	B
0.8944	= NORM.DIST(51,50,0.8,TRUE)
0.1056	= NORM.DIST(49,50,0.8,TRUE)
0.7887	= A1 - A2

Figure 7.15 Calculations for $P(49 \le \overline{X} \le 51)$.

SECTION 7.2 EXERCISES

Concept Check

7.26 True/False   The sample mean varies from sample to sample.

7.27 Fill in the Blank   Let $\overline{X}$ be the mean of observations in a random sample of size n from a population with mean μ and variance σ^2.
   a. The mean of $\overline{X}$ is _____.
   b. The variance of $\overline{X}$ is _____.

7.28 True/False   The distribution of $\overline{X}$ is never exactly normal.

7.29 True/False   If the underlying population is not normal, the central limit theorem says the distribution of $\overline{X}$ approaches a normal distribution as n increases.

7.30 True/False   As n increases, the distribution of the sum of the observations approaches a normal distribution.

7.31 True/False   As n increases, the variance of $\overline{X}$ also increases.

Note: To find the distribution of $\overline{X}$, or any random variable, you need to describe the distribution, by name if possible, and provide the values of the parameters that characterize the distribution. For example, $\overline{X}$ is normally distributed with mean $\mu = 27$ and variance $\sigma_{\overline{X}}^2 = 0.35$, or $\overline{X} \sim N(27, 0.35)$, completely describes the distribution of $\overline{X}$.

Practice

7.32 Consider a normally distributed population with mean $\mu = 10$ and standard deviation $\sigma = 2.5$. Suppose a random sample of size n is selected from this population. Find the distribution of $\overline{X}$ and the indicated probability in each of the following cases.
  a. $n = 7$, $P(\overline{X} \le 9)$.
  b. $n = 12$, $P(\overline{X} > 11.5)$.
  c. $n = 15$, $P(9.5 \le \overline{X} \le 10.5)$.
  d. $n = 25$, $P(\overline{X} \ge 10.25)$.
  e. $n = 100$, $P(\overline{X} \le 9.8 \cup \overline{X} \ge 10.2)$.

7.33 Suppose X is a normal random variable with mean $\mu = 17.5$ and standard deviation $\sigma = 6$. A random sample of size $n = 24$ is selected from this population.
  a. Find the distribution of $\overline{X}$.
  b. Carefully sketch a graph of the probability density functions for X and $\overline{X}$ on the same coordinate axes.
  c. Find $P(X \le 14)$ and $P(\overline{X} < 14)$.
  d. Find $P(15 < X < 19)$ and $P(15 < \overline{X} < 19)$.

7.34 Suppose X is a random variable with mean $\mu = 50$ and variance $\sigma^2 = 49$. A random sample of size $n = 38$ is selected from this population.
  a. Find the approximate distribution of $\overline{X}$. Why is the central limit theorem necessary here?
  b. Find $P(\overline{X} < 49)$.
  c. Find $P(\overline{X} \ge 52)$.
  d. Find $P(49.5 \le \overline{X} \le 51.5)$.
  e. Find a value c such that $P(\overline{X} > c) = 0.15$.

7.35 Suppose X is a random variable with mean $\mu = 1000$ and standard deviation $\sigma = 100$. A random sample of size $n = 36$ is selected from this population.
  a. Find the approximate distribution of $\overline{X}$. Carefully sketch a graph of the probability density function.
  b. Find $P(\overline{X} > 975)$.
  c. Find $P(\overline{X} \le 1030)$.
  d. Find $P(\mu_{\overline{X}} - \sigma_{\overline{X}} \le \overline{X} \le \mu_{\overline{X}} + \sigma_{\overline{X}})$.
  e. Find a symmetric interval about the mean $\mu_{\overline{X}}$ such that $P(\mu_{\overline{X}} - c \le \overline{X} \le \mu_{\overline{X}} + c) = 0.95$.

7.36 Suppose X is a random variable with mean $\mu = 30$ and standard deviation $\sigma = 50$. A random sample of size $n = 40$ is selected from this population.
  a. Find the approximate distribution of $\overline{X}$. Carefully sketch a graph of the probability density function.
  b. Find $P(\overline{X} \ge 38)$.
  c. Find $P(20 \le \overline{X} \le 40)$.
  d. Find $P(\overline{X} < 15)$.
  e. Find a value of c such that $P(\overline{X} \le c) = 0.001$.

7.37 The figure below shows the graphs of the probability density functions for the random variable X, the random variable $\overline{X}$ for $n = 5$, and the random variable $\overline{X}$ for $n = 15$.

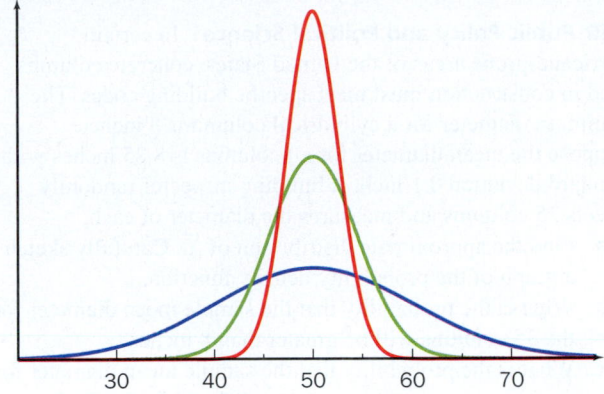

Identify each probability density function.

7.38. The figure below shows graphs of the probability density function for the random variable X and the approximate density functions for the random variable $\overline{X}$ for $n = 5$ and the random variable $\overline{X}$ for $n = 15$.

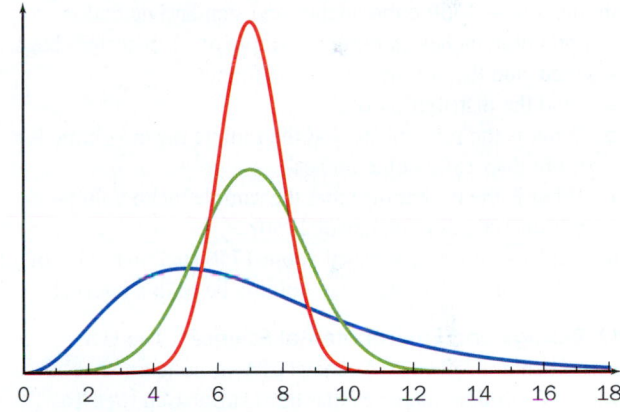

Identify each probability density function.

Applications

7.39 Public Health and Nutrition One measure of general health is your body mass index (BMI). Adults with a BMI between 18.5 and 24.9 are generally considered to have an ideal body weight for their height. Despite a reputation as a food-loving nation, the mean BMI for adults in France is 24.5.[11] Suppose the BMI for adults in France is normally distributed with standard deviation 1.3.
  a. Suppose one adult from France is selected at random. What is the probability that the person's BMI is more than 25.5?
  b. Suppose 10 adults from France are selected at random. What is the probability that the sample mean BMI is greater than 25.5?
  c. What is the probability that the sample mean BMI is less than 24?

d. A certain health spa claims its clients have BMI between 23.5 and 24. Suppose eight health spa clients are selected at random. What is the probability that the sample mean will be in this interval?

7.40 Public Policy and Political Science   In certain hurricane-prone areas of the United States, concrete columns used in construction must meet specific building codes. The minimum diameter for a cylindrical column is 8 inches. Suppose the mean diameter for all columns is 8.25 inches with standard deviation 0.1 inch. A building inspector randomly selects 35 columns and measures the diameter of each.

a. Find the approximate distribution of $\overline{X}$. Carefully sketch a graph of the probability density function.

b. What is the probability that the sample mean diameter for the 35 columns will be greater than 8 inches?

c. What is the probability that the sample mean diameter for the 35 columns will be between 8.2 and 8.4 inches?

d. Suppose the standard deviation is 0.15 inch. Answer parts (a), (b), and (c) using this value of σ.

7.41 Manufacturing and Prodcut Development   A large part of the luggage market is made up of overnight bags. These bags vary by weight, exterior appearance, material, and size. Suppose the volume of overnight bags is normally distributed with mean $\mu = 1750$ cubic inches and standard deviation $\sigma = 250$ cubic inches. A random sample of 15 overnight bags is selected, and the volume of each is found.

a. Find the distribution of $\overline{X}$.

b. What is the probability that the sample mean volume is more than 1800 cubic inches?

c. What is the probability that the sample mean volume is within 100 cubic inches of 1750?

d. Find a symmetric interval about 1750 such that 95% of all values of the sample mean volume lie in this interval.

7.42 Biology and Environmental Science   The U.S. Environmental Protection Agency (EPA) is concerned about pollution caused by factories that burn sulfur-rich fuel. To decrease the impact on the environment, factory chimneys must be high enough to allow pollutants to dissipate over a larger area. Assume the mean height of chimneys in these factories is 100 meters (an EPA-acceptable height) with standard deviation 12 meters. A random sample of 40 chimney heights is obtained.

a. What is the probability that the sample mean height for the 40 chimneys is greater than 102 meters?

b. What is the probability that the sample mean height is between 101 and 103 meters?

c. Suppose the sample mean is 98.5 meters. Is there any evidence to suggest that the true mean height for chimneys is less than 100 meters? Justify your answer.

7.43 Manufacturing and Product Development   The manager at an HEB grocery store in Corpus Christi, Texas, suspects a supplier is systematically underfilling 12-ounce bags of potato chips. To check the manufacturer's claim, a random sample of 100 bags of potato chips is obtained, and each bag is carefully weighed. The sample mean is 11.9 ounces. Assume $\sigma = 0.3$ ounces.

a. Find the probability that the sample mean is 11.9 ounces or less.

b. How can this probability in part (a) be so small when 11.9 is so *close* to the population mean $\mu = 12$?

c. From your answer to part (a), is there any evidence to suggest that the mean weight of bags of potato chips is less than 12 ounces?

7.44 Psychology and Human Behavior   In attempting to flee from police, criminals often barricade themselves inside a building and create a police standoff. The criminal is usually armed with a dangerous weapon and may hold hostages. Suppose the mean length of a police standoff, ending with some sort of resolution, is 6.5 hours with standard deviation 4 hours. A random sample of 35 police standoffs is selected.

a. Find the distribution of the sample mean.

b. What is the probability that the sample mean for the 35 police standoffs will be greater than 7 hours?

c. A new psychological technique was used in negotiations with the criminals involved in the 35 police standoffs. Suppose the sample mean for the 35 police standoffs is $\overline{x} = 5.1$ hours. Is there any evidence to suggest that the mean police standoff time is lower when this new technique is used?

7.45 Biology and Environmental Science   During monsoon season in India, the mean amount of rainfall is 89 centimeters.[12] Suppose the standard deviation is 5 centimeters and 30 monsoon seasons are selected at random.

a. What is the probability that the sample mean rainfall is less than 87 centimeters?

b. What is the probability that the sample mean rainfall is greater than 91.5 centimeters?

c. Find a symmetric interval about the mean, 89, such that the probability that the sample mean lies in this interval is 0.90.

7.46 Biology and Environmental Science   Carbon dioxide (CO_2) is one of the primary gases contributing to the greenhouse effect and global warming. The mean amount of CO_2 in the atmosphere for March 2013 was 397.34 parts per million (ppm).[13] Suppose 40 atmospheric samples are selected at random in May 2013 and the standard deviation for CO_2 in the atmosphere is $\sigma = 20$ ppm.

a. Find the probability that the sample mean CO_2 level is less than 393 ppm.

b. Find the probability that the sample mean CO_2 level is between 395 and 403 ppm.

c. Suppose the sample mean CO_2 level is 400. Is there any evidence to suggest that the population mean CO_2 level has increased? Justify your answer.

7.47 Public Health and Nutrition   Typhoid fever is more common in developing countries with poor sanitation and greater incidence of food contamination. The mean duration of this illness is approximately 4 weeks.[14] Periodically, a team of doctors randomly selects 10 patients with typhoid fever and carefully measures the duration of the illness. If the mean duration is greater than 4.5 weeks, then a health alert is issued. Suppose the duration of this illness is normally distributed with standard deviation one week.

a. Find the probability that the sample mean duration is less than 4.5 weeks.

b. Find the probability that a health alert is issued.

c. Suppose the true mean duration has increased to 4.3 weeks. What is the probability that a health alert will be issued?

7.48 Biology and Environmental Science   There are many regulations for catching lobsters off the coast of New England, including required permits, allowable gear, and size prohibitions. The Massachusetts Division of Marine Fisheries requires a minimum carapace length measured from a rear eye socket to the center line of the body shell. Any lobster measuring less than $3\frac{1}{4}$ inches must be returned to the ocean. For all lobster caught, suppose the carapace length is normally distributed with mean $4\frac{1}{8}$ inches and standard deviation 1 inch. A random sample of 15 lobsters is obtained.

a. Find the distribution of the sample mean carapace length.

b. What is the probability that the sample mean carapace length is more than $4\frac{1}{2}$ inches?

c. What is the probability that the sample mean carapace length is between $3\frac{7}{8}$ and $4\frac{1}{8}$ inches?

d. If the sample mean carapace length is less than $3\frac{1}{2}$ inches, a lobsterman will look for other places to set his traps. What is the probability that a lobsterman will be looking for a different location?

7.49 Sports and Leisure   One measure of an athlete's ability is the height of his or her vertical leap. Many professional basketball players are known for their remarkable vertical leaps, which lead to amazing dunks. DJ Stephens is the current vertical-leap record holder at 46 inches. However, the mean vertical leap of all NBA players is 28 inches.[15] Suppose the standard deviation is 7 inches and 36 NBA players are selected at random.

a. What is the probability that the mean vertical leap for the 36 players will be less than 26 inches?

b. What is the probability that the mean vertical leap for the 36 players will be between 27.5 and 28.5 inches?

c. A high-priced athletic trainer has been hired to work with a group of NBA players to improve their hip flexibility, which should improve their vertical leap. After one month of training, the mean vertical leap for 50 of these players selected at random was 29.75 inches. Is there any evidence to suggest this flexibility program has increased the mean vertical leap (from $\mu = 28$ inches)?

7.50 Physical Sciences   The ozone hole is a region in the Southern Hemisphere that has increased since 1979. The mean ozone hole area (in million km^2) in 2012 was 17.9[16] with standard deviation 5.6. Suppose 33 days are selected at random and the ozone hole is measured each day.

a. What is the probability that the sample mean ozone hole area is less than 15.5 million km^2?

b. What is the probability that the sample mean ozone hole area is between 16 and 18 million km^2?

c. Some researchers suggest that increased use of chemicals has caused the ozone hole area to increase. Suppose the sample mean for the 33 days is 20.7 million km^2. Is there any evidence to suggest that the mean ozone hole area has increased? Justify your answer.

Extended Applications

7.51 Public Health and Nutrition   Recent research studies suggest that lycopene, found in tomatoes, may reduce the risk of certain cancers. According to the USDA/NCC Carotenoid Database for U.S. Foods, the mean amount of lycopene in a medium-sized fresh tomato is 3.7 milligrams (mg). Suppose the amount of lycopene is a normal random variable with standard deviation 1 mg. Consider the sample mean based on 5 observations, $\overline{X}_5$, and the sample mean based on 20 observations, $\overline{X}_{20}$.

a. Find the distributions for $\overline{X}_5$ and $\overline{X}_{20}$.

b. Carefully sketch the probability density functions for X, $\overline{X}_5$, and $\overline{X}_{20}$ on the same coordinate axes.

c. Find $P(X < 3)$, $P(\overline{X}_5 < 3)$, and $P(\overline{X}_{20} < 3)$. Explain why these values are different.

d. Find $P(3.6 \le X \le 3.8)$, $P(3.6 \le \overline{X}_5 \le 3.8)$, and $P(3.6 \le \overline{X}_{20} \le 3.8)$. Explain why these values are different.

7.52 Sports and Leisure   A health club recently added tanning booths in addition to a weight room, pool, and racquetball courts. For insurance purposes, an employee maintains careful records of the length of time each member spends in a tanning booth. The mean length of time spent tanning is 15 minutes with standard deviation 2 minutes. Consider a random sample of 35 members who use a tanning booth.

a. Find the distribution of the total time spent tanning by the 35 members, T.

b. Find the probability that the total time spent tanning is between 8 and 9 hours.

c. If the total time spent tanning is more than 9.2 hours, an employee must work overtime. What is the probability that the employee works overtime on a day when 35 members tan?

d. Find a value t such that $P(T \ge t) = 0.01$.

7.53 Sports and Leisure   *The Tonight Show* with Johnny Carson lasted for 30 years and featured movie stars, comedians, animal acts, and Carnac the Magnificent. The first part of the show was reserved for Johnny's monologue. The mean length of a monologue was 12 minutes with standard deviation 45 seconds. Suppose 40 shows are selected at random and the length of each monologue is recorded.

a. Find the distribution of the total monologue time for the 40 shows, T.

b. Find $P(T > 500)$.

c. Find $P(470 \le T \le 490)$.

d. After his retirement, Johnny Carson's company started to sell DVDs of his monologues. Suppose one DVD holds 7.9 hours of recording. What is the probability that the company will be able to fit 40 randomly selected monologues onto one DVD?

7.54 Manufacturing and Product Development   A (destructive) tensile test is standard procedure for testing a cross-wire weld. Suppose a certain weld is designed to withstand a force with mean 0.8 kilonewtons (kN). To check the quality of welds, a manufacturer randomly selects 25 welds (every hour) and performs a tensile test on each weld. If the mean force required (for the 25 welds) is between 0.75 and 0.85 kN, the process is allowed to continue. Otherwise, it is shut down. Suppose the force distribution is normal with standard deviation 0.1 kN.

  a. If the true mean is 0.8 kN, what is the probability that the process will be shut down? This probability represents the chance of making one kind of error.

  b. If the true mean is 0.82 kN, what is the probability that the process will be allowed to continue? What if $\mu = 0.84$ kN? These probabilities represent the chance of making a different kind of error.

  c. Suppose the process is allowed to continue if the mean force required is between 0.76 and 0.84 kN. Answer parts (a) and (b) with this new interval.

7.55 Manufacturing and Product Development Shetland wool is considered some of the finest in the world because it is soft, durable, and easy to spin. The mean fleece weight from a typical sheep is 3.25 pounds with a standard deviation of 0.4 pound. Suppose a farmer has 100 sheep ready to be sheared.

  a. Find the distribution for the total weight of fleece from the 100 sheep, T.

  b. What is the probability that the total fleece weight is less than 323 pounds?

  c. What is the probability that the total fleece weight is between 330 and 340 pounds?

  d. Find a value t such that $P(T \geq t) = 0.15$.

7.56 Sports and Leisure A wingsuit is a specially designed jumpsuit with added surface area to provide more lift. Extreme-sports enthusiasts wear this suit and jump from planes or a fixed object (BASE jumping, where BASE stands for Building, Antenna, Span, Earth). One of the most popular BASE environments is Troll Wall in Norway. Suppose the mean flight time for a wingsuit jump at Troll Wall is 22.5 seconds with standard deviation 2.3 seconds. Thirty-two jumpers are selected at random to test a new wingsuit at Troll Wall.

  a. What is the probability that the sample mean flight time is less than 22 seconds?

  b. Suppose the sample mean flight time for these jumpers is 23.15 seconds. Is there any evidence to suggest that this new wingsuit increases the mean flight time?

  c. Find a value w such that $P(\overline{X} \leq w) = 0.005$.

Challenge

7.57 Fuel Consumption and Cars Companies receiving large shipments of raw materials of any product often use a specific plan for accepting the entire shipment. An *acceptance sampling plan* usually includes the sample size for close inspection, the *acceptance* criterion, and the *rejection* criterion. For a given plan, the *operating characteristic* (OC) *curve* shows the probability of accepting the entire lot as a function of the actual quality level.

Standard clip-on weights for steel rims on automobiles (used when tires are balanced) are available in $\frac{1}{4}$-ounce to 6-ounce sizes. Suppose an automobile garage receives a large shipment of 2-ounce weights. A garage mechanic will select a random sample of 30 weights and weigh each one on a precise scale. If the sample mean is within 0.05 ounce of the printed weight (of 2 ounces), then the shipment is accepted. Otherwise, the entire shipment is rejected and returned to the manufacturer. Suppose the population standard deviation is 0.13 ounce.

  a. Find the probability of accepting the entire shipment if the true population mean weight is 1.86, 1.88, 1.90, 1.92, 1.94, 1.96, 1.98, 2.00, 2.02, 2.04, 2.06, 2.08, 2.10, 2.12, of 2.14 ounces.

  b. Plot the probability of accepting the entire shipment versus the true population mean weight. The resulting graph is the OC curve for the given acceptance sampling plan.

7.58 Normal Approximation to the Binomial Distribution Consider a binomial experiment with n trials and probability of success p. If we assign a 0 to each failure and a 1 to each success, then the binomial random variable can be defined as a sum. By the central limit theorem, as n increases the distribution of X (the sum) approaches a normal distribution with mean np and variance $np(1 - p)$. Suppose X is a binomial random variable with $n = 30$ and probability of success $p = 0.5$.

  a. Construct a probability histogram for the binomial random variable X. Find $P(12 \leq X \leq 16)$ using the binomial distribution.

  b. Find the approximate normal distribution for X. Find $P(12 \leq X \leq 16)$ using the normal distribution for X.

  c. Compare the probabilities found in parts (a) and (b).

  d. Find $P(11.5 \leq X \leq 16.5)$ using the normal distribution for X. Compare this answer with the probability in part (a). Why do you think this is a much better approximation?

7.59 Manufacturing and Product Development A hardware store has 20 interior plantation shutter sets in stock. The width (in inches) of each set is given in the table below.

SHUTTER

| 30 | 30 | 28 | 34 | 36 | 28 | 34 | 36 | 24 | 35 |
| 28 | 30 | 32 | 30 | 44 | 34 | 22 | 32 | 20 | 30 |

  a. Find the (population) mean width of the 20 shutters.

  b. Use technology to generate at least 100 random samples of size five (without replacement). Find the sample mean width for each sample.

  c. Construct a histogram of the sample means found in part (a). Describe the distribution.

7.3 The Distribution of the Sample Proportion

In Section 7.2 the sampling distribution of the sample mean was introduced. The central limit theorem helps if the underlying population is not normal. Knowing the distribution of $\overline{X}$, we can use the sample mean to make an inference about the population mean μ. Similarly, we are often interested in drawing a conclusion about the population proportion

p (the probability of a success). For example, a politician might want to estimate the proportion of voters in a district in favor of a certain highway bill, or a quality-control supervisor might need to estimate the true proportion of defective parts in a large shipment.

It seems reasonable to use a value of the sample proportion, $\hat{p}$, to make an inference concerning the population proportion p. Therefore, knowledge of the **sampling distribution of the sample proportion** is necessary. We need to completely characterize the variability of this statistic.

Consider a sample of n individuals or objects (or trials) and let X be the number of successes in the sample. The sample proportion (introduced in Section 3.1) is defined to be

$$\hat{P} = \frac{X}{n} = \frac{\text{the number of successes in the sample}}{\text{the sample size}} \tag{7.1}$$

The sample proportion is simply the proportion of successes in the sample, or a relative frequency.

An approach similar to the one in Section 7.2 can be used to approximate the distribution of $\hat{P}$: Generate lots of sample proportions, construct a histogram, and try to characterize the distribution in terms of shape, center, and variability. The sampling distribution is summarized below.

The Sampling Distribution of $\hat{P}$

Let $\hat{P}$ be the sample proportion of successes in a sample of size n from a population with true proportion of success p.

1. The mean of $\hat{P}$ is the true population proportion.

   In symbols: $\mu_{\hat{P}} = p$.

2. The variance of $\hat{P}$ is $\sigma_{\hat{P}}^2 = \dfrac{p(1-p)}{n}$.

   The standard deviation of $\hat{P}$ is $\sigma_{\hat{P}} = \sqrt{\dfrac{p(1-p)}{n}}$.

3. If n is large and both $np \geq 5$ and $n(1-p) \geq 5$, then the distribution of $\hat{P}$ is approximately normal.

   In symbols: $\hat{P} \overset{\bullet}{\sim} N(p, p(1-p)/n)$.

As n increases, the distribution of $\hat{P}$ approaches a normal distribution. There is no threshold value for n. The larger the value of n and the closer p is to 0.5, the better is the approximation.

A CLOSER LOOK

1. It may be a little surprising to learn that $\hat{P}$ is approximately normal. However, the sample proportion can be written as a sample mean. Assign 0 to each failure and 1 to each success. X, the total number of successes, is a sum (of 0s and 1s). Therefore, the sample proportion, X/n, is really a sample mean. Remember, the central limit theorem says that the sample mean is approximately normal for n sufficiently large.

2. Because the mean of $\hat{P}$ is the true population proportion, $\hat{P}$ is an unbiased estimator for p.

3. A *large* sample isn't enough for normality. The two products np and $n(1-p)$ must *both* be greater than or equal to 5. This is called the *nonskewness criterion*. It guarantees that the distribution of $\hat{P}$ is approximately symmetric, that is, centered far enough away from 0 or 1.

4. To compute a probability involving the sample proportion, treat $\hat{P}$ like any other normal random variable: Standardize and use cumulative probability where appropriate.   ●

The following example illustrates the properties of $\hat{P}$ and the technique for computing probabilities associated with this random variable.

Solution Trail 7.10

KEYWORDS

- 110 people
- 45%
- Sample proportion

TRANSLATION

- $n = 110$ trials
- Probability of a success (changing policies) is $p = 0.45$
- $\hat{P}$

CONCEPTS

- Distribution of $\hat{P}$

VISION

This question involves the random variable $\hat{P}$. Use the properties of $\hat{P}$ to find the distribution, translate each question into a probabiity statement, and write one equivalent statement involving Z.

Example 7.10 Discount Double Check

In early 2013, the number of people looking to change auto insurance companies reached a record low. However, for the population of those people who were looking for a better deal, 45% did change companies.[17] Suppose 110 people looking for a better deal on auto insurance are selected at random and the number who actually switch policies is determined.

a. Find the distribution of the sample proportion of people who switch policies, $\hat{P}$. Carefully sketch the probability density function for this random variable.

b. What is the probability that the sample proportion (for the 110 people selected) is greater than 0.50?

c. Find the probability that the sample proportion will be between 0.37 and 0.47.

SOLUTION

a. For $n = 110$ and $p = 0.45$, check the nonskewness criteria.

$$np = (110)(0.45) = 49.5 \geq 5 \qquad \text{and} \qquad n(1 - p) = (110)(0.55) = 60.5 \geq 5$$

Both inequalities are satisfied. The distribution of $\hat{P}$ is approximately normal with

$$\mu_{\hat{p}} = p = 0.45 \qquad \text{and} \qquad \sigma_{\hat{p}}^2 = \frac{p(1 - p)}{n} = \frac{(0.45)(0.55)}{120} = 0.00225$$

In symbols: $\qquad \hat{P} \overset{\bullet}{\sim} N(0.45, 0.00225) \qquad \sigma_{\hat{p}} = \sqrt{0.00225}.$

Figure 7.16 shows a graph of the probability density function and the associated probability.

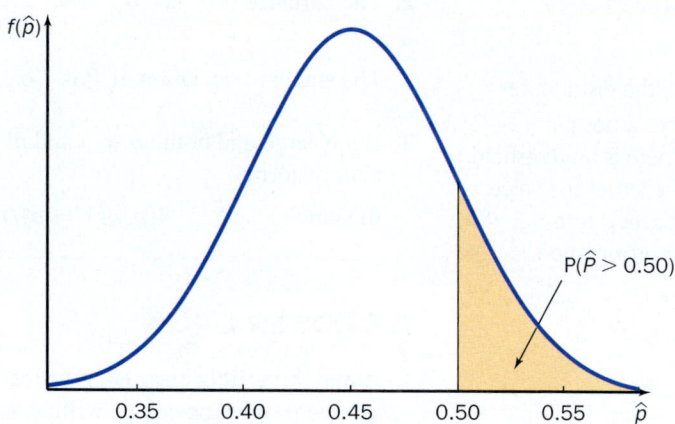

$f(\hat{p})$

$P(\hat{P} > 0.50)$

0.35   0.40   0.45   0.50   0.55 $\hat{p}$

Figure 7.16 The probability density function for $\hat{P}$. The shaded area represents $P(\hat{P} > 0.50)$.

b. $P(\hat{P} > 0.50)$ *Probability that the sample proportion is greater than 0.50.*

$$= P\left(\frac{\hat{P} - 0.45}{\sqrt{0.00225}} > \frac{0.50 - 0.45}{\sqrt{0.00225}}\right) \qquad \textit{Standardize.}$$

$$= P(Z > 1.05) \qquad \textit{Equation 6.8; simplify.}$$

$$= 1 - P(Z \leq 1.05) \qquad \textit{Use cumulative probability.}$$

$$= 1 - 0.8531 \qquad \textit{Use Table III in the Appendix.}$$

$$= 0.1469$$

The probability that the sample proportion is greater than 0.50 is 0.1469.

Standardization illustrated part (c):

P(0.37 ≤ $\hat{P}$ ≤ 0.47)

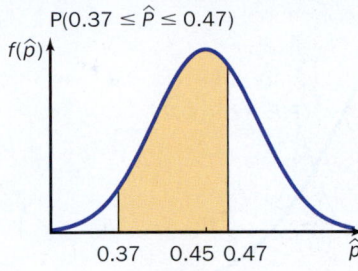

P(−1.69 ≤ Z ≤ 0.42)

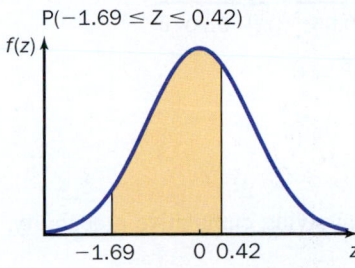

Remember, it is very important for the sample to be selected *randomly* from the underlying population. Otherwise, the results are not valid.

Solution Trail 7.11a

KEYWORDS

- 150 CEOs
- 60% of all CEOs
- Sample proportion

TRANSLATION

- $n = 150$ trials
- Probability of a success (a belief in too many regulations) is $p = 0.60$
- $\hat{P}$

CONCEPTS

- Distribution of $\hat{P}$

VISION

Knowing the distribution of $\hat{P}$, we can work backward to find the value r. Draw a picture to illustrate the probability statement, standardize, and use (inverse) cumulative probability.

Another way to ask this question: Find the third quartile of the $\hat{P}$ distribution.

---

c. P(0.37 ≤ $\hat{P}$ ≤ 0.47)   Probability the sample proportion is between 0.37 and 0.47.

$$= P\left(\frac{0.37 - 0.45}{\sqrt{0.00225}} \le \frac{\hat{P} - 0.45}{\sqrt{0.00225}} \le \frac{0.47 - 0.45}{\sqrt{0.00225}}\right)$$ Standardize.

$$= P(-1.69 \le Z \le 0.42)$$ Equation 6.8; simplify.

$$= P(Z \le 0.42) - P(Z \le -1.69)$$ Use cumulative probability.

$$= 0.6628 - 0.0455$$ Use Table III in the Appendix.

$$= 0.6173$$

Figure 7.17 and 7.18 show technology solutions to parts (b) and (c).

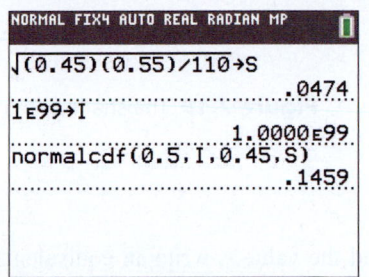

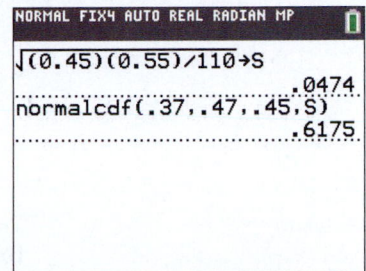

Figure 7.17 P($\hat{P}$ > 0.50): right-tail probability.

Figure 7.18 P(0.37 ≤ $\hat{P}$ ≤ 0.47).

TRY IT NOW   GO TO EXERCISE 7.70

The following example illustrates an inverse cumulative probability problem and an inference associated with the random variable $\hat{P}$.

Example 7.11 Too Many Regulations

Company executives often complain that there are too many government regulations. Stifling rules and endless bureaucracy may limit creativity, new-product research, and corporate profits. Suppose 60% of all CEOs believe there are too many government regulations for business. A random sample of 150 CEOs is obtained, and each is asked whether he or she believes there are too many government regulations.

a. Find a value r such that the probability the sample proportion is greater than r is 0.25.

b. In recent years, big business has lobbied politicians to relax regulations to stimulate the economy. Suppose the sample proportion for the 150 CEOs is 0.56. Is there any evidence to suggest that the true proportion of CEOs who believe there are too many regulations has decreased?

SOLUTION

For $n = 150$ and $p = 0.60$, check the nonskewness criterion.

$$np = (150)(0.60) = 90 \ge 5 \quad \text{and} \quad n(1 - p) = (150)(0.40) = 60 \ge 5$$

Both inequalities are satisfied. The distribution of $\hat{P}$ is approximately normal with

$$\mu_{\hat{P}} = p = 0.60 \quad \text{and} \quad \sigma_{\hat{P}}^2 = \frac{p(1 - p)}{n} = \frac{(0.60)(0.40)}{150} = 0.0016$$

In symbols: $\hat{P} \overset{\bullet}{\sim} N(0.60, 0.0016)$ $\sigma_{\hat{P}} = \sqrt{0.0016} = 0.04$.

a. Find a value r such that P($\hat{P}$ > r) = 0.25. Figure 7.18 illustrates this probability statement.

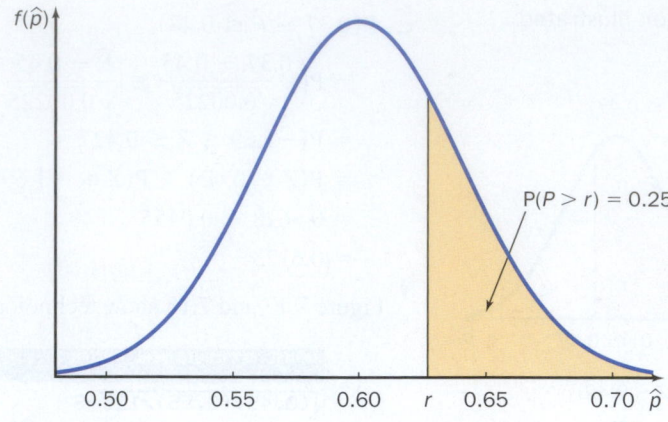

Figure 7.19 The shaded area represents $P(\hat{P} > r)$.

To find the value r, write an equivalent expression involving cumulative probability. Standardize and work backward in Table III.

$$P(\hat{P} \le r) = P\left(\frac{\hat{P} - 0.60}{0.04} \le \frac{r - 0.60}{0.04}\right) \qquad \text{Standardize.}$$

$$= P\left(Z \le \frac{r - 0.60}{0.04}\right) = 0.75 \qquad \text{Equation 6.8.}$$

There is no further simplification, but the resulting probability statement involves Z and is cumulative probability. Find a value in the body of Table III in the Appendix that is as close to 0.75 as possible. Set the corresponding z value equal to $\left(\frac{r - 0.60}{0.04}\right)$, and solve for r.

$$\frac{r - 0.60}{0.04} = 0.675 \qquad \text{Table III in the Appendix; interpolation.}$$

$$r - 0.60 = 0.027 \qquad \text{Multiply both sides by 0.04.}$$

$$r = 0.627 \qquad \text{Add 0.60 to both sides.}$$

The probability that the sample proportion is greater than 0.627 is 0.25. See Figure 7.19 for a technology solution.

b. Follow the usual four-step inference procedure and consider a tail probability as a measure of likelihood.

Claim: 60% of all CEOs believe there are too many government regulations.

$$p = 0.60 \rightarrow \hat{P} \overset{\bullet}{\sim} N(0.60, 0.04).$$

Experiment: The proportion of CEOs who believe there are too many government regulations is $\hat{p} = 0.56$.

Likelihood: Because $\hat{p} = 0.56$ is to the left of the mean ($p = 0.60$), consider a left-tail probability as a measure of likelihood.

$$P(\hat{P} \le 0.56) = P\left(\frac{\hat{P} - 0.60}{0.04} \le \frac{0.56 - 0.60}{0.04}\right) \qquad \text{Standardize.}$$

$$= P(Z \le -1.00) \qquad \text{Equation 6.8; simplify.}$$

$$= 0.1587 \qquad \text{Table III in the Appendix.}$$

NORMAL FIX4 AUTO REAL RADIAN MP

invNorm(0.75,0.6,0.04)
.6270

Figure 7.19 A technology solution: Use `invNorm` to solve an inverse cumulative probability problem.

 Solution Trail 7.11b

KEYWORDS
- Is there any evidence?

TRANSLATION
- Use available evidence to draw a reasonable conclusion.

CONCEPTS
- Inference procedure

VISION

To decide whether the observed sample proportion is reasonable, we need to follow the four-step inference procedure. Consider the claim, the experiment, and the likelihood of the experimental outcome.

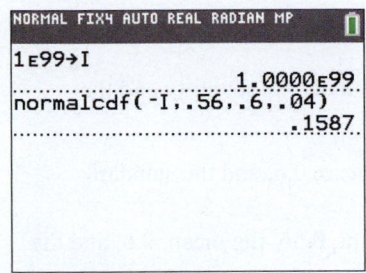

Figure 7.21 $\hat{P}$ is approximately normal, so use `normalcdf`.

Note: Even though the sample proportion can not be less than 0, −1E99 was used in this technology solution because a normal distribution is defined for all real numbers. Most of the time, when the approximation is good, using 0 (or 1 as a right bound) will not change the probability.

Standardization illustrated part (b):

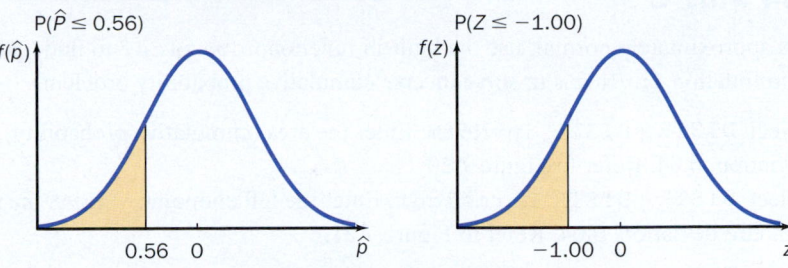

In statistical inference problems, any probability less than or equal to 0.05 is considered small.

Conclusion: This probability is larger than 0.05, so it is reasonable to observe a sample proportion of 0.56 or smaller. There is no evidence to suggest that the claim of $p = 0.60$ is wrong.

Figure 7.21 shows a technology solution.

TRY IT NOW   GO TO EXERCISE 7.75

Technology Corner

Procedure: Solve probability questions involving the sample proportion $\hat{P}$.
Reconsider: Example 7.11, solutions, and interpretations.

CrunchIt!

If $\hat{P}$ is approximately normal, use the Distribution Calculator as described in Section 6.2 to find probabilities and to solve inverse cumulative probability problems.

1. Select Distribution Calculator; Normal. Enter the mean, 0.60, and the standard deviation, 0.04. Under the Quantile tab, enter the cumulative probability and click Calculate. See Figure 7.22.
2. Select Distribution Calculator; Normal. Enter the mean, 0.60, and the standard deviation, 0.04. Under the Probability tab, select ≤ and enter 0.56 for the endpoint. See Figure 7.23.

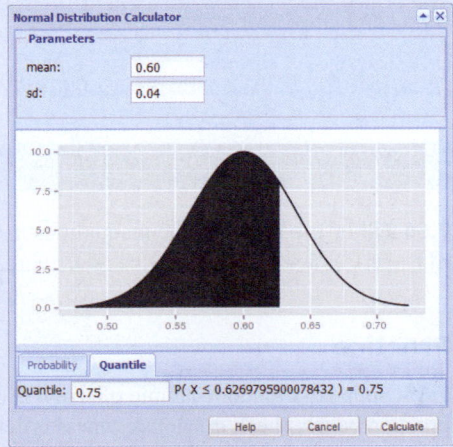

Figure 7.22 A solution to $P(\hat{P} \leq r) = 0.75$.

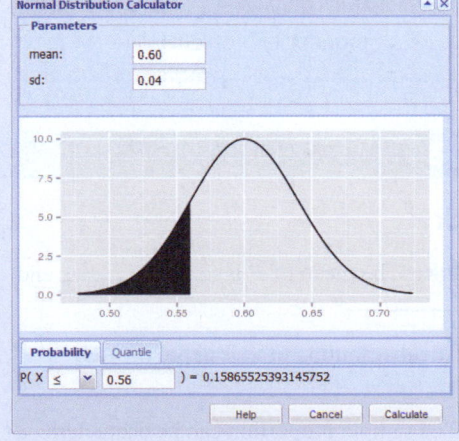

Figure 7.23 $P(\hat{P} \leq 0.56)$.

TI-84 Plus C

If $\widehat{P}$ is approximately normal, use the built-in function `normalcdf` to find (calculate) cumulative probability and the built-in function `invNorm` to solve inverse cumulative probability problems.

1. Select DISTR; DISTR; invNorm. Enter the area (cumulative probability), 0.75, the mean, 0.6, and the standard deviation, 0.04. Refer to Figure 7.20.
2. Select DISTR; DISTR; normalcdf. Enter the left endpoint, −1E99, the right endpoint, 0.56, the mean, 0.6, and the standard deviation, 0.04. Refer to Figure 7.21.

Minitab

As in Chapter 6 for normal random variables, use the built-in functions to find cumulative probability or inverse cumulative probability. These functions may be accessed through a graphical input window: Calc; Probability Distributions; Normal, Graph; Probability Distribution Plot; or by using the command language.

1. Use the function INVCDF to solve inverse cumulative probability problems. See Figure 7.24.
2. Use the function CDF to find cumulative probability. See Figure 7.25.

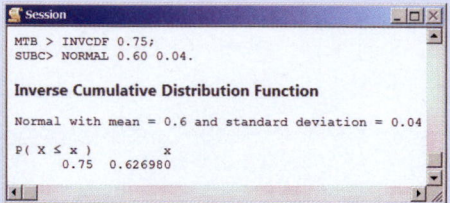

Figure 7.24 Use INVCDF to solve inverse cumulative probability problems.

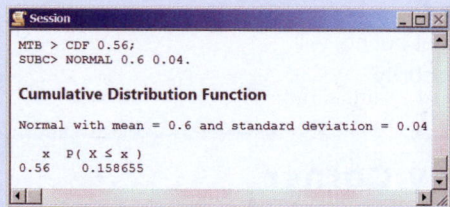

Figure 7.25 Use CDF to find cumulative probability.

Excel

As in Chapter 6 for normal random variables, use the built-in function `NORM.DIST` to compute cumulative probability associated with a normal random variable, and the function `NORM.INV` to solve inverse cumulative probability problems.

1. Use the function `NORM.INV` to find the value r such that $P(\widehat{P} \leq r) = 0.75$. See Figure 7.26.
2. Use the function `NORM.DIST` to find cumulative probability. See Figure 7.27.

A	B
0.6270	= NORM.INV(0.75,0.6,0.04)

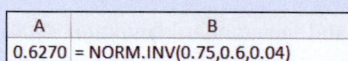

Figure 7.26 Use `NORM.INV` to solve inverse cumulative probability problems.

A	B
0.1587	= NORM.DIST(0.56,0.6,0.04,TRUE)

Figure 7.27 Use `NORM.DIST` to find cumulative probability.

SECTION 7.3 EXERCISES

Concept Check

7.60 Fill in the Blank The mean of $\widehat{P}$ is __________ and the variance is __________.

7.61 True/False The distribution of $\widehat{P}$ is approximately normal for any sample size n.

7.62 True/False The sample proportion can be characterized as a sample mean.

7.63 True/False The sample proportion is an unbiased estimator for p.

7.64 Fill in the Blank The inequalities $np \geq 5$ and $n(1-p) \geq 5$ are called the __________.

Practice

7.65 Suppose a random sample of size n is obtained. In each problem below, check the nonskewness criterion, and find the distribution of the sample proportion $\widehat{P}$.

a. $n = 100$, $p = 0.25$. **b.** $n = 150$, $p = 0.90$.
c. $n = 100$, $p = 0.75$. **d.** $n = 1000$, $p = 0.85$.
e. $n = 5000$, $p = 0.006$.

7.66 Suppose a random sample of size $n = 200$ is obtained from a population with probability of success $p = 0.40$. Find each of the following probabilities.

   a. $P(\hat{P} \leq 0.37)$.

   b. $P(\hat{P} > 0.45)$.

   c. $P(0.38 \leq \hat{P} \leq 0.42)$.

   d. $P(\hat{P} < 0.33)$ or $P(\hat{P} > 0.47)$.

7.67 Suppose a random sample of size $n = 500$ is obtained from a population with probability of success $p = 0.50$. Find each of the following probabilities.

   a. $P(0.52 > \hat{P})$.   b. $P(\hat{P} \geq 0.47)$.

   c. $P(0.44 \leq \hat{P} < 0.49)$.   d. $P(0.45 \leq \hat{P} < 0.55)$.

7.68 Suppose a random sample of size $n = 80$ is obtained from a population with probability of success $p = 0.35$.

   a. Find a value a such that $P(\hat{P} \leq a) = 0.10$.

   b. Find a value b such that $P(\hat{P} > b) = 0.01$.

   c. Find a value c such that $P(0.35 - c \leq \hat{P} \leq 0.35 + c) = 0.95$.

7.69 Suppose a random sample of size $n = 1000$ is obtained from a population with probability of success $p = 0.25$.

   a. Find a value a such that $P(\hat{P} \leq a) = 0.05$.

   b. Find a value b such that $P(\hat{P} > b) = 0.005$.

   c. Find a value c such that $P(0.25 - c \leq \hat{P} \leq 0.25 + c) = 0.99$.

   d. Find the quartiles of the distribution of $\hat{P}$.

Applications

7.70 Public Health and Nutrition Approximately 70% of American adults monitor a specific health indicator, for example weight, diet, or exercise.[18] Suppose a random sample of 250 American adults is obtained, and all were asked if they monitor their health in some way.

   a. Find the distribution of the sample proportion of American adults who monitor their health.

   b. Find the probability that the sample proportion is less than 0.66.

   c. Find the probability that the sample proportion is more than 0.71.

   d. Find the probability that the sample proportion is between 0.68 and 0.78.

7.71 Technology and the Internet According to a report from Microsoft, 24% of PCs worldwide are not adequately protected by antivirus software.[19] Suppose 200 PCs from around the world are selected at random.

   a. Find the distribution of the sample proportion of PCs that are not adequately protected.

   b. Find the probability that the sample proportion is less than 0.20.

   c. Find the probability that the sample proportion is more than 0.29.

   d. Find a value v such that the probability the sample proportion is less than v is 0.01.

7.72 Marketing and Consumer Behavior Movie trailers are designed to attract large audiences. However, according to a recent survey, 32% of Americans believe movie trailers give away too many of a movie's best scenes, that is, they reveal too much.[20] Suppose 250 Americans are selected at random and asked if they believe movie trailers reveal too much.

   a. Find the distribution of the sample proportion of Americans who believe movie trailers give away too much.

   b. Find the probability that the sample proportion is more than 0.35.

   c. Find the probability that the sample proportion is less than 0.25.

   d. Find a symmetric interval about the mean ($p = 0.32$) such that the probability that the sample proportion is in this interval is 0.90.

7.73 Sports and Leisure Anyone who bets money in a casino is classified as a winner if he or she wins more than he or she loses. Casino operators in Atlantic City, New Jersey, believe the proportion of all players who *go home* a winner is 0.46. Suppose 75 Atlantic City gamblers are selected at random.

   a. Find the sampling distribution of the proportion of gamblers who go home winners.

   b. Find the probability that the sample proportion is less than 0.40.

   c. Find the probability that the sample proportion is more than 0.45.

   d. Find a symmetric interval about the mean ($p = 0.46$) such that the probability that the sample proportion is in this interval is 0.99.

7.74 Marketing and Consumer Behavior According to a recent study, approximately 69% of consumers order food online using a mobile device, for example, a smartphone or tablet.[21] Consumers use mobile applications to check restaurant menus, consider fast-food options, and even to order a pizza. Suppose 220 consumers are randomly selected.

   a. Find the probability that the sample proportion of consumers who order food online using a mobile device is less than 0.65.

   b. Find the probability that the sample proportion of consumers who order food online using a mobile device is more than 0.72.

   c. Find a value t such that $P(\hat{P} < t) = 0.95$.

7.75 Public Health and Nutrition The U.S. Centers for Disease Control announced that skin allergies have risen among U.S. children to 12.5%.[22] Serious allergies can affect a child's education, sleep, and ordinary daily activities. Suppose 320 U.S. children are randomly selected and tested for skin allergies.

   a. Find the probability that the sample proportion of children with skin allergies is less than 0.10.

   b. Find the probability that the sample proportion of children with skin allergies is between 0.08 and 0.16.

   c. Suppose 150 U.S. children aged 5 to 9 are randomly selected and each is tested for skin allergies. Nine children are found to have skin allergies. Is there any evidence to suggest that children in this age group have a lower incidence rate of skin allergies? Justify your answer.

7.76 Education and Child Development Admissions offices at universities and colleges keep careful records of acceptance and yield rates. The yield rate is the percentage of admitted students who decide to accept an offer of admission. Historical data help them to plan for the incoming classes and guide some admission decisions. The yield rate for Yale University is 64.1%.[23] Suppose 100 students who were accepted to Yale are randomly selected.

 a. Find the probability that the sample proportion of those who enroll is less than 0.55.
 b. Find the probability that the sample proportion of those who enroll is between 0.60 and 0.65.
 c. Suppose the actual sample proportion of those who enroll is 0.70. Is there any evidence to suggest that the yield has increased? Justify your answer.

7.77 Business and Management A certain philanthropic organization funds one out of every ten grant proposals. Suppose a random sample of 300 grant proposals is obtained.

 a. Find the probability that the sample proportion of funded grant proposals is less than 0.075.
 b. Find the probability that the sample proportion of funded grant proposals is between 0.11 and 0.15.
 c. If the funding rate increases, the board of directors may become concerned about resources being depleted. Suppose the actual sample proportion of funded grant proposals is 0.16. Is there any evidence to suggest that the funding rate has increased? Justify your answer.

7.78 Sports and Leisure The fielding percentage for a major league baseball team is a measure of how well defensive players handle a batted ball (without error) and is based on total chances (putouts + assists + errors). One quarter into the 2013 baseball season, the Boston Red Sox fielding percentage was 0.984.[24] Suppose 150 chances are randomly selected.

 a. Find the distribution of the sample proportion of chances handled without error.
 b. Find the probability that the sample proportion is less than 0.97.
 c. Find a value f such that $P(\hat{P} \leq f) = 0.05$.
 d. If the team fielding percentage for these random chances is greater than 0.99, then everyone on the team receives a bonus. What is the probability of a team bonus?

7.79 Psychology and Human Behavior Many Americans own a DVR or subscribe to a special TV recording service. For example, the Hopper from DISH network allows users to record up to six HD programs at once. Despite this increased recording flexibility, Motorola Mobility recently announced that for U.S. TV viewers, 41% of the recorded content on DVRs is never watched.[25] Suppose 425 programs recorded on DVRs in the United States are randomly selected.

 a. Find the probability that the sample proportion of programs never watched is less than 0.37.
 b. Find the probability that the sample proportion is between 0.42 and 0.43.
 c. Find a value w such that $P(\hat{P} \geq w) = 0.01$.

7.80 Biology and Environmental Science Ships dumping garbage and ordinary beachgoers contribute to the increasing amount of trash that washes onto the shore and collects in the oceans all over the world. In 2012 the International Coastal Cleanup collected more than 9 million pounds of garbage from the world's oceans. Common debris items collected included food wrappers, bottle caps, and beverage bottles. For the United States, cigarettes accounted for 28.1% of all debris items.[26] A random sample of 430 debris items from a clean-up along a Texas beach was obtained.

 a. Find the probability that the proportion of cigarette debris items is greater than 0.31.
 b. Find the probability that the proportion of cigarette debris items is between 0.25 and 0.30.
 c. Suppose 94 of the debris items are cigarettes. Is there any evidence to suggest that the true proportion of cigarette debris items is different from 0.281? Justify your answer.

Extended Applications

7.81 Manufacturing and Product Development Low-quality coffee shipped from various locations in Europe tends to contain a high proportion of defective beans (beans composed of foreign matter; moldy, black, unripe, or fermented beans; or those known as stinkers). Suppose a shipper claims the proportion of defective beans is 0.07. A U.S. packaging company receives a huge shipment of coffee beans and randomly select 1000 beans. If the sample proportion of defective beans is more than 0.09, then the entire shipment will be returned to the supplier in Europe.

 a. If the true proportion of defective beans is 0.07 (as claimed), what is the probability that the shipment will be sent back?
 b. If the true proportion of defective beans is 0.08, what is the probability that the shipment will be accepted?

7.82 Manufacturing and Product Development McGuckin Hardware in Boulder, Colorado, routinely receives shipments of 4×8 foot sheets of $\frac{1}{2}$-inch-thick plywood. A plywood sheet may contain defects, for example, a knot, a split, or a deviation in wood structure. Defective sheets reduce profits because they are sold at a lower price. The manufacturer claims the proportion of all plywood sheets that are defective is 0.05. Suppose a large shipment of plywood sheets is received and 200 are randomly selected for inspection. If the sample proportion of defective sheets is more than 0.09, the entire shipment will be sent back to the supplier.

 a. If the true proportion of defective plywood sheets is 0.05, what is the probability that the entire shipment will be sent back?
 b. If the true proportion of defective plywood sheets is 0.03, what is the probability that the entire shipment will be sent back?
 c. If the true proportion of defective plywood sheets is 0.10, what is the probability that the shipment will be accepted?

7.83 Demographics and Population Statistics According to National Geographic, 28% of Canadians are of British

descent.[27] Suppose 150 Canadians in Alberta are randomly selected and 185 Canadians in Ontario are randomly selected.
   a. What is the probability that the sample proportion of those Canadians from Alberta of British descent is less than 0.23?
   b. What is the probability that the sample proportion of those Canadians from Ontario of British descent is less than 0.23?
   c. What is the probability that the sample proportion of those Canadians from both provinces is greater than 0.35?

Challenge

7.84 Manufacturing and Product Development   Suppose a company is receiving a large shipment of peel-and-stick vinyl floor tiles. The manufacturer claims the proportion of defective floor tiles is 0.05. The company will select a random sample of 200 floor tiles, carefully inspect each, and determine whether it

is defective. The acceptance sampling plan states: Accept the entire shipment if the sample proportion of defective tiles is 0.08 or less; otherwise, reject the entire shipment.
   a. Find the probability of accepting the entire shipment if the true proportion of defective floor tiles is 0.01, 0.02, 0.03, 0.04, 0.05, 0.06, 0.07, 0.08, 0.09, 0.10, 0.15.
   b. Plot the probability of accepting the entire shipment (on the y axis) versus the true proportion of defective floor tiles (on the x axis). The resulting graph is the OC curve for the given acceptance sampling plan.

7.85 Maximum Variance   For a fixed sample size n, find the value of p that maximizes the variance of the sample proportion, $\sigma_{\hat{P}}^2 = \dfrac{p(1-p)}{n}$.

Hint: Compute the value of the variance for several different values of p. Plot the variance (on the y axis) versus the value of p (on the x axis).

CHAPTER 7 SUMMARY

Concept	Page	Notation / Formula / Description
Parameter	296	A numerical descriptive measure of a population.
Statistic	296	Any quantity computed from values in a sample.
Sampling distribution	297	The probability distribution of a statistic.
Properties of the sample mean $\overline{X}$	306	$\mu_{\overline{X}} = \mu, \quad \sigma_{\overline{X}}^2 = \dfrac{\sigma^2}{n}$. If the underlying population is normal, then $\overline{X}$ is normal.
Central limit theorem	308	As the sample size n increases, the sampling distribution of $\overline{X}$ will increasingly approximate a normal distribution, with mean μ and variance σ^2/n, regardless of the shape of the underlying population distribution.
Distribution of the sample proportion of successes $\hat{P}$	319	$\mu_{\hat{P}} = p, \quad \sigma_{\hat{P}}^2 = \dfrac{p(1-p)}{n}$. If n is large and both $np \geq 5$ and $n(1-p) \geq 5$, then $\hat{P}$ is approximately normal.

CHAPTER 7 EXERCISES

APPLICATIONS

7.86 Manufacturing and Product Development   Up until the early 1900s, people who colored their hair used only herbs and natural dyes. Today, hair-coloring products contain two main ingredients: hydrogen peroxide and ammonia. The makers of a Clairol hair-color product claim the mean amount of hydrogen peroxide in each bottle is 0.10 mg/m³. Assume the standard deviation is 0.05 mg/m³. A random sample of these hair-color products was obtained, and the amount of hydrogen peroxide in each bottle was measured.   **HAIRDYE**
   a. Suppose the manufacturer's claim is true. Find the distribution of the sample mean.

   b. Is there any evidence to suggest the manufacturer is including too much hydrogen peroxide in the product? Justify your answer.

7.87 Physical Sciences   A floor slip tester is used to measure the safety of a floor by comparing the measured coefficient of static friction with accepted standards and guidelines. Several factors can affect floor safety, for example, dampness, polishes, and maintenance chemicals. A marble floor is consider safe if the coefficient of static friction is no greater than 0.5. A random sample of 50 rainy days was selected, and the coefficient of static friction for the marble floor was measured on each day. The resulting sample mean was 0.6. Is there any evidence to suggest the marble floor is unsafe on rainy days? Assume the underlying population standard deviation is 0.2 and justify your answer.

7.88 Marketing and Consumer Behavior In the fresh fruits and vegetables section of a Kroger grocery store, customers can purchase any desired amount (by placing the food in a plastic bag to be weighed and/or priced at the checkout line). For those people who purchase cucumbers, the probability distribution for the number purchased is given in the table below. ▦ CUCUMBER

x	1	2	3	4	5
$p(x)$	0.10	0.50	0.20	0.15	0.05

a. Suppose two customers who purchase cucumbers are selected at random. Find the exact probability distribution for the sample mean number of cucumbers purchased, $\overline{X}$.
b. Find the mean, variance, and standard deviation of $\overline{X}$.

7.89 Marketing and Consumer Behavior A drive-in movie theater charges viewers by the carload, but keeps careful records of the number of people in each car. The probability distribution for the number of people in each car entering the drive-in is given in the table below. ▦ DRIVEIN

x	1	2	3	4	5	6
$p(x)$	0.02	0.30	0.10	0.30	0.20	0.08

a. Suppose two cars entering the drive-in are selected at random. Find the exact probability distribution for the maximum number of people in either one of the cars, M.
b. Find the mean, variance, and standard deviation of M.

7.90 Demographics and Population Statistics Many clubs and companies around the country offer hot-air balloon rides. Most impose strict safety regulations and take at most four adults at one time, plus a pilot. The mean weight for an adult male in the United States is 195.5 pounds.[28] Suppose the distribution is normal with a standard deviation of 30 pounds.

a. If a hot-air balloon pilot (an adult male) takes three adult males for a ride, what is the distribution for the total weight aboard, T? Carefully sketch the probability distribution for T.
b. If the total weight is less than 900 pounds, the pilot will have enough fuel to extend the ride by a few minutes. What is the probability of an extended ride?
c. If the total weight is over 975 pounds, then the balloon will not be able to take off. What is the probability that the balloon will not be able to take off?

7.91 Biology and Environmental Science A typical houseplant produces oxygen (from carbon dioxide) in varying amounts, depending on the amount of light and water. Suppose a medium Norfolk Island pine produces 7.5 ml per hour of oxygen when exposed to normal sunlight. Thirty-five Norfolk Island pines are selected at random, and the oxygen output is carefully measured for each. Assume $\sigma = 1.75$ ml/h.

a. What is the probability that the sample mean oxygen produced is less than 7 ml/h?

b. What is the probability that the sample mean oxygen produced is between 7.25 and 7.5 ml/h?
c. Suppose each plant is exposed to a new high-intensity grow light and the sample mean oxygen produced for the 35 plants is 8.1 ml/h. Is there any evidence that the new lamp has increased oxygen output?
d. Answer parts (a), (b), and (c) if $\sigma = 3.75$ ml/h.

7.92 Marketing and Consumer Behavior Seafood restaurants along the coast of New England offer a variety of entrees, but lobster is the most popular meal. At Newick's Seafood Restaurant in Dover, New Hampshire, 37% of all diners order lobster. Suppose 120 customers are selected at random, and the meal ordered by each is recorded.

a. Find the distribution of the sample proportion of diners who order lobster, and carefully sketch the probability distribution.
b. Find the probability that the sample proportion of diners who order lobster is less than 0.30.
c. Find the probability that the sample proportion of diners who order lobster is between 0.35 and 0.40.
d. The manager of Newick's is concerned that more customers might be ordering lobster. This would require a change in restaurant ordering and a shift in kitchen staff. Suppose the actual proportion of diners who order lobster is 0.42. Is there any evidence to suggest that the proportion of diners who order lobster has increased? Justify your answer.

7.93 Physical Sciences Tropical rainforests are located in areas near the Equator. The temperature in a typical rainforest ranges from 20 to 30°C, and the humidity is usually between 77% and 88%.[29] The mean amount of rain per year in any one rainforest is $\mu = 155$ inches with standard deviation $\sigma = 35$ inches. Suppose 30 rainforests are selected at random and the amount of rain per year is recorded for each.

a. What is the probability that the mean rainfall per year for the 30 rainforests is more than 170 inches?
b. What is the probability that the mean rainfall is between 140 and 150 inches?
c. Find a value r such that the probability that the mean rainfall is less than r is 0.001.

7.94 Technology and the Internet An office manager has several computers running distributed programs. Because of the demands on the system, the machines may crash at various times during the day and require a hard reset. The probability distribution for the number of times a randomly selected machine crashes during a day, X, is given in the table below. ▦ COMPUTER

x	0	1	2	3	4	5
$p(x)$	0.50	0.30	0.10	0.07	0.02	0.01

a. Find the mean, variance, and standard deviation for the number of crashes by a single machine during a day.
b. Suppose $n = 2$ machines are selected at random. Find the sampling distribution of the statistic T, the total number of crashes for the two machines.

c. Find the mean, variance, and standard deviation of T.

d. Verify the relationships $\mu_T = 2\mu_X$ and $\sigma_T^2 = 2\sigma_X^2$.

7.95 Parameter Versus Statistic   In each of the following statements, identify the **boldface** number as the value of a population parameter or a sample statistic.

a. Some political observers claim that state senators spend too much time addressing colleagues about pending legislation. The mean length for a random sample of speeches was **23.7** minutes.

b. A consulting firm prepared a report from census data and concluded that the proportion of single-family homes in a certain county is **0.52**.

c. In a random sample of adults, the mean number of blinks per day was **22,037**.

d. A ballet instructor found the variance of the number of round-trip miles (to practice) traveled by all students was **150.76**.

e. A random sample of people who snowboard at least five times per year found the mean number of injuries per person per year to be **3.4**.

f. In a survey conducted by the U.S. Postal Service, **80%** of respondents supported the proposed new six-day package, five-day delivery schedule.

7.96 Medicine and Clinical Studies   The stirrup bone, located in the middle ear, is the smallest bone in the human body, with a mean length of 3 mm.[30] Suppose the standard deviation of the length is 0.16 mm. This bone, located in the ear, is part of a leverage system that can affect hearing. There is some speculation that high noise levels can affect the development of this bone and inhibit hearing. In a random sample of adults who have lived in a large city their entire lives, the length of the right-ear stirrup bone was obtained for each. ▐▐▐ STIRRUP

a. Find the distribution of the sample mean. Carefully sketch the probability density function.

b. Is there any evidence to suggest the length of the stirrup bone for lifetime *city* adults is different from 3 mm? Justify your answer.

7.97 Sampling Distribution   Each of the following graphs shows the probability distribution for an underlying distribution and for the sampling distribution of the sample mean of n observations (drawn from the underlying distribution). Identify each probability distribution function.

a.

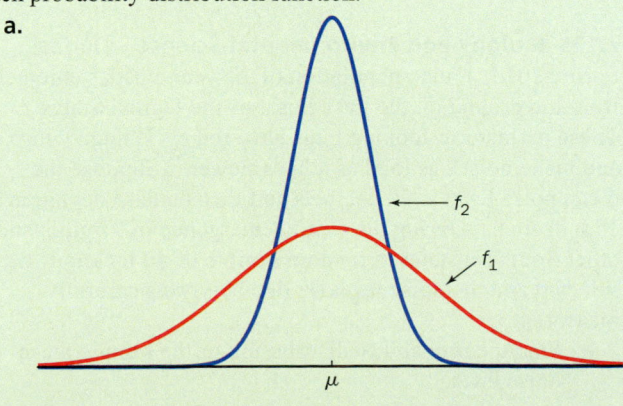

b.

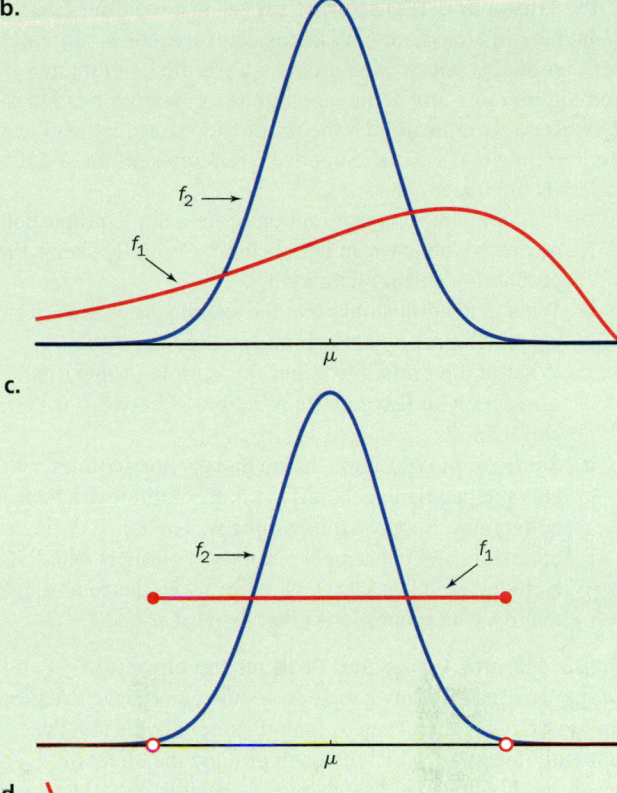

c.

d.

7.98 Manufacturing and Product Development   Above-ground swimming pools are very popular in some parts of the country. The vinyl liners used in these pools vary in thickness, but most pools are sold with a 15- or 16-mil-thick liner.[31] A certain company sells special liners and claims the mean thickness is 20 mil. Suppose the thickness of these liners is normally distributed with standard deviation 0.63. A random sample of these vinyl liners was obtained and the thickness of each was carefully measured. The resulting data are given in the following table. ▐▐▐ LINERS

20.73	18.31	20.19	19.55	19.35
20.42	19.59	20.42	20.76	20.03

a. Suppose the company's claim is true. Find the distribution of the mean thickness of these liners.

b. Use the data to determine whether there is any evidence to suggest that the company's claim is false.

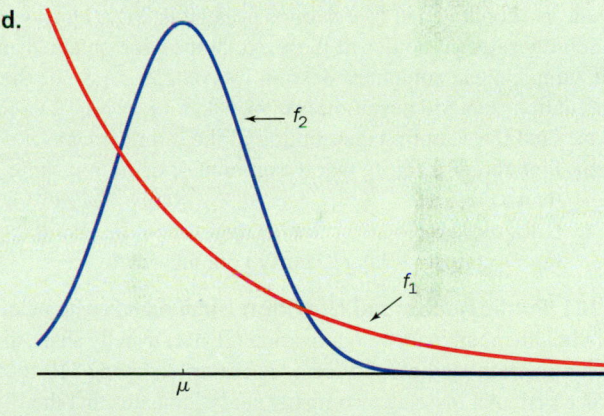

7.99 Travel and Transportation   The Greenline Taxi Company in New York City keeps careful records of items left behind by riders. Most items are claimed, but many remain in a lost-and-found area at the company's headquarters. Records indicate that the proportion of riders who leave an item in a taxi is 0.12. Suppose a random sample of 250 riders is obtained.

a. Find the sampling distribution of the sample proportion of riders who leave an item behind. Carefully sketch the probability density function.

b. What is the probability that the sample proportion of riders who leave an item behind is more than 0.15?

c. What is the probability that the sample proportion of riders who leave an item behind is between 0.11 and 0.115?

d. Some people speculate that in bad economic times, riders (and people in general) are more careful with their belongings. Suppose this sample was obtained during a recession, and the sample proportion of riders who left an item behind was 0.09. Is there any evidence to suggest that the true proportion is less than 0.12?

7.100 Manufacturing and Product Development   Eyedrops are used by many people to soothe and relieve irritation and to lubricate their eyes. A manufacturer claims that the amount of dextran, which helps to prolong the effect of eyedrops, contained in its eyedrops product is 70%. Thirty-six randomly selected bottles of these eye drops were obtained and the sample mean amount of dextran was 68.25%. Assume the population standard deviation is $\sigma = 5\%$.

a. Find the sampling distribution of the sample mean.

b. Find the probability that the amount of dextran is more than 71%.

c. Does the sample mean found suggest that the manufacturer's claim is wrong? Justify your answer.

7.101 Public Health and Nutrition   Individuals who belong to a health maintenance organization (HMO) usually share the cost of certain medical services, for example, paying $10 for an office visit. An insurance company study indicates that the proportion of all people who belong to an HMO who have a copayment of more than $10 for an office visit is 0.65. Suppose 1000 people who belong to an HMO are randomly selected, and the office copayment amount is recorded.

a. Find the distribution of the sample proportion of people who have a copayment of more than $10 for an office visit. Verify the nonskewness criterion.

b. Find the probability that the sample proportion is more than 0.66.

c. Find the probability that the sample proportion is between 0.64 and 0.67.

d. Find a value h such that the probability that the sample proportion is less than h is 0.01.

EXTENDED APPLICATIONS

7.102 Physical Sciences   Carlinville, Illinois, has just started an ambitious recycling program. Special trucks collect recyclable products once a week, sorted into barrels of paper, glass, and plastic. The amount of glass recycled per week by a single household is normally distributed with mean $\mu = 27$ pounds and standard deviation $\sigma = 7$ pounds. Suppose 12 households are randomly selected.

a. What is the probability that the total glass collected for the 12 homes, T, will be less than 350 pounds?

b. Find a value g such that $P(T \geq g) = 0.05$.

c. If the total glass collected for the 12 homes is more than 400 pounds, the recycling plant will make a profit. Find a value of μ such that the probability of making a profit is 0.10.

7.103 Sports and Leisure   The Hotel Association of Canada reported that, because of the high cost of airline tickets and gasoline, more people are taking a *staycation*, a vacation close to their hometown. According to a survey, 31% of travelers stayed in a hotel near home, visited local attractions, restaurants, and shopped.[32] Suppose 250 Canadians planning to take a vacation are selected at random and asked whether they will take a staycation.

a. What is the probability that the sample proportion who will take a staycation is between 0.30 and 0.35?

b. What is the probability that the sample proportion is more than 0.37?

c. Find a value of the sample size n such that the probability that the sample proportion is less than 0.40 is 0.95.

7.104 Manufacturing and Product Development   A manufacturer of ice pops fills each plastic container with a fruity liquid, leaving enough room so that consumers can freeze the product. A filling machine is set so that the amount of liquid in each ice pop is normally distributed with mean 8.00 ounces and standard deviation 0.25 ounce. Suppose 16 ice pops are randomly selected.

a. Find the probability distribution of the sample mean number of ounces in each container, $\overline{X}$.

b. Find the probability that the sample mean is less than 7.9 ounces.

c. Find the probability that the sample mean is more than 8.15 ounces.

d. Suppose the filling machine operator can fine-tune the process by controlling the standard deviation of the fill. Find a value for σ such that the probability that the sample mean is more than 8.05 ounces is 0.05.

7.105 Biology and Environmental Science   During Spring 2013, a huge population of 17-year cicadas emerged from the ground on the East Coast of the United States. These prehistoric-looking bugs have red eyes, huge wings, and make noises as loud as a lawnmower.[33] Suppose the mean noise level is 90 decibels and the standard deviation is 15.6 decibels. During the cicada emergence in Virginia and other Southern states, a random sample of 40 locations was selected and the noise level (in decibels) was carefully measured.

a. Find the probability distribution of the sample mean noise level.

b. Find the probability that the sample mean is less than 60 decibels, the level of moderate conversational speech.
c. Find the probability that the sample mean is more than 110 decibels, the level of a passing train.
d. Find a value of n such that the probability that the sample mean is less than 93 decibels with probability 0.95.

7.106 Sports and Leisure A large sporting goods company has just received a shipment of 100,000 table tennis balls. USA Table Tennis tournament regulations specify that the diameter of the ball must be 40 millimeters (mm). Suppose the distribution of the diameter is normal with standard deviation 0.4 mm. Twenty-five table tennis balls will be selected at random, and the diameter of each will be carefully measured. If the mean diameter is within 0.2 mm of 40 mm, then the shipment will be accepted. Otherwise the entire shipment will be returned to the manufacturer.

a. Suppose the true mean diameter of the table tennis balls is 40 mm. What is the probability that the entire shipment will be sent back to the manufacturer?

b. Suppose the true mean diameter of the table tennis balls is 40.4 mm. What is the probability that the shipment will be accepted by the sporting goods store?
c. Suppose the true mean diameter of the table tennis balls is 39.4 mm. What is the probability that the shipment will be accepted by the sporting goods store?

LAST STEP

7.107 How much was the typical foreclosure check? The checks sent to Americans in 2013 were in the amount of $300 to $125,000. Suppose a spokesperson from the U.S. Treasury claimed that the mean check amount was $800 with standard deviation $125. A random sample of 40 Americans compensated under this program was obtained. The amount of each check was recorded and the sample mean check amount for these Americans was $755.

a. Find the distribution of the sample mean.
b. Is there any evidence to suggest that the true mean check amount was less than $800?

8 Confidence Intervals Based on a Single Sample

Is the concentration of ultrafine particles dangerous?

Ultrafine particles in the air we breathe are nano-sized, less than 100 nm. These particles result from combustion, friction, or some natural process in the air or water. Several centers in the United States are studying the effect of ultrafine particles on health. There is some evidence to suggest that high levels of these particles can cause cardiovascular or lung disease because they can easily penetrate the body's natural defense mechanisms.

The U.S. Environmental Protection Agency sets National Ambient Air Quality Standards for pollutants that are considered harmful to the environment and to the public health. For ultrafine particles the EPA standard is 35 μg/m^3 over a 24-hour period. New York City and Los Angeles have some of the highest levels of ultrafine particles in the United States. In addition, tunnels, highways, and urban areas in general also may have very high levels of ultrafine particles.

In Kansas City, Missouri, suppose 15 24-hour periods are selected at random and the concentration of ultrafine particles is carefully measured over each period. The summary statistics are $\bar{x} = 10.76$ μg/m^3 and $s = 4.7$ μg/m^3. The concepts presented in this chapter will be used to construct an interval in which we are fairly certain the true mean level of ultrafine particles lies. This interval of numbers can also be used to make an inference concerning the population mean.

CONTENTS

Alan Marsh/age fotostock

8.1 Point Estimation

A **point estimate** of a population parameter is a single number computed from a sample, which serves as a guess for the parameter. Using the terminology introduced in Chapter 7:

1. An **estimator** is a statistic of interest and is, therefore, a random variable. An estimator has a distribution, a mean, a variance, and a standard deviation.

2. An **estimate** is simply a specific value of an estimator.

Definition

An **estimator** (statistic) is a rule used to produce a point estimate of a population parameter.

Suppose we need to estimate a population parameter θ and there are many different statistics (rules) available. Which one should we use? Figure 8.1 shows the sampling distributions of three different statistics for estimating θ. These graphs suggest some properties of a *good* statistic.

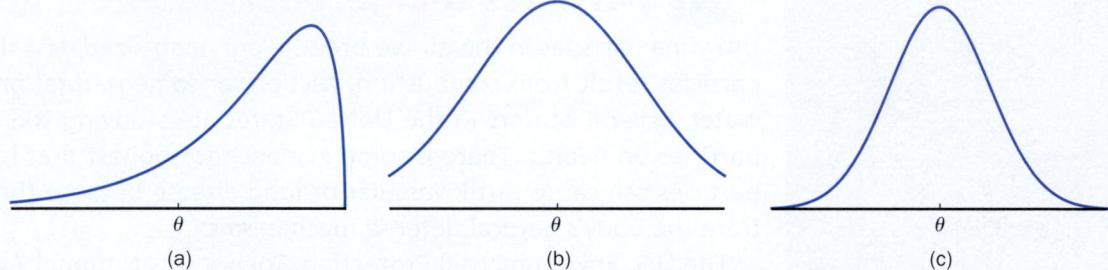

Figure 8.1 The sampling distributions for three different statistics for estimating θ: (a) a skewed statistic; (b) a statistic with large variance; (c) a statistic with small variance. The horizontal axis represents all possible values of $\hat{\theta}$.

The statistic in (a) is unlikely to produce a value close to θ. The sampling distribution is skewed to the left, and most of the values of the statistic are to the right of θ. The statistic in (b) is centered at θ. On average (in the long run), this statistic will produce θ (that's good!). However, this statistic has large variance (that's bad!). Even though the sampling distribution is centered at the true value of the population parameter, specific estimates will probably be *far away* from θ. The statistic in (c) exhibits two very desirable properties. It is centered at the true value of the population parameter, and it has small variance. These observations suggest two rules for selecting a statistic.

Definition

A statistic $\hat{\theta}$ is an **unbiased estimator** of a population parameter θ if $E(\hat{\theta}) = \theta$, the mean of $\hat{\theta}$ is θ.

If $E(\hat{\theta}) \neq \theta$, then the statistic $\hat{\theta}$ is a **biased estimator** of θ.

Figure 8.2 illustrates the sampling distribution of an unbiased estimator for θ. The distribution of the statistic is centered at θ; the value of the statistic is, on average, θ. Figure 8.2 also shows the sampling distribution of a biased estimator for θ. On average, the value of this statistic is greater than θ.

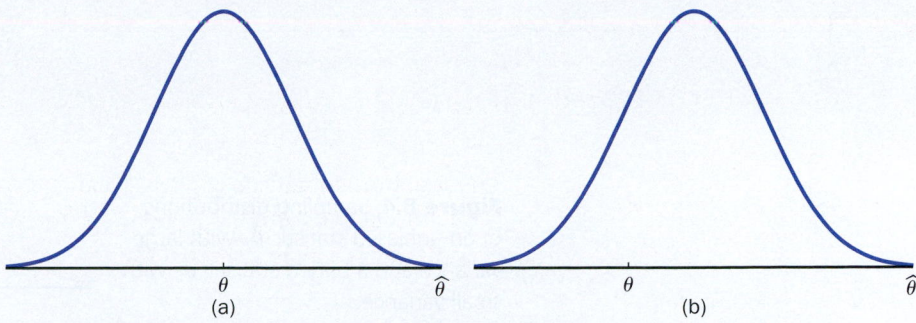

Figure 8.2 The sampling distribution for (a) an unbiased estimator for θ, and (b) a biased estimator for θ. The horizontal axis represents all possible values of $\hat{\theta}$.

We have already worked with several unbiased estimators in previous chapters:

1. The sample mean, $\overline{X}$, is an unbiased statistic for estimating the population mean μ, because $E(\overline{X}) = \mu$.

2. The sample proportion, $\hat{P}$, is an unbiased statistic for estimating the population proportion p, because $E(\hat{P}) = p$.

3. The sample variance, S^2, is an unbiased statistic for estimating the population variance σ^2, because $E(S^2) = \sigma^2$.

Even though S^2 is an unbiased estimator for σ^2, the sample standard deviation S is a *biased* estimator for the population standard deviation σ.

The expected-value operation does not pass freely through the square-root symbol.

$$E(S) = E(\sqrt{S^2}) \neq \sqrt{E(S^2)} = \sqrt{\sigma^2} = \sigma$$

And, even though S is biased, it is still important in statistical inference.

If there are several statistics available, it seems reasonable to use one that is unbiased. Therefore, the first rule for choosing a statistic is that the sampling distribution should be centered at θ: The estimator should be unbiased.

The second rule for choosing a statistic is that, of all unbiased statistics, the best statistic to use is the one with the smallest variance. The point estimate produced using this statistic will, on average, be *close* to the true value of the population parameter. Figure 8.3 illustrates this second rule for choosing a statistic.

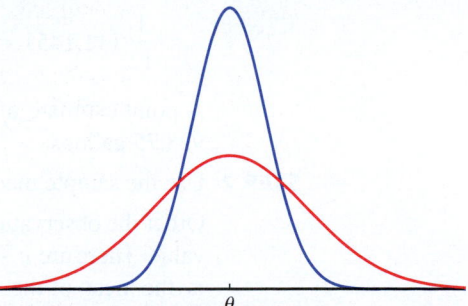

Figure 8.3 Sampling distributions of two unbiased statistics for estimating θ. Use the statistic with smaller variance.

Suppose there are two statistics to choose from, $\hat{\theta}_1$ and $\hat{\theta}_2$, for estimating θ, as shown in Figure 8.4. The statistic $\hat{\theta}_1$ is unbiased but has large variance; $\hat{\theta}_2$ is slightly biased but has small variance. The choice of an estimator is a difficult decision, and there is no definitive answer.

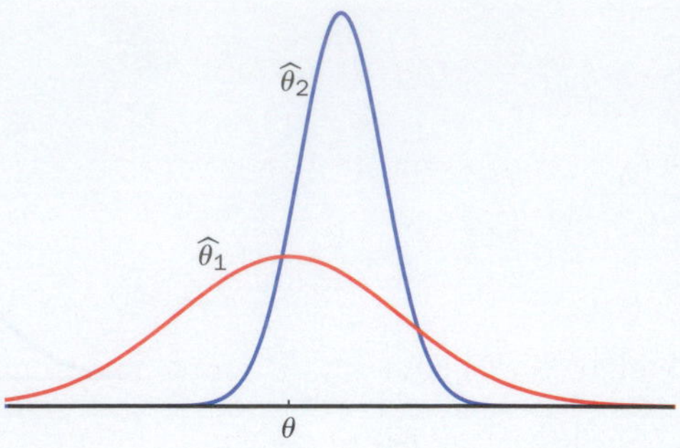

Figure 8.4 Sampling distributions of an unbiased statistic $\hat{\theta}_1$ with large variance and a biased statistic $\hat{\theta}_2$ with small variance.

For a given parameter θ, a MVUE may not exist.

Suppose we need to estimate the population parameter θ, and there are several unbiased statistics from which to choose. If one of these statistics has the smallest possible variance, it is called the MVUE (minimum-variance unbiased estimator). If the underlying population is normal, the sample mean, $\overline{X}$, is the MVUE for estimating μ. So, if the population is normal, the sample mean is a *really good* statistic to use for estimating μ. $\overline{X}$ is unbiased, and it has the smallest variance of all possible unbiased estimators for μ.

DATA SET

SOLVENTS

Example 8.1 Storing Petroleum Solvents

A dry-cleaning company in the United States must comply with many government regulations, for example, the federal Clean Water Act and the Oil Pollution Act. Some regulations are based on the aboveground storage capacity of petroleum solvents. A random sample of dry-cleaning companies in Tacoma, Washington, was obtained. The petroleum-solvent storage capacity (in gallons) for each is given in the table below.

770	875	850	1000	830	980
800	950	940	1125	925	1100

Find point estimates for the population mean petroleum-solvent storage capacity and for the population median petroleum-solvent storage capacity.

SOLUTION

STEP 1 Use the sample mean to estimate the population mean.

$$\overline{x} = \frac{1}{12}(770 + 875 + \cdots + 925 + 1100)$$

$$= \frac{1}{12}(11{,}145) = 928.75$$

A point estimate of the population mean petroleum-solvent storage capacity is 928.75 gallons.

STEP 2 Use the sample median to estimate the population median.

Order the observations from smallest to largest. The sample median is the *middle* value. There are $n = 12$ observations, so the middle value is in position 6.5, that is, the mean of the values in positions 6 and 7.

Ordered observations:

770 800 830 850 875 925 940 950 980 1000 1100 1125
↑

$$\tilde{x} = \tfrac{1}{2}(925 + 940) = 932.50$$

A point estimate of the population median petroleum-solvent storage capacity is 932.50 gallons. ●

TRY IT NOW GO TO EXERCISE 8.12

Technology Corner

Procedure: Compute point estimates.
Reconsider: Example 8.1, solution, and interpretations.

CrunchIt!

CrunchIt! has a built-in function to find certain descriptive statistics, including the sample mean and the sample median.

1. Enter the data into a column.
2. Select Statistics; Descriptive Statistics. Choose the appropriate column and click the Calculate button. See Figure 8.5.

Figure 8.5 CrunchIt! descriptive statistics. Note that the output is rounded in the tenths place.

Results - Descriptive Statistics									
Export ▾									
	n	Sample Mean	Standard Deviation	Min	Q1	Median	Q3	Max
Var1	12	928.8	111.5	770	840	932.5	990	1125

TI-84 Plus C

There are several built-in functions to compute summary statistics. STAT ; CALC ; 1-Var Stats may be used to find several summary statistics at once. The LIST ; MATH menu contains several functions that return single point estimates.

1. Enter the data into list L1.
2. Use LIST ; MATH; mean to find the sample mean and LIST ; MATH; median to find the sample median. See Figure 8.6.

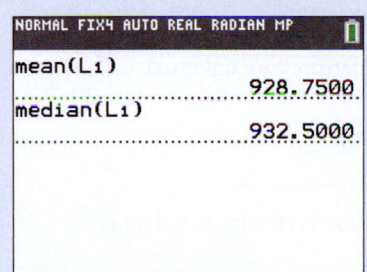

```
NORMAL FIX4 AUTO REAL RADIAN MP

mean(L1)
                          928.7500
median(L1)
                          932.5000
```

Figure 8.6 The sample mean and the sample median computed using LIST ; MATH functions.

Minitab

There are several built-in functions to compute summary statistics, in a session window, using a graphical input window, or the Calculator.

1. Enter the data into column C1.
2. Select Stat; Basic Statistics; Display Descriptive Statistics.
3. Enter C1 in the Variables input window. Use the Statistics options button to select the desired estimates. See Figure 8.7.

Figure 8.7 Descriptive statistics, selected in a graphical input window.

```
Session

Descriptive Statistics: Capacity

Variable    N    Mean   Median
Capacity   12   928.8    932.5
```

Excel

There are several built-in functions to compute summary statistics and under the Data tab, Data Analysis; Descriptive Statistics returns several summary statistics at once.

1. Enter the data into column A.

2. Use the function AVERAGE to find the sample mean and the function MEDIAN to find the sample median. See Figure 8.8.

B	C
928.75	= AVERAGE(A1:A12)
932.50	= MEDIAN(A1:A12)

Figure 8.8 Built-in functions to find the sample mean and sample median.

SECTION 8.1 EXERCISES

Concept Check

8.1 True/False   An estimator is a random variable.

8.2 Fill in the Blank   A statistic $\hat{\theta}$ is an unbiased estimator of θ if _____.

8.3 Short Answer   Name two characteristics of a good estimator.

8.4 Short Answer   If two statistics are unbiased, which one would you use?

8.5 Fill in the Blank   An unbiased statistic with the smallest possible variance is called _____.

8.6 Short Answer   Would you ever want to use a biased statistic? If so, why?

Practice

8.7 The graph below shows the probability density functions for three different statistics that could be used to estimate a population parameter θ. Which statistic would you use, and why?

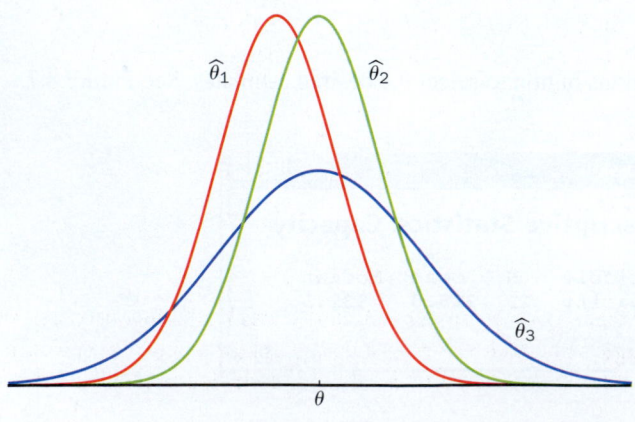

8.8 The graph below shows the probability density functions for three different statistics that could be used to estimate a population parameter θ. Which statistic would you use, and why?

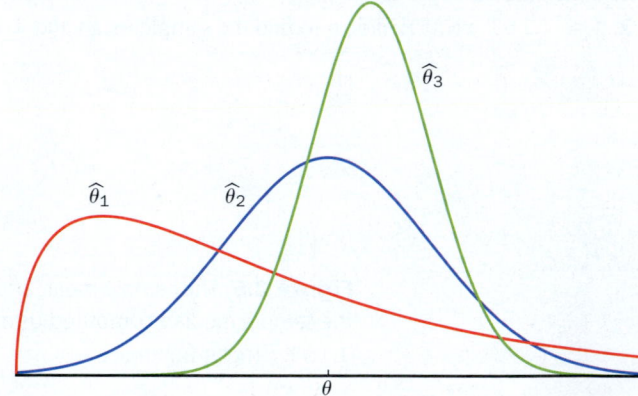

8.9 Why is an unbiased estimator for θ better than a biased estimator for θ?

8.10 Suppose there are several unbiased estimators for θ. What criterion would you use for selecting one of these estimators, and why?

Applications

8.11 Business and Management   A recent survey asked employees in technology-related jobs what perks they would like their employer to provide. Out of 1200 randomly selected workers polled, 975 said they would like ongoing training paid for by their employer. Find a point estimate for the population proportion of all workers in technology-related jobs who would like ongoing training.

8.12 Biology and Environmental Science   A *hand* is a traditional unit used to measure the height of a horse. One hand is four inches, and the height of a horse is measured from the ground to the horse's shoulder. A random sample of horses sold at auctions around the country revealed the following heights (in hands). ▫ HORSEHT

15.8	13.7	11.0	17.1	19.3	14.6	14.4	13.8
18.7	16.5	12.2	17.9	16.8	16.3	12.8	18.7
10.4	15.6	11.7	12.8				

a. Find a point estimate for the population mean height of all horses sold at auctions.
b. Find a point estimate for the population median height of all horses sold at auctions.
c. Find a point estimate for the population variance of the height of all horses sold at auctions.
d. Any horse with a height less than 14.5 hands is considered *short*. Find a point estimate for the true proportion of *short* horses sold at auctions.

8.13 Biology and Environmental Science   During the first few weeks of January every year, biologists associated with the U.S. Fish and Wildlife Service and the Maryland Department of Natural Resources physically count ducks, geese, and swans near the Chesapeake Bay. In 2013 these scientists counted 175,500 ducks, of which 33,100 were Mallards.[1] Find a point estimate for the proportion of Mallards that were observed during the 2013 winter duck count.

8.14 Biology and Environmental Science   Cowrie shells have been used as money in many parts of the world, for example, China and Africa. Today, they are used in decorations and jewelry, and cowrie-shell bracelets and necklaces are popular near seaside resorts. A jeweler recently purchased several hundred cowrie shells to use in making earrings. A random sample of the finished earrings was obtained, and the weight (in grams) of each is given in the table below. ▫ COWRIE

7.3	7.2	7.2	7.9	7.3	7.3	7.0	7.0	7.4	7.4
7.4	7.2	7.5	7.7	7.1	7.0	7.2	7.2	7.7	6.9
7.4	7.8	7.7	7.4	7.5	7.7	7.5	7.3	7.6	7.3

a. Find a point estimate for the first quartile and a point estimate for the third quartile.
b. The smallest (in weight) 20% of all cowrie-shell earrings are sold at a discount. Find a point estimate for the 20th percentile of the cowrie-shell earring weight distribution.

8.15 Manufacturing and Product Development   A company that manufactures a centrifugal pump for golf-

course sprayers would like to rate the pressure (in psi) developed by this unit. Thirty pumps were randomly selected and tested. ▫ PUMPS

a. Find a point estimate for the minimum pressure developed by this pump.
b. Find a point estimate for the maximum pressure developed by this pump.
c. Use your answers to parts (a) and (b) to construct an *interval* estimate for the pressure developed by this pump.

8.16 Travel and Transportation   The San Francisco Bay Area Rapid Transit (BART) system conducted a survey to learn if bikes on board would have any effect on riders. A random sample of riders was selected and asked if there was enough room for bikes and passengers. Of the 1720 riders selected, 671 indicated there was enough room and 327 said it would be too crowded.[2]

a. Find a point estimate for the proportion of all BART riders who believe there is enough room for passengers and bikes.
b. Find a point estimate for the proportion of all BART riders who believe it is too crowded with bikes on board.
c. Let p_d denote the difference in population proportions between all BART riders who believe there is enough room for passengers and bikes and all BART riders who believe it is too crowded. Find a point estimate for p_d.

8.17 Medicine and Clinical Studies   Merck recently released an experimental insomnia drug, suvorexant. This drug temporarily blocks chemical messengers that work to keep people alert and awake.[3] The new pill seems to work but does have some potential harmful side effects, for example, extended daytime drowsiness. A random sample of patients taking this experimental drug was obtained. The dosage in milligrams was recorded for each patient. ▫ SLEEP

a. Find a point estimate for the mean dosage.
b. Find a point estimate for the variance of the dosage.
c. Find a point estimate for the first quartile and a point estimate for the third quartile.

8.18 Ambulance Response Time   In March 2013, the 17 ambulances in Manatee County, Florida, responded to 3645 calls.[4] The response time is measured from the time the ambulance is notified of the call until the crew arrives on the scene. A random sample of the March response times was obtained. ▫ AMBRESP

a. Find a point estimate for the minimum response time.
b. Find a point estimate for the maximum response time.
c. Find a point estimate for the interquartile range of response times.

8.2 A Confidence Interval for a Population Mean When σ Is Known

In Section 8.1 we discovered that a *good* estimator is unbiased and has small variance. An estimator produces only a single value that serves as a best guess for a population parameter. In this section we use this single value to produce a **confidence interval**. This interval of values is constructed so that we can be reasonably sure the true value of the population parameter lies within it.

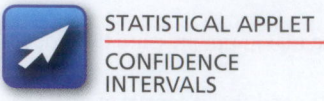

STATISTICAL APPLET

CONFIDENCE
INTERVALS

Definition

A **confidence interval** (CI) for a population parameter is an interval of values constructed so that, with a specified degree of confidence, the value of the population parameter lies within it.

The **confidence coefficient** is the probability that the CI encloses the population parameter in repeated samplings.

The **confidence level** is the confidence coefficient expressed as a percentage.

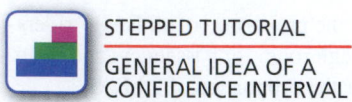

STEPPED TUTORIAL

GENERAL IDEA OF A
CONFIDENCE INTERVAL

A CLOSER LOOK

1. A confidence interval is usually expressed as an *open* interval, for example, (10.5, 15.8), where 10.5 is the left endpoint, or lower bound, and 15.8 is the right endpoint, or upper bound. The interval extends all the way to, but does not include, the endpoints.

2. Typical confidence coefficients are 0.95 and 0.99.

3. Typical confidence levels are, therefore, 95% and 99%.

The following steps provide background for the construction of a confidence interval for a population mean μ.

1. Suppose either (a) the underlying population is normal, or (b) the sample size n is large, or both, and the population standard deviation σ is known.

2. Using the properties of $\overline{X}$ and the central limit theorem (if necessary): The sample mean $\overline{X}$ is (approximately) normal with mean μ and variance σ^2/n.

   In symbols, $\overline{X} \sim N(\mu, \sigma^2/n)$.

3. Using the empirical rule, approximately 95% of all values of the sample mean lie within two standard deviations of the population mean. Figure 8.9 shows an example of a single value, or point estimate, $\bar{x}$ within two standard deviations of the mean.

4. Even though we know the distribution of $\overline{X}$ is centered at μ, we do not know the true value of μ. To *capture* μ it seems reasonable to *step* two standard deviations from an estimate $\bar{x}$ in both directions. The resulting (rough) 95% confidence interval $(\bar{x} - 2\sigma/\sqrt{n}, \bar{x} + 2\sigma/\sqrt{n})$ is illustrated in Figure 8.10.

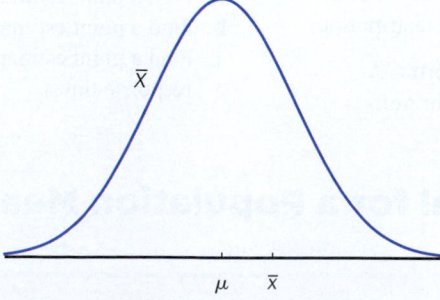

Figure 8.9 The sampling distribution of $\overline{X}$ and a typical value.

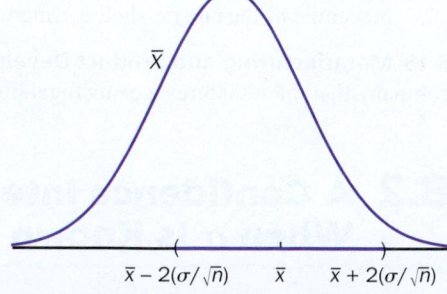

Figure 8.10 A rough 95% confidence interval that probably captures the true value μ.

To construct a more accurate 95% confidence interval for μ, using the same assumptions, begin with the standardized random variable Z.

1. $\bar{X} \sim N(\mu, \sigma^2/n)$ $\rightarrow$ $Z = \dfrac{\bar{X} - \mu}{\sigma/\sqrt{n}} \sim N(0, 1)$

2. Find a symmetric interval about 0 such that the probability that Z lies in this interval is 0.95.

$P(-1.96 < Z < 1.96) = 0.95$ Use Table III in the Appendix; see Figure 8.11.

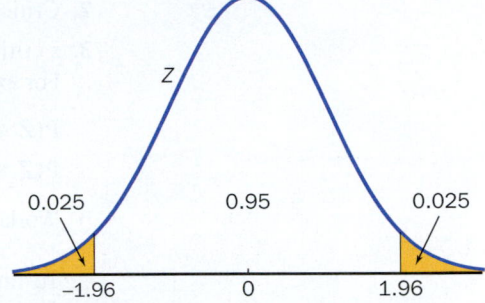

Figure 8.11 A symmetric interval about 0 such that $P(-1.96 < Z < 1.96) = 0.95$.

3. Substitute for Z and manipulate the interval inside the probability statement so that μ is *caught* in the middle.

▶ $P\left(-1.96 < \dfrac{\bar{X} - \mu}{\sigma/\sqrt{n}} < 1.96\right) = 0.95$ Substitute for Z.

$P\left((-1.96)\dfrac{\sigma}{\sqrt{n}} < \bar{X} - \mu < (1.96)\dfrac{\sigma}{\sqrt{n}}\right) = 0.95$ Multiply all three parts by $\sigma/\sqrt{n}$.

$P\left(-\bar{X} - (1.96)\dfrac{\sigma}{\sqrt{n}} < -\mu < -\bar{X} + (1.96)\dfrac{\sigma}{\sqrt{n}}\right) = 0.95$ Subtract $\bar{X}$ from all three parts.

$P\left(\bar{X} + 1.96\dfrac{\sigma}{\sqrt{n}} > \mu > \bar{X} - 1.96\dfrac{\sigma}{\sqrt{n}}\right) = 0.95$ Multiply all three parts by -1. Multiplying by -1 changes the directions of the inequalities.

$P\left(\bar{X} - 1.96\dfrac{\sigma}{\sqrt{n}} < \mu < \bar{X} + 1.96\dfrac{\sigma}{\sqrt{n}}\right) = 0.95$ Rewrite the expressions in increasing order. ◀

The last expression includes a formula for a 95% confidence interval for μ: *Step* exactly 1.96 (not 2) standard deviations from a specific value $\bar{x}$ in both directions. This interval is shown in Figure 8.12.

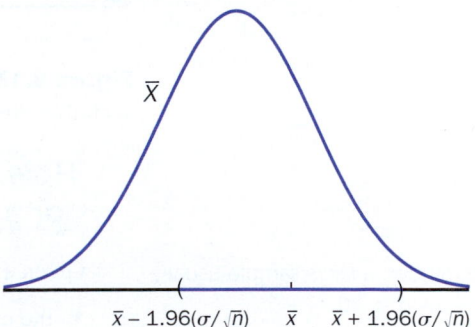

Figure 8.12 An exact 95% confidence interval for μ.

$\bar{x} - 1.96(\sigma/\sqrt{n})$ $\bar{x}$ $\bar{x} + 1.96(\sigma/\sqrt{n})$

The last step is to find a more general $100(1 - \alpha)\%$ confidence interval for μ, using the same assumptions. α is usually small. For example, if $\alpha = 0.05$, the resulting confidence level is $100(1 - 0.05)\% = 95\%$. Usually, the confidence level is given and we need to work backward to find α. The following definition is necessary in order to start with a probability statement involving Z.

In this definition, the subscript on z could be *any variable*, or letter. For example,

$$P(Z \geq z_c) = c$$

Definition

$z_{\alpha/2}$ is a **critical value**. It is a value on the measurement axis in a **standard normal distribution** such that $P(Z \geq z_{\alpha/2}) = \alpha/2$.

A CLOSER LOOK

1. $z_{\alpha/2}$ is simply a z value such that there is $\alpha/2$ of the area (probability) to the right of $z_{\alpha/2}$. $-z_{\alpha/2}$ is just the negative critical value.

2. Critical values are *always* defined in terms of right-tail probability.

3. z critical values are easy to find by using the complement rule and working backward. For example,

$$P(Z \geq z_{\alpha/2}) = \alpha/2 \qquad \text{Definition of critical value.}$$
$$P(Z \leq z_{\alpha/2}) = 1 - \alpha/2 \qquad \text{The complement rule.}$$

Work backward in Table III to find $z_{\alpha/2}$.

To find a general confidence interval for μ, start once again in the Z world. Find a symmetric interval about 0 such that the probability that Z lies in this interval is $1 - \alpha$ (Figure 8.13).

$$P(-z_{\alpha/2} < Z < z_{\alpha/2}) = 1 - \alpha \qquad (8.1)$$

Manipulate Equation 8.1 to obtain the probability statement.

$$P\left(\overline{X} - z_{\alpha/2}\frac{\sigma}{\sqrt{n}} < \mu < \overline{X} + z_{\alpha/2}\frac{\sigma}{\sqrt{n}}\right) = 1 - \alpha$$

STEPPED TUTORIAL

CONFIDENCE INTERVAL FOR A MEAN

Figure 8.14 illustrates this interval for a specific value $\bar{x}$.

These derivations lead to the following general result.

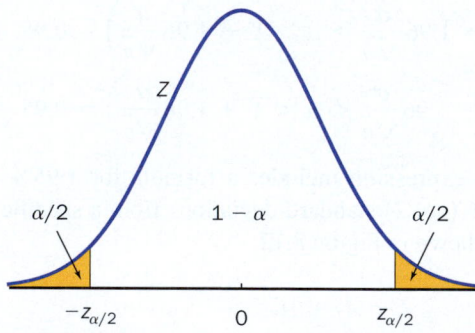

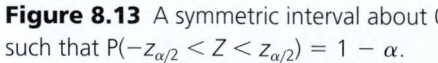

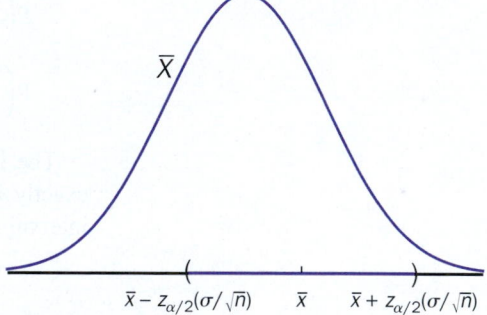

Figure 8.13 A symmetric interval about 0 such that $P(-z_{\alpha/2} < Z < z_{\alpha/2}) = 1 - \alpha$.

Figure 8.14 A $100(1 - \alpha)\%$ confidence interval for μ.

How to Find a $100(1 - \alpha)\%$ Confidence Interval for a Population Mean When σ Is Known

Reminder: a *large* sample usually means $n \geq 30$.

Given a random sample of size n from a population with mean μ, if

1. the underlying population distribution is normal and/or n is large, and
2. the population standard deviation σ is known, then

a $100(1 - \alpha)\%$ confidence interval for μ has as endpoints the values

$$\bar{x} \pm z_{\alpha/2}\frac{\sigma}{\sqrt{n}} \qquad (8.2)$$

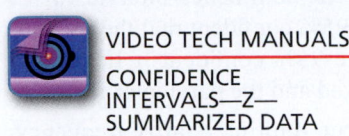

VIDEO TECH MANUALS

CONFIDENCE
INTERVALS—Z—
SUMMARIZED DATA

Solution Trail 8.2

KEYWORDS

- 95% CI for the true mean
- Normally distributed
- Standard deviation 1.25 pounds
- Random sample

TRANSLATION

- 95% CI for μ
- Underlying population is normal
- $\sigma = 1.25$

CONCEPTS

- A $100(1 - \alpha)$% CI for a population mean when σ is known

VISION

We need a 95% CI for μ. The population is normal and σ is known. Find the appropriate critical value and use Equation 8.2.

A CLOSER LOOK

1. Equation 8.2 can be used *only* if σ is known.

2. If n is large and σ is unknown, some statisticians suggest using the sample standard deviation s in Equation 8.2. in place of σ. This produces an *approximate* confidence interval for μ. The next section presents an *exact* confidence interval for μ when σ is unknown.

3. As the confidence coefficient increases (with σ and n constant), the critical value $z_{\alpha/2}$ increases. Therefore, the confidence interval is wider.

Example 8.2 Tire Weight

The total weight of an air-filled tire can dramatically affect the performance and safety of an automobile. Some transportation officials argue that mechanics should check the tire weights of every vehicle as part of an annual inspection. Suppose the weight of a 185/60/14 tire filled with air is normally distributed with standard deviation 1.25 pounds. In a random sample of 15 filled tires, the sample mean weight was $\bar{x} = 18.75$ pounds. Find a 95% confidence interval for the true mean weight of 185/60/14 tires.

SOLUTION

STEP 1 $\bar{x} = 18.75$ $\sigma = 1.25$ $n = 15$ Given.

$1 - \alpha = 0.95 \ \Rightarrow \ \alpha = 0.05 \ \Rightarrow \ \alpha/2 = 0.025$ Find $\alpha/2$.

$P(Z \geq z_{\alpha/2}) = P(Z \geq z_{0.025}) = 0.025$ Definition of critical value.

$P(Z \leq z_{0.025}) = 1 - 0.025 = 0.975$ The complement rule.

$z_{0.025} = 1.96$ Use Table III in the Appendix.

STEP 2 Use Equation 8.2.

$$\bar{x} \pm z_{\alpha/2} \frac{\sigma}{\sqrt{n}}$$ Equation 8.2.

$$= \bar{x} \pm z_{0.025} \frac{\sigma}{\sqrt{n}}$$ Use the value of α.

$$= 18.75 \pm (1.96)\frac{1.25}{\sqrt{15}}$$ Use summary statistics and values for σ and $z_{0.025}$.

$$= 18.75 \pm 0.63$$ Simplify.

$$= (18.12, 19.38)$$ Compute endpoints.

(18.12, 19.38) is a 95% confidence interval for the true mean weight (in pounds) of 185/60/14 tires.

Figure 8.15 and 8.16 together show a technology solution.

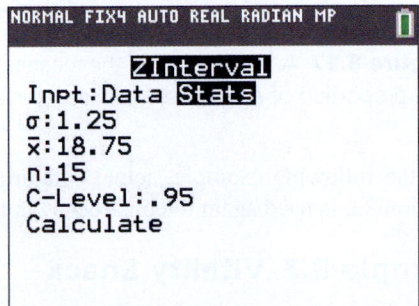

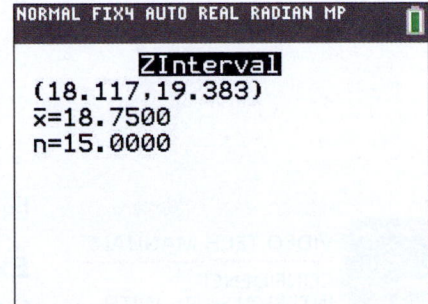

Figure 8.15 ZInterval input screen.

Figure 8.16 Resulting confidence interval.

TRY IT NOW   GO TO EXERCISE 8.31

There are two important ideas to remember when constructing a confidence interval.

1. The population parameter, in this case μ, is *fixed*. The confidence interval *varies* from sample to sample. It is correct to say, "We are 95% confident that the interval *captures* the true mean μ." The statement, "We are 95% confident μ lies in the interval," (incorrectly) implies that the interval is fixed and the parameter μ varies.

2. The confidence coefficient, a probability, is a long-run limiting relative frequency. In repeated samples, the proportion of confidence intervals that capture the true value of μ approaches 0.95. Figure 8.17 illustrates this concept. We cannot be certain about any one specific confidence interval. The confidence is in the long-run process.

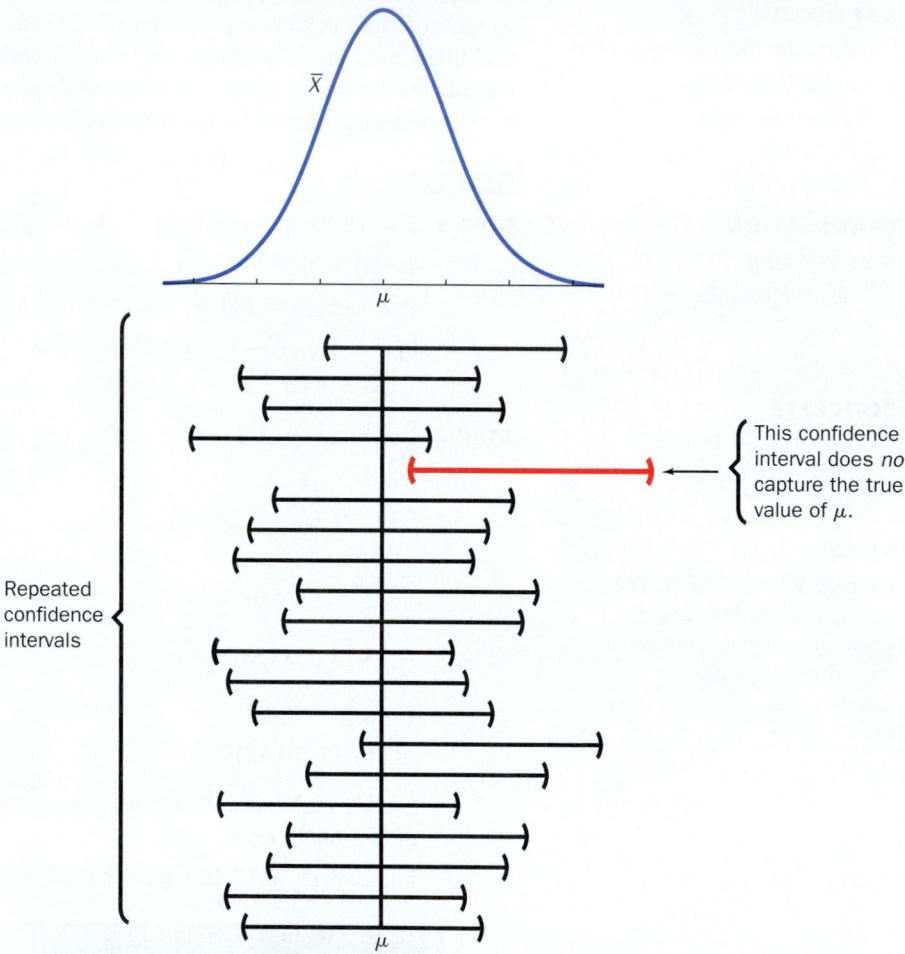

Repeated confidence intervals

This confidence interval does *not* capture the true value of μ.

Figure 8.17 An illustration of the meaning of confidence coefficient. In repeated CIs, the proportion of all 95% confidence intervals that capture the true value of μ is 0.95.

In the following example, actual data are presented, rather than summary statistics. Equation 8.2 is used again to construct a confidence interval for a population mean.

Example 8.3 Vitality Snack

In May 2013, Mamma Chia introduced a new drink in four flavors called the Chia Squeeze Vitality Snack. The product was advertised as a good source of fiber, containing fruit and vegetables, no added sugar, gluten-free and vegan, and only 70 calories. The drink had a smoothie-like texture, and the company claimed each squeeze packet contained 1200 mg of omega-3.[5] A random sample of Green Magic Vitality Snacks was obtained and the

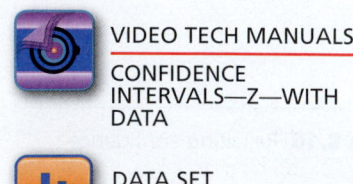

VIDEO TECH MANUALS

CONFIDENCE INTERVALS—Z—WITH DATA

DATA SET

VITALITY

Solution Trail 8.3a

KEYWORDS

- 99% CI for the mean
- $\sigma = 12$ mg
- Random sample

TRANSLATION

- 99% CI for μ
- Known standard deviation

CONCEPTS

- A $100(1 - \alpha)\%$ CI for a population mean when σ is known

VISION

Construct a 99% CI for μ. No information is given about the shape of the underlying population distribution. However, $n = 40$ (≥ 30), so the central limit theorem applies, and we can use Equation 8.2.

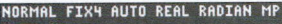

```
NORMAL FIX4 AUTO REAL RADIAN MP
        ZInterval
(1194.6,1204.4)
x̄=1199.4750
Sx=11.4107
n=40.0000
```

Figure 8.18 TI-84 Plus C confidence interval.

Solution Trail 8.3b

KEYWORDS

- Is there any evidence?

TRANSLATION

- Inference procedure

CONCEPTS

- A $100(1 - \alpha)\%$ CI for a population mean when σ is known

VISION

A 99% CI for μ is an interval in which we are 99% confident that the true value of μ lies. If the CI in part (a) captures, or includes, 1200, then there is no evidence to suggest that μ is different from 1200. Use the four-step inference procedure.

amount of omega-3 was carefully measured in each. The data are given in the table below. Assume that $\sigma = 12$ mg.

1192	1200	1207	1185	1198	1194	1210	1197	1212	1209
1189	1202	1194	1196	1179	1191	1214	1197	1213	1213
1211	1193	1204	1187	1196	1194	1220	1193	1194	1194
1177	1181	1217	1213	1204	1197	1221	1198	1210	1183

a. Find a 99% confidence interval for the mean amount of omega-3 in each Green Magic Vitality Snacks.

b. Using the confidence interval in part (a), is there any evidence to suggest that the population mean amount of omega-3 is different from 1200 mg?

SOLUTION

a. $\sigma = 12$ $n = 40$ *Given.*

$$\bar{x} = \frac{1}{40}(1192 + \cdots + 1183) = 1199.47$$ *Compute the sample mean.*

$1 - \alpha = 0.99 \Rightarrow \alpha = 0.01 \Rightarrow \alpha/2 = 0.005$ *Find $\alpha/2$.*

$P(Z \geq z_{\alpha/2}) = P(Z \geq z_{0.005}) = 0.005$ *Definition of critical value.*

$P(Z \leq z_{0.005}) = 1 - 0.005 = 0.995$ *The complement rule.*

$z_{0.005} = 2.5758$ *Use Table III in the Appendix.*

Use Equation 8.2.

$$\bar{x} \pm z_{\sigma/2} \frac{\sigma}{\sqrt{n}}$$ *Equation 8.2.*

$$= \bar{x} \pm z_{0.005} \frac{\sigma}{\sqrt{n}}$$ *Use the value of α.*

$$= 1199.47 \pm (2.5758) \frac{12}{\sqrt{40}}$$ *Use summary statistics and values for σ and $z_{0.005}$.*

$$= 1199.47 \pm 4.89$$ *Simplify.*

$$= (1194.58, 1204.36)$$ *Compute endpoints.*

$(1194.58, 1204.36)$ is a 99% confidence interval for the true mean amount of omega-3 in a Green Magic Vitality Snack. Figure 8.18 shows a technology solution.

b. Claim: $\mu = 1200$.

Experiment: $\bar{x} = 1199.47$.

Likelihood: The likelihood in this case is expressed as a 99% confidence interval, an interval of likely values for μ: $(1194.57, 1204.37)$.

Conclusion: This CI includes 1200, so there is no evidence to suggest that μ is different from 1200.

TRY IT NOW   GO TO EXERCISE 8.35

Here is one more example involving a data set instead of summary statistics.

Example 8.4 Health Canada

In April 2013, the second set of biomonitoring data from the Canadian Health Measures Survey was released. This report contained information on 91 chemicals and was designed to track levels over time in order to assess certain regulations and health-risk management actions. 100% of Canadians tested in this survey, aged 3 to 79, had lead in their blood. However, almost all had blood lead levels lower than the current intervention level.[6] A

DATA SET

HEALTHCA

random sample of 50 Canadians was obtained and the blood lead level in each was recorded, in μg/dL. The data are given in the following table. Assume that $\sigma = 0.3\ \mu$g/dL.

1.6	1.3	1.6	1.3	1.3	1.2	1.2	1.3	1.0	1.3	1.1	0.7	1.2
1.3	1.3	1.5	0.9	1.3	1.1	1.0	1.5	1.7	1.3	1.6	0.8	1.1
1.5	1.3	1.4	0.9	1.0	1.5	1.2	1.3	1.5	1.3	1.6	1.4	1.9
0.8	0.9	1.6	1.4	1.3	1.2	1.0	0.7	1.2	1.5	0.8		

Find a 98% confidence interval for the true mean blood lead level in Canadians.

SOLUTION

STEP 1 $\sigma = 0.3$ $n = 50$ Given.

$$\bar{x} = \frac{1}{50}(1.6 + \cdots + 0.8) = 1.254$$ Compute the sample mean.

$$1 - \alpha = 0.98 \ \Rightarrow\ \alpha = 0.02 \ \Rightarrow\ \alpha/2 = 0.01$$ Find $\alpha/2$.

$$P(Z \geq z_{\alpha/2}) = P(Z \geq z_{0.01}) = 0.01$$ Definition of critical value.

$$P(Z \leq z_{0.01}) = 1 - 0.01 = 0.99$$ The complement rule.

$$z_{0.01} = 2.3263$$ Use Table III in the Appendix.

STEP 2   Use Equation 8.2.

$$\bar{x} \pm z_{\alpha/2}\frac{\sigma}{\sqrt{n}}$$ Equation 8.2.

$$= \bar{x} \pm z_{0.01}\frac{\sigma}{\sqrt{n}}$$ Use the value of α.

$$= 1.254 \pm (2.3263)\frac{0.3}{\sqrt{50}}$$ Use summary statistics and values for σ and $z_{0.01}$.

$$= 1.254 \pm 0.0987$$ Simplify.

$$= (1.1553, 1.3527)$$ Compute endpoints.

(1.1553, 1.3527) is a 98% confidence interval for the true mean blood lead level in Canadians. Figure 8.19 shows a technology solution.

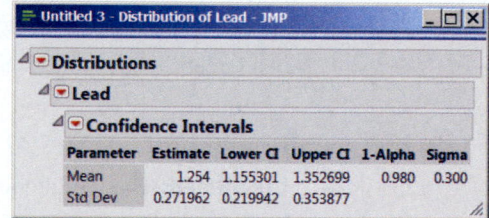

Figure 8.19 JMP confidence interval.

Many confidence intervals have endpoint formulas that are of the same general form. Suppose the statistic $\hat{\theta}$ is used to estimate the population parameter θ. The general form of a confidence interval is

(Point estimate using $\hat{\theta}$) $\pm$ (critical value) $\cdot$ [(estimate of) standard deviation of $\hat{\theta}$].

The critical value may be from the standard normal distribution or some other reference distribution. If the actual standard deviation of the statistic is not known (a likely situation), then an estimate is used.

The (two-sided) confidence interval derived above is the most common CI. However, there are other two-sided confidence intervals, and also one-sided confidence intervals consisting of either only a lower bound or only an upper bound. The derivation of several one-sided confidence intervals is outlined in the Challenge problems.

Solution Trail 8.4

KEYWORDS

- 98% CI for the true mean
- $\sigma = 0.3$
- Random sample

TRANSLATION

- 98% CI for μ
- Known standard deviation

CONCEPTS

- A $100(1 - \alpha)$% CI for a population mean when σ is known

VISION

Construct a 98% CI for μ. No information is given about the shape of the underlying distribution. Because $n = 50\ (\geq 30)$, the central limit theorem applies, and we can use Equation 8.2.

The error of estimation is the step, the distance from the sample mean to the left and to the right endpoint of the confidence interval. We would like to find a sample size n so that the step is no bigger than B. Therefore, B is called a bound on the error of estimation.

Most of the time, as in the examples above, there is no control over the sample size. The data or summary statistics are presented, and a confidence interval is constructed. However, a certain desired accuracy may be expressed by the width of a confidence interval. Given σ and the confidence level, a sample size n can be computed such that the resulting confidence interval has a certain desired width.

Suppose n is large (but unknown), σ is known, and the confidence level is $100(1 - \alpha)\%$. If the desired width is W, let $B = W/2$, called the **bound on the error of estimation**. Half the width of the confidence interval is given by the *step* (in each direction) from $\bar{x}$. The endpoints of the confidence interval for μ are (from Equation 8.2)

$$\bar{x} \pm \underbrace{z_{\alpha/2} \frac{\sigma}{\sqrt{n}}}_{B \,=\, \text{bound}}$$

Let $B = z_{\alpha/2}(\sigma/\sqrt{n})$, and solve for n. The resulting formula for n is given by

$$n = \left(\frac{\sigma z_{\alpha/2}}{B} \right)^2 \tag{8.3}$$

A CLOSER LOOK

In symbols,

$\sigma \uparrow \;\Rightarrow\; n \uparrow$

$z_{\alpha/2} \uparrow \;\Rightarrow\; n \uparrow$

$B \downarrow \;\Rightarrow\; n \uparrow$

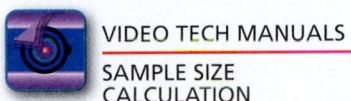

VIDEO TECH MANUALS

SAMPLE SIZE
CALCULATION

Solution Trail 8.5

KEYWORDS

- 95% CI for the mean
- Bound of 5 feet, standard deviation 15 feet
- How large a sample?

TRANSLATION

- $B = 5, \sigma = 15$
- Find a value of n

CONCEPTS

- Equation 8.3 for n

VISION

The researcher is interested in a 95% CI for a population mean with a certain bound on the error of estimation. There is a good estimate for the population standard deviation. Find the appropriate critical value, and use Equation 8.3 to determine a lower bound for the sample size.

1. ▶ Consider the effect on n as one value (on the right-hand side of Equation 8.3) changes while the other two remain constant.

   As σ increases (with $z_{\alpha/2}$ and B constant), the fraction is larger, the square is larger, and n is larger. This result makes sense, because a larger underlying variance would require a larger sample size to preserve a certain bound on the error of estimation.

   As $z_{\alpha/2}$ increases (equivalently, as the confidence level increases, with σ and B constant), again the fraction is larger, the square is larger, and n is larger. More confidence requires a larger n to maintain the same bound on the error of estimation.

   As B decreases (with σ and $z_{\alpha/2}$ constant), the denominator is smaller, which makes the fraction larger, the square larger, and n larger. A smaller bound on the error of estimation (i.e., a smaller CI) means more information (a larger sample size) is needed to maintain the same confidence level. ◀

2. When Equation 8.3 is applied, it is likely that the value of n will *not* be an integer. *Always* round up. This guarantees a $100(1 - \alpha)\%$ confidence interval with a bound on the error of estimation of at most B. Any larger sample size will also be sufficient.

3. It is unlikely that σ, needed in Equation 8.3, is known. However, often an experimenter can make a very good guess at σ from previous experience or a small preliminary study.

The following example illustrates the use of Equation 8.3.

Example 8.5 Geyser Height

Yellowstone National Park, which opened in 1872, includes magnificent scenery, historic sites, attractions of scientific interest, and recreational areas. The Cliff Geyser is located in the Black Sands Basin on Iron Creek. The time between eruptions varies considerably, from a few minutes to several hours.[7] A National Park Service researcher would like to find a 95% confidence interval for the mean height of the Cliff Geyser eruption with a bound on the error of estimation of five feet. Previous experience suggests that the population standard deviation for the height is approximately 15 feet. How large a sample is necessary to achieve this accuracy?

SOLUTION

STEP 1 $\sigma = 15$ $B = 5$ Given.

$1 - \alpha = 0.95 \;\Rightarrow\; \alpha/2 = 0.25 \;\Rightarrow\; z_{0.025} = 1.96$ Find the critical value.

STEP 2 Use Equation 8.3.

$$n = \left(\frac{\sigma\, z_{\alpha/2}}{B}\right)^2$$ Equation 8.3.

$$= \left[\frac{(15)(1.96)}{5}\right]^2$$ Substitute given values.

$$= (5.88)^2 = 34.57$$ Computation.

The necessary sample size is $n \geq 35$ (always round up). This will guarantee that a 95% confidence interval for the population mean height will have a bound on the error of estimation no greater than five.   ◼

TRY IT NOW   GO TO EXERCISE 8.39

Technology Corner

Procedure: Construct a confidence interval for a population mean when σ is known.
Reconsider: Example 8.2, solution, and interpretations.

CrunchIt!

Use the built-in function z 1-Sample. Input is either summary statistics or a column containing the sample data.

1. Select Statistics; z; 1-sample.
2. Under the Summarized tab, enter values for n, the Sample Mean, the Standard Deviation, and the Confidence Interval Level (Figure 8.20). Click Calculate. The results are displayed in a new window (Figure 8.21).

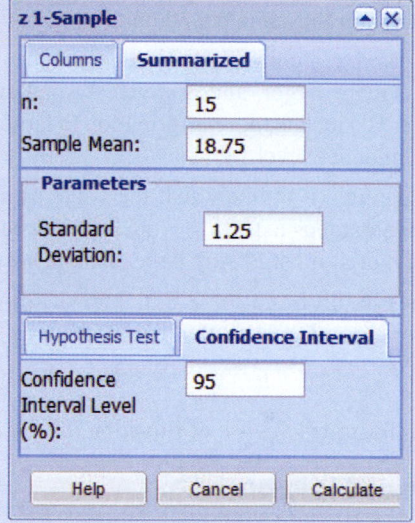

Figure 8.20 z 1-sample input screen.

Results - z 1-Sample	
Export ▾	
n:	15
Sample Mean:	18.75
Standard Deviation:	1.250
95% ConfInt:	(18.12, 19.38)

Figure 8.21 CrunchIt! results screen: a confidence interval for the population mean.

TI-84 Plus C

Use the built-in function ZInterval. Input is either summary statistics or a list containing the data.

1. Select STAT ; TESTS; ZInterval.
2. Highlight Stats (for summary statistics). Enter values for σ, $\bar{x}$, n, and the C-Level (the confidence coefficient). See Figure 8.15.
3. Highlight Calculate and press ENTER . See Figure 8.16.

Minitab

Use the built-in function 1-Sample Z. Input is either summary statistics or a column containing data.

1. Select Stat; Basic Statistics; 1-Sample Z.
2. Choose Summarized data and enter the Sample size, Mean, and Standard deviation.
3. Select the Options button and enter the confidence level. Click OK. The results are displayed in the session window. See Figure 8.22.

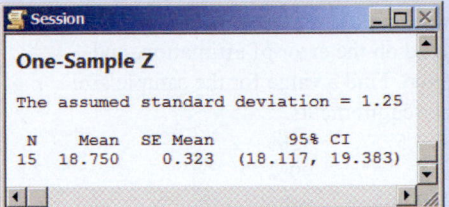

Figure 8.22 Minitab confidence interval for the population mean.

```
Session                                    _□×

One-Sample Z

The assumed standard deviation = 1.25

 N    Mean   SE Mean       95% CI
15   18.750    0.323   (18.117, 19.383)
```

Excel

There is no built-in function to compute the endpoints of a confidence interval for a population mean when σ is known. Use the function CONFIDENCE.NORM to find the step ($z_{\alpha/2} \cdot \sigma/\sqrt{n}$) in each direction from the mean, and the function AVERAGE to find the sample mean, if necessary.

1. Use the command CONFIDENCE.NORM to find the step. The arguments are α, σ, and n.
2. Compute the left and right endpoints of the confidence interval. See Figure 8.23.

Figure 8.23 Excel calculations to construct the confidence interval.

	A	B	C
1	0.6326	= CONFIDENCE.NORM(0.05,1.25,15)	The step
2	18.1174	= 18.75 - A1	Left endpoint
3	19.3826	= 18.75 + A1	Right endpoint

SECTION 8.2 EXERCISES

Concept Check

8.19 True/False A confidence interval for a population mean is guaranteed to contain the true value of μ.

8.20 True/False A critical value is always defined in terms of right-tail probability.

8.21 True/False Many different two-sided confidence intervals and one-sided confidence intervals can be constructed based on the same information/data.

8.22 Fill in the Blank

$P(Z \geq z_{\alpha/2}) =$ _____.

$P(Z \leq z_{\alpha/2}) =$ _____.

$P(Z \leq -z_{\alpha/2}) =$ _____.

8.23 Fill in the Blank As the confidence coefficient increases, the confidence interval becomes _____.

8.24 Fill in the Blank To find a $100(1 - \alpha)\%$ confidence interval for a population mean when σ is known, either the underlying population must be _____ or _____.

8.25 Short Answer Explain the meaning of a confidence coefficient.

Practice

8.26 Find each of the following critical values.

a. $z_{0.10}$ **b.** $z_{0.05}$ **c.** $z_{0.025}$ **d.** $z_{0.01}$

e. $z_{0.005}$ **f.** $z_{0.001}$ **g.** $z_{0.0005}$ **h.** $z_{0.0001}$

8.27 In each of the following problems, the sample mean, the sample size, the population standard deviation, and the confidence level are given. Assume that the underlying population is normally distributed. Find the associated confidence interval for the population mean.

a. $\bar{x} = 15.6$, $\quad n = 12$, $\quad \sigma = 3.7$, $\quad 95\%$
b. $\bar{x} = 6322$, $\quad n = 17$, $\quad \sigma = 225$, $\quad 90\%$
c. $\bar{x} = -45.78$, $\quad n = 9$, $\quad \sigma = 12.35$, $\quad 80\%$
d. $\bar{x} = 0.0795$, $\quad n = 24$, $\quad \sigma = 0.006$, $\quad 99\%$
e. $\bar{x} = 37.68$, $\quad n = 27$, $\quad \sigma = 2.2$, $\quad 99.9\%$

8.28 In each of the following problems, the sample mean, the sample size, the population standard deviation, and the confidence level are given. Find the associated confidence interval for the population mean.

a. $\bar{x} = 17.6$, $\quad n = 32$, $\quad \sigma = 10.27$, $\quad 95\%$
b. $\bar{x} = 136.8$, $\quad n = 45$, $\quad \sigma = 25.44$, $\quad 99\%$
c. $\bar{x} = 335.7$, $\quad n = 65$, $\quad \sigma = 125.3$, $\quad 90\%$
d. $\bar{x} = -6.7$, $\quad n = 52$, $\quad \sigma = 2.25$, $\quad 98\%$
e. $\bar{x} = 20.11$, $\quad n = 37$, $\quad \sigma = 1.76$, $\quad 99.99\%$

8.29 Two statisticians are given the same data from an experiment. Each uses these data to construct a confidence interval for the true population mean μ. The resulting CIs are (8.55, 10.85) and (8.40, 11.0).

    a. What is the value of the sample mean $\bar{x}$?

    b. One CI has a confidence level of 95%, and the other has a confidence level of 99.9%. Match the confidence level with the confidence interval, and justify your answer.

8.30 In each of the following problems, the population standard deviation, the bound on the error of estimation, and the confidence level are given. Find a value for the sample size n necessary to satisfy these requirements.

    a. $\sigma = 7.9$, $B = 2.5$,   95%
    b. $\sigma = 10.77$, $B = 5$,   99%
    c. $\sigma = 0.55$, $B = 0.001$,   98%
    d. $\sigma = 35.97$, $B = 3.5$,   95%
    e. $\sigma = 55$, $B = 2$,   99.9%

Applications

8.31 Physical Science The number of active coal mines in Utah has steadily decreased since 1960. However, peak production occurred in 2001.[8] One measure of mine productivity is tons/employee hour. In a random sample of 40 days, the mean tons/employee hour was 4.99. Assume $\sigma = 1.14$. Find a 95% confidence interval for the true mean tons/employee hour.

8.32 Public Health and Nutrition After a ham is cured, it may be smoked to add flavor or to ensure it lasts longer. Typical grocery-store hams are smoked for a short period of time, whereas gourmet hams are usually smoked for at least one month. A random sample of 36 grocery-store hams was obtained, and the length of the smoking time was recorded for each. The mean was $\bar{x} = 140$ hours. Assume $\sigma = 8$ hours.

    a. Find a 99% confidence interval for the mean amount of time a grocery-store ham is smoked. Write a Solution Trail for this problem.

    b. What assumptions did you make in order to construct the confidence interval in part (a)?

8.33 Sports and Leisure In May 2013 the manager of Chicago's Navy Pier set the world record for the longest ferris wheel ride. Clinton Shepherd spent more than 2 days and 384 revolutions on the wheel in a specially designed gondola.[9] During normal park hours, a random sample of ferris wheel riders was obtained and the length of each ride was measured (in minutes). Assume $\sigma = 0.5$. **LONGRIDE**

    a. Find a 95% confidence interval for the mean amount of time on this ride.

    b. Find a 99% confidence interval for the mean amount of time on this ride.

    c. Explain why the interval in part (b) is wider.

8.34 Marketing and Consumer Behavior The owner of a small tailoring shop keeps careful records of all alterations. Two common alterations to men's clothing are lengthening the

inseam on pants and letting out a sports coat around the waist. A random sample of each type of alteration was obtained, and the summary statistics are given in the table below. Measurements are in inches.

Alteration	Sample size	Sample mean	Assumed σ
Pants inseam	33	0.74	0.22
Sports coat waist	42	1.05	0.37

    a. Find a 95% confidence interval for the mean alteration of the inseams on men's pants.

    b. How large a sample is necessary for the bound on the error of estimation to be 0.05 for the confidence interval in part (a)?

    c. Find a 90% confidence interval for the mean alteration of the waists of sports coats.

    d. How large a sample is necessary for the bound on the error of estimation to be 0.07 for the confidence interval in part (c)?

8.35 Physical Sciences An iceberg consists of frozen freshwater that has broken away from a glacier, and could be more than 15,000 years old. Icebergs in the Grand Banks area are estimated to weigh between 100,000 and 200,000 tons.[10] Suppose a team of scientists aboard a research vessel operating in the Grand Banks area selects a random sample of 49 icebergs. The height is carefully measured for each, and the sample mean is $\bar{x} = 102$ meters. Assume $\sigma = 25$ meters.

    a. Find a 99% confidence interval for the mean height of icebergs in the Grand Banks area.

    b. Some experts suggest that global warming has caused there to be more icebergs of greater height. Using the interval in part (a), is there any evidence to suggest that the mean height of icebergs in the Grand Banks area is more than 95 meters? Justify your answer. Write a Solution Trail for this problem.

8.36 Travel and Transportation Lighthouses are constructed to guide ships traveling in rocky waters and to allow sailing at night. There are many ways to report the size of a lighthouse. However, the height is usually measured from the base of the tower to the top of the ventilator ball. A random sample of 18 lighthouses in France and England was obtained, and the height of each was recorded. The sample mean was $\bar{x} = 33.75$ meters. Assume the distribution of lighthouse heights is normal and $\sigma = 5.4$ meters.

    a. Find a 95% confidence interval for the mean height of all lighthouses in France and England.

    b. How large a sample is necessary to ensure that the width of the resulting 95% confidence interval is 2 meters?

    c. Why is the confidence interval constructed in part (a) valid even though the sample size $n = 18$ is less than 30?

8.37 Biology and Environmental Science Koalas are not really bears, but are related to the wombat and the kangaroo and live in eastern Australia. These cuddly-looking creatures grow to a size of 23.5 to 33.5 inches and are a threatened species.[11]

A random sample of 35 koalas was obtained, and each was carefully weighed. The mean weight was $\bar{x} = 20.75$ pounds. Assume $\sigma = 3.05$ pounds.

a. Find a 99% confidence interval for the mean weight of all koalas.

b. Researchers believe that if the true mean weight is less than 23 pounds, these animals may be suffering from malnutrition. Is there any evidence to suggest that the koala population is suffering from malnutrition? Justify your answer.

8.38 Technology and the Internet Many companies are increasingly concerned about computer security and employees using their computers for personal use. A random sample of 50 employees of Liberty Mutual Insurance Company was asked whether they use the Internet for nonbusiness (personal) purposes during the day. The number of hours for each employee is given on the text website. Assume $\sigma = 0.5$ hours. **INTERNET**

a. Find a 95% confidence interval for the mean number of hours all employees at Liberty Mutual use the Internet for personal reasons.

b. Using the confidence interval in part (a), is there any evidence to suggest that the mean time spent using the Internet for personal reasons is more than 1 hour? Justify your answer.

c. How large a sample is necessary for the bound on the error of estimation to be 0.1 hours?

d. Do you think the underlying distribution of time spent on the Internet for personal use is normal? Explain your answer.

8.39 Sports and Leisure A random sample of professional wrestlers was obtained, and the annual salary (in dollars) for each was recorded. The summary statistics were $\bar{x} = 47,500$ and $n = 18$. Assume the distribution of annual salary is normal with $\sigma = 8,500$.

a. Find a 90% confidence interval for the true mean annual salary for all professional wrestlers.

b. How large a sample is necessary for the bound on the error of estimation to be 3000?

c. How large a sample is necessary for the bound on the error of estimation to be 1000?

8.40 Travel and Transportation The Boeing 747-8 is a technologically advanced aircraft with a new wing design and GEnx-2B engines, and is capable of holding more passengers and payload than previous versions of the 747. In addition, this plane has better fuel efficiency than the 747-400 and the A380, which means lower carbon emissions.[12] A random sample of twelve 747-8 test flights was obtained and the carbon emission was measured on each (in grams CO_2 per seat-km). The sample mean was 72. Assume the carbon emission distribution is normal and $\sigma = 8.7$.

a. Find a 99% confidence interval for the true mean carbon emission of all 747-8 planes.

b. The carbon emission for the A380 is 80 g CO_2 per seat-km. Is there any evidence to suggest that the mean carbon emission for the 747-8 is lower than that of the A380? Justify your answer.

8.41 Public Policy and Political Science Many chimney fires are caused by a buildup of creosote, a highly flammable material that forms in flues due to the condensation of certain gases. In an effort to promote fireplace and wood-stove safety, a town in Vermont has started a new inspection program. Forty homes with wood stoves were selected at random, and the amount of creosote buildup one foot from the top of the chimney was carefully measured. The resulting sample mean thickness was $\bar{x} = 0.131$ inch. Assume $\sigma = 0.02$ inch.

a. Find a 95% confidence interval for the true mean thickness of creosote buildup one foot from the top of chimneys.

b. One-eighth of an inch of creosote buildup is considered safe. If there is evidence to suggest that the true mean thickness is greater than $\frac{1}{8}$ inch, the town will embark on an extensive safety program. Using the interval constructed in part (a), should the town stress greater safety? Justify your answer.

8.42 Physical Sciences According to the U.S. National Weather Service, at any given moment of any day, approximately 2000 thunderstorms are occurring worldwide. Many of these storms include lightning strikes. Sensitive electronic equipment is used to record the number of lightning strikes worldwide every day. Twelve days were selected at random, and the number of lightning strikes on each day was recorded. The sample mean was $\bar{x} = 8.6$ million. Assume the distribution of the number of lightning strikes per day is normal with $\sigma = 0.35$ million.

a. Find a 99% confidence interval for the mean number of lightning strikes per day worldwide.

b. Do you think the normality assumption in this problem is reasonable? Why or why not?

8.43 Technology and the Internet In early 2013, Cree, Inc., introduced a new, cheap LED light that uses much less electricity and is designed to last for years. These new bulbs are replacements for traditional incandescent bulbs, and the company claims that its 60-watt warm white bulb has a brightness of 800 lumens.[13] A random sample of 45 60-watt warm white bulbs was obtained and the brightness of each was carefully measured. The sample mean was $\bar{x} = 798$ lumens. Assume $\sigma = 12.2$ lumens.

a. Find a 95% confidence interval for the true mean brightness for the 60-watt warm white bulb.

b. Using your answer in part (a), is there any evidence to suggest that the mean brightness is less than 800 lumens? Justify your answer.

c. Suppose $\sigma = 5.2$. Find a 95% confidence interval for the true mean brightness for the 60-watt warm white bulb. In this case, is there any evidence to suggest that the true mean brightness is less than 800 lumens? Justify your answer.

Extended Applications

8.44 Biology and Environmental Science Peregrine falcons were placed on the U.S. Endangered Species list in the 1970s. Their population dwindled primarily as a result of the introduction of new chemicals and pesticides. Although it may seem like a peculiar nesting area, there are many falcon pairs in the New York City area. The tall buildings and bridges provide lots of open space for hunting. These birds have a wingspan of 3.3 to 3.6 feet and weigh between 18.8 and 56.5 ounces. A random sample of 35 falcons was obtained and each was carefully weighed. The sample mean was $\bar{x} = 47.3$ ounces. Assume $\sigma = 6.2$ ounces.

a. Find a 95% confidence interval for the true mean weight of peregrine falcons.

b. Using your answer from part (a), is there any evidence to suggest that the mean weight is less than 48 ounces?

8.45 Economics and Finance   Representative Bill Cassidy has become concerned about the disparity in home prices in Louisiana's 6th district (Monroe). A random sample of homes sold within the past year in each of two parishes was obtained. The summary statistics are given in the table below. Prices are in dollars.

Parish	Sample size	Sample mean	Assumed σ
East Feliciana	30	125,200	5,750
Iberville	36	155,900	25,390

a. Find a 99% confidence interval for the mean selling price of all homes in East Feliciana Parish.

b. Find a 99% confidence interval for the mean selling price of all homes in Iberville Parish.

c. Using the confidence intervals in parts (a) and (b), is there any evidence to suggest that the mean selling price is different for the two parishes? Justify your answer.

8.46 Sports and Leisure   There are three main types of exercises: range-of-motion (flexibility), strengthening, and endurance. A random sample of people who exercise regularly was obtained. The type of exercise and length (in minutes) of each workout was recorded. The table below summarizes the information obtained.

Exercise type	Sample size	Sample mean	Assumed σ
Range-of-motion	65	25.2	5.2
Strengthening	32	73.6	10.7
Endurance	40	82.2	12.5

a. Construct a 99% confidence interval for the mean workout length for each of the three types of exercises.

b. A group of exercise scientists claims the length of a workout for range-of-motion exercises is the same as for the other two types. Based on the confidence intervals in part (a), is there any evidence to refute this claim?

8.47 Psychology and Human Behavior   A random sample of male college athletes was obtained and their coping skills levels were measured by means of an extensive psychological profile. Each athlete was also classified by sport. The table below summarizes the information obtained.

Sport	Sample size	Sample mean	Assumed σ
Football	35	65.77	14.07
Basketball	30	53.90	12.50
Hockey	32	68.45	10.25

a. Find a 95% confidence interval for the true mean coping skills level for all male athletes in each sport.

b. Use your answers to part (a) to determine whether there is any evidence to suggest that the mean coping skills level for male basketball players is different from that of male football players. Justify your answer.

c. How large a sample is necessary for the bound on the error of estimation for each confidence interval in part (a) to be 2?

8.48 Sports and Leisure   Two competing ski slopes in Colorado advertise their powder base each day in an effort to attract more skiers. A random sample of the powder base depth (in inches) was obtained for each ski resort on days during a recent winter. Assume $\sigma_B = 2.5$ and $\sigma_V = 2.7$.   **POWDER**

a. Find a 99% confidence interval for the true mean powder depth at Breckenridge Ski Resort.

b. Find a 99% confidence interval for the true mean powder depth at Vail Ski Resort.

c. Use the results in parts (a) and (b) to determine whether there is any evidence to suggest a difference in the true mean powder depths at these two ski resorts.

d. What assumptions were necessary to construct the confidence intervals in parts (a) and (b)?

8.49 Public Health and Nutrition   The nutritional value of every food product sold in the United States is listed on the package in terms of protein, fat, etc. A researcher randomly selected 50 one-ounce samples of various nuts and carefully measured the amount of protein (in grams) in each sample. The sample mean amount of protein and the assumed standard deviation for each type of nut are given in the table below.[14]

Nut	Sample mean	Assumed σ
Cashew	5.17	0.40
Filbert	4.24	0.60
Pecan	2.60	0.95

a. Find a 95% confidence interval for the true mean amount of protein in each type of nut. Based on these intervals, is there any evidence to suggest that the mean amount of protein is different for cashews and pecans? How about filberts and pecans?

b. Answer part (a) using a sample size for each nut of 18.

8.50 Business and Management   Snail farming, or heliciculture, has become more popular and lucrative in many parts of the world. In April 2013, John Morton, head of U.S. Immigration and Customs Enforcement, resigned with plans to open a snail farm.[15] The *Helix pomatia*, or escargot snail, grows to full size in 2 to 3 years and weighs 15 to 25 grams. A random sample of snails of this type was obtained from a farm near the town of Targovishte, Bulgaria. Each snail was carefully weighed (in ounces). Assume $\sigma = 2.4$ grams.   **SNAILS**

a. Find a 99% confidence interval for the true mean weight of these snails.

b. Using your answer in part (a), is there is any evidence to suggest that the population mean weight is more than 20 grams?

c. How large a sample size is necessary for the bound on the error of estimation to be 0.5 gram?

Challenge

8.51 One-Sided Confidence Intervals   Given a random sample of size n from a population with mean μ, assume the underlying population is normal or n is large, and the population standard deviation σ is known.

a. Manipulate the probability statement

$$P\left(\frac{\overline{X} - \mu}{\sigma/\sqrt{n}} > -z_\alpha\right) = 1 - \alpha$$

to find a *one-sided* $100(1 - \alpha)\%$ confidence interval for μ. That is, find an upper bound for the population mean μ.

b. Manipulate a similar probability statement to find a one-sided $100(1 - \alpha)\%$ confidence interval for μ bounded below.

c. First-class mail in the United States includes personal correspondence and all kinds of bills. Each piece of first-class mail must weigh less than 13 ounces. In a random sample of 25 first-class letters, the sample mean weight was $\bar{x} = 2.2$ ounces. Assume the population is normal and $\sigma = 0.75$. Find a 95% one-sided confidence interval bounded above for the population mean weight of first-class letters.

8.52 Sports and Leisure   Steeplechase horseraces are run on courses that include obstacles such as brush fences, stone walls, timber walls, and water jumps. A random sample of 17 winning times (in seconds) in the $2\frac{3}{8}$-mile Saratoga Steeplechase Race

was obtained. The sample mean was $\bar{x} = 259.79$. Assume and $\sigma = 7.5$ the population distribution of winning times is normal, and find a one-sided 99% confidence interval, bounded below, for the true mean winning time.

8.53 The "Best" CI   There are actually *infinitely* many different confidence intervals for a population mean μ. For example, suppose that, in a random sample of size n from a population with mean μ, the underlying population is normal or n is large, and the population standard deviation σ is known. One could start with a probability statement of the form

$$P\left(-z_{3\alpha/4} < \frac{\overline{X} - \mu}{\sigma/\sqrt{n}} < z_{\alpha/4}\right) = 1 - \alpha$$

to find a $100(1 - \alpha)\%$ confidence interval for μ. Why is the traditional two-sided confidence interval (presented in this chapter) the *best* CI for a parameter?

8.54 The Real Meaning of a 95% CI   In the *Confidence Intervals* applet, select a 95% Confidence Level (C). Click the Sample 50 button to generate 50 simple random samples. The applet will construct the confidence interval associated with each sample and display a visual representation of each CI with a line centered at the sample mean. Lines in red do not capture the true population mean.

a. Describe any pattern in the confidence intervals.

b. Click the Sample 50 button repeatedly and observe the Percent hit. This indicates the percentage of CIs that captured the true population mean. Describe the behavior of the percent hit as the total CIs generated increases. Justify your answer.

8.3   A Confidence Interval for a Population Mean When σ Is Unknown

The symbol T is used in notation to represent the sample total (in Chapter 7) and a random variable with a t distribution. The context in which the notation is used implies the appropriate concept.

Okay, here is the probability density function for a t random variable:

$$f(x) = \frac{1}{\sqrt{\pi\nu}} \cdot \frac{\Gamma\left(\frac{\nu + 1}{2}\right)}{\Gamma\left(\frac{\nu}{2}\right)} \cdot \left(1 + \frac{x^2}{\nu}\right)^{-(\nu+1)/2}$$

$-\infty < x < \infty$ $\nu = df$

In Section 8.2, a confidence interval for a population mean μ was presented. However, Equation 8.2, based on a standard normal, or Z, distribution, is valid only if σ is known (and either the underlying population is normal or the sample size is large). It is unrealistic to assume that the population standard deviation is known. A more practical and useful approach to constructing a confidence interval for μ is presented in this section.

Recall that if σ is known and either the underlying population is normal or the sample size is large, then the expression

$$Z = \frac{\overline{X} - \mu}{\sigma/\sqrt{n}}$$

has a standard normal distribution. This expression was used to derive the confidence interval in Section 8.2. Note that only one component of this expression, $\overline{X}$, contributes to the variability.

If σ is unknown, a similar *standardization* is used:

$$T = \frac{\overline{X} - \mu}{S/\sqrt{n}} \tag{8.4}$$

However, the distribution of this random variable is not normal. It is reasonable to believe that the distribution of T is centered at 0, but there is more variability in this expression, with contributions from two sources, $\overline{X}$ and S. The most common confidence interval for a population mean μ is based on Equation 8.4, and the t distribution introduced below.

The t distribution is closely related to the standard normal distribution. Most continuous random variables are defined by a probability density function, but it is only important here to understand the properties of a t distribution.

Properties of a t Distribution

The t distribution was derived by William Gosset in 1908. Working for Guinness Breweries, he published his result using the pseudonym *Student*. For this reason, the distribution is still often called "Student's t distribution."

The symbol ν is the Greek letter "nu." It is often used to represent the number of degrees of freedom associated with a random variable.

1. A t distribution is completely determined (characterized) by only one parameter, ν, called the number of degrees of freedom (df). ν must be a positive integer, $\nu = 1, 2, 3, 4, \ldots$, and there is a different t distribution corresponding to each value of ν.

2. If T (a random variable) has a t distribution with ν degrees of freedom, denoted $T \sim t_\nu$, then

$$\mu_T = 0 \quad \text{and} \quad \sigma_T^2 = \frac{\nu}{\nu - 2} \quad (\nu \geq 3) \tag{8.5}$$

3. The density curve for every t distribution is bell-shaped and centered at 0, but more spread out than the density curve for a standard normal random variable, Z. As ν increases, the density curve for T becomes more compact and closer to the density curve for Z. See Figure 8.24.

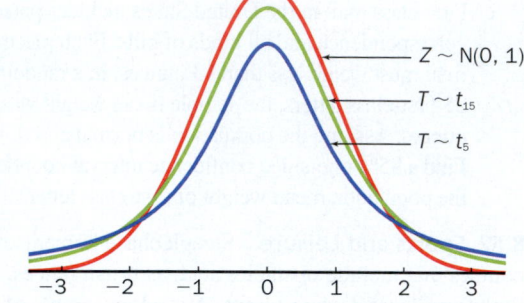

Figure 8.24 A comparison of density curves for two t distributions with the density curve for a standard normal random variable.

To construct a confidence interval for μ based on a t distribution, the following definition for a ***t* critical value** is necessary.

Definition

$t_{\alpha,\nu}$ is a **critical value** related to a ***t* distribution** with ν degrees of freedom. If T has a t distribution with ν degrees of freedom, then $P(T \geq t_{\alpha,\nu}) = \alpha$.

A CLOSER LOOK

Remember, the α in $t_{\alpha,\nu}$ is simply a *placeholder*. Any symbol could be used here to represent a right-tail probability.

$$P(T \geq t_{\alpha,\nu}) = \alpha$$

1. For any t distribution, $t_{\alpha,\nu}$ is simply a t value (a value on the measurement axis) such that there is α of the area (probability) to the right of $t_{\alpha,\nu}$. The negative critical value is $-t_{\alpha,\nu}$. Because the t distribution is symmetric, $P(T \leq -t_{\alpha,\nu}) = P(T \geq t_{\alpha,\nu}) = \alpha$. See Figure 8.25.

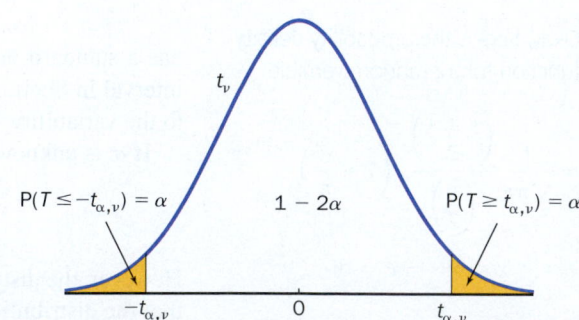

Figure 8.25 An illustration of the definition of a t critical value.

2. Remember, critical values are always defined in terms of right-tail probability.

Table V in the Appendix presents selected critical values associated with various t distributions. Right-tail probabilities are in the top row and degrees of freedom are listed in the left column. In the body of the table, $t_{\alpha,\nu}$ is at the intersection of the α column and the ν row. The following example illustrates the use of this table for finding critical values associated with a t distribution.

Example 8.6 Critical Value Look-ups

Find each critical value: (a) $t_{0.05,12}$, (b) $t_{0.01,21}$.

SOLUTION

a. Using the following portion of Table V, find the intersection of the $\alpha = 0.05$ column and the $\nu = 12$ row.

ν	0.20	0.10	0.05	0.025	0.01	0.005	0.001	0.0005	0.0001
⋮	⋮	⋮	⋮	⋮	⋮	⋮	⋮	⋮	⋮
10	.8791	1.3722	1.8125	2.2281	2.7638	3.1693	4.1437	4.5869	5.6938
11	.8755	1.3634	1.7959	2.2010	2.7181	3.1058	4.0247	4.4370	5.4528
12	.8726	1.3562	1.7823	2.1788	2.6810	3.0545	3.9296	4.3178	5.2633
13	.8702	1.3502	1.7709	2.1604	2.6503	3.0123	3.8520	4.2208	5.1106
14	.8681	1.3450	1.7613	2.1448	2.6245	2.9768	3.7874	4.1405	4.9850
⋮	⋮	⋮	⋮	⋮	⋮	⋮	⋮	⋮	⋮

The top of the α columns is labeled α.

Therefore, $t_{0.05,12} = 1.7823$, and if T has a t distribution with $\nu = 12$, then $P(T \geq 1.7823) = 0.05$, as illustrated in Figure 8.26.

Figure 8.28 shows a technology solution.

b. Similarly, using Table V, find the intersection of the $\alpha = 0.01$ column and the $\nu = 21$ row. $t_{0.01,21} = 2.5176$, and if T has a t distribution with $\nu = 21$, then $P(T \geq 2.5176) = 0.01$, as illustrated in Figure 8.27.

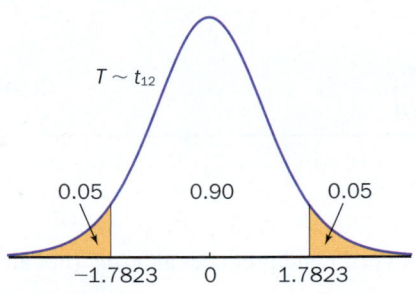

Figure 8.26 Visualization of $t_{0.05,12} = 1.7823$.

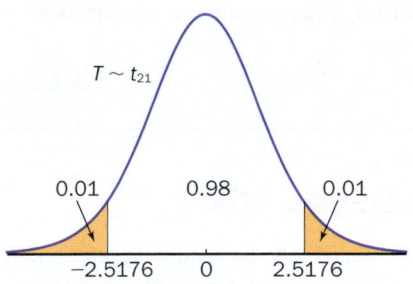

Figure 8.27 Visualization of $t_{0.01,21} = 2.5176$.

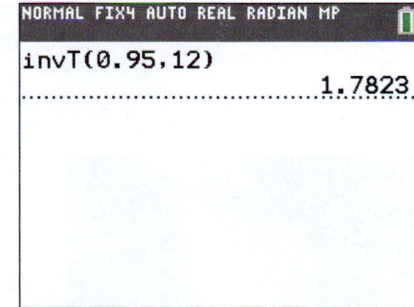

Figure 8.28 Use the `invT` function to find critical values associated with a t distribution.

TRY IT NOW   GO TO EXERCISE 8.61

A CLOSER LOOK

1. Table V is *very* limited. However, a calculator or computer can find almost any critical value needed.

2. In Table V, as ν increases, the t critical values approach the corresponding Z critical values. This numerical observation is analogous to the graphical comparison in Figure 8.24. The density curve for T approaches the density curve for Z as ν increases.

Even for $\nu = 40$, $t_{\alpha,\nu} \approx z_\alpha$.

▶ A confidence interval for a population mean when σ is unknown is based on the following theorem.

8.1 Theorem

Let $\overline{X}$ be the mean of a random sample of size n from a normal distribution with mean μ. The random variable

$$T = \frac{\overline{X} - \mu}{S/\sqrt{n}} \tag{8.6}$$

has a t distribution with $n - 1$ degrees of freedom.

Even though n is in the denominator, $n - 1$ degrees of freedom is correct!

This theorem is used to construct a confidence interval for μ. As in Section 8.2, start in the appropriate t world. Find a symmetric interval about 0 such that the probability T lies in this interval is $1 - \alpha$.

$$\text{P}(-t_{\alpha/2,n-1} < T < t_{\alpha/2,n-1}) = 1 - \alpha$$

$$\text{P}\left(-t_{\alpha/2,n-1} < \frac{\overline{X} - \mu}{S/\sqrt{n}} < t_{\alpha/2,n-1}\right) = 1 - \alpha \tag{8.7}$$

Manipulate Equation 8.7 to obtain the probability statement

$$\text{P}\left(\overline{X} - t_{\alpha/2,n-1}\frac{S}{\sqrt{n}} < \mu < \overline{X} + t_{\alpha/2,n-1}\frac{S}{\sqrt{n}}\right) = 1 - \alpha \tag{8.8}$$

This probability statement leads to the following general result. ◀

How to Find a 100(1 − α)% Confidence Interval for a Population Mean When σ Is Unknown

Given a random sample of size n with sample standard deviation s from a population with mean μ; if the underlying population distribution is normal, a $100(1 - \alpha)\%$ confidence interval for μ has as endpoints the values

$$\overline{x} \pm t_{\alpha/2,n-1}\frac{s}{\sqrt{n}} \tag{8.9}$$

A CLOSER LOK

1. Equation 8.9 can be used with *any* sample size n (≥ 2) and produces an *exact* (not an approximate) confidence interval for μ.

2. This confidence interval for μ (Equation 8.9) is valid *only* if the underlying population is normal. ■

Example 8.7 Orange Blossom Perfume

Distillation is a process for separating and collecting substances according to their reaction to heat. When heat is applied to a mixture, the substance that evaporates and is collected as it cools is the distillate. The unevaporated portion of the mixture is the residue. Oil obtained from orange blossoms through distillation is used in perfume. Suppose the oil yield is normally distributed. In a random sample of 11 distillations, the sample mean oil yield was $\overline{x} = 980.2$ grams with standard deviation $s = 27.6$ grams. Find a 95% confidence interval for the true mean oil yield per batch.

Emilio Ereza/AgeFotostock

Solution Trail 8.7

KEYWORDS

- 95% CI for the true mean
- Normally distributed
- $s = 27.6$
- Random sample

TRANSLATION

- 95% CI for μ
- Underlying population is normal
- σ is unknown

CONCEPTS

- A $100(1 - \alpha)$% CI for a population mean when σ is unknown

VISION

We need a 95% CI for μ. The population is normal, and σ is unknown. Find the appropriate t critical value, and use Equation 8.9.

SOLUTION

STEP 1 $\bar{x} = 980.2 \qquad s = 27.6 \qquad n = 11$ — Given.

$1 - \alpha = 0.95 \Rightarrow \alpha = 0.05 \Rightarrow \alpha/2 = 0.025$ — Find $\alpha/2$.

$t_{\alpha/2,n-1} = t_{0.025,10} = 2.2281$ — Use Table V in the Appendix with $\nu = 10$.

STEP 2 Use Equation 8.9.

$$\bar{x} \pm t_{\alpha/2,n-1}\frac{s}{\sqrt{n}}$$ — Equation 8.9.

$$= \bar{x} \pm t_{0.025,10}\frac{s}{\sqrt{n}}$$ — Use the values of α and n.

$$= 980.2 \pm (2.2281)\frac{27.6}{\sqrt{11}}$$ — Use summary statistics and value for $t_{0.025,10}$.

$$= 980.2 \pm 18.54$$ — Simplify.

$$= (961.66, 998.74)$$ — Compute endpoints.

(961.66, 998.74) is a 95% confidence interval for the true mean oil yield (in grams) through distillation in a typical batch.

Figures 8.29 and 8.30 together show a technology solution.

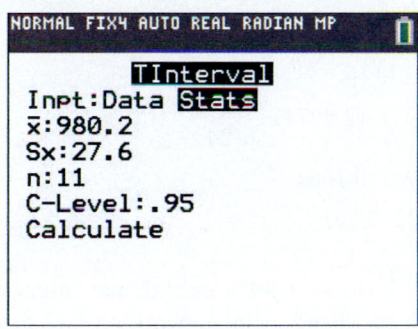

Figure 8.29 TInterval input screen.

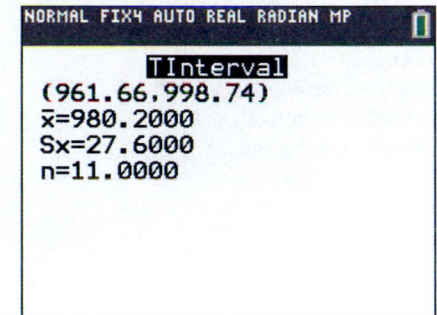

Figure 8.30 Resulting confidence interval.

TRY IT NOW GO TO EXERCISE 8.68

The following example illustrates the same process using actual data rather than summary statistics. The resulting confidence interval is used to draw a conclusion about the value of the population mean.

Example 8.8 Social Networking

DATA SET
SOCIAL

The use of social networks, for example, Facebook and Twitter, has grown dramatically all over the world. In a recent survey among social network users, it was reported that Indonesians and Saudi Arabians spend the most time per day using social networks. American social network users reportedly spend over 3 hours per day social networking from a computer, tablet, and/or mobile phone.[16] A random sample of 24 American social network users was obtained and each was asked for the amount of time spent (in hours) social networking each day. The data are given in the following table.

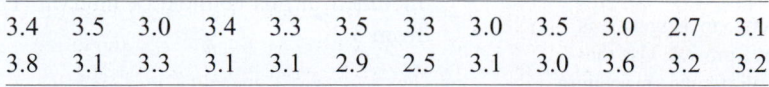

| 3.4 | 3.5 | 3.0 | 3.4 | 3.3 | 3.5 | 3.3 | 3.0 | 3.5 | 3.0 | 2.7 | 3.1 |
| 3.8 | 3.1 | 3.3 | 3.1 | 3.1 | 2.9 | 2.5 | 3.1 | 3.0 | 3.6 | 3.2 | 3.2 |

a. Find a 99% confidence interval for the true mean amount of time Americans spend social networking each day. Assume that the underlying population is normal.

b. Using the confidence interval constructed in part (a), is there any evidence to suggest that the true mean amount of time spent social networking each day is different from 3.2 hours? Justify your answer.

SOLUTION

a. $n = 24$ — Sample size.

$$\bar{x} = \frac{1}{24}(3.4 + \cdots + 3.2) = 3.19$$ — Compute the sample mean.

$$s^2 = \frac{1}{23}[(3.4 - 3.19)^2 + \cdots + (3.2 - 3.19)^2] = 0.0843$$

$$s = \sqrt{0.0843} = 0.2903$$ — Compute the sample standard deviation.

$$1 - \alpha = 0.99 \implies \alpha = 0.01 \implies \alpha/2 = 0.005$$ — Find $\alpha/2$.

$$t_{\alpha/2,n-1} = t_{0.005,23} = 2.8073$$ — Use Table V with $\nu = 22$.

Use Equation 8.9.

$$\bar{x} \pm t_{\alpha/2,n-1}\frac{s}{\sqrt{n}}$$ — Equation 8.9.

$$= \bar{x} \pm t_{0.005,23}\frac{s}{\sqrt{n}}$$ — Use the values of α and n.

$$= 3.19 \pm (2.8073)\frac{0.2903}{\sqrt{24}}$$ — Use summary statistics and value for $t_{0.005,23}$.

$$= 3.19 \pm 0.1664$$ — Simplify.

$$= (3.02, 3.36)$$ — Compute endpoints.

(3.02, 3.36) is a 99% confidence interval for the true mean amount of time Americans spend social networking each day. Figure 8.31 and 8.32 show technology solutions.

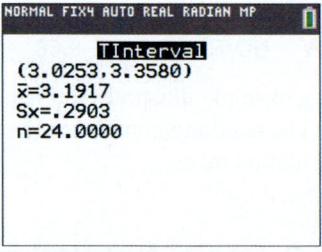

Figure 8.31 TI-84 Plus C TInterval.

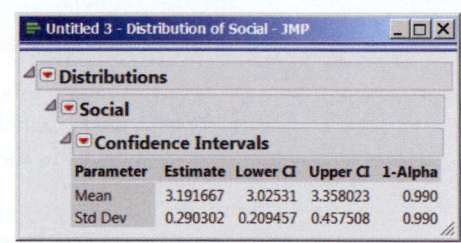

Figure 8.32 JMP confidence interval.

b. Claim: $\mu = 3.2$.

Experiment: $\bar{x} = 3.19$.

Likelihood: The likelihood is expressed as a 99% confidence interval, an interval of likely values for μ: (3.02, 3.36), from part (a).

Conclusion: The claimed amount of time spent social networking, 3.2 hours, is included in this confidence interval. There is no evidence to suggest μ is different from 3.2.

TRY IT NOW GO TO EXERCISE 8.71

A CLOSER LOOK

1. Technology solutions can find the appropriate t critical value for any sample size and any confidence level. If you must use Table V to find a critical value for a value of ν and/or a value of α not listed, use linear interpolation.

   Recall that interpolation is a method of approximation. It is often used to estimate a value at a position between two given values in a table. Linear interpolation assumes the two known values lie on a straight line and was discussed in Chapter 6.

2. Sample size calculation is more complicated in this case. σ is unknown and the critical value, $t_{\alpha/2,n-1}$, depends on n. Equation 8.3 is often used with an estimate for σ and the assumption that $z_{\alpha/2} \approx t_{\alpha/2,n-1}$. ●

Technology Corner

Procedure: Construct a confidence interval for a population mean when σ is unknown.
Reconsider: Example 8.8, solution, and interpretations.

CrunchIt!

Use the function t; 1-sample. Input is either summary statistics or a column containing data.

1. Enter the data in column Var1.
2. Select Statistics; t; 1-sample.
3. Under the Columns tab, enter the name of the column containing the data.
4. Under the Confidence Interval tab, enter the confidence level. Click Calculate. The confidence interval is displayed in a new window. See Figure 8.33.

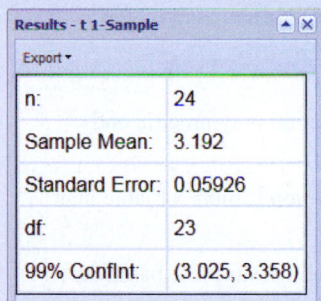

Figure 8.33 CrunchIt! t 1-Sample confidence interval.

Results - t 1-Sample	
Export ▾	
n:	24
Sample Mean:	3.192
Standard Error:	0.05926
df:	23
99% ConfInt:	(3.025, 3.358)

TI-84 Plus C

Use the calculator function `TInterval`. Input is either summary statistics or a list containing data.

1. Enter the data into list `L1`.
2. Select `STATS`; TESTS; TInterval.
3. Highlight `Data`. Enter the name of the list `L1`, set the frequency to 1, and enter the C-Level (the confidence coefficient).
4. Highlight `Calculate` and press `ENTER`. The confidence interval and summary statistics are displayed on the Home Screen. See Figure 8.31.

Minitab

Use the function 1-Sample t. Input is either summary statistics or a column containing data.

1. Enter the data into column `C1`.
2. Select Stat; Basic Statistics; 1-Sample t.
3. Choose One or more samples, each in a column, and enter `C1` in the entry window.

4. Select the Options option button and enter the confidence level. The results are displayed in a session window. See Figure 8.34.

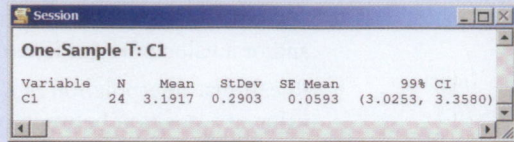

Figure 8.34 Minitab 1-Sample t confidence interval.

Excel

There is no built-in function to compute the endpoints of a confidence interval for a population mean when σ is unknown. Use the function CONFIDENCE.T to find the bound on the confidence interval and the functions AVERAGE and STDEV.S if necessary.

1. Enter the data into column A.
2. Compute the sample mean and the sample standard deviation.
3. Find the bound on the confidence interval using the function CONFIDENCE.T. This is the value $t_{\alpha/2,n-1}\dfrac{s}{\sqrt{n}}$, half the width of the CI. The arguments are α, $\bar{x}$, and n.
4. Compute the left and right endpoints of the confidence interval. See Figure 8.35.

B	C	D
3.1917	= AVERAGE(A1:A24)	Sample mean
0.2903	= STDEV.S(A1:A24)	Sample standard deviation
0.1664	= CONFIDENCE.T(0.01,B2,24)	CI bound
3.0253	= B1 - B3	Left endpoint
3.3580	= B1 + B3	Right endpoint

Figure 8.35 Calculations to construct a confidence interval.

SECTION 8.3 EXERCISES

Concept Check

8.55 True/False Every t distribution is symmetric and centered at 0.

8.56 True/False Every t distribution is more variable than the Z distribution.

8.57 True/False The confidence interval for μ based on a t distribution is valid for any underlying distribution.

8.58 True/False The confidence interval for μ based on a t distribution is used only for sample sizes less than or equal to 30.

8.59 Fill in the Blank A critical value associated with a t distribution depends on _____ and _____.

8.60 Short Answer What are the endpoints for a $100(1 - \alpha)\%$ CI for a population mean μ when σ is unknown?

Practice

8.61 Find each of the following critical values.
a. $t_{0.10,5}$ b. $t_{0.20,24}$ c. $t_{0.005,19}$
d. $t_{0.025,7}$ e. $t_{0.005,15}$ f. $t_{0.001,6}$
g. $t_{0.0005,23}$ h. $t_{0.0001,3}$ i. $t_{0.05,11}$

8.62 In each of the following problems, the sample size and the confidence level are given. Assume σ is unknown. Find the appropriate t critical value for use in constructing a confidence interval for the population mean.
a. $n = 15$,   95%
b. $n = 21$,   98%
c. $n = 31$,   99%
d. $n = 17$,   99.9%
e. $n = 12$,   99.99%
f. $n = 4$,   98%

8.63 Use Table V in the Appendix and linear interpolation (if necessary) to approximate each of the following critical values. Verify each approximation using technology.
a. $t_{0.15,10}$ b. $t_{0.07,23}$ c. $t_{0.0025,20}$
d. $t_{0.01,35}$ e. $t_{0.005,42}$ f. $t_{0.025,75}$
g. $t_{0.02,45}$ h. $t_{0.003,52}$ i. $t_{0.05,45}$

8.64 In each of the following problems, the sample mean, the sample standard deviation, the sample size, and the confidence level are given. Assume the underlying population is normally distributed. Find the associated confidence interval for the population mean.
a. $\bar{x} = 211.2$, $s = 44.37$, $n = 27$,   95%
b. $\bar{x} = 74.42$, $s = 31.8$, $n = 10$,   98%
c. $\bar{x} = 138.9$, $s = 22.3$, $n = 28$,   99%
d. $\bar{x} = -28.3$, $s = 41.33$, $n = 20$,   95%
e. $\bar{x} = 1014.5$, $s = 67.9$, $n = 17$,   99.9%

8.65 In each of the following problems, the sample mean, the sample standard deviation, the sample size, and the confidence level are given. Assume the underlying population is normally distributed. Find the associated confidence interval for the population mean.

a. $\bar{x} = 0.234$, $\quad s = 0.081$, $\quad n = 16$, $\quad 95\%$
b. $\bar{x} = 259.6$, $\quad s = 76.9$, $\quad n = 26$, $\quad 99\%$
c. $\bar{x} = 22.85$, $\quad s = 7.19$, $\quad n = 27$, $\quad 99\%$
d. $\bar{x} = 380.9$, $\quad s = 28.4$, $\quad n = 21$, $\quad 95\%$
e. $\bar{x} = 88.1$, $\quad s = 17.45$, $\quad n = 19$, $\quad 99.9\%$

8.66 In each of the following problems, put the values in order from smallest to largest. (*Note*: There is no need to use a table or technology here. Use your knowledge of the t distribution and the Z distribution.)

a. $t_{0.01,5}, t_{0.01,27}, z_{0.01}, t_{0.01,17}$
b. $t_{0.025,13}, t_{0.025,11}, t_{0.025,45}, z_{0.025}$
c. $t_{0.05,15}, t_{0.001,15}, t_{0.02,15}, t_{0.025,15}$
d. $t_{0.0001,21}, t_{0.1,21}, t_{0.005,21}, t_{0.05,21}$
e. $t_{0.10,6}, t_{0.001,26}, t_{0.05,17}, z_{0.0001}$

Applications

8.67 Manufacturing and Product Development During the manufacture of certain commercial windows and doors, hot steel ingots are passed through a rolling mill and flattened to a prescribed thickness. The machinery is set to produce a steel section 0.25 inch thick. Fourteen steel sections were selected at random and the thickness of each was recorded. The data are given in the table below.

0.213	0.298	0.236	0.324	0.254	0.271	0.204
0.252	0.307	0.297	0.301	0.291	0.222	0.246

Assume the underlying distribution of section thickness is normal.

a. Find a 99% confidence interval for the true mean steel section thickness.
b. As the rollers erode, the machine begins to produce steel sections that are too thick. Using your answer to part (a), is there any evidence to suggest the true mean steel section thickness is more than 0.25 inch? Justify your answer.
c. Use the methods described in Section 6.3, page 272, to check for any evidence of non-normality.

8.68 Manufacturing and Product Development A typical washing machine has several different cycles, including soak, wash, and rinse. The energy consumption of a washing machine is linked to the length of each cycle. A random sample of 21 washing machines was obtained, and the length (in minutes) of each wash cycle was recorded. The summary statistics are $\bar{x} = 37.8$ and $s = 5.9$. Assume the underlying distribution of main wash cycle times is normal.

a. Find a 90% confidence interval for the true mean wash cycle time. Write a Solution Trail for this problem.

b. Interpret the confidence interval found in part (a).
c. Suppose a 95% confidence interval for the true mean wash cycle time is constructed. Would this interval be smaller or larger than the interval in part (a)? Justify your answer.

8.69 Biology and Environmental Science For the week ending 5/29/13, the Iowa Department of Agriculture reported the mean weight of barrows and gilts (young male and female hogs) as 275.4 pounds.[17] To check this claim, a random sample of hogs was obtained and each was carefully weighed. The weight (in pounds) of each is given in the following table. **HOGS**

269.7	253.2	277.3	264.1	264.2	266.7
267.4	268.3	268.2	272.4	259.7	263.3

a. Assuming the underlying distribution of barrow and gilt weights is normal, find a 95% confidence interval for the true mean weight of barrows and gilts.
b. Using your answer in part (a), is there any evidence to suggest that the claim ($\mu = 275.4$) is wrong? Justify your answer.

8.70 Physical Sciences The Earth is structured in layers: crust, mantle, and core. A recent study was conducted to estimate the mean depth of the upper mantle in a specific farming region in California. Twenty-six sample sites were selected at random, and the depth to the upper mantle was measured using changes in seismic velocity and density. The summary statistics are $\bar{x} = 127.5$ km and $s = 21.3$ km. Suppose the depth of the upper mantle is normally distributed. Find a 95% confidence interval for the true mean depth of the upper mantle in this farming region, and interpret your result.

8.71 Fuel Consumption and Cars In many rural areas, newspaper carriers deliver morning papers using their automobiles because the length of the route prohibits walking. In a random sample of 28 carriers who use their automobiles, the sample mean route length was $\bar{x} = 16.7$ miles with $s = 3.4$.

a. If the distribution of route lengths is normal, find a 95% confidence interval for the true mean route length of newspaper carriers who use their automobiles.
b. If the mean length of the routes is over 20 miles, the circulation department becomes concerned that papers will not be delivered by 7:00 A.M. Using your answer to part (a), is there any evidence to suggest the true mean route length is over 20 miles? Justify your answer. Write a Solution Trail for this problem.

8.72 Sports and Leisure The popular sport of mountain biking involves riding bikes off-road, often over very rough terrain. Trek is a leading manufacturer of mountain bikes, specially designed for durability and performance. A random sample of Trek mountain bikes was obtained and the weight (in kg) of each is given in the following table.[18] ▎▎ MTNBIKE

11.74	12.32	12.06	13.58	12.19	12.29	10.84	12.04
13.01	11.37	9.01	7.34	7.19	7.45	7.69	

a. Assume the underlying distribution is normal. Find a 99% confidence interval for the true mean weight of Trek mountain bikes.

b. Suppose Trek claims to have the lightest mountain bikes on the market, with a mean weight of 10 kg. Using your answer in part (a), is there any evidence to suggest this claim is false? Justify your answer.

8.73 Technology and the Internet A computer supply store sells a wide variety of generic replacement ink cartridges for printers. A consumer group is concerned that the cartridges may not contain the specified amount of ink (30 ml). A random sample of 17 black replacement cartridges was obtained, and the amount of ink (in ml) in each is given in the table below. ▎▎ INK

30.27	29.70	29.35	29.08	29.74	29.26	29.50
29.12	29.68	28.54	30.01	29.87	30.61	29.80
29.33	29.21	28.84				

Assume the underlying distribution of ink amount is normal.

a. Find a 95% confidence interval for the true mean amount of ink in each black cartridge.

b. Is there any evidence to suggest the cartridges are underfilled? Justify your answer.

8.74 Travel and Transportation Some municipal managers complain a city bus route is the most difficult service to fulfill. Traffic jams and roadwork are often the cause of unreliable service. In an effort to analyze current city bus transportation in Atlanta, Georgia, a random sample of routes and stops was obtained. The number of minutes the bus was late (compared with the posted route times) was recorded. The resulting data are given in the table below. A value of 0.00 means the bus arrived on time. ▎▎ CITYBUS

10.29	0.00	0.00	9.96	2.10	0.00	9.52
1.83	2.37	1.47	0.00	7.98	5.19	0.40
3.21	0.00	6.91	0.00	5.02		

a. Assume the underlying distribution of late times is normal. Find a 90% confidence interval for the true mean number of minutes a city bus in Atlanta is late.

b. Use the methods described in Section 6.3, page 272, to determine whether there is any evidence to suggest that the distribution of late times is non-normal.

c. If city buses in Atlanta consistently arrive more than five minutes late, the mayor receives a large number of phone calls from irate citizens. Using your answer to part (a), is there any evidence to suggest the mayor will be receiving nasty phone calls? Justify your answer.

8.75 Business and Management Recently the Health Services Purchasing Organization started a patient-focused funding (PFF) program in British Columbia. One component of this program included financial incentives for hospitals that operate more efficiently, which is associated with the patient average length of stay (ALOS). In a random sample of 25 patients with congestive heart failure at Vancouver Coastal Health, the resulting mean ALOS (in days) was 9.9.[19] The standard deviation was $s = 3.7$ days and assume that the underlying distribution is normal.

a. Find a 99% confidence interval for the true mean ALOS for patients with congestive heart failure.

b. Suppose the Health Services Purchasing Organization will award hospitals performance funding money if the mean ALOS for these patients is less than 12 days. Is there any evidence to suggest that Vancouver Coastal Health will receive performance funding money? Justify your answer.

8.76 Sports and Leisure One factor in rating a National Hockey League team is the mean weight of its players. A random sample of players from five teams was obtained. The weight (in pounds) of each player was carefully measured, and the resulting data are given in the following table.[20]

Team	Sample size	Sample mean	Sample standard deviation
Boston	15	200.0	10.5
Detroit	18	201.2	11.8
NY Rangers	12	200.3	10.9
Philadelphia	10	202.7	12.1
San Jose	15	210.7	14.2

a. Assuming normality, find a 95% confidence interval for the true mean weight of players for each hockey team.

b. The sample size for Boston and San Jose is the same. Why is the 95% confidence interval for the weight of San Jose players wider?

c. If the true mean weight for two teams is different, then it is likely that there will be a more physical game when the two teams meet. Is there any evidence to suggest that the true mean player weights are different from Boston and San Jose? Justify your answer.

8.77 Marketing and Consumer Behavior According to a recent survey, Americans are keeping their cars longer.[21]

Some of the reasons may include the poor economy and better-built vehicles. A random sample of new- and used-car owners was asked how long they usually kept their car (in months). **CARS**

a. Find a 99% confidence interval for the true mean time people keep a new car.

b. Find a 99% confidence interval for the true mean time people keep a used car.

c. Using your answers in parts (a) and (b), do you think new-car owners and used-car owners keep their cars for different lengths of times? Justify your answer.

8.78 Economics and Finance   Along with gold and silver, platinum is also considered a precious metal and is traded on the commodities market. A random sample of the price of platinum (in dollars per troy ounce) was obtained from jewelers around the world. The resulting summary statistics were $\bar{x} = 664.50$, $s = 5.25$, and $n = 55$. Assume the underlying distribution of the price is normal.

a. Find a 95% confidence interval for the current true mean price per troy ounce of platinum.

b. One month ago, a brokerage firm reported the true mean price of platinum to be $660.79 per troy ounce, with a recommendation to buy. Is there any evidence to suggest the true mean price has increased? Justify your answer.

Extended Applications

8.79 Psychology and Human Behavior   The ambient temperature in which humans are comfortable varies with culture, activity, metabolic rate, psychological state, environment, and season. For most people in the United States, the *comfort zone* is 68 to 78°F. During a recent winter, a random sample of homeowners was selected from two different parts of the country. The thermostat temperature setting (in °F) was recorded for each home, and the summary statistics are given in the following table.

Region	Sample size	Sample mean	Sample standard deviation
New England	14	70.2	2.75
South	11	72.1	1.55

a. Find a 95% confidence interval for the true mean thermostat setting for New England homeowners during winter.

b. Find a 95% confidence interval for the true mean thermostat setting for Southern homeowners during winter.

c. What assumption(s) did you make in constructing these two confidence intervals?

d. Using your answers to parts (a) and (b), is there any evidence to suggest the New England mean thermostat setting is different from the Southern thermostat setting during winter? Justify your answer.

8.80 Public Policy and Political Science   Juvenile courts in all states maintain careful records for cases, including demographics, charges, and dispositions. One variable of interest for repeat offenders is the number of days since the last offense (or arrest). A random sample of repeat offenders appearing in juvenile court was obtained for three states. The number of days since the last offense was recorded for each, and the summary statistics are given in the table below.

State	Sample size	Sample mean	Sample standard deviation
Ohio	17	180.6	37.8
California	29	162.7	25.2
Massachusetts	22	115.3	17.6

a. Assume the underlying distribution of days since the last offense is normal for each state. Find a 99% confidence interval for the true mean number of days since the last offense for repeat offenders in each state.

b. Using your answers to part (a), is there any evidence to suggest the true mean number of days since the last offense in Ohio and California is different? How about California and Massachusetts? Justify each answer.

8.81 Medicine and Clinical Studies   Arrhythmia, or an irregular heart beat, may be caused by heart disease or by environmental factors such as stress, caffeine, tobacco, or even cold medicine. One type of arrhythmia is atrial flutter, in which the heart beats very fast, at over 250 beats per minute. In a recent research study, randomly selected patients identified to have atrial flutter were carefully monitored. The number of beats per minute for the most recent flutter for each patient, by sex, is given on the text website. Assume the underlying distribution of beats per minute in atrial flutter patients is normal. **FLUTTER**

a. Find a 98% confidence interval for the true mean beats per minute in male atrial flutter patients.

b. Find a 98% confidence interval for the true mean beats per minute in female atrial flutter patients.

c. Using your answers to parts (a) and (b), is there any evidence to suggest the true mean beats per minute in atrial flutter patients is different for men and women? Justify your answer.

d. Do you think the normality assumption is reasonable in this case? Why or why not?

8.82 Sports and Leisure   The Wimbledon Championships tennis tournament, where women players curtsy to the Queen and fans eat strawberries and cream, finishes in the first week of July. Each match is played on a grass court, as opposed to a clay or hard court. Some people believe the grass court is more challenging and speeds up the game. Random samples of match lengths (in minutes) from Wimbledon and from the U.S. Clay Court Championships were obtained. The data are summarized in the table below.

Court type	Sample size	Sample mean	Sample standard deviation
Grass	18	65.7	25.3
Clay	12	83.2	35.8

Assume the underlying distribution of match lengths is normal on grass and on clay.

- **a.** Find a 99% confidence interval for the true mean match length for grass courts at Wimbledon.
- **b.** Find a 99% confidence interval for the true mean match length for clay courts at the U.S. Clay Court Championships.
- **c.** Is there any evidence to suggest the mean match time for grass courts and clay courts is different? Justify your answer.

8.83 Public Health and Nutrition   During a medical emergency, people who dial 911 expect an ambulance to arrive quickly and personnel to provide vital care. Health insurance companies believe there is a marked difference in the response time for rural areas versus cities because of differences in coverage area and number of qualified paramedics. Random samples of ambulance response times (in minutes) were obtained for rural areas and cities. **PARAMED**

- **a.** Assuming normality, find a 99% confidence interval for the true mean response time for rural-area ambulances.
- **b.** Assuming normality, find a 99% confidence interval for the true mean response time for city ambulances.
- **c.** Use the methods described in Section 6.3, page 272, to determine whether there is any evidence to suggest either response time distribution is non-normal.
- **d.** Is there any evidence to suggest the mean response time is different for rural areas and cities? Justify your answer.

8.84 Sports and Leisure   In May 2013 the *New York Post* published an article about a scheme by wealthy Manhattan mothers to avoid long wait times at Disney World.[22] Although the wait times can be long and the Disney VIP service is expensive, the plot to cut in line has been difficult to prove. A random sample of wait times (in minutes) for the Magic Kingdom, Epcot, and Animal Kingdom was obtained.[23] Assume each underlying distribution of wait times is normal. **WAITTIME**

- **a.** Find a 95% confidence interval for the true mean wait time for each park.
- **b.** Is there any evidence that any of the parks have different true mean wait times? Justify your answer.

8.85 Sports and Leisure   The Jackpine Gypsies Motorcycle Club holds the Sturgis Rally in the Black Hills of South Dakota every summer. There are motocross, hill climb, and short-track races, and plenty of parties and bikes. In an attempt to learn

more about the approximately 1 million participants, a random sample of bike enthusiasts was obtained. Some of the summary data are given in the table below.

Variable	Sample size	Sample mean	Sample standard deviation
Age (males, years)	60	38.9	7.9
Age (females, years)	40	35.6	4.5
Distance traveled (miles)	75	257.5	56.8

Assume the underlying distributions are normal.

- **a.** Find a 95% confidence interval for the true mean age of men attending the rally and a 95% confidence interval for the true mean age of women attending the rally.
- **b.** Is there any evidence to suggest the mean age for men is different from the mean age for women? Justify your answer.
- **c.** Find a 99% confidence interval for the true mean distance traveled to the rally. Interpret this result.

8.86 Biology and Environmental Science   There have been many studies that suggest global climate change is affecting the population of polar bears. One measure of the health of a polar bear population is weight, and the mean weight for a male polar bear is approximately 900 pounds. A recent study suggests that the polar bears in the Chukchi Sea are doing very well, despite changes in the climate.[24] A random sample of 14 male polar bears from the Chukchi Sea was obtained from a capture-and-release program. The sample mean weight was $\bar{x} = 1016$ pounds. Assume the underlying weight distribution is normal and that $s = 78$.

- **a.** Find a 99% confidence interval for the true mean weight of male polar bears in the Chukchi Sea.
- **b.** Using your answer from part (a), is there any evidence to suggest that the mean weight of polar bears in the Chukchi Sea is greater than 900 pounds? Justify your answer.

Challenge

8.87 Public Health and Nutrition   Given a random sample of size n from a normal population with mean μ (and σ unknown):

- **a.** Use the method outlined in Exercise 8.51 to find a one-sided $100(1 - \alpha)\%$ confidence interval for μ bounded above, and another bounded below.
- **b.** A recent study suggests that drinks made with broccoli sprouts may filter out harmful chemicals from our bodies.[25] A new "super broccoli" has been developed that contains two to three times more glucoraphanin, which may help to reduce the risk of certain kinds of cancer. A random sample of the amount of glucoraphanin in 17 super broccoli seeds was obtained. The sample mean was 21.6 μmol/g with a standard deviation of 3.2. Assume the underlying population distribution is normal, and find a one-sided 99% confidence interval, bounded below, for the true mean amount of glucoraphanin in a super broccoli seed.

8.4 A Large-Sample Confidence Interval for a Population Proportion

Many surveys and experiments are conducted to estimate a population proportion, the true fraction of individuals or objects that exhibit a specific characteristic. For example, pollsters routinely estimate the proportion of Americans who favor a particular candidate for office. Food companies use randomly selected consumers to test-market new products, because the true proportion of shoppers who will purchase a product is important for predicting profit. And insurance agencies constantly analyze data to estimate the proportion of 18-year-old drivers who will be in an accident during the next year.

Let p = the true population proportion, the fraction of individuals or objects with a specific characteristic (the probability of a success). As in Section 7.3, it is reasonable to use a value of the sample proportion, $\hat{p}$, to construct a confidence interval for p. In a sample of n individuals or objects, let X be the number of individuals with the characteristic (or the number of successes in the sample). Recall that the sample proportion is the proportion of individuals with a specific characteristic in the sample, or a relative frequency.

$$\hat{P} = \frac{X}{n} = \frac{\text{the number of individuals with the characteristic}}{\text{the sample size}} \quad (8.10)$$

From Section 7.3, if n is large and both $np \geq 5$ and $n(1 - p) \geq 5$, then the random variable $\hat{P}$ is approximately normal with mean p and variance $p(1 - p)/n$: $\hat{P} \overset{\bullet}{\sim} N(p, p(1 - p)/n)$. $\hat{P}$ is approximately normal, so we can standardize to obtain

$$Z = \frac{\hat{P} - p}{\sqrt{\dfrac{p(1 - p)}{n}}} \overset{\bullet}{\sim} N(0, 1) \quad (8.11)$$

▶ As in Section 8.2, start with an appropriate probability statement involving the random variable Z in Equation 8.11. Find a symmetric interval about 0 such that the probability Z lies in this interval is $1 - \alpha$.

$$P(-z_{\alpha/2} < Z < z_{\alpha/2}) = 1 - \alpha$$

$$P\left(-z_{\alpha/2} < \frac{\hat{P} - p}{\sqrt{\dfrac{p(1 - p)}{n}}} < z_{\alpha/2}\right) = 1 - \alpha \quad (8.12)$$

A Challenge exercise in this section asks you to find a confidence interval for p without using $\hat{p}$ in the denominator.

Trying to sandwich p in Equation 8.12 is a little tricky because p appears in both the numerator and the denominator (inside a square root). Instead, because n is large, we usually use the sample proportion, $\hat{p}$, as a good estimate of p *in the denominator*. Manipulating the inequality (inside the probability statement) leads to the following general result. ◀

How to Find a Large-Sample 100(1 − α)% Confidence Interval for a Population Proportion

Use $\hat{p}$ as an estimate of p to check the *nonskewness criterion*.

Given a random sample of size n. If n is large and both $n\hat{p} \geq 5$ and $n(1 - \hat{p}) \geq 5$, a large-sample $100(1 - \alpha)\%$ confidence interval for p, the true population proportion, has as endpoints the values

$$\hat{p} \pm z_{\alpha/2}\sqrt{\frac{\hat{p}(1 - \hat{p})}{n}} \quad (8.13)$$

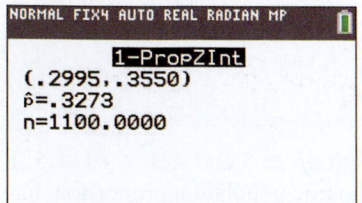

```
NORMAL FIX4 AUTO REAL RADIAN MP

       1-PropZInt
(.2995,.3550)
p̂=.3273
n=1100.0000
```

Figure 8.36 TI-84 Plus C confidence interval for p.

Example 8.9 July 4th Travelers

The period around the July 4 holiday in the United States is usually the busiest travel time of the summer. According to the AAA, over 40 million Americans take a trip of 50 miles or more during this holiday period.[26] To avoid crowded highways, many people begin their holiday by traveling on July 3. In a recent survey, 1100 Independence Day travelers were randomly selected, and 360 indicated they were leaving on July 3.

a. Find a 95% confidence interval for the true proportion of holiday travelers who will leave on July 3.

b. The AAA claims that 32% of travelers will leave on July 3. Is there any evidence to suggest that the percentage of travelers leaving on July 3 is different from this value? Justify your answer.

SOLUTION

a. The given information:

Sample size: $n = 1100$.

Number of people with the specific characteristic, July 3 travelers: $x = 360$.

Compute the sample proportion and check the nonskewness criterion.

$$\hat{p} = \frac{x}{n} = \frac{360}{1100} = 0.3273.$$

$$n\hat{p} = (1100)(0.3273) = 360 \geq 5$$

$$n(1 - \hat{p}) = (1100)(0.6727) = 740 \geq 5$$

Both inequalities are satisfied, so $\widehat{P}$ is approximately normal, and Equation 8.13 can be used to construct a confidence interval for p.

$$1 - \alpha = 0.95 \implies \alpha = 0.05 \implies \alpha/2 = 0.025 \qquad \text{Find } \alpha/2.$$

$$z_{\alpha/2} = z_{0.025} = 1.96 \qquad \text{Use Table III in the Appendix.}$$

Use Equation 8.13.

$$\hat{p} \pm z_{\alpha/2}\sqrt{\frac{\hat{p}(1 - \hat{p})}{n}} \qquad \text{Equation 8.13.}$$

$$= \hat{p} \pm z_{0.025}\sqrt{\frac{\hat{p}(1 - \hat{p})}{n}} \qquad \text{Use the value of } \alpha.$$

$$= 0.3273 \pm 1.96\sqrt{\frac{(0.3273)(0.6727)}{1100}} \qquad \text{Use the values for } z_{0.025} \text{ and } \hat{p}.$$

$$= 0.3273 \pm 0.0277 \qquad \text{Simplify.}$$

$$= (0.2996, 0.3550) \qquad \text{Compute endpoints.}$$

(0.2996, 0.3550) is a 95% confidence interval for the true proportion, p, of July 3 travelers. Figure 8.36 shows a technology solution.

b. Claim: $p = 0.32$.

Experiment: $\hat{p} = 0.3273$.

Likelihood: The likelihood is expressed as a 95% confidence interval, an interval of likely values for p: (0.2996, 0.3550).

Conclusion: The claimed value, 0.32, lies in the confidence interval. There is no evidence to suggest p is different from 0.32. ●

TRY IT NOW GO TO EXERCISE 8.100

Here is one more example involving a confidence interval for a population proportion.

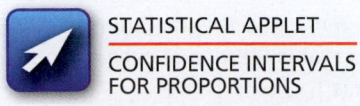

STATISTICAL APPLET

CONFIDENCE INTERVALS FOR PROPORTIONS

JAXA/NASA

Example 8.10 Solar Flare Activity

Solar flares on the sun are violent explosions that release enormous amounts of energy and radiation. These emissions can harm spacecraft and satellites in the Earth's orbit and disrupt communication on the ground. In July 2013 the sun was near the maximum point of the 11-year solar flare activity cycle.[27] In an effort to plan for these possible disruptions, a study was conducted over several years. On 86 of 350 randomly selected days there was at least one solar flare. Find a 99% confidence interval for the true proportion of days on which there is at least one solar flare.

SOLUTION

STEP 1 $n = 350$, $x = 86$. Given.

STEP 2 The sample proportion: $\hat{p} = \dfrac{86}{350} = 0.2457$.

$$n\hat{p} = (350)(0.2457) = 86 \geq 5$$
$$n(1 - \hat{p}) = (350)(0.7543) = 264 \geq 5$$

Both inequalities are satisfied. $\hat{P}$ is approximately normal.

STEP 3 $1 - \alpha = 0.99 \implies \alpha = 0.01 \implies \alpha/2 = 0.005$ Find $\alpha/2$.

$z_{\alpha/2} = z_{0.005} = 2.5758$ Use Table III in the Appendix.

STEP 4 Use Equation 8.13.

$$\hat{p} \pm z_{\alpha/2}\sqrt{\frac{\hat{p}(1 - \hat{p})}{n}}$$ Equation 8.13.

$$= \hat{p} \pm z_{0.005}\sqrt{\frac{\hat{p}(1 - \hat{p})}{n}}$$ Use the value of α.

$$= 0.2457 \pm 2.5758\sqrt{\frac{(0.2457)(0.7543)}{350}}$$ Use the values for $z_{0.005}$ and $\hat{p}$.

$$= 0.2457 \pm 0.0593$$ Simplify.

$$= (0.1864, 0.3050)$$ Compute endpoints.

(0.1864, 0.3050) is a 99% confidence interval for the true proportion, p, of days on which there is at least one solar flare. Figures 8.37 and 8.38 show technology solutions.

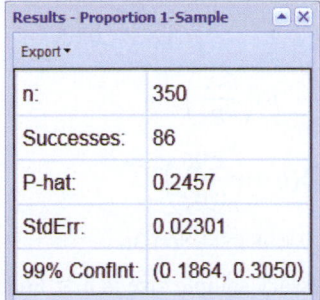

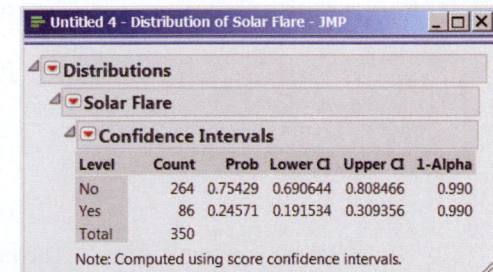

Figure 8.37 CrunchIt! confidence interval for p.

Figure 8.38 JMP confidence interval for p using the Wilson score interval.

TRY IT NOW GO TO EXERCISE 8.109

Recall from Section 8.2 that a sample size can be determined such that the resulting confidence interval for μ has certain properties. Similarly, given the confidence level, a sample size n can be computed such that the resulting confidence interval for p has a desired width.

Suppose n is large, the confidence level is $100(1 - \alpha)\%$, and B is the bound on the error of estimation. Half the width of the confidence interval is the *step* in each direction from $\hat{p}$. Therefore, the bound on the error of estimation is

$$B = z_{\alpha/2} \cdot \sqrt{\frac{\hat{p}(1 - \hat{p})}{n}} \tag{8.14}$$

Solving for n yields

$$n = \hat{p}(1 - \hat{p})\left(\frac{z_{\alpha/2}}{B}\right)^2 \tag{8.15}$$

Look carefully at this result. Although the mathematics is correct, there is a real problem with Equation 8.15. The critical value, $z_{\alpha/2}$, and the bound on the error of estimation, B, can be specified. But $\hat{p}$ *is unknown*. We do not know $\hat{p}$ until we have the sample size n (and x). There are two solutions to this problem.

1. Use a reasonable estimate for $\hat{p}$ from previous experience. Researchers often conduct many similar experiments over time and can make very realistic guesses for $\hat{p}$.

For given $z_{\alpha/2}$ and B, Equation 8.15 is greatest (maximized) when $\hat{p} = 0.5$.

2. If no prior information is available, use $\hat{p} = 0.5$ in Equation 8.15. This produces a very *conservative*, large value for n.

In either case, it is unlikely that the value of n will be an integer. *Always* round up. This guarantees the resulting confidence interval will have a bound on the error of estimation of at most B. Any larger sample size will also suffice. The following example illustrates the use of Equation 8.15.

Example 8.11 Time for a New Car

Drivers in the United States are keeping their cars longer. This could be due to better-built automobiles, longer warranties, and the state of the economy. To plan production and sales events, a company plans to estimate the proportion of automobiles more than 10 years old that are still on the road. A 95% confidence interval for p with bound on the error of estimation of 0.02 is needed. How large a sample size is necessary in each of the following cases?

a. Prior experience suggests $\hat{p} \approx 0.1$.

b. There is no prior information about the value of $\hat{p}$.

SOLUTION

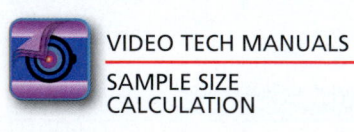
VIDEO TECH MANUALS

SAMPLE SIZE
CALCULATION

a. Use $\hat{p} = 0.1$ and $B = 0.02$ in Equation 8.15.

$1 - \alpha = 0.95 \implies \alpha/2 = 0.05 \implies z_{0.025} = 1.96$ Find the critical value.

$n = \hat{p}(1 - \hat{p})\left(\frac{z_{\alpha/2}}{B}\right)^2$ Equation 8.15.

$= (0.1)(0.9)\left(\frac{1.96}{0.02}\right)^2$ Substitute given values.

$= (0.09)(98)^2 = 864.36$ Computation.

The necessary sample size is $n \geq 865$.

b. With no prior information, use $\hat{p} = 0.5$, $B = 0.02$, and $z_{0.025} = 1.96$ in Equation 8.15.

$n = \hat{p}(1 - \hat{p})\left(\frac{z_{\alpha/2}}{B}\right)^2$ Equation 8.15.

$= (0.5)(0.5)\left(\frac{1.96}{0.02}\right)^2$ Substitute given values.

$= (0.25)(98)^2 = 2401$ Computation.

The necessary sample size in this case is $n \geq 2401$.   ●

TRY IT NOW   GO TO EXERCISE 8.106

Technology Corner

Procedure: Construct a confidence interval for a population proportion.
Reconsider: Example 8.9, solution, and interpretations.

CrunchIt!

Use the built-in function Proportion 1-Sample. Input is either summary statistics or a single column of values in which one is designated as a success.

1. Select Statistics; Proportion; 1-sample.
2. Under the Summarized tab, enter values for n, Successes, and the Confidence Interval Level. Click Calculate. The results are displayed in a new window (Figure 8.37).

TI-84 Plus C

Use the built-in function 1-PropZInt.

1. Select STATS ; TESTS; 1-PropZInt.
2. Enter values for x, the number of success, n, the total number of trials, and C-Level, the confidence coefficient.
3. Highlight Calculate and press ENTER . The confidence interval and summary statistics are displayed on the Home Screen. See Figure 8.36.

Minitab

Use the built-in function 1 Proportion. Input is either a sample in a column or summarized data.

1. Select Stat; Basic Statistics; 1 Proportion.
2. Choose Summarized data and enter the Number of events (the number of successes), and the Number of trials.
3. Select the Options option button. Enter the Confidence level and select Method: Normal appoximation. Click OK. The results are displayed in a session window. See Figure 8.39.

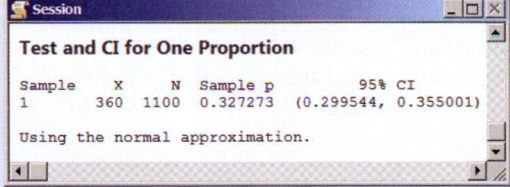

Figure 8.39 Minitab confidence interval for a population proportion.

Excel

There is no built-in function to compute the endpoints of a confidence interval for a population proportion. Use the function NORM.S.INV to find the critical value.

1. Enter values for x, the number of successes, n, the number of trials, and α.
2. Use the function NORM.S.INV to find the critical value.
3. Compute the left and right endpoints of the confidence interval. See Figure 8.40.

	A	B	C
1	360	= x	Number of successes
2	1100	= n	Number of trials
3	0.3273	= A1/A2	Sample proportion
4	0.05	= alpha	Alpha
5	1.9600	= NORM.S.INV(1-A4/2)	Critical Value
6	0.2995	= A3-A5*SQRT(A3*(1-A3)/A2)	Left endpoint
7	0.3550	= A3+A5*SQRT(A3*(1-A3)/A2)	Right endpoint

Figure 8.40 Excel calculations to construct a confidence interval for a population proportion.

SECTION 8.4 EXERCISES

Concept Check

8.88 Short Answer   What two conditions must be true in order to construct a large sample confidence interval for a population proportion?

8.89 Short Answer   Explain how a large sample confidence interval for a population proportion can be used in statistical inference.

8.90 True/False   A large sample confidence interval for a population proportion will always include 0.50.

8.91 True/False   When computing the sample size necessary to construct a confidence interval for a population proportion of a certain width, round n to the nearest 100.

8.92 True/False   When computing the sample size necessary to construct a confidence interval for a population proportion of a certain width, if no prior information is available, use a value for $\hat{p}$ close to 0 or close to 1.

Practice

8.93   In each of the following problems, the sample size and the number of individuals or objects with a specified characteristic are given. Check the nonskewness criterion to determine whether the distribution of $\hat{P}$ is approximately normal.

  a. $n = 105$, $x = 85$ **b.** $n = 1750$, $x = 1645$
  c. $n = 225$, $x = 220$ **d.** $n = 183$, $x = 3$
  e. $n = 377$, $x = 350$ **f.** $n = 480$, $x = 478$

8.94   In each of the following problems, the sample size, the number of individuals or objects with a specified characteristic, and the confidence level are given. Find the associated confidence interval for the population proportion.

  a. $n = 150$, $x = 70$,   95%
  b. $n = 225$, $x = 65$,   98%
  c. $n = 500$, $x = 468$,   90%
  d. $n = 95$, $x = 63$,   99%
  e. $n = 2450$, $x = 986$,   99.9%

8.95   In each of the following problems, the sample size, the number of individuals or objects with a specified characteristic, and the confidence level are given. Find the associated confidence interval for the population proportion.

  a. $n = 1336$, $x = 1001$,   99%
  b. $n = 775$, $x = 680$,   95%
  c. $n = 85$, $x = 41$,   95%
  d. $n = 335$, $x = 290$,   98%
  e. $n = 566$, $x = 47$,   99.9%

8.96   In each of the following problems, the confidence level, the bound on the error of estimation, and an estimate for $\hat{p}$ are given. Find the sample size necessary to produce a confidence interval for p with this bound and confidence.

  a. 95%, $B = 0.05$, $\hat{p} \approx 0.45$
  b. 98%, $B = 0.07$, $\hat{p} \approx 0.32$
  c. 99%, $B = 0.10$, $\hat{p} \approx 0.14$
  d. 90%, $B = 0.001$, $\hat{p} \approx 0.057$
  e. 99.9%, $B = 0.03$, $\hat{p} \approx 0.22$

8.97   In each of the following problems, the confidence level and the bound on the error of estimation are given. Find the sample size necessary to produce a confidence interval for p with this bound and confidence level.

  a. 99%, $B = 0.06$
  b. 95%, $B = 0.10$
  c. 98%, $B = 0.002$
  d. 90%, $B = 0.05$
  e. 99.9%, $B = 0.2$

8.98   Three factors affect the width of a large-sample confidence interval for p: the confidence level, the sample size, and the sample proportion, $\hat{p}$. In each of the following problems, determine whether the width of the resulting confidence interval for p increases or decreases.

  a. Confidence level and $\hat{p}$ are constant, and n increases.
  b. Sample size and $\hat{p}$ are constant, and confidence level decreases.
  c. Sample size and confidence level are constant, and $\hat{p}\,(> 0.5)$ increases.

8.99   Three factors affect the size of the sample necessary to produce a confidence interval for p with certain properties: the confidence level, the bound on the error of estimation, and the estimate of $\hat{p}$. In each of the following problems, determine whether the necessary sample size increases or decreases.

  a. Confidence level and estimate of $\hat{p}$ are constant, and B is smaller.
  b. B and estimate of $\hat{p}$ are constant, and confidence level increases.
  c. B and confidence level are constant, and estimate of $\hat{p}$ is closer to 0.
  d. B and confidence level are constant, and estimate of $\hat{p}$ is closer to 1.

Applications

8.100 Public Health and Nutrition   Many restaurants are now offering "*healthy*" items on their menus. Research suggests that these items may increase sales and profits because more Americans claim to be eating healthier foods. In a recent survey by the NPD Group, only 25% of adults say they eat healthy foods when they dine at restaurants.[28] A random sample of 260 adults was asked if they eat healthy foods when they dine out. Seventy adults indicated they eat healthy foods at restaurants.

  a. Find a 95% confidence interval for the true proportion of adults who eat healthy foods when they dine out.
  b. Is there any evidence to suggest that the claim of $p = 0.25$ is wrong? Justify your answer.

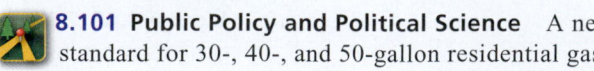 **8.101 Public Policy and Political Science**   A new standard for 30-, 40-, and 50-gallon residential gas

water heaters includes a flame arrester. This device helps to prevent flashback fires from flammable liquid vapor nearby. The Consumer Product Safety Commission would like to estimate the proportion of homes affected by this safety standard.

a. In a random sample of 575 homes, 235 had gas water heaters. Find a 90% confidence interval for the true proportion of homes with gas water heaters. Write a Solution Trail for this problem.

b. Prior research suggests that the proportion of homes with gas water heaters is approximately 0.40. How large a sample is necessary for the bound on the error of estimation to be 0.03 for a 95% confidence interval?

8.102 **Marketing and Consumer Behavior** A successful company usually has high brand-name and logo recognition among consumers. For example, Coca-Cola products are available to 98% of all people in the world, and may have the highest logo recognition of any company. A software firm developing a product would like to estimate the proportion of people who recognize the Linux penguin logo. Of the 952 randomly selected consumers surveyed, 132 could identify the product associated with the penguin.

a. Is the distribution of the sample proportion, $\hat{P}$, approximately normal? Justify your answer.

b. Find a 95% confidence interval for the true proportion of consumers who recognize the Linux penguin.

c. The company will market a Linux version of their new software if the true proportion of people who recognize the logo is greater than 0.10. Is there any evidence to suggest the true proportion of people who recognize the logo is greater than 0.10? Justify your answer.

8.103 **Travel and Transportation** To advertise appropriate vacation packages, Best Bets Travel would like to learn more about families planning overseas trips. In a random sample of 125 families planning a trip to Europe, 15 indicated France as their travel destination.

a. For those families planning vacations to Europe, find a 98% confidence interval for the true proportion traveling to France.

b. Suppose no prior estimate of $\hat{p}$ is known. How large a sample is necessary for the bound on the error of estimation to be 0.05 with the confidence level in part (a)?

8.104 **Public Policy and Political Science** In a recent survey, a random sample of 650 Americans were asked several questions regarding television broadcasting and programming. The table below lists three proposals and the number of Americans who responded in favor of each.

Proposal	Number in favor
To limit the number of commercials shown during each children's show	566
To provide more adult-education programs	530
To make children's shows commercial-free	468

a. Find a 95% confidence interval for the true proportion of Americans in favor of limiting the number of commercials shown during each children's show.

b. Find a 95% confidence interval for the true proportion of Americans in favor of providing more adult-education programs.

c. Find a 95% confidence interval for the true proportion of Americans in favor of making all children's programming commercial-free.

d. Which of the above confidence intervals is the narrowest? Why?

8.105 **Marketing and Consumer Behavior** In the CRFA 2013 Canadian Chef Survey, 350 professional chefs were randomly selected and asked to rate the popularity of several items, including cooking methods, hot trends, and perennial favorites.[29] In the Hot Trends section, suppose 209 chefs selected slow cooking as the most popular preparation method, 107 chefs selected gluten-free as the most popular culinary theme, and 98 selected bite-size desserts as the most popular dessert item.

a. Find a 95% confidence interval for the true proportion of Canadian chefs who believe slow cooking is the most popular preparation method.

b. Find a 95% confidence interval for the true proportion of Canadian chefs who believe gluten-free is the most popular culinary theme.

c. Find a 95% confidence interval for the true proportion of Canadian chefs who believe bite-size desserts are the most popular dessert item.

8.106 **Civic Engagement** The Hyde Park Neighborhood Association Outreach Committee recently conducted a survey to learn more about the issues and concerns of the residents. In a random sample of 124 residents, 20 indicated that the main barrier to involvement in community events is scheduling conflicts.[30]

a. Find a 90% confidence interval for the true proportion of residents who believe the main barrier to involvement in community events is scheduling conflicts.

b. Suppose no prior estimate of $\hat{p}$ is known. How large a sample is necessary for the bound on the error of estimation to be 0.04 with confidence level 90%?

8.107 **Economics and Finance** When unemployment rises, high school students face more competition from college students and adult workers for summer jobs. In a random sample of 188 high school students looking for summer work, 61 said they were able to find a job.

a. Find a 95% confidence interval for the true proportion of high school students who were able to find a summer job.

b. In a similar study the previous year, the sample proportion of high school students who were able to find a job was 0.25. Use this estimate to find the sample size necessary for the bound on the error of estimation to be 0.025 with confidence level 95%.

8.108 **Marketing and Consumer Behavior** Most walk-behind lawn mowers have three options for disposal of grass clippings: by bagging, by mulching, or by side discharge. The manager at an Aubuchon Hardware store conducted a survey to determine which disposal method is most common. The results are given in the table below, classified by area mowed. A small area is less than $\frac{1}{2}$ acre, a medium area is $\frac{1}{2}$ to 1 acre, and a large area is over 1 acre.

		Disposal method		
Area	Sample size	Bagging	Mulching	Side discharge
Small	125	85	35	5
Medium	157	70	40	47
Large	144	42	45	57

a. For people with small yards, find a 95% confidence interval for the true proportion who dispose of grass clippings by bagging.

b. For people with medium yards, find a 95% confidence interval for the true proportion who dispose of grass clippings by mulching.

c. For people with large yards, find a 95% confidence interval for the true proportion who dispose of grass clippings by side discharge.

8.109 Business and Management A U.S. textile company is interested in the proportion of orders shipped to another country. In a random sample of 1560 clothing orders placed to U.S. companies, 500 were exported.

a. Find a 99% confidence interval for the true proportion of clothing orders shipped to other countries.

b. In the previous year, the true proportion of clothing orders shipped to other countries was believed to be 0.30. Using your answer to part (a), is there any evidence to suggest the true proportion has changed? Justify your answer. Write a Solution Trail for this problem.

8.110 Business and Management Many dairy farms have experienced bankruptcy over the past decade as a result of wild fluctuations in conventional milk prices. However, organic farms, those that do not treat cows with antibiotics or hormones and use hay grown without chemicals, have remained solvent and even expanded. In a random sample of 1400 New England dairy farms, 90 are certified as organic.

a. Find a 99% confidence interval for the true proportion of New England dairy farms certified as organic.

b. Five years ago, an extensive census reported that 3% of all New England dairy farms were organic. Is there any evidence to suggest this proportion has changed? Justify your answer.

8.111 Marketing and Consumer Behavior In a recent survey of home-buyer preferences, consumers were asked about desired characteristics, such as a wood-burning fireplace, a den/library, and flooring. In particular, each person was asked whether a separate dining room is essential. The sample size and the number who responded "*Yes*" to this question are given in the table below by geographic region.

Geographic region	Sample size	Number who responded *Yes*
Northeast	225	180
Midwest	276	224
South Central	301	232
South Atlantic	454	377
West	366	304

a. Find a 99% confidence interval for the true proportion of home-buyers who believe a separate dining room is essential in each geographic region.

b. Which confidence interval in part (a) is the largest? Why?

8.112 Travel and Transportation The U.S. Transportation Security Administration (TSA) screens approximately 1.8 million travelers in the nation's airports each day. All carry-on and checked baggage is carefully screened, and every passenger must pass through a metal detector or an advanced imaging-technology scanner. The agency claims that approximately 3% of all passengers are patted down.[31] In a random sample of 478 airline passengers, 31 were patted down.

a. Find a 99% confidence interval for the true proportion of airline passengers who are patted down.

b. Is there any evidence to suggest the TSA claim is false? Justify your answer.

Extended Applications

8.113 Business and Management The 2013 Global Management Education Graduate Survey asked students in 159 business schools from 33 countries several questions about their job search. Approximately 60% of job seekers reported that they had an offer of employment.[32] A random sample of graduate students in their final year at business schools in the Northeast was asked whether they had a job offer. The resulting data are given in the table below, by degree program.

Sample size	Number with a job offer (x)	Degree program
260	160	MBA
380	288	MAcc
310	125	MFin

a. Find a 99% confidence interval for the true proportion of graduate business students who have a job offer for each degree program.

b. Is there any evidence to suggest that the true proportion of MFin graduate business students who have a job offer is different from that in either of the other two degree programs? Justify your answer.

8.114 Psychology and Human Behavior The Drivers Technology Association in the United Kingdom recently studied the behavior of drivers with and without radar detectors. A random sample of 550 users and 562 non-users was obtained. In the past three years, 108 users and 68 non-users have had an accident.

a. Find a 99% confidence interval for the true proportion of radar-detector users who have had an accident in the past three years.
b. Find a 99% confidence interval for the true proportion of radar-detector non-users who have had an accident in the past three years.
c. Is there any evidence to suggest the two true proportions are different? Justify your answer.

8.115 Education and Child Development   Many families that must find care for their children have turned to an au pair. An au pair usually provides more than just child care, for example, help with meals, house cleaning, and even participation in children's activities. Some become an extended family member. According to Aupair World, the majority of au pairs are from Europe. However, approximately 8% are from North America.[33] A random sample of au pairs in the United States and Canada was asked for their home country. The resulting data are given in the table below.

Sample size	Number from North America (x)	Country
450	41	United States
506	51	Canada

a. Find a 95% confidence interval for the true proportion of au pairs in the United States that come from North America.
b. Find a 95% confidence interval for the true proportion of au pairs in Canada that come from North America.
c. Is there any evidence to suggest that the two true proportions are different? Justify your answer.

8.116 Economics and Finance   The first question on the U.S. IRS Income Tax Form 1040 asks taxpayers whether they want $3 to go to the Presidential Election Campaign Fund. A large Washington political-action committee obtained a random sample of registered voters and asked them if they checked "*Yes*" on this question on the past year's return. The results are given in the table below, by party affiliation.

Political party	Sample size	Number who checked *Yes*
Democrat	237	70
Republican	388	184
Independent	155	23

a. Find a 95% confidence interval for the true proportion of registered voters in each political party who checked "*Yes*" on the Presidential Election Campaign Fund question.
b. Is there any evidence to suggest the proportion of Independent voters who checked "*Yes*" is different from either the proportion of Democrat or Republican voters who checked "*Yes*"? Justify your answer.

c. Suppose the results obtained in this study are preliminary and a larger survey is planned. How large a sample is necessary for each political party for the bound on the error of estimation to be 0.02 with confidence level 95%?

8.117 Medicine and Clinical Studies   A new prescription medication is designed to ease the pain of arthritis. In a clinical trial, both treatment and placebo groups were studied to determine whether there were any adverse reactions. The table below lists the number of patients in each group who experienced each adverse reaction. Assume each group represents a random sample.

Adverse reaction	Treatment group ($n = 465$)	Placebo group ($n = 154$)
Headache	61	15
Rash	31	9

a. Find a 95% confidence interval for the true proportion of people who suffered a headache in each of the groups.
b. Is there any evidence to suggest the true proportion of people who suffered a headache is different for the two groups? Justify your answer.
c. Find a 98% confidence interval for the true proportion of people who experienced a rash in each of the groups.
d. Is there any evidence to suggest the true proportion of people who experienced a rash is different for the two groups? Justify your answer.

8.118 Public Policy and Political Science   A recent survey of Canadians on privacy-related issues indicated that approximately two-thirds of all Canadians are concerned about privacy protection. In addition, 70% of Canadians believe their personal information is less protected than it was 10 years ago.[34] Suppose the table below lists the number of respondents concerned about privacy protection by province.

Province	Sample size	Number concerned about privacy protection
Ontario	580	401
British Columbia	617	381
Nova Scotia	478	320
Alberta	436	292

a. Find a 95% confidence interval for the true proportion of Canadians who are concerned about privacy protection for each province.
b. Is there any evidence to suggest that the true proportion of Canadians who are concerned about privacy protection is different from 67% in any province? Justify your answer.
c. Is there any evidence to suggest that the true proportion of Canadians who are concerned about privacy protection is different for any two provinces? Justify your answer.

Challenge

8.119 Technology and the Internet   Given a random sample of size n, suppose n is large and both $n\hat{p} \geq 5$ and $n(1 - \hat{p}) \geq 5$.

a. Use the method outlined in Exercise 8.51 to find a one-sided $100(1 - \alpha)\%$ confidence interval for p bounded above, and another bounded below.

b. Many Internet users download and share illegal copies of songs (and movies). A random sample of 260 Internet users who regularly download music was obtained, and 171 indicated they did not care if they were violating copyright laws. Find a one-sided 95% confidence interval, bounded above, for the population proportion of Internet users who download music who do not care about violating copyright laws.

8.120 The Wilson CI   Consider the probability statement used to construct a confidence interval for a population proportion (Equation 8.12):

$$P\left(-z_{\alpha/2} < \frac{\hat{P} - p}{\sqrt{\dfrac{p(1 - p)}{n}}} < z_{\alpha/2} \right) = 1 - \alpha$$

Manipulate the inequality (without substituting $\hat{p}$ for p in the denominator) to obtain a $100(1 - \alpha)\%$

confidence interval for p (the Wilson interval, with endpoints given below).

$$\frac{n\hat{p} + z_{\alpha/2}^2/2}{n + z_{\alpha/2}^2} \pm \frac{z_{\alpha/2}\sqrt{n}}{n + z_{\alpha/2}^2}\sqrt{\hat{p}(1 - \hat{p}) + z_{\alpha/2}^2/(4n)} \quad (8.16)$$

a. In a random sample of size 100, suppose the sample proportion is $\hat{p} = 0.60$. Find a 95% confidence interval for p using Equation 8.13. Find a 95% confidence interval for p using the Wilson interval (Equation 8.16). Which is wider? Why?

b. Let $n = 120, 140, 160, \ldots, 500$. For each value of n, let $\hat{p} = 0.60$ and compute both CIs for p. What happens to the Wilson CI as n increases? Why?

8.121 Necessary Sample Size   To find the sample size necessary to construct a $100(1 - \alpha)\%$ confidence interval for a population proportion with bound on the error of estimation B, Equation 8.15 is used:

$$n = \hat{p}(1 - \hat{p})\left(\frac{z_{\alpha/2}}{B}\right)^2$$

For a 95% confidence interval and $B = 0.10$, let $\hat{p} = 0.05, 0.10, 0.15, \ldots, 0.95$, and find the sample size necessary for each value of $\hat{p}$. Plot the values of n versus $\hat{p}$ (n on the vertical axis and $\hat{p}$ on the horizontal axis). Describe the pattern. When is n largest?

8.5 A Confidence Interval for a Population Variance or Standard Deviation

Many real-world problems involve estimation of variability to find out whether measurements are clustered around a central value or spread out over a wide range. For example, quality-control specialists continuously monitor production processes for increases or changes in range or variability, and most scientists claim increased quantities of carbon dioxide in the atmosphere have contributed to greater climate variability. It seems reasonable to use the sample variance, S^2, as an estimator for the population variance, σ^2, a measure of variability. A confidence interval for a population variance, σ^2, is based on S^2, a new *standardization*, and a **chi-square distribution** introduced below.

S^2 is a *good* estimator for σ^2. It is unbiased: $E(S^2) = \sigma^2$.

A **chi-square** (abbreviated χ^2) **distribution** has positive probability only for non-negative values. The probability density function for a χ^2 random variable (details given in the margin) is 0 for $x < 0$. Focus on the properties of a chi-square distribution and the method for finding critical values associated with this distribution.

Properties of a Chi-Square Distribution

1. A chi-square distribution is completely determined by one parameter, the number of degrees of freedom (ν). The degrees of freedom must be a positive integer, $\nu = 1, 2, 3, 4, \ldots$. There is a different chi-square distribution corresponding to each value of ν.

The probability density function for a chi-square random variable, X, with ν degrees of freedom is given by

$$f(x) = \begin{cases} \dfrac{e^{-x/2}x^{(\nu/2)-1}}{2^{\nu/2} \cdot (\nu/2)} & x \geq 0 \\ 0 & \text{otherwise} \end{cases}$$

2. If X has a chi-square distribution with ν degrees of freedom, denoted $X \sim \chi_\nu^2$, then

$$\mu_X = \nu \quad \text{and} \quad \sigma_X^2 = 2\nu \quad (8.17)$$

The mean of X is ν, the number of degrees of freedom, and the variance is 2ν, twice the number of degrees of freedom.

3. Suppose $X \sim \chi_\nu^2$. The density curve for X is positively skewed (*not* symmetric), and as x increases it gets closer and closer to the x axis but never touches it. As ν increases, the density curve becomes flatter and actually looks more normal. See Figure 8.41.

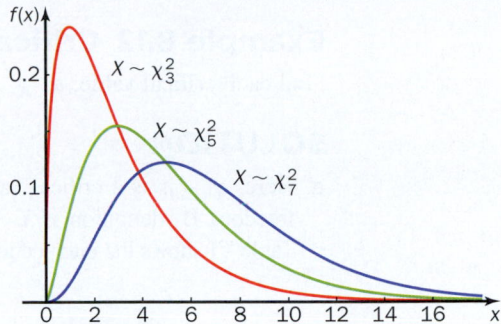

Figure 8.41 Density curves for several chi-square distributions.

The definition and notation for a χ^2 **critical value** is analogous to those for a Z and a t critical value.

Definition

$\chi_{\alpha,\nu}^2$ is a **critical value** related to a **chi-square distribution** (a χ^2 **critical value**) with ν degrees of freedom. If $X \sim \chi_\nu^2$, then $P(X \geq \chi_{\alpha,\nu}^2) = \alpha$.

A CLOSER LOOK

1. $\chi_{\alpha,\nu}^2$ is a value on the measurement axis in a chi-square world with ν degrees of freedom such that there is α of the area (probability) to the right of $\chi_{\alpha,\nu}^2$. There is *no* symmetry in chi-square critical values (as there was for Z and t critical values).

2. Critical values are always defined in terms of right-tail probability. The notation is just a little trickier here because a chi-square distribution is *not* symmetric. It will be necessary to find critical values denoted $\chi_{1-\alpha,\nu}^2$ (with *large* values for $1 - \alpha$). By definition, $P(X \geq \chi_{1-\alpha,\nu}^2) = 1 - \alpha$, and by the complement rule, $P(X \leq \chi_{1-\alpha,\nu}^2) = \alpha$. See Figure 8.42.

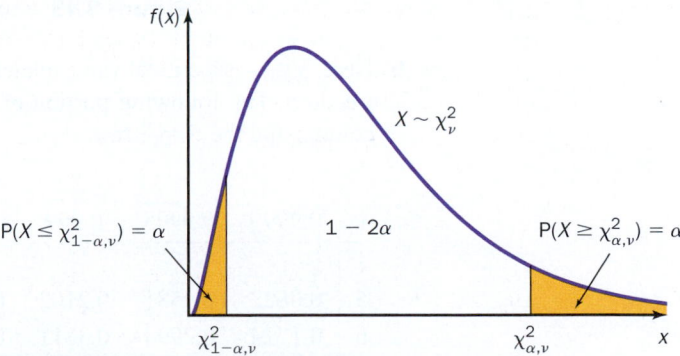

Figure 8.42 An illustration of chi-square critical values.

Table VI in the Appendix presents selected critical values associated with various chi-square distributions. Right-tail probabilities are in the top row and degrees of freedom are listed in the left column. In the body of the table, $\chi_{\alpha,\nu}^2$ is at the intersection of the α column and the ν row. The first part of this table presents *left-tail critical values* corresponding to *large* right-tail probabilities. The second half contains *right-tail critical values* corresponding to *small* right-tail probabilities. The following example illustrates the use of this table for finding critical values associated with a chi-square distribution.

Example 8.12 Critical Value Look-ups

Find each critical value: **a.** $\chi^2_{0.05,10}$, **b.** $\chi^2_{0.99,7}$.

SOLUTION

a. Here, $\chi^2_{0.05,10}$ is a critical value related to a chi-square distribution with 10 degrees of freedom. By definition, if $X \sim \chi^2_{10}$, then $P(X \geq \chi^2_{0.05,10}) = 0.05$. The following portion of Table VI shows the intersection of the $\alpha = 0.05$ column and the $\nu = 10$ row.

					α			
ν	0.10	0.05	0.025	0.01	0.005	0.001	0.0005	0.0001
$\vdots$	$\vdots$	$\vdots$	$\vdots$	$\vdots$	$\vdots$	$\vdots$	$\vdots$	$\vdots$
8	13.3616	15.5073	17.5345	20.0902	21.9550	26.1245	27.8680	31.8276
9	14.6837	16.9190	19.0228	21.6660	23.5894	27.8772	29.6658	33.7199
10	15.9872	18.3070	20.4832	23.2093	25.1882	29.5883	31.4198	35.5640
11	17.2750	19.6751	21.9200	24.7250	26.7568	31.2641	33.1366	37.3670
12	18.5493	21.0261	23.3367	26.2170	28.2995	32.9095	34.8213	39.1344
$\vdots$	$\vdots$	$\vdots$	$\vdots$	$\vdots$	$\vdots$	$\vdots$	$\vdots$	$\vdots$

Therefore, $\chi^2_{0.05,10} = 18.3070$, and if $X \sim \chi^2_{10}$, then $P(X \geq 18.3070) = 0.05$, as illustrated in Figure 8.43.

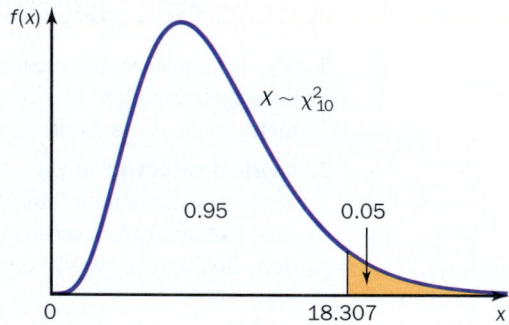

Figure 8.43 Visualization of $\chi^2_{0.05,10} = 18.3070$.

b. Here, $\chi^2_{0.99,7}$ is a critical value related to a chi-square distribution with seven degrees of freedom. The following portion of Table VI shows the intersection of the $\alpha = 0.99$ column and the $\nu = 7$ row.

					α			
ν	0.9999	0.9995	0.999	0.995	0.99	0.975	0.95	0.90
$\vdots$	$\vdots$	$\vdots$	$\vdots$	$\vdots$	$\vdots$	$\vdots$	$\vdots$	$\vdots$
5	0.0822	0.1581	0.2102	0.4117	0.5543	0.8312	1.1455	1.6103
6	0.1724	0.2994	0.3811	0.6757	0.8721	1.2373	1.6354	2.2041
7	0.3000	0.4849	0.5985	0.9893	1.2390	1.6899	2.1673	2.8331
8	0.4636	0.7104	0.8571	1.3444	1.6465	2.1797	2.7326	3.4895
9	0.6608	0.9717	1.1519	1.7349	2.0879	2.7004	3.3251	4.1682

$\chi^2_{0.99,7} = 1.2390$ and if $X \sim \chi^2_7$, then $P(X \geq 1.2390) = 0.99$ and $P(X \leq 1.2390) = 0.01$, as illustrated in Figure 8.44.

Figure 8.45 shows a technology solution.

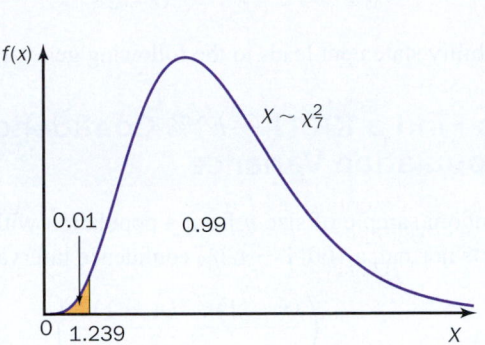

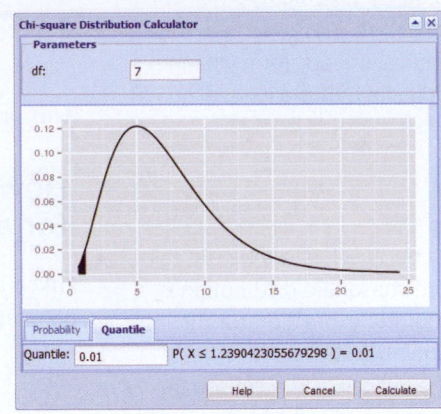

Figure 8.44 Visualization of $\chi^2_{0.99,7} = 1.2390$.

Figure 8.45 Use the Quantile option in the Chi-square Distribution Calculator of CrunchIt! to find critical values associated with a chi-square distribution.

TRY IT NOW   GO TO EXERCISE 8.127

Table VI is limited. There are only a handful of values for α, and $\nu = 1-40, 50, 60, 70, 80, 90, 100$. However, a calculator or computer can find almost any critical value needed.

A confidence interval for a population variance is based on the following theorem.

Theorem

Let S^2 be the sample variance of a random sample of size n from a normal distribution with variance σ^2. The random variable

$$X = \frac{(n-1)S^2}{\sigma^2} \tag{8.18}$$

has a chi-square distribution with $n-1$ degrees of freedom.

This is another kind of *standardization*, a transformation to a chi-square distribution.

This theorem is used to construct a confidence interval for σ^2.

▶ Let $X \sim \chi^2_{n-1}$ and find an interval that captures $1 - \alpha$ in the *middle* of this chi-square distribution (see Figure 8.46).

$$P(\chi^2_{1-\alpha/2,n-1} < X < \chi^2_{\alpha/2,n-1}) = 1 - \alpha \tag{8.19}$$

$$P\left(\chi^2_{1-\alpha/2,n-1} < \frac{(n-1)S^2}{\sigma^2} < \chi^2_{\alpha/2,n-1}\right) = 1 - \alpha \tag{8.20}$$

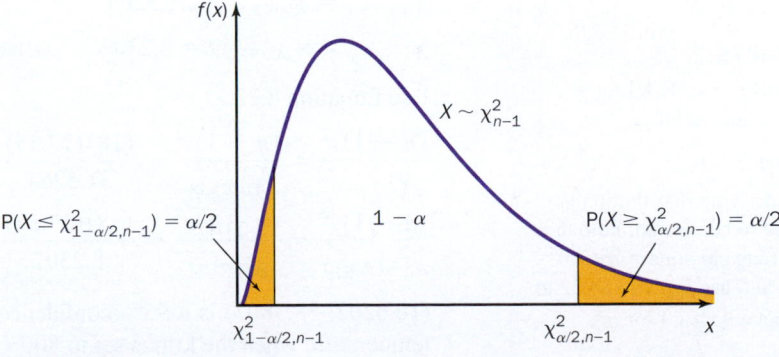

Figure 8.46 An illustration of the probability statement in Equation 8.20.

Manipulate here means: Inside the probability statement, divide each term by $(n - 1)S^2$ and take the reciprocal of each term (change the direction of the inequality).

Manipulate Equation 8.20 to obtain the probability statement

$$P\left(\frac{(n-1)S^2}{\chi^2_{\alpha/2,n-1}} < \sigma^2 < \frac{(n-1)S^2}{\chi^2_{1-\alpha/2,n-1}}\right) = 1 - \alpha \tag{8.21}$$

This probability statement leads to the following general result. ◀

How to Find a $100(1 - \alpha)$% Confidence Interval for a Population Variance

The critical value in the right tail of the chi-square distribution is part of the expression for the left endpoint, and the critical value in the left tail of the chi-square distribution is part of the expression for the right endpoint.

Given a random sample of size n from a population with variance σ^2; if the underlying population is normal, a $100(1 - \alpha)$% confidence interval for σ^2 is given by

$$\left(\frac{(n-1)s^2}{\chi^2_{\alpha/2,n-1}}, \frac{(n-1)s^2}{\chi^2_{1-\alpha/2,n-1}}\right) \tag{8.22}$$

A CLOSER LOOK

1. This confidence interval for σ^2 is valid *only* if the underlying population is normal.
2. Take the square root of each endpoint of Equation 8.22 to find a $100(1 - \alpha)$% confidence interval for the population standard deviation σ.

The following example illustrates the use of Equation 8.22 to construct a confidence interval for σ^2 and to answer an inference question.

Example 8.13 Kiln-Fired Dishes

Earthenware dishes are made from clay and are fired, or exposed to heat, in a large kiln. Large fluctuations in the kiln temperature can cause cracks, bumps, or other flaws (and increase cost). With the kiln set at 800°C, a random sample of 19 temperature measurements (in °C) was obtained. The sample variance was $s^2 = 17.55$.

a. Find a 95% confidence interval for the true population variance in temperature of the kiln when set to 800°C. Assume the underlying distribution is normal.

b. Quality-control engineers have determined that the maximum variance in temperature during firing should be 16°C. Using the confidence interval constructed in part (a), is there any evidence to suggest the true temperature variance is greater than 16°C? Justify your answer.

SOLUTION

a. $s^2 = 17.55$, $n = 19$, $\nu = 19 - 1 = 18$ Given.

$1 - \alpha = 0.95$ $\Rightarrow$ $\alpha = 0.05$ $\Rightarrow$ $\alpha/2 = 0.025$ Find $\alpha/2$.

$\chi^2_{\alpha/2,n-1} = \chi^2_{0.025,18} = 31.5264$

$\chi^2_{1-\alpha/2,n-1} = \chi^2_{0.975,18} = 8.2307$ Use Table VI in the Appendix with $\nu = 18$ to find the critical values.

Use Equation 8.22.

$\dfrac{(n-1)s^2}{\chi^2_{\alpha/2,n-1}} = \dfrac{(n-1)s^2}{\chi^2_{0.025,18}} = \dfrac{(18)(17.55)}{31.5264} = 10.0202$ Left endpoint.

$\dfrac{(n-1)s^2}{\chi^2_{1-\alpha/2,\,n-1}} = \dfrac{(n-1)s^2}{\chi^2_{0.975,18}} = \dfrac{(18)(17.55)}{8.2307} = 38.3807$ Right endpoint.

(10.0202, 38.3807) is a 95% confidence interval for the true population variance in temperature when the kiln is set to 800°C.

Solution Trail 8.13a

KEYWORDS

- 95% CI for the true population variance
- Sample variance $s^2 = 17.55$
- Random sample

TRANSLATION

- 95% CI for σ^2

CONCEPTS

- A $100(1 - \alpha)$% CI for a population variance

VISION

The underlying distribution is assumed to be normal. Find the appropriate chi-square critical values, and use Equation 8.22 to construct a 95% CI for σ^2.

$(\sqrt{10.0202}, \sqrt{38.3807}) = (3.1655, 6.1952)$ is a 95% confidence interval for the population standard deviation.

Figure 8.47 shows a technology solution.

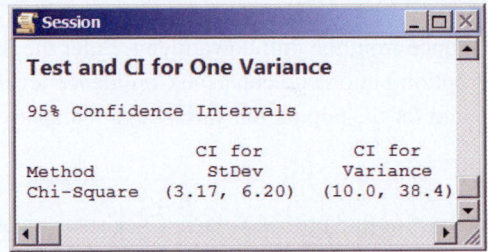

Figure 8.47 Confidence intervals for the population standard deviation and population variance.

Solution Trail 8.13b

KEYWORDS
- Is there any evidence?

TRANSLATION
- Inference procedure

CONCEPTS
- A $100(1 - \alpha)\%$ CI for a population variance

VISION

The CI constructed in part (a) is an interval in which we are 95% confident the true value of σ^2 lies. If 16 lies in this interval, there is no evidence to suggest σ^2 is different from 16. Use the four-step inference procedure.

b. Claim: $\sigma^2 = 16$.

Experiment: $s^2 = 17.55$.

Likelihood: The likelihood is expressed as a 95% confidence interval, an interval of likely values for σ^2: $(10.0202, 38.3807)$, from part (a).

Conclusion: The required kiln temperature variance, 16°C, is included in this confidence interval. There is no evidence to suggest σ^2 is different from 16.

Note: The confidence interval for σ^2 is *not* symmetric about the point estimate s^2 and is quite large (see Figure 8.48). The confidence interval will be *narrow* only for very large values of n.

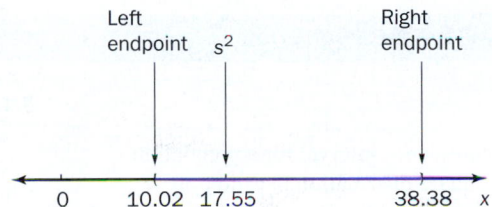

Figure 8.48 A 95% confidence interval for the population variance σ^2.

TRY IT NOW   GO TO EXERCISE 8.136

Technology Corner

Procedure: Construct a confidence interval for a population variance.
Reconsider: Example 8.13, solution, and interpretations.

CrunchIt!

There is no built-in function to construct a confidence interval for a population variance. The Chi-square Distribution Calculator may be used to find critical values.

TI-84 Plus C

There is no built-in function to compute critical values associated with a chi-square distribution nor to construct a confidence interval for a population variance. However, the EQUATION SOLVER, or equivalently, the solve function, may be used to find critical values, and other confidence interval calculations can be completed on the Home screen.

Minitab

Use the built-in function 1 Variance. Input is either summary statistics or a column containing data.

1. Select Stat; Basic Statistics; 1 Variance.
2. Choose Sample variance from the pull down menu. Enter the Sample size and the Sample variance.
3. Select the Options option button and enter the Confidence level. The results (a confidence interval for the population standard deviation and for the population variance) are displayed in a session window. See Figure 8.47.

Excel

There are built-in functions to find critical values. Use these values in the appropriate formula to construct a confidence interval for a population variance.

1. Use the function CHISQ.INV to find the left tail critical values and the function CHISQ.INV.RT to find the right tail critical value. The arguments are tail probability and degrees of freedom.
2. Compute the left and right endpoints of the confidence interval. See Figure 8.49.

	A	B	C
1	8.2307	= CHISQ.INV(0.025,18)	Left tail critical value
2	31.5264	= CHISQ.INV.RT(0.025,18)	Right tail critical value
3	10.0202	= 18*17.55/A2	Left endpoint
4	38.3805	= 18*17.55/A1	Right endpoint

Figure 8.49 Excel calculations to construct a confidence interval for a population variance.

SECTION 8.5 EXERCISES

Concept Check

8.122 True/False A confidence interval for a population variance based on a chi-square distribution is valid only if the underlying population is normal.

8.123 True/False A confidence interval for a population variance based on a chi-square distribution is symmetric about the sample variance.

8.124 True/False A confidence interval for a population variance based on a chi-square distribution is based on $n - 1$ degrees of freedom, where n is the sample size.

8.125 True/False Chi-square critical values satisfy the equation $\chi^2_{1-\alpha/2,\nu} = -\chi^2_{\alpha/2,\nu}$.

8.126 True/False A confidence interval for a population standard deviation can be constructed by taking the square root of each endpoint of the confidence interval for the population variance.

Practice

8.127 Find each of the following critical values.
a. $\chi^2_{0.10,5}$
b. $\chi^2_{0.001,31}$
c. $\chi^2_{0.05,16}$
d. $\chi^2_{0.025,21}$
e. $\chi^2_{0.99,11}$
f. $\chi^2_{0.95,15}$
g. $\chi^2_{0.975,23}$
h. $\chi^2_{0.995,9}$

8.128 Find each of the following critical values.
a. $\chi^2_{0.90,12}$
b. $\chi^2_{0.01,15}$
c. $\chi^2_{0.9999,22}$
d. $\chi^2_{0.005,3}$
e. $\chi^2_{0.9995,19}$
f. $\chi^2_{0.005,34}$
g. $\chi^2_{0.999,26}$
h. $\chi^2_{0.0001,40}$

8.129 In each of the following problems, the sample size and the confidence level are given. Find the appropriate chi-square critical values for use in constructing a confidence interval for the population variance.
a. $n = 22$, 95%
b. $n = 37$, 99%
c. $n = 11$, 98%
d. $n = 31$, 90%
e. $n = 5$, 95%
f. $n = 37$, 99.9%

8.130 In each of the following problems, the sample variance, the sample size, and the confidence level are given. Assume the underlying population is normally distributed. Find the associated confidence interval for the population variance.
a. $s^2 = 5.65$, $n = 35$, 95%
b. $s^2 = 45.62$, $n = 26$, 98%
c. $s^2 = 50.41$, $n = 6$, 80%
d. $s^2 = 7.68$, $n = 37$, 90%
e. $s^2 = 32.22$, $n = 5$, 99%
f. $s^2 = 70.67$, $n = 28$, 99.9%

8.131 In each of the following problems, the sample variance, the sample size, and the confidence level are given. Assume the underlying population is normally distributed. Find the associated confidence intervals for the population variance and the population standard deviation.

a. $s^2 = 3.08$, $n = 14$,   99%
b. $s^2 = 64.10$, $n = 11$,   95%
c. $s^2 = 59.07$, $n = 6$,   80%
d. $s^2 = 7.35$, $n = 27$,   99.98%
e. $s^2 = 31.38$, $n = 22$,   95%
f. $s^2 = 12.39$, $n = 18$,   99%

8.132 Use Table VI in the Appendix and linear interpolation to approximate each critical value. Verify each approximation using technology.

a. $\chi^2_{0.05,45}$ b. $\chi^2_{0.005,65}$
c. $\chi^2_{0.01,72}$ d. $\chi^2_{0.025,56}$
e. $\chi^2_{0.95,85}$ f. $\chi^2_{0.999,52}$
g. $\chi^2_{0.90,66}$ h. $\chi^2_{0.975,75}$

Applications

8.133 Business and Management A personnel manager at Inserra Supermarkets is concerned about the pattern of overtime hours claimed by employees. Although there is a special budget to pay for overtime hours, large fluctuations may cause cash-flow problems. In a random sample of 25 employees, the sample variance for the number of overtime hours claimed was 4.25. Assume the distribution of overtime hours is normal.

a. Find a 95% confidence interval for the population variance of overtime hours.
b. Find a 95% confidence interval for the population standard deviation of overtime hours.

8.134 Got Milk? One measure of milk production in the United States is the amount (in gallons) produced per cow. Milk production is affected by weather, feed, and other environmental factors. A random sample of the amount of milk produced per cow per quarter is given on the text website.[35] **MILK**

a. Find a 95% confidence interval for the true population variance of milk produced per cow per quarter. Write a Solution Trail for this problem.
b. What assumption(s) did you make in constructing this confidence interval?
c. Find a 95% confidence interval for the true population standard deviation of milk produced per cow per quarter.
d. The U.S. Department of Agriculture claims the standard deviation of milk produced per cow per quarter is 400 gallons. Is there any evidence to refute this claim? Justify your answer.

8.135 Technology and the Internet Everyone who owns or has access to a computer probably has several passwords for various programs. These codes are supposed to be a secret, random combination of characters. However, some psychologists believe passwords are predictable based on personality traits, and many computer hackers claim that any password can be *acquired*. To study computer security at Spectaguard Acquisition, a special program was developed and used to systematically try various passwords for randomly selected user accounts. The time (in hours) needed to obtain each password is given in the table below. **CODES**

1.88	1.71	2.09	6.60	2.28	3.52	2.64	4.94
1.78	4.13	2.55	1.66	0.28	4.02	4.47	0.68
5.67	0.03						

Assume the distribution of times is normal.
a. Find a 98% confidence interval for the true population variance of the time needed to acquire someone's password.
b. Use the methods described in Section 6.3, page 272, to check for any evidence of non-normality.

8.136 Physical Sciences St. Simons Island, Georgia, is a popular vacation destination. Attractions include a salt-marsh tour, a dolphin watch, a golf course, and the art and antiques trail. In a sample of high-tide observations (in feet) from June 2013, $\bar{x} = 7.07$, $s = 0.702$, and $n = 30$.[36]

a. Assuming normality, find a 99% confidence interval for the true variance in high-tide level.
b. Historical evidence suggests the variance in high tide is 0.50. Is there any evidence to suggest the variance in high tide has decreased? Justify your answer. Write a Solution Trail for this problem.

8.137 Physical Sciences June temperatures in Phoenix, Arizona, routinely soar above 100°F, but residents claim that it is "dry" heat. The low humidity makes outside physical activity easier and attracts tourists, especially golfers, and retirees. The following table contains relative humidity data (percentages) for several days in June in Phoenix.[37] **HUMID**

16	19	22	23	18	17	18	16	11	8
8	7	9	10	12	18	19	16	12	10
11	9	9	7	6	9	10	12	15	15

a. Find a 95% confidence interval for the true variance in relative humidity in Phoenix.
b. Is there any evidence to suggest the data are from a non-normal distribution? Justify your answer.

8.138 Manufacturing and Product Development In residential and commercial buildings, the thickness of window glass varies according to the width of the window. Wider windows require thicker window panes. As part of quality control, a manufacturer randomly samples windows and carefully measures the width of each (in millimeters). Consider the data in the following table.

Window width	Sample size	Sample mean	Sample variance
< 250	15	6.02	0.003
250–1100	18	7.95	0.04
1100–2250	22	10.12	0.5

a. Find a 99% confidence interval for the true variance of window thickness for each window width category. Assume normality.

b. Suppose a process is considered "in control" if the variance is 0.02 mm (or less). Is there any evidence to suggest any of the three processes is out of control?

8.139 Fuel Consumption and Cars   New hybrid automobiles run on both gasoline and electricity. Toyota claims the Prius is capable of traveling 15 miles on electricity alone, and GM reports that the all-electric range of the 2013 Volt is 38 miles.[38] A random sample of each automobile was obtained. All cars were fully charged, and the all-electric range for each was carefully measured. The data are given in the following table.

Automobile	Sample size	Sample variance
Volt	12	8.31
Prius	8	41.50

a. Find a 95% confidence interval for the true population variance in all-electric miles for the Volt.
b. Find a 95% confidence interval for the true population variance in all-electric miles for the Prius.
c. Is there any evidence to suggest that the two population variances are different? Justify your answer.

8.140 Sports and Leisure   Some of the stages of the Tour de France bicycling championships are over 200 km. In a random sample of 22 riders on Stage 4 of the 2012 Tour de France, Abbeville to Rouen, 214.5 km, times were measured (in minutes), and the sample variance was 4.81 minutes.[39] Assuming normality, find a 99% confidence interval for the true population variance in times for Stage 4 of the Tour de France.

8.141 Sports and Leisure   During the National Football League preseason, coaches appraise new players and tested veterans. Suppose a rookie is trying to unseat a veteran for the starting tailback position. A random sample of preseason runs is selected for each player, and the number of yards gained on each play is recorded. **RUNS**
a. Find a 99% confidence interval for the variance of yardage gained per run for the veteran tailback.
b. Find a 99% confidence interval for the variance of yardage gained per run for the rookie tailback.
c. Suppose the coach will start the more consistent and dependable tailback. Which player do you think will get the starting position, and why?

8.142 Travel and Transportation   Airline travelers are often plagued by long delays on the tarmac waiting to take off. In April 2013 there were six flights in the United States with tarmac times of more than 3 hours; four were international flights.[40] A random sample of 25 planes at the Charlotte Douglas International Airport was obtained, and the tarmac time (in minutes) was recorded for each. The sample variance was 181.55.
a. Find a 95% confidence interval for the population variance in tarmac times. Assume normality.
b. Airlines would like to reduce tarmac times, of course, and lower the variance. Is there any evidence to suggest that the population variance is less than 100 minutes? Justify your answer.

8.143 Physical Sciences   The well depth in certain locations is important for agriculture and for the development and maintenance of residential and commercial areas. A random sample of the depth to water (in centimeters) at the Upper Gordon Gulch Ground Water Well from 2011 to 2013 is given in the following table.[41] **GULCH**

1177	1021	989	984	981	979	974	974	973	975
970	925	877	959	962	937	960	985	984	970

a. Find a 98% confidence interval for the population variance in depth to water at this site.
b. Find a 98% confidence interval for the population standard deviation in depth to water at this site.
c. Use the methods described in Section 6.3, page 272, to test for any evidence of non-normality.

8.144 Fuel Consumption and Cars   One measure of a vehicle's front-end alignment is the caster angle—the relationship between the upper and lower ball joints. The specifications for a certain sports car include a caster angle of 2.8° with variance 0.40. Twenty-two new sports cars were randomly selected, and the caster angle was measured for each. The sample variance was 0.55.
a. Find a 95% confidence interval for the population variance in caster angle. Assume normality.
b. Is there any evidence to suggest the population variance in caster angle is different from 0.40? Justify your answer.

8.145 Business and Management   Even though the budget for a movie may be close to $300 million, worldwide distribution can provide a huge net income for a studio. The following table includes a sample of movies, the budget (in dollars) for each, and the U.S. gross income (in dollars) for each.[42] **MOVIES**

Movie	Budget	Gross
Avatar	237,000,000	760,507,625
Superman Returns	232,000,000	200,120,000
Quantum of Solace	230,000,000	169,368,427
Green Lantern	200,000,000	116,601,172
Wrath of the Titans	150,000,000	36,030,512
Die Another Day	142,000,000	160,942,139
Lethal Weapon 4	140,000,000	130,444,603
Sahara	145,000,000	68,671,925
I Am Legend	150,000,000	256,393,010
Transformers	151,000,000	319,246,193
Cowboys and Aliens	163,000,000	100,240,551
Iron Man 2	170,000,000	312,433,331
Wild Wild West	175,000,000	113,805,681
Salt	130,000,000	118,311,368
Bad Boys II	130,000,000	138,540,870

a. Find a 95% confidence interval for the population variance in gross income.
b. Compute the ratio of U.S. revenue to budget (ratio = gross/budget) for each movie. Find a 95% confidence

interval for the population variance in ratio of revenue to budget.

c. If the variance in the ratio is greater than 1, it is considered *risky* to invest in a new film. Is there any evidence to suggest investing in a new movie is risky? Justify your answer.

8.146 Manufacturing and Product Development
Silver bars used in trading on the bullion market are manufactured to certain height, width, and length specifications. In addition, there is variation in the silver content. Five randomly selected silver bars were carefully analyzed for silver content (in ounces). The sample variance was 65.5. Assume the distribution of silver content is normal, and find a 99% confidence interval for the population variance in silver content.

8.147 Psychology and Human Behavior
The Great Wall of China was started in 214 B.C. and designed as a defense against nomadic tribes. This 1500-mile-long wall, built from earth and stone, is visible from space and is one of the most famous structures in the world. Suppose a random sample of midpoints between guard towers along the Great Wall was obtained, and the height of the wall at each point was measured (in feet). The data are given in the following table.[43] **GRWALL**

26.9	22.2	21.8	26.6	19.2	29.3	21.9	19.0
23.5	26.7	18.3	27.7	18.8	25.0	24.9	21.1
24.1	24.3	19.5	20.8	27.1			

a. Find a 95% confidence interval for the population variance in height at the midpoints between guard towers of the Great Wall.

b. Using the confidence interval constructed in part (a), is there any evidence to suggest the true population variance in the height at midpoints between guard towers is less than 12 feet? Justify your answer.

Extended Applications

8.148 Medicine and Clinical Studies
Many researchers believe moderate physical activity, such as walking, will help prevent weight gain. To study this claim, doctors in Colorado had randomly selected patients wear a pedometer to measure the distance walked (in miles) per week. The table below presents data for men and women.

	Sample size	Sample variance
Men	32	5.75
Women	28	7.66

a. Find a 95% confidence interval for the true variance in distance walked per week for men.

b. Find a 95% confidence interval for the true variance in distance walked per week for women.

c. Is there any evidence to suggest the true variance is different for men and women? Justify your answer.

d. What assumption(s) did you make in constructing the confidence intervals above?

8.149 Biology and Environmental Science
Rice is a staple food for much of the world's population. The height of a rice plant (usually 0.4–5 mm) depends on the rice variety and the environment. Genetic engineering is currently underway to protect rice plants from disease and to decrease the variability in plant height. In a controlled experiment, plant heights were measured (in millimeters); 30 natural rice plants had a sample variance in height of 1.5 mm, and 22 genetically engineered rice plants had a sample variance in height of 0.89 mm. Assume normality.

a. Find a 95% confidence interval for the true variance in height for natural rice plants.

b. Find a 95% confidence interval for the true variance in height for genetically engineered rice plants.

c. Is there any evidence to suggest the population variance is different for natural and genetically engineered rice plants?

8.150 Public Health and Nutrition
In June 2013 the Agriculture Department banned all high-calorie sports drinks and candy bars from school vending machines.[44] Many beverage companies had already added sports drinks to replace high-calorie sodas. The new rules stipulate that sports drinks must contain 60 calories or less in a 12-ounce serving. A random sample of sports drinks was obtained, and the calories in each was recorded.[45] **DRINKS**

a. Find a 95% confidence interval for the population variance in calories for sports drinks.

b. Find a 95% confidence interval for the population standard deviation in calories for sport drinks.

8.151 Physical Sciences
A microwave radiometer is used to measure the column water vapor and the infrared brightness (IB) temperatures in clouds. Clouds were randomly selected, and a weather station in Coffeyville, Kansas, collected the following summary statistics.

Cloud type	Column water vapor (cm)		IB Temperature (°C)	
	Sample size	Sample variance	Sample size	Sample variance
Cirrus	11	0.06	17	201.7
Cumulus	21	0.08	28	225.6

a. Find a 99% confidence interval for the population variance in column water vapor and in infrared brightness temperature for cirrus clouds.

b. Find a 99% confidence interval for the population variance in column water vapor and in infrared brightness temperature for cumulus clouds.

c. Is there any evidence to suggest the variance in column water vapor or infrared brightness temperature is different in cirrus and cumulus clouds? Justify your answers.

8.152 Medicine and Clinical Studies
A new medicinal spray has been developed to help ease the itch and burn associated with

poison ivy and poison oak. People who suffer from these poisons want and need immediate relief. A research study was conducted to measure the time (in minutes) from application of this spray to relief from itching. The following summary statistics were reported for a random sample of children and adults.

Population	Sample size	Sample variance
Children	11	1.57
Adults	25	2.38

a. Find a 95% confidence interval for the population variance in time to relief for children.

b. Find a 95% confidence interval for the population variance in time to relief for adults.

c. Is there any evidence to suggest the variance in time to relief is different in children and adults? Justify your answer.

8.153 Electric Shock During the first half of 2013, it was discovered that British Columbia Hydro was dramatically overcharging customers in a Burnaby condominium complex. New "smart" meters had been installed but not connected to units, and the company was estimating electricity use based on very poor comparisons.[46] Suppose a random sample of bimonthly electricity bills for these condominium owners was obtained before and after the new meters were installed. **METERS**

a. Find a 99% confidence interval for the population variance in electricity bills before the new meters were installed.

b. Find a 99% confidence interval for the population variance in electricity bills after the new meters were installed.

c. Is there any evidence to suggest that the variance is different before and after the new meters were installed? Justify your answer.

Challenge

8.154 Sports and Leisure Given a random sample of size n from a normal population with variance σ^2:

a. Use the method outlined in Exercise 8.51 to find a one-sided $100(1 - \alpha)\%$ confidence interval for σ^2 bounded above, and another bounded below.

b. A random sample of soccer stadiums from around the world was obtained, and the seating capacity for each is given on the text website. Find a one-sided 95% confidence interval, bounded above, for the population variance in seating capacity for soccer stadiums. **SEATCAP**

8.155 Normal Approximation to the Chi-Square Distribution Given a random sample of size n from a normal population with variance σ^2, a $100(1 - \alpha)\%$ confidence interval for σ^2 (based on a chi-square distribution) is given by Equation 8.22:

$$\frac{(n-1)s^2}{\chi^2_{\alpha/2}} < \sigma^2 < \frac{(n-1)s^2}{\chi^2_{1-\alpha/2}}$$

If n is large ($n > 30$), then the chi-square random variable $(n-1)S^2/\sigma^2$ is approximately normal with mean $n - 1$ and variance $2(n - 1)$. Therefore, for large n,

$$P\left(-z_{\alpha/2} < \frac{\frac{(n-1)S^2}{\sigma^2} - (n-1)}{\sqrt{2(n-1)}} < z_{\alpha/2}\right) = 1 - \alpha \quad (8.23)$$

a. Manipulate Equation 8.23 to obtain an approximate $100(1 - \alpha)\%$ confidence interval for a population variance.

b. The thickness of pavement on roads and highways depends on the predicted weight and volume of vehicular traffic. The thickness of the pavement (in mm) was measured at 51 random locations along Route 95. The sample variance was $s^2 = 16.25$.

i. Find a 95% confidence interval for the population variance in pavement thickness using Equation 8.22.

ii. Find an approximate 95% confidence interval for the population variance in pavement thickness using the equation derived in part (a).

iii. Which of these two intervals is wider? Why?

CHAPTER 8 SUMMARY

Concept	Page	Notation / Formula / Description
Estimator	334	A statistic, or rule, used to produce a point estimate of a population parameter.
Unbiased estimator	334	A statistic $\hat{\theta}$ is an unbiased estimator of θ if $E(\hat{\theta}) = \theta$.
Biased estimator	334	A statistic $\hat{\theta}$ is a biased estimator of θ if $E(\hat{\theta}) \neq \theta$.
MVUE	336	Minimum variance unbiased estimator.
Confidence interval (CI)	340	An interval of values constructed so that with a specified degree of confidence, the value of the population parameter lies in this interval.
Confidence coefficient	340	The probability the confidence interval encloses the population parameter in repeated samplings.
Confidence level	340	The confidence coefficient expressed as a percentage.
z_α	342	z critical value; a value such that $P(Z \geq z_\alpha) = \alpha$.

$t_{\alpha,\nu}$ 354 t critical value; a value such that $P(T \geq t_{\alpha,\nu}) = \alpha$ where T has a t distribution with ν degrees of freedom.

$\chi^2_{\alpha,\nu}$ 375   Chi-square critical value; a value such that $P(X \geq \chi^2_{\alpha,\nu}) = \alpha$, where X has a chi-square distribution with ν degrees of freedom.

Summary of confidence intervals

Parameter	Assumptions	$100(1-\alpha)\%$ confidence interval
μ	n large, σ known, or normality, σ known	$\bar{x} \pm z_{\alpha/2} \dfrac{\sigma}{\sqrt{n}}$
μ	normality, σ unknown	$\bar{x} \pm t_{\alpha/2,n-1} \dfrac{s}{\sqrt{n}}$
p	n large, nonskewness	$\hat{p} \pm z_{\alpha/2}\sqrt{\dfrac{\hat{p}(1-\hat{p})}{n}}$
σ^2	normality	$\left(\dfrac{(n-1)s^2}{\chi^2_{\alpha/2,n-1}}, \dfrac{(n-1)s^2}{\chi^2_{1-\alpha/2,n-1}} \right)$

Common sample-size calculations

Parameter	Estimate	Sample size
μ	$\bar{x}$	$n = \left(\dfrac{\sigma z_{\alpha/2}}{B} \right)^2$
p	$\hat{p}$	$n = \hat{p}(1-\hat{p})\left(\dfrac{z_{\alpha/2}}{B} \right)^2$

CHAPTER 8 EXERCISES

APPLICATIONS

8.156 Medicine and Clinical Studies Most patients in need of a kidney transplant are placed on a dialysis machine. There is some evidence to suggest that the longer patients remain on dialysis, the worse they fare following a transplant. The mean waiting time for a kidney transplant is five years (60 months). Suppose a random sample of kidney transplant patients was obtained, and the wait time (in months) was recorded for each. The data are given in the table below. **KIDNEY**

58.1	63.9	74.4	57.8	85.5	68.5	53.5	100.8
87.6	83.7	78.4	49.7	76.6	40.7	63.6	50.3
75.5	76.8	101.4	64.0	43.7	64.3	48.8	76.2

  a. Find a 95% confidence interval for the true mean wait time for a kidney transplant.

  b. Find a 95% confidence interval for the true variance in wait time for a kidney transplant.

  c. Construct a normal probability plot for the wait-time data. Is there any evidence to suggest non-normality?

  d. Is there any evidence to suggest that the waiting time is different from 60 months? Justify your answer.

8.157 Biology and Environmental Science The amount of chlorophyll near Stearns Wharf in Santa Barbara, California,

varies due to the seasons, the entry of fresh water into the area, and the salt water from the ocean. A random sample of 35 days was selected and the chlorophyll near Stearns Wharf was measured (in μg/L) for each.[47] The sample mean chlorophyll concentration was 2.93 μg/L. Assume the population standard deviation is $\sigma = 2.12$.

  a. Find a 98% confidence interval for the true mean chlorophyll concentration near Stearns Wharf.

  b. Find the sample size necessary to construct a 98% confidence interval for the true mean chlorophyll concentration with a bound on the error of estimation $B = 0.5$.

8.158 Marketing and Consumer Behavior Parrot Jungle Island is a roadside attraction in Miami, Florida, featuring tropical birds, crocodiles, and over 2000 varieties of plants and flowers. A new advertising campaign was developed to attract more out-of-state visitors. In a random sample of 270 visitors, 189 were area residents.

  a. Find the sample proportion of visitors who were area residents. Check the nonskewness criterion.

  b. Find a 99% confidence interval for the true proportion of visitors who were area residents.

  c. Suppose no prior knowledge of the proportion of visitors who were area residents is available. Find the sample size necessary for a 99% confidence interval with a bound on the error of estimation of 0.05.

8.159 Sports and Leisure   Most Major League baseball parks have a device to measure the speed of every pitch. The results are often displayed on a scoreboard and tracked by coaches. A random sample of pitches made during the first inning of baseball games around the country was selected. The speed of each pitch (in mph) is given in the table below. ▓ **PITCH**

85	90	82	86	83	88	87	90	92	84	92
90	89	90	89	89	84	87	92	89	86	89

a. Assume normality. Find a 95% confidence interval for the population mean speed of pitches in the first inning of Major League baseball games.
b. Find a 99% confidence interval for the population variance in speed of pitches in the first inning of Major League baseball games.
c. The mean speed of all pitches made in the first inning of games during the previous season was 90.225. Is there any evidence to suggest the mean speed has changed?

8.160 Economics and Finance   ATM machines have made it easier and faster for customers to check account balances, transfer money, and obtain cash. Many banks are now considering the next generation of ATM machines, which will feature news headlines, full-motion video, and tickets to events. In a random sample of 500 customers, 280 said ATM machines should offer postage stamps.

a. Find the sample proportion of customers who believe ATM machines should offer postage stamps, and check the nonskewness criterion.
b. Find a 99% confidence interval for the true proportion of customers who believe ATM machines should offer postage stamps.
c. A bank official claims the proportion of customers who believe ATM machines should offer postage stamps is 0.60. Is there any evidence to refute this claim?

8.161 Manufacturing and Product Development   A 0.2-kiloton "bunker buster" missile is designed to destroy enemy bunkers approximately 70 feet deep. To prevent fallout, the missile must penetrate 120 feet below the ground. In tests conducted by the military, eight missiles were fired and the penetration depth (in feet) of each was recorded. The data are given in the table below. ▓ **MISSILE**

122	117	119	119	121	124	119	120

a. Assume normality. Find a 99% confidence interval for the population mean penetration depth.
b. Is there any evidence to suggest the population mean penetration depth is less than 120 feet?

8.162 Medicine and Clinical Studies   Passengers on long airline flights may develop deep vein thrombosis (DVT), potentially dangerous blood clots. Although most blood clots in the bloodstream dissolve naturally, travelers on long flights have three times the risk of developing a blood clot. In a research study of people who traveled frequently as part of their job, 22 of 8755 developed a blood clot within eight weeks of a long flight.

a. Find a 90% confidence interval for the true proportion of passengers on long flights who develop DVT within eight weeks of a long flight.
b. Is there any evidence to suggest this proportion is greater than 0.5%? Justify your answer.

8.163 Psychology and Human Behavior   People who watch multiple DVDs or several online TV episodes in a row are indulging in *binge-watching*. Most experts agree that this is mostly a harmless, enjoyable addiction. In a recent survey of 2693 U.S. adults with Internet access, 1670 watch multiple TV episodes back to back.[48]

a. Find a 99% confidence interval for the true proportion of adults with Internet access who watch multiple TV episodes back to back.
b. Find the sample size necessary to construct a 99% confidence interval with a bound on the error of estimation of 0.02.

8.164 Travel and Transportation   An English racing team called the Bloodhound Project is testing a rocket car that they hope will break the current land speed record. They expect this hybrid vehicle, which runs on concentrated peroxide and a synthetic rubber, to reach speeds over 1000 mph.[49] Suppose a random sample of test runs was obtained, and the speed of the rocket car was recorded for each. ▓ **ROCKET**

a. Find a 99% confidence interval for the population mean speed of the rocket car.
b. Find a 99% confidence interval for the population variance in speed of the rocket car.
c. What assumptions did you make in constructing these two confidence intervals?
d. The Bloodhound Team members will attempt to break the 1000 mph barrier on the Hakskeen Pan desert in South Africa in 2016. Using the test speed data, is there any evidence to suggest the mean speed of the rocket car will be great than 1000 mph? Justify your answer.

8.165 Business and Management   Many businesses rent a limousine to chauffeur important clients to/from an airport, hotel, or office. A random sample of the cost (in dollars) of renting a limousine for the entire day in Cincinnati was obtained. For $n = 25$, the mean was $\bar{x} = 410.25$ and the standard deviation was $s = 35.07$. Assume the cost distribution is normal.

a. Find a 95% confidence interval for the population mean cost of renting a limousine for the entire day.
b. Find a 95% confidence interval for the population variance in the cost of renting a limousine for the entire day.

c. Find a 95% confidence interval for the standard deviation in the cost of renting a limousine for the entire day.

8.166 Psychology and Human Behavior According to a recent survey, approximately one of every five people hides money from their spouse. Women are slightly more likely to hide money than men, but when men stash cash, they hide more of it. The report claims that the mean amount hidden is $10,000.[50] Suppose a random sample of 17 people who hide money from their spouse was obtained. The mean amount hidden was $\bar{x} = 10,460$ and the standard deviation was $s = 325$. Assume the distribution of hidden cash is normal.

a. Find a 99% confidence interval for the mean amount of hidden cash.

b. Suppose divorce lawyers always assume a spouse has hidden cash of approximately $10,000. Is there any evidence to suggest that the mean amount of stashed cash is more than $10,000?

c. Find a 99% confidence interval for the variance in the amount of hidden cash.

d. A large variance in the amount of hidden cash means divorce lawyers need to hire more forensic accountants. Is there any evidence to suggest that the variance in hidden cash is greater than $55,000?

8.167 Physical Sciences The Quabbin Reservoir in Massachusetts is 18 miles long and has a capacity of 412 billion gallons. This reservoir serves approximately 2.5 million people, with a daily yield of 300 million gallons of water. The depth of the reservoir is carefully monitored and measured at certain locations and times each day. In a random sample of 18 summer days, the depth (in feet) was measured at location S-1 at noon. The mean depth was 75.4. Assume the population standard deviation is $\sigma = 7.58$.

a. Find a 99% confidence interval for the depth of the reservoir at this location.

b. Historical records indicate the mean depth at this location for the previous 10 years is 78.4 feet. Is there any evidence to suggest this population mean has changed?

8.168 Public Health and Nutrition Rising costs caused many Americans to do without health insurance. A random sample of adults of various ethnicities was obtained and asked whether they had health insurance. The resulting data are given in the table below.

Group	Sample size	Number without health insurance
White	1220	166
African American	1080	205
Hispanic	1156	384

a. Find a 95% confidence interval for the proportion of adults without health insurance for each group.

b. A government publication claims that the proportion of people without health insurance for the general population is 0.146. Using the confidence intervals in part (a), is there evidence that any group has a different population proportion without health insurance?

EXTENDED APPLICATIONS

8.169 Medicine and Clinical Studies The drugs Ritalin and Adderall are designed to stimulate the central nervous system and are widely prescribed for children diagnosed with attention deficit disorder (ADD). Suppose a random sample of girls and boys in 12th grade was obtained and the following results were obtained.

Group	Sample size	Number taking Ritalin/Adderall
Girls	375	24
Boys	480	39

a. Find a 95% confidence interval for the true proportion of 12th-grade girls who are taking Ritalin/Adderall.

b. Find a 95% confidence interval for the true proportion of 12th grade boys who are taking Ritalin or Adderall.

c. Is there any evidence to suggest that the proportion of girls in this age group taking Ritalin or Adderall is different from the proportion of boys? Justify your answer.

8.170 Public Health and Nutrition Tannin is a general term for certain nonvolatile phenolic substances in many fruits that provide an astringent sensation, in apple cider for example. There is some evidence to suggest the tannin level in apples is affected by the fertilizer regimen. A random sample of apples was obtained from both fertilized and unfertilized trees. The percentage of tannin was measured in each apple, and the data (sample size, sample mean, and sample standard deviation) are reported in the table below.

Apples	Sample size	Sample mean	Sample standard deviation
Fertilized trees	48	0.30	0.058
Unfertilized trees	55	0.35	0.077

a. Find a 95% confidence interval for the true mean tannin percentage in apples from fertilized trees.

b. Find a 95% confidence interval for the true mean tannin percentage in apples from unfertilized trees.

c. Is there any evidence to suggest the percentage of tannin is different in apples from fertilized and unfertilized trees?

8.171 Sports and Leisure Most high schools in the United States offer a wide variety of sports for both boys and girls.

However, the participation rate has historically been higher for boys. A random sample of high school students from across the country was obtained, and the students were asked whether they participate in a high school sport. The results are given in the following table.

Group	Sample size	Number participating
Boys	1250	583
Girls	1475	494

a. Find a 95% confidence interval for the true proportion of boys participating in a high school sport.
b. Find a 95% confidence interval for the true proportion of girls participating in a high school sport.
c. Is there any evidence to suggest the proportion of boys and girls participating in a high school sport is different? Justify your answer.

8.172 Public Health and Nutrition Elevated levels of mercury can cause hair loss, fatigue, and memory lapses. The concentration of mercury in a person's blood can be greatly affected by diet, specifically, the amount of fish consumed. Random samples of adults who eat fish two to three times per week and of adults who never eat fish were obtained. The mercury concentration in the blood of each person was measured (in micrograms per liter of blood), and the summary data are reported in the table below.

Group	Sample size	Sample mean	Sample standard deviation
Fish	18	4.662	0.298
No fish	27	2.079	0.309

Assume both distributions are normal.
a. Find a 95% confidence interval for the population mean mercury concentration in the blood of adults who eat fish regularly.
b. Find a 95% confidence interval for the population mean mercury concentration in the blood of adults who never eat fish.
c. The safe level of mercury (set by the U.S. Environmental Protection Agency) is five micrograms per liter of blood. Is there any evidence to suggest either group is over the safe limit of mercury concentration?
d. Use the confidence intervals in parts (a) and (b) to determine whether the two groups have different mean mercury concentration levels.

8.173 Biology and Environmental Science There is growing concern that Americans are generating too much trash. More zoning permits are being sought for landfills in rural areas, and trash haulers move garbage out of state and even out to the ocean. In a recent study, 65 households were randomly selected, and the amount of trash (in pounds)

generated by each in one week was recorded. The sample mean was $\bar{x} = 52.3$ with a standard deviation of $s = 10.75$. Assume normality.
a. Find a 98% confidence interval for the true mean amount of trash generated by an American household per week.
b. Find an *approximate* 98% confidence interval for the true mean amount of trash generated by an American household per week (based on a Z distribution rather than a t distribution). Compare your answers in parts (a) and (b). How do they differ?
c. Find a 95% confidence interval for the variance in the amount of trash generated by an American household per week.

8.174 Medicine and Clinical Studies In a major health study, a random sample of adult males was obtained and each was tested for symptoms of heart disease yearly for a decade. Subjects were divided into those who regularly donated blood and those who did not. The results are given in the table below.

Group	Sample size	Number with heart disease
Donate blood	145	51
Do not donate blood	527	210

a. Find a 95% confidence interval for the population proportion of males who donate blood who have heart disease.
b. Find a 95% confidence interval for the population proportion of males who do not donate blood who have heart disease.
c. Is there any evidence to suggest the proportion of males with heart disease is different for those who donate blood and those who do not?

8.175 Manufacturing and Product Development Teflon-coated pots and pans are designed to be nonstick and to make cooking and clean-up easier. A recent study compared the thickness of the Teflon layer in new, factory-coated pans with recoated pans. A random sample of pans was obtained for each group, and the Teflon thickness was carefully measured (in inches). **|III| TEFLON**
a. Find a 95% confidence interval for the population mean Teflon thickness in factory-coated pans.
b. Find a 95% confidence interval for the population mean Teflon thickness in recoated pans.
c. Is there any evidence to suggest these two population means are different? Justify your answer.

8.176 Public Health and Nutrition Most people understand that highly processed, carbohydrate-heavy foods are not very healthy. However, we often justify a carbohydrate-rich meal by vowing to "eat healthy" next time. A new study suggests that those who eat a high-carbohydrate meal start to crave even more of these foods shortly afterwards.[51] An experiment was conducted to test this claim. Researchers gave two groups of adults similar test meals, one of which was rich in carbohydrates. After four hours, brain scans were used to determine if there was activity related to reward, craving, and addictive behavior. The following results were obtained.

Group	Sample size	Number with craving
Low-carbohydrate meal	255	87
High-carbohydrate meal	303	143

a. Find a 99% confidence interval for the true proportion of people who exhibit addictive behavior after eating a low-carbohydrate meal.
b. Find a 99% confidence interval for the true proportion of people who exhibit addictive behavior after eating a high-carbohydrate meal.

c. Is there any evidence to suggest that the proportion of people who exhibit addictive behavior is different after eating a high- or low-carbohydrate meal?

LAST STEP

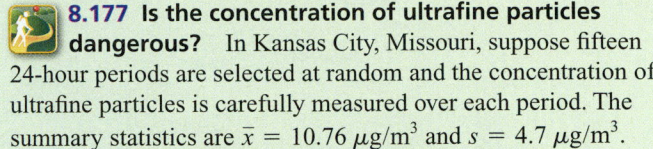

8.177 Is the concentration of ultrafine particles dangerous? In Kansas City, Missouri, suppose fifteen 24-hour periods are selected at random and the concentration of ultrafine particles is carefully measured over each period. The summary statistics are $\bar{x} = 10.76\ \mu g/m^3$ and $s = 4.7\ \mu g/m^3$.
a. Find a 95% confidence interval for the true mean level of ultrafine particles in a 24-hour period in Kansas City.
b. Is there any evidence to suggest that the mean level of ultrafine particles in a 24-hour period in Kansas City is over the EPA standard of 35 $\mu g/m^3$?

9 Hypothesis Tests Based on a Single Sample

Is it OK to pad an insurance claim?

Most automobile and home insurance policies include a deductible amount that applies to each claim. A deductible is the portion of a covered loss that the policyholder must pay. Most insurance companies simply subtract the deductible from the total amount of a claim. Typical home insurance deductibles are $500 or $1000, and generally, a larger deductible will lower the overall cost of a policy.

The results of a recent survey revealed that 24% of Americans believe it is acceptable to increase an insurance claim by a small amount to make up for deductibles that they are required to pay.[1] Even though a very high percentage of Americans believes that insurance fraud contributes to higher premiums for everyone, almost one in four thinks claim padding is acceptable. In addition, 18% believe it is acceptable to increase a claim to recoup high premiums in previous years. The survey also revealed that Americans aged 18–34, especially males in this age range, are more likely to view insurance claim padding as acceptable.

Suppose Liberty Mutual Insurance Company has hired an independent agency to conduct a survey of its automobile policyholders. The concepts presented in this chapter will be used to construct a formal hypothesis test to determine whether there is any evidence to suggest that more than 24% of policyholders view claim padding as acceptable.

CONTENTS

Marcelo Santos/Stone/Getty Images

9.1 The Parts of a Hypothesis Test and Choosing the Alternative Hypothesis

You have probably heard the word **hypothesis** used in many different contexts. An engineer might have a hypothesis concerning gas mileage on a certain car. She might claim that a new hybrid engine design will significantly improve the miles-per-gallon rating. Or a biologist might hypothesize that a special combination of nutrients will significantly increase the growth rate of yellow corn.

Definition

In statistics, a **hypothesis** is a declaration, or claim, in the form of a mathematical statement, about the value of a specific population parameter (or about the values of several population characteristics).

Here are some examples of statistical hypotheses.

1. $\mu = 14.5$

   where μ is the population mean time (in minutes) it takes for an adult's pupils to dilate after treatment with phenylephrine.

2. $p > 0.70$

   where p is the population proportion of all people over 65 who will need long-term care services at some point in their lifetime.

3. $\sigma^2 \neq 30.5$

   where σ^2 is the population variance in the amount (in gallons) of coal tar in a five-gallon bucket.

A hypothesis is a claim about a *population* parameter, *not* about a sample statistic. For example, $\mu = 5$ and $p = 0.27$ are valid hypotheses, but $\bar{x} = 27$ and $s = 32.5$ are not.

There are four parts to every hypothesis test, and it is important to identify each part in every test.

Four Parts of a Hypothesis Test

H_0 is read as "H sub zero," and H_a is read as "H sub a."

1. The **null hypothesis**, denoted H_0.

   This is the claim (about a population parameter) assumed to be true, what is believed to be true, or the hypothesis to be tested. Sometimes referred to as the *no change* hypothesis, this claim usually represents the status quo or existing state. There is an implied inequality in H_0, however, the null hypothesis is written in terms of a single value (with an equal sign), for example, $\theta = 5$. Although it may seem strange, we usually try to *reject* the null hypothesis.

2. The **alternative hypothesis**, denoted H_a.

   This statement identifies other possible values of the population parameter, or simply a possibility not included in the null hypothesis. H_a indicates the possible values of the parameter if H_0 is false. Experiments are often designed to determine whether there is evidence in favor of H_a. The alternative hypothesis represents change in the current standard or existing state.

3. The **test statistic**, denoted TS.

   This statistic is a rule, related to the null hypothesis, involving the information in a sample. The *value* of the test statistic will be used to determine which hypothesis is more likely to be true, H_0 or H_a.

4. The **rejection region** or **critical region,** denoted RR or CR.

This is an interval or set of numbers specified such that if the value of the test statistic lies in the rejection region, then the null hypothesis is rejected. There is also a corresponding *nonrejection region*; if the value of the test statistic lies in this set, then we *cannot reject H_0.*

A CLOSER LOOK

STATISTICAL APPLET

THE REASONING OF A STATISTICAL TEST

1. The test of a statistical hypothesis is a procedure by which we decide whether there is evidence to suggest that the **alternative hypothesis**, H_a, is true. The ultimate objective of a hypothesis test is to use the information in a sample to decide which hypothesis is more likely to be true, H_0 or H_a. Usually, we are trying to reject the **null hypothesis**.

2. The **rejection region** and the nonrejection region divide the world (values of the **test statistic**) into parts. Figure 9.1 illustrates this concept in terms of a parameter. The cutoff point, or dividing line, is determined by considering likely values for the test statistic if H_0 is true, and is included in one of the regions.

The value of the parameter must lie in one of the regions. In Section 9.3, we'll see how easy it is to specify a rejection region.

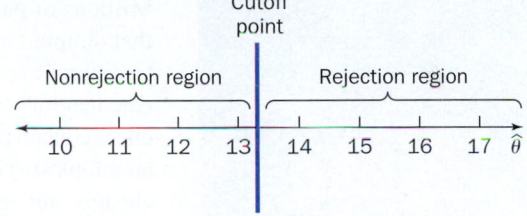

Figure 9.1 An illustration of a rejection region and a nonrejection region, where θ is a population parameter.

3. The hypothesis test procedure is very prescriptive. Once the four parts are identified, the sample data are used to compute a value of the test statistic. There are only two possible conclusions.

   a. If the value of the test statistic lies in the rejection region, then we reject H_0. There is evidence to suggest the alternative hypothesis is true.

   b. If the value of the test statistic does not lie in the rejection region, then we cannot reject H_0. There is no evidence to suggest the alternative hypothesis is true.

   We never say that we *accept H_0*. A hypothesis test is designed to prove the alternative hypothesis. If there is no evidence in favor of H_a, this does *not* imply H_0 is true.

A hypothesis test can only provide support in favor of H_a. If the value of the test statistic lies in the rejection region, reject the null hypothesis. There is evidence to suggest H_a is true. If the value of the test statistic does not lie in the rejection region, do not reject H_0. There is no evidence to suggest H_a is true. Watch the wording: We **never** *accept* the null hypothesis. Rather, the value of the test statistic *does not lie in the rejection region.*

4. This formal hypothesis test procedure is analogous to the four-step inference procedure used in the previous chapters. In fact, many of the concepts presented in earlier chapters are combined here to produce this traditional, well-established hypothesis test procedure.

The **claim** corresponds to H_0, a claim about a population parameter. The **experiment** is equivalent to a value of the test statistic. **Likelihood** is expressed in terms of the nonrejection region (likely values of the test statistic) and the rejection region (unlikely values of the test statistic). The **conclusion** is completely determined by the region in which the value of the test statistic lies. If the value is in the rejection region, we reject H_0; otherwise, we cannot reject H_0.

Suppose a hypothesis test is conducted concerning the population parameter θ, and θ_0 is a specific value of θ. The null hypothesis is always stated in terms of a single value. And there are only three possible alternative hypotheses.

$$H_0 : \theta = \theta_0$$

$$\left. \begin{array}{l} H_a : \theta > \theta_0 \\ \quad\ \theta < \theta_0 \end{array} \right\} \text{ one-sided alternatives}$$

$$\theta \neq \theta_0 \ \} \ \text{two-sided alternative}$$

Only one alternative hypothesis is selected. H_a answers the question, "What is the experimenter trying to prove, or detect, about θ?" It takes a little practice to decide which H_a is appropriate. The same specific value of the parameter (θ_0, above) always appears in H_0 and H_a.

A valid set of null and alternative hypotheses must include a statement similar to H_0 above and the relevant alternative, one of the three given above. For example, H_0: $\mu = 17$; H_a: $\mu < 17$ is a valid set of null and alternative hypotheses. H_0: $\tilde{\mu} \neq 25$; H_a: $\tilde{\mu} = 26$ is not. The following examples focus on identifying H_0 and the relevant alternative hypothesis.

> There are two one-sided alternatives and one two-sided alternative.

Example 9.1 Back Pain

Millions of people suffer from back pain, and some experience disc problems so severe that simple tasks, such as driving, sitting in a chair, or even sleeping, are painful. The traditional remedy for a damaged disc is surgery: spinal fusion. Historical records indicate that 65% of all patients who endure this costly, complicated surgery actually experience reduced pain and greater mobility. A new treatment (IDET, intradiscal electrothermal annuloplasty) has been developed, and researchers claim this procedure is more effective, cheaper, and less painful. An experiment is conducted to determine whether IDET is more effective than spinal fusion. What null and alternative hypotheses should be used?

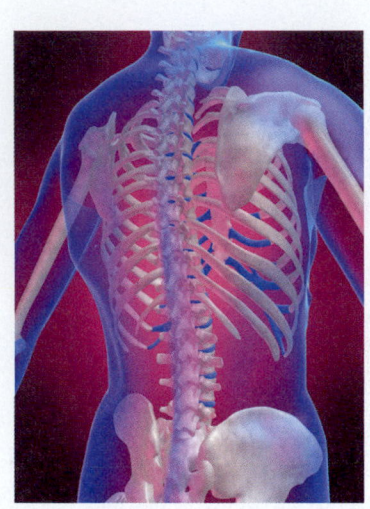

SOLUTION

STEP 1 The claim assumed to be true involves the population proportion of patients who experience reduced pain and greater mobility with the existing surgical technique. It is assumed that 65% of all patients experience relief. Therefore, the null hypothesis, stated in terms of a single value for p, is

$$H_0: p = 0.65$$

STEP 2 The existing proportion of patients who experience relief is 0.65. The experiment is designed to detect an increase in this proportion, i.e., to answer the question, "Is the new procedure more effective?" Researchers hope to find evidence that the proportion of patients who experience relief with IDET is greater than 0.65. Therefore, the alternative hypothesis is

$$H_a: p > 0.65$$

TRY IT NOW   GO TO EXERCISE 9.12

Example 9.2 A Big Place, Mate

As in so many places around the world, motorists in Australia can easily be caught in rush-hour traffic. Even though the roads are crowded, it was reported that the mean number of kilometers driven per year by each Australian driver is 15,300.[2] A new national advertising campaign is designed to encourage people to drive less and to use buses and trains instead. An observational study is conducted to determine whether the mean number of kilometers driven each year has decreased. State the appropriate null and alternative hypotheses.

SOLUTION

STEP 1 The claim assumed to be true involves the population mean number of kilometers driven by all Australians. The null hypothesis, the status quo, is given in terms of the parameter μ.

$$H_0: \mu = 15{,}300$$

STEP 2 We are searching for evidence in favor of a smaller mean number of kilometers driven. The advertising campaign is designed with the hope that the mean number of miles driven is less than the current mean. Therefore,

$$H_a: \mu < 15{,}300$$

TRY IT NOW   GO TO EXERCISE 9.14

Example 9.3 Recycled Paper

The thickness (measured in inches) of recycled printer paper is important, because sheets that are too thick will clog the machine, and paper that is too thin will rip and bleed toner. The variance in thickness for 20-pound printer paper at a manufacturing plant is known to be 0.0007. A new process is developed that uses more recycled fiber, and an experiment is conducted to detect any difference in the variance in paper thickness. State the appropriate null and alternative hypotheses.

SOLUTION

STEP 1 The assumption, or existing state, involves the population variance in thickness of recycled printer paper. The null hypothesis is given in terms of σ^2.

$$H_0: \sigma^2 = 0.0007$$

STEP 2 The experiment is designed to detect *any* difference in the population variance. This suggests a two-sided alternative.

$$H_a: \sigma^2 \neq 0.0007$$

TRY IT NOW   GO TO EXERCISE 9.17

SECTION 9.1 EXERCISES

Concept Check

9.1 True/False A statistical hypothesis is always stated in terms of a population parameter.

9.2 True/False In a test of a statistical hypothesis, we attempt to find evidence in favor of the null hypothesis.

9.3 True/False In a test of a statistical hypothesis, there may be more than one alternative hypothesis.

9.4 Short Answer State the four parts to every hypothesis test.

9.5 Short Answer State the only two possible conclusions in a test of a statistical hypothesis.

Practice

9.6 For each of the following statements, determine whether it is a valid hypothesis. Classify each valid hypothesis as a null hypothesis or an alternative hypothesis. Justify your answers.

  a. $\mu = 0.355$ **b.** $\hat{p} < 0.42$ **c.** $s > 3.5$
  d. $\bar{x} \neq 16$ **e.** $\mu > 22.66$ **f.** $p \neq 0.15$
  g. $\tilde{x} < 47.5$ **h.** $\tilde{\mu} = 12$

9.7 For each of the following statements, determine whether it is a valid hypothesis. Classify each valid hypothesis as a null hypothesis or an alternative hypothesis. Justify your answers.

  a. $\sigma^2 = 49.55$ **b.** IQR $= 25$ **c.** $\mu \neq 17$
  d. $\bar{y} = 100.7$ **e.** $Q_3 = 7.65$ **f.** $\mu < 33.79$
  g. $\sigma \neq 8.95$ **h.** $p = 0.77$

9.8 In each of the following problems, determine whether each pair of statements is a valid set of null and alternative hypotheses. Justify your answers.

  a. $H_0: p = 0.55;$ $H_a: p < 0.55$
  b. $H_0: \mu = 9.7;$ $H_a: \mu \geq 9.7$
  c. $H_0: \sigma^2 = 98.6;$ $H_a: \sigma^2 = 101$
  d. $H_0: \tilde{\mu} = 38.9;$ $H_a: \tilde{\mu} < 38.9$

9.9 In each of the following problems, determine whether each pair of statements is a valid set of null and alternative hypotheses. Justify your answers.

   a. $H_0: \mu = 30;$ $H_a: \mu \neq 30$

   b. $H_0: \sigma = 3.55;$ $H_a: \sigma > 3.55$

   c. $H_0: p \leq 0.32;$ $H_a: p > 0.32$

   d. $H_0: \bar{x} = 78.5;$ $H_a: \bar{x} \neq 78.5$

9.10 In each of the following problems, determine whether each pair of statements is a valid set of null and alternative hypotheses. Justify your answers.

   a. $H_0: p = 0.50;$ $H_a: p \neq 0.50$

   b. $H_0: \mu = 25.6;$ $H_a: \mu < 25.6$

   c. $H_0: \mu < 35.9;$ $H_a: \mu \geq 35.9$

   d. $H_0: \sigma^2 = 95;$ $H_a: \sigma^2 > 95$

9.11 In each of the following problems, a conclusion to a hypothesis test is presented. Determine whether each statement is permissible.

   a. The value of the test statistic does not lie in the rejection region. Therefore, we accept the null hypothesis.

   b. The value of the test statistic lies in the rejection region. Therefore, there is evidence to suggest the null hypothesis is not true.

   c. The value of the test statistic does not lie in the rejection region. Therefore, there is evidence to suggest the null hypothesis is true.

   d. The value of the test statistic lies in the nonrejection region. Therefore, there is no evidence to suggest the alternative hypothesis is true.

   e. The value of the test statistic does not lie in the rejection region. Therefore, there is no evidence to suggest the alternative hypothesis is true.

   f. The value of the test statistic lies in the rejection region. Therefore, there is evidence to suggest the alternative hypothesis is true.

Applications

9.12 Education and Child Development The College Board reported that the mean cumulative SAT score (Critical Reading, Mathematics, and Writing) for 2013 college-bound seniors was 1498.[3] Officials from the Pennsylvania State System of Higher Education would like to know whether the students enrolled for fall 2013 classes have a mean cumulative SAT score greater than 1498. A random sample of students from across the system is obtained, each cumulative SAT score is recorded, and the mean cumulative score is recorded. State the null and alternative hypotheses.

9.13 Marketing and Consumer Behavior A marketing research study conducted in 2013 indicated that 11% of all households in the United States have at least one telescope. During May 2013, Venus, Jupiter, and Mercury were visible in a near-perfect triangle formation. A company that manufactures telescopes believes the increased media attention caused more people to buy telescopes in order to see this spectacular sight. State the null and alternative hypotheses in terms of the population proportion of households that have at least one telescope.

9.14 Biology and Environmental Science During the summer months, wildfires in the western United States pose a great hazard to people, residential and commercial buildings, and animals. Previous records indicate that the mean number of acres burned during a wildfire is 17,060. The most recent summer was unusually wet, and firefighting officials would like to know whether the mean number of acres burned during wildfires was any less. State the null and alternative hypotheses.

9.15 Biology and Environmental Science Tourism officials in Palm Beach County, Florida, claim that the mean diameter of sand dollars found on Delray Beach is 4.25 centimeters. Some scientists claim that warmer water off the coast has inhibited growth of all sea life and that the diameter of sand dollars has decreased. State the null and alternative hypothesis.

9.16 Marketing and Consumer Behavior The stereotypical video-game player is a male teenager. However, recent studies suggest that at least 45% of all players are women.[4] A software company has decided to develop and market a sophisticated stock-market video game if the mean age of all video-game players is greater than 25. A random sample of players will be obtained, and the resulting data will be used to test the relevant hypothesis. Let μ represent the mean age of all video-game players. Which of the following sets of null and alternative hypotheses is appropriate? Justify your answer.

   a. $H_0: \mu = 25$ versus $H_a: \mu > 25$

   b. $H_0: \mu = 25$ versus $H_a: \mu < 25$

   c. $H_0: \mu = 25$ versus $H_a: \mu \neq 25$

9.17 Physical Sciences There are approximately 16 fiber-optic lines under the ocean off the Florida coast. These lines carry telephone and Internet communications between Florida and Europe, Latin America, and the Caribbean. During storms, these lines sway dramatically and often damage coral or get caught in anchors. Technical reports including the distance the line swayed indicate that the variance in sway during storms is 32 feet. A study will be conducted to determine whether a new, heavier cable housing will decrease the variance in sway. State the relevant null and alternative hypotheses in terms of σ^2, the variance in cable sway.

9.18 Technology and the Internet Many limousines now offer the latest in high-tech gadgets. DVD players, satellite radio, and wireless Internet connections are some of the high-priced accessories. Suppose Boston Coach will install wireless Internet in all of its vehicles if more than 75% of all clients desire this service. A random sample of clients will be selected, and the resulting information will be used to test the relevant hypothesis. Let p represent the proportion of all clients who would like wireless Internet in a limousine. State the null and alternative hypotheses in terms of p.

9.19 Sports and Leisure The mean duration of a Major League baseball game from the first pitch to the last out over the first eight weeks of the 2013 season was 2 hours and 58 minutes.[5] During the second half of the season, Major League umpires tried various techniques to speed up the game. A random sample of second-half games will be selected, and the resulting game durations will be used to determine whether games take less time to complete. Let μ be the mean duration of a baseball game. State the relevant null and alternative hypotheses in terms of μ.

9.20 Travel and Transportation DATTCO, a school bus company in New Britain, Connecticut, will install seat belts on all buses if more than 50% of all parents favor this change. A random sample of parents in the school district will be obtained, and the responses will be used to test the relevant hypothesis. If p is the true proportion of parents who favor seat-belt installation, which of the following sets of null and alternative hypotheses is appropriate? Justify your answer.
   a. $H_0: p = 0.50$ versus $H_a: p \neq 0.50$
   b. $H_0: p = 0.50$ versus $H_a: p < 0.50$
   c. $H_0: p = 0.50$ versus $H_a: p > 0.50$

9.21 Public Policy and Political Science Suppose Colby Coash, a state senator from Lancaster County, Nebraska, is considering a run for governor. He will enter the campaign if there is evidence to suggest that more than 65% of all state residents favor his candidacy. A random sample of likely voters is obtained, and the resulting survey data will be used to test the relevant hypothesis. Let p be the proportion of all voters who favor his candidacy. State the null and alternative hypotheses in terms of p.

9.22 Education and Child Development Students at Stetson University plan to ask administrators to build more on-campus housing on the DeLand campus if there is any evidence that the median monthly rent for off-campus housing is more than $350 per person. A random sample of students living off campus will be obtained, and the resulting rent data will be used to test the relevant hypothesis. Let $\tilde{\mu}$ be the median monthly rent for students. State the null and alternative hypotheses in terms of the population median $\tilde{\mu}$.

9.23 Travel and Transportation Commercial airline pilots often begin as flight engineers or first officers for smaller, regional airlines. Historically, newly hired pilots at major airline companies have approximately 2000 hours of flight experience. The FAA is conducting a study to determine if the mean flight experience of newly hired pilots has decreased. State the null and alternative hypotheses in terms of the population mean μ.

9.24 Medicine and Clinical Studies A new surgical procedure has been developed to remove cataracts; it involves a smaller incision. This more expensive procedure will be implemented at a major hospital only if there is evidence that the standard deviation in time to recovery is less than seven days. A random sample of patients will receive the new procedure, and the resulting recovery times will be used to test the relevant hypothesis. Let σ be the population standard deviation in recovery time. State the null and alternative hypotheses in terms of σ.

9.25 Public Policy and Political Science Officials in Dexter, a small town in upstate New York, have decided to install more fire hydrants in order to decrease insurance rates for many businesses. Prior to installing the new fire hydrants, the mean distance to a fire hydrant for downtown buildings was 525 feet. After the new installations, a random sample of downtown buildings will be obtained, and the distance to the nearest fire hydrant will be recorded. The data will be used to determine whether there is evidence that the mean distance to a fire hydrant has decreased. State the null and alternative hypotheses in terms of μ, the population mean distance to a fire hydrant.

9.26 Public Policy and Political Science The U.S. Health Insurance Portability and Accountability Act of 1996 (HIPAA) was designed to protect personal privacy. However, this law has created mountains of paperwork and may even increase the cost of medical care. The federal government has decided to consider repealing this law if more than 60% of all hospitals are experiencing increased costs due to this regulation. A random sample of hospitals will be obtained, and the resulting information will be used to test the relevant hypothesis. State the null and alternative hypotheses in terms of p, the population proportion of hospitals experiencing increased costs as a result of HIPAA.

9.27 Public Policy and Political Science The mean time from receipt of application to a final decision for a routine citizenship request in Canada is 25 months.[6] In an effort to shorten this lengthy process, the government of Canada has installed a sophisticated electronic record system and hired more staff. A study will be conducted to determine whether these actions have been effective. State the null and alternative hypothesis in terms of μ, the mean time from receipt of application to a final decision.

9.28 Public Policy and Political Science The Sparks City Council in Nevada will build a new dock at the Sparks Marina if more than 80% of the residents favor the plan. A random sample of residents will be obtained, and their responses will be used to test the relevant hypothesis. Let p be the population proportion of residents who favor a new marina. Which of the following sets of null and alternative hypotheses is appropriate? Justify your answer.
   a. $H_0: p = 0.80$ versus $H_a: p \neq 0.80$
   b. $H_0: p = 0.80$ versus $H_a: p < 0.80$
   c. $H_0: p = 0.80$ versus $H_a: p > 0.80$

9.29 Business and Management The Savannah Sugar Refinery in Louisiana will invest in new energy-saving devices if there is evidence to suggest that the mean amount of energy (in kilowatts) used per day will decrease from 1925. Suppose μ is the population mean amount of energy used per day, and an experiment is designed to test the new equipment. State the null and alternative hypotheses in terms of μ.

9.30 Economics and Finance Suppose U.S. monetary policy will be set by the Federal Reserve Board by examining the median consumer price, $\tilde{\mu}$, for a large fixed set of common commodities. The Board will raise the interest rate if there is evidence that the median price is less than $125.50. A random sample of counties will be obtained, and the cost of these goods in each county will be used to test the relevant hypothesis. State H_0 and H_a in terms of $\tilde{\mu}$.

9.31 Public Health and Nutrition The typical American adult male consumes approximately 2666 calories per day. New research suggests that those people who stay up late lose more sleep and consume more calories.[7] A random sample of 125 American adult males who routinely stay up late was obtained, and the number of calories per day consumed by each was carefully recorded. Let μ represent the mean number of calories consumed per day by an American adult male. State the relevant null and alternative hypotheses in terms of μ.

9.2 Hypothesis Test Errors

To conduct a hypothesis test, we use the information in a sample to reach a decision about the value of a population parameter. The sample data lead to a value of the test statistic and the ultimate decision (reject or do not reject H_0). However, there is always a chance of making a mistake (the wrong decision). Even a simple random sample is only a (usually small) portion of the entire population. This limited information could lead to the wrong conclusion about the population parameter. To fully understand the structure of a hypothesis test, we need to examine what could possibly go wrong.

The water in Lake Jean, located in Ricketts Glen State Park, Pennsylvania, is clear and cool, sometimes *very* cool. Park officials have decided to post an advisory warning for swimmers if there is any evidence to suggest that the mean water temperature, μ, is less than 62°F. Otherwise, the lifeguards will post no signs and allow swimming as usual. Each day, lifeguards measure the temperature of the water (in °F) at 15 randomly selected locations around the lake. They use the sample mean water temperature, $\bar{x}$, to test the hypotheses

In this hypothesis test, $\overline{X}$ is the test statistic.

$$H_0: \mu = 62 \qquad \text{versus} \qquad H_a: \mu < 62$$

The lifeguards have decided to use a *cutoff point* of 61.5°F. They cannot measure the temperature of the water at every location in the lake (and hence compute the population mean), and remember, the sample mean varies around the population mean. The lifeguards have decided that a sample mean of $\bar{x} \leq 61.5$ is far enough away from 62 that it cannot be attributed to ordinary variation about the population mean.

The rejection region is any value of $\bar{x}$ less than or equal to 61.5 (see Figure 9.2). If the value of $\bar{x}$ is 61.5 or less, then H_0 is rejected, and an advisory warning is posted. If the value of $\bar{x}$ is greater than 61.5, the lifeguards cannot reject H_0. There is no evidence to suggest that the mean water temperature is less than 62°F.

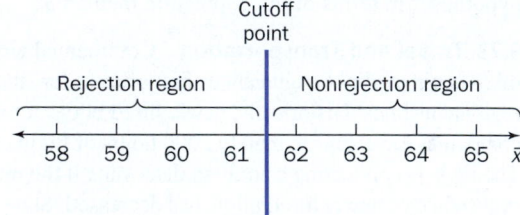

Figure 9.2 The rejection region and nonrejection region for the Lake Jean water-temperature hypothesis test.

Here is what can happen when the sample data are collected and the hypothesis is tested.

This is all very theoretical. You never really know whether you made a mistake in a hypothesis test.

1. Suppose the true population mean μ is 62°F or greater; the water is warm enough for swimming.

   a. If $\bar{x} > 61.5°F$, then there is no evidence to reject H_0. This conclusion is *correct*.

   b. If $\bar{x} \leq 61.5°F$, then there is evidence to reject H_0. An advisory warning is posted. This conclusion is *incorrect*. The water really is warm enough for swimming.

2. Suppose the true population mean μ is less than 62°F; the water is too cold for swimming.

   a. If $\bar{x} > 61.5°F$, then there is no evidence to reject H_0. This conclusion is *incorrect*. The water is really too cold for swimming.

   b. If $\bar{x} \leq 61.5°F$, then there is evidence to reject H_0. An advisory warning is posted. This conclusion is *correct*.

This example illustrates the two possible errors in a hypothesis test.

Definition

1. The value of the test statistic may lie in the rejection region, but the null hypothesis is true. If we reject H_0 when H_0 is really true, this is called a **type I error**. The probability of a type I error is called the *significance level* of the hypothesis test and is denoted by α: P(type I error) = α.

2. The value of the test statistic may not lie in the rejection region, but the alternative hypothesis is true. If we do not reject the null hypothesis when H_a is really true, this is called a **type II error**. The probability of a type II error is denoted by β: P(type II error) = β.

The following table illustrates the decisions and errors in a hypothesis test.

		Decision	
		Reject H_0	Do not reject H_0
Truth	H_0	Type I error	Correct decision
	H_a	Correct decision	Type II error

Each example below describes a specific hypothesis test. The type I error and the type II error are described in context, and the real-world consequences of a wrong decision are given.

Example 9.4 Defective Solar Panels

Many utility companies in the United States have constructed new power plants using solar panels. As a result of increased competition, the cost of solar panels has decreased significantly since 2009. However, now heavily subsidized solar panel manufacturers in China are under extreme pressure to cut costs and substitute less expensive material. There is little quality control. One company in Shanghai has a known defect rate of 0.078.[8] Florida Power and Light company has ordered a large number of solar panels from this company and is willing to accept the known defect rate or lower.

A random sample of solar panels will be obtained and carefully inspected. The information in the sample will be used to test the hypotheses

$$H_0: p = 0.078 \qquad \text{versus} \qquad H_a: p > 0.078$$

There are two possible truth assumptions, as indicated in the table above: H_0 is true or H_a is true.

If H_0 is rejected, the entire shipment of solar panels will be sent back to the company in Shanghai. Discuss the consequences of the decision to reject or not to reject for each truth assumption.

SOLUTION

STEP 1 Suppose H_0 is true: The proportion of defective solar panels is really 0.078 (or less).

   a. If H_0 is rejected, the entire solar panel shipment will be sent back. This is a type I error, and in this case the company in Shanghai will not be happy. The solar panels are of acceptable quality, but the entire shipment is returned.

   b. If H_0 is not rejected, then the solar panel shipment is accepted. Everyone is happy in this case. The hypothesis test indicated no evidence of a higher proportion of defective solar panels.

STEP 2 Suppose H_a is true: The proportion of defective solar panels is *greater than* 0.078.

   a. If H_0 is rejected, the entire solar panel shipment will be returned. This is the correct conclusion. The hypothesis test indicated the solar panel shipment contained a higher proportion of defectives, and Florida Power and Light will be glad to return a bad batch.

b. If H_0 is not rejected, then the solar panel shipment will be accepted. This is a type II error, and in this case Florida Power and Light will not be happy. Too many solar panels are defective, yet Florida Power and Light will accept a shipment of poor quality.

TRY IT NOW   GO TO EXERCISE 9.43

Example 9.5 Take Two Aspirin

Approximately 50 million people in the United States take an aspirin daily to treat or prevent heart disease.[9] Suppose Walgreens receives a large shipment of Bayer aspirin, in which each tablet should contain 80 mg of acetylsalicylic acid, the active ingredient in aspirin. A random sample of tablets will be obtained and chemically analyzed. The information in the sample will be used to test the hypotheses

$$H_0: \mu = 80 \qquad \text{versus} \qquad H_a: \mu \neq 80$$

where μ is the mean amount of acetylsalicylic acid in each tablet. If H_0 is rejected, the entire aspirin shipment will be returned to the Bayer corporation. Discuss the consequences of the decision to reject or not to reject for each truth assumption.

SOLUTION

STEP 1 Suppose H_0 is true: The mean amount of the active ingredient is 80 mg.

    a. If H_0 is rejected, the aspirin shipment is returned to Bayer. This is a type I error, and Bayer will not like this decision. The aspirin has the specified amount of acetylsalicylic acid, but the entire shipment is returned.

    b. If H_0 is not rejected, then the aspirin bottles are placed on the shelves at Walgreens. Everyone is happy. The aspirin contains the correct amount of the active ingredient, and the hypothesis test did not indicate otherwise.

STEP 2 Suppose H_a is true: The mean amount of active ingredient is not 80 mg.

    a. If H_0 is rejected, the aspirin shipment is returned to Bayer. This is the correct conclusion. The hypothesis test indicated the mean amount of acetylsalicylic acid is different from 80 mg, and Walgreens is happy to return a flawed shipment.

    b. If H_0 is not rejected, the aspirin bottles are placed on the shelves at Walgreens, ready for sale. This is a type II error, and in this case Walgreens may encounter some very unhappy customers. The mean amount of the active ingredient is not as specified, yet the aspirin is placed in store stock and sold.

TRY IT NOW   GO TO EXERCISE 9.45

Example 9.6 Cold and Rolled

Remember, H_0 is stated in terms of an equality, but there is an implied inequality in the null hypothesis. The manufacturing process will be stopped only if there is evidence that the population variance is greater than 0.04. The status quo or no-change state is really $\sigma^2 \leq 0.04$.

The Nucor Corporation manufactures steel in cold-rolled sheets for use in a wide variety of products. The steel is produced to have a certain thickness, but the focus in the manufacturing process is on the variance in thickness. Each hour, a random sample of cold-rolled sheets of 10-gauge steel is obtained and the thickness of each sheet (in inches) is carefully measured. The sample variance is used to test the hypotheses

$$H_0: \sigma^2 = 0.04 \qquad \text{versus} \qquad H_a: \sigma^2 > 0.04$$

where σ^2 is the population variance in sheet metal thickness. If H_0 is rejected, the entire manufacturing process is shut down for inspection. Discuss the consequences of the decision to reject or not to reject for each truth assumption.

SOLUTION

STEP 1 Suppose H_0 is true: The population variance is 0.04 inch (or less).

  a. If H_0 is rejected, the manufacturing process is shut down. This is a type I error. The null hypothesis is true, but H_0 is rejected. This error is bad for Nucor. The process is stopped unnecessarily, and production time is lost.

  b. If H_0 is not rejected, then the facility continues to manufacture metal sheets. This is a correct decision.

STEP 2 Suppose H_a is true: The population variance in thickness is greater than 0.04 inch.

  a. If H_0 is rejected, the correct decision has been made. The variance is too high and the hypothesis test procedure suggested that the manufacturing process should be shut down for inspection.

  b. If H_0 is not rejected, the facility continues to hum along. This is an incorrect decision, a type II error. H_a is true, but the null hypothesis is not rejected. In this case, there is a good chance the metal sheets are being made with too much variability in thickness.

TRY IT NOW   GO TO EXERCISE 9.52

A *perfect* hypothesis test would have the probability of a type I error and the probability of a type II error both equal to 0.

The *efficiency* or *goodness* of a hypothesis test is often measured in terms of α and β. Intuitively, an effective test should have the probability of both types of errors small. In fact, because no one wants to make a mistake, it would be ideal to have $\alpha = \beta = 0$. However, this is impossible, because any decision in a hypothesis test is made using information in a *sample*. We can never examine the entire population, so there is always a chance of making a mistake.

α and β are inversely related, but not by any set formula. We know only that as α decreases, β increases; and as β decreases, α increases.

▶ In a hypothesis test, we usually control the value of α. We do not have any direct control over β, but we can often compute the probability of a type II error for a specific alternative value of the parameter. It seems reasonable to set α as small as possible, but for a fixed sample size, α and β are inversely related. For a fixed n, making α smaller forces β to be larger. The only way to decrease both α and β (simultaneously) is to increase the sample size. More information (larger n) means less of a chance of making a mistake. The most common values for α are 0.05 and 0.01. The probability of a type II error actually *depends* on the specific alternative value of the population parameter being tested. Therefore, the probability of a type II error is usually written as a *function* of the population parameter. For example, $\beta(\mu_a)$ represents the probability of a type II error if the true value of μ is μ_a.

Consider a hypothesis test with $H_0: \theta = \theta_0$ and $H_a: \theta \neq \theta_0$. As the alternative value of the parameter, θ_a, moves farther away from the hypothesized value, θ_0, the probability of a type II error, $\beta(\theta_a)$, decreases. This seems reasonable because for values of θ_a far away from θ_0, we have a better chance of detecting the difference between the hypothesized value and the alternative value of the population parameter.

Later in this chapter, we will visualize and learn how to compute the probability of a type II error for a specific alternative value of the population characteristic. ◀

SECTION 9.2 EXERCISES

Concept Check

9.32 Fill in the Blank
  a. If we reject H_0 when H_0 is really true, this is called a _____.

  b. If we do not reject the null hypothesis when H_a is true, this is called a _____.

9.33 True/False   Under certain circumstances, the probability of a type I error and the probability of a type II error can be 0.

9.34 True/False   In a hypothesis test, we always know if we made an error in the conclusion.

9.35 True/False   In a hypothesis test, we never really know if we have made the correct decision.

9.36 True/False   In a hypothesis test, for a fixed sample size, α and β are inversely related.

Practice

9.37 Consider a hypothesis test with

$$H_0: \mu = 180 \quad \text{and} \quad H_a: \mu < 180$$

Determine whether each of the following decisions is correct or in error. Identify each error as type I or type II.
  a. The true value of μ is 180 and H_0 is rejected.
  b. The true value of μ is 179 and H_0 is rejected.
  c. The true value of μ is 160 and H_0 is not rejected.
  d. The true value of μ is 182 and H_0 is rejected.

9.38 Consider a hypothesis test with

$$H_0: p = 0.44 \quad \text{and} \quad H_a: p \neq 0.44$$

Determine whether each of the following decisions is correct or in error. Identify each error as type I or type II.
  a. The true value of p is 0.44 and H_0 is not rejected.
  b. The true value of p is 0.41 and H_0 is not rejected.
  c. The true value of p is 0.45 and H_0 is not rejected.
  d. The true value of p is 0.42 and H_0 is rejected.

9.39 Consider a hypothesis test with

$$H_0: \sigma^2 = 26.5 \quad \text{and} \quad H_a: \sigma^2 > 26.5$$

Determine whether each of the following decisions is correct or in error. Identify each error as type I or type II.
  a. The true value of σ^2 is 26.0 and H_0 is rejected.
  b. The true value of σ^2 is 27.0 and H_0 is not rejected.
  c. The true value of σ^2 is 26.4 and H_0 is not rejected.
  d. The true value of σ^2 is 26.5 and H_0 is not rejected.

9.40 Recall that the probability of a type II error depends on the alternative specific value of the population parameter. Consider a hypothesis test with

$$H_0: \mu = 10 \quad \text{and} \quad H_a: \mu > 10$$

  a. For a fixed sample size, how do $\beta(11)$ and $\beta(15)$ compare? Are these two values approximately the same or different? If they are different, which is smaller, and why?
  b. What happens to $\beta(\mu_a)$ as μ_a increases (gets farther and farther away from 10)?

9.41 Why is there always a chance of making a mistake (an incorrect decision) in any hypothesis test?

9.42 Because we usually control the value of α, one could simply set the probability of a type I error to a very small value, say 0.0001. What's wrong with this strategy?

Applications

9.43 Travel and Transportation The 30-mile I-287 corridor near Tarrytown, New York, is heavily traveled and is a major interstate transportation link. The Tappan Zee Bridge is part of this road network and is in need of structural repairs. Approximately 140,000 vehicles cross this bridge every day.[10] Transportation officials have decided to conduct a hypothesis test and will raise tolls to fund planned repairs if there is

evidence to suggest that the mean number of cars per day using this bridge has increased.
  a. Write the null and alternative hypotheses about μ, the mean number of cars per day that cross the Tappan Zee Bridge.
  b. For the hypotheses in part (a), describe the type I and type II errors.
  c. If a type I error is committed, who is more angry, the transportation officials or drivers, and why?
  d. If a type II error is committed, who is more angry, the transportation officials or drivers, and why?

9.44 Public Policy and Political Science The Avenal State Prison in California has thousands of files on former inmates, an estimated 20 million pages of documents. A proposal has been made to archive many of these old files to compact discs. The state will release money for this project only if prison officials present evidence to suggest the true mean age of all files is greater than 10 years. A random sample of files will be obtained, and the information in the sample will be used to test the hypotheses

$$H_0: \mu = 10 \quad \text{versus} \quad H_a: \mu > 10$$

  a. Describe a type I error and a type II error in this context.
  b. From the warden's perspective, which error is more serious? Why?
  c. From a state senator's point of view, which error is more serious? Why?

9.45 Physical Sciences The annual Waikiki Roughwater Swim contest is held over a 2.4-mile course and ends near the Hilton Rainbow Tower. The 2013 winner was Rhys Mainstone in 48 minutes, 10 seconds.[11] Before the race, a random sample of the current velocity (in knots) along the race course is obtained, and the resulting information is used to determine whether the race should be canceled. A mean current velocity, μ, of more than 0.65 knots is considered unsafe.
  a. State the null and alternative hypotheses.
  b. Describe type I and type II errors in this context.
  c. Which error is more serious for the swimmers? Why?
  d. Which error is more serious for the race organizers? Why?

9.46 Education and Child Development A Harris Interactive Poll for Tylenol PM indicated that 8% of American students have missed a test because they overslept. Suppose officials at a state college decide to apply a new academic policy regarding exams. If a student arrives late for a test, then the student receives a 0 for the test and cannot take a make-up test at a later time.
  a. What hypotheses should be tested, in terms of p, the true proportion of students who have missed a test because they overslept, in order for college officials to prove that the new academic policy is causing fewer students to be late for exams?
  b. Which error, type I or type II, is more serious for college officials? Why?
  c. Which error, type I or type II, is more serious for students? Why?

9.47 Medicine and Clinical Studies Recently, researchers have speculated that eating a moderate amount of dark chocolate may increase the level of antioxidants, compounds that protect us against free radicals, which can cause heart disease and cancer.[12] A study was designed, and the antioxidant concentration in patients who ate dark chocolate regularly was measured. This information was used to test the hypotheses

$$H_0: \mu = 0.4 \qquad \text{versus} \qquad H_a: \mu > 0.4$$

where μ is the true mean percentage of antioxidants in the bloodstream.

a. Describe a type I error and a type II error in this context.

b. The Hershey Company, maker of Hershey's chocolates, is very interested in the results of this study. Which error is more serious for this company? Why?

9.48 Public Policy and Political Science The Dallas, Texas, city council is going to consider a zoning variance for the Estates on Frankford apartment complex so that the developer may extend the structure closer to the nearest road. Some council members are concerned about safety, but they will vote for the measure if more than 60% of all city residents favor the variance. A random sample of residents will be obtained and asked whether they favor the zoning variance. Let p be the true proportion of residents who favor the extended structure.

a. State the null and alternative hypotheses in terms of p.

b. Describe a type I error and a type II error in this context.

c. Which error is more serious for the developer? Why?

d. Which error is more serious for the city council members? Why?

9.49 Manufacturing and Product Development The National Fire Protection Association maintains an extensive list of codes and standards that address equipment, extinguishing systems, and fire-protective gear.[13] A firefighter's helmet should withstand a temperature of 1200°F for 30 seconds. The Seattle Fire Department is ordering helmets from Kidde Fire Fighting. They will accept the shipment only if there is no evidence to suggest the helmets fail to meet the standard. Let μ be the mean temperature the helmets can withstand.

a. State the null and alternative hypothesis.

b. Describe a type I error and a type II error in the context of this problem.

9.50 Economics and Finance A shadow chief secretary to the treasury claims that first-time home buyers in the South Downs area (near Worthing in Sussex, England) are paying (on average) more than £1367 in stamp duty. The shadow chief secretary believes this steep tax is prohibiting families from buying a first home in this area. A random sample of first-time home buyers will be obtained, and the stamp duty paid by each family will be recorded. These data will be used to determine whether there is any evidence the mean stamp duty for first-time home buyers, μ, is more than £1367.

a. State the null and alternative hypotheses in terms of μ.

b. Describe a type I error and a type II error in this context.

9.51 Psychology and Human Behavior Recently a panel of Canadian hotel industry IT executives reported on hotel technology trends. Although 22–30% of guests shop for a room using a mobile device, only 2% actually book a room using this technology.[14] Many shoppers still feel insecure completing a transaction with a credit card via a mobile device. Executives at Alt Hotels have decided to conduct a survey to determine whether hotel guests are becoming more comfortable booking a room using a mobile device. If the proportion ($p = 0.02$) has increased, they intend to invest more money in this technology.

a. State the null and the alternative hypothesis.

b. Describe a type I error and a type II error in this context.

9.52 Medicine and Clinical Studies An electroencephalogram (EEG) is often used to measure brain, or neural, activity expressed as electrical voltage. Research suggests that transcendental meditation (TM) produces a simplified state of rest and relaxation resulting in an EEG with smaller than normal variance. Suppose normal brain activity has variance $\sigma^2 = 15$ volts^2. A random sample of patients who can achieve TM will be selected, and their brain activity will be measured during TM. The resulting information will be used to test the research theory.

a. State the null and alternative hypotheses in terms of σ^2.

b. Describe a type I and a type II error in this context.

c. If there is evidence to suggest TM decreases the variance in brain activity, then the National Science Foundation (NSF) will commit more money for TM research. Which error is more serious for the NSF? Why? Which error is more serious for TM researchers? Why?

Extended Applications

9.53 Marketing and Consumer Behavior Many families have decided to use a satellite dish instead of traditional cable TV. Satellite-dish companies offer a wide variety of premium channels and popular sports packages. The local cable TV provider in Neodesha, Kansas, Cable ONE, is concerned about losing market share and is going to conduct a hypothesis test to determine whether more advertising is needed. A random sample of homes in the city will be obtained, and the data will be used to determine whether there is any evidence that the true proportion of homes with satellite dishes is greater than 0.25.

a. State the null and alternative hypotheses in terms of p.

b. Describe type I and type II errors in this context.

c. For a fixed sample size, which is smaller, $\beta(0.27)$ or $\beta(0.35)$? Justify your answer.

9.54 Sports and Leisure South Carolina Department of Natural Resources personnel enforce hunting and fishing regulations and conduct routine safety checks on recreational boats. Past experience indicates that 15% of all boats inspected have at least one safety violation. Recent accidents on lakes in South Carolina have prompted calls for more extensive inspections. A random sample of recreational boats will be obtained and inspected. If there is evidence that the true proportion of boats with safety violations, p, is more than 15%, a methodical inspection of every boat launched at popular sites will be started.

a. State the null and alternative hypotheses in terms of p.

b. Describe a type I error and a type II error in this context.

c. What happens to the probability of a type I error as the value of p approaches 0.15 from the right (that is, 0.20, 0.19, . . .)?

9.55 Physical Sciences Civil engineers are going to test a highway bridge outside of Washington, D.C., for structural integrity. A random sample of locations along the bridge will be selected, and the concrete stress (in pounds per square inch, psi) will be measured. These data will be used to determine the safety of the bridge.

a. Suppose 6400 psi is considered a safe concrete stress level. Which pair of hypotheses should be tested?

$H_0: \mu = 6400$ versus $H_a: \mu > 6400$

or

$H_0: \mu = 6400$ versus $H_a: \mu < 6400$

b. If you regularly drive over the bridge being tested, would you prefer $\alpha = 0.1$ or $\alpha = 0.01$? Why?

9.56 Medicine and Clinical Studies The failure rate of hip implants in men is approximately $p = 0.019$, and this proportion is even higher for women.[15] A new implant design by researchers at the Missouri Bone & Joint Research Foundation uses a thinner plastic and will hopefully cause fewer complications and further surgeries. A long-term study was conducted to examine the failure rate of the new implant design in men.

a. State the null and alternative hypotheses in terms of p.
b. Describe a type I error and a type II error in this context.
c. For a fixed sample size, which is smaller, $\beta(0.015)$ or $\beta(0.010)$?
d. If you have invested in the company that manufactures this new implant, would you prefer $\alpha = 0.1$ or $\alpha = 0.01$? Why?

9.3 Hypothesis Tests Concerning a Population Mean When σ Is Known

In the previous sections and chapters, all of the statistics tools have been provided to construct a formal hypothesis test. Recall, every hypothesis test consists of four parts, and the null hypothesis, an equality stated in terms of a population parameter, represents the current state, i.e., what is assumed to be true. The following example is used to develop a hypothesis test about a population mean μ when σ is known.

Example 9.7 Patient Triage

Hospital emergency rooms across the country are experiencing shortages of doctors and nurses, and have too few beds. These constraints make it difficult to treat patients in a timely manner. University Hospital in Syracuse, New York, which treats approximately 58,000 patients in its emergency room each year, decided to address this issue by moving into the waiting room to treat patients, similar to a MASH unit. Prior to this experiment, the mean time to treat *very ill* patients (as opposed to critically ill patients or those with a minor injury) entering the emergency room was 20 minutes (with standard deviation $\sigma = 5.0$ minutes). During the waiting-room experiment, a random sample of 36 very ill patients was selected and the time to treatment for each was recorded. The sample mean time was $\bar{x} = 16.1$ minutes. Conduct a hypothesis test to determine whether there is any evidence to suggest the waiting-room experiment reduced the mean time to treatment for very ill patients. Use $\alpha = 0.05$.

An Army MASH unit is a mobile surgical hospital, usually deployed close to battle lines. It is designed to provide early surgical procedures and to stabilize critically wounded soldiers.

SOLUTION

STEP 1 Let μ be the mean time to treatment for very ill patients entering the emergency room. The current state, or what is assumed to be true, involves time to treatment prior to the waiting-room experiment. Therefore, the null hypothesis is

$H_0: \mu = 20$

STEP 2 The chairman of the Department of Emergency Medicine for University Hospital hopes doctors going directly to patients in the waiting room will *decrease* time to treatment for very ill patients. The alternative hypothesis is

$H_a: \mu < 20$

STEP 3 We need to know whether 16.1 minutes is a reasonable observation under the null hypothesis, or whether it is *too far away* from the assumed mean, 20 minutes. Because H_0 is assumed to be true, and $n = 36 \geq 30$, by the central limit theorem, $\overline{X}$ is approximately normal with mean $\mu = 20$ and standard deviation $\sigma/\sqrt{n}$. Although $\overline{X}$ could be used as the test statistic, it's easier to identify unlikely values of a standard normal random variable. Therefore, the appropriate test statistic is obtained by standardizing,

$$\text{TS: } Z = \frac{\bar{X} - 20}{\sigma/\sqrt{n}}$$

If H_0 is true, then Z is approximately standard normal.

STEP 4 The alternative hypothesis is one-sided, left-tailed, so unusual values of the test statistic, Z, are in the left tail of the distribution. We should reject the null hypothesis if the value of Z is to the left of some cutoff value, or endpoint. The cutoff value is determined so that the probability of a type I error is 0.05. Select the cutoff value such that if H_0 is true,

$$P(Z \le \text{cutoff value}) = 0.05$$

Using the definition of a Z critical value (from Section 8.2),

$$P(Z \le -z_{0.05}) = 0.05 \quad \text{and} \quad -z_{0.05} = -1.6449 \qquad \text{Table III in the Appendix.}$$

> In interval notation, the rejection region is $(-\infty, -1.6449]$.

The rejection region is written as

$$\text{RR: } Z \le -z_{0.05} = -1.6449$$

If the value of Z is less than or equal to the critical value -1.6449, then the observed value of $\bar{X}$ is considered rare, and we reject the null hypothesis. Figure 9.3 illustrates the critical value and the rejection region for this hypothesis test.

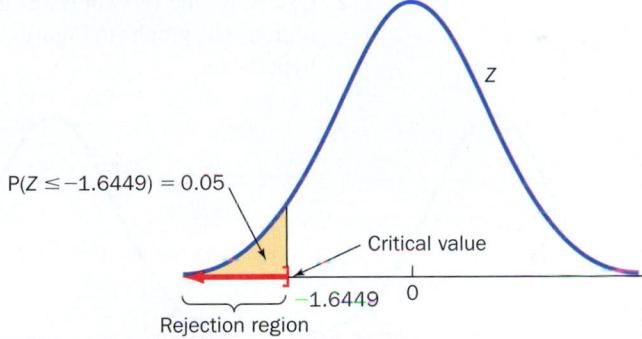

$P(Z \le -1.6449) = 0.05$

> The critical value divides the z measurement axis into two parts: the rejection region and the nonrejection region.

Figure 9.3 The critical value and the rejection region for the emergency-room hypothesis test.

STEP 5 The value of the test statistic is

> The small letters here (z and $\bar{x}$) represent actual values.

$$z = \frac{\bar{x} - 20}{\sigma/\sqrt{n}} = \frac{16.1 - 20}{5.0/\sqrt{36}} = -4.68$$

The value $\bar{x} = 16.1$ is 4.68 standard deviations to the left of the mean—a very unusual observation if the null hypothesis is really true. This means either 16.1 is an incredibly lucky observation or the assumption ($\mu = 20$) is wrong. As usual, we discount the lucky possibility.

More formally, because $z = -4.68$ (≤ -1.6449) lies in the rejection region, we reject the null hypothesis at the $\alpha = 0.05$ level of significance. There is evidence to suggest the true population mean time to treatment, μ, is less than 20 minutes. ●

In any hypothesis test, as in the example above, we assume H_0 is true and consider the likelihood of the sample outcome (expressed as a single value of the test statistic).

1. If the value of the test statistic is reasonable under the null hypothesis, then we cannot reject H_0.

2. If the value of the test statistic is unlikely under the null hypothesis, then we reject H_0.

As presented in Section 9.1, there are three possible alternative hypotheses: one two-sided alternative and two one-sided alternatives. The rejection region depends on the alternative hypothesis. The general procedure for a hypothesis test concerning μ is summarized below.

Hypothesis Tests Concerning a Population Mean When σ Is Known

Use this as a *template* for a hypothesis test about a population mean when σ is known.

Given a random sample of size n from a population with mean μ, assume

1. The underlying population is normal or n is large, and
2. The population standard deviation σ is known.

A hypothesis test about the population mean μ with significance level α has the form:

$H_0: \mu = \mu_0$

$H_a: \mu > \mu_0, \qquad \mu < \mu_0, \qquad$ or $\qquad \mu \neq \mu_0$

The rejection region *always* includes the endpoint of the (infinite) interval.

TS: $Z = \dfrac{\bar{X} - \mu_0}{\sigma/\sqrt{n}}$

RR: $Z \geq z_\alpha, \qquad Z \leq -z_\alpha, \qquad$ or $\qquad |Z| \geq z_{\alpha/2}$

A CLOSER LOOK

1. μ_0 is a fixed, hypothesized value of the population mean μ.
2. Use only one (appropriate) alternative hypothesis and the corresponding rejection region. The graphs in Figures 9.4–9.6 illustrate the rejection region for each alternative hypothesis.

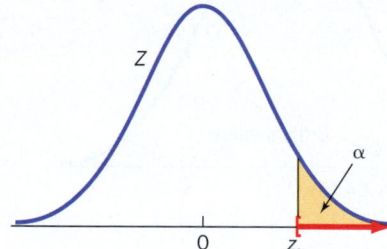

Figure 9.4 Rejection region for $H_a: \mu > \mu_0$.

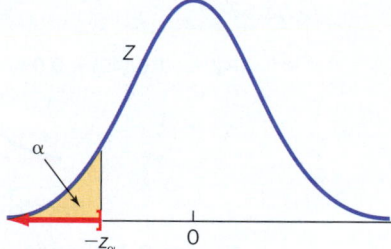

Figure 9.5 Rejection region for $H_a: \mu < \mu_0$.

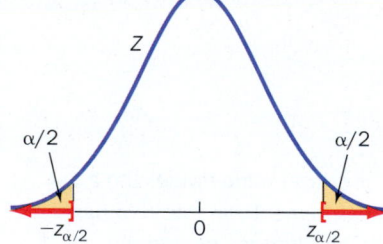

Figure 9.6 Rejection region for $H_a: \mu \neq \mu_0$.

3. For a two-sided alternative hypothesis, the rejection region $|Z| \geq z_{\alpha/2}$ (written using the absolute value of Z) is simply a shorthand way to write $Z \geq z_{\alpha/2}$ or $Z \leq -z_{\alpha/2}$.
4. For a given significance level, the corresponding critical value is found using Table III in the Appendix, backward. This procedure was presented in Chapter 8. Common values for α are 0.05, 0.025, and 0.01.
5. The hypothesis test procedure described above can be used *only* if σ is known. If n is large and σ is unknown, some statisticians suggest using the sample standard deviation s in place of σ. This produces an *approximate* test statistic. Section 9.5 presents an *exact* test procedure concerning μ when σ is unknown. ●

The following examples illustrate this hypothesis test procedure.

Example 9.8 The Dead Zone

The biological dessert in the Gulf of Mexico called the Dead Zone is a region in which there is very little or no oxygen. Most marine life in the Dead Zone dies or leaves the region. The area of this region varies and is affected by agriculture, fertilizer runoff, and

Solution Trail 9.8

KEYWORDS

- Is there any evidence?
- Greater than the long-term mean
- Standard deviation 1850
- Random sample

TRANSLATION

- Conduct a one-sided, right-tailed test about a population mean μ
- $\mu_0 = 5960$
- $\sigma = 1850$

CONCEPTS

- Hypothesis test concerning a population mean when σ is known

VISION

Use the template for a one-sided, right-tailed test about μ. The underlying population distribution is unknown, but n is large and σ is known. Determine the appropriate alternative hypothesis and the corresponding rejection region, find the value of the test statistic, and draw a conclusion.

weather. The long-term mean area of the Dead Zone is 5960 square miles.[16] As a result of recent flooding in the Midwest and subsequent runoff from the Mississippi River, researchers believe that the Dead Zone area will increase. A random sample of 35 days was obtained, and the sample mean area of the Dead Zone was 6759 mi^2. Is there any evidence to suggest that the current mean area of the Dead Zone is greater than the long-term mean? Assume that the population standard deviation is 1850 and use $\alpha = 0.025$.

SOLUTION

STEP 1 The current state, or assumed mean, is $\mu = 5960 \; (= \mu_0)$.

The sample size is $n = 35$; $\sigma = 1850$ and $\alpha = 0.025$.

We are looking for evidence that the current mean area of the Dead Zone is *greater* than the long-term mean. Therefore, the alternative hypothesis is one-sided, right-tailed.

STEP 2 The four parts of the hypothesis test are

H_0: $\mu = 5960$

H_a: $\mu > 5960$

TS: $Z = \dfrac{\overline{X} - \mu_0}{\sigma/\sqrt{n}}$

RR: $Z \geq z_\alpha = z_{0.025} = 1.96$

STEP 3 The value of the test statistic is

$$z = \frac{\overline{x} - \mu_0}{\sigma/\sqrt{n}} = \frac{6759 - 5960}{1850/\sqrt{35}} = 2.555 \geq 1.96$$

STEP 4 Because 2.555 lies in the rejection region, we reject the null hypothesis at the $\alpha = 0.025$ level. There is evidence to suggest the current mean area of the Dead Zone is greater than 5960 mi^2.

Figures Figure 9.7–Figure 9.9 show a technology solution.

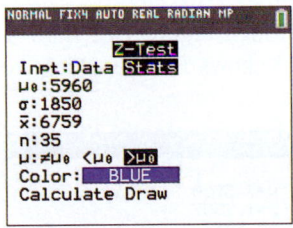

Figure 9.7 TI-84 Plus C Z-Test input screen.

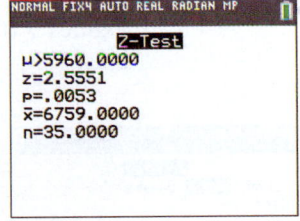

Figure 9.8 TI-84 Plus C Z-Test Calculate results.

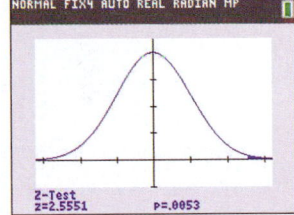

Figure 9.9 TI-84 Plus C Z-Test Draw results.

VIDEO TECH MANUALS

ONE MEAN TEST - *z* - SUMMARIZED DATA

DATA SET

WHTCELLS

VIDEO TECH MANUALS

ONE MEAN TEST - *z* - WITH DATA

TRY IT NOW   GO TO EXERCISE 9.74

Example 9.9 Natural Defense

White blood cells are the body's natural defense mechanism against disease and infection. The mean white blood cell count in healthy adults, measured as part of a CBC (complete blood count), is approximately $7.5 \times 10^3/\mu$l.[17] A company developing a new drug to treat arthritis pain must check for any side effects. A random sample of patients using the new drug was selected, and the white blood cell count of each patient was measured. The results are given in the table below ($\times 10^3/\mu$l).

6.50	8.69	6.85	6.76	6.58	8.84	8.44	8.28	7.65	6.95	10.12
8.74	8.00	8.84	7.93	7.65	7.00	6.70	9.20	6.45	7.66	

Assume the distribution of white blood cell counts is normal and $\sigma = 1.1$. Conduct a hypothesis test to determine whether there is any change in the mean white blood cell count due to the arthritis drug. Use $\alpha = 0.01$.

SOLUTION

STEP 1   The assumed mean is $\mu = 7.5 \,(= \mu_0)$; the sample size is $n = 21$; $\sigma = 1.1$; and $\alpha = 0.01$.

The company is looking for *any* change in the mean white blood cell count. Therefore, the relevant alternative hypothesis is two-sided.

The sample size is small ($n < 30$), but the population is assumed to be normal. The hypothesis test concerning a population mean when σ is known can be used.

STEP 2   The four parts of the hypothesis test are

H_0: $\mu = 7.5$

H_a: $\mu \neq 7.5$

TS: $Z = \dfrac{\overline{X} - \mu_0}{\sigma/\sqrt{n}}$

RR: $|Z| \geq z_{\alpha/2} = z_{0.005} = 2.5758$ ($Z \leq -2.5758$ or $Z \geq 2.5758$)

STEP 3   The sample mean is

$$\overline{x} = \frac{1}{21}(6.50 + 8.69 + \cdots + 7.66) = 7.8014$$

The value of the test statistic is

$$z = \frac{\overline{x} - \mu_0}{\sigma/\sqrt{n}} = \frac{7.8014 - 7.5}{1.1/\sqrt{21}} = 1.2556$$

STEP 4   The value of the test statistic, $z = 1.2556$, does *not* lie in the rejection region. We do not reject the null hypothesis at the $\alpha = 0.01$ level of significance. There is no evidence to suggest the new arthritis drug has changed the mean white blood cell count.

Figures 9.10 through Figure 9.12 show a technology solution using the TI-84 plus; Figure 9.13 shows JMP Test Mean results.

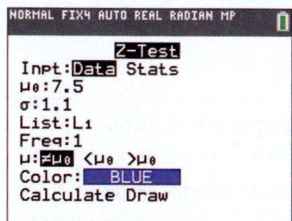

Figure 9.10 TI-84 Plus C input screen.

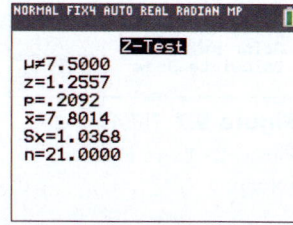

Figure 9.11 TI-84 Plus C Z-Test Calculate results.

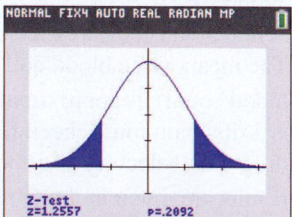

Figure 9.12 TI-84 Plus C Z-Test Draw results.

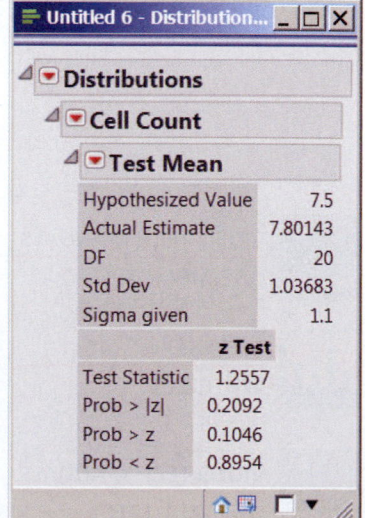

Figure 9.13 JMP Test Mean results.

TRY IT NOW   GO TO EXERCISE 9.86

▶ Recall that the probability of a type II error depends on the alternative value of the parameter under investigation. The example below presents a method for computing β in a hypothesis test about a population mean μ when σ is known. ◀

Example 9.10 The Blind Side

UPI/Robert Cornforth/Landov

Michael Oher, an offensive lineman for the Baltimore Ravens, helped to glamorize and increase the value of tackles in the National Football League (NFL). These lineman protect the quarterback's blind side, and teams draft prospects specifically for this purpose. The mean weight of offensive tackles at the NFL scouting combine from 2008 to 2012 was 315 pounds.[18] A random sample of 36 tackles at the 2013 NFL combine is selected and each is weighed. A hypothesis test is conducted to determine whether there is any evidence that the mean weight of tackles at the 2013 combine is less than 315 pounds. Let $\alpha = 0.025$, assume $\sigma = 29$, and find the probability of a type II error if the true mean weight of the tackles at the combine is 300 pounds.

SOLUTION

STEP 1 The assumed mean is $\mu = 315 \; (= \mu_0)$; $n = 36$, $\sigma = 29.0$, and $\alpha = 0.025$.

The NFL scouts are looking for evidence of a smaller mean weight in offensive tackles. This is a one-sided, left-tailed test.

The underlying population distribution is unknown, but the sample size is large ($n = 36 \geq 30$). The hypothesis test concerning a population mean μ when σ is known is relevant.

STEP 2 The four parts of the hypothesis test are

$$H_0: \mu = 315.0$$
$$H_a: \mu < 315.0$$
$$\text{TS: } Z = \frac{\overline{X} - \mu_0}{\sigma/\sqrt{n}} = \frac{\overline{X} - 315.0}{29.0/\sqrt{36}}$$
$$\text{RR: } Z \leq -z_\alpha = -z_{0.025} = -1.96$$

STEP 3 To compute the probability of a type II error, you must visualize the rejection region in terms of $\overline{X}$. Start by writing the definition of a type I error, and work backward to the $\overline{X}$ world.

$$P(Z \leq -1.96) = 0.025 \qquad \text{Definition of type I error.}$$

$$P\left(\frac{\overline{X} - 315.0}{29.0/\sqrt{36}} \leq -1.96\right) = 0.025 \qquad \text{Use the definition of the test statistic.}$$

$$P(\overline{X} \leq 305.5) = 0.025 \qquad \begin{array}{l}\text{Isolate } \overline{X}. \text{ Within the probability expression, multiply} \\ \text{both sides by } 29.0/\sqrt{36}, \text{ and add 315.0 to both sides.}\end{array}$$

$305.5 = \overline{X}_c$ is the *critical value in the* $\overline{X}$ *world* (see Figure 9.14). If the sample mean weight of the tackles is less than or equal to 305.5, reject the null hypothesis.

STEP 4 $\beta(300)$ is the probability of *not* rejecting the null hypothesis if the real mean is 300 (often denoted $\mu_a = 300$; the a in the subscript stands for **alternative mean**). Write a probability statement for $\beta(300)$ using the critical value 305.5, and standardize.

$$\beta(\mu_a) = P(\overline{X} > \overline{X}_c) \qquad \text{Definition of a type II error.}$$

$$\beta(300) = P(\overline{X} > 305.5) \qquad \text{Use values for } \mu_a \text{ and } \overline{X} > \overline{X}_c.$$

$$= P\left(\frac{\overline{X} - 300}{29.0/\sqrt{36}} > \frac{305.5 - 300}{29.0/\sqrt{36}}\right) \qquad \begin{array}{l}\text{Standardize using } \mu_a = 300. \text{ Assume } \sigma = 29.0 \\ \text{is unchanged even if the true mean is 300.}\end{array}$$

$$= P(Z > 1.14) \qquad \text{Equation 6.8.}$$

$$= 1 - P(Z \leq 1.14) \qquad \text{Use cumulative probability.}$$

$$= 1 - 0.8729 = 0.1271 \qquad \text{Use Table III in the Appendix.}$$

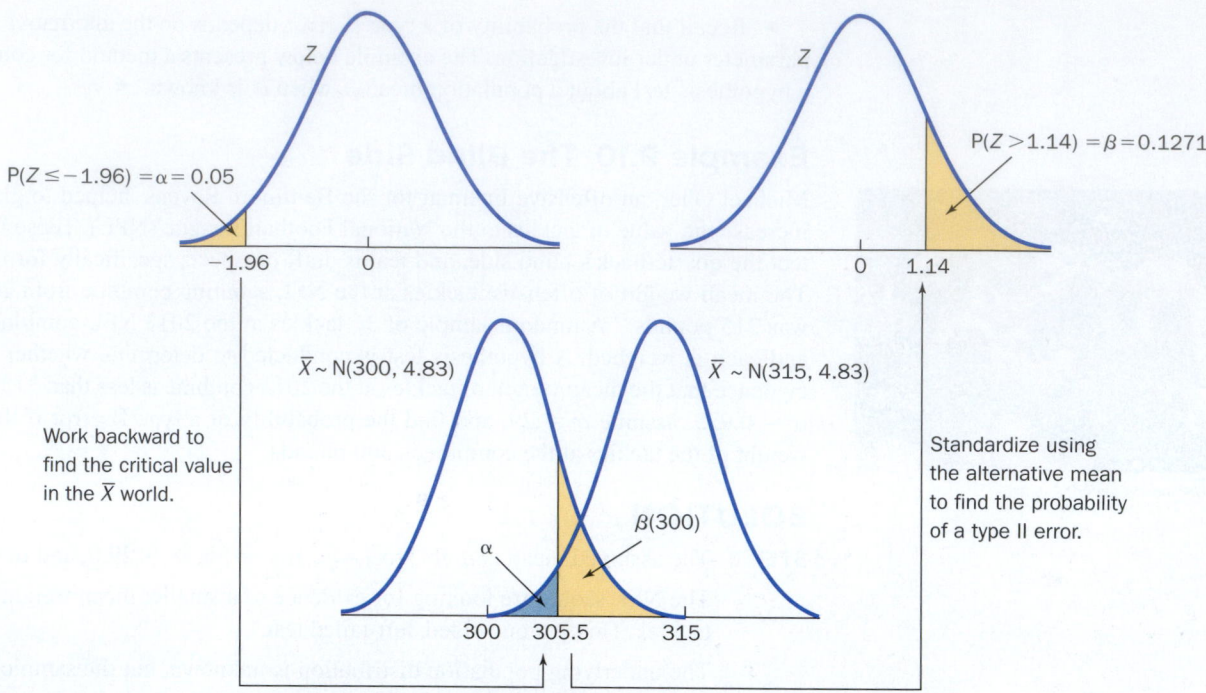

Figure 9.14 Visualization of the calculations for $\beta(300)$, the probability of a type II error for $\mu_a = 300$.

If the true mean weight of tackles at the combine is 300 pounds, the probability of a type II error is 0.1271, at the $\alpha = 0.025$ significance level.

TRY IT NOW   GO TO EXERCISE 9.88

A CLOSER LOOK

1. By referring to Figure 9.14, one can visualize the inverse relationship between α and β. For example, if α increases, $\overline{X}_c$ moves to the right, and the area of the region corresponding to β decreases. Exercise 9.88 at the end of this section is designed to explore and confirm this concept numerically.

2. The logic and method for finding the probability of a type II error for the other alternative hypotheses is the same:

   a. Work backward to find $\overline{X}_c$.

   b. Write a probability statement for $\beta(\mu_a)$ in terms of $\overline{X}$ and $\overline{X}_c$.

   c. Standardize to find the probability.

3. Rather than trying to memorize a formula for the probability of a type II error corresponding to each alternative hypothesis, use the definition, a drawing, and the three-step procedure described above. This conceptual approach will help when we consider the hypothesis test procedure concerning a population proportion.

A little educational philosophy sneaks in here.

Technology Corner

Procedure: Hypothesis test concerning a population mean when σ is known.
Reconsider: Example 9.9, solution, and interpretations.

CrunchIt!

Use the built-in function z 1-Sample. Input is either summary statistics or a column containing the sample data.

1. Enter the data in column Var1.
2. Select Statistics; z; 1-sample.
3. Under the Columns tab, select Var1 for the Sample and enter the standard deviation. Under the Hypothesis Test tab, enter the value for μ_0 and select the appropriate Alternative. (Figure 9.15).
4. Click Calculate. The results are displayed in a new window (Figure 9.16).

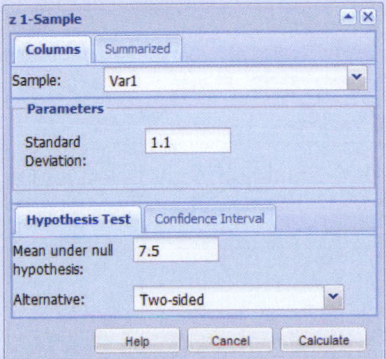
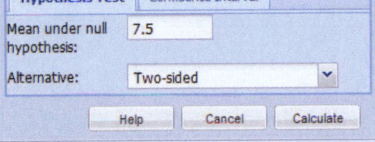

Figure 9.15 1-Sample input screen.   **Figure 9.16** 1-Sample test results.

TI-84 Plus C

Use the calculator function Z-Test. The input for this function is either data in a list or summary statistics.

1. Enter the data into list L1.
2. Select STAT; TESTS; Z-Test.
3. The data are given in this example; highlight Data. Enter μ_0, σ, the name of the list containing the data, the frequency of occurrence of each observation, and highlight the appropriate alternative hypothesis. See Figure 9.10.
4. In the last row of the input screen, select Calculate and press ENTER.
5. The p value associated with the hypothesis test is given on the Z-Test output screen. This topic is covered in Section 9.4. See Figure 9.11.
6. The Z-Test input screen includes an option to Draw, or visualize, the results of the hypothesis test. The calculator will automatically determine a suitable WINDOW, sketch a standard normal density curve, display the computed z and p values, and shade the area under the curve corresponding to the p value. See Figure 9.12.

Minitab

Use the function 1-Sample Z. The input is either data in a column (Samples in columns) or summary statistics (Summarized data).

1. Enter the data into column C1.
2. Choose Stat; Basic Statistics; 1-Sample Z.
3. In this example, select One or more samples, each in a column and enter C1. Enter the Standard deviation, check the box to Perform hypothesis test, and enter the Hypothesized mean (Figure 9.17).
4. Choose Options. Enter a Confidence level and select the Alternative hypothesis (Figure 9.18).
5. Summary statistics, the confidence interval, the value of the test statistic, and the p value are displayed in the Session window (Figure 9.19).

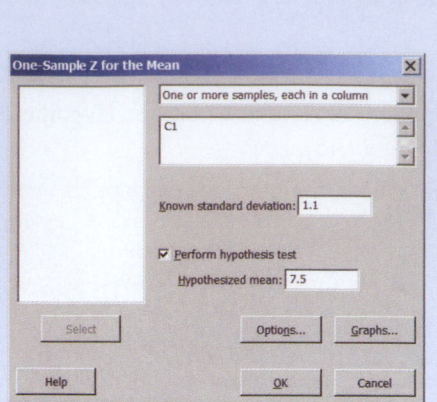

Figure 9.17 1-Sample Z input screen.

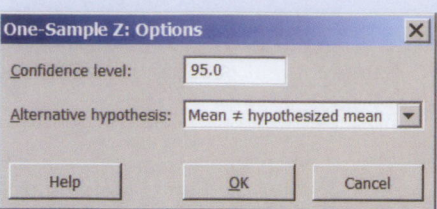

Figure 9.18 1-Sample Z Options screen.

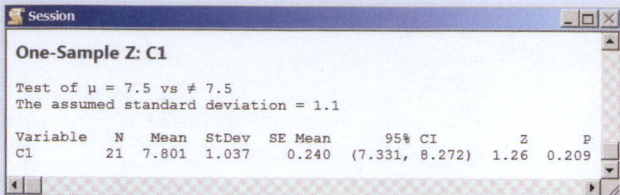

Figure 9.19 1-Sample Z hypothesis test results.

Excel

Use the function Z.TEST. The input is a column (array) of data.

1. Enter the data into column A.
2. The arguments for Z.TEST are the array, μ_0, and σ. The function returns the probability that the sample mean is greater than the hypothesized mean. Other calculations may be necessary to find the desired p value (discussed in Section 9.4). See Figure 9.20.

Figure 9.20 Excel function Z.TEST.

	B	C
	0.1046	= Z.TEST(A1:A21,7.5,1.1)
	0.2092	= 2*B1

SECTION 9.3 EXERCISES

Concept Check

9.57 True/False The hypothesis test concerning a population mean when σ is known can only be used when the underlying population is normal.

9.58 True/False In general, we reject the null hypothesis in a test concerning a population mean when σ is known when the value of the test statistic is *large* in magnitude.

9.59 True/False In a hypothesis test concerning a population mean when σ is known, the probability of a type II error does not depend on the true value of the population mean.

9.60 True/False In a hypothesis test concerning a population mean when σ is known, there are only three possible alternative hypotheses.

9.61 True/False In a hypothesis test concerning a population mean when σ is known, the conclusion is dependent on whether the value of the test statistic lies in the rejection region.

Practice

9.62 Consider a hypothesis test concerning a population mean with $H_0: \mu = 170$, $H_a: \mu < 170$, $n = 38$, and $\sigma = 15$.
   a. Write the appropriate test statistic.
   b. Write the rejection region corresponding to each value of α:
      i. $\alpha = 0.01$ **ii.** $\alpha = 0.025$ **iii.** $\alpha = 0.05$
      iv. $\alpha = 0.10$ **v.** $\alpha = 0.001$ **vi.** $\alpha = 0.0001$

9.63 Consider a hypothesis test concerning a population mean from a normal population with $H_0: \mu = 45.6$, $H_a: \mu > 45.6$, $n = 16$, and $\sigma = 15$.
   a. Write the appropriate test statistic.
   b. Write the rejection region corresponding to each value of α:
      i. $\alpha = 0.01$ **ii.** $\alpha = 0.025$ **iii.** $\alpha = 0.05$
      iv. $\alpha = 0.10$ **v.** $\alpha = 0.005$ **vi.** $\alpha = 0.0005$

9.64 Consider a hypothesis test concerning a population mean from a normal population with $H_0: \mu = -11$, $H_a: \mu \neq -11$, $n = 21$, and $\sigma = 4.5$.

a. Write the appropriate test statistic.

b. Write the rejection region corresponding to each value of α:

    i. $\alpha = 0.01$ **ii.** $\alpha = 0.2$ **iii.** $\alpha = 0.05$

    iv. $\alpha = 0.10$ **v.** $\alpha = 0.001$ **vi.** $\alpha = 0.0002$

9.65 Consider a hypothesis test concerning a population mean with H_0: $\mu = 3.55$, H_a: $\mu < 3.55$, $n = 49$, and $\sigma = 6.2$. Find the significance level (α) for each rejection region:

    a. $Z \leq -1.6449$ **b.** $Z \leq -2.5758$ **c.** $Z \leq -2.0537$

    d. $Z \leq -2.3263$ **e.** $Z \leq -3.0902$ **f.** $Z \leq -3.7190$

9.66 Consider a hypothesis test concerning a population mean with H_0: $\mu = 7.6$, H_a: $\mu \neq 7.6$, $n = 37$, and $\sigma = 4.506$. Find the significance level (α) for each rejection region:

    a. $|Z| \geq 1.96$ **b.** $|Z| \geq 1.6449$ **c.** $|Z| \geq 2.8070$

    d. $|Z| \geq 3.2905$ **e.** $|Z| \geq 1.2816$ **f.** $|Z| \geq 2.3263$

9.67 Consider a hypothesis test concerning a population mean from a normal population with H_0: $\mu = 98.6$, H_a: $\mu > 98.6$, $n = 10$, and $\sigma = 1.2$. Find the significance level (α) for each rejection region:

    a. $Z \geq 3.7190$ **b.** $Z \geq 0.8416$ **c.** $Z \geq 2.3263$

    d. $Z \geq 1.6449$ **e.** $Z \geq 3.2905$ **f.** $Z \geq 2.8782$

9.68 Consider a random sample of size 25 from a normal population with hypothesized mean 212 and $\sigma = 2.88$.

a. Write the four parts for a one-sided, right-tailed hypothesis test concerning the population mean with $\alpha = 0.01$.

b. What assumptions are made in order for the test in part (a) to be appropriate?

c. Suppose the sample mean is $\bar{x} = 213.5$. Find the value of the test statistic, and draw a conclusion about the population mean.

9.69 Consider a random sample of size 32 from a population with hypothesized mean 3.14 and $\sigma = 6.8$. **EX9.69**

a. Write the four parts for a one-sided, left-tailed hypothesis test concerning the population mean with $\alpha = 0.001$.

b. What assumptions are made in order for the test in part (a) to be appropriate?

c. Compute the sample mean, and find the value of the test statistic. Draw a conclusion about the population mean.

9.70 Consider a random sample of size 48 from a population with hypothesized mean 365.25 and $\sigma = 22.3$.

a. Write the four parts for a two-sided hypothesis test concerning the population mean with $\alpha = 0.05$.

b. What assumptions are necessary in order for the test in part (a) to be appropriate?

c. Suppose the sample mean is $\bar{x} = 360.0$. Find the value of the test statistic, and draw a conclusion about the population mean.

9.71 Consider a one-sided, right-tailed hypothesis test concerning the mean, μ_0, from a normal population with σ known, sample size n, and $\alpha = 0.05$. Explain the error in each of the following statements.

a. The rejection region is RR: $Z \leq 1.96$.

b. The test statistic is $Z = \dfrac{\mu_0 - \bar{X}}{\sigma/\sqrt{n}}$.

c. The value of the test statistic does not lie in the rejection region. We accept the null hypothesis and conclude that $\mu = \mu_0$.

d. The null hypothesis is H_0: $\mu > \mu_0$.

e. The probability of a type II error **when** $\mu = \mu_a$ is $1 - \alpha = 1 - 0.05 = 0.95$.

Applications

9.72 **Business and Management**   The mean income per year of employees who produce internal and external newsletters and magazines for corporations was reported to be $51,500. These editors and designers work on corporate publications but not on marketing materials. As a result of poor economic conditions and oversupply, these corporate communications workers may be experiencing a decrease in salary. A random sample of 38 corporate communications workers revealed that $\bar{x} = 49,762$. Conduct a hypothesis test to determine whether there is any evidence to suggest that the mean income per year of corporate communications workers has decreased. Assume $\sigma = 3750$ and use $\alpha = 0.01$.

9.73 **Biology and Environment Science**   Motor vehicles contribute approximately 15% of all carbon dioxide released into the atmosphere each year. Carbon dioxide emissions are often used to determine if a motor vehicle is efficient, or *green*. A random sample of 40 2013 automobiles was obtained and the CO_2 emissions for each was measured (in g/mi).[19] The mean CO_2 emissions for all automobiles tested by the U.S. Environmental Protection Agency in 2012 was 335 g/mi. Is there any evidence that the mean CO_2 emissions has decreased? Assume $\sigma = 65$ and use $\alpha = 0.05$. Write a Solution Trail for this problem. **GREENCAR**

9.74 **Marketing and Consumer Behavior**   After analyzing a database of over 2 million telephone calls, Vonage reported that the mean length of all international calls was 295 seconds. Following an extensive advertising campaign, the company expected the length of international calls to increase. A few months after the ads first appeared, a random sample of 48 calls was obtained. The sample mean was $\bar{x} = 306.3$ seconds. Assume that the population standard deviation is 52 seconds. Is there any evidence to suggest the advertising campaign has been successful? Use $\alpha = 0.01$.

9.75 **Physical Sciences**   The U.S. Geological Survey collects water-level measurement data from various locations. Data from the La Pine Basin, Oregon, are being used to study changes in the water table. Twelve days were randomly selected, and the water table in feet below land surface on each day is given in the following table. **WTRTABLE**

12.30	12.29	12.30	12.35	12.37	12.40
12.38	12.37	12.38	12.42	12.46	12.52

Is there any evidence to suggest the mean water table is less than 12.4 feet? Assume the underlying distribution is normal, $\sigma = 0.07$ and use $\alpha = 0.025$.

9.76 Sports and Leisure The length of a motorboat is traditionally measured in two ways. The distance along the centerline from the outside of the front hull to the rear is the length overall (LOA). The length of waterline, or load waterline (LWL), is the length of the boat on the line where the boat meets the water. Members of the Community Association in the lakeside city of Vermilion, Ohio, are concerned that residents are using bigger boats, contributing to more noise and water pollution. Past registration records indicate the mean LOA for boats allowed on the lake is 35 feet. A random sample of 41 boats on the lake was obtained, and each LOA was carefully measured. The sample mean LOA was $\bar{x} = 36.22$ feet. Assume $\sigma^2 = 5.7$ ft^2.

a. Is there any evidence to suggest the mean LOA has increased? Use $\alpha = 0.01$.
b. Does your answer to part (a) change if $\alpha = 0.1$? Why or why not?

9.77 Economics and Finance The price of coffee has been steadily declining since October 2012. The International Coffee Organization (ICO) reported that the mean price for coffee in May 2013 was 126.96 U.S. cents/lb.[20] A random sample of 34 days was obtained and the composite indicator for Brazilian Natural coffee was recorded for each. The sample mean was $\bar{x} = 130.29$. Assume the population standard deviation is 8.6 U.S. cents/lb. Is there any evidence to suggest that the mean composite indicator for Brazilian Natural is greater than 126.96? Use $\alpha = 0.05$. Write a Solution Trail for this problem.

9.78 Public Health and Nutrition The mean daily energy requirement for eight-year-old boys is 2200 calories. An education researcher believes many students in this group do poorly in school because they have an inadequate diet and therefore not enough energy. A random sample of academically at-risk eight-year-old boys was obtained, and their caloric intake was carefully measured for one day. The summary statistics were $n = 37, \bar{x} = 2089$. Assume $\sigma = 358$.

a. Is there any evidence this group of students has a mean caloric intake below the daily energy requirement? Use $\alpha = 0.05$.
b. How would your answer to part (a) change if $\alpha = 0.01$?

9.79 Biology and Environment Science An average lawn has a mean of 21 blades of grass per square inch. A garden store sells an expensive fertilizer designed to transform an average lawn into a lush, thick carpet within three weeks. To test the claim, a random sample of average lawn plots was obtained. Each was treated with the fertilizer according to the instructions on the package. Three weeks later the density of each plot was measured by counting the blades of grass per square inch. The summary statistics were $n = 32, \bar{x} = 22.4$. Is there any evidence to suggest the fertilizer improves the thickness of an average lawn? Assume $\sigma = 2.7$ and use $\alpha = 0.005$.

9.80 Technology and the Internet Based on test results, Speedtest.net publishes an Internet download index for countries all over the world. In July 2013, the mean download speed for the entire world was 13.91 Mbps.[21] A random sample of states was obtained and the download index for each was recorded. Assume the underlying population is normal and that $\sigma = 3.4$. Is there any evidence to suggest that the download speed for the United States is greater than the world population mean? Use $\alpha = 0.01$. **DOWNLOAD**

9.81 Public Policy and Political Science The U.S. Department of Health and Human Services defines "response time" as the time from receipt of a report of child neglect to the initial investigation. Once a call to the agency is received and logged in, the clock starts. Face-to-face contact with the alleged victim marks the initial investigation. There is some concern that workload and poor organization have contributed to a significant increase in the long-term mean response time of 14.0 hours. A random sample of cases was selected, and the response time for each was recorded. Assume $\sigma = 3.2$ and conduct a hypothesis test to determine whether there is any evidence that the mean response time has increased. Use $\alpha = 0.001$. **HHSCASES**

9.82 Manufacturing and Product Development A major cause of injuries in highway work zones is weak construction barriers. One federal government road-barrier specification involves the velocity of a front-seat passenger immediately following impact. This impact velocity must be less than 12 m/s. TSS GmbH, a German company, has just developed a new high-impact road barrier with special absorbing material. In controlled tests the impact velocity was measured for 12 randomly selected crashes. The sample mean was 11.85 m/s. Assume the distribution of impact velocities is normal and $\sigma = 0.26$.

a. Conduct a hypothesis test to determine whether there is sufficient evidence to suggest the true mean impact velocity is less than 12 m/s. Use $\alpha = 0.05$.
b. What is your conclusion if $\alpha = 0.01$?

9.83 Marketing and Consumer Behavior Many people live under a lot of stress and various time constraints. Consequently, some people run out of the house without the keys, and are locked out. Rather than break into one's own home, most people in these circumstances call a locksmith. In 2013, the mean service charge for a locksmith was $68.[22] A random sample of 26 locksmith service charges in Black Springs, Arkansas, was obtained. The sample mean was $72.35. Assume the underlying population distribution is normal and that $\sigma = 15.5$. Is there any evidence to suggest the mean locksmith service charge in this city is greater than $68? Use $\alpha = 0.05$.

9.84 Manufacturing and Product Development The Brunton Optimus Crux Camping Stove, a lightweight camping stove, is designed to accept a tri-blend fuel made of propane, iso-butane, and butane. The cartridges for this stove are sold with camping equipment, and the label indicates 225 grams of fuel. To check this claim, 52 cartridges were randomly selected and the amount of fuel in each was carefully measured. The sample mean was $\bar{x} = 224.2$ grams.

a. Is there any evidence that the true mean amount of fuel in these cartridges is less than the advertised amount? Assume $\sigma = 3.6$ and use $\alpha = 0.05$.
b. What assumptions did you make in order to conduct the hypothesis test in part (a)?

9.85 Public Health and Nutrition Egg Beaters are egg substitutes with no fat or cholesterol. The Original Egg Beaters are produced to have 5 grams of protein, one less gram than a regular egg.[23] To check the protein claim, 18 Egg Beaters were randomly selected and the amount of protein in each was measured. The resulting data are given in the following table. **EGGSUB**

5.54	3.83	3.01	5.83	4.64	6.26	4.45	5.71	4.84
4.41	5.62	5.49	5.51	4.11	6.46	6.26	4.95	4.40

Is there any evidence to suggest the mean amount of protein in an Egg Beater is less than 5 grams? Assume the distribution of protein is normal and $\sigma = 0.95$. Use $\alpha = 0.02$.

9.86 Public Policy and Political Science Residential mailboxes in Des Moines, Iowa, should be installed such that the bottom of the mailbox is 42 inches above the ground. This rule is designed for safety and to accommodate short mail carriers. A random sample of 75 mailboxes in the city was selected. The height of each was carefully measured, and the sample mean was 43.22 inches. Assume $\sigma = 7.6$ inches and use $\alpha = 0.05$. Is there any evidence to suggest the true mean height of mailboxes in Des Moines is different from 42 inches?

9.87 Public Health and Nutrition Iodine is an important nutrient for the human body, especially in hormone development. Milk is a natural source of iodine, and the mean amount in cow's milk is 88 μg/250ml.[24] Milk from organic farms is reported to have a lower concentration of elements such as zinc, iodine, and selenium. A random sample of 35 gallons of organic milk from Humboldt County, California, was obtained and the iodine concentration was measured in each. The sample mean was $\bar{x} = 86.5$. Assume $\sigma = 3.4$. Is there any evidence to suggest that the mean iodine concentration in organic milk is less than 88? Use $\alpha = 0.01$.

Extended Applications

9.88 Type II Errors The four parts of a hypothesis test concerning a population mean from a normal population are shown below.

H_0: $\mu = 50$

H_a: $\mu > 50$

TS: $Z = \dfrac{\bar{X} - \mu_0}{\sigma/\sqrt{n}}$

RR: $Z \geq z_\alpha$

Assume the sample size is $n = 25$; $\sigma = 7.5$ and $\alpha = 0.01$.
  a. Find the probability of a type II error for the alternative mean $\mu_a = 54$; that is, find $\beta(54)$.
  b. Find $\beta(55)$ and $\beta(56)$.
  c. Repeat parts (a) and (b) for $\alpha = 0.025$.

9.89 Fuel Consumption and Cars Biodiesel fuels are made from vegetable oils or animal fats and may be used instead of conventional diesel fuel. One advantage to using biodiesel is a possible decrease in regulated emissions, specifically total hydrocarbons (HC). Using the heavy-duty transient Federal Test Procedure (FTP), the mean HC emission is 0.23 g/hp-hr (grams per horsepower-hour) with standard deviation $\sigma = 0.07$. A random sample of heavy-duty engines was obtained, and each

was tested with biodiesel fuel. The resulting HC measurements are given in the table below. **BIODIES**

0.17	0.19	0.10	0.21	0.15	0.23	0.06	0.21
0.18	0.01	0.24	0.29	0.23	0.14	0.18	0.10
0.02	0.16	0.13	0.10	0.14	0.30	0.24	0.24
0.18	0.06						

  a. Suppose the underlying population is normal. Is there any evidence to suggest the use of biodiesel has decreased the mean level of HC emissions? Use $\alpha = 0.01$.
  b. Suppose a company is thinking about building a $20 million biodiesel fuel production facility. The company will invest the money only if there is overwhelming evidence in favor of decreased HC emissions. Which error (type I or type II) is more important to the fuel company? Would the company prefer a smaller or a larger significance level? Justify your answer.

9.90 Manufacturing and Product Development The left outside panel of an apartment-size, frost-free refrigerator is designed to have width $23\frac{5}{8}$ inches. Each hour, 10 such panels are randomly selected from the assembly line and carefully measured. If there is any evidence that the mean width is different from $23\frac{5}{8}$ inches, the assembly line is shut down for cleaning and inspection. Suppose the distribution of panel widths is normal and $\sigma = 0.15$.
  a. During a specific hour of operation, $\bar{x} = 23.7$ inches. Should the assembly line be shut down? Use $\alpha = 0.05$.
  b. Using $\alpha = 0.05$, find the critical values in the $\bar{x}$ world. (*Note:* There are two critical values, because this is a two-sided hypothesis test.)

9.91 Manufacturing and Product Development A manufacturer claims the weight of a package of their unsalted pretzels is (at least) 15.5 ounces. To test this claim, a consumer group randomly selected 250 packages and carefully weighed the contents of each. The sample mean was $\bar{x} = 15.45$ ounces. **PRETZEL**
  a. Conduct a one-sided, left-tailed hypothesis test to see whether there is any evidence that the true mean weight of the pretzel packages is less than 15.5 ounces. Use $\alpha = 0.01$ and assume $\sigma = 0.26$.
  b. If your conclusion in part (a) is to reject the null hypothesis, explain how this conclusion is possible even though the sample mean, 15.45, is so close to the hypothesized mean, 15.5. If your conclusion in part (a) is to not reject H_0, check those calculations one more time.

9.92 Biology and Environmental Science Lake Vostok is Antarctica's largest and deepest subsurface lake. It is buried under approximately 2 miles of ice, and scientists believe it contains evidence of thousands of tiny organisms and fish. The mean depth of the lake is believed to be 344 meters.[25] A random sample of the depth of this lake was obtained. Assume the underlying distribution is normal and that $\sigma = 30$ meters. **VOSTOK**
  a. Conduct a hypothesis test to determine whether there is any evidence to suggest the mean depth is less than 344 meters. Use $\alpha = 0.01$.

b. Find the probability of a type II error if the true mean depth of the lake is 315 meters; that is, find $\beta(315)$.

c. Carefully sketch a graph illustrating the probability found in part (b) using density curves for $\overline{X}$.

d. Find $\beta(310)$.

9.93 Marketing and Consumer Behavior   The water-treatment facility owners in Lake Havasu City, Arizona, recently proposed changes to all user fees. There is a treatment capacity fee, a connection fee, and a monthly minimum fee. All prices have been set assuming the mean monthly usage is 714 cubic feet per household. To check this claim, a random sample of 16 homes was obtained and the monthly usage for each home was recorded. The sample mean was 601.2 cubic feet. Assume the distribution of monthly water usage per household is normal with standard deviation 283 cubic feet.

a. Conduct a two-sided hypothesis test to determine whether there is any evidence the mean monthly water usage is different from 714 cubic feet. Use $\alpha = 0.05$.

b. If your conclusion in part (a) is to not reject the null hypothesis, explain how this conclusion is possible even though the sample mean, 601.2, is so far away from the hypothesized mean, 714. If your conclusion in part (a) is to reject the null hypothesis, try checking your calculations one more time.

9.94 Marketing and Consumer Behavior   The Carpet Corner in Gladstone, Missouri, offers a variety of carpets, wood flooring, and tiles. Sales records indicate that the mean amount of carpet installed in a wall-to-wall carpeted residential home by crews from this store is 1250 square feet. The store manager, Frank Vida, believes that when there is a sale, customers translate the savings into carpeting a larger area. A random sample of 45 wall-to-wall carpet purchases was selected during a sale. The sample mean was $\overline{x} = 1305$ square feet; assume that $\sigma = 155$ square feet.

a. Conduct a one-sided, right-tailed test of $H_0\colon \mu = 1250$ versus $H_a\colon \mu > 1250$. Is there any evidence to suggest the true mean square footage is larger during a sale? Use $\alpha = 0.01$.

b. Find the probability of a type II error if the true mean square footage during a sale is $\mu_a = 1330$; that is, find $\beta(1330)$.

c. Carefully sketch a graph that includes the distribution of $\overline{X}$ if $\mu = 1250$ and if $\mu = 1330$. Shade in the areas that correspond to the probability of a type I error (0.01) and to the probability of a type II error found in part (b).

9.95 Physical Sciences   Kiln-dried solid grade A teak wood should have a moisture content of no more than 12%. A furniture company recently purchased a large shipment of this wood for use in constructing dining room sets. Thirty-seven pieces of teak wood were randomly selected and carefully measured for moisture content. The sample mean moisture content was $\overline{x} = 12.3\%$. Assume $\sigma = 1.25$.

a. Is there any evidence to suggest the true population mean moisture content is greater than 12%? Use $\alpha = 0.01$.

b. Find the probability of a type II error if the true population mean moisture content is 12.2%; that is, find $\beta(12.2)$.

9.96 Fuel Consumption and Cars   The joint venture ContiTech-Jiebao Power Transmission Systems Ltd. produces many makes and models of automobile drive belts. One popular poly V-belt is designed to have length 1050 mm. To ensure that the belts meet this specification, 18 are randomly selected from the assembly line every hour and carefully measured. If there is any evidence that the population mean length is different from 1050 mm, the entire line is shut down for inspection.   ▮▮▮ **FUEL**

a. Assume the distribution of belt lengths is normal and $\sigma = 3.7$ mm. In a two-sided hypothesis test of $H_0\colon \mu = 1050$ versus $H_a\colon \mu \neq 1050$, find the critical values in the $\overline{X}$ world. Use $\alpha = 0.01$.

b. In a sample of 18 belts, suppose $\overline{x} = 1049$. Should the assembly line be shut down? Justify your answer using the Z distribution and the appropriate $\overline{X}$ distribution.

9.97 Manufacturing and Product Development   The 2013 Kawasaki Jet Ski Ultra 300LX is supercharged and intercooled, has a four-stoke inline-four engine, and is designed to have a load capacity of 496 pounds.[26] If the passenger and gear weight exceed this capacity, the jet ski will not function properly and there will be a severe strain on the engine. To determine if customers are following the load capacity recommendation, a random sample of these jet skis is obtained from marinas around the country and each is carefully weighed as owners prepare for a ride. The summary statistics were $n = 33$ and $\overline{x} = 498.42$. Assume $\sigma = 46.7$.

a. Is there any evidence to suggest that the true mean weight of passengers and gear is greater than 496 pounds? Use $\alpha = 0.01$.

b. Suppose the engineers have determined that the watercraft will overheat rapidly if the load is 525 pounds. Find the probability of a type II error if the true mean weight of passengers and gear is 525 pounds; that is, find $\beta(525)$.

Challenge

9.98 The Power of a Test   The probability of a type II error, β, represents the likelihood of accepting the null hypothesis when the alternative is true. The *power* of a statistical test is $\pi = 1 - \beta$, the probability of (correctly) rejecting the null hypothesis, of detecting a difference in the hypothesized value of the population parameter. Try the *Statistical Power* applet on the text website.

A hot torsion test is used to determine the workability of a metal. Suppose a carbon steel rod is designed to fail (break) with mean axial load of 800 newtons (N) (under certain temperature and speed conditions). A random sample of 25 rods is obtained, and the axial load failure is measured for each. Suppose the underlying distribution of axial load is normal and $\sigma = 50$.

a. Consider a hypothesis test of $H_0\colon \mu = 800$ versus $H_a\colon \mu < 800$ with $\alpha = 0.025$. Find the probability of a type II error and the power of this test for $\mu = 775$; that is, find $\beta(775)$ and $\pi(775)$.

b. Use a calculator or computer to compute the power, $\pi(\mu_a)$, for $\mu_a = 730, 735, 740, \ldots, 800$.

c. Use the values from part (b) to carefully sketch a plot of $\pi(\mu_a)$ versus μ_a. The resulting plot is called a *power curve*.

9.4 *p* Values

The last piece of a hypothesis test, the rejection region, establishes a firm decision rule based on the value of the test statistic. If the value of the test statistic lies in the rejection region, we reject H_0. If not, we cannot reject the null hypothesis. Using a fixed rejection region associated with a specific value of α (the significance level) can lead to a peculiar dilemma. Consider the following example.

Example 9.11 Sleep Deprivation

Research suggests that most people need at least 7 hours of sleep each night in order to function well the next day. Some research suggests that lack of sleep can adversely affect health, memory, learning, creativity, productivity, emotional stability, and even life expectancy.[27] Suppose adults need 7.0 hours of sleep per night. A random sample of 32 adults was obtained, and the sleeping time for each was recorded. The sample mean was $\bar{x} = 6.8$ hours and assume $\sigma = 0.67$ hours. Is there any evidence to suggest that the mean sleeping time for adults is less than 7 hours?

SOLUTION

STEP 1 The assumed mean is $\mu = 7.0 \; (= \mu_0)$; $n = 32$ and $\sigma = 0.67$.

We are searching for any evidence to suggest that the mean sleeping time for adults is less than 7.0 hours. This is a one-sided, left-tailed test.

The underlying population distribution is unknown, but the sample size is large ($n = 32 \geq 30$). The hypothesis test concerning a population mean μ when σ is known, is relevant.

STEP 2 No value of α is given. The first three parts of the hypothesis test are

$$H_0: \mu = 7$$
$$H_a: \mu < 7$$
$$\text{TS: } Z = \frac{\bar{X} - \mu_0}{\sigma/\sqrt{n}} = \frac{\bar{X} - 7}{0.67/\sqrt{32}}$$

The value of the test statistic is

$$z = \frac{6.8 - 7.0}{0.67/\sqrt{32}} = -1.6886$$

STEP 3 The following table shows the resulting rejection region and conclusion for various values of α.

α	Rejection region	Conclusion
0.10	$Z \leq -z_{0.10} = -1.2816$	$z = -1.6886$ lies in the rejection region. We **reject** the null hypothesis at the $\alpha = 0.10$ level.
0.05	$Z \leq -z_{0.05} = -1.6449$	$z = -1.6886$ lies in the rejection region. We **reject** the null hypothesis at the $\alpha = 0.05$ level.
0.025	$Z \leq -z_{0.025} = -1.9600$	$z = -1.6886$ does *not* lie in the rejection region. We **cannot reject** the null hypothesis at the $\alpha = 0.025$ level.
0.01	$Z \leq -z_{0.01} = -2.3263$	$z = -1.6886$ does *not* lie in the rejection region. We **cannot reject** the null hypothesis at the $\alpha = 0.01$ level.

STEP 4 In this example, the decision to reject or not reject H_0 depends on the value of α. To avoid this dilemma, an alternative method for reporting the result of a hypothesis test is often used. This technique involves computing a tail probability, or *p* **value**. ●

Most statistics software packages report the *p* value associated with a hypothesis test.

The symbol *p* is used to represent the population proportion of successes and the *p* value. The context in which the notation is used implies the appropriate concept.

Definition

The *p* **value**, denoted *p*, for a hypothesis test is the smallest significance level (value of α) for which the null hypothesis, H_0, can be rejected.

The *p* value is simply a tail probability, and the tail is determined by the alternative hypothesis. Consider a hypothesis test concerning a population mean μ when σ is known. Suppose either the underlying population is normal or *n* is large, and let *z* be the value of the test statistic. The following table presents the probability definition of a *p* value for each alternative hypothesis.

Alternative hypothesis	Probability definition	
$H_a: \mu > \mu_0$	$p = P(Z \geq z)$	Figure 9.21
$H_a: \mu < \mu_0$	$p = P(Z \leq z)$	Figure 9.22
$H_a: \mu \neq \mu_0$	$p/2 = P(Z \geq z)$ if $z \geq 0$	Figure 9.23
	$p/2 = P(Z \leq z)$ if $z < 0$	Figure 9.24

The *p* value conveys the strength of the evidence in favor of the alternative hypothesis. For a small *p* value, the value of the test statistic lies far out in a tail of the distribution, the more unlikely the value of the test statistic, and the more evidence in favor of H_a. *Small values of p are usually* $p \leq 0.05$.

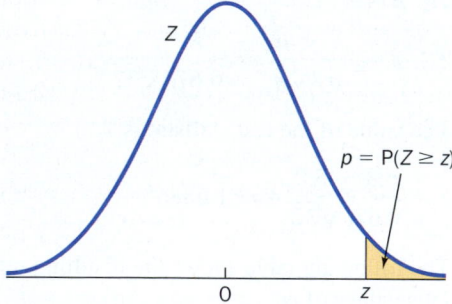

Figure 9.21 Illustration of the *p* value for $H_a: \mu > \mu_a$.

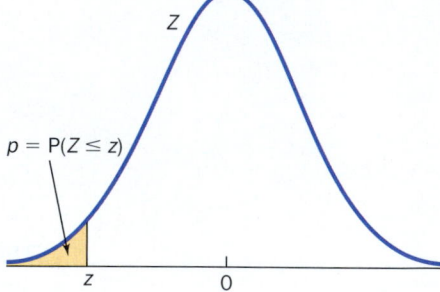

Figure 9.22 Illustration of the *p* value for $H_a: \mu < \mu_a$.

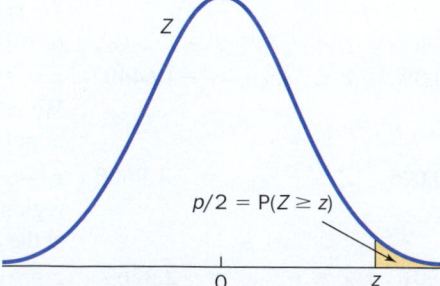

Figure 9.23 Illustration of the *p* value for $H_a: \mu \neq \mu_a$, $z \geq 0$.

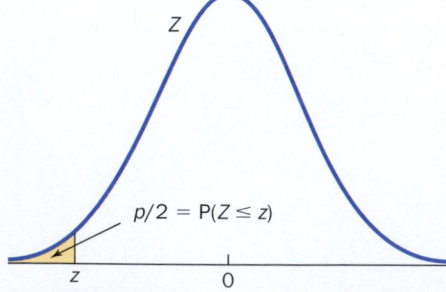

Figure 9.24 Illustration of the *p* value for $H_a: \mu \neq \mu_a$, $z < 0$.

If α is the significance level of the hypothesis test, the conclusion is usually written in one of two ways.

1. If $p \leq \alpha$: Reject the null hypothesis. There is evidence to suggest H_a is true at the p = (observed *p* value) level of significance.

2. If $p > \alpha$: Do not reject the null hypothesis. There is no evidence to suggest the null hypothesis is false.

Consider a one-sided, right-tailed hypothesis test concerning a population mean μ when σ is known (H_a: $\mu > \mu_0$). Suppose either the underlying population is normal or n is large, α is the significance level, and let z be the value of the test statistic. Recall the definition of a critical value: $\alpha = P(Z \geq z_\alpha)$. Figures 9.25 and 9.26 demonstrate the relationship between the *p* value and a fixed significance level. These figures also illustrate the definition of a *p* value, that is, the *smallest* significance level for which H_0 can be rejected. If the significance level (α) is less than p, then $z < z_\alpha$, and we cannot reject H_0.

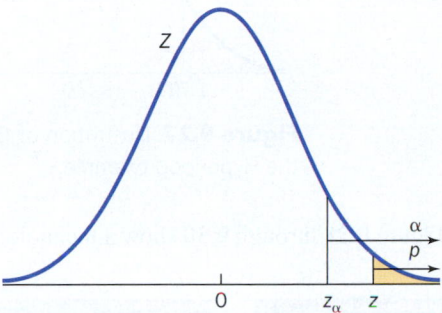

Figure 9.25 If $p \leq \alpha$, then $z \geq z_\alpha$.
Conclusion: Reject H_0.

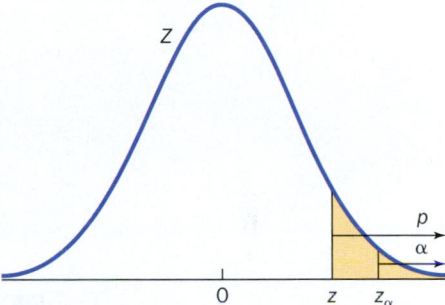

Figure 9.26 If $p > \alpha$, then $z < z_\alpha$.
Conclusion: Do not reject H_0.

Example 9.12 The Hyperloop

Elon Musk, the billionaire founder of PayPal, Tesla, and SpaceX, has been developing a transportation system called the Hyperloop, an alternative to bullet trains. Passengers would ride in a capsule and travel at a speed of approximately 620 mph in a vacuum tube, or tunnel.[28] A three-mile test tunnel has been constructed in California. Suppose 20 random tests are selected and the speed of the capsule is carefully measured for each. The sample mean is $\bar{x} = 610.1$. Assume the distribution of capsule speed is normal with $\sigma = 25$. Conduct a one-sided, left-tailed hypothesis test with $\alpha = 0.05$ and compute the *p* value. Is there any evidence to suggest that the true mean speed is less than 620 mph?

SOLUTION

STEP 1 The assumed mean is $\mu = 620$ ($= \mu_0$); $n = 20$ and $\sigma = 25$.

This is a one-sided, left-tailed test. The underlying population is assumed to be normal. The hypothesis test concerning a population mean μ when σ is known, is relevant.

STEP 2 The value of the test statistic will be used to compute a *p* value. The first three parts of the hypothesis test are

H_0: $\mu = 620$

H_a: $\mu < 620$

TS: $Z = \dfrac{\bar{X} - \mu_0}{\sigma/\sqrt{n}} = \dfrac{\bar{X} - 620}{25/\sqrt{20}}$

The value of the test statistic is

$$z = \frac{610.1 - 620}{25/\sqrt{20}} = -1.7710$$

STEP 3 The p value is a left-tail probability (see Figure 9.27).

$$p = P(Z \le -1.7710)$$

$$= 0.0384$$

Definition of p value for H_a: $\mu < \mu_0$.

Use Table III in the Appendix.

Because $p = 0.0384 \le 0.05 \ (= \alpha)$, we reject the null hypothesis. There is evidence to suggest that the true mean capsule speed is less than 620 mph at the $p = 0.0384$ level of significance.

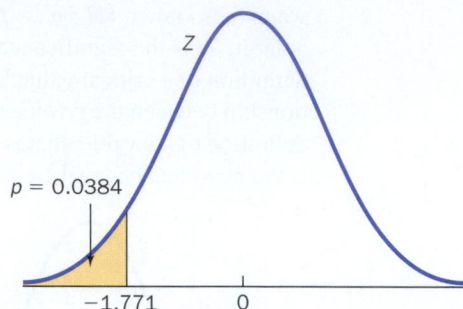

$p = 0.0384$

-1.771 0

Figure 9.27 Illustration of the p value for the Hyperloop example.

Figure 9.28 through 9.30 show a technology solution.

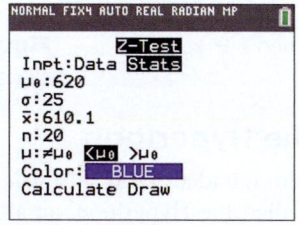

Figure 9.28 TI-84 Plus C Z-Test input screen.

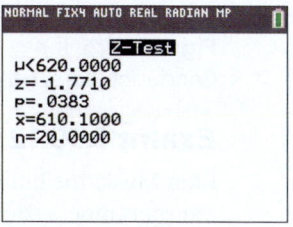

Figure 9.29 TI-84 Plus C Z-Test Calculate results.

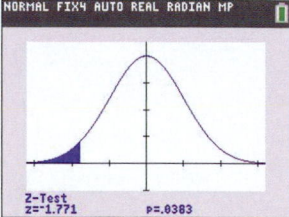

Figure 9.30 TI-84 Plus C Z-Test Draw results.

TRY IT NOW   GO TO EXERCISE 9.111

The following example involves a two-sided hypothesis test.

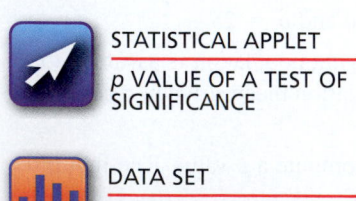

STATISTICAL APPLET

p VALUE OF A TEST OF SIGNIFICANCE

DATA SET

FLOODING

Example 9.13 Flash Flooding

On July 8, 2013, a very heavy rainstorm over Toronto swamped commuter trains, caused power outages, and clogged storm sewers. Approximately 4 inches (101 mm) of rain fell near the heart of the city and almost 5 inches fell at Pearson International Airport, breaking a 1954 record.[29] Toronto city officials considering changes to the storm drainage system examined 24-hour rain totals. A random sample of days in which there was measurable precipitation was selected. The daily rainfall total (in mm) was recorded for each, and the data are given in the following table.[30]

0.1	0.8	13.7	0.2	1.7	0.1	1.0	7.7	23.1
4.2	4.2	0.7	29.0	23.5	1.3	3.6	5.0	13.8
11.1	0.2	28.2	14.1	1.8	11.7	1.6	1.1	

Assume the underlying population is normal and that $\sigma = 10$ mm. Is there any evidence that the true mean daily rainfall total in Toronto is different from 5 mm? Use $\alpha = 0.05$ and the p value associated with the test statistic to justify your answer.

SOLUTION

STEP 1 The assumed mean is $\mu = 5.0\ (= \mu_0)$; $n = 26$ and $\sigma = 10.0$.

The phrase "is different" means this is a two-sided test. The distribution of 24-hour rainfall total is assumed to be normal. The hypothesis test concerning a population mean μ when σ is known, is relevant.

STEP 2 The first three parts of the hypothesis test are

$$H_0: \mu = 5.0$$
$$H_a: \mu \neq 5.0$$
$$\text{TS: } Z = \frac{\overline{X} - \mu_0}{\sigma/\sqrt{n}} = \frac{\overline{X} - 5.0}{10.0/\sqrt{26}}$$

STEP 3 The sample mean is

$$\bar{x} = \frac{1}{26}(0.1 + 4.2 + \cdots + 1.1) = 7.8269$$

The value of the test statistic is

$$z = \frac{\bar{x} - \mu_0}{\sigma/\sqrt{n}} = \frac{7.8269 - 5.0}{10.0/\sqrt{26}} = 1.4414$$

STEP 4 Since this is a two-sided test and the value of the test statistic ($z = 1.4414$) is positive, $p/2$ is a right-tail probability (see Figure 9.31).

$$p/2 = P(Z \geq 1.4414)$$
$$= 0.0747$$
$$p = 2(0.0747) = 0.1494$$

Definition of *p* value for $H_a: \mu \neq \mu_0$.

Use Table III in the Appendix.

Solve for *p*.

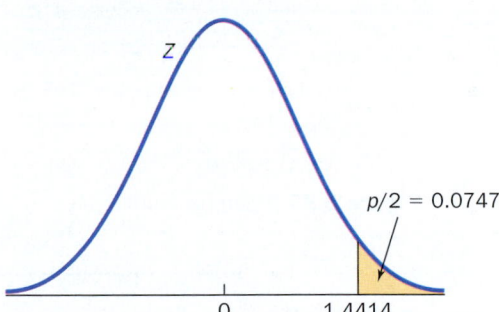

Figure 9.31 Illustration of the *p* value for the Toronto 24-hour rainfall example.

Figures 9.32 through 9.34 show a technology solution.

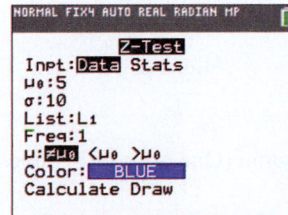

Figure 9.32 TI-84 Plus C Z-Test input screen.

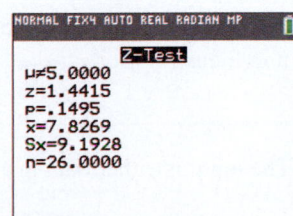

Figure 9.33 TI-84 Plus C Z-Test Calculate results.

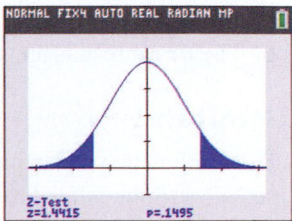

Figure 9.34 TI-84 Plus C Z-Test Draw results.

Because $p = 0.1494 > 0.05 \, (= \alpha)$, we do not reject the null hypothesis. There is no evidence to suggest that the mean 24-hour rainfall total is different from 5 mm.

TRY IT NOW   GO TO EXERCISE 9.113

Technology Corner

Procedure: Compute the p value in a hypothesis test concerning a population mean when σ is known.
Reconsider: Example 9.13, solution, and interpretations.

CrunchIt!

Use the built-in function z 1-Sample. Input is either summary statistics or a column containing the sample data.

1. Enter the data in column Var1. Rename the column if desired.
2. Select Statistics; z; 1-sample.
3. Under the Columns tab, select the appropriate column name for the Sample and enter the standard deviation. Under the Hypothesis Test tab, enter the value for μ_0 and select the appropriate Alternative. (Figure 9.35).
4. Click Calculate. The results are displayed in a new window (Figure 9.36).

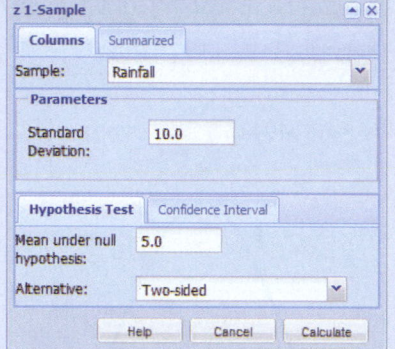

Figure 9.35 1-Sample input screen.

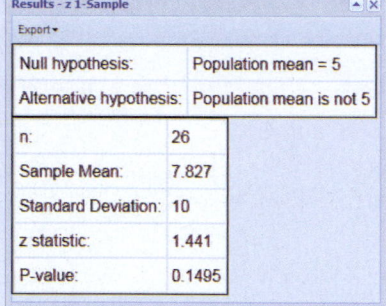

Figure 9.36 1-Sample test results.

TI-84 Plus C

Use the calculator function Z-Test. The input for this function is either data in a list or summary statistics.

1. Enter the data into list L1.
2. Select STAT; TESTS; Z-Test.
3. The data are given in this example; highlight Data. Enter μ_0, σ, the name of the list containing the data, the frequency of occurrence of each observation, and highlight the appropriate alternative hypothesis. See Figure 9.31.
4. In the last row of the input screen, select Calculate and press ENTER.
5. The p value associated with the hypothesis test is given on the Z-Test output screen. See Figure 9.32.
6. The Draw results are shown in Figure 9.33.

Minitab

Use the function 1-Sample Z. The input is either data in a column (One or more samples, each in a column) or summary statistics (Summarized data).

1. Enter the data into column C1.
2. Choose Stat; Basic Statistics; 1-Sample Z.

3. In this example, select One or more samples, each in a column and enter C1. Enter the Standard deviation, check the box to Perform hypothesis test, and enter the Hypothesized mean (Figure 9.37).
4. Choose Options. Enter a Confidence level and select the Alternative hypothesis (Figure 9.38).
5. Summary statistics, the confidence interval, the value of the test statistic, and the *p* value are displayed in the Session window (Figure 9.39).

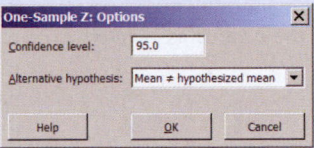

Figure 9.38 1-Sample Z Options screen.

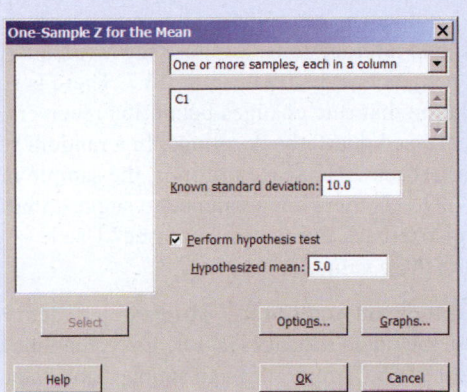

Figure 9.37 1-Sample Z input screen.

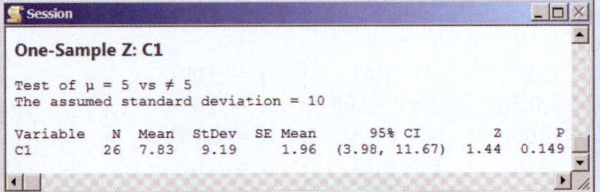

Figure 9.39 1-Sample Z hypothesis test results.

Excel

Use the function Z.TEST. The input is a column (array) of data.

1. Enter the data into column A.
2. The arguments for Z.TEST are the array, μ_0, and σ. The function returns the probability that the sample mean is greater than the hypothesized mean. Use this probability to compute the *p* value. See Figure 9.40.

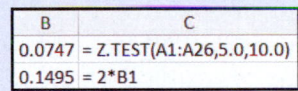

Figure 9.40 Excel function Z.TEST.

SECTION 9.4 EXERCISES

Concept Check

9.99 True/False   In a two-sided hypothesis test, the *p* value may be greater than 1.

9.100 True/False   A small *p* value suggests more evidence in favor of H_a.

9.101 True/False   The *p* value is the smallest significance level for which the null hypothesis can be rejected.

9.102 True/False   The decision to reject or not reject H_0 depends on the value of α.

9.103 Short Answer   If α is the significance level of a hypothesis test and $p \leq \alpha$, what is the conclusion?

Practice

9.104   For each *p* value and significance level, determine whether the null hypothesis would be rejected or not rejected.

   a. $p = 0.067$, $\alpha = 0.05$ **b.** $p = 0.0043$, $\alpha = 0.01$
   c. $p = 0.159$, $\alpha = 0.05$ **d.** $p = 0.0260$, $\alpha = 0.025$
   e. $p = 0.001$, $\alpha = 0.05$ **f.** $p = 0.1770$, $\alpha = 0.01$

9.105   Consider a hypothesis test concerning a population mean with σ known, *n* large, and alternative hypothesis $H_a: \mu > \mu_0$. Find the *p* value associated with each value of the test statistic.

   a. $z = 1.87$ **b.** $z = 2.55$ **c.** $z = 1.20$
   d. $z = 0.57$ **e.** $z = 3.88$ **f.** $z = -1.14$

9.106 Consider a hypothesis test concerning a population mean with σ known, underlying population normal, and alternative hypothesis H_a: $\mu < \mu_0$. Find the p value associated with each value of the test statistic.

   a. $z = -2.05$ **b.** $z = -1.43$ **c.** $z = -3.22$
   d. $z = -0.67$ **e.** $z = -4.58$ **f.** $z = 0.25$

9.107 Consider a hypothesis test concerning a population mean with σ known, underlying population normal, and alternative hypothesis H_a: $\mu \neq \mu_0$. Find the p value associated with each value of the test statistic.

   a. $z = -1.77$ **b.** $z = 1.43$ **c.** $z = 2.58$
   d. $z = -0.37$ **e.** $z = 3.58$ **f.** $z = 0.85$

9.108 Consider a hypothesis test concerning a population mean with σ known and n large. For each alternative hypothesis, value of the test statistic, and significance level, find the p value and determine whether H_0 is rejected or not rejected.

   a. H_a: $\mu > 12.5$, $z = 1.43$, $\alpha = 0.001$
   b. H_a: $\mu < -0.56$, $z = -2.05$, $\alpha = 0.05$
   c. H_a: $\mu \neq 1200$, $z = 1.75$, $\alpha = 0.10$
   d. H_a: $\mu > 37.7$, $z = 3.11$, $\alpha = 0.01$
   e. H_a: $\mu < 52.68$, $z = -1.16$, $\alpha = 0.025$
   f. H_a: $\mu \neq 46.68$, $z = -2.35$, $\alpha = 0.001$

9.109 Consider a hypothesis test concerning a population mean with σ known and underlying population normal. For each alternative hypothesis, value of the test statistic, and significance level, find the p value and determine whether H_0 is rejected or not rejected.

   a. H_a: $\mu > 3.14$, $z = 2.52$, $\alpha = 0.05$
   b. H_a: $\mu > 9.80$, $z = 1.39$, $\alpha = 0.05$
   c. H_a: $\mu < 186,000$, $z = -2.28$, $\alpha = 0.01$
   d. H_a: $\mu < 4.135$, $z = 0.17$, $\alpha = 0.01$
   e. H_a: $\mu \neq 1.62$, $z = -1.63$, $\alpha = 0.001$
   f. H_a: $\mu \neq 0.671$, $z = 2.96$, $\alpha = 0.001$

9.110 Consider a hypothesis test concerning a population mean with σ known and underlying population normal. For each hypothesis test find the p value and determine whether H_0 is rejected or not rejected.

	H_0	H_a	$\bar{x}$	σ	n	α
a.	$\mu = 10$	$\mu > 10$	11.50	7.56	18	0.05
b.	$\mu = 2.718$	$\mu < 2.718$	2.60	0.56	21	0.01
c.	$\mu = 57.72$	$\mu \neq 57.72$	56.42	1.58	14	0.01
d.	$\mu = -16.18$	$\mu > -16.18$	2.35	21.23	8	0.001
e.	$\mu = 273$	$\mu < 273$	275.80	17.80	15	0.05
f.	$\mu = 6.63$	$\mu \neq 6.63$	7.17	1.08	27	0.05

Applications

9.111 Public Health and Nutrition High levels of the chemical dioxin, ingested through food and air, can cause diabetes, immune disorders, and cancer. The U.S. EPA has set a maximum acceptable level of dioxin in blood at 10 ppt (parts per trillion). Residents living near an old incinerator known to produce dioxin air pollution have complained of many health problems. A random sample of people living within a 10-mile radius of the facility was obtained, and the dioxin level (in ppt) in the blood of each was measured. Is there any evidence to suggest that the mean dioxin level in the blood of residents living near the incinerator is greater than 10 ppt? Assume $\sigma = 2.66$, use $\alpha = 0.05$, and compute the p value.   **DIOXIN**

9.112 Sports and Leisure The National Football League rates each quarterback after every game according to a formula that involves pass attempts, passing yards, touchdown passes, and interceptions. Suppose the mean quarterback rating per game is 52.1 with $\sigma = 16.7$. There is some speculation that rule changes benefiting receivers have also increased quarterback ratings. In a random sample of 34 quarterback 2012 season ratings, the sample mean was $\bar{x} = 57.09$.[31] Is there any evidence to suggest that the true mean quarterback rating has increased? Use $\alpha = 0.05$ and compute the p value.

9.113 Medicine and Clinical Studies The active ingredient in antiseptic liquid bandages is 8-hydroxyquinoline (8h). This chemical can be harmful through simple skin contact, by swallowing, or inhalation. Suppose the list of ingredients on a 3M Nexcare liquid bandage indicates that the volume of 8h is 1%. A random sample of eight bottles of this product was obtained, and each was analyzed to obtain the volume of 8h. The sample mean was $\bar{x} = 1.025\%$. Assume the underlying population is normal and $\sigma = 0.04$. Is there any evidence to suggest that the true mean percentage of 8h is different from 1%? Use $\alpha = 0.01$ and compute the p value.

9.114 Fuel Consumption and Cars Jack Keef General Motors, an automobile dealer in Lincoln, Nebraska, advertised a \$9.99 40-point safety checkup for any car and claimed that the service department would finish the job in less than 30 minutes. A local newspaper reporter selected a random sample of 58 customers who took advantage of the dealer's offer and found the sample mean time to complete the safety checkup was $\bar{x} = 32.2$ minutes. Assume $\sigma = 5.7$ and use $\alpha = 0.01$.

   a. Is there any evidence to suggest that the true mean time to complete the safety checkup is greater than 30 minutes?
   b. Find the p value for the hypothesis test in part (a).

9.115 Sports and Leisure The Der Dachstein mountain in Austria has several hiking trails, lots of options for skiers, and a spectacular sky walk that extends out from a ledge to offer a view of the Alps. A new cable car takes passengers to approximately 2700 meters above sea level in approximately 5.5 minutes.[32] A random sample of cable car rides was selected and the time (in minutes) to reach the top was recorded for each. The data are given in the following table.   **CBLCAR**

5.7	6.4	6.1	5.2	8.7	5.4	5.9	4.7
7.1	6.1	8.9	6.2	6.8	6.0	4.5	6.1

Assume the underlying population is normal and $\sigma = 1.3$. Is there any evidence to suggest that the mean cable car time is

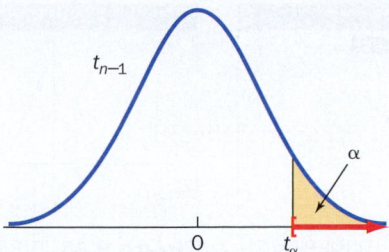

Figure 9.41 Rejection region for $H_a: \mu > \mu_0$.

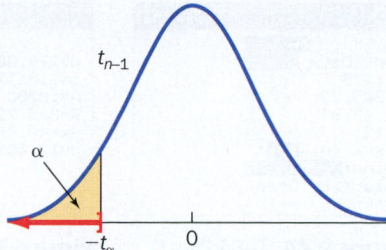

Figure 9.42 Rejection region for $H_a: \mu < \mu_0$.

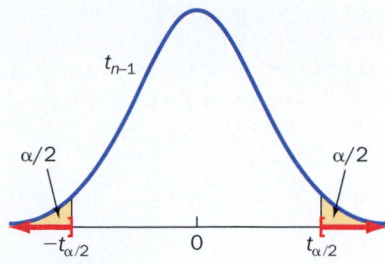

Figure 9.43 Rejection region for $H_a: \mu \neq \mu_0$.

The following example illustrates this hypothesis test procedure.

Solution Trail 9.14

KEYWORDS

- Is there any evidence?
- Mean residential bill
- Greater than 279
- $s = 40.05$
- Population is normal
- Random sample

TRANSLATION

- Conduct a one-sided, right-tailed test about μ.

CONCEPTS

- Hypothesis test concerning a population mean when σ is unknown

VISION

Use the template for a one-sided, right-tailed t test about μ. The underlying population is assumed to be normal, and σ is unknown. Determine the appropriate alternative hypothesis and the corresponding rejection region, find the value of the test statistic, and draw a conclusion.

Example 9.14 Residential Electric Bill

U.S. residential electric bills were expected to decrease slightly during the summer months of 2013, compared with 2012. This was partly due to decreased demand and weather patterns. Electric bills vary by region, and the mean residential bill for July 2013 in the Pacific region was estimated to be \$279.[36] A random sample of 12 homes in Oregon were selected and the electric bill for July 2013 was obtained for each. The summary statistics were $\bar{x} = 305.72$ dollars and $s = 40.05$ dollars. Is there any evidence to suggest that the true mean July residential electric bill in the Pacific region is greater than \$279? Assume the underlying distribution is normal, and use $\alpha = 0.05$.

SOLUTION

STEP 1 The assumed mean is $\mu = 279 \, (= \mu_0)$; $n = 12$, $s = 40.05$, and $\alpha = 0.05$.

We are looking for evidence to suggest that the mean residential electric bill is *greater* than the estimate. The relevant alternative hypothesis is one-sided, right-tailed.

The underlying population is assumed to be normal, but σ is unknown. A t test is appropriate.

STEP 2 The four parts of the hypothesis test are

$$H_0: \mu = 279$$
$$H_a: \mu > 279$$
$$\text{TS: } T = \frac{\bar{X} - \mu_0}{S/\sqrt{n}}$$
$$\text{RR: } T \geq t_{\alpha, n-1} = t_{0.05, 11} = 1.7959$$

STEP 3 The value of the test statistic is

$$t = \frac{\bar{x} - \mu_0}{s/\sqrt{n}} = \frac{305.72 - 279}{40.05/\sqrt{12}} = 2.3111 \geq 1.7959$$

STEP 4 Because 2.3111 lies in the rejection region, we reject the null hypothesis at the $\alpha = 0.05$ level. There is evidence to suggest that the mean July residential electric bill in the Pacific region is greater than 279 dollars.

Figures 9.44–9.46 show a technology solution. Figures 9.46 and 9.47 illustrate the p value associated with this test.

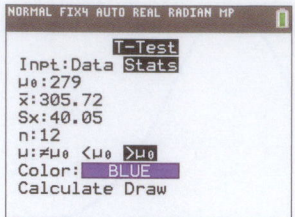

Figure 9.44 TI-84 Plus C T-Test input screen.

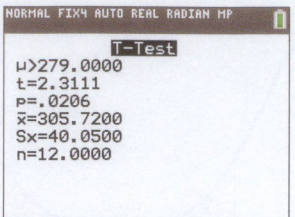

Figure 9.45 TI-84 Plus C T-Test Calculate results.

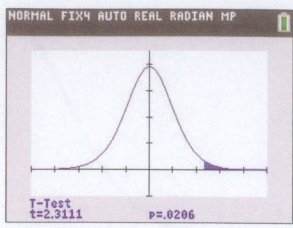

Figure 9.46 TI-84 Plus T-Test Draw results.

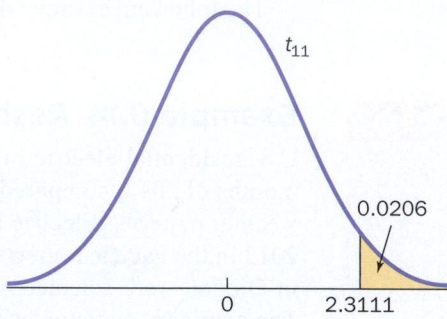

Figure 9.47 p-Value illustration: $p = P(T \geq 2.3111) = 0.0206 \leq 0.05 = \alpha$

TRY IT NOW GO TO EXERCISE 9.143

Example 9.15 Water Park Water

DATA SET

WTRPARK

Mayangsari/Dreamstime.com

Most water parks use a recycling system so that the only water loss is due to evaporation and splashing. Despite a sophisticated recycling system, the Schlitterbahn Waterpark in New Braunfels, Texas, has informed the city water department of their need for 250,000 gallons of water per day. The city water department selected a random sample of days, and the park's water usage (in thousands of gallons) on each day is given in the following table.

140.1	270.1	234.3	242.3	233.3	337.4	248.6	210.4
244.6	269.6	292.3	263.0	229.0	264.0	260.5	236.9
244.1	201.0	205.7	219.0	303.5			

Is there any evidence to suggest the mean water usage is different from 250,000 gallons? Assume the underlying population is normal and use $\alpha = 0.01$.

SOLUTION

STEP 1 The assumed mean is 250 ($= \mu_0$) thousand gallons, $n = 21$, and $\alpha = 0.01$.

The water department is looking for water usage *different* from 250,000. Therefore, the relevant alternative hypothesis is two-sided.

The underlying population is assumed to be normal, and σ is unknown. A t test is appropriate.

STEP 2   The four parts of the hypothesis test are

$H_0: \mu = 250$

$H_a: \mu \neq 250$

TS: $T = \dfrac{\bar{X} - \mu_0}{S/\sqrt{n}}$

RR: $|T| \geq t_{\alpha/2, n-1} = t_{0.005, 20} = 2.8453$ $(T \leq -2.8453 \text{ or } T \geq 2.8453)$

STEP 3   The sample mean is

$$\bar{x} = \frac{1}{21}(140.1 + 270.1 + \cdots + 303.5) = 245.2238$$

Find the sample standard deviation.

$$s^2 = \frac{1}{20}\left(1{,}296{,}137.59 - \frac{1}{21}(5149.7)^2\right) = 1665.4269$$

$$s = \sqrt{1665.4269} = 40.8096$$

The value of the test statistic is

$$t = \frac{\bar{x} - \mu_0}{s/\sqrt{n}} = \frac{245.2238 - 250}{40.8096/\sqrt{21}} = -0.5363$$

STEP 4   The value of the test statistic, $t = -0.5363$, does *not* lie in the rejection region. We cannot reject the null hypothesis. There is no evidence to suggest park water usage is different from 250,000 gallons per day.

Figure 9.48 shows a technology solution, and Figure 9.49 illustrates the p value associated with this test.

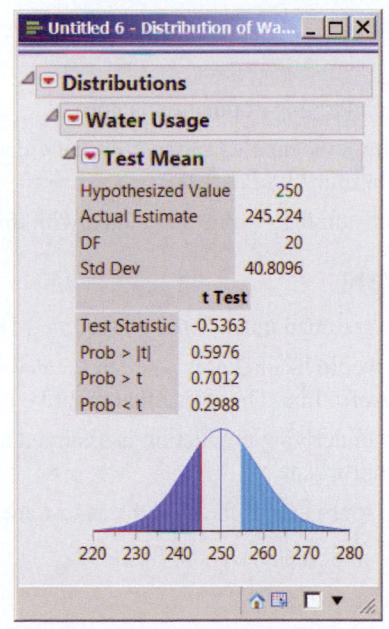

Figure 9.48 JMP Test Mean results.

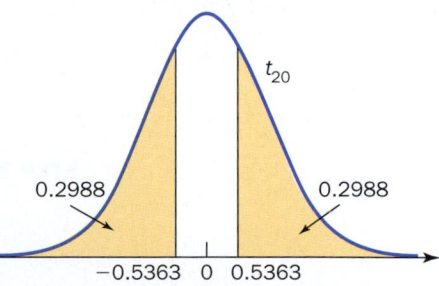

Figure 9.49 p-Value illustration:

$p = 2P(T \leq -0.5363)$

$\quad = 0.5976 > 0.01 = \alpha$

TRY IT NOW   GO TO EXERCISE 9.149

The table listing critical values for various t distributions is very limited. Using this table, the best we can do is *bound* the p value associated with a hypothesis test.

How to Bound the p Value for a t Test

Suppose t is the value of the test statistic in a one-sided hypothesis test.

1. Select the row in Table V in the Appendix that corresponds to $n - 1$, the number of degrees of freedom associated with the test.
2. Place $|t|$ in this ordered list of critical values.
3. To compute p:
    a. If $|t|$ is between two critical values in the $n - 1$ row, then the p value is bounded by the corresponding significance levels.
    b. If $|t|$ is greater than the largest critical value in the $n - 1$ row, then $p < 0.0001$ (the smallest significance level in the table).
    c. If $|t|$ is less than the smallest critical value in the $n - 1$ row, then $p > 0.20$ (the largest significance level in the table).

If $t < 0$ for a right-tailed test, or $t > 0$ for a left-tailed test, then $p > 0.5$. If the hypothesis test is two-sided, this method produces a bound on $p/2$.

Example 9.16 Delivery Time

DATA SET

DELIVERY

A certain daily delivery route for Hostess breads and snack cakes includes eight grocery stores and four convenience stores. The historical mean time to complete these deliveries (to the 12 stores) and return to the distribution center is 6.5 hours. A new driver has been assigned to this route, and a random sample of his route completion times (in hours) was obtained. The data are given in the following table.

6.61	6.25	6.40	6.57	6.35	5.95	6.53	6.29

Assume the underlying population is normal.

a. Is there any evidence to suggest the new driver has been able to shorten the route completion time? Use $\alpha = 0.01$.

b. Find bounds on the p value associated with this hypothesis test.

SOLUTION

STEP 1 The assumed mean is 6.5 ($= \mu_0$); $n = 8$ and $\alpha = 0.01$.

We would like to know whether the new driver has been able to shorten the mean delivery time. The relevant alternative hypothesis is one-sided, left-tailed.

The underlying population is assumed to be normal, and σ is unknown. A t test is appropriate.

STEP 2 The four parts of the hypothesis test are

$H_0: \mu = 6.5$

$H_a: \mu < 6.5$

TS: $T = \dfrac{\overline{X} - \mu_0}{S/\sqrt{n}}$

RR: $T \leq -t_{\alpha, n-1} = -t_{0.01,7} = -2.9980$

STEP 3 The sample mean is

$$\overline{x} = \frac{1}{8}(6.61 + 6.25 + \cdots + 6.29) = 6.3688$$

Find the sample standard deviation.

$$s^2 = \frac{1}{7}\left(324.8095 - \frac{1}{8}(50.95)^2\right) = 0.0460$$

$$s = \sqrt{0.0460} = 0.2144$$

The value of the test statistic is

$$t = \frac{\bar{x} - \mu_0}{s/\sqrt{n}} = \frac{6.3688 - 6.5}{0.2144/\sqrt{8}} = -1.7308$$

The value of the test statistic, $t = -1.7308$, does *not* lie in the rejection region. We cannot reject the null hypothesis. There is no evidence to suggest the new driver has been able to shorten the mean delivery time for this route.

STEP 4 $|t| = |-1.7308| = 1.7308$

In Table V in the Appendix A, row $n - 1 = 8 - 1 = 7$, place 1.7308 in the ordered list of critical values.

$$1.4149 \leq 1.7308 \leq 1.8946$$

$$t_{0.10,7} \leq 1.7308 \leq t_{0.05,7}$$

Therefore, $\quad 0.05 \leq \quad p \quad \leq 0.10$ $\qquad$ (See Figure 9.50.)

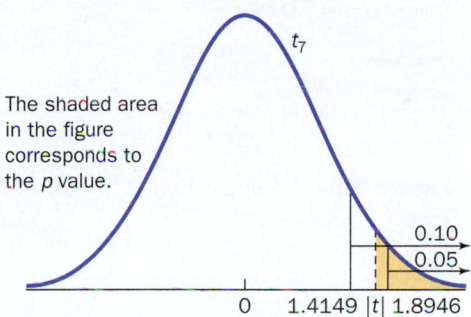

The shaded area in the figure corresponds to the *p* value.

0.10
0.05

0   1.4149 |t| 1.8946

Figure 9.50 Visualization of the bounds on the *p* value.

The *p* value associated with this hypothesis test is between 0.05 and 0.10. Because the smallest possible value, 0.05, is greater than the significance level, 0.01, we cannot reject the null hypothesis.

Figures 9.51 through 9.53 show a technology solution.

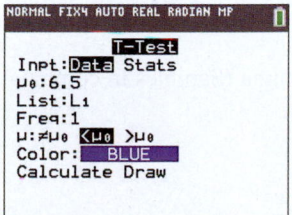

Figure 9.51 TI-84 Plus C T-Test input screen.

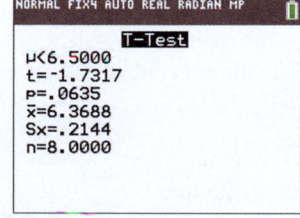

Figure 9.52 TI-84 Plus C T-Test Calculate results.

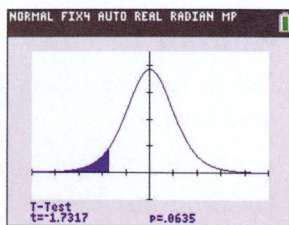

Figure 9.53 TI-84 Plus C T-Test Draw results.

TRY IT NOW   GO TO EXERCISE 9.153

Technology Corner

Procedure: Hypothesis test concerning a population mean when σ is unknown.
Reconsider: Example 9.16, solution, and interpretations.

CrunchIt!

Use the built-in function t 1-Sample. Input is either summary statistics or a column containing the sample data.

1. Enter the data in column Var1.
2. Select Statistics; t; 1-sample.
3. Under the Columns tab, select Var1 for the Sample. Under the Hypothesis Test tab, enter the value for μ_0 and select the appropriate Alternative. (Figure 9.54).
4. Click Calculate. The results are displayed in a new window (Figure 9.55).

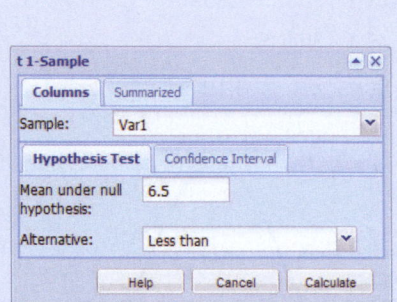

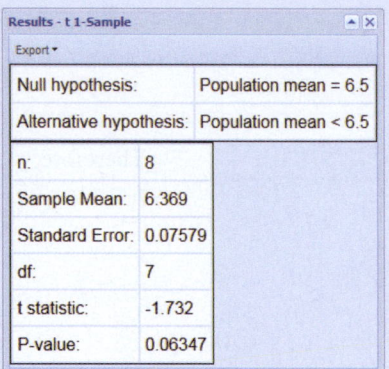

Figure 9.54 1-Sample input screen.

Figure 9.55 1-Sample test results.

TI-84 Plus C

Use the calculator function T-Test. The input for this function is either data in a list or summary statistics.

1. Enter the data into list L1.
2. Select STAT ; TESTS; T-Test.
3. The data are given in this example; highlight Data. Enter μ_0, the name of the list containing the data, the frequency of occurrence of each observation, and highlight the appropriate alternative hypothesis. Refer to Figure 9.50.
4. In the last row of the input screen, select Calculate and press ENTER .
5. The alternative hypothesis, value of the test statistic, p value, and summary statistics are displayed on the output screen. Refer to Figure 9.51.
6. The Draw results are shown in Figure 9.52.

Minitab

Use the function 1-Sample t. The input is either data in a column (Samples in columns) or summary statistics (Summarized data).

1. Enter the data into column C1.
2. Choose Stat; Basic Statistics; 1-Sample t.
3. In this example, select One or more samples, each in a column, and enter C1. Check the box to Perform a hypothesis test, and enter the Hypothesized mean (Figure 9.56).
4. Choose Options. Enter a Confidence level and select the Alternative hypothesis (Figure 9.57).
5. Summary statistics, the confidence interval, the value of the test statistic, and the p value are displayed in the Session window (Figure 9.58).

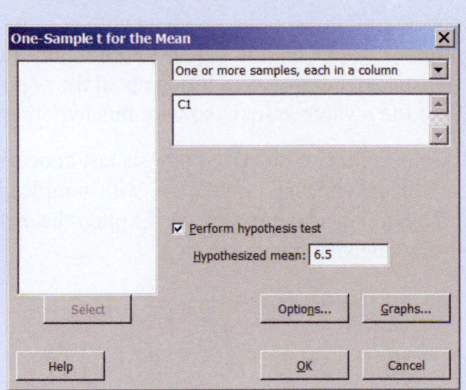

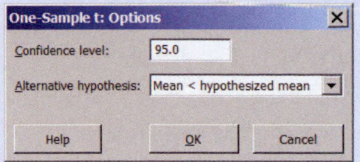

Figure 9.57 1-Sample t Options screen.

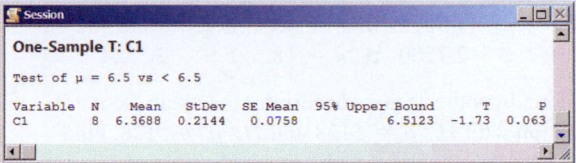

Figure 9.56 1-Sample t input screen.

Figure 9.58 1-Sample t hypothesis test results.

Excel

There is no built-in function for a one-sample t test. Compute the summary statistics and use the appropriate equation to find the value of the test statistic.

1. Enter the data into column A.
2. Compute the sample mean, sample standard deviation, and value of the test statistic. Use the appropriate Excel function to compute the p value. See Figure 9.59.

Figure 9.59 Excel computations for the value of the test statistic and the p value.

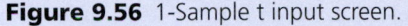

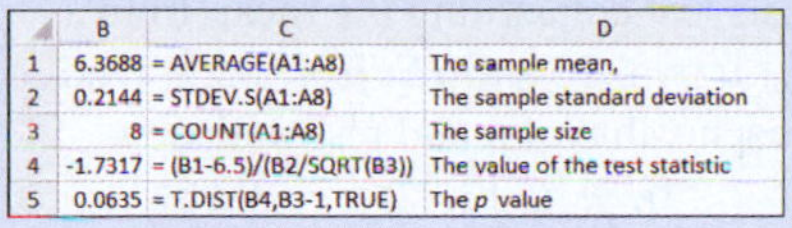

	B	C	D
1	6.3688	= AVERAGE(A1:A8)	The sample mean,
2	0.2144	= STDEV.S(A1:A8)	The sample standard deviation
3	8	= COUNT(A1:A8)	The sample size
4	-1.7317	= (B1-6.5)/(B2/SQRT(B3))	The value of the test statistic
5	0.0635	= T.DIST(B4,B3-1,TRUE)	The p value

SECTION 9.5 EXERCISES

Concept Check

9.124 True/False In a hypothesis test concerning a population mean when σ is unknown, the test statistic is based on a t distribution with $n - 1$ degrees of freedom.

9.125 True/False The hypothesis test concerning a population mean when σ is unknown can only be used when the underlying population is normal.

9.126 True/False In a hypothesis test concerning a population mean when σ is unknown, there are only three possible alternative hypotheses.

9.127 True/False In a hypothesis test concerning a population mean when σ is unknown, the p value can never be determined.

9.128 True/False In a hypothesis test concerning a population mean when σ is unknown, the test statistic is based on the standard normal distribution when the sample size is greater than 30.

Practice

9.129 Consider a hypothesis test concerning a population mean from a normal population with $H_0: \mu = 10.5$ and $H_a: \mu > 10.5$.

a. Write the appropriate test statistic.
b. Find the rejection region corresponding to each value of n and α.
   i. $n = 6,$ $\alpha = 0.01$ **ii.** $n = 23,$ $\alpha = 0.025$
   iii. $n = 17,$ $\alpha = 0.05$ **iv.** $n = 29,$ $\alpha = 0.10$
   v. $n = 10,$ $\alpha = 0.001$ **vi.** $n = 9,$ $\alpha = 0.0001$

9.130 Consider a hypothesis test concerning the mean from a normal population with $H_0: \mu = 22.41$ and $H_a: \mu < 22.41$.

a. Write the appropriate test statistic.
b. Find the rejection region corresponding to each value of n and α.
   i. $n = 15,$ $\alpha = 0.01$ **ii.** $n = 11,$ $\alpha = 0.0005$
   iii. $n = 21,$ $\alpha = 0.05$ **iv.** $n = 24,$ $\alpha = 0.10$
   v. $n = 5,$ $\alpha = 0.001$ **vi.** $n = 31,$ $\alpha = 0.0001$

9.131 Consider a hypothesis test concerning the mean from a normal population with $H_0: \mu = 1.67$ and $H_a: \mu \neq 1.67$.

a. Write the appropriate test statistic.
b. Find the rejection region corresponding to each value of n and α.
   i. $n = 12,$ $\alpha = 0.01$ **ii.** $n = 19,$ $\alpha = 0.20$
   iii. $n = 26,$ $\alpha = 0.05$ **iv.** $n = 28,$ $\alpha = 0.10$
   v. $n = 7,$ $\alpha = 0.001$ **vi.** $n = 56,$ $\alpha = 0.02$

9.132 Consider a hypothesis test concerning the mean from a normal population with $H_0: \mu = 0.082$ and $H_a: \mu > 0.082$.

Find the significance level (α) corresponding to each value of n and rejection region.

  a. $n = 27$, $T \geq 2.0555$ **b.** $n = 9$, $T \geq 4.5008$
  c. $n = 16$, $T \geq 2.6025$ **d.** $n = 20$, $T \geq 3.5794$

9.133 Consider a hypothesis test concerning the mean from a normal population with H_0: $\mu = -15.76$ and H_a: $\mu < -15.76$. Find the significance level (α) corresponding to each value of n and rejection region.

  a. $n = 23$, $T \leq -1.3212$ **b.** $n = 4$, $T \leq -10.2145$
  c. $n = 31$, $T \leq -2.7500$ **d.** $n = 18$, $T \leq -2.5524$

9.134 Consider a hypothesis test concerning the mean from a normal population with H_0: $\mu = 5128$ and H_a: $\mu \neq 5128$. Find the significance level (α) corresponding to each value of n and rejection region.

  a. $n = 30$, $|T| \geq 1.6991$ **b.** $n = 6$, $|T| \geq 4.0321$
  c. $n = 17$, $|T| \geq 3.6862$ **d.** $n = 14$, $|T| \geq 5.1106$

9.135 Consider a hypothesis test concerning the mean from a normal population with H_0: $\mu = 2.53$ and H_a: $\mu > 2.53$. Find bounds on the p value for each value of n and test statistic t.

  a. $n = 24$, $t = 2.35$ **b.** $n = 3$, $t = 8.55$
  c. $n = 12$, $t = 5.68$ **d.** $n = 8$, $t = 1.52$

9.136 Consider a hypothesis test concerning the mean from a normal population with H_0: $\mu = 6.28$ and H_a: $\mu < 6.28$. Find bounds on the p value for each value of n and test statistic t.

  a. $n = 25$, $t = -1.97$ **b.** $n = 13$, $t = -0.63$
  c. $n = 18$, $t = -2.28$ **d.** $n = 27$, $t = -3.58$

9.137 Consider a hypothesis test concerning the mean from a normal population with H_0: $\mu = 1.414$ and H_a: $\mu \neq 1.414$. Find bounds on the p value for each value of n and test statistic t.

  a. $n = 10$, $t = 2.04$ **b.** $n = 23$, $t = -3.14$
  c. $n = 14$, $t = -5.52$ **d.** $n = 11$, $t = 1.75$

9.138 Consider a random sample of size 20 from a normal population with hypothesized mean 1.618.

  a. Write the four parts for a one-sided, left-tailed hypothesis test concerning the population mean with $\alpha = 0.05$.
  b. Suppose $\bar{x} = 1.5$ and $s = 0.45$. Find the value of the test statistic, and draw a conclusion about the population mean.
  c. Find the p value associated with this hypothesis test.

9.139 Consider a random sample of 11 observations, given in the following table, from a normal population with hypothesized mean 57.71. ▮▮ EX9.139

59.94	58.93	59.41	60.66	59.00	60.98
58.85	55.21	59.02	61.14	59.25	

  a. Write the four parts for a one-sided, right-tailed hypothesis test concerning the population mean with $\alpha = 0.01$.
  b. Compute the sample mean, the sample standard deviation, and the value of the test statistic. Draw a conclusion about the population mean.
  c. Find the p value associated with this hypothesis test.

9.140 Consider a random sample of size 28 from a normal population with hypothesized mean $\mu_0 = 9.96$.

  a. Write the four parts for a two-sided hypothesis test concerning the population mean with $\alpha = 0.002$.
  b. Suppose $\bar{x} = 9.04$ and $s = 1.20$. Find the value of the test statistic, and draw a conclusion about the population mean.
  c. Find the p value associated with this hypothesis test.

9.141 Consider a two-sided hypothesis test concerning the mean, μ_0, from a normal population, with sample size $n = 25$, standard deviation s, and $\alpha = 0.05$. Explain the error in each of the following statements.

  a. The test statistic is $T = \dfrac{\bar{X} - \mu_0}{\sigma/\sqrt{25}}$.
  b. The test statistic is $T = \dfrac{\bar{X} - \mu_0}{S/\sqrt{24}}$.
  c. The rejection region is RR: $T \geq 1.7109$.
  d. If the value of the test statistic is $t = 2.6732$, then $p \leq 0.005$.

Applications

9.142 Biology and Environmental Science   The Atlantic bluefin tuna is the largest and most endangered of the tuna species. They are found throughout the North Atlantic Ocean and the mean weight is 550 pounds.[37] The Center for Biological Diversity is concerned that this species has been overfished and that the mean weight has decreased. Suppose a random sample of 12 Atlantic bluefin tuna was obtained from commercial fishing boats and weighed. The summary statistics were $\bar{x} = 535.7$ and $s = 37.8$. Conduct a hypothesis test to determine whether there is any evidence that the mean weight of Atlantic bluefin tuna is less than 550 pounds. Assume the distribution of weight is normal and use $\alpha = 0.05$. Write a Solution Trail for this problem.

9.143 Travel and Transportation   During a routine commercial airline flight, much of the time in the aircraft is spent waiting to take off and taxiing to an arrival gate. However, airlines also keep careful records of the actual air time of each flight. A random sample of the air times (in minutes) of US Airways flights from Philadelphia to Las Vegas during December 2012 was obtained, and the data are given in the following table.[38] ▮▮ FLIGHTS

297	301	299	306	307	300	308	327
316	300	323	313	348	303	316	

The air time for this flight is affected by the prevailing west-to-east winds and weather systems and normally takes 303 minutes. Is there any evidence to suggest the true mean air time is greater than this 303 minutes? Assume the underlying distribution is normal and use $\alpha = 0.025$.

9.144 Biology and Environmental Science   The historical mean yield of soybeans from farms in Indiana is 31.9 bushels per acre. Following a recent dry summer, a random sample of 26 farms across the state was obtained. The mean yield in bushels per acre was $\bar{x} = 30.088$ with $s = 4.433$. Is there any evidence to suggest the lack of rain adversely affected the soybean yield in Indiana? Assume the underlying population is normal and use $\alpha = 0.01$.

9.145 Public Policy and Political Science In 2013, the mayor of North Olmsted, Ohio, began a mayor's court designed to hear cases involving traffic citations and misdemeanor charges. The purpose was to address the case backlog and to add to the city's revenue. During the first quarter of 2013, the mean fine was $64.98.[39] Later in the year, a random sample of 14 fines from the mayor's court was obtained. The summary statistics were $\bar{x} = 70.45$ and $s = 15.59$. Is there any evidence to suggest that the mean fine has increased? Assume the underlying distribution is normal and use $\alpha = 0.01$. Write a Solution Trail for this problem.

9.146 Business and Management The 40-hour work week did not become a U.S. standard until 1940. Today, many white-collar employees work more than 40 hours per week because management demands longer hours or offers large monetary incentives. A random sample of white-collar employees was obtained, and the number of hours each worked during the last week are given in the following table. WORK

44.7	42.0	45.8	43.0	42.8	50.9	47.0
41.9	49.3	45.6	45.7	39.4	39.0	44.4

Is there any evidence the true mean number of hours worked by white-collar employees is greater than 40? Use $\alpha = 0.01$. What assumption(s) did you make in order to complete this hypothesis test?

9.147 Marketing and Consumer Behavior Kennebunkport, Maine, is a popular tourist town, but businesses suffer during the winter months, especially January. The mean hotel room occupation rate per day, a measure of tourist activity, is 23.1% during winter months. A new advertising campaign was launched to attract more tourists to this town during the winter. Following the campaign, a random sample of nine winter days was selected. The mean hotel room occupation rate was $\bar{x} = 24.6\%$ and $s = 2.1\%$. Is there any evidence to suggest the mean hotel room occupation rate has increased? Assume normality and use $\alpha = 0.01$. If this test is significant, can you conclude the ad campaign caused the increase?

9.148 Economics and Finance Economic growth in China has contributed to increased demand for crude oil and rising gasoline prices in several countries. In May 2013, the mean crude oil imported per day in China was 5.64 million barrels.[40] A random sample of 17 days in the last quarter of 2013 was obtained and the crude oil imported was recorded for each (in million barrels per day, bpd). The summary statistics were $\bar{x} = 5.869$ million bpd and $s = 1.16$ million bpd. Is there any evidence to suggest that the mean crude oil imported per day in China has increased? Use $\alpha = 0.05$ and assume normality.

9.149 Medicine and Clinical Studies According to the Organization for Economic Co-operation and Development (OECD), the mean density of physicians per 1000 people worldwide is 3.2.[41] A random sample of countries was obtained

and the physician density per 1000 people was recorded for each. The data are given in the following table. OECD

1.6	3.6	3.5	3.3	3.3	3.3	3.8	6.1	3.0	3.5	2.7
3.3	4.1	2.2	2.0	3.0	2.2	3.0	2.6	3.7	2.2	

Is there any evidence to suggest that the mean physician density is different from 3.2? Use $\alpha = 0.05$ and assume normality.

9.150 Medicine and Clinical Studies Over time, certain metabolic changes can cause the lens of the eye to become opaque, which leads to loss of vision. In this case, many adults elect to have cataract surgery in which the natural lens of the eye is removed and an artificial lens implant is inserted. A random sample of cataract surgeries was obtained and the length of each surgery was recorded (in hours).[42] Past records indicate that the mean time for this type of surgery is 0.75 hour. Assume the underlying population is normal. CATRACT
   a. Is there any evidence to suggest that the mean time has decreased? Use $\alpha = 0.01$.
   b. Find bounds on the p value associated with this test.

9.151 Business and Management In June 2013 the largest building in the world opened in Chengdu, China. The Century Global Center has an ice-skating rink, shopping mall, 14-screen IMAX theater, and an artificial beach.[43] Some researchers believe that as the economy improves, larger buildings are being constructed. A random sample of buildings from around the world with large floor space was obtained and the floor area (in million square feet) was measured for each. BIGBLDG
   a. Is there any evidence to suggest that the mean floor area of very large buildings is greater than 9 million square feet? Assume the underlying population is normal and use $\alpha = 0.05$.
   b. Find bounds on the p value.

9.152 Marketing and Consumer Behavior As the use of the Internet has increased, companies are spending more money advertising online. A sample of leading national advertisers was obtained, and the Internet advertising spending (in millions of dollars) for the year was recorded for each. The data are given in the following table. ADVERT

Company	Spending
Procter & Gamble Co.	80.6
Verizon Communications	189.1
Time Warner	98.4
Walt Disney Co.	141.4
General Electric Co.	78.9
Toyota Motor Corp.	57.3
Sony Corp.	71.5
Kraft Foods	36.3
Macy's	11.1
PepsiCo	28.1
Pfizer	24.2
McDonald's Corp.	27.2

Suppose the mean spending on Internet advertising for these companies the previous year was $55.5 million. Is there any evidence to suggest the true mean spending on Internet advertising has increased? Use $\alpha = 0.05$.

Extended Applications

9.153 Demographics and Population Statistics   Police reports and insurance claims indicate that the mean loss during a residential break-in is $1381. Actuaries working for Eagle Pacific Insurance Company need to check whether the mean loss has changed during the past year in order to determine new policy rates. A random sample of 17 residential break-ins was obtained, and the loss due to each burglary was recorded. The sample mean loss (in dollars) was 1857 and the standard deviation was $s = 786$. Assume the loss distribution is normal and use $\alpha = 0.05$.

   a. Is there any evidence to suggest the true mean loss as a result of a residential break-in has changed?

   b. Find bounds on the p value associated with this hypothesis test.

   c. Carefully sketch a graph illustrating the p value and the value of the test statistic.

9.154 Sports and Leisure   It's legal to bet on sports in Nevada, and some research suggests that Americans wage at least $500 billion each year on sporting events. However, only about 1% of this betting is legal. In 2013, New Jersey filed a suit to win the right to allow betting on all the major sports. The purpose was to increase state revenue from this billion-dollar business. According to *Baseball Press*, the mean amount bet on each Major League baseball game is $8.85 million.[44] A random sample of Major League baseball games is obtained, and the amount legally bet in Nevada on each game is given in the following table (in millions of dollars).   ▬ **BETS**

9.52	13.27	12.08	10.38	13.73	14.23	10.27	10.97
16.35	10.47	6.23	5.52	14.02	11.76	10.32	5.49

   a. Is there any evidence to suggest that the mean amount bet on Major League baseball games has increased? Assume the distribution is normal and use $\alpha = 0.05$.

   b. Find bounds on the p value associated with this hypothesis test.

9.155 Fuel Consumption and Cars   Coalbed methane is an important source of natural gas in the United States. The Powder River coalfield has approximately 2500 wells, each producing on average 159,350 cubic feet of gas per day. To maintain sufficient storage facilities, the mean well output is carefully monitored. A random sample of 11 wells was obtained, and the daily methane output for each was recorded. The sample mean was $\bar{x} = 163,288$ cubic feet and the standard deviation was $s = 8792$ cubic feet. Assume the distribution of methane output per well per day is normal, and use $\alpha = 0.01$.

   a. Is there any evidence to suggest the mean methane output per well per day has increased?

   b. Find bounds on the p value associated with this hypothesis test.

9.156 Marketing and Consumer Behavior   The manager of a Piggly Wiggly grocery store claims a membership card will save consumers (through automatic discounts and extra coupons) at least $15.00 per week on average. To check this claim, a random sample of 11 shoppers with membership cards was obtained and their weekly grocery bills were inspected. The sample mean savings was $\bar{x} = \$14.35$ and $s = \$3.75$. Assume the distribution of savings is normal.

   a. Is there any evidence to refute the manager's claim? Use $\alpha = 0.025$.

   b. If you rejected the null hypothesis, how can you explain this conclusion when $14.35 is so close to $15.00? If you did not reject the null hypothesis, how can you explain this conclusion when $14.35 is certainly less than $15.00?

   c. Find bounds on the p value associated with this hypothesis test.

9.157 Biology and Environmental Science   Lionfish are native to Indo-Pacific waters but were introduced into the Atlantic in the 1990s. These spiney fish have huge appetites, feed on native species, and have a mean length of 13.5 inches.[45] Recently, invasive lionfish were found 300 feet below the water's surface off the coast of Florida. There was some speculation that these fish were larger than previously observed. A random sample of 24 lionfish was obtained near the sunken *Bill Boyd* cargo ship close to Fort Lauderdale, and the length of each was recorded (in inches). The summary statistics were $\bar{x} = 16.005$ and $s = 4.039$.

   a. Is there any evidence to suggest that the true mean length of lionfish in this area near Fort Lauderdale is greater than 13.5 inches? Use $\alpha = 0.01$ and assume the underlying population is normal.

   b. Find bounds on the p value associated with this test.

9.158 Sports and Leisure   Triathlons involve swimming, cycling, and running in succession over various distances. The July 2013 Sydenham Lakeside Olympic Triathlon near North Kingston, Ontario, featured a 1500-m swim, 40-km cycling course, and a 10-km run. A random sample of racers was obtained, and their finish times are given in the table below (in hours).[46]   ▬ **TRIATH**

2.07	2.21	2.32	2.36	2.51	2.54
2.64	2.65	2.71	2.92	3.09	3.50

Food vendors at the finish line planned to stay for 2.5 hours. Is there any evidence to suggest that the mean time to finish this triathlon was more than 2.5 hours? Assume the underlying distribution is normal and use $\alpha = 0.01$.

9.159 Biology and Environmental Science A quahog is a chewy Atlantic hard-shell clam with a blue-gray shell. Quahogs have different names, depending on size; for example, the width of a littleneck clam is under 5 centimeters (across the shell). The historical mean width of a littleneck is 4.75 cm. Following a recent oil spill off the coast of Maine, a random sample of 46 littlenecks was obtained and the width of each was recorded. The summary statistics were $\bar{x} = 4.66$ cm and $s = 0.25$ cm.

a. Use a one-sided, left-tailed hypothesis test to show that there is evidence to suggest the mean width of littlenecks has decreased. Assume the population of littleneck widths is normal and use $\alpha = 0.01$.

b. Explain why there is statistical evidence that the population mean has decreased even though the sample mean (4.66) is so close to the historical mean (4.75).

c. Find bounds on the p value associated with this hypothesis test.

9.160 Physical Sciences The Palmer Drought Severity Index (PDSI) is a measure of prolonged abnormal dryness or wetness. It indicates general conditions and is not affected by local variations. An index value of -4 indicates extreme drought, and $+4$ represents very wet conditions. A random sample of the PDSI for selected regions in the Eastern United States for the week ending July 6, 2013, was obtained, and the data are given in the following table.[47] **DROUGHT**

0.56	2.51	2.41	1.75	-0.07	-1.34	1.79
2.08	-0.42	2.62	1.74	2.61	5.11	1.53
1.29	1.30	-0.28	-1.14	-0.85	3.63	2.10

a. Is there any evidence to suggest the true mean PDSI is different from 0? Assume the underlying distribution is normal and use $\alpha = 0.01$. Find bounds on the p value associated with this test.

b. Explain your conclusion in part (a) in terms of farming conditions and reservoir levels.

9.161 Physical Sciences Alaska, California, and Hawaii are the states with the most earthquakes, and over 50% of the earthquakes in the United States occur in Alaska. A random sample of earthquakes around the world during July 2013 was obtained, and the magnitude of each is given in the following table.[48] **QUAKE**

1.5	0.4	0.6	0.5	0.3	0.6	1.1	1.0	1.7	1.4
1.9	1.6	1.9	1.2	1.5	1.6	5.3	4.5	0.5	1.9
1.3	0.5	0.3	2.0	2.4	0.5	1.8	0.6		

a. An earthquake with magnitude of 1.0 to 4.0 is considered weak and generally causes minor or no damage. Is there any evidence to suggest the true mean magnitude of earthquakes is greater than 1.0? Assume the underlying distribution is normal and use $\alpha = 0.05$.

b. Find bounds on the p value associated with this test.

9.162 Public Health and Nutrition Fluoride is added to public water supplies and many dental products in order to help prevent tooth decay. Sodium fluoride is a common additive in toothpaste, and the concentration is usually measured in parts per million by weight (ppmF). The manufacturer of Pepsodent toothpaste claims the concentration of fluoride in every tube is 1000 ppmF. A random sample of toothpaste tubes was obtained, and the fluoride concentration in each was determined. Some research suggests that high concentrations of fluoride can be toxic and may cause brain damage, immune disorders, and changes in bone structure and strength. **FLUOR**

a. Test the relevant hypothesis concerning the mean fluoride concentration per tube of toothpaste. Assume normality and use $\alpha = 0.05$.

b. Find bounds on the p value associated with this hypothesis test.

9.163 Public Health and Nutrition A recent study suggested that artificially sweetened soft drinks actually have a negative impact on weight and other health issues. Aspartame is the most common sweetener used in diet sodas. A random sample of diet sodas was obtained and the amount of aspartame in an 8-ounce can was recorded for each (in mg). The data are given in the following table.[49] **ASPRTME**

125	125	0	58	118	118	83
123	57	50	50	19	66	0

Assume the underlying population is normal.

a. Is there any evidence to suggest that the mean amount of aspartame in an 8-ounce can of diet soda is greater than 65 mg? Use $\alpha = 0.01$.

b. Find bounds on the p value associated with this test.

9.164 Marketing and Consumer Behavior Red Rocks Park in Morrison, Colorado, is a naturally formed, rock, open-air amphitheater 6450 feet above sea level. A random sample of events from Summer 2013 was obtained and the attendance for each was recorded. The summary statistics were $\bar{x} = 8722$, $s = 460.4$, and $n = 11$. Assume the distribution of attendance is normal and use $\alpha = 0.01$.

a. The amphitheater operators believe that 9000 people is the optimal attendance in terms of comfort and profit. Is there any evidence to suggest that the true mean attendance is different from 9000?

b. Find bounds on the p value associated with this test.

c. How large a sample size would be necessary for this test to be significant (assuming $\bar{x}$ and s remain the same)?

Challenge

9.165 Psychology and Human Behavior A small-sample (exact) hypothesis test concerning the mean of a Poisson random variable, λ, is constructed in the following manner. Suppose the random variable X is a count, modeled by a Poisson random variable. The four parts of the hypothesis test are

$H_0: \lambda = \lambda_0$

$H_a: \lambda > \lambda_0, \quad \lambda < \lambda_0, \quad \text{or} \quad \lambda \neq \lambda_0$

TS: X = the number of events that occur in the specified interval

RR: $X \geq x_\alpha, \quad X \leq x'_\alpha, \quad \text{or} \quad X \leq x_{\alpha/2} \quad \text{or} \quad X \geq x'_{\alpha/2}$

The critical values $x_\alpha, x'_\alpha, x_{\alpha/2}$, and $x'_{\alpha/2}$ are obtained from the Poisson distribution with parameter λ_0 to yield the desired significance level α.

At a large hotel in Los Angeles, on average four people per day forget to take their room key and, therefore, lock themselves out. In an effort to lower the number of these service calls, the hotel staff has placed special signs on the back of every hotel room door reminding visitors to take their key. On a randomly selected day, two people were locked out of their rooms.

   a. Write the four parts of a small-sample, one-sided, left-tailed test concerning the mean number of people locked out of their room, based on a Poisson distribution. Use $\alpha = 0.05$.

   b. Is there any evidence to suggest the mean has decreased?

   c. Find the p value associated with this hypothesis test.

9.166 Public Policy and Political Science A large-sample test concerning the mean of a Poisson random variable is based on a normal approximation. If X has a Poisson distribution with (large) mean λ, then X is approximately normal with mean λ and standard deviation $\sqrt{\lambda}$.

The four parts of the large-sample hypothesis test are

$H_0: \lambda = \lambda_0$

$H_a: \lambda > \lambda_0, \quad \lambda < \lambda_0, \quad \text{or} \quad \lambda \neq \lambda_0$

TS: $Z = \dfrac{X - \lambda_0}{\sqrt{\lambda_0}}$

   X = the number of events that occur in the specified *interval*

RR: $Z \geq z_\alpha, \quad Z \leq -z_\alpha, \quad \text{or} \quad |Z| \geq z_{\alpha/2}$

Postal workers who deliver mail are at high risk for dog bites. The number of dog bites to postal employees in the United States per (fiscal) year peaked during the mid-1980s, but with increased public awareness and employee training the number steadily decreased to less than 3000 per year. However, in fiscal year 2002, the number of dog bites began to rise again, especially during the "dog days" of summer. In 2012 the mean number of dog bites to postal workers per day was 19.[50] A random day in July 2013 was obtained and the number of dog bites to postal workers was 24.

   a. Write the four parts of a large-sample, one-sided, right-tailed test concerning the mean number of dog bites to postal employees per day, based on a normal approximation to the Poisson distribution. Use $\alpha = 0.01$.

   b. Is there any evidence to suggest the mean number of dog bites to postal employees per day has increased?

   c. Find the p value associated with this hypothesis test.

9.6 Large-Sample Hypothesis Tests Concerning a Population Proportion

Many experiments and observational studies are conducted in order to draw a conclusion about a population proportion. For example, a quality control inspector may need to decide whether the proportion of defective products in a delivery is greater than 0.05. Based on a random sample, a decision is made either to accept the delivery or to send the entire shipment back. Medical researchers routinely assess the proportion of patients who recover from various illnesses. This information may be used to determine whether there is evidence a new drug performs better than an existing treatment, or whether the proportion of patients who recover is greater than some threshold value.

These decisions involve p, the true population proportion, the fraction of individuals or objects with a specific characteristic, or the probability of a success. As in Sections 7.3 and 8.4, it is reasonable to use the sample proportion, $\hat{p}$, as an estimate of the population proportion p. In a sample of n individuals or objects, let X be the number of individuals with the relevant characteristic. Recall the definition of the random variable $\hat{P}$, the sample proportion:

$$\hat{P} = \frac{X}{n} = \frac{\text{the number of individuals with the characteristic}}{\text{the sample size}} \qquad (9.1)$$

From Section 7.3, if n is large and both $np \geq 5$ and $n(1 - p) \geq 5$, then the random variable $\hat{P}$ is approximately normal with mean p and variance $p(1 - p)/n$: $\hat{P} \overset{\cdot}{\sim} \mathrm{N}(p, p(1 - p)/n)$. These results concerning the distribution of $\hat{P}$ are used to construct a general procedure for a hypothesis test concerning p.

Large-Sample Hypothesis Tests Concerning a Population Proportion

Given a random sample of size n, a large-sample hypothesis concerning the population proportion p with significance level α has the form

$H_0: p = p_0$

$H_a: p > p_0, \quad p < p_0, \quad$ or $\quad p \neq p_0$

$$\text{TS: } Z = \frac{\hat{P} - p_0}{\sqrt{\dfrac{p_0(1 - p_0)}{n}}}$$

$\text{RR: } Z \geq z_\alpha, \quad Z \leq -z_\alpha, \quad$ or $\quad |Z| \geq z_{\alpha/2}$

This is the template for a hypothesis test about a population proportion. The hypothesized value is p_0.

As usual, use only one (appropriate) alternative hypothesis and the corresponding rejection region.

A CLOSER LOOK

1. This test is valid as long as $np_0 \geq 5$ and $n(1 - p_0) \geq 5$ (the nonskewness criterion).
2. The critical values for this test are from the standard normal distribution, Z (as in a hypothesis test about a population mean when σ is known).

The following example illustrates this hypothesis test procedure.

Example 9.17 Online Diagnosers

It's hard to believe, but doctors used to make house calls! Today, many people first consult the Internet when they are sick, starting with a search engine and then a specialized health site such as WebMD. According to a Pew Research Center survey, 35% of U.S. adults have used the Internet specifically to diagnose a certain medical condition.[51] A medical insurance company is concerned that the percentage of online diagnosers is increasing in response to rising health care costs. A random sample of 2400 U.S. adults was obtained, and 871 indicated that they diagnose online. Is there any evidence to suggest that the proportion of online diagnosers has increased? Use $\alpha = 0.01$.

SOLUTION

STEP 1 The given information:

Sample size: $n = 2400$.

Number of people with the specific characteristic, online diagnoser: $x = 871$.

The sample proportion: $\hat{p} = x/n = 871/2400 = 0.3629$.

The assumed value of the population proportion is $0.35 \, (= p_0)$, and the significance level is $\alpha = 0.01$.

STEP 2 Check the nonskewness criterion.

$np_0 = (2400)(0.35) = 840 \geq 5$

$n(1 - p_0) = (2400)(0.65) = 1560 \geq 5$

Both inequalities are satisfied, so $\hat{P}$ is approximately normal, and the large-sample hypothesis test concerning a population proportion can be used.

VIDEO TECH MANUALS

ONE PROPORTION INFERENCE CI - SUMMARIZED DATA

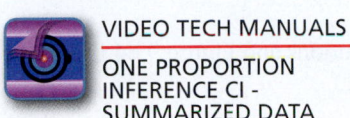

Solution Trail 9.17

KEYWORDS

- Is there any evidence?
- Proportion increased
- 2400 U.S. adults
- 871 online diagnosers
- 35%
- Random sample

TRANSLATION

- Conduct a one-sided, right-tailed test about p
- $n = 2400$
- $x = 871$
- $p_0 = 0.35$

CONCEPTS

- Large-sample hypothesis test concerning a population proportion

VISION

Check the nonskewness criterion. Use the template for a one-sided, right-tailed test about p. Use $\alpha = 0.01$ to find the critical value, compute the value of the test statistic, and draw a conclusion.

STEP 3 The four parts of the hypothesis test are

$$H_0: p = 0.35$$

$$H_a: p > 0.35$$

$$\text{TS: } Z = \frac{\hat{P} - p_0}{\sqrt{\dfrac{p_0(1 - p_0)}{n}}}$$

$$\text{RR: } Z \geq z_\alpha = z_{0.01} = 2.3263$$

STEP 4 The value of the test statistic is

$$z = \frac{\hat{p} - p_0}{\sqrt{\dfrac{p_0(1 - p_0)}{n}}} = \frac{0.3629 - 0.35}{\sqrt{\dfrac{(0.35)(0.65)}{2400}}} = 1.3250$$

STEP 5 The value of the test statistic does *not* lie in the rejection region. Equivalently, $p = 0.0923 > 0.01$ as shown in the figures below. We do not reject the null hypothesis. There is no evidence to suggest the true population proportion of online diagnosers is greater than 0.35. There is no evidence to suggest the true proportion of online diagnosers has increased.

Figures 9.60 through 9.63 show technology solutions and Figure 9.64 illustrates the p value associated with the hypothesis test.

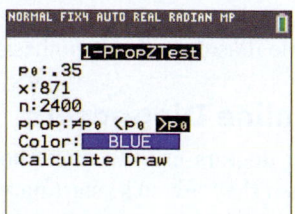

Figure 9.60 TI-84 Plus C `1-PropZTest` input screen.

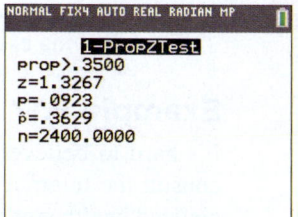

Figure 9.61 TI-84 Plus C `1-PropZTest Calculate` results.

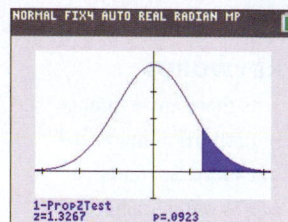

Figure 9.62 TI-84 Plus C `1-PropZTest Draw` results.

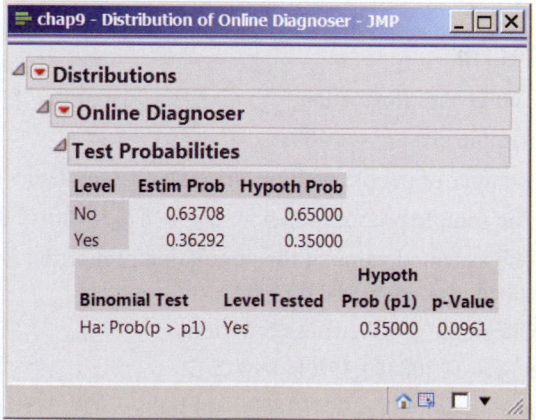

Figure 9.63 JMP hypothesis test results.

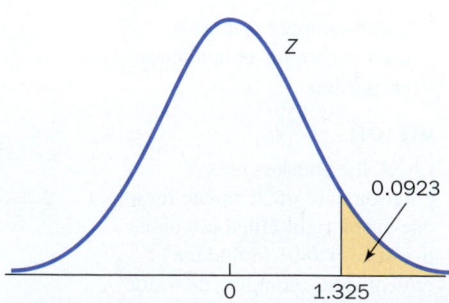

Figure 9.64 p-Value illustration:
$p = P(Z \geq 1.325)$
$\quad = 0.0923 > 0.01 = \alpha$

Just a reminder: The p in H_0 and H_a represents the population proportion. The symbol p is also used later in this problem to represent the p value associated with the hypothesis test.   ●

TRY IT NOW   GO TO EXERCISE 9.179

Because a large-sample test concerning a population proportion is based on a standard normal distribution, Z, the p value is computed the same way as in a hypothesis test concerning a population mean when σ is known (which is also based on a standard normal distribution). The following example includes a p-value computation.

Example 9.18 Sick or Tired

A recent survey indicated that 54% of employed Canadians admit that they have taken a sick day from work when they really were not sick.[52] The survey results suggest that there are marked differences in this percentage among provinces. A random sample of 400 employed Canadians from Quebec was selected, and 204 admitted that they have faked an illness.

a. Is there any evidence to suggest that the proportion of employed Canadians in Quebec who have faked an illness is different from 0.54? Use $\alpha = 0.05$.

b. Find the p value associated with this hypothesis test.

SOLUTION

a. The sample size is $n = 400$ and $x = 204$.

The sample proportion is $\hat{p} = x/n = 204/400 = 0.51$.

The assumed value of the population proportion is $p_0 = 0.54$ and $\alpha = 0.05$.

Check the nonskewness criterion:

$np_0 = (400)(0.54) = 216 \geq 5$

$n(1 - p_0) = (400)(0.46) = 184 \geq 5$

Both inequalities are satisfied. $\hat{P}$ is approximately normal, and the large-sample hypothesis test concerning a population proportion can be used.

We are looking for *any* difference from $p_0 = 0.54$, so this is a two-sided hypothesis test:

$H_0: p = 0.54$

$H_a: p \neq 0.54$

TS: $Z = \dfrac{\hat{P} - p_0}{\sqrt{\dfrac{p_0(1 - p_0)}{n}}}$

RR: $|Z| \geq z_{\alpha/2} = z_{0.025} = 1.96$ ($Z \leq -1.96$ or $Z \geq 1.96$)

The value of the test statistic is

$z = \dfrac{\hat{p} - p_0}{\sqrt{\dfrac{p_0(1 - p_0)}{n}}} = \dfrac{0.51 - 0.54}{\sqrt{\dfrac{(0.54)(0.46)}{400}}} = -1.2039$

The value of the test statistic does *not* lie in the rejection region. We do not reject the null hypothesis. There is no evidence to suggest the proportion of employed Canadians in Quebec who fake an illness is different from 0.54.

b. Because this is a two-sided test and the value of the test statistic ($z = -1.2039$) is negative, $p/2$ is a left-tail probability.

$$p/2 = P(Z \le -1.2039)$$ Definition of p value for $H_a: p \ne p_0$.

$$= 0.1143$$ Use Table III in the Appendix.

$$p = 2(0.1143) = 0.2286 \ (> 0.05 = \alpha)$$ Solve for p.

Because $p > \alpha$, we do not reject the null hypothesis.

Figure 9.65 through 9.67 show a technology solution, and Figure 9.68 illustrates the p value associated with the hypothesis test.

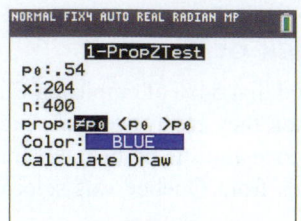

Figure 9.65 TI-84 Plus C `1-PropZTest` input screen.

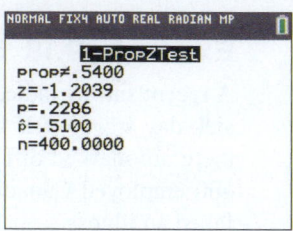

Figure 9.66 TI-84 Plus C `1-PropZTest` `Calculate` results.

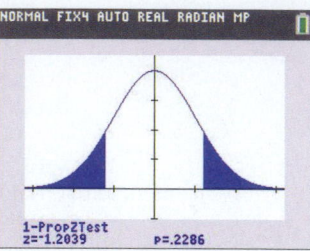

Figure 9.67 TI-84 Plus C `1-PropZTest` `Draw` results.

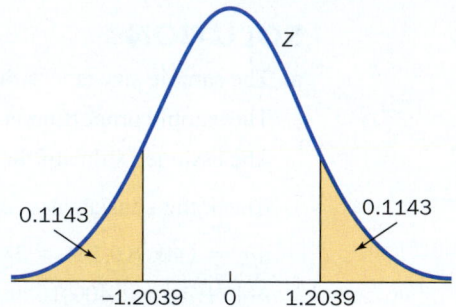

Figure 9.68 p-Value illustration:
$p = 2P(Z \le -1.2039)$
$= 0.2286 > 0.05 = \alpha$

TRY IT NOW   GO TO EXERCISE 9.180

Technology Corner

Procedure: Hypothesis test concerning a population proportion.
Reconsider: Example 9.18, solution, and interpretations.

CrunchIt!

Use the built-in function Proportion 1-Sample. Input is either summary statistics or a column containing the sample data.

1. Select Statistics; Proportion; 1-sample.
2. Under the Summarized tab, enter the value for n and Successes (x). Under the Hypothesis Test tab, enter the value for p_0 and select the appropriate Alternative. (Figure 9.69).
3. Click Calculate. The results are displayed in a new window (Figure 9.70).

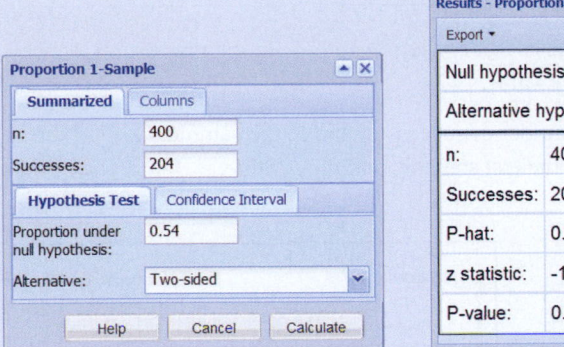

Figure 9.69 Proportion 1-Sample input screen.

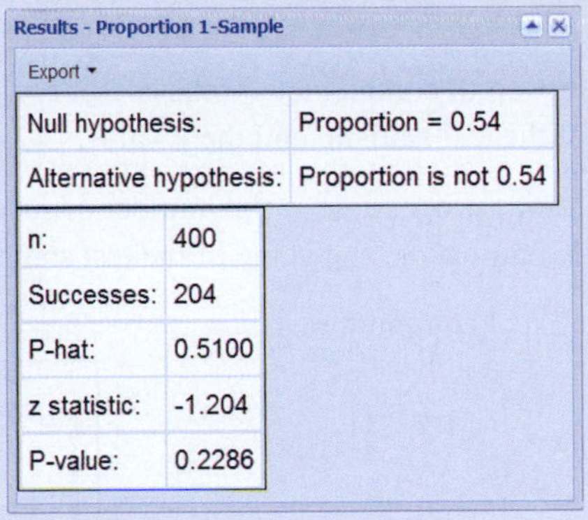

Figure 9.70 Proportion 1-Sample test results.

TI-84 Plus C

Use the calculator function 1-PropZTest. The input is the value of p_0, the number of successes, and the number of trials.

1. Select STAT; TESTS; 1-PropZTest.
2. Enter p_0, the hypothesized proportion, x, the number of successes, and n, the sample size.
3. Choose the appropriate alternative hypothesis and Calculate. Press ENTER. See Figure 9.65.
4. The 1-PropZTest output includes the alternative hypothesis, the value of the test statistic, the p value, the sample proportion, and the sample size. Refer to Figure 9.66.
5. The Draw results are shown in Figure 9.67.

Minitab

Use the function 1 Proportion. The input is either data in a column (Samples in columns) or summary statistics (Summarized data).

1. Choose Stat; Basic Statistics; 1 Proportion.
2. In this example, select Summarized data. Enter the Number of events (x) and the Number of trials (n). Check the box to Perform a hypothesis test, and enter the Hypothesized proportion. See Figure 9.71.
3. Choose Options. Enter a Confidence level, select the Alternative hypothesis, and select the Normal approximation Method. See Figure 9.72.
4. Summary statistics, the confidence interval, the value of the test statistic, and the p value are displayed in the Session window. See Figure 9.73.

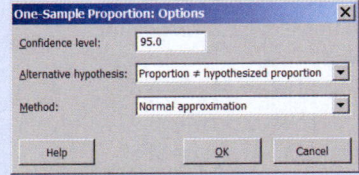

Figure 9.72 1 Proportion Options screen.

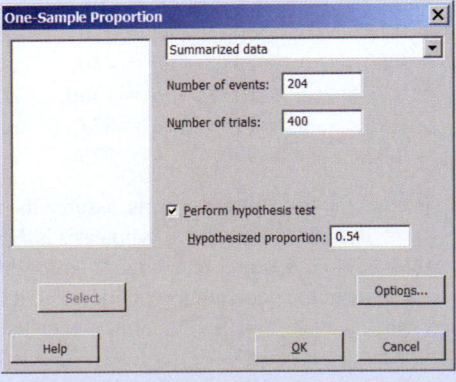

Figure 9.71 1 Proportion input screen.

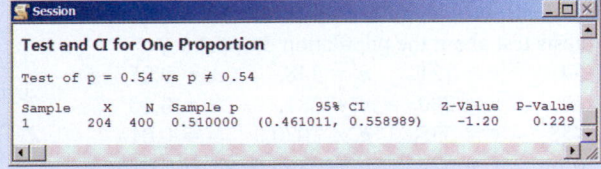

Figure 9.73 1 Proportion hypothesis test results.

Excel

There is no built-in function to conduct a hypothesis test concerning a population proportion. However, Excel can be used to compute the value of the test statistic and the p value.

1. Enter p_0, the hypothesize proportion, x, the number of successes, and n, the sample size.
2. Compute the sample proportion, the value of the test statistic, and the p value.

Figure 9.74 shows the Excel computations.

	A	B	C
1	0.5400	= p_0	The hypothesized proportion
2	204	= x	The number of successes
3	400	= n	The sample size
4	0.5100	= A2/A3	The sample proportion
5	-1.2039	= (A4-A1)/SQRT(A1*(1-A1)/A3)	The z value
6	0.2286	= 2*NORM.S.DIST(A5,TRUE)	The p value

Figure 9.74 Excel computations.

SECTION 9.6 EXERCISES

Concept Check

9.167 True/False   A large-sample hypothesis test concerning a population proportion is based on a standard normal distribution.

9.168 True/False   In a large-sample hypothesis test concerning a population proportion, the hypothesized proportion, p_0, cannot be close to 0 or 1.

9.169 True/False   In a large-sample hypothesis test concerning a population proportion, the number of trials must be at least 250.

9.170 True/False   In a large-sample hypothesis test concerning a population proportion, there are only two possible alternative hypotheses.

9.171 Short Answer   In a large-sample hypothesis test concerning a population proportion, what happens if the nonskewness criterion is not satisfied?

Practice

9.172   In each of the following problems, the sample size and p_0 are given for a large-sample hypothesis test concerning a population proportion. Check the two nonskewness inequalities and determine whether this test is appropriate.

a. $n = 276$, $p_0 = 0.30$ b. $n = 1158$, $p_0 = 0.60$
c. $n = 645$, $p_0 = 0.03$ d. $n = 159$, $p_0 = 0.97$
e. $n = 322$, $p_0 = 0.38$ f. $n = 443$, $p_0 = 0.82$

9.173   In each of the following problems, assume the null hypothesis is $H_0: p = p_0$ and the alternative hypothesis is $H_a: p > p_0$. Use the values of p_0, x, n, and α to conduct a large-sample hypothesis test about the population proportion.

a. $p_0 = 0.34$, $x = 121$, $n = 348$, $\alpha = 0.05$
b. $p_0 = 0.67$, $x = 200$, $n = 281$, $\alpha = 0.10$
c. $p_0 = 0.488$, $x = 535$, $n = 1020$, $\alpha = 0.01$
d. $p_0 = 0.02$, $x = 52$, $n = 2366$, $\alpha = 0.025$
e. $p_0 = 0.85$, $x = 570$, $n = 667$, $\alpha = 0.01$

9.174   In each of the following problems, assume the null hypothesis is $H_0: p = p_0$ and the alternative hypothesis is $H_a: p < p_0$.

Use the values of p_0, x, n, and α to conduct a large-sample hypothesis test about the population proportion.

a. $p_0 = 0.14$, $x = 40$, $n = 317$, $\alpha = 0.01$
b. $p_0 = 0.275$, $x = 98$, $n = 404$, $\alpha = 0.05$
c. $p_0 = 0.52$, $x = 250$, $n = 546$, $\alpha = 0.025$
d. $p_0 = 0.78$, $x = 2710$, $n = 3580$, $\alpha = 0.001$
e. $p_0 = 0.605$, $x = 1102$, $n = 1862$, $\alpha = 0.05$

9.175   In each of the following problems, assume the null hypothesis is $H_0: p = p_0$ and the alternative hypothesis is $H_a: p \neq p_0$. Use the values of p_0, x, n, and α to conduct a large-sample hypothesis test about the population proportion.

a. $p_0 = 0.28$, $x = 88$, $n = 377$, $\alpha = 0.025$
b. $p_0 = 0.46$, $x = 130$, $n = 243$, $\alpha = 0.02$
c. $p_0 = 0.337$, $x = 120$, $n = 414$, $\alpha = 0.05$
d. $p_0 = 0.71$, $x = 865$, $n = 1250$, $\alpha = 0.005$
e. $p_0 = 0.93$, $x = 1515$, $n = 1600$, $\alpha = 0.01$

9.176   In each of the following problems, assume the null hypothesis is $H_0: p = p_0$ and the alternative hypothesis is $H_a: p > p_0$. Use the values of p_0, x, n, and α to conduct a large-sample hypothesis test about the population proportion. Find the p value associated with each test, and use it to draw a conclusion.

a. $p_0 = 0.15$, $x = 60$, $n = 356$, $\alpha = 0.10$
b. $p_0 = 0.62$, $x = 298$, $n = 450$, $\alpha = 0.05$
c. $p_0 = 0.743$, $x = 1035$, $n = 1360$, $\alpha = 0.05$
d. $p_0 = 0.94$, $x = 795$, $n = 825$, $\alpha = 0.01$
e. $p_0 = 0.32$, $x = 960$, $n = 2750$, $\alpha = 0.001$

9.177   In each of the following problems, assume the null hypothesis is $H_0: p = p_0$ and the alternative hypothesis is $H_a: p < p_0$. Use the values of p_0, x, n, and α to conduct a large-sample hypothesis test about the population proportion. Find the p value associated with each test, and use it to draw a conclusion.

a. $p_0 = 0.54$, $x = 145$, $n = 301$, $\alpha = 0.05$
b. $p_0 = 0.39$, $x = 180$, $n = 460$, $\alpha = 0.005$
c. $p_0 = 0.07$, $x = 35$, $n = 566$, $\alpha = 0.01$
d. $p_0 = 0.64$, $x = 449$, $n = 747$, $\alpha = 0.025$
e. $p_0 = 0.47$, $x = 395$, $n = 925$, $\alpha = 0.005$

9.178 In each of the following problems, assume the null hypothesis is $H_0: p = p_0$ and the alternative hypothesis is $H_a: p \neq p_0$. Use the values of p_0, x, n, and α to conduct a large-sample hypothesis test about the population proportion. Find the p value associated with each test, and use it to draw a conclusion.

   a. $p_0 = 0.50$,　$x = 418$,　$n = 882$,　$\alpha = 0.01$
   b. $p_0 = 0.19$,　$x = 158$,　$n = 700$,　$\alpha = 0.05$
   c. $p_0 = 0.90$,　$x = 445$,　$n = 520$,　$\alpha = 0.001$
   d. $p_0 = 0.75$,　$x = 1095$,　$n = 1400$,　$\alpha = 0.025$
   e. $p_0 = 0.45$,　$x = 2386$,　$n = 5525$,　$\alpha = 0.01$

Applications

9.179 Public Health and Nutrition　In a recent research study concerning nutrition, it was reported that only 2% of children in the United States ages 6–10 eat the recommended number of servings for all five major food groups each day. A certain state conducted a nutrition campaign through elementary schools, newspapers, and television. Following the campaign, a random sample of 500 children in this age group was obtained, and 16 were found to eat the recommended number of servings each day.
   a. Identify n, x, and p_0.
   b. Check the nonskewness criterion. Is a large-sample hypothesis test about p appropriate?
   c. Conduct the appropriate hypothesis test to determine whether the proportion of children who eat the recommended number of servings each day has increased. Use $\alpha = 0.05$.
   d. Compute the p value associated with this test.

9.180 Psychology and Human Behavior　It is very common to complain about one's boss. There are many examples in TV, movies, and real life of bullying and harassment. In a recent survey it was reported that 20% of workers say that a manager has hurt their career.[53] A random sample of 500 workers in the retail industry was obtained, and 115 indicated that a boss had hurt their career.
   a. Identify n, x, and p_0.
   b. Check the nonskewness criterion. Is a large-sample hypothesis test about p appropriate?
   c. Conduct the appropriate hypothesis test to determine whether the proportion of workers in the retail industry who believe a boss has hurt their career is different from p_0. Use $\alpha = 0.05$.
   d. Compute the p value associated with this test.

9.181 Education and Child Development　Homeschooling has become very popular in the United States, and many colleges try to attract students from this group. Evidence suggests that approximately 90% of all homeschooled children attend college. To check this claim, a random sample of 225 home-schooled children was obtained and 189 of them were found to have attended college.
   a. Identify n, x, and p_0.
   b. Check the nonskewness criterion. Is a large-sample hypothesis test about p appropriate?
   c. Conduct the appropriate test to determine whether the proportion of homeschooled children who attend college is different from 0.90. Use $\alpha = 0.05$.
   d. Compute the p value associated with this test.

9.182 Public Policy and Political Science　It has always been difficult to attract general physicians to rural areas. Small-town doctors do not have the opportunity to take much time off, and the salary is relatively low. However, the quality of life is often appealing. Past records indicate that 52% of all physicians in rural areas leave after one year. The federal government has decided to offer more incentives for doctors to stay in rural areas, for example, loan forgiveness and housing allowances. A few years after this program was implemented, a random sample of rural-area physician positions showed 62 of 130 left after one year.
   a. Identify n, x, and p_0.
   b. Check the nonskewness criterion. Is a large-sample hypothesis test about p appropriate?
   c. Conduct the appropriate test to determine whether the turnover rate of rural doctors has decreased. Use $\alpha = 0.01$.
   d. Compute the p value associated with this test.

9.183 Medicine and Clinical Studies　According to a recent study, taller women have a significantly higher risk of developing some form of cancer.[54] The American Cancer Society reports that approximately 38% of all women develop some form of cancer in their lifetime. A long-term Canadian study was conducted involving 1250 tall women between the ages of 50 and 79, and 502 developed some form of cancer. Is there any evidence to suggest that the proportion of taller women who develop cancer is greater than 0.38? Use $\alpha = 0.01$. Write a Solution Trail for this problem.

9.184 Public Policy and Political Science　Soon after Edward Snowden disclosed information about the surveillance programs of the U.S. National Security Agency, a survey of Americans revealed that 56% were worried the United States would go too far in violating privacy rights.[55] Aides to Senator Roy Blunt selected a random sample of 575 Missouri residents, and 358 said they were worried about violations of their privacy rights. Is there any evidence to suggest that the true proportion of Missouri residents worried about violations of privacy rights is greater than 0.58? Use $\alpha = 0.05$ and use the p value to draw a conclusion.

9.185 Psychology and Human Behavior　There are many urban legends involving a full moon and human behavior. Research at the University of Basel in Switzerland suggests that sleep is associated with the lunar cycle.[56] In a new sleep study, 120 random adults were selected and studied during a full moon phase. Melatonin levels were used to determine whether each person experienced a deep sleep and 76 experienced low levels, and therefore, trouble sleeping, during the full moon. Is there any evidence to suggest that more than half of all people have trouble sleeping during a full moon? Use $\alpha = 0.05$.

9.186 Marketing and Consumer Behavior　A recent study suggests that Canadians are getting richer but they are among the most indebted people in the world.[57] The Bank of Canada suggests that no household should have debt in excess of 160% of annual disposable income. A random sample of 1500 households in Canada's wealthiest cities, Vancouver, Calgary, and Toronto, was obtained, and 235 were found to be too much in debt. Is there any evidence to suggest that the true proportion of Canadian households that are too much in debt is greater than 14%? Use $\alpha = 0.05$. Write a Solution Trail for this problem.

9.187 Public Policy and Political Science Ron Littlefield is considering a campaign for mayor of Chattanooga. He will enter the race if there is evidence to suggest that less than 40% of all residents are satisfied with the local government. A random sample of 375 residents is obtained, and 127 indicate they are satisfied with the local government. Do you think this politician will enter the race for mayor? Justify your answer. (Use $\alpha = 0.01$.)

9.188 Physical Sciences It's known as the Hum, a steady, droning sound heard in places all around the world. It is a low-frequency, annoying, throbbing, unexplained sound. It seems to be louder at night and indoors, and approximately 2% of the people living in any given Hum-prone area can hear the sound.[58] To determine if more people are affected by the Hum, a random sample of 1300 adults in Taos, New Mexico, was obtained, and 24 reported that they regularly hear the noise. Is there any evidence to suggest that the proportion of adults in this area who hear the Hum has changed? Use $\alpha = 0.05$ and use the p value associated with this hypothesis test to draw a conclusion.

9.189 Public Policy and Political Science A local planning board must consider whether to require all new projects with five or more apartments to designate some of the units as rent-controlled. A random sample of 100 apartments in Cheyenne, Wyoming, was obtained, and 9 were found to be rent-controlled. Is there any evidence to suggest the proportion of rent-controlled apartments is less than 10%? Let $\alpha = 0.05$, and use the p value associated with this hypothesis test to draw a conclusion.

9.190 Public Policy and Political Science Early in 2013, a political report indicated that 86% of Americans disapprove of the job Congress is doing. In a poll of 1000 Americans during Summer 2013 by NBC News, 830 said they disapprove of the job Congress is doing.[59] Is there any evidence to suggest that the proportion of Americans who disapprove of the job Congress is doing has decreased? Use $\alpha = 0.01$.

9.191 Technology and the Internet Technology is constantly improving, increasing the speed of communication, and many software programs on tablets, laptops, and phones now anticipate our next words or phrases. Docmail, a UK-based printing and mailing company, conducted a study and concluded that one of every three adults had not been required to produce something in handwriting for more than a year.[60] To check this claim in the United States, a random sample of 656 adults was obtained, and 235 said they had not been required to produce something in handwriting for more than a year. Is there any evidence to suggest that the proportion of Americans not using handwriting is different in the United States? Use $\alpha = 0.05$ and find the p value associated with this test.

Extended Applications

9.192 Education and Child Development College study-abroad programs have great educational value, compel students to become more globally aware, and invite participants to learn about different nations and cultures. Some colleges sponsor several programs and have high participation rates. However, nationally, only 14% of all college students participate in a study-abroad program.[61] Recent world events may have made more students leery of travel and life in another country. In a random sample of 1200 graduating college students, 150 said they had participated in a study-abroad program. Is there any evidence to suggest the proportion of students participating in these programs has decreased? Use $\alpha = 0.05$.

9.193 Physical Sciences Interstate Batteries guarantees their lawn tractor battery will last at least three years. To increase sales, the company is planning to offer a new replacement warranty, as long as 95% of all batteries do indeed last at least three years. A random sample of 200 customers was contacted, and 183 had batteries that lasted at least three years.

a. Is there any evidence to suggest the proportion of tractor batteries that last at least three years is less than 0.95? Use $\alpha = 0.05$.

b. Find the p value associated with this hypothesis test.

c. Based on your results, would you recommend implementation of the new replacement warranty? Justify your answer.

9.194 Medicine and Clinical Studies A certain puzzle task is designed to measure spatial reasoning performance. Forty-five percent of all people attempting the puzzle complete the task within the allotted time. A researcher decided to test the theory that classical music increases brain activity and improves the ability to perform such tasks. A random sample of 400 people was obtained, and each listened to 15 minutes of classical music and attempted the task. Two hundred and eleven completed the task within the allotted time.

a. Is there any evidence to suggest the proportion of people who complete the puzzle within the allotted time has increased? Use $\alpha = 0.001$.

b. Compute the p value associated with this test.

c. Carefully sketch a graph indicating the critical value from part (a) and the p value from part (b).

9.195 Location, Location, Location The 2013 Profile of International Home Buying Activity reported the percentage of total foreign purchases of U.S. homes by country of origin. Although buyers from China, India, and Mexico accounted for a large portion of all foreign purchases, Canadians purchased approximately twice as many homes as the second largest group of foreign purchasers, the Chinese.[62] In 2014, a random sample of 347 foreign purchases was obtained, and 63 were purchased by Canadians.

a. Is there any evidence to suggest that the proportion of foreign purchases of U.S. homes by Canadians has decreased from the 2013 proportion (0.23)? Use $\alpha = 0.05$.

b. Compute the p value associated with this test.

c. Carefully sketch a graph that shows the critical value from part (a) and the p value from part (b).

Challenge

9.196 Public Policy and Political Science The high cost of health care has forced some people to choose between medical treatment and other necessities such as food, heat, or electricity. A recent survey concluded that 43% of America's working-age adults did not see a doctor or access other medical services during the past year because of the cost.[63] Following a recent advertising campaign urging adults to seek medical assistance when needed, physicians' groups hope this percentage has decreased. A random sample of 1000 working Americans was obtained, and 380 did not seek medical assistance when needed, because of the cost.

    a. Is there any evidence to suggest the proportion of American workers who did not seek medical assistance has decreased? Use $\alpha = 0.01$.

    b. Find the probability of a type II error in this hypothesis test for $p_a = 0.38$; that is, find $\beta(0.38)$.

    c. Find a value of p_a such that the probability of a type II error is 0.1; that is, solve $\beta(p_a) = 0.1$ for p_a.

    d. Find the sample size necessary such that the probability of a type II error for $p_a = 0.35$ is 0.025; that is, solve $\beta(0.35) = 0.025$ for n.

9.197 Demographics and Population Statistics A large-sample hypothesis test concerning a population proportion is based on a normal approximation and the nonskewness criterion. If n is small (and the nonskewness criterion fails), the test statistic in an exact hypothesis test concerning p and is based on the number of successes in the sample, X. If H_0: $p = p_0$ is true, then X has a binomial distribution with n trials and probability of success p_0.

The four parts of the hypothesis test are

$$H_0: p = p_0$$

$$H_a: p > p_0, \quad p < p_0, \quad \text{or} \quad p \neq p_0$$

TS: $X =$ the number of successes in n trials

RR: $X \geq x_\alpha, \quad X \leq x'_\alpha \quad \text{or} \quad X \leq x_{\alpha/2} \quad \text{or} \quad X \geq x'_{\alpha/2}$

The critical values x_α, x'_α, $x_{\alpha/2}$, and $x'_{\alpha/2}$ are obtained from the binomial distribution with parameters n and p_0 to yield the desired significance level α. For example, in a one-sided, right-tailed test, the critical value x'_α is found such that $P(X \geq x'_\alpha) \approx \alpha$.

A recent poll indicated that 51% of all students admit to cheating.[64] A sociologist conducting research believes this percentage is much lower. In a random sample of 25 students from Long Island public schools, 9 admitted to cheating.

    a. Write the four parts of a small-sample (exact) one-sided, left-tailed test concerning the population proportion of Long Island students who admit to cheating, based on a binomial distribution. Use $\alpha = 0.01$.

    b. Is there any evidence to suggest the true proportion is less than 0.51?

    c. Find the p value associated with this hypothesis test.

9.7 Hypothesis Tests Concerning a Population Variance or Standard Deviation

Many real-world, practical decisions involve variability, or a population variance. For example, road inspectors make sure asphalt pavement mix meets design specifications and that variability in asphalt properties is small. This improves the quality, safety, and lifetime of the road surface. The dose of an anticancer drug is usually determined by the patient's body surface area (BSA). However, there is still large variability in drug exposure, the amount of the drug absorbed into the bloodstream. Researchers continue to search for better methods of calculating the dose in order to decrease the variability in exposure from patient to patient.

As in Section 8.5, the sample variance, S^2, is used as an estimator for the population variance, σ^2. The hypothesis test procedure is based on Theorem 8.5, a standardization, or transformation, to a chi-square random variable. Recall that if S^2 is the sample variance of a random sample of size n from a normal distribution with variance σ^2, then the random variable $(n-1)S^2/\sigma^2$ has a chi-square distribution with $n-1$ degrees of freedom. This result is used to construct a general procedure for a hypothesis test concerning σ^2.

This is the template for a hypothesis test about a population variance. The hypothesized value is σ_0^2.

Use only one (appropriate) alternative hypothesis and the corresponding rejection region. Remember, the two-sided alternative rejection region can *not* be written with an absolute value symbol. A chi-square distribution is not symmetric (about zero).

Hypothesis Test Concerning a Population Variance

Given a random sample of size n from a normal population with variance σ^2, a hypothesis test concerning the population variance σ^2 with significance level α has the form

$$H_0: \sigma^2 = \sigma_0^2$$

$$H_a: \sigma^2 > \sigma_0^2, \quad \sigma^2 < \sigma_0^2, \quad \text{or} \quad \sigma^2 \neq \sigma_0^2$$

$$\text{TS: } X^2 = \frac{(n-1)S^2}{\sigma_0^2}$$

$$\text{RR: } X^2 \geq \chi_{\alpha, n-1}^2, \quad X^2 \leq \chi_{1-\alpha, n-1}^2, \quad \text{or} \quad X^2 \leq \chi_{1-\alpha/2, n-1}^2 \quad \text{or} \quad X^2 \geq \chi_{\alpha/2, n-1}^2$$

A CLOSER LOOK

1. X (a Greek capital chi) is a random variable. χ (a Greek lowercase chi) is a specific value.

2. This test is valid for any sample size, as long as the underlying population is normal.

3. Table VI in the Appendix presents selected critical values associated with various chi-square distributions. These critical values were used to construct confidence intervals in Section 8.5.

4. Figures 9.75, 9.76, and 9.77 illustrate the rejection region for each alternative hypothesis.

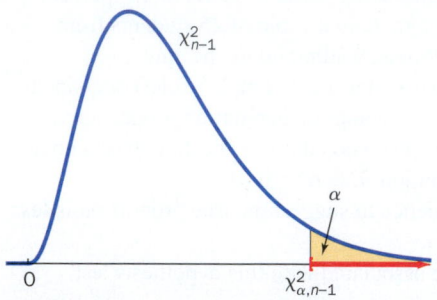

Figure 9.75 Rejection region for $H_a\colon \sigma^2 > \sigma_0^2$.

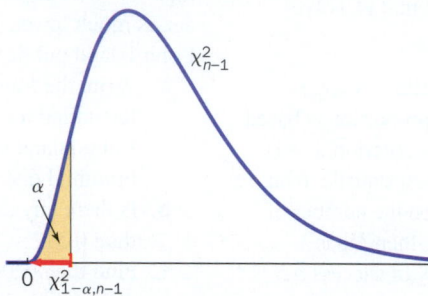

Figure 9.76 Rejection region for $H_a\colon \sigma^2 < \sigma_0^2$.

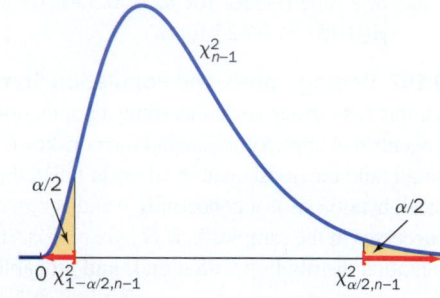

Figure 9.77 Rejection region for $H_a\colon \sigma^2 \neq \sigma_0^2$.

Solution Trail 9.19

KEYWORDS

- Is there any evidence?
- Population variance
- Greater than 0.002
- 15 locations
- Sample variance 0.0025
- Random sample

TRANSLATION

- Conduct a one-sided, right-tailed test about σ^2
- $n = 14$
- $s^2 = 0.0025$

CONCEPTS

- Hypothesis test concerning a population variance

VISION

There is no mention of normality. We assume the underlying distribution is normal. Use the template for a one-sided, right-tailed test about σ^2. Use $\alpha = 0.05$ to find the critical value. Compute the value of the test statistic, and draw a conclusion. Note: $\bar{x}$ is not used in the calculation of the test statistic.

Example 9.19 Permafrost Measurements

One measure of permafrost is the volumetric liquid water content (VWC), a unitless quantity. This measurement is used to study climate changes in the Arctic ice cap, greenhouse gases, and vegetation. Along the northern part of the Trans-Alaskan Pipeline, variation in permafrost was an important consideration in planning and construction. Suppose a random sample of 15 locations was obtained, and the VWC was measured at each location during the winter months. The sample mean was $\bar{x} = 0.225$ and the sample variance was $s^2 = 0.0025$. Is there any evidence to suggest the population variance is greater than 0.002? Use $\alpha = 0.05$.

SOLUTION

STEP 1 The given information:

Sample size: $n = 15$.

The sample variance: $s^2 = 0.0025$.

The assumed value of the population variance is 0.002 ($= \sigma_0^2$), and the significance level is $\alpha = 0.05$.

STEP 2 The four parts of the hypothesis test are

$$H_0\colon \sigma^2 = 0.002$$

$$H_a\colon \sigma^2 > 0.002$$

$$\text{TS:}\ X^2 = \frac{(n-1)S^2}{\sigma_0^2}$$

$$\text{RR:}\ X^2 \geq \chi_{\alpha,n-1}^2 = \chi_{0.05,14}^2 = 23.6848$$

STEP 3 The value of the test statistic is

$$\chi^2 = \frac{(n-1)s^2}{\sigma_0^2} = \frac{(14)(0.0025)}{0.002} = 17.5$$

STEP 4 The value of the test statistic does *not* lie in the rejection region. We do not reject the null hypothesis. There is no evidence to suggest the true population variance in volumetric liquid water content is greater than 0.002.

Figure 9.78 shows a technology solution.

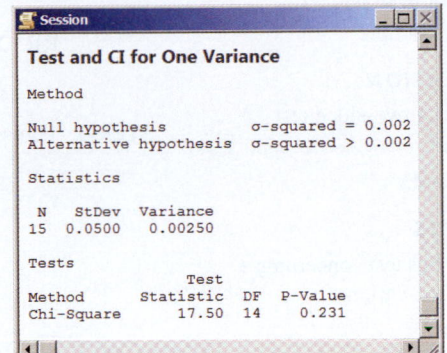

Figure 9.78 Minitab 1 Variance output.

TRY IT NOW   GO TO EXERCISE 9.215

A CLOSER LOOK

1. Neither the TI-84 Plus nor Excel has a built-in command to conduct a hypothesis test concerning a population variance. However, both can be used to find the appropriate critical value and associated *p* value.

2. The table listing critical values for various chi-square distributions is very limited. Using this table, the best we can do is bound the *p* value associated with a hypothesis test (as for a *t* test).

Example 9.20 Don't Forget the Sprinkles

DATA SET

ICECREAM

The Ice Cream Store in Rehoboth Beach, Delaware, has over 133 flavors including Boston Cream Pie, Cake Batter Cookie Dough, and Fruity Pebbles. The mean number of gallons of ice cream sold during a summer day is 354 gallons with a standard deviation of 25.7 gallons. This information is used to plan production schedules and order supplies. A random sample of summer days was obtained, and the amount of ice cream sold each day (in gallons) is given in the following table.

360	347	346	347	338	370	362	356	330	339
372	358	387	359	343	372	369	349	344	334

a. Is there any evidence to suggest that the true population variance in ice cream purchased per day is different from 660.49 gallons^2? Assume normality and use $\alpha = 0.01$.

b. Find bounds on the *p* value associated with this test.

SOLUTION

STEP 1 The assumed population variance is $25.7^2 = 660.49 (= \sigma_0^2)$; $n = 20$ and $\alpha = 0.01$.

We would like to know whether the population variance is different from 660.49. The relevant alternative hypothesis is two-sided.

The underlying population is assumed to be normal. A hypothesis test concerning a population variance can be used.

Solution Trail 9.20

KEYWORDS

- Is there any evidence?
- Population variance
- Different from 660.49
- Random sample

TRANSLATION

- Conduct a two-sided test about σ^2
- Find n and s^2

CONCEPTS

- Hypothesis test concerning a population variance

VISION

We can assume that the underlying distribution is normal. Use the template for a two-sided test about σ^2. Use $\alpha = 0.01$ to find the critical values. Compute the value of the test statistic, and draw a conclusion. Use Table VI in the Appendix to find bounds on the p value.

The initial bounds are on $p/2$, because the hypothesis test is two-sided.

STEP 2 The four parts of the hypothesis test are

$$H_0: \sigma^2 = 660.49$$
$$H_a: \sigma^2 \neq 660.49$$
$$\text{TS: } X^2 = \frac{(n-1)S^2}{\sigma_0^2}$$

RR: $X^2 \leq \chi^2_{1-\alpha/2, n-1} = \chi^2_{0.995, 19} = 6.8440$ or
$X^2 \geq \chi^2_{\alpha/2, n-1} = \chi^2_{0.005, 19} = 38.5823$

STEP 3 The sample variance is

$$s^2 = \frac{1}{n-1}\left[\sum x_i^2 - \frac{1}{n}(\sum x_i)^2\right]$$ Computational formula, sample variance.

$$= \frac{1}{19}\left[2,511,964 - \frac{1}{20}(7082)^2\right] = 222.52$$ Use given data.

The value of the test statistic is

$$\chi^2 = \frac{(n-1)s^2}{\sigma_0^2} = \frac{(19)(222.52)}{660.49} = 6.4011$$

The value of the test statistic lies in the rejection region ($\chi^2 = 6.4011 \leq 6.8840$). We reject the null hypothesis. There is evidence to suggest the true population variance is different from 660.49. Figure 9.79 shows a technology solution.

STEP 4 $\chi^2 = 6.4011$.

Using Table VI in the Appendix, row $n - 1 = 20 - 1 = 19$, place 6.4011 in the ordered list of critical values.

$$5.4068 \leq 6.4011 \leq 6.8440$$
$$\chi^2_{0.999, 19} \leq 6.4011 \leq \chi^2_{0.995, 19}$$
$$\chi^2_{1-0.001, 19} \leq 6.4011 \leq \chi^2_{1-0.005, 19}$$

Therefore, $0.001 \leq p/2 \leq 0.005$
and $0.002 \leq p \leq 0.010$ See Figure 9.80.

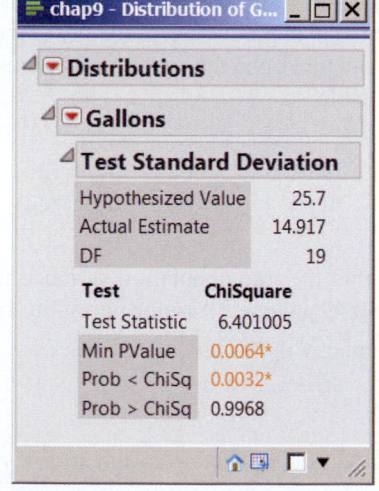

Figure 9.79 JMP Hypothesis test concerning a population standard deviation.

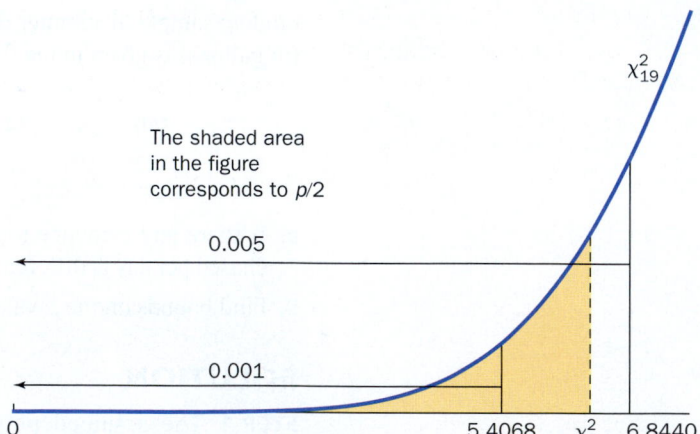

The shaded area in the figure corresponds to $p/2$

Figure 9.80 The left tail of a chi-square distribution with 19 degrees of freedom. This illustrates the computations to find bounds on the p value.

TRY IT NOW   GO TO EXERCISE 9.227

Technology Corner

Procedure: Hypothesis test concerning a population variance or standard deviation.
Reconsider: Example 9.20, solution, and interpretations.

CrunchIt!

There is no built-in function to conduct a hypothesis test concerning a population variance or standard deviation. However, CrunchIt! may be used to compute the p value.

1. Select Distribution Calculator; Chi-square.
2. Enter the degrees of freedom associated with the test statistic (df).
3. Under the Probability tab, select the appropriate inequality and enter the value of the test statistic. Click Calculate. A graph and the tail probability are displayed. See Figure 9.81. Note that the probability displayed in this example is $p/2$.

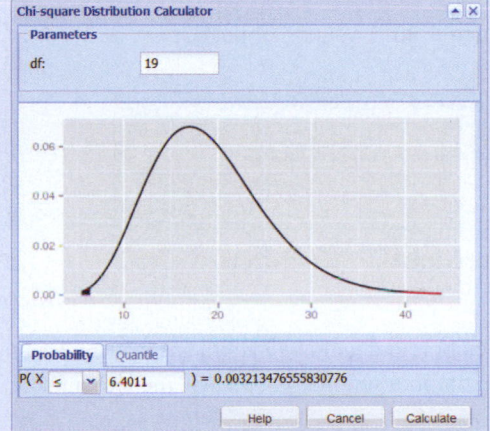

Figure 9.81 Left-tail probability calculation, $p/2$.

TI-84 Plus C

There is no built-in function to conduct a hypothesis test concerning a population variance or standard deviation. However, the calculator may be used to compute summary statistics, the test statistic, and the p value.

1. Enter the data into list `L1`.
2. Use `LIST`; MATH; `stdDev` to compute the sample standard deviation.
3. Compute the value of the test statistic.
4. Use `DISTR`; DISTR; χ2cdf to compute the left tail probability, and multiply by 2 to obtain the p value. See Figure 9.82.

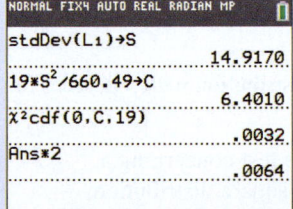

Figure 9.82 Standard deviation, test statistic, and p-value calculations.

Minitab

Use the function 1 Variance to conduct a hypothesis test concerning a population variance or standard deviation. The input may be either Samples in columns or Summarized data, and the test may be conducted in term of the variance or standard deviation.

1. Enter the data into column `C1`.
2. Select Stat; Basic Statistics; 1 Variance.

3. Choose One or more samples, each in a column, and enter C1. Check the box to Perform a hypothesis test, and enter the Hypothesized variance. See Figure 9.83.

4. Choose Options. Enter a Confidence level and select the Alternative hypothesis. See Figure 9.84.

5. Click OK. Summary statistics, the confidence interval, the value of the test statistic, and the p value are displayed in the Session window. See Figure 9.85.

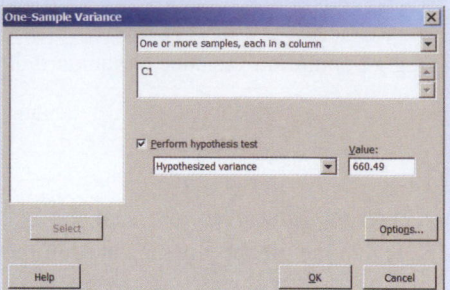

Figure 9.83 1 Variance input screen.

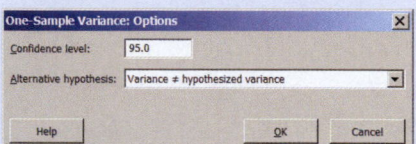

Figure 9.84 1 Variance Options screen.

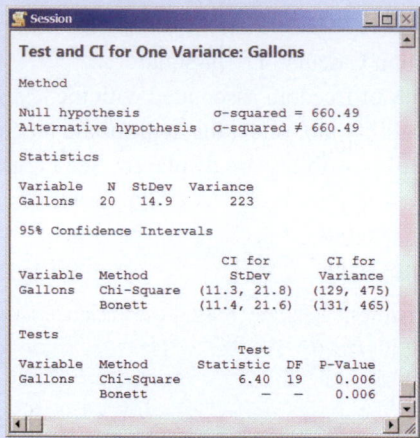

Figure 9.85 Hypothesis test results.

Excel

There is no built-in function to conduct a hypothesis test concerning a population variance or standard deviation. However, Excel may be used to compute summary statistics, the test statistic, and the p value.

1. Enter the data into column A.

2. Compute the sample variance and the value of the test statistic.

3. Use CHISQ.DIST to find the left tail probability and calculate the p value. See Figure 9.86.

Figure 9.86 Excel computations to conduct a hypothesis test concerning a population variance.

	B	C	D
1	660.49	= σ_0^2	The hypothesized variance
2	20	= COUNT(A1:A20)	The sample size
3	222.5158	= VAR.S(A1:A20)	The sample variance
4	6.4010	= (B2-1)*B3/B1	The value of the test statistic
5	0.0064	= 2*CHISQ.DIST(B4,B2-1,TRUE)	The p value

SECTION 9.7 EXERCISES

Concept Check

9.198 True/False A chi-square distribution with ν degrees of freedom is symmetric about ν.

9.199 True/False In a hypothesis test concerning a population variance based on a chi-square distribution, there is no assumption about the distribution of the underlying population.

9.200 True/False In a hypothesis test concerning a population variance, there are $n - 1$ degrees of freedom associated with the test statistic.

9.201 True/False In a hypothesis test concerning a population variance based on a chi-square distribution, the test statistic can always be adjusted to use a right-tail rejection region.

9.202 True/False The hypothesis test concerning a population variance based on a chi-square distribution can also be used to conduct a hypothesis test concerning a population standard deviation.

Practice

9.203 Consider a hypothesis test concerning a population variance from a normal population with $H_0: \sigma^2 = 27.2$ and $H_a: \sigma^2 > 27.2$.

  a. Write the appropriate test statistic.

  b. Find the rejection region corresponding to each value of n and α.

    i. $n = 12$, $\alpha = 0.05$ **ii.** $n = 19$, $\alpha = 0.025$

   iii. $n = 23$, $\alpha = 0.01$ **iv.** $n = 29$, $\alpha = 0.10$

    v. $n = 6$, $\alpha = 0.001$ **vi.** $n = 15$, $\alpha = 0.0001$

9.204 Consider a hypothesis test concerning a population variance from a normal population with $H_0: \sigma^2 = 352.98$ and $H_a: \sigma^2 < 352.98$.
   a. Write the appropriate test statistic.
   b. Find the rejection region corresponding to each value of n and α.
      i. $n = 17$, $\alpha = 0.001$ **ii.** $n = 13$, $\alpha = 0.05$
      iii. $n = 37$, $\alpha = 0.0001$ **iv.** $n = 27$, $\alpha = 0.10$
      v. $n = 33$, $\alpha = 0.05$ **vi.** $n = 8$, $\alpha = 0.005$

9.205 Consider a hypothesis test concerning a population variance from a normal population with $H_0: \sigma^2 = 43.8$ and $H_a: \sigma^2 \neq 43.8$.
   a. Write the appropriate test statistic.
   b. Find the rejection region corresponding to each value of n and α.
      i. $n = 40$, $\alpha = 0.05$ **ii.** $n = 31$, $\alpha = 0.01$
      iii. $n = 22$, $\alpha = 0.001$ **iv.** $n = 16$, $\alpha = 0.02$
      v. $n = 29$, $\alpha = 0.002$ **vi.** $n = 5$, $\alpha = 0.20$

9.206 Consider a hypothesis test concerning the variance from a normal population with $H_0: \sigma^2 = 1.28$ and $H_a: \sigma^2 > 1.28$. Find the significance level (α) corresponding to each value of n and rejection region.
   a. $n = 21$, $X^2 \geq 31.4104$ b. $n = 33$, $X^2 \geq 56.3281$
   c. $n = 14$, $X^2 \geq 29.8195$ d. $n = 29$, $X^2 \geq 59.3000$

9.207 Consider a hypothesis test concerning the variance from a normal population with $H_0: \sigma^2 = 48.92$ and $H_a: \sigma^2 < 48.92$. Find the significance level (α) corresponding to each value of n and rejection region.
   a. $n = 25$, $X^2 \leq 7.4527$ b. $n = 36$, $X^2 \leq 18.5089$
   c. $n = 12$, $X^2 \leq 3.8157$ d. $n = 51$, $X^2 \leq 27.9907$

9.208 Consider a hypothesis test concerning the variance from a normal population with $H_0: \sigma^2 = 15.667$ and $H_a: \sigma^2 \neq 15.667$. Find the significance level (α) corresponding to each value of n and rejection region.
   a. $n = 9$, $X^2 \leq 1.3444$ or $X^2 \geq 21.9550$
   b. $n = 21$, $X^2 \leq 8.2604$ or $X^2 \geq 37.5662$
   c. $n = 38$, $X^2 \leq 15.0202$ or $X^2 \geq 73.3512$
   d. $n = 18$, $X^2 \leq 7.5642$ or $X^2 \geq 30.1910$

9.209 Consider a hypothesis test concerning the variance from a normal population with $H_0: \sigma^2 = 11.4$ and $H_a: \sigma^2 > 11.4$. Find bounds on the p value for each value of n and test statistic.
   a. $n = 32$, $\chi^2 = 50.05$ b. $n = 7$, $\chi^2 = 11.99$
   c. $n = 19$, $\chi^2 = 38.62$ d. $n = 24$, $\chi^2 = 60.15$

9.210 Consider a hypothesis test concerning the variance from a normal population with $H_0: \sigma^2 = 404.7$ and $H_a: \sigma^2 < 404.7$. Find bounds on the p value for each value of n and test statistic.
   a. $n = 11$, $\chi^2 = 1.36$ b. $n = 17$, $\chi^2 = 1.97$
   c. $n = 31$, $\chi^2 = 14.05$ d. $n = 51$, $\chi^2 = 32.85$

9.211 Consider a hypothesis test concerning the variance from a normal population with $H_0: \sigma^2 = 232$ and $H_a: \sigma^2 \neq 232$. Find bounds on the p value for each value of n and test statistic.
   a. $n = 25$, $\chi^2 = 41.67$ b. $n = 28$, $\chi^2 = 8.12$
   c. $n = 5$, $\chi^2 = 0.5005$ d. $n = 16$, $\chi^2 = 38.88$

9.212 Consider a random sample of size 21 from a normal population with hypothesized variance 16.7.
   a. Write the four parts for a one-sided, right-tailed hypothesis test concerning the population variance with $\alpha = 0.01$.
   b. Suppose $s^2 = 28$. Find the value of the test statistic, and draw a conclusion about the population variance.
   c. Find bounds on the p value associated with this hypothesis test, and carefully sketch a graph to illustrate this value.

9.213 Consider a random sample of 16 observations, given in the following table, from a normal population with hypothesized variance 36.8. **EX9.213**

233.1	226.1	220.3	247.6	232.9	232.8	235.9
232.4	249.4	207.4	231.8	232.1	220.7	229.6
242.5	229.3					

   a. Write the four parts for a two-sided hypothesis test concerning the population variance with $\alpha = 0.05$.
   b. Compute the sample variance and the value of the test statistic. Draw a conclusion about the population variance.
   c. Find bounds on the p value associated with this hypothesis test.

9.214 Consider a random sample of size 40 from a normal population with hypothesized variance 75.6.
   a. Write the four parts for a one-sided, left-tailed hypothesis test concerning the population variance with $\alpha = 0.001$.
   b. Suppose $s^2 = 48.5$. Find the value of the test statistic, and draw a conclusion about the population variance.
   c. Find bounds on the p value associated with this hypothesis test.

Applications

9.215 Physical Sciences A water-droplet generator is used to produce simulated rain or fog, and, for example, to test emission drift rates from nuclear-power-plant cooling towers. A piezoelectric water-droplet generator is designed to produce 10-microliter (μl) mist with a variance of 0.25. A random sample of 35 water drops was obtained, the amount of water in each (in μl) was measured, and the summary statistics were $\bar{x} = 10.004$ and $s = 0.558$. Is there any evidence to suggest the variance is larger than specified? Assume normality and use $\alpha = 0.05$.

9.216 Physical Sciences Energy companies assess potential basin and formations for recoverable shale oil and gas resources. This information is used to forecast production levels and plan further exploration. A random sample of world basins and formations was obtained, and the technically recoverable shale gas (in trillion cubic feet, tcf) for each site is given in the following table.[65] **SHALE**

94	158	50	24	23	89	19	25	235	22
202	35	51	144	141	37	25	17	15	10
176	56	30	65	211	82	215	287	57	101

Is there any evidence to suggest that the population variance in technically recoverable shale gas is greater than 5800 tcf²?

Assume normality and use $\alpha = 0.01$. Write a Solution Trail for this problem.

9.217 Manufacturing and Product Development A manufactured glass rod is tested for the stress required for fracture. Experiments suggest that a flame-polished rod has a higher mean fracture stress than a rod with an abraded surface. In both cases, the variance in stress is designed to be at most 324 MPa^2. A random sample of 12 polished glass rods was obtained, and the fracture stress for each was measured. The sample standard deviation in fracture stress was $s = 21.56$. Is there any evidence to refute the manufacturer's claim? Assume the underlying distribution is normal and use $\alpha = 0.01$.

9.218 Public Health and Nutrition The Italian Peoples Bakery in Pennington, New Jersey, advertises an original-recipe cream puff packed with 3 ounces of filling. To produce a consistent product, the variance in cream filling is carefully monitored. In a random sample of 23 cream puffs, the variance in filling was $s^2 = 0.105$ ounce^2. Is there any evidence to suggest the variance in filling is greater than 0.09 ounce^2? Assume normality and use $\alpha = 0.025$.

9.219 Manufacturing and Product Development In an effort to improve efficiency, Samuel Adams brewing company would like to produce yeast slurry with variance no greater than 62.5. A new delivery pump was installed, and a random sample of 10 slurry mixtures was obtained. Each mixture was measured in billion cells/ml and the sample variance was $s^2 = 70.1$. Is there any evidence to suggest the variance is greater than the desired value? Assume normality and use $\alpha = 0.01$. Write a Solution Trail for this problem.

9.220 Sports and Leisure The specified weight of a 22-mm die used at casinos in Las Vegas, Nevada, is 10.4 g. The variability in die weight must be negligible (at most 0.04 g^2) in order for patrons to believe that games involving dice are fair. A random sample of 25 new 22-mm dice was obtained and each die was carefully weighed. The sample mean weight was $\bar{x} = 10.38$ g and the sample standard deviation was $s = 0.244$ g. Conduct the relevant hypothesis test to determine whether there is any evidence that the variance in die weight is greater than 0.04. Use $\alpha = 0.01$ and assume normality.

9.221 Manufacturing and Product Development SoLux produces special light bulbs for museums, photography studios, automotive paint finishing, and machine vision. The SoLux 4100 Kelvin 50-watt 12-volt 10° bulb is designed to produce a beam diameter of 1.75 feet 10 feet away.[66] In a random sample of 26 bulbs from the production line, the standard deviation was 0.342 feet. Is there any evidence to suggest that the standard deviation in beam diameter at 10 feet is greater than 0.3? Assume normality and use $\alpha = 0.05$.

9.222 Business and Management Many farmers craft unique corn mazes in their fields after the growing season. These huge puzzles challenge both children and adults, and increase revenue for farmers. Suppose a farmer has constructed a corn maze so that the mean time for completion is 1.5 hours with variance 0.57 hour^2. A random sample of 18 people entering the maze was selected, and the time (in hours) to complete the puzzle was recorded for each. The sample standard deviation was $s = 0.34$. Is there any evidence the completion-time variance is different from the intended variance? Use $\alpha = 0.05$ and assume the underlying distribution is normal.

9.223 Sports and Leisure Bull riding has become a very popular rodeo sport—the action is fast and dangerous. A bull ride is scored by two judges; one actually rates the bull, while the other evaluates the rider. The variance in bull-riding time tends to be large, approximately 22.5 seconds^2. A random sample of 37 bull rides was selected over an entire rodeo season. The sample variance in riding times was $s^2 = 15.6$ seconds^2. Smaller variability in bull-riding time translates to a more monotonous, unexciting rodeo. Is there any evidence that bull riding has become less exciting? Assume normality and use $\alpha = 0.05$.

Extended Applications

9.224 Public Health and Nutrition The Obama administration worked to make health care more affordable and accountable. However, there is considerable variation across the country in provider charges for common medical services. A random sample of hospitals was obtained and the charge for outpatient service 300 was recorded for each.[67] **HEALTH**
  a. Is there any evidence to suggest that the standard deviation in outpatient charges for this service is greater than 1000? Assume normality and use $\alpha = 0.05$.
  b. Find bounds on the p value associated with this hypothesis test, and carefully sketch a graph to illustrate this value.

9.225 Manufacturing and Product Development Bittersharp apple cider is made from apples that contain more than 0.45% malic acid. It is important for the variance in malic acid content to be small, because large variability may cause the resulting cider to be too sharp. In preparation for making a batch of cider, a random sample of 20 Foxwhelp Bittersharp apples was obtained, and the malic acid content in each was measured. The sample variance was $s^2 = 0.42$.
  a. Is there any evidence to suggest the population variance in malic acid is greater than 0.36? Assume the underlying distribution is normal and use $\alpha = 0.05$.
  b. Find bounds on the p value associated with this hypothesis test.

9.226 Business and Management Researchers studying the European Union have suggested that small variability in minimum wage among countries leads to more economic stability and growth. A random sample of European countries was obtained and the minimum wage (in euros) was recorded for each.[68] **MINWAGE**
  a. Is there any evidence to suggest that the standard deviation in minimum wage is greater than 400 euros? Assume normality and use $\alpha = 0.05$.
  b. Find bounds on the p value associated with this hypothesis test.

9.227 Manufacturing and Product Development Cell phone companies attract customers by offering a wide coverage area and a consistent, strong signal. Signal strength is measured in dBm, decibels above or below 1.0 milliwatt. A random sample of locations in a company's coverage area was obtained, and the signal strength (in dBm) at each spot was measured. **CPHONE**

a. Suppose a consistent signal means a variance of at most 230 mw^2. Is there any evidence of an inconsistent signal in this company's coverage area? Use $\alpha = 0.05$ and assume normality.

b. Find bounds on the p value associated with this hypothesis test.

9.228 Manufacturing and Product Development The Estes Pro Series II Ventris model rocket weighs approximately 16 ounces and can reach altitudes of over 2000 feet.[69] Suppose the company would like the standard deviation in height to be approximately 50 feet. If the standard deviation is any larger, then more rockets are lost (because they soar too high and drift away) or customers become angry because more launches do not achieve the advertised height. A random sample of rockets was obtained, and the height (in feet) achieved on each launch was measured. **ROCKET**

a. Is there any evidence to suggest the population variance in height is greater than the company's desired value? Assume normality and use $\alpha = 0.01$.

b. Find bounds on the p value for this hypothesis test.

9.229 Biology and Environmental Science A monarch butterfly may fly as many as 2000 miles during its migration to Mexico. These amazing creatures actually vary dramatically in weight, color, and even flying behavior. Previous research suggests the mean wingspan for a monarch butterfly is 50 mm with a standard deviation of 2.75 mm. There is some speculation that changes in the environment (for example, pollution and climate) have caused greater variability in the wingspan. A random sample of monarch butterflies was obtained, and the wingspan of each (in mm) is given in the following table. **MONARCH**

49.3	52.4	43.3	51.2	55.1	48.4	41.6	54.9
51.9	51.6	47.0	50.7	47.2	55.7	50.3	48.3
52.9	47.8	49.4	56.8	47.0	51.5		

a. Is there any evidence that the variability in wingspan of the monarch butterfly has increased? Assume normality and use $\alpha = 0.05$.

b. Find bounds on the p value associated with this hypothesis test.

9.230 Manufacturing and Product Development Following a recent night-time helicopter crash, the air ambulance service in Ontario is considering night-vision goggles for all pilots. One measure of the effectiveness of these goggles is the resolution, in lp/mm (line pairs per millimeter). A random sample of 18 ATN PVS7-CGT high-performance night-vision goggles was obtained and the resolution was measured for each. The sample variance was $s^2 = 5.230$.

a. Is there any evidence to suggest that the standard deviation in the resolution for these goggles is different from 2? Assume normality and use $\alpha = 0.01$.

b. Find bounds on the p value associated with this hypothesis test.

9.231 Manufacturing and Product Development Oxford Paper Company manufactures boxboard for folding cartons made entirely from recycled material. One specific product is designed to have thickness 375 micrometers with standard deviation 7 micrometers. A random sample of 21 pieces of boxboard was obtained from the assembly line.

a. Assume the thickness distribution is normal and let $\alpha = 0.05$. Consider a two-sided hypothesis test to determine whether the boxboard is being manufactured with the designed variance in thickness.

b. Find the critical values in terms of the sample variance. Work backward to determine two values such that if $S^2 \leq s_L^2$ or $S^2 \geq s_H^2$, then the null hypothesis is rejected.

c. Suppose $s^2 = 56$. Use the critical values from part (b) to draw a conclusion about the population variance.

d. Suppose $s^2 = 15.6$. Use the critical values from part (b) to draw a conclusion about the population variance.

Challenge

9.232 Economics and Finance An approximate test concerning a population variance is based on a normal approximation. Recall that if S^2 is the sample variance of a random sample of size n from a normal distribution with variance σ^2, the random variable $X = (n - 1)S^2/\sigma^2$ has a chi-square distribution with $n - 1$ degrees of freedom. If n is large, the random variable X is approximately normal with mean $n - 1$ and variance $2(n - 1)$. An approximate hypothesis test is based on standardizing X to a Z random variable.

The four parts of the hypothesis test are

H_0: $\sigma^2 = \sigma_0^2$.

H_a: $\sigma^2 > \sigma_0^2$, $\qquad \sigma^2 < \sigma_0^2$, $\qquad$ or $\qquad \sigma^2 \neq \sigma_0^2$

TS: $Z = \dfrac{S^2 - \sigma_0^2}{\sqrt{2}\sigma_0^2/\sqrt{n - 1}}$

RR: $Z \geq z_\alpha$, $\qquad Z \leq -z_\alpha$, $\qquad$ or $\qquad |Z| \geq z_{\alpha/2}$

The historic variance in the exchange rate for the Japanese yen against the U.S. dollar is approximately 1.56. A random sample of 48 closing exchange rates was obtained, and the summary statistics were $\bar{x} = 103.74$ and $s^2 = 2.2$.

a. Write the four parts of a large-sample, two-sided test concerning the variance in exchange rate, based on a normal approximation. Use $\alpha = 0.05$.

b. Is there any evidence to suggest the variance in exchange rate has changed?

c. Find the p value associated with this hypothesis test.

d. Conduct an exact hypothesis test based on the chi-square distribution. Compute the p value, and compare it with your answer to part (c).

CHAPTER 9 **SUMMARY**

Concept	Page	Notation / Formula / Description
Hypothesis	392	A claim about the value of a specific population parameter.
Null hypothesis	392	H_0, the claim about a parameter assumed to be true, or the hypothesis to be tested.
Alternative hypothesis	392	H_a, the possible values of the parameter if H_0 is false.
Test statistic	392	A rule related to the null hypothesis, involving information in the sample. The value of the test statistic is used to determine which hypothesis, H_0 or H_a, is more likely.
Rejection region	393	An interval or set of numbers determined such that if the value of the test statistic lies in the rejection region, then the null hypothesis is rejected.
One-sided alternatives	394	$H_a: \theta > \theta_0$ (right-tailed), $H_a: \theta < \theta_0$ (left-tailed).
Two-sided alternative	394	$H_a: \theta \neq \theta_0$.
Type I error	399	H_0 is rejected (because the value of the test statistic lies in the rejection region), but H_0 is really true.
Type II error	399	H_0 is not rejected (because the value of the test statistic does not lie in the rejection region), but H_a is really true.
Significance level	399	The probability of a type I error, denoted by P(type I error) $= \alpha$.
p value	418	The smallest significance level for which the null hypothesis can be rejected.

Parameter	Assumptions	Alternative hypothesis	Test statistic	Rejection region
μ	n large, σ known, or normality, σ known	$\mu > \mu_0$ $\mu < \mu_0$ $\mu \neq \mu_0$	$Z = \dfrac{\bar{X} - \mu_0}{\sigma/\sqrt{n}}$	$Z \geq z_\alpha$ $Z \leq -z_\alpha$ $\lvert Z \rvert \geq z_{\alpha/2}$
μ	normality, σ unknown	$\mu > \mu_0$ $\mu < \mu_0$ $\mu \neq \mu_0$	$T = \dfrac{\bar{X} - \mu_0}{S/\sqrt{n}}$	$T \geq t_{\alpha,n-1}$ $T \leq -t_{\alpha,n-1}$ $\lvert T \rvert \geq t_{\alpha/2,n-1}$
p	n large, nonskewness criteria	$p > p_0$ $p < p_0$ $p \neq p_0$	$Z = \dfrac{\hat{P} - p_0}{\sqrt{\dfrac{p_0(1 - p_0)}{n}}}$	$Z \geq z_\alpha$ $Z \leq -z_\alpha$ $\lvert Z \rvert \geq z_{\alpha/2}$
σ^2	normality	$\sigma^2 > \sigma_0^2$ $\sigma^2 < \sigma_0^2$ $\sigma^2 \neq \sigma_0^2$	$X^2 = \dfrac{(n-1)S^2}{\sigma_0^2}$	$X^2 \geq \chi_{\alpha,n-1}^2$ $X^2 \leq \chi_{1-\alpha,n-1}^2$ $X^2 \leq \chi_{1-\alpha/2,n-1}^2$ or $X^2 \geq \chi_{\alpha/2,n-1}^2$

CHAPTER 9 **EXERCISES**

9 **APPLICATIONS**

9.233 Manufacturing and Product Development Vacuum-packed coffee stays fresh longer, but the package tends to be bumpy and unappealing. In a nitrogen-flushed package, all the oxygen is pushed out by heavier nitrogen, producing a smoother, more attractive package that also stays fresh. A machine used to produce a nitrogen-flushed package should dispense 1.6 moles of nitrogen for a 1-pound package of coffee. In a random sample of 23 packings, the sample mean amount of nitrogen dispensed was $\bar{x} = 1.78$ moles. Assume the underlying distribution is normal, with $\sigma = 0.5$ moles, and use $\alpha = 0.01$.

a. Is there any evidence to suggest the machine is malfunctioning?

b. Compute the p value associated with this hypothesis test.

9.234 Manufacturing and Product Development Shanghai Creative Material, Co. Ltd., makes plating tape for use on circuit boards, designed to be 4 mil thick. (A mil is equal to 0.001 inch: a milli-inch.) As part of quality control, every hour a random sample of 40 pieces of tape are carefully measured. If there is any evidence the thickness is different from 4 mil, then the entire process is stopped and the machinery checked and cleaned. Assume $\sigma = 0.05$ mil and use $\alpha = 0.01$.

   a. Suppose the sample mean is $\bar{x} = 4.014$. Should the process be stopped?

   b. Suppose the sample mean is $\bar{x} = 3.979$. Should the process be stopped?

   c. Comment on the statistical and practical differences between the decisions in parts (a) and (b).

9.235 Physical Sciences CubeSats are tiny satellites that began as an educational experiment but are now included in many commercial and government space launches. A CubeSat is a 10-cm cube that weighs up to 1.33 kg. A random sample of CubeSats set for launch in 2013 was obtained and the weight of each was measured. **CUBESAT**

   a. Each CubeSat has optimal launch characteristics if the weight is close to 1 kg. Is there any evidence to suggest that the mean weight of CubeSats scheduled for launch is different from 1? Assume the distribution is normal, $\sigma = 0.1$, and use $\alpha = 0.05$.

   b. Compute the p value for this hypothesis test.

9.236 Biology and Environmental Science A special water channel is being constructed, connected to the Campbell River in British Columbia, to provide more area for chinook salmon spawning. The plans call for the channel to be 23 meters wide. Following completion, a random sample of 18 locations along the channel was selected, and the width at each location was measured. The sample mean was $\bar{x} = 24.6$ meters. Assume the distribution of channel widths is normal and $\sigma = 2.28$ meters.

   a. Is there any evidence to suggest the mean width of the channel is greater than 23 meters? Use $\alpha = 0.01$.

   b. Find the p value associated with this hypothesis test.

9.237 Manufacturing and Product Development A piston in a particular 12-cylinder diesel engine is manufactured to have diameter 13 mm. Any larger or smaller diameter will cause immediate, costly damage to the engine. Every half-hour, 10 finished pistons are selected and the diameter of each (in mm) is carefully measured. If there is any evidence that the mean diameter is different from 13 mm, the manufacturing process is stopped and the machinery inspected. The quality control inspector uses a significance level of $\alpha = 0.05$.

   a. Suppose the sample mean is $\bar{x} = 12.89$ and the sample standard deviation is $s = 0.96$. Should the manufacturing process be stopped?

   b. Suppose the sample mean is $\bar{x} = 13.04$ and the sample standard deviation is $s = 0.045$. Should the manufacturing process be stopped?

   c. If you were buying these pistons, would you like the manufacturer to use a smaller or a larger significance level? Justify your answer.

9.238 Medicine and Clinical Studies According to a recent study, doctors are spending less time with each patient, approximately eight minutes each day with each patient.[70] A random sample of residents was obtained and the time spent with a random patient was recorded for each. The data are given in the following table (in minutes). **DOCTIME**

| 6.8 | 9.0 | 7.2 | 8.2 | 7.4 | 4.8 | 7.8 | 5.7 | 9.0 |
| 7.6 | 6.2 | 7.5 | 6.8 | 5.7 | 6.4 | 11.6 | 8.0 | |

   a. Is there any evidence to suggest that the true population mean time spent with each patient is less than eight minutes? Assume normality and use $\alpha = 0.05$.

   b. Find bounds on the p value associated with this hypothesis test.

9.239 Business and Management According to a report issued by the management consulting firm McKinsey & Company, 20% of all companies require genetic or family medical history information from employees or job applicants. A random sample of 1500 companies was obtained, and 345 said they require this information from employees or job applicants.

   a. Identify p_0 and n, and compute $\hat{p}$.

   b. Check the nonskewness criterion.

   c. Is there any evidence to suggest the true proportion of companies that require this information is greater than 0.20? Use $\alpha = 0.01$.

   d. Compute the p value associated with this hypothesis test.

9.240 Public Policy and Political Science A candidate for district attorney claims that 75% of all residents in Elko County, Nevada, favor granting police additional powers to tap telephone lines. To check this claim, a random sample of 560 residents was selected, and 392 said they were in favor of more phone taps.

   a. Is there any evidence to suggest the true proportion of residents who favor additional power to tap phones is less than 0.75? Use $\alpha = 0.01$.

   b. Compute the p value associated with this hypothesis test.

9.241 Public Policy and Political Science The sequester—potentially harmful, automatic budget cuts—can affect jobs, services, education, and public safety. In 2013, the U.S. Congress was unable to reach a compromise on the federal budget, and on March 1, the sequester began. According to a poll conducted during the Summer, 22% of Americans said they had been significantly affected by resulting cuts.[71] Early in the Fall, in a new survey, 55 of 310 randomly selected Americans said they were significantly affected by the resulting cuts.

   a. Is there any evidence to suggest that the proportion of Americans affected by the sequester changed? Use $\alpha = 0.05$.

   b. Find the p value associated with this hypothesis test.

9.242 Public Health and Nutrition Many remember actor Michael J. Fox as conservative Republican Alex Keaton on the TV show *Family Ties*. After more than 10 years, he returned to television during the Fall of 2013 as a news anchor diagnosed with Parkinson's disease. The show reflects much of his frustrating experience. Most cases of Parkinson's disease are linked to environmental or genetic factors. It has been reported that approximately 15% of people with Parkinson's disease have a family history of the disease.[72] A random sample of 420 Parkinson's patients was obtained, and 78 had a family history of the disease. Is there any evidence to suggest that the proportion of Parkinson's patients with a family history of the disease is different from 0.15? Use $\alpha = 0.05$.

9.243 Travel and Transportation In July 2013 a train carrying 72 tank cars filled with oil exploded and rolled off the tracks in Lac-Mégantic, Quebec. This was one of the worst train accidents in Canadian history, and the disaster prompted many to discuss improved regulations for the transportation of hazardous materials. The Association of American Railroads reported that the safety record for moving hazardous materials is outstanding, and that 74% of railroad crude-oil spills involve less than five gallons.[73] Over the next year, a random sample of 129 railroad crude-oil spills was obtained, and 85 involved spills of less than five gallons.

a. Is there any evidence to suggest that the proportion of crude-oil spills that involve less than five gallons has decreased? Use $\alpha = 0.05$.

b. Explain the practical implications of the hypothesis test results in part (a).

c. Find the p value associated with this hypothesis test.

9.244 Medicine and Clinical Studies The diameter of a virus is approximately 0.3 μm (micrometers). There is some speculation that new virus strains exhibit greater variability in diameter. A medical research lab obtained a random sample of 15 new virus strains and measured the diameter (in μm) of each. The sample mean diameter was 0.323 and the sample variance was $s^2 = 0.0026$. Is there any evidence to suggest the true population variation in diameter of viruses has increased from 0.0015? Assume normality and use $\alpha = 0.05$.

9.245 Manufacturing and Product Development The radial shrinkage in paper birch wood from green to oven dry is approximately 6.3%. A new process has been developed to decrease the variability in shrinkage. In a random sample of 21 pieces of paper birch, the sample variance in shrinkage was 0.39. Is there any evidence to suggest the population variance in shrinkage is less than 0.50? Assume normality and use $\alpha = 0.10$.

9.246 Medicine and Clinical Studies The normal blood platelet count for an adult ranges from 150,000 to 400,000, with a standard deviation of approximately 62,500. A researcher has speculated that increased exposure to pollutants has increased the variability in numbers of blood platelets. A random sample of 37 adults was obtained, and the blood platelet count was measured in each person. The sample standard deviation was $s = 65{,}268$. Is there any evidence to suggest the population variance in blood platelet count has increased? Assume normality and use $\alpha = 0.001$.

9.247 Public Policy and Political Science With over $18 billion in debt, in July 2013 the city of Detroit declared bankruptcy. In many parts of the city there are few public services, roads in need of repair, and approximately 40% of all streetlights do not work at night. In addition, the mean police response time is 58 minutes.[74] A random sample of calls to police

was obtained and the response time (in minutes) for each was recorded. The data are given in the following table. **DETROIT**

52	53	56	50	48	51	56	60	49	57	60	59	61
59	60	44	57	61	66	54	55	59	53	54	62	

a. The chief of police claims that response time has improved. Is there any evidence to suggest that the mean police response time has decreased? Assume normality and use $\alpha = 0.05$.

b. Find bounds on the p value for this hypothesis test.

EXTENDED APPLICATIONS

9.248 Physical Sciences Workers at the Daivik diamond mine in northern Canada extract approximately 1800 DMT (dry metric tons) of ore per day, which is sifted and examined for diamonds. New machinery has just been installed that is designed to increase the amount of ore extracted per day. A random sample of 36 days was selected, and the amount of ore extracted each day was recorded. The sample mean was $\bar{x} = 1852$ DMT. Assume $\sigma = 202$ DMT.

a. Is there any evidence to suggest the new machinery has improved production? Use $\alpha = 0.05$.

b. What is the probability of a type II error if the true mean amount of ore extracted has changed to 1875 DMT; that is, find $\beta(1875)$? Find the probability of a type II error if the true mean is 1925 DMT.

9.249 Biology and Environmental Science A random sample of days in July 2013 was obtained and the highest significant wave heights that occurred in July 2013 near Southeast Queensland, Australia, was obtained. The data are given in the following table (in meters).[75] **WAVES**

0.9	0.4	2.4	0.9	1.7	1.5	2.0
1.3	2.1	2.1	2.0	1.3	2.1	

a. Is there any evidence to suggest that the mean significant wave height is greater than 1 meter? Assume normality and use $\alpha = 0.01$.

b. Find bounds on the p value for this hypothesis test.

9.250 Physical Sciences The design specifications for a new gymnasium at a local high school call for the lights to produce at least 40 footcandles (a measure of brightness). Before the new facility was opened to students, the contractor collected a sample of 23 brightness measurements at random locations in the gym. The sample mean was $\bar{x} = 38.63$ footcandles and the sample standard deviation was $s = 5.6$ footcandles.

a. Is there any evidence to suggest the mean brightness in the gym is less than the design specification? Assume normality and use $\alpha = 0.01$.

b. Find bounds on the p value associated with this hypothesis test.

9.251 Business and Management The largest above-ground storage tanks in Bayonne, New Jersey, have the capacity to hold 6,000,000 gallons of oil. For safety reasons, managers prefer the mean amount stored at any given time to be no greater than 4,500,000 gallons. A random sample of nine large tanks was selected, and the amount of oil stored in each (in gallons) was recorded. The summary statistics were $\bar{x} = 4{,}675{,}250$ and $s = 482{,}556$.

  a. Is there any evidence to suggest the mean amount of oil stored in the large tanks is above the safety level? Assume normality and use $\alpha = 0.025$.

  b. Find bounds on the p value associated with this hypothesis test.

9.252 Public Health and Nutrition   Cast iron is an extremely durable cookware material, good for searing and blackening foods. However, a cast-iron pan, for example, can be full of bacteria and can lend unwanted flavors to food. A company trying to promote alternative ceramic cookware asked members of a community to bring their favorite cast-iron cookware to their store for bacteria testing. The company manager claimed that at least 60% of all cast-iron cookware contains harmful bacteria. A random sample of 120 pans was selected, and 57 were found to contain harmful bacteria.

  a. Is there any evidence to refute the manager's claim? Use $\alpha = 0.01$.

  b. Find the p value associated with this hypothesis test.

  c. Do you believe the sample is really random? Why or why not?

9.253 Marketing and Consumer Behavior   An assisted-living home is an alternative to a nursing home and a bridge between a skilled-care facility and a patient's residence. A recent report indicated 92% of all patients in assisted-living homes are satisfied with the facility and the care. An insurance company believes that this percentage is actually much lower (due to health-care violations and strict for-profit motives). A random sample of 5000 patients in assisted-living homes around the country was obtained, and 4576 said they were satisfied with the facility and the care.

  a. Is there any evidence to suggest the true proportion of assisted-living patients who are satisfied is less than 0.92? Use $\alpha = 0.01$.

  b. Find the p value associated with this hypothesis test.

  c. Carefully sketch a graph illustrating the critical value, value of the test statistic, and p value.

9.254 Biology and Environmental Science   The Castaic Lake reservoir provides water for the northern portion of the Greater Los Angeles Area. A random sample of days during Summer 2013 was obtained and the capacity of the lake on each day (in thousands of acre feet) was measured.[76] **WATER**

  a. Suppose the historic mean capacity during the summer is 277 thousand acre feet. Is there any evidence to suggest that the true population mean capacity has decreased? Assume normality and use $\alpha = 0.05$.

  b. Find the p value associated with this hypothesis test.

9.255 Manufacturing and Product Development   The Liberty Stone Vista standard unit is manufactured to weigh 70 pounds.[77] This double-sided stone is part of a wall system for residential and commercial retaining walls. Large variability in the weight of these stones could cause structural deficiencies in a wall. A random sample of 17 standard units was obtained and the sample variance was 7.75 pounds^2.

  a. Is there any evidence to suggest that the population variance is greater than 4 pounds^2? Assume normality and use $\alpha = 0.05$.

  b. Find bounds on the p value for this hypothesis test.

9.256 Business and Management   According to Catalyst, a nonprofit organization working to provide more opportunities for women and business, women currently hold 4.6% of Fortune 1000 CEO positions.[78] Some economic researchers believe that women own a larger proportion of small businesses. A random sample of 550 small businesses in the United States was obtained, and 37 were owned by women.

  a. Is there any evidence to suggest that the proportion of small businesses owned by women is greater than 4.6%? Use $\alpha = 0.01$.

  c. Find the p value for this hypothesis test.

LAST STEP

9.257 Is it OK to pad an insurance claim?   The results of a recent survey revealed that 24% of Americans believe that it is acceptable to increase an insurance claim by a small amount to make up for deductibles that they are required to pay. Suppose Liberty Mutual Insurance Company has hired an independent agency to conduct a survey of its automobile policy holders. Liberty Mutual plans to increase all policies by a fixed rate if there is evidence to suggest that more than 24% of policy holders view claim padding as acceptable. Of the 355 policy holders contacted, 96 said they view claim padding as acceptable. Is there any evidence to suggest that the proportion of policy holders that view claim padding as acceptable is greater than 0.24? Use $\alpha = 0.05$.

10 Confidence Intervals and Hypothesis Tests Based on Two Samples or Treatments

◀ **Looking Back**

■ Recall the formal, four-part hypothesis test process.
■ Remember the specific inference procedures concerning a single population parameter: μ, p, or σ^2.

▶ **Looking Forward**

■ Adapt and extend the single-sample hypothesis test procedures.
■ Construct confidence intervals to estimate the difference between two population parameters.
■ Conduct hypothesis tests to compare two population parameters.

Are people who live at higher altitudes slimmer?

A recent study suggests that people who live at higher altitudes tend to be thinner than those who live in low-lying areas. This does not mean we should all pack up and move to the mountains. There are certainly other factors affecting obesity. However, the data suggest that Americans who live near sea level are more likely to be obese, compared with people who live at higher altitudes, for example, in Colorado.

A possible explanation for a difference in obesity rates is that altitude can affect appetite hormones. In addition, altitude can also affect how many calories the body burns in ordinary daily activities. People who live at higher altitudes also tend to drink more water, which may help weight loss.

To examine this trend more closely, random samples were obtained for 125 adults living in Denver, Colorado (the Mile High City), and 150 adults living in New Orleans, Louisiana (where the mean elevation is zero feet above sea level). Using body mass index as a measure of obesity, 38 of those living in Denver and 61 of those living in New Orleans were classified as obese.

The hypothesis test procedures presented in this chapter will be used to compare parameters (means or variances, for example) from two different populations. In this case, we will compare the population proportion of obese adults in Denver with the population proportion of obese adults in New Orleans. These tests are constructed using methods of *standardization* similar to those in Chapter 9.

CONTENTS

Katja Kreder/AWL Images/Getty Images

Notation

To conduct a hypothesis test to compare two (similar) population parameters, we will simply modify the single-sample procedures presented in the previous chapter. Perhaps the most tricky aspect of these procedures is the notation. The following table summarizes the notation used to represent similar parameters associated with two different populations.

	Population parameters			
	Mean	Variance	Standard deviation	Proportion
Population 1	μ_1	σ_1^2	σ_1	p_1
Population 2	μ_2	σ_2^2	σ_2	p_2

The following table summarizes the notation used to represent *values* of summary statistics associated with samples from two different populations.

Note: We do not necessarily use every summary statistic associated with a sample in every problem. For example, we may only need the sample size and proportion in one case, but use the sample size, mean, and standard deviation in another problem.

	Sample statistics				
	Sample size	Mean	Variance	Standard deviation	Proportion
Sample from population 1	n_1	$\bar{x}_1$	s_1^2	s_1	$\hat{p}_1$
Sample from population 2	n_2	$\bar{x}_2$	s_2^2	s_2	$\hat{p}_2$

To compare two population parameters to see whether there is any evidence that they are different, we often consider a difference. For example, to compare two population means, μ_1 and μ_2, we consider the difference $\mu_1 - \mu_2$. In searching for evidence that p_1 is larger than p_2, we look at the difference $p_1 - p_2$.

There are two reasons to consider a difference.

1. A typical relationship between two population parameters can be written in terms of a difference. For example, suppose we need to compare the means from two populations, μ_1 and μ_2.

Standard notation		Difference notation
$\mu_1 = \mu_2$	is equivalent to	$\mu_1 - \mu_2 = 0$
$\mu_1 > \mu_2$	is equivalent to	$\mu_1 - \mu_2 > 0$
$\mu_1 < \mu_2$	is equivalent to	$\mu_1 - \mu_2 < 0$

Therefore, a statistical test with null hypothesis $H_0: \mu_1 - \mu_2 = 0$ corresponds to a test of $H_0: \mu_1 = \mu_2$. And $H_a: \mu_1 - \mu_2 > 0$ is equivalent to $H_a: \mu_1 > \mu_2$. The hypothesized difference between the two means may be nonzero. The null hypothesis $H_0: \mu_1 = \mu_2 + 5$ written using a difference is equivalent to $H_0: \mu_1 - \mu_2 = 5$.

2. In addition, a difference (for example, $\mu_1 - \mu_2$) is itself a *single* population parameter. A natural, intuitive statistic, $\bar{X}_1 - \bar{X}_2$, may be used to estimate the value of this parameter. The properties of $\bar{X}_1 - \bar{X}_2$ will be used to develop a test statistic.

As in the statistical tests presented in the last chapter, in any two-sample hypothesis test we usually make certain assumptions. The assumptions associated with the hypothesis tests in this chapter include a statement concerning the selection of individuals or objects from *two* different populations.

Definition

1. Two samples are **independent** if the process of selecting individuals or objects in sample 1 has no effect on, or no relation to, the selection of individuals or objects in sample 2. If the samples are not independent, they are **dependent**.

2. A **paired** data set is the result of matching each individual or object in sample 1 with a *similar* individual or object in sample 2. A common experiment in which paired data are obtained involves a *before* and *after* measurement on each individual or object. Each *before* observation is matched, or paired, with an *after* observation.

Similar means the individuals or objects share some common, fundamental characteristic. They may even be the same individual or object!

The notation, the idea of using differences, and the extra assumptions are all used in the following sections to construct hypothesis tests for comparing various characteristics of two populations.

10.1   Comparing Two Population Means Using Independent Samples When Population Variances Are Known

As in Chapters 8 and 9, the first hypothesis test presented here, for comparing two population means, is instructive but not very realistic. Because $\overline{X}_1$ is a good estimator for μ_1 and $\overline{X}_2$ is a good estimator for μ_2, it is reasonable to use the estimator $\overline{X}_1 - \overline{X}_2$ to estimate the parameter $\mu_1 - \mu_2$. To develop a hypothesis test, we need to know the properties of the estimator, or the distribution of the random variable, $\overline{X}_1 - \overline{X}_2$.

Properties of $\overline{X}_1 - \overline{X}_2$

Suppose

1. $\overline{X}_1$ is the mean of a random sample of size n_1 from a population with mean μ_1 and variance σ_1^2.

2. $\overline{X}_2$ is the mean of a random sample of size n_2 from a population with mean μ_2 and variance σ_2^2.

3. The samples are independent.

If the distributions of both populations are normal, then the random variable $\overline{X}_1 - \overline{X}_2$ has the following properties.

1. $\mathrm{E}(\overline{X}_1 - \overline{X}_2) = \mu_{\overline{X}_1 - \overline{X}_2} = \mu_1 - \mu_2$.

   $\overline{X}_1 - \overline{X}_2$ is an unbiased estimator of the parameter $\mu_1 - \mu_2$. The distribution is centered at $\mu_1 - \mu_2$.

2. $\mathrm{Var}(\overline{X}_1 - \overline{X}_2) = \sigma_{\overline{X}_1 - \overline{X}_2}^2 = \dfrac{\sigma_1^2}{n_1} + \dfrac{\sigma_2^2}{n_2}$ and the standard deviation is

   $\sigma_{\overline{X}_1 - \overline{X}_2} = \sqrt{\dfrac{\sigma_1^2}{n_1} + \dfrac{\sigma_2^2}{n_2}}$.

3. The distribution of $\overline{X}_1 - \overline{X}_2$ is normal.

   If the underlying distributions are not known, but both n_1 and n_2 are large, then $\overline{X}_1 - \overline{X}_2$ is approximately normal (by the central limit theorem).

Can you see the standardization coming?

Because the distribution of $\overline{X}_1 - \overline{X}_2$ is (approximately) normal, the usual standardization can be used to obtain a Z random variable. The resulting hypothesis test has a very typical form.

Hypothesis Test Concerning Two Population Means When Population Variances Are Known

Given two independent random samples, the first of size n_1 from a population with mean μ_1 and the second of size n_2 from a population with mean μ_2, assume that:

For reference, we'll call these the two-sample Z-test assumptions.

1. The underlying populations are normal and/or both sample sizes are large, and
2. The population variances, σ_1^2 and σ_2^2, are known.

A hypothesis test concerning two population means, in terms of the difference in means $\mu_1 - \mu_2$, with significance level α, has the form

This is the template for a hypothesis test concerning two population means when variances are known, sometimes called a two-sample Z test.

$H_0: \mu_1 - \mu_2 = \Delta_0$

$H_a: \mu_1 - \mu_2 > \Delta_0, \qquad \mu_1 - \mu_2 < \Delta_0, \qquad \text{or} \qquad \mu_1 - \mu_2 \neq \Delta_0$

$$\text{TS: } Z = \frac{(\overline{X}_1 - \overline{X}_2) - \Delta_0}{\sqrt{\dfrac{\sigma_1^2}{n_1} + \dfrac{\sigma_2^2}{n_2}}}$$

RR: $Z \geq z_\alpha, \qquad Z \leq -z_\alpha, \qquad \text{or} \qquad |Z| \geq z_{\alpha/2}$

A CLOSER LOOK

Δ is the uppercase Greek letter delta.

1. The value Δ_0 is the fixed, hypothesized difference in means. Usually $\Delta_0 = 0$, that is, the means are assumed equal. The null hypothesis is then $H_0: \mu_1 - \mu_2 = 0$, which is equivalent to $H_0: \mu_1 = \mu_2$. However, Δ_0 may be some nonzero value. For example, two population means may historically differ by 12 so that $H_0: \mu_1 - \mu_2 = 12 \, (= \Delta_0)$. We may want to conduct a test to see whether there is any change in this difference, with $H_a: \mu_1 - \mu_2 \neq 12$.

2. Just a reminder: Use only one (appropriate) alternative hypothesis and the corresponding rejection region. The z critical values are from the standard normal distribution.

3. This hypothesis test procedure can be used *only* if both population variances are known. If they are unknown but both sample sizes are large, some statisticians substitute s_1^2 for σ_1^2 and s_2^2 for σ_2^2. This produces an *approximate* test statistic. Section 10.2 presents an *exact* test procedure for comparing population means (under certain assumptions) when the population variances are unknown.

The following example illustrates this hypothesis test procedure.

Example 10.1 Turning off the TV

Some social science researchers believe that with increased usage of social media, mobile devices, and other communication devices, younger people are watching less TV. Independent random samples of people in two age groups were obtained, and the weekly time spent watching TV (in hours) was recorded for each. The summary statistics and known variances are given in the following table.[1]

Age group	Sample size	Sample mean	Population variance
18–24 (1)	$n_1 = 18$	$\bar{x}_1 = 23.4$	$\sigma_1^2 = 44.89$
25–34 (2)	$n_2 = 24$	$\bar{x}_2 = 28.9$	$\sigma_2^2 = 65.61$

Is there any evidence to suggest that the mean weekly time spent watching TV for 18–24 year olds is less that the mean weekly time spent watching TV for 25–34 year olds? Use

Solution Trail 10.1

KEYWORDS

- Is there any evidence?
- Less than
- Known variances
- Independent random samples
- Each underlying distribution is normal

TRANSLATION

- Conduct a one-sided, left-tailed test to compare μ_1 and μ_2.

CONCEPTS

- Hypothesis test concerning two population means when variances are known

VISION

Use the template for this hypothesis test. The samples are random and independent, the underlying populations are normal, and the population variances are known. Use a one-sided alternative hypothesis and the corresponding rejection region, find the value of the test statistic, and draw a conclusion.

$\alpha = 0.01$ and assume that each underlying distribution of weekly time spent watching TV is normal.

SOLUTION

STEP 1 Arbitrarily, let the 18–24-year-old group be population 1, and the 25–34-year-old group be population 2.

The current state, or assumption, is that the two population mean weekly times spent watching TV are equal:

$$\mu_1 = \mu_2 \Rightarrow \mu_1 - \mu_2 = 0 \ (= \Delta_0).$$

The sample sizes, sample means, and population variances are given.

We are trying to find evidence that the 18–24-year-old group has *smaller* mean weekly time spent watching TV: $\mu_1 < \mu_2$, which is the same as $\mu_1 - \mu_2 < 0$. Therefore, the alternative hypothesis is one-sided, left-tailed.

STEP 2 The four parts of the hypothesis test are

$$H_0: \mu_1 - \mu_2 = 0$$
$$H_a: \mu_1 - \mu_2 < 0$$
$$\text{TS: } Z = \frac{(\overline{X}_1 - \overline{X}_2) - 0}{\sqrt{\dfrac{\sigma_1^2}{n_1} + \dfrac{\sigma_2^2}{n_2}}}$$

RR: $Z \leq -z_\alpha = -z_{0.01} = -2.3263$

STEP 3 The value of the test statistic is

$$z = \frac{(\bar{x}_1 - \bar{x}_2) - 0}{\sqrt{\dfrac{\sigma_1^2}{n_1} + \dfrac{\sigma_2^2}{n_2}}} = \frac{23.4 - 28.9}{\sqrt{\dfrac{44.89}{18} + \dfrac{65.61}{24}}} = -2.4055 \ (\leq -2.3263)$$

STEP 4 Because -2.4055 lies in the rejection region, we reject the null hypothesis at the $\alpha = 0.01$ significance level. There is evidence to suggest that the mean weekly time spent watching TV for 18–24-year-olds is less than the mean weekly time spent watching TV for 25–34-year-olds.

The p value for this hypothesis test is

$$p = P(Z \leq -2.4055) = 0.0081 \ (\leq 0.05)$$ Use Table III in the Appendix.

Because $p \leq \alpha$, we reject the null hypothesis. (See Figure 10.1.)

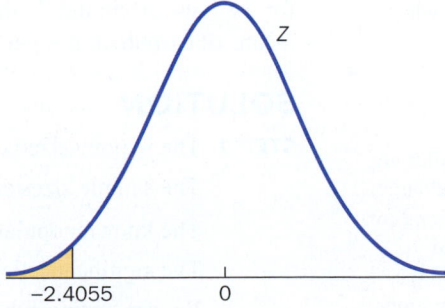

Figure 10.1 p-Value illustration:
$p = P(Z \leq -2.4055)$
$= 0.0081 \leq 0.05 = \alpha$

Figures 10.2 through 10.4 show a technology solution.

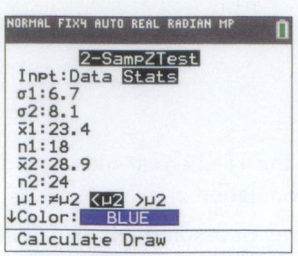

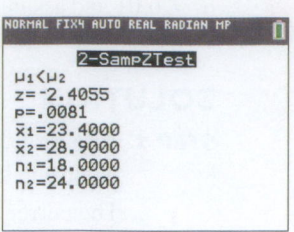

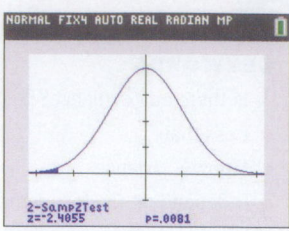

Figure 10.2
2-SampZTest input screen.

Figure 10.3
2-SampZTest Hypothesis test results.

Figure 10.4
2-SampZTest Draw results.

TRY IT NOW   GO TO EXERCISE 10.14

The following example involves a hypothesis test with a nonzero value for the hypothesized difference in means, Δ_0.

Example 10.2   Low-Carb Ice Cream

Low-carbohydrate foods are very popular as many Americans try to avoid this sugar and starch combination that many believe causes weight gain. An advertisement for a low-carb ice cream claims that the product has 16 fewer grams of carbohydrates per serving than the leading store brand. To check this claim, independent random samples of each type of ice cream were obtained, and the amount of carbohydrates in each serving was measured. The data are given (in grams) in the following table.

Store brand (1)

15.4	20.4	21.0	24.3	23.3	18.7	19.8	22.5	18.9	22.8
25.4	25.1	20.3	24.1	16.6	22.6	22.1	19.4	16.6	24.4
17.8	18.6	14.9	24.6	19.1	17.9	18.7	20.1	26.3	18.4
21.8	17.1	21.5	19.6	22.9	22.2	21.5	18.3		

Low-carb brand (2)

3.7	3.9	4.5	4.3	3.2	3.6	3.7	3.6	3.7	4.0
4.1	3.1	4.3	3.4	3.4	3.5	4.4	4.9	3.7	3.8
4.1	4.7	3.7	4.2	3.1	4.4	4.2	3.4	4.8	3.6
3.2	3.4	4.2	3.0	3.9					

The variance in carbohydrates per serving is known to be 8.5 for the store brand and 0.253 for the low-carb brand. Is there any evidence to suggest that the difference in population means of carbohydrates per serving is not 16 grams? Use $\alpha = 0.01$.

SOLUTION

STEP 1   The hypothesized difference is $\mu_1 - \mu_2 = 16 \, (= \Delta_0)$.

The sample sizes are $n_1 = 38$ and $n_2 = 35$.

The known population variances are $\sigma_1^2 = 8.5$ and $\sigma_2^2 = 0.253$.

The significance level is $\alpha = 0.01$.

We are testing for *any* difference in population means other than 16 grams of carbohydrates. This is a two-sided test.

The samples are random and independent, and the population variances are known. The underlying population distributions are unknown, but both sample sizes are large (≥ 30). A hypothesis test concerning two population means when variances are known is relevant.

STEP 2 The four parts of the hypothesis test are

$$H_0: \mu_1 - \mu_2 = 16$$

$$H_a: \mu_1 - \mu_2 \neq 16$$

$$\text{TS: } Z = \frac{(\bar{X}_1 - \bar{X}_2) - 16}{\sqrt{\dfrac{\sigma_1^2}{n_1} + \dfrac{\sigma_2^2}{n_2}}}$$

RR: $|Z| \geq z_{\alpha/2} = z_{0.005} = 2.5758$

STEP 3 The sample means are

$$\bar{x}_1 = \frac{1}{38}(15.4 + 20.4 + \cdots + 18.3) = 20.6579$$

$$\bar{x}_2 = \frac{1}{35}(3.7 + 3.9 + \cdots + 3.9) = 3.8486$$

The value of the test statistic is

$$z = \frac{(\bar{x}_1 - \bar{x}_2) - 16}{\sqrt{\dfrac{\sigma_1^2}{n_1} + \dfrac{\sigma_2^2}{n_2}}} = \frac{(20.6579 - 3.8486) - 16}{\sqrt{\dfrac{8.5}{38} + \dfrac{0.253}{35}}} = 1.6842$$

STEP 4 The value of the test statistic, $z = 1.6842$, does not lie in the rejection region. We do not reject the null hypothesis. There is no evidence to suggest that the difference in population mean carbohydrates is different from 16 grams at the $\alpha = 0.01$ significance level.

This is a two-sided test and the value of the test statistic is positive, so $p/2$ is a right-tail probability.

$p/2 = P(Z \geq 1.6842)$	Definition of p value for a two-sided test.
$= 1 - P(Z \leq 1.6842)$	The complement rule.
$= 1 - 0.9539 = 0.0461$	Use Table III in the Appendix.
$p = 2(0.0461) = 0.0922$	Solve for p.

Because $p = 0.0922 > 0.01 (= \alpha)$, we do not reject the null hypothesis. See Figure 10.5.

Figure 10.6 shows a technology solution.

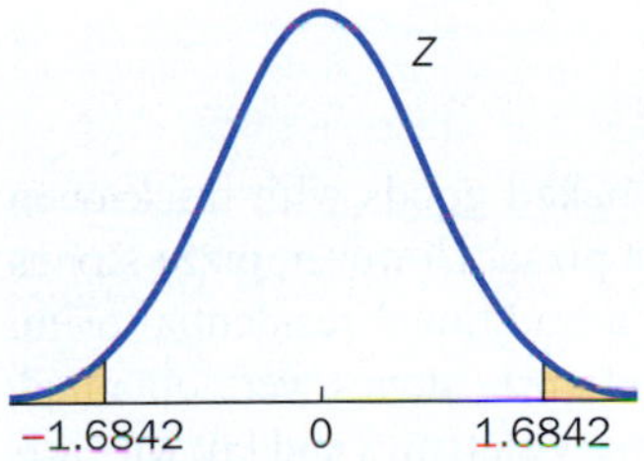

Figure 10.5 p-Value illustration:
$p = 2P(Z \geq 1.6842)$
$= 0.0922 > 0.01 = \alpha$

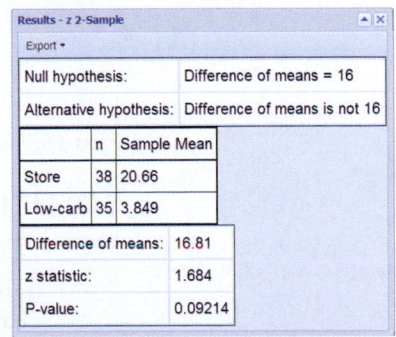

Figure 10.6 CrunchIt! z 2-Sample Hypothesis test results.

Results – z 2-Sample		
Export ▾		
Null hypothesis:	Difference of means = 16	
Alternative hypothesis:	Difference of means is not 16	
	n	Sample Mean
Store	38	20.66
Low-carb	35	3.849
Difference of means:	16.81	
z statistic:	1.684	
P-value:	0.09214	

TRY IT NOW   GO TO EXERCISE 10.16

Given the two-sample Z test assumptions and the properties of the random variable $\bar{X}_1 - \bar{X}_2$, we can construct a confidence interval (CI) for the (difference) parameter $\mu_1 - \mu_2$.

▶ As usual, to find a general CI, start with an appropriate symmetric interval about 0 such that the probability Z lies in this interval is $1 - \alpha$.

$$P\left[-z_{\alpha/2} < \frac{(\bar{X}_1 - \bar{X}_2) - (\mu_1 - \mu_2)}{\underbrace{\sqrt{\dfrac{\sigma_1^2}{n_1} + \dfrac{\sigma_2^2}{n_2}}}_{Z}} < z_{\alpha/2}\right] = 1 - \alpha \qquad (10.1)$$

Manipulate the inequality in Equation 10.1 to *sandwich* the parameter $\mu_1 - \mu_2$. We obtain the following probability statement:

$$P\left[(\bar{X}_1 - \bar{X}_2) - z_{\alpha/2}\sqrt{\frac{\sigma_1^2}{n_1} + \frac{\sigma_2^2}{n_2}} < \mu_1 - \mu_2 < (\bar{X}_1 - \bar{X}_2) + z_{\alpha/2}\sqrt{\frac{\sigma_1^2}{n_1} + \frac{\sigma_2^2}{n_2}}\right] = 1 - \alpha$$

This leads to the following general result. ◀

How to Find a 100(1 − α)% Confidence Interval for $\mu_1 - \mu_2$ When Variances Are Known

Given the two-sample Z test assumptions, a $100(1 - \alpha)\%$ confidence interval for $\mu_1 - \mu_2$ has as endpoints the values

$$(\bar{x}_1 - \bar{x}_2) \pm z_{\alpha/2}\sqrt{\frac{\sigma_1^2}{n_1} + \frac{\sigma_2^2}{n_2}} \qquad (10.2)$$

Example 10.3 No Anchovies

Tracy Hornbrook/Dreamstime.com

Pizza stones designed for home use help cooks produce baked goods with brick-oven qualities, for example, a *crusty* loaf of bread or crispy-crust pizza. However, pizza stones can be very heavy and can also take up a lot of space in a traditional residential oven. Independent random samples of two similar types of round pizza stones were obtained, and the weight (in pounds) of each was recorded. The summary statistics and known variances are given in the following table.

Pizza stone	Sample size	Sample mean	Population variance
Kitchen Depot (1)	$n_1 = 35$	$\bar{x}_1 = 6.21$	$\sigma_1^2 = 2.1$
Head Chef (2)	$n_2 = 31$	$\bar{x}_2 = 7.08$	$\sigma_2^2 = 3.5$

Find a 95% confidence interval for the difference in population mean pizza-stone weights.

SOLUTION

STEP 1 Sample sizes, sample means, and known variances are given.

The underlying weight distributions are unknown, but the sample sizes are both large (≥ 30).

$1 - \alpha = 0.95 \quad \Rightarrow \quad \alpha = 0.05 \quad \Rightarrow \quad \alpha/2 = 0.025$ Find $\alpha/2$.

$z_{\alpha/2} = z_{0.025} = 1.960$ Find the z critical value.

STEP 2 Use Equation 10.2.

$$(\bar{x}_1 - \bar{x}_2) \pm z_{\alpha/2}\sqrt{\frac{\sigma_1^2}{n_1} + \frac{\sigma_2^2}{n_2}}$$ Equation 10.2.

$$= (6.21 - 7.08) \pm (1.96)\sqrt{\frac{2.1}{35} + \frac{3.5}{31}}$$ Use summary statistics and critical value.

$$= -0.87 \pm 0.8150$$ Simplify.

$$= (-1.6850, -0.0550)$$ Compute endpoints.

$(-1.6850, -0.0550)$ is a 95% confidence interval for the difference in population mean weights (in pounds) of the pizza stones, $\mu_1 - \mu_2$. This interval represents a set of very plausible values for the difference in population mean weights.

Figures 10.7 and 10.8 together show a technology solution.

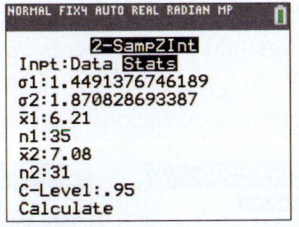 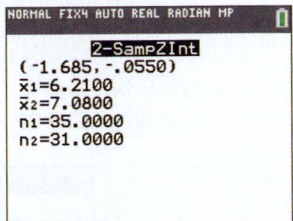

Figure 10.7
2-SampZInt input screen.

Figure 10.8 Resulting confidence interval.

TRY IT NOW   GO TO EXERCISE 10.17

Technology Corner

Procedure: Hypothesis tests and confidence intervals concerning two population means when the population variances are known.

Reconsider: Example 10.2, solution, and interpretations.

CrunchIt!

Use the function z 2-Sample to conduct a hypothesis test concerning two population means.

1. Enter the store-brand data into column Var1 and the low-carb brand data into column Var2.
2. Select Statistics; z; 2-Sample. Under the Columns tab, select Var1 for Sample 1 and Var2 for Sample 2. Enter the standard deviation for each group.
3. Under the Hypothesis Test tab, enter the Difference of means under null hypothesis, 16 ($= \Delta_0$). Choose the appropriate Alternative (hypothesis).
4. Click Calculate. The results are shown in Figure 10.6.

TI-84 Plus C

Use the calculator functions 2-SampZTest and 2-SampZInt. Input is either summary statistics or data in lists.

1. Enter the store brand data into list L1 and the low-carb brand data into list L2.
2. Subtract $\Delta_0 = 16$ from each observation in list L1 and store the results in list L1.
3. Select STAT; TESTS; 2-SampZTest. Highlight Data. Enter σ_1, σ_2, List1, and List2. Set each frequency to 1. Highlight the alternative hypothesis. See Figure 10.9.
4. Highlight Calculate and press ENTER. The results are displayed on the Home screen. See Figure 10.10. The Draw results are shown in Figure 10.11.

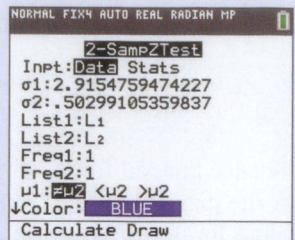

Figure 10.9
2-SampZTest input screen.

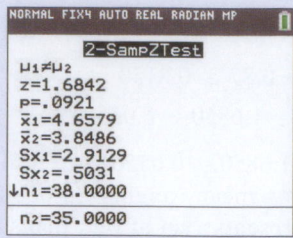

Figure 10.10
2-SampZTest Hypothesis
test results.

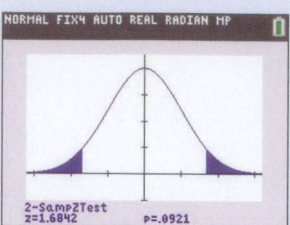

Figure 10.11
2-SampZTest Draw
results.

5. To construct a confidence interval for the difference in population means, use the original data, in list L1.
6. Select STAT; TESTS; 2-SampZInt. Highlight Data. Enter σ_1, σ_2, List1, and List2. Set each frequency to 1 and enter the C-Level. See Figure 10.12.
7. Highlight Calculate and press ENTER. The resulting confidence interval is displayed on the Home screen. See Figure 10.13.

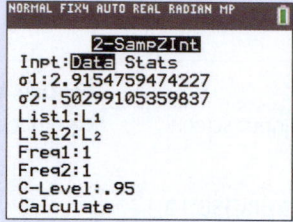

Figure 10.12
2-SampZInt input screen.

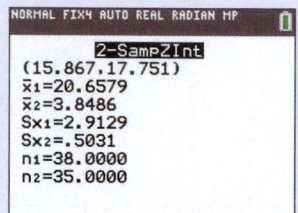

Figure 10.13 Resulting
95% confidence interval.

Minitab

There is no built-in function to conduct hypothesis tests and construct confidence intervals concerning two population means when the population variances are known. Remember, this is an instructive situation, not very realistic. It is pretty unlikely we would know the population variances, but not the population means.

Excel

Use the built-in function z-test: Two Sample for Means.

1. Enter the store brand data into column A and the low-carb brand data into column B.
2. Under the Data tab, select Data Analysis; z-test: Two Sample for Means.
3. Enter the Variable 1 Range and Variable 2 Range, the Hypothesized Mean Difference, and the known population variances. Enter the value for Alpha and choose an Output option. See Figure 10.14. Click OK.
4. Summary statistics along with the value of the test statistic, critical values, and p values are displayed. See Figure 10.15.

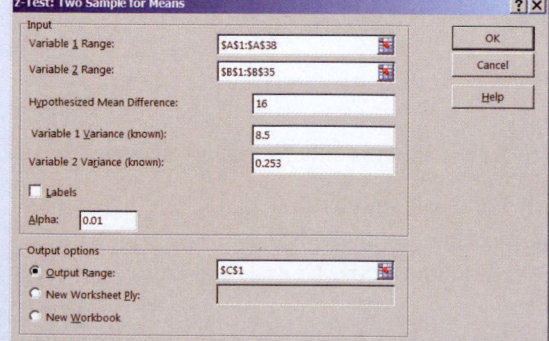

Figure 10.14 z-Test: Two Sample for Means input
screen.

z-Test: Two Sample for Means		
	Variable 1	Variable 2
Mean	20.6579	3.8486
Known Variance	8.5	0.253
Observations	38	35
Hypothesized Mean Difference	16	
z	1.6842	
P(Z<=z) one-tail	0.0461	
z Critical one-tail	2.3263	
P(Z<=z) two-tail	0.0921	
z Critical two-tail	2.5758	

Figure 10.15 Hypothesis test
results.

SECTION 10.1 EXERCISES

Concept Check

10.1 True/False A paired data set often involves a before-and-after measurement on each individual or object.

10.2 True/False In a two-sample Z test, the hypothesized difference in means must be 0.

10.3 True/False The two-sample Z test can be used only if both population variances are known.

10.4 True/False In a two-sample Z test, both sample sizes must be large.

10.5 True/False In a two-sample Z test, the observations may be dependent.

10.6 Short Answer Given the two-sample Z test assumptions, a $100(1 - \alpha)\%$ confidence interval for $\mu_1 - \mu_2$ has as endpoints the values _____.

Practice

10.7 In each of the following problems, rewrite the standard-notation hypothesis concerning two population means in terms of a difference, $\mu_1 - \mu_2$.

a. $\mu_1 = \mu_2$
b. $\mu_1 < \mu_2$
c. $\mu_1 \neq \mu_2 + 7$
d. $\mu_1 > \mu_2 - 4$
e. $\mu_1 \neq \mu_2$
f. $\mu_1 - 10 = \mu_2$

10.8 In each of the following problems, $\mu_1, \mu_2, \sigma_1, \sigma_2, n_1,$ and n_2 are given. Assume the underlying distributions are normal. Find the mean, variance, and standard deviation of the random variable $\bar{X}_1 - \bar{X}_2$, and carefully sketch the probability density function.

a. $\mu_1 = 12, \mu_2 = 9, \sigma_1 = 3, \sigma_2 = 7, n_1 = 15, n_2 = 11$
b. $\mu_1 = 25.6, \mu_2 = 37.8, \sigma_1 = 7.5, \sigma_2 = 10.5,$
$n_1 = 10, n_2 = 25$
c. $\mu_1 = 125.3, \mu_2 = 250.6, \sigma_1 = 15.6, \sigma_2 = 25.6,$
$n_1 = 8, n_2 = 12$
d. $\mu_1 = 3.1, \mu_2 = 2.2, \sigma_1 = 0.50, \sigma_2 = 0.75,$
$n_1 = 21, n_2 = 21$

10.9 Given the two-sample Z test assumptions, consider the following table of sample sizes, sample means, and known standard deviations.

Group	Sample size	Sample mean	Population standard deviation
One	18	17.5	1.5
Two	26	16.2	2.6

a. Write the four parts of a hypothesis test of
$H_0: \mu_1 - \mu_2 = 0$ versus $H_a: \mu_1 - \mu_2 > 0$. Use $\alpha = 0.05$.
b. Compute the value of the test statistic and draw a conclusion.
c. Find the p value associated with this hypothesis test.

10.10 Given the two-sample Z test assumptions, consider the following table of sample sizes, sample means, and known variances.

Group	Sample size	Sample mean	Population variance
One	25	186	14.7
Two	24	190	23.8

a. Write the four parts of a hypothesis test of
$H_0: \mu_1 - \mu_2 = 2$ versus $H_a: \mu_1 - \mu_2 < 2$. Use $\alpha = 0.01$.
b. Compute the value of the test statistic and draw a conclusion.
c. Carefully sketch a graph to illustrate the p value associated with this hypothesis test. Compute the p value.

10.11 Given the two-sample Z test assumptions, consider the following table of sample sizes, sample means, and known variances.

Group	Sample size	Sample mean	Population variance
One	37	1025.6	225.3
Two	42	1031.3	107.6

a. Write the four parts of a hypothesis test of
$H_0: \mu_1 - \mu_2 = 0$ versus $H_a: \mu_1 - \mu_2 \neq 0$. Use $\alpha = 0.001$.
b. Compute the value of the test statistic and draw a conclusion.
c. Is the normality assumption necessary in order to conduct this hypothesis test? Justify your answer.

10.12 Two random samples were obtained independently and the resulting data are given on the text website. Assume both populations are normal with $\sigma_1 = 8$ and $\sigma_2 = 12$. 📊 **EX10.12**

a. Find a 95% confidence interval for the true difference in means, $\mu_1 - \mu_2$.
b. Using the confidence interval in part (a), is there any evidence to suggest that the two population means are different? Justify your answer.

10.13 Suppose a random sample of size 15 is taken from a normal population with mean 25 and standard deviation 5; and a second, independent random sample of size 21 is taken from a normal population with mean 10 and standard deviation 4.

a. Describe the distribution of the difference in sample means, $\bar{X}_1 - \bar{X}_2$ (in terms of type of distribution, mean, variance, and standard deviation).
b. Carefully sketch the probability distribution for $\bar{X}_1 - \bar{X}_2$.
c. Find $P(\bar{X}_1 - \bar{X}_2 \geq 17)$.
d. Find $P(13.5 < \bar{X}_1 - \bar{X}_2 < 14.5)$.
e. Find $P(\bar{X}_1 < \bar{X}_2 + 14)$.

Applications

10.14 Manufacturing and Product Development The efficiency of an electric toothbrush is often judged by the rotation speed, in revolutions per minute (rpm). Two brands were selected for comparison, and independent random samples of each electric toothbrush were obtained. The rotation speed for each toothbrush was measured, and the summary statistics are given in the following table.

Electric toothbrush	Sample size	Sample mean	Population variance
Sonicare Elite	23	7992.2	1260.25
Oral-B	25	7988.2	1697.44

a. Is there any evidence to suggest that the Sonicare Elite has a greater population mean rotation speed than the Oral-B? Assume normality and use $\alpha = 0.05$.

b. Find the p value associated with this hypothesis test.

10.15 Business and Management Gift cards have become a popular present. Retailers like these cards because they are easier to process than paper gift certificates and more difficult to forge. Customers appreciate the convenience; the cards make great stocking stuffers and are easy to mail. Independent random samples of credit card–type gift certificates from two merchants were obtained, and the purchased value (in dollars) of each was recorded. The summary statistics and known variances are given in the following table.

Store	Sample size	Sample mean	Population variance
Nordstrom	41	24.07	16.81
Macy's	38	26.61	10.24

Is there any evidence to suggest that the true mean Nordstrom gift-certificate purchased value is different from the true mean Macy's gift-certificate purchase value? Use $\alpha = 0.01$. Write a Solution Trail for this problem.

10.16 Manufacturing and Product Development The energy rating, water consumption, and noise level of an electric dishwasher are all-important selling features. Suppose the makers of the Hotpoint DF55 claim that this model has a lower noise-level rating than any other comparable dishwasher. Independent random samples of the Hotpoint DF55 and of a similar Maytag dishwasher were obtained, and the noise level (in decibels) was measured for each. Is there any evidence to suggest that the population mean noise level for the Hotpoint dishwasher is less than the population mean noise level for the Maytag? Assume the underlying distributions are normal, with $\sigma_1 = 3.75$ and $\sigma_2 = 4.14$. Use $\alpha = 0.05$. **DISHWASH**

10.17 Copper Thieves As a result of increasing commodity prices, the theft of certain metals has increased dramatically. Because copper is used in so many items and is difficult to trace, thieves across the United States have become more brazen. A major target for thieves has been copper wiring in electrical power substations and utility poles. The leading states for thefts are Ohio, Texas, and Georgia.[2] Independent random samples of copper theft reports were obtained from two states, and the estimated dollar amount was recorded for each. The summary statistics and known variances are given in the following table.

State	Sample size	Sample mean	Population variance
Ohio (1)	16	427.40	8500
Texas (2)	24	419.50	7400

a. Assume the underlying distributions are normal and find a 95% confidence interval for the true difference in population mean amounts stolen.

b. Use the confidence interval to determine whether there is any evidence that the mean dollar amount of copper stolen differs for the two states.

10.18 Marketing and Consumer Behavior A new advertising program involves placing small screens on the back of taxi front seats in order to run several advertisements continuously. The theory is that riders give their undivided attention to these ads during the entire trip. However, advertisers worry that their ad may not be viewed during a typical ride. Independent random samples of taxi ride times (in minutes) in two cities were obtained. Is there any evidence to suggest that the mean taxi ride time is different in San Diego and Phoenix? Assume normality, with $\sigma_1 = 6.2$ and $\sigma_2 = 4.9$, and use $\alpha = 0.01$. Write a Solution Trail for this problem. **TAXIRIDE**

10.19 Manufacturing and Product Development The total weight (with the case) of a portable sewing machine is an important consideration. Suppose Singer claims to have the lightest machine by five pounds. Independent random samples of a Singer machine and a comparable Simplicity machine were obtained, and the weight (in pounds) of each was recorded. The summary statistics and known variances are given in the following table.

Sewing machine	Sample size	Sample mean	Population variance
Simplicity	42	17.99	2.89
Singer	38	13.26	2.25

a. Is there any evidence to refute the claim made by Singer? Use $\alpha = 0.01$.

b. Find the p value associated with this hypothesis test.

c. Is the normality assumption necessary in this problem? Why or why not?

10.20 Medicine and Clinical Studies Many people consume protein shakes to help build muscle mass and eliminate body fat. In a recent study, the amount of protein in two competing drinks was compared. Independent random samples were obtained, and the protein content (in grams) in each drink was measured. The summary statistics and known variances are given in the following table.

Protein drink	Sample size	Sample mean	Population variance
Met-Rx	12	39.38	5.06
Pure Gro	24	39.01	6.01

Is there any evidence to suggest that the mean amount of protein is different in these two products? Use $\alpha = 0.01$ and assume normality.

10.21 Manufacturing and Product Development Several factors determine how well a ceiling fan cools a room, including blade pitch, height from the ceiling, and revolutions per minute. Independent random samples of two types of ceiling fans were obtained, and the revolutions per minute (on high) for each was measured. The summary statistics and known variances are given in the following table.

Ceiling fan	Sample size	Sample mean	Population variance
Hampton	34	295.05	11.55
Altura	35	300.38	6.25

a. Find a 99% confidence interval for the true mean difference in revolutions per minute, $\mu_1 - \mu_2$.
b. Using the interval in part (a), is there any evidence to suggest that the mean revolutions per minute for the Altura ceiling fan is greater than the mean revolutions per minute for the Hampton ceiling fan? Justify your answer.

10.22 Travel and Transportation The recommended tire pressure for an off-road bicycle depends on the weight of the rider. As you would expect, the greater the rider's weight, the greater the recommended tire pressure. At a well-used bicycle trail in the Black River State Forest in Wisconsin, independent random samples were obtained from two different weight groups. The front tire pressure (in psi) was measured for each person, and the summary statistics and known variances are given in the following table.

Weight group	Sample size	Sample mean	Population variance
≈ 150 pounds	18	38.91	2.25
≈ 180 pounds	23	41.99	6.25

Is there any evidence to suggest that the difference between the 180-pound riders' mean tire pressure and the 150-pound riders' mean tire pressure is greater than 3 psi? Assume normality and use $\alpha = 0.05$.

10.23 Manufacturing and Product Development Although they are called leaf blowers, these hand-held machines are used to sweep patios, clean driveways, and even move light snow. Leaf blowers are often compared using weight, noise, and airspeed. Independent random samples of two types of leaf blowers were obtained, and the airspeed (in mph) was measured for each. The summary statistics and the known variances are given in the following table.

Leaf blower	Sample size	Sample mean	Population variance
Craftsman	18	200.28	24.5
Echo	19	196.74	35.7

a. Assume the underlying distributions are normal. Is there any evidence to suggest that the population mean airspeeds are different? Use $\alpha = 0.05$.
b. Find the p value associated with this hypothesis test.

10.24 Medicine and Clinical Studies The time it takes for general anesthesia to work (time to induction) is an important consideration during an emergency and for scheduled surgeries. Recently, a study was conducted to compare the mean induction time of similar drugs administered via inhalation and intravenously. Independent random samples of patients requiring general anesthesia were obtained, and the induction times (in minutes) were measured. Assume the variance in induction time for inhalation administration is 0.0625 and for intravenous administration is 0.1225. Is there any evidence to suggest that the mean time to induction for intravenous administration is less than the mean time to induction for inhalation administration? Use $\alpha = 0.05$. 🔊 ANESTH

Extended Applications

10.25 Public Health and Nutrition Magnesium is used by every cell in your body, is required for over 300 biochemical reactions, and helps muscles and nerves function properly. According to the U.S. Department of Agriculture National Nutritional Database, $\frac{1}{2}$ cup of vegetarian baked beans and one medium baked potato without the skin contain the same amount of magnesium (40 milligrams). To check this claim, independent random samples of baked beans and potatoes were obtained, and the amount of magnesium in each serving was recorded (in milligrams). The summary statistics and known variances are given in the following table.

Food	Sample size	Sample mean	Population variance
Vegetarian baked beans (1)	18	39.58	2.47
Medium potato (2)	18	40.12	0.87

a. Assume the underlying distributions are normal. Is there any evidence to refute the claim? Use $\alpha = 0.01$.
b. Suppose that, instead, the sample sizes are $n_1 = n_2 = 38$. Now, is there any evidence to refute the claim? Find the p value for this hypothesis test.
c. How large would the sample sizes ($n_1 = n_2$) have to be for the hypothesis test to be significant at the $\alpha = 0.01$ level?

10.26 Physical Sciences The manufacturer of a Kenmore residential stove can order parts from two different suppliers: The Repair Clinic and The Parts Pros. The small burners, or elements, are designed to produce 7.4 kilowatts at 240 volts. To decide which supplier to use, independent random samples from each supplier were obtained, and each element's output (in kilowatts) was carefully measured. For The Repair Clinic, $n_1 = 12$, $\bar{x}_1 = 7.361$, and $\sigma_1^2 = 0.81$; for The Parts Pros, $n_2 = 15$, $\bar{x}_2 = 7.307$, and $\sigma_2^2 = 0.64$. Assume the underlying distributions are normal and use $\alpha = 0.05$ for the following.

a. Is there any evidence to suggest that the population mean output of elements from The Repair Clinic is different from 7.4?

b. Is there any evidence to suggest that the population mean output of elements from The Parts Pros is different from 7.4?

c. Is there any evidence to suggest that μ_1 is different from μ_2?

d. Using the results from parts (a), (b), and (c), which supplier should the manufacturer use?

10.27 Marketing and Consumer Behavior The Press Association Mediapoint recently released a housing market report for England that included how long it would take a typical first-time buyer to save for a deposit in their local area.[3] Independent random samples of first-time buyers were obtained in two areas, and the time (in years) needed for a couple to save for the deposit was recorded for each. The summary statistics are given in the following table.

Area	Sample size	Sample mean	Population variance
Yorkshire & The Humer	60	4.5	1.56
East Midlands	75	4.8	3.24

a. Is there any evidence to suggest that the mean time needed for a couple to save for a deposit is different in these two areas? Use $\alpha = 0.05$.

b. Find the p value associated with this hypothesis test.

c. The sample means, 4.5 and 4.8, seem close together. Can you find a value $n = n_1 = n_2$ such that the hypothesis test is significant at the $\alpha = 0.05$ level?

10.28 Flood Insurance According to the U.S. National Flood Insurance Program, the cost of policies in New Orleans continued to rise for several years after Hurricane Katrina.

Florida still has the most flood insurance policies, followed by Texas, then Louisiana. A random sample of flood insurance policies in these three states was obtained, and the premium in dollars for each was recorded. The summary statistics are given in the following table.[4]

State	Sample size	Sample mean	Population standard deviation
Louisiana	18	716.15	250
Florida	22	498.71	275
Texas	26	560.50	300

Assume the underlying populations are normal. Is there any evidence to suggest that any pairs of population mean flood insurance premiums are different? That is, conduct three separate hypothesis tests to consider $\mu_1 - \mu_2$, $\mu_1 - \mu_3$, and $\mu_2 - \mu_3$. Use $\alpha = 0.05$ in each case.

Challenge

10.29 Sample Size Calculation Suppose a $100(1 - \alpha)\%$ confidence interval is needed for the difference in two population means, $\mu_1 - \mu_2$. In addition, suppose the underlying populations are normal, the population variances, σ_1^2 and σ_2^2, are known, and the samples sizes are equal, $n_1 = n_2 = n$.

a. Find an expression for the sample size necessary (from each population) in order for the resulting confidence interval to have a bound on the error of estimation B (half the width of the confidence interval).

b. How large a sample size is necessary if $\sigma_1 = 12.7$, $\sigma_2 = 9.5$, $B = 5$, and the confidence level is 95%?

c. Use the sample size in part (b) with $\bar{x}_1 = 57.3$ and $\bar{x}_2 = 48.6$ to construct a 95% confidence interval for $\mu_1 - \mu_2$. Compute the exact bound on the error of estimation. How does this compare with $B = 5$?

10.2 Comparing Two Population Means Using Independent Samples from Normal Populations

In Section 10.1, the hypothesis tests concerning the difference between two population means (or for comparing two population means) were based on the standard normal, or Z, distribution. These tests are valid *only* if both population variances are known (and with normality and/or large samples, and independent random samples). It is unrealistic to assume that the population variances are known. As in Chapter 9, we will assume that the underlying populations are normal. But one additional assumption is necessary to construct a similar *two-sample t* test.

Suppose that

For reference, these are the two-sample t test assumptions.

A test of equality of population variances will be discussed in Section 10.5.

1. $\bar{X}_1$ is the mean of a random sample of size n_1 from a normal population with mean μ_1.

2. $\bar{X}_2$ is the mean of a random sample of size n_2 from a normal population with mean μ_2.

3. The samples are independent.

4. The two population variances are *unknown* but *equal*. The common variance is denoted $\sigma^2 (= \sigma_1^2 = \sigma_2^2)$.

The last assumption is new and implies we are comparing populations with the same variability. If we do not assume equal variances, there is no *nice* test procedure. More on this later.

Properties of $\overline{X}_1 - \overline{X}_2$

If the two-sample t test assumptions are true, then the estimator $\overline{X}_1 - \overline{X}_2$ has the following properties.

1. $E(\overline{X}_1 - \overline{X}_2) = \mu_{\overline{X}_1 - \overline{X}_2} = \mu_1 - \mu_2$

   $\overline{X}_1 - \overline{X}_2$ is still an unbiased estimator of the parameter $\mu_1 - \mu_2$.

2. $\mathrm{Var}(\overline{X}_1 - \overline{X}_2) = \sigma^2_{\overline{X}_1 - \overline{X}_2} = \dfrac{\sigma_1^2}{n_1} + \dfrac{\sigma_2^2}{n_2} = \dfrac{\sigma^2}{n_1} + \dfrac{\sigma^2}{n_2} = \sigma^2\left(\dfrac{1}{n_1} + \dfrac{1}{n_2}\right)$ and the standard

   deviation is $\sigma_{\overline{X}_1 - \overline{X}_2} = \sqrt{\sigma^2\left(\dfrac{1}{n_1} + \dfrac{1}{n_2}\right)}$.

3. Both underlying populations are normal, so the distribution of $\overline{X}_1 - \overline{X}_2$ is also normal.

In the previous section, we used the known population variances, standardized, and constructed a test based on the Z distribution. Here, an estimate of the common variance σ^2 is necessary. The appropriate *standardization* results in a t distribution.

S_1^2 and S_2^2 are separate estimators for the common variance, but using only one of these means ignoring additional, useful information. Because σ^2 is the variance for both underlying populations, an estimator for this common variance should depend on both samples. However, it also seems reasonable for the estimator to rely more on the larger sample. Therefore an estimate of the common variance uses both S_1^2 and S_2^2 in a *weighted average*.

Definition

The **pooled estimator** for the common variance σ^2, denoted S_p^2, is

$$S_p^2 = \frac{(n_1 - 1)S_1^2 + (n_2 - 1)S_2^2}{n_1 + n_2 - 2} \tag{10.3}$$

$$= \left(\frac{n_1 - 1}{n_1 + n_2 - 2}\right)S_1^2 + \left(\frac{n_2 - 1}{n_1 + n_2 - 2}\right)S_2^2$$

The pooled estimator for the common standard deviation σ is $S_p = \sqrt{S_p^2}$.

A CLOSER LOOK

λ is the lowercase Greek letter lambda and represents a constant.

1. S_p^2 is indeed a weighted average. This estimator can be written in the form

   $$S_p^2 = \lambda S_1^2 + (1 - \lambda)S_2^2 \qquad \text{where} \qquad 0 \le \lambda \le 1$$

   If $n_1 = n_2$, then $\lambda = \frac{1}{2}$ and $S_p^2 = \frac{1}{2}S_1^2 + \frac{1}{2}S_2^2$. If $n_1 \ne n_2$, then more *weight* is given to the larger sample.

2. The constants in Equation 10.3 are related to the number of degrees of freedom. S_1^2 contributes $n_1 - 1$ degrees of freedom and S_2^2 contributes $n_2 - 1$ degrees of freedom. Consequently, there are a total of $(n_1 - 1) + (n_2 - 1) = n_1 + n_2 - 2$ degrees of freedom associated with the estimator S_p^2.

The hypothesis test procedure is based on the following theorem.

Theorem

If the two-sample t test assumptions are true, then the random variable

$$T = \frac{(\bar{X}_1 - \bar{X}_2) - (\mu_1 - \mu_2)}{\sqrt{S_p^2\left(\dfrac{1}{n_1} + \dfrac{1}{n_2}\right)}}$$

has a t distribution with $n_1 + n_2 - 2$ degrees of freedom.

As in a two-sample Z test, the null and alternative hypotheses are stated in terms of the difference $\mu_1 - \mu_2$. The critical values are from the appropriate t distribution.

Hypothesis Tests Concerning Two Population Means When Variances Are Unknown but Equal

Given the two-sample t test assumptions, a hypothesis test concerning two population means in terms of the difference in means $\mu_1 - \mu_2$, with significance level α, has the form

$H_0: \mu_1 - \mu_2 = \Delta_0$

This is the template for a hypothesis test concerning two population means when variances are unknown but equal: a two-sample t test, with pooled variance.

$H_a: \mu_1 - \mu_2 > \Delta_0, \qquad \mu_1 - \mu_2 < \Delta_0, \qquad$ or $\qquad \mu_1 - \mu_2 \neq \Delta_0$

TS: $T = \dfrac{(\bar{X}_1 - \bar{X}_2) - \Delta_0}{\sqrt{S_p^2\left(\dfrac{1}{n_1} + \dfrac{1}{n_2}\right)}}$

RR: $T \geq t_{\alpha, n_1+n_2-2}, \qquad T \leq -t_{\alpha, n_1+n_2-2}, \qquad$ or $\qquad |T| \geq t_{\alpha/2, n_1+n_2-2}$

Example 10.4 Surgical Wait Times

Frequently, patients must wait a long time for elective surgery. For those with chronic pain, the wait can be unbearable. Suppose the wait time for patients needing a knee replacement at two hospitals in British Columbia was investigated. Independent random samples of patients were obtained, and the wait time for each (in weeks) was recorded. The resulting summary statistics are given in the following table.[5]

Hospital	Sample size	Sample mean	Sample variance
Abbotsford Regional (1)	15	17.4	34.81
Earl Ridge (2)	17	12.1	46.24

a. Is there any evidence to suggest that there is a difference in the population mean waiting time for a knee replacement between Abbotsford Regional Hospital and Earl Ridge Hospital? Use $\alpha = 0.05$ and assume the underlying distributions are normal, with equal variances.

b. Find bounds on the p value associated with this hypothesis test.

SOLUTION

STEP 1 Let Abbotsford Regional Hospital be population 1 and Earl Ridge Hospital be population 2.

The null hypothesis is that the two population means are equal—with the same waiting time for a knee replacement: $\mu_1 = \mu_2 \Rightarrow \mu_1 - \mu_2 = 0 \ (= \Delta_0)$.

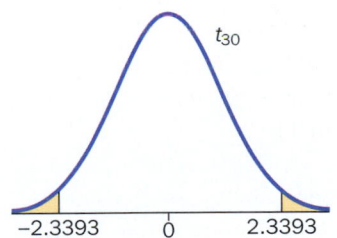

Solution Trail 10.4

KEYWORDS

- Is there any evidence?
- Difference in population mean
- Underlying distributions are normal, with equal variances
- Independent random samples

TRANSLATION

- Conduct a two-sided test to compare μ_2 and μ_2
- Variances are unknown but assumed equal

CONCEPTS

- Hypothesis test concerning two population means when variances are unknown but equal

VISION

Use the template for this hypothesis test. The samples are random and independent, the underlying distributions are normal, and the population variances are unknown but assumed equal. Use the two-sided alternative hypothesis and the corresponding rejection region.

The summary statistics are given, and the samples were obtained independently. The population variances are unknown but assumed equal. A two-sample t test is appropriate.

We are looking for *any* difference in population means, so this is a two-sided test.

STEP 2   The four parts of the hypothesis test are

H_0: $\mu_1 - \mu_2 = 0$

H_a: $\mu_1 - \mu_2 \neq 0$

TS: $T = \dfrac{(\bar{X}_1 - \bar{X}_2) - 0}{\sqrt{S_p^2\left(\dfrac{1}{n_1} + \dfrac{1}{n_2}\right)}}$

RR: $|T| \geq t_{\alpha/2, n_1 + n_2 - 2} = t_{0.025,30} = 2.0423$

STEP 3   The pooled estimate of the common population variance is

$$s_p^2 = \frac{(n_1 - 1)s_1^2 + (n_2 - 1)s_2^2}{n_1 + n_2 - 2} = \frac{(14)(34.81) + (16)(46.24)}{30} = 40.906$$

The value of the test statistic is

$$t = \frac{(\bar{x}_1 - \bar{x}_2) - 0}{\sqrt{s_p^2\left(\dfrac{1}{n_1} + \dfrac{1}{n_2}\right)}} = \frac{17.4 - 12.1}{\sqrt{(40.906)\left(\dfrac{1}{15} + \dfrac{1}{17}\right)}} = 2.3393 \ (\geq 2.0423)$$

The value of the test statistic, $t = 2.3393$, lies in the rejection region, hence we reject the null hypothesis at the $\alpha = 0.05$ significance level. There is evidence to suggest the mean waiting time for a knee replacement is different at Abbotsford Regional Hospital and Earl Ridge Hospital.

STEP 4   Recall, because of the nature of the table of critical values for t distributions, we can only bound the p value.

$$|t| = |2.3393| = 2.3393$$

In Table V in the Appendix, row $n_1 + n_2 - 2 = 15 + 17 - 2 = 30$, place 2.3393 in the ordered list of critical values.

$$2.0423 \leq 2.3393 \leq 2.4573$$

$$t_{0.025,30} \leq 2.3444 \leq t_{0.01,30}$$

Therefore, $0.01 \leq p/2 \leq 0.025$

And $0.02 \leq p \leq 0.05$

See Figure 10.16.

Figures 10.17 through 10.19 together show a technology solution.

Figure 10.16 *p* Value illustration:
$p = 2P(T \geq 2.3393)$
$= 0.0262 \leq 0.05 = \alpha$

VIDEO TECH MANUALS

TWO SAMPLE MEAN
INFERENCE - *t* -
SUMMARIZED DATA

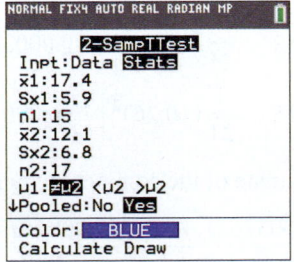

Figure 10.17
2 - SampTTest input
screen.

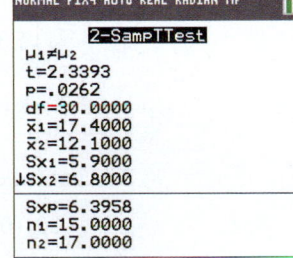

Figure 10.18
2 - SampTTest Hypothesis
test results.

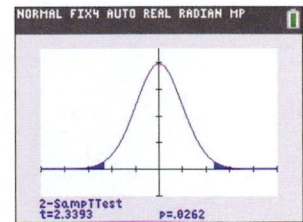

Figure 10.19
2 - SampTTest Draw
results.

TRY IT NOW   GO TO EXERCISE 10.45

DATA SET

ALUMCAN

Why do you suppose a small significance level is important here?

Solution Trail 10.5

KEYWORDS

- Is there any evidence?
- Smaller population mean
- Populations are normal, with equal variances
- Independent random samples

TRANSLATION

- Conduct a one-sided test to compare μ_1 and μ_2
- Variances are unknown but equal

CONCEPTS

- Hypothesis test concerning two population means when variances are unknown but equal

VISION

Use the template for this hypothesis test. The samples are random and independent, the underlying distributions are normal, and the population variances are unknown but assumed equal. Use a one-sided alternative hypothesis and the corresponding rejection region.

Example 10.5 Weight of Aluminum Cans

Aluminum cans are made from huge solid ingots pressed under high-pressure rollers and are cut like cookies from thin sheets. Aluminum is ideal for cans because it is lightweight, strong, and recyclable. A company claims that a new manufacturing process decreases the amount of aluminum needed to make a can, and therefore, decreases the weight. Independent random samples of aluminum cans made by the old and new processes were obtained, and the weight (in ounces) of each is given in the following table.

Old process (1)

0.52	0.49	0.47	0.47	0.48	0.52	0.55	0.49	0.52	0.50	0.50
0.50	0.51	0.51	0.50	0.53	0.49	0.51	0.52	0.51	0.51	

New process (2)

0.51	0.51	0.50	0.48	0.47	0.49	0.46	0.46	0.52	0.50	0.48
0.51	0.50	0.48	0.51	0.44	0.48	0.47	0.50	0.51	0.48	

Is there any evidence that the new-process aluminum cans have a smaller population mean weight? Assume the populations are normal, with equal variances, and use $\alpha = 0.01$.

SOLUTION

STEP 1 The null hypothesis is that the two mean weights are the same: $\mu_1 - \mu_2 = 0$. We are looking for evidence that the new-process cans have a smaller mean weight. The alternative hypothesis is $\mu_1 - \mu_2 > 0$.

The underlying populations are assumed normal with equal variances, and the samples were obtained independently. A two-sample t test is relevant.

STEP 2 The four parts of the hypothesis test are

$$H_0: \mu_1 - \mu_2 = 0$$
$$H_a: \mu_1 - \mu_2 > 0$$
$$\text{TS: } T = \frac{(\bar{X}_1 - \bar{X}_2) - 0}{\sqrt{S_p^2\left(\frac{1}{n_1} + \frac{1}{n_2}\right)}}$$

RR: $T \geq t_{\alpha, n_1 + n_2 - 2} = t_{0.01, 40} = 2.4233$

STEP 3 The summary statistics are

$$\bar{x}_1 = \frac{1}{21}(0.52 + 0.49 + \cdots + 0.51) = 0.5048$$

$$\bar{x}_2 = \frac{1}{21}(0.51 + 0.51 + \cdots + 0.48) = 0.4886$$

$$s_1^2 = \frac{1}{20}\left[5.3580 - \frac{1}{21}(10.6)^2\right] = 0.0003762$$

$$s_2^2 = \frac{1}{20}\left[5.0216 - \frac{1}{21}(10.26)^2\right] = 0.0004429$$

The pooled estimate of the common population variances is

$$s_p^2 = \frac{(20)(0.0003762) + (20)(0.0004429)}{40} = 0.0004095$$

The value of the test statistic is

$$t = \frac{(\bar{x}_1 - \bar{x}_2) - 0}{\sqrt{s_p^2\left(\frac{1}{n_1} + \frac{1}{n_2}\right)}} = \frac{0.5048 - 0.4886}{\sqrt{(0.0004095)\left(\frac{1}{21} + \frac{1}{21}\right)}} = 2.5941$$

STEP 4 The value of the test statistic lies in the rejection region ($t = 2.5941 \geq 2.4233$; $p = 0.0066 \leq .01$, see Figure 10.20). We reject the null hypothesis at the $\alpha = 0.01$ significance level. There is evidence to suggest that new-process aluminum cans have a smaller mean weight.

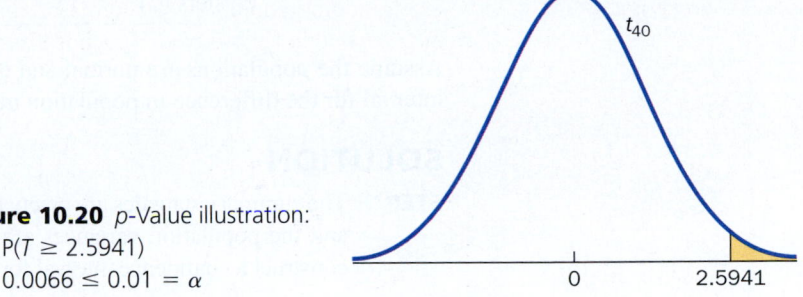

Figure 10.20 *p*-Value illustration:
$p = P(T \geq 2.5941)$
$= 0.0066 \leq 0.01 = \alpha$

Figures 10.21 and 10.22 show technology solutions.

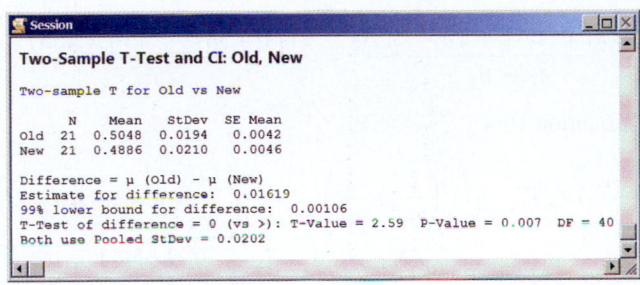

Figure 10.21 Minitab hypothesis test (and confidence interval) results.

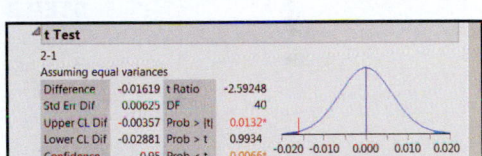

Figure 10.22 JMP two-sample *t* test.

TRY IT NOW   GO TO EXERCISE 10.47

This methodology has been used several times, beginning in Chapter 8.

Using the assumptions presented in this section and the technique presented in Section 10.1, a confidence interval for $\mu_1 - \mu_2$ can be derived. Start with a symmetric interval about 0 such that the probability T lies in this interval is $1 - \alpha$. Manipulate the inequality to *sandwich* the parameter $\mu_1 - \mu_2$.

How to Find a 100(1 − α)% Confidence Interval for $\mu_1 - \mu_2$ When Variances Are Unknown but Equal

Given the two-sample *t* test assumptions, a $100(1 - \alpha)\%$ confidence interval for $\mu_1 - \mu_2$ has as endpoints the values

$$(\bar{x}_1 - \bar{x}_2) \pm t_{\alpha/2, n_1 + n_2 - 2} \sqrt{s_p^2 \left(\frac{1}{n_1} + \frac{1}{n_2} \right)} \qquad (10.4)$$

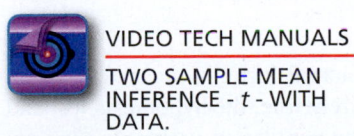

VIDEO TECH MANUALS

TWO SAMPLE MEAN INFERENCE - *t* - WITH DATA.

Example 10.6 Iron Man

Iron is an essential mineral. It is used by the body to carry oxygen, and even a slight deficiency can cause fatigue and weakness. Certain kinds of mollusks are very high in iron content, for example, clams, mussels, and oysters. Independent random samples of 3-ounce servings of clams and oysters were obtained, and the iron content (in mg) was measured in each.

The summary statistics are given in the following table.

Mollusk	Sample size	Sample mean	Sample standard deviation
Clams (1)	12	23.17	4.38
Oysters (2)	15	24.19	3.35

Assume the populations are normal and the variances are equal. Find a 99% confidence interval for the difference in population mean iron content.[6]

SOLUTION

STEP 1 The summary statistics are given, the underlying distributions are assumed normal, and the population variances are assumed equal. Equation 10.4 can be used to construct a confidence interval for the difference $\mu_1 - \mu_2$.

$$1 - \alpha = 0.99 \quad \Rightarrow \quad \alpha = 0.01 \quad \Rightarrow \quad \alpha/2 = 0.005$$ *Find $\alpha/2$.*

$$t_{\alpha/2, n_1+n_2-2} = t_{0.005,25} = 2.7874$$ *Find the t critical value.*

STEP 2 Find the pooled estimate of the common variance.

$$s_p^2 = \frac{(n_1 - 1)s_1^2 + (n_2 - 1)s_2^2}{n_1 + n_2 - 2} = \frac{(11)(4.38)^2 + (14)(3.35)^2}{25} = 14.7257$$

STEP 3 Use Equation 10.4.

$$(\bar{x}_1 - \bar{x}_2) \pm t_{\alpha/2} \cdot \sqrt{s_p^2 \left(\frac{1}{n_1} + \frac{1}{n_2} \right)}$$ *Equation 10.4.*

$$= (23.17 - 24.19) \pm (2.7874)\sqrt{(14.7257)\left(\frac{1}{12} + \frac{1}{15} \right)}$$

Use summary statistics and critical values.

$$= -1.02 \pm 4.1427$$ *Simplify.*

$$= (-5.1627, 3.1227)$$ *Compute endpoints.*

$(-5.1627, 3.1227)$ is a 99% confidence interval for the difference (in mg) in population mean iron content, $\mu_1 - \mu_2$. Note that because 0 is included in, or captured by, this interval, there is no evidence to suggest the mean iron content is different.

Figures 10.23 and 10.24 together show a technology solution.

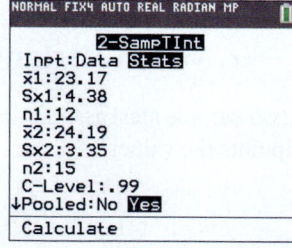

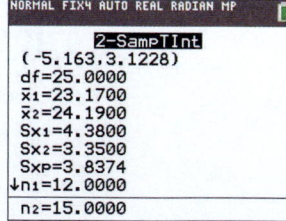

Figure 10.23
2-SampTInt input screen.

Figure 10.24 Resulting
99% confidence interval.

TRY IT NOW   GO TO EXERCISE 10.52

The underlying distributions might only be approximately normal, or the population variances might not be exactly the same.

The hypothesis test procedure and the confidence interval formula presented in this section are *robust*. That is, if the assumptions aren't entirely true, the hypothesis test and the confidence interval are still very reliable. Even if the population variances are very different, as long as the underlying populations are normal and $n_1 = n_2$, the results are still very reliable.

Nice means a reasonable *standardization* to produce a common random variable.

If the underlying populations are normal, the population variances are unequal, and the sample sizes are different, there is no *nice* test procedure concerning $\mu_1 - \mu_2$ (or confidence interval for $\mu_1 - \mu_2$). It is reasonable to use each sample variance as an approximation for the corresponding population variance. However, the resulting logical standardization produces only an *approximate* test statistic. If the sample sizes are small and the underlying populations are not normal, then a nonparametric test must be used.

Hypothesis Tests and Confidence Interval Concerning Two Population Means When Variances Are Unknown and Unequal

This is the template for an *approximate* two-sample t test.

Given the *modified* two-sample t test assumptions (population variances unknown and assumed unequal), an *approximate* hypothesis test concerning two population means in terms of the difference, $\mu_1 - \mu_2$, with significance level α, has the form

$H_0: \mu_1 - \mu_2 = \Delta_0$

$H_a: \mu_1 - \mu_2 > \Delta_0, \qquad \mu_1 - \mu_2 < \Delta_0, \qquad \text{or} \qquad \mu_1 - \mu_2 \neq \Delta_0$

$$\text{TS: } T' = \frac{(\overline{X}_1 - \overline{X}_2) - \Delta_0}{\sqrt{\dfrac{S_1^2}{n_1} + \dfrac{S_2^2}{n_2}}}$$

The formula for ν is the Satterthwaite approximation for the number of degrees of freedom.

$$\text{RR: } T' \geq t_{\alpha,\nu}, \qquad T' \leq -t_{\alpha,\nu}, \qquad \text{or} \qquad |T'| \geq t_{\alpha/2,\nu}$$

$$\text{where } \nu \approx \frac{\left(\dfrac{s_1^2}{n_1} + \dfrac{s_2^2}{n_2}\right)^2}{\dfrac{(s_1^2/n_1)^2}{n_1 - 1} + \dfrac{(s_2^2/n_2)^2}{n_2 - 1}}$$

An approximate $100(1 - \alpha)\%$ confidence interval for $\mu_1 - \mu_2$ has as endpoints the values

$$(\overline{x}_1 - \overline{x}_2) \pm t_{\alpha/2,\nu} \sqrt{\frac{s_1^2}{n_1} + \frac{s_2^2}{n_2}} \tag{10.5}$$

A CLOSER LOOK

1. The random variable T' has an approximate t distribution with ν degrees of freedom.

2. It is likely that the value of ν will *not* be an integer. To be conservative, always round down (to the nearest integer).

3. A test for equality of population variances is presented in Section 10.5. This hypothesis test is often used to determine whether equal population variances is a reasonable assumption.

Example 10.7 Poker Chip Weights

DATA SET

POKER

Clay-composite poker chips used in Las Vegas and Atlantic City weigh between 8.5 and 10 grams each, and last between 3 and 6 years. In a recent study of poker-chip weights, a casino obtained independent random samples of $100 and $500 chips. The weight of each chip (in grams) is given in the following table.

Solution Trail 10.7

KEYWORDS

- Is there any evidence?
- Population mean is different
- Both populations are normal
- Independent random samples

TRANSLATION

- Conduct a two-sided test to compare μ_1 and μ_2
- Variances are unknown and assumed unequal

CONCEPTS

- Hypothesis test concerning two population means when variances are unknown and unequal

VISION

Use the template for this hypothesis test. The samples are random and independent, the underlying distributions are normal, and the population variances are unknown and unequal. Use the two-sided alternative hypothesis and the corresponding rejection region, the test statistic, and draw a conclusion.

$100 chips (1)

9.17	9.21	9.25	9.29	9.16	9.08	9.39	9.23	9.15	9.14
9.34	9.26	9.08	9.11						

$500 chips (2)

9.37	9.98	9.04	8.74	9.58	9.45	9.08	9.96	9.69

Is there any evidence to suggest that the population mean weight of $100 chips is different from that of $500 chips? Assume both populations are normal, and use $\alpha = 0.05$.

SOLUTION

STEP 1 The null hypothesis is that the two mean weights are the same, and the alternative is two-sided. The underlying populations are assumed normal and the samples were obtained independently. However, there is no assumption of equal variances. The approximate two-sample t test is appropriate.

STEP 2 The summary statistics are

$$\bar{x}_1 = \frac{1}{14}(9.17 + 9.21 + \cdots + 9.11) = 9.2043$$

$$\bar{x}_2 = \frac{1}{9}(9.37 + 9.98 + \cdots + 9.69) = 9.4322$$

$$s_1^2 = \frac{1}{13}\left[1186.1804 - \frac{1}{14}(128.86)^2\right] = 0.008934$$

$$s_2^2 = \frac{1}{8}\left[802.1295 - \frac{1}{9}(84.89)^2\right] = 0.1785$$

The approximate number of degrees of freedom are

$$\nu \approx \frac{\left(\dfrac{0.008934}{14} + \dfrac{0.1785}{9}\right)^2}{\dfrac{(0.008934/14)^2}{13} + \dfrac{(0.1785/9)^2}{8}} = 8.5177$$

We round ν down to 8.

STEP 3 The four parts of the hypothesis test are

H_0: $\mu_1 - \mu_2 = 0$

H_a: $\mu_1 - \mu_2 \neq 0$

TS: $T' = \dfrac{(\bar{X}_1 - \bar{X}_2) - 0}{\sqrt{\dfrac{S_1^2}{n_1} + \dfrac{S_2^2}{n_2}}}$

RR: $|T'| \geq t_{\alpha/2,\nu} = t_{0.025,8} = 2.3060$

STEP 4 The value of the test statistic is

$$t' = \frac{(\bar{x}_1 - \bar{x}_2) - 0}{\sqrt{\dfrac{s_1^2}{n_1} + \dfrac{s_2^2}{n_2}}} = \frac{9.2043 - 9.4322}{\sqrt{\dfrac{0.008934}{14} + \dfrac{0.1785}{9}}} = -1.5928$$

The value of the test statistic does not lie in the rejection region. Equivalently, $p = 0.1475 > 0.05$, illustrated in Figure 10.25. We do not reject the null hypothesis at the $\alpha = 0.05$ significance level. There is no evidence to suggest that the mean weight of $100 chips is different from the mean weight of $500 chips.

Figure 10.26 shows a technology solution.

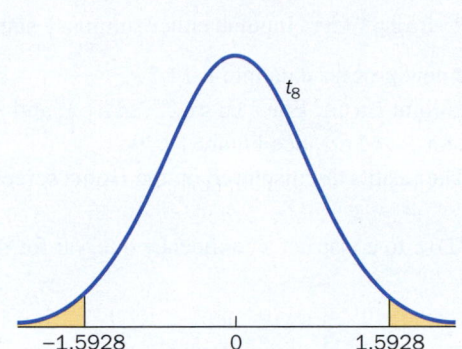

Figure 10.25 *p* Value illustration:
$$p = 2P(T \geq 1.5928)$$
$$= 0.1475 > 0.05 = \alpha$$

	C	D	E
t-Test: Two-Sample Assuming Unequal Variances			
		Variable 1	Variable 2
Mean		9.2043	9.4322
Variance		0.0089	0.1785
Observations		14	9
Hypothesized Mean Difference		0	
df		9	
t Stat		-1.5930	
P(T<=t) one-tail		0.0728	
t Critical one-tail		1.8331	
P(T<=t) two-tail		0.1456	
t Critical two-tail		2.2622	

Figure 10.26 Excel hypothesis test results.

TRY IT NOW   GO TO EXERCISE 10.55

Technology Corner

Procedure: Hypothesis tests and confidence intervals concerning two population means when the population variances are unknown.

Reconsider: Example 10.5, solution, and interpretations.

CrunchIt!

Use the built-in function t 2-Sample. Input is either summary statistics or data in columns.

1. Enter the old process data into column Var1 and the new process data into column Var2.
2. Select Statistics; t; 2-sample. Using the pull-down menus, select Var1 for Sample 1 and Var2 for Sample 2. Check the Pooled Variance box.
3. Under the Hypothesis tab, enter the Difference of means under null hypothesis (0) and select the appropriate Alternative (Greater than). See Figure 10.27.
4. Click Calculate. The results are displayed in a new window (Figure 10.28).
5. Use the Confidence Interval tab to construct a confidence interval for the difference of two population means.

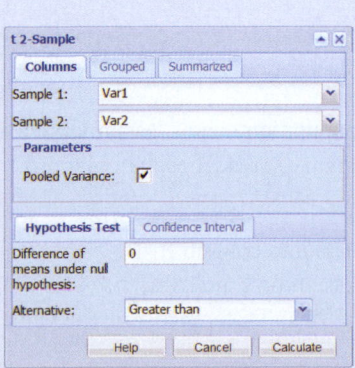

Figure 10.27 t 2-Sample input screen.

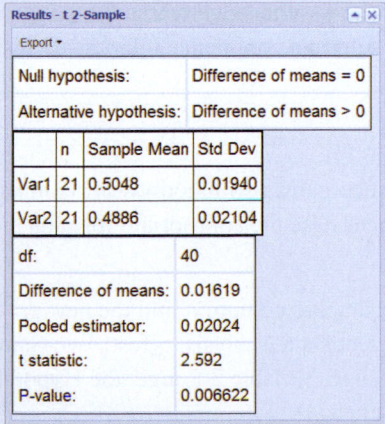

Figure 10.28 t 2-Sample results.

TI-84 Plus C

Use the built-in functions 2-SampTTest and 2-SampTInt. Input is either summary statistics or data in lists.

1. Enter the old process data into list L1 and the new process data into list L2.
2. Select STAT ; TESTS; 2-SampTTest. Highlight Data. Enter List1, List2, and set each frequency to 1. Highlight the alternative hypothesis and Yes for Pooled. See Figure 10.29.
3. Highlight Calculate and press ENTER . The results are displayed on the Home screen. See Figure 10.30.
4. The Draw results are shown in Figure 10.31.
5. Use the function STAT ; TESTS; 2-SampTInt to construct a confidence interval for the difference of two population means.

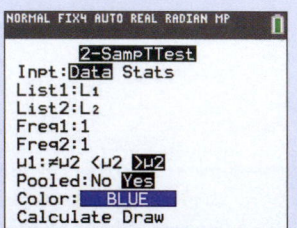

Figure 10.29
2-SampTTest input screen.

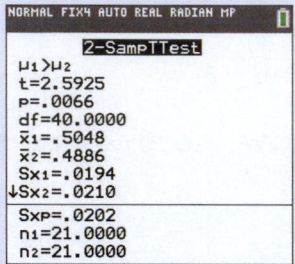

Figure 10.30
2-SampTTest hypothesis test results.

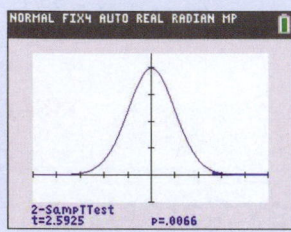

Figure 10.31
2-SampTTest Draw results.

Minitab

Use the built-in function 2-Sample t to conduct a hypothesis test and to construct a confidence interval. Input is either data in one or two columns (data in one column requires a subscript, or group-identifying, column) or summarized data.

1. Enter the old process data into column C1 and the new process data into column C2.
2. Select Stat; Basic Statistics; 2-Sample t.
3. Choose Each sample is in its own column and enter C1 in the Sample 1 input window and C2 in the Sample 2 input window.
4. Choose the Options option button. Enter a Confidence level, the hypothesized Test difference, and choose the appropriate Alternative, check the Assume equal variances box.
5. The hypothesis test results and confidence interval are displayed in a session window. Refer to Figure 10.21.

Excel

The Data Analysis toolkit contains two functions for comparing population means, assuming equal variances and assuming unequal variances. Use the appropriate formula and ordinary spreadsheet calculations to find the endpoints of a confidence interval.

1. Enter the old process data into column A and the new process data into column B.
2. Under the Data tab, select Data Analysis; t-Test: Two-Sample Assuming Equal Variances.
3. Enter the Variable 1 Range, Variable 2 Range, the Hypothesized Mean Difference, and the value for α. Choose an Output option and click OK.
4. Summary statistics along with the value of the test statistic, critical values, and p values are displayed. See Figure 10.32.

	C	D	E
t-Test: Two-Sample Assuming Equal Variances			
		Variable 1	Variable 2
Mean		0.5048	0.4886
Variance		0.0004	0.0004
Observations		21	21
Pooled Variance		0.0004	
Hypothesized Mean Difference		0	
df		40	
t Stat		2.5925	
P(T<=t) one-tail		0.0066	
t Critical one-tail		2.4233	
P(T<=t) two-tail		0.0132	
t Critical two-tail		2.7045	

Figure 10.32 Excel hypothesis test results.

SECTION 10.2 EXERCISES

Concept Check

10.30 Short Answer   State the two-sample t test assumptions.

10.31 Short Answer   Under the two-sample t test assumptions, $\mathrm{Var}(\overline{X}_1 - \overline{X}_2) =$ _____.

10.32 True/False   In a two-sample t test, the pooled estimator for the common variance is the sample variance associated with the larger sample.

10.33 Short Answer   The two-sample t test is robust. Explain what this means in practice.

10.34 Short Answer   If the population variances are unequal, there is no nice test procedure to compare the population means. Explain what *nice* means in terms of statistics.

10.35 True/False   Consider a hypothesis test or confidence interval concerning two population means when the variances are unknown and unequal. If the approximate number of degrees of freedom is a decimal, round down to the nearest integer.

10.36 True/False   In a hypothesis test or confidence interval concerning two population means when the variances are unknown and unequal, the hypothesized difference in means must always be 0.

Practice

10.37   Given the two-sample t test assumptions, consider the following table of summary statistics.

Group	Sample size	Sample mean	Sample variance
One	14	49.6	134.56
Two	16	50.2	243.36

a. Conduct a hypothesis test of $H_0: \mu_1 - \mu_2 = 0$ versus $H_a: \mu_1 - \mu_2 < 0$. Use $\alpha = 0.05$.
b. Find bounds on the p value associated with this test.

10.38   Given the two-sample t test assumptions, consider the following table of summary statistics.

Group	Sample size	Sample mean	Sample standard deviation
One	10	156.5	26.5
Two	11	132.6	21.5

a. Conduct a hypothesis test of $H_0: \mu_1 - \mu_2 = 0$ versus $H_a: \mu_1 - \mu_2 > 0$. Use $\alpha = 0.01$.
b. Find bounds on the p value associated with this test.

10.39   Given the two-sample t test assumptions, consider the independent random samples from two different populations. 📊 EX10.39
a. Conduct a hypothesis test of $H_0: \mu_1 - \mu_2 = 0$ versus $H_a: \mu_1 - \mu_2 \neq 0$. Use $\alpha = 0.05$.
b. Find bounds on the p value associated with this test.

10.40   Given the two-sample t test assumptions, consider the following table of summary statistics.

Group	Sample size	Sample mean	Sample standard deviation
One	23	49.03	9.24
Two	23	49.57	8.15

a. Find a 95% confidence interval for the difference in population means, $\mu_1 - \mu_2$.
b. Using the confidence interval in part (a), is there any evidence to suggest that the two population means are different? Justify your answer.

10.41 In each of the following problems, n_1, n_2, s_1, and s_2 are given. Assume normal underlying distributions, independent random samples, and unknown, *unequal* variances. Find the approximate number of degrees of freedom, v, in the critical value of an approximate two-sample t test.

a. $n_1 = 12$, $\quad n_2 = 15$, $\quad s_1 = 11.7$, $\quad s_2 = 16.7$
b. $n_1 = 8$, $\quad n_2 = 23$, $\quad s_1 = 5.46$, $\quad s_2 = 6.78$
c. $n_1 = 18$, $\quad n_2 = 26$, $\quad s_1 = 57.8$, $\quad s_2 = 49.9$
d. $n_1 = 32$, $\quad n_2 = 34$, $\quad s_1 = 5.51$, $\quad s_2 = 5.03$

10.42 Consider the following table of summary statistics.

Group	Sample size	Sample mean	Sample variance
One	8	173.9	320.41
Two	9	150.3	655.36

Assume normal underlying distributions, independent random samples, and unknown, unequal variances.

a. Conduct a hypothesis test of H_0: $\mu_1 - \mu_2 = 0$ versus H_a: $\mu_1 - \mu_2 > 0$. Use $\alpha = 0.05$.
b. Find bounds on the p value associated with this test.

10.43 Consider the following table of summary statistics.

Group	Sample size	Sample mean	Sample standard deviation
One	7	76.83	3.30
Two	16	66.80	14.00

Assume the underlying distributions are normal and the random samples were obtained independently.

a. Suppose the population variances are assumed equal. Conduct a hypothesis test of H_0: $\mu_1 - \mu_2 = 0$ versus H_a: $\mu_1 - \mu_2 \neq 0$. Use $\alpha = 0.05$.
b. Suppose the population variances are assumed unequal. Conduct a hypothesis test of H_0: $\mu_1 - \mu_2 = 0$ versus H_a: $\mu_1 - \mu_2 \neq 0$. Use $\alpha = 0.05$.
c. Which of the two tests do you think is more appropriate here? Justify your answer.

10.44 Independent random samples from two normal populations were obtained. The summary statistics were $n_1 = 17$, $\bar{x}_1 = 32.3$, $s_1 = 12.9$, $n_2 = 19$, $\bar{x}_2 = 43.8$, $s_2 = 14.9$.

a. Assume the population variances are unequal. Find a 99% confidence interval for the difference in population means, $\mu_1 - \mu_2$.
b. Using the confidence interval in part (a), is there any evidence to suggest the two population means are different? Justify your answer.

Applications

10.45 Fuel Consumption and Cars The durability and flatness of the front rotors on an automobile are important for braking and for a smooth ride. The flatness of a rotor can be determined by a special optical measuring device that measures the largest deviation from perfect flatness in microinches.

Suppose independent random samples of rotors from two different manufacturers were obtained. The largest deviation from perfect flatness of each rotor was measured, and the resulting summary statistics are given in the following table.

Manufacturer	Sample size	Sample mean	Sample standard deviation
Tire Rack	11	26.74	8.31
JC Whitney	14	29.53	6.85

Assume the underlying distributions are normal and the population variances are equal.

a. Is there any evidence to suggest that there is a difference in population mean deviations from perfect flatness for these two rotor brands? Use $\alpha = 0.05$.
b. Find bounds on the p value associated with this hypothesis test.

10.46 Manufacturing and Product Development The mean weight of an ordinary key is an important consideration, as most Americans carry a pocketful of keys. A manufacturer claims that a new process produces a lighter and more durable key. Independent random samples of both types of keys were obtained, and each key was carefully weighed and its weight (in ounces) was recorded. The resulting summary statistics are given in the following table, with the sample means and sample variances.

Key type	Sample size	Sample mean	Sample variance
Old process	10	0.321	0.0137
New process	10	0.199	0.0202

Assume the underlying distributions are normal and the population variances are equal.

a. Is there any evidence that the population mean weight of a new-process key is less than the population mean weight of an old-process key? Use $\alpha = 0.05$.
b. Find bounds on the p value associated with this hypothesis test.

10.47 Biology and Environmental Science The tiny zebra and quagga mussels have invaded at least 600 bodies of water in the United States and are causing many problems. These mollusks disrupt the natural food chain, clog pipes, cling to machinery, and foul water-delivery systems.[7] A random sample of quagga mussels was obtained from Lake Texoma and Lake Mead, and each was carefully measured. The size of each (in cm) is given on the text website. Is there any evidence to suggest that the population mean size of quagga mussels is larger in Lake Texoma than in Lake Mead? Use $\alpha = 0.05$, and assume the underlying distributions are normal with equal population variances. Write a Solution Trail for this problem. **MUSSELS**

10.48 Manufacturing and Product Development Shelf Safe Milk does not need to be refrigerated until it is opened. Although it is convenient, there is some concern that this Grade A milk contains less protein than regular milk. Independent random samples of 8-ounce servings of Shelf Safe Milk and regular milk were obtained, and the amount of protein (in grams) in each was measured. The summary statistics are given in the following table.

Milk	Sample size	Sample mean	Sample standard deviation
Shelf Safe	25	13.95	3.93
Regular	23	19.09	5.91

Is there any evidence to suggest that the population mean amount of protein in Shelf Safe Milk is less than the population mean amount of protein in regular milk? Use $\alpha = 0.01$, and assume the underlying distributions are normal, with equal variances.

10.49 Medicine and Clinical Studies A study was conducted to determine standard reference values for musculoskeletal ultrasonography in healthy adults. Independent random samples of men and women were obtained, and the sagittal diameter (in mm) of the biceps tendon was measured in each subject. The resulting summary statistics are given in the following table.

Group	Sample size	Sample mean	Sample standard deviation
Women	54	2.5	0.49
Men	48	2.8	0.49

Assume the underlying populations are normal, with equal variances.

a. Is there any evidence to suggest that the population mean sagittal diameter of women's biceps tendons is different from that of men's biceps tendons? Use $\alpha = 0.01$.
b. Construct a 95% confidence interval for the difference in population mean sagittal diameters, $\mu_1 - \mu_2$.

10.50 Manufacturing and Product Development A company that produces hospital furniture has two assembly lines dedicated to cutting and drilling wood for medical cabinets. Each computer-controlled process is designed to drill holes in a certain cabinet part with depth 12.7 mm. Independent random samples of drilled holes were obtained from the two assembly lines, the resulting hole depths (in mm) were recorded. Assume the underlying populations are normal, with equal variances. **CABINET**

a. Is there any evidence to suggest that Line 2 is producing holes with a greater population mean depth than Line 1? Use $\alpha = 0.05$.
b. Find bounds on the p value associated with this hypothesis test.

10.51 Travel and Transportation During Summer 2013, American Airlines introduced a new method for passengers to board a narrowbody aircraft. The new procedure affected passengers traveling light, those carrying one item that fits under the seat. The system was designed to decrease the total boarding time, and improve on-time performance.[8] Independent random samples of American and US Airways flights were obtained and

the boarding time (in minutes) for each was recorded. The summary statistics are given in the following table.

Airline	Sample size	Sample mean	Sample standard deviation
American	21	44.5	12.3
US Airways	26	50.7	15.5

Is there any evidence to suggest that the new boarding procedure has decreased the population mean boarding time? Use $\alpha = 0.05$, and assume the populations are normal with equal variances.

10.52 Manufacturing and Product Development A recent study was conducted to determine the curing efficiency (time to harden) of dental composites (resins for the restoration of damaged teeth) using two different types of lights. Independent random samples of lights were obtained and a certain composite was cured for 40 seconds. The depth of each cure (in mm) was measured using a penetrometer. The summary statistics for the Halogen light were $n_1 = 10$, $\bar{x}_1 = 5.35$, and $s_1 = 0.7$. The summary statistics for the LuxOMax light were $n_2 = 10$, $\bar{x}_2 = 3.90$, and $s_2 = 0.8$. Assume the underlying populations are normal, with equal variances.

a. The maker of the Halogen light claims that they produce a larger cure depth after 40 seconds than LuxOMax lights. Is there any evidence to support this claim? Use $\alpha = 0.01$.
b. Construct a 99% confidence interval for the difference in population mean cure depths.

10.53 Manufacturing and Product Development Certain masonry ties used in residential construction receive a hot-dipped galvanized finish for strength and protection against moisture. Independent random samples of masonry ties from two competing companies were obtained. The amount of coating on one side of each tie was measured (in g/m^2). The resulting summary statistics are given in the following table.

Company	Sample size	Sample mean	Sample variance
Fero	10	331.4	201.64
Cintex	18	298.7	1190.25

Assume the underlying populations are normal.

a. Managers at Fero claim that their product has a larger mean coating than Cintex. Is there any evidence to support this claim? Use $\alpha = 0.01$. Write a Solution Trail for this problem.
b. Find a 95% confidence interval for the difference in population mean coatings.

10.54 Sports and Leisure The curve in a hockey stick is measured by first placing the face of the blade against a flat surface. The curvature of the stick is restricted so that the perpendicular distance from any point at the heel to the end of the blade is at most $\frac{3}{4}$ inch.[9] Independent random samples of hockey sticks used by players on the Toronto Maple Leafs and

Montreal Canadiens teams were obtained. The curve in each stick was measured (in inches), and the resulting data are summarized in the following table.

Team	Sample size	Sample mean	Sample standard deviation
Toronto	10	0.361	0.122
Montreal	20	0.425	0.051

Assume the underlying distributions are normal, with unequal variances. Is there any evidence to suggest that the mean curve in Toronto sticks is different from the mean curve in Montreal sticks? Use $\alpha = 0.001$.

10.55 Manufacturing and Product Development   The tear strength, tensile strength, backing, and thickness all contribute to the durability of vinyl wallpaper. A new company (Aries Wallcoverings) claims to sell the thickest vinyl wallpaper of any currently on the market. Independent random samples of Aries wallpaper and all others were obtained. The thickness of each wallpaper (in inches) was measured. Assume the underlying distributions are normal, with unequal variances. Is there any evidence to support the claim made by Aries Wallcoverings? Use $\alpha = 0.01$.   ▉ **WALLCVR**

10.56 Manufacturing and Product Development   Avid video-game players are always searching for the best graphics card. Overall performance is improved and everything just runs more smoothly with a better graphics card, creating a more enjoyable game experience. Independent random samples of two video cards were obtained and the frame rate (in fps) was measured for each. The summary statistics are given in the following table.[10]

Card	Sample size	Sample mean	Sample standard deviation
GeForce GTX 680 SLI	15	54.7	10.75
Radeon HD 7970 CrossFire	18	58.8	12.00

Assume the underlying populations are normal and the population variances are equal. Is there any evidence to suggest that the mean frame rate for the two cards is different? Use $\alpha = 0.05$.

10.57 Biology and Environmental Science   Over the last few years, the bee population has declined by approximately one-third as a result of mites, fungus, and colony collapse disorder.[11] Some states have been more affected than others. To compare the effect on honey production, independent random samples of hives in Kentucky and Missouri were selected and the amount of honey harvested (in pounds) from each was recorded. Assume the underlying populations are normal and the population variances are equal.   ▉ **BEES**

   a. Is there any evidence to suggest that the mean honey harvest per hive is different in the two states? Use $\alpha = 0.05$.

   b. Find bounds on the p value for the hypothesis test in part (a).

10.58 Medicine and Clinical Studies   An abdominal aortic aneurysm (AAA) is often signaled by inflammation. A cardiovascular magnetic resonance study was conducted to identify wall edema as a marker for inflammation. Independent random samples of AAA and normal patients were obtained, and the MR-STIR intensity values were recorded for each.[12] Assume the underlying populations are normal with unequal variances.   ▉ **EDEMA**

   a. Is there any evidence to suggest that the population mean intensity values are different for the two patient groups? Use $\alpha = 0.05$.

   b. Find bounds on the p value for this hypothesis test.

Extended Applications

10.59 Physical Sciences   Tinted residential windows have become popular because they help a home absorb solar energy, keep out harmful ultraviolet rays, and add privacy. Two independent random samples of tinted windows were obtained, each produced by applying a thin film of a specified color and density. The shading coefficient of each tinted window (a unitless quantity) was measured, and the summary statistics are given in the following table.

Tinted window	Sample size	Sample mean	Sample standard deviation
Silver	8	0.601	0.113
Neutral	11	0.741	0.077

Assume the underlying populations are normal and the population variances are equal.

   a. Is there any evidence to suggest the population mean shading coefficients are different? Use $\alpha = 0.01$.

   b. Construct a 99% confidence interval for the difference in population mean shading coefficients, $\mu_1 - \mu_2$.

   c. Use the confidence interval in part (b) to determine whether there is any evidence to suggest the shading coefficients are different. Does your answer agree with part (a)? If so, why? If not, why not?

10.60 Economics and Finance   The U.S. Bureau of Engraving and Printing produces $1, $5, $10, $20, $50, and $100 bills. The $2 banknote is still legal tender but is currently not in production. Each bill is designed to have the same width, but many people perceive larger-denomination bills to be larger in size. Independent random samples of newly minted $1 and $20 bills were obtained, and the width of each (in mm) was recorded. The summary statistics are given in the following table.

Bill	Sample size	Sample mean	Sample variance
$1	23	66.5990	0.0132
$20	24	66.6924	0.0057

Assume the underlying distributions are normal.

a. If the population variances are assumed equal, is there any evidence to suggest the mean width of a $20 bill is greater than the mean width of a $1 bill? Use $\alpha = 0.01$.

b. If the population variances are assumed unequal, is there any evidence to suggest the mean width of a $20 bill is greater than the mean width of a $1 bill? Use $\alpha = 0.01$.

c. Why do both tests lead to the same conclusion (with very similar p values)?

10.61 Marketing and Consumer Behavior Many home-owners use TIKI torches for outside decoration and to burn special oil to repel insects. Independent random samples of two types of oil were obtained, and the burn time for 3 ounces of each was recorded (in hours). The summary statistics are given in the following table.

Oil	Sample size	Sample mean	Sample variance
Citronella Torch Fuel	21	6.25	1.04
Black Flag Mosquito Control	28	5.98	0.77

Assume the underlying populations are normal.

a. Do you think the assumption of equal variances is reasonable? Why or why not?

b. Based on your answer to part (a), conduct the appropriate hypothesis test to determine whether there is any evidence that the mean burn time is different for these two brands. Use $\alpha = 0.01$.

c. Find bounds on the p value associated with the hypothesis test in part (b).

10.62 Physical Sciences A pressure-relief valve (PRV) is installed on a residential hot-water heater to protect against overheating and, of course, high pressure. Independent random samples of PRVs from different companies were obtained. Each value was tested by recording the pressure (in psi) required to cause the valve to open. The summary statistics are given in the following table.

Company	Sample size	Sample mean	Sample variance
Delta	30	147.6	7.09
Gamma	35	147.8	13.70

Assume the underlying populations are normal, with unequal variances.

a. Is there any evidence to suggest that the mean pressure required to open each valve is different? Use $\alpha = 0.05$.

b. Find a 95% confidence interval for the difference in population mean pressure required to open each valve. Is this confidence interval consistent with the results in part (a)? Explain.

10.63 Physical Sciences The lifetime of a fuel rod in a commercial light-water nuclear reactor is related to the internal pressure. Typically, a fuel rod lasts for 36 months, and one-third of all fuel rods are replaced each year during a plant shutdown. A new type of fuel rod includes a gas-relief capsule and is

designed to last longer. Independent random samples of the two types of fuel rods were obtained, and the lifetime of each (in months) was recorded. The summary statistics are given in the following table.

Fuel-rod design	Sample size	Sample mean	Sample standard deviation
Old	11	34.91	3.20
New	11	39.55	3.55

Assume the underlying populations are normal and the variances are equal.

a. Conduct the relevant hypothesis test to determine whether the new fuel rod does last longer. Use $\alpha = 0.01$.

b. Construct a 99% confidence interval for the difference in population mean lifetimes. Does this confidence interval support the hypothesis test conclusion in part (a)? Explain.

10.64 Manufacturing and Product Development Root beer was originally made using the sarsaparilla root. However, the oil from this root was shown to be carcinogenic (cancer-causing). Since then, many varieties are now made with cane sugar, herbs, spices, and vanilla. Independent random samples of 12-ounce cans of A&W root beer and Barq's root beer were obtained, and the amount of sugar (in grams) was measured in each. Assume the underlying populations are normal, with unequal variances. **ROOTBEER**

a. Is there any evidence to suggest that the population mean amount of sugar in A&W root beer is greater than in Barq's root beer? Use $\alpha = 0.05$.

b. Construct a 95% confidence interval for the difference in the population mean sugar amounts. Does this confidence interval support the hypothesis test conclusion in part (a)? Explain.

Challenge

10.65 Robust Statistics The two-sample t test for comparing population means when the variances are equal is a robust statistical procedure. If the population variances are different, as long as the underlying populations are normal and the sample sizes are equal, then the hypothesis test is still very reliable.

Generate a random sample of size 25 from a normal distribution with mean $\mu_1 = 100$ and standard deviation $\sigma_1 = 5$. Generate a second random sample of size 25 from a normal distribution with mean $\mu_2 = 100$ and standard deviation $\sigma_2 = 5$. Conduct a two-sided, two-sample t test for comparing population means assuming the population variances are equal and with $\alpha = 0.05$. Do this 100 times and record the number of times you reject the null hypothesis.

Repeat the same procedure but use $\sigma_2 = 7$. Record the number of times you reject the null hypothesis. Repeat the same procedure for $\sigma_2 = 10, 15, 20, 25, 30, 50$. Use your results to explain the robust nature of this hypothesis test.

10.3 Paired Data

When comparing population means in the previous two sections, one necessary assumption was that the samples were obtained independently. The n_1 observations from the first population and the n_2 observations from the second population were unrelated. Many experiments, however, involve only n individuals, or objects, where two observations are made of each individual. A classic example involves a diet-and-exercise program designed to help people lose weight. A random sample of n individuals is selected and each is weighed. Each person follows the regimen for a specified time period and is weighed again at the end of the experiment. There are two observations of each individual, a *before* weight and an *after* weight. The data are used to determine whether the diet-and-exercise program is effective.

Trial-and-error learning or memory experiments in animals present another good example. In a typical psychology experiment, a random sample of rats is obtained and each is timed as it maneuvers through a maze. After several weeks of training, each rat is timed again. This produces two observations of each animal, a *before* time and an *after* time. This experiment might be designed to determine whether animals can *learn* the correct path through a maze and hence decrease the mean time needed to traverse the course.

The difference between the experiments described above and those in the previous two sections is that here, the *paired* observations are *dependent*. We are still interested in comparing population means (the *before* mean μ_1 and the *after* mean μ_2), and, therefore, still interested in the difference $\mu_1 - \mu_2$. However, the sample means, $\overline{X}_1$ and $\overline{X}_2$, are *not* independent. The standardizations used previously are not applicable, because the variance of $\overline{X}_1 - \overline{X}_2$ is more complicated. Therefore, another method is necessary, one that addresses the dependence and yet still considers the difference $\mu_1 - \mu_2$.

The variance of $\overline{X}_1 - \overline{X}_2$ must account for the dependence.

Suppose that

For reference, these are the two-sample paired t test assumptions.

1. There are n individuals or objects, or n pairs of individuals or objects, that are related in an important way or share a common characteristic; and
2. There are two observations of each *individual*. The population of first observations is normal, and the population of second observations is also normal.

Even through the word individual is used, this means individual or object.

▶ Let X_1 represent a randomly selected first observation and let X_2 represent the corresponding second observation on the same individual. Consider the random variable $D = X_1 - X_2$, the difference in the observations, and the n observed differences $d_i = (x_1)_i - (x_2)_i$, $i = 1, 2, \ldots, n$. X_1 and X_2 are both normal, so D is also normal. More important, the differences are independent. A hypothesis test concerning $\mu_1 - \mu_2$ is based on the sample mean of the differences, $\overline{D}$. This random variable has the following properties.

Properties of $\overline{D}$

1. $E(\overline{D}) = \mu_1 - \mu_2$. $\overline{D}$ is an unbiased estimator for the difference in means $\mu_1 - \mu_2$.
2. The variance of $\overline{D}$ is unknown, but it can be estimated using the sample variance of the differences.
3. Both underlying populations are normal, so D is normal, and hence, $\overline{D}$ is also normal. ◀

Here's what all of these results mean for us. To compare population means, μ_1 and μ_2, when the data are paired, we focus on the difference $\mu_1 - \mu_2$. As in earlier two-sample tests, the null hypothesis $H_0: \mu_1 = \mu_2$ is equivalent to $H_0: \mu_1 - \mu_2 = 0$. A test to determine whether the underlying population means of two paired samples are equal is equivalent to a test to determine whether the population mean of the paired differences is zero. We compute the differences, $d_1, d_2, \ldots, d_n$, and conduct a one-sample t test (with $n - 1$ degrees of freedom) using the differences.

Hypothesis Tests Concerning Two Population Means When Data Are Paired

This is the template for a hypothesis test concerning two population means when data are paired: a paired t test.

Given the two-sample paired t test assumptions, a hypothesis test concerning the two population means in terms of the difference $\mu_D = \mu_1 - \mu_2$, with significance level α, has the form

H_0: $\mu_D = \mu_1 - \mu_2 = \Delta_0$

H_a: $\mu_D > \Delta_0$, $\mu_D < \Delta_0$,   or $\mu_D \neq \Delta_0$

TS: $T = \dfrac{\overline{D} - \Delta_0}{S_D/\sqrt{n}}$

where S_D is the sample standard deviation of the differences.

RR: $T \geq t_{\alpha,n-1}$, $T \leq -t_{\alpha,n-1}$,   or $|T| \geq t_{\alpha/2,n-1}$

A CLOSER LOOK

1. Δ_0 is the hypothesized difference in the population means. Usually $\Delta_0 = 0$: The null hypothesis is that the two population means are equal. However, Δ_0 may be nonzero. For example, the null hypothesis H_0: $\mu_D = \mu_1 - \mu_2 = 5 = \Delta_0$ specifies that the difference in population means is 5.

2. A paired t test is valid even if the underlying population variances are unequal, that is, even if $\sigma_1^2 \neq \sigma_2^2$. The sample variance of the differences, S_D^2, is a good estimator of $\text{Var}(\overline{X}_1 - \overline{X}_2)$ when the observations are paired.

3. If a paired t test is appropriate, the test statistic is based on $n - 1$ degrees of freedom. A two-sample t test (incorrect here) would be based on a test statistic with $n + n - 2 = 2n - 2$ degrees of freedom. Therefore, the correct analysis is based on a distribution with greater variability and is more conservative. ●

Example 10.8 Relaxing Music

DATA SET

MUSIC

The common characteristic is patient identity.

There is no direct scientific measure of stress, but some physical properties of the body that are believed to be related to stress include pulse rate, blood pressure, breathing rate, brain waves, muscle tension, skin resistance, and body temperature. Some researchers claim that music can be relaxing and, therefore, reduce stress. Twelve patients who claim to be suffering from job-related stress were selected at random. An initial resting pulse rate (in beats per minute, bpm) was obtained, and each person participated in a month-long music-listening, relaxation-therapy program. A final resting pulse rate was taken at the end of the experiment. The data are given in the following table.

Subject	1	2	3	4	5	6
Initial pulse rate	67	71	67	83	70	75
Final pulse rate	61	72	70	76	58	61
Difference	6	−1	−3	7	12	14

Subject	7	8	9	10	11	12
Initial pulse rate	71	68	72	88	78	70
Final pulse rate	74	59	61	64	71	77
Difference	−3	9	11	24	7	−7

Is there any evidence to suggest that the music-listening, relaxation-therapy program reduced the mean pulse rate and, therefore, the stress level? Assume the underlying distributions of initial and final pulse rate are normal, and use $\alpha = 0.05$.

Solution Trail 10.8

KEYWORDS
- Initial pulse rate
- Final pulse rate
- Is there any evidence?
- Reduce the mean
- Random sample

TRANSLATION
- *Before* and *after* measurements on the same individual
- Conduct a one-sided, right-tailed test to compare the *before* and *after* mean pulse rates.

CONCEPTS
- Hypothesis test concerning two population means when data are paired

VISION
The data are certainly paired—there is a *before* and an *after* measurement on each individual—and each population is assumed normal. Compute the differences, use a one-sided alternative hypothesis and the corresponding rejection region, find the value of the test statistic, and draw a conclusion.

Try to find bounds for the *p* value associated with this hypothesis test.

SOLUTION

STEP 1 Traditionally, the *before* measurements are population 1 and the *after* measurements are population 2.

The null hypothesis is that the two population means are equal (i.e., the music-listening, relaxation-therapy program has no effect): $\mu_1 = \mu_2 \Rightarrow \mu_1 - \mu_2 = \mu_D = 0 \, (= \Delta_0)$.

Each population is assumed normal, and there are two observations on each individual. A paired *t* test is appropriate.

The therapy program is designed to reduce stress, so the alternative hypothesis is $\mu_1 > \mu_2 \Rightarrow \mu_1 - \mu_2 = \mu_D > 0$. This is a one-sided, right-tailed test.

STEP 2 The four parts of the hypothesis test are

$$H_0: \mu_D = 0$$
$$H_a: \mu_D > 0$$
$$\text{TS: } T = \frac{\overline{D} - \Delta_0}{S_D/\sqrt{n}}$$

RR: $T \geq t_{\alpha, n-1} = t_{0.05, 11} = 1.7959$

STEP 3 In anticipation of a paired *t* test, the differences are given in the table above.

The sample mean of the differences is

$$\overline{d} = \frac{1}{12}[6 + (-1) + \cdots + (-7)] = 6.3333$$

The sample variance of the differences is

$$s_D^2 = \frac{1}{11}\left[1320 - \frac{1}{12}(76)^2\right] = 76.2424$$

The sample standard deviation of the differences is

$$s_D = \sqrt{76.2424} = 8.7317$$

The value of the test statistic is

$$t = \frac{\overline{d} - 0}{s_d/\sqrt{n}} = \frac{6.3333}{8.7317/\sqrt{12}} = 2.5126 \, (\geq 1.7959)$$

STEP 4 The value of the test statistic, $t = 2.5126$, lies in the rejection region, so we reject the null hypothesis at the $\alpha = 0.05$ significance level. There is evidence to suggest that the music-listening, relaxation-therapy program does reduce a person's resting pulse rate (and therefore the stress level).

Figure 10.33 shows a technology solution.

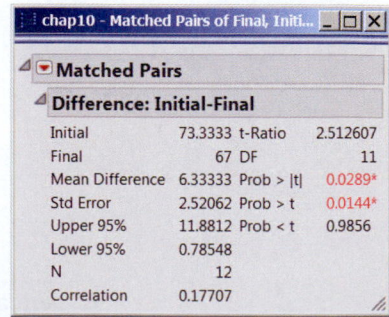

Figure 10.33 JMP Matched Pairs output.

TRY IT NOW GO TO EXERCISE 10.79

DATA SET

LEAKAGE

Example 10.9 Radiation Leakage

In August 2013, radioactive water from the damaged Fukushima Daiichi nuclear plant leaked into the Pacific Ocean. The contaminated groundwater seeped through an underground barrier created through chemical injections, and the Japanese Nuclear Regulatory Authority declared an emergency. In response, Tepco (the plant's operator) began pumping out more groundwater to ease the spill into the ocean. Concentrations of cesium-137 (in Bq/1) in the seawater 10–30 kilometers off the Japanese coast were taken at eight different locations one week apart, before and after the additional pumping. The data are given in the following table.

Location	1	2	3	4	5	6	7	8
August 23	16.0	12.0	14.0	18.0	12.0	12.5	10.5	15.5
August 30	16.0	9.0	11.0	16.0	14.0	8.5	6.5	14.5

Is there any evidence to suggest that the population mean cesium-137 level decreased from August 23 to August 30? Assume the underlying populations are normal, use $\alpha = 0.05$, and find bounds on the p value associated with this hypothesis test.

Solution Trail 10.9

KEYWORDS

- August 23, August 30
- Is there any evidence?
- Mean level decreased

TRANSLATION

- Two measurements from each location
- Conduct a one-sided test to compare the cesium-137 levels at different times.

CONCEPTS

- Hypothesis test concerning two population means when data are paired

VISION

The data are paired, and each population is assumed normal. Compute the differences, use the one-sided, right-tailed alternative hypothesis and the corresponding rejection region, find the value of the test statistic, and draw a conclusion.

SOLUTION

STEP 1 Let population 1 be the cesium-137 concentrations on August 23, and let population 2 be the cesium-137 concentrations on August 30.

The null hypothesis is that the two population means are equal and the cesium-137 concentrations are the same on both days:

$$\mu_1 = \mu_2 \implies \mu_1 - \mu_2 = \mu_D = 0$$

Each population is assumed normal and there are two observations at each location. A paired t test is appropriate.

We would like to determine if the additional pumping helped to reduce the cesium-137 concentrations. Therefore, the alternative hypothesis is

$$\mu_1 > \mu_2 \implies \mu_1 - \mu_2 > \mu_D = 0$$

STEP 2 The four parts of the hypothesis test are

$$H_0: \mu_D = 0$$

$$H_a: \mu_D > 0$$

$$\text{TS: } T = \frac{\overline{D} - \Delta_0}{S_D/\sqrt{n}}$$

RR: $T \geq t_{\alpha,n-1} = t_{0.05,7} = 1.8946$

STEP 3 The summary statistics for the differences are

$$\overline{d} = 1.875, \qquad s_D = 2.1002$$

The value of the test statistic is

$$t = \frac{\overline{d} - 0}{s_d/\sqrt{n}} = \frac{1.875}{2.1002/\sqrt{8}} = 2.5251 \; (\geq 1.8946)$$

STEP 4 The value of the test statistic, $t = 2.5251$, lies in the rejection region. Therefore, we reject the null hypothesis. There is evidence to suggest that the population mean cesium-137 concentration was less on August 30.

STEP 5 Using Table V in the Appendix, we can only bound the p value. In Table V, row $n - 1 = 8 - 1 = 7$, place 2.5251 in the ordered list of critical values.

$$2.4469 \leq 2.5251 \leq 3.1427$$

$$t_{0.025,7} \leq 2.5251 \leq t_{0.01,7}$$

Therefore, $\qquad 0.01 \leq \quad p \quad \leq 0.025$

See Figure 10.34.

Figure 10.35 shows a technology solution.

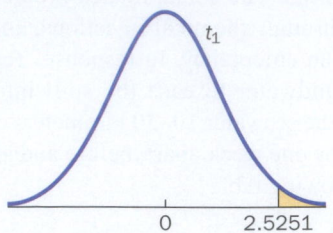

Figure 10.34 p-Value illustration:
$p = P(T \geq 2.5251)$
$= 0.0198 \leq 0.05 = \alpha$

Figure 10.35 Minitab Paired t (Test and Confidence Interval) results.

TRY IT NOW   GO TO EXERCISE 10.81

The random variable T here is

$$T = \frac{\overline{D} - \mu_D}{S_D/\sqrt{n}}$$

The usual technique can be used to construct a confidence interval for the difference in means, $\mu_D = \mu_1 - \mu_2$, when the observations are paired. Start with a symmetric interval about 0 such that the probability T lies in this interval is $1 - \alpha$. Manipulate the inequality to sandwich μ_D.

How to Find a 100(1 − α)% Confidence Interval for μ_D

Given the paired t test assumptions, a $100(1 - \alpha)\%$ confidence interval for μ_D has as endpoints the values

$$\overline{d} \pm t_{\alpha/2,n-1}\frac{s_D}{\sqrt{n}} \tag{10.6}$$

Example 10.10 Improved Mileage

A local automotive repair shop advertises a special maintenance package, including tire balancing, new spark plugs, engine oil additive, and a front-end alignment, that will certainly improve gas mileage. To check this claim, 18 cars (and drivers) were randomly selected. Each car was driven on a specially designed route and the miles per gallon for each car was recorded. Following the maintenance package, each driver took the same route, and the miles per gallon were measured again. The summary statistics for the differences (before maintenance mpg − after maintenance mpg) were $\overline{d} = -1.28$, $s_D = 5.62$. Assuming normality, find a 99% confidence interval for the true difference in mean miles per gallon.

SOLUTION

STEP 1 The sample size and summary statistics are given, and the underlying distributions (before maintenance mpg and after maintenance mpg) are assumed normal.

The observations are paired, so we can use Equation 10.6.

$1 - \alpha = 0.99 \;\Rightarrow\; \alpha = 0.01 \;\Rightarrow\; \alpha/2 = 0.005$ Find $\alpha/2$.

$t_{\alpha/2,n-1} = t_{0.005,17} = 2.8982$ Find the t critical value with $\nu = 17$.

STEP 2 Use Equation 10.6.

$$\overline{d} \pm t_{\alpha/2,n-1}\frac{s_D}{\sqrt{n}}$$ Equation 10.6.

$$= -1.28 \pm (2.8982)\frac{5.62}{\sqrt{18}}$$ Use summary statistics and critical value.

$$= -1.28 \pm 3.8391$$ Simplify.

$$= (-5.1191, 2.5591)$$ Compute endpoints.

$(-5.1191, 2.5591)$ is a 99% confidence interval for the true mean difference in miles per gallon, μ_D.

Note that because 0 is included in this confidence interval, there is no evidence to suggest that μ_D is different from 0, and there is no evidence to suggest that the maintenance program improves mileage.

Figures 10.36 and 10.37 together show a technology solution.

VIDEO TECH MANUALS

PAIRED SAMPLES INFERENCE

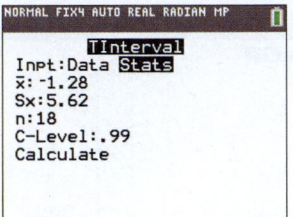

Figure 10.36 TInterval input screen.

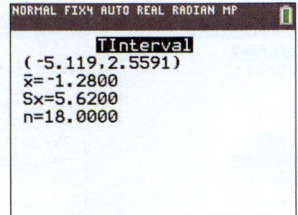

Figure 10.37 Resulting confidence interval.

TRY IT NOW GO TO EXERCISE 10.93

Technology Corner

Procedure: Hypothesis tests and confidence intervals concerning paired data.
Reconsider: Example 10.9, solution, and interpretations.

CrunchIt!

Use the built-in function Statistics; t; Paired. There are tabs to conduct a hypothesis test and construct a confidence interval.

1. Enter the data from the first date into column Var1 and the data from the second date into column Var2.
2. Select Statistics; t; Paired. Choose the First and Second Variables from the drop-down menus. Under the Hypothesis Test tab, enter the Mean difference under the null hypothesis and select the appropriate alternative. See Figure 10.38.
3. Click Calculate. The results are displayed in a separate window. See Figure 10.39.

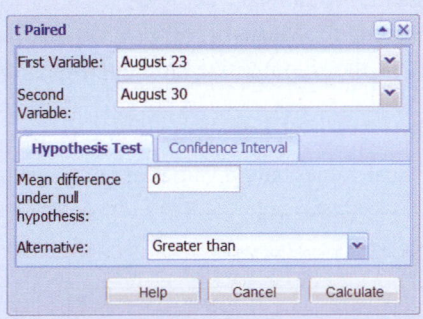

Figure 10.38 CrunchIt! t Paired input screen.

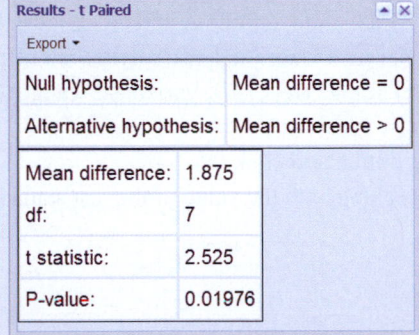

Figure 10.39 CrunchIt! t Paired hypothesis test results.

TI-84 Plus C

Compute the differences if necessary. Use the built-in functions T-Test and TInterval. Input is either summary statistics or data in lists.

1. Enter the August 23 data into list L1 and the August 30 data into list L2.
2. Find the paired differences on the Home screen, and store them in the list L3.

3. Select STAT ; TESTS; T-Test. Highlight Data. Enter μ_0, the hypothesized difference in means and the list containing the paired differences. Set the frequency to 1 and highlight the alternative hypothesis. See Figure 10.40.

4. Highlight Calculate and press ENTER . The results are displayed on the Home screen. See Figure 10.41. The Draw results are shown in Figure 10.42.

5. Use STAT ; TESTS; TInterval to construct a confidence interval for the paired differences. This procedure is described in the Technology Corner in Section 8.3. A 95% confidence interval for the paired differences in shown in Figure 10.43.

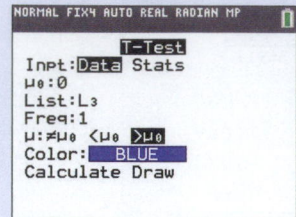 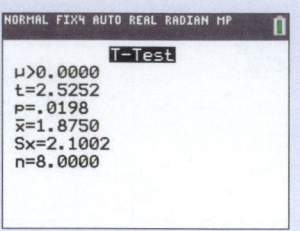 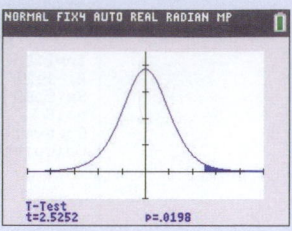 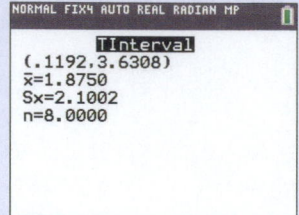

Figure 10.40 T-Test input screen.   **Figure 10.41** T-Test hypothesis test results.   **Figure 10.42** T-Test Draw results.   **Figure 10.43** A 95% confidence interval for the paired differences.

Minitab

Use the built-in function Paired t to conduct a hypothesis test and to construct a confidence interval. Input is either paired data in two columns or summarized data for the paired differences.

1. Enter the August 23 data into column C1 and the August 30 data into column C2.
2. Select Stat; Basic Statistics; Paired t.
3. Choose Each sample is in a column, and enter C1 in the Sample 1 input window and C2 in the Sample 2 input window.
4. Choose the Options option button. Enter a Confidence level, the Hypothesized difference (Δ_0), and the Alternative hypothesis.
5. The hypothesis test results and the confidence interval are displayed in a session window. Refer to Figure 10.35.

Excel

Use the Data Analysis toolkit function t-Test: Paired Two Sample for Means. If only summarized data for the paired differences are given, use the function CONFIDENCE.T to find the bound on the confidence interval and compute the endpoints of the CI as described in the Technology Corner in Section 8.3.

1. Enter the August 23 data into column A and the August 30 data into column B.
2. Under the Data tab, select Data Analysis; t-Test: Paired Two Sample for Means.
3. Enter the Variable 1 Range (A1:A8), Variable 2 Range (B1:B8), the Hypothesized Mean Difference (Δ_0), and a value for Alpha.
4. Choose an Output option and click OK.
5. Summary statistics along with the value of the test statistic, critical values, and p values are displayed. See Figure 10.44.

t-Test: Paired Two Sample for Means		
	Variable 1	Variable 2
Mean	13.8125	11.9375
Variance	6.3527	13.5313
Observations	8	8
Pearson Correlation	0.8345	
Hypothesized Mean Difference	0	
df	7	
t Stat	2.5252	
P(T<=t) one-tail	0.0198	
t Critical one-tail	1.8946	
P(T<=t) two-tail	0.0395	
t Critical two-tail	2.3646	

Figure 10.44 Excel paired *t* test results.

SECTION 10.3 EXERCISES

Concept Check

10.66 Short Answer State the two-sample paired t test assumptions.

10.67 True/False In a two-sample paired t test, the before and after sample sizes must be the same.

10.68 True/False In a two-sample paired t test, the sample mean of the differences, $\overline{D}$, is an unbiased estimator for $\mu_1 - \mu_2$.

10.69 True/False A paired t test is valid only if the underlying population variances are equal.

10.70 Short Answer Explain how a confidence interval for μ_D can be used to make an inference concerning μ_1 and μ_2.

10.71 Short Answer Suppose a paired t test is appropriate. How does this reference distribution compare to the distribution based on a two-sample t test?

Practice

10.72 In each experiment, determine whether the data are obtained independently or are paired. If the data are independent, indicate the two distinct populations. If the data are paired, indicate the common characteristic.

a. School board members believe adding a teacher's aide to each K–4 class will improve classroom management and increase instruction time. Twenty-six elementary classrooms are selected at random, and the daily instruction time for each is recorded. A teacher's aide is then added to each classroom, and the daily instruction time is recorded again. The data will be used to determine whether there is any evidence that adding a teacher's aide increases the mean daily instruction time.

b. A researcher investigating home-insurance costs obtained a random sample of homes in the Northeast and another random sample of homes in the South. The yearly insurance cost for each home was recorded. The data will be used to determine whether there is any difference in the mean yearly home-insurance costs between the Northeast and the South.

c. Officials at the transit authority of a large city would like to compare the route times during the morning and evening rush hours. Eleven routes are selected at random. A morning and an evening route completion time are recorded for each. The data will be used to determine whether the mean evening route time is less than the mean morning route time.

d. A supplementary health-insurance provider is investigating changes in claim patterns. A random sample of 32 policyholders was selected. The total amounts claimed in 2013 and in 2014 were recorded for each person. The data will be used to determine whether there is any evidence that the mean amount claimed has increased from 2013 to 2014.

e. Random samples of 45 new home sites in Kansas and 52 new home sites in upstate New York were selected. The *flatness coefficient* (a unitless quantity between 0 and 1) of each lot was measured. The data will be used to determine whether the mean flatness coefficient for new home sites in Kansas is less than the mean flatness coefficient for new home sites in upstate New York.

10.73 In each experiment, determine whether the data are obtained independently or are paired. If the data are independent, indicate the two distinct populations. If the data are paired, indicate the common characteristic.

a. A surgeon is investigating the effect of physical therapy on patients who have had rotator-cuff injuries. Twenty-eight patients were selected at random. The range of motion in the affected arm was measured in each patient prior to starting physical therapy. After three weeks of consistent therapy, the range of motion in each patient was measured again. The data will be used to determine whether physical therapy has increased the mean range of motion.

b. A woodwork manufacturer has several lathes used for shaving parts. The accuracy of each is determined by measuring the production width of a designed 1-mm wood strip. Nine lathes were selected at random. The accuracy of each was measured. Then the gib screws on each lathe were adjusted, and the width of another 1-mm wood strip was measured. The data will be used to determine if the mean width before adjustment is different from the mean width after adjustment.

c. A random sample of sixty 20-year-old males and a random sample of forty-five 70-year-old males were obtained. The length of each person's right ear was measured (in inches). The data will be used to determine whether the mean ear length of a 70-year-old male is greater than the mean ear length of a 20-year-old male.

d. Modular homes are built in a closed, factory setting, indoors, and are not subject to adverse weather conditions. The quality control is often better than for on-site construction, but the cost may also differ. Ten different styles of homes were selected, and the cost of each built as a modular home and on-site was estimated. The data will be used to determine whether the mean cost of modular homes is greater than for on-site construction.

e. Random samples of 25 frequent flyers on United Airlines and on Delta Airlines were obtained. The total number of accumulated frequent-flyer miles for each person was recorded. The data will be used to determine whether the mean number of frequent-flyer miles is greater for United Airlines passengers than for Delta Airlines passengers.

10.74 In each experiment, determine whether the data are obtained independently or are paired. If the data are independent, indicate the two distinct populations. If the data are paired, indicate the common characteristic.

a. A random sample of soccer players on the Brazil national team and a random sample of soccer players on the

Argentina national team was obtained. The amount of time (in minutes) each played during a World Cup match was obtained. The data will be used to determine whether the mean time is different for the two teams.

b. A random sample of bananas from Columbia and Costa Rica was obtained. The amount of fiber in the peel of each banana was carefully measured and recorded. The data will be used to determine whether there is a difference in the mean amount of fiber in the banana peel by country.

c. A new stent has been designed for use in patients with diseased arteries. A random sample of patients in need of a stent was selected. The intrasaccular pressure of each was carefully measured, and then a stent was surgically placed in the diseased artery. The intrasaccular pressure was measured again following surgery. The data will be used to determine whether the stent placement reduced intrasaccular pressure.

d. An automobile manufacturer claims that using a lighter weight engine oil can actually increase a car's miles per gallon (of gasoline). A random sample of cars (and drivers) was selected, all who currently use a heavy weight oil in their engines. The miles per gallon for each car was recorded. The engine oil was drained, a lighter weight oil was used, and the miles per gallon was recorded again. The data will be used to determine if the mean miles per gallon has increased.

e. A new process has been developed that theoretically improves the nutritional value of barley for use in fish feed.[*] A random sample of traditional barley and another random sample of the new barley was obtained. The percentage of protein in each grain was carefully measured. The data will be used to determine whether the new barley has a greater mean percentage of protein.

10.75 The following summary statistics were obtained in a paired-data study: $\bar{d} = 15.68$, $s_D = 33.55$, and $n = 17$. Assume normality and conduct a test of $H_0: \mu_D = 0$ versus $H_a: \mu_D > 0$. Use $\alpha = 0.05$.

10.76 Consider the following paired data. ▥ **EX10.76**

Subject	1	2	3	4	5
Before treatment	332.5	289.3	288.2	268.0	278.0
After treatment	317.3	302.5	312.9	325.4	267.3

a. Assume normality and conduct a test of $H_0: \mu_D = 0$ versus $H_a: \mu_D < 0$. Use $\alpha = 0.01$.

b. Find bounds on the p value associated with this hypothesis test.

10.77 Consider the paired data provided. ▥ **EX10.77**

a. Assume normality and conduct a test of $H_0: \mu_D = 0$ versus $H_a: \mu_D \neq 0$. Use $\alpha = 0.01$.

b. Find bounds on the p value associated with this hypothesis test.

[*]US Department of Agriculture, accessed on July 15, 2014, http://www.ars.usda.gov/is/pr/2014/140714.htm.

Applications

10.78 Technology and the Internet Twenty-one computer programmers from IT firms around the country were selected at random. Each was asked to write code in C++ and in Java for a specific application. The runtime (in seconds) for each program, by computer language, was recorded. ▥ **RUNTIME**

a. What is the common characteristic that makes these data paired?

b. Assume normality. Conduct the appropriate hypothesis test to determine whether there is any evidence that the mean runtime for Java programs is greater than the mean runtime for C++ programs. Use $\alpha = 0.001$.

c. Find bounds on the p value associated with this hypothesis test.

10.79 Physical Sciences A consultant working for a State Police barracks contends that service weapons will fire with a higher muzzle velocity if the barrel is properly cleaned. A random sample of Glock 9-mm handguns was obtained, and the muzzle velocity (in feet per second) of a single shot from each gun was measured. Each gun was professionally cleaned, and the muzzle velocity of a second shot (with the same bullet type) was measured. The data are given in the following table. ▥ **GLOCK**

Gun	1	2	3	4	5	6
Before	1505	1419	1504	1494	1510	1506
After	1625	1511	1459	1441	1472	1521

a. What is the common characteristic that makes these data paired?

b. Assume normality. Conduct the appropriate hypothesis test to determine whether there is any evidence that a clean gun fires with a higher muzzle velocity. Use $\alpha = 0.01$.

c. Find bounds on the p value associated with this hypothesis test.

10.80 Sports and Leisure In 1969, the height of Major League Baseball pitching mounds was lowered from 15 to 10 inches. This decreased a pitcher's leverage and presumably the speed of a typical fastball. Fifteen Major League pitchers were selected at random, and each threw his best fastball from a 15-inch mound and from a 10-inch mound. The speed of each pitch (in mph) was recorded. ▥ **FASTBALL**

a. Assume normality. Construct a 99% confidence interval for the true mean difference in fastball speeds from a 15-inch mound and a 10-inch mound.

b. Using the confidence interval in part (a), is there any evidence to suggest that pitching speed is, on average, faster from a higher mound? Justify your answer.

10.81 Public Health and Nutrition A filtration system made for small businesses is designed to remove particulate matter from the air. To test the new device, a random sample of businesses was obtained. The concentration of particulate matter was measured (in $\mu g/m^3$). The filtration system was then allowed to run for 24 hours, and the concentration of particulate matter was measured again. The differences (before filtration – after filtration) were recorded. Assume normality. Is there any evidence to suggest that the new filtration system improves air quality by removing particulate matter? Use $\alpha = 0.05$. Write a Solution Trail for this problem. **FILTER**

10.82 Biology and Environmental Science The best hurricane forecasting models use global data and take hours to run on the world's fastest supercomputers. A random sample of 26 tropical storms or hurricanes that passed within 25 miles of Miami were selected. The European Center for Medium-Range Weather Forecasting (ECMWF) and the Global Forecast System (GFS) models were used to predict storm surge (in feet). The summary statistics for the differences (ECMWF – GFS) were $\bar{d} = 1.4923$ and $s_D = 2.8097$.

  a. Is there any evidence to suggest that the ECMWF model predicts higher storm surges? Assume normality and use $\alpha = 0.05$.

  b. Find bounds on the p value associated with this hypothesis test.

10.83 Physical Sciences It is important to maintain a low ammonia-ion concentration in freshwater aquariums to ensure healthy fish (and plants). An ammonia neutralizer is advertised to almost instantly detoxify ammonia (i.e., reduce the concentration of ammonia ions) in order to protect fish. Fourteen untreated 20-gallon aquariums were selected at random, and the ammonia-ion concentration (in ppm) in each was measured. One hour after the directed amount of the neutralizer was used, the ammonia ion concentration was measured again. Assuming normality, is there any evidence to suggest that the neutralizer decreases the mean ammonia-ion concentration? Use $\alpha = 0.025$. **AQUARIUM**

10.84 Public Health and Nutrition Beef boullion generally has a high salt content, which can cause health problems. A food columnist for a local newspaper suggested simmering boullion with slices of raw potato to remove salt. To check this claim, 10 different boullion brands were selected at random and the salt content in each was measured (in mg/cup of water). Five potato slices were then added to each broth and the mixtures were left to simmer for 15 minutes. Following this procedure, the salt content was measured again. The difference between the initial and the final salt content was computed for each boullion brand, and the data are given in the following table. **BOULLION**

−169	−222	431	110	−168
353	−207	68	25	203

Assume normality. Is there evidence to suggest that simmering with raw potatoes decreases the mean salt content in beef boullion? Use $\alpha = 0.05$.

10.85 Psychology and Human Behavior A new study suggests that the speed of a person's step is related to the genre of music being listened to while walking.[13] A random sample of 18 adults was obtained. Each walked in a circular path while listening to two different genres of music, hard rock and

ballads. The steps per minute for each person for each genre were recorded. **WALKING**

  a. What is the common characteristic that makes these data paired?

  b. Assume normality. Conduct the appropriate hypothesis to determine whether there is any evidence to suggest that the population mean steps per minute listening to hard rock is greater than the population mean steps per minute listening to ballads. Use $\alpha = 0.01$.

10.86 Public Health and Nutrition The cost of long-term health care continues to rise each year, but it varies considerable from state to state. Many older adults who must enter a nursing home may select either a semi-private or private room. Several states were selected at random and the cost of each type of room was recorded.[14] **ROOMS**

  a. Assume normality and conduct the appropriate hypothesis test to determine whether the population mean cost of a private room is $15 greater than the population mean cost of a semi-private room. Use $\alpha = 0.001$. Write a Solution Trail for this problem.

  b. What characteristic of the differences suggests that the hypothesis test in part (a) will be significant?

  c. Find bounds on the p value associated with this hypothesis test.

10.87 Manufacturing and Product Development The porosity of a concrete block is a measure (as a percentage) of the amount of empty space in the block. In residential homes with concrete-block foundations, a larger porosity leads to damper, colder basements. A contractor recommends pretreatment of concrete blocks with a product designed to decrease the porosity. A random sample of concrete blocks was obtained, and the porosity of each was measured. The clear, paintlike product was applied to each block, and the porosity was measured again. The differences (before treatment – after treatment) in porosities were recorded. Assume normality. Is there any evidence to suggest the new product decreases the mean porosity of concrete blocks? Use $\alpha = 0.001$. **CONCRETE**

10.88 Psychology and Human Behavior Each employee hired at an electronics parts assembly line in Edmonton, Alberta, is given a general intelligence test. To determine which method of training is more effective, eight pairs of new hires were matched according to their exam scores. One set of employees was asked to read appropriate training manuals, while the other group watched interactive training videos. Each employee was then asked to assemble a part used in a locater-beacon transmitter, and the time (in minutes) to completion was recorded. The data are given in the following table. **TRAINING**

Employee pair	1	2	3	4
Written manual	4.9	4.6	5.3	4.9
Interactive video	3.1	4.1	4.4	4.9

Employee pair	5	6	7	8
Written manual	4.9	5.4	5.5	5.0
Interactive video	3.6	3.9	6.5	5.3

a. What is the common characteristic that makes these data paired?

b. Is there any evidence to suggest the true mean time difference, μ_D, is different from 0? Assume normality and use $\alpha = 0.05$.

10.89 Public Health and Nutrition   Americans love hamburgers, but the high fat content in some cooked patties presents a severe health threat. Certain electric grills are designed to drain fat away from the patty, resulting in a healthier, although perhaps less tasty, meal. A random sample of ground beef packages was obtained (with various fat contents). Two patties were made from each package. One was cooked in an electric grill, while the other was prepared in a frying pan on top of a stove. The fat content (as a percentage) in each cooked patty was measured. **HAMBURG**

a. Conduct the appropriate hypothesis test to determine whether the true mean fat content in hamburgers cooked on an electric grill is less than the true mean fat content of hamburgers cooked in a frying pan. Assume normality and use $\alpha = 0.001$.

b. Find bounds on the p value associated with this hypothesis test.

Extended Applications

10.90 Medicine and Clinical Studies   A new drug designed to reduce fever (and relieve aches and pains) is being tested for efficacy and side effects. Ten patients entering a hospital with a high fever were selected at random. The temperature (in °F) of each patient was measured, the drug was administered, and two hours later the temperature was measured again. The data are given in the following table. **FEVER**

Patient	1	2	3	4	5
Before drug	102.6	99.2	102.3	101.1	102.7
After drug	99.8	98.8	97.5	100.3	99.6

Patient	6	7	8	9	10
Before drug	102.6	100.5	103.5	105.7	104.3
After drug	102.8	99.0	101.8	97.1	99.2

a. What is the common characteristic that makes these data paired?

b. Assume normality. Conduct the appropriate hypothesis test to determine whether there is any evidence that the new drug reduces the mean patient temperature after two hours. Use $\alpha = 0.05$.

c. Find bounds on the p value associated with this hypothesis test.

d. What characteristic of the differences suggests that a hypothesis test will be significant?

10.91 Fuel Consumption and Cars   Biodiesel fuel has a cloud point, the temperature at which the fuel becomes cloudy, of approximately 13°C. This clouding can lead to poor engine performance and can even cause an engine to stop completely. An industrial chemical company produces an additive designed

to lower the cloud point of this type of fuel. A random sample of six different biodiesel fuels was obtained and the cloud point was measured for each. One ounce of the chemical additive was mixed in with every fuel sample and the cloud point was measured again. The resulting data are given in the following table (temperatures in °C). **CLOUDING**

Fuel	1	2	3	4	5	6
Before additive	11.7	12.9	14.2	12.7	11.3	12.4
After additive	10.3	10.7	14.1	10.0	11.2	12.1

a. Assume normality, and conduct the appropriate hypothesis test to determine whether the additive lowers the mean cloud point in biodiesel fuel. Use $\alpha = 0.05$.

b. Conduct an inappropriate two-sample t test to compare the population mean cloud point before treatment with the population mean cloud point after treatment. Assume the population variances are unequal and use $\alpha = 0.05$.

c. Compare the conclusions in parts (a) and (b). How are the test statistics the same, and how do they differ?

10.92 Biology and Environmental Science   In August 2013, Vancouver Coastal Health warned swimmers that the coliform count at East False Creek was approximately twice the safe level.[15] The contamination was attributed to boats, birds, geese, and warm weather. A random sample of locations along the creek was obtained, and the coliform count (bacteria per 100 ml of water) was measured in early August and again following several rain storms. **COLIFORM**

a. What is the common characteristic that makes these data paired?

b. Use a one-sided paired t test to determine whether the rain caused a decrease in the coliform count. Assume normality and use $\alpha = 0.001$.

c. Find the safe level of coliform. Do you think it was safe to swim in the creek after the rain? Why or why not?

10.93 Travel and Transportation   For anyone planning to travel, either for vacation or on business, the pricing plans of airlines remain a guarded mystery. A study by CheapAir.com suggested that the cheapest fares are found 49 days before a flight. However, according to Travelers Today, the best prices are offered 21 days before a flight.[16] The Washington, D.C.-to-Los Angeles route was selected as a test case. A random sample of days was selected, and the best price for a one-way ticket was recorded 21 and 49 days prior to the flight. **AIRPRICE**

a. Assume normality. Is there any evidence to suggest that the price of a ticket on this route differs if purchased 21 or 49 days in advance? Use $\alpha = 0.05$.

b. Find bounds on the p value associated with this hypothesis test.

c. Find a 95% confidence interval for the difference in population mean cost per ticket. Does this confidence interval support your conclusion in part (a)? Why or why not?

10.4 Comparing Two Population Proportions Using Large Samples

The methods presented in this section can be used to compare two population proportions. For example, a social scientist may conduct an experiment to determine whether the true proportion of men who favor legalized gambling is the same as the true proportion of women. Or an advertising agency might be interested in comparing the true proportions of children who saw a certain television commercial in two different regions of the country.

Here is a quick review of the notation (presented at the beginning of this chapter) associated with populations 1 and 2 and samples 1 and 2.

Population proportion of successes:	p_1, p_2
Sample size:	n_1, n_2
Number of successes:	x_1, x_2
Corresponding random variables:	X_1, X_2
Sample proportion of successes:	$\hat{p}_1 = x_1/n_1$, $\hat{p}_2 = x_2/n_2$
Corresponding random variables:	$\hat{P}_1 = X_1/n_1$, $\hat{P}_2 = X_2/n_2$

The general null hypothesis is stated (as usual) in terms of a difference, $H_0: p_1 - p_2 = \Delta_0$. However, there are two cases to consider: (1) $\Delta_0 = 0$ and (2) $\Delta_0 \neq 0$. In both cases, a reasonable estimator for $p_1 - p_2$ is the difference between the sample proportions, $\hat{P}_1 - \hat{P}_2$. The following properties are used to construct a hypothesis test (and confidence interval) concerning the difference between two population proportions.

Properties of the Sampling Distribution of $\hat{P}_1 - \hat{P}_2$

1. The mean of $\hat{P}_1 - \hat{P}_2$ is the true difference between population proportions, $p_1 - p_2$. That is,

$$E(\hat{P}_1 - \hat{P}_2) = \mu_{\hat{P}_1 - \hat{P}_2} = p_1 - p_2$$

2. The variance of $\hat{P}_1 - \hat{P}_2$ is

$$\text{Var}(\hat{P}_1 - \hat{P}_2) = \sigma^2_{\hat{P}_1 - \hat{P}_2} = \frac{p_1(1 - p_1)}{n_1} + \frac{p_2(1 - p_2)}{n_2}$$

The standard deviation of $\hat{P}_1 - \hat{P}_2$ is

$$\sigma_{\hat{P}_1 - \hat{P}_2} = \sqrt{\frac{p_1(1 - p_1)}{n_1} + \frac{p_2(1 - p_2)}{n_2}}$$

3. If

(a) both n_1 and n_2 are large,

(b) $n_1 p_1 \geq 5$ and $n_1(1 - p_1) \geq 5$, and

Items (b) and (c) are the nonskewness criterion.

(c) $n_2 p_2 \geq 5$ and $n_2(1 - p_2) \geq 5$,

then the distribution of $\hat{P}_1 - \hat{P}_2$ is approximately normal.

In symbols: $\hat{P}_1 - \hat{P}_2 \overset{\bullet}{\sim} N\left[p_1 - p_2, \frac{p_1(1 - p_1)}{n_1} + \frac{p_2(1 - p_2)}{n_2}\right]$

The appropriate standardization will result in an approximate Z distribution. The estimate of the standard deviation, $\sigma_{\hat{P}_1 - \hat{P}_2}$, is determined by the value of Δ_0.

Case 1: $H_0: p_1 - p_2 = 0,$ or $p_1 = p_2 (\Delta_0 = 0)$

If this null hypothesis is true, there is one common value for the two population proportions, denoted $p\ (= p_1 = p_2)$. The variance of $\hat{P}_1 - \hat{P}_2$ becomes

$$\sigma^2_{\hat{P}_1 - \hat{P}_2} = \frac{p(1-p)}{n_1} + \frac{p(1-p)}{n_2} = p(1-p)\left(\frac{1}{n_1} + \frac{1}{n_2}\right) \tag{10.7}$$

Using the properties of $\hat{P}_1 - \hat{P}_2$, the random variable

$$Z = \frac{(\hat{P}_1 - \hat{P}_2) - 0}{\sqrt{p(1-p)\left(\dfrac{1}{n_1} + \dfrac{1}{n_2}\right)}} \tag{10.8}$$

is approximately standard normal. As for the common variance in Section 10.2, an estimator for the *common* proportion, p, is obtained by using information from both samples. The pooled or **combined estimate of the common population proportion** is

$\hat{P}_c$ is another weighted average.

$$\hat{P}_c = \frac{X_1 + X_2}{n_1 + n_2} = \left(\frac{n_1}{n_1 + n_2}\right)\hat{P}_1 + \left(\frac{n_2}{n_1 + n_2}\right)\hat{P}_2 \tag{10.9}$$

The general hypothesis test procedure is based on the standardization in Equation 10.8 with $\hat{p}_c$ as an estimate of p.

Hypothesis Tests Concerning Two Population Proportions When $\Delta_0 = 0$

Given two random samples of sizes n_1 and n_2, a large-sample hypothesis test concerning two population proportions in terms of the difference $p_1 - p_2$ (with $\Delta_0 = 0$) with significance level α has the form

This is the template for a large-sample hypothesis test concerning two population proportions when $\Delta_0 = 0$.

$H_0: p_1 - p_2 = 0$

$H_a: p_1 - p_2 > 0,$ $p_1 - p_2 < 0,$ or $p_1 - p_2 \neq 0$

$$\text{TS: } Z = \frac{\hat{P}_1 - \hat{P}_2}{\sqrt{\hat{P}_c(1 - \hat{P}_c)\left(\dfrac{1}{n_1} + \dfrac{1}{n_2}\right)}}$$

Again, use only one (appropriate) alternative hypothesis and the corresponding rejection region.

RR: $Z \geq z_\alpha,$ $Z \leq -z_\alpha,$ or $|Z| \geq z_{\alpha/2}$

There is no confidence interval for the difference in population proportions in this case. If we assume $p_1 = p_2$, then there is no reason to construct a confidence interval for the difference $p_1 - p_2 = 0$.

A CLOSER LOOK

1. This test is valid as long as the nonskewness criterion holds for both samples. Use the estimates $\hat{p}_1$ and $\hat{p}_2$ to check the inequalities.

2. Just as a reminder, the z critical values for this test are from the standard normal distribution.

3. Remember, we can also determine whether to reject or not to reject the null hypothesis by comparing the p value associated with the value of the test statistic to the significance level α.

The following example illustrates this hypothesis test procedure.

Solution Trail 10.11

KEYWORDS

- Is there any evidence?
- Greater than the true proportion
- Random sample

TRANSLATION

- Conduct a one-sided hypothesis test about $p_1 - p_2$
- $\Delta_0 = 0$.

CONCEPTS

- Large-sample hypothesis test concerning two population proportions when $\Delta_0 = 0$

VISION

Check the large-sample assumptions. Use the template for a one-sided, right-tailed test concerning $p_1 - p_2$ when $\Delta_0 = 0$. Use $\alpha = 0.05$ to find the critical value, compute the value of the test statistic, and draw a conclusion.

Example 10.11 Living with Parents

A high percentage of the millennial generation, adults approximately 18 to 31 years old, are living in their parents' homes. This may be due to unemployment, lower marriage rates, and/or the recession.[17] In a random sample of 275 male millennials, 110 lived with their parents, and in a random sample of 300 female millennials, 96 lived with their parents. Is there any evidence to suggest that the true proportion of male millennials living with their parents is greater than the true proportion of female millennials living with their parents? Use $\alpha = 0.05$.

SOLUTION

STEP 1 This is a one-sided test in which we are looking for evidence that a greater proportion of males than females are living with their parents. Therefore, $\Delta_0 = 0$, and case 1 is appropriate. Arbitrarily, let male millennials be population 1 and female millennials be population 2. The given information:

	Males	Females
Sample size	$n_1 = 275$	$n_2 = 300$
Number of successes	$x_1 = 110$	$x_2 = 96$
Sample proportion	$\hat{p}_1 = 110/275 = 0.40$	$\hat{p}_2 = 96/300 = 0.32$

STEP 2 Check the nonskewness criterion using estimates for p_1 and p_2.

$$n_1\hat{p}_1 = (275)(0.40) = 110 \geq 5 \qquad n_1(1 - \hat{p}_1) = (275)(0.60) = 165 \geq 5$$
$$n_2\hat{p}_2 = (300)(0.32) = 96 \geq 5 \qquad n_2(1 - \hat{p}_2) = (300)(0.68) = 204 \geq 5$$

All of the inequalities are satisfied, so $\hat{P}_1 - \hat{P}_2$ is approximately normal, and the large-sample hypothesis test concerning population proportions can be used.

STEP 3 The four parts of the hypothesis test are

$$H_0: p_1 - p_2 = 0$$
$$H_a: p_1 - p_2 > 0$$

$$\text{TS: } Z = \frac{\hat{P}_1 - \hat{P}_2}{\sqrt{\hat{P}_c(1 - \hat{P}_c)\left(\dfrac{1}{n_1} + \dfrac{1}{n_2}\right)}}$$

RR: $Z \geq z_\alpha = z_{0.05} = 1.6449$

STEP 4 The estimate of the common population proportion is

$$\hat{p}_c = \frac{x_1 + x_2}{n_1 + n_2} = \frac{110 + 96}{275 + 300} = 0.3583$$

The value of the test statistic is

$$z = \frac{\hat{p}_1 - \hat{p}_2}{\sqrt{\hat{p}_c(1 - \hat{p}_c)\left(\dfrac{1}{n_1} + \dfrac{1}{n_2}\right)}} = \frac{0.40 - 0.32}{\sqrt{(0.3583)(0.6417)\left(\dfrac{1}{275} + \dfrac{1}{300}\right)}} = 1.9985$$

STEP 5 The value of the test statistic lies in the rejection region. Equivalency, $p = 0.0228 \leq 0.05$, as illustrated in Figure 10.45. We reject the null hypothesis at the $\alpha = 0.05$ significance level. There is evidence to suggest that the true proportion of male millennials living with their parents is greater than the true proportion of female millennials living with their parents.

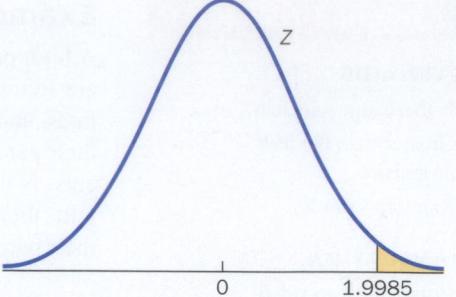

Figure 10.45 p-Value illustration:
$$p = P(Z \geq 1.9985)$$
$$= 0.0228 \leq 0.05 = \alpha$$

Figures 10.46 through 10.48 together show a technology solution.

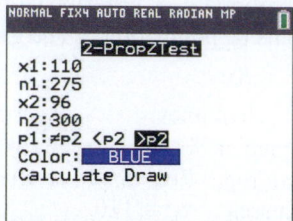

Figure 10.46 2-Prop ZTest input screen.

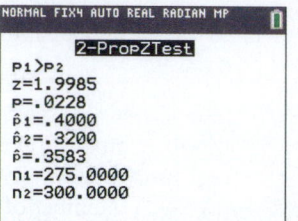

Figure 10.47 2-PropZ Test hypothesis test results.

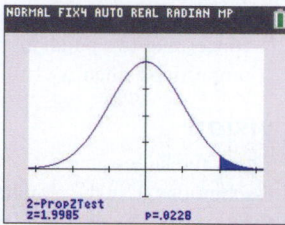

Figure 10.48 2-PropZ Test Draw results. ●

TRY IT NOW   GO TO EXERCISE 10.106

Case 2: $H_0\!: p_1 - p_2 = \Delta_0 \neq 0$

This case, with $\Delta_0 \neq 0$, is less common. Because p_1 and p_2 are assumed unequal, there is no hypothesized, *common* population proportion. The hypothesis test follows routinely from the properties of $\widehat{P}_1 - \widehat{P}_2$.

Hypothesis Tests Concerning Two Population Proportions When $\Delta_0 \neq 0$

Given two random samples of sizes n_1 and n_2, a large-sample hypothesis test concerning two population proportions in terms of the difference $p_1 - p_2$ (with $\Delta_0 \neq 0$) with significance level α has the form

This is the template for a large-sample hypothesis test concerning two population proportions when $\Delta_0 \neq 0$. The nonskewness criterion must also be met.

$H_0\!: p_1 - p_2 = \Delta_0$

$H_a\!: p_1 - p_2 > \Delta_0, \qquad p_1 - p_2 < \Delta_0, \qquad$ or $\qquad p_1 - p_2 \neq \Delta_0$

$$\text{TS: } Z = \frac{(\widehat{P}_1 - \widehat{P}_2) - \Delta_0}{\sqrt{\dfrac{\widehat{P}_1(1 - \widehat{P}_1)}{n_1} + \dfrac{\widehat{P}_2(1 - \widehat{P}_2)}{n_2}}}$$

RR: $Z \geq z_\alpha, \quad Z \leq -z_\alpha, \quad$ or $\quad |Z| \geq z_{\alpha/2}$

Suppose two random samples of sizes n_1 and n_2 are obtained, and the nonskewness criterion is satisfied. Using the properties of $\widehat{P}_1 - \widehat{P}_2$, a confidence interval for $p_1 - p_2$ can be derived (p_1 and p_2 are assumed unequal). Start with a symmetric interval about 0 such that the probability Z lies in this interval is $1 - \alpha$. As usual, manipulate the inequality to sandwich the parameter $p_1 - p_2$.

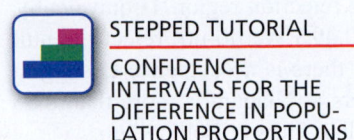

STEPPED TUTORIAL

CONFIDENCE INTERVALS FOR THE DIFFERENCE IN POPULATION PROPORTIONS

How to Find a $100(1 - \alpha)$% Confidence Interval for $p_1 - p_2$

Given two (large) random samples of sizes n_1 and n_2, a $100(1 - \alpha)$% confidence interval for $p_1 - p_2$ has as endpoints the values

$$(\hat{p}_1 - \hat{p}_2) \pm z_{\alpha/2} \sqrt{\frac{\hat{p}_1(1 - \hat{p}_1)}{n_1} + \frac{\hat{p}_2(1 - \hat{p}_2)}{n_2}} \qquad (10.10)$$

Solution Trail 10.12

KEYWORDS

- Is there any evidence?
- True proportion is more than 0.05 greater than
- Random sample

TRANSLATION

- Conduct a one-sided hypothesis test about $p_1 - p_2$
- $\Delta_0 = 0.05$

CONCEPTS

- Large-sample hypothesis test concerning two population proportions when $\Delta_0 \neq 0$

VISION

Check the large-sample assumptions. Use the template for a one-sided test concerning $p_1 - p_2$ when $\Delta_0 \neq 0$. Use $\alpha = 0.01$ to find the critical value, compute the value of the test statistic, and draw a conclusion.

Example 10.12 Make Room for Canadian Fliers

More Canadians are using smaller U.S. airports near the border when they travel to cities in the United States. Reasons include a strong Canadian dollar and higher taxes and fees on air travel in Canada.[18] In a random sample of 500 fliers at Buffalo Niagara International Airport, 245 were Canadian, and in a random sample of 400 fliers at Bellingham (Washington) International Airport, 160 were Canadian.

a. Conduct the appropriate hypothesis test to determine whether there is evidence that the true proportion of Canadian fliers at Niagara is more than 0.05 greater than the true proportion of Canadian fliers at Bellingham. Use $\alpha = 0.01$.

b. Find the p value associated with this hypothesis test.

SOLUTION

STEP 1 Let Niagara fliers be population 1 and Bellingham fliers be population 2. We are looking for evidence that the difference $p_1 - p_2$ is greater than $0.05 = \Delta_0 \neq 0$. Therefore, case 2 is appropriate.

The given information is presented here.

	Niagara fliers	Bellingham fliers
Sample size	$n_1 = 500$	$n_2 = 400$
Number of successes	$x_1 = 245$	$x_2 = 160$
Sample proportion	$\hat{p}_1 = 245/500 = 0.49$	$\hat{p}_2 = 160/400 = 0.40$

Check the nonskewness criterion using estimates for p_1 and p_2.

$n_1\hat{p}_1 = (500)(0.49) = 245 \geq 5 \qquad n_1(1 - \hat{p}_1) = (500)(0.51) = 255 \geq 5$

$n_2\hat{p}_2 = (400)(0.40) = 160 \geq 5 \qquad n_2(1 - \hat{p}_2) = (400)(0.60) = 240 \geq 5$

All of the inequalities are satisfied, so $\hat{P}_1 - \hat{P}_2$ is approximately normal, and the large-sample hypothesis test concerning population proportions can be used.

STEP 2 The four parts of the hypothesis test are

$H_0: p_1 - p_2 = 0.05$

$H_a: p_1 - p_2 > 0.05$

$$\text{TS: } Z = \frac{(\hat{P}_1 - \hat{P}_2) - 0.05}{\sqrt{\dfrac{\hat{P}_1(1 - \hat{P}_1)}{n_1} + \dfrac{\hat{P}_2(1 - \hat{P}_2)}{n_2}}}$$

RR: $Z \geq z_\alpha = z_{0.01} = 2.3263$

STEP 3 The value of the test statistic is

$$z = \frac{(\hat{p}_1 - \hat{p}_2) - 0.05}{\sqrt{\dfrac{\hat{p}_1(1 - \hat{p}_1)}{n_1} + \dfrac{\hat{p}_2(1 - \hat{p}_2)}{n_2}}} = \frac{(0.49 - 0.40) - 0.05}{\sqrt{\dfrac{(0.49)(0.51)}{500} + \dfrac{(0.40)(0.60)}{400}}} = 1.2062$$

The value of the test statistic does *not* lie in the rejection region. (Equivalently, $p = 0.1139 > 0.05$, as illustrated in Figure 10.49.) We do not reject the null hypothesis. At the $\alpha = 0.01$ significance level, there is no evidence to suggest that the population proportion of Canadian fliers at Niagara is more than 0.05 greater than the population proportion of Canadian fliers at Bellingham.

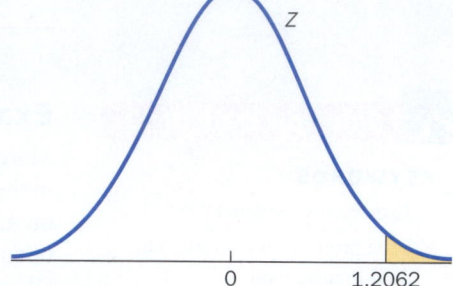

Figure 10.49 *p*-Value illustration:
$p = P(Z \geq 1.2062)$
$= 0.1139 > 0.05 = \alpha$

STEP 4 This is a one-sided, right-tailed test, so the *p* value is a right-tail probability.

$p = P(Z \geq 1.2062)$ Definition of p value for H_a: $p_1 - p_2 > 0.05$.

$= 1 - P(Z \leq 1.2062)$ The complement rule.

$= 1 - 0.8861 = 0.1139 \, (> 0.05 = \alpha)$ Use Table III in the Appendix.

Figure 10.50 shows a technology solution.

Figure 10.50 The 2 Proportions hypothesis test results and (one-sided) confidence interval from Minitab.

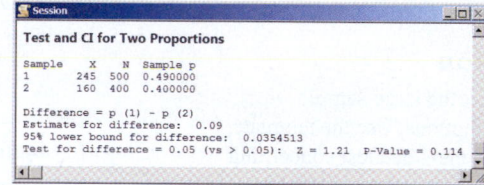

TRY IT NOW   GO TO EXERCISE 10.121

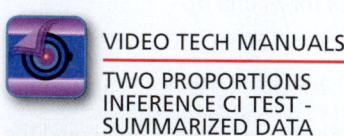

VIDEO TECH MANUALS

TWO PROPORTIONS
INFERENCE CI TEST –
SUMMARIZED DATA

Carolina K. Smith MD/Shutterstock

The computations for a confidence interval for $p_1 - p_2$ are illustrated in the next example.

Example 10.13 Storm Watch

The Weather Channel (TWC) is one of the most popular cable TV networks. Who hasn't seen Jim Cantore in the middle of some wild weather? However, the number of viewers varies greatly according to geographic region and the current weather conditions. Random samples of cable TV viewers in the Northeast (population 1) and in the West (population 2) were obtained. The number of viewers who watched TWC in the past week was recorded. The data are given in the following table.

	Northeast	West
Sample size	$n_1 = 1000$	$n_2 = 1500$
Number of successes	$x_1 = 446$	$x_2 = 303$
Sample proportion	$\hat{p}_1 = 446/1000 = 0.4460$	$\hat{p}_2 = 303/1500 = 0.2020$

Construct a 99% confidence interval for the true difference in proportions of cable TV viewers who watched TWC in the past week.

SOLUTION

STEP 1 The sample sizes, number of successes, and sample proportions are given. Check the nonskewness criterion using estimates for p_1 and p_2.

$$n_1 \hat{p}_1 = (1000)(0.4460) = 446 \geq 5 \qquad n_1(1 - \hat{p}_1) = (1000)(0.5540) = 540 \geq 5$$
$$n_2 \hat{p}_2 = (1500)(0.2020) = 303 \geq 5 \qquad n_2(1 - \hat{p}_2) = (1500)(0.7980) = 1197 \geq 5$$

All of the inequalities are satisfied, so the distribution of the difference in sample proportions is approximately normal. A large-sample confidence interval is appropriate.

STEP 2 Find the critical value.

$$1 - \alpha = 0.99 \quad \Rightarrow \quad \alpha = 0.01 \quad \Rightarrow \quad \alpha/2 = 0.005 \qquad \text{Find } \alpha/2.$$

$$z_{\alpha/2} = z_{0.005} = 2.5758 \qquad \text{Common critical value.}$$

STEP 3 Use Equation 10.10.

$$(\hat{p}_1 - \hat{p}_2) \pm z_{\alpha/2} \sqrt{\frac{\hat{p}_1(1 - \hat{p}_1)}{n_1} + \frac{\hat{p}_2(1 - \hat{p}_2)}{n_2}} \qquad \text{Equation 10.10.}$$

$$= (0.4460 - 0.2020) \pm (2.5758)\sqrt{\frac{(0.4460)(0.5540)}{1000} + \frac{(0.2020)(0.7980)}{1500}}$$

Use summary statistics and critical value.

$$= 0.2440 \pm 0.0485 \qquad \text{Simplify.}$$
$$= (0.1955, 0.2925) \qquad \text{Compute endpoints.}$$

(0.1955, 0.2925) is a 99% confidence interval for the difference in the proportion of cable TV viewers who watched TWC in the past week in the Northeast and in the West, $p_1 - p_2$. Note that because 0 is not included in this interval, there is evidence to suggest that the two proportions are different.

Figures 10.51 and 10.52 together show a technology solution.

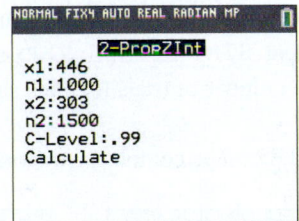

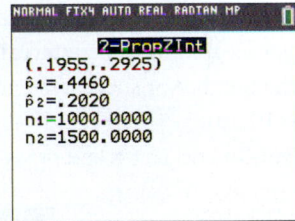

Figure 10.51 2-PropZ Int input screen.

Figure 10.52 Resulting confidence interval.

TRY IT NOW   GO TO EXERCISE 10.122

Technology Corner

Procedure: Hypothesis tests and confidence intervals concerning two population proportions.
Reconsider: Example 10.11, solution, and interpretations.

CrunchIt!

Use Proportion 2-sample to conduct a hypothesis test and concerning two population proportions, $\Delta_0 = 0$ or $\Delta_0 \neq 0$, and to construct a confidence interval for the difference in population proportions.

1. Select Statistics; Proportion; 2-sample.
2. Under the Summarized tab, enter n and the number of successes for each sample.
3. Under the Hypothesis Test tab, enter the Difference of proportions under the null hypothesis (Δ_0), and select the appropriate Alternative hypothesis. See Figure 10.53.
4. Click Calculate. The results are displayed in a separate window. See Figure 10.54.
5. To construct a confidence interval, under the Confidence Interval tab, enter the Confidence Interval Level, and click Calculate. See Figure 10.55.

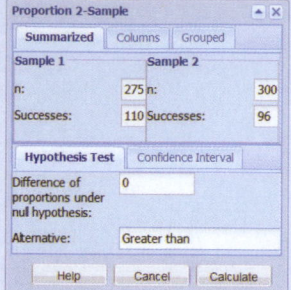

Figure 10.53 Proportion 2-Sample input screen.

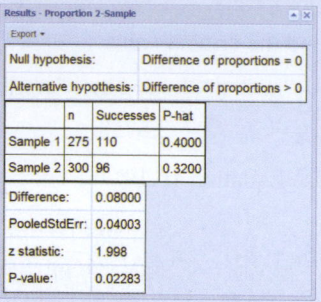

Figure 10.54 Proportion 2-Sample hypothesis test results.

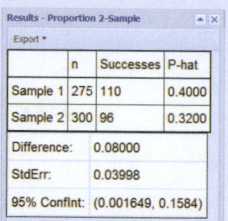

Figure 10.55 Proportion 2-Sample confidence interval.

TI-84 Plus C

Use 2-PropZTest to conduct a hypothesis test concerning two population proportions, $\Delta_0 = 0$, and 2-PropZInt to construct a confidence interval for the difference in population proportions. There is no built-in function to conduct a hypothesis test if $\Delta_0 \neq 0$.

1. Select STAT ; TESTS; 2-PropZTest.
2. Enter the number of successes and the number of trials for each sample: x_1, n_1, x_2, n_2. Highlight the appropriate alternative hypothesis. See Figure 10.46.
3. The Calculate and Draw results are shown in Figures 10.47 and 10.48.
4. To construct a confidence interval, select STAT ; TESTS; 2-PropZInt.
5. Enter the number of successes and the number of trials for each sample, x_1, n_1, x_2, n_2, and the confidence level (Figure 10.56).
6. Highlight Calculate and press ENTER . The confidence interval is displayed on the Home Screen. See Figure 10.57.

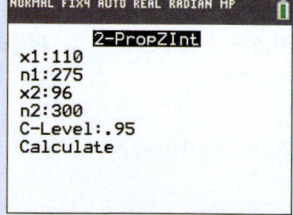

Figure 10.56 2-PropZ Int input screen.

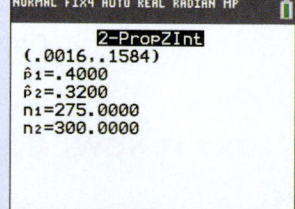

Figure 10.57 Resulting confidence interval.

Minitab

Use the function 2 Proportions to conduct a hypothesis test and to construct a confidence interval. Input is samples in one column, samples in different columns, or summarized data.

1. Select Stat; Basic Statistics; 2 Proportions.
2. Choose Summarized data and enter the number of successes (Events) and number of trials for each sample.

3. Choose the Options option button. Enter a Confidence level, the (hypothesized) Test difference (Δ_0), and select the appropriate Alternative hypothesis. If $\Delta_0 = 0$, use the Test method: Use the pooled estimate of the proportion. If $\Delta_0 \neq 0$, use Estimate the proportions separately.
4. The hypothesis test results and confidence interval are displayed in a session window. See Figure 10.58.

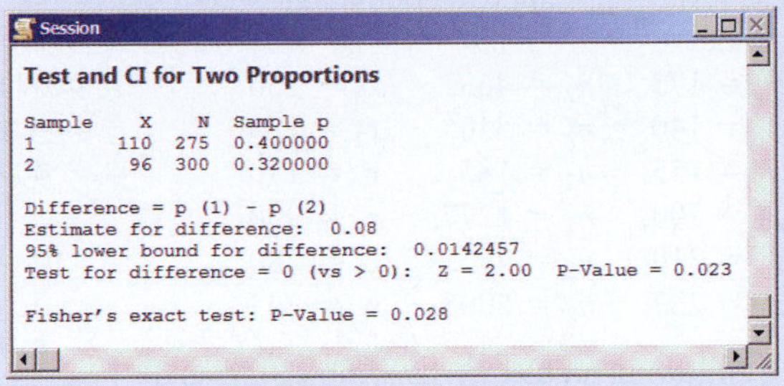

Figure 10.58 The 2 Proportions hypothesis test results and (one-sided) confidence interval from Minitab.

```
Session                                                    _ □ X

Test and CI for Two Proportions

Sample    X    N    Sample p
1       110  275   0.400000
2        96  300   0.320000

Difference = p (1) - p (2)
Estimate for difference:  0.08
95% lower bound for difference:  0.0142457
Test for difference = 0 (vs > 0):  Z = 2.00   P-Value = 0.023

Fisher's exact test: P-Value = 0.028
```

Excel

There are no built-in functions to conduct a hypothesis test concerning two population proportions or to construct a confidence interval for the difference in population proportions. However, functions associated with the standard normal distribution may be used to find critical values and p values. Use ordinary spreadsheet calculations where necessary.

1. Enter the number of successes and number of trials for each sample.
2. Compute $\hat{p}_1$, $\hat{p}_2$, and $\hat{p}_c$.
3. Compute the value of the test statistic, z, and use the function NORM.S.DIST to find the p value.
4. To construct the confidence interval, compute the difference, $\hat{p}_1 - \hat{p}_2$, use the function NORM.S.INV to find the critical value, and compute $\sqrt{\dfrac{\hat{p}_1(1 - \hat{p}_1)}{n_1} + \dfrac{\hat{p}_2(1 - \hat{p}_2)}{n_2}}$.
5. Find the left endpoint and the right endpoint of the confidence interval. See Figure 10.59.

Figure 10.59 Excel calculations to conduct a hypothesis test and construct a confidence interval concerning two population proportions.

	A	B	C	D
1	110	= x_1	96	= x_2
2	275	= n_1	300	= n_2
3	0.40	= A1/A2 = phat_1	0.32	= C1/C2 = phat_2
4	0.3583	= (A1+C1)/(A2+C2) = phat_c		
5	1.9985	= (A3-C3)/(SQRT(A4*(1-A4)*(1/A2+1/C2))) = z		
6	0.0228	= 1-NORM.S.DIST(A5,TRUE) = p value		
7				
8	0.08	= A3-C3 = phat_1 - phat_2		
9	1.9600	= NORM.S.INV(0.975) = critical value		
10	0.0400	= SQRT(A3*(1-A3)/A2+C3*(1-C3)/C2)		
11	0.0016	= A8-A9*A10 = left endpoint		
12	0.1584	= A8+A9*A10 = right endpoint		

SECTION 10.4 EXERCISES

Concept Check

10.94 Short Answer State the nonskewness criterion.

10.95 True/False The estimate of $\sigma_{\hat{p}_1 - \hat{p}_2}$ is determined by the value of Δ_0.

10.96 True/False The pooled estimate of the common population proportion is based on the largest sample.

10.97 True/False If p_1 and p_2 are assumed unequal, there is no hypothesized common population proportion.

10.98 True/False In a hypothesis test concerning two population proportions, Δ_0 can be negative.

10.99 True/False A confidence interval for $p_1 - p_2$ can be used to draw a conclusion about the equality of the two population proportions.

Practice

10.100 In each of the following problems, the size and the number of individuals or objects with a certain characteristic are given for samples from two populations. Check the two nonskewness inequalities for both samples, and determine whether a large-sample test concerning two population proportions is appropriate.

  a. $n_1 = 303$, $x_1 = 175$, $n_2 = 463$, $x_2 = 250$
  b. $n_1 = 560$, $x_1 = 140$, $n_2 = 530$, $x_2 = 125$
  c. $n_1 = 160$, $x_1 = 155$, $n_2 = 185$, $x_2 = 170$
  d. $n_1 = 1020$, $x_1 = 700$, $n_2 = 1277$, $x_2 = 950$
  e. $n_1 = 842$, $x_1 = 319$, $n_2 = 755$, $x_2 = 280$
  f. $n_1 = 4375$, $x_1 = 237$, $n_2 = 5005$, $x_2 = 245$

10.101 In each of the following problems, the sample sizes and population proportions are given. Find the mean, variance, and standard deviation of the estimator $\hat{P}_1 - \hat{P}_2$, and compute each probability.

  a. $n_1 = 645$, $p_1 = 0.24$, $n_2 = 650$, $p_2 = 0.26$,
    $P(\hat{P}_1 - \hat{P}_2 \geq 0.045)$
  b. $n_1 = 250$, $p_1 = 0.37$, $n_2 = 270$, $p_2 = 0.33$,
    $P(\hat{P}_1 - \hat{P}_2 \leq -0.04)$
  c. $n_1 = 144$, $p_1 = 0.87$, $n_2 = 156$, $p_2 = 0.86$,
    $P(-0.05 < \hat{P}_1 - \hat{P}_2 < 0.05)$
  d. $n_1 = 520$, $p_1 = 0.65$, $n_2 = 480$, $p_2 = 0.72$,
    $P(\hat{P}_1 - \hat{P}_2 > -0.10)$
  e. $n_1 = 1200$, $p_1 = 0.73$, $n_2 = 1150$, $p_2 = 0.85$,
    $P(\hat{P}_1 - \hat{P}_2 < -0.06)$
  f. $n_1 = 500$, $p_1 = 0.645$, $n_2 = 525$, $p_2 = 0.604$,
    $P(-0.02 \leq \hat{P}_1 - \hat{P}_2 \leq 0.10)$

10.102 A hypothesis test concerning two population proportions is described in each of the following problems. Identify each population, and determine the appropriate null and alternative hypotheses in terms of p_1 and p_2.

  a. A study was conducted to determine whether there is any difference in the proportion of people who listen to satellite radio in California versus Tennessee.
  b. Two random samples, of men and of women who tried to talk their way out of a traffic ticket, were obtained. The number of each who said they missed a street sign was recorded. The data will be used to determine whether the proportion of women who say they missed a street sign is greater than the proportion of men who say so.
  c. Random samples of teens ages 13–19 from two different school districts were obtained. The number of teens who own a cell phone was recorded. The data will be used to determine whether there is any difference in the proportion of teens who own a cell phone in the two districts.
  d. Random samples of Americans who received an income tax refund were obtained. All were classified by income level (low versus high) and asked whether they intended to use their refund to pay outstanding bills. The data will be used to determine whether the proportion of high-income Americans who pay bills with tax refunds is 0.10 greater than the proportion of low-income Americans who pay bills with tax refunds.

  e. Independent random samples of videos on U.S. and Canadian newspaper websites were obtained. The number of videos that were advertisements was recorded. The data will be used to determine whether the proportion of videos that are advertisements in the United States is higher than in Canada.

10.103 In each of the following problems, use the given data to conduct the appropriate hypothesis test concerning two population proportions, find the p value, and state your conclusion.

  a. $n_1 = 500$, $x_1 = 400$, $n_2 = 525$, $x_2 = 405$,
    $H_0: p_1 - p_2 = 0$, $H_a: p_1 - p_2 > 0$, $\alpha = 0.05$
  b. $n_1 = 646$, $x_1 = 280$, $n_2 = 680$, $x_2 = 330$,
    $H_0: p_1 - p_2 = 0$, $H_a: p_1 - p_2 < 0$, $\alpha = 0.01$
  c. $n_1 = 255$, $x_1 = 81$, $n_2 = 266$, $x_2 = 110$,
    $H_0: p_1 - p_2 = 0$, $H_a: p_1 - p_2 \neq 0$, $\alpha = 0.025$
  d. $n_1 = 1440$, $x_1 = 907$, $n_2 = 1562$, $x_2 = 970$,
    $H_0: p_1 - p_2 = 0$, $H_a: p_1 - p_2 \neq 0$, $\alpha = 0.001$

10.104 In each of the following problems, use the given data to conduct the appropriate hypothesis test concerning two population proportions, find the p value, and state your conclusion.

  a. $n_1 = 200$, $x_1 = 100$, $n_2 = 300$, $x_2 = 165$,
    $H_0: p_1 - p_2 = 0.05$, $H_a: p_1 - p_2 < 0.05$, $\alpha = 0.01$
  b. $n_1 = 480$, $x_1 = 384$, $n_2 = 490$, $x_2 = 367$,
    $H_0: p_1 - p_2 = 0.02$, $H_a: p_1 - p_2 > 0.02$, $\alpha = 0.05$
  c. $n_1 = 610$, $x_1 = 450$, $n_2 = 675$, $x_2 = 470$,
    $H_0: p_1 - p_2 = 0.10$, $H_a: p_1 - p_2 \neq 0.10$, $\alpha = 0.01$
  d. $n_1 = 2500$, $x_1 = 710$, $n_2 = 3100$, $x_2 = 770$,
    $H_0: p_1 - p_2 = 0.07$, $H_a: p_1 - p_2 \neq 0.07$, $\alpha = 0.001$

10.105 In each of the following problems, use the given data and confidence level to construct a confidence interval for the difference of two population proportions, $p_1 - p_2$.

  a. $n_1 = 388$, $x_1 = 230$, $n_2 = 402$, $x_2 = 250$,   95%
  b. $n_1 = 528$, $x_1 = 475$, $n_2 = 530$, $x_2 = 497$,   95%
  c. $n_1 = 180$, $x_1 = 92$, $n_2 = 194$, $x_2 = 100$,   99%
  d. $n_1 = 2300$, $x_1 = 1705$, $n_2 = 2404$, $x_2 = 1690$,   90%

Applications

10.106 Public Health and Nutrition Over the last decade, many Americans have been able to stop smoking. However, a recent survey suggests that asthmatic children are more likely to be exposed to second-hand smoke than children without asthma.[19] In a random sample of 300 children without asthma, 132 were regularly exposed to second-hand smoke, and in a random sample of 325 children with asthma, 177 were regularly exposed to second-hand smoke. Is there any evidence to suggest that the proportion of children with asthma who are exposed to second-hand smoke is greater than the proportion for children without asthma? Use $\alpha = 0.01$.

10.107 Public Policy and Political Science A recent survey suggested that more than one-quarter of registered U.S. voters believe an armed revolution might be necessary to protect our liberties.[20] In a random sample of 250 voters in Western states, 73 indicated that an armed revolution might be necessary, and in a random sample of 275 voters in Eastern states, 85 said that an armed revolution might be necessary. Is there any evidence to suggest that the proportion of voters in the West who believe that an armed revolution might be necessary is different from the proportion of voters in the East? Use $\alpha = 0.05$. Write a Solution Trail for this problem.

10.108 **Technology and the Internet** In its Millennium Development Goals Report, the United Nations suggested that by the end of 2013 there will be 6.8 billion cell-phone subscription plans, and approximately 2.7 billion people will be connected to the Internet.[21] A random sample of households in Brazil and Russia was obtained, and the number of people connected to the Internet was recorded for each. The data are given in the following table.

Country	Sample size	Number connected to the Internet
Brazil	326	161
Russia	387	210

Is there any evidence to suggest that the proportion of households connected to the Internet is greater in Russia than in Brazil? Use $\alpha = 0.01$.

10.109 **Marketing and Consumer Behavior** Many critics have been known to say, "They sure don't make movies like they used to." To assess Americans' opinions of movies, a random sample of people was obtained and each was asked about the quality of movies. The data are given in the following table.

Age group	Sample size	Number who said movies are getting better
18–29	347	238
30–49	387	221

a. Conduct the appropriate hypothesis test to determine whether there is any evidence to suggest that the true proportion of 18–29-year-olds who believe movies are getting better is greater than the proportion of 30–49-year-olds. Use $\alpha = 0.001$.

b. Find the p value associated with the hypothesis test in part (a).

10.110 **Public Health and Nutrition** Several years ago, most doctors believed that it was not necessary to take any dietary supplement. Now, because many Americans do not eat a healthy, balanced diet, many physicians recommend a once-a-day multivitamin. A random sample of people was obtained and asked whether they regularly take a multivitamin. The data are given in the following table.

Group	Sample size	Number who take a multivitamin
Men	490	181
Women	428	214

Is there any evidence that the proportion of women who take a multivitamin is greater than the proportion of men? Use $\alpha = 0.005$. Write a Solution Trail for this problem.

10.111 **Travel and Transportation** Many people who commute to work by car in New York City every day use either the George Washington Bridge or the Lincoln Tunnel. A random sample of commuters who use one of these two routes was obtained, and each was asked whether they carpooled to work. The data are given in the following table.

Commuting route	Sample size	Number who carpool
Bridge	1055	530
Tunnel	1663	825

a. Verify that the nonskewness criterion inequalities are satisfied.

b. Is there any evidence to suggest that the proportion of carpoolers crossing the George Washington Bridge is greater than the proportion of carpoolers using the Lincoln Tunnel? Use $\alpha = 0.01$.

c. Find the p value associated with the hypothesis test in part (b).

10.112 **Conspiracy Theory** In a recent survey, Americans were asked about 20 popular conspiracy theories. Thirty-seven percent of voters believe global warming is a hoax, 21% are certain a UFO crashed in Roswell, New Mexico, and 7% believe the moon landings were faked.[22] (You might consider investigating the theory about lizard people who control our society.) The survey results were often very different according to political affiliation. A random sample of Democrats and Republicans were asked if they believe pharmaceutical companies invent new diseases to make money. The resulting data are given in the following table.

Political affiliation	Sample size	Number who believe pharmaceutical companies invent diseases
Democrats	788	137
Republicans	866	105

Is there any evidence to suggest that the proportion of voters who believe pharmaceutical companies invent diseases to make money is different for Democrats and Republicans? Use $\alpha = 0.01$.

10.113 **Medicine and Clinical Studies** According to the National Institute of Allergy and Infectious Diseases, approximately 54.6% of all U.S. citizens test positive to one or more allergens. Between 9% and 16% suffer from hay fever. A random sample of people who suffer from hay fever was obtained, and each was treated with either a conventional antihistamine or butterbur extract. The number of subjects who experienced relief from hay fever was recorded for each group. The resulting data are given in the following table.

Treatment	Sample size	Number who experienced relief
Antihistamine	255	71
Butterbur extract	237	55

a. Compute the sample proportion of people who experienced relief for each treatment.

b. Conduct the appropriate hypothesis test to determine whether the proportion of people who experience relief due to the antihistamine is different from the proportion of people who experience relief from butterbur extract. Use $\alpha = 0.01$.

10.114 Public Health and Nutrition   A survey was conducted concerning physical activity of adults in two states. Random samples of adults were obtained from Arizona and from West Virginia, and they were all asked whether they consider themselves physically inactive. The data are given in the following table.

State	Sample size	Number who are physically inactive
Arizona	1122	163
West Virginia	1181	205

Is there any evidence to suggest that the proportion of adults who consider themselves physically inactive is greater in West Virginia than in Arizona? Use $\alpha = 0.001$ and find the p value.

10.115 Manufacturing and Product Development   Blenko Specs has two different processes for the manufacture of optical lenses supplied to the military. A random sample of finished lenses was obtained from each process, and each lens was carefully inspected for defects. Of the 106 lenses from Process A, eight were defective, and 12 of the 121 lenses from Process B were defective.

a. Compute the sample proportion of defectives for each process.

b. Check the nonskewness criterion and verify that the inequalities are satisfied.

c. Conduct a hypothesis test to determine whether there is any evidence that the sample proportion of defective lenses is different for Process A and Process B. Use $\alpha = 0.05$.

10.116 Sports and Leisure   A major league sports franchise can contribute a great deal to the local economy and unite an entire region. In a recent survey, a random sample of adults in the Portland, Oregon, area were asked if they would support a National Football League team. The data are given in the following table.[23]

County	Sample size	Number who would support an NFL team
Clackamas	469	117
Multnomah	1985	337

Is there any evidence to suggest that the proportion of residents who would support an NFL team is different in the two counties? Use $\alpha = 0.01$.

10.117 Psychology and Human Behavior   In a recent survey, residents of Reston, Virginia, were asked about their quality of life, characteristics of the community, child care, and crime.

Respondents were asked to rate how safe or unsafe they felt in their neighborhood. The data are given in the following table.[24]

District	Sample size	Number who felt safe in their neighborhood after dark
Hunter Woods	213	141
Lake Anne	218	155

Is there any evidence to suggest that the true proportion of residents who feel safe after dark is different in the two districts? Use $\alpha = 0.05$.

10.118 Technology and the Internet   The percentage of Canadians who use technology is very high, but a recent survey suggests that they greatly overrate their tech savviness. Approximately 60% of Canadians rated themselves as B or better for tech savviness, but a large proportion of respondents could not explain roaming, data usage, or online security.[25] A random sample of Canadians from two regions was obtained and asked to rate their tech savviness. The results are given in the following table.

Region	Sample size	Number who rated their tech savviness B or better
Edmonton	566	345
Thunder Bay	617	330

Is there any evidence to suggest that the true proportion of Canadians who rate their tech savviness as B or better is greater in Edmonton than in Thunder Bay? Use $\alpha = 0.01$.

Extended Applications

10.119 Marketing and Consumer Behavior   Americans have many sources for daily news, for example, local television shows, public radio, or national newspapers. A random sample of Americans was obtained and classified by age. Each person was asked whether he or she obtained news every day from three specific sources. The data are given in the following table.

News source	Age group 18- to 29-year-olds Sample size	Number who obtained news every day	30- to 49-year-olds Sample size	Number who obtained news every day
Nightly network news	570	103	462	120
Cable news networks	450	108	520	182
Internet	546	197	568	239

a. Conduct the appropriate hypothesis test to determine whether there is evidence that the true proportion of

18- to 29-year-olds who obtain news every day from nightly network news shows is less than the true proportion of 30- to 49-year-olds who obtain news every day from nightly network news shows. Use $\alpha = 0.05$. Find the p value associated with this hypothesis test.

b. Conduct the appropriate hypothesis test to determine whether there is evidence that the true proportion of 18- to 29-year-olds who obtain news every day from cable news networks is less than the true proportion of 30- to 49-year-olds who obtain news every day from cable news networks shows. Use $\alpha = 0.01$. Find the p value associated with this hypothesis test.

c. Conduct the appropriate hypothesis test to determine whether there is evidence that the true proportion of 18- to 29-year-olds who obtain news every day from the Internet is different from the true proportion of 30- to 49-year-olds who obtain news every day from the Internet. Use $\alpha = 0.005$. Find the p value associated with this hypothesis test.

10.120 Manufacturing and Product Development Two different machines in a manufacturing facility are designed to fill cans with 280 grams of Tang orange-flavored drink mix. A random sample of filled cans from each machine was obtained, and each can was carefully weighed. Of the 134 cans from machine A, 10 were underfilled, and 7 of 114 cans from machine B were underfilled.

a. Compute the sample proportion of underfilled cans for each machine.
b. Verify the nonskewness criterion.
c. Find a 95% confidence interval for the true difference in the proportion of underfilled cans for machines A and B.
d. Using the confidence interval in part (c), is there any evidence to suggest that the proportion of underfilled cans is different for the two machines? Justify your answer.

10.121 Psychology and Human Behavior Historically, the three most popular home-improvement projects are interior decorating, landscaping, and expansion (respectively). A random sample of homeowners and condominium owners was

obtained asked whether they planned any landscaping within the next year. The data are given in the following table.

Residence	Sample size	Number who are planning to landscape
Homeowner	261	90
Condominium owner	303	65

Is there any evidence to suggest that the proportion of homeowners planning a landscaping project is more than 0.10 greater than the proportion of condominium owners planning a landscaping project? Use $\alpha = 0.01$.

10.122 Medicine and Clinical Studies Young children usually get 5–10 colds each year. To ease cold symptoms, for example, a runny nose or sore throat, some parents give their children over-the-counter cough and cold medicines. However, many of these medicines are not effective and can cause serious side effects in young children. A random sample of parents of young children was obtained and asked whether they give their children cough or cold medicine. The data are given in the following table.

Parent	Sample size	Number who give cold medicine
Male	376	142
Female	428	183

a. Is there any evidence to suggest that the true population proportion of males who give their children cough medicine is different from the true proportion of females? Use $\alpha = 0.05$.
b. Find the p value associated with this hypothesis test.
c. Find a 95% confidence interval for the difference in the proportion of males who give their children cold medicine and the proportion of females who give their children cold medicine. Does this confidence interval support your conclusion in part (a)? Why or why not?
d. What must be true of the respondents in order for the hypothesis test in part (a) to be valid?

10.5 Comparing Two Population Variances or Standard Deviations

Many practical business decisions are based on a comparison of variability, or manufacturing precision. For example, a company that produces a certain drug via fermentation would like to maintain a very small variability in yield. One fermentation process may be more reliable and less variable than another. A hardware store wants very little variability in paint color from gallon to gallon. One paint mixer may be more precise (less variable) than another. Even food manufacturers strive for small differences in product taste from one batch to the next. And we may compare population variances to decide which two-sample t test is appropriate, the pooled test or the approximate test.

S_1^2 and S_2^2 are good (unbiased) estimators for the population variances σ_1^2 and σ_2^2, respectively. However, a hypothesis test for comparing σ_1^2 and σ_2^2 is based on a new standardization and an F distribution, introduced below.

An F distribution has positive probability only for non-negative values. The probability density function for an F random variable is 0 for $x < 0$. Once again, it is important to focus on the properties of an F distribution and the method for finding critical values associated with this distribution.

Properties of an F Distribution

The numerator and denominator designations will make more sense as you read on.

1. An F distribution is completely determined by two parameters, the number of degrees of freedom in the numerator and the number of degrees of freedom in the denominator, given in that order. Both values must be positive integers (1, 2, 3, . . .) and there is, of course, a different F distribution for every combination.

2. If X has an F distribution with v_1 and v_2 degrees of freedom, $(X \sim F_{v_1,v_2})$, then

Why are these restrictions on v_2 necessary? What do you suppose the mean is if $v_2 = 2$?

$$\mu_X = \frac{v_2}{v_2 - 2},\ v_2 \geq 3 \quad \text{and} \quad \sigma_X^2 = \frac{2v_2^2(v_1 + v_2 - 2)}{v_1(v_2 - 2)^2(v_2 - 4)},\ v_2 \geq 5 \quad (10.11)$$

3. Suppose $X \sim F_{v_1,v_2}$. The density curve for X is positively skewed (*not* symmetric), and gets closer and closer to the x axis but never touches it. As both degrees of freedom increase, the density curve becomes taller and more compact. See Figure 10.60.

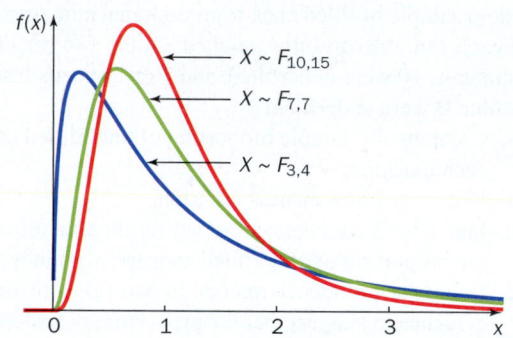

$X \sim F_{10,15}$
$X \sim F_{7,7}$
$X \sim F_{3,4}$

Figure 10.60 Density curves for several F distributions.

The definition and notation for an F *critical value* are analogous to those for Z, t, and χ^2 critical values.

Definition

F_{α,v_1,v_2} is a critical value related to an F distribution with v_1 and v_2 degrees of freedom. If $X \sim F_{v_1,v_2}$, then $P(X \geq F_{\alpha,v_1,v_2}) = \alpha$.

Yes, there are three subscripts, but don't panic. The notation looks more complicated than it is. F_{α,v_1,v_2} is simply consistent, concise notation to represent a critical value related to an F distribution.

A CLOSER LOOK

1. F_{α,v_1,v_2} is a value on the measurement axis in an F world such that there is α of the area (probability) to the right of F_{α,v_1,v_2}. Remember, as for a chi-square distribution, there is no symmetry in F critical values.

2. Critical values are defined in terms of right-tail probability, and the F distribution is not symmetric, so the notation here is similar to that for chi-square critical values. It will be necessary to find critical values denoted $F_{1-\alpha,v_1,v_2}$, where $1 - \alpha$ is large. By definition, $P(X \geq F_{1-\alpha,v_1,v_2}) = 1 - \alpha$, and by the complement rule, $P(X \leq F_{1-\alpha,v_1,v_2}) = \alpha$. See Figure 10.61.

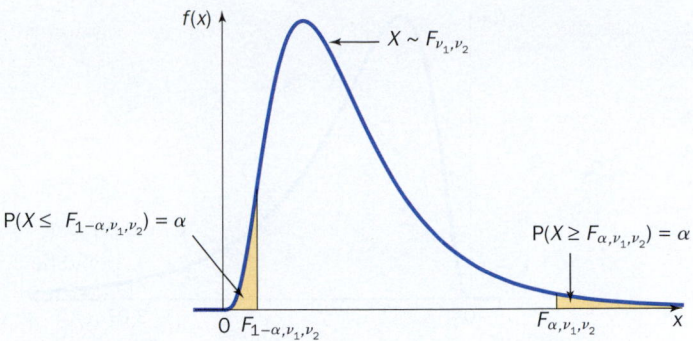

Figure 10.61 An illustration of F distribution critical values.

Notice how the degrees of freedom switch positions.

3. F critical values are related according to the following equation:

$$F_{1-\alpha,v_1,v_2} = \frac{1}{F_{\alpha,v_2,v_1}}.$$ (10.12)

Table VII in the Appendix presents selected critical values associated with various F distributions and right-tail probabilities. The degrees of freedom in the numerator are given in the top row and the degrees of freedom in the denominator are given in the left column. In the body of the table, F_{α,v_1,v_2} is at the intersection of the appropriate row and column. Left-tail probabilities are found using Equation 10.12. The following example illustrates the use of Table VII in the Appendix for finding critical values associated with an F distribution.

Example 10.14 Critical Value Look-ups

Find each critical value: **a.** $F_{0.05,8,10}$ and   **b.** $F_{0.99,9,15}$.

SOLUTION

a. $F_{0.05,8,10}$ is a critical value related to an F distribution with 8 and 10 degrees of freedom. By definition, if $X \sim F_{8,10}$, then $P(X \geq F_{0.05,8,10}) = 0.05$. Using Table VII in the Appendix, for $\alpha = 0.05$, find the intersection of the $v_1 = 8$ column and the $v_2 = 10$ row.

$\alpha = 0.05$

			v_1				
v_2	6	7	8	9	10	
$\vdots$	$\vdots$	$\vdots$	$\vdots$	$\vdots$	$\vdots$	$\vdots$	$\vdots$
8	$\cdots$	3.58	3.50	3.44	3.39	3.35	$\cdots$
9	$\cdots$	3.37	3.29	3.23	3.18	3.14	$\cdots$
10	$\cdots$	3.22	3.14	3.07	3.02	2.98	$\cdots$
11	$\cdots$	3.09	3.01	2.95	2.90	2.85	$\cdots$
12	$\cdots$	3.00	2.91	2.85	2.80	2.75	$\cdots$
$\vdots$	$\vdots$	$\vdots$	$\vdots$	$\vdots$	$\vdots$	$\vdots$	$\vdots$

Therefore, $F_{0.05,8,10} = 3.07$ and if $X \sim F_{8,10}$, then $P(X \geq 3.07) = 0.05$, as illustrated in Figure 10.62.

b. $F_{0.99,9,15}$ is a critical value related to an F distribution with 9 and 15 degrees of freedom. By definition, if $X \sim F_{9,15}$, then $P(X \geq F_{0.99,9,15}) = 0.99$. Because $F_{0.99,9,15}$ is in the left tail of the distribution, use Equation 10.12.

$$F_{0.99,9,15} = F_{1-0.01,9,15} = \frac{1}{F_{0.01,15,9}}$$

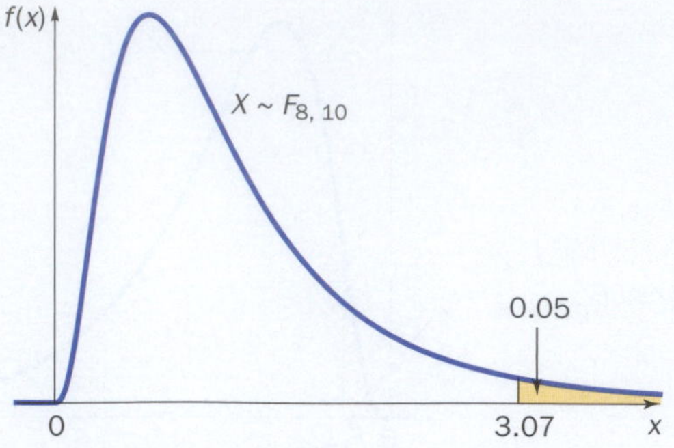

Figure 10.62 Visualization of $F_{0.05,8,10}$.

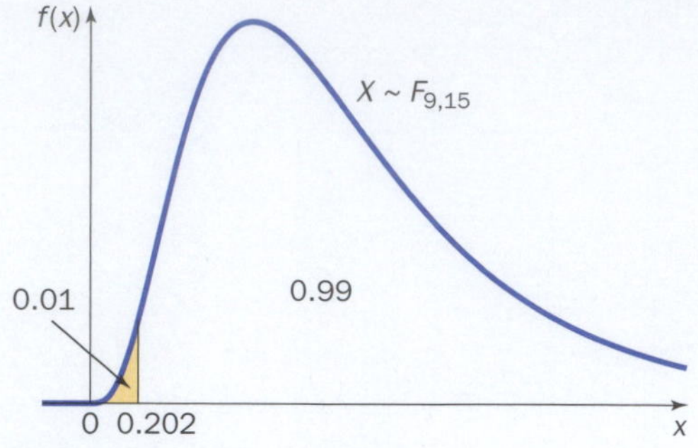

Figure 10.63 Visualization of $F_{0.99,9,15}$.

Using Table VII in the Appendix, for $\alpha = 0.01$, find the intersection of the $v_1 = 15$ column and the $v_2 = 9$ row.

$\alpha = 0.01$

		v_1					
v_2		9	10	15	20	30	
⋮	⋮	⋮	⋮	⋮	⋮	⋮	⋮
7	⋯	6.72	6.62	6.31	6.16	5.99	⋯
8	⋯	5.91	5.81	5.52	5.36	5.20	⋯
9	⋯	5.35	5.26	4.96	4.81	4.65	⋯
10	⋯	4.94	4.85	4.56	4.41	4.25	⋯
11	⋯	4.63	4.54	4.25	4.10	3.94	⋯
⋮	⋮	⋮	⋮	⋮	⋮	⋮	⋮

A	B
3.0717	= F.INV.RT(0.05,8,10)
0.2015	= F.INV.RT(0.99,9,15)

Figure 10.64 Use the Excel function `F.INV.RT` to find F critical values.

Therefore, $F_{0.99,9,15} = 1/4.96 = 0.202$. If the random variable $X \sim F_{9,15}$ then $P(X \geq 0.202) = 0.99$ and $P(X \leq 0.202) = 0.01$, as illustrated in Figure 10.63.

TRY IT NOW GO TO EXERCISE 10.128

A CLOSER LOOK

1. Table VII in the Appendix is very limited. There are only three values for α and a limited number of values for v_1 and v_2. The TI-84 Plus C does *not* have a built-in function for finding F critical values. However, the `SOLVE` feature may be used to find a critical value related to any F distribution.
2. Minitab and Excel may also be used to find a critical value related to any F distribution. Minitab uses inverse cumulative probability and Excel uses right-tail probability. For example, to find $F_{0.99,9,15}$ using Minitab, let the random variable $X \sim F_{9,15}$.

   $P(X \geq F_{0.99,9,15}) = 0.99$ — Definition of F critical value.

   $P(X \leq F_{0.99,9,15}) = 1 - 0.99 = 0.01$ — The complement rule.

   $F_{0.99,9,15} = 0.2015$ — Use Minitab.

Hypothesis tests concerning two population variances and a confidence interval for the ratio of two population variances are based on the following results.

For reference, we'll call these the two-sample F test assumptions.

Let S_1^2 be the sample variance of a random sample of size n_1 from a normal distribution with variance σ_1^2, let S_2^2 be the sample variance of a random sample of size n_2 from a normal distribution with variance σ_2^2, and suppose the samples are independent.

This is yet another kind of standardization, a transformation to an F distribution.

1. The random variable

$$F = \frac{S_1^2/\sigma_1^2}{S_2^2/\sigma_2^2}$$
(10.13)

has an F distribution with $n_1 - 1$ (from the numerator) and $n_2 - 1$ (from the denominator) degrees of freedom.

If σ_1^2 and σ_2^2 are equal, they cancel out.

2. If the null hypothesis is H_0: $\sigma_1^2 = \sigma_2^2$, then the random variable simplifies to $F = \dfrac{S_1^2/\sigma_1^2}{S_2^2/\sigma_2^2} = S_1^2/S_2^2$. This simple ratio is the test statistic for comparing two population variances.

Hypothesis Tests Concerning Two Population Variances

Given the two-sample F test assumptions, a hypothesis test concerning two population variances with significance level α has the form

This is the template for a hypothesis test concerning two population variances, sometimes called a two-sample F test.

H_0: $\sigma_1^2 = \sigma_2^2$

H_a: $\sigma_1^2 > \sigma_2^2$, $\sigma_1^2 < \sigma_2^2$,   or $\sigma_1^2 \neq \sigma_2^2$

TS: $F = S_1^2/S_2^2$

RR: $F \geq F_{\alpha, n_1-1, n_2-1}$, $F \leq F_{1-\alpha, n_1-1, n_2-1}$ or

$F \leq F_{1-\alpha/2, n_1-1, n_2-1}$ or $F \geq F_{\alpha/2, n_1-1, n_2-1}$

Using the same assumptions, a confidence interval for the ratio of two population variances can be derived. Let $X \sim F_{n_1-1, n_2-1}$ and find an interval that captures $1 - \alpha$ in the *middle* of this F distribution. Manipulate the inequality to sandwich the ratio σ_1^2/σ_2^2.

How to Find a $100(1 - \alpha)$% Confidence Interval for the Ratio of Two Population Variances

Given the two-sample F test assumptions, a $100(1 - \alpha)$% confidence interval for σ_1^2/σ_2^2 is given by

$$\left(\frac{s_1^2}{s_2^2} \frac{1}{F_{\alpha/2, n_1-1, n_2-1}}, \frac{s_1^2}{s_2^2} \frac{1}{F_{1-\alpha/2, n_1-1, n_2-1}} \right)$$
(10.14)

Using Equation 10.12, the confidence interval can be written as

$$\left(\frac{s_1^2}{s_2^2} \frac{1}{F_{\alpha/2, n_1-1, n_2-1}}, \frac{s_1^2}{s_2^2} F_{\alpha/2, n_2-1, n_1-1} \right)$$
(10.15)

The hypothesis test procedure described above can also be used to compare two population standard deviations. And, you can take the square root of each endpoint of Equation 10.14 to find a $100(1 - \alpha)$% confidence interval for the ratio of two population standard deviations. The following example illustrates the hypothesis test procedure.

Example 10.15 Long-Term-Care Cost

DATA SET

CARECOST

The cost of long-term care in a nursing home varies considerably by region and may be as much as $50,000 per year. Two independent samples of nursing homes in Connecticut and in Colorado were obtained, and the cost-per-day for each was recorded. The data are given in the following table.

Connecticut (1)

270	294	174	180	314	274	160	210	255	187	271

Colorado (2)

161	150	164	109	168	172	133	148	120	157	138	94
166	116	98	168	153	118	138	116	120			

a. Conduct the appropriate hypothesis test to determine whether there is any evidence that the population variance in cost per day is different in Connecticut and Colorado. Assume the costs per day underlying populations are normal and use $\alpha = 0.02$.

b. Find bounds on the p value for the hypothesis test in part (a).

SOLUTION

STEP 1   The null hypothesis is that the two population variances are equal. We are looking for *any difference* in the variances, so the alternative hypothesis is two-sided. The underlying populations are assumed normal and the samples were obtained independently. A two-sample F test is appropriate.

In this case, $n_1 = 11$ and $n_2 = 21$; $\alpha/2 = 0.01$ and $1 - \alpha/2 = 0.99$.

STEP 2   The four parts of the hypothesis test are

$$H_0: \sigma_1^2 = \sigma_2^2$$
$$H_a: \sigma_1^2 \neq \sigma_2^2$$
$$\text{TS: } F = S_1^2/S_2^2$$
$$\text{RR: } F \leq F_{1-\alpha/2,n_1-1,n_2-1} = F_{0.99,10,20} = 1/4.41 = 0.2268 \qquad \text{or}$$
$$F \geq F_{\alpha/2,n_1-1,n_2-1} = F_{0.01,10,20} = 3.37$$

STEP 3   The summary statistics are

$$s_1^2 = \frac{1}{10}\left[638,819 - \frac{1}{11}(2589)^2\right] = 2946.25$$

$$s_2^2 = \frac{1}{20}\left[414,601 - \frac{1}{21}(2907)^2\right] = 609.46$$

The value of the test statistic is

$$f = \frac{s_1^2}{s_2^2} = \frac{2946.25}{609.46} = 4.83 \; (\geq 3.37)$$

The value of the test statistic lies in the rejection region. Therefore, we reject the null hypothesis at the $\alpha = 0.02$ significance level. There is evidence to suggest that the two population variances are different.

STEP 4   Because the tables of critical values for F distributions are very limited, we can only bound the p value. Place the value of the test statistic, $f = 4.83$, in an ordered list of critical values with df 10 and 20.

$$3.37 \leq 4.83 \leq 5.08$$
$$F_{0.01,10,20} \leq 4.83 \leq F_{0.001,10,20}$$

Therefore, $0.001 \leq p/2 \leq 0.01$

and, $0.002 \leq p \leq 0.02$

The exact p value is illustrated in Figure 10.65.

Figures 10.66 through 10.69 show technology solutions.

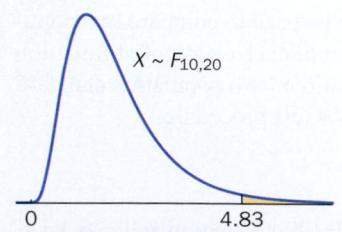

Figure 10.65 *p*-Value illustration:
$p = 2P(X \geq 4.83)$
$= 0.0027 \leq 0.02 = \alpha$

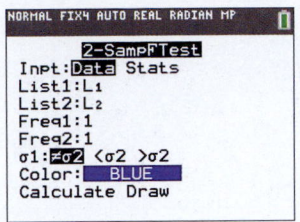

Figure 10.66
2-SampFTest input screen.

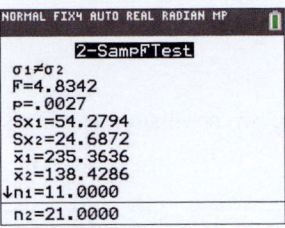

Figure 10.67
2-SampFTest hypothesis test results.

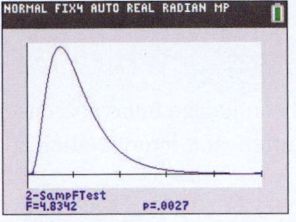

Figure 10.68
2-SampFTest Draw results.

Figure 10.69 JMP tests for equality of two population variances.

TRY IT NOW   GO TO EXERCISE 10.138

The following example involves constructing a confidence interval for the ratio of two population variances.

Example 10.16 Importance of Coral Reefs

Coral reefs are diverse and valuable ecosystems. These vast ocean environments support approximately 4000 species of fish, lead to medical advances, and contribute to local economies through tourism. Independent random samples of brain corals were obtained from two Caribbean reefs. The diameter of each polyp was carefully measured (in mm). The data are summarized in the following table.

Navassa Island	$n_1 = 16$,	$s_1^2 = 0.267$
U.S. Virgin Islands	$n_2 = 21$,	$s_2^2 = 0.172$

Construct a 90% confidence interval for the ratio of population variances in diameter of brain coral polyps.

SOLUTION

STEP 1 The samples are independent; the sample sizes and sample variances are given. A confidence interval for the ratio of two population variances is appropriate.

STEP 2 Find the critical values.

$$1 - \alpha = 0.90 \;\Rightarrow\; \alpha = 0.10 \;\Rightarrow\; \alpha/2 = 0.05 \qquad \text{Find } \alpha/2.$$

$$F_{\alpha/2, n_1-1, n_2-1} = F_{0.05, 15, 20} = 2.20 \qquad \text{Critical value, left endpoint. Use Table VII in the Appendix.}$$

$$F_{\alpha/2, n_2-1, n_1-1} = F_{0.05, 20, 15} = 2.33 \qquad \text{Critical value, right endpoint. Use Table VII in the Appendix.}$$

STEP 3 Use Equation 10.15.

$$\left(\frac{s_1^2}{s_2^2} \frac{1}{F_{\alpha/2, n_1-1, n_2-1}}, \; \frac{s_1^2}{s_2^2} F_{\alpha/2, n_2-1, n_1-1} \right) \qquad \text{Equation 10.15.}$$

$$= \left(\frac{0.267}{0.172} \frac{1}{2.20}, \; \frac{0.267}{0.172}(2.33) \right) \qquad \text{Use sample variances and critical values.}$$

$$= (0.7056, 3.6169) \qquad \text{Simplify.}$$

(0.7056, 3.6169) is a 90% confidence interval for the ratio of the population variances.

TRY IT NOW   GO TO EXERCISE 10.141

Suppose a two-sample t test will be used to compare two population means. The hypothesis test presented in this section is often used first to compare the population variances. The results and conclusion suggest the appropriate hypothesis test concerning population means from Section 10.2, according to whether or not there is evidence that the two population variances are unequal.

Technology Corner

Procedure: Hypothesis tests and confidence intervals concerning two population variances.
Reconsider: Example 10.15, solution, and interpretations.

CrunchIt!

There is no built-in function to conduct hypothesis tests nor to construct a confidence interval concerning two population variances. However, the F Distribution Calculator can be used to find critical values.

TI-84 Plus C

Use the built-in function 2-SampFTest to conduct a hypothesis test concerning two population variances. Input is either data in lists or summary statistics. There is no built-in function to construct a confidence interval for the ratio of population variances.

1. Enter the Connecticut data into list L1 and the Colorado data into list L2.
2. Select STAT; TESTS; 2-SampFTest.
3. Highlight Data and enter the two lists. Set each frequency to 1 and highlight the appropriate alternative hypothesis. See Figure 10.66.
4. Highlight Calculate and press ENTER to display the hypothesis test results. See Figure 10.67. The Draw results are shown in Figure 10.68.

Minitab

Use the built-in function 2 Variances to conduct a hypothesis test concerning two population variances and to construct a confidence interval for the ratio of the population variances. Input is samples in one column (with subscripts), samples in different columns, or summarized data. Several graph options are also available.

1. Enter the Connecticut data into column C1 and the Colorado data into column C2.
2. Select Stat; Basic Statistics; 2 Variances.
3. Choose Each sample is in its own column, and enter C1 in the Sample 1 input window and C2 in the Sample 2 input window.
4. Choose the Options option button. Select the Ratio of sample variances, enter a Confidence level, enter a Hypothesized ratio (and value, usually 1), and an Alternative hypothesis.
5. Click OK and the results are displayed in a session and graph window. See Figures 10.70 and 10.71.

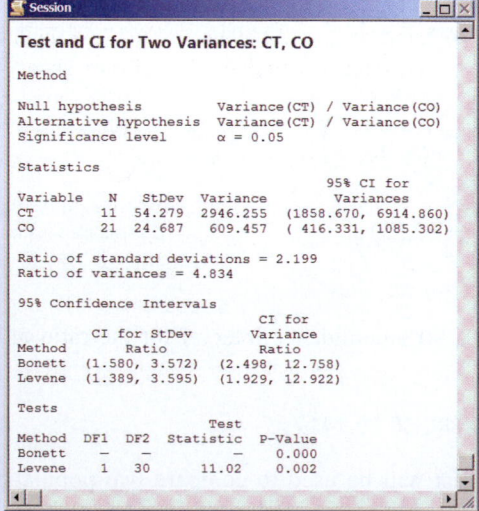

Figure 10.70 The 2 Variances output from Minitab.

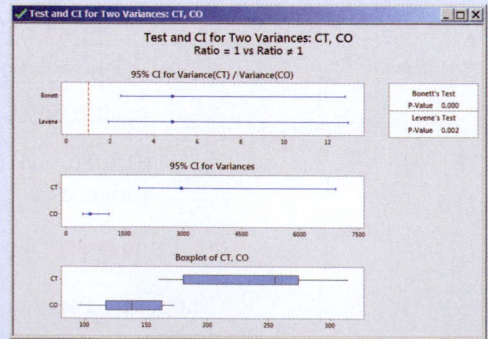

Figure 10.71 The 2 Variances graph results from Minitab.

Excel

Use the built-in function F-Test Two-Sample for Variances to conduct a hypothesis test concerning two population variances. There is no built-in function to construct a confidence interval for the ratio of population variances.

1. Enter the Connecticut data into column A and the Colorado data into column B.
2. Under the Data tab, select Data Analysis; F-Test Two-Sample for Variances.
3. Enter the Variable 1 range, Variable 2 Range, a value for Alpha, and specify an Output option. There is no alternative hypothesis option.
4. Click OK to view the summary statistics and the hypothesis test results. See Figure 10.72. The p value displayed is for a one-sided hypothesis test. Double this value for a two-sided test.

Figure 10.72 Excel Hypothesis test results.

	C	D	E
F-Test Two-Sample for Variances			
		Variable 1	Variable 2
Mean		235.36	138.43
Variance		2946.25	609.46
Observations		11	21
df		10	20
F		4.8342	
P(F<=f) one-tail		0.0013	
F Critical one-tail		2.9148	

SECTION 10.5 EXERCISES

Concept Check

10.123 True/False Every F distribution is symmetric about its mode.

10.124 True/False An F distribution has positive probability only for non-negative values.

10.125 True/False $F_{\alpha,v_1,v_2} = F_{1-\alpha,v_2,v_1}$

10.126 True/False A hypothesis test to compare two population variances can also be used to test for equality of population standard deviations.

10.127 Short Answer S_1^2 and S_2^2 are unbiased estimators for the population variances σ_1^2 and σ_2^2, respectively, Explain what this means.

Practice

10.128 Find each of the following critical values.
a. $F_{0.05,7,19}$
b. $F_{0.05,30,25}$
c. $F_{0.01,6,19}$
d. $F_{0.001,40,40}$
e. $F_{0.95,17,15}$
f. $F_{0.95,12,10}$
g. $F_{0.99,21,30}$
h. $F_{0.999,11,8}$

10.129 Find each of the following critical values.
a. $F_{0.01,20,60}$
b. $F_{0.01,15,19}$
c. $F_{0.05,6,8}$
d. $F_{0.001,8,24}$
e. $F_{0.99,12,9}$
f. $F_{0.99,6,8}$
g. $F_{0.95,10,10}$
h. $F_{0.999,23,20}$

10.130 In each of the following problems, the null hypothesis is $H_0: \sigma_1^2 = \sigma_2^2$. The alternative hypothesis, the sample sizes,

and the value of the test statistic are given. Find bounds on the p value associated with each hypothesis test.
a. $H_a: \sigma_1^2 > \sigma_2^2$, $n_1 = 16$, $n_2 = 17$, $f = 2.29$
b. $H_a: \sigma_1^2 < \sigma_2^2$, $n_1 = 11$, $n_2 = 16$, $f = 0.34$
c. $H_a: \sigma_1^2 \neq \sigma_2^2$, $n_1 = 7$, $n_2 = 10$, $f = 7.36$
d. $H_a: \sigma_1^2 > \sigma_2^2$, $n_1 = 31$, $n_2 = 26$, $f = 4.26$

10.131 Consider independent random samples of sizes $n_1 = 31$ and $n_2 = 25$ from normal populations.
a. Write the four parts for a one-sided, right-tailed hypothesis test concerning the population variances with $\alpha = 0.05$.
b. Suppose $s_1^2 = 44.89$ and $s_2^2 = 17.64$. Find the value of the test statistic, and draw a conclusion about the population variances.
c. Find bounds on the p value associated with this hypothesis test and carefully sketch a graph to illustrate this value.

10.132 Consider the two independent random samples from normal distributions given in the following table. **EX10.132**

Sample 1

89.6	61.1	83.7	74.2	60.6	50.4	82.4	79.0
56.5	72.2	72.4	77.1	58.2	72.3	71.6	70.3
76.1	73.2						

Sample 2

73.7	37.8	76.2	64.7	74.8	75.8	67.9	61.6
76.2	82.7	88.9	60.3	34.4	74.5	68.0	100.2
73.7	41.5	76.9	55.0	76.0			

a. Write the four parts for a one-sided, left-tailed hypothesis test concerning the two population variances with $\alpha = 0.05$.
b. Compute each sample variance, find the value of the test statistic, and draw a conclusion.
c. Find bounds on the p value associated with this hypothesis test.

10.133 Consider independent random samples of sizes $n_1 = 10$ and $n_2 = 16$ from normal populations.
a. Write the four parts for a two-sided hypothesis test concerning the population variances with $\alpha = 0.01$.
b. Suppose $s_1^2 = 426.42$ and $s_2^2 = 88.36$. Find the value of the test statistic, and draw a conclusion about the population variances.
c. Find bounds on the p value associated with this hypothesis test.

10.134 In each of the following problems, the sample sizes and the confidence level are given. Find the appropriate F critical values for use in constructing a confidence interval for the ratio of the population variances.
a. $n_1 = 10$, $n_2 = 10$, 90%
b. $n_1 = 21$, $n_2 = 31$, 98%
c. $n_1 = 9$, $n_2 = 7$, 98%
d. $n_1 = 41$, $n_2 = 31$, 99.8%

10.135 In each of the following problems, the sample sizes, the sample variances, and the confidence level are given. Assume the underlying populations are normal and the samples were obtained independently. Find the associated confidence interval for the ratio of the population variances.
a. $n_1 = 10$, $s_1^2 = 17.2$, $n_2 = 9$, $s_2^2 = 15.6$, 90%
b. $n_1 = 16$, $s_1^2 = 54.1$, $n_2 = 16$, $s_2^2 = 32.6$, 98%
c. $n_1 = 16$, $s_1^2 = 3.35$, $n_2 = 31$, $s_2^2 = 4.59$, 98%
d. $n_1 = 31$, $s_1^2 = 126.8$, $n_2 = 41$, $s_2^2 = 155.3$, 99.8%

10.136 Use Table VII in the Appendix and linear interpolation to approximate each critical value. Verify each approximation using technology.
a. $F_{0.05,25,15}$
b. $F_{0.99,20,32}$
c. $F_{0.01,10,56}$
d. $F_{0.025,15,20}$
e. $F_{0.995,10,7}$
f. $F_{0.05,35,35}$

Applications

10.137 Biology and Environmental Science In a recent study conducted by the NOAA, the aerosol light absorption coefficient was measured (in Mm^{-1}) at randomly selected locations in Africa and in South America. The resulting data are summarized in the following table.

Country	Sample size	Sample variance
Africa	10	243.36
South America	21	51.84

Is there any evidence that the population variance in aerosol light absorption coefficient is greater in Africa than in South America? Use $\alpha = 0.05$ and assume normality.

10.138 Medicine and Clinical Studies A study in the *British Medical Journal* suggested that children who received a diagnostic CT scan using ionizing radiation were more likely to develop a cancer 10 years after radiation exposure. Newer CT scanners use less radiation. Therefore, the increased risk for children today may be decreased.[26] Independent random samples of hospital CT scans in the United States and England were obtained, and the amount of radiation for each was recorded (in mSV). For the United States, $n_1 = 10$ and $s_1^2 = 1.075$; for England $n_2 = 16$ and $s_2^2 = 2.786$. Is there any evidence to suggest that the variability in CT radiation per scan is different for these two countries? Use $\alpha = 0.05$ and assume normality.

10.139 Travel and Transportation Most airlines now charge passengers to check a bag and impose a surcharge if the weight of the bag is over 50 pounds. Independent random samples of checked luggage on Delta and American flights were obtained, and the weight (in pounds) of each was recorded. For Delta, $n_1 = 25$, and $s_1^2 = 96.23$; for American, $n_2 = 21$ and $s_2^2 = 194.02$. Is there any evidence to suggest that the variability in checked baggage weight is different for these two airlines? Use $\alpha = 0.05$ and assume normality.

10.140 Sports and Leisure Many basketball purists believe that the three-point shot (a shot from behind the three-point line, 22 feet from the basket) has dramatically changed the game, for the worse. Independent random samples of attempted shots from National Basketball Association games played in 1975 (prior to the three-point shot) and in 2013 were obtained. The shot distance (in feet) was recorded for each attempt. The data are given in the following table.

Year	Sample size	Sample variance
1975	61	12.25
2013	61	26.01

Is there any evidence to suggest that the variability in shot distance is greater in the year 2013 than it was in 1975? Use $\alpha = 0.01$ and assume normality. (Why do you suppose there is greater variability in shot distance with a three-point shot?)

10.141 Sports and Leisure A Laurel Downs racetrack official believes there is less variability in winning times for a race in which the purse is at least $10,000, called a stakes race. Independent random samples of ordinary races and stakes races were obtained, and the winning time (in seconds) for each race was recorded. The summary statistics were as follows: ordinary race, $n_1 = 26$ and $s_1^2 = 110.25$; stakes race, $n_2 = 26$ and $s_2^2 = 38.44$.
a. Write the four parts for a hypothesis test to check for evidence of the official's assertion. Use $\alpha = 0.01$, assume normality, and find the critical value using technology. Conduct the hypothesis test and draw a conclusion.
b. Construct a 98% confidence interval for the ratio of population variances.

10.142 Sports and Leisure A study was conducted to compare the variability in times for men and women involved in collegiate swimming events. Independent random samples of 800-meter freestyle competitors were obtained, and the time (in minutes) was recorded for each swimmer. The data are summarized in the following table.

Group	Sample size	Sample variance
Men	11	0.1025
Women	12	0.1241

a. Find the critical values necessary to construct a 95% confidence interval for the ratio of population variances.
b. Construct the confidence interval.

10.143 Take the Stairs The World Summit Wing Hotel in China is Beijing's tallest hotel and hosts the Vertical Run. There are 81 floors, 330 meters, and 2041 steps to reach the roof. Independent random samples of men's and women's times (in minutes) from the 2013 run were obtained.[27] **VERTRUN**

a. Is there any evidence of a difference in variability of times for men and women who finished this vertical run? Use $\alpha = 0.05$ and assume normality.
b. Find bounds on the p value associated with this hypothesis test.

10.144 Manufacturing and Product Development A sailboat manufacturer has two machines for constructing main mast poles with diameter designed to be 76.2 mm. Small variability in production is very important to ensure boat control and safety. Independent random samples of mast poles produced on each machine were obtained, and the diameter of each was carefully measured. The summary statistics were as follows: Machine A, $n_1 = 7$, $s_1^2 = 0.0231$; Machine B, $n_2 = 8$, $s_2^2 = 0.0096$. Conduct the appropriate hypothesis test to determine whether there is any evidence of a difference in variability of mast-pole diameter between the two machines. Assume normality, use $\alpha = 0.05$, find the p value associated with this test, and use this value to draw a conclusion.

10.145 Manufacturing and Product Development Tower cranes along a city skyline often indicate the success of economic development efforts. In early January 2013, more than 50 tower crane permits were in use in Washington, D.C. Independent random samples of items lifted by cranes at two different sites were obtained, and the weight (in tons) of each item was recorded for each. The summary statistics are given in the following table.

Site	Sample size	Sample variance
New York Avenue	31	109.86
Jefferson at Market Place	31	339.43

Is there any evidence to suggest that the variability in item weight at Jefferson is greater than at New York Avenue? Use $\alpha = 0.01$ and assume normality.

10.146 Sports and Leisure The Pikes Peak International Hill Climb is also known as The Race to the Clouds. This 12.42-mile race up Pikes Peak in Colorado features over 156 turns, grades of 7%, and a finish line at 14,110 feet. Independent random samples of two classes of cars in the 2013 race were obtained, and the speed (in mph) for each was recorded. The summary statistics are given in the following table.[28]

Class	Sample size	Sample variance
Exhibition Powersports	9	44.05
Heavyweight Supermoto	9	9.76

Is there any evidence to suggest that the variability in speed for the Exhibition Powersports class is greater than the variability in speed for the Heavyweight Supermoto class? Use $\alpha = 0.01$ and assume normality.

10.147 Manufacturing and Product Development The Akashi-Kaikyo bridge in Japan is the longest suspension bridge in the world, with a main span of 1991 meters. Two million workers took 10 years to construct this bridge using 181,000 tons of steel and 1.4 million cubic meters of concrete. Independent random samples of suspension bridges in China and the United States were obtained, and the span (in meters) of each was recorded. For China, $n_1 = 14$ and $s_1^2 = 66,096.8$; and for the United States, $n_2 = 10$ and $s_2^2 = 59,524.5$. Is there any evidence to suggest a difference in the variability of the span of suspension bridges in China and the United States? Use $\alpha = 0.05$ and assume normality.

Extended Applications

10.148 Physical Sciences Crude oil pumped from ocean wells contains salt that must be removed before the oil is refined. Otherwise, equipment would erode quickly. Independent random samples of unrefined crude oil from two ocean wells were obtained, and the percentage of salt in each sample was recorded. The data are given in the following table.

Oil well	Sample size	Sample variance
North Sea	21	56.40
Antarctica	31	82.42

a. Conduct the appropriate test to determine whether there is evidence of any difference in variability of salt content between these two wells. Use $\alpha = 0.10$ and assume normality.
b. Use Table VII in the Appendix to find bounds on the p value for this hypothesis test. Use technology to find an exact p value.

10.149 Public Health and Nutrition Saccharin is a low-calorie sweetener used in sugar-free foods and beverages. According to the U.S. Food and Drug Administration, the acceptable daily intake (ADI) of saccharin is 5 mg for a person with a body weight of 60 kg. If saccharin is used as an additive, it must be included on the food label and cannot exceed certain limits. Independent random samples of

12-ounce bottles of iced tea from two different manufacturers were obtained, and the amount of saccharin in each drink was measured (in mg). The summary statistics were as follows: Fishing Creek, $n_1 = 20$, $s_1^2 = 7.84$; Honest Tea, $n_2 = 15$, $s_2^2 = 2.89$.

a. Conduct the appropriate test to determine whether there is any difference in the population variance of saccharin amounts. Use $\alpha = 0.02$.

b. Find bounds on the p value associated with this hypothesis test.

10.150 Fuel Consumption and Cars The length of time brake pads last in an automobile varies depending on driving style and the type of car. Brake pads are made from organic, semimetallic, metallic, or synthetic materials, and typically last between 30,000 and 70,000 miles. Independent random samples of cars in for an inspection at dealerships and private garages were obtained and the width (in mm) of the brake pad on the front driver's side was measured for each. The data are given in the following table.

Location	Sample size	Sample variance
Dealership	41	1.056
Private garage	41	2.771

a. Is there any evidence to suggest that the variance in brake-pad widths of cars in for inspection is greater at private garages than at dealerships? Use $\alpha = 0.01$ and assume normality.

b. Find bounds on the p value associated with this hypothesis test.

Challenge

10.151 Physical Sciences When we are comparing two population variances, if both sample sizes, n_1 and n_2, are large and

the null hypothesis, $H_0: \sigma_1^2 = \sigma_2^2$, is true, then the test statistic $F = S_1^2/S_2^2$ is approximately normal with

$$\mu_F = \frac{n_2 - 1}{n_2 - 3} \quad \text{and} \quad \sigma_F^2 = \frac{2(n_2 - 1)^2(n_1 + n_2 - 4)}{(n_1 - 1)(n_2 - 3)^2(n_2 - 5)}$$

An approximate hypothesis test is based on standardizing F to a Z random variable.

The four parts of the hypothesis test are

$$H_0: \sigma_1^2 = \sigma_2^2$$

$$H_a: \sigma_1^2 > \sigma_2^2, \quad \sigma_1^2 < \sigma_2^2, \quad \text{or} \quad \sigma_1^2 \neq \sigma_2^2$$

$$\text{TS: } Z = \frac{(S_1^2/S_2^2) - [(n_2 - 1)/(n_2 - 3)]}{\sqrt{\dfrac{2(n_2 - 1)^2(n_1 + n_2 - 4)}{(n_1 - 1)(n_2 - 3)^2(n_2 - 5)}}}$$

$$\text{RR: } Z \geq z_\alpha, \quad Z \leq -z_\alpha \quad \text{or} \quad |Z| \geq z_{\alpha/2}$$

The National Wind Energy Assessment contains data from 975 stations and includes measurements of wind speed and wind power density. Suppose independent random samples of wind power density (in watts/m^2) during the winter were obtained from two stations. The data are summarized in the following table.

Location	Sample size	Sample mean	Sample variance
Chanute	31	207	95.35
Dodge City	31	283	53.68

a. Write the four parts of a large-sample, two-sided, approximate test based on the standard normal distribution to determine whether there is any evidence to suggest that the two population variances are different. Conduct the test using $\alpha = 0.05$.

b. Conduct an *exact* hypothesis test based on the F distribution. Compare your answer to part (a).

CHAPTER 10 SUMMARY

Concept	Page	Notation / Formula / Description
Independent samples	463	Two samples are independent if the process of selecting individuals or objects in sample 1 has no effect on the selection of individuals or objects in sample 2.
Paired data set	463	The result of matching each individual or object in sample 1 with a similar individual or object in sample 2.
Pooled estimator for the common variance	475	$S_p^2 = \dfrac{(n_1 - 1)S_1^2 + (n_2 - 1)S_2^2}{n_1 + n_2 - 2}$
Combined estimate of the common population proportion	502	$\hat{P}_c = \dfrac{X_1 + X_2}{n_1 + n_2}$

Summary of confidence intervals

Parameter	Assumptions	$100(1-\alpha)\%$ Confidence interval
$\mu_1 - \mu_2$	n_1, n_2 large, independence, σ_1^2, σ_2^2 known, or normality, independence, σ_1^2, σ_2^2 known.	$(\bar{x}_1 - \bar{x}_2) \pm z_{\alpha/2} \sqrt{\dfrac{\sigma_1^2}{n_1} + \dfrac{\sigma_2^2}{n_2}}$
$\mu_1 - \mu_2$	Normality, independence, σ_1^2, σ_2^2 unknown but equal.	$(\bar{x}_1 - \bar{x}_2) \pm t_{\alpha/2, n_1+n_2-2} \sqrt{s_p^2 \left(\dfrac{1}{n_1} + \dfrac{1}{n_2} \right)}$ $s_p^2 = \dfrac{(n_1-1)s_1^2 + (n_2-1)s_2^2}{n_1 + n_2 - 2}$
$\mu_1 - \mu_2$	Normality, independence, σ_1^2, σ_2^2 unknown and unequal.	$(\bar{x}_1 - \bar{x}_2) \pm t_{\alpha/2, v} \sqrt{\dfrac{s_1^2}{n_1} + \dfrac{s_2^2}{n_2}}$ $v = \dfrac{\left(\dfrac{s_1^2}{n_1} + \dfrac{s_2^2}{n_2} \right)^2}{\dfrac{(s_1^2/n_1)^2}{n_1 - 1} + \dfrac{(s_2^2/n_2)^2}{n_2 - 1}}$
$\mu_D = \mu_1 - \mu_2$	Normality, n pairs, dependence.	$\bar{d} \pm t_{\alpha/2, n-1} \dfrac{s_D}{\sqrt{n}}$
$p_1 - p_2$	n_1, n_2 large, nonskewness, independence.	$(\hat{p}_1 - \hat{p}_2) \pm z_{\alpha/2} \sqrt{\dfrac{\hat{p}_1(1-\hat{p}_1)}{n_1} + \dfrac{\hat{p}_2(1-\hat{p}_2)}{n_2}}$
$\dfrac{\sigma_1^2}{\sigma_2^2}$	Normality, independence.	$\left(\dfrac{s_1^2}{s_2^2} \dfrac{1}{F_{\alpha/2, n_1-1, n_2-1}}, \dfrac{s_1^2}{s_2^2} \dfrac{1}{F_{1-\alpha/2, n_1-1, n_2-1}} \right)$

Summary of hypothesis tests

Null hypothesis	Assumptions	Alternative hypothesis	Test statistic	Rejection region
$\mu_1 - \mu_2 = \Delta_0$	n_1, n_2 large, independence, σ_1^2, σ_2^2 known, or normality, independence, σ_1^2, σ_2^2 known.	$\mu_1 - \mu_2 > \Delta_0$ $\mu_1 - \mu_2 < \Delta_0$ $\mu_1 - \mu_2 \neq \Delta_0$	$Z = \dfrac{(\bar{X}_1 - \bar{X}_2) - \Delta_0}{\sqrt{\dfrac{\sigma_1^2}{n_1} + \dfrac{\sigma_2^2}{n_2}}}$	$Z \geq z_\alpha$ $Z \leq -z_\alpha$ $|Z| \geq z_{\alpha/2}$
$\mu_1 - \mu_2 = \Delta_0$	Normality, independence, σ_1^2, σ_2^2 unknown, $\sigma_1^2 = \sigma_2^2$.	$\mu_1 - \mu_2 > \Delta_0$ $\mu_1 - \mu_2 < \Delta_0$ $\mu_1 - \mu_2 \neq \Delta_0$	$T = \dfrac{(\bar{X}_1 - \bar{X}_2) - \Delta_0}{\sqrt{S_p^2 \left(\dfrac{1}{n_1} + \dfrac{1}{n_2} \right)}}$ $S_p^2 = \dfrac{(n_1-1)S_1^2 + (n_2-1)S_2^2}{n_1 + n_2 - 2}$	$T \geq t_{\alpha, n_1+n_2-2}$ $T \leq -t_{\alpha, n_1+n_2-2}$ $|T| \geq t_{\alpha/2, n_1+n_2-2}$

Null hypothesis	Assumptions	Alternative hypothesis	Test statistic	Rejection region
$\mu_1 - \mu_2 = \Delta_0$	Normality, independence, σ_1^2, σ_2^2 unknown, $\sigma_1^2 \neq \sigma_2^2$.	$\mu_1 - \mu_2 > \Delta_0$ $\mu_1 - \mu_2 < \Delta_0$ $\mu_1 - \mu_2 \neq \Delta_0$	$T' = \dfrac{(\bar{X}_1 - \bar{X}_2) - \Delta_0}{\sqrt{\dfrac{S_1^2}{n_1} + \dfrac{S_2^2}{n_2}}}$ $v = \dfrac{\left(\dfrac{s_1^2}{n_1} + \dfrac{s_2^2}{n_2}\right)^2}{\dfrac{(s_1^2/n_1)^2}{n_1 - 1} + \dfrac{(s_2^2/n_2)^2}{n_2 - 1}}$	$T' \geq t_{\alpha,v}$ $T' \leq -t_{\alpha,v}$ $|T'| \geq t_{\alpha/2,v}$
$\mu_D = \Delta_0$	Normality, n pairs, dependence.	$\mu_D > \Delta_0$ $\mu_D < \Delta_0$ $\mu_D \neq \Delta_0$	$T = \dfrac{\bar{D} - \Delta_0}{S_D/\sqrt{n}}$	$T \geq t_{\alpha,n-1}$ $T \leq -t_{\alpha,n-1}$ $|T| \geq t_{\alpha/2,n-1}$
$p_1 - p_2 = 0$	n_1, n_2 large, nonskewness, independence.	$p_1 - p_2 > 0$ $p_1 - p_2 < 0$ $p_1 - p_2 \neq 0$	$Z = \dfrac{\hat{P}_1 - \hat{P}_2}{\sqrt{\hat{P}_c(1 - \hat{P}_c)\left(\dfrac{1}{n_1} + \dfrac{1}{n_2}\right)}}$ $\hat{P}_c = \dfrac{X_1 + X_2}{n_1 + n_2}$	$Z \geq z_\alpha$ $Z \leq -z_\alpha$ $|Z| \geq z_{\alpha/2}$
$p_1 - p_2 = \Delta_0$	n_1, n_2 large, nonskewness, independence.	$p_1 - p_2 > \Delta_0$ $p_1 - p_2 < \Delta_0$ $p_1 - p_2 \neq \Delta_0$	$Z = \dfrac{(\hat{P}_1 - \hat{P}_2) - \Delta_0}{\sqrt{\dfrac{\hat{P}_1(1 - \hat{P}_1)}{n_1} + \dfrac{\hat{P}_2(1 - \hat{P}_2)}{n_2}}}$	$Z \geq z_\alpha$ $Z \leq -z_\alpha$ $|Z| \geq z_{\alpha/2}$
$\sigma_1^2 = \sigma_2^2$	Normality, independence.	$\sigma_1^2 > \sigma_2^2$ $\sigma_1^2 < \sigma_2^2$ $\sigma_1^2 \neq \sigma_2^2$	$F = \dfrac{S_1^2}{S_2^2}$	$F \geq F_{\alpha,n_1-1,n_2-1}$ $F \leq F_{1-\alpha,n_1-1,n_2-1}$ $F \leq F_{1-\alpha/2,n_1-1,n_2-1}$ or $F \geq F_{\alpha/2,n_1-1,n_2-1}$

CHAPTER 10 EXERCISES

10 APPLICATIONS

10.152 Travel and Transportation The U.S. Department of Transportation requires vehicles transporting hazardous materials to use special placards indicating the type of cargo. There are many other regulations involving containers, separation of various materials, and gross weight. Independent random samples of trucks carrying corrosive materials were stopped on highways in North Carolina and in Virginia, and the weight (in kg) of the hazardous material was recorded. The summary statistics and known variances are given in the following table.

State	Sample size	Sample mean	Population variance
North Carolina	22	835.6	3192.25
Virginia	25	884.2	3956.41

Is there any evidence to suggest that the mean amount of corrosive material carried by trucks in North Carolina is different from the mean amount of corrosive material carried by trucks in Virginia? Use $\alpha = 0.01$ and assume each underlying distribution of weight is normal.

10.153 Taser Accuracy Tasers are nonlethal weapons used by police to subdue dangerous people. This electroshock weapon uses electrical current to disrupt control of an individual's muscles. The police forces in England maintain careful records concerning the use of Tasers. Records from 2010–2012 indicate the following.[29]

Police force	Lancashire	West Mercia
Number of Taser uses	186	138
Number of chest hits	120	62

Is there any evidence to suggest that the true proportion of chest hits is greater in Lancashire than in West Mercia? Use $\alpha = 0.01$.

10.154 Biology and Environmental Science Soybeans are an important source of oil and protein and are also used to produce many food additives. The leading producers of soybeans are the United States, Brazil, Argentina, and China. The first genetically modified (GM) soybeans were grown in the United States in 1996, and now GM soybeans are grown in at least nine countries. Independent random samples of soybean farmers in the United States and Brazil were obtained, and the number growing GM soybeans was recorded. The data are given in the following table.

Country	Sample size	Number of GM soybean farmers
United States	238	202
Brazil	162	104

a. Find the sample proportion of GM soybean farmers for each country. Verify the nonskewness criterion.

b. Conduct the appropriate hypothesis test to determine whether there is any evidence that the true proportion of GM soybean farmers in the United States is 0.15 greater than in Brazil. Use $\alpha = 0.05$.

c. Find the p value associated with this hypothesis test.

10.155 Biology and Environmental Science   Maple syrup producers in New York and Vermont collect sweet-water sap from sugar maples and black maples in early spring. It takes approximately 30–50 gallons of sap to yield, through boiling and evaporation, 1 gallon of maple syrup. Independent random samples of maple trees in both states were obtained, and the amount of sap collected from each tree was recorded. Assume the underlying populations are normal, with equal variances. Is there any evidence to suggest that the population mean amount of sap from trees in New York is different from the population mean amount of sap from trees in Vermont? Use $\alpha = 0.01$. **SAP**

10.156 Physical Sciences   Recycling of aluminum, glass, newspapers, and magazines is good for the environment and the economy. In 2013, San Francisco had the highest recycling rate in the United States[30] (recycling rate = tons collected for recycling/tons of all waste generated). Despite efforts to make the process easier, many people still do not recycle. Independent random samples of residents in Ohio and in Florida were obtained and asked whether they recycle newspapers. Of the 909 Ohio residents, 700 said they recycled newspapers, and 691 of the 923 Florida residents said they recycled newspapers.

a. Is there any evidence to suggest that the population proportion of residents in Ohio who recycle newspapers is greater than the population proportion of residents in Florida? Use $\alpha = 0.01$.

b. Find the p value for this hypothesis test.

10.157 Sports and Leisure   Archery target shooters use a variety of arrows made from wood, carbon, aluminum, or even platinum. One measure of the quality of an arrow (and bow) is the speed of the arrow when shot. A random sample of archers was obtained, and each was asked to shoot a carbon arrow and a similarly made aluminum arrow. The speed (in feet per second) of each arrow was measured. **ARCHERY**

a. What is the common characteristic that makes these data paired?

b. Assume normality. Conduct the appropriate hypothesis test to determine whether there is any evidence that the aluminum arrow flies faster. Use $\alpha = 0.05$.

c. Find bounds on the p value associated with this hypothesis test.

10.158 Manufacturing and Product Development
Raytheon Aircraft is now manufacturing business jets with a molded carbon fiber fuselage instead of aluminum. This reduces the overall weight of the plane, speeds production time, and increases cabin space. The total wall thickness of a carbon fiber fuselage is 0.81 inch versus 3 inches for aluminum, and the variability in thickness is theoretically much smaller also. Independent random samples of the two fuselage types were obtained, and the thickness was measured (in inches) on each. The data are given in the following table.

Fuselage type	Sample size	Sample variance
Aluminum	9	0.0196
Carbon fiber	11	0.0025

Is there any evidence to suggest that the variability in fuselage thickness is less for carbon fiber fuselages? Use $\alpha = 0.01$.

10.159 Biology and Environmental Science   Piers on public beaches are usually supported by widely spread piles or pillars and can extend a thousand feet into the ocean. Many piers are extensions of boardwalks, and visitors frequently fish or simply sightsee along these walkways. Longer piers tend to be more susceptible to wind and storm damage. Independent random samples of concrete and wooden piers on public beaches along the California and Florida coasts were obtained and the length (in feet) of each was recorded. Assume the underlying populations are normal, with equal variances. Is there any evidence to suggest that the population mean pier length in California is greater than the population mean pier length in Florida? Use $\alpha = 0.01$. **PIERS**

10.160 Public Health and Nutrition   In case you missed it, the United Nations declared 2008 as the International Year of the Potato. Seriously, potatoes are a good source of carbohydrates, protein, fiber, and potassium. However, the amount of each element varies depending on where the potato is grown. Independent random samples of medium-sized potatoes from Russia and China were obtained, and the amount of potassium (in mg) was measured in each. The data are summarized in the following table.

Location	Sample size	Sample mean	Sample standard deviation
Russia	25	896.8	92.9
China	30	866.0	120.0

Assume normality and equal population variances. Is there any evidence to suggest that the population mean potassium level is different for a medium-sized potato in Russia and China? Use $\alpha = 0.05$.

10.161 Biology and Environmental Science The moisture content in bulk grain is important, because high values can encourage the development of fungi. Potential buyers want to know how much water they are buying along with their grain. Two direct methods for measuring the moisture content are by means of a chemical reaction (with iodine in the presence of sulfur dioxide) and by distillation. A random sample of bulk grain was obtained, and the moisture content of each grain sample was measured as a percentage of water using each method. Assuming normality, conduct the appropriate hypothesis test to determine whether there is any difference in the population mean moisture content of bulk grain measured by chemical reaction and by distillation. Use $\alpha = 0.05$. **GRAIN**

10.162 Psychology and Human Behavior Two recent studies suggest that people who drive really nice cars exhibit some very bad habits. In one study, as a car approached a crosswalk, a person stepped into the road, and the driver's reaction was recorded. In another, similar study, independent random samples of drivers were selected, and their behavior was observed at a four-way intersection. For luxury-car drivers, $n_1 = 217$ and 130 cut ahead in the usual four-way rotation. For ordinary-car drivers, $n_2 = 182$ and 82 violated the four-way-intersection rotation rule. Is there any evidence to suggest that the proportion of luxury-car drivers with insufferable driving habits is greater than the proportion of ordinary-car drivers with similar habits? Use $\alpha = 0.01$. Note: The largest group of driving-rule etiquette violators were men, ages 35–50, with blue BMWs.[31]

10.163 Public Policy and Political Science California law requires fuel outlets to install special catch basins designed to contain gasoline leaks in underground storage tanks. Owners who do not comply can face stiff fines and other penalties. Independent random samples of gasoline stations around Los Angeles and around San Francisco were obtained, and each station was inspected for catch basins. Sixteen of 140 stations near Los Angeles had no catch basins, and 12 of 126 in San Francisco were not complying with the law.

a. Find the sample proportion of stations without catch basins near each city. Verify the nonskewness criterion.
b. Is there any evidence that the population proportion of stations in noncompliance with the law is different near Los Angeles and near San Francisco? Use $\alpha = 0.01$.

10.164 Manufacturing and Product Development During Summer 2013, Procter and Gamble (P&G) recalled 30 different types of cat and dog food because they may have been contaminated with *Salmonella*. While pets can become ill from eating contaminated foods, the Centers for Disease Control and Prevention also reminded people to wash their hands thoroughly after handling pet food. Independent random samples of P&G cat foods and dog foods were obtained, and each was tested for *Salmonella*. For cat food, $n_1 = 1250$ and 50 were contaminated, and for dog food,

$n_2 = 1448$ and 87 were contaminated. Is there any evidence to suggest that the true proportion of contaminated cat food is different from the true proportion of contaminated dog food? Use $\alpha = 0.05$.

EXTENDED APPLICATIONS

10.165 Economics and Finance The U.S. Internal Revenue Service estimates that the average taxpayer takes approximately six hours to complete Form 1040. A study was conducted to examine the amount of time it takes to complete this dreaded form, by income level. Independent random samples of federal filers in two income ranges were obtained, and the length of time (in hours) to complete Form 1040 was recorded for each. The summary statistics are given in the following table.

Income level (in dollars)	Sample size	Sample mean	Sample variance
50,000–<100,000	17	4.56	1.5625
100,000–<200,000	14	6.58	15.0544

Assume the underlying populations are normal.
a. Conduct an F test to determine whether there is any evidence that the two population variances are different. Use $\alpha = 0.02$.
b. Using your conclusion from part (a), conduct the appropriate test for evidence that the mean time to complete Form 1040 for the lower-income level is less than the mean time for the higher-income level. Use $\alpha = 0.05$. State your conclusion and find bounds on the p value.

10.166 Public Policy and Political Science In many states, lawyers are encouraged to do *pro bono* work by both their firms and judicial advisory councils. However, in recent years lawyers have been devoting more time to paying clients and less time to *pro bono* legal aid. Independent random samples of lawyers from two large firms were obtained, and the number of *pro bono* hours for the past year was recorded for each lawyer. The summary statistics are given in the following table.

Law firm	Sample size	Sample mean	Sample variance
Dewey, Cheatum, & Howe	26	75.1	5.92
Fine, Howard, & Fine	26	80.9	5.65

Assume the underlying populations are normal, with equal variances.
a. Is there any evidence to suggest the mean number of yearly *pro bono* hours is different at these two law firms? Use $\alpha = 0.01$.
b. Construct a 99% confidence interval for the difference in mean *pro bono* hours.
c. Does the confidence interval in part (b) support your conclusion in part (a)? Explain.

10.167 Economics and Finance   Online investing has grown with the Internet and with companies like E*TRADE and Ameritrade. Independent random samples of investors were obtained and asked whether they traded online within the past year. The data are given in the following table, by portfolio size.

Portfolio	Sample size	Number of online traders
Less than $100,000	348	132
At least $100,000	226	65

a. Compute the sample proportions and verify the nonskewness criterion.
b. Construct a 95% confidence interval for the difference in population proportions of online investors.
c. Using the interval in part (b), is there any evidence to suggest that the population proportion of online investors is different for these two portfolio classifications? Justify your answer.

10.168 Look, in the Sky   The American Meteor Society maintains a running fireball-tracking system. All reports are analyzed and grouped according to several variables. In 2012, there were 2302 fireball reports, and 590 of these were confirmed by 2–5 witnesses. As of August 2013, there were 507 reports and 142 were confirmed by 2–5 witnesses.[32] Assume the samples are independent. Is there any evidence to suggest that the proportion of fireball reports that are confirmed by 2–5 witnesses is different for these two time periods? Use $\alpha = 0.05$.

10.169 Biology and Environmental Science   Benzene, toluene, ethylbenzene, m-, p-xylenes, and o-xylene are volatile organic compounds that are found in residential environments and can cause severe health problems, for example, dizziness, tremors, eye, ear, and throat irritation. The Canadian Health Measures Survey was administered to more than 5000 individuals in order to predict the presence of these compounds. Suppose a random sample of these respondents was selected, and the concentration of toluene was measured in each residence (in $\mu g/m^3$). The results are given in the following table.[33]

Dwelling	Sample size	Sample mean	Sample standard deviation
Single detached	15	8.68	6.32
Double/duplex	15	13.16	9.20
Apartment	16	23.70	16.52

Assume all three underlying populations are normal, with equal variances. Conduct the appropriate hypothesis tests to determine which pairs of population mean toluene levels are different. Use $\alpha = 0.01$ in each test.

10.170 Physical Sciences   Independent random samples of ore taken from two high-grade gold mines were obtained, and the gold value (in grams/tonne) was measured for each. The summary statistics were

El Aguila mine	$n_1 = 8$	$\bar{x}_1 = 15.6$	$s_1 = 5.2$
Dolaucothi mine	$n_2 = 11$	$\bar{x}_2 = 26.8$	$s_2 = 21.6$

Assume normality and unequal variances.
a. Conduct the appropriate hypothesis test to determine whether there is any evidence that the population mean gold value at the El Aguila mine is less than at the Dolaucothi mine. Use $\alpha = 0.05$.
b. Your conclusion in part (a) should be that there is no evidence to suggest a difference. Explain why this result is correct even though the sample means are *very far apart*.

LAST STEP

10.171 Are people who live at higher altitudes slimmer?   Random samples were obtained for 125 adults living in Denver, Colorado (the Mile High City), and 150 adults living in New Orleans, Louisiana (where the mean elevation is zero feet above sea level). Using body mass index as a measure of obesity, 38 of those living in Denver and 61 of those living in New Orleans were classified as obese. Is there any evidence to suggest that the people living in Denver are thinner? That is, is there any evidence to suggest that the proportion of adults classified as obese in Denver is less than the proportion of adults classified as obese in New Orleans? Use $\alpha = 0.05$.

11 The Analysis of Variance

Do the hours worked by farm laborers differ by region?

Farm workers make a huge contribution to the welfare of the country. They are involved in planting, cultivating, and harvesting many different types of foods for the United States and the world. Their work is tiring and grueling, yet farm workers have a very low income level, frequently below the poverty line.

Wages vary by region and by the number of years a person has worked for the same employer. In addition, some farm workers are paid based on a piece rate, that is, by the number of baskets or crates they pick of the crop being harvested. There is no overtime pay and little job security.

Some research suggests farm workers work longer hours in different regions of the United States. Random samples of farm workers from four regions were obtained, and the number of hours worked during a spring planting week was recorded for each. The summary statistics are given in the following table.[1]

Region	Sample size	Sample mean	Sample standard deviation
Northeast II	15	38.6	4.370
Cornbelt I	15	39.5	4.042
Mountain I	15	43.9	4.597
California	15	42.8	3.651

The statistical techniques presented in this chapter can be used to determine whether *any* two population mean number of hours worked per week are different. If at least two means are different, then we need to identify which pairs of means are contributing to an overall difference.

CONTENTS

11.1 One-Way ANOVA

A one-way, or single-factor, ANOVA (ANalysis Of VAriance) involves the analysis of data sampled from more than two populations. The only difference among the populations is a single factor. For example, consider a study in which random samples of the amount of carbon dioxide in underground train tunnels are obtained in four different cities. The data may be used to determine whether there is any difference in the mean amount of carbon dioxide in train tunnels among the four cities. The single factor that varies among the populations is the city. Or suppose a researcher is investigating techniques for controlling the amount of electricity lost during transmission over utility lines. Experimental results may be used to determine whether there is any difference in the mean amount of electricity lost for five differently designed lines. The single factor here is the design of the electricity line. In theory, everything else is the same among the five populations, for example, the initial amount of electricity transmitted, the distance the electricity is transmitted, the weather conditions, etc.

Suppose three random samples are obtained, and a histogram is constructed using *all* of the data. The resulting graph is shown in Figure 11.1. Analysis of variance is used to determine whether the data came from a single population, or whether at least two samples came from populations with different means.

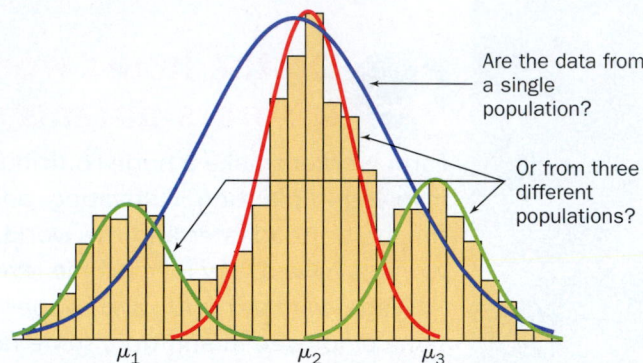

Figure 11.1 A visualization of a typical ANOVA problem.

The notation used in a one-way ANOVA is similar to and is an extension of the notation used in Chapter 10.

ANOVA Notation

k = the number of populations under investigation.

Population	1	2	$\cdots$	i	$\cdots$	k
Population mean	μ_1	μ_2	$\cdots$	μ_i	$\cdots$	μ_k
Population variance	σ_1^2	σ_2^2	$\cdots$	σ_i^2	$\cdots$	σ_k^2
Sample size	n_1	n_2	$\cdots$	n_i	$\cdots$	n_k
Sample mean	$\bar{x}_{1.}$	$\bar{x}_{2.}$	$\cdots$	$\bar{x}_{i.}$	$\cdots$	$\bar{x}_{k.}$
Sample variance	s_1^2	s_2^2	$\cdots$	s_i^2	$\cdots$	s_k^2

$$n = n_1 + n_2 + \cdots + n_k$$

= the total number of observations in the *entire* data set

The null and alternative hypotheses are stated in terms of the population means.

$H_0: \mu_1 = \mu_2 = \cdots = \mu_k$ (All k population means are equal.)

$H_a: \mu_i \neq \mu_j$ for some $i \neq j$ (At least two of the k population means differ.)

The assumptions for this test procedure are similar to those for a two-sample t test.

For reference, these are the one-way ANOVA assumptions.

1. The k population distributions are normal.
2. The k population variances are equal; that is, $\sigma_1^2 = \sigma_2^2 = \cdots = \sigma_k^2$.
3. The samples are selected randomly and independently from the respective populations.

To denote observations, we use a single letter with two subscripts. The first subscript indicates the sample number and the second subscript denotes the observation number within that sample. In general,

$$x_{ij} = \text{the } j\text{th measurement taken from the } i\text{th population.}$$
$$X_{ij} = \text{the corresponding random variable.}$$

A comma is placed between i and j if there is ambiguity. For example, $x_{1,23}$ is the 23rd observation in the first sample, but $x_{12,3}$ is the 3rd observation in the 12th sample.

The mean of the observations in the ith sample is

$$\bar{x}_{i.} = \frac{1}{n_i} \sum_{j=1}^{n_i} x_{ij} = \frac{1}{n_i}(x_{i1} + x_{i2} + \cdots + x_{in_i}) \tag{11.1}$$

The dot in the second subscript is used to indicate a sum over that subscript (j) while the first subscript (i) is held fixed.

The mean of *all* the observations, called the **grand mean**, is the sum of all the observations divided by n.

$$\bar{x}_{..} = \frac{1}{n} \sum_{i=1}^{k} \sum_{j=1}^{n_i} x_{ij} \tag{11.2}$$

There is just a little more notation that will make some of the calculations easier.

A more theoretical development of a one-way ANOVA test includes the corresponding *random variables* $\bar{X}_{i.}$, $\bar{X}_{..}$, $T_{i.}$, and $T_{..}$.

$$t_{i.} = \sum_{j=1}^{n_i} x_{ij} \quad = \text{sum of the observations in the } i\text{th sample.}$$

$$t_{..} = \sum_{i=1}^{k} \sum_{j=1}^{n_i} x_{ij} = \text{sum of } all \text{ the observations.}$$

Here's where the analysis of *variance* plays a role. The total variation in the data (the total sum of squares) is decomposed into a sum of *between-samples* variation (the sum of squares due to factor) and *within-samples* variation (the sum of squares due to error). The total variation is the variability of individual observations from the grand mean. The between-samples (or between-factor) variation is the variability in the sample means; this tells us how different the sample means are from each other. The within-samples variation is the variability of the observations from their sample mean; this is just like the sample variance we have already been using. The three sums of squares are defined in the following fundamental identity, which shows the decomposition of the total variation in the data.

One-Way ANOVA Identity

Let SST = total sum of squares, SSA = sum of squares due to factor, and SSE = sum of squares due to error.

$$\underbrace{\sum_{i=1}^{k} \sum_{j=1}^{n_i} (x_{ij} - \bar{x}_{..})^2}_{\text{SST}} = \underbrace{\sum_{i=1}^{k} n_i(\bar{x}_{i.} - \bar{x}_{..})^2}_{\text{SSA}} + \underbrace{\sum_{i=1}^{k} \sum_{j=1}^{n_i} (x_{ij} - \bar{x}_{i.})^2}_{\text{SSE}} \tag{11.3}$$

A CLOSER LOOK

1. SSA is used to denote the sum of squares due to factor instead of SSF because in a two-way ANOVA there is a factor A and a factor B.
2. The sample size is used as a *weight* in the expression for SSA.

If the null hypothesis is true, then each observation comes from the same population with mean μ and variance σ^2. Therefore, the sample means, the $\bar{x}_i$'s, should all be about the same, should all be close to the grand mean, $\bar{x}_{..}$, and SSA should be (relatively) *small*. If at least two population means are different, then at least two $\bar{x}_i$'s should be very different, and these values will be far away from $\bar{x}_{..}$. In this case, SSA should be (relatively) *large*.

The one-way ANOVA test statistic is based on two separate estimates for the common variance, σ^2, computed using SSA and SSE.

Definition

The **mean square due to factor, MSA**, is SSA divided by $k-1$:

$$MSA = \frac{SSA}{k-1} \tag{11.4}$$

The **mean square due to error, MSE**, is SSE divided by $n-k$:

$$MSE = \frac{SSE}{n-k} \tag{11.5}$$

If the null hypothesis is true, then (the random variable) MSA is an unbiased estimator of σ^2. If H_a is true, then MSA tends to overestimate σ^2. The mean square due to error, MSE, is an unbiased estimator of the common variance σ^2 whether H_0 or H_a is true.

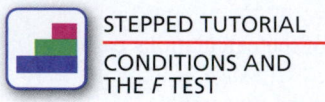

STEPPED TUTORIAL

CONDITIONS AND THE *F* TEST

One-way ANOVA does not mean a one-sided statistical test, but indicates that there is one factor.

Consider the ratio $F = MSA/MSE$. If the value of F is close to 1, then the two estimates of σ^2, or sources of variation, are approximately the same. There is no evidence to suggest that the population means are different. If the value of F is much greater than 1, then the variation *between* samples is greater than the variation *within* samples. This suggests that the alternative hypothesis is true.

If the one-way ANOVA assumptions are satisfied and H_0 is true, then the statistic $F = MSA/MSE$ has an F distribution with $k-1$ and $n-k$ degrees of freedom. Because *large* values of F suggest that H_a is true, the rejection region is *only* in the right tail of the distribution.

One-Way ANOVA Test Procedure

Given the one-way ANOVA assumptions, the test procedure with significance level α is

H_0: $\mu_1 = \mu_2 = \cdots = \mu_k$

H_a: $\mu_i \neq \mu_j$ for some $i \neq j$

TS: $F = \dfrac{MSA}{MSE}$

RR: $F \geq F_{\alpha, k-1, n-k}$

If the value of F is smaller than the critical value, then there is no evidence from the data to reject H_0. If the value of F is in the rejection region, we say that the F test is significant and that there is a statistically significant difference among the population means. Recall that we can also use the p value to conduct an equivalent test: If $p \leq \alpha$, then we reject H_0.

Remember that if any of the one-way ANOVA assumptions are violated, then the conclusion is unreliable. Several methods to check for normality were presented in Section 6.3, and there are statistical procedures for testing equality of variances. The samples must also be selected randomly and independently from the appropriate populations.

The following *computational formulas* (rather than the definitions) are used to find the sums of squares, and then the mean squares.

These equations are used for the same reasons the computational formula for the sample variance is used: They are easier, faster, and more accurate.

Computational Formulas

$$\text{SST} = \underbrace{\sum_{i=1}^{k}\sum_{j=1}^{n_i}(x_{ij} - \bar{x}_{..})^2}_{\text{definition}} = \underbrace{\left(\sum_{i=1}^{k}\sum_{j=1}^{n_i}x_{ij}^2\right) - \frac{t_{..}^2}{n}}_{\text{computational formula}}$$

$$\text{SSA} = \underbrace{\sum_{i=1}^{k}n_i(\bar{x}_{i.} - \bar{x}_{..})^2}_{\text{definition}} = \underbrace{\left(\sum_{i=1}^{k}\frac{t_{i.}^2}{n_i}\right) - \frac{t_{..}^2}{n}}_{\text{computational formula}}$$

$$\text{SSE} = \underbrace{\sum_{i=1}^{k}\sum_{j=1}^{n_i}(x_{ij} - \bar{x}_{i.})^2}_{\text{definition}} = \underbrace{\text{SST} - \text{SSA}}_{\text{computational formula}}$$

One last detail: One-way ANOVA calculations are often presented in an **analysis of variance table**, or ANOVA table. The values included in this table are associated with the three sources of variation and the calculation of the F statistic.

One-way ANOVA summary table

Source of variation	Sum of squares	Degrees of freedom	Mean square	F	p Value
Factor	SSA	$k - 1$	$\text{MSA} = \dfrac{\text{SSA}}{k - 1}$	$\dfrac{\text{MSA}}{\text{MSE}}$	p
Error	SSE	$n - k$	$\text{MSE} = \dfrac{\text{SSE}}{n - k}$		
Total	SST	$n - 1$			

The following example illustrates the computations involved in a one-way ANOVA and the process of making an inference based on the value of the test statistic.

DATA SET

SODIUM

Solution Trail 11.1

KEYWORDS

- Is there any evidence?
- At least two of the population means are different
- Four categories
- Independent random samples

TRANSLATION

- Statistical inference
- Analysis of variance

CONCEPTS

- One-way ANOVA test procedure

VISION

Compute the summary statistics necessary to complete the ANOVA summary table. Find the critical value, and draw the appropriate conclusion.

Example 11.1 Take with a Grain of Salt

A recent research study suggests that a high-salt diet in older women increases the risk of breaking a bone.[2] Although the biological mechanism is still unclear, there does appear to be an association between excessive sodium intake and bone fragility. One measure of bone health is the vitamin D blood level. As determined by food questionnaires, independent random samples of older women in four different salt intake categories were obtained. The vitamin D blood level (in nmol/L) was measured in each. The data are given in the following table.

Sample	Observations				
Very high	91.5	77.5	94.5	77.5	92.0
High	89.0	92.0	98.2	80.0	86.7
Moderate	92.5	100.7	94.0	93.3	106.3
Low	100.1	98.0	99.1	103.9	97.6

Is there any evidence to suggest that at least two of the population mean vitamin D blood levels are different? Use $\alpha = 0.05$.

SOLUTION

STEP 1 Assume the one-way ANOVA assumptions are true. There are $k = 4$ samples or groups and $n = 5 + 5 + 5 + 5 = 20$ total observations. Some summary statistics are given in the following table.

The summary statistics table suggests a possible violation of the equal-variances assumption. There are formal statistical procedures for testing equality of (several) variances, but the ANOVA test statistic is robust. That is, even if the population variances are a little different, the test still provides reliable results. In addition, because the sample sizes are small, it is more difficult to detect a real difference in population variances.

Sample	Sample size	Sample total	Sample mean	Sample variance
Very high	$n_1 = 5$	$t_{1.} = 433.0$	$\bar{x}_{1.} = 86.60$	$s_1^2 = 70.30$
High	$n_2 = 5$	$t_{2.} = 445.9$	$\bar{x}_{2.} = 89.18$	$s_2^2 = 44.94$
Moderate	$n_3 = 5$	$t_{3.} = 486.8$	$\bar{x}_{3.} = 97.36$	$s_3^2 = 35.62$
Low	$n_4 = 5$	$t_{4.} = 498.7$	$\bar{x}_{4.} = 99.74$	$s_4^2 = 6.36$

The sum of all the observations, or grand total, is

$$t_{..} = 433.0 + 445.9 + 486.8 + 498.70 = 1864.4$$

STEP 2 The four parts of the hypothesis test are

$H_0: \mu_1 = \mu_2 = \mu_3 = \mu_4$ (all four population means are equal)

$H_a: \mu_i \neq \mu_j$ for some $i \neq j$ (at least two population means differ)

TS: $F = \dfrac{\text{MSA}}{\text{MSE}}$

RR: $F \geq F_{\alpha, k-1, n-k} = F_{0.05, 3, 16} = 3.24$

STEP 3 Find the total sum of squares.

$$\text{SST} = \left(\sum_{i=1}^{k} \sum_{j=1}^{n_i} x_{ij}^2 \right) - \frac{t_{..}^2}{n}$$ Computational formula for SST.

$$= (91.5^2 + 77.5^2 + \cdots + 97.6^2) - \frac{1864.4^2}{20}$$ Apply the formula.

$$= 175{,}027.24 - 173{,}799.368 = 1227.872$$ Simplify.

Find the sum of squares due to factor.

$$\text{SSA} = \left(\sum_{i=1}^{k} \frac{t_{i.}^2}{n_i} \right) - \frac{t_{..}^2}{n}$$ Computational formula for SSA.

$$= \left(\frac{433.0^2}{5} + \frac{445.9^2}{5} + \frac{486.8^2}{5} + \frac{498.7^2}{5} \right) - \frac{1864.4^2}{20}$$ Use sample totals and grand total.

$$= 174{,}398.348 - 173{,}799.368 = 598.98$$ Simplify.

Use these two values to find the sum of squares due to error.

$$\text{SSE} = \text{SST} - \text{SSA} = 1227.872 - 598.98 = 628.892$$

STEP 4 Compute MSA and MSE.

MSA = SSA/$(k - 1)$ Definition.

$= 598.98/(4 - 1) = 199.66$ Use SSA and $k = 4$ groups.

MSE = SSE/$(n - k)$ Definition.

$= 628.892/(20 - 4) = 39.3058$ Use SSE, $n = 20$ total observations, and $k = 4$.

The next reasonable question is, "Which pairs of means are contributing to this *overall* difference?" We'll address this issue in Section 11.2.

STEP 5 The value of the test statistic is

$$f = \frac{\text{MSA}}{\text{MSE}} = \frac{199.66}{39.3058} = 5.08 \ (\geq 3.24)$$

The value of the test statistic lies in the rejection region. Reject the null hypothesis. There is evidence to suggest that at least two population means are different.

STEP 6 Recall that because the tables of critical values for F distributions are limited, we can only bound the p value. Place the value of the test statistic, $f = 5.08$, in an ordered list of critical values with degrees of freedom 3 and 16.

$$3.24 \quad \leq 5.08 \leq 5.29$$

$$F_{0.05, 3, 16} \leq 5.08 \leq F_{0.01, 3, 16}$$

Therefore, $0.01 \quad \leq \ p \ \leq 0.05$

The exact p value is illustrated in Figure 11.2.

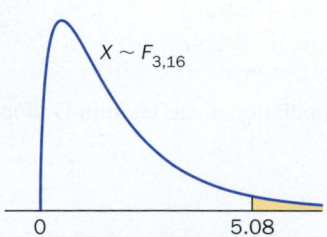

Figure 11.2 *p*-Value illustration:

$p = P(X \geq 5.08)$

$= 0.0117 \leq 0.05 = \alpha$

STEP 7   Here's how all of these calculations are presented in an ANOVA table.

ANOVA summary table

Source of variation	Sum of squares	Degrees of freedom	Mean square	F	p Value
Factor	598.980	3	199.66	5.08	0.0117
Error	628.892	16	39.31		
Total	1227.870	19			

The p value in this table is from a technology solution.

Figure 11.3 shows a technology solution.

Results - One-Way ANOVA					
Export ▾					
Source	Sum of Squares	df	Mean Square	F-value	P-value
Treatment	599.0	3	199.7	5.080	0.01166
Error	628.9	16	39.31		
Total	1228	19			

Figure 11.3 ANOVA one-way results.

VIDEO TECH MANUALS
ONE-WAY ANOVA

STATISTICAL APPLET
ONE-WAY ANOVA

DATA SET
MERCURY

TRY IT NOW   GO TO EXERCISE 11.15

In the next example, fewer detailed calculations are shown. However, the ANOVA table is presented, with the focus on the inference process.

Example 11.2 Mercury Rising

Although mercury is useful in many ways, recent research suggests that overexposure can cause severe health problems. The use of mercury has increased worldwide in industry and in mining, and traces of mercury are emitted during power generation. The National Atmospheric Deposition Program maintains a long-term record of total mercury concentration in precipitation in the United States and Canada. Independent random samples from five sites were obtained and the mercury concentrations (in ng/L) are given in the following table.[3]

Site	Observations					
CA 20	6.59	2.78	7.85	5.99	4.27	5.25
CA 75	4.48	6.82	3.44	7.66	2.76	5.85
CA 94	6.08	8.81	3.11	3.16	2.57	3.87
CO 96	2.62	4.92	4.78	7.45	3.43	3.01
NV 99	3.75	2.45	3.31	8.62	4.16	3.12

Is there any evidence to suggest that there is a difference in population mean mercury concentration in precipitation at these five sites? Use $\alpha = 0.05$.

SOLUTION

STEP 1   Assume the one-way ANOVA assumptions are true. There are $k = 5$ groups and $n = 6 + 6 + 6 + 6 + 6 = 30$ total observations. We would like to know whether the data suggest that any two of the population mean mercury concentrations are different.

The four parts of the hypothesis test are

$H_0: \mu_1 = \mu_2 = \mu_3 = \mu_4 = \mu_5$

$H_a: \mu_i \neq \mu_j$ for some $i \neq j$

TS: $F = \dfrac{\text{MSA}}{\text{MSE}}$

RR: $F \geq F_{\alpha,4,25} = F_{0.05,4,25} = 2.76$

Solution Trail 11.2

KEYWORDS

- Is there any evidence?
- Difference in population means
- Five sites
- Independent random samples

TRANSLATION

- Statistical inference
- Analysis of variance

CONCEPTS

- One-way ANOVA test procedure

VISION

Compute the summary statistics necessary to complete the ANOVA summary table. Find the critical value, and draw the appropriate conclusion.

STEP 2 Find the total sum of squares.

$$\text{SST} = \left(\sum_{i=1}^{k} \sum_{j=1}^{n_i} x_{ij}^2 \right) - \frac{t_{..}^2}{n} = \cdots = 110.5687$$

Find the sum of squares due to factor.

$$\text{SSA} = \left(\sum_{i=1}^{k} \frac{t_{i.}^2}{n_i} \right) - \frac{t_{..}^2}{n} = \cdots = 6.6255$$

Use these two values to find the sum of squares due to error.

$$\text{SSE} = \text{SST} - \text{SSA} = 110.5687 - 6.6255 = 103.9432$$

STEP 3 Compute MSA and MSE.

$$\text{MSA} = \text{SSA}/(k - 1) \qquad \qquad \text{Definition.}$$
$$= 6.6255/(5 - 1) = 1.6564 \qquad \text{Use SSA and } k = 5 \text{ groups.}$$
$$\text{MSE} = \text{SSE}/(n - k) \qquad \qquad \text{Definition.}$$
$$= 103.9432/(30 - 5) = 4.1577 \quad \text{Use SSE, } n = 30 \text{ total observations, and } k = 5.$$

STEP 4 The value of the test statistic is

$$f = \frac{\text{MSA}}{\text{MSE}} = \frac{1.6564}{4.1577} = 0.3984$$

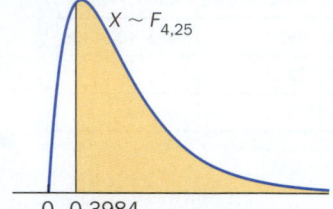

Figure 11.4 p-Value illustration:
$p = P(X \geq 0.3984)$
$= 0.8079 > 0.05 = \alpha$

The value of the test statistic does not lie in the rejection region. Equivalently, $p = 0.8079 > 0.05$ as illustrated in Figure 11.4. Do not reject the null hypothesis. There is no evidence to suggest that any two population mean mercury concentrations are different.

STEP 5 Here is the ANOVA summary table.

ANOVA summary table

Source of variation	Sum of squares	Degrees of freedom	Mean square	F	p Value
Factor	6.6255	4	1.6564	0.39	0.8079
Error	103.9432	25	4.1577		
Total	110.5687	29			

The p value in this table is from a technology solution.

Figure 11.5 shows a technology solution.

Analysis of Variance

Source	DF	Sum of Squares	Mean Square	F Ratio	Prob > F
Site	4	6.62548	1.65637	0.3984	0.8079
Error	25	103.94327	4.15773		
C. Total	29	110.56875			

Figure 11.5 JMP analysis of variance table.

TRY IT NOW   GO TO EXERCISE 11.20

Technology Corner

Procedure: Conduct a one-way analysis of variance test.
Reconsider: Example 11.1, solution, and interpretations.

CrunchIt!

Use the ANOVA; one-way function.

1. Enter the data from sample 1 into column Var1, from sample 2 into column Var2, from sample 3 into column Var3, and from sample 4 into column Var4.
2. Select Statistics; ANOVA; One-way.
3. Under the Columns tab, check the column names containing the data.
4. Click Calculate. The results are displayed in a new window. Refer to Figure 11.3.

TI-84 Plus C

Use the built-in function ANOVA.

1. Enter the data from sample 1 into list L1, from sample 2 into list L2, from sample 3 into list L3, and from sample 4 into list L4.
2. Select STAT; TESTS; ANOVA.
3. Enter the lists containing the data as function arguments. See Figure 11.6.
4. Press ENTER. The results are displayed on the Home Screen. See Figure 11.7.

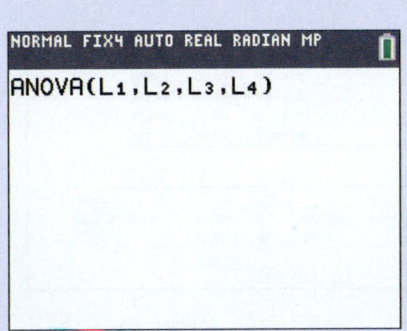

Figure 11.6 ANOVA function with the data lists as arguments.

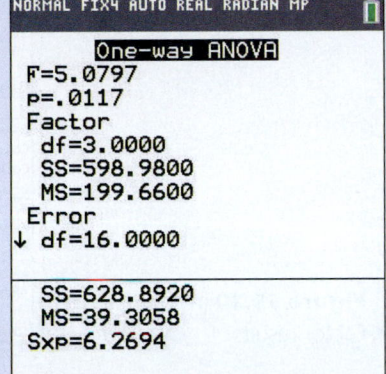

Figure 11.7 One-way ANOVA results.

Minitab

Minitab has several built-in functions to conduct an analysis of variance test, depending on the model. Use <u>O</u>ne-Way for the model presented in this section.

1. Enter the data from sample 1 into column C1, from sample 2 into column C2, from sample 3 into column C3, and from sample 4 into column C4. Each sample is in a separate column.
2. Select <u>S</u>tat; <u>A</u>NOVA; <u>O</u>ne-Way.
3. Select Response data are in a separate column for each factor level. Enter the Responses: the columns containing the data.
4. Select Options. Check the box to Assume equal variances and select an appropriate Confidence level and Type of confidence interval.
5. Click OK. The output is displayed in a session window and an Interval Plot is constructed in a graph window. See Figure 11.8 and 11.9.

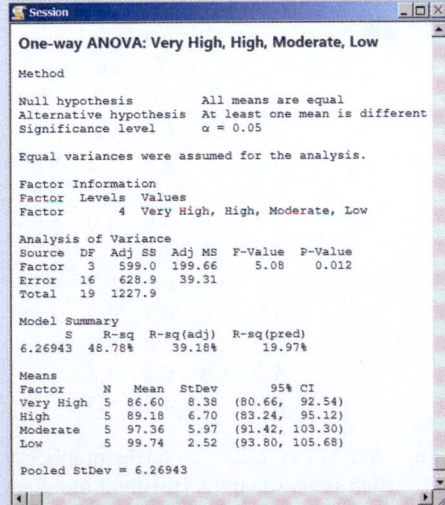

Figure 11.8 Minitab ANOVA one-way results.

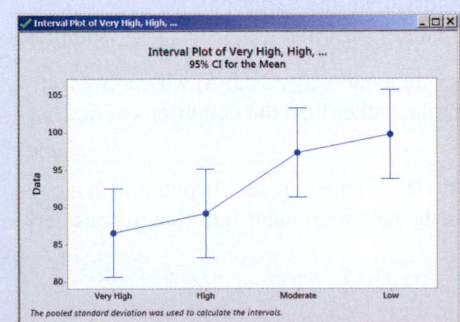

Figure 11.9 Interval plot associated with one-way ANOVA results.

Excel

Use the Data Analysis Toolkit function Anova: Single Factor.

1. Enter the data from sample 1 into column A, from sample 2 into column B, from sample 3 into column C, and from sample 4 into column D.
2. Under the Data tab, select Data Analysis; Anova: Single Factor.
3. Enter the Input Range, choose Grouped By Columns, and select an Output option. Click OK to display the results. See Figure 11.10.

Anova: Single Factor

SUMMARY

Groups	Count	Sum	Average	Variance
Column 1	5	433.0	86.60	70.300
Column 2	5	445.9	89.18	44.942
Column 3	5	486.8	97.36	35.618
Column 4	5	498.7	99.74	6.363

ANOVA

Source of Variation	SS	df	MS	F	P-value	F crit
Between Groups	598.980	3	199.66	5.08	0.0117	3.24
Within Groups	628.892	16	39.31			
Total	1227.872	19				

Figure 11.10 Anova: Single Factor results.

SECTION 11.1 EXERCISES

Concept Check

11.1 True/False In a one-way ANOVA, the k population variances are assumed equal.

11.2 True/False In a one-way ANOVA, all sample sizes must be the same in order for the test statistic to be valid.

11.3 True/False If we reject the null hypothesis in a one-way ANOVA, there is evidence to suggest all pairs of population means are unequal.

11.4 True/False If we reject the null hypothesis in a one-way ANOVA, there is evidence to suggest the underlying population variances are unequal.

11.5 Short Answer If the null hypothesis is true in a one-way ANOVA, then MSA is an unbiased estimator for _____.

11.6 Short Answer In a one-way ANOVA, if the value of F is much greater than 1, then the variation _____ samples is greater than the variation _____ samples.

11.7 Short Answer In a one-way ANOVA, why do we use the computation formulas rather than the definitions to find SSA, SSE, and SST?

11.8 Short Answer If we reject the null hypothesis in a one-way ANOVA, what is the next reasonable question to consider?

Practice

11.9 Consider the following data in the context of an ANOVA test. EX11.9

Group	Observations
1	33 27 27 32 27 31 23 26 34
2	27 35 32 28 35 39 33
3	30 36 33 35 33 28

a. Find n_i, $i = 1, 2, 3$ (the number of observations in each sample), and n (the total number of observations).
b. Find t_i, $i = 1, 2, 3$ (the total for each sample), and $t_{..}$ (the grand total).
c. Find $\sum_{i=1}^{3}\sum_{j=1}^{n_i} x_{ij}^2$ (the sum of all the squared observations).

11.10 Consider the following data in the context of an ANOVA test. EX11.10

1	2	3	4	5	6
2.8	4.8	6.3	6.8	6.4	6.3
5.6	4.2	3.9	3.3	5.8	4.7
3.7	4.4	5.4	5.7	6.8	5.9
4.4	2.8	4.4	4.0	5.7	3.4
5.8	5.1	3.0	6.1	5.7	5.7
6.1	3.2	6.3	7.5	4.3	2.3
5.0	4.6	6.9	2.9	2.5	4.5
5.9	2.3	2.2	5.1	6.8	4.9
5.2	6.3	6.8	3.8		3.6
6.8		3.9	5.5		5.6

a. Find n_i, $i = 1, 2, \ldots, 6$ (the number of observations in each sample), and n (the total number of observations).
b. Find t_i, $i = 1, 2, \ldots, 6$ (the total for each sample), and $t_{..}$ (the grand total).

c. Find $\displaystyle\sum_{i=1}^{6}\sum_{j=1}^{n_i} x_{ij}^2$ (the sum of all the squared observations).

11.11 Consider the following data in the context of an ANOVA test. ⬛ EX11.11

Factor	Observations				
1	162	155	157	144	157
2	168	147	163	131	136
3	135	138	155	172	168
4	144	162	150	140	157

a. Find $t_{i.}$ for $i = 1, 2, 3, 4$. Find $t_{..}$.

b. Find $\displaystyle\sum_{i=1}^{4}\sum_{j=1}^{n_i} x_{ij}^2$.

c. Find SST (the total sum of squares), SSA (the sum of squares due to factor), and SSE (the sum of squares due to error).

d. Find MSA (the mean square due to factor), and MSE (the mean square due to error).

e. Compute the value of the test statistic.

11.12 Consider the data on the text website in the context of an ANOVA test. ⬛ EX11.12

a. Find SST, SSA, and SSE.

b. Find MSA and MSE.

c. Compute the value of the test statistic.

d. If $\alpha = 0.05$, would you reject the null hypothesis? Justify your answer.

11.13 Complete the following ANOVA table.

ANOVA summary table

Source of variation	Sum of squares	Degrees of freedom	Mean square	F	p Value
Factor		4			
Error	12,062.1				
Total	12,646.2	51			

11.14 Complete the following ANOVA table.

ANOVA summary table

Source of variation	Sum of squares	Degrees of freedom	Mean square	F	p Value
Factor			4.522		
Error		62	0.988		
Total		65			

a. State the null and the alternative hypothesis in terms of the population means.

b. Find the rejection region for $\alpha = 0.01$.

c. Conduct the test. Is there any evidence to suggest that at least two population means are different? Justify your answer.

Applications

11.15 Public Health and Nutrition   A study was conducted to compare the amount of salt in potato chips. Random samples

of four varieties were obtained and the amount of salt in each one-ounce portion of potato chips was recorded (in mg of sodium). The data are given in the following table. ⬛ CHIPS

Variety	Observations					
BBQ	338	155	239	184	185	261
Cheese-Flavored	235	238	251	229	233	232
Olestra-Based	164	197	136	214	148	230
Baked	290	343	294	373	306	357

Conduct an analysis of variance test to determine whether there is any evidence that the population mean amount of salt per serving is different for at least two varieties. Use $\alpha = 0.05$.

a. State the four parts of the hypothesis test.

b. Complete an ANOVA table.

c. Draw the appropriate conclusion.

11.16 Manufacturing and Product Development   Breitling sells men's gold, silver, and titanium watchbands. A random sample of each type was obtained (in similar styles), and the weight of each watchband (in grams) was recorded. The data are given in the following table. ⬛ WATCHES

Watchband	Observations							
Gold	7.9	7.2	7.8	8.1	7.9	8.3	9.9	
Silver	9.5	7.0	8.7	7.6	7.5	9.3	7.3	6.9
Titanium	6.7	7.1	6.5	7.1	5.5	6.7	4.9	3.9

a. Conduct an analysis of variance test to determine whether there is any evidence that the mean weights of any two watchband types are different. Include an ANOVA table. Use $\alpha = 0.01$. Write a Solution Trail for this problem.

b. Compute the sample mean weight for each sample. Given your conclusion in part (a), which pair(s) of population means do you think are different?

11.17 Travel and Transportation   According to a recent study, U.S. drivers spend approximately 38 hours each year just sitting in traffic. Los Angeles drivers experience the worst traffic congestion; for example, on Interstate 405, drivers traveled just 14 miles per hour.[4] Random drivers in some of the most congested cities were asked for the time (in minutes) it normally takes them to get through the traffic congestion during morning rush hour. The cities and summary statistics are given in the following table.

City	Sample size	Sample total	Sum of squared observations
Chicago	14	463	16,185
Honolulu	18	620	22,458
New York	20	506	13,686
Seattle	15	356	8,964
Washington, D.C.	19	586	18,840

Conduct an analysis of variance test to determine whether there is any evidence that the population mean times through some pair of traffic bottlenecks are different. Use $\alpha = 0.05$.

11.18 Sports and Leisure   A study was conducted to determine the tension needed by various types of guitar strings in order to produce the proper frequency. A high E was used for comparison, and tension was measured in newtons. The guitar string brands and summary statistics are given in the following table.

String brand	Sample size	Sample total	Sum of squared observations
Darco Acoustic	8	458.5	26,347.8
Ernie Ball	9	508.8	28,947.4
Martin	9	554.5	34,242.7
Gibson	8	523.9	34,373.9

Conduct the appropriate test to check the hypothesis of no difference in population mean tensions. Use $\alpha = 0.01$.

11.19 Marketing and Consumer Behavior   The deli department in a Publix Supermarket conducted a survey to compare orders on certain items. A random sample of sliced ham, roast beef, and turkey orders was obtained, and the weight of each order (in pounds) was recorded. Conduct an analysis of variance test to determine whether there is any evidence to suggest that at least two of the population mean weights are different. Use $\alpha = 0.01$. **DELI**

11.20 Public Health and Nutrition   Nondairy creamers (for coffee and tea) contain vegetable fat, corn-syrup solids, casein, and other ingredients. Independent random samples of various nondairy creamers were obtained, and the percentage of fat in a serving size was measured. Is there any evidence to suggest that at least two nondairy creamers have a different population mean percentage of fat per serving? Use $\alpha = 0.05$. Write a Solution Trail for this problem. **CREAMER**

11.21 Biology and Environmental Science   Sulfur dioxide is a toxic gas, has a distinct, pungent odor, and contributes to acid rain and changes in climate. It is usually released into the atmosphere by volcanoes, but sulfur dioxide is also emitted from coal-fired electricity-generating plants. Independent random times at four coal-fired power plants in the United States were selected, and the sulfur dioxide emission rate was measured (in lb/MWh) for each. Summary statistics are given in the following table. Use this information with $\alpha = 0.01$ to test the null hypothesis of no difference in population mean sulfur dioxide emission rates.[5]

Coal-fired plant	Sample size	Sample total	Sum of squared observations
Warrick	18	574.0	18,731.76
Portland	21	584.1	16,493.11
Shawville	15	409.0	11,169.80
Keystone	16	421.4	11,455.32

11.22 Medicine and Clinical Studies   The sweetener sorbitol has fewer calories than sucrose, adds texture to food, is non-carcinogenic, and is very useful for people with diabetes. Surprisingly, this sweetener is also found in children's cough

syrups. Independent random samples of four cough syrups were obtained, and the amount of sorbitol per teaspoon (in grams) was measured. Conduct the appropriate test to determine whether there is any evidence that at least two population mean levels of sorbitol are different. Use $\alpha = 0.01$ and include an ANOVA table. **SORBITOL**

11.23 Fuel Consumption and Cars   Biodiesel is a specific type of biofuel. It is renewable, burns cleanly, and helps to reduce our dependence on imported diesel fuel. Biodiesel is made from a mixture of agricultural oils, cooking oil, and animal fats. Independent random samples of daily production (in gallons) was obtained from five facilities near Houston, Texas. Use these data to test the hypothesis of no difference in the population mean daily production of biodiesel at these five plants. Use $\alpha = 0.05$. **BIOFUEL**

11.24 Manufacturing and Product Development   Lumber is an important part of the Canadian economy. Over 60,000 people are employed in this industry that involves shipping and manufacturing lumber. Independent random samples of monthly hardwood shipments (in cubic meters) from three provinces were obtained. The summary statistics are given in the following table.[6] Conduct an analysis of variance test to determine whether there is evidence that some pair of population mean monthly shipments are different. Use $\alpha = 0.01$ and include an ANOVA table.

Province	Sample size	Sample total	Sum of squared observations
New Brunswick	5	992.6	198,016.20
Ontario	6	1,275.5	272,977.23
Nova Scotia	6	425.7	30,443.45
British Columbia Coast	5	1,285.8	336,878.00

11.25 Biology and Environmental Science   A study was conducted to examine the thaw depths in late August to early September in the floodplain sites of the Tanana River southwest of Fairbanks, Alaska. The data in the following table are a subset of the thaw depth measurements (in cm) from four locations. **THAW**

Study site	Observations					
253	45	47	44	45	50	38
1133	47	58	46	50	55	46
254	69	34	37	39	40	62
1113	44	36	38	41	43	55

Is there any evidence to suggest that at least two population mean thaw depths are different? Use $\alpha = 0.05$.

11.26 Biology and Environmental Science   Dissolved organic compounds (DOCs) consist of a variety of organic material in water systems. These dissolved particles are mostly the result of decaying organic matter and highly organic soils. The DOC concentration is one measure of the quality of a water system. Independent random samples of water in the Clackamas River in northwestern Oregon at four locations were obtained, and the DOC was measured for each (in mg/L).[7] Is there any evidence to suggest at least two population mean DOC concentrations are different? Use $\alpha = 0.01$. **DOCS**

11.27 Kerosene Purity   Many people who live in northern states supplement their main heating system with kerosene heaters. These devices are generally inexpensive, but a kerosene flame emits a noticeable ash and the fumes must be well ventilated. The flash point of kerosene is related to purity, ash, fumes, and safety. A low flash point is a fire hazard and may be an indication of contamination. Independent random samples of kerosene from five dealers in Maine were obtained and the flash point of each was measured (in °C). Is there any evidence to suggest that at least two of the population mean flash points are different? Use $\alpha = 0.05$. **KEROSENE**

11.28 Expense IQ   A recent publication by Concur provided a summary of travel expenses for both small and medium-sized businesses (SMBs) and large-scale companies.[8] A random sample of business trips to some of the most visited international cities was obtained, and the amount spent on ground transportation was recorded (in dollars). Is there any evidence to suggest that at least two of the population mean amounts spent on ground transportation are different? Use $\alpha = 0.01$. **EXPENSE**

Extended Applications

11.29 Manufacturing and Product Development   Many cities, towns, and college campuses use a rotary broom to sweep debris and snow from sidewalks. The pressure created by the broom is one measure of how effective the machine will be in removing debris. Random samples of rotary brooms from four manufacturers were obtained, and the pressure of each broom was measured (in psi). The data are given in the following table. **BROOMS**

Company	Observations			
Ditch Witch	2081	1980	2210	2297
	2204	2765	2327	
Schwarze	2567	1799	2422	2437
	2367	2244	2245	
Elgin	2228	2581	2364	2375
	2066	2091	2543	
Holder	2905	2695	2503	2931
	2657	2591	2138	

a. Conduct the appropriate test to determine whether there is any evidence that at least two population mean pressures are different. Use $\alpha = 0.05$.
b. Which manufacturer would you recommend and why?

11.30 Fuel Consumption and Cars   Automobile service clubs offer free maps, trip-interruption protection, payment for some legal fees, and roadside assistance. Independent random samples of roadside service calls were selected for three different clubs. The time required (in minutes) for a tow truck to arrive was recorded for each call. The summary statistics are given in the following table.

Club	Sample size	Sample mean	Sample variance
AAA	15	36.8	10.18
Discover	17	43.8	9.97
Executive	20	34.8	12.15

a. Use the definitions to find SSA and SSE. Use these two values to find SST.
b. Complete an ANOVA table, and use this information to determine whether there is any evidence to suggest that at least two population mean waiting times are different. Use $\alpha = 0.01$.

11.31 Physical Sciences   During the summer months, grocery stores, convenience stores, and gas stations sell bags of ice. Independent random sample of bags were obtained from various locations, and the weight (in pounds) of each was recorded. The summary statistics are given in the following table.

Location	Sample size	Sample total	Sum of squared observations
Giant	10	80.8	654.38
Sheetz	10	88.8	789.20
Star Market	14	119.9	1028.31
Unimart	12	104.1	903.57
Weis	15	139.7	1303.41

a. Do the data suggest that the population mean weight of bags of ice is the same at these five locations? Use $\alpha = 0.001$.
b. If each store sells these bags of ice for approximately the same price, where would you make your purchase? Justify your answer.

11.32 Sports and Leisure   The ESPN Home Run Tracker computes a variety of measurements for every home run hit in Major League Baseball, including true distance, speed off bat, elevation angle, and apex. Independent random samples of home runs during the 2013 season were obtained from six different ballparks, and the true distance was recorded for each.[9] **HOMERUN**

a. Is there any evidence to suggest that at least two population mean home-run distances are different? Use $\alpha = 0.05$.
b. Do you think there are any violations in the one-way ANOVA assumptions? If so, why?
c. Do the data suggest that population mean home-run distance is greatest in one ballpark? Why do you suppose the distances tend to be greater in that ballpark?

Challenge

11.33 Public Policy and Political Science   Every city has design specifications for streets, including minimum right of way, minimum vertical grade, and minimum centerline radii on curves. Independent random samples of city streets were obtained from Washington, D.C., and New York City, and the width (in feet) of a randomly selected section was recorded. The data are given in the following table. **STREETS**

City	Observations					
New York City	28.2	32.9	34.6	31.5	31.6	29.5
	30.3	29.2	25.6	28.6	28.8	31.5
Washington, D.C.	32.5	36.3	34.3	33.0	31.0	36.5
	29.8	30.0	30.2	34.7	31.9	30.0

a. Conduct a two-sample t test to determine whether there is any evidence to suggest that the population mean street widths are different. Assume the population variances are equal and use $\alpha = 0.05$. Find the exact p value associated with this test.

b. Conduct a one-way analysis of variance test to determine whether there is any difference among the $k = 2$ population mean street widths due to city. Use $\alpha = 0.05$. Find the exact p value associated with this test.

c. What is the relationship between the value of the test statistic in part (a) and the value of the test statistic in part (b)? How are the p values related? Why do these relationships make sense?

11.34 The Effect Size Generate a random sample of size $n_1 = 20$ from a normal distribution with mean $\mu_1 = 50$ and

variance $\sigma_1^2 = 100$. Generate a second random sample of size $n_2 = 20$ from a normal distribution with mean $\mu_2 = 50$ and variance $\sigma_2^2 = 100$. Generate a third random sample of size $n_3 = 20$ from a normal distribution with mean $\mu_3 = 52$ and variance $\sigma_3^2 = 100$. Conduct a one-way ANOVA test with $\alpha = 0.05$ to determine whether there is any evidence to suggest that the population means differ.

Repeat this process 100 times and record the proportion of times you reject the null hypothesis, p_r.

Let $\mu_T = (\mu_1 + \mu_2 + \mu_3)/3$ and compute the *effect size e,*

$$e = \sqrt{\frac{\sum_{i=1}^{3}(\mu_i - \mu_T)^2/3}{100}}$$

Plot the point (e, p_r).

Repeat this process for various values of μ_1, μ_2, and μ_3, and therefore, various effect sizes. Plot the points (e, p_r).

Explain the resulting graph. What concept related to an ANOVA test does this graph illustrate?

11.2 Isolating Differences

If we reject H_0, then we would like to know which means are different.

If we fail to reject the null hypothesis in a one-way ANOVA, there is no evidence to suggest any difference among population means. The statistical analysis stops there. However, if the null hypothesis is rejected, there is evidence to suggest an *overall* difference among means. The next logical step is to try and isolate the difference(s), to determine which pair(s) of means are contributing to the overall (significant) difference. There are several **multiple comparison procedures** for isolating differences. Two will be presented in this section.

If two population means are being compared, then a t test (or Z test) is usually appropriate. However, if there are three or more population means to compare, the analysis requires a little more finesse. Here's why. Suppose that, in a one-way ANOVA with three groups, we reject H_0. It seems reasonable to conduct a test on every possible pair of means (μ_1 versus μ_2, μ_1 versus μ_3, and μ_2 versus μ_3). However, we cannot simply set the significance level in each *individual* hypothesis test. The probability α of making a type I error (a mistake) is set in each test under the assumption that *only one* test is conducted per experiment. Therefore, the more tests that are conducted, the greater is the chance of making an error.

Recall, a type I error means rejecting H_0 when H_0 is true.

Think about a waiter totaling a customer's bill. Suppose the probability that the waiter makes a mistake on any single bill is 0.10. If the waiter must total three bills, the probability that he makes a mistake on *at least* one bill is 0.2710. The more bills and totals, the greater the chance of making at least one error. The same principle applies to hypothesis tests. Suppose three hypothesis tests (using data from the same experiment) are conducted, each with significance level 0.05. The probability of making at least one mistake is 0.1426. The more hypothesis tests (or the more comparisons) are conducted, the more likely we are to make at least one error. As we have seen, the probability of making at least one error is more than two times as big as α.

X = number of bills in which there is a mistake.
$X \sim B(3, 0.10)$.
$P(X \geq 1)$
$= 1 - P(X = 0)$
$= 1 - 0.7290 = 0.2710$

We would really like to control the (overall) probability of making at least one mistake. For example, we might want the probability of making at least one mistake in three hypothesis tests to be 0.10. We typically set this overall error probability and work backward to compute the individual error probabilities associated with each individual test.

The procedures presented here provide methods for capping the probability of making at least one mistake in all of the comparisons.

Although we could actually conduct hypothesis tests, usually we construct multiple confidence intervals for the difference between population means. Recall that, if a confidence interval for $\mu_1 - \mu_2$ contains 0, there is no evidence to suggest that the population means are different. However, if 0 is not included in the confidence interval, there is evidence to suggest that the two population means are different. If we want to find a $100(1 - \alpha)\%$ confidence interval for all possible paired comparisons (i.e., so that the overall probability is α), we must make the intervals much wider than those for individual differences of means.

In a one-way ANOVA with three groups, suppose there is evidence to suggest that at least one pair of means is different and a multiple comparison procedure produces the following confidence intervals.

Difference	Confidence interval
$\mu_1 - \mu_2$	$(-1.21, \quad 9.00)$
$\mu_1 - \mu_3$	$(\quad 3.09, 13.31)$
$\mu_2 - \mu_3$	$(-0.81, \quad 9.41)$

Zero is included in the confidence intervals for the differences $\mu_1 - \mu_2$ and $\mu_2 - \mu_3$. There is no evidence to suggest that these pairs of population means are different. The confidence interval for $\mu_1 - \mu_3$ does not include 0. There is evidence to suggest that μ_1 is different from μ_3.

The general form of each **Bonferroni confidence interval** is very similar to a confidence interval for the difference between two means based on a t distribution using a pooled estimate of the common variance (introduced in Section 10.2). Here, we use the mean square due to error (MSE) as an estimate of the common variance. And a t critical value is used to achieve a *simultaneous*, or *familywise*, *confidence level* of $100(1 - \alpha)\%$.

The Bonferroni Multiple Comparison Procedure

In a one-way analysis of variance, suppose there are k groups, $n = n_1 + n_2 + \cdots + n_k$ total observations, and H_0 is rejected.

1. There are $c = \dbinom{k}{2} = \dfrac{k(k-1)}{2}$ pairs of population means to compare.

2. The c simultaneous $100(1 - \alpha)\%$ **Bonferroni confidence intervals** have as endpoints the values

$$(\bar{x}_{i.} - \bar{x}_{j.}) \pm t_{\alpha/(2c), n-k} \sqrt{\mathrm{MSE}} \sqrt{\frac{1}{n_i} + \frac{1}{n_j}} \qquad \text{for all } i \neq j \qquad (11.6)$$

Example 11.3 Small-Scale Farming

DATA SET

FARM

Many people believe most farms are huge, corporate-like enterprises covering thousands of acres. However, in Canada, most of the country's farms are small-scale, family-owned and operated. Only a small portion of the farms in Canada earn over $1 million in revenue. Independent random samples of farms were obtained from four Canadian provinces, with 10 observations in each group. The size (in acres) was recorded for each farm.[10] The resulting sample means and the ANOVA table are shown below.

Group number	Province (factor)	Sample mean
1	Prince Edward Island	402.3
2	New Brunswick	421.1
3	Quebec	326.1
4	British Columbia	314.3

ANOVA summary table

Source of variation	Sum of squares	Degrees of freedom	Mean square	F	p Value
Factor	86,185.9	3	28,728.6	6.26	0.0016
Error	165,086.0	36	4,585.7		
Total	251,271.9	39			

The ANOVA test is significant at the $p = 0.0016$ level. There is evidence to suggest that at least one pair of population means is different (an overall difference). Construct the Bonferroni 95% confidence intervals and use them to isolate the pair(s) of means contributing to this overall experiment difference.

SOLUTION

STEP 1 The number of pairwise comparisons needed is

$$c = \frac{4 \cdot 3}{2} = 6$$

$$95\% = 100(1 - \alpha)\% \quad \Rightarrow \quad \alpha = 0.05 \quad \Rightarrow \quad \frac{\alpha}{2c} = \frac{0.05}{2 \cdot 6} = 0.0042$$

This (infrequent) right-tail probability is not specified in Table V in the Appendix; however, one can use linear interpolation or technology, with $n - k = 40 - 4 = 36$ degrees of freedom, to find $t_{\alpha/(2c),36} = t_{0.0042,36} = 2.7888$.

STEP 2 The Bonferroni confidence interval for the difference $\mu_1 - \mu_2$ is

$$(\bar{x}_{1.} - \bar{x}_{2.}) \pm t_{\alpha/(2c),n-k} \sqrt{\text{MSE}} \sqrt{\frac{1}{n_1} + \frac{1}{n_2}}$$

$$= (\bar{x}_{1.} - \bar{x}_{2.}) \pm t_{0.0042,36} \sqrt{\text{MSE}} \sqrt{\frac{1}{n_1} + \frac{1}{n_2}}$$

$$= (402.3 - 421.1) \pm (2.7888)\sqrt{4585.7} \sqrt{\frac{1}{10} + \frac{1}{10}}$$

$$= (-103.3, 65.7)$$

STEP 3 The remaining five confidence intervals are found in the same manner. Each Bonferroni confidence interval and its corresponding conclusion are shown in the following table.

Difference	Bonferroni confidence interval	Conclusion
$\mu_1 - \mu_2$	$(-103.3, \ 65.7)$	0 in CI. μ_1 and μ_2 are not significantly different.
$\mu_1 - \mu_3$	$(-8.3, \ 160.7)$	0 in CI. μ_1 and μ_3 are not significantly different.
$\mu_1 - \mu_4$	$(3.5, \ 172.5)$	0 **not** in CI. Evidence to suggest that $\mu_1 \neq \mu_4$.
$\mu_2 - \mu_3$	$(10.5, \ 179.5)$	0 **not** in CI. Evidence to suggest that $\mu_2 \neq \mu_3$.
$\mu_2 - \mu_4$	$(22.3, \ 191.3)$	0 **not** in CI. Evidence to suggest that $\mu_2 \neq \mu_4$.
$\mu_3 - \mu_4$	$(-72.7, \ 96.3)$	0 in CI. μ_3 and μ_4 are not significantly different.

Finally, the initial ANOVA test indicates that there is an overall difference among the four population means. The simultaneous 95% Bonferroni confidence intervals suggest that this overall difference is due to a difference between μ_1 and μ_4, μ_2 and μ_3, and μ_2 and μ_4. Figure 11.11 shows a technology solution.

Ordered Differences Report						
Level	- Level	Difference	Std Err Dif	Lower CL	Upper CL	p-Value
2	4	106.8000	30.28439	22.1989	191.4011	0.0012*
2	3	95.0000	30.28439	10.3989	179.6011	0.0034*
1	4	88.0000	30.28439	3.3989	172.6011	0.0062*
1	3	76.2000	30.28439	-8.4011	160.8011	0.0165*
2	1	18.8000	30.28439	-65.8011	103.4011	0.5387
3	4	11.8000	30.28439	-72.8011	96.4011	0.6991

Figure 11.11 JMP Bonferroni confidence intervals.

There is another common, compact, graphical method for summarizing the results of a multiple comparison procedure. Write the sample means in order from smallest to largest. Use the results from a multiple comparison procedure to draw a horizontal line under the groups of means that are *not* significantly different.

Here is the graphical summary for Example 11.3.

1. The sample means in order:

	$\bar{x}_{4.}$	$\bar{x}_{3.}$	$\bar{x}_{1.}$	$\bar{x}_{2.}$
Sample mean	314.3	326.1	402.3	421.1

2. There is no significant difference between μ_1 and μ_2. Draw a horizontal line under the sample means from population 1 and 2.

	$\bar{x}_{4.}$	$\bar{x}_{3.}$	$\bar{x}_{1.}$	$\bar{x}_{2.}$
Sample mean	314.3	326.1	402.3	421.1

3. There is no significant difference between μ_1 and μ_3. Draw a horizontal line under the sample means from population 1 and 3. There is no significant difference between μ_3 and μ_4. Draw a horizontal line under the sample means from population 3 and 4.

	$\bar{x}_{4.}$	$\bar{x}_{3.}$	$\bar{x}_{1.}$	$\bar{x}_{2.}$
Sample mean	314.3	326.1	402.3	421.1

Those pairs of means *not* connected by a horizontal line are significantly different.

TRY IT NOW   GO TO EXERCISE 11.51

Tukey's multiple comparison procedure also yields simultaneous confidence intervals for all pairwise differences. The form of the confidence intervals is similar to Equation 11.6, but it uses a Q critical value from the **Studentized range distribution**. This distribution is completely characterized by two parameters, the degrees of freedom in the numerator and denominator, m and ν, respectively.

Using the usual notation, let $Q_{\alpha,m,\nu}$ denote the right-tail critical value of the Studentized range distribution with m and ν degrees of freedom. Table VIII in the Appendix presents selected critical values associated with various Studentized range distributions. The degrees of freedom in the numerator are given in the top row and the degrees of freedom in the denominator are given in the left column. In the body of the table, $Q_{\alpha,m,\nu}$ is at the intersection of column m and row ν.

Tukey's Multiple Comparison Procedure

In a one-way analysis of variance, suppose there are k groups, $n = n_1 + n_2 + \cdots + n_k$ total observations, and H_0 is rejected. The set of $c = \binom{k}{2}$ simultaneous $100(1 - \alpha)\%$ confidence intervals have as endpoints the values

$$(\bar{x}_{i.} - \bar{x}_{j.}) \pm \frac{1}{\sqrt{2}} Q_{\alpha,k,n-k} \sqrt{\text{MSE}} \sqrt{\frac{1}{n_i} + \frac{1}{n_j}} \qquad \text{for all } i \neq j \qquad (11.7)$$

If all pairwise comparisons are considered, the Bonferroni procedure produces wider confidence intervals than the Tukey procedure. However, if only a subset of all pairwise comparisons is needed, then the Bonferroni method may be better. There are also other methods for comparing population means following an ANOVA test. No single comparison method is uniformly best.

In the following example, Tukey's procedure is used to isolate pairwise differences contributing to an overall significant ANOVA test.

Bill O'Leary/The Washington Post/Getty Images

Example 11.4 People on the Go - Go to Wawa

A recent study suggests that convenience store shopping is affected by geographic region.[11] For example, Northeast shoppers are more likely to frequent a convenience store during late night hours, and Westerners are the most likely to utilize a DVD rental kiosk. Independent random samples of shoppers at convenience stores were selected in four regions, and the total amount spent was recorded (in dollars). The summary statistics and the ANOVA table are shown below.

Group number	1	2	3	4
Region	Northeast	South	Midwest	West
Sample size	10	11	12	11
Sample mean	15.64	13.99	9.50	12.98

ANOVA summary table

Source of variation	Sum of squares	Degrees of freedom	Mean square	F	p Value
Factor	226.87	3	75.62	6.13	0.0016
Error	493.41	40	12.34		
Total	720.28	43			

The ANOVA test is significant at the $p = 0.0016$ level. There is evidence to suggest that at least one pair of means is different (an overall difference). Construct the Tukey 95% simultaneous confidence intervals and use them to isolate the pair(s) of means contributing to this overall experiment difference.

SOLUTION

STEP 1 The number of pairwise comparisons needed is $\binom{4}{2} = 6$ and $\alpha = 0.05$.

The Studentized range critical value (Table VIII in the Appendix) is $Q_{\alpha,k,n-k} = Q_{0.05,4,40} = 3.791$.

STEP 2 The first confidence interval, for the difference $\mu_1 - \mu_2$, using Tukey's procedure is

$$(\bar{x}_{1.} - \bar{x}_{2.}) \pm \frac{1}{\sqrt{2}} Q_{0.05,k,n-k} \sqrt{\text{MSE}} \sqrt{\frac{1}{n_1} + \frac{1}{n_2}}$$

$$= (15.64 - 13.99) \pm \frac{1}{\sqrt{2}}(3.791)\sqrt{12.34} \sqrt{\frac{1}{10} + \frac{1}{11}}$$

$$= 1.65 \pm 4.11 = (-2.47, 5.75)$$

STEP 3 The remaining confidence intervals and conclusions are given in the following table.

Difference	Tukey confidence interval	Conclusion
$\mu_1 - \mu_2$	$(-2.47, \ 5.75)$	0 in CI. μ_1 and μ_2 are not significantly different.
$\mu_1 - \mu_3$	$(\ 2.11, 10.17)$	0 **not** in CI. Evidence to suggest that $\mu_1 \neq \mu_3$.
$\mu_1 - \mu_4$	$(-1.45, \ 6.77)$	0 in CI. μ_1 and μ_4 are not significantly different.
$\mu_2 - \mu_3$	$(\ 0.56, \ 8.42)$	0 **not** in CI. Evidence to suggest that $\mu_2 \neq \mu_3$.
$\mu_2 - \mu_4$	$(-3.01, \ 5.03)$	0 in CI. μ_2 and μ_4 are not significantly different.
$\mu_3 - \mu_4$	$(-7.41, \ 0.45)$	0 in CI. μ_3 and μ_4 are not significantly different.

Here are the results presented in graphical form.

	$\bar{x}_{3.}$	$\bar{x}_{4.}$	$\bar{x}_{2.}$	$\bar{x}_{1.}$
Sample mean	9.50	12.98	13.99	15.64

The simultaneous 95% confidence intervals constructed using Tukey's procedure suggest that the overall significance is due to a difference between μ_1 and μ_3, and μ_2 and μ_3.

TRY IT NOW GO TO EXERCISE 11.52

VIDEO TECH MANUALS

ONE-WAY ANOVA -
FOLLOW-UP ANALYSIS:
BONFERRONI, TUKEY

The Bonferroni multiple comparison procedure is conservative. This means the *overall* resulting confidence level is probably higher than the *specified* value. To compute Bonferroni intervals, the overall confidence level is simply divided evenly among all pairwise comparisons. If we are only interested in some of the $\binom{k}{2} = c$ pairwise comparisons, the overall confidence level is divided accordingly. In this case, the resulting confidence level may be closer to the specified value. Tukey's procedure is based on a special distribution and tends to result in more accurate confidence levels. Therefore, if all pairwise comparisons are necessary, Tukey's multiple comparison procedure is usually better.

Consider the Bonferroni confidence intervals for Example 11.4.

Difference	Bonferroni confidence interval	Conclusion
$\mu_1 - \mu_2$	$(-2.61, \ 5.91)$	0 in CI. μ_1 and μ_2 are not significantly different.
$\mu_1 - \mu_3$	$(\ 1.97, 10.31)$	0 **not** in CI. Evidence to suggest that $\mu_1 \neq \mu_3$.
$\mu_1 - \mu_4$	$(-1.60, \ 6.92)$	0 in CI. μ_1 and μ_4 are not significantly different.
$\mu_2 - \mu_3$	$(\ 0.42, \ 8.56)$	0 **not** in CI. Evidence to suggest that $\mu_2 \neq \mu_3$.
$\mu_2 - \mu_4$	$(-3.14, \ 5.16)$	0 in CI. μ_2 and μ_4 are not significantly different.
$\mu_3 - \mu_4$	$(-7.55, \ 0.59)$	0 in CI. μ_3 and μ_4 are not significantly different.

Notice that each confidence interval is wider than the corresponding Tukey confidence interval. A wider, more conservative, Bonferroni confidence interval could result in a different conclusion concerning the difference between two population means. For example, compare the Tukey and Bonferroni confidence intervals for $\mu_2 - \mu_3$; the left endpoint of the Bonferroni CI is closer to 0.

Technology Corner

Procedure: Construct simultaneous confidence intervals to isolate the pair(s) of means contributing to an overall experiment difference detected using an analysis of variance test.

Reconsider: Example 11.3, solution, and interpretations.

CrunchIt!, the TI-84 Plus C, and Excel do not have built-in functions to construct Bonferroni or Tukey confidence intervals. In Excel, one can use ordinary spreadsheet functions to construct the Bonferroni confidence intervals.

Minitab

There is an option to One-Way Analysis of Variance to construct the Tukey confidence intervals. There is no option for Bonferroni confidence intervals.

1. The data may be entered in a one column for all factor levels or in a separate column for each factor level. In this example, enter the data from samples 1, 2, 3, and 4 into columns C1, C2, C3, and C4, respectively.
2. Select Stat; ANOVA; One-Way. Choose Response data are in a separate column for each factor level.
3. Enter the Responses: the columns containing the data.
4. Choose the Options button. Check the box to Assume equal variances. Select an appropriate Confidence level and Type of confidence interval.
5. Choose the Comparisons options button. Enter an Error rate for comparisons which is $1 -$ confidence level, and check Tukey. Check Interval plot for the differences of means, Grouping information, and Tests as desired.
6. Click OK. The output is displayed in a session window. See Figure 11.12. Note that the mean differences are reversed.

```
Session                                                      _|□|×|

Tukey Pairwise Comparisons

Tukey Simultaneous Tests for Differences of Means
                              Difference    SE of
Difference of Levels          of Means   Difference      95% CI       T-Value
New Brunswic - Prince Edwar       18.8       30.3  ( -62.8, 100.4)       0.62
Quebec - Prince Edwar            -76.2       30.3  (-157.8,   5.4)      -2.52
British Colu - Prince Edwar      -88.0       30.3  (-169.6,  -6.4)      -2.91
Quebec - New Brunswic            -95.0       30.3  (-176.6, -13.4)      -3.14
British Colu - New Brunswic     -106.8       30.3  (-188.4, -25.2)      -3.53
British Colu - Quebec            -11.8       30.3  ( -93.4,  69.8)      -0.39

                              Adjusted
Difference of Levels           P-Value
New Brunswic - Prince Edwar      0.925
Quebec - Prince Edwar            0.074
British Colu - Prince Edwar      0.030
Quebec - New Brunswic            0.017
British Colu - New Brunswic      0.006
British Colu - Quebec            0.980

Individual confidence level = 98.93%
```

Figure 11.12 Minitab Tukey confidence intervals.

Excel

There is no built-in function to construct simultaneous confidence intervals. However, you can use the function T.INV, the ANOVA summary table, and ordinary spreadsheet calculations to construct the Bonferroni confidence intervals.

1. Enter the data from samples 1, 2, 3, and 4 into columns A, B, C, and D, respectively.
2. Under the Data tab, select Data Analysis; Anova: Single Factor.
3. Enter the Input Range, choose Grouped by Columns, enter a value for Alpha, and select an Output option. Click OK to display the results.
4. Use T.INV to find the critical value, and compute $\sqrt{MSE}$. Use these values, the sample means, and the sample sizes in Equation 11.6 to construct the confidence intervals. See Figure 11.13.

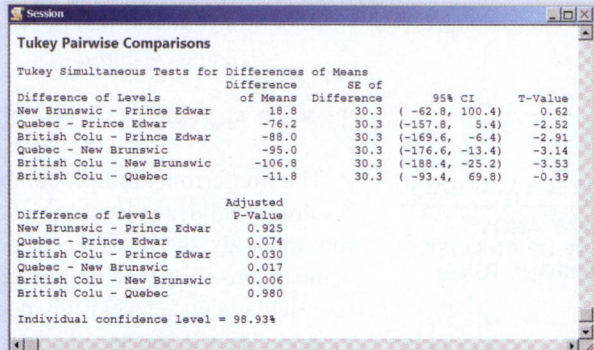

	E	F	G	H	I	J	K	L
	Anova: Single Factor							
	SUMMARY							
	Groups	Count	Sum	Average	Variance			
	Prince Edward Island	10	4023	402.3	5833.8			
	New Brunswick	10	4211	421.1	3286.3			
	Quebec	10	3261	326.1	2240.8			
	British Columbia	10	3143	314.3	6982.0			
	ANOVA							
	Source of Variation	SS	df	MS	F	P-value	F crit	
	Between Groups	86185.9	3	28728.6	6.26	0.0016	2.87	
	Within Groups	165086.0	36	4585.7				
	Total	251271.9	39					
	Multiple Comparisons	Left	Right					
	$\mu_1 - \mu_2$	-103.3	65.7		2.7888	= T.INV(0.9958,36) = $t_{.0042,36}$		
	$\mu_1 - \mu_3$	-8.3	160.7		67.7180	= SQRT(H14) = SQRT(MSE)		
	$\mu_1 - \mu_4$	3.5	172.5					
	$\mu_2 - \mu_3$	10.5	179.5					
	$\mu_2 - \mu_4$	22.3	191.3					
	$\mu_3 - \mu_4$	-72.7	96.3					

Figure 11.13 Excel Bonferroni confidence intervals.

SECTION 11.2 EXERCISES

Concept Check

11.35 True/False The Bonferroni multiple comparison procedure is conservative.

11.36 True/False The Bonferroni multiple comparison procedure can be used even if we are only interested in some of the pairwise comparisons.

11.37 True/False For Tukey's multiple comparison procedure to be valid, all sample sizes must be the same.

11.38 True/False In a one-way ANOVA with k groups, suppose we reject the null hypothesis. Using a multiple comparison procedure, there must be at least one pairwise comparison in which there is evidence to suggest the population means are significantly different.

11.39 Short Answer In a one-way ANOVA with k groups, suppose we reject the null hypothesis. What is the total number of possible pairwise comparisons?

11.40 Short Answer Explain why we use a multiple comparison procedure rather than several individual hypothesis tests or confidence intervals.

11.41 Short Answer In a one-way ANOVA, suppose the null hypothesis is not rejected. Explain why there is no need for a multiple comparison procedure.

Practice

11.42 In each of the following problems, the number of groups, k, the total number of observations, n, and the overall confidence level for the Bonferroni multiple comparison procedure are given. Find the total number of comparisons, c, and the t critical value used in the calculation of each Bonferroni confidence interval.

a. $k = 3$, $n = 30$, 95%
b. $k = 3$, $n = 43$, 99%
c. $k = 4$, $n = 32$, 99%
d. $k = 5$, $n = 55$, 90%
e. $k = 6$, $n = 30$, 95%

11.43 In each of the following problems, the number of observations in each group and the overall confidence level for Tukey's multiple comparison procedure are given. Find the critical value from the Studentized range distribution to be used in the calculation of each Tukey confidence interval.

a. $n_1 = 6$, $n_2 = 8$, $n_3 = 7$, 95%
b. $n_1 = n_2 = 11$, $n_3 = 12$, $n_4 = 10$, 95%
c. $n_1 = 14$, $n_2 = 12$, $n_2 = 10$, $n_4 = 18$, 99%
d. $n_1 = n_2 = n_3 = n_4 = n_5 = 11$, 99%
e. $n_1 = n_2 = n_3 = 5$, $n_4 = n_5 = 6$, $n_6 = 9$, 99.9%

11.44 In each of the following problems, use the summary statistics and the Bonferroni confidence intervals to construct the corresponding graph indicating which pairs of means are and are not significantly different.

a.

Sample means	Difference	Confidence interval
$\bar{x}_{1.} = 52.8$	$\mu_1 - \mu_2$	(−6.00, 17.85)
$\bar{x}_{2.} = 46.9$	$\mu_1 - \mu_3$	(−9.50, 14.35)
$\bar{x}_{3.} = 50.4$	$\mu_2 - \mu_3$	(−15.42, 8.42)

b.

Sample means	Difference	Confidence interval
$\bar{x}_{1.} = 4.82$	$\mu_1 - \mu_2$	(−5.69, −1.13)
$\bar{x}_{2.} = 8.23$	$\mu_1 - \mu_3$	(−4.49, 0.07)
$\bar{x}_{3.} = 7.03$	$\mu_2 - \mu_3$	(−1.08, 3.48)

c.

Sample means	Difference	Confidence interval
$\bar{x}_{1.} = 16.09$	$\mu_1 - \mu_2$	(−12.16, −1.07)
$\bar{x}_{2.} = 22.71$	$\mu_1 - \mu_3$	(−7.98, 3.11)
$\bar{x}_{3.} = 18.53$	$\mu_1 - \mu_4$	(−1.37, 9.72)
$\bar{x}_{4.} = 16.33$	$\mu_2 - \mu_3$	(−5.78, 5.31)
	$\mu_2 - \mu_4$	(0.83, 11.92)
	$\mu_3 - \mu_4$	(−3.34, 7.75)

d.

Sample means	Difference	Confidence interval
$\bar{x}_{1.} = 201.7$	$\mu_1 - \mu_2$	(9.98, 56.20)
$\bar{x}_{2.} = 168.6$	$\mu_1 - \mu_3$	(13.20, 59.42)
$\bar{x}_{3.} = 165.4$	$\mu_1 - \mu_4$	(−19.89, 26.33)
$\bar{x}_{4.} = 219.7$	$\mu_2 - \mu_3$	(−41.10, 5.12)
	$\mu_2 - \mu_4$	(−74.19, −27.98)
	$\mu_3 - \mu_4$	(−77.41, −31.20)

11.45 In each of the following problems, use the summary statistics and the Tukey confidence intervals to construct the corresponding graph indicating which pairs of means are and are not significantly different.

a.

Sample means	Difference	Confidence interval
$\bar{x}_{1.} = -33.44$	$\mu_1 - \mu_2$	(−37.57, 0.36)
$\bar{x}_{2.} = -14.83$	$\mu_1 - \mu_3$	(−52.88, −14.95)
$\bar{x}_{3.} = 0.48$	$\mu_1 - \mu_4$	(−56.71, −18.77)
$\bar{x}_{4.} = 4.30$	$\mu_2 - \mu_3$	(−34.28, 3.66)
	$\mu_2 - \mu_4$	(−38.10, −0.17)
	$\mu_3 - \mu_4$	(−22.79, 15.14)

b.

Sample means	Difference	Confidence interval
$\bar{x}_{1.} = 1.62$	$\mu_1 - \mu_2$	(0.086, 0.320)
$\bar{x}_{2.} = 1.41$	$\mu_1 - \mu_3$	(0.197, 0.432)
$\bar{x}_{3.} = 1.30$	$\mu_1 - \mu_4$	(−0.006, 0.229)
$\bar{x}_{4.} = 1.50$	$\mu_2 - \mu_3$	(0.001, 0.235)
	$\mu_2 - \mu_4$	(−0.203, 0.320)
	$\mu_3 - \mu_4$	(−0.314, −0.079)

c.

Sample means	Difference	Confidence interval
$\bar{x}_{1.} = 64.35$	$\mu_1 - \mu_2$	$(\ \ 0.91,\ 19.38)$
$\bar{x}_{2.} = 54.21$	$\mu_1 - \mu_3$	$(-5.68,\ 12.79)$
$\bar{x}_{3.} = 60.80$	$\mu_1 - \mu_4$	$(\ \ 3.20,\ 21.66)$
$\bar{x}_{4.} = 51.92$	$\mu_1 - \mu_5$	$(-9.73,\ \ 8.73)$
$\bar{x}_{5.} = 64.85$	$\mu_2 - \mu_3$	$(-15.83,\ \ 2.64)$
	$\mu_2 - \mu_4$	$(\ -6.95,\ 11.52)$
	$\mu_2 - \mu_5$	$(-19.88,\ -1.41)$
	$\mu_3 - \mu_4$	$(\ -0.36,\ 18.11)$
	$\mu_3 - \mu_5$	$(-13.29,\ \ 5.18)$
	$\mu_4 - \mu_5$	$(-22.16,\ -3.70)$

11.46 In each of the following problems, the results from a multiple comparison procedure are shown graphically. Use each illustration to identify all pairs of population means that are significantly different.

a.

	$\bar{x}_{3.}$	$\bar{x}_{1.}$	$\bar{x}_{2.}$	$\bar{x}_{4.}$
Sample mean	12.69	14.64	15.94	16.21

b.

	$\bar{x}_{1.}$	$\bar{x}_{2.}$	$\bar{x}_{3.}$	$\bar{x}_{4.}$
Sample mean	29.98	32.59	32.62	37.25

c.

	$\bar{x}_{1.}$	$\bar{x}_{2.}$	$\bar{x}_{3.}$	$\bar{x}_{4.}$
Sample mean	39.01	41.40	50.83	51.62

d.

	$\bar{x}_{2.}$	$\bar{x}_{3.}$	$\bar{x}_{5.}$	$\bar{x}_{1.}$	$\bar{x}_{4.}$
Sample mean	4.67	6.08	6.23	6.30	7.20

11.47 Suppose data collected from an experiment resulted in a significant ANOVA test. Using the summary statistics and MSE = 29.83, construct the Bonferroni 99% confidence intervals for all pairwise comparisons. Use these intervals to identify the significantly different population means.

Group	1	2	3	4
Sample size	20	20	20	20
Sample mean	19.59	21.58	16.50	25.76

11.48 Suppose data collected from an experiment resulted in a significant ANOVA test. Using the summary statistics and MSE = 8111.8, construct the Tukey 95% confidence intervals for all pairwise comparisons. Use these intervals to identify the significantly different population means.

Group	1	2	3	4
Sample size	16	18	21	9
Sample mean	476.3	450.2	698.2	597.6

Applications

11.49 Medicine and Clinical Studies Some people believe that unequal Medicare spending across the country is due to waste and unnecessary tests. However, recent research suggests that health care costs do indeed differ by region.[12] Independent random samples of adults in three states were obtained, and the Medicare payments to each were recorded. The resulting

ANOVA test was significant at the $p = 0.0190$ level. The summary statistics are given below. If MSE = 136,146, construct the Bonferroni 95% confidence intervals to isolate the population mean payments that are significantly different.

State	Sample size	Sample mean
Connecticut	18	3489.22
Georgia	22	3774.50
Mississippi	25	3794.76

11.50 Demographics and Population Statistics A study was conducted to compare the population mean age of women at the time of their first marriage. Independent random samples of size 11 were obtained from weddings in three different years. The resulting ANOVA test was significant at the $p = 0.003$ level. The sample means are given below. If MSE = 17.68, construct the Tukey 95% confidence intervals to isolate the population mean ages that are significantly different.

Year	1985	1995	2003
Sample mean	23.3	24.5	25.3

11.51 Manufacturing and Product Development An experiment was conducted to compare the true cost of microwave popcorn brands by examining the percentage of popped kernels. Independent random samples of six packages of buttered microwave popcorn (all weighing the same) were obtained for each of four different brands. Each bag was popped, and the percentage of popped kernels was measured. An ANOVA test resulted in the test statistic $f = 7.69$ and MSE = 2.72. The sample means are given in the following table.

Microwave popcorn	Pop Secret	Best Choice	Act II	Orville Redenbacher's
Sample mean	90.8	86.9	90.9	89.2

a. Find bounds on the p value associated with this test statistic. State the conclusion of the ANOVA test.

b. Construct the Bonferroni 99% simultaneous confidence intervals to determine which pairs of population mean percentages of popped kernels are significantly different.

11.52 Biology and Environmental Science A person's metabolic rate certainly depends on level of activity. However, a recent study suggests that metabolic rate is also related to personality type. For example, the performer, people oriented and living for the moment, might have a higher resting metabolic rate, and the artist, quiet, sensitive, and kind, could have a lower resting metabolic rate. Suppose independent random samples of different personality types were obtained and the resting metabolic rate (in kcal/day) was measured for each using a metabolic chamber. An ANOVA test was conducted to determine whether there was any difference in population mean resting metabolic rates ($f = 5.83$ and MSE = 52,025.4). The summary statistics by personality type are given in the following table.

Personality type	Sample size	Sample mean
Mechanic	14	1962.7
Nurturer	16	1757.4
Protector	20	2057.8
Idealist	18	2019.5

a. Show that there is evidence of an overall difference in population mean resting metabolic rate. Use $\alpha = 0.01$ and justify your answer.

b. Construct the Tukey 99% confidence intervals and indicate which pairs of means are contributing to the overall difference.

11.53 Biology and Environmental Science   Soil permeability, the rate at which water can flow through the soil, is an important measure prior to construction. Four different potential county development sites in Wisconsin were selected for testing. Independent random samples of locations within each development were selected, and the soil permeability was measured at each location (in inches per hour). An ANOVA test was conducted to determine whether there was any difference in population mean permeability rates ($f = 9.10$ and MSE $= 0.02504$). The summary statistics are given in the following table.

Development	Sample size	Sample mean
Grant	10	1.3164
Green	12	0.2567
Dane	11	0.5601
Rock	11	0.9206

a. Find bounds on the p value associated with this (significant) F statistic.

b. Construct the Bonferroni 95% confidence intervals for all pairwise comparisons and draw a graph to represent the results.

11.54 Sports and Leisure   There is a 35-second shot clock for men's NCAA basketball games. A team must shoot before the shot clock expires, or they turn the ball over to the opposing team. A team rarely uses the entire 35 seconds, and some teams shoot as quickly as possible. A study was conducted to compare the mean time to take a shot in five different athletic conferences. Independent random samples of possessions were obtained, and the time (in seconds) to take a shot was recorded for each. An ANOVA test was conducted to determine whether there was any overall difference in population mean shot times ($f = 6.33$ and MSE $= 65.201$). The summary statistics are given in the following table.

Conference	Sample size	Sample mean
ACC	20	17.27
Big East	26	27.90
Pac-12	23	19.14
Sun Belt	25	22.97
WAC	26	20.05

Construct the Bonferroni 99% confidence intervals for all pairwise comparisons and draw a graph to represent the results. Identify the pairs of athletic conference population means that are significantly different.

11.55 Biology and Environmental Science   Most honey comes from European honey bee colonies cultivated in the United States. Africanized killer bees tend to be aggressive and produce less honey. A study was conducted to compare the mean amount of honey produced by colonies in four parts of the country. Independent random samples of colonies were obtained in each area, and the amount of honey (in pounds) per year per colony was recorded. **HONEY**

a. Conduct an analysis of variance test to show that there is evidence of an overall difference in population mean honey production. Use $\alpha = 0.05$.

b. Construct the Bonferroni 95% confidence intervals to isolate the pairs of means contributing to the overall difference.

c. Draw a graph to represent the results of the multiple comparison procedure in part (b).

11.56 Olive Oil Purity   Olive oil is used in a variety of cooking methods and has been shown to provide some health benefits. However, the chemical composition and purity of olive oil can also vary and depends on the supplier region and climate. Independent random samples of domestic regular olive oils were obtained and the peroxide value (in mEq O_2/kg oil) was measured for each. **OLIVEOIL**

a. Conduct an analysis of variance test to show that there is evidence of an overall difference in population mean peroxide value. Use $\alpha = 0.01$.

b. Construct the Tukey 99% confidence intervals to isolate the pairs of means contributing to the overall difference.

c. Draw a graph to represent the results of the multiple comparison procedure in part (b).

d. Based on these data, which olive oil appears to be the purest? Why?

11.57 Demographics and Population Statistics   A stun gun delivers a high-voltage, low-amperage electrical shock to an attacker. In certain states, cities, and countries the use of this self-protection device is restricted or even illegal. Independent random samples of stun gun owners in cities, suburban areas, and rural areas were obtained. Each stun gun was examined, and the voltage (in thousands of volts) on the device was carefully measured in a single test. **STUNGUN**

a. Conduct an analysis of variance test to show that there is evidence of an overall difference in population mean stun gun voltage. Use $\alpha = 0.05$.

b. Construct the Tukey 95% confidence intervals to isolate the pairs of means contributing to the overall difference.

c. Draw a graph to represent the results of the multiple comparison procedure in part (b).

11.58 Public Health and Nutrition   A simple gelatin is known today as "America's most famous dessert." Although it was patented in 1845, Jell-O sales were minimal until the early 1900s. While primarily made from processed collagen, gelatin also contains sugar, artificial flavors, and coloring. Independent random samples of three gelatin brands were obtained, and the amount of sugar (in grams) in one serving was measured. **GELATIN**

a. Conduct an analysis of variance test to show that there is evidence of an overall difference in population mean sugar content in gelatin servings. Use $\alpha = 0.01$.

b. Construct the Bonferroni 99% confidence intervals to isolate the pairs of means contributing to the overall difference.

c. Draw a graph to represent the results of the multiple comparison procedure in part (b).

11.59 Sports and Leisure   The outside dimensions and playing lines of a tennis court are standard and well established. However, the distance between courts, or sidelines, varies. Many construction guidelines recommend 24 feet between courts, but to conserve space, builders plan for only at least 12 feet. Independent random samples of public courts were obtained in five different cities, and the distance (in feet) between courts was recorded. The data are given in the following table. **TENNIS**

City	Observations				
Atlanta	14.0	14.2	13.0	15.2	15.0
Boston	14.2	16.8	18.6	15.5	16.6
Los Angeles	14.3	14.9	16.5	15.1	14.4
Miami	14.3	17.3	17.3	14.9	16.4
New York	22.0	18.3	19.3	20.5	18.5

a. Conduct an analysis of variance test to show that there is evidence of an overall difference in population mean distances between tennis courts in public parks. Use $\alpha = 0.01$.

b. Construct the Bonferroni 95% confidence intervals to isolate the pairs of means contributing to the overall difference.

c. Draw a graph to represent the results of the multiple comparison procedure in part (b).

Extended Applications

11.60 Biology and Environmental Science   Hatcheries around the country provide chicks to poultry farms for growing into eventual roasters and broilers. Independent random samples of chicks from different hatcheries were obtained, and the shipping weight (in grams) of each chick was measured. The data are given in the following table. **CHICKS**

Hatchery	Observations				
Bedwell Farms	47.1	48.8	52.6	49.1	53.3
Clinton Chicks	63.0	52.0	56.4	56.2	55.3
Sunny Creek	51.6	52.9	54.9	52.8	54.6
Wild Wings	57.4	57.4	55.2	56.0	56.0

a. Conduct an analysis of variance test to show that there is evidence of an overall difference in population mean chick weights. Use $\alpha = 0.05$.

b. Construct the Bonferroni 95% confidence intervals to isolate the pairs of means contributing to the overall difference.

c. Construct the Tukey 95% confidence intervals to isolate the pairs of means contributing to the overall difference.

d. Are your answers to parts (b) and (c) the same? If so, did you expect this to happen? If not, why not?

11.61 Public Health and Nutrition   Butter and margarine contain saturated fat, which can increase the "bad" cholesterol in your blood. In order to advise clients, a dietitian examined four types of margarine. A random sample of each type was obtained, and the amount of saturated fat (in grams) per serving was recorded. **SATFAT**

a. Conduct an analysis of variance test to show that there is evidence of an overall difference in population mean saturated fat per serving. Use $\alpha = 0.01$.

b. Construct the Tukey 99% confidence intervals to isolate the pairs of means contributing to the overall difference.

c. Are the differences found in part (b) consistent with the percentage of fat indicated on each product label? If not, can you explain any discrepancy?

11.62 Washer Water Usage   According to the U.S. Environmental Protection Agency, a typical American family washes about 400 loads of laundry each year. Front-loading washing machines have become popular because they have a small carbon footprint, use less electricity, and tend to get clothes cleaner. Independent random samples of front-loading washing machines were obtained and the amount of water used (in gallons) in a regular load was measured for each. **LAUNDRY**

a. Conduct an analysis of variance test to show that there is evidence of an overall difference in population mean water usage. Use $\alpha = 0.05$.

b. Construct the Bonferroni 95% confidence intervals to isolate the pairs of means contributing to the difference.

c. Construct the Tukey 95% confidence intervals to isolate the pairs of means contributing to the overall difference.

d. Are the results the same in parts (b) and (c)? If so why? If not, why not?

Challenge

11.63 Manufacturing and Product Development   An experiment was conducted to compare the acoustic properties of a control plastic (styrene) with four alternative treatment plastics. Independent random samples were selected, and the attenuation value of each piece was measured (in dB/mm at 5 MHz). **ACOUSTIC**

a. Conduct a one-way analysis of variance to test for an overall difference among the five population means. Use $\alpha = 0.05$.

b. Construct 95% Bonferroni confidence intervals *only* for the differences between each treatment plastic population mean and the control, styrene, population mean. Distribute the confidence level among the four comparisons (1 versus 2, 1 versus 3, 1 versus 4, and 1 versus 5) so that the simultaneous confidence level is 95%. Which treatment plastic attenuation means are significantly different from the control?

11.3 Two-Way ANOVA

In the previous two sections, we considered the effect of a single factor on a response variable. In this section, we will consider experiments in which two factors may contribute to the overall variability in response. A two-way ANOVA is designed to compare the means of populations that can be classified in two different ways. For example, suppose we are interested in the miles on a charge for electric cars. The miles may vary by the motor and the battery configuration.

Without as much detail as Section 11.1, suppose there are a levels of factor A, b levels of factor B, and n observations for each combination of levels, for a total of abn observations. Using the electric car example, there could be $a = 5$ levels of factor A, 5 different motors; and $b = 3$ levels of factor B, 3 different battery configurations; and $n = 6$ observations for each combination of motor and battery configuration, for a total of $abn = (5)(3)(6) = 90$ observations.

In Table 11.1 there are $a = 3$ levels of factor A, $b = 4$ levels of factor B, and $n = 2$ observations per cell. Let x_{ijk} represent the kth observation for the ith level of factor A and the jth level of factor B. For example, in Table 11.1, $x_{132} = 71$, the 2nd observation for the 1st level of factor A and the 3rd level of factor B.

Table 11.1 Presentation of data in a two-way ANOVA

		Factor B						
		1		**2**		**3**		**4**	
	1	83	84	84	77	84	71	100	102
Factor A	**2**	71	68	78	66	114	90	108	114
	3	78	89	95	104	115	119	119	126

An interaction effect is significant if one level of factor A and one level of factor B interact differently, or inconsistently, from other factor combinations.

The total variation in the data, sum of squares (**SST**), is decomposed into the sum of squares due to factor A (**SSA**), the sum of squares due to factor B (**SSB**), the sum of squares due to interaction [**SS(AB)**], and the sum of squares due to error (**SSE**).

Consistent with previous notation, dots in the subscript of $\bar{x}$ and t indicate the mean and the sum of x_{ijk}, respectively, over the appropriate subscript(s), for example,

$$\bar{x}_{.j.} = \frac{1}{an}\sum_{i=1}^{a}\sum_{k=1}^{n}x_{ijk} \quad \text{and} \quad t_{...} = \sum_{i=1}^{a}\sum_{j=1}^{b}\sum_{k=1}^{n}x_{ijk}$$

Here are the fundamental identity, the definitions, and the computational formulas for a two-way ANOVA.

Two-Way ANOVA Identity

Let SST = total sum of squares, SSA = sum of squares due to factor A, SSB = sum of squares due to factor B, SS(AB) = sum of squares due to interaction, and SSE = sum of squares due to error.

$$\text{SST} = \underbrace{\sum_{i=1}^{a}\sum_{j=1}^{b}\sum_{k=1}^{n}(x_{ijk} - \bar{x}_{...})^2}_{\text{definition}} = \underbrace{\left(\sum_{i=1}^{a}\sum_{j=1}^{b}\sum_{k=1}^{n}x_{ijk}^2\right) - \frac{t_{...}^2}{abn}}_{\text{computational formula}}$$

$$\text{SSA} = \underbrace{bn\sum_{i=1}^{a}(\bar{x}_{i..} - \bar{x}_{...})^2}_{\text{definition}} = \underbrace{\frac{\sum_{i=1}^{a}t_{i..}^2}{bn} - \frac{t_{...}^2}{abn}}_{\text{computational formula}}$$

$$SSB = \underbrace{an \sum_{j=1}^{b} (\bar{x}_{.j.} - \bar{x}_{...})^2}_{\text{definition}} = \underbrace{\frac{\sum_{j=1}^{b} t_{.j.}^2}{an} - \frac{t_{...}^2}{abn}}_{\text{computational formula}}$$

$$SS(AB) = \underbrace{n \sum_{i=1}^{a} \sum_{j=1}^{b} (\bar{x}_{ij.} - \bar{x}_{i..} - \bar{x}_{.j.} + \bar{x}_{...})^2}_{\text{definition}}$$

$$= \underbrace{\frac{\sum_{i=1}^{a} \sum_{j=1}^{b} t_{ij.}^2}{n} - \frac{\sum_{i=1}^{a} t_{i..}^2}{bn} - \frac{\sum_{j=1}^{b} t_{.j.}^2}{an} + \frac{t_{...}^2}{abn}}_{\text{computational formula}}$$

$$SSE = \underbrace{\sum_{i=1}^{a} \sum_{j=1}^{b} \sum_{k=1}^{n} (x_{ijk} - \bar{x}_{ij.})^2}_{\text{definition}} = \underbrace{SST - SSA - SSB - SS(AB)}_{\text{computational formula}}$$

$$SST = SSA + SSB + SS(AB) + SSE$$

The assumptions for a two-way ANOVA are stated in terms of each *cell*, which is considered a *population*.

For reference, these are the two-way ANOVA assumptions.

1. The ab population distributions are normal.

2. The ab population variances are equal.

3. The samples are selected randomly and independently from the respective populations.

Two-way ANOVA calculations are also usually presented in a summary table. The values in this table are associated with the five sources of variation, and the F statistics are used to conduct appropriate hypothesis tests.

Two-way ANOVA summary table

Source of variation	Sum of squares	Degrees of freedom	Mean square	F	p Value
Factor A	SSA	$a - 1$	$MSA = \dfrac{SSA}{a - 1}$	$F_A = \dfrac{MSA}{MSE}$	p_A
Factor B	SSB	$b - 1$	$MSB = \dfrac{SSB}{b - 1}$	$F_B = \dfrac{MSB}{MSE}$	p_B
Interaction	SS(AB)	$(a - 1)(b - 1)$	$MS(AB) = \dfrac{SS(AB)}{(a - 1)(b - 1)}$	$F_{AB} = \dfrac{MS(AB)}{MSE}$	p_{AB}
Error	SSE	$ab(n - 1)$	$MSE = \dfrac{SSE}{ab(n - 1)}$		
Total	SST	$abn - 1$			

Check to make sure that the expressions for the number of degrees of freedom do sum to $abn - 1$.

There are three hypothesis tests associated with a two-way ANOVA.

Two-Way ANOVA Hypothesis Tests

Test 1. Test for an interaction effect.

H_0: There is no interaction effect.

H_a: There is an effect due to interaction.

TS: $F_{AB} = \dfrac{\text{MS(AB)}}{\text{MSE}}$

RR: $F_{AB} \geq F_{\alpha,(a-1)(b-1),ab(n-1)}$

Test 2. Test for an effect due to factor A.

H_0: There is no effect due to factor A.

H_a: There is an effect due to factor A.

TS: $F_A = \dfrac{\text{MSA}}{\text{MSE}}$

RR: $F_A \geq F_{\alpha,a-1,ab(n-1)}$

Test 3. Test for an effect due to factor B.

H_0: There is no effect due to factor B.

H_a: There is an effect due to factor B.

TS: $F_B = \dfrac{\text{MSB}}{\text{MSE}}$

RR: $F_B \geq F_{\alpha,b-1,ab(n-1)}$

The hypothesis test for an interaction effect is usually considered first. An interaction effect is present when the relationship between the two factors is not *linear*, or *additive*, for at least one combination of levels. For example, suppose the total production of corn depends on two factors, the amount of water and the amount of fertilizer. Intuitively, one might expect the total production of corn to increase as the amount of water increases and/or as the amount of fertilizer increases. Each change in water and/or fertilizer results in a predictable, additive change in the total amount of corn produced. However, consider a very high level of water and a high level of fertilizer. The additive model predicts a huge total production in corn. However, there could be an interaction or inconsistent effect associated with these two levels. Too much water and too much fertilizer might combine to produce very little corn production. An interaction plot is a scatter plot of each cell sample mean versus factor A level. Connect the points in the graph corresponding to the same factor B levels. Parallel lines suggest no evidence of interaction.

Case 1. If the null hypothesis is *not* rejected, then the other two hypothesis tests can be conducted as usual, to see whether there are effects due to either (or both) factors.

Case 2. If the null hypothesis is rejected, then there is evidence of a significant interaction. Interpretation of the other two hypothesis tests is tricky.

   1. If we reject a null hypothesis of no effect due to factor A (and/or factor B), then the effect due to factor A (and/or factor B) is probably significant.

   2. If we do not reject a null hypothesis of no effect due to factor A (and/or factor B), then the effect due to factor A (and/or factor B) is inconclusive.

DATA SET

SATELTV

Example 11.5 Satellite TV Quality

The quality of the picture on a home satellite TV depends on the strength of the signal. Professional installation is often necessary to properly align a dish and tune in a satellite.

A study was conducted to determine whether the signal strength was related to the satellite company and/or the geographic region. For each company and region, a random sample of satellite TV users was obtained, and the signal strength of each system was measured (in dBμV). The data are given in the following table.

		Geographic region			
		Northeast	**Southeast**	**Midwest**	**West**
Company	**DIRECTTV**	65.9 73.9 70.9 74.9	64.0 61.4 55.0 52.1	55.9 64.2 67.2 74.3	54.7 69.9 62.2 65.5
	Dish Network	71.3 76.8 71.4 65.2	64.3 64.0 61.3 68.2	69.4 68.6 61.2 64.9	65.6 67.8 55.2 66.0
	Echostar	64.0 64.2 76.7 65.1	60.8 62.9 57.0 69.6	58.3 64.4 69.3 73.1	62.4 62.9 64.5 71.3

Conduct a two-way analysis of variance to determine whether signal strength is affected by company and/or geographic region. Use $\alpha = 0.05$.

SOLUTION

STEP 1 Assume the two-way ANOVA assumptions are true. There are $a = 3$ levels of factor A (satellite TV company), $b = 4$ levels of factor B (geographic region), and $n = 4$ observations in each cell. Sample totals are given in the following table.

		Geographic region				
		Northeast	**Southeast**	**Midwest**	**West**	
Company	**DIRECTTV**	$t_{11.} = 285.6$	$t_{12.} = 232.5$	$t_{13.} = 261.6$	$t_{14.} = 252.3$	$t_{1..} = 1032.0$
	Dish Network	$t_{21.} = 284.7$	$t_{22.} = 257.8$	$t_{23.} = 264.1$	$t_{24.} = 254.6$	$t_{2..} = 1061.2$
	Echostar	$t_{31.} = 270.0$	$t_{32.} = 250.3$	$t_{33.} = 265.1$	$t_{34.} = 261.1$	$t_{3..} = 1046.5$
		$t_{.1.} = 840.3$	$t_{.2.} = 740.6$	$t_{.3.} = 790.8$	$t_{.4.} = 768.0$	$t_{...} = 3139.7$

STEP 2 Find the sums of squares.

$$\text{SST} = \left(\sum_{i=1}^{3}\sum_{j=1}^{4}\sum_{k=1}^{4} x_{ijk}^2\right) - \frac{t_{...}^2}{(3)(4)(4)}$$

Computational formula.

$$= (65.9^2 + 73.9^2 + \cdots + 71.3^2) - \frac{3139.7^2}{48}$$

Apply the formula.

$$= 206{,}990.75 - 205{,}369.09 = 1621.66$$

Simplify.

$$\text{SSA} = \frac{\sum_{i=1}^{3} t_{i..}^2}{(4)(4)} - \frac{t_{...}^2}{(3)(4)(4)}$$

Computational formula.

$$= \frac{1032.0^2 + 1061.2^2 + 1046.5^2}{16} - \frac{3139.7^2}{48}$$

Apply the formula.

$$= 205{,}395.73 - 205{,}369.09 = 26.64$$

Simplify.

$$\text{SSB} = \frac{\sum_{j=1}^{4} t_{.j.}^2}{(3)(4)} - \frac{t_{...}^2}{(3)(4)(4)}$$

Computational formula.

$$= \frac{840.3^2 + 740.6^2 + 790.8^2 + 768.0^2}{12} - \frac{3139.7^2}{48}$$

Apply the formula.

$$= 205{,}815.09 - 205{,}369.09 = 446.00$$

Simplify.

$$SS(AB) = \frac{\sum_{i=1}^{3}\sum_{j=1}^{4} t_{ij.}^2}{4} - \frac{\sum_{i=1}^{3} t_{i..}^2}{(4)(4)} - \frac{\sum_{j=1}^{4} t_{.j.}^2}{(3)(4)} + \frac{t_{...}^2}{(3)(4)(4)}$$ Computational formula.

$$= \frac{285.6^2 + 232.5^2 + \cdots + 261.1^2}{4}$$

$$-205,395.73 - 205,815.09 + 205369.09$$ Apply the formula.

$$= 108.19$$ Simplify.

$$SSE = SST - SSA - SSB - SS(AB)$$ Computational formula.

$$= 1621.66 - 26.64 - 446.00 - 108.19 = 1040.83$$ Apply the formula.

STEP 3 Check for an interaction effect first.

H_0: There is no interaction effect.

H_a: There is an effect due to interaction.

TS: $F_{AB} = \dfrac{MS(AB)}{MSE}$

RR: $F_{AB} \geq F_{\alpha,(a-1)(b-1),ab(n-1)} = F_{0.05,6,36} = 2.37$

$$MS(AB) = \frac{SS(AB)}{(a-1)(b-1)} = \frac{108.19}{6} = 18.0317$$

$$MSE = \frac{SSE}{ab(n-1)} = \frac{1040.83}{36} = 28.9119$$

The value of the test statistic is

$$f_{AB} = \frac{MS(AB)}{MSE} = \frac{18.0317}{28.9119} = 0.6237$$

The value of the test statistic does not lie in the rejection region. Equivalently, $p = 0.7101 > 0.05$, as illustrated in Figure 11.14. There is no evidence of an interaction effect. The tests for factor effects can be conducted as usual.

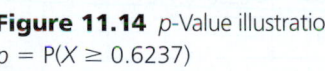

$X \sim F_{6,36}$

0   0.6237

Figure 11.14 *p*-Value illustration: $p = P(X \geq 0.6237)$ $= 0.7101 > 0.05 = \alpha$

STEP 4 Check for an effect due to satellite company (factor A).

H_0: There is no effect due to factor A.

H_a: There is an effect due to factor A.

TS: $F_A = \dfrac{MSA}{MSE}$

RR: $F_A \geq F_{\alpha,\, a-1,ab(n-1)} = F_{0.05,\, 2,36} = 3.26$

The value of the test statistic is

$$f_A = \frac{MSA}{MSE} = \frac{SSA/(a-1)}{SSE/ab(n-1)} = \frac{26.64/2}{1040.83/36} = 0.4607$$

The value of the test statistic does not lie in the rejection region. Equivalently, $p = 0.6345 > 0.05$, as illustrated in Figure 11.15. There is no evidence to suggest that satellite company has an effect on the signal strength.

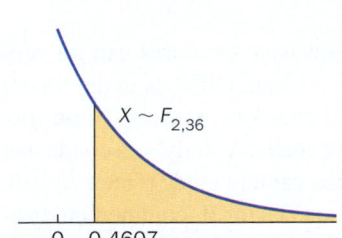

$X \sim F_{2,36}$

0   0.4607

Figure 11.15 *p*-Value illustration: $p = P(X \geq 0.4607)$ $= 0.6345 > 0.05 = \alpha$

STEP 5 Check for an effect due to geographic region (factor B).

H_0: There is no effect due to factor B.

H_a: There is an effect due to factor B.

TS: $F_B = \dfrac{MSB}{MSE}$

RR: $F_B \geq F_{\alpha,b-1,ab(n-1)} = F_{0.05,\, 3,36} = 2.87$

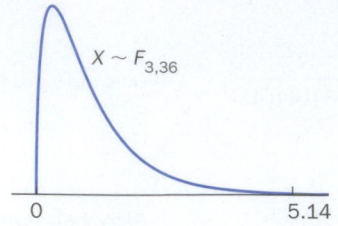

Figure 11.16 *p*-Value illustration:
$p = P(X \geq 5.14)$
$= 0.0046 \leq 0.05 = \alpha$

The *p* values in this table were found using technology and are slightly different from those given below due to round-off error in computing f_A, f_B, and f_{AB}.

The commands to produce this output are detailed in the Technology Corner at the end of this section. Note that SST is not displayed in this output.

The value of the test statistic is

$$f_B = \frac{\text{MSB}}{\text{MSE}} = \frac{\text{SSB}/(b-1)}{\text{SSE}/ab(n-1)} = \frac{446.00/3}{1040.83/36} = 5.14 \; (\geq 2.87)$$

The value of the test statistic lies in the rejection region. Equivalently, $p = 0.0046 \leq 0.05$, as illustrated in Figure 11.16. There is evidence to suggest that signal strength is affected by geographic region.

STEP 6 And finally, here is the ANOVA summary table.

Source of variation	Sum of squares	Degrees of freedom	Mean square	F	*p* Value
Factor A	26.64	2	13.32	0.46	0.6349
Factor B	446.00	3	148.67	5.14	0.0046
Interaction	108.19	6	18.03	0.62	0.7129
Error	1040.83	36	28.91		
Total	1621.66	47			

Figure 11.17 shows a technology solution.

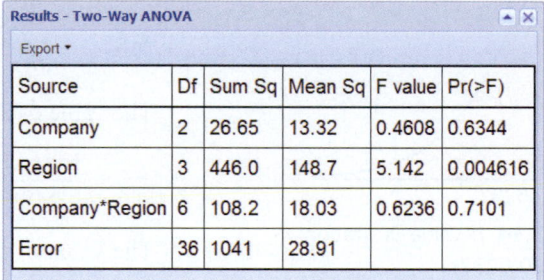

Figure 11.17 Two-way ANOVA results.

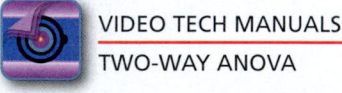

VIDEO TECH MANUALS
TWO-WAY ANOVA

Robert C Nunnington/Oxford Scientific/Getty Images

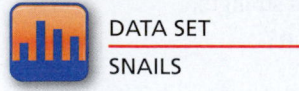

DATA SET
SNAILS

TRY IT NOW GO TO EXERCISE 11.17

Example 11.6 Snail Invasion

Giant African land snails have invaded Florida. These invasive creatures can be very destructive, eating plants, stucco, and plaster in search of calcium. Officials in the Miami area have started a new plan to eradicate the snails using Labrador retrievers. These specially trained dogs are able to detect and help capture these snails. A study was conducted to determine whether the size of these snails is related to the capture mode (factor A: bait, chemical treatments, trap, Labrador, human) and/or location (factor B: commercial, residential). For each of the five capture modes ($a = 5$) and locations ($b = 2$), independent random samples of size $n = 3$ were obtained, and the length of each shell (in centimeters) was measured. The following ANOVA summary table was obtained.

Source of variation	Sum of squares	Degrees of freedom	Mean square	F	*p* Value
Factor A	44.690	4	11.172	1.08	0.3929
Factor B	122.533	1	122.533	11.84	0.0026
Interaction	24.383	4	6.096	0.59	0.6744
Error	206.973	20	10.349		
Total	398.579	29			

a. Is there any evidence of interaction? Use $\alpha = 0.05$.

b. Is there any evidence that capture mode affects the length of the shell? Use $\alpha = 0.05$.

c. Is there any evidence that location affects the length of the shell? Use $\alpha = 0.05$.

SOLUTION

a. Check for an interaction effect.

H_0: There is no interaction effect.

H_a: There is an effect due to interaction.

TS: $F_{AB} = \dfrac{MS(AB)}{MSE}$

RR: $F_{AB} \geq F_{\alpha,(a-1)(b-1),ab(n-1)} = F_{0.05,4,20} = 2.87$

Using the ANOVA summary table, the value of the test statistic is

$$f_{AB} = \frac{MS(AB)}{MSE} = \frac{6.096}{10.349} = 0.59$$

The value of the test statistic does not lie in the rejection region. Equivalently, $p = 0.6744 > 0.05$, as illustrated in Figure 11.18. Therefore, we do not reject the null hypothesis. There is no evidence of an interaction effect.

b. Check for an effect due to capture mode (factor A).

H_0: There is no effect due to factor A.

H_a: There is an effect due to factor A.

TS: $F_A = \dfrac{MSA}{MSE}$

RR: $F_A \geq F_{\alpha,a-1,ab(n-1)} = F_{0.05,4,20} = 2.87$

Using the ANOVA summary table, the value of the test statistic is

$$f_A = \frac{MSA}{MSE} = \frac{11.172}{10.349} = 1.08$$

The value of the test statistic does not lie in the rejection region. Equivalently, $p = 0.3929 > 0.05$, as illustrated in Figure 11.19. Therefore, we do not reject the null hypothesis. There is no evidence of an effect on snail shell size due to capture mode.

c. Check for an effect due to gender (factor B).

H_0: There is no effect due to factor B.

H_a: There is an effect due to factor B.

TS: $F_B = \dfrac{MSB}{MSE}$

RR: $F_B \geq F_{\alpha,b-1,ab(n-1)} = F_{0.05,1,20} = 4.35$

Using the ANOVA summary table, the value of the test statistic is

$$f_B = \frac{MSB}{MSE} = \frac{122.533}{10.349} = 11.84$$

The value of the test statistic lies in the rejection region. Equivalently, $p = 0.0026 \leq 0.05$, as illustrated in Figure 11.20. Therefore, we reject the null hypothesis. There is evidence to suggest an effect on snail shell size due to location. Figure 11.21 shows a technology solution.

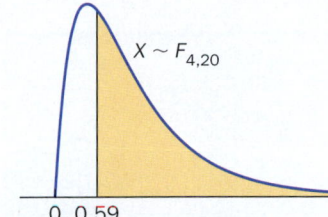

Figure 11.18 *p*-Value illustration:

$p = P(X \geq 0.59)$
$= 0.6744 > 0.05 = \alpha$

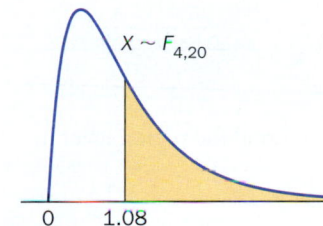

Figure 11.19 *p*-Value illustration:

$p = P(X \geq 1.08)$
$= 0.3929 > 0.05 = \alpha$

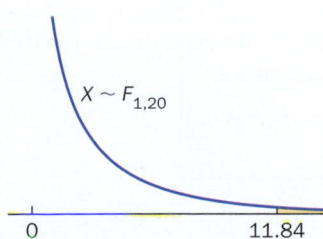

Figure 11.20 *p*-Value illustration:

$p = P(X \geq 11.84)$
$= 0.0026 \leq 0.05 = \alpha$

Effect Tests					
Source	Nparm	DF	Sum of Squares	F Ratio	Prob > F
Capture	4	4	44.68989	1.0796	0.3929
Location	1	1	122.53323	11.8405	0.0026*
Capture*Location	4	4	24.38255	0.5890	0.6744

Figure 11.21 JMP tests for interaction and treatment effects.

There is no built-in TI-84 Plus C function to conduct a two-way analysis of variance.

TRY IT NOW GO TO EXERCISE 11.80

Technology Corner

Procedure: Two-way analysis of variance.
Reconsider: Example 11.5, solution, and interpretations.

CrunchIt!

Use the Two-Way ANOVA function.

1. Enter the signal data in column Var3. Enter the corresponding company level in column Var1 and the region level in column Var2.
2. Select Statistics; ANOVA; Two-way.
3. Select the Factor 1 column (Var1), the Factor 2 column (Var2), and the Values column (Var3).
4. Click Calculate. The results are displayed in a new window. Refer to Figure 11.17.

Minitab

Use the Balanced ANOVA function.

1. Enter the signal data in a single worksheet column (C3). Enter the company level in column C1 and the region level in column C2.
2. Select Stat; ANOVA; Balanced ANOVA.
3. Enter the Response column (C3) and specify the model: C1 C2 C1*C2. The last expression is the interaction term.
4. Click OK. The results are displayed in a session window. See Figure 11.22.

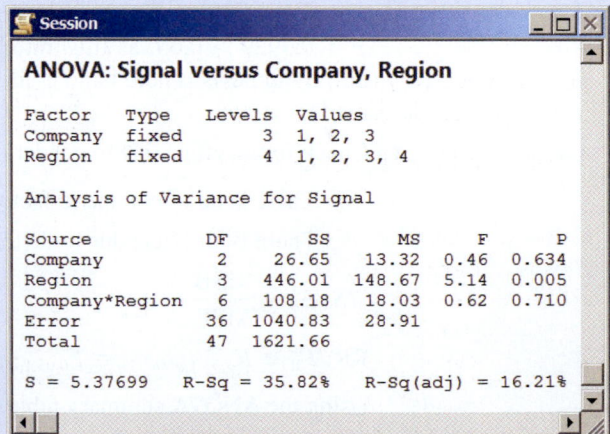

```
Session                                              _ □ ×

ANOVA: Signal versus Company, Region

Factor    Type    Levels  Values
Company   fixed       3   1, 2, 3
Region    fixed       4   1, 2, 3, 4

Analysis of Variance for Signal

Source          DF       SS      MS      F      P
Company          2    26.65   13.32   0.46  0.634
Region           3   446.01  148.67   5.14  0.005
Company*Region   6   108.18   18.03   0.62  0.710
Error           36  1040.83   28.91
Total           47  1621.66

S = 5.37699   R-Sq = 35.82%   R-Sq(adj) = 16.21%
```

Figure 11.22 Minitab ANOVA Two-way results.

Excel

Use the Data Analysis function Anova: Two-Factor With Replication.

1. Enter the data in an array as shown in Figure 11.23.
2. Under the Data tab, select Data Analysis; Anova: Two-Factor With Replication.

3. Enter the Input Range, including the row and column headings, the Rows per sample (4), and the Alpha level (0.05).
4. Choose an Output option and click OK.
5. The analysis of variance table and summary statistics by row and column are displayed. The Anova summary table is show in Figure 11.24.

	A	B	C	D	E
		Northeast	Southwest	Midwest	West
DIRECTTV		65.9	64.0	55.9	54.7
		70.9	55.0	67.2	62.2
		73.9	61.4	64.2	69.9
		74.9	52.1	74.3	65.5
Dish Network		71.3	64.3	69.4	65.6
		71.4	61.3	61.2	55.2
		76.8	64.0	68.6	67.8
		65.2	68.2	64.9	66.0
Echostar		64.0	60.8	58.3	62.4
		76.7	57.0	69.3	64.5
		64.2	62.9	64.4	62.9
		65.1	69.6	73.1	71.3

Figure 11.23 Data input array for Anova: Two-Factor With Replication.

ANOVA

Source of Variation	SS	df	MS	F	P-value	F crit
Sample	26.6454	2	13.3227	0.46	0.6344	3.26
Columns	446.0056	3	148.6685	5.14	0.0046	2.87
Interaction	108.1812	6	18.0302	0.62	0.7101	2.36
Within	1040.8325	36	28.9120			
Total	1621.6648	47				

Figure 11.24 Excel Anova summary table: Two-Factor With Replication.

SECTION 11.3 EXERCISES

Remember to use technology wherever possible to obtain numerical results. Focus on interpretation of results.

Concept Check

11.64 True/False In a two-way ANOVA, a significant interaction means the relationship between the two factors is not additive.

11.65 True/False In a two-way ANOVA, it is only possible to identify one of two factors as significant.

11.66 True/False If there is evidence of significant interaction, the analysis stops.

11.67 Short Answer In a two-way ANOVA, list the degrees of freedom associated with the F critical value for each test.
a. Test for an interaction effect.
b. Test for an effect due to factor A.
c. Test for an effect due to factor B.

11.68 Short Answer In a two-way ANOVA, explain why this analysis fails if the number of observations per cell is $n = 1$.

Practice

11.69 Consider the following data in the context of a two-way ANOVA test. EX11.69

	Factor B								
Factor A		**1**			**2**			**3**	
1	5	7	9	8	9	14	14	12	11
2	8	13	7	10	11	11	16	17	15

a. Find a, the number of levels for factor A, b, the number of levels for factor B, and n, the number of observations for each combination of levels.
b. Find $t_{ij.}$ for $i = 1, 2$ and $j = 1, 2, 3$.
c. Find $t_{i..}$ for $i = 1, 2$. Find $t_{.j.}$ for $j = 1, 2, 3$. Find $t_{...}$.

11.70 Consider the following data in the context of a two-way ANOVA test. EX11.70

	Factor B							
Factor A		**1**				**2**		
1	1.5	2.5	0.7	1.8	3.1	4.4	4.1	3.9
2	3.6	2.8	2.8	3.4	5.7	4.7	7.0	6.0
3	1.0	2.1	1.5	1.6	4.6	3.2	2.0	4.5
4	3.8	2.6	3.3	1.3	4.4	4.1	3.6	4.8

a. Find $t_{i..}$ for $i = 1, 2, 3, 4$, the sample total for each level of factor A. Find $t_{.j.}$ for $j = 1, 2$, the sample total for each level of factor B. Find $t_{ij.}$, the sample total for each *cell*. And, find $t_{...}$, the grand total.

b. Find $\sum_{i=1}^{4}\sum_{j=1}^{2}\sum_{k=1}^{4}x_{ijk}^2$.

c. Compute SST, SSA, SSB, SS(AB), and SSE.

11.71 Consider the following data in the context of a two-way ANOVA test. ▮▮ EX11.71

Factor B

		1		2		3	
Factor A	**1**	11.4 9.6	16.4	4.9 12.6	9.6	17.7 16.8	14.9
	2	8.9 11.7	8.5	8.6 8.5	12.7	6.5 9.0	17.3
	3	8.7 11.2	6.4	9.4 9.2	6.6	8.4 11.7	10.0
	4	14.8 9.3	13.9	13.8 15.6	11.7	19.8 16.2	15.3

a. Compute SST, SSA, SSB, SS(AB), and SSE.
b. Compute MSA, MSB, MS(AB), and MSE.
c. Compute f_A, f_B, and f_{AB}.
d. Conduct the hypothesis tests to check for effects due to interaction, factor A, and factor B. Use $\alpha = 0.05$ for each test. State your conclusions.

11.72 Consider the following data in the context of a two-way ANOVA test. ▮▮ EX11.72

Factor B

		1		2		3	
Factor A	**1**	34.2 30.4	33.3 27.9	36.3 28.5	25.9 34.7	29.7 38.9	37.6 35.6
	2	45.5 38.3	39.7 35.9	37.5 42.4	37.9 37.5	33.2 40.1	45.0 44.5
	3	31.0 29.5	35.1 39.1	38.3 36.3	35.7 33.5	41.4 39.0	39.8 36.7

a. Complete an ANOVA summary table.
b. Conduct the hypothesis tests to check for effects due to interaction, factor A, and factor B. Use $\alpha = 0.01$ for each test. State your conclusions.

11.73 Complete the following ANOVA table.

Source of variation	Sum of squares	Degrees of freedom	Mean square	F	p Value
Factor A		4	40.66		
Factor B		2			
Interaction	144.23				
Error	1206.87	75			
Total	1670.28	89			

Conduct the hypothesis tests to check for effects due to interaction, factor A, and factor B. Use $\alpha = 0.05$ for each test. State your conclusions.

11.74 Complete the following ANOVA table.

Source of variation	Sum of squares	Degrees of freedom	Mean square	F	p Value
Factor A	121.25			1.57	
Factor B	91.12	4			
Interaction					
Error	1446.41	75			
Total	1833.33	99			

Conduct the hypothesis tests to check for effects due to interaction, factor A, and factor B. Use $\alpha = 0.01$ for each test. State your conclusions.

Applications

11.75 Public Health and Nutrition   William Soeltz, the food editor for *The Boston Herald*, conducted a study to examine the quality of wine (measured by age only) at local restaurants. Independent random samples of three types of wines (red, white, and rosé) were obtained from four exclusive restaurants (The Federalist, Top of the Hub, Capital Grille, and Azure). The age (in years) of each wine was recorded. Consider the following partial ANOVA summary table.

Source of variation	Sum of squares	Degrees of freedom	Mean square	F	p Value
Type	90.14	2			
Restaurant	266.49	3			
Interaction	56.59				
Error		48			
Total	1128.82	59			

a. Complete the ANOVA table.
b. What was the total number of observations?
c. Conduct the hypothesis test for any effect due to interaction. Use $\alpha = 0.05$. What does your conclusion imply about the other two hypothesis tests?
d. Conduct the hypothesis tests for factor effects. Use $\alpha = 0.05$. State your conclusions.

11.76 Business and Management   A temp agency recently conducted a study to determine the effects of job type and sex on length of employment. Independent random samples of employees who worked in service, technology, sales, security, and labor were obtained. The length (in weeks) of each assignment was recorded. Consider the following partial ANOVA summary table.

Source of variation	Sum of squares	Degrees of freedom	Mean square	F	p Value
Sex	16.33	1			
Job type	184.39	4			
Interaction		4			
Error	202.42	50			
Total	422.48				

a. Complete the ANOVA table.
b. Is there any evidence of interaction? Use $\alpha = 0.01$.
c. Is there any evidence that gender or job type affects the length of employment? Use $\alpha = 0.01$.

11.77 Biology and Environmental Science According to a recent study, many factors affect feed intake of chickens and, therefore, egg production.[13] For example, lighting, noise, and water may all affect the efficiency of poultry production. A study was conducted to determine whether the feed intake of chickens (in kg/day) is affected by flock size and/or temperature. **FEED**
a. Construct a two-way ANOVA summary table.
b. Use $\alpha = 0.05$ to test the following hypotheses: (i) there is no effect due to interaction; (ii) there is no effect due to flock size; and (iii) there is no effect due to temperature.

11.78 Sports and Leisure Tailgate parties before college football games have become long and lavish, and some present an added security risk. A study was conducted to determine whether the length of a tailgate party is affected by the outside temperature and/or by the college team. The data presented in the following table are the lengths (in hours) of randomly selected tailgate parties (before a game) by school and temperature (C, cold; M, moderate; H, hot). **TAILGATE**

School

Temperature		Michigan		Miami		Ohio State		UCLA	
C		5.1	5.0	4.5	7.0	3.4	4.6	2.9	2.5
		6.0	2.2	6.6	2.8	0.2	3.6	3.1	3.6
M		5.9	5.5	4.7	3.4	3.1	5.7	6.7	7.4
		4.4	4.9	5.4	7.9	5.0	7.2	5.3	5.8
H		4.8	2.8	6.5	5.1	6.3	5.8	4.4	5.8
		4.7	3.4	7.5	5.3	8.0	8.5	9.8	7.2

Use $\alpha = 0.05$ to test the following hypotheses: (a) there is no effect due to interaction; (b) there is no effect due to temperature; and (c) there is no effect due to school. Include the ANOVA summary table.

11.79 Public Policy and Political Science The city of Austin, Texas, recently introduced new procedures to streamline the review of commercial and residential building proposals. A random sample of proposals was obtained and classified by type of development and by the city office conducting the review. The length of the review process (in business days) for each proposal is given in the following table. **BUILDING**

Type of development

Office		Commercial				Residential			
1		5	18	32	32	30	27	27	30
2		20	23	23	28	31	26	31	30
3		33	24	24	20	31	31	26	34

Use $\alpha = 0.05$ to test the following hypotheses: (a) there is no effect due to interaction; (b) there is no effect due to type of development; and (c) there is no effect due to office. Include the ANOVA summary table.

11.80 Travel and Transportation Many "snowbirds," people who live in the Northeast during the summer and in Florida during the winter, ship their vehicles back and forth. However, automobile transportation costs are affected by a number of factors. In a study conducted by Corsia Logistics, a random sample of automobiles shipped from Boston to Miami was obtained for each combination of type of car (compact, midsize, full-size, SUV, pickup) and transport type (open, enclosed). The cost for each (in dollars) was recorded. Conduct an analysis of variance with $\alpha = 0.01$ to test for interaction and factor effects. State your conclusions. **AUTOTRAN**

11.81 Biology and Environmental Science Wetlands, sometimes called the "nurseries of life," are home to thousands of species of plants and animals and are found on every continent expect Antarctica. Canada has approximately 130 million hectares of wetlands, which is about 25% of all the wetlands on our planet.[14] Suppose that in a recent Department of Agriculture study, a random sample of wetlands was obtained for each combination of state and cover type. The cover type classifications are defined by the U.S. Fish and Wildlife service and are determined by the representative plant species. The area covered by water was measured (in hectares), and the data are given in the following table. **WETLAND**

State

Cover type		Louisiana		Mississippi		California		Arkansas	
AB3		0.8	1.1	1.2	0.7	1.2	1.8	3.3	0.2
FO1		2.0	1.7	0.5	0.1	2.4	3.0	1.4	1.5
OW		2.6	0.6	2.7	1.6	1.3	1.8	2.7	1.8
SS1		2.7	2.3	1.1	0.8	1.3	2.4	1.8	1.3

a. Construct the ANOVA summary table. Is there any evidence of an effect due to interaction of state and cover type? Use $\alpha = 0.01$.
b. Is there any evidence of an effect due to state or cover type? Use $\alpha = 0.01$ for both hypothesis tests. State your conclusions.

11.82 Sports and Leisure The Genesis Diving Institute of Florida certifies scuba divers using a variety of different systems of education. This organization recently studied the time spent underwater by scuba divers exploring caves and in open water. Each diver was also classified by age group. Independent dives were selected at random for each combination of age group and dive type. The data (in minutes) are given in the following table. **DIVERS**

Dive type

Age group		Cave				Open water			
20–30		39	38	41	39	42	40	40	39
		41	42	40	40	37	45	36	40
30–40		43	40	38	36	45	39	43	43
		41	46	40	37	47	50	46	44
40–50		42	41	50	47	37	40	40	38
		43	38	46	47	34	45	34	41
≥50		38	34	35	36	33	37	42	38
		40	38	42	38	30	39	41	39

Construct the two-way ANOVA summary table. Interpret the results. Use $\alpha = 0.001$ for each hypothesis test.

Extended Applications

11.83 Travel and Transportation The Swedish Highway Department conducted a study to examine the speed of cars on major highways. Three regions were selected and rated using four different quality scores (based on retroreflective properties) for road markings. (The four classes of road markings in Sweden are K1, certainly approved; K2, probably approved; K3, probably rejected; and K4, certainly rejected.) The speed of randomly selected cars (in km/hr) was recorded, the sample totals for each cell are given in the following table ($n = 4$), and SST $= 1680.66$. **SWEHWY**

	Road marking quality			
Region	K1	K2	K3	K4
North	402.4	393.4	386.6	353.3
Mälardalen	398.6	418.3	381.6	374.2
Mitt	419.4	425.3	385.2	386.3

Use $\alpha = 0.05$ to test the following hypotheses: (a) there is no effect due to interaction; (b) there is no effect due to region; and (c) there is no effect due to road-marking quality.

11.84 Psychology and Human Behavior The FBI studied the amount of money stolen (in thousands of dollars) during randomly selected bank robberies at different types of banks and in different locations. The sample totals for each combination of bank type and location are given in the following table ($n = 3$) and $\sum_{i=1}^{3}\sum_{j=1}^{4}\sum_{k=1}^{3} x_{ijk}^2 = 2514.28$. **BANKROB**

	Bank type			
Location	Commercial	Savings	Savings & Loan	Credit Union
City	25.2	23.7	26.8	24.6
Suburban	32.2	20.2	24.4	31.6
Rural	20.5	22.5	16.0	22.0

Use $\alpha = 0.01$ to test the following hypotheses: (a) there is no effect due to interaction; (b) there is no effect due to location; and (c) there is no effect due to bank type. Construct the ANOVA summary table.

11.85 Medicine and Clinical Studies A low hemoglobin level may be an indication of anemia, an iron deficiency, or even kidney disease. A high hemoglobin level could indicate lung disease or extreme physical exercise. In a study conducted by a new testing facility, a random sample of adults was obtained for each combination of sex and altitude. The hemoglobin level (in grams per liter) for each person was recorded. **HEMOGLOB**

a. Construct the summary ANOVA table. Is there any evidence of an effect due to interaction of gender and altitude? Use $\alpha = 0.05$.

b. Is there any evidence of an effect due to gender or altitude? Use $\alpha = 0.05$ for both hypothesis tests.

11.86 Public Health and Nutrition Cerner Corporation recently presented their vision of future hospital rooms: more technology to improve medical efficiency and decrease the chance of human error. In addition to a system that scans staff IDs and a monitor that displays all relevant patient medical information, another consideration is the size of the room. Suppose a random sample of existing hospital rooms was obtained, and the amount of square feet per patient in each room was measured. Each room was classified by location and type, and the data are given in the following table. **ROOMSIZE**

		Room type					
		Private		Semiprivate two beds		Semiprivate four beds	
Location	City	206	190	210	244	182	183
		263	212	233	201	142	240
		217	205	185	213	177	176
	Suburban	206	203	228	192	143	200
		198	191	204	216	170	195
		197	179	178	200	124	187

a. Construct the ANOVA summary table. Is there any evidence of an effect due to interaction of location and hospital room type? Use $\alpha = 0.05$.

b. Is there any evidence of an effect due to location or hospital room type? Use $\alpha = 0.05$ for both hypothesis tests. State your conclusions.

c. Does this analysis suggest there is one combination of location and room type which is *on average* the smallest in square feet per patient? If so, what is the location and room type?

11.87 Medicine and Clinical Studies A recent study in Nova Scotia, Canada, suggested that several factors affect the age at diagnosis of autism spectrum disorders (ASD).[15] Suppose an additional study was conducted to determine whether age at ASD diagnosis is affected by county and a diagnosis of attention-deficit hyperactivity disorder (ADHD). A random sample of newly diagnosed ASD children was obtained for each combination of county and ADHD diagnosis (1 = no; 2 = yes), and the age of each child was recorded. **ADHD**

a. Construct the ANOVA summary table. Is there any evidence of an effect due to interaction of county and ADHD diagnosis? Use $\alpha = 0.05$.

b. Is there any evidence of an effect due to county or ADHD diagnosis? Use $\alpha = 0.05$ for both hypothesis tests. State your conclusions.

c. Early diagnosis of ASD is important in order to begin intervention services. Is diagnosis of ADHD associated with a decrease or increase in age at diagnosis for ASD? Justify your answer.

CHAPTER 11 SUMMARY

One-Way ANOVA

Assumptions

1. The k population distributions are normal.

2. The k population variances are equal; that is, $\sigma_1^2 = \sigma_2^2 = \cdots = \sigma_k^2$.

3. The samples are selected randomly and independently from the respective populations.

A one-way, or single-factor, analysis of variance is a statistical technique used to determine whether there is any difference among k population means.

The sums of squares

The fundamental identity is SST = SSA + SSE.

$$\text{SST} = \text{total sum of squares} = \sum_{i=1}^{k} \sum_{j=1}^{n_i} (x_{ij} - \bar{x}_{..})^2 = \left(\sum_{i=1}^{k} \sum_{j=1}^{n_i} x_{ij}^2 \right) - \frac{t_{..}^2}{n}$$

$$\text{SSA} = \text{sum of squares due to factor} = \sum_{i=1}^{k} n_i (\bar{x}_{i.} - \bar{x}_{..})^2 = \left(\sum_{i=1}^{k} \frac{t_{i.}^2}{n_i} \right) - \frac{t_{..}^2}{n}$$

$$\text{SSE} = \text{sum of squares due to error} = \sum_{i=1}^{k} \sum_{j=1}^{n_i} (x_{ij} - \bar{x}_{i.})^2 = \text{SST} - \text{SSA}$$

ANOVA summary table

Source of variation	Sum of squares	Degrees of freedom	Mean square	F	p Value
Factor	SSA	$k - 1$	$\text{MSA} = \dfrac{\text{SSA}}{k - 1}$	$\dfrac{\text{MSA}}{\text{MSE}}$	p
Error	SSE	$n - k$	$\text{MSE} = \dfrac{\text{SSE}}{n - k}$		
Total	SST	$n - 1$			

Hypothesis test

H_0: $\mu_1 = \mu_2 = \cdots = \mu_k$ (all k population means are equal)

H_a: $\mu_i \neq \mu_j$ for some $i \neq j$ (at least two population means differ)

TS: $F = \dfrac{\text{MSA}}{\text{MSE}}$

RR: $F \geq F_{\alpha, k-1, n-k}$

Bonferroni multiple comparison procedure

The c simultaneous $100(1 - \alpha)\%$ Bonferroni confidence intervals have as end points

$$(\bar{x}_{i.} - \bar{x}_{j.}) \pm t_{\alpha/(2c), n-k} \sqrt{\text{MSE}} \sqrt{\frac{1}{n_i} + \frac{1}{n_j}} \qquad \text{for all } i \neq j$$

Tukey multiple comparison procedure

The c simultaneous $100(1 - \alpha)\%$ confidence intervals have as endpoints

$$(\bar{x}_{i.} - \bar{x}_{j.}) \pm \frac{1}{\sqrt{2}} Q_{\alpha, k, n-k} \sqrt{\text{MSE}} \sqrt{\frac{1}{n_i} + \frac{1}{n_j}} \qquad \text{for all } i \neq j$$

Two-Way ANOVA
A two-way analysis of variance is a statistical procedure used to determine the effect of two factors on a response variable.

The sums of squares

The fundamental identity is SST = SSA + SSB + SS(AB) + SSE.

SST = total sum of squares

$$= \sum_{i=1}^{a}\sum_{j=1}^{b}\sum_{k=1}^{n}(x_{ijk}-\bar{x}_{...})^2 = \left(\sum_{i=1}^{a}\sum_{j=1}^{b}\sum_{k=1}^{n}x_{ijk}^2\right) - \frac{t_{...}^2}{abn}$$

SSA = sum of squares due to factor A

$$= bn\sum_{i=1}^{a}(\bar{x}_{i..}-\bar{x}_{...})^2 = \frac{\sum_{i=1}^{a}t_{i..}^2}{bn} - \frac{t_{...}^2}{abn}$$

SSB = sum of squares due to factor B

$$= an\sum_{j=1}^{b}(\bar{x}_{.j.}-\bar{x}_{...})^2 = \frac{\sum_{j=1}^{b}t_{.j.}^2}{an} - \frac{t_{...}^2}{abn}$$

SS(AB) = sum of squares due to interaction

$$= n\sum_{i=1}^{a}\sum_{j=1}^{b}(\bar{x}_{ij.}-\bar{x}_{i..}-\bar{x}_{.j.}+\bar{x}_{...})^2 = \frac{\sum_{i=1}^{a}\sum_{j=1}^{b}t_{ij.}^2}{n} - \frac{\sum_{i=1}^{a}t_{i..}^2}{bn} - \frac{\sum_{j=1}^{b}t_{.j.}^2}{an} + \frac{t_{...}^2}{abn}$$

SSE = sum of squares due to error

$$= \sum_{i=1}^{a}\sum_{j=1}^{b}\sum_{k=1}^{n}(x_{ijk}-\bar{x}_{ij.})^2 = \text{SST} - \text{SSA} - \text{SSB} - \text{SS(AB)}$$

ANOVA summary table

Source of variation	Sum of squares	Degrees of freedom	Mean square	F	p Value
Factor A	SSA	$a-1$	$\text{MSA} = \frac{\text{SSA}}{a-1}$	$F_A = \frac{\text{MSA}}{\text{MSE}}$	p_A
Factor B	SSB	$b-1$	$\text{MSB} = \frac{\text{SSB}}{b-1}$	$F_B = \frac{\text{MSB}}{\text{MSE}}$	p_B
Interaction	SS(AB)	$(a-1)(b-1)$	$\text{MS(AB)} = \frac{\text{SS(AB)}}{(a-1)(b-1)}$	$F_{AB} = \frac{\text{MS(AB)}}{\text{MSE}}$	p_{AB}
Error	SSE	$ab(n-1)$	$\text{MSE} = \frac{\text{SSE}}{ab(n-1)}$		
Total	SST	$abn-1$			

Hypothesis tests

1. Test for an interaction effect

H_0: There is no interaction effect.

H_a: There is an effect due to interaction.

TS: $F_{AB} = \dfrac{\text{MS(AB)}}{\text{MSE}}$

RR: $F_{AB} \geq F_{\alpha,(a-1)(b-1),ab(n-1)}$

2. Test for an effect due to factor A

H_0: There is no effect due to factor A.

H_a: There is an effect due to factor A.

TS: $F_A = \dfrac{\text{MSA}}{\text{MSE}}$

RR: $F_A \geq F_{\alpha, a-1, ab(n-1)}$

3. Test for an effect due to factor B

H_0: There is no effect due to factor B.

H_a: There is an effect due to factor B.

TS: $F_B = \dfrac{\text{MSB}}{\text{MSE}}$

RR: $F_B \geq F_{\alpha, b-1, ab(n-1)}$

CHAPTER 11 EXERCISES

11

APPLICATIONS

11.88 Sports and Leisure Backing a trailer down a boat ramp to launch a boat can be awkward and tricky. A steep ramp angle often makes this task even more intimidating, especially for new boat owners. Independent random samples of boat ramps in various regions of Florida were obtained, and the angle of each was measured (in degrees). Consider the following partial ANOVA table.

ANOVA summary table

Source of variation	Sum of squares	Degrees of freedom	Mean square	F	p Value
Factor		4			
Error	470.355	45			
Total	502.650				

a. Complete the ANOVA summary table.

b. How many regions in Florida were considered? How many total observations were obtained?

c. Is there any evidence that the population mean boat-ramp angle differs among these populations? Use $\alpha = 0.05$.

11.89 Physical Sciences It can get very hot for actors on stage. Not only do they have to remember all of their lines, theater spotlights are usually bright and intense. Independent random samples of spotlights were obtained from four different Broadway theaters (Cort Theater, Imperial Theater, Majestic Theater, and the New Amsterdam Theater). The wattage of each light was measured, and a partial ANOVA summary table is shown below.

ANOVA summary table

Source of variation	Sum of squares	Degrees of freedom	Mean square	F	p Value
Factor	106,568	3			
Error		20			
Total	258,565				

Complete the table and use this information to determine whether there is any evidence that the population mean wattage of spotlights is different among the four theaters. Use $\alpha = 0.05$.

11.90 Manufacturing and Product Development A new section of the San Francisco–Oakland Bay Bridge opened on September 2, 2013. The bridge took over 25 years to complete and cost approximately $6.4 billion. It is open, spacious, and even has a bicycle path, but the "Case of the Corrupt Bolts" lingers.[16] Independent random samples of bolts were obtained, and each was subjected to a motion simulating an earthquake. The horizontal displacement of the attached panel was measured for each (in inches). Construct the one-way ANOVA table and test the hypothesis of no difference in population mean horizontal displacement due to bolt type. Use $\alpha = 0.01$. **BOLTS**

11.91 Manufacturing and Product Development Microlithography is a process that includes baking a semiconductor wafer. An experiment was conducted to study the temperature of a 200-mm-diameter wafer at different locations on the wafer during a new baking process. Independent random samples of wafers were selected, and a temperature sensor was placed on each wafer at one of four distances from the center of the wafer. The temperature was recorded (in °C) 80 seconds into the process. The summary statistics are given in the following table.

Location	T1	T2	T3	T4
Sample size	10	10	10	10
Sample mean	106.7	104.6	104.3	98.6

A one-way analysis of variance test was significant at the $p = 0.009$ level. Use MSE $= 26.92$ to find the Bonferroni 95% confidence intervals for all pairwise differences. Draw a graph to represent the results.

11.92 Technology and the Internet An experiment was conducted to compare the accuracy of five different computer algorithms for translating Chinese into English. Each algorithm was tested 25 times on randomly selected passages, and an accuracy score (between 0 and 1) was recorded for each trial.

A one-way analysis of variance test was significant at the $p = 0.003$ level. Use MSE $= 0.0276$ and the summary statistics below to find the Tukey 95% confidence intervals for all pairwise differences. Draw a graph to represent the results.

Algorithm	OO	CP	SLCP	AC	T
Sample mean	0.5566	0.7020	0.6190	0.7023	0.7115

11.93 Manufacturing and Product Development   Using a new atomic imaging technique, a study was conducted to measure randomly selected thin films of NaCl on a metal substrate. Four different growth modes were used, and the step height was measured (in nm) in each case. **NACLFILM**

a. Conduct a one-way analysis of variance test to determine whether there is an overall difference in population mean step heights. Use $\alpha = 0.05$.

b. Construct the Bonferroni 99% confidence intervals to isolate any population means contributing to an overall difference, and draw a graph to represent the results.

11.94 Public Health and Nutrition   A root canal is an endodontic procedure to remove the infected or damaged nerve tissue of a tooth. The American Dental Association conducted a study to determine whether the major canal diameter of a tooth is related to the tooth type and/or the patient's race. Independent random samples of adult male root canal patients were selected for each combination of tooth type and race. The canal diameters (in mm) are given in the following table. **TOOTH**

		Tooth type		
		Maxillary	**Mandibular**	
	White	0.97 1.00	1.09 1.39	
		0.75 1.06	1.17 1.19	
Group	**African American**	1.17 1.00	1.21 1.48	
		1.09 0.88	0.94 1.16	
	Native American	0.86 1.03	1.10 1.45	
		0.77 0.88	1.58 1.18	
	Hispanic	1.13 0.81	0.90 1.01	
		0.78 0.86	1.57 1.42	

Construct the two-way ANOVA summary table. Interpret the results. Use $\alpha = 0.05$ for each hypothesis test.

11.95 Travel and Transportation   Many factors affect the number of automobile accidents on major highways, for example, increased traffic, weather conditions, and driver fatigue. A recent study was conducted to determine whether the frequency of median crashes is related to the number of lanes and/or the differential elevation or opposite travel lanes. Independent random samples of U.S. highways were obtained for each combination of lanes and differential elevation. The number of median crashes over a 6-month time period was recorded for each. Construct a two-way ANOVA table. Interpret the results. Use $\alpha = 0.05$ for each hypothesis test. **CRASH**

EXTENDED APPLICATIONS

11.96 Physical Sciences   Approximately 437 nuclear power plants are in operation around the world, with an estimated power production of 372,210 MWh per year.[17] Independent random samples of nuclear power plants in three countries were obtained, and the net generating capacity per year (in MW) of each was obtained. Summary statistics are given in the following table.

Country	Sample size	Sample total	Sum of squared observations
United States	5	5,848	6,861,942
France	6	6,470	7,623,250
Germany	5	5,212	5,726,814

Conduct the appropriate test to check the hypothesis of no difference in population mean net generating capacity due to country. Use $\alpha = 0.01$.

11.97 Sports and Leisure   The biggest catfish caught in the Catfish Chasers Tournament at Milford Reservoir in June 2013 was a record 82.5 pounds.[18] In preparation for the next catfish tournament, a fisherman (familiar with statistics) randomly selected anglers from five different locations and recorded the weight (in pounds) of their last catfish caught. The data (in pounds) are given in the following table. **CATFISH**

Location	Observations					
Hill's Landing	22.7	14.1	9.6	5.9	21.5	28.7
	45.7	17.4				
Eagle's Nest	38.7	44.6	34.6	19.9	59.3	23.2
	28.3	35.5	38.8	34.5	39.1	
Santee	37.5	53.2	47.0	31.8	19.9	35.9
	50.0	27.6	30.6	22.4		
Campground	23.7	32.1	38.9	31.1	34.2	37.2
	49.8	56.6	52.2	34.2	44.5	38.9
Rock Hill	42.0	41.6	63.2	41.6	25.5	28.4
	33.8					

a. Conduct a one-way analysis of variance test to determine whether there is an overall difference in population mean catfish weights. Use $\alpha = 0.01$.

b. Construct the Tukey 99% confidence intervals to isolate any population means contributing to an overall difference, and draw a graph to represent the results. Which site would you recommend for the fisherman to have a good chance at winning the tournament? Why?

11.98 Biology and Environmental Science   We all expect hospitals to be clean, almost spotless. However, according to the U.S. Centers for Disease Control and Prevention, approximately one million people contract an infection directly from a hospital visit each year. A recent study suggested that hospital lobbies may have high levels of airborne particles, including bacteria and fungi.[19] Independent random samples of hospital lobbies were obtained for each combination of season (1, winter; 2, spring; 3, summer; 4, fall) and lobby size (1, small; 2, medium; 3, large). The level of airborne bacteria was measured in each (in CFU/m^3). **BACTERIA**

a. Construct the two-way ANOVA summary table. Interpret the results. Use $\alpha = 0.05$ for each hypothesis test.

b. Suppose more airborne bacteria in the lobby means there is a greater chance of contracting an infection. Is there one season in which we should avoid going to a hospital? Why?

c. Similarly, is there a specific size of hospital lobby to avoid? Why?

11.99 Marketing and Consumer Behavior A real-estate agent conducted a study to compare the effect of location and season on weekly time-share costs (in dollars). The sample totals for each combination of island (A, Aruba; M, Martinique; SK, St. Kitts; SL, St. Lucia) and season are given in the following table ($n = 6$) and SST = 487,902,980.64.

		Season			
		Spring	Summer	Fall	Winter
Island	A	18,045	12,168	19,894	26,214
	M	14,495	1,925	13,890	17,538
	SK	18,075	32,887	24,398	18,834
	SL	25,457	20,757	27,505	36,428

Use $\alpha = 0.01$ to test the following hypotheses: (a) there is no effect due to interaction; (b) there is no effect due to island; and (c) there is no effect due to season.

11.100 Biology and Environmental Science A random sample of male baby California sea lions was captured, tagged, and monitored for several years. The weight of each sea lion was measured (in pounds) once, and each sea lion was classi-fied by species ID and age (in years). The sample totals are given in the following table ($n = 4$) and SST = 103,152.38.

		Age		
		1	2	3
Species ID	50011	1768	1735	1752
	50023	1794	1563	1517
	50037	1750	1742	1328

Use $\alpha = 0.05$ to test the following hypotheses: (a) there is no effect due to interaction; (b) there is no effect due to species; and (c) there is no effect due to age.

11.101 Free Radicals Glutathione is an antioxidant found in every cell and is important to our health because it neutralizes free radicals. This antioxidant is believed to contribute to our immune system and is found naturally in many foods. However, some research suggests that common, over-the-counter pain medica-tions may lower the body's glutathione level. Independent random samples of healthy male adults and of those taking medication for ordinary aches were obtained. The glutathione level in each was measured (in micrograms/10(10) erythrocytes). **ANTIOX**

a. Conduct a one-way analysis of variance test to determine whether there is an overall difference in population mean glutathione levels. Use $\alpha = 0.05$.

b. Construct the Tukey 95% confidence intervals to isolate any population means contributing to an overall difference. Draw a graph to represent the results.

c. Because we are interested mainly in whether pain relievers affect glutathione levels in healthy males, find the Bonferroni 95% confidence intervals only for the differences between each pain reliever mean and the healthy adult mean. Based on these results, do you believe pain relievers lower glutathione levels? Justify your answer.

11.102 Public Health and Nutrition A consumer group recently studied the amount of partially hydrogenated oils (trans fat) in various peanut butter brands. Four brands were selected in smooth and chunky varieties. The amount of trans fat was measured (in grams) per serving (two tablespoons) in each randomly selected jar, and the data are given in the table below. **PEANUT**

a. Construct the two-way ANOVA summary table. Interpret the results. Use $\alpha = 0.05$ for each hypothesis test.

b. If a consumer wants to avoid trans fat as much as possible, which combination of brand and style would you recommend? Why?

	Peanut butter brand							
	Jif		**Peter Pan**		**Skippy**		**Smucker's**	
Smooth	0.51	0.72	0.89	0.69	0.75	0.85	0.73	0.71
Chunky	0.46	0.63	0.88	0.81	0.76	0.69	0.73	0.76

LAST STEP

11.103 Do the hours worked by farm laborers differ by region? Farm worker wages vary by region and by the number of years a person has worked for the same employer. In addition, some farm workers are paid based on a piece rate, that is, by the number of baskets or crates they pick of the crop being harvested. Some research suggests farm workers work longer hours in some regions of the United States. Random samples of farm workers from four regions were obtained, and the number of hours worked during a spring planting week was recorded for each. The summary statistics are given in the following table.[20] **FARMHR**

Region	Sample size	Sample mean	Sample standard deviation
Northeast II	15	38.6	4.370
Cornbelt I	15	39.5	4.042
Mountain I	15	43.9	4.597
California	15	42.8	3.651

Is there any evidence to suggest that there is a difference in the mean number of hours worked per week due to region?

12 Correlation and Linear Regression

Are windmills too noisy?

Modern windmills are quickly becoming a more efficient, clean alternative for producing energy in the United States and other countries. In a typical three-blade system, the length of a blade is approximately 50–60 feet with diameter 12 feet, and each blade is angled. While a cost-effective windmill can utilize low wind velocity, the ideal locations for wind farms are along ocean coastlines or in mountainous areas, where there is a consistent wind speed of at least 15 mph.

Despite favorable wind conditions, windmills cannot be constructed near many towns because of noise-level regulations. For example, a typical windmill produces approximately 55 dB 100 meters away.[1] This is softer than the sound of human speech (which is about 70 dB). Suppose a small coastal community is considering the construction of a windmill to generate electricity for the town hall. An experiment is conducted to measure the windmill's noise level (in dB) at various distances (in meters) from the proposed site. The data are summarized in the following table.

Distance	10	50	75	120	150	160	200	250	400	500
Noise level	75	110	73	52	58	77	56	57	28	4

The techniques presented in this chapter can be used to determine whether there is a significant linear relationship between distance and noise level. We will see how regression analysis can be used to predict a value of the noise level for a given distance from the windmill.

CONTENTS

Eugene Everett/Dreamstime.com

12.1 Simple Linear Regression

Remember function notation from your algebra class? $f(x)$ is read as "f of x" and is the rule for determining y for a value of x.

Remember $y = mx + b$? This is the same idea, rewritten. Here, the Greek letter β is not related to a type II error, as discussed in Section 9.3.

There is often a need to predict the value of the dependent variable for a given value of the independent variable. For example, if the price is $110, predict the number of tickets that will be sold. Simple linear regression allows us to find the best prediction equation.

A *deterministic* relationship between two variables x and y is one in which the value of y is completely determined by the value of x. In general, $y = f(x)$ is a deterministic relationship between x and y. The value of y *depends* on the value of x. The *independent* variable is x and the *dependent* variable is y. We are free to choose x, but y depends on the value selected. For example, $y = x^2 + 2x + 5$ is a deterministic relationship between x and y. If $x = 1$, then y is completely determined and $y = 1^2 + 2(1) + 5 = 8$.

One of the simplest deterministic relationships between x and y is a *linear* relationship: $y = \beta_0 + \beta_1 x$, where β_0 and β_1 are constants. The set of ordered pairs (x, y) such that $y = \beta_0 + \beta_1 x$ forms a straight line with slope β_1 and y intercept β_0. For example, the graph of $y = 3 + 7x$ is a straight line with slope 7 and y intercept 3. (Could you sketch this line?)

An extension of a deterministic relationship is a *probabilistic model*. For a fixed value x, the value of the second variable is randomly distributed. For example, suppose we are investigating the relationship between one-way airfare (in dollars) from New York City to Los Angeles and the number of tickets sold on the 10:00 A.M. flight. If we select $x = 99$, then the number of tickets sold (for $x = 99$) is a random variable Y. On one particular day, the observed value of Y associated with $x = 99$ may be $y = 135$ tickets.

The *independent variable*, fixed by the experimenter, is usually denoted by x. For a fixed value x, the second variable is randomly distributed. This random variable is the *dependent variable* and is usually denoted by Y. Consistent with previous notation, an observed value of Y is denoted by y.

In a general *additive probabilistic model* there is a deterministic part and a random part. The value of Y differs from $f(x)$ (the deterministic part) by a random amount. The model can be written as

$$Y = (\text{deterministic function of } x) + (\text{random deviation})$$
$$= f(x) + E \tag{12.1}$$

where E is a random variable, called the *random error*.

A value of the random variable E is denoted by e. Here, e is not the base of the natural logarithm.

Suppose we fix a value of x, say $x = x_0$, and observe a value of Y for this value of x, denoted y. If the value of the random variable E is positive, then $y > f(x_0)$. Similarly, if $e < 0$, then $y < f(x_0)$. And if $e = 0$, then $y = f(x_0)$. Geometrically, the value y lies either above, below, or on the graph of $y = f(x)$. Figure 12.1 illustrates the case where $e > 0$.

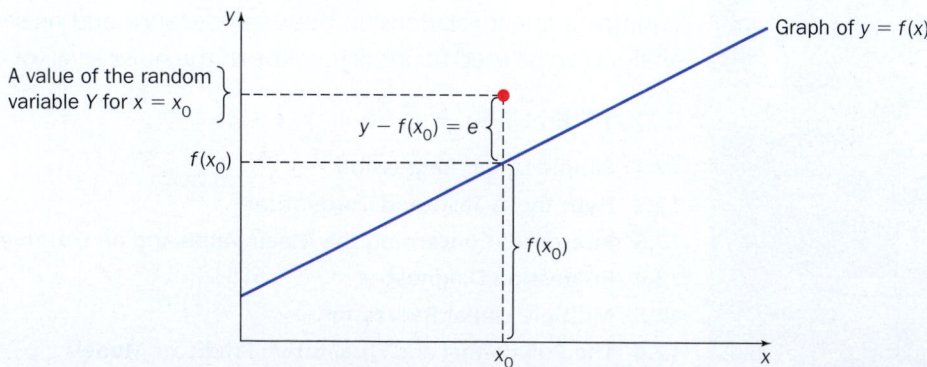

Figure 12.1 An illustration of a probabilistic model. An observed value y is composed of a deterministic part, $f(x_0)$, and a random error, e.

Suppose there are n observations on specific, fixed values of the independent variable. Here is the notation we use.

1. The observed values of the independent variable are denoted $x_1, x_2, \ldots, x_n$.

2. Y_i and y_i are the random variable and the observed value of the random variable associated with x_i, for $i = 1, 2, \ldots, n$. This is tricky. For each x_i, there is a corresponding random variable Y_i. So there are really n random variables in these problems.

3. The data set consists of n ordered pairs: $(x_1, y_1), (x_2, y_2), \ldots, (x_n, y_n)$.

Example 12.1   Fore!

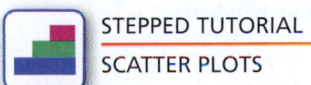

Golf courses in Arizona are watered regularly to keep the fairways and greens pleasing and playable. However, high levels of dissolved salts in the water can harm the course. A groundskeeper is investigating the relationship between the amount of sodium in the water and the total dissolved salts (TDS). A random sample of Arizona golf courses was obtained and a water sample from each was used to measure the amount of sodium (in mEq/liter) and TDS (in mg/L). The data are given in the following table.[2]

Sodium, x	4.1	5.4	4.0	6.2	8.6	9.6	14.3	10.1	12.3	13.4	11.3
TDS, y	450	672	740	860	1307	891	1190	1676	1512	2700	2398

a. Identify the independent and dependent variables.

b. List the ordered pairs in the data set.

c. Construct a scatter plot for these data. What does the plot suggest about the relationship between the variables?

SOLUTION

a. The independent variable is sodium, and the dependent variable is TDS. The groundskeeper believes that there is a relationship between these two variables and hopes to be able to predict the TDS as a function of sodium.

b. The data set consists of 11 ordered pairs: $(4.1, 450), (5.4, 672), \ldots, (11.3, 2398)$. For each (fixed) value of sodium, there is a corresponding observation on a random variable for TDS.

c. Figure 12.2 is a *scatter plot* of TDS (y) versus sodium (x). This plot suggests that as sodium increases, TDS also increases, and therefore the relationship might be positive linear. Figure 12.3 shows a technology solution.

DATA SET
GOLF

STEPPED TUTORIAL
SCATTER PLOTS

VIDEO TECH MANUALS
SCATTER PLOTS

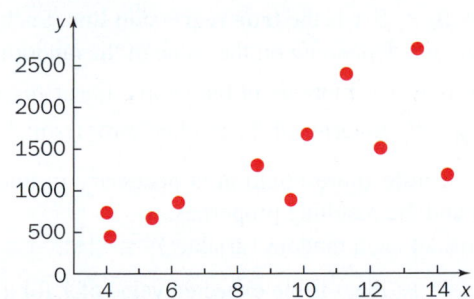

Figure 12.2 A scatter plot of TDS versus sodium level.

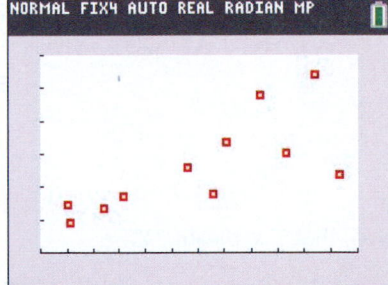

Figure 12.3 A TI-84 Plus C scatter plot of TDS versus sodium level.

TRY IT NOW   GO TO EXERCISE 12.12

In a **simple linear regression model**, the deterministic function $f(x)$ in Equation 12.1 is assumed to be linear [i.e., $f(x) = \beta_0 + \beta_1 x$]. The graph of this regression equation is a straight line that describes how a dependent variable y changes as an independent variable x changes. Figure 12.4 shows the graph of a possible deterministic straight line added to the scatter plot. Notice that the data points lie *close* to the line; the vertical distance between a point and the line depends on the value of the random error. And the line has positive slope, which conveys the relationship between the two variables: as the sodium level rises, so does the TDS.

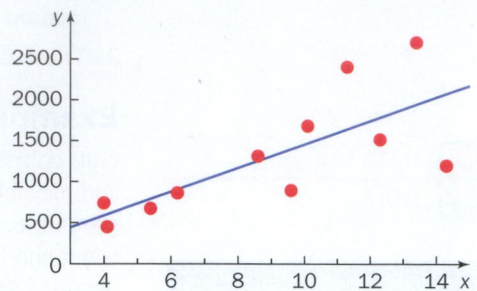

Figure 12.4 TDS versus sodium level scatter plot with a (deterministic) line added.

In just a few pages, we'll define what is meant by *best*.

This section presents a method for finding the *best* deterministic straight line for a given set of data. Hypothesis tests will be used to determine whether there is a *significant* linear relationship between two variables. As with all other hypothesis tests, certain assumptions are required for the related statistical procedures to be valid.

Simple Linear Regression Model

Let $(x_1, y_1), (x_2, y_2), \ldots, (x_n, y_n)$ be n pairs of observations such that y_i is an observed value of the random variable Y_i. We assume that there exist constants β_0 and β_1 such that

$$Y_i = \beta_0 + \beta_1 x_i + E_i$$

where $E_1, E_2, \ldots, E_n$ are independent, normal random variables with mean 0 and variance σ^2. That is:

1. The E_i's are normally distributed (which implies that the Y_i's are normally distributed).
2. The expected value of E_i is 0 [which implies that $E(Y_i) = \beta_0 + \beta_1 x_i$].
3. $\text{Var}(E_i) = \sigma^2$ [which implies that $\text{Var}(Y_i) = \sigma^2$].
4. The E_i's are independent (which implies that the Y_i's are independent).

A CLOSER LOOK

1. The E_i's are the **random deviations** or **random error terms**.
2. $y = \beta_0 + \beta_1 x$ is the **true regression line**. Each point (x_i, y_i) lies *near* the true regression line, depending on the value of the random error term e_i.
3. The four assumptions in the simple linear regression model definition can be stated compactly in terms of the random error term: $E_i \overset{\text{ind}}{\sim} N(0, \sigma^2)$.

Just a little more notation is necessary to understand the simple linear regression model and the resulting properties.

Recall, that $Y|x_i$ means "Y given x_i."

Consider each random variable $Y_i = Y|x_i$.

$\mu_{Y|x_i} = E(Y|x_i)$ is the expected value of Y for a fixed value x_i, and

$\sigma^2_{Y|x_i}$ is the variance of Y for a fixed value x_i.

The simple linear regression model assumptions imply that

$\mu_{Y|x_i} = E(\beta_0 + \beta_1 x_i + E_i) = \beta_0 + \beta_1 x_i + E(E_i) = \beta_0 + \beta_1 x_i$

$\sigma^2_{Y|x_i} = \text{Var}(\beta_0 + \beta_1 x_i + E_i) = \sigma^2$

$Y|x_i$ is normal.

Therefore, the mean value of Y is a linear function of x. The true regression line passes through the *line of mean values*.

The variability in the distribution of Y is the *same* for every value of x (this is called **homogeneity of variance**).

Figure 12.5 illustrates the model assumptions and resulting properties. Each Y_i has a normal distribution, centered at $\beta_0 + \beta_1 x_i$. All the distributions have the same width, or variance.

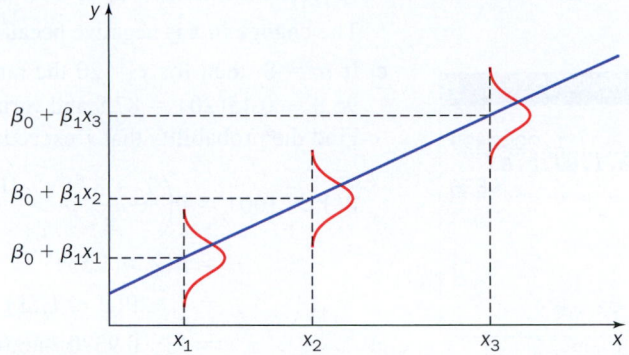

Figure 12.5 The true regression line connects the mean values $\beta_0 + \beta_1 x_i$.

Example 12.2 Get the Lead Out

A recent study suggests that exposure to high levels of lead may lead to more violent crime, more cases of ADHD, and even lower IQs.[3] For six-year-old children, suppose IQ level (y) is related to blood lead level (x, measured in μg/dL) and that the true regression line is $y = 96.8 - 0.45x$.

a. Find the expected IQ for a six-year-old when the blood lead level is 30 μg/dL.

b. How much change in IQ is expected if the blood lead level increases by 10 μg/dL? What if it decreases by 20 μg/dL?

c. Suppose $\sigma = 8$ μg/dl. Find the probability that an observed IQ is greater than 100 when the blood lead level is 20 μg/dL.

SOLUTION

a. We need the expected value of Y for the value $x = 30$.

$$E(Y|x) = \beta_0 + \beta_1 x \qquad \text{Simple linear regression model implication.}$$
$$E(Y|30) = 96.8 - 0.45(30) \qquad \text{Use the true regression line with } x = 30.$$
$$= 83.3 \qquad \text{Simplify.}$$

The expected IQ for a six-year-old child with a blood lead level of 30 μg/dL is 83.3. See Figure 12.6.

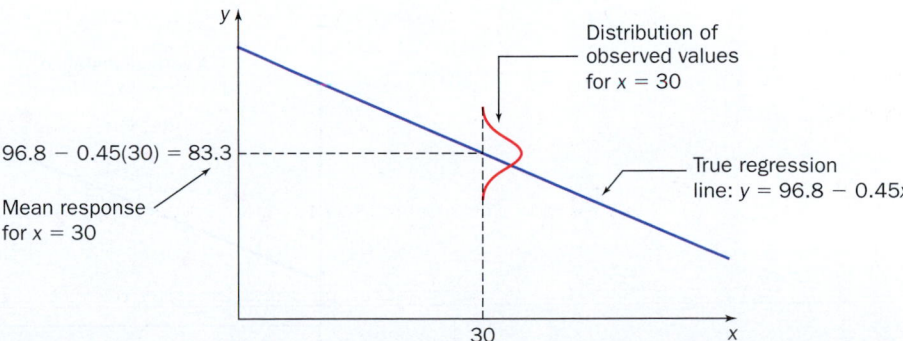

Figure 12.6 The true regression line, the mean response for $x = 30$, and the distribution of observed values around the mean response.

b. The slope of the true regression line, $\beta_1 = -0.45$, is the change in IQ associated with a 1 μg/dL change in blood lead level.

If the blood lead level increases by 10 μg/dL, the expected change in IQ is $(\beta_1)(\text{change in } x) = (-0.45)(10) = -4.5$.

If the blood lead level decreases by 20 μg/dL, the expected change in IQ is $(\beta_1)(\text{change in } x) = (-0.45)(-20) = 9.0$.

The change in x is negative because the blood lead level is *decreasing*.

c. If $\sigma = 8$, then for $x = 20$ the random variable Y is normally distributed with mean $96.8 - 0.45(20) = 87.8$ and variance $\sigma^2 = 8^2 = 64$: $Y \sim N(87.8, 64)$. Find the probability that Y exceeds 100.

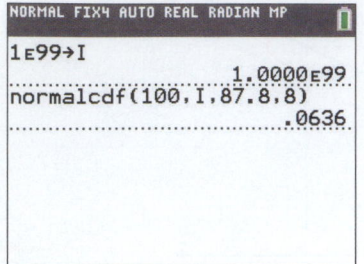

Figure 12.7 Probability calculation using the TI-84 Plus C.

$$P(Y > 100) = P\left(\frac{Y - 87.8}{8} > \frac{100 - 87.8}{8}\right) \qquad \text{Standardize.}$$

$$= P(Z > 1.53) \qquad \text{Equation 6.8, simplification.}$$

$$= 1 - P(Z \le 1.53) \qquad \text{Use cumulative probability.}$$

$$= 1 - 0.9370 = 0.0630 \qquad \text{Table III in the Appendix.}$$

Figure 12.7 shows a technology calculation. For a blood lead level of 20 μg/dL, the probability of observing an IQ greater than 100 in a six-year-old child is 0.0630. See Figure 12.8.

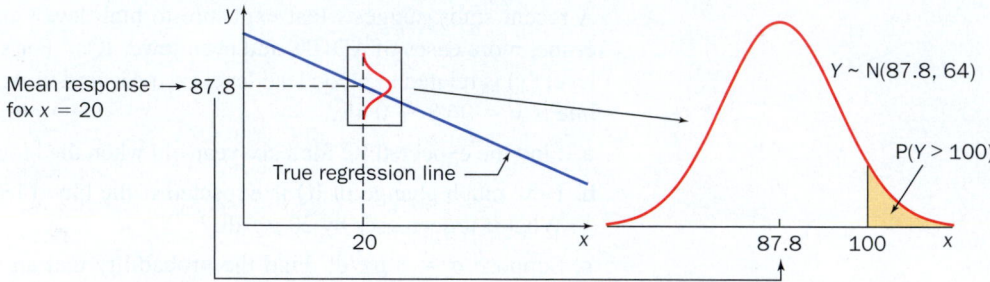

Figure 12.8 The distribution of Y for $x = 20$.

TRY IT NOW   GO TO EXERCISE 12.18

Suppose two variables are related via a simple linear regression model. The parameters β_0 and β_1 are usually unknown. However, if we assume that the observations (x_1, y_1), $(x_2, y_2), \ldots, (x_n, y_n)$ are independent, then these sample data can be used to estimate the model parameters β_0 and β_1.

The **line of best fit**, or **estimated regression line**, is obtained using the **principle of least squares**. Figure 12.9 illustrates this concept: Minimize the sum of the squared deviations, or vertical distances from the observed points to the line. Consider the

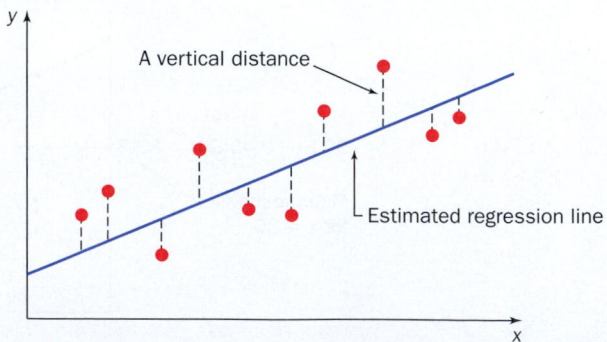

Figure 12.9 An illustration of the principle of least squares.

vertical distances from the points $(x_1, y_1), (x_2, y_2), \ldots, (x_n, y_n)$ to the line. The principle of least squares produces an estimated regression line such that the sum of all squared vertical distances is a minimum.

The focus in the remainder of this section is on the interpretations associated with finding the line of best fit. In Section 12.2, several hypothesis tests utilizing these various statistics will be introduced.

Least-Squares Estimates

$\hat{\beta}_0$ and $\hat{\beta}_1$ are estimates of β_0 and β_1, respectively, but $\hat{\beta}_1$ is found first, because its value is used in the calculation of $\hat{\beta}_0$.

The least-squares estimates of the y intercept (β_0) and the slope (β_1) of the true regression line are

$$\hat{\beta}_1 = \frac{n\sum x_i y_i - (\sum x_i)(\sum y_i)}{n\sum x_i^2 - (\sum x_i)^2} \qquad (12.2)$$

and

$$\hat{\beta}_0 = \frac{\sum y_i - \hat{\beta}_1 \sum x_i}{n} = \bar{y} - \hat{\beta}_1 \bar{x} \qquad (12.3)$$

The estimated regression line is $y = \hat{\beta}_0 + \hat{\beta}_1 x$.

A CLOSER LOOK

1. Before using these equations, always consider a scatter plot and compute the sample correlation coefficient (presented in Section 12.2) to make sure a linear model is reasonable. For example, in Figure 12.10 a linear model seems reasonable; the relationship between x and y appears to be negative linear. In Figure 12.11, a linear model is not reasonable; the relationship between the variables appears to be quadratic.

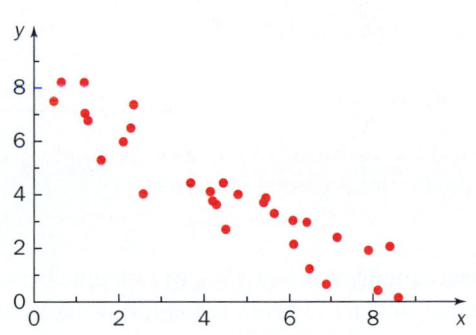

Figure 12.10 A scatter plot of data in which the relationship appears linear.

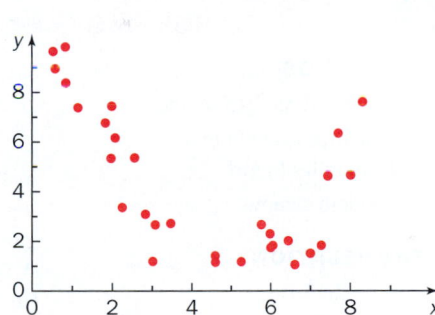

Figure 12.11 A scatter plot of data in which the relationship does not appear to be linear.

The predictor variable is the independent variable, and the response variable is the dependent variable.

2. If x^* is a specific value of the independent, or *predictor*, variable x, let $y^* = \hat{\beta}_0 + \hat{\beta}_1 x^*$.
   a. y^* is an estimate of the *mean* value of Y for $x = x^*$, denoted $\mu_{Y|x^*}$.
   b. y^* is also an estimate of an *observed* value of Y for $x = x^*$.

Example 12.3 Growth Charts

DATA SET

GROWTH

Doctors often use growth charts to compare the heights and weights of children of similar ages. The clinical growth charts of the U.S. Centers for Disease Control and Prevention show that the weight of infant girls is approximately linear between the ages of 12 and 36 months.

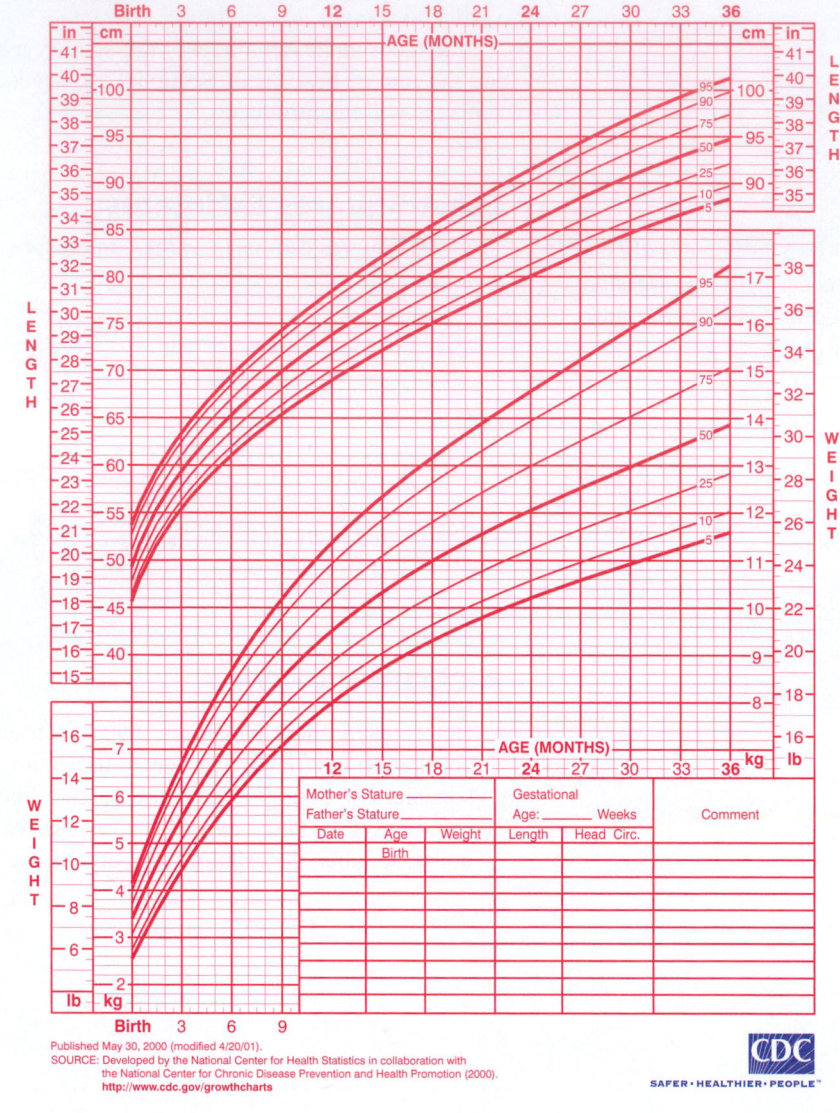

Birth to 36 months: Girls
Length-for-age and Weight-for-age percentiles

NAME _____

RECORD # _____

Published May 30, 2000 (modified 4/20/01).
SOURCE: Developed by the National Center for Health Statistics in collaboration with
the National Center for Chronic Disease Prevention and Health Promotion (2000).
http://www.cdc.gov/growthcharts

CDC
SAFER · HEALTHIER · PEOPLE™

Solution Trail 12.3

KEYWORDS

- Estimated regression line
- True mean weight of a 22-month-old girl
- Random sample

TRANSLATION

- $y = \hat{\beta}_0 + \hat{\beta}_1 x$
- $E(Y|22)$

CONCEPTS

- Least-squares estimates
- Estimate of the mean value of Y for $x = x^*$

VISION

Find the necessary summary statistics. Use Equation 12.2 to find $\hat{\beta}_1$ and Equation 12.3 to find $\hat{\beta}_0$. Compute $y^* = \hat{\beta}_0 + \hat{\beta}_1(22)$ as an estimate of the mean value of Y for $x = 22$.

A random sample of infant girls was obtained, and the weight (in kilograms) and age (in months) for each is given in the following table.

Age	33	32	14	20	15	16	30	17	21	23
Weight	12.9	13.8	8.2	12.2	8.5	12.9	13.7	11.2	11.9	10.4

a. Find the estimated regression line.

b. Estimate the true mean weight of a 22-month-old girl.

SOLUTION

a. Age (x) is the independent, or predictor, variable, and weight (y) is the dependent, or response, variable. Find the necessary summary statistics for the $n = 10$ pairs of observations, and use Equations 12.2 and 12.3.

$$\sum x_i = 221.0 \qquad \sum y_i = 115.7 \qquad \sum x_i y_i = 2650.5$$

$$\sum x_i^2 = 5349.0 \qquad \sum y_i^2 = 1374.5$$

$$\hat{\beta}_1 = \frac{n\sum x_i y_i - (\sum x_i)(\sum y_i)}{n\sum x_i^2 - (\sum x_i)^2}$$ Equation 12.2.

$$= \frac{(10)(2650.5) - (221.0)(115.7)}{(10)(5349.0) - (221.0)^2}$$ Use summary statistics.

$$= \frac{935.3}{4649} = 0.2012$$ Simplify.

$$\hat{\beta}_0 = \frac{\sum y_i - \hat{\beta}_1 \sum x_i}{n}$$ Equation 12.3.

$$= \frac{115.7 - (0.2012)(221.0)}{10}$$ Use summary statistics and $\hat{\beta}_1$.

$$= 7.123$$ Simplify.

VIDEO TECH MANUALS
LINEAR REGRESSION

The estimated regression line is $y = 7.123 + 0.2012x$. Figure 12.12 shows a scatter plot of the data and the graph of the estimated regression line.

b. The estimated true mean weight of a 22-month-old girl is found by substituting $x^* = 22$ into the estimated regression line equation.

$$y = 7.123 + 0.2012x$$ Estimated regression line equation.

$$= 7.123 + (0.2012)(22) = 11.55$$ Substitute and simplify.

An estimate for the expected weight of a 22-month-old girl is 11.55 kilograms. See Figure 12.13.

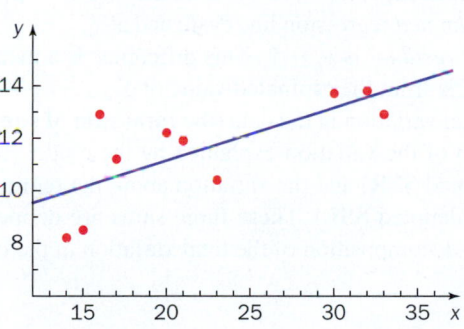

Figure 12.12 A scatter plot of the data and the graph of the estimated regression line.

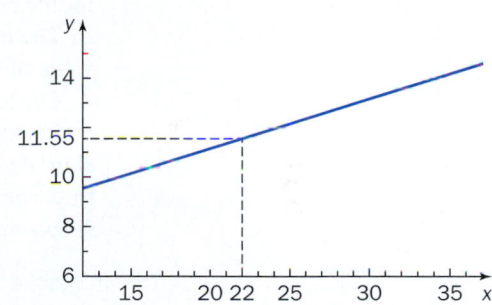

Figure 12.13 The expected weight for a 22-month-old girl.

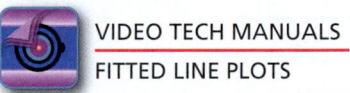
VIDEO TECH MANUALS
FITTED LINE PLOTS

Figure 12.14 shows a technology solution.

Figure 12.14 The estimated regression line and relevant hypothesis tests concerning the regression coefficients (explanations to follow in Section 12.2).

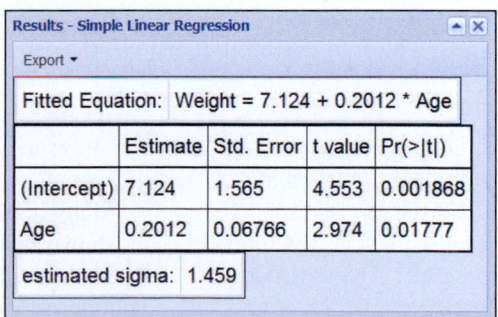

Results - Simple Linear Regression

Export ▾

Fitted Equation: Weight = 7.124 + 0.2012 * Age

	Estimate	Std. Error	t value	Pr(>|t|)
(Intercept)	7.124	1.565	4.553	0.001868
Age	0.2012	0.06766	2.974	0.01777

estimated sigma: 1.459

TRY IT NOW   GO TO EXERCISE 12.24

The variance σ^2 is a measure of the underlying variability in the simple linear regression model. A *large* σ^2 means the data will vary widely from the true regression line. A *small* σ^2 implies that the observed values will lie close to the true regression line.

An estimate of σ^2 is used to conduct hypothesis tests and to construct confidence intervals related to simple linear regression. The estimate is based on the deviations of the observed values from the estimated regression line. More on hypothesis tests and confidence intervals in Section 12.2.

One method to assess the accuracy of a simple linear regression model involves an analysis of variance table, similar to those constructed in Chapter 11. To explain the ANOVA table entries and to make computations easier, a little more notation is necessary.

For computational purposes, define

$$S_{xx} = \underbrace{\sum(x_i - \bar{x})^2}_{\text{definition}} = \underbrace{\sum x_i^2 - \frac{1}{n}(\sum x_i)^2}_{\text{computational formula}}$$

$$S_{yy} = \underbrace{\sum(y_i - \bar{y})^2}_{\text{definition}} = \underbrace{\sum y_i^2 - \frac{1}{n}(\sum y_i)^2}_{\text{computational formula}}$$

$$S_{xy} = \underbrace{\sum(x_i - \bar{x})(y_i - \bar{y})}_{\text{definition}} = \underbrace{\sum x_i y_i - \frac{1}{n}(\sum x_i)(\sum y_i)}_{\text{computational formula}}$$

The *ith predicted*, or *fitted*, *value*, denoted $\hat{y}_i$, is $\hat{y}_i = \hat{\beta}_0 + \hat{\beta}_1 x_i$. This is simply the equation for the *estimated* regression line evaluated at x_i.

The *ith residual* is $y_i - \hat{y}_i$. This difference is a measure of how far away the observed value of Y is from the estimated value of Y.

The total variation in the data (the **total sum of squares**, denoted **SST**) is decomposed into a sum of the variation explained by the model (the **sum of squares due to regression**, denoted **SSR**) and the variation about the regression line (the **sum of squares due to error**, denoted **SSE**). These three sums are defined in the following identity, which shows the decomposition of the total variation in the data.

Sum of Squares

$$\underbrace{\sum(y_i - \bar{y})^2}_{\text{SST}} = \underbrace{\sum(\hat{y}_i - \bar{y})^2}_{\text{SSR}} + \underbrace{\sum(y_i - \hat{y}_i)^2}_{\text{SSE}} \tag{12.4}$$

Here are the *computational formulas* for these sums of squares.

$$SST = S_{yy} \qquad SSR = \hat{\beta}_1 S_{xy} \qquad SSE = SST - SSR$$

You can probably guess that a large F statistic suggests a significant regression. The formal hypothesis test is presented in Section 12.2.

Many regression computations are often summarized in an analysis of variance table, as shown below. As in Chapter 11, **mean squares** are corresponding sums of squares divided by the corresponding degrees of freedom. The F test for a significant regression and the associated p value are discussed in Section 12.2.

ANOVA summary table for simple linear regression

Source of variation	Sum of squares	Degrees of freedom	Mean square	F	p Value
Regression	SSR	1	$MSR = \dfrac{SSR}{1}$	$\dfrac{MSR}{MSE}$	p
Error	SSE	$n - 2$	$MSE = \dfrac{SSE}{n - 2}$		
Total	SST	$n - 1$			

Coefficient of Determination

The **coefficient of determination**, denoted r^2, is a measure of the proportion of the variation in the data that is explained by the regression model, and is defined by

$$r^2 = SSR/SST \tag{12.5}$$

$r^2 = 1$ implies a perfect fit. What does $r^2 = 0$ imply?

Because $0 \le SSR \le SST$, the **coefficient of determination**, r^2, is always a number between 0 and 1 (inclusive). The higher r^2 is, the better the model is. Many statistical software packages report $100r^2$, the percentage of variation explained by the regression model.

The following example illustrates the computations necessary to produce the estimated regression line and the ANOVA table.

Example 12.4 Turn Down the Noise

DATA SET

NOISE

A recent study suggests that noise pollution as well as air pollution can increase the risk of cardiovascular disease.[4] Suppose a random sample of adults who have lived in the same home for at least five years was obtained. The nighttime noise level was measured in each home (in dB) and an Agatston score for each person was found using electron-beam computed tomography. This score is an indication of coronary calcification, and a higher Agatston score suggests a greater cardiovascular risk. The data are given in the following table.

Noise	58	95	102	105	23	68	58	89	54
Agatston	72	135	82	188	108	192	40	61	89

Noise	15	105	21	61	20	72	113	80	27
Agatston	23	137	38	106	108	75	170	69	40

a. Find the estimated regression line and explain the meaning of the estimated coefficient $\hat{\beta}_1$.

b. Complete the ANOVA table (without the p value), and find and interpret the coefficient of determination.

Solution Trail 12.4

KEYWORDS

- Estimated regression line
- ANOVA table
- Coefficient of determination
- Random sample

TRANSLATION

- $y = \hat{\beta}_0 + \hat{\beta}_1 x$
- ANOVA summary table
- r^2

CONCEPTS

- Least-squares estimates
- ANOVA summary table for simple linear regression
- The proportion of variation in the data explained by the regression model

VISION

Find the necessary summary statistics. Use Equation 12.2 to find $\hat{\beta}_1$ and Equation 12.3 to find $\hat{\beta}_0$. Compute the ANOVA summary table and compute r^2.

SOLUTION

a. Noise level (x) is the independent variable and Agatston score (y) is the dependent variable. The summary statistics for the $n = 18$ pairs of observations are

$$\sum x_i = 1{,}166 \qquad \sum y_i = 1{,}733 \qquad \sum x_i y_i = 128{,}564$$

$$\sum x_i^2 = 94{,}066 \qquad \sum y_i^2 = 211{,}775$$

$$\hat{\beta}_1 = \frac{n\sum x_i y_i - (\sum x_i)(\sum y_i)}{n\sum x_i^2 - (\sum x_i)^2} \qquad \text{Equation 12.2.}$$

$$= \frac{(18)(128{,}564) - (1{,}166)(1{,}733)}{(18)(94{,}066) - (1{,}166)^2} \qquad \text{Use summary statistics.}$$

$$= \frac{293{,}474}{333{,}632} = 0.8796 \qquad \text{Simplify.}$$

$$\hat{\beta}_0 = \frac{\sum y_i - \hat{\beta}_1 \sum x_i}{n} \qquad \text{Equation 12.3.}$$

$$= \frac{1{,}733 - (0.8796)(1{,}166)}{18} \qquad \text{Use summary statistics and } \hat{\beta}_1.$$

$$= 39.3 \qquad \text{Simplify.}$$

The estimated regression line is $y = 39.3 + 0.8796x$. The value $\hat{\beta}_1 = 0.8796$ suggests that an increase of 1 dB leads to an increase of approximately 0.9 in the Agatston score. Figure 12.15 shows a scatter plot of the data and the graph of the estimated regression line.

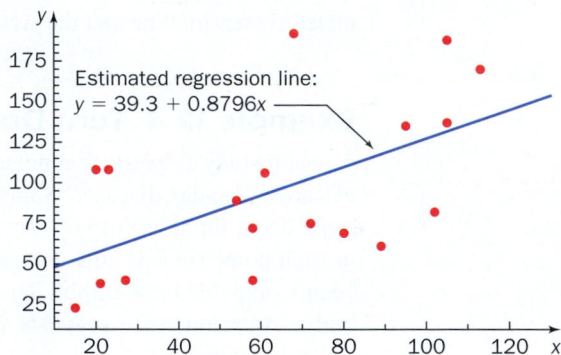

Figure 12.15 A scatter plot of the data and the graph of the estimated regression line.

b. Complete the ANOVA table, except for the p value ($n = 18$).

$$\text{SST} = S_{yy} = \sum y_i^2 - \frac{1}{18}(\sum y_i)^2 \qquad \text{Computational formula.}$$

$$= 211{,}775 - \frac{1}{18}(1{,}733)^2 = 44{,}925.6 \qquad \text{Use summary statistics.}$$

$$\text{SSR} = \hat{\beta}_1 S_{xy} = \hat{\beta}_1 \left[\sum x_i y_i - \frac{1}{18}(\sum x_i)(\sum y_i) \right] \qquad \text{Computational formula.}$$

$$= 0.8796 \left[128{,}564 - \frac{1}{18}(1{,}166)(1{,}733) \right] \qquad \text{Use summary statistics.}$$

$$= 14{,}341.1 \qquad \text{Simplify.}$$

$$\text{SSE} = \text{SST} - \text{SSR} \qquad \text{Computational formula.}$$

$$= 44{,}925.6 - 14{,}341.1 = 30{,}584.5$$

Compute the mean squares, MSR and MSE.

$$MSR = SSR/1$$

Definition.

$$= 14,341.1/1 = 14,341.1$$

Use SSR.

$$MSE = SSE/(n - 2)$$

Definition.

$$= 30,584.5/16 = 1,911.5$$

Use SSE and $n = 18$.

The value of the test statistic is

You can probably guess that we will reject the null hypothesis if the value of the test statistic is large. The formal hypothesis test is presented in Section 12.2.

$$f = \frac{MSR}{MSE} = \frac{14,341.1}{1,911.5} = 7.50$$

Here are all of these calculations presented in an ANOVA table.

ANOVA summary table for simple linear regression

Source of variation	Sum of squares	Degrees of freedom	Mean square	F
Regression	14,341.1	1	14,341.1	7.50
Error	30,584.5	16	1,911.5	
Total	44,925.6	17		

The coefficient of determination is

$$r^2 = SSR/SST = 14,341.1/44,925.6 = 0.3192$$

Approximately 0.3192, or 32%, of the variation in the data is explained by the regression model.

Figures 12.16 and 12.17 show technology solutions.

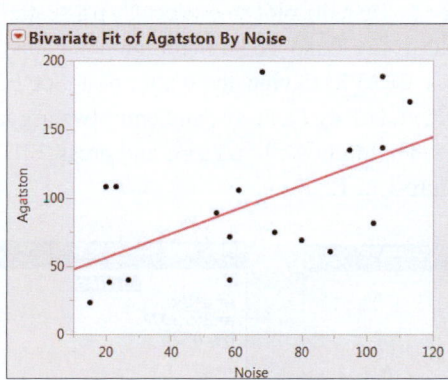

Figure 12.16 JMP scatter plot and estimated regression line.

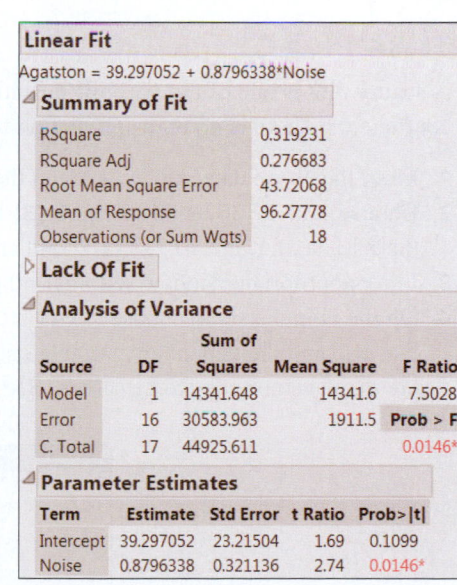

Figure 12.17 JMP Linear Fit results.

TRY IT NOW   GO TO EXERCISE 12.26

Technology Corner

Procedure: Construct a scatter plot, find the estimated regression line, and complete an ANOVA table.
Reconsider: Example 12.4, solution, and interpretations.

CrunchIt!

A Scatterplot option is available under the Graphics menu, and the Simple linear regression command is used to find the estimated regression line and construct the ANOVA table.

1. Enter the Noise data into column Var1 and the Agatston score data into column Var2. Rename the columns if desired.
2. Choose Graphics; Scatterplot. Enter the independent variable (X) and the dependent variable (Y).
3. In the Parameters section, select Points, and optionally enter a Title, X Label, Y Label, Slope of line, and Intercept of line. Click Calculate to produce the scatter plot (Figure 12.18).
4. Choose Statistics; Regression; Simple Linear. Enter the Dependent Variable, the Independent Variable, and in the Parameters section, select the desired Display. Click Calculate. The estimated regression line and coefficient statistics are displayed. See Figure 12.19.

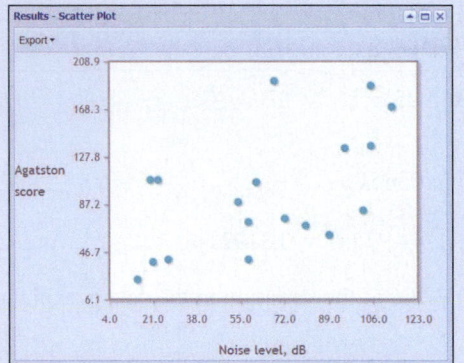

Figure 12.18 CrunchIt! scatter plot.

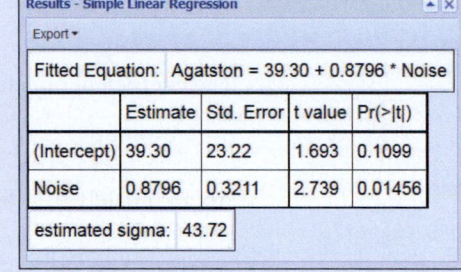

Figure 12.19 The estimated regression line.

TI-84 Plus C

A scatter plot is one of the six built-in statistical plots. Use the function `LinReg(a+bx)` to compute the regression coefficients. There is no built-in function to produce the ANOVA table.

1. Enter the Noise data into list `L1` and the Agatston score data into list `L2`.
2. Choose `STATPLOT`; STAT PLOTS; Plot1. Turn the plot On, select Type scatter plot (the first graph icon), enter the Xlist, L1, the Ylist, L2, and choose a Mark and Color (for the points on the graph).
3. Enter appropriate window settings and press `GRAPH` to view the scatter plot. See Figure 12.20.
4. On the Home screen, choose `STAT`; CALC; LinReg(a+bx), and enter two arguments: the independent variable list, L1, and the dependent variable list, L2. Highlight Calculate and press `ENTER`. The regression coefficients are displayed on the Home screen. See Figure 12.21.

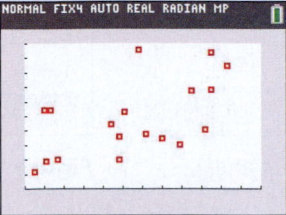

Figure 12.20 TI-84 Plus C scatter plot.

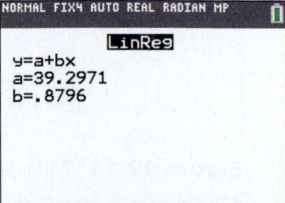

Figure 12.21 The regression coefficients.

Note that the function `LinReg(a+bx)` has an optional argument, a function variable, for example `Y1`, for storing the regression equation. There are other calculator functions that will also produce the regression coefficients, for example, `LinReg(ax+b)` and `LinRegTTest`.

Minitab

Minitab has a Scatterplot option in the Graph menu, and there are several built-in functions to compute the regression coefficients and the ANOVA table, in the Stat; Regression menu.

1. Enter the Noise data into column C1 and the Agatston score data into column C2.
2. Select Graph; Scatterplot. Choose a Simple scatter plot.
3. Enter C2 under Y variables and C1 under X variables. Click OK. The scatter plot is displayed in a graph window. Adjust the labels and axes as necessary. See Figure 12.22.
4. Select Stat; Regression; Regression; Fit Regression Model. Enter the Response variable, C2, and the Continuous predictors, C1. Click OK. The regression coefficients and the ANOVA table are displayed in a Session window. See Figure 12.23.

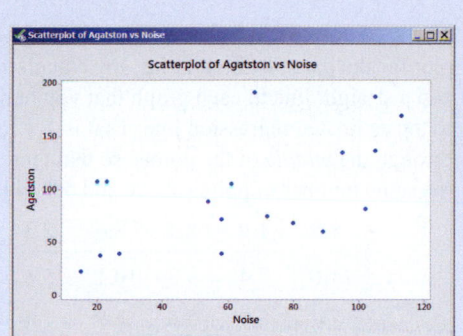

Figure 12.22 Minitab scatter plot.

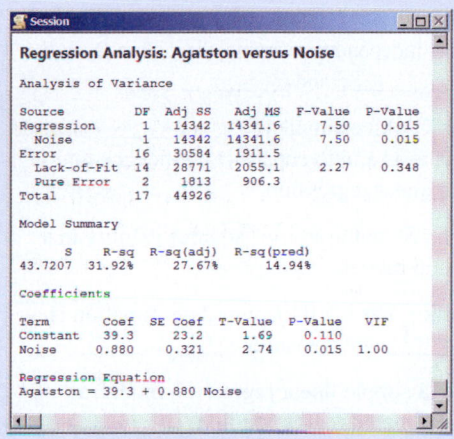

Figure 12.23 Minitab output: regression coefficients and analysis of variance table.

Excel

Use the Scatter chart option to construct a scatter plot and the Data Analysis Toolkit function Regression to compute the regression coefficients and ANOVA table.

1. Enter the Noise data into column A and the Agatston score data into column B.
2. Highlight, or select, the data range, A1:B18. Under the Insert tab, select Scatter; Scatter with only Markers. The scatter plot is displayed in the current worksheet. Edit the plot options as necessary. See Figure 12.24.
2. Under the Data tab, select Data Analysis; Regression. Enter the Y Range (B1:B19), the X Range, (A1:A19), and the Output Range. Click OK to display the regression coefficients and ANOVA table. See Figure 12.25.

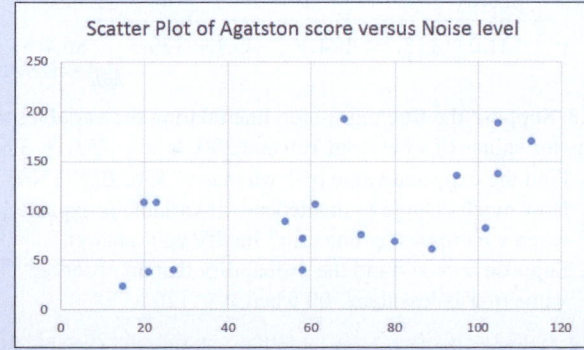

Figure 12.24 Excel scatter plot.

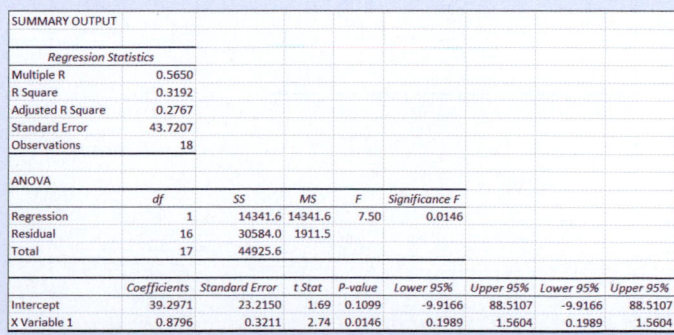

Figure 12.25 Estimated regression coefficients and ANOVA table.

SECTION 12.1 EXERCISES

Concept Check

12.1 Short Answer Name the two parts in a general additive probabilistic model.

12.2 Short Answer Name the four assumptions in a simple linear regression model.

12.3 Fill in the Blank In a simple linear regression model, the variability in the distribution of Y is the same for every value of x. This property is called _____.

12.4 Fill in the Blank The principle of least squares produces an estimated regression line such that _____ is a minimum.

12.5 Fill in the Blank In a simple linear regression model, if x^* is a specific value of the independent variable, then $y^* = \widehat{\beta}_0 + \widehat{\beta}_1 x^*$ is an estimate of _____ and _____.

12.6 Fill in the Blank An estimate of _____ is used to conduct hypothesis tests and to construct confidence intervals related to simple linear regression.

12.7 Short Answer State the sum-of-squares identity in a simple linear regression model.

12.8 Fill in the Blank The coefficient of determination is defined as _____ and measures _____.

12.9 True/False In a simple linear regression model, all of the points lie on the estimated regression line.

12.10 Short Answer Explain why a scatter plot should be considered prior to finding an estimated regression line.

Practice

12.11 Decide whether a simple linear regression model is appropriate for each of the following graphs. If it is, indicate whether the slope ($\widehat{\beta}_1$) of the estimated regression line is positive, negative, or zero. If it is not, state why.

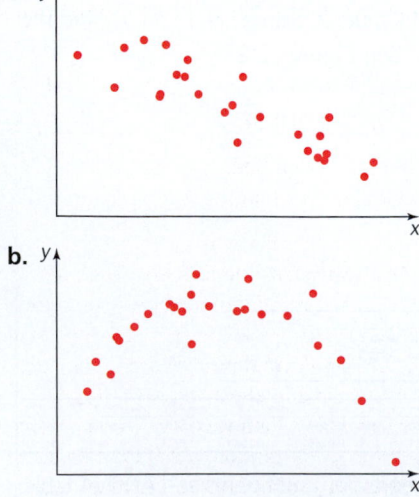

a. y

b. y

c. y

d. y

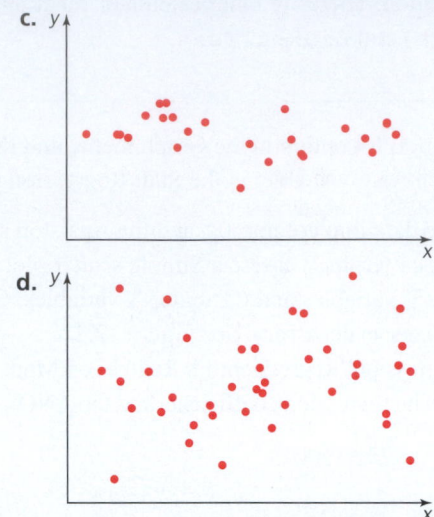

12.12 In each of the following problems, construct a scatter plot for the data. Without doing any calculations for $\widehat{\beta}_0$ and $\widehat{\beta}_1$, add a straight line to each graph that you believe would be *close* to the estimated regression line. That is, try to draw a line through the *middle* of the points, so that there are approximately the same number of points above and below the line.

a.

x	8.3	1.9	8.4	8.6	2.3	4.6	2.4
y	11.6	3.4	9.3	10.1	5.6	7.6	4.3
x	7.7	8.0	5.6	6.9	1.8	4.8	9.4
y	10.5	11.5	6.6	9.8	2.9	6.8	12.8
x	6.8	4.9	7.8	5.3	1.1	0.3	
y	10.2	4.9	9.4	8.2	3.6	3.2	

EX12.12A

b.

x	98.2	56.4	52.9	99.3	73.5	82.7
y	88.7	167.0	185.5	72.5	149.5	113.0
x	54.1	68.2	70.3	71.2	53.4	90.2
y	163.0	135.4	135.8	130.1	168.4	82.0
x	53.8	99.4	85.5	72.6	92.7	89.6
y	184.2	90.1	130.3	132.9	93.2	103.6
x	80.9	76.4	54.5	97.3	58.7	92.6
y	121.0	115.7	174.1	79.1	159.3	86.4

EX12.12B

12.13 Suppose the true regression line relating the variables x and y, for values of x between 100 and 200, is $y = 75.0 + 3.6x$.
   a. Find the expected value of Y when $x = 150$, $E(Y|150)$.
   b. How much change in the dependent variable is expected when x increases by one unit? Justify your answer.
   c. Suppose $\sigma = 6$. Find the probability that an observed value of Y is less than 500 when $x = 120$.

12.14 Suppose the true regression line relating the variables x and y, for values of x between 25 and 35, is $y = -35.1 - 7.2x$.

a. Find the expected value of Y when $x = 27.4$.

b. How much change in the dependent variable is expected when x decreases by 5 units? Justify your answer.

c. Suppose $\sigma = 1.5$. Find the probability that an observed value of Y is between -250 and -252 when $x = 30$.

12.15 Suppose a simple linear regression model is used to explain the relationship between x and y. A random sample of $n = 12$ values for the independent variable was selected, and the corresponding values of the dependent variable were observed. The summary statistics were

$$\sum x_i = 460.53 \qquad \sum y_i = -6,349.7$$
$$\sum x_i^2 = 17,875.1 \qquad \sum y_i^2 = 3,421,892$$
$$\sum x_i y_i = -246,677$$

a. Find the estimated regression line.

b. Estimate the true mean value of Y when $x = 41$.

12.16 Suppose a simple linear regression model is used to explain the relationship between x and y. A random sample of $n = 15$ values for the independent variable was selected, and the corresponding values of the dependent variable were observed. The data are given in the following table. ▥ EX12.16

x	4.9	9.9	0.3	3.8	9.8	4.6	9.8	5.4
y	49.1	53.4	21.7	40.4	78.6	60.4	73.9	37.4
x	3.1	3.0	4.2	4.0	1.7	1.2	7.0	
y	41.6	29.6	46.7	28.5	23.8	15.2	72.3	

a. Construct a scatter plot for these data. Is a simple linear regression model reasonable? Justify your answer.

b. Find the estimated regression line. Add a graph of this line to your scatter plot.

c. What is the predicted value of Y when $x = 5.7$?

12.17 Suppose a simple linear regression model is used to explain the relationship between x and y. A random sample of $n = 21$ values for the independent variable was selected, and the corresponding values of the dependent variable were observed. ▥ EX12.17

a. Construct a scatter plot for these data. Is a simple linear regression model reasonable? Justify your answer.

b. Find the estimated regression line. Add a graph of this line to your scatter plot.

c. Estimate the true mean value of Y when $x = 62$.

d. Construct an ANOVA summary table, except for the p value.

Applications

12.18 Biology and Environmental Science Agricultural research suggests that the final corn yield in bushels per acre (y) is linearly related to the number of inches between rows (x). Suppose the true regression line is $y = 197.5 - 6.1x$.

a. Find the expected yield when there are 15 inches between rows.

b. How much change in yield is expected if the distance between rows decreases by 2 inches?

c. Suppose $\sigma = 3.2$ bushels per acre. Find the probability that an observed value of yield is between 90 and 95 bushels per acre when rows are 17 inches apart.

12.19 Physical Sciences Once a PVC pipe joint is assembled using solvent cement, a certain amount of time must be allowed for the new joint to set. For 4- to 8-inch-diameter pipes, suppose the setting time (y, measured in hours) is linearly related to the ambient temperature (x, measured in °F). The true regression line is assumed to be $y = 8.3 - 0.09x$.

a. Find an estimate of an observed setting time for an ambient temperature of 65°F.

b. How much less time does a joint take to set if the temperature rises by 10°F?

c. If $\sigma = 15$ minutes, find the probability that an observed set time is less than one hour if the ambient temperature is 80°F.

12.20 Sports and Leisure A recent study investigated the association between heading a soccer ball and subclinical evidence of traumatic brain injury.[5] Amateur soccer players were selected, and each completed a questionnaire to determine the number of times (x) they headed the ball in the last 12 months. In addition, each player was subject to diffusion-tensor magnetic resonance to measure the fractional anisotropy (FA, y). Suppose the estimated regression line was $y = -0.00026x + 0.6877$.

a. Find the expected value of y if the number of headers is 1000.

b. Find an estimate of an observed y value for $x = 1200$.

c. Suppose $\sigma = 0.06$. Find the probability that an observed y value is more than 0.5 when the number of headers is 750.

12.21 Manufacturing and Product Development As soon as a bottle of soda is opened, it begins to lose its carbonation. Fourteen 12-ounce bottles of cola were obtained, and each was assigned a randomly selected time period (in hours). Each bottle was opened and allowed to stand at room temperature. The carbonation (y) in each bottle was measured (in volumes) after the prescribed time period (x). The summary statistics are given below.

$$\sum x_i = 8.6 \qquad \sum y_i = 37.4 \qquad \sum x_i y_i = 19.61$$
$$\sum x_i^2 = 7.26 \qquad \sum y_i^2 = 110.24$$

a. Find the estimated regression line.

b. Estimate the true mean carbonation after 1 hour and 15 minutes.

12.22 Biology and Environmental Science Some research suggests that higher levels of carbon dioxide (CO_2) increase the rate at which plants and trees grow.[6] In fact, even without ideal conditions (for example, normal precipitation and temperature), some trees grow more in elevated levels of CO_2. Suppose that, in a research study, data were collected on oak trees in northern U.S. forests. For randomly selected levels of CO_2 (x, measured in parts per million, ppm), the most recent tree-ring growth (y, in cm) was measured. The summary statistics are given below ($n = 10$).

$$\sum x_i = 5,035.0 \qquad \sum y_i = 13.7 \qquad \sum x_i y_i = 7,012.5$$
$$\sum x_i^2 = 2,607,575 \qquad \sum y_i^2 = 19.25$$

a. Find the estimated regression line.
b. Find an estimate of an observed value of tree-ring growth if the level of CO_2 is 600 ppm.

12.23 Business and Management   A recent study considered the effects of innovation on employment in Latin America.[7] It seems reasonable that as more firms produce new products, they would need more workers, and employment would rise. For small firms in Argentina, let y be the yearly percentage of employment growth and x be the percentage of small firms that are product or process innovators. Assume the estimated regression line is $y = -5.399 + 5.790x$.

a. Find the expected percentage in employment growth if the percentage of product or process innovators is 2%.
b. Find an estimate of an observed value y for $x = 3.5$.
c. Suppose $\sigma = 1.5$. Find the probability that an observed y value is more than 19 when $x = 4$.
d. Suppose $x = 0.9\%$. Find the expected value of y and interpret this result.

12.24 Sports and Leisure   Sailboat enthusiasts believe that the wind speed (x, in miles per hour) is linearly related to (downwind) boat speed (y, in knots). For wind speeds between 10 and 30 mph and Hobie catamarans, the following data were recorded. 📊 EX12.24

x	26.2	11.8	27.3	24.3	20.4	24.6	20.7	14.0	15.5	20.3
y	16.3	7.7	17.3	18.2	16.9	10.3	12.7	13.3	10.0	6.8

a. Construct a scatter plot for these data.
b. Find the estimated regression line. Add a graph of the estimated regression line to the scatter plot in part (a).
c. Estimate the mean speed of a Hobie catamaran for a wind speed of 18 mph.

12.25 Psychology and Human Behavior   A development officer at a large university believes that the amount of money donated to the general fund by alumni each year is linearly related to the football team's winning percentage. A random sample of Division I football teams was selected. The winning percentage (x) and subsequent alumni donation amount (y, in millions of dollars) are given in the following table. 📊 EX12.25

x	92	67	92	25	83	83	58	8	75
y	15.8	11.1	13.5	7.0	17.4	10.8	11.2	3.9	23.7

x	100	92	58	92	17	25	75	92	33
y	27.8	23.3	11.3	6.6	7.3	10.3	15.8	12.0	6.8

a. Construct a scatter plot for these data.
b. Find the estimated regression line. Add a graph of the estimated regression line to the scatter plot in part (a).

c. Find an estimate of an observed value of the donation amount for a football team with a winning percentage of 75%.

Extended Applications

12.26 Public Health and Nutrition   A recent study suggests that the number of steps walked per day is strongly associated with good health.[8] A person's heart rate indicates how hard the heart is working to circulate blood throughout the body and is a measure of health. A lower resting heart rate may reduce the risk of heart attack and stroke, and also increase endurance. A random sample of adults between 35 and 40 years old was obtained, and each person wore a pedometer for a day. The number of steps taken, x, was recorded, and the next day the resting pulse rate (beats/minute), y, for each person was also measured.

a. Find the estimated regression line and explain the meaning of the estimated coefficient $\hat{\beta}_1$.
b. Complete the ANOVA table (without the p value), and find and interpret the coefficient of determination. 📊 WALKING

12.27 Medicine and Clinical Studies   Many automobile accidents are caused by tired drivers. A typical test for alertness involves measuring the speed and degree to which a person's eyes respond to certain stimuli.[9] A random sample of 25 drivers was obtained, and the oscillations in pupil size (x, in millimeters per second) was measured using a pupillograph. Each person's tiredness (y) was also recorded using the pupil unrest index (PUI). The summary statistics are given below.

$$\sum x_i = 7.1 \qquad \sum y_i = 192.0 \qquad \sum x_i y_i = 49.22$$
$$\sum x_i^2 = 2.1064 \qquad \sum y_i^2 = 2094.0$$

a. Find the estimated regression line.
b. Find the expected PUI for $x = 0.3$ millimeters per second.
c. Suppose a driver is considered too tired to drive if the PUI score is 15 (or higher). What value of x yields an expected PUI score of 15?

12.28 Biology and Environment Science   The Virginia Cooperative Extension Service provides yield data and other performance statistics on barley as well as a summary of the growing season. A random sample of barley varieties was obtained and the 2013 yield (y, in bushels per acre) and the date headed (x, number of days since the beginning of the calendar year at which 50% of the heads have emerged from the plants) are given in the following table.[10] 📊 BARLEY

Date headed	115	111	114	116	115
Yield	88	87	87	84	82

Date headed	120	119	113	112	119
Yield	82	81	78	82	81

a. Find the estimated regression line.
b. Complete the ANOVA table (without the p value).

c. Do you believe that the date headed is a good predictor of yield? Why or why not?

12.29 Medicine and Clinical Studies Ultrasound measurements are often used to predict the birth weight of a newborn. However, there is some evidence to suggest that pre-pregnancy body mass index (BMI) is related to birth weight.[11] A random sample of Caucasian pregnant women experiencing no complications was obtained. The BMI pre-pregnancy (x, in kg/m^2) of each was measured, and the newborn weight (y, in grams) was also recorded. **NEWBORN**

a. Construct a scatter plot for these data.
b. Find the estimated regression line.
c. Estimate the expected weight of a newborn if the mother's pre-pregnancy BMI is 25 kg/m^2.
d. If ultrasound is not available, do you believe that BMI is a good predictor of birth weight? Why or why not?

12.30 Physical Sciences Deep-water (>300 m) wave forecasts are important for large cargo ships. One method of prediction suggests that the wind speed (x, in knots) is linearly related to the wave height (y, in feet). A random sample of buoys was obtained, and the wind speed and wave height was measured at each. The data are given in the following table.[12] **WAVES**

Wind speed	9	11	10	10	11	9	9	6
Wave height	2.9	1.4	1.7	0.9	1.2	1.0	1.5	0.7

Wind speed	9	5	8	9	12	9	12	9
Wave height	1.9	0.1	2.0	2.6	3.0	1.7	2.1	1.5

Wind speed	12	8	7	13	9	8	6	8
Wave height	3.1	2.7	0.4	2.5	1.7	0.6	0.7	1.4

a. Find the estimated regression line.
b. Complete the ANOVA table (without the p value).
c. Find the coefficient of determination. Interpret this value.
d. Suppose a 10-foot wave is considered to be the storm threshold. What wind speed yields an expected storm threshold?

12.31 Manufacturing and Product Development There is good evidence to suggest that the depth of a bounce on a certain circular trampoline is linearly related to the stiffness of the springs around the edges, or spring constant k. A random sample of production trampolines was obtained and the spring constant of each was recorded (in lb/in). The bounce was measured (in feet) with a testing weight of 200 pounds released from a height of 2 feet. The data are given in the following table. **BOUNCE**

Spring constant	0.43	0.77	0.63	0.73	0.55	0.42
Bounce	1.1	2.6	2.3	3.0	3.0	1.9

Spring constant	0.38	0.39	0.20	0.73	0.56	0.57
Bounce	1.4	1.4	1.3	2.2	2.0	1.2

a. Find the estimated regression line.
b. Complete the ANOVA table (without the p value).
c. Find the coefficient of determination. Interpret this value.
d. Suppose the trampoline sits 4 feet above the ground. For what value of k can a 200-pound person expect to hit the ground?

12.32 Public Health and Nutrition According to Map the Meal Gap 2013, there are Americans living at risk of hunger in every county in the United States.[13] A random sample of counties was obtained, and the cost of a typical meal (x, in dollars) and the food insecurity (y, a percentage of the population, adults and children, who are food insecure) was recorded for each. **HUNGER**

a. Construct a scatter plot for these data. Based on this plot, do you believe there is a significant linear relationship between the cost of a typical meal and food insecurity? Why or why not?
b. Find the estimate regression line.
c. Complete the ANOVA table (without the p value).
d. Does the value of the F statistic in the ANOVA table support your conclusion in part (a)? Why or why not?

Challenge

12.33 Medicine and Clinical Studies Some physicians use the cholesterol ratio (CR = total cholesterol/HDL cholesterol) as a measure of a patient's risk of heart disease. In addition, the triglyceride concentration (TG) is associated with coronary artery disease in many patients. In a study of the relationship between these two variables, a random sample of adults was obtained, and the triglyceride level (x_1, mg/dL) and cholesterol ratio (y) was obtained for each person. **CHOLEST**

a. Carefully sketch a scatter plot of these data (y versus x_1). The relationship does not appear to be linear. Can you describe this relationship between y and x_1?
b. Compute the natural logarithm of each difference $(x_1 - 129)$. That is, find the values of a new predictor variable, $x_2 = \ln(x_1 - 129)$.
c. Carefully sketch a scatter plot of y versus the new predictor variable x_2. Describe this relationship.
d. Find the estimated regression line, and complete the ANOVA table using x_2 as the predictor variable.

12.2 Hypothesis Tests and Correlation

Suppose there is theoretical or empirical evidence that two variables are linearly related. The mean squares (in the regression ANOVA table) are used to determine whether the linear relationship is statistically significant. The null hypothesis states that the variation in Y is completely random and is independent of the value of x;

Even though a true regression line is shown in Figure 12.27, there is really no significant linear relationship. The true regression line is of absolutely no use for predicting or estimating values of Y.

Remember that β_1 is the slope of the true regression line. A horizontal line has slope, β_1, equal to 0.

knowing the value of x provides no additional information about the value of Y. In this case, a scatter plot would have no discernible pattern, with points scattered randomly in the plane (Figure 12.26).

For simple linear regression, a test of significance is equivalent to testing the hypothesis $H_0: \beta_1 = 0$. If H_0 is true, then the model assumptions imply that the mean value of Y for any value of x is the same. That is, $\mu_{Y|x_i} = \beta_0$. Therefore, the values of Y vary around the horizontal line $y = \beta_0$, and knowing the value of x adds no additional information. See Figure 12.27.

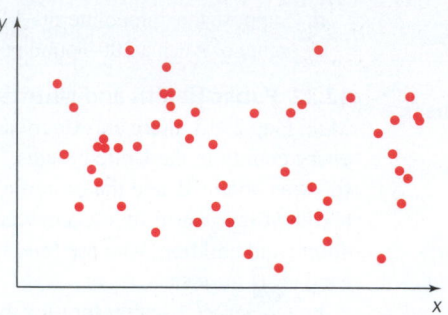

Figure 12.26 Scatter plot of data showing no linear pattern.

Figure 12.27 If $\beta_1 = 0$, the values of Y vary randomly around the true regression line $y = \beta_0$.

Here is a summary of an F test for a significant regression with significance level α.

Hypothesis Test for a Significant Linear Regression

H_0: There is no significant linear relationship ($\beta_1 = 0$).

H_a: There is a significant linear relationship ($\beta_1 \neq 0$).

TS: $F = \dfrac{\text{MSR}}{\text{MSE}}$

RR: $F \geq F_{\alpha, 1, n-2}$

A CLOSER LOOK

1. The null hypothesis is rejected only for large values of the test statistic. The associated p value is a right-tail probability. The F ratio will be larger when $\beta_1 \neq 0$ than when $\beta_1 = 0$.

2. This is often called a **model utility test**. In general, if H_0 is rejected, then the value of r^2 is usually large.

3. Alternatively, we can conduct this hypothesis test by using the p value. Recall that if H_0 is true, the p value is the probability of obtaining a value of the test statistic at least as large as the observed value. If $p \leq \alpha$, we reject the null hypothesis. If $p > \alpha$, we do not reject the null hypothesis.

Recall that a specific value of B_1 is $\hat{\beta}_1$.

Consider the random variable B_1, an estimator for β_1, and $S^2 = \text{MSE} = \text{SSE}/(n - 2)$, an estimator for the underlying variance σ^2. If the simple linear regression assumptions are true, then S^2 is an unbiased estimator for σ^2, and the estimator B_1 has the following properties.

1. B_1 is an unbiased estimator for β_1: $E(B_1) = \mu_{B_1} = \beta_1$.

2. The variance of B_1 is $\text{Var}(B_1) = \sigma_{B_1}^2 = \sigma^2/S_{xx}$. If we use s^2 as an estimate of σ^2, then an estimate of the variance of B_1 is $s_{B_1}^2 = s^2/S_{xx}$.

3. The random variable B_1 has a normal distribution.

The hypothesis test procedure concerning β_1 is based on the following theorem.

Theorem

If the simple linear regression assumptions are true, then the random variable

$$T = \frac{B_1 - \beta_1}{S/\sqrt{S_{xx}}} = \frac{B_1 - \beta_1}{S_{B_1}}$$

has a t distribution with $n - 2$ degrees of freedom.

No proof is given here that the two tests are equivalent, but there is some numerical evidence in Example 12.5.

β_{10} is the hypothesized value of β_0.

The null and alternative hypotheses are stated in terms of the parameter β_1. The most common test has $H_0: \beta_1 = 0$. If $\beta_1 = 0$, there is no significant linear relationship between the two variables. The true regression line is $y = \beta_0$ (a horizontal line), and knowing the value of x is of no use in predicting the value of $Y \mid x$. For simple linear regression, the following hypothesis test (with $\beta_{10} = 0$) is equivalent to an F test for a significant regression with significance level α.

Hypothesis Test and Confidence Interval Concerning β_1

$H_0: \beta_1 = \beta_{10}$

$H_a: \beta_1 > \beta_{10}, \qquad \beta_1 < \beta_{10}, \qquad \text{or} \qquad \beta_1 \neq \beta_{10}$

TS: $T = \dfrac{B_1 - \beta_{10}}{S_{B_1}}$

RR: $T \geq t_{\alpha, n-2}, \qquad T \leq -t_{\alpha, n-2}, \qquad \text{or} \qquad |T| \geq t_{\alpha/2, n-2}$

VIDEO TECH MANUALS

LINEAR REGRESSION: INFERENCE FOR SLOPE, CI FOR SLOPE

A $100(1 - \alpha)\%$ confidence interval for β_1 has as endpoints the values

$$\hat{\beta}_1 \pm t_{\alpha/2, n-2} \cdot s_{B_1} \qquad (12.6)$$

There are similar properties, a hypothesis test procedure, and confidence interval concerning the simple linear regression parameter β_0. The estimator B_0 has the following properties.

1. B_0 is an unbiased estimator for β_0: $E(B_0) = \mu_{B_0} = \beta_0$.

2. The variance of B_0 is $\text{Var}(B_0) = \sigma_{B_0}^2 = \dfrac{\sigma^2 \sum x_i^2}{n S_{xx}}$. If we use s^2 as an estimate of σ^2, then

an estimate of the variance of B_0 is $s_{B_0}^2 = \dfrac{s^2 \sum x_i^2}{n S_{xx}}$.

3. The random variable B_0 has a normal distribution.

Theorem

If the simple linear regression assumptions are true, then the random variable

$$T = \frac{B_0 - \beta_0}{S\sqrt{\sum x_i^2 / n S_{xx}}} = \frac{B_0 - \beta_0}{S_{B_0}}$$

has a t distribution with $n - 2$ degrees of freedom.

Hypothesis Test and Confidence Interval Concerning β_0

$H_0: \beta_0 = \beta_{00}$

β_{00} is the hypothesized value of β_0.

$H_a: \beta_0 > \beta_{00}, \qquad \beta_0 < \beta_{00}, \qquad \text{or} \qquad \beta_0 \neq \beta_{00}$

What does $\beta_0 = 0$ imply about the true regression line?

TS: $T = \dfrac{B_0 - \beta_{00}}{S_{B_0}}$

RR: $T \geq t_{\alpha, n-2}, \qquad T \leq -t_{\alpha, n-2}, \qquad \text{or} \qquad |T| \geq t_{\alpha/2, n-2}$

A $100(1 - \alpha)\%$ confidence interval for β_0 has as endpoints the values

$$\hat{\beta}_0 \pm t_{\alpha/2,\, n-2} \cdot s_{B_0} \tag{12.7}$$

Example 12.5 Ready to Operate

DATA SET

MEDTOOLS

Solution Trail 12.5

KEYWORDS

- Estimated regression line
- ANOVA table
- t test (concerning β_1) for significant regression
- Random sample

TRANSLATION

- $y = \hat{\beta}_0 + \hat{\beta}_1 x$
- ANOVA summary table
- Hypothesis test with $\beta_{10} = 0$

CONCEPTS

- Least-squares estimates
- ANOVA summary table for simple linear regression
- Hypothesis test concerning β_1

VISION

Find the necessary summary statistics. Use Equation 12.2 to find $\hat{\beta}_1$ and Equation 12.3 to find $\hat{\beta}_0$. Complete the ANOVA summary table. Use the template for a hypothesis test concerning β_1 with $\beta_{10} = 0$.

The Pulsar corporation sells a large sterilizer with four extendable shelves for medical tools. Company engineers believe that the time to reach operating temperature from a cold start (y, measured in minutes) is linearly related to the thickness of insulation (x, in inches). A random sample of $n = 12$ thicknesses was selected, and the time to reach operating temperature was recorded for each. The data and the summary statistics are given below.

x	1.3	1.8	0.9	1.6	2.6	1.5	2.1	3.0	0.8	2.4	2.5	2.6
y	8.0	6.9	8.1	7.0	6.3	6.5	6.4	5.8	8.3	8.3	6.6	6.6

$\sum x_i = 23.1$ $\sum y_i = 84.8$ $\sum x_i y_i = 158.5$

$\sum x_i^2 = 50.13$ $\sum y_i^2 = 607.66$

a. Find the estimated regression line.

b. Complete the ANOVA table and conduct an F test for a significant regression. Use a significance level of 0.05.

c. Conduct a t test (concerning β_1) for a significant regression. Use a significance level of 0.05.

d. Interpret your results.

SOLUTION

a. Use the summary statistics to find $\hat{\beta}_0$ and $\hat{\beta}_1$.

$$\hat{\beta}_1 = \frac{n\sum x_i y_i - (\sum x_i)(\sum y_i)}{n\sum x_i^2 - (\sum x_i)^2} \qquad \text{Equation 12.2.}$$

$$= \frac{(12)(158.5) - (23.1)(84.8)}{(12)(50.13) - (23.1)^2} \qquad \text{Use summary statistics.}$$

$$= \frac{-56.88}{67.95} = -0.8371 \qquad \text{Simplify.}$$

$$\hat{\beta}_0 = \frac{\sum y_i - \hat{\beta}_1 \sum x_i}{n} \qquad \text{Equation 12.3.}$$

$$= \frac{84.8 - (-0.8371)(23.1)}{12} \qquad \text{Use summary statistics and } \hat{\beta}_1.$$

$$= 8.6781 \qquad \text{Simplify.}$$

The estimated regression line is $y = 8.6781 - 0.8371x$.

b. Here are the calculations for the ANOVA table and the F test for a significant regression.

$$\text{SST} = S_{yy} = \sum y_i^2 - \frac{1}{n}(\sum y_i)^2 \qquad \text{Computational formula.}$$

$$= 607.66 - \frac{1}{12}(84.8)^2 = 8.4067 \qquad \text{Use summary statistics.}$$

$$\text{SSR} = \hat{\beta}_1 S_{xy} = \hat{\beta}_1 \left[\sum x_i y_i - \frac{1}{n}(\sum x_i)(\sum y_i) \right] \qquad \text{Computation formula.}$$

$$= -0.8371 \left[158.5 - \frac{1}{12}(23.1)(84.8) \right] \qquad \text{Use summary statistics.}$$

$$= 3.9679 \qquad \text{Simplify.}$$

$$\text{SSE} = \text{SST} - \text{SSR} \qquad \text{Computational formula.}$$

$$= 8.4067 - 3.9679 = 4.4388$$

Compute the mean squares.

$$\text{MSR} = \text{SSR}/1 = 3.9679/1 = 3.9679$$

$$\text{MSE} = \text{SSE}/(n-2) = 4.4388/10 = 0.4439$$

The F test for a significant regression with significance level 0.05 is

H_0: There is no significant linear relationship.

H_a: There is a significant linear relationship.

TS: $F = \text{MSR}/\text{MSE}$

RR: $F \geq F_{\alpha,1,n-2} = F_{0.05,1,10} = 4.96$

The value of the test statistic is

$$f = \text{MSR}/\text{MSE} = 3.9679/0.4439 = 8.9387 \; (\geq 4.96)$$

Because f lies in the rejection region, there is evidence to suggest that insulation thickness is linearly related to time to reach operating temperature.

Recall that we can use the tables of critical values for F distributions to bound the p value. Place the value of the test statistic in an ordered list of critical values with $v_1 = 1$ and $v_2 = 10$.

$$4.96 \quad \leq 8.9387 \leq 10.04$$

$$F_{0.05,1,10} \leq 8.9387 \leq F_{0.01,1,10}$$

Therefore, $0.01 \quad \leq \quad p \quad \leq 0.05$.

Figure 12.28 shows the calculation and an illustration of the exact p value.

Here are all the calculations presented in the ANOVA table, with some help from technology to find the exact p value.

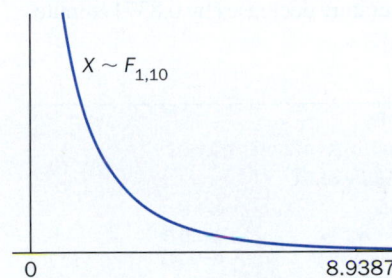

Figure 12.28 p-Value illustration:

$$p = P(F \geq 8.9387)$$
$$= 0.0136 \leq 0.05 = \alpha$$

ANOVA summary table for simple linear regression

Source of variation	Sum of squares	Degrees of freedom	Mean square	F	p Value
Regression	3.9679	1	3.9679	8.9387	0.0136
Error	4.4388	10	0.4439		
Total	8.4067	11			

c. The t test for a significant regression (concerning β_1) is two-sided and has $\beta_{10} = 0$. Use a significance level of 0.05.

H_0: $\beta_1 = 0$

H_a: $\beta_1 \neq 0$

TS: $T = \dfrac{B_1}{S_{B_1}}$

RR: $|T| \geq t_{\alpha/2,n-2} = t_{0.025,10} = 2.2281$

Remember that $s^2 = \text{MSE}$.

$$S_{xx} = \sum x_i^2 - \frac{1}{n}\left(\sum x_i\right)^2$$

Computational formula.

$$= 50.13 - \frac{1}{12}(23.1)^2 = 5.6625$$

Use summary statistics.

$$s_{B_1}^2 = s^2/S_{xx} = 0.4439/5.6625 = 0.0784$$

and the value of the test statistic is

$$t = \frac{\hat{\beta}_1}{s_{B_1}} = \frac{-0.8371}{\sqrt{0.0784}} = -2.9896; \qquad |-2.9896| = 2.9896 \geq 2.2281$$

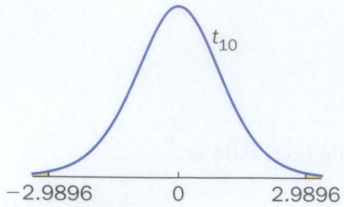

Figure 12.29 *p*-Value illustration:

$p = 2P(T \leq -2.9896)$
$= 0.0136 \leq 0.05 = \alpha$

Why have two hypothesis tests that are essentially the same? In multiple linear regression there may be many (more than one) hypothesized explanatory variables. An overall *F* test for a significant regression is conducted first. If the regression is significant, then a *t* test is conducted for each variable to see whether it contributes significantly to the variability in *y*.

Because $|t|$ lies in the rejection region, there is evidence to suggest that $\beta_1 \neq 0$, the regression is significant. The tables of critical values for *t* distributions can be used to bound the *p* value. Using technology,

$$p/2 = P(T \geq 2.9896) = 0.0068 \quad \Rightarrow \quad p = 0.0136$$

Figure 12.29 shows the calculation and an illustration of the exact *p* value.

Note: Comparing the two tests for a significant regression, we notice that $t^2 \approx f$. In fact, for simple linear regression $t^2 = f$ in every case, subject to round-off error. The *p* values associated with these two tests are also always the same. It seems reasonable that these two tests should always lead to the same conclusion, because they are both overall tests of a significant regression.

d. This analysis suggests that insulation thickness is linearly related to the time to reach operating temperature. Because $\hat{\beta}_1 = -0.8371 < 0$, this suggests that for each 1-inch increase in thickness, the time to reach operating temperature decreases by 0.8371 minute.

Figures 12.30 and 12.31 show technology solutions.

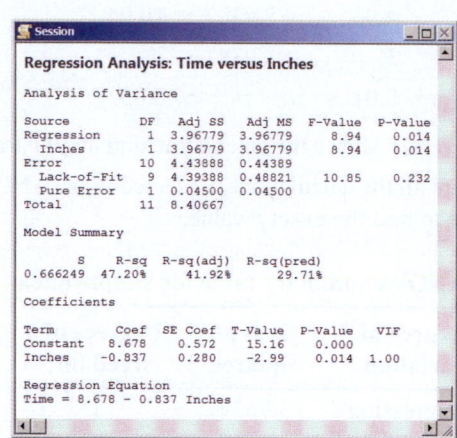

Figure 12.30 Minitab regression analysis output.

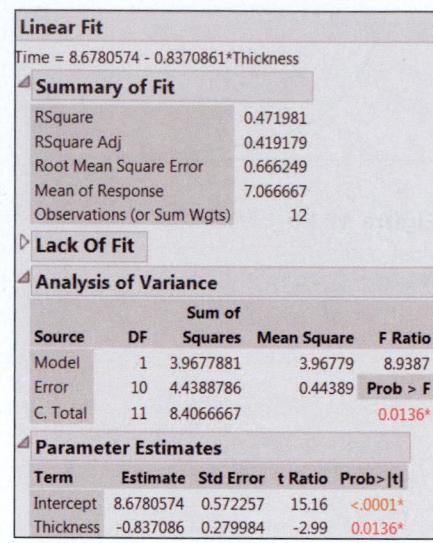

Figure 12.31 JMP Linear Fit results.

TRY IT NOW   GO TO EXERCISE 12.48

Remember that correlation, or an association, between two variables, does not imply causation.

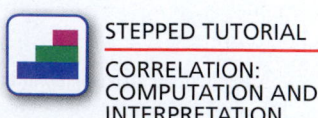

STEPPED TUTORIAL

CORRELATION: COMPUTATION AND INTERPRETATION

Correlation is a statistical term indicating a relationship between two variables. For example, the temperature is correlated with the number of cars that will not start in the morning. Or amount of lead exposure as a child is correlated with IQ level. In each case, a change in one variable is associated with a steady, consistent change in the other variable. As the temperature decreases for example, the number of cars that will not start in the morning increases.

The **sample correlation coefficient** is a measure of the strength of a linear relationship between two continuous variables, *x* and *y*. Suppose there are *n* pairs of observations $(x_1, y_1), (x_2, y_2), \ldots, (x_n, y_n)$. If large values of *x* are associated with large values of *y*, or, as *x* increases, the corresponding value of *y* tends to increase, then *x* and *y* are positively related. If small values of *x* are associated with large values of *y*, or, as *x* increases, the corresponding value of *y* tends to decrease, then *x* and *y* are negatively related.

To understand the formula for the sample correlation coefficient, consider the scatter plots in Figures 12.32 and 12.33 and the quantity $S_{xy} = \sum (x_i - \bar{x})(y_i - \bar{y})$. The horizontal line $y = \bar{y}$ and the vertical line $x = \bar{x}$ divide the plane region into four parts. The signs, plus (+) or minus (−), on each graph indicate the sign of the product $(x_i - \bar{x})(y_i - \bar{y})$

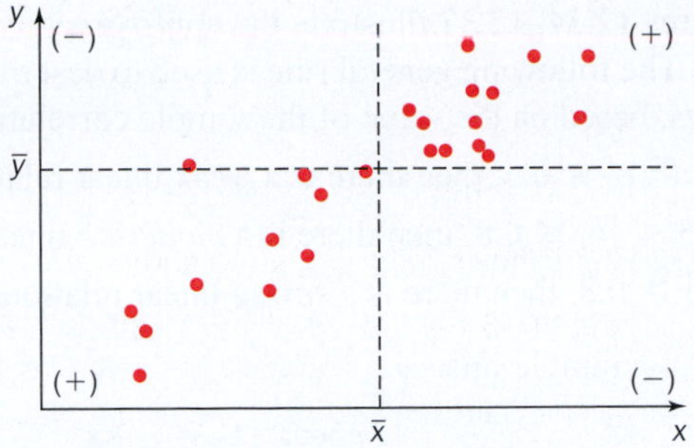

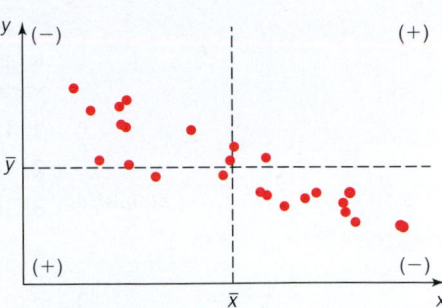

Figure 12.32 If x and y have a positive linear relationship, then most of the products $(x_i - \bar{x})(y_i - \bar{y})$ are positive.

Figure 12.33 If x and y have a negative linear relationship, then most of the products $(x_i - \bar{x})(y_i - \bar{y})$ are negative.

in each part. For example, for any ordered pair in the top right corner of Figure 12.32, $x_i > \bar{x}$ and $y_i > \bar{y}$. Therefore, $x_i - \bar{x} > 0$ and $y_i - \bar{y} > 0$, and the product is positive. Similarly, consider any ordered pair in the bottom right corner of Figure 12.33, $x_i > \bar{x}$ and $y_i < \bar{y}$. Therefore, $x_i - \bar{x} > 0$ and $y_i - \bar{y} > 0$, and the product is negative.

If x and y are positively related, as in Figure 12.32, then most of the products $(x_i - \bar{x})(y_i - \bar{y})$ are positive. And if x and y are negatively related, as in Figure 12.33, then most of the products $(x_i - \bar{x})(y_i - \bar{y})$ are negative. Therefore, it seems reasonable to find the sum of all of these products, S_{xy}, and use this as a measure of the linear relationship between the two variables. Large positive values of S_{xy} should indicate a positive linear relationship, and large negative values should indicate a negative linear relationship.

This approach is intuitive, but it must be modified slightly. The magnitude of S_{xy} depends on the units of x and y. For example, if we multiply every value of x by 10, the inherent linear relationship between x and y should not change, but the value of S_{xy} increases. The sample correlation coefficient adjusts S_{xy} so that it is unit-independent.

The sample correlation coefficient, r, is an estimate of the population correlation coefficient, ρ.

Sample Correlation Coefficient

Suppose there are n pairs of observations $(x_1, y_1), (x_2, y_2), \ldots, (x_n, y_n)$. **The sample correlation coefficient** for these n pairs is

$$r = \frac{S_{xy}}{\sqrt{S_{xx}S_{yy}}} = \frac{\sum x_i y_i - \frac{1}{n}(\sum x_i)(\sum y_i)}{\sqrt{\left[\sum x_i^2 - \frac{1}{n}(\sum x_i)^2\right]\left[\sum y_i^2 - \frac{1}{n}(\sum y_i)^2\right]}} \qquad (12.8)$$

A CLOSER LOOK

1. The value of r does not depend on the order of the variables and is independent of units, or is unitless.

2. The value of r is always between -1 and $+1$, that is, $-1 \le r \le +1$. r is exactly $+1$ if and only if all of the ordered pairs lie on a straight line with positive slope. r is exactly -1 if and only if all of the ordered pairs lie on a straight line with negative slope.

3. The square of the sample correlation coefficient is the coefficient of determination in a simple linear regression model. Because $-1 \le r \le +1$, $0 \le r^2 \le 1$.

4. r is a measure of the strength of a *linear* relationship. If r is near 0, there is no evidence of a linear relationship, but x and y may be related in another way.

5. Suppose there is a horizontal line ($y = \beta_0$) with zero slope, and all the data points lie very close to this line. There is no association between the variables; the correlation is close to 0. Intuitively, some of the products $(x_i - \bar{x})(y_i - \bar{y})$ are positive and some are negative, and they tend to cancel out one another.

Figures 12.34–12.37 illustrate the approximate value of *r* corresponding to each scatter plot. The following general rule is used to describe the linear relationship between two variables, based on the value of the sample correlation coefficient.

1. If $0 \le |r| \le 0.5$, then there is a *weak* linear relationship.
2. If $0.5 < |r| \le 0.8$, then there is a *moderate* linear relationship.
3. If $|r| > 0.8$, then there is a *strong* linear relationship.

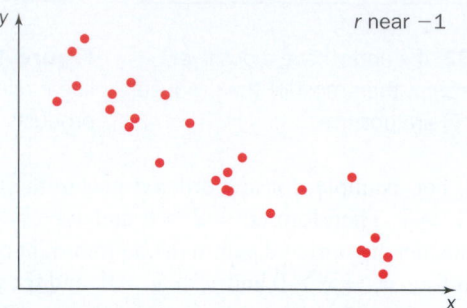

Figure 12.34 The variables *x* and *y* are (strongly) negatively related. The points all fall near a straight line with negative slope. The sample correlation coefficient is near −1.

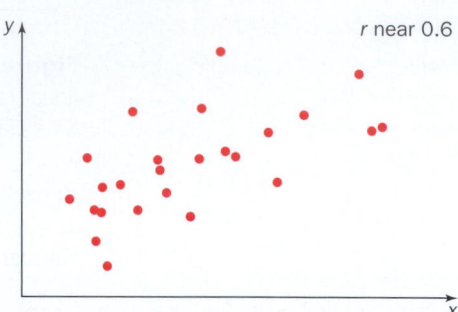

Figure 12.35 The variables *x* and *y* are (moderately) positively related. As the values of *x* increase, the values of *y* tend to increase. The sample correlation coefficient is near 0.6.

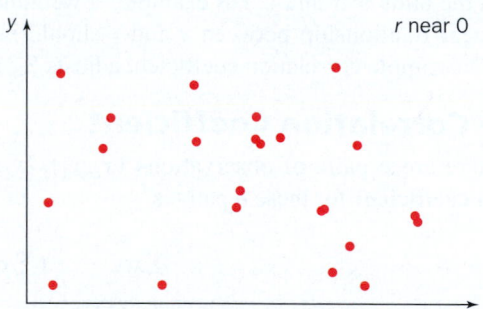

Figure 12.36 The scatter plot shows no linear relationship between the variables *x* and *y*. The sample correlation coefficient is near 0.

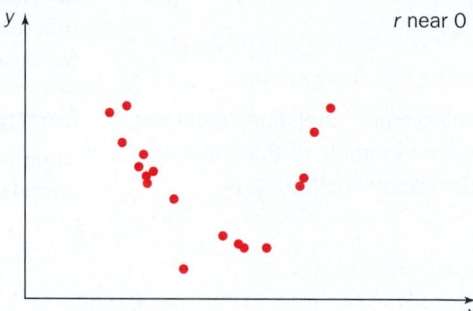

Figure 12.37 The variables *x* and *y* appear to be related but not linearly. There is a pattern in the scatter plot, but because the sample correlation coefficient measures linear association, *r* is near 0.

Example 12.6 Parking Fines

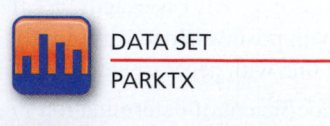

DATA SET

PARKTX

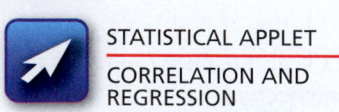

STATISTICAL APPLET

CORRELATION AND REGRESSION

The city of Toronto has many public parking lots downtown, but most are very small. This forces drivers to park on nearby streets, where they frequently receive parking tickets. A random sample of Toronto public parking lots was obtained. The capacity and the number of parking tickets issued per month within 100 meters of each lot was recorded. The data are given in the following table.[14]

Capacity, *x*	20	43	24	144	184	40	83	38	40	18
Tickets, *y*	167	394	109	291	165	247	212	81	51	239

Find the sample correlation coefficient between capacity and number of tickets issued, and interpret this value.

SOLUTION

STEP 1 Compute the summary statistics needed to find the sample correlation coefficient.

$$S_{xx} = \sum x_i^2 - \frac{1}{n}(\sum x_i)^2 = 69{,}274 - \frac{1}{10}(634)^2 = 29{,}078.4$$

$$S_{yy} = \sum y_i^2 - \frac{1}{n}(\sum y_i)^2 = 479{,}148 - \frac{1}{10}(1{,}956)^2 = 96{,}554.4$$

$$S_{xy} = \sum x_i y_i - \frac{1}{n}(\sum x_i)(\sum y_i)$$

$$= 132{,}058 - \frac{1}{10}(634)(1{,}956) = 8{,}047.6$$

STEP 2 Using Equation 12.8, the sample correlation coefficient is

$$r = \frac{S_{xy}}{\sqrt{S_{xx}S_{yy}}} = \frac{8{,}047.6}{\sqrt{(29{,}078.4)(96{,}554.4)}} = 0.1519$$

Because $r = 0.1519 < 0.5$, there is a weak positive linear relationship between parking-lot capacity and the number of tickets issued each month within 100 meters. This suggests that more tickets are issued close to larger parking lots. Figure 12.38 shows a technology solution. ●

TRY IT NOW   GO TO EXERCISE 12.52

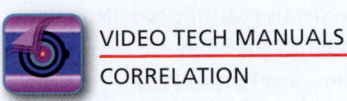

VIDEO TECH MANUALS

CORRELATION

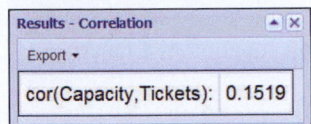

Figure 12.38 The sample correlation coefficient computed using CrunchIt!.

Remember that r is a measure of *association*, not *causation*. Two variables may be strongly related, but that does not mean one variable causes the other. In children, for example, shoe size and height are associated, but probably both are caused by a third variable, age.

Technology Corner

Procedure: Find the estimated regression line, complete the ANOVA table, and conduct a t test (concerning β_1) for a significant regression.
Reconsider: Example 12.5, solution, and interpretations.

CrunchIt!

Use the Simple Linear Regression command. CrunchIt! does not display the ANOVA table.

1. Enter the thickness of insulation values into column Var1 and the time to reach operating temperature into column Var2.
2. Select Statistics; Regression; Simple Linear. Using the drop-down menus, select the Dependent Variable, y, the Independent Variable, x; and in the Parameters section, select Numeric Results as the Display option.
3. Click Calculate. The results are displayed in a new window. See Figure 12.39.

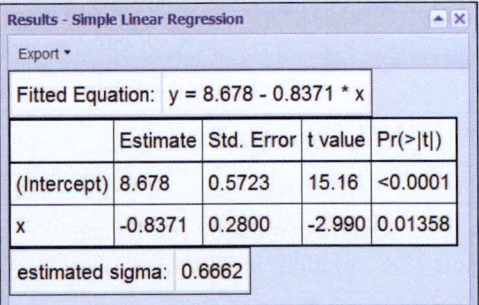

Figure 12.39 CrunchIt! Statistics; Regression; Simple Linear results.

TI-84 Plus C

Execute `DiagnosticOn` on the Home screen to display the coefficient of determination and sample correlation coefficient with the estimated regression line.

1. Enter the thickness of insulation values into list L1 and the times into list L2.

2. Select $\boxed{\text{STAT}}$; CALC; `LinReg(a+bx)`. Enter the name of the list containing the observed values of the independent variable, `L1`, and the name of the list containing the observed values of the dependent variable, `L2`. Highlight `Calculate` and press $\boxed{\text{ENTER}}$. The estimated regression coefficients are displayed on the Home screen. See Figure 12.40.

3. To test for a significant regression in terms of β_1, select $\boxed{\text{STAT}}$; TESTS; `LinRegTTest`. Enter the name of the list containing the observed values of the independent variable, `L1`, and the name of the list containing the observed values of the dependent variable, `L2`, set `Freq` to 1, highlight $\neq 0$, and optionally store the regression equation in a function variable. See Figure 12.41.

4. Highlight `Calculate` and press $\boxed{\text{ENTER}}$. The output is displayed on the Home screen. See Figure 12.42.

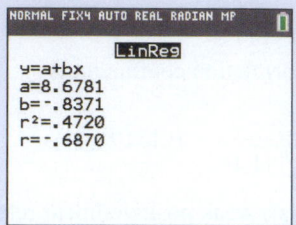

Figure 12.40 `LinReg` results.

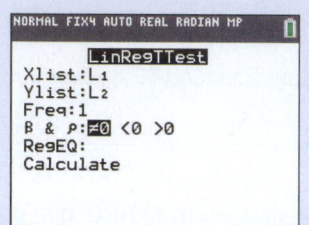

Figure 12.41 `LinRegTTest` setup screen.

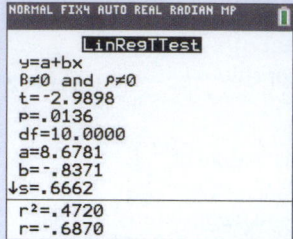

Figure 12.42 `LinRegTTest` results.

Minitab

There are several ways to find the estimated regression line in Minitab, in a Session window and in the <u>Stat</u>; <u>R</u>egression menu.

1. Enter the thickness of insulation values into column `C1` and the times into column `C2`.

2. Select <u>Stat</u>; <u>R</u>egression; <u>R</u>egression; <u>F</u>it Regression Model. Enter the Response variable, `C2`, and the Continuous predictor variable, `C1`. Click OK. The results are displayed in a session window. Refer to Figure 12.30.

Excel

Regression analysis is part of the Data Analysis tool pack.

1. Enter the thickness of insulation values into column A and the times into column B.

2. Under the Data tab, select Data Analysis; Regression. Enter the Y Range, X Range, and the Output Range. Click OK. The output is displayed in the specified range. See Figure 12.43

	C	D	E	F	G	H	I	J	K
SUMMARY OUTPUT									
Regression Statistics									
Multiple R	0.6870								
R Square	0.4720								
Adjusted R Square	0.4192								
Standard Error	0.6662								
Observations	12								
ANOVA									
	df	*SS*	*MS*	*F*	*Significance F*				
Regression	1	3.9678	3.9678	8.94	0.0136				
Residual	10	4.4389	0.4439						
Total	11	8.4067							
	Coefficients	*Standard Error*	*t Stat*	*P-value*	*Lower 95%*	*Upper 95%*	*Lower 95.0%*	*Upper 95.0%*	
Intercept	8.6781	0.5723	15.1646	0.0000	7.4030	9.9531	7.4030	9.9531	
X Variable 1	-0.8371	0.2800	-2.9898	0.0136	-1.4609	-0.2132	-1.4609	-0.2132	

Figure 12.43 Regression analysis output.

Procedure: Compute the sample correlation coefficient.

Reconsider: Example 12.6, solution, and interpretations.

CrunchIt!

Use the command Correlation to find the sample correlation coefficient.

1. Enter the capacity values into column Var1 and the number of tickets into column Var2.

2. Select Statistics; Correlation. Check the appropriate columns and click Calculate. The sample correlation coefficient is displayed in a new window. Refer to Figure 12.38.

TI-84 Plus C

Use the built-in function `LinReg(a+bx)` to find the sample correlation coefficient. Execute `DiagonisticOn` on the Home screen to display the coefficient of determination and sample correlation coefficient with the estimated regression line.

1. Enter the values for capacity into list `L1` and the values for tickets into list `L2`.
2. Select STAT ; CALC; `LinReg(a+bx)`. Enter the name of the list containing the observed values of the independent variable, `L1`, and the name of the list containing the observed values of the dependent variable, `L2`. Press ENTER .
3. The sample correlation coefficient, r, is displayed at the bottom of the screen. See Figure 12.44.

Minitab

Minitab has a built-in function to compute the sample correlation coefficient, in the <u>S</u>tat; <u>B</u>asic Statistics menu.

1. Enter the values for capacity into column `C1` and the values for tickets into column `C2`.
2. Select <u>S</u>tat; <u>B</u>asic Statistics; <u>C</u>orrelation.
3. Select the Variables (columns) containing the data. Select the Method and optionally display the p-values. Click OK. The sample correlation coefficient is displayed in a Session window. See Figure 12.45.

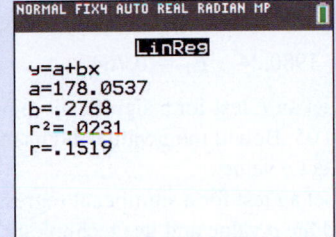

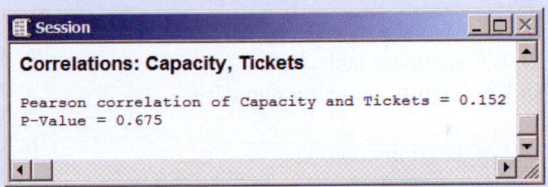

Figure 12.44 Using `DiagnosticOn`, the sample correlation coefficient is displayed at the bottom of the `LinReg(a+bx)` output screen.

Figure 12.45 The sample correlation coefficient computed using Minitab.

Excel

Use the built-in function CORREL to compute the sample correlation coefficient.

1. Enter the values for capacity into column A and the values for tickets into column B.
2. Use the function CORREL with the two arrays. See Figure 12.46.

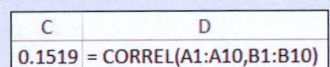

Figure 12.46 The sample correlation coefficient computed using Excel.

SECTION 12.2 EXERCISES

Concept Check

12.34 Short Answer If the variation in Y is independent of the value of x, describe the resulting scatter plot.

12.35 Short Answer For simple linear regression, state an equivalent test to the F test for a significant regression.

12.36 Short Answer Using the model utility test for a significant regression, the test statistic is _____.

12.37 True/False Using the model utility test for a significant linear regression, we will reject the null hypothesis for small and large values of the test statistic.

12.38 True/False The sample correlation coefficient is a measure of the strength of a linear relationship between two continuous variables.

12.39 Short Answer The sample correlation coefficient is always between what two numbers?

12.40 Short Answer Suppose there is a strong positive linear relationship between the variables x and y. Describe the products $(x_i - \bar{x})(y_i - \bar{y})$.

Practice

12.41 Consider the following (partial) ANOVA summary table from a simple linear regression analysis.

ANOVA summary table for simple linear regression

Source of variation	Sum of squares	Degrees of freedom	Mean square	F	p Value
Regression	11,691.9				
Error		23			
Total	116,064.0				

a. Complete the ANOVA summary table.
b. Conduct an F test for a significant regression. Use $\alpha = 0.05$.
c. Find the coefficient of determination.
d. Use your answer in part (c) to find the sample correlation coefficient.

12.42 Consider the following (partial) ANOVA summary table from a simple linear regression analysis.

ANOVA summary table for simple linear regression

Source of variation	Sum of squares	Degrees of freedom	Mean square	F	p Value
Regression	2,772.93				
Error	12,988.70				
Total		35			

a. Complete the ANOVA summary table.
b. Conduct an F test for a significant regression. Use $\alpha = 0.01$.
c. Find the coefficient of determination.
d. Use your answer in part (c) to find the sample correlation coefficient. Is the relationship between these two variables positive or negative? Justify your answer.

12.43 Consider the following summary statistics for data obtained in a simple linear regression analysis.

$$n = 23 \qquad MSE = 561.088$$
$$S_{xx} = 151.086 \qquad \hat{\beta}_1 = 4.4285$$

a. Conduct a hypothesis test for a significant regression based on the value of β_1 (H_0: $\beta_1 = 0$ versus H_a: $\beta_1 \neq 0$). Use a significance level of 0.05.
b. Find a 95% confidence interval for the slope of the true regression line, β_1.

c. Using your confidence interval in part (b), is there any evidence to suggest that $\beta_1 \neq 0$? Justify your answer. How does your conclusion compare with part (a)?

12.44 Consider the following data obtained in a simple linear regression study. ▥ **EX12.44**

x	3.27	1.26	4.55	0.86	4.07	4.79	3.25
y	16.67	19.93	14.65	17.48	18.18	13.58	15.70

a. Find the estimated regression line, and complete the ANOVA table.
b. Conduct the hypothesis test H_0: $\beta_0 = 0$ versus H_a: $\beta_0 \neq 0$ with significance level 0.01. Is there any evidence to suggest that the true regression line does not pass through the origin?
c. Find a 99% confidence interval for β_0.
d. Interpret the value of $\hat{\beta}_0$.

12.45 Consider the following summary statistics for data obtained in a simple linear regression analysis.

$$n = 16 \qquad SSR = 1155.9 \qquad SSE = 3912.82$$
$$S_{xx} = 1980.24 \qquad \hat{\beta}_1 = 0.7640$$

a. Conduct an F test for a significant regression. Use $\alpha = 0.05$. Bound the p value and use technology to find the exact p value.
b. Conduct a t test for a significant regression. Use $\alpha = 0.05$. Bound the p value and use technology to find the exact p value.
c. Square the value of the test statistic in part (a). Compare this value with the test statistic in part (b).
d. How do the exact p values compare in parts (a) and (b)? Explain this result.

12.46 Consider the following summary statistics for data obtained in a study of the linear relationship between two variables.

$$S_{xx} = 199.418 \qquad S_{yy} = 81.430 \qquad S_{xy} = 111.774$$

a. Find the sample correlation coefficient.
b. Use the value of r from part (a) to describe the relationship between the two variables.

12.47 Consider the following data obtained in a study of the relationship between two variables. ▥ **EX12.47**

x	40.4	44.8	40.7	31.7	41.3	38.1
y	-320.3	-303.9	-264.1	-197.0	-311.8	-280.3

a. Find the estimated regression line and complete the ANOVA table.
b. Conduct an F test for a significant regression. Use $\alpha = 0.05$.
c. Find the coefficient of determination.
d. Find the sample correlation coefficient. Use this value to describe the relationship between the two variables. How does the estimate b_1 support this answer?
e. Square the sample correlation coefficient. Verify that this value is the coefficient of determination.

Applications

12.48 Sports and Leisure Many factors affect the length of a professional football game. A study was conducted to determine the relationship between the total number of penalty yards (x) and the time required to complete a game (y, in hours). The following data were obtained. **FOOTBALL**

x	196	164	167	35	111	78	150	121	40
y	4.2	4.1	3.5	3.2	3.2	3.6	4.0	3.1	1.9

a. Find the estimated regression line, and complete the ANOVA table.
b. Conduct a t test (concerning β_1) for a significant regression. Use a significance level of 0.05.
c. What proportion of the observed variation in game length can be explained by this simple linear regression model?

12.49 Biology and Environmental Science In a recent study, the weight of an orange (x, in pounds) was compared with the amount of fresh-squeezed juice from the orange (y, in ounces). A random sample of $n = 15$ oranges was obtained. Each was carefully weighed and squeezed. The following summary statistics were obtained.

$$\Sigma x_i = 9.09 \qquad \Sigma y_i = 50.44 \qquad \Sigma x_i y_i = 30.72$$
$$\Sigma x_i^2 = 5.73 \qquad \Sigma y_i^2 = 173.21$$

a. Find the estimated regression line, and complete the ANOVA table. Interpret your results.
b. Conduct an F test for a significant regression. Use a significance level of 0.01.
c. Find a 95% confidence interval for the regression parameter β_0. Does this interval suggest that β_0 is different from 0?

12.50 Physical Sciences The temperature of the upper layer of ocean water is affected by sunlight and wind. There is often a very sharp difference in temperature between the surface zone and the more stationary deep zone. The thermocline layer marks the abrupt drop-off in temperature. The following data were obtained in a study of temperature (x, measured in °C) versus depth (y, measured in meters) above the thermocline layer in the Mediterranean Sea. **OCEAN**

x	76	54	146	7	91	130	131	117
y	11.0	16.9	8.0	24.7	17.4	16.0	11.1	16.5

a. Complete the ANOVA summary table for simple linear regression.
b. Conduct an F test for a significant regression and find bounds on the p value associated with this test. Interpret your results.
c. Find a 95% confidence interval for the true value of β_1.

12.51 Physical Sciences Permafrost is soil or rock that remains at or below 0°C for at least two years. Suppose an Arctic study analyzed the observed changes in mean annual air temperature and the depth of the freezing layer. The depth of the upper surface of the permafrost layer (y, in feet) was measured for $n = 26$ mean annual air temperatures (x, in °F). The summary statistics are given below.

$$S_{xx} = 8.6786 \qquad S_{yy} = 2.2771 \qquad S_{xy} = -2.8600$$

a. Find the sample correlation coefficient.
b. Use the value of r from part (a) to describe the relationship between mean annual temperature and the depth of the permafrost layer.

12.52 Public Health and Nutrition Crimini mushrooms are more common than white mushrooms, and they contain a high amount of copper, which is an essential element according to the U.S. Food and Drug Administration. A study was conducted to determine whether the weight of a mushroom is linearly related to the amount of copper it contains. A random sample of crimini mushrooms was obtained, and the weight (in grams) and the total copper content (in mg) was measured for each. **CRIMINI**

a. Construct a scatter plot of the data.
b. Find the sample correlation coefficient.
c. Use your results in parts (a) and (b) to describe the relationship between crimini mushroom weight and copper content.
d. If a simple linear regression analysis were conducted, identify the independent and dependent variables that would be used.

12.53 Sports and Leisure A recent study suggests that, for NASCAR racers, yearly winnings are related to average finish.[15] A random sample of NASCAR racers was obtained, and the average finish (x) and total winnings (y, in millions of dollars) through 31 races in 2013 were recorded for each.[16] **NASCAR**

a. Construct a scatter plot of the data. Describe the relationship between average finish and winnings.
b. Find the sample correlation coefficient. Is the value of r consistent with your interpretation in part (a)? Why or why not?
c. Find the estimated regression line, and complete the ANOVA table.
d. Using your results from part (c), what would the average finish have to be in order to win at least $6 million through 31 races?

12.54 Fuel Consumption and Cars An investigative reporter believes that certain automobile service stations that offer state vehicle inspections routinely charge for unnecessary repair work. Preliminary data suggest that the cost of the repair work may be related to the age of the car. A random sample of automobiles inspected at these stations was obtained, and the age (in years) along with the cost of the repairs (in dollars) were recorded for each vehicle. The data are given in the following table. **REPAIR**

Age	Repair cost	Age	Repair cost
9.1	1882	3.2	17
3.8	193	3.3	1268
5.0	368	6.7	1126
1.7	1047	3.9	646
1.9	315	9.7	955
5.3	1631	2.0	801
5.9	652	5.4	973
6.4	475		

a. Find the sample correlation coefficient, and describe the linear relationship.

b. Find the estimated regression line for age (x) and repair cost (y). Conduct an F test for a significant repression, and find the bounds on the p value for this test.

c. Using your results in parts (a) and (b), do you believe the reporter's claim? Justify your answer.

12.55 Physical Sciences   As the temperature of air increases, it has a greater capacity to hold moisture. However, humidity can only increase if there is a supply of moisture. For the area of southern Florida, a random sample of days was obtained and the relative humidity (x, a percentage) and daily temperature (y, in °F) were recorded for each.[17] **HUMID**

a. Construct a scatter plot of the data.

b. Find the sample correlation coefficient.

c. Based on the value of r in part (b), predict the sign of $\hat{\beta}_1$ in the estimated regression line.

d. Find the estimated regression line. Do the results support your answer to part (c)? Why or why not?

12.56 Biology and Environmental Science   A recent study suggests that airplanes flying into and out of Australia and New Zealand create the most amount of aircraft pollution in the form of ozone.[18] An area over the Pacific Ocean, about 1000 km to the east of the Solomon Islands, is very sensitive to aircraft emissions. A random sample of aircraft flying from Australia was obtained, and the amount of aircraft emissions (x, oxides, in kg) and the amount of ozone created (y, in kg) was measured for each. **OZONE**

a. Construct a scatter plot of these data. Describe the relationship.

b. Compute the sample correlation coefficient, and use this value to describe the linear relationship.

c. Find the estimated regression line. Conduct an F test for a significant regression.

d. Conduct the appropriate test to determine if there is any evidence that the coefficient β_0 is different from 0. Do your results seem practical? Why or why not?

12.57 Biology and Environmental Science   The Arctic Oscillation (AO) has two distinct phases (positive and negative) that vary in intensity. The phase and intensity affect the weather patterns of North America and especially Eastern North America. There is some research that suggests the snow cover in the month of October in Siberia/Eurasia is associated with the AO, and therefore, weather patterns in North America. The text website presents a data set of Eurasian snow cover (x, in sq km) and AO anomaly (y).[19] **ARCTIC**

a. Construct a scatter plot of these data. Describe the relationship between the variables x and y.

b. Compute the sample correlation coefficient. Interpret this value.

c. Suppose high AO anomalies are associated with fierce, cold winters. Find the estimated regression line and use it to predict the AO anomaly for this winter.

Extended Applications

12.58 Business and Management   The owner of a small ice cream stand believes that total weekly revenue (y, in dollars) during the summer months is related to money spent on advertising (x, dollars per week). A random sample of summer weeks was selected, and the resulting data are given in the following table. **ICECREAM**

x	30	300	380	275	350	190	85
y	957	1125	1202	1028	1134	1124	1062

a. Find the estimated regression line, and complete the ANOVA table.

b. What proportion of the observed variation in weekly revenue is explained by this regression model?

c. Find a 99% confidence interval for the regression parameter β_1.

d. Conduct a hypothesis test of $H_0: \beta_0 = 0$ versus $H_a: \beta_0 > 0$. Interpret the results. (What happens if the owner spends nothing on advertising in a week?)

12.59 Public Health and Nutrition   Research suggests that salt intake is related to fluid intake in adults and soft drink consumption in children and adolescents.[20] Suppose a random sample of 18 children and adolescents was obtained and the amount of salt consumed (x, g/week) and the amount of sugar-sweetened soft drinks consumed (y, g/week) was recorded for each. The following summary statistics were obtained.

$$\sum x_i = 430 \qquad \sum y_i = 17{,}512 \qquad \sum x_i y_i = 441{,}608$$
$$\sum x_i^2 = 11{,}138 \qquad \sum y_i^2 = 18{,}197{,}768$$

a. Find the estimated regression line, and complete the ANOVA table.

b. Conduct an F test for a significant regression. Find bounds on the p value for this test.

c. Conduct the hypothesis test $H_0: \beta_1 = 25$ versus $H_a: \beta_1 \neq 25$. Use a significance level of 0.05.

d. Find a 99% confidence interval for the true value of β_1.

12.60 Biology and Environmental Science   The water quality in many lakes and rivers is routinely monitored by tracking the turbidity, a measure of the total amount of suspended solids in the water. High turbidity measured in nephelometric turbidity units, or NTUs, suggests murky water. Heavy rains tend to raise river levels and increase turbidity. A sample of weekly rainfall total (x, measured in inches) and turbidity (y, in NTUs) was obtained for the Wide Waters site at Owasco Lake in New York. The data are given in the following table.[21] **RAINFALL**

x	0.67	0.08	0.64	0.98	0.34	0.37	0.85
y	5.29	3.21	5.68	1.77	3.66	3.38	3.19

x	0.75	0.07	0.68	0.64	0.26	0.09	1.44
y	2.45	2.22	2.44	2.26	2.55	1.66	8.41

a. Find the estimated regression line.

b. Conduct a test of $H_0: \beta_1 = 0$ versus $H_a: \beta_1 \neq 0$ with a significance level of 0.05.

c. Find an estimate of the expected turbidity if 0.55 inch of rain has fallen within the past week.

d. What proportion of the observed variation in turbidity is explained by this regression model? How do you think this model could be improved?

12.61 Economics and Finance   Stock brokers and even casual investors are always searching for better methods to predict the movement in the price of stocks. The "skirt-length theory" suggests that if women's skirts are short, then the markets will rise. If skirts are long, then the markets will be headed down. To test this theory, a random sample of years was obtained, and the length of women's skirts was measured (*x*, in inches) for a typical fashion model. The change in the S&P 500 market index (*y*) was also recorded for that year. The data are given in the following table. ▥ STOCKS

x	24	15	24	16	16	16	19	15
y	7	−7	3	36	−35	23	−35	79

x	19	25	15	18	17	22	16	19
y	−5	−55	−29	8	42	7	−9	−45

a. Find the estimated regression line, and complete the ANOVA table.

b. Conduct an *F* test for a significant regression. Use a significance level of 0.01.

c. Construct a scatter plot of the data, and find the sample correlation coefficient.

d. Using your answers to parts (b) and (c), do you think the skirt-length theory is worth using? Justify your answer.

12.62 Physical Sciences   The rate of evaporation at the surface of the water in a swimming pool (kg/h) is believed to be related to the air velocity (m/s) or the relative humidity (measured as a percentage). ▥ POOL

a. Find the estimated regression line, and complete the ANOVA table for evaporation rate (*y*) and air velocity (*x*).

b. Find the estimated regression line and complete the ANOVA table for evaporation rate (*y*) and relative humidity (*x*).

c. Which of these two models do you think is better? Justify your answer.

12.63 Physical Sciences   The U.S. Geological Survey recently completed a national assessment of geologic carbon dioxide storage resources.[22] A random sample of formations was obtained, and the area of each formation (*x*, in thousands of

acres) and the depth from the surface (*y*, in thousands of feet) was obtained for each. ▥ GEOLCO2

a. Construct a scatter plot of these data. Describe the relationship between these two variables.

b. Compute the sample correlation coefficient. Does this value support your answer in part (a)? Why or why not?

c. Using the results in parts (a) and (b), do you believe the relationship between these two variables is significant? Why or why not?

d. Find the estimated regression line and conduct an *F* test for a significant regression. Use $\alpha = 0.05$. Does your conclusion support your answer in part (c)?

Challenge

12.64 Sports and Leisure   Let ρ be the population correlation coefficient between two variables. A hypothesis test for a correlation different from zero is based on the sample correlation coefficient, *R* (the random variable), and involves the *t* distribution.

$H_0: \rho = 0$

$H_a: \rho > 0, \qquad \rho < 0, \qquad \text{or} \qquad \rho \neq 0$

$$\text{TS: } T = \frac{R\sqrt{n-2}}{\sqrt{1-R^2}}$$

RR: $T \geq t_{\alpha, n-2}, \qquad T \leq -t_{\alpha, n-2}, \qquad \text{or} \qquad |T| \geq t_{\alpha/2, n-2}$

A new mountain climber believes that there is a linear relationship between the diameter of a single rope and its dry weight. A random sample of climbing ropes was obtained. The diameter of each was measured (*x*, in mm) and the weight of each (*y*, in grams per meter) was also recorded. The data are given in the following table. ▥ MTNCLIMB

x	10.07	9.54	10.34	10.85	9.85	9.45	9.95
y	71.2	64.5	67.2	73.5	64.5	65.2	68.2

x	10.63	9.73	10.34
y	70.9	67.8	72.0

a. Find the sample correlation coefficient, and conduct a two-sided hypothesis test with $H_0: \rho = 0$. Use a significance level of 0.05.

b. Find the estimated regression line, and conduct a test of $H_0: \beta_1 = 0$ versus $H_a: \beta_1 \neq 0$. Use a significance level of 0.05.

c. Explain the similarities between the values of the test statistics in parts (a) and (b). Explain why this relationship makes sense.

12.3 Inferences Concerning the Mean Value and an Observed Value of *Y* for *x* = *x**

Recall from Section 12.1: Suppose *x** is a specific value of the independent, or predictor, variable *x* and $y = \hat{\beta}_0 + \hat{\beta}_1 x$ is the estimated regression line. The value $y^* = \hat{\beta}_0 + \hat{\beta}_1 x^*$ is

1. An estimate of the *mean* value of *Y* for *x* = *x**, and

2. An estimate of an *observed* value of *Y* for *x* = *x**.

Remember that $E(Y|x^*)$ is the mean of *all* values of Y for which $x = x^*$. And y^* is a single observation for $x = x^*$.

The error in estimating the *mean* value of Y is less than the error in estimating an *observed* value of Y. In the first case, y^* is used to estimate a *single value*, the mean. However, an estimate of an *observed* value of Y is a guess at the next value selected from an entire distribution; the error in estimation must be greater. In this section, we will first consider a hypothesis test and confidence interval concerning the mean value of Y for $x = x^*$. Then we will consider a **prediction interval** for an observed value of Y if $x = x^*$.

Suppose the simple linear regression model assumptions are true. For $x = x^*$, the random variable $B_0 + B_1x^*$ has a normal distribution with expected value

$$E(B_0 + B_1x^*) = \beta_0 + \beta_1x^* \tag{12.9}$$

and variance

$$\text{Var}(B_0 + B_1x^*) = \sigma^2\left[\frac{1}{n} + \frac{(x^* - \bar{x})^2}{S_{xx}}\right] \tag{12.10}$$

The standard deviation is the square root of the expression in Equation 12.10, and an estimate of the standard deviation is obtained by using s as an estimate for σ.

The numerator, $(x^* - \bar{x})^2$, is 0 when $x = \bar{x}$.

The variance of $B_0 + B_1x^*$ is smallest when $x = \bar{x}$. The further x^* is from $\bar{x}$, the greater the squared difference $(x^* - \bar{x})^2$ and the greater the variance. Therefore, the estimator, $B_0 + B_1x^*$, for the mean value of Y is most precise near $\bar{x}$. And a confidence interval for the mean value of Y would be narrower for values of x^* near $\bar{x}$. Intuitively, think about predicting the weather. The further into the future we try to predict the weather, the more inaccurate will be the forecast.

The following theorem is used to conduct a hypothesis test and construct a confidence interval concerning the mean value of Y given $x = x^*$.

Theorem

If the simple linear regression assumptions are true, then the random variable

$$T = \frac{(B_0 + B_1x^*) - (\beta_0 + \beta_1x^*)}{S\sqrt{(1/n) + [(x^* - \bar{x})^2/S_{xx}]}}$$

has a t distribution with $n - 2$ degrees of freedom.

The null and alternative hypotheses are stated in terms of the parameter $y^* = \beta_0 + \beta_1x^*$.

Hypothesis Test and Confidence Interval Concerning the Mean Value of Y for $x = x^*$

y_0^* can be any constant. We can conduct a test for evidence that the mean value of Y for $x = x^*$ is different from any value.

▶ H_0: $y^* = y_0^*$

H_a: $y^* > y_0^*$, $y^* < y_0^*$,   or $y^* \neq y_0^*$

TS: $T = \dfrac{(B_0 + B_1x^*) - y_0^*}{S\sqrt{(1/n) + [(x^* - \bar{x})^2/S_{xx}]}}$

RR: $T \geq t_{\alpha,n-2}$, $T \leq -t_{\alpha,n-2}$,   or $|T| \geq t_{\alpha/2,n-2}$ ◀

A $100(1 - \alpha)\%$ confidence interval for $\mu_{Y|x^*}$, the mean value of Y for $x = x^*$, has as endpoints the values

$$(\hat{\beta}_0 + \hat{\beta}_1x^*) \pm t_{\alpha/2,n-2}\,s\sqrt{\frac{1}{n} + \frac{(x^* - \bar{x})^2}{S_{xx}}} \tag{12.11}$$

Example 12.7 Health Care Expenditures

DATA SET

HEALTH

Research suggests that a country's total expenditures on health care are related to the gross domestic product (GDP). A random sample of countries was obtained and the GDP per capita, x, and health expenditures per capita, y (each in thousands of US dollars), was obtained for each. The data are given in the following table.

x	5.523	67.039	49.686	22.431	4.636	12.594	16.542	30.523	59.581	43.865	36.874	31.820
y	0.225	5.939	5.280	1.723	0.262	1.121	1.075	2.123	6.648	4.875	3.666	3.027

Solution Trail 12.7

KEYWORDS

■ Hypothesis test

■ 30 thousand dollars

■ Mean expenditures greater than

■ Random sample

TRANSLATION

■ Conduct a one-sided, right-tailed, hypothesis test concerning the mean when $x = 30$

CONCEPTS

■ Hypothesis test concerning the mean value of *Y* when $x = x^*$

VISION

The hypothesized value of the mean expenditures for a GDP of 30 is $y_0^* = 2.6$. Use this value and the summary statistics to conduct a one-sided, right-tailed hypothesis test concerning the mean value of *Y* for $x = x^* = 30$.

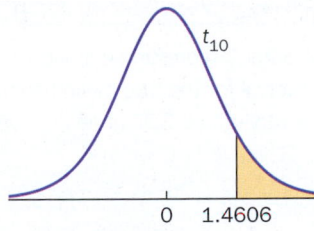

Figure 12.47 *p*-Value illustration:

$p = P(T \geq 1.4606)$

$= 0.0874 > 0.05 = \alpha$

The estimated regression line is $y = -0.4201 + 0.1076x$, and the summary ANOVA table is as follows.

ANOVA summary table for simple linear regression

Source of variation	Sum of squares	Degrees of freedom	Mean square	*F*	*p* Value
Regression	53.5304	1	53.5304	222.41	0.0000
Error	2.4068	10	0.2407		
Total	55.9372	11			

In addition, $s = \sqrt{MSE} = \sqrt{0.2407} = 0.4906$, $\bar{x} = 31.7595$, and $S_{xx} = 4624.22$.

a. ▶ For $x = 30$ thousand dollars per capita, conduct a hypothesis test to determine whether there is any evidence that the mean expenditures for health care per capita is greater than 2.6 thousand dollars. ◀

b. Construct a 95% confidence interval for the true mean health care expenditures in thousands of dollars for a country with a GDP of 48 thousand dollars per capita.

SOLUTION

a. ▶ The hypothesis test is one-sided, right-tailed with $y_0^* = 2.6$. Use a significance level of 0.05.

$H_0: y^* = 2.6$

$H_a: y^* > 2.6$

TS: $T = \dfrac{(B_0 + B_1 x^*) - y_0^*}{S\sqrt{(1/n) + [(x^* - \bar{x})^2/S_{xx}]}}$

RR: $T \geq t_{\alpha, n-2} = t_{0.05, 10} = 1.8125$

The value of the test statistic is

$$t = \frac{(\hat{\beta}_0 + \hat{\beta}_1 x^*) - y_0^*}{s\sqrt{(1/n) + [(x^* - \bar{x})^2/S_{xx}]}}$$

$$= \frac{[-0.4201 + 0.1076(30)] - 2.6}{0.4906\sqrt{(1/12) + [(30 - 31.7595)^2/4624.22]}} = 1.4606$$

The value of the test statistic does not lie in the rejection region. Equivalently, $p = 0.0874 > 0.05$ as shown in Figure 12.47. At the $\alpha = 0.05$ significance level, there is no evidence to suggest that the mean health care expenditures per capita is greater than 2.6 thousand dollars for a country with GDP 30 thousand dollars per capita. ◀

b. To construct the confidence interval, first find the appropriate critical value.

$1 - \alpha = 0.95 \implies \alpha = 0.05 \implies \alpha/2 = 0.025$ Find $\alpha/2$.

$t_{\alpha/2, n-2} = t_{0.025, 10} = 2.2281$ Use Table V in the Appendix to find the critical value.

Use Equation 12.11.

$$(\hat{\beta}_0 + \hat{\beta}_1 x^*) \pm t_{\alpha/2, n-2} \, s\sqrt{\frac{1}{n} + \frac{(x^* - \bar{x})^2}{S_{xx}}}$$

Equation 12.11.

$$= (\hat{\beta}_0 + \hat{\beta}_1 x^*) \pm t_{0.025, 10} \, s\sqrt{\frac{1}{n} + \frac{(x^* - \bar{x})^2}{S_{xx}}}$$

Use the values of α and n.

$$= [-0.4201 + 0.1076(48)] \pm (2.2281)(0.4906)\sqrt{\frac{1}{12} + \frac{(48 - 31.7595)^2}{4624.22}}$$

Use summary statistics, values for s and $t_{0.025, 10}$.

$$= 4.7447 \pm 0.4095$$

Simplify.

$$= (4.3352, 5.1542)$$

Compute endpoints.

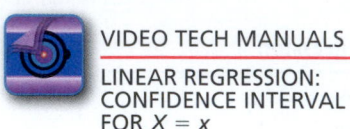

VIDEO TECH MANUALS

LINEAR REGRESSION:
CONFIDENCE INTERVAL
FOR $X = x$

(4.3352, 5.1542) is a 95% confidence interval for the true mean health care expenditures in a country with GDP 48 thousand dollars per capita.

Note: Figure 12.48 shows a scatter plot of the data, the estimated regression line, and the 95% **confidence bands** for the true mean health expenditures for each GDP. The confidence bands allow us to visualize 95% confidence intervals for the true mean value of Y for any value of $x = x^*$. The top confidence band connects the right, or upper, endpoint of each 95% confidence interval, and the bottom confidence band connects the left, or lower, endpoint of each 95% confidence interval. To estimate the 95% confidence interval for the true mean value of y for $x = 48$ from the graph, draw a vertical line at $x = 48$. The intersection of the line and the confidence bands yields the endpoints of the interval. As x^* moves away from $\bar{x}$, the width of the confidence interval increases.

Figure 12.49 shows a technology solution.

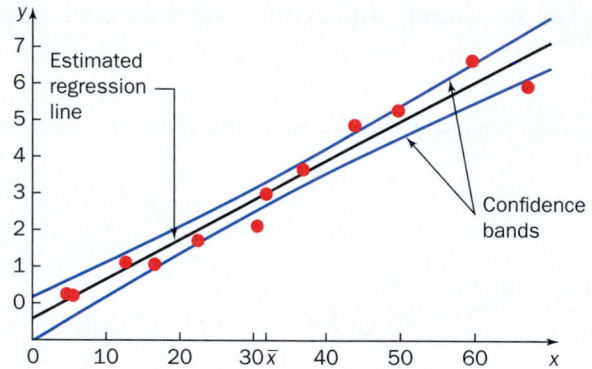

Figure 12.48 Scatter plot, estimated regression line, and confidence bands for the health care expenditure–GDP data.

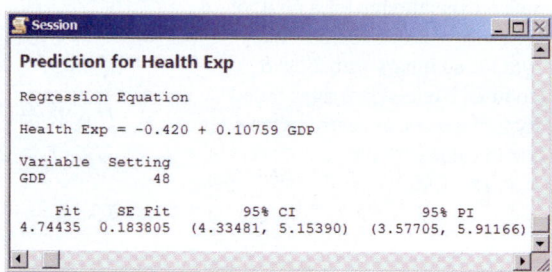

Figure 12.49 Minitab output: regression equation and a 95% confidence interval for the true mean health care expenditures in a country with a GDP of 48.

TRY IT NOW GO TO EXERCISE 12.75

An investigator may be interested in constructing an interval of possible values for an *observed* value of Y if $x = x^*$. For example, suppose temperature is used to predict the amount of expansion (in inches) of a certain type of vinyl siding. We may need an interval of possible values of expansion when the temperature is 90°F.

Note that we construct a *confidence interval* for the mean value of Y for $x = x^*$, and we construct a *prediction interval* for an observed value of Y when $x = x^*$.

An observed value of Y (for $x = x^*$) is a value of a random variable, not a fixed parameter. This helps to explain why the error of estimation for an observed value is larger than the error of estimation for a single mean value of Y. The interval of possible values is called a **prediction interval**.

The random variable $(B_0 + B_1 x^*) - (\beta_0 + \beta_1 x^* + E^*)$ is used to derive a prediction interval for Y. Using the properties of this random variable, a prediction interval can be found.

The only difference between this prediction interval and the confidence interval in Equation 12.10 is the extra 1 underneath the square-root symbol. This extra 1 is reasonable because the random variable here includes an extra term, E^*, with variance s^2.

Prediction Interval for an Observed Value of Y

A $100(1 - \alpha)\%$ prediction interval for an observed value of Y when $x = x^*$ has as endpoints the values

$$(\hat{\beta}_0 + \hat{\beta}_1 x^*) \pm t_{\alpha/2, n-2} \, s \sqrt{1 + \frac{1}{n} + \frac{(x^* - \bar{x})^2}{S_{xx}}} \qquad (12.12)$$

Example 12.8 Health Care Expenditures (Continued)

Use the health care expenditure–GDP data presented in Example 12.7. Suppose a country with a GDP of 50 thousands dollars per capita is selected at random. (This was the approximate U.S. GDP in 2012.) Find a 90% prediction interval for the health care expenditures per capita.

SOLUTION

STEP 1 Find the appropriate critical value.

$$1 - \alpha = 0.90 \implies \alpha = 0.10 \implies \alpha/2 = 0.05 \qquad \text{Find } \alpha/2.$$

$$t_{\alpha/2, n-2} = t_{0.05, 10} = 1.8125 \qquad \text{Use Table V in the Appendix to find the critical value.}$$

STEP 2 Use Equation 12.12.

$$(\hat{\beta}_0 + \hat{\beta}_1 x^*) \pm t_{\alpha/2, n-2} \, s \sqrt{1 + \frac{1}{n} + \frac{(x^* - \bar{x})^2}{S_{xx}}} \qquad \text{Equation 12.12.}$$

$$= (\hat{\beta}_0 + \hat{\beta}_1 x^*) \pm t_{0.05, 10} \, s \sqrt{1 + \frac{1}{n} + \frac{(x^* - \bar{x})^2}{S_{xx}}} \qquad \text{Use the values of } \alpha \text{ and } n.$$

$$= [-0.4201 + 0.1076(50)] \pm (1.8125)(0.4906) \sqrt{1 + \frac{1}{12} + \frac{(50 - 31.7595)^2}{4624.22}}$$

Use summary statistics, values for s and $t_{0.05, 10}$.

$$= 4.9599 \pm 0.9558 \qquad \text{Simplify.}$$

$$= (4.0041, 5.9157) \qquad \text{Compute endpoints.}$$

(4.0041, 5.9157) is a 90% prediction interval for an observed value of health care expenditures per capita in a country with a GDP of 50 thousand dollars per capita. Figure 12.50 shows a technology solution.

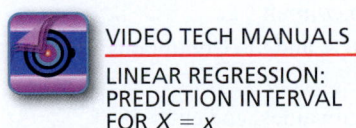
VIDEO TECH MANUALS

LINEAR REGRESSION: PREDICTION INTERVAL FOR X = x

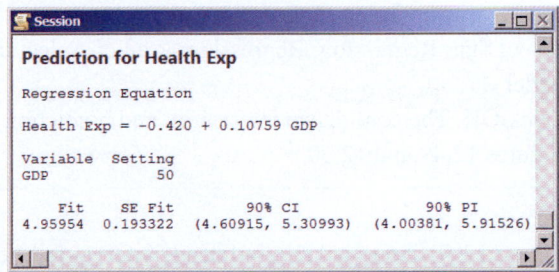

Figure 12.50 Minitab output: a 90% prediction interval (and a 90% confidence interval).

A CLOSER LOOK

1. A confidence interval for the mean value of Y and a prediction interval for an observed value of Y (for $x = x^*$) are centered at the same value. Compare the endpoints for these two intervals in Equations 12.11 and 12.12. The only difference is a 1 underneath the radical in the prediction interval.

2. As x^* moves farther away from $\bar{x}$, the width of the corresponding prediction interval increases.

Can you explain why the prediction interval is wider the further x^* is from $\bar{x}$? Look carefully at Equation 12.12.

TRY IT NOW   GO TO EXERCISE 12.78

Technology Corner

Procedure: Construct a confidence interval for the mean value of Y and a prediction interval for an observed value of Y when $x = x^*$.

Reconsider: Examples 12.7 and 12.8, solution, and interpretations.

CrunchIt!

Use Simple Linear Regression options to construct a prediction interval.

1. Enter the GDP data into column Var1 and the health care expenditure data into column Var2. Rename the columns if desired.
2. Choose Statistics; Regression; Simple Linear. Enter the Dependent Variable, the Independent Variable. In the Parameters section, select the desired Display, enter 50 in the Predict box, and the Prediction Interval Level. Click Calculate. The estimated regression line, coefficient statistics, and the prediction interval are displayed. See Figure 12.51. Note: CrunchIt! does not have an option for a confidence interval for Y when $x = x^*$.

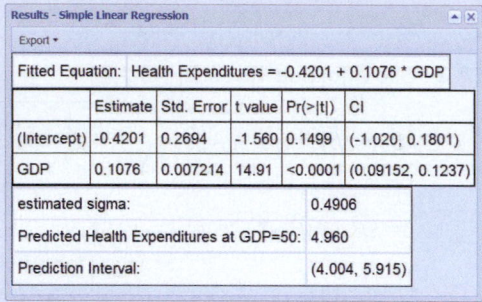

Figure 12.51 CrunchIt! output; the prediction interval is given in the last row.

Minitab

Use regression options to construct a confidence interval and a prediction interval.

1. Enter the values for GDP into column C1 and the values for health care expenditures into column C2.
2. Select Stat; Regression; Regression; Fit Regression Model. Enter the Response column, C2, and the Continuous predictor column, C1. Click OK.
3. Select Stat; Regression; Regression; Predict. Select the Response variable, enter individual values, and enter 48 under C1.
4. Click OK. The confidence interval(s), and prediction interval(s) are displayed in a session window. Refer to Figures 12.49 and 12.50.

Note

1. Figure 12.52 shows a 95% confidence interval for the mean value of Y and a 90% prediction interval for an observed value of Y when $x = 48$ using JMP.

GDP	Health Exp	Lower 95% Mean Health Exp	Upper 95% Mean Health Exp	Lower 90% Indiv Health Exp	Upper 90% Indiv Health Exp
48	.	4.3348	5.1539	3.7948	5.6939

Figure 12.52 JMP confidence interval and prediction interval.

2. The TI-84 Plus and Excel do not have a built-in function for constructing a confidence interval for Y nor a prediction interval for y when $x = x^*$.

SECTION 12.3 **EXERCISES**

Concept Check

12.65 Short Answer The value $y^* = \hat{\beta}_0 + \hat{\beta}_1 x^*$ is used to estimate _____ and _____.

12.66 True/False For $x = x^*$ and a fixed confidence level, a prediction interval for an observed value Y is wider than a confidence interval for the mean value of Y.

12.67 True/False For a fixed confidence level, the width of a confidence interval for the mean value of Y is the same for any value of x^*.

12.68 True/False For $x = x^*$, a confidence interval for the mean value of Y and a prediction interval for an observed value of Y are centered at the same value.

12.69 Short Answer When is the variance of the estimator for the mean value of Y the smallest?

Practice

12.70 Consider the following summary statistics for data obtained in a simple linear regression analysis.

$$\hat{\beta}_0 = -34.38 \qquad n = 18 \qquad MSE = 103.27$$
$$\hat{\beta}_1 = 3.38 \qquad \bar{x} = 15.367 \qquad S_{xx} = 138.14$$

a. Conduct a hypothesis test to determine whether there is any evidence that the true mean value of Y for $x = 16.2$ is greater than 20. Use a significance level of 0.05.
b. Conduct a hypothesis test to determine whether there is any evidence that the true mean value of Y for $x = 11.5$ is different from 5. Use a significance level of 0.01.

12.71 Consider the following summary statistics for data obtained in a simple linear regression analysis.

$$\hat{\beta}_0 = 38.86 \qquad n = 21 \qquad MSE = 21.23$$
$$\hat{\beta}_1 = -4.318 \qquad \bar{x} = 30.891 \qquad S_{xx} = 6.298$$

a. Find a 95% confidence interval for the true mean value of Y when $x = 31.5$. Find the width of the resulting interval.
b. Find a 95% confidence interval for the true mean value of Y when $x = 31.9$. Find the width of the resulting interval.
c. Explain why the width of the confidence interval in part (b) is greater than the width of the confidence interval in part (a).

12.72 Consider the following summary statistics for data obtained in a simple linear regression analysis.

$$\hat{\beta}_0 = 23.69 \qquad n = 7 \qquad MSE = 277.0$$
$$\hat{\beta}_1 = 1.452 \qquad \bar{x} = 19.1 \qquad S_{xx} = 15.131$$

a. Find a 99% prediction interval for an observed value of Y when $x = 19.25$. Find the width of the resulting interval.
b. Find a 99% prediction interval for an observed value of Y when $x = 18.10$. Find the width of the resulting interval.
c. Explain why the width of the prediction interval in part (b) is greater than the width of the prediction interval in part (a).

12.73 Consider the following data obtained in a simple linear regression analysis. 📊 EX12.73

x	5.6	5.1	7.6	3.9	6.5	6.7	5.1	2.5
y	5.7	2.8	10.5	1.6	7.7	8.8	3.4	6.1

a. Find the estimated regression line, and complete the summary ANOVA table. Conduct an F test for a significant regression.
b. Find s, an estimate of the standard deviation.
c. Conduct a hypothesis test to determine whether there is any evidence that the true mean value of Y for $x = 6$ is greater than 4. Use a significance level of 0.05.
d. Find a 99% confidence interval for the true mean value of Y when $x = 8.5$. Give an interpretation of this interval.

12.74 Consider the data given on the website, obtained in a simple linear regression analysis. 📊 EX12.74
a. Find the estimated regression line, and complete the summary ANOVA table. Conduct an F test for a significant regression. Use a significance level of 0.01.
b. Find s, an estimate of the standard deviation.
c. Find a 95% prediction interval for an observed value of Y when $x = 37$. Based on this interval, do you think it is likely that an observed value of Y (for $x = 37$) will be greater than 170? Justify your answer.

Applications

12.75 Psychology and Human Behavior The Linguistics Department at the University of Massachusetts at Amherst recently studied the relationship between external facial movements and the acoustics of speech sounds. Facial motions were captured using infrared markers and summarized using a unitless measure (x). A special unidirectional microphone was used for acoustic recording and to measure the sound level of speech $(y$, in dB). A random sample of individuals was selected, and each was asked to read a certain word. The facial expression and sound level were recorded for each person, and the summary statistics are given below.

$$\hat{\beta}_0 = 11.669 \qquad n = 16 \qquad MSE = 65.42$$
$$\hat{\beta}_1 = 28.009 \qquad \bar{x} = 0.4269 \qquad S_{xx} = 0.8615$$

a. Find an estimate of the true mean decibel level for a facial expression value of $x = 0.40$.
b. Find a 95% confidence interval for the true mean decibel level for a facial expression value of $x = 0.40$.
c. Suppose the facial-expression value is 0.55. Conduct a hypothesis test to determine whether there is any evidence that the true mean decibel level is greater than 30. Use a significance level of 0.01.

12.76 Physical Sciences A new solar collector is being tested for use in charging batteries that can provide electricity for an entire home. A random sample of days was selected and the amount of solar radiation was measured (x, in langleys) for each. The total battery charge was measured as a proportion (y, between 0 and 1). The summary statistics are given below.

$$\hat{\beta}_0 = 0.2007 \qquad n = 21 \qquad \text{MSE} = 0.06135$$
$$\hat{\beta}_1 = 0.00446 \qquad \bar{x} = 103.095 \qquad S_{xx} = 12335.8$$

a. What proportion of a charge can you expect the batteries to take if the amount of solar radiation is 100 langleys?
b. Find a 95% confidence interval for the true mean battery charge proportion if the amount of solar radiation is 130 langleys.
c. A value of 80 langleys indicates a typical cloudy day. On a typical cloudy day, is there any evidence to suggest that the true mean charge proportion is greater than 0.06 (the proportion needed to ensure a home will have sufficient energy until the next day)? Use a significance level of 0.01.

12.77 Fuel Consumption and Cars   An automobile mechanic believes the weight of a tire is related to the overall diameter of the tire. A random sample of automobile tires was obtained and the diameter (x, inches) and weight (y, pounds) were measured for each. The estimated regression line was $y = -37.8316 + 2.46657x$ with $n = 19$, SSR = 87.1673, SSE = 258.8706, $\bar{x} = 25.3474$, $\sum x_i^2 = 12221.6504$, and $S_{xx} = 14.3274$.

a. Complete the ANOVA summary table, and conduct an F test for a significant regression. Explain the relationship between tire weight and tire diameter.
b. Find an estimate of the observed tire weight for a tire diameter of 25.2 inches.
c. Find a 99% prediction interval for an observed tire weight if the tire diameter is 24.8 inches.
d. Conduct the hypothesis test $H_0: \beta_0 = 0$ versus $H_a: \beta_0 \neq 0$ with significance level 0.01. Is there any evidence to suggest that the true regression line does not pass through the origin? Does this result make practical sense? Why or why not?

12.78 Biology and Environmental Science   A study was conducted to investigate the relationship between asthma prevalence and annual rainfall in countries around the world. A random sample of 11 countries was obtained. The total annual rainfall (x, in mm) was recorded for each country, as well as the percentage of adults, 22–44 years old, treated for asthma (y). The following summary statistics were reported.

$$S_{xx} = 178,661 \qquad S_{yy} = 49.2655 \qquad S_{xy} = 380.955$$
$$\bar{x} = 792.636 \qquad \bar{y} = 5.26$$

a. Find the estimated regression line.
b. Complete the ANOVA summary table, and conduct an F test for a significant regression. Does annual rainfall help to explain the variation in asthma prevalence? Justify your answer.
c. Find a 95% prediction interval for an observed asthma prevalence if the total annual rainfall is 1000 mm. Comment on anything odd about this interval.

12.79 Public Health and Nutrition   The European Food Safety Authority recently issued a scientific opinion on the public health risks related to mechanically separated meat (MSM). The analysis suggested that calcium could be used to distinguish between MSM and non-MSM products.[23] A random sample of MSM poultry was obtained and the deboner head pressure (x, lb/in^2) and the amount of calcium (y, ppm) was measured for each. The data are given in the following table.   **POULTRY**

Pressure	Calcium	Pressure	Calcium
51	573	112	577
95	654	76	600
104	581	143	666
143	709	93	616
77	560	87	514
109	629	70	586
102	623	49	584
72	560	142	634
120	598	132	632

a. Find the estimated regression line.
b. Find an estimate for the true mean calcium for a pressure of 100 lb/in^2.
c. Suppose a deboner is set at a pressure of 120 lb/in^2. Is there any evidence to suggest that the true mean calcium is greater than 625 ppm? Find the p value associated with this test.
d. Suppose certain hand-deboned poultry has on average 600 ppm calcium. What should the pressure be for the mean calcium level for MSM poultry to be the same value?

12.80 Travel and Transportation   Highway engineers have long argued that roads designed with high skid resistance help to prevent accidents, especially in wet conditions. A random sample of two-lane highways was selected from across the United States, and the skid resistance was measured (in skid numbers) using a Skid Resistance Tester (SRT). The accident rate (per 10,000 vehicles) was computed for 25-mile sections of each highway during wet conditions.   **SKIDTEST**

a. Identify the independent variable and the dependent variable.
b. Construct a scatter plot for these data, and describe the relationship between the two variables.
c. Find the estimated regression line.
d. Suppose the skid resistance is 0.50. Conduct a hypothesis test to determine whether there is any evidence that the true mean accident rate is less than 0.60. Use a significance level of 0.05.

12.81 Biology and Environmental Science   A recent study suggests that the onset of Alzheimer's disease (AD) is related to exposure to microorganisms.[24] A greater exposure to microorganisms may actually protect against AD. A random sample of countries was obtained and a measure of contemporary parasite stress (x) and AD burden (y) was recorded for each. The data are given in the following table. ALZHEIM

Parasite stress	AD burden	Parasite stress	AD burden
3.81	5.14	4.99	3.48
3.84	4.59	4.91	3.59
4.45	4.27	3.79	5.06
4.86	4.46	4.36	4.76
3.95	4.28	4.07	4.31
4.19	4.40	3.89	4.33
4.05	4.49	4.59	4.80
4.46	4.43	3.54	4.77
4.42	4.28	4.51	4.96
4.27	4.99	4.18	5.44

a. Find the estimated regression line. Explain the relevance of the sign on the estimate of β_1 in the context of this problem.
b. Conduct an F test for a significant regression. Find the p value associated with this test.
c. Suppose the United States has a parasite stress of approximately 5.6. Find a 95% confidence interval for the true mean AD burden in the United States.

12.82 Psychology and Human Behavior Several studies suggest that the number of violent crimes is related to temperature. Suppose the FBI investigated this relationship between temperature and the number of violent crimes in several large cities in the United States. A random sample of days was obtained, and the *average* temperature (x, in °F) and the number of violent crimes per 100,000 people (y) were recorded for each day. CRIME
a. Construct a scatter plot for these data, and describe the relationship between the two variables.
b. Find the estimated regression line.
c. Find a 95% prediction interval for an observed number of violent crimes if the average temperature is 80°F.
d. Find a 95% prediction interval for an observed number of violent crimes if the average temperature is 60°F.

12.83 Demographics and Population Statistics A recent study suggests that the linguistic diversity in a community is related to the number of traffic accidents.[25] A random sample of U.S. cities was obtained and the annual road fatalities per 1000 people (y) and the linguistic diversity (x, Greenberg diversity index) was recorded for each. LINGDIV
a. Construct a scatter plot for these data, and describe the relationship between the two variables. Does your answer support the conclusions of the recent study? Why or why not?
b. Find the estimated regression line. Explain how the sign of $\hat{\beta}_1$ supports the conclusions of the recent study.
c. The Greenburg diversity index in the United States is approximately 0.333. Find a 95% interval for the true mean number of accidents per 1000 people in the United States.

Extended Applications

12.84 Technology and the Internet A recent study suggests that there is a relationship between Twitter use and TV ratings.[26] A random sample of prime-time television shows was obtained. During the month of September, the Twitter volume (x, in millions) of tweets on the day of each show and the Nielsen TV rating (y) was recorded for each show. The summary statistics are

$$\hat{\beta}_0 = -0.54 \qquad n = 25 \qquad MSE = 0.2042$$
$$\hat{\beta}_1 = 0.006795 \qquad \bar{x} = 407.6 \qquad S_{xx} = 21612.0$$

a. Find an estimate of the true mean TV rating if there are 400 million tweets on the day of the show.
b. Suppose the number of tweets during a day is 450 million. Is there any evidence that the true mean TV rating will exceed 2.3? Use a significance level of 0.05.
c. Construct a 99% confidence interval for the true mean TV rating when $x = 375$. On the basis of this interval, do you think the mean TV rating is less than 2.3? Justify your answer.

12.85 Public Policy and Political Science In a global study by Transparency International, it was found that a country's environmental performance is related to political corruption. A random sample of 15 countries was selected. The Transparency International Corruption Perceptions Index (CPI, x), a measure of perceived corruption among public officials and politicians, was computed for each country. In addition, the Environmental Sustainability Index (ESI, y), a measure of overall progress toward environmental sustainability, was computed for each country. The estimated regression line was $y = 26.432 + 4.546x$, with SSR = 853.50, SSE = 1234.23, $\bar{x} = 5.4333$, and $S_{xx} = 41.2933$.
a. Complete the ANOVA summary table, and conduct an F test for a significant regression. Explain the relationship between CPI and ESI.
b. Find an estimate of the observed ESI for a CPI score of 6.7.
c. Find a 95% prediction interval for an observed ESI if the CPI is 8.2. Based on this interval, do you think a randomly selected country with CPI 8.2 will have an ESI of 90 or greater? Justify your answer.

12.86 Physical Sciences The frequency of air exchange is believed to influence the indoor climate in unheated historic buildings. This is important because many unheated historic buildings hold collections of artifacts, and the indoor climate is related to preservation of these items. Skokloster Castle in Sweden is a masonry building without active climate control that houses many artifacts.[27] A random sample of days during the year was obtained. The air changes per hour (x, ACH) and the climate fluctuations transmittance related to relative humidity (y, CFT) was measured in similar rooms. CLIMATE
a. Construct a scatter plot for these data, and describe the relationship between the two variables.
b. Find the estimated regression line. Does the estimate of β_1 agree with your description in part (a)? Why or why not?

c. Conduct an F test for a significant regression. Find the p value associated with this test.

d. Find a 99% confidence interval for the true mean CFT for an ACH of 0.5.

e. Using your confidence interval in part (d), is there any evidence that the mean CFT is different from 0.75?

12.87 Medicine and Clinical Studies   Golden Rule medical insurance company recently investigated the relationship between the number of patients per registered nurse in a hospital and the patient's length of stay. A random sample of hospitals was selected, and the number of patients per registered nurse was computed (x). A patient was randomly selected for each hospital, and the length of stay was recorded (y, in hours). ▙▐▌ **PATIENT**

a. Construct a scatter plot for these data, and describe the relationship between the two variables.

b. Find the estimated regression line.

c. Find a 99% prediction interval for an observed length of stay if the number of patients per nurse is 3.7.

d. Find a 99% prediction interval for an observed length of stay if the number of patients per nurse is 3.3.

e. Why is the prediction interval in part (c) wider than the prediction interval in part (d)?

12.88 Physical Sciences   Doxylamine succinate (DS) is an over-the-counter antihistamine and is used in many nighttime sleep-aid products. Unfortunately, because this drug is so accessible, there are frequent cases of overdoses. To treat these patients properly, it is important to know the amount of DS ingested. However, many overdose patients are unconscious when they arrive at a hospital emergency room. A study was conducted to determine if the DS amount ingested could be reliably predicted from the plasma drug concentration upon arrival at the hospital. A random sample of DS overdose patients was obtained from various hospitals. The plasma drug concentration (x, in μg/mL) was measured for each upon arrival at the hospital. Following recovery, each patient was interviewed and the amount of DS ingested (y, in mg) was also recorded. ▙▐▌ **ANTIHIST**

a. Construct a scatter plot for these data, and describe the relationship between the plasma concentration and the amount ingested.

b. Find the estimated regression line.

c. Find a 95% confidence interval for the true mean amount of DS ingested for a plasma concentration of 6 μg/mL.

d. Suppose the lethal dose of DS for a typical teenager is 2000 mg and that an overdose patient has a plasma concentration of 5 μg/mL. Do you believe that this patient has taken a lethal dose of DS? Why or why not?

12.4 Regression Diagnostics

Recall that the assumptions in a simple linear regression model are stated in terms of the random deviations, E_i, $i = 1, 2, \ldots, n$. It is assumed that the E_i's are independent, normal random variables with mean 0 and (constant) variance σ^2. If any one of these assumptions is violated, then the results and subsequent inferences are in doubt.

If the true regression line were known, the set of actual random errors,

$$e_1 = y_1 - (\beta_0 + \beta_1 x_1)$$
$$e_2 = y_2 - (\beta_0 + \beta_1 x_2)$$
$$\vdots$$
$$e_n = y_n - (\beta_0 + \beta_1 x_n)$$

could be computed and used to check the assumptions. However, we usually do not know the values for β_0 and β_1, so the **residuals**, or deviations from the estimated regression line,

$$\hat{e}_1 = y_1 - (\hat{\beta}_0 + \hat{\beta}_1 x_1)$$
$$\hat{e}_2 = y_2 - (\hat{\beta}_0 + \hat{\beta}_1 x_2)$$
$$\vdots$$
$$\hat{e}_n = y_n - (\hat{\beta}_0 + \hat{\beta}_1 x_n)$$

are used to check for assumption violations. In practice, these estimates of the random errors are used in a variety of diagnostic checks. This section presents several preliminary graphical procedures used to reveal assumption violations.

Recall from Section 6.3 that a normal probability plot is a scatter plot of each observation versus its corresponding expected value from a Z distribution, or normal score. For observations from a normal distribution, the points will fall along a straight line. The data axis can be horizontal or vertical. If the scatter plot is nonlinear, there is evidence to suggest that the data did not come from a normal distribution.

A normal probability plot of the residuals may be used to check the normality assumption. A simple histogram or stem-and-leaf plot may also be used to reveal departures from normality.

A scatter plot of the residuals versus the independent variable values is also used to check the simple linear regression assumptions. The ordered pairs in this plot are $(x_i, \hat{e}_i)$. Figure 12.53 illustrates the definition of a residual. This is a scatter plot of y versus x with the estimated regression line. Figure 12.54 shows a scatter plot of the resulting residuals versus the independent variable.

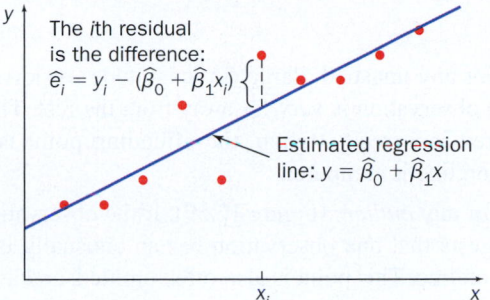

Figure 12.53 An illustration of the definition of a residual.

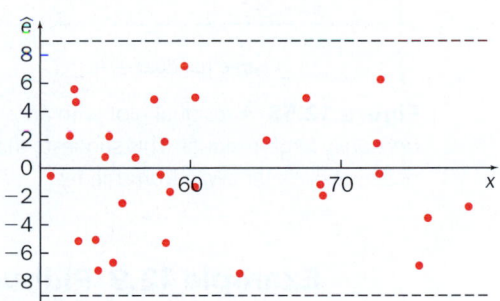

Figure 12.54 A scatter plot of the residuals versus the independent variable.

If there are no violations in assumptions, a scatter plot of the residuals versus the independent variable should look like a horizontal band around zero with randomly distributed points and no discernible pattern. For example, see Figure 12.55. There should be no obvious relation or pattern between the residuals and the predictor variable.

Figure 12.55 A desirable residual plot: no discernible pattern in a horizontal band.

Here is a list of *patterns* in a residual plot that indicate a possible violation in assumptions.

1. Check for a distinct *curve* in the plot, either mound- or bowl-shaped (parabolic). See Figure 12.56. A curved plot suggests that an additional (or different) predictor variable may be necessary. A *linear* model is not appropriate.

2. Check for a *nonconstant* spread. See Figure 12.57. If there is not a uniform horizontal band, or if the spread of the residuals varies outside this band, this suggests that the variance is not constant.

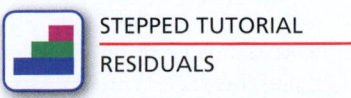

STEPPED TUTORIAL

RESIDUALS

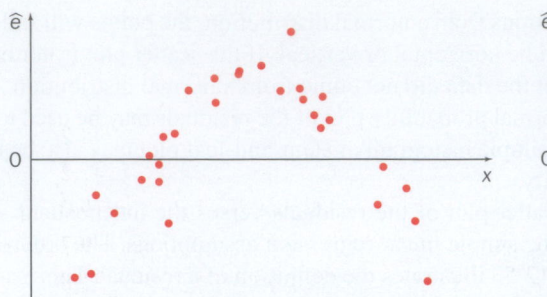

Figure 12.56 A *curved* residual plot. This suggests that a linear model is not appropriate.

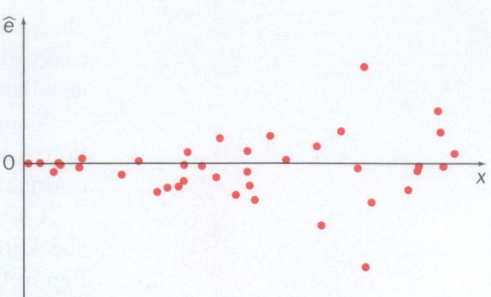

Figure 12.57 A residual plot with *nonconstant* spread. This suggests that the variance is not the same for each value of *x*.

3. Check for any unusually large (in magnitude) residual (Figure 12.58). This suggests that one observation is very far away from the rest. The data may have been recorded or entered incorrectly. Often, the offending point is omitted and a new estimated regression line is computed.

4. Check for any *outliers* (Figure 12.59). If the observation is correct, an outlying residual suggests that one observation has an unusually large influence on the estimated regression line. This point is also often omitted, and a new line is computed.

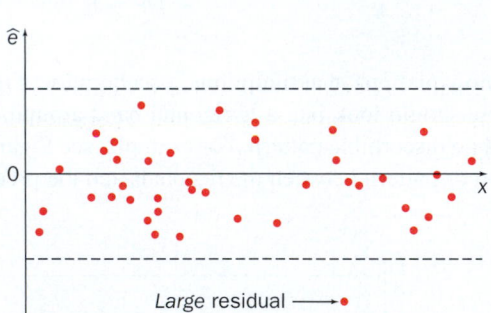

Large residual ⟶ •

Figure 12.58 A residual plot with an *unusually large* residual. This suggests that an observation is far away from the test.

Possible influential observation

Figure 12.59 A residual plot with an *outlying* residual. This observation is very influential when the estimated regression line is computed.

Example 12.9 Pillbugs and Red Clover

DATA SET

PILLBUG

Ryan Pike/Feature Pics

Isopods, or pillbugs, use moss, ferns, and even poison ivy for shelter. However, some scientists believe that the preferred natural shelter is red clover, and that the density of pillbugs can be predicted from the density of red clover. A random sample of fields was obtained. The number of red clover plants per square meter (x) and the number of pillbugs per square meter (y) were measured for each field. The data are given below, and the estimated regression line is $y = -12.360 + 14.213x$. Compute the residuals, construct a normal probability plot of the residuals, and carefully sketch a graph of the residuals versus the predictor variable values.

SOLUTION

STEP 1 Use the estimated regression line to find the predicted value, $\hat{y}_i$, for each x value: $\hat{y}_i = -12.360 + 14.213x_i$. Then, compute each residual: $\hat{e}_i = y_i - \hat{y}_i$. The normal scores for $n = 20$ are given in Table IV in the Appendix.

Percho/Dreamstime

Reminder: The residuals should be placed in order from smallest to largest when the normal scores are assigned.

x_i	y_i	$\hat{y}_i$	$\hat{e}_i$	Normal score
5.9	108	71.5	36.5	0.92
4.4	12	50.2	−38.2	−0.92
3.6	86	38.8	47.2	1.13
9.1	136	117.0	19.0	0.45
8.8	103	112.7	−9.7	−0.19
6.3	99	77.2	21.8	0.59
11.7	136	153.9	−17.9	−0.31
10.9	94	142.6	−48.6	−1.13
6.3	9	77.2	−68.2	−1.87
5.7	145	68.7	76.3	1.87
5.0	4	58.7	−54.7	−1.40
8.4	124	107.0	17.0	0.31
4.5	68	51.6	16.4	0.19
13.3	177	176.7	0.3	0.06
4.9	91	57.3	33.7	0.74
14.2	259	189.5	69.5	1.40
10.4	101	135.5	−34.5	−0.74
10.2	101	132.6	−31.6	−0.45
5.0	56	58.7	−2.7	−0.06
3.6	7	38.8	−31.8	−0.59

STEP 2 The normal probability plot of the residuals is shown in Figure 12.60. The points lie along an approximate straight line. There is no evidence to suggest that the normality assumption is violated.

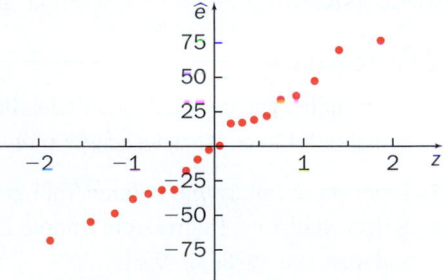

Figure 12.60 The normal probability plot of the residuals.

STEP 3 Figure 12.61 shows the residual plot ($\hat{e}_i$ versus x_i). There is no discernible pattern in this graph. One point in the upper right, corresponding to $x = 14.2$, may be an outlier, but the graphical evidence is not convincing enough. These two graphs suggest that there is no violation in the simple linear regression assumptions.

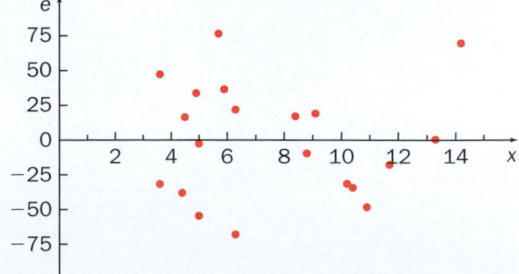

Figure 12.61 The residual plot of residuals versus predictor variable values.

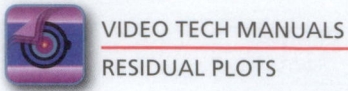

VIDEO TECH MANUALS

RESIDUAL PLOTS

Figures 12.62 and 12.63 together show a technology solution.

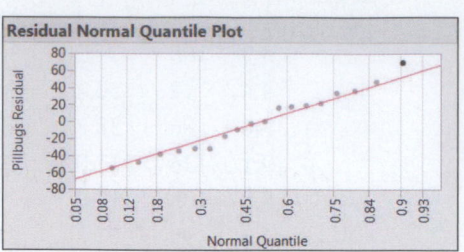

Figure 12.62 JMP normal probability plot of the residuals.

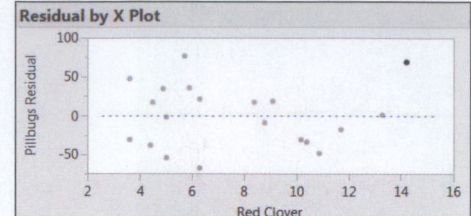

Figure 12.63 JMP scatter plot of the residuals versus the predictor variable values.

TRY IT NOW   GO TO EXERCISE 12.106

A CLOSER LOOK

1. The sum of the residuals is always 0, subject to round-off error. We'll check a few examples empirically.

2. Other diagnostic plots include a plot of the residuals versus the predicted values $\hat{y}_i$. Any distinct pattern suggests a violation in the simple linear regression assumptions.

3. Other types of residuals are also used to check for violations in assumptions. **Standardized residuals** utilize the standard deviation of each residual and are useful for identifying residuals with large magnitudes. **Studentized residuals** are also standardized but by using a model without the current observation.

Technology Corner

Procedure: Construct a normal probability plot of the residuals and a scatter plot of the residuals versus the predictor variable values.

Reconsider: Example 12.9, solution, and interpretations.

CrunchIt!

In the Simple Linear Regression dialog box, use the Residuals QQ Plot option to produce a normal probability plot and the Residuals Plot to construct a scatter plot of the residuals versus the predictor variable.

1. Enter the x values into column Var1 and the observed y values into column Var2.
2. Select Statistics; Regression; Simple Linear. Using the pull-down menus, select the Dependent Variable, Var2, and the Independent Variable, Var1.
3. In the Parameters section, select Residuals QQ Plot from the pull-down menu for Display and click Calculate. See Figure 12.64.
4. To construct a scatter plot of the residuals versus the predictor variable values, select Residuals Plot from the pull-down menu for Display and click Calculate. See Figure 12.65.

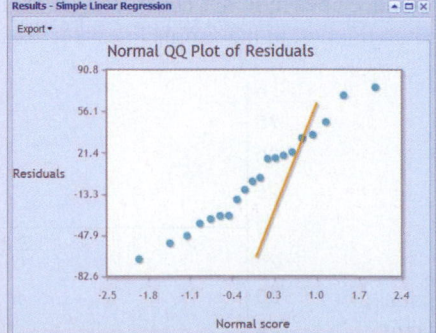

Figure 12.64 Normal probability plot of the residuals.

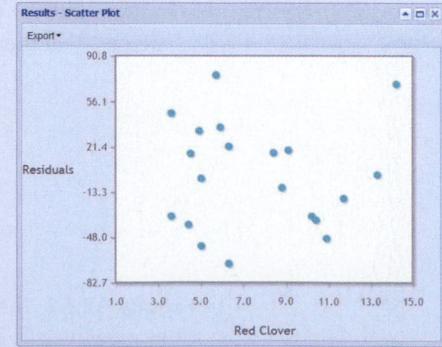

Figure 12.65 Scatter plot of the residuals versus the predictor variable values.

TI-84 Plus C

A normal probability plot and a scatter plot are built-in statistical plots.

1. Enter the x values into list L1 and the observed y values into list L2.
2. Select STAT ; CALC; LinReg(a+bx). Enter the name of the list containing the values of the independent variable, L1, and the name of the list containing the observed values of the dependent variable, L2. When this command is executed, the residuals are automatically stored in the named list RESID.
3. Construct a normal probability plot for the residuals (in the list RESID as described in Section 6.3. See Figure 12.66.
4. Construct a scatter plot of the residuals versus the predictor variable values as described in Section 12.1. See Figure 12.67.

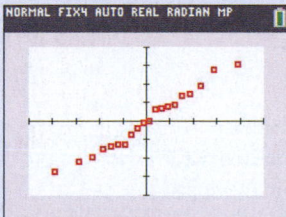   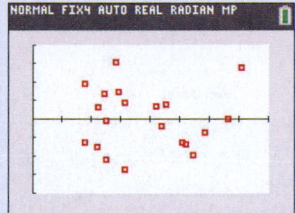

Figure 12.66 Normal probability plot of the residuals.

Figure 12.67 Scatter plot of the residuals versus the predictor variable values.

Minitab

Use the Graphs option to construct a scatter plot of the residuals and a scatter plot of the residuals versus the predictor variable values.

1. Enter the x values into column C1 and the observed y values into column C2.
2. Select Stat; Regression; Regression; Fit Regression Model. Enter the Response and Continuous predictor columns.
3. In the Graphs options, use Regular residuals. Check Normal probability plot of residuals and enter C1 under Residuals versus the variables.
4. Click OK to conduct the regression analysis and display the graphs. See Figures 12.68 and 12.69.

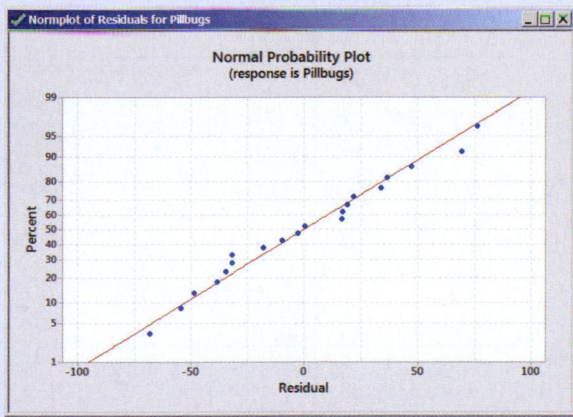   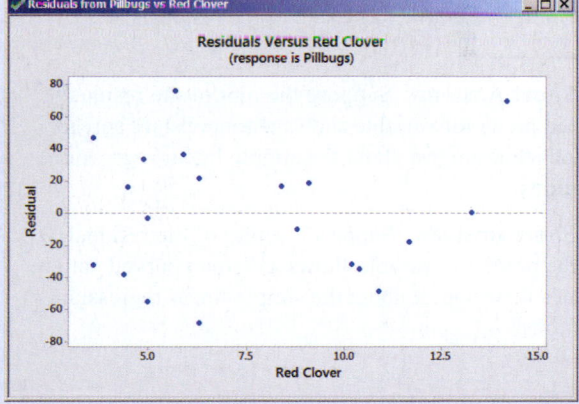

Figure 12.68 Minitab normal probability plot of the residuals.

Figure 12.69 Scatter plot of the residuals versus the predictor variable values.

Excel

Compute the normal scores for the residuals as described in Section 6.3. Construct scatter plots as described in Section 12.1.

1. Enter the x values into column A and the observed y values into column B.
2. Under the Data tab, select Data Analysis; Regression. Enter the Y Range, the X Range, and an Output Range. Check Residuals to save the residuals and Residual Plots to construct a plot of the residuals versus the predictor variable values.

3. Click OK. The scatter plot of the residuals versus predictor variable values is shown in Figure 12.70.
4. Compute the normal scores and construct the normal probability plot as described in Section 6.3. See Figure 12.71.

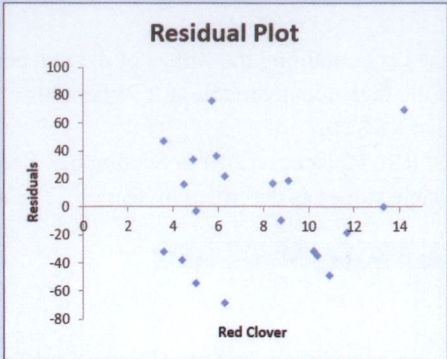

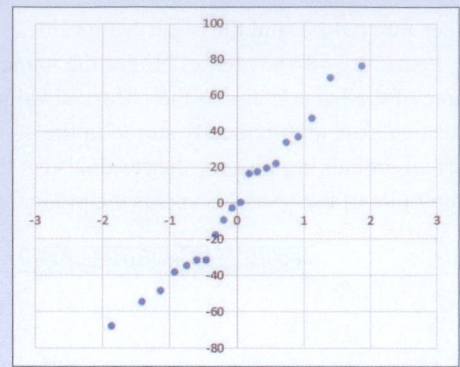

Figure 12.70 Scatter plot of the residuals versus the predictor variable values.

Figure 12.71 Normal probability plot of the residuals.

SECTION 12.4 EXERCISES

Concept Check

12.89 Short Answer Write a mathematical expression for the ith residual, $\hat{e}_i$.

12.90 Short Answer What are the simple linear regression assumptions in terms of the random deviations, E_i, $i = 1, 2, \ldots, n$.

12.91 Short Answer If the simple linear regression assumptions are satisfied, describe a normal probability plot of the residuals.

12.92 Fill in the Blank The sum of the residuals is

_____ .

12.93 Short Answer Suppose the plot of the residuals versus the predictor variable shows a nonconstant spread. What does this suggest about the simple linear regression assumptions?

12.94 Short Answer Suppose the plot of the residuals versus the predictor variable shows a distinct curved pattern. What does this suggest about the simple linear regression assumptions?

Practice

12.95 Consider the following data used in a simple linear regression analysis. EX12.95

x	11.9	11.3	10.4	12.4	18.5	12.5	15.7	10.2
y	45.2	33.1	34.4	25.1	−4.4	37.6	27.5	52.6

The estimated regression line is $y = 97.63 - 5.15x$.
   a. Find the residuals.
   b. Find the sum of the residuals.

12.96 For the data on the text website, the estimated regression line is $y = 4.7245 + 3.3717x$. EX12.96
   a. Find the residuals.
   b. Find the sum of the residuals.

12.97 Examine each normal probability plot of residuals. Is there any evidence to suggest that the random error terms are not normal?

a.

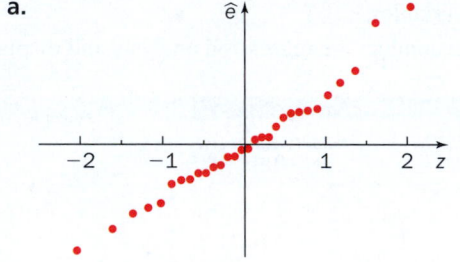

b.

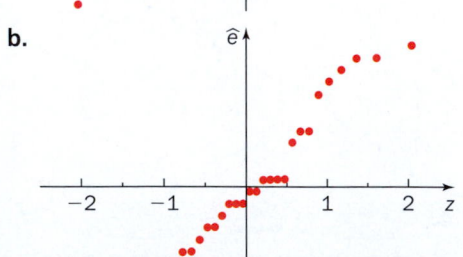

c.

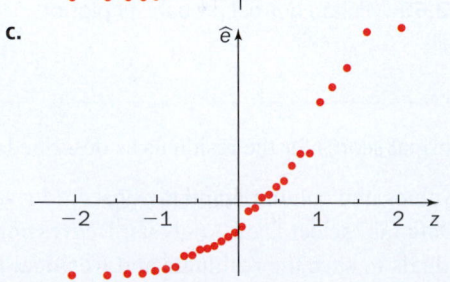

d.
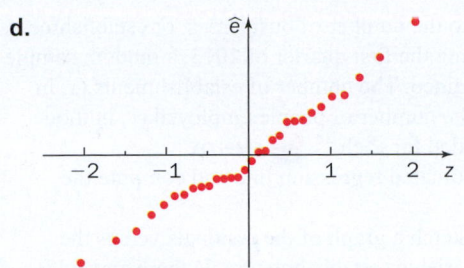

12.98 The following residuals were obtained in a simple linear regression analysis. [ili] **EX12.98**

−22.2	−67.9	−1.1	−6.3	−12.4	−33.0	−24.2
35.0	−13.2	33.7	42.3	14.2	13.6	−10.7
−1.6	23.7	38.0	0.4	−20.3	12.1	

a. Construct a normal probability plot for these residuals.
b. Is there any evidence to suggest that the random errors are not normal? Justify your answer.

12.99 The residuals on the text website were obtained in a simple linear regression analysis. [ili] **EX12.99**
a. Construct a normal probability plot for these residuals.
b. Is there any evidence to suggest that the random errors are not normal? Justify your answer.

12.100 Examine each residual plot (residuals versus predictor variable). Is there any evidence to suggest a violation in the simple linear regression model assumptions? If the graph suggests a violation, indicate which assumption is in doubt.

a.
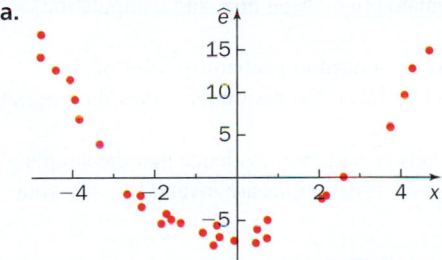

b.
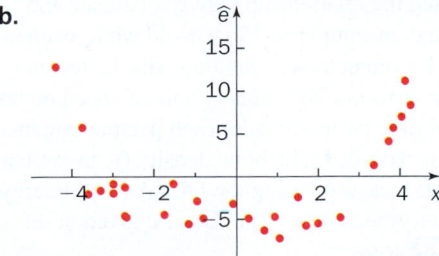

c.
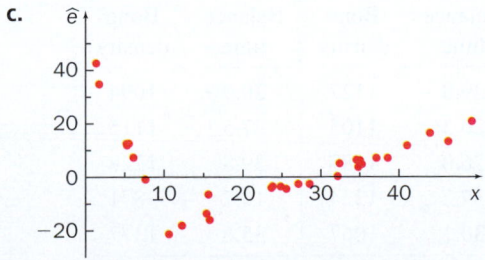

12.101 The data on the text website were used in a simple linear regression analysis. The estimated regression line is $y = 91.74 − 3.7384x$. [ili] **EX12.101**
a. Find the residuals.
b. Carefully sketch a graph of the residuals versus the predictor variable, x. Is there any evidence of a violation in the regression assumptions? Justify your answer.

12.102 The data on the text website were used in a simple linear regression analysis. [ili] **EX12.102**
a. Find the estimated regression line.
b. Compute the residuals.
c. Carefully sketch a graph of the residuals versus the predictor variable, x. Is there any evidence of a violation in the regression assumptions? Justify your answer.

Applications

12.103 Physical Sciences The Atlantic Elevator Company conducted a study to examine the relationship between the total weight of passengers on an elevator (x, in pounds) and the total energy (y, in kW) required to lift the passengers one floor in an office building. A simple linear regression analysis was conducted, and the following residuals were obtained. [ili] **ELEVAT**

−1.2	−1.1	1.5	−8.5	−13.1	−3.3	−17.4
7.5	−6.0	8.2	15.6	6.9	−4.3	−19.2
18.9	−9.4	16.3	0.8	−3.0	10.9	

a. Carefully sketch a normal probability plot for the residuals.
b. Is there any evidence that the random error terms are not normal? Justify your answer.

12.104 Manufacturing and Product Development Suppose the research department for Century Tire Company is developing a snow tire based on a new formulation. A study was conducted to examine the relationship between the amount of the new additive per liter (x, in mL) and the amount of wear after 50,000 indoor test miles (y, in mm). A simple linear regression analysis was conducted, and the following residuals were obtained. [ili] **SNOTIRE**

0.7	−0.2	0.5	−1.1	−0.8	−0.2	0.2
−0.1	0.7	0.3	−0.4	0.5	0.3	0.5
0.8	0.2	0.6	0.1	−0.8	−0.3	0.4
−0.7	−0.4	−0.7	−0.1			

d.
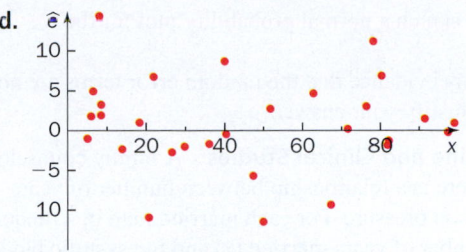

a. Carefully sketch a normal probability plot for the residuals.

b. Is there any evidence that the random error terms are not normal? Justify your answer.

12.105 Medicine and Clinical Studies   A family counselor believes that there is a relationship between number of years married and blood pressure. For each married man in a random sample, the number of years married (x) and the systolic blood pressure (y, in mmHg) were recorded. The data are given in the following table. ▮▮ MARRIED

Years married	Blood pressure	Years married	Blood pressure
4.9	133	3.4	94
10.3	157	3.1	114
11.9	129	12.0	144
13.0	141	12.9	147
13.0	166	7.1	124

The estimated regression line is $y = 97.95 + 4.034x$.

a. Compute the residuals, and verify that they sum to 0.

b. Carefully sketch a normal probability plot for the residuals. Is there any evidence that the random error terms are not normal? Justify your answer.

12.106 Manufacturing and Product Development   An automotive component manufacturer believes that the speed of the milling machine is related to the surface roughness in aluminum components. A random sample of aluminum components was obtained. Each piece had a 50×50 mm cross section and was 100 mm in length. The speed of the milling machine (x, in rpm) and the surface roughness (y, in μm) was measured for each piece. ▮▮ MILLING

a. Find the estimated regression line and compute the residuals.

b. Carefully sketch a normal probability plot of the residuals and a plot of the residuals versus the predictor variable.

c. Do these graphs provide any evidence that the simple linear regression assumptions are invalid? Justify your answer.

12.107 Biology and Environmental Science   Agriculture researchers recently investigated the relationship between wheat yield and wind speed. A random sample of wheat-field plots was obtained. The mean daily wind speed over the growing period (x, in m/s) was recorded for each field, along with the yield (y, in bushels per acre). The estimated regression line is $y = 50.052 - 1.103x$. ▮▮ WHEAT

a. Compute the residuals, and verify that they sum to 0.

b. Carefully sketch a graph of the residuals versus the predictor variable (wind speed). Is there any evidence that the simple linear regression assumptions are violated? Justify your answer.

12.108 Business and Management   Many economists believe that the number of people employed in an area is strongly related to the number of businesses, or establishments, in that area. During the first quarter of 2013, a random sample of states was obtained. The number of establishments (x, in thousands) and the number of people employed (y, in thousands) was recorded for each.[28] ▮▮ EMPLOY

a. Find the estimated regression line and compute the residuals.

b. Carefully sketch a graph of the residuals versus the predictor variable (establishments). Is there any evidence that the simple linear regression assumptions are violated? Justify your answer.

12.109 Physical Sciences   The solar wind, or the flow of charged particles from the Sun, affects communications, navigation, and the Earth's static pressure (the force exerted by the atmosphere on the surface of the Earth). A random sample of days was selected, and the solar wind density was measured (x, in protons/cm^3) using a special NASA satellite. For each day selected, the static pressure (y, in MPa) was also recorded. The data are given in the following table. ▮▮ SOLAR

Solar wind density	Static pressure	Solar wind density	Static pressure
5.6	0.099	9.9	0.117
7.4	0.080	9.4	0.110
8.0	0.115	11.0	0.138
8.7	0.110	6.7	0.104
10.1	0.110	6.9	0.094

a. Find the estimated regression line, and compute the residuals.

b. Carefully sketch a normal probability plot of the residuals and a plot of the residuals versus the predictor variable.

c. Do these graphs provide any evidence that the simple linear regression assumptions are invalid? Justify your answer.

12.110 Medicine and Clinical Studies   A health researcher recently investigated the relationship between balance and bone density. A random sample of 25-year-old white women was obtained. Each woman took a standing balance test in which she held her arms out horizontally and balanced on one foot. The length of time (x, in seconds) each person remained in that position was recorded. The bone density (y, in mg/cm^2) for each person was measured using the DEXA (dual-energy X-ray absorptiometry) technique. The data are given in the following table. ▮▮ BONE

Balance time	Bone density	Balance time	Bone density
29.8	1127	20.9	1094
26.3	1105	27.5	1115
26.0	1187	39.8	1228
37.2	1334	10.7	871
30.1	1067	35.6	1377

a. Find the estimated regression line, and compute the residuals.
b. Carefully sketch a normal probability plot of the residuals and a plot of the residuals versus the predictor variable.
c. Do these graphs provide any evidence that the simple linear regression assumptions are invalid? Justify your answer.

12.111 Medicine and Clinical Studies   A coal miner's job is extremely hazardous and can cause serious health problems, for example, black lung disease and vibration-induced white finger. A study was conducted to examine the hearing loss of coal miners exposed to years of vibration and cool temperatures underground. A random sample of coal miners was selected, and the length of time on the job (x, in years) was recorded for each person. Each miner took a test in which the hearing threshold level (HTL, y, in dB) was measured.   MINER
a. Find the estimated regression line, and compute the residuals.
b. Carefully sketch a normal probability plot of the residuals and a plot of the residuals versus the predictor variable.
c. Do these graphs provide any evidence that the simple linear regression assumptions are invalid? Justify your answer. How do you think this regression model could be improved?

Extended Applications

12.112 Sports and Leisure   There have been many studies to determine the most important characteristic for golfers to produce long drives. A recent investigation considered grip strength. A random sample of golfers was obtained and a hand-grip dynamometer was used to measure grip strength (x, in Newtons) on each person. Each golfer then drove a ball, and the length of the drive was measured (y, in yards).   GOLFER
a. Find the estimated regression line, and compute the residuals.
b. Conduct an F test for a significant regression. Use a significance level of 0.01.
c. Carefully sketch a normal probability plot of the residuals and a plot of the residuals versus the predictor variable.
d. Do these graphs provide any evidence that the simple linear regression assumptions are invalid? Justify your answer. How do you think this regression model could be improved?

12.113 Business and Management   Suppose the Human Resources director at NVR, a large construction company, believes that the number of sick hours per year taken by an employee is related to the commuting distance. A random sample of employees was obtained, and each person's travel distance to work (x, in miles) was recorded. Personnel records were used to obtain the number of hours off for sickness in a year (y) for each employee included in the study.   SICKHRS

a. Find the estimated regression line, and compute the residuals.
b. Conduct an F test for a significant regression. Use a significance level of 0.10. Do you believe that there is a relationship between commuting distance and sick hours? Why or why not?
c. Carefully sketch a normal probability plot of the residuals and a plot of the residuals versus the predictor variable.
d. Do these graphs provide any evidence that the simple linear regression assumptions are invalid? Justify your answer.

12.114 Physical Sciences   Researchers at a marine institute investigated the relationship between the dissolved barium concentration and the silicate/nitrate ratio in the North Pacific Ocean. The silicate/nitrate ratio (x) and the dissolved barium concentration (y, in nmol/kg) were measured at randomly selected locations.   NPOCEAN
a. Find the estimated regression line, and compute the residuals.
b. Conduct an F test for a significant regression. Use a significance level of 0.01.
c. Carefully sketch a normal probability plot of the residuals and a plot of the residuals versus the predictor variable.
d. Do these graphs provide any evidence that the simple linear regression assumptions are invalid? Justify your answer.

12.115 Biology and Environmental Science   Research suggests that the electrical conductivity (EC) of soil is related to plant growth. Nutrients in the soil are in the form of ions. These ions in the soil become dissolved in water and carry an electrical charge, and this determines the soil EC. Too little EC indicates a lack of nutrients, and a high EC level suggests a salinity problem. A random sample of corn fields in Iowa was obtained and the soil EC was measured in each (x, in μS/cm). The resulting crop yield was also recorded (y, in bushels per acre). The data are given in the following table.   SOIL

EC	Yield	EC	Yield	EC	Yield
1087	193	711	165	534	90
399	104	672	159	661	163
709	165	931	171	789	152
916	134	636	122	482	155
941	151	921	165	841	157
877	191	707	131	533	131
687	177	1045	140	737	176

a. Find the estimated regression line.
b. Conduct an F test for a significant regression ($\alpha = 0.05$). Explain your results in the context of this problem. Find bounds on the p value associated with this test. Use technology to verify your answer.
c. Carefully sketch a normal probability plot of the residuals and a plot of the residuals versus the predictor variable. Do

these graphs provide any evidence that the simple linear regression assumptions are invalid? Justify your answer.

12.116 Sports and Leisure   In a study conducted in Scandinavia, certain physiological parameters of endurance horses were found to be related to performance.[29] A random sample of horses and races was obtained. The total fluid intake pre-race (x, in liters) and the speed of the horse during the race (y, in km/h) was measured for each.   **HORSES**

a. Find the estimated regression line, and compute the residuals.
b. Conduct an F test for a significant regression. Explain your results in the context of this problem.

c. Carefully sketch a plot of the residuals versus the predictor variable and a normal probability plot of the residuals. Do these graphs provide any evidence that the simple linear regression assumptions are invalid? Justify your answer.

Challenge

12.117 Sum of the Residuals   In a simple linear regression analysis, show that the sum of the residuals is always 0. *Hint*: In the definition of the ith residual, use the definition of $\hat{y}_i$, and then the formula for $\hat{\beta}_0$.

12.5 Multiple Linear Regression

Although simple linear regression analysis has practical uses, many real-world applications involve a model with a dependent variable Y and at least two independent variables, $x_1, x_2, \ldots, x_k$. For example, the number of days it takes to complete construction of a new home (Y) might be predicted by the square footage of the home (x_1) and the total linear feet of wiring in the home (x_2). Or the temperature of an automobile engine (Y) might be modeled (predicted) by the amount of coolant (x_1), the amount of engine oil (x_2), and the rate of air flow (x_3). The purpose of this section is to extend the simple linear regression model to cases involving k (≥ 2) predictor variables. The formal model and assumptions are given below. The procedure is similar to that used for the simple linear regression model.

Multiple Linear Regression Model

Let $(x_{11}, x_{21}, \ldots, x_{k1}, y_1), (x_{12}, x_{22}, \ldots, x_{k2}, y_2), \ldots, (x_{1n}, x_{2n}, \ldots, x_{kn}, y_n)$ be n sets of observations such that y_i is an observed value of the random variable Y_i. We assume that there exist constants $\beta_0, \beta_1, \ldots, \beta_k$ such that

$$Y_i = \beta_0 + \beta_1 x_{1i} + \beta_2 x_{2i} + \cdots + \beta_k x_{ki} + E_i \tag{12.13}$$

where $E_1, E_2, \ldots, E_n$ are independent, normal random variables with mean 0 and variance σ^2. That is,

1. The E_i's are normally distributed (which means that the Y_i's are normally distributed).
2. The expected value of E_i is 0 [which implies that $E(Y_i) = \beta_0 + \beta_1 x_{1i} + \beta_2 x_{2i} + \cdots + \beta_k x_{ki}$].
3. $Var(E_i) = \sigma^2$ [which implies that $Var(Y_i) = \sigma^2$].
4. The E_i's are independent (which implies that the Y_i's are independent).

A CLOSER LOOK

1. The double subscript notation on x is necessary to indicate both the variable and the observation. For example, x_{21} is the value of the variable x_2 that corresponds to the observed value of y_1. Similarly, x_{12} is the value of the variable x_1 that corresponds to the observed value of y_2.

2. In the multiple linear regression model, the E_i's again represent the **random deviations** or **random error terms**.

3. The **true regression equation** is $y = \beta_0 + \beta_1 x_1 + \beta_2 x_2 + \cdots + \beta_k x_k$. Notice that this equation is a *linear* function of the unknown parameters $\beta_0, \beta_1, \ldots, \beta_k$. The graph of the true regression equation is, in general, a *surface*, not a line. However, we often refer to this equation as the true regression *line*.

4. The unknown constants $\beta_0, \beta_1, \ldots, \beta_k$ are called partial regression coefficients. Recall that in Section 12.1, β_1 represented the mean amount of change in y for every one-unit increase in x_1. In a multiple linear regression model, β_i represents the mean change in y for every increase of one unit in x_i if the values of all the other predictor variables are kept fixed.

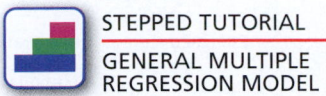

STEPPED TUTORIAL

GENERAL MULTIPLE REGRESSION MODEL

The focus of this section is on the method for finding the best deterministic linear model, that is, finding estimates of the unknown parameters $\beta_0, \beta_1, \ldots, \beta_k$. Hypothesis tests will be used to determine whether the overall model explains a significant amount of the variability in the dependent variable and to evaluate the contribution of each independent variable. The principle of least squares can again be used to minimize the sum of the squared deviations between the observations and the estimated values (the sum of squares due to error).

Although hand calculations are possible, it is much more efficient to use technology to compute the estimates $\hat{\beta}_0, \hat{\beta}_1, \ldots, \hat{\beta}_k$ for the true regression parameters, or coefficients, $\beta_0, \beta_1, \ldots, \beta_k$. For specific values of the independent variables, $(x_1^*, x_2^*, \ldots, x_k^*) = x^*$, let $y^* = \hat{\beta}_0 + \hat{\beta}_1 x_1^* + \hat{\beta}_2 x_2^* + \cdots + \hat{\beta}_k x_k^*$. The value y^* is an estimate of the mean value of Y for $x = x^*$, denoted $\mu_{Y|x^*}$, and also an estimate of an observed value of Y for $x = x^*$.

DATA SET

BASEBALL

Example 12.10 MLB Ticket Prices

The cost of a Major League Baseball (MLB) ticket has increased rapidly over the last decade. Add in the price of a few soft drinks, hot dogs, a program, and parking, and the total cost can exceed $200 for a family of four. A sports writer believes that the cost of a typical ticket is affected by the team's opening-day payroll and the cost of a hot dog in the stadium. To test this theory, data were obtained for each of the 30 MLB teams during the 2013 season.[30] Some of the data are given in the following table, with payroll in millions of dollars.

Team	Ticket	Payroll	Hot Dog
Boston Red Sox	53.38	173.19	4.50
New York Yankees	51.55	197.96	3.00
Los Angeles Dodgers	22.37	95.14	5.00
St. Louis Cardinals	33.11	110.30	4.25
Chicago Cubs	44.55	88.20	4.50
⋮	⋮	⋮	⋮

a. Find the estimated regression line.

b. Estimate the true mean ticket price for a payroll of 100 million dollars and a 5-dollar hot dog.

SOLUTION

a. Ticket price (y) is the dependent, or response, variable. Payroll (x_1) and hot dog price (x_2) are the independent, or predictor, variables. Use technology to find the estimated regression coefficients. The results from JMP are shown in Figure 12.72.

Solution Trail 12.10

KEYWORDS

- Estimated regression line
- True mean ticket price

TRANSLATION

- Multiple linear regression
- Find the expected ticket price for a 100-million-dollar payroll and a 5-dollar hot dog

CONCEPTS

- Least-square estimates
- Estimate the mean value of Y for $x = x^*$

VISION

Use technology to find the estimates of the regression coefficients. Compute $y^* = \hat{\beta}_0 + \hat{\beta}_1(100) + \hat{\beta}_2(5)$ as an estimate of the mean value of Y for $x = (100, 5)$.

Parameter Estimates				
Term	Estimate	Std Error	t Ratio	Prob>|t|
Intercept	8.2214071	5.563581	1.48	0.1511
Payroll	0.1811181	0.034502	5.25	<.0001*
Hot Dog	0.3584748	1.065904	0.34	0.7392

Figure 12.72 Estimated regression coefficients.

The sign of $\hat{\beta}_i$ represents the way in which the surface is tilted along the x_i axis.

The estimated regression equation is $y = 8.2214 + 0.1811x_1 + 0.3585x_2$. That is, $\hat{\beta}_0 = 8.2214$, $\hat{\beta}_1 = 0.1811$, and $\hat{\beta}_2 = 0.3585$. Note that the estimated regression coefficients on both payroll and the price of a hot dog are positive. This suggests that as payroll and the price of a hot dog increase, ticket prices also increase. The signs on both estimated regression coefficients seem reasonable.

b. The estimated true mean ticket price is found by substituting $x_1 = x_1^* = 100$ and $x_2 = x_2^* = 5$ into the estimated regression equation.

$y = 8.2214 + 0.1811x_i + 0.3585x_2$ Estimated regression equation.

$= 8.2214 + 0.1811(100) + 0.3585(5)$ Substitute.

$= 28.12$ Simplify.

An estimate of the expected ticket price for a 100-million-dollar payroll and a 5-dollar hot dog is approximately $28.12.

In this example, the graph of $y = 8.2214 + 0.1811x_1 + 0.3585x_2$ is a plane in three dimensions. Figure 12.73 is a three-dimensional scatter plot of the data and the graph of the estimated regression equation. Note that the remaining points are behind the plane and are not visible in this particular view. Figures 12.74, 12.75, and 12.76 illustrate how y depends on each predictor variable separately. In addition to the graph of the estimated regression equation, Figure 12.74 includes the graphs of two lines (in three dimensions): one for $x_1 = 130$ held constant, and one for $x_2 = 3.25$ held constant. Figures 12.75 and 12.76 each represent a *slice* through the three-dimensional plane at a particular value of the other predictor variable. In Figure 12.75, $\hat{\beta}_1 = 0.1811$ is the slope of the line when $x_2 = 3.25$, and in Figure 12.76, $\hat{\beta}_2 = 0.3585$ is the slope of the line when $x_2 = 130$.

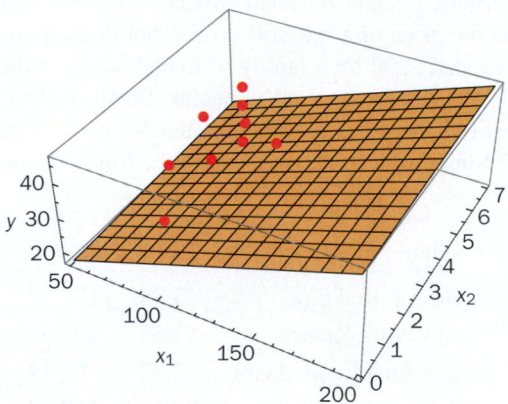

Figure 12.73 A scatter plot of the data and the graph of the estimated regression equation.

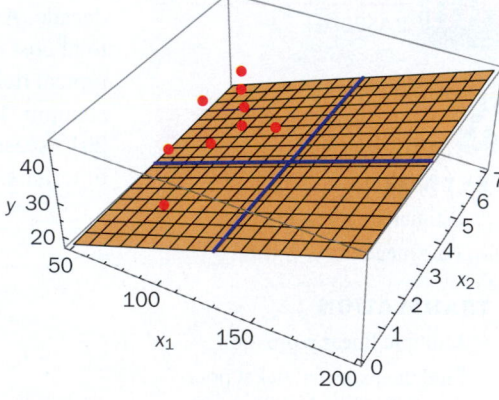

Figure 12.74 A scatter plot of the data, the graph of the estimated regression equation, and two lines.

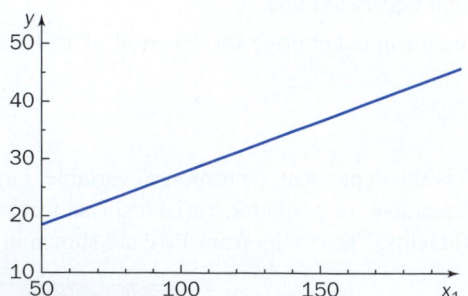

Figure 12.75 A graph of y versus x_1 when $x_2 = 3.25$.

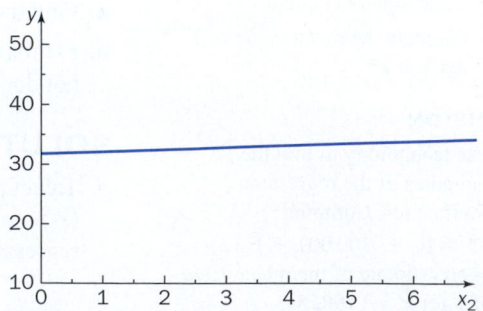

Figure 12.76 A graph of y versus x_2 when $x_1 = 130$.

TRY IT NOW   GO TO EXERCISE 12.142

The following concepts from simple linear regression also apply to multiple linear regression.

1. The variance σ^2 is a measure of the underlying variability in the model. An estimate of σ^2 is used to conduct hypothesis tests and to construct confidence intervals related to multiple linear regression.

2. The ith predicted value is $\hat{y}_i = \hat{\beta}_0 + \hat{\beta}_1 x_{1i} + \cdots + \hat{\beta}_k x_{ki}$.

3. The ith residual is $y_i - \hat{y}_i$.

4. The total sum of squares, the sum of squares due to regression, and the sum of squares due to error have the same definitions. The sum of squares identity, Equation 12.4, is also true. The total variation in the data (SST) is decomposed into a sum of the variation explained by the model (SSR) and the variation about the regression equation (SSE).

An analysis of variance table is used to summarize the computations in multiple linear regression. The differences are in the degrees of freedom associated with the sum of squares due to regression and the sum of squares due to error. As usual, the mean squares are corresponding sums of squares divided by the corresponding degrees of freedom.

ANOVA summary table for multiple linear regression

Source of variation	Sum of squares	Degrees of freedom	Mean square	F	p Value
Regression	SSR	k	$\text{MSR} = \dfrac{\text{SSR}}{k}$	$\dfrac{\text{MSR}}{\text{MSE}}$	p
Error	SSE	$n - k - 1$	$\text{MSE} = \dfrac{\text{SSE}}{n - k - 1}$		
Total	SST	$n - 1$			

The coefficient of determination, $r^2 = \text{SSR}/\text{SST}$, is a measure of the proportion of variation in the data that is explained by the regression model. For multiple linear regression, a test of significance is equivalent to testing the hypothesis that all regression coefficients (except the constant term) are zero. That is, the null hypothesis is $H_0: \beta_1 = \beta_2 = \cdots = \beta_k = 0$, and none of the predictor variables help to explain any variation in the dependent variable. Suppose $k = 2$ and H_0 is true. Graphically, this means the values of Y vary around the plane (in three dimensions) $y = \beta_0$ with no discernible pattern.

A test for a significant multiple linear regression model is based on the ratio of the mean square due to regression and the mean square due to error. Here is a summary of an F test for a significant regression with significance level α.

This is the model utility test to determine whether the multiple linear regression model is appropriate/useful (similar to the simple linear regression case).

Hypothesis Test for a Significant Multiple Linear Regression

$H_0: \beta_1 = \beta_2 = \cdots = \beta_k = 0$
(None of the predictor variables helps to explain the variation in y.)

$H_a: \beta_i \neq 0$ for at least one i
(At least one predictor variable helps to explain the variation in y.)

TS: $F = \dfrac{\text{MSR}}{\text{MSE}}$

RR: $F \geq F_{\alpha, k, n-k-1}$

Solution Trail 12.11

KEYWORDS

- The effects of . . .
- F test
- r^2

TRANSLATION

- Three predictor variables
- f test for a significant regression
- Coefficient of determination

CONCEPTS

- Multiple linear regression
- Summary ANOVA table
- Hypothesis test for a significant regression
- Computation of r^2

VISION

Use the sums of squares, number of predictor variables, and number of observations to complete the ANOVA summary table for multiple linear regression. Use the template for a hypothesis test for a significant regression with $\alpha = 0.05$. The sum of squares due to regression and the total sum of squares are used to compute r^2.

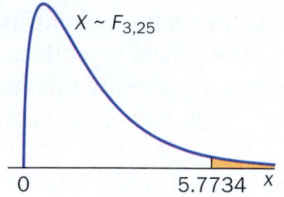

$X \sim F_{3,25}$

0   5.7734 X

p-Value illustration:
$p = P(X \geq 5.7734)$
$\quad = 0.0038 \leq 0.05 = \alpha$

Example 12.11 Price at the Pump

A study was conducted to investigate the effects of the cost of crude oil (x_1, in dollars per barrel), the state tax on gasoline (x_2, in dollars per gallon), and the number of publicly owned automobiles in a state (x_3, in thousands) on the price of gasoline (in dollars per gallon). Data from 29 randomly selected days were used to produce the following multiple linear regression equation:

$$y = 2.59 + 0.008x_1 + 0.191x_2 + 0.0146x_3.$$

In addition, SSR = 0.3517 and SSE = 0.5085.

a. Complete the summary ANOVA table and conduct an F test for a significant regression. Use a significance level of 0.05.

b. Compute r^2 and interpret this value.

SOLUTION

a. Use the sum of squares identity to compute the total sum of squares.

SST = SSR + SSE = 0.3517 + 0.5085 = 0.8602   Use Equation 12.4.

There are $n = 29$ observations and $k = 3$ predictor variables. Use these values to compute the mean squares.

MSR = SSR/k = 0.3517/3 = 0.1172

MSE = SSE/($n - k - 1$) = SSE/(29 − 3 − 1) = 0.5085/25 = 0.0203

The F test for a significant regression with $\alpha = 0.05$ is

$H_0: \beta_1 = \beta_2 = \beta_3 = 0$

$H_a: \beta_i \neq 0$ for at least one i

TS: $F = \dfrac{\text{MSR}}{\text{MSE}}$

RR: $F \geq F_{\alpha,k,n-k-1} = F_{0.05,3,25} = 2.99$

The value of the test statistic is

$f = \text{MSR/MSE} = 0.1172/0.0203 = 5.7734\ (\geq 2.99)$

Because f lies in the rejection region, equivalently, $p = 0.0038 \leq 0.05$, there is evidence to suggest that at least one of the regression coefficients is different from zero. At least one of the predictor variables can be used to explain a significant amount of variation in the price per gallon of gasoline. The next reasonable step is to determine which of the three predictors is really contributing to this overall significant regression.

Recall that using tables of critical values for F distributions, we can only bound the p value. Place the value of the test statistic in an ordered list of critical values with $v_1 = 3$ and $v_2 = 25$.

$4.68 \quad \leq 5.7734 \leq 7.45$

$F_{0.01,3,25} \leq 5.7734 \leq F_{0.001,3,25}$

Therefore, $0.001 \quad \leq \quad p \quad \leq 0.01$

Here are all the calculations presented in the ANOVA table, with some help from technology to find the exact p value.

ANOVA summary table for multiple linear regression

Source of variation	Sum of squares	Degrees of freedom	Mean square	F	p Value
Regression	0.3517	3	0.1172	5.7734	0.0038
Error	0.5085	25	0.0203		
Total	0.8602	28			

b. The coefficient of determination is

$$r^2 = \text{SSR}/\text{SST} = 0.3517/0.8602 = 0.4089$$

Approximately 0.4089, or 41%, of the variation in the price per gallon of gasoline is explained by this regression model.

TRY IT NOW   GO TO EXERCISE 12.143

If the overall multiple linear regression F test is significant, then there is evidence to suggest that at least one of the independent variables can be used to predict the value of Y. Hypothesis tests (or confidence intervals) can be used to determine whether x_i helps to predict the value of $Y|x$, or, equivalently, whether β_i is different from 0.

Consider the random variable B_i, an estimator for β_i, and $S^2 = \text{MSE} = \text{SSE}/(n - k - 1)$, an estimator for the underlying variance σ^2. If the multiple linear regression assumptions are true, then S^2 is an unbiased estimator for σ^2 and is used in constructing a hypothesis test concerning β_i.

<table>
<tr><td>

β_{i0} can be any constant. We can conduct a test for evidence that the regression coefficient β_i is different from any value. However, usually $\beta_{i0} = 0$ and $H_a: \beta_i \neq 0$. This means we are looking for any evidence to suggest that the value of x_i helps to predict the value of $Y|x$.

</td></tr>
</table>

Hypothesis Test and Confidence Interval Concerning β_i

$H_0: \beta_i = \beta_{i0}$

$H_a: \beta_i > \beta_{i0}, \qquad \beta_i < \beta_{i0}, \qquad$ or $\qquad \beta_i \neq \beta_{i0}$

TS: $T = \dfrac{B_i - \beta_{i0}}{S_{B_i}}$

RR: $T \geq t_{\alpha, n-k-1}, \qquad T \leq -t_{\alpha, n-k-1}, \qquad$ or $\qquad |T| \geq t_{\alpha/2, n-k-1}$

A $100(1 - \alpha)\%$ confidence interval for β_i has as endpoints the values

$$\widehat{\beta}_i \pm t_{\alpha/2, n-k-1} \, s_{B_i} \tag{12.14}$$

Example 12.12 Silk Versus Milk

DATA SET

SOYMILK

In a recent study of soy milk, a random sample of 1-cup servings from various brands was obtained and the number of calories was measured in each (y). In addition, the total fat (x_1, in grams) and sugar (x_2, in grams) were also measured.[31] Multiple linear regression was used to investigate the effects of total fat and sugar on the number of calories. The resulting Minitab output is shown in Figure 12.77.

a. Verify that the multiple linear regression is significant at the $\alpha = 0.01$ level.

b. Conduct separate hypothesis tests to determine whether each predictor variable contributes to the overall significant regression. Use $\alpha = 0.05$ in each test.

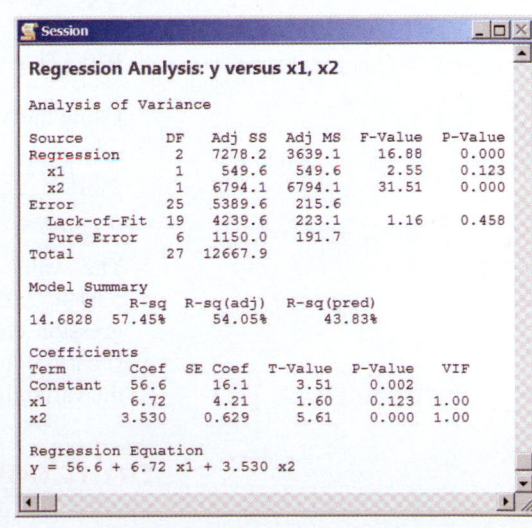

Figure 12.77 Minitab regression analysis.

Notice that Minitab rounds the value of the test statistic, f, to two decimal places.

Notice again that Minitab rounds the value of the test statistic, t, to two decimal places.

SOLUTION

a. There are $k = 2$ predictor variables and, from the ANOVA table, there are $n = 28$ observations ($n - 1 = 27$).

The F test for a significant regression with $\alpha = 0.05$ is

H_0: $\beta_1 = \beta_2 = 0$.

H_a: $\beta_i \neq 0$ for at least one i.

TS: $F = $ MSR/MSE

RR: $F \geq F_{\alpha,k,n-k-1} = F_{0.05,2,25} = 3.39$

As given in the ANOVA table, the value of the test statistic is

$f = $ MSR/MSE $= 3639.1/215.6 = 16.88 \ (\geq 3.39)$

Because f lies in the rejection region (or, equivalently, $p < 0.001$), there is evidence to suggest that at least one of the regression coefficients is different from 0.

b. To test whether x_1, total fat, is a significant predictor, conduct a hypothesis test concerning the regression coefficient β_1.

H_0: $\beta_1 = 0$

H_a: $\beta_1 \neq 0$

TS: $T = \dfrac{B_1 - 0}{S_{B_i}}$

RR: $|T| \geq t_{\alpha/2,n-k-1} = t_{0.025,25} = 2.0595$

Using the Minitab output, the value of the test statistic is

$$t = \frac{\hat{\beta}_1 - 0}{s_{B_1}} = \frac{6.72}{4.21} = 1.60$$

The value of the test statistic does not lie in the rejection region, $|t| = |1.60| = 1.60 < 2.0595$ (equivalently, $p = 0.123 > 0.05$). There is no evidence to suggest that the predictor total fat is contributing to the overall regression effect nor to the variability in y.

Conduct a similar hypothesis test concerning the regression coefficient β_2.

H_0: $\beta_2 = 0$

H_a: $\beta_2 \neq 0$

TS: $T = \dfrac{B_2 - 0}{S_{B_2}}$

RR: $|T| \geq t_{\alpha/2,n-k-1} = t_{0.025,25} = 2.0595$

Using the Minitab output, the value of the test statistic is

$$t = \frac{\hat{\beta}_2 - 0}{s_{B_2}} = \frac{3.530}{0.629} = 5.61$$

The value of the test statistic lies in the rejection region, $|t| = |5.61| = 5.61 \geq 2.0595$ (equivalently, $p < 0.001$). There is evidence to suggest that the regression coefficient β_2 is different from 0. This implies that the predictor variable sugar is contributing to the overall regression effect and is contributing significantly to the variability in y.   ●

TRY IT NOW   GO TO EXERCISE 12.145

A CLOSER LOOK

Recall that if a confidence interval for β_i contains 0, there is no evidence to suggest that the *i*th predictor variable is significant. However, if 0 is not included in the confidence interval, there is evidence to suggest that the *i*th regression coefficient is different from 0.

1. The Minitab output in Example 12.12 also includes the results of a hypothesis test concerning the constant term, β_0, with $H_0: \beta_0 = 0$. If we fail to reject this null hypothesis, a model without a constant term might be more appropriate.

2. If a model utility test is significant in a multiple linear regression model, then there are $k \geq 2$ hypothesis tests to consider in order to isolate those variables contributing to the overall effect. Recall from Chapter 11 that we cannot set the significance level in each *individual* hypothesis test. The probability α of making a type I error (a mistake) is set in each test under the assumption that *only one* test is conducted per experiment. Therefore, the more tests that are conducted, the greater is the chance of making an error. To control the probability of making at least one mistake, we can use the Bonferroni technique (presented in Chapter 11) applied to simultaneous hypothesis tests or confidence intervals.

    a. If k hypothesis tests are conducted, then the significance level in each case is α/k.

    b. The k simultaneous $100(1 - \alpha)\%$ confidence intervals have as endpoints the values $\hat{\beta}_i \pm t_{\alpha/(2k),n-k-1} \, s_{B_i}$.

Consider additional reading about a partial *F* test, stepwise, and best subsets regression.

3. Many different statistical procedures can be used to select the *best* regression model. For now, the most reasonable method is simply to keep only those variables in the model that have regression coefficients significantly different from zero. Eliminate the others, and calculate a new, *reduced* model for prediction.

Two very useful inferences in multiple linear regression involve an estimate of the *mean* value of Y for $x = x^*$ and an estimate of an *observed* value of Y for $x = x^*$. Recall from the simple linear regression case that the error in estimating the mean value of Y is less than the error in estimating an observed value of Y. The difference is due to estimating a single value versus estimating the next value from an entire distribution.

If the multiple linear regression assumptions are true, then a hypothesis test and confidence interval concerning the mean value of Y for $x = x^*$ is based on the t distribution. The by-hand calculation of the standard deviation is taxing. An appropriate symbol is used below, and we will rely on technology to provide the necessary calculations. The null and alternative hypotheses are stated in terms of the parameter $y^* = \beta_0 + \beta_1 x_1^* + \beta_2 x_2^* + \cdots + \beta_k x_k^*$. The random variable Y^* is used as an estimate of y^*.

Recall that we can conduct a test of $y^* = y_0^*$ using either a formal hypothesis test or a confidence interval. Most statistical software packages produce a confidence interval rather than conduct a hypothesis test.

Hypothesis Test and Confidence Interval Concerning the Mean Value of Y for $x = x^*$

$H_0: y^* = y_0^*$

$H_a: y^* > y_0^*, \qquad y^* < y_0^*, \qquad \text{or} \qquad y^* \neq y_0^*$

$$\text{TS: } T = \frac{(B_0 + B_1 x_1^* + \cdots + B_k x_k^*) - y_0^*}{S_{Y^*}}$$

$\text{RR: } T \geq t_{\alpha,n-k-1}, \qquad \text{or} \qquad T \leq -t_{\alpha,n-k-1}, \qquad |T| \geq t_{\alpha/2,n-k-1}$

A $100(1 - \alpha)\%$ confidence interval for $\mu_{Y|x^*}$, the mean value of Y for $x = x^*$, has as endpoints the values

$$(\hat{\beta}_0 + \hat{\beta}_1 x_1^* + \cdots + \hat{\beta}_k x_k^*) \pm t_{\alpha/2,n-k-1} s_{Y^*} \qquad (12.15)$$

Prediction Interval for an Observed Value of Y

A $100(1 - \alpha)\%$ prediction interval for an observed value of Y when $x = x^*$ has as endpoints the values

Recall that $s = \sqrt{\text{MSE}}$.

$$(\hat{\beta}_0 + \hat{\beta}_1 x_1^* + \cdots + \hat{\beta}_k x_k^*) \pm t_{\alpha/2,n-k-1}\sqrt{s^2 + s_{Y^*}^2} \qquad (12.16)$$

Example 12.13 Shear Stress

The total weight of an automobile contributes to fuel consumption and, therefore, to CO_2 emissions. Recent automobile design research has focused on lighter materials to reduce total weight. However, the strength and durability of any new product must also be considered. Data were obtained from several new chassis designs and used to fit a multiple linear regression equation of the form $y = \beta_0 + \beta_1 x_1 + \beta_2 x_2 + \beta_3 x_3$, where

y = shear stress, in MPa, STRESS

x_1 = thickness of web, in mm, WEB

x_2 = thickness of upper flange, in mm, UPPER

x_3 = thickness of lower flange, in mm, LOWER

a. Construct a 95% confidence interval for the mean chassis shear stress when $x_1 = 3$, $x_2 = 5$, and $x_3 = 7$. Use this confidence interval to determine whether there is any evidence that the mean chassis shear stress for these values is different from 125 MPa.

b. Construct a 95% prediction interval for an observed value of chassis shear stress when $x_1 = 3$, $x_2 = 5$, and $x_3 = 7$.

SOLUTION

a. The Fit Model option in JMP is used to produce a summary of fit, the analysis of variance table, and the parameter estimates. The Save Columns option is used with specific values for x_1, x_2, and x_3 to produce a confidence interval and a prediction interval. The relevant output is shown in Figures 12.78–12.80.

Analysis of Variance

Source	DF	Sum of Squares	Mean Square	F Ratio
Model	3	3918.0967	1306.03	9.6254
Error	26	3527.8485	135.69	**Prob > F**
C. Total	29	7445.9451		0.0002*

Parameter Estimates

Term	Estimate	Std Error	t Ratio	Prob>|t|
Intercept	207.35391	17.46776	11.87	<.0001*
Web	-7.796625	1.805966	-4.32	0.0002*
Upper	-5.665126	1.885751	-3.00	0.0058*
Lower	-5.040539	2.266411	-2.22	0.0350*

Figure 12.78 Analysis of variance table and parameter estimates.

Summary of Fit

RSquare	0.526205
RSquare Adj	0.471537
Root Mean Square Error	11.64845
Mean of Response	117.0513
Observations (or Sum Wgts)	30

Figure 12.79 Summary of fit.

Lower 95% Mean Stress	Upper 95% Mean Stress	Lower 95% Indiv Stress	Upper 95% Indiv Stress
108.64	132.07	93.70	147.01

Figure 12.80 Confidence interval and prediction interval.

The estimated regression equation is

$$y = 207.35 - 7.797 x_1 - 5.665 x_2 - 5.041 x_3$$

The model utility test (using the ANOVA table) is significant at the $p = 0.0002$ level, and each predictor variable is also significant. These results together with $r^2 = 0.5262$ suggest that the model is effective in predicting the chassis shear stress.

Using the JMP output, a 95% confidence interval for the true mean chassis shear stress when $x_1 = 3$, $x_2 = 5$, and $x_3 = 7$ is (108.64, 132.07). Because 125 is included in this interval, there is no evidence to suggest that the mean shear stress is different from 125.

b. Using the JMP output, a 95% prediction interval for a single observation of the chassis shear stress when $x_1 = 3$, $x_2 = 5$, and $x_3 = 7$ is (93.70, 147.01). Notice that this 95% prediction interval is larger (wider) than the corresponding confidence interval.

TRY IT NOW   GO TO EXERCISE 12.148

The assumptions in a multiple linear regression model are also given in terms of the random deviations, E_i, $i = 1, 2, \ldots, n$. As in a simple linear regression model, it is assumed that the E_i's are independent, normal random variables with mean 0 and constant variance σ^2. If any of these assumptions are violated, then the resulting analysis is unreliable. The residuals, $\hat{e}_i = y_i - (\hat{\beta}_0 + \hat{\beta}_1 x_{1i} + \cdots + \hat{\beta}_k x_{ki})$, $i = 1, 2, \ldots, n$, are estimates of the random errors and are used to check the regression assumptions.

A residual, $\hat{e}_i$, is the difference between the observed value of Y_i and the predicted value for Y_i.

The following graphical procedures, developed for a simple linear regression model, can be extended to a multiple linear regression model.

1. Construct a histogram, stem-and-leaf plot, scatter plot, and/or normal probability plot of the residuals. These graphs are all used to check the normality assumption.

2. Construct a scatter plot of the residuals versus *each* independent variable. For example, the first scatter plot has the ordered pairs $(x_{1i}, \hat{e}_i)$, the second has the ordered pairs $(x_{2i}, \hat{e}_i)$, etc. If there are no violations in assumptions, each scatter plot should appear as a horizontal band around 0. There should be no recognizable pattern. Typical patterns in a residual plot that suggest a violation in assumptions include a distinct curve, nonconstant spread, an unusually large residual, or an outlier.

Example 12.14 Hurricane Damage

DATA SET

DAMAGE

Recent research suggests that the building damage due to a hurricane is related to several vulnerability indicators.[32] Suppose several buildings damaged due to hurricanes in 2012 and 2013 were selected at random. The following data were recorded for each building: the damage ratio, the quotient of the insurance payout and the building appraised value times 100, y; the maximum sustained wind speed, in m/s, x_1; the building age, in years, x_2; and the distance from the shoreline, in 1000 m, x_3. The estimated regression equation is shown in Figure 12.81. Construct a normal probability plot of the residuals and sketch a graph of the residuals versus each predictor variable. Discuss any indication of violations in regression assumptions.

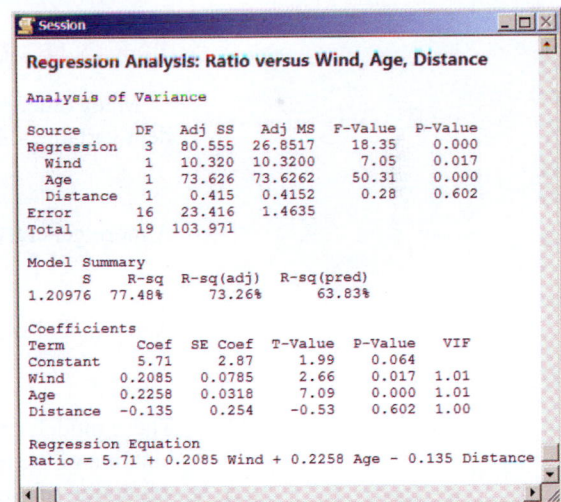

Figure 12.81 Minitab regression analysis.

SOLUTION

STEP 1 Using technology, the normal probability plot of the residuals is shown in Figure 12.82. The points appear to lie along an approximate straight line. There is no overwhelming evidence to suggest that the normality assumption is violated.

STEP 2 Figures 12.83–12.85 show the residual plots: $\hat{e}_i$ versus x_{1i}, $\hat{e}_i$ versus x_{2i}, and $\hat{e}_i$ versus x_{3i}. There is no obvious pattern in any of the these graphs.

Together, these four graphs suggest that there is no violation in the multiple linear regression model assumptions.

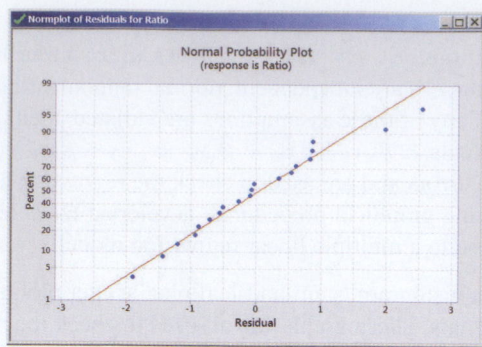

Figure 12.82 Minitab normal probability plot.

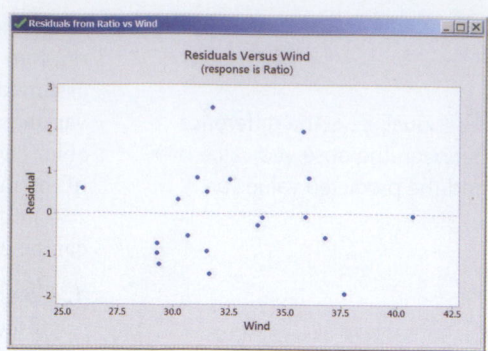

Figure 12.83 Plot of the residuals versus the first predictor.

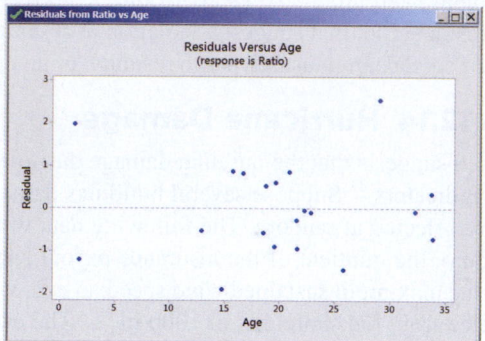

Figure 12.84 Plot of the residuals versus the second predictor.

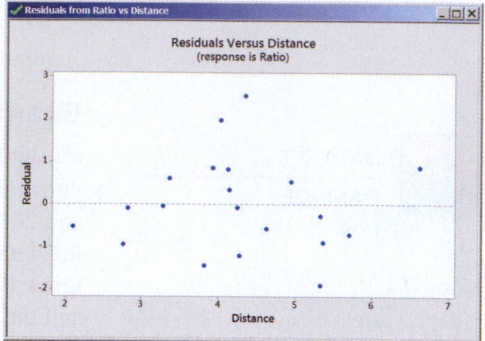

Figure 12.85 Plot of the residuals versus the third predictor.

TRY IT NOW GO TO EXERCISE 12.146

Many other possible models can also be considered to explain the variation in a dependent variable y. For example, a quadratic model with one predictor variable has the form

$$Y_i = \beta_0 + \beta_1 x_i + \beta_2 x_i^2 + E_i$$

A more general kth-degree polynomial model with one predictor has the form

$$Y_i = \beta_0 + \beta_1 x_i + \beta_2 x_i^2 + \cdots + \beta_k x_i^k + E_i$$

And a model with two predictor variables and an interaction term has the form

$$Y_i = \beta_0 + \beta_1 x_{1i} + \beta_2 x_{2i} + \beta_3 x_{1i} x_{2i} + E_i$$

These models are still considered *linear* regression models because each is a linear combination of the regression coefficients, that is, a sum of terms of the form $\beta_i v_i$, where v_i is a function of one or more predictor variables.

There are other models that do not appear to be linear but are *intrinsically* linear. These models can be transformed into linear models, and the techniques presented in this chapter can be used to estimate the regression coefficients. For example, the exponential model

$$Y_i = \beta_0 e^{\beta_1 x_i} E_i$$

is intrinsically linear. However, the general growth model

$$Y_i = \beta_0 + \beta_1 e^{\beta_2 x_i} + E_i$$

cannot be made into a linear model. The power model and the general logistic model are other common models. No matter what model you decide is best, technology should be used to estimate the unknown parameters.

Technology Corner

Procedure: Multiple Linear Regression analysis, normal probability plot of the residuals, and plot of the residuals versus values of a predictor variable.
Reconsider: Example 12.14, solution, and interpretations.

CrunchIt!

Use Multiple Linear regression options to produce a plot of the residuals versus the dependent variable and a normal probability plot of the residuals.

1. Enter the data into columns Var1, Var2, Var3, and Var4. Rename the columns if desired.
2. Select Statistics; Regression; Multiple Linear. Choose the Dependent Variable from the pull-down menu and check the Independent Variables. Select Numeric Results from the Display pull-down menu to view the estimated regression line and hypothesis test results concerning the coefficients. The CI Level box is used to obtain a confidence interval for each coefficient. Click Calculate to display the results. See Figure 12.86.

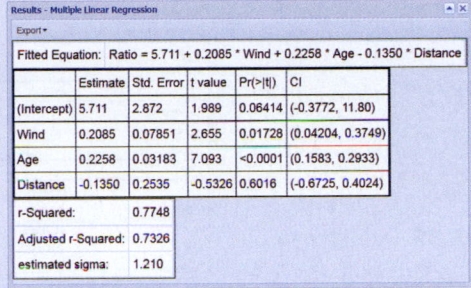

Figure 12.86 CrunchIt! Multiple Linear Regression summary output.

3. Construct a normal probability plot of the residuals by using the Residuals QQ Plot option from the Display pull-down menu. See Figure 12.87. Construct a scatter plot of the residuals versus the dependent variable by using the Residuals Plot option from the Display pull-down menu. See Figure 12.88.

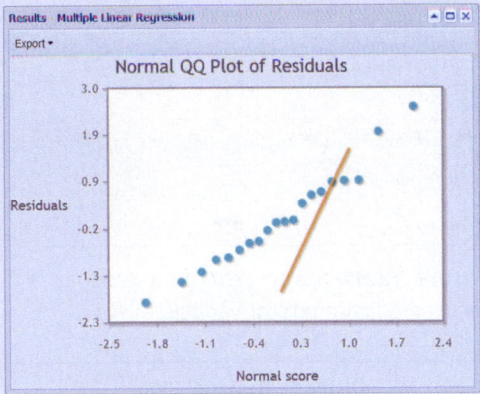

Figure 12.87 Normal probability plot of the residuals.

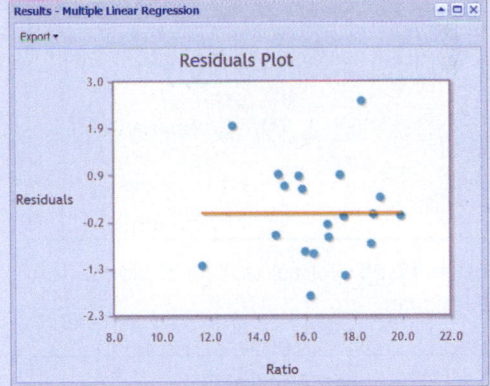

Figure 12.88 Scatter plot of the residuals versus the dependent variable.

Minitab

There are several ways to conduct a multiple linear regression analysis in Minitab, in a Session window and in the Stat; Regression menu.

1. Enter values of the dependent variable into column C1, and values for the predictor variables into columns C2, C3, and C4. Name the columns if desired.
2. Select Stat; Regression; Regression. Enter the Response column, C1, and the Predictor columns, C2, C3, and C4.

3. Select the Storage button and check Residuals to store the residuals in a Worksheet column. Click OK to display the regression analysis. See Figure 12.81.
4. Construct a normal probability plot of the residuals as described in Section 12.4 (Figures 12.82). Construct scatter plots of the residuals versus values of each predictor variable as described in Section 12.4 (Figures 12.83–12.85).

Excel

Regression analysis is included in the Data Analysis tool pack.

1. Enter values of the dependent variable into column A, and values for the predictor variables into columns B, C, and D.
2. Under the Data tab, select Data Analysis; Regression. Enter the Y Range, X Range, and the Output Range. Check Residuals Plot to construct a plot of the residuals versus each predictor variable. Check Residuals to save the residuals. The summary output is shown in Figure 12.89.

SUMMARY OUTPUT

Regression Statistics	
Multiple R	0.8802
R Square	0.7748
Adjusted R Square	0.7326
Standard Error	1.2098
Observations	20

ANOVA

	df	SS	MS	F	Significance F
Regression	3	80.5550	26.8517	18.35	2.0E-05
Residual	16	23.4164	1.4635		
Total	19	103.9714			

	Coefficients	Standard Error	t Stat	P-value	Lower 95%	Upper 95%	Lower 95.0%	Upper 95.0%
Intercept	5.7108	2.8718	1.9885	0.0641	-0.3772	11.7988	-0.3772	11.7988
Wind	0.2085	0.0785	2.6555	0.0173	0.0420	0.3749	0.0420	0.3749
Age	0.2258	0.0318	7.0928	0.0000	0.1583	0.2933	0.1583	0.2933
Distance	-0.1350	0.2535	-0.5326	0.6016	-0.6725	0.4024	-0.6725	0.4024

Figure 12.89 Excel regression analysis summary output.

Compute the normal scores and construct a normal probability plot of the residuals as described in Section 6.3. The normal probability plot is shown in Figure 12.90, and the other residual plots are shown in Figures 12.91–12.93.

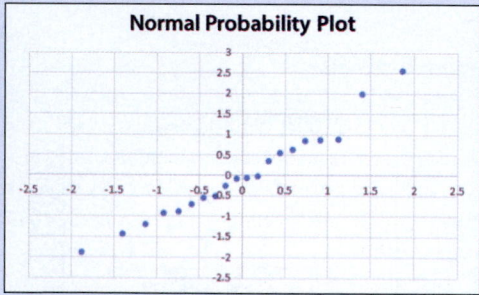

Figure 12.90 Normal probability plot of the residuals.

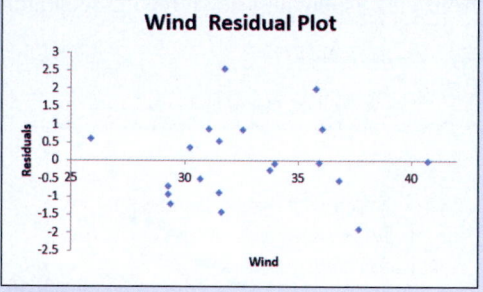

Figure 12.91 Scatter plot of the residuals versus the first predictor variable.

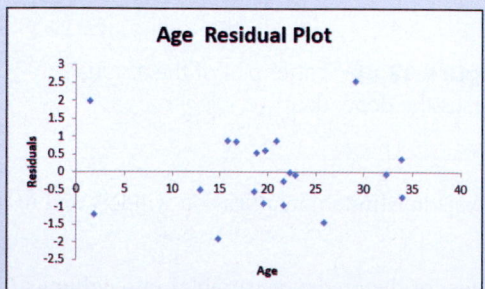

Figure 12.92 Scatter plot of the residuals versus the second predictor variable.

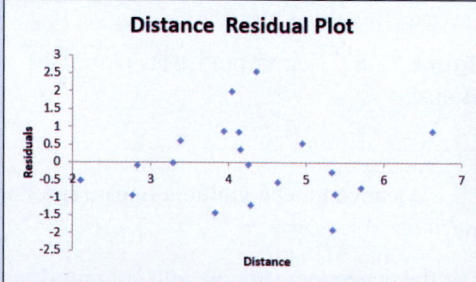

Figure 12.93 Scatter plot of the residuals versus the third predictor variable.

SECTION 12.5 EXERCISES

Concept Check

12.118 Short Answer   State the four assumptions in a multiple linear regression model in terms of the error terms.

12.119 Fill in the Blank   The true regression equation $y = \beta_1 + \beta_1 x_1 + \beta_2 x_2 + \cdots + \beta_k x_k$ is a linear function of _____.

12.120 Fill in the Blank   In the regression equation $y = \beta_1 + \beta_1 x_1 + \beta_2 x_2 + \cdots + \beta_k x_k$, β_i represents the mean change in y for every increase of _____ if the values of all the other predictor variables held constant.

12.121 Fill in the Blank   In a multiple linear regression model, _____ is used to minimize the sum of squares due to error.

12.122 True/False   There can be no more than five predictor variables in a multiple linear regression model.

12.123 True/False   The graph of the estimated regression equation $y = \hat{\beta}_0 + \hat{\beta}_1 x_1 + \hat{\beta}_2 x_2$ is a line in space.

12.124 Short Answer   A test for a significant multiple linear regression model is based on what ratio?

12.125 Short Answer   In a hypothesis test for a significant multiple linear regression, suppose we reject the null hypothesis. What does this suggest?

12.126 True/False   Suppose a test for a significant multiple linear regression model is significant. Every predictor in the model helps to determine the value of the dependent variable.

12.127 Short Answer   In a multiple linear regression model with k predictor variables, we cannot set the significance level in each individual hypothesis test concerning β_i. Why?

12.128 Short Answer   Suppose a test for a significant multiple linear regression model is significant. What is the most reasonable way to select the best model?

12.129 Fill in the Blank   In a multiple linear regression model, the value of Y for $x = x^*$ can be used as an estimate of _____ and _____.

12.130 Short Answer   Describe two type of plots used to check the assumptions in a multiple linear regression model.

Practice

12.131   Suppose the true regression equation relating the variables x_1, x_2, and y, for values of x_1 between 50 and 100 and for values of x_2 between 4.5 and 11, is $y = 32.0 + 7.5x_1 - 5.3x_2$.
   a. Find the expected value of Y when $x_1 = 65$ and $x_2 = 7$.
   b. Use the sign of the coefficient of x_1 to explain the relationship between the x_1 and y.
   c. How much change in the dependent variable is expected when x_2 increases by 1 unit?

d.   Suppose $\sigma = 3.7$. Find the probability an observed value of Y is less than 554 when $x_1 = 77$ and $x_2 = 9.8$.

12.132   Suppose the true regression equation relating the variables x_1, x_2, x_3, x_4, and y is $y = -10.7 + 5x_1 - 14x_2 - 23x_3 + 6.7x_4$.
   a. Find the expected value of Y when $x_1 = 1$, $x_2 = 2$, $x_3 = 2$, and $x_4 = 4.5$ [or $x = (1, 2, 2, 4.5)$].
   b. How much change in the dependent variable is expected when x_2 increases by 2 units and x_4 decreases by 5 units?
   c. Suppose $\sigma = 24$. Find the probability an observed value of Y is between -460 and 400 when $x = (10, 12.5, 15, 7)$.

12.133   Suppose a multiple linear regression model is used to explain the relationship between y, x_1, and x_2. A random sample of $n = 12$ pairs for the independent variables were selected, and the corresponding values of the dependent variable were observed. The data are given in the following table. 🔳 **EX12.133**

y	196.7	203.4	227.0	221.3	154.8	185.1
x_1	7.8	8.7	4.0	2.6	1.9	3.1
x_2	22.5	24.0	27.4	27.8	19.0	21.8
y	198.7	200.9	220.8	193.6	206.2	214.1
x_1	8.3	7.8	2.2	2.4	3.9	6.4
x_2	22.1	23.4	26.6	24.6	24.3	25.6

   a. Find the estimated regression equation.
   b. What is the predicted value of Y when $x_1 = 3.3$ and $x_2 = 20$?

12.134   An experiment resulted in observations on a single dependent variable and three independent variables. 🔳 **EX12.134**
   a. Construct three separate scatter plots of y versus x_1, y versus x_2, and y versus x_3. Describe the relationship in each plot.
   b. Estimate the regression coefficients in the model $Y_i = \beta_0 + \beta_1 x_{1i} + \beta_2 x_{2i} + \beta_3 x_{3i} + E_i$. Explain the relationship between the sign of each estimated regression coefficient and the scatter plots in part (a).
   c. Estimate the true mean value of Y when $x_1 = -10$, $x_2 = 1.35$, and $x_3 = 52.6$.

12.135   Data from an observational study were used to fit a multiple linear regression model and the following summary statistics were obtained:

$$n = 23 \quad SSE = 80.75 \quad SST = 152.5 \quad MSE = 4.75$$

   a. Complete the ANOVA summary table.
   b. How many predictor variables are in the multiple linear regression model?
   c. Conduct a model utility test with $\alpha = 0.05$. Find bounds on the p value associated with this test. Use technology to find the exact p value and to support your answer.

d. Find the value of r^2 and explain the meaning of this value.

12.136 An experiment was conducted and the data obtained were used to fit the multiple linear regression model $Y_i = \beta_0 + \beta_1 x_{1i} + \beta_2 x_{2i} + \beta_3 x_{3i} + \beta_4 x_{4i} + E_i$. Minitab was used to estimate the regression coefficients, and the output is shown below.

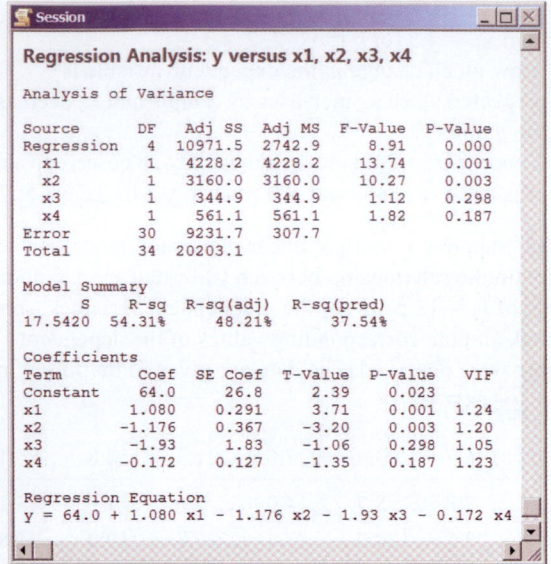

a. Is the overall regression significant? Conduct the appropriate hypothesis test and justify your answer.

b. Conduct four hypothesis tests with $H_0: \beta_i = 0$ for $i = 1, 2, 3, 4$ and $\alpha = 0.05$ in each test. Which regression coefficients are significantly different from 0, and therefore, which predictor variables are significant?

c. Conduct four hypothesis tests with $H_0: \beta_i = 0$ for $i = 1, 2, 3, 4$ with an *overall* type I error of $\alpha = 0.05$. Using these results, which predictor variables are significant? Compare your answers in parts (b) and (c).

12.137 An experiment resulted in the following observations on a single dependent variable and three independent variables. **EX12.137**

y	x_1	x_2	x_3
-31.56	3.76	0.94	16.5
-44.48	4.47	0.23	17.2
-26.28	1.97	0.52	15.7
-27.54	1.16	0.45	15.8
-22.68	0.73	0.48	16.7
-20.34	0.82	0.63	16.5
-46.19	5.84	0.27	18.2
-29.00	2.69	0.61	16.4
-31.60	3.08	0.46	16.3
-34.31	5.38	0.73	17.5

a. Estimate the regression coefficients in the model $Y_i = \beta_0 + \beta_1 x_{1i} + \beta_2 x_{2i} + \beta_3 x_{3i} + E_i$.

b. Conduct an F test for a significant regression. Use $\alpha = 0.05$. Use technology to find the exact p value.

c. Find the coefficient of determination.

d. Which predictor variable(s) is(are) significant in explaining the variation in the dependent variable? Conduct the appropriate hypothesis tests.

e. Construct a 95% confidence interval for the *constant* regression coefficient, β_0. Use this interval to determine whether there is evidence to suggest the constant regression coefficient is different from 0.

12.138 An experiment resulted in observations on a single dependent variable and four independent variables. **EX12.138**

a. Estimate the regression coefficients in the model $Y_i = \beta_0 + \beta_1 x_{1i} + \beta_2 x_{2i} + \beta_3 x_{3i} + \beta_4 x_{4i} + E_i$. Find the coefficient of determination.

b. Conduct the appropriate hypothesis tests to determine which predictor variables are significant (equivalently, which regression coefficients are significantly different from 0).

c. Using the results from part (b), write the *reduced* multiple linear regression model (including only the significant predictor variables). Estimate the regression coefficients and find the coefficient of determination in this *reduced* model.

d. Compare the value of r^2 in the two models. Which model do you think is better? Why?

12.139 An experiment was conducted to determine whether the freezing point of a certain solution can be predicted from the concentrations of three chemicals. Twenty combinations of chemical concentrations were studied, and the data were recorded. **FREEZE**

a. Estimate the regression coefficients in the model $Y_i = \beta_0 + \beta_1 x_{1i} + \beta_2 x_{2i} + \beta_3 x_{3i} + E_i$.

b. Compute the residuals and construct a normal probability plot for the residuals. Is there any evidence to suggest the random error terms are not normal? Justify your answer.

c. Construct a graph of the residuals versus each predictor variable. Is there any evidence of a violation in the regression assumptions? If so, what modification(s) could you make to improve the model?

12.140 The amount of sludge buildup in an automobile engine can severely affect performance. In an article in the *Vanagon Maintenance Guide*, a technical editor discussed the advantages of synthetic lubricants. Suppose an experiment was conducted to determine whether the sludge buildup in an automobile engine can be predicted by the oil viscosity, oxidation inhibitors, detergents, dispersants, and/or anti-wear additives. The data obtained were used to fit the multiple linear regression model $Y_i = \beta_0 + \beta_1 x_{1i} + \beta_2 x_{2i} + \beta_3 x_{3i} + \beta_4 x_{4i} + \beta_5 x_{5i} + E_i$.

Minitab was used to estimate the regression coefficients, and the output is shown below.

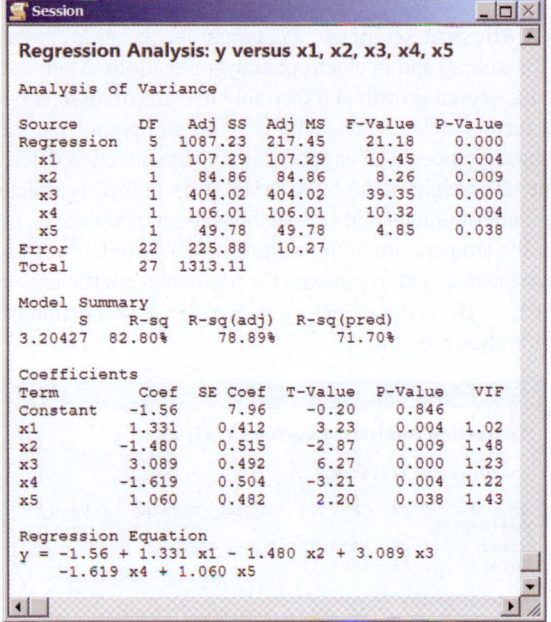

Consider the value $x^* = (6.1, 5.5, 8.3, 7.2, 6.5)$ and suppose an estimate of the standard deviation of Y^* is $s_{Y*} = 1.973$.

a. Find a 95% confidence interval for the true mean value of Y when $x = x^*$. Give an interpretation of this interval.

b. Find a 95% prediction interval for an observed value of Y when $x = x^*$. Give an interpretation of this interval.

Applications

12.141 Economics and Finance A financial analyst believes that the U.S. 6-month treasury yield (y) can be predicted from the crude oil price per barrel (x_1), the price of gold (x_2), and the M1 money supply (x_3, in billions of dollars). A random sample of days was selected and the data were recorded.[33] USTREAS

a. Find the estimated regression line. Conduct an F test for a significant regression with $\alpha = 0.05$.

b. Find an estimate of the true mean 6-month treasury yield for $x_1 = 100$, $x_2 = 1200$, and $x_3 = 2600$.

c. Which of these variables do you think is the most important predictor for the 6-month treasury yield? Why?

12.142 Business and Management A study was conducted to determine if the rental price of an apartment in a college town is related to the distance from campus and the apartment size. A random sample of apartments in State College, Pennsylvania, was obtained, and the monthly rental price (y, in dollars), distance from campus (x_1, in miles), and square feet (x_2) was recorded for each. RENTAL

a. Estimate the regression coefficients in the model $Y_i = \beta_0 + \beta_1 x_{1i} + \beta_2 x_{2i} + E_i$. Interpret the value of each estimated coefficient.

b. Do these predictor variables explain a significant amount of variation in Y? Conduct the appropriate model utility test using $\alpha = 0.05$.

c. Find an estimate of an observed value of Y for $x_1 = 1.5$ and $x_2 = 1200$.

12.143 Physical Sciences An experiment was conducted to test the effect of temperature (x_1, °F), contact area (x_2, cm²), and wood density (x_3, g/cm³) on the bonding strength (y) of a certain (wood) glue. Each piece of wood was glued to the same control block and the entire fixture was placed in an industrial oven. The bonding strength was measured by recording the time (in minutes) until the test piece of wood separated from the control block. The multiple linear regression model equation is $Y_i = \beta_0 + \beta_1 x_{1i} + \beta_2 x_{2i} + \beta_3 x_{3i} + E_i$. Minitab was used to analyze the data, and a portion of the output is shown below.

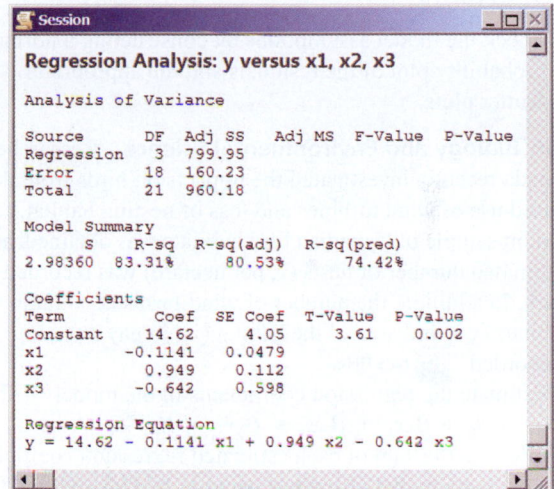

a. Complete the ANOVA table and conduct a model utility test.

b. Interpret the sign of each estimated regression coefficient.

c. Conduct three hypothesis tests with $H_0: \beta_i = 0$ and $\alpha = 0.05$ in each test. Find bounds on the p value associated with each test. Use these results to determine which predictor variables are the most important in determining the bonding strength of glue.

12.144 Physical Sciences A Daniel cell uses zinc and copper solutions to produce electricity. An experiment was conducted to determine whether voltage (y) is affected by the temperature of the solutions (x_1, °F), the concentration of the solutions (x_2, M), and/or the surface area of the electrodes (x_3, cm²). DANIEL

a. Estimate the regression coefficients in the model $Y_i = \beta_0 + \beta_1 x_{1i} + \beta_2 x_{2i} + \beta_3 x_{3i} + E_i$.

b. Conduct a model utility test and the other appropriate tests to determine which variables are the most important predictors of voltage.

c. Construct a normal probability plot of the residuals. Use this plot to determine whether there is any evidence of violation in multiple linear regression assumptions.

12.145 Physical Sciences To keep golf courses attractive in the arid southwest, groundskeepers carefully monitor salt content in water used for irrigation. It is believed that high salt concentration can damage plants and grass. A random sample of golf courses in the southwest was obtained and a sample of water used for irrigating was obtained for each. The following variables were measured for each water sample: dissolved salt (y, dS/m), total dissolved solids (x_1, mg/L), sodium, calcium magnesium, and bicarbonate (x_2, x_3, x_4, and x_5, all in milliequivalents/liter).[34] **GROUNDS**

a. Estimate the regression coefficients in the model
$Y_i = \beta_0 + \beta_1 x_{1i} + \beta_2 x_{2i} + \beta_3 x_{3i} + \beta_4 x_{4i} + \beta_5 x_{5i} + E_i$.

b. Construct the ANOVA table and conduct the model utility test.

c. Conduct the necessary hypothesis tests to determine the most important variables in predicting the time to failure. What do you think is the single most important predictor variable? Why?

d. Check the model assumptions by constructing a normal probability plot of the residuals and the appropriate scatter plots.

12.146 Biology and Environmental Science Researchers in Canada recently investigated the impacts on birds from the increased use of wind turbines and loss of nesting habitat.[35] A random sample of Canadian bird habitats was obtained, and the estimated number of nests (y, per hectare) was recorded for each. In addition, the number of wind farms (x_1), number of turbines (x_2), and area of the habitat (x_3, in ha) was also recorded. **NESTING**

a. Estimate the regression coefficients in the model
$Y_i = \beta_0 + \beta_1 x_{1i} + \beta_2 x_{2i} + \beta_3 x_{3i} + E_i$.

b. Interpret the sign of each estimated regression coefficient.

c. Conduct a model utility test using $\alpha = 0.05$ and find the value of the coefficient of determination.

d. Which predictor variable do you believe is most important? Why?

e. Check the model assumptions by constructing a normal probability plot of the residuals and the appropriate scatter plots.

Extended Applications

12.147 Biology and Environmental Science There is some evidence to suggest that the variation in the size of penguin colonies is related to sea ice extent. Suppose a study was conducted to investigate the effect of sea ice extent and stormy weather on the size of penguin colonies along the west coast of the Antarctic Peninsula. Fifteen years were selected at random, and the data were recorded: **PENGUIN**

y = size of the penguin colony population.
x_1 = sea ice extent (as a percentage)
x_2 = number of stormy days (yearly total)

a. Estimate the regression coefficients in the model
$Y_i = \beta_0 + \beta_1 x_{1i} + \beta_2 x_{2i} + E_i$.

b. Conduct the model utility test. Use $\alpha = 0.01$.

c. Interpret the sign of each estimated regression coefficient.

d. Find the value of r^2 and interpret this value.

e. Estimate the mean value of Y for a sea ice extent of 12.5% and 35 stormy days.

12.148 Physical Sciences Potassium ferricyanide crystals are used for etching and in eloctroplating applications. Many factors affect the crystal growth and therefore the size of these crystals. These factors include the solubility of the compound and the evaporation process. An experiment was conducted to determine whether the weight of the final crystal (y, in grams) is affected by the amount of initial solid (x_1, g), the amount of water (x_2, ml), and/or the temperature of the water (x_3, °F). Minitab was used to analyze the data and to estimate the regression coefficients in the model $Y_i = \beta_0 + \beta_1 x_{1i} + \beta_2 x_{2i} + \beta_3 x_{3i} + E_i$. A portion of the output is shown below.

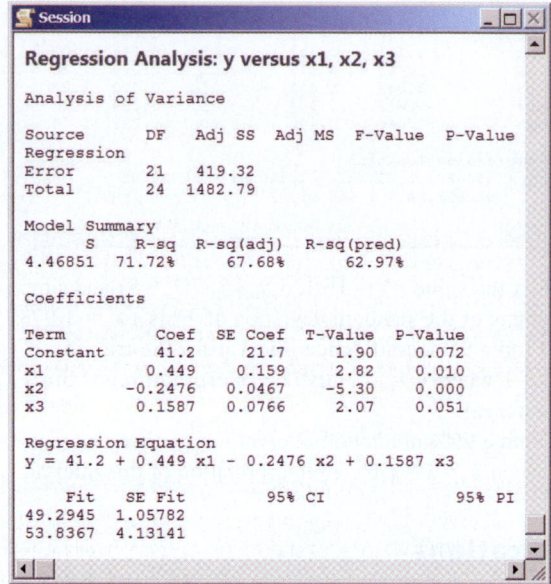

a. Complete the ANOVA table and conduct a model utility test.

b. Use the output to determine which of the three predictor variables contributes to the overall significant regression.

c. Suppose $x_1^* = (93, 200, 100)$ and $x_2^* = (100, 220, 140)$. The predicted value for each is given in the Minitab output, along with the standard error of each estimate (SE Fit). For example, $s_{Y_1^*} = 1.05782$. Find a 95% confidence interval for the true mean value of Y when $x = x_1^*$. Find a 95% confidence interval for the true mean value of Y when $x = x_2^*$.

d. Find a 95% prediction interval for an observed value of Y when $x = x_1^*$. Find a 95% prediction interval for an observed value of Y when $x = x_2^*$.

e. Why do you suppose $s_{Y_1^*} < s_{Y_2^*}$?

12.149 Biology and Environmental Science In many rural areas in countries around the world, it is very costly to construct an efficient water supply system. Therefore, groundwater is the principal source of water for drinking and other activities. Prior to use, the groundwater must be tested for safety and quality. Chloride in water affects the taste of many foodstuffs, and large amounts may lead to hypertension. However, little is known about the long-term effects. A recent multiple linear regression model was used to explain the variation in chloride in groundwater (y). The model included five predictor variables. The independent variables are x_1, total dissolved solids (mg/L); x_2, iron (mg/L); x_3, nitrate (mg/L); x_4, total hardness (ppm); x_5, depth of sample (feet).

A portion of the analysis associated with this model follows.

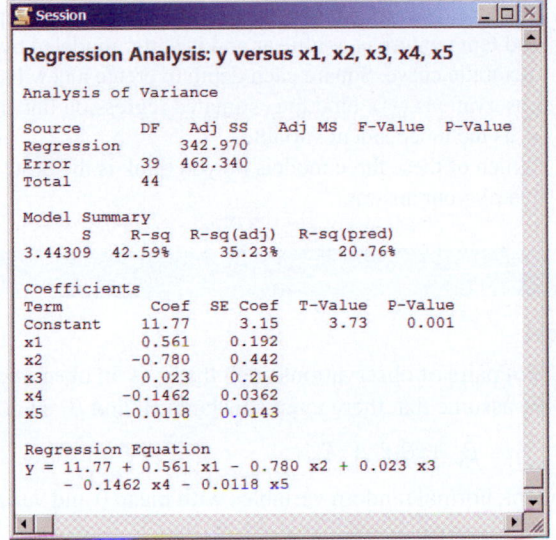

Regression Analysis: y versus x1, x2, x3, x4, x5

Analysis of Variance

Source	DF	Adj SS	Adj MS	F-Value	P-Value
Regression		342.970			
Error	39	462.340			
Total	44				

Model Summary

S	R-sq	R-sq(adj)	R-sq(pred)
3.44309	42.59%	35.23%	20.76%

Coefficients

Term	Coef	SE Coef	T-Value	P-Value
Constant	11.77	3.15	3.73	0.001
x1	0.561	0.192		
x2	-0.780	0.442		
x3	0.023	0.216		
x4	-0.1462	0.0362		
x5	-0.0118	0.0143		

Regression Equation

y = 11.77 + 0.561 x1 - 0.780 x2 + 0.023 x3 - 0.1462 x4 - 0.0118 x5

a. Complete the ANOVA table and conduct a model utility test. Use technology to find the p value associated with this test. Find the value of r^2. Use these results to described the overall significance of the model.

b. Find the value of the test statistic and the p value associated with each hypothesis test for a significant regression coefficient. Use these results to determine the most important predictor variables in the model. For each of these variables, explain the effect on chloride in groundwater. Justify your answers.

12.150 Economics and Finance A recent study investigated the variables that can be used to predict economic growth in China, as measured by gross domestic product (GDP, y, in hundred millions).[36] A sample of years was obtained and the following measurements were recorded: capital stock (x_1, in hundred millions); labor force (x_2, in ten thousands); and energy consumption (x_3, in ten thousand tons standard coal). **GDP**

a. Estimate the regression coefficients in the model $Y_i = \beta_0 + \beta_1 x_{1i} + \beta_2 x_{2i} + \beta_3 x_{3i} + E_i$.

b. Conduct the model utility test. Use $\alpha = 0.05$.

c. Find the value of r^2 and interpret this value. Explain this remarkable value in the context of this problem.

d. Conduct the appropriate hypothesis tests to determine which regression coefficient(s) is(are) significantly different from 0. Use $\alpha = 0.05$ in each test.

e. Do you believe there is any way to improve this model? Why or why not?

12.151 Public Health and Nutrition A recent experiment was designed to determine whether the percentage of lactic acid

(y) in cultured buttermilk (produced from whole milk) can be predicted by the temperature to which the milk is heated (x_1, °C), the temperature to which the milk is cooled before the started culture is added (x_2, °C), the percentage of started culture added (x_3), and/or the fermentation time (x_4, in hours). **MILK**

a. Estimate the regression coefficients in the model $Y_i = \beta_0 + \beta_1 x_{1i} + \beta_2 x_{2i} + \beta_3 x_{3i} + \beta_4 x_{4i} + E_i$.

b. Conduct a model utility test using $\alpha = 0.05$ and find the value of the coefficient of determination.

c. Conduct the appropriate hypothesis tests to determine which predictor(s) is(are) significant.

d. Conduct a hypothesis test to determine whether there is any evidence to suggest β_3 is less than 0.93. Use $\alpha = 0.05$.

e. Find a 95% confidence interval for the true mean amount of lactic acid in cultured buttermilk when $x_1 = 95$, $x_2 = 22$, $x_3 = 1.5$, and $x_4 = 20$.

12.152 The Rainy Season Hawaii is a popular tourist destination during the winter months, and Honolulu offers many outdoor attractions. However, the months of December through February are the rainy season, and records indicate that the most rain falls during the month of January. Using data from random years,[37] a multiple linear regression model was used to explain the total January rainfall (y) that consisted of four predictor variables, one of which was a *dummy* variable (a variable with a value of either 0 or 1). The independent variables were monthly rainfall total in the previous December, January, and February (x_1, x_2, and x_3, respectively) and a dummy variable (x_4) which indicated whether there was a moderate to weak El Niño during the previous year. Rainfall totals were recorded in inches. **HAWAII**

a. Construct the ANOVA table and conduct a model utility test. Use technology to find the p value associated with this test. Find the value of r^2. Use these results to describe the overall significance of the model.

b. Find the value of the test statistic and the p value associated with each hypothesis test for a significant regression coefficient. Use these results to determine the most important predictor variables in this model.

c. Consider a reduced model with the two most important predictor variables. Conduct a model utility test. Describe the significance of this model compared to the full model in part (a).

d. Find the most recent rainfall totals for Honolulu and records on El Niño. Use this information to find a 95% prediction interval for the most recent January rainfall total in Honolulu. Does the actual value lie in this interval? What does this suggest about the reduced model?

Challenge

12.153 Geothermal Gradient The temperature naturally increases as one digs deeper into the continental crust. This phenomenon is called the geothermal gradient. A random sample of continental crust depths (x, in km) was obtained, and the temperature at each depth was recorded (y, in °C). The data are given in the following table. **GEOGRAD**

Depth	192	306	75	375	96	120	399
Temperature	212	682	15	855	93	49	943

Depth	469	439	209	224	313	379	296
Temperature	1456	1156	271	336	604	911	606

Depth	37	316	41	242	21	230
Temperature	35	576	31	462	10	274

a. Carefully sketch a scatter plot of these data.
b. Find the estimated regression line, and complete the ANOVA summary table.

c. There is some evidence to suggest that the rate of change in temperature is different according to two depth zones. For example, at small depths, the temperature changes only slightly for each kilometer. However, at large depths, the temperature change is much greater for each kilometer. Use your scatter plot from part (a) to divide the data into two depth zones. Find the estimated regression line and the ANOVA summary table for each zone.
d. Some researchers believe the relationship between depth and temperature is nonlinear and is better modeled by a quadratic curve. Square each depth to create a new list of observations (x^2). Find the estimated regression line using x^2 as the independent variable.
e. Which of these three models do you think is the best? Justify your answer.

CHAPTER 12 SUMMARY

Simple Linear Regression Model

Let $(x_1, y_1), (x_2, y_2), \ldots, (x_n, y_n)$ be n pairs of observations such that y_i is an observed value of the random variable Y_i. We assume that there exist constants β_0 and β_1 such that

$$Y_i = \beta_0 + \beta_1 x_i + E_i$$

where $E_1, E_2, \ldots, E_n$ are independent, normal random variables with mean 0 and variance σ^2. That is,

1. The E_i's are normally distributed (which implies that the Y_i's are normally distributed).
2. The expected value of E_i is 0 [which implies that $E(Y_i) = \beta_0 + \beta_1 x_i$].
3. $\text{Var}(E_i) = \sigma^2$ [which implies that $\text{Var}(Y_i) = \sigma^2$].
4. The E_i's are independent (which implies that the Y_i's are independent).

The E_i's are the random deviations or random error terms. $y = \beta_0 + \beta_1 x$ is the true regression line.

Principle of least squares

The estimated regression line is obtained by minimizing the sum of the squared deviations, or vertical distances from the observed points to the line.

Least-squares estimates

$$\hat{\beta}_1 = \frac{n\Sigma x_i y_i - (\Sigma x_i)(\Sigma y_i)}{n\Sigma x_i^2 - (\Sigma x_i)^2} \qquad \hat{\beta}_0 = \frac{\Sigma y_i - \hat{\beta}_1 \Sigma x_i}{n} = \bar{y} - \hat{\beta}_1 \bar{x}$$

The estimated regression line is $y = \hat{\beta}_0 + \hat{\beta}_1 x$.

The ith predicted (fitted) value is $\hat{y}_i = \hat{\beta}_0 + \hat{\beta}_1 x_i$ ($i = 1, 2, \ldots, n$).

The ith residual is $\hat{e}_i = y_i - \hat{y}_i$.

The sum of squares

$$\underbrace{\Sigma(y_i - \bar{y})^2}_{\text{SST}} = \underbrace{\Sigma(\hat{y}_i - \bar{y})^2}_{\text{SSR}} + \underbrace{\Sigma(y_i - \hat{y}_i)^2}_{\text{SSE}}$$

Computation formulas

$$\text{SST} = S_{yy} = \Sigma(y_i - \bar{y})^2 = \Sigma y_i^2 - \frac{1}{n}(\Sigma y_i)^2$$

$$\text{SSR} = \hat{\beta}_1 S_{xy} = \hat{\beta}_1 \Sigma(x_i - \bar{x})(y_i - \bar{y}) = \hat{\beta}_1 \left[\Sigma x_i y_i - \frac{1}{n}(\Sigma x_i)(\Sigma y_i) \right]$$

$$\text{SSE} = \text{SST} - \text{SSR}$$

Analysis of variance for simple linear regression

Source of variation	Sum of squares	Degrees of freedom	Mean square	F	p Value
Regression	SSR	1	$MSR = \dfrac{SSR}{1}$	$\dfrac{MSR}{MSE}$	p
Error	SSE	$n-2$	$MSE = \dfrac{SSE}{n-2}$		
Total	SST	$n-1$			

Coefficient of determination

$$r^2 = SSR/SST$$

Estimate of variance

$$s^2 = SSE/(n-2)$$

F test for a significant linear regression

H_0: There is no significant linear relationship ($\beta_1 = 0$).

H_a: There is a significant linear relationship ($\beta_1 \neq 0$).

TS: $F = \dfrac{MSR}{MSE}$

RR: $F \geq F_{\alpha,1,n-2}$

Hypothesis test and confidence interval concerning β_1

H_0: $\beta_1 = \beta_{10}$

H_a: $\beta_1 > \beta_{10}$, $\quad \beta_1 < \beta_{10}$, $\quad$ or $\quad \beta_1 \neq \beta_{10}$

TS: $T = \dfrac{B_1 - \beta_{10}}{S/\sqrt{S_{xx}}} = \dfrac{B_1 - \beta_{10}}{S_{B_1}}$

RR: $T \geq t_{\alpha,n-2}$, $\quad T \leq -t_{\alpha,n-2}$, $\quad |T| \geq t_{\alpha/2,n-2}$

A $100(1-\alpha)\%$ confidence interval for β_1 has as endpoints $\hat{\beta}_1 \pm t_{\alpha/2,n-2}\, s_{B_1}$.

Hypothesis test and confidence interval concerning β_0

H_0: $\beta_0 = \beta_{00}$

H_a: $\beta_0 > \beta_{00}$, $\quad \beta_0 < \beta_{00}$, $\quad$ or $\quad \beta_0 \neq \beta_{00}$

TS: $T = \dfrac{B_0 - \beta_{00}}{S\sqrt{\sum x_i^2/nS_{xx}}} = \dfrac{B_0 - \beta_{00}}{S_{B_0}}$

RR: $T \geq t_{\alpha,n-2}$, $\quad T \leq -t_{\alpha,n-2}$, $\quad$ or $\quad |T| \geq t_{\alpha/2,n-2}$

A $100(1-\alpha)\%$ confidence interval for β_0 has as endpoints $\hat{\beta}_0 \pm t_{\alpha/2,n-2}s_{B_0}$

Sample correlation coefficient

$$r = \frac{S_{xy}}{\sqrt{S_{xx}S_{yy}}} = \frac{\sum x_i y_i - \dfrac{1}{n}(\sum x_i)(\sum y_i)}{\sqrt{\left[\sum x_i^2 - \dfrac{1}{n}(\sum x_i)^2\right]\left[\sum y_i^2 - \dfrac{1}{n}(\sum y_i)^2\right]}}$$

Hypothesis test and confidence interval concerning the mean value of Y for $x = x^*$

H_0: $y^* = y_0^*$

H_a: $y^* > y_0^*$, $\quad y^* < y_0^*$, $\quad$ or $\quad y^* \neq y_0^*$

TS: $T = \dfrac{(B_0 + B_1 x^*) - y_0^*}{S\sqrt{(1/n) + [(x^* - \bar{x})^2/S_{xx}]}}$

RR: $T \geq t_{\alpha,n-2}$, $\quad T \leq -t_{\alpha,n-2}$, $\quad$ or $\quad |T| \geq t_{\alpha/2,n-2}$

A $100(1 - \alpha)\%$ confidence interval for $\mu_{Y|x^*}$, the mean value of Y for $x = x^*$, has as endpoints the values

$$(\widehat{\beta}_0 + \widehat{\beta}_1 x^*) \pm t_{\alpha/2, n-2} s \sqrt{\frac{1}{n} + \frac{(x^* - \bar{x})^2}{S_{xx}}}$$

Prediction interval for an observed value of Y

A $100(1 - \alpha)\%$ prediction interval for an observed value of Y when $x = x^*$ has as endpoints the values

$$(\widehat{\beta}_0 + \widehat{\beta}_1 x^*) \pm t_{\alpha/2, n-2} s \sqrt{1 + \frac{1}{n} + \frac{(x^* - \bar{x})^2}{S_{xx}}}$$

Regression diagnostics

1. Normal probability plot of residuals. The points should lie along an approximately straight line.

2. Scatter plot of residuals versus predictor variable. There should be no distinct pattern.

Multiple Linear Regression Model

Let $(x_{11}, x_{21}, \ldots, x_{k1}, y_1), (x_{12}, x_{22}, \ldots, x_{k2}, y_2), \ldots, (x_{1n}, x_{2n}, \ldots, x_{kn}, y_n)$ be n sets of observations such that y_i an observed value of the random variable Y_i. We assume that there exist constants $\beta_0, \beta_1, \ldots, \beta_k$ such that

$$Y_i = \beta_0 + \beta_1 x_{1i} + \beta_2 x_{2i} + \cdots + \beta_k x_{ki} + E_i$$

where $E_1, E_2, \ldots, E_n$ are independent, normal random variables with mean 0 and variance σ^2; that is,

1. The E_i's are normally distributed (which means that the Y_i's are normally distributed).

2. The expected value of E_i is 0 [which implies that $E(Y_i) = \beta_0 + \beta_1 x_{1i} + \beta_2 x_{2i} + \cdots + \beta_k x_{ki}$].

3. $\text{Var}(E_i) = \sigma^2$ [which implies that $\text{Var}(Y_i) = \sigma^2$].

4. The E_i's are independent (which implies that the Y_i's are independent).

The E_i's are the random deviations or random error terms.

$y = \beta_0 + \beta_1 x_1 + \beta_2 x_2 + \cdots + \beta_k x_k$ is the true regression line.

Principle of least squares

The estimated regression equation is obtained by minimizing the sum of the squared deviations between the observations and the estimated values.

The estimated regression equation is $y = \widehat{\beta}_0 + \widehat{\beta}_1 x_1 + \widehat{\beta}_2 x_2 + \cdots + \widehat{\beta}_k x_k$.

The ith predicted (fitted) value is $\widehat{y}_i = \widehat{\beta}_0 + \widehat{\beta}_1 x_{1i} + \widehat{\beta}_2 x_{2i} + \cdots + \widehat{\beta}_k x_{ki}$ $(i = 1, 2, \ldots, n)$.

The ith residual is $\widehat{e}_i = y_i - \widehat{y}_i$.

The sum of squares

$$\underbrace{\sum (y_i - \bar{y})^2}_{\text{SST}} = \underbrace{\sum (\widehat{y}_i - \bar{y})^2}_{\text{SSR}} + \underbrace{\sum (y_i - \widehat{y}_i)^2}_{\text{SSE}}$$

Analysis of variance for multiple linear regression

Source of variation	Sum of squares	Degrees of freedom	Mean square	F	p Value
Regression	SSR	k	$\text{MSR} = \dfrac{\text{SSR}}{k}$	$\dfrac{\text{MSR}}{\text{MSE}}$	p
Error	SSE	$n - k - 1$	$\text{MSE} = \dfrac{\text{SSE}}{n - k - 1}$		
Total	SST	$n - 1$			

Coefficient of determination

$$r^2 = \text{SSR}/\text{SST}$$

Estimate of variance

$$s^2 = \text{SSE}/(n - k - 1)$$

F test for a significant multiple linear regression

H_0: $\beta_1 = \beta_2 = \cdots = \beta_k = 0$ (None of the predictor variables helps to predict y.)

H_a: $\beta_i \neq 0$ for at least one i (At least one predictor variable helps to predict y.)

TS: $F = \dfrac{\text{MSR}}{\text{MSE}}$

RR: $F \geq F_{\alpha,k,n-k-1}$

Hypothesis test and confidence interval concerning β_i

H_0: $\beta_i = \beta_{i0}$

H_a: $\beta_i > \beta_{i0}$, $\qquad \beta_i < \beta_{i0}$, $\qquad$ or $\qquad \beta_i \neq \beta_{i0}$

TS: $T = \dfrac{B_i - \beta_{i0}}{S_{B_i}}$

RR: $T \geq t_{\alpha,n-k-1}$, $\qquad T \leq -t_{\alpha,n-k-1}$, $\qquad$ or $\qquad |T| \geq t_{\alpha/2,n-k-1}$

A $100(1 - \alpha)\%$ confidence interval for β_i has as endpoints the values

$$\hat{\beta}_i \pm t_{\alpha/2,n-k-1}\, s_{B_i}$$

Hypothesis test and confidence interval concerning the mean value of Y for $x = x^*$

H_0: $y^* = y_0^*$

H_a: $y^* > y_0^*$, $\qquad y^* < y_0^*$, $\qquad$ or $\qquad \text{y}^* \neq y_0^*$

TS: $T = \dfrac{(B_0 + B_1 x_1^* + \cdots + B_k x_k^*) - y_0^*}{S_{y^*}}$

RR: $T \geq t_{\alpha,n-k-1}$, $\qquad T \leq -t_{\alpha,n-k-1}$, $\qquad$ or $\qquad |T| \geq t_{\alpha/2,n-k-1}$

A $100(1 - \alpha)\%$ confidence interval for $\mu_{Y|x^*}$, the mean value of Y for $x = x^*$, has as endpoints the values

$$(\hat{\beta}_0 + \hat{\beta}_1 x_1^* + \cdots + \hat{\beta}_k x_k^*) \pm t_{\alpha/2,n-k-1}\, s_{Y^*}$$

Prediction interval for an observed value of Y

A $100(1 - \alpha)\%$ prediction interval for an observed value of Y when $x = x^*$ has as endpoints the values

$$(\hat{\beta}_0 + \hat{\beta}_1 x_1^* + \cdots + \hat{\beta}_k x_k^*) \pm t_{\alpha/2,n-k-1}\sqrt{s^2 + s_{y^*}^2}$$

Regression diagnostics

1. Construct a histogram, stem-and-leaf plot, scatter plot, and/or normal probability plot of the residuals. These graphs are all used to check the normality assumption.

2. Construct a scatter plot of the residuals versus *each* independent variable. If there are no violations in assumptions, each scatter plot should appear as a horizontal band around 0. There should be no recognizable pattern.

CHAPTER 12 EXERCISES

12 **APPLICATIONS**

12.154 Medicine and Clinical Studies Officials at the U.S. Environmental Protection Agency worked with hospital physicians in Anaheim, California, to examine the relationship between environmental pollution and illness. A random sample of summer days was selected, and at 2:00 P.M. on each day the air quality was assessed by measuring the concentration of carbon monoxide (x, in ppm). Illness was measured by counting the number of new patients (y) seen for respiratory problems at the hospital that day. Suppose the true regression line is $y = -5.1 + 0.9x$.

a. Find the expected number of respiratory patients at the hospital on a day when the carbon monoxide concentration is 12 ppm.

b. What is the expected change in the number of patients per day if the carbon monoxide concentration decreases by 5 ppm?

c. Suppose $\sigma = 1.2$. Find the probability that an observed number of patients is between 5 and 7 when the carbon monoxide concentration is 13 ppm.

12.155 Economics and Finance An economist believes that a credit card company sets the spending limits for customers based on loyalty, that is, the number of years the customer has been a cardholder. A random sample of 22 cardholders was selected. The number of years since obtaining the card (x) was recorded as well as the spending limit (y, in thousands of dollars) on the card. The summary statistics are given below.

$$\Sigma x_i = 155.2 \qquad \Sigma y_i = 392.0$$
$$\Sigma x_i^2 = 1508.8 \qquad \Sigma y_i^2 = 10{,}090.0$$
$$\Sigma x_i y_i = 3644.2$$

a. Find the estimated regression line.

b. Estimate the true mean spending limit for a customer who has had this credit card for ten years.

c. Find an estimate of an observed value of the spending limit for a customer who has had this credit card for two years.

12.156 Biology and Environmental Science Farmers in northern Sweden sell one of the most expensive cheeses in the world, which is made from moose milk. Suppose a random sample of female moose was obtained, and the weight of each was measured (x, in kilograms). The amount of milk produced by each moose in one day was also measured (y, in liters). **MOOSE**

a. Construct a scatter plot for these data.

b. Find the estimated regression line. Add a graph of the estimated regression line to the scatter plot in part (a).

c. Complete the ANOVA summary table, and conduct an F test for a significant regression. Use a significance level of 0.05.

d. Remove the first observation, a possible outlier, from the data set. Find the new estimated regression line. Is this new regression significant at the 0.05 level?

12.157 Public Health and Nutrition Fragments of protein waste, beta amyloids, may be a reliable marker in Alzheimer's disease (AD) patients. Recent studies suggest that these protein fragments accumulate in the brain and form hard plaques, which are found in many AD patients. However, as omega-3 intake increases, there is some evidence to suggest that the beta amyloid blood level decreases.[38] Suppose a study was conducted to investigate this relationship. A random sample of adults was obtained, the amount of omega-3 that each consumed in a day (x, in mg) was estimated, and the beta amyloids 40 level (y, in pg/ml) was measured for each. **PLAQUE**

a. Find the estimated regression line, and complete the ANOVA table.

b. Conduct an F test for a significant regression. Use $\alpha = 0.01$.

c. Find the coefficient of determination, r^2.

d. Interpret your results. Do you believe that omega-3 intake is a good predictor of beta amyloid blood level? Why or why not?

12.158 Technology and the Internet An economist at the National Bureau of Economic Research (NBER) used government data to conclude that as Americans spend more time online, they spend less time on many routine duties and tasks, that is, less time working.[39] A random sample of adults ages 25–55 was obtained, and each completed a survey concerning work and leisure activities. The amount of time spent online each day (x, in hours) and the amount of time spent working each day (y, in hours) for each are given in the following table. **NBER**

Time online	Time working	Time online	Time working
3.1	7.6	5.8	7.0
5.9	10.1	0.0	9.9
2.2	11.9	5.8	8.7
1.0	11.1	5.0	10.2
3.9	4.0	3.0	7.3
2.8	11.0	2.2	8.5
3.9	11.4	1.0	14.5
0.5	10.0	5.6	9.2
2.7	10.3	5.0	10.2
2.4	11.9	5.6	6.1
0.5	13.9	0.8	13.0
5.2	10.0	0.9	8.9

a. Carefully sketch a scatter plot of these data.

b. Find the sample correlation coefficient, and describe the relationship between time spent online and time spent working.

c. Find the estimated regression line. Explain why the estimated regression coefficient $\hat{\beta}_1$ supports your answer in part (b).

12.159 Biology and Environmental Science In a study by the U.S. Forestry Service, the relationship between certain chemical elements and oak tree defoliation was studied. A random sample of oak trees was obtained, and the concentration of phosphorus (x, in g/kg) in leaves was measured for each tree. The percentage of defoliation (y) for each tree was carefully estimated at the end of the growing season. 📊 OAKTREE

a. Carefully sketch a scatter plot of these data. Describe the relationship between these two variables.
b. Find the estimated regression line. Is this regression significant? Conduct an appropriate test at the 0.05 level. Add a graph of this line to your scatter plot in part (a).
c. Find the residuals, and sketch a graph of the residuals versus the predictor variable.
d. How could this model be improved? Justify your answer.

12.160 Medicine and Clinical Studies In a study of hair regeneration, 30 men over 45 years old who had used minoxidil for six months were randomly selected. The daily dose of minoxidil (x, in mg) and the hair density (y, in hairs/mm²) were recorded for each man. The following summary statistics were reported.

$$S_{xx} = 21423.0 \quad S_{yy} = 80.119 \quad S_{xy} = 165.593$$
$$\bar{x} = 49.033 \quad \bar{y} = 2.207$$

a. Find the estimated regression line.
b. Complete the ANOVA table, and conduct an F test for a significant regression. Does the dosage of minoxidil help to explain variation in hair density? Justify your answer.
c. Find a 95% confidence interval for the mean hair density for a man taking a daily 50-mg dose of minoxidil.

12.161 Economics and Finance A statistician (and cigar smoker) recently studied the relationship between *Cigar Aficionado*'s blind ratings of cigars and price. The purpose of this study was to determine whether premium cigars are really more expensive. A random sample of 50 cigars was selected, and the rating (x) and price per cigar (y, in dollars) were recorded for each. The following summary statistics were obtained.

$$S_{xx} = 185.725 \quad S_{yy} = 1481.83 \quad S_{xy} = 111.877$$
$$\bar{x} = 5.6084 \quad \bar{y} = 11.7654$$

a. Find the estimated regression line.
b. Complete the ANOVA table.
c. Conduct a test of $H_0: \beta_1 = 0$ versus $H_a: \beta_1 \neq 0$ with a significance level of 0.05.
d. Do you believe the saying, "You get what you pay for" applies to cigars? Justify your answer.

12.162 Manufacturing and Product Development Consumer Federation of America recently studied the relationship between tensile strength and amount of nickel in household stainless steel products. The percentage of nickel by weight (x) and the tensile strength (y, in MPa) were measured for each product. The data are given in the following table. 📊 NICKEL

x	5.9	2.8	5.3	4.6	4.6	3.5	3.8
y	948	859	921	909	915	876	828

x	5.0	5.3	4.2
y	964	964	900

a. Find the estimated regression line, and complete the ANOVA table.
b. Compute the residuals.
c. Carefully sketch a normal probability plot of the residuals, and interpret the graph.
d. Carefully sketch a scatter plot of the residuals versus the predictor variable. Is there any indication of a violation in the regression model assumptions?

12.163 Marketing and Consumer Behavior "Black Friday," the Friday after Thanksgiving, is traditionally the beginning of the Christmas shopping season, and is the busiest shopping day of the year. An economist believes there is a relationship between the time of day at which a consumer starts shopping on Black Friday and the amount of money spent that day. Suppose a random sample of Black Friday shoppers was obtained and the shopping start time (x, in hours after 6:00 A.M.) and the total amount spent (y, in hundreds of dollars) was recorded for each. 📊 SHOPPER

a. Find the estimated regression line, and complete the ANOVA table.
b. Describe the relationship between shopping start time and total amount spent on Black Friday. What would you recommend to retailers to increase business?
c. Find a 95% prediction interval for an observed value of shopping start time of 8:00 A.M.
d. Suppose the shopping start time is 9:00 A.M. Is there any evidence to suggest that the true mean amount spent is greater than $725? Use a significance level of 0.05.

12.164 Economics and Finance Scientists at the U.S. Mint believe that the number of years a quarter is in circulation is linearly related to the condition of the coin. To test this theory, a random sample of quarters was obtained. The number of years in circulation (x) was recorded for each coin, and each quarter was assessed for wear using the Official American Numismatic Association Grading Standards for U.S. Coins (y, a 70-point scale). A simple linear regression analysis was performed, and the following partial ANOVA table was obtained.

ANOVA summary table

Source of variation	Sum of squares	Degrees of freedom	Mean square	F	p Value
Regression					
Error	7671.0				
Total	11,148.4	54			

a. Complete the ANOVA table.
b. Find an estimate of the variance of the random error terms.

12.165 Biology and Environmental Science Every 13 or 17 years a brood of cicadas, sometimes in Biblical proportions, emerges from underground in the northeastern part of the United States. Although harmless to humans, cicadas cause damage to small trees and shrubs, and they make a piercing, irritating sound. Some research suggests that the density of cicadas is related to the density of moles, natural predators of cicada. Plots were randomly selected, and the density of moles per acre (x) and the density of cicadas per acre (y, in millions) were carefully estimated. The data are given in the following table. ▪▪▪ CICADA

x	4	6	3	0	2	4	0	5	4
y	1.5	0.8	1.2	1.4	1.5	1.5	1.7	1.2	1.1

a. Carefully sketch a scatter plot of the data. Describe the relationship between the density of moles and the density of cicadas.
b. Find the estimated regression line. Conduct a hypothesis test of H_0: $\beta_1 = 0$ versus H_a: $\beta_1 \neq 0$. Use a significance level of 0.05.
c. Construct a normal probability plot of the residuals and a scatter plot of the residuals versus the predictor variable. Is there any evidence that the regression assumptions are not satisfied?

12.166 Travel and Transportation Since airlines began charging for checked luggage, more passengers use carry-on bags to avoid the extra fee. For those who check luggage, the wait at the baggage carousel while jostling with other passengers can be long and infuriating. A study was conducted to determine whether there is a relationship between the flight time (x, in hours) and the amount of time for a bag to arrive from a plane to the carousel (y, in minutes). A random sample of arriving flights at the Denver International Airport was obtained and the flight time and baggage time were recorded for each. ▪▪▪ LUGGAGE
a. Construct a scatter plot for these data. Describe the relationship between flight time and the amount of time for a bag to arrive from a plane to the carousel.
b. Find the estimate regression line, complete the ANOVA table, and conduct an F test for a significant regression. Use $\alpha = 0.05$.
c. Based on your results in part (b), does flight time affect the time it takes for a bag to arrive at the carousel? What are some other variables that might affect this time?

12.167 Manufacturing and Product Development In structural tests of material to be used in an airplane's fuselage, rivets are cycled (tapped) until they crack. The number of cycles (x, in thousands) is recorded along with the detectable crack length (y, in mm). The data are given in the following table. ▪▪▪ RIVETS

Cycles	Length	Cycles	Length	Cycles	Length
128.0	0.193	118.1	0.165	199.0	0.191
246.4	0.167	100.7	0.191	100.8	0.177
188.8	0.209	205.9	0.208	131.1	0.207
152.3	0.180	107.6	0.153	145.8	0.193
129.9	0.153	182.5	0.169	229.1	0.191
101.5	0.172	132.8	0.187		

a. Carefully sketch a scatter plot of these data. Does there appear to be a linear relationship between the number of cycles and the length of the crack? Explain.

b. Find the estimated regression line, and complete the ANOVA table.
c. How does the F test for a significant regression support your answer to part (a)?

12.168 Medicine and Clinical Studies Twenty-one adults were selected at random for a study of the effect of castor oil on the immune system. Participants took a certain amount of castor oil (x, in milliliters) each day for three months. At the end of this regimen, the white blood cell count was measured (y, thousands of cells per microliter of blood) for each person. Analyze these data using any appropriate methods to determine whether there is a significant linear relationship between the amount of castor oil consumed and the number of white blood cells. Check the relevant assumptions. Assume that the more white blood cells, the stronger the immune system, and draw a conclusion about the effect of castor oil on the immune system. ▪▪▪ CASTOR

12.169 Sports and Leisure Despite the rising cost of equipment, the number of fishing licenses issued in many states continues to increase. Officials are trying to predict the number of participants in freshwater fishing in order to hire and deploy the appropriate number of fish and game wardens. The data given on the website list the percentage of freshwater (x) and the freshwater fishing participation rate (y) by state.[40] Note: No data are given for Hawaii. Analyze these data using any appropriate methods to determine whether there is a significant linear relationship between the percentage of freshwater and the participation rate. Check the relevant assumptions. Does the sign of $\widehat{\beta}_1$ seem appropriate? Why or why not? ▪▪▪ FISHING

EXTENDED APPLICATIONS

12.170 Manufacturing and Product Development A study was conducted to determine the effect of thread count (x_1) and the percentage of cotton (x_2) on the lifetime of standard twin bed sheets (y, in years). A machine was constructed to simulate continued usage and to allow a measurement of the lifetime. ▪▪▪ SHEETS
a. Estimate the regression coefficients in the model $Y_i = \beta_0 + \beta_1 x_{1i} + \beta_2 x_{2i} + E_i$.
b. Construct an ANOVA table and conduct a model utility test. Use technology to find the p value associated with this test. Find the value of r^2. Explain the meaning of this value.
c. Conduct the necessary hypothesis tests to determine whether both predictor variables are significant. Use an *overall* significance level of $\alpha = 0.05$.
d. Find a 99% prediction interval for an observed value of y for $x_1 = 320$ and $x_2 = 100$.
e. Suppose a certain brand of bed sheets is sold with 50% cotton and various thread counts. What thread count would guarantee an estimate of the mean lifetime of at least 15 years?

12.171 Travel and Transportation Many of the highways in Pennsylvania are paved with concrete rather than tar. Suppose an experiment was conducted to determine whether the compression strength in cured concrete pavement (y, psi) is affected by the water-to-cement ratio (x_1), the sand-to-cement ratio (x_2), the percentage of fly ash (x_3), and/or the ambient temperature during curing (x_4, °F). **PAHWY**

a. Estimate the regression coefficients in the model $Y_i = \beta_0 + \beta_1 x_{1i} + \beta_2 x_{2i} + \beta_3 x_{3i} + \beta_4 x_{4i} + E_i$. Use the sign of each estimated regression coefficient to explain the effect of each predictor variable on the compression strength of concrete.

b. Construct an ANOVA table and conduct a model utility test. Use technology to find the p value associated with this test.

c. Conduct the appropriate hypothesis tests to determine the significant predictor variables.

d. Construct a normal probability plot of the residuals and the plots of the residuals versus each predictor variable. Is there any evidence of a violation in the multiple linear regression assumptions?

12.172 Public Health and Nutrition There is some evidence to suggest that exposure to airborne fungal products may be associated with adverse health effects, for example, respiratory tract infections.[41] A random sample of Canadian elementary schools was obtained and the following data was obtained for each: mesophilic fungi (y, in cfu/m^3), building age (x_1, in years), air exchange rate (x_2, per hour), indoor CO_2 (x_3, ppm), outdoor temperature (x_4, °C), and a dummy variable, signs of moisture in the room (x_5, 0 for no signs, 1 for signs). **FUNGI**

a. Estimate the regression coefficients in the model $Y_i = \beta_0 + \beta_1 x_{1i} + \beta_2 x_{2i} + \beta_3 x_{3i} + \beta_4 x_{4i} + \beta_5 x_{5i} + E_i$.

b. Determine the most important predictor variables. Find the estimated regression line for this reduced model.

c. Use the sign of each estimated regression coefficient in the reduced model to explain the effect of each predictor variable on the volume of mesophilic fungi.

d. Check the relevant assumptions for the reduced model.

LAST STEP

12.173 Are windmills too noisy? An experiment was conducted to measure the windmill noise level (in dB) at various distances (in meters) from a proposed site. The data are summarized in the following table.

Distance	10	50	75	120	150	160	200	250	400	500
Noise level	75	110	73	52	58	77	56	57	28	4

a. Find the estimated regression line, and complete the ANOVA table.

b. Find the coefficient of determination and interpret this value.

c. Suppose a windmill is constructed 350 meters from your home. Find an estimate for the mean noise level and a 95% prediction interval for the noise level.

d. Construct a normal probability plot of the residuals and a scatter plot of the residuals versus distance. Is there any evidence of non-normality? Justify your answer.

Exit ④⑤ Grand Central Ese⋯
⑥ Ⓢ Terminal ups⋯

← **5 Avenue**
Times Square

13 Categorical Data and Frequency Tables

◀ **Looking Back**

- Remember how univariate categorical data are identified: non-numerical observations that may be placed in categories.
- Recall how bivariate categorical data are obtained and characterized: two non-numerical observations on each individual or object.
- Think about the natural summary measures for categorical data: frequency and relative frequency.

▶ **Looking Forward**

- Learn the background, computations, and interpretations of a goodness-of-fit test concerning the true population proportions.
- Conduct a test for homogeneity or independence.

Is public transportation use the same for all New York boroughs?

A recent study concluded that New York City residents spend the most time commuting when compared to all workers in the United States.[1] According to the Partnership for New York City, a typical New Yorker spends 48 minutes getting to work, and, therefore, almost two hours each day commuting. Millions of people commute between the city's five boroughs. However, the public transportation options are limited in some boroughs, and there are plans to improve subway, bus, and ferry service.

A random sample of New York City commuters was obtained and each was classified by resident borough and the type of transportation used to get to work. The resulting frequencies, or counts, are given in the following two-way, or contingency, table.

		Type of transportation			
		Bus	Subway	Ferry	Drive
Borough	Bronx	40	45	19	15
	Brooklyn	45	52	14	8
	Manhattan	30	54	17	11
	Queens	47	48	15	13
	Staten Island	30	43	20	14

The techniques presented in this chapter will be used to determine whether two categorical variables are dependent. In the context of this problem, we would like to know whether residents in all boroughs have the same public transportation patterns.

CONTENTS

© David M. Grossman/The Image Works

13.1 Univariate Categorical Data, Goodness-of-Fit Tests

Categorical data are often displayed in a frequency distribution. In this section, we will focus only on the number of observations in each category, and display these results in a one-way frequency table. This table lists each possible category and the number of times each category occurred (the observed count or frequency for each category). For example, a recent article in *Forbes* suggested that there are four types of employees.[2] Suppose the Human Resources department at a large corporation selects 200 employees at random and classifies each. The total in each category is given in the following one-way frequency table.

Employee type	Motivated	Indifferent	Disgruntled	Con-artist
Frequency	70	55	45	30

The hypothesis test procedure presented in this section is designed to compare a set of *hypothesized* proportions with the set of *true* proportions, to check the **goodness of fit (GOF)**. For example, the HR director might use the data above to determine whether there is any evidence that the true proportions are different from the following hypothesized proportions: 0.4 (for motivated employees), 0.3 (for indifferent employees), 0.2 (for disgruntled employees), and 0.1 (for con-artist employees).

The notation used here is an extension of the notation used to represent sample and population proportions. Suppose each observation falls into one of k categories.

	True proportion	Hypothesized proportion
Category 1	p_1	p_{10}
Category 2	p_2	p_{20}
$\vdots$	$\vdots$	$\vdots$
Category i	p_i	p_{i0}
$\vdots$	$\vdots$	$\vdots$
Category k	p_k	p_{k0}

> The subscript in the hypothesized proportion p_{10} is read as "one zero," not "ten." The first number in the subscript (1) denotes the category, and the second number (0) indicates that this is a hypothesized value.

The true proportion and the hypothesized proportion are both population proportions. The sum of the hypothesized proportions (like the sum of the true proportions) must be 1. That is, $p_{10} + p_{20} + \cdots + p_{k0} = 1$.

The goodness-of-fit test is done to determine whether there is any evidence (from the observed, or sample, counts) that the true population proportions differ from the hypothesized population proportions. The null hypothesis and the alternative hypothesis are stated in terms of the true and hypothesized category proportions.

> H_0 is a composite null hypothesis. It involves several parameters and is true only if all the equalities hold.

$H_0: p_1 = p_{10}, p_2 = p_{20}, \ldots, p_k = p_{k0}.$
(Each true category proportion is equal to a specified hypothesized value.)

$H_a: p_i \neq p_{i0}$ for at least one i.
(There is at least one true category proportion that is not equal to the corresponding specified hypothesized value.)

> In this context, a *cell* is simply a category.

Suppose a random sample of size n is selected; let n_i ($i = 1, 2, \ldots, k$) be the number of observations falling into each category. To decide whether the sample data *fit* the hypothesized proportions, *observed* cell counts (the n_i's) are compared with *expected* cell counts. If the null hypothesis is true, then the expected frequency, or count, for category 1, or in cell 1, is $e_1 = np_{10}$. For cell 2, the expected frequency is $e_2 = np_{20}$, etc. For example, if the sample size is $n = 100$ and $p_{10} = 0.25$, then we expect the count for category 1 to be, on average, $np_{10} = (100)(0.25) = 25$.

The test statistic is a measure of how far away the observed cell counts are from the expected cell counts. If the null hypothesis is true, then the random variable

$$X^2 = \sum_{i=1}^{k} \frac{(\text{observed cell count} - \text{expected cell count})^2}{\text{expected cell count}} = \sum_{i=1}^{k} \frac{(n_i - e_i)^2}{e_i}$$

The X in X^2 is the uppercase Greek letter chi (χ is the lowercase form).

has approximately a chi-square distribution with $k - 1$ degrees of freedom. This approximation is good if $e_i = np_{i0} \geq 5$ for all i, that is, if all expected cell counts are at least 5.

If the observed cell counts are *close* to the expected cell counts, then the value of X^2 will be small. If the observed cell counts are considerably different from the expected cell counts, then the value of X^2 will be large. Therefore, the null hypothesis is rejected only for large values of the test statistic. Now we have all of the pieces for the complete hypothesis test.

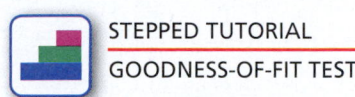

Goodness-of-Fit Test

Let n_i be the number of observations falling into the ith category ($i = 1, 2, \ldots, k$), and let $n = n_1 + n_2 + \cdots + n_k$. A hypothesis test about the true category population proportions with significance level α has the form

H_0: $p_1 = p_{10}, p_2 = p_{20}, \ldots, p_k = p_{k0}$

H_a: $p_i \neq p_{i0}$ for at least one i.

TS: $X^2 = \sum_{i=1}^{k} \dfrac{(n_i - e_i)^2}{e_i}$ where $e_i = np_{i0}$

RR: $X^2 \geq \chi^2_{\alpha, k-1}$

DATA SET

BEES

A reminder: This test is appropriate if all expected cell counts are at least 5 ($e_i = np_{i0} \geq 5$ for all i). The chi-square distribution was introduced in Section 8.5, and a hypothesis test based on this distribution was discussed in Section 9.7.

Example 13.1 Sweet as Tupelo Honey

Solution Trail 13.1

KEYWORDS

- Test the hypothesis
- Four possible bee purchases occur with equal frequency

TRANSLATION

- Statistical inference
- True category population proportions are equal

CONCEPTS

- Goodness-of-fit test

VISION

Find the expected cell counts, and use the goodness-of-fit test procedure to determine whether there is any evidence that any one of the true population proportions differs from 0.25.

For beekeepers looking to harvest more honey, there are four possibilities for obtaining more bees: package bees, nucs, established colonies, or swarms. A university agricultural sciences department obtained a random sample of recent bee purchases, and each was classified into one of the four categories. Use the following one-way frequency table to test the hypothesis that the four possible bee purchases occur with equal frequency. Use $\alpha = 0.05$.

Bee purchase	Package bees	Nucs	Colonies	Swarms
Frequency	31	36	26	20

SOLUTION

STEP 1 If the bee purchase types are equally likely, the proportion of purchases falling into each category is $1/k = 1/4 = 0.25$.

The four parts of the hypothesis test are

H_0: $p_1 = 0.25, p_2 = 0.25, p_3 = 0.25, p_4 = 0.25$

H_a: $p_i \neq p_{i0}$ for at least one i.

TS: $X^2 = \sum_{i=1}^{k} \dfrac{(n_i - e_i)^2}{e_i}$

RR: $X^2 \geq \chi^2_{\alpha, k-1} = \chi^2_{0.05, 3} = 7.8147$ From Table VI in the Appendix.

STEP 2 There are $n = 31 + 36 + 26 + 20 = 113$ total observations. The expected counts are given in the following table.

Cell	Category	Observed cell count	Expected cell count
1	Package bees	31	$e_1 = np_{10} = (113)(0.25) = 28.25$
2	Nucs	36	$e_2 = np_{20} = (113)(0.25) = 28.25$
3	Colonies	26	$e_3 = np_{30} = (113)(0.25) = 28.25$
4	Swarms	20	$e_4 = np_{40} = (113)(0.25) = 28.25$

All four expected cell counts are greater than 5. The chi-square goodness-of-fit test is appropriate.

STEP 3 The value of the test statistic is

$$\chi^2 = \sum_{i=1}^{4} \frac{(n_i - e_i)^2}{e_i}$$

$$= \frac{(31 - 28.25)^2}{28.25} + \frac{(36 - 28.25)^2}{28.25} + \frac{(26 - 28.25)^2}{28.25} + \frac{(20 - 28.25)^2}{28.25}$$

Use observed and expected cell counts.

$$= 0.2677 + 2.1261 + 0.1792 + 2.4093 = 4.9823 \ (< 7.8147)$$ Simplify.

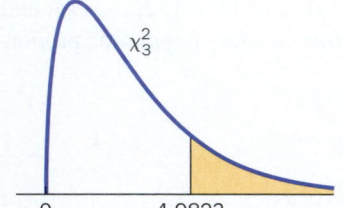

Figure 13.1 p-Value illustration:
$p = P(X \geq 4.9823)$
$= 0.1731 > 0.05 = \alpha$

Recall that, using Table VI in the Appendix, we can only *bound* the p value for this hypothesis test. CrunchIt!, Minitab, or the TI-84 can be used to find the exact p value for this test.

STEP 4 The value of the test statistic does not lie in the rejection region or equivalently, $p = 0.1731 > 0.05$. Figure 13.1 shows the calculation and an illustration of the exact p value. At the $\alpha = 0.05$ significance level, there is no evidence to suggest that any one of the true population proportions differs from 0.25.

Figure 13.2 shows a technology solution.

Test Probabilities

Level	Estim Prob	Hypoth Prob
Colonies	0.23009	0.25000
Nucs	0.31858	0.25000
Package bees	0.27434	0.25000
Swarms	0.17699	0.25000

Test	ChiSquare	DF	Prob>Chisq
Likelihood Ratio	5.0838	3	0.1658
Pearson	4.9823	3	0.1731

Figure 13.2 JMP goodness-of-fit test results.

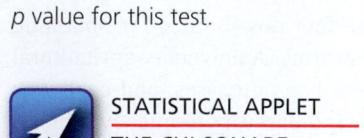

STATISTICAL APPLET

THE CHI-SQUARE
GOODNESS-OF-FIT TEST

DATA SET

THANKS

TRY IT NOW   GO TO EXERCISE 13.12

Example 13.2 Thanksgiving Traditions

On Thanksgiving, many families traditionally gather for a wonderful meal, lively conversation, and in some cases a spirited street hockey game. A random sample of adults over age 18 was obtained and asked to name their favorite Thanksgiving food. The data and the proportions from a previous survey[3] are given in the following table.

Favorite food	Frequency	Previous proportions
Turkey	250	0.38
Stuffing	148	0.26
Mashed potatoes	98	0.17
Yams	55	0.10
Green bean casserole	30	0.05
Cranberry sauce	42	0.04

Is there evidence to suggest that any of the true cell proportions differ from the previous proportions? Use $\alpha = 0.05$.

SOLUTION

STEP 1 A goodness-of-fit test is appropriate to determine whether there is evidence that any of the true category population proportions have changed. There are $k = 6$ categories.

The four parts of the hypothesis test are

H_0: $p_1 = 0.38, p_2 = 0.26, p_3 = 0.17, p_4 = 0.10, p_5 = 0.05, p_6 = 0.04$

H_a: $p_i \neq p_{i0}$ for at least one i.

TS: $X^2 = \sum_{i=1}^{k} \frac{(n_i - e_i)^2}{e_i}$

RR: $X^2 \geq \chi^2_{\alpha, k-1} = \chi^2_{0.05, 5} = 11.0705$ Table VI in the Appendix.

STEP 2 There are $n = 250 + 148 + 98 + 55 + 30 + 42 = 623$ total observations. The expected cell counts are given in the following table.

Cell	Favorite food	Observed cell count	Expected cell count
1	Turkey	250	$e_1 = np_{10} = (623)(0.38) = 236.74$
2	Stuffing	148	$e_2 = np_{20} = (623)(0.26) = 161.98$
3	Mashed potatoes	98	$e_3 = np_{30} = (623)(0.17) = 105.91$
4	Yams	55	$e_4 = np_{40} = (623)(0.10) = 62.30$
5	Green bean casserole	30	$e_5 = np_{50} = (623)(0.05) = 31.15$
6	Cranberry sauce	42	$e_6 = np_{60} = (623)(0.04) = 24.92$

All expected cell counts are greater than 5. The chi-square goodness-of-fit test is appropriate.

STEP 3 The value of the test statistic is

$$\chi^2 = \sum_{i=1}^{6} \frac{(n_i - e_i)^2}{e_i}$$

$$= \frac{(250 - 236.74)^2}{236.74} + \frac{(148 - 161.98)^2}{161.98} + \frac{(98 - 105.91)^2}{105.91}$$

$$+ \frac{(55 - 62.30)^2}{62.30} + \frac{(30 - 31.15)^2}{31.15} + \frac{(42 - 24.92)^2}{24.92}$$

Use observed and expected cell counts.

$$= 0.74 + 1.21 + 0.59 + 0.86 + 0.04 + 11.71$$

$$= 15.15 \, (\geq 11.0705)$$ Simplify.

STEP 4 The value of the test statistic ($\chi^2 = 15.15$) lies in the rejection region or equivalently, $p = 0.0097 \leq 0.05$. At the $\alpha = 0.05$ significance level, there is evidence to suggest that at least one population proportion has changed from its previous value.

Note: we can use Table VI in the Appendix to bound the p value for this hypothesis test. In row $k - 1 = 5$ (degrees of freedom), place 15.15 in the ordered list of critical values.

$$15.0863 \leq 15.15 \leq 16.7496$$

$$\chi^2_{0.01, 5} \leq 15.15 \leq \chi^2_{0.005, 5}$$

Therefore, $0.005 \leq p \leq 0.01$ See Figure 13.3.

Figure 13.3 shows the calculation and an illustration of the exact p value.

Jon Feingersh/Blend Images/Getty Images

Which food has gained or lost most in popularity? How do you know?

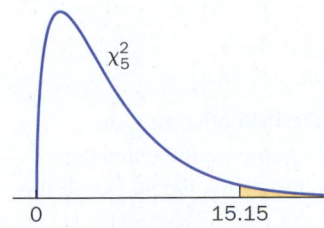

Figure 13.3 p-Value illustration:
$p = P(X \geq 15.15)$
$= 0.0097 \leq 0.05 = \alpha$

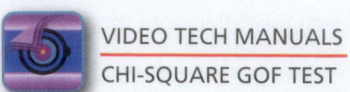

VIDEO TECH MANUALS

CHI-SQUARE GOF TEST

Figure 13.4 shows a technology solution.

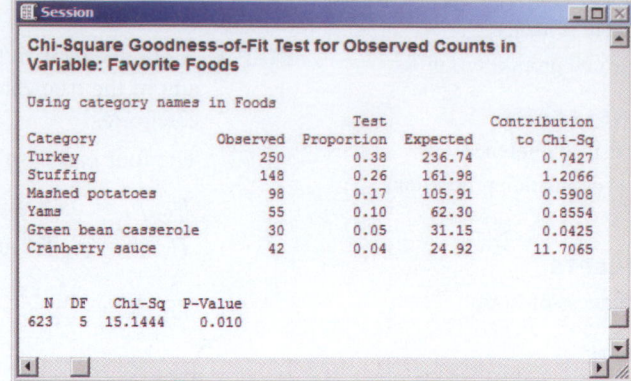

Figure 13.4 Minitab goodness-of-fit test.

TRY IT NOW   GO TO EXERCISE 13.17

Technology Corner

Procedure: Chi-square goodness-of-fit test.
Reconsider: Example 13.2, solution, and interpretations.

CrunchIt!

Use the built-in function Chi-squared Goodness-of-fit.

1. Enter the observed frequencies into column Var1. If all categories are assumed to be equally likely, continue with Step 2. Otherwise, enter the corresponding expected probabilities into column Var2.
2. Select Statistics; Chi-squared Goodness-of-fit. Choose the Observed Frequencies column and optionally the Expected Probabilities column.
3. Click Calculate. The results are displayed in a separate window. See Figure 13.5.

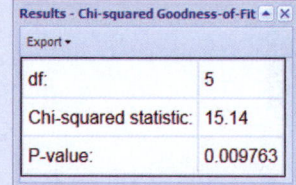

Figure 13.5 CrunchIt! Chi-squared Goodness-of-fit results.

TI-84 Plus C

Use the built-in function χ^2GOF-Test.

1. Enter the observed frequencies into list L1 and the expected frequencies into list L2.
2. Select STAT; TESTS; χ^2GOF-Test. Enter the list containing the observed frequencies, the list containing the expected frequencies, and the degrees of freedom associated with the test statistics. Note: To compute the expected frequencies, enter the hypothesized proportions into a list. Compute n (623 in this example) times this list and store the results (expected frequencies) in a new list.
3. Highlight Calculate and press ENTER. The hypothesis test results are displayed on the Home screen. See Figure 13.6. The Draw results are shown in Figure 13.7.

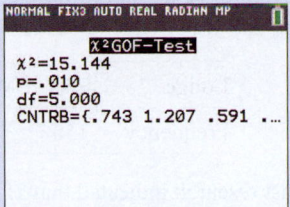

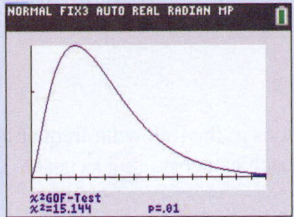

Figure 13.6 χ^2GOF-Test hypothesis test results.

Figure 13.7 χ^2GOF-Test Draw results.

Minitab

The Chi-square goodness-of-fit test is in the Stat; Tables menu.

1. Enter the category names into column C1 (optional), the hypothesized proportions into column C2, and the observed counts into column C3.
2. Select Stat; Tables; Chi-Square Goodness-of-Fit Test (One variable). Enter the Observed counts column, the Category names column, and the Specific proportions column.
3. Click OK. The results are displayed in a session window. Refer to Figure 13.4. This Minitab procedure also automatically produces a chart of observed and expected values by category, and a chart of the contribution to the value of the chi-square statistic by category.

Excel

Use the built-in function CHISQ.TEST.

1. Enter the observed frequencies into column B. Compute and enter the expected frequencies into column D.
2. Use the function CHISQ.TEST to find the p value associated with the goodness-of-fit test and the function CHISQ.INV to find the value of the test statistic if desired. See Figure 13.8.

	A	B	C	D
1	Favorite	Observed	Hypothesized	Expected
2	food	frequencies	proportions	frequencies
3	Turkey	250	0.38	236.74
4	Stuffing	148	0.26	161.98
5	Mashed potatoes	98	0.17	105.91
6	Yams	55	0.10	62.30
7	Green bean casserole	30	0.05	31.15
8	Cranberry sauce	42	0.04	24.92
9	Total	623	1.00	623.00
10				
11	p value		0.0098	= CHISQ.TEST(B3:B8,D3:D8)
12	Test statistic		15.14	= CHISQ.INV(1-B11,5)

Figure 13.8 Excel template for a goodness-of-fit test.

SECTION 13.1 EXERCISES

Concept Check

13.1 True/False In a goodness-of-fit test the hypothesized proportions are always assumed to be equal.

13.2 Short Answer Explain how to compute each expected cell count in a goodness-of-fit test.

13.3 Fill in the Blank If the observed cell counts are close to the expected cell counts in a goodness-of-fit test, then the value of X^2 should be _____.

13.4 Fill in the Blank In a goodness-of-fit test with k categories, the test statistic is compared to a critical value from a chi-square distribution with _____ degrees of freedom.

13.5 Fill in the Blank The goodness-of-fit test is appropriate if all expected cell counts are _____ .

Practice

13.6 Use the hypothesized proportions in the following frequency table to find the expected count for each category. EX13.6

Category	1	2	3	4
Frequency	42	58	62	56
p_{i0}	0.25	0.40	0.20	0.15

13.7 Use the hypothesized proportions in the following frequency table to find the expected count for each category. EX13.7

Category	1	2	3	4	5
Frequency	125	150	300	220	205
p_{i0}	0.15	0.14	0.31	0.23	0.17

13.8 Consider the following one-way frequency table. EX13.8

Category	1	2	3	4
Frequency	115	85	70	30

a. Suppose the hypothesized proportions are $p_{10} = 0.4$, $p_{20} = 0.3$, $p_{30} = 0.2$, and $p_{40} = 0.1$. Find the four parts of the appropriate goodness-of-fit test. Use $\alpha = 0.01$.
b. Find the expected count for each cell. Verify that each is at least 5.
c. Find the value of the test statistic. State your conclusion, and justify your answer.

13.9 Consider the following one-way frequency table. EX13.9

Category	1	2	3	4	5	6
Frequency	90	82	75	95	96	36

a. Find the four parts of a goodness-of-fit test with $p_{10} = 0.175$, $p_{20} = 0.171$, $p_{30} = 0.162$, $p_{40} = 0.225$, $p_{50} = 0.202$, and $p_{60} = 0.065$. Use $\alpha = 0.05$.
b. Find the value of the test statistic. State your conclusion and justify your answer.
c. Find bounds on the p value.

13.10 Consider the following one-way frequency table. EX13.10

Category	1	2	3	4	5
Frequency	140	135	155	152	168

Conduct a goodness-of-fit test to check the hypothesis that all five categories are equally likely. Use $\alpha = 0.05$. Find bounds on the p value.

Applications

13.11 Psychology and Human Behavior A random sample of adult males was obtained and asked whether they belonged to a fraternal organization. The results are given in the following one-way frequency table. EX13.11

Lodge	Elks	Moose	Raccoon	None
Frequency	150	60	60	230

Past research indicated that 25% of all men belonged to an Elks lodge, 15% belonged to a Moose lodge, and 10% belonged to a Raccoon lodge. Conduct a goodness-of-fit test to determine whether there is evidence that at least one of the population proportions associated with the four categories above has changed. Use $\alpha = 0.05$.

13.12 Marketing and Consumer Behavior In a random sample of Caribou coffee-shop customers who purchased flavor shots, 185 bought almond, 180 purchased vanilla, 230 chose raspberry, 220 opted for caramel, and 185 went with hazelnut. Use a goodness-of-fit test to determine whether there is any evidence that customers who purchase flavor shots prefer one over the others. Use $\alpha = 0.05$.

13.13 Economics and Finance The AAII Sentiment Survey is administered every week to determine the percentage of investors who are bullish, bearish, or neutral on stock prices in the next six months. The long-term, or historical, percentages are bullish 39.0%, neutral 30.5%, and bearish 30.5%.[4] The following data were obtained in a random sample of investors for the week ending 12/11/2013. STOCK

Sentiment	Bullish	Neutral	Bearish
Frequency	413	337	250

Is there any evidence that the percentages during this week are different from the long-term percentages? Use $\alpha = 0.01$.

13.14 Biology and Environmental Science In a recent study of adult eye color, the hypothesized distribution was given as blue 25%, green 10%, brown 50%, and black 15%. A random sample of adults was obtained, and the resulting eye colors are summarized in the following one-way frequency table. EYES

Eye color	Blue	Green	Brown	Black
Frequency	121	65	242	72

Conduct a goodness-of-fit test to determine whether these data provide any evidence that the true proportions differ from the hypothesized proportions. Use $\alpha = 0.05$.

13.15 Marketing and Consumer Behavior Original Designs sells home plans in five different styles. A random sample of purchases was obtained, and the number falling into each category was recorded. The data are given in the following table. HOMES

Style	A-frame	Cape Cod	Colonial	Log home	Ranch
Frequency	75	180	268	58	385

Conduct a goodness-of-fit test to determine whether there is evidence that any of the true proportions differ from the hypothesized values 0.10, 0.20, 0.25, 0.05, and 0.40. Use $\alpha = 0.05$.

13.16 Marketing and Consumer Behavior A random sample of West Bay Yacht Club members was obtained and asked to identify their dinner event preference. The resulting data along with the historical proportions are given in the following table.[5] **DINNER**

Dinner event preference	Frequency	p_{i0}
Pot luck	120	0.33
Catered $8	95	0.18
Catered $12	100	0.26
Catered $15	36	0.10
Catered $18	61	0.13

Conduct a goodness-of-fit test to determine whether there is evidence that any of the true proportions differs from the historical value. Use $\alpha = 0.05$.

13.17 Marketing and Consumer Behavior The manager at a Publix grocery store is trying to determine which types of lettuce to order and sell to customers. A random sample of shoppers was obtained and asked to select their favorite lettuce type. The data are given in the following table. **LETTUCE**

Lettuce	Frequency
Crisphead	45
Butterhead	17
Romaine	130
Leafy	90
Celtuce	28

Test the fit of these data to the hypothesized proportions 0.14, 0.06, 0.37, 0.30, and 0.13. Use $\alpha = 0.05$.

13.18 Public Health and Nutrition The U.S. Centers for Disease Control and Prevention collects data on the number of reported flu cases. The following table shows the number of reported H1N1 flu cases by region for the week ending December 15, 2013, and the hypothesized proportions from 2009.[6] **H1N1**

Surveillance region	Frequency	p_{i0}
1	40	0.014
2	125	0.047
3	145	0.061
4	501	0.210
5	345	0.132
6	380	0.151
7	140	0.051
8	501	0.221
9	164	0.066
10	110	0.047

Conduct a goodness-of-fit test to determine whether there is evidence that any of this year's true population proportions differs from the 2009 proportion. Use $\alpha = 0.05$.

13.19 Business and Management A random sample of employed adults in Canada was obtained and classified by occupation. The resulting data and the hypothesized proportions are given in the following table.[7] **EMPLOYED**

Occupation	Frequency	p_{i0}
Management	1503	0.09
Business, finance, administration	3157	0.18
Natural and applied sciences	1357	0.07
Health	1183	0.07
Social science, education, gov, and rel	1659	0.09
Art, culture, recreation, sport	595	0.03
Sales and service	4262	0.24
Trades, transport, eq operators	2634	0.15
Unique to primary	557	0.03
Unique to processing, man, util	788	0.05

Conduct a goodness-of-fit test to determine whether there is evidence that any of the true proportions differs from the hypothesized value. Use $\alpha = 0.01$.

13.20 Public Health and Nutrition Prescription medication can be very expensive, in part because of the high cost of research and development. The most expensive drug in the world is Soliris, which is used to treat a rare blood disorder. Many insurance companies have adopted a tiered plan that allows the user to select a generic or preferred drug at a reduced cost or copay instead of the named or specialty drug at a higher copay. Medicare Blue RX has four tiers for prescription medication, each with different copays. Past records indicate that the proportions of users who select each tier are as follows: Tier 1, 0.35; Tier 2, 0.28; Tier 3, 0.12; Tier 4, 0.25. Suppose a random sample of policyholders who purchased medication was obtained and the frequency associated with each tier is given in the following table. **SOLIRIS**

Tier	1	2	3	4
Frequency	152	107	36	76

a. Use a goodness-of-fit test to show that there is a shift in the proportion of users by tiers. Use $\alpha = 0.05$.
b. Use your results in part (a) to explain the shift in the proportions, that is, which tiers are more/less utilized.

13.21 Sports and Leisure As soon as the gates at an amusement park open, there is a mad rush to the most popular rides. A random sample of people waiting in line was obtained and asked which ride they were headed to first. The results and the hypothesized proportions are given in the following table. **RIDES**

Ride	Frequency	p_{i0}
Tower of Terror	63	0.40
Rockin' Roller Coaster	35	0.30
Star Tours	16	0.15
Studios Backlot Tour	14	0.15

Conduct a goodness-of-fit test to determine whether any of the true population proportions differ from the hypothesized proportions. Use $\alpha = 0.01$.

13.22 Marketing and Consumer Behavior According to a Rasmussen Report, 58% of American adults dine out at least once per week.[8] Consequently, customers are quick to voice dissatisfaction when their dining experience is unpleasant. A random sample of diners was obtained and asked to name their most common restaurant complaint. The results and the hypothesized proportions are given in the following table. 📊 COMPLAINT

Complaint	Frequency	p_{i0}
Food quality	45	0.21
Cleanliness	32	0.19
Service speed	50	0.28
Prices	66	0.32

Conduct a goodness-of-fit test to determine whether any of the true population proportions differs from the hypothesized proportion. Use $\alpha = 0.05$.

13.23 Marketing and Consumer Behavior In a recent survey by the National Association of Realtors, participants were asked to select their housing type preference.[9] A random sample of adults in Florida was obtained and asked to select their housing type preference. The results are given in the following table. 📊 HOUSING

Housing type	Florida frequencies	Realtors proportions
Single-family detached home, large yard	144	0.52
Single-family detached home, small yard	101	0.24
Apartment or condominium	74	0.14
Single-family home or townhouse	25	0.06
Something else	21	0.04

a. Conduct a goodness-of-fit test to show that there is evidence the true proportions in Florida differ from those found in the national survey. Use $\alpha = 0.05$.

b. Use technology to find the exact p value associated with this hypothesis test.

c. Use your results in part (a) to explain which proportions are different. Why would you expect these proportions to be different in Florida?

Extended Applications

13.24 Psychology and Human Behavior Berkeley Breathed stopped writing his Pulitzer Prize–winning comic strip *Bloom County* in 1989. Since then, he has written several books and animations. Recently, Mr. Breathed has been involved in writing a children's movie and two sequel comic strips, *Outland* and *Opus*, featuring some of the old *Bloom County* gang. A random sample of people who read the original *Bloom County* comic strip was obtained and asked to name their favorite character. The results are given in the following table. 📊 BLOOM

Character	Frequency
Opus	250
Michael Binkley	210
Oliver Wendell Jones	205
Milo Bloom	190
Bill the Cat	260
Cutter John	195
Steve Dallas	201
Portnoy	206
Hodge Podge	185
Rosebud	204

Is there any evidence to suggest that one (or more) character(s) are more popular than the others? Find bounds on the p value associated with this test.

13.25 Biology and Environmental Science Maine is the largest producer of wild blueberries in the United States and in the world.[10] Blueberry farms use several methods of pest management and pruning practices to increase the harvest. The following table contains the total number of farms for each pruning practice in 2013 and for a random sample in 2014. 📊 PRUNING

Pruning practice	2013 totals	2014 sample
Straw burn	33	44
Oil burn	44	40
Mow	87	110
Prune every other year	93	94

a. Compute the proportion associated for each pruning practice based on the 2013 totals.

b. Conduct a goodness-of-fit test to determine whether there is any evidence that the true 2013 proportions of pruning practice have changed in 2014. Use $\alpha = 0.01$.

13.26 Business and Management Big banks and cable television providers traditionally receive very low customer service ratings. Many of us simply hate dealing with banks and cable companies. In a recent survey by MSN Money, each participant was asked to select the one item that matters the most in customer service.[11] A random sample of consumers was obtained shortly after the 2014 holiday shopping season, and each was also asked the same question. The results are given in the following table. **SERVICE**

Customer service item	2014 sample	MSN survey
Knowledgeable staff	150	0.30
Service after the sale	130	0.18
Friendly staff	85	0.16
Flexible policies for returns/exchanges	75	0.13
Readily available staff	47	0.09
The product is all that matters	45	0.08
Not sure	34	0.06

Conduct a goodness-of-fit test to determine whether there is evidence that the true 2014 proportions of the most important customer service items have changed since the MSN survey. Use $\alpha = 0.05$.

13.27 Travel and Transportation Capital Bikeshare is a Washington, D.C., organization sponsored by several agencies that offers short-term use of over 1650 bicycles to registered members. There are approximately 175 bicycle stations in the District of Columbia, Arlington County, and the City of Alexandria, Virginia.[12] To distribute bicycles appropriately to stations, a survey is conducted each year to determine the home location and work location of members. The following table contains the historical proportions associated with each home location and a summary of a random sample of members in 2014. **BIKES**

Home location	2014 sample	Historical proportions
District of Columbia	3750	0.78
Arlington County (VA)	530	0.11
Montgomery County (MD)	185	0.04
Fairfax County (VA)	101	0.02
Prince Georges County (MD)	52	0.01
Alexandria City (VA)	98	0.02
Other	92	0.02

Conduct a goodness-of-fit test to determine whether there is evidence that any of the historical proportions have changed. Use $\alpha = 0.05$.

13.28 Public Policy and Political Science The Metropolitan Council routinely conducts a survey to learn the opinions of Twin City residents about the region's quality of life and key regional problems and possible solutions. One key question involves the most important problem in the Twin Cities metro area, grouped by major categories. Suppose the following table

contains the 2012 responses and a random sample of responses from residents in 2014. **METRO**

Category	2012 responses	2014 sample
Transportation	127	135
Growth	127	124
Crime	120	110
Economy	80	89
Government	67	62
Taxes	40	43
Social	33	39
Housing	39	45
Other	33	36

a. Compute the proportion associated with each category for the 2012 responses.

b. Conduct a goodness-of-fit test to determine whether there is any evidence that the true 2012 population proportions have changed in 2014. Use $\alpha = 0.05$.

Challenge

13.29 Business and Management In Britain, a small baker does not have a fully automatic plant and sells the majority of his or her production on-site or from vehicles. According to the Birmingham City Council trading standards, the law states that the average weight of one loaf type must be 400 grams. Suppose the Small Baker's Association (SBA) claims that the weight of each loaf of bread of this type is approximately normal with mean 400 grams and standard deviation 15 grams. To check this claim, a random sample of small-baker loaves was obtained, and each was carefully weighed. The observed frequency of weights in each specified interval (in grams) is summarized in the following one-way table.

Interval	Frequency
<370	18
370–385	67
385–400	175
400–415	184
415–430	75
≥430	14

a. Assume the SBA claim is true. Find the probability that a randomly selected loaf of bread falls into each interval.

b. Use the probabilities computed in part (a) as the hypothesized population proportions associated with each interval. Conduct a goodness-of-fit test to determine whether the observed weights fit the hypothesized distribution. Use $\alpha = 0.05$. State your conclusion.

Note: The goodness-of-fit test provides a formal test for normality. It complements the methods used to check for normality in Section 6.3.

13.2   Bivariate Categorical Data, Tests for Homogeneity and Independence

If *any* two observations are made on an individual or object, the data set is bivariate. For example, one observation might be categorical and the other numerical, or both observations could be numerical.

If two categorical observations are made on the same individual or object, the data set is *bivariate*. This type of data arises in two common ways.

1. Random samples are obtained from two or more populations, and each individual is classified by values of a categorical variable.

2. Suppose there are two categorical variables of interest. In a (single) random sample, a value of each variable is recorded for each individual.

The test for homogeneity applies to the first kind of data (samples from two or more populations), and the test for independence applies to the second kind of data (data from a single sample, with two categorical variables).

Let's focus on the first type of bivariate categorical data. Suppose random samples of products traded on futures exchanges are obtained from three brokerage firms, and the type of futures product is recorded for each. The brokerage firms are the *populations* and the futures type is the *categorical variable*. The data is *bivariate* because there are two values (firm and futures type) for, or associated with, each trade. The data may be recorded in the following manner.

Trade	Firm	Futures type
1	Fidelity	Energy
2	Vanguard	Agriculture
3	WellsTrade	Financials
4	Vanguard	Metals
⋮	⋮	⋮
300	Fidelity	Metals

Two-way tables are described by the number of rows and the number of columns. For example, a 2 × 6 contingency table has 2 rows and 6 columns.

The natural summary for this type of bivariate data set (in which each observation is categorical) is the number of observations in each category *combination*. For example, compute the number of futures trades (or frequency) by Fidelity *and* involving agriculture, the number of futures trades by Fidelity *and* involving energy, etc. This summary information can be displayed in a 3 × 4 *two-way frequency table*, or *contingency table*, as shown in Table 13.1. Each cell in this table contains the number of observations (observed count or frequency) in a category combination, or pairing.

Table 13.1 A two-way frequency table for the data obtained from futures trades

| | | Futures type | | |
		Agriculture	Energy	Financials	Metals
Firm	Fidelity	15	25	30	22
	Vanguard	22	24	15	30
	WellsTrade	32	25	20	40

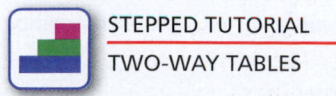

STEPPED TUTORIAL

TWO-WAY TABLES

The columns of the two-way table in Table 13.1 correspond to futures type, and the rows correspond to brokerage firms. Each *cell* in the body of the table contains a frequency, or observed cell count. For example, there were 15 trades by Vanguard involving financials, and there were 40 trades by WellsTrade involving metals.

Bar charts were introduced in Chapter 2.

If we consider a *single* firm (population), say Fidelity, then a bar chart may be used to represent the distribution of futures type (Figure 13.9). A bar chart may be constructed for each firm. For example, Figure 13.10 is another bar chart associated with these data, a summary of futures type for Vanguard.

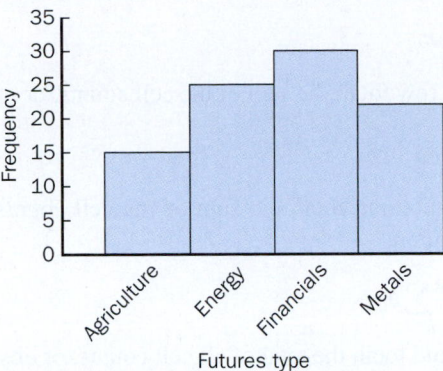

Figure 13.9 A bar chart showing the frequency of futures type for trades by Fidelity.

Figure 13.10 A bar chart showing the frequency of futures type for trades by Vanguard.

VIDEO TECH MANUALS

TWO-WAY TABLES - STACKED/SEGMENTED BAR CHARTS

A side-by-side or stacked bar chart may be used to compare categorical data from two or more sources, or populations. Figure 13.11 shows a side-by-side bar chart of futures type grouped by brokerage firm. Figure 13.12 shows a stacked bar chart of futures type grouped by brokerage firm.

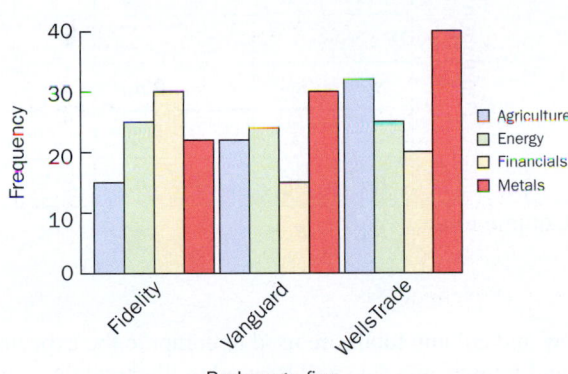

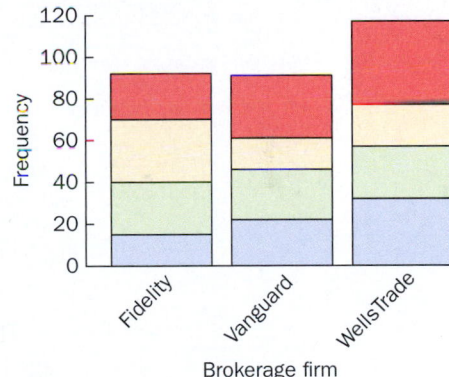

Figure 13.11 A side-by-side bar chart showing the frequency of occurrence of each futures type, by firm.

Figure 13.12 A stacked bar chart showing the frequency of occurrence of each futures type, by brokerage firm.

Although these graphs are useful for suggesting possible differences among populations, there is a more precise statistical test. The practical problem to consider is: Are all of the true category proportions the same for each population? This is a test for **homogeneity** of populations. Homogeneity is the state of having identical properties of values; in this case, it refers to populations having identical true category proportions. In the firm and futures type example, it seems reasonable to ask whether the proportion of futures type trades is the same for each firm. The statistical procedure used to analyze this problem is based on observed and expected cell counts (as in Section 13.1). Under the null hypothesis that the populations have the same category proportions, the test statistic also has a chi-square distribution.

Suppose there are I rows and J columns in a two-way frequency table. The notation associated with this table includes the *dot* notation in a subscript (introduced in Chapter 11) to indicate a sum over that subscript while the other subscript is held fixed.

n_{ij} = observed cell count, or frequency, in the (ij) cell (the intersection of the ith row and the jth column).

$$n_{i.} = \sum_{j=1}^{J} n_{ij}$$

= ith row total, the sum of the cell counts, or observed frequencies, in the ith row.

$$n_{.j} = \sum_{i=1}^{I} n_{ij}$$

= jth column total, the sum of the cell counts, or observed frequencies, in the jth column.

$$n = \sum_{i=1}^{I} \sum_{j=1}^{J} n_{ij}$$

= grand total, the total of all cell counts, or observed frequencies.

Table 13.2 is a visualization of these symbols in an $I \times J$ two-way frequency table.

Table 13.2 Notation used in an $I \times J$ two-way frequency table.

		Category 1	2	$\cdots$	j	$\cdots$	J	Row total
Population	1	n_{11}	n_{12}	$\cdots$	n_{1j}	$\cdots$	n_{1J}	$n_{1.}$
	2	n_{21}	n_{22}	$\cdots$	n_{2j}	$\cdots$	n_{2J}	$n_{2.}$
	$\vdots$	$\vdots$	$\vdots$	$\vdots$	$\vdots$	$\vdots$	$\vdots$	$\vdots$
	i	n_{i1}	n_{i2}	$\cdots$	n_{ij}	$\cdots$	n_{iJ}	$n_{i.}$
	$\vdots$	$\vdots$	$\vdots$	$\vdots$	$\vdots$	$\vdots$	$\vdots$	$\vdots$
	I	n_{I1}	n_{I2}	$\cdots$	n_{Ij}	$\cdots$	n_{IJ}	$n_{I.}$
Column total		$n_{.1}$	$n_{.2}$	$\cdots$	$n_{.j}$	$\cdots$	$n_{.J}$	n

The row and column totals are used to compute the expected cell counts. The brokerage firm and futures type data will be used to illustrate these calculations. Table 13.3 is a modified two-way table containing the observed cell counts, the row and column totals, and the grand total.

Table 13.3 A modified two-way frequency table including the row and column totals and the grand total for the trade data

		Futures type Agriculture	Energy	Financials	Metals	Row total
Firm	Fidelity	15	25	30	22	92
	Vanguard	22	24	15	30	91
	WellsTrade	32	25	20	40	117
	Column total	69	74	65	92	300

Here, as in Section 13.1, the expected cell count may not be an integer.

The e here is used to denote expected frequency and is not connected in any way with the random errors or residuals in Chapter 12.

There were 300 futures trades in this study, and 69 involve agriculture. The proportion of all agriculture trades in the data set is $69/300 = 0.23$. Suppose there is no difference in the proportion of agriculture trades among firms. Then we expect 23% of the futures trades by Fidelity to involve agriculture. This expected frequency in the (11) cell is denoted e_{11} and is computed by

$$e_{11} = 0.23 \times 92 = \frac{69}{300} \times 92 = \frac{(92)(69)}{300}$$

$$= \frac{(\text{1st row total})(\text{1st column total})}{\text{grand total}} = \frac{n_{1.} \times n_{.1}}{n} = 21.16$$

Similarly, we expect 23% of the futures trades by Vanguard to involve agriculture. The expected cell count in the (21) cell is

$$e_{21} = 0.23 \times 91 = \frac{69}{300} \times 91 = \frac{(91)(69)}{300}$$

$$= \frac{(\text{2nd row total})(\text{1st column total})}{\text{grand total}} = \frac{n_{2.} \times n_{.1}}{n} = 20.93$$

The expected counts in column 2 are computed in a similar manner. There are 74 futures trades that involve energy. The proportion of all futures trades involving energy is $74/300 = 0.2467$. If there is no difference in the proportion of energy trades, then we expect 24.67% of the trades, or

$$e_{12} = 0.2467 \times 92 = \frac{74}{300} \times 92 = \frac{(92)(74)}{300}$$

$$= \frac{(\text{1st row total})(\text{2nd column total})}{\text{grand total}} = \frac{n_{1.} \times n_{.2}}{n} = 22.69$$

to be the energy trades by Fidelity. We continue in the same manner to compute all the expected counts. Table 13.4 shows each expected count in parentheses beneath the corresponding observed count.

Table 13.4 A modified two-way frequency table including the row and column totals, the grand total, and the expected cell counts for the trade data

	Futures type				
	Agriculture	Energy	Financials	Metals	Row total
Fidelity	15 (21.16)	25 (22.69)	30 (19.93)	22 (28.21)	92
Vanguard	22 (20.93)	24 (22.45)	15 (19.72)	30 (27.91)	91
WellsTrade	32 (26.91)	25 (28.86)	20 (25.35)	40 (35.88)	117
Column total	69	74	65	92	300

(Firm)

The computations in this example suggest an easy formula for finding the expected frequencies. In an $I \times J$ two-way table, the expected count, or frequency, in the (ij) cell can be written as

$$e_{ij} = \frac{(i\text{th row total})(j\text{th column total})}{\text{grand total}} = \frac{n_{i.} \times n_{.j}}{n}$$

The test statistic is a measure of how far away the observed cell counts are from the expected cell counts. If there is no difference in category proportions among populations, the random variable

$$X^2 = \sum_{\text{All cells}} \frac{(\text{observed cell count} - \text{expected cell count})^2}{\text{expected cell count}} = \sum_{i=1}^{I} \sum_{j=1}^{J} \frac{(n_{ij} - e_{ij})^2}{e_{ij}}$$

has approximately a chi-square distribution with $(I-1)(J-1)$ degrees of freedom. This approximation is good if $e_{ij} \geq 5$ for all i and j, that is, if all expected cell counts are at least 5.

If the observed cell counts are *close* to the expected cell counts, then the value of X^2 will be small. If the observed cell counts are considerably different from the expected cell counts, then the value of X^2 will be large. As in Section 13.1, the null hypothesis is rejected only for large values of the test statistic.

For the trade data, the value of the test statistic is

$$\chi^2 = \sum_{i=1}^{3} \sum_{j=1}^{4} \frac{(n_{ij} - e_{ij})^2}{e_{ij}}$$
$$= \frac{(15 - 21.16)^2}{21.16} + \frac{(25 - 22.69)^2}{22.69} + \cdots + \frac{(40 - 35.88)^2}{35.88} = 13.009$$

This test is significant at the $\alpha = 0.05$ level, because the critical value is

$$\chi^2_{\alpha,(I-1)(J-1)} = \chi^2_{0.05,(2)(3)} = \chi^2_{0.05,6} = 12.5916.$$

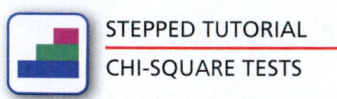

STEPPED TUTORIAL

CHI-SQUARE TESTS

Test for Homogeneity of Populations

In an $I \times J$ two-way frequency table, let n_{ij} be the observed count in the (ij) cell and let e_{ij} be the expected count in the (ij) cell. A hypothesis test for homogeneity of populations with significance level α has the following form:

H_0: The true category proportions are the same for all populations (homogeneity of populations).

H_a: The true category proportions are not the same for all populations.

TS: $X^2 = \sum_{i=1}^{I} \sum_{j=1}^{J} \frac{(n_{ij} - e_{ij})^2}{e_{ij}}$ where $e_{ij} = \frac{(i\text{th row total})(j\text{th column total})}{\text{grand total}} = \frac{n_{i.} \times n_{.j}}{n}$

RR: $X^2 \geq \chi^2_{\alpha,(I-1)(J-1)}$

A reminder: This test is appropriate if all expected cell counts are at least 5 ($e_{ij} \geq 5$ for all i and j). Note that this procedure is called a test of homogeneity and this property is stated in the null hypothesis. However, we are really testing for evidence of inhomogeneity. We cannot *prove* homogeneity, we can only *test* for evidence of inhomogeneity.

DATA SET

HOSPITAL

Hero Images/Getty Images

Example 13.3 Hospital Readmission

One measure of the quality of health care is patient readmissions. In particular, some children with chronic illnesses are readmitted to a hospital frequently. A random sample of children admitted to a Pediatric Health Information System hospital was obtained. The number of times each child was readmitted within one year was recorded. The observed frequencies are given in the table below.[13] Is there any evidence to suggest that the true category proportions of readmittance are different for males and females? Use $\alpha = 0.05$.

		Number of times readmitted				
		0	1	2	3	≥ 4
Gender	Males	346	56	19	15	24
	Females	454	66	28	18	16

Solution Trail 13.3

KEYWORDS

- Is there any evidence?
- True category proportions . . . are different

TRANSLATION

- Statistical inference
- Population proportions are different

CONCEPTS

- Test for homogeneity of populations

VISION

We need to compare the proportion of the number of readmits for males and females. Find the expected cell counts, and use the test for homogeneity of populations procedure to compute the value of the test statistic and draw the appropriate conclusion.

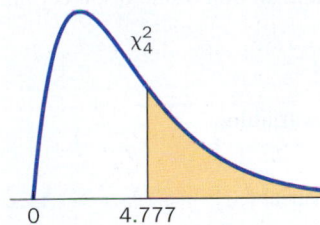

Figure 13.13 *p*-Value illustration:
$p = P(X \geq 4.777)$
$= 0.3109 > 0.05 = \alpha$

Contingency Table

			Readmitted			
Count	>=4	0	1	2	3	
Total %						
Col %						
Row %						
Men	24	346	56	19	15	460
	2.30	33.21	5.37	1.82	1.44	44.15
	60.00	43.25	45.90	40.43	45.45	
	5.22	75.22	12.17	4.13	3.26	
Women	16	454	66	28	18	582
	1.54	43.57	6.33	2.69	1.73	55.85
	40.00	56.75	54.10	59.57	54.55	
	2.75	78.01	11.34	4.81	3.09	
	40	800	122	47	33	1042
	3.84	76.78	11.71	4.51	3.17	

Tests

N	DF	-LogLike	RSquare (U)
1042	4	2.3731801	0.0027

Test	ChiSquare	Prob>ChiSq
Likelihood Ratio	4.746	0.3143
Pearson	4.777	0.3109

Figure 13.14 JMP chi-square test results.

SOLUTION

STEP 1 This is a test for homogeneity of populations. There are two populations, males and females, and five categories for readmittance ($I = 2, J = 5$). We would like to know whether the true proportions of readmittance categories are the same for each population.

STEP 2 The four parts of the hypothesis test are

H_0: The true readmission category proportions are the same for males and females.

H_a: The true readmission category proportions are not the same.

TS: $X^2 = \sum_{i=1}^{2} \sum_{j=1}^{5} \frac{(n_{ij} - e_{ij})^2}{e_{ij}}$

RR: $X^2 \geq \chi^2_{\alpha,(I-1)(J-1)} = \chi^2_{0.05,4} = 9.4877$

STEP 3 Find each expected cell count. Here is one calculation.

$$e_{11} = \frac{n_{1.} \times n_{.1}}{n} = \frac{(460)(800)}{1042} = 353.17$$

All of the expected cell counts are given in Table 13.5.

Table 13.5 The two-way table for the hospital readmission example, including the row and column totals, the grand total, and the expected cell counts

		Number of times readmitted					
		0	1	2	3	≥ 4	Row total
Gender	Males	346 (353.17)	56 (53.86)	19 (20.75)	15 (14.57)	24 (17.66)	460
	Females	454 (446.83)	66 (68.14)	28 (26.25)	18 (18.43)	16 (22.34)	582
		800	122	47	33	40	1042

STEP 4 The value of the test statistic is

$$\chi^2 = \sum_{i=1}^{2} \sum_{j=1}^{5} \frac{(n_{ij} - e_{ij})^2}{e_{ij}}$$

$$= \frac{(346 - 353.17)^2}{353.17} + \frac{(56 - 53.86)^2}{53.86} + \cdots + \frac{(16 - 22.34)^2}{22.34}$$

$$= 4.777 \ (< 9.4877)$$

STEP 5 The value of the test statistic does not lie in the rejection region or equivalently, $p = 0.3109 > 0.05$. Figure 13.13 shows the calculation and illustration of the exact p value. At the $\alpha = 0.05$ significance level, there is no evidence to suggest that the true readmission category proportions are different for males and females. Figure 13.14 shows a technology solution.

TRY IT NOW GO TO EXERCISE 13.42

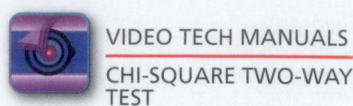

VIDEO TECH MANUALS

CHI-SQUARE TWO-WAY TEST

Suppose bivariate data arise from a single random sample in which the values of two categorical variables are recorded for each individual or object. For example, in a random sample of home burglaries, the type of item stolen and the method of entry might be recorded. Or suppose a random sample of employed people is obtained, and the occupation and any job-related injury is recorded for each person. In each instance, the data are bivariate, and there are observations on two categorical variables.

Given this type of bivariate data, it seems reasonable to ask whether the values of one variable affect the values of the other (are the variables dependent?). Or are the two variables independent? Then, knowing the value of one variable suggests nothing special about the value of the other. For example, suppose the proportion of all employed people who suffer from job-related respiratory illnesses is 0.15 for farm workers, 0.12 for industry and construction workers, 0.03 for service workers, and 0.04 for all other occupations. Now, suppose we know an employed person suffered a job-related broken bone. Does this change the proportions? If so, then the variables (occupation and job-related injury) are dependent. If knowing the occupation does not alter the job-related injury proportions, then the variables are independent.

We use the same notation as in the test for homogeneity. In an $I \times J$ two-way table, there are I categories for the first variable (instead of I populations), and J categories for the second variable. The test for independence is based on observed and expected counts once again. The test statistic is exactly the same as in the test for homogeneity. Here's why.

As in the test for homogeneity, this is actually a test for dependence. We cannot prove independence, we can only test for evidence of dependence.

Recall from Chapter 4 that if two events A and B are independent, then the probability of A and B is the product of the corresponding probabilities; that is,

$$P(A \cap B) = P(A) \cdot P(B) \qquad \text{if } A \text{ and } B \text{ are independent}$$

Suppose the two categorical variables are independent and an individual or object falls into the (ij) cell. Consider the following probability.

Remember that *and* means intersection.

$P[\text{an individual falls into the } (ij) \text{ cell}]$

$= P\left(\begin{array}{l}\text{an individual responds with the } i\text{th value of the first variable} \\ \text{and the } j\text{th value of the second variable}\end{array}\right)$

$= P[(i\text{th value for first variable}) \cap (j\text{th value for second variable})]$ *And* means intersection.

$= P(i\text{th value for first variable}) \cdot P(j\text{th value for second variable})$

Independent events; multiply corresponding probabilities.

$= \dfrac{n_{i.}}{n} \cdot \dfrac{n_{.j}}{n}$ Using the notation introduced in this section, the probability of falling into the ith row times the probability of falling into the jth column.

Because there is a total of n individuals, the expected count, or frequency, in the (ij) cell is

$e_{ij} = \text{expected count in the } (ij) \text{ cell}$

$= \left(\begin{array}{c}\text{sample} \\ \text{size}\end{array}\right) \cdot \left[\begin{array}{l}\text{probability an individual falls} \\ \text{into the } (ij) \text{ cell}\end{array}\right]$

$= n \cdot \left(\dfrac{n_{i.}}{n} \cdot \dfrac{n_{.j}}{n}\right)$

$= \dfrac{n_{i.} \times n_{.j}}{n}$ Simplify: cancel an n.

$= \dfrac{(i\text{th row total})(j\text{th column total})}{\text{grand total}}$ Symbol translation.

This is identical to the expression we used before, so the test statistic is the same as in the test for homogeneity. The test statistic is again a measure of how far away the observed cell counts are from the expected cell counts. The formal hypothesis test follows.

Test for Independence of Two Categorical Variables

In a random sample of n individuals, suppose the values of two categorical variables are recorded. In the resulting $I \times J$ two-way frequency table, let n_{ij} be the observed count in the (ij) cell and let e_{ij} be the expected count in the (ij) cell. A hypothesis test for independence of the two categorical variables with significance level α has the form

H_0: The two variables are independent.

H_a: The two variables are dependent.

TS: $X^2 = \displaystyle\sum_{i=1}^{I} \sum_{j=1}^{J} \frac{(n_{ij} - e_{ij})^2}{e_{ij}}$ where $e_{ij} = \dfrac{(i\text{th row total})(j\text{th column total})}{\text{grand total}} = \dfrac{n_{i.} \times n_{.j}}{n}$

RR: $X^2 \geq \chi^2_{\alpha,(I-1)(J-1)}$

This test is appropriate if all expected cell counts are at least 5 ($e_{ij} \geq 5$ for all i and j).

DATA SET
RESTSTOP

Example 13.4 Rest Stop Preferences

The Pilot Travel Plaza in Pennsylvania is located at the intersection of Routes 487 and 80 so that travelers on each road have easy-on/easy-off access in both directions. Research is being conducted to summarize food preferences and to attract vendors. A random sample of people who purchased food at this plaza was obtained, and the traveling direction and the food vendor were recorded for each person. The observed frequencies are given in the table below. Is there any evidence to suggest that traveling direction and food vendor are dependent? Use $\alpha = 0.01$.

Solution Trail 13.4

KEYWORDS
- Is there any evidence?
- Traveling direction and food vendor are dependent

TRANSLATION
- Statistical inference
- Are these two categorical variables dependent?

CONCEPTS
- Test for independence of two categorical variables

VISION
We need to determine whether there is any evidence that traveling direction and food vendor are dependent. Compute the expected cell counts, and use the test for independence of two categorical variables procedure to find the value of the test statistic and draw the appropriate conclusion.

		Food vendor				
		Aunt Annie's	Pizza Hut	Taco Bell	Mrs. Fields	Hot Dog Company
Traveling direction	North	25	30	17	38	56
	South	40	22	25	45	41
	East	34	24	20	43	48
	West	28	27	25	31	32

SOLUTION

STEP 1 This is a test for independence of two categorical variables: traveling direction and food vendor. There are four possible values for traveling direction and five possible responses for food vendor ($I = 4$ and $J = 5$).

STEP 2 The four parts of the hypothesis test are

H_0: Traveling direction and food vendor are independent.

H_a: Traveling direction and food vendor are not independent.

TS: $X^2 = \displaystyle\sum_{i=1}^{4} \sum_{j=1}^{5} \frac{(n_{ij} - e_{ij})^2}{e_{ij}}$

RR: $X^2 \geq \chi^2_{\alpha,(I-1)(J-1)} = \chi^2_{0.01,12} = 26.2170$

STEP 3 Find each expected cell count. Here is one calculation.

$$e_{11} = \frac{n_{1.} \times n_{.1}}{n} = \frac{(166)(127)}{651} = 32.38$$

All of the expected cell counts are given in Table 13.6.

Table 13.6 The two-way table for the travel plaza example, including the row and column totals, the grand total, and the expected cell counts

		Food vendor					
		Aunt Annie's	Pizza Hut	Taco Bell	Mrs. Fields	Hot Dog Company	Row total
Traveling direction	North	25 (32.38)	30 (26.26)	17 (22.18)	38 (40.03)	56 (45.13)	166
	South	40 (33.75)	22 (27.37)	25 (23.12)	45 (41.72)	41 (47.04)	173
	East	34 (32.97)	24 (26.74)	20 (22.59)	43 (40.76)	48 (45.95)	169
	West	28 (27.90)	27 (22.63)	25 (19.11)	31 (34.49)	32 (38.88)	143
Column total		127	103	87	157	177	651

STEP 4 The value of the test statistic is

$$\chi^2 = \sum_{i=1}^{4} \sum_{j=1}^{5} \frac{(n_{ij} - e_{ij})^2}{e_{ij}}$$

$$= \frac{(25 - 32.38)^2}{32.38} + \frac{(30 - 26.26)^2}{26.26} + \cdots + \frac{(32 - 38.88)^2}{38.88}$$

$$= 14.598 \ (< 26.2170)$$

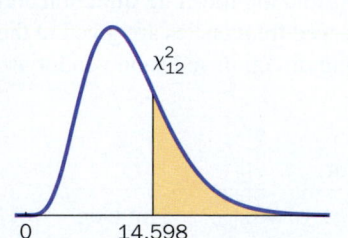

Figure 13.15 *p*-Value illustration:
$p = P(X \geq 14.598)$
$= 0.2642 > 0.01 = \alpha$

STEP 5 The value of the test statistic does not lie in the rejection region or equivalently, $p = 0.2642 > 0.01$. Figure 13.15 shows the calculation and illustration of the exact p value. At the $\alpha = 0.01$ significance level, there is no evidence to suggest that the two categorical variables are dependent.

Figure 13.16 shows a technology solution.

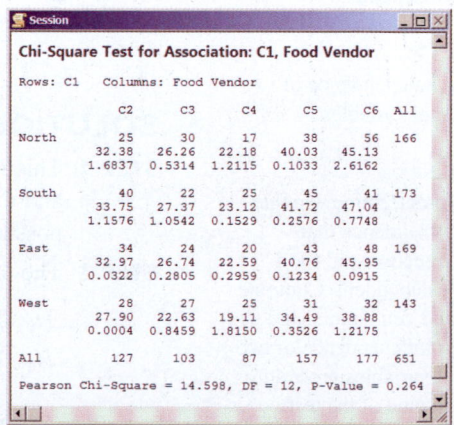

Figure 13.16 Minitab chi-square test results.

TRY IT NOW GO TO EXERCISE 13.48

Technology Corner

Procedure: Chi-square test for homogeneity or independence.
Reconsider: Example 13.4, solution, and interpretations.

CrunchIt!

Use the built-in function Contingency Table; with counts.

1. Enter the row numbers in Var1, the column numbers in Var2, and the corresponding counts in Var3.
2. Choose Statistics; Contingency Table; with counts. Select the Row Variable, the Column Variable, and the Counts (variable).
3. Click Calculate. CrunchIt! computes and displays a table with counts, row percentages, column percentages, and total percentages in addition to the value of the test statistic. See Figure 13.17.

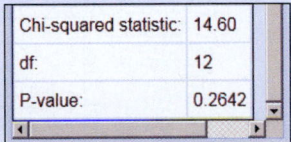

Figure 13.17 CrunchIt! chi-square test results. The table with counts and percentages is not shown.

TI-84 Plus C

Use the built-in function χ^2-Test.

1. Enter the observed cell counts into the matrix [A]. See Figure 13.18.
2. Select STAT; TESTS; χ^2-Test. Enter the matrix containing the observed cell counts, [A], and specify a matrix for the expected frequency, [B]. See Figure 13.19.

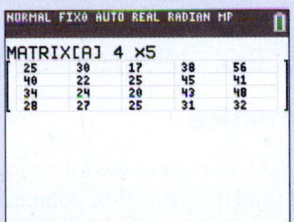

Figure 13.18 Observed cell counts.

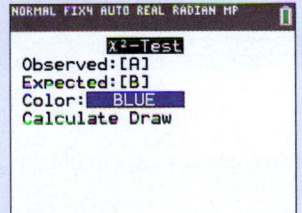

Figure 13.19 χ^2-Test input screen.

3. Highlight Calculate and press ENTER. The expected frequencies are shown in Figure 13.20 and the chi-square test results are shown in Figure 13.21. The Draw results are shown in Figure 13.22.

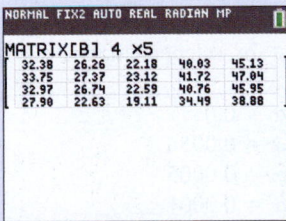

Figure 13.20 Expected frequencies.

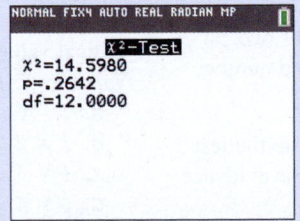

Figure 13.21 χ^2-Test results.

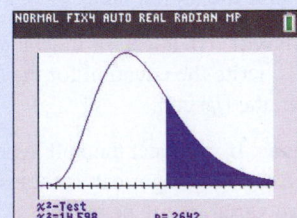

Figure 13.22 χ^2-Test Draw results.

Minitab

Use the built-in function Chi-Square Test for Association in the <u>Stat</u>; <u>T</u>ables menu.

1. Enter the observed category frequencies into the columns C1–C5.
2. Select <u>Stat</u>; <u>T</u>ables; Chi-<u>S</u>quare for Association.
3. Select Summarized data in a two-way table. Enter the columns containing the table.
4. Click OK. The chi-square hypothesis test results are displayed in a sessions window. See Figure 13.16.

Excel

Use the built-in function CHISQ.TEST.

1. Enter the observed cell counts in a rectangular array (matrix) and the expected cell counts in another rectangular array.
2. Use the function CHISQ.TEST to find the p value associated with the chi-square test and the function CHISQ.INV to find the value of the test statistic if desired. See Figure 13.23.

	A	B	C	D	E	F	G
1				Observed counts			
2		Aunt Annie	Pizza Hut	Taco Bell	Mrs. Fields	Hot Dog Co	Total
3	North	25	30	17	38	56	166
4	South	40	22	25	45	41	173
5	East	34	24	20	43	48	169
6	West	28	27	25	31	32	143
7	Total	127	103	87	157	177	651
8				Expected counts			
9		Aunt Annie	Pizza Hut	Taco Bell	Mrs. Fields	Hot Dog Co	
10	North	32.38	26.26	22.18	40.03	45.13	
11	South	33.75	27.37	23.12	41.72	47.04	
12	East	32.97	26.74	22.59	40.76	45.95	
13	West	27.90	22.63	19.11	34.49	38.88	
14							
15		p value		0.2642	= CHISQ.TEST(B3:F6,B10:F13)		
16		Test stat		14.598	= CHISQ.INV(1-C15,12)		

Figure 13.23 Rectangular arrays and the chi-square test results.

SECTION 13.2 EXERCISES

Concept Check

13.30 Short Answer   Explain two common ways to obtain bivariate data.

13.31 Short Answer   Name two types of graphs that may be used to compare categorical data from two or more groups.

13.32 Short Answer   State the null hypothesis in a test for homogeneity of populations.

13.33 True/False   If we reject the null hypothesis in the test for homogeneity of populations, there is evidence to suggest the category proportions are all the same.

13.34 Short Answer   In a test for independence of two categorical variables, write the equation for the expected number of observations in the (ij) cell.

13.35 True/False   If we reject the null hypothesis in the test for independence of two categorical variables, there is evidence to suggest the two variables are dependent.

13.36 True/False   In a test for independence of two categorical variables, we reject the null hypothesis only for large values of the test statistic.

Practice

13.37 In each of the following problems, the number of rows (I) and the number of columns (J) for a two-way frequency table are given. Use the value of α to determine the critical value in a test for homogeneity of populations.
   a. $I = 3$, $J = 4$, $\alpha = 0.05$
   b. $I = 2$, $J = 6$, $\alpha = 0.01$
   c. $I = 4$, $J = 3$, $\alpha = 0.025$
   d. $I = 5$, $J = 3$, $\alpha = 0.001$

13.38 In each of the following problems, the number of rows (I) and the number of columns (J) for a two-way frequency table are given. Use the value of α to determine the critical value in a test for independence of two categorical variables.
   a. $I = 6$, $J = 4$, $\alpha = 0.05$
   b. $I = 2$, $J = 8$, $\alpha = 0.005$
   c. $I = 3$, $J = 7$, $\alpha = 0.0005$
   d. $I = 4$, $J = 6$, $\alpha = 0.0001$

13.39 Find the missing observed cell counts and row and column totals in the following two-way frequency table.

	Category				Row total
	1	2	3	4	
Population 1	18	14	18		65
Population 2	25			12	
Population 3		33	26	28	119
Column total		68	60		258

13.40 Consider the following two-way frequency table with three populations and three values of a categorical variable. **EX13.40**

	Category		
	1	2	3
Population 1	58	62	33
Population 2	65	55	40
Population 3	70	60	25

a. Find the row and column totals and the grand total.
b. Suppose the true category proportions are the same for each population. Find the expected count for each cell.
c. Conduct a test for homogeneity of populations. Use $\alpha = 0.025$. State your conclusion.

13.41 Consider the following two-way frequency table, in which the values of two categorical variables were recorded for each individual. **EX13.41**

	Variable 2			
	1	2	3	4
Variable 1 1	235	267	245	386
2	241	264	280	305
3	228	254	270	394
4	219	235	263	363

Conduct a test for independence of the two categorical variables. Use $\alpha = 0.05$. State your conclusion.

Applications

13.42 Sports and Leisure Random samples of gamblers at four Las Vegas casinos were obtained, and each gambler was asked which game he or she played most. The results are given in the following two-way frequency table. **CASINO**

	Game			
Casino	Blackjack	Poker	Roulette	Slots
Bellagio	22	20	38	66
Caesar's	30	38	22	68
Golden Nugget	28	25	21	81
Harrah's	38	25	29	84

Conduct a test for homogeneity of populations. Is there any evidence to suggest that the true proportion of gamblers at each game is not the same for all casinos? Use $\alpha = 0.05$.

13.43 Marketing and Consumer Behavior A marketing manager obtained random samples of shoppers from three different grocery stores and asked each shopper to name his or her favorite Tastykake product. The results are summarized in the following two-way frequency table. **TASTY**

	Product			
Grocery store	Krimpets	Cupcakes	Kandy Kakes	Creamies
Giant	90	80	95	92
Shaw	81	66	87	56
Weis	94	83	92	55

Is there any evidence to suggest that the true proportion of each favorite is not the same for all populations? Use $\alpha = 0.01$.

13.44 Marketing and Consumer Behavior Random samples of customers at two different office-supply stores were obtained, and each customer was asked which type of writing implement he or she prefers. The results are summarized in the following two-way frequency table. **WRITING**

	Writing implement		
Store	Traditional pencil	Mechanical pencil	Pen
Office Max	183	164	480
Staples	130	202	420

Conduct a test for homogeneity of populations. Use $\alpha = 0.01$. State your conclusion, justify your answer, and find bounds on the p value associated with this test.

13.45 Public Policy and Political Science CBC/Radio Canada is the nation's public broadcaster and airs regional and cultural programs in several languages. A survey was conducted concerning the funding level for this broadcasting corporation. A random sample of Canadians was selected, and each was asked whether funding should be increased, decreased, or maintained at the current levels. The data are summarized in the following two-way table. **RADIO**

	Funding for CBC			
Region	Decrease	Maintain	Increase	Unsure
Atlantic	17	58	36	9
Quebec	27	167	65	37
Ontario	59	191	79	26
West	68	212	106	44

Conduct a test for homogeneity of populations. Use $\alpha = 0.05$. State your conclusion, and justify your answer.

13.46 Education and Child Development Random samples of boys and girls in elementary and secondary schools were obtained, and their parents were asked how often they helped with homework. The data are summarized in the following two-way frequency table. **HOMEWORK**

Number of times spent helping with homework per week			
	Less than once	1 or 2 times	3 or 4 times	5 or more times
Girls	259	369	254	118
Boys	274	335	262	129

Is there any evidence that the number of times spent helping with homework is different for boys and girls? Use $\alpha = 0.01$. Find bounds on the p value associated with this test.

13.47 Psychology and Human Behavior   While food-and-wine pairing is subjective and an inexact science, traditionally red wine goes with red meat, and white wine goes with fish and poultry. A random sample of diners at four-star restaurants was obtained, and each diner was classified according to the food and wine ordered. Here is the resulting two-way frequency table. ▦ **PAIRING**

	Wine	
Food	Red	White
Red meat	86	46
Fish or poultry	50	64

Is there any evidence that food and wine are dependent? Test the relevant hypothesis with $\alpha = 0.05$. Do these data suggest that diners are still following the traditional food-and-wine pairings?

13.48 Sports and Leisure   The Whitewater Ski Resort in British Columbia offers downhill skiing, cross-country skiing, snowboarding, and snow tubing. The manager is planning to develop a new, targeted advertising campaign. A random sample of customers was obtained, and each was classified by age group and activity. The data are summarized in the following two-way table. ▦ **RESORT**

	Resort activity			
Age group	Downhill skiing	Cross-country skiing	Snow-boarding	Snow tubing
<16	24	35	23	48
16–20	50	25	40	32
20–30	21	27	27	32
30–40	29	39	29	29
≥40	26	37	35	30

Is there any evidence to suggest that resort activity is dependent on age group? Conduct the appropriate hypothesis test with $\alpha = 0.01$.

13.49 Sports and Leisure   Amid the hype and media coverage, the National Football League draft takes place every spring. NFL teams jostle for position in an attempt to draft a college player who will help their team win a Super Bowl. A random sample of players selected in the draft was obtained, and each was classified by round and position (offense, defense, special teams).[14] The data are summarized in the following two-way table. ▦ **NFL**

	Position		
Round	Offense	Defense	Special teams
1	462	338	6
2	426	360	11
3	441	355	30
4	443	358	26

a. Is there any evidence to suggest that draft round is dependent on position? Conduct the appropriate hypothesis test with $\alpha = 0.01$.

b. Interpret your results in part (a). For example, when is a special teams player most likely to be drafted?

13.50 Medicine and Clinical Studies   Some research suggests that college student athletes who are subject to high stress levels (due to academic demands, personal problems, or other sources) are more likely to be injured during a game. A random sample of student athletes was obtained, and a questionnaire was used to determine the stress level of each prior to a game. Following the game, the injury status of each athlete was also recorded. The data are summarized in the following table. ▦ **STRESS**

	Stress level		
Injury	Low	Medium	High
Yes	12	17	25
No	335	292	288

Is there any evidence to suggest that stress level and injury are dependent? Conduct the appropriate hypothesis test with $\alpha = 0.05$.

13.51 Public Health and Nutrition   The National Advisory Committee on Immunization in Canada provides the public with specific health recommendations and summary reports. A random sample of patients who tested positive for influenza during the 2012–2013 flu season was obtained. Each person was classified by age group and influenza type. The data are summarized in the following table.[15] ▦ **FLU**

	Influenza type			
Age group	A/H1N1	A/H3N2	A unsub	B
<5	207	849	1592	617
5–19	66	630	754	895
20–44	319	1207	1617	512
45–64	300	1196	1946	510
65+	120	3599	5860	597

Is there any evidence to suggest that the age group and influenza type are dependent? Conduct the appropriate hypothesis test with $\alpha = 0.01$.

Extended Applications

13.52 Public Policy and Political Science   Public swimming pools are routinely inspected by city health officials. In a random sample of pool code violations, the type of pool and the violation class were recorded.[16] The resulting two-way frequency table is shown. ▦ **POOLS**

	Violation		
Pool type	Serious, pool closed	Water chemistry	Policy/ management
Hotel/motel	1,525	5,462	2,914
Condominium/ apartments	3,282	13,227	7,542
School/university	90	550	390
Private club	367	1,508	863
Child care	6	32	20
Water park	33	111	62
Hospital	18	72	35
Municipal	84	368	259
Campground	32	134	91
Camp	20	227	192

Is there any evidence to suggest that the type of violation and the type of pool are dependent? Conduct the appropriate hypothesis test with $\alpha = 0.01$ and find bounds on the p value associated with this test.

13.53 Psychology and Human Behavior Most municipalities provide the public with a variety of crime statistics, including types of crimes and class to the police. A random sample of gunshot incidents in Oakland, California, was obtained, and each was classified by time of day and day of week. The data are summarized in the two-way frequency table below.[17] **CRIME**

	Time of day		
Day	Overnight	Daytime	Evening
Mon	17	10	11
Tue	32	4	9
Wed	16	7	17
Thu	14	7	8
Fri	24	6	17
Sat	49	3	12
Sun	46	7	15

a. Is there any evidence that day of the week and time of day are dependent? Use $\alpha = 0.05$.
b. Are the assumptions for the hypothesis test in part (a) met? Why or why not?
c. Use your results in part (a) to suggest the day and time of day when gunshot incidents are much different than expected.

13.54 Public Policy and Political Science The United Nations is almost 70 years old. However, many people in countries all over the world are still unfamiliar with the UN and do not understand its function. A survey concerning familiarity with the United Nations was conducted in six nations. The data are summarized in the two-way frequency table. Is there any evidence to suggest that the

familiarity with the UN and the country are dependent? Conduct the appropriate hypothesis test with $\alpha = 0.01$. **UNITED**

	Response			
Country	Very familiar	Somewhat familiar	Not that familiar	Not at all familiar
U.S.	340	1020	553	213
G. Brit.	33	402	457	196
France	31	261	669	84
Italy	32	547	400	74
Spain	60	423	403	121
Germany	21	238	610	165

13.55 Psychology and Human Behavior Homeopathic medicine is based on the theory that the body has the ability to heal itself. However, homeopathic health practitioners do use pills or liquids with a small amount of an active ingredient to treat diseases. These treatments are loosely regulated, and homeopathic practitioners hold various degrees. A random sample of homeopathic practitioners was obtained in the United States and Canada and the academic degree of each was recorded. The data are summarized in the following table. **MEDICINE**

	Country	
Degree	United States	Canada
MD	78	30
ND	22	31
RN	15	14
DO	12	5
NP	11	4
DVM	10	18
Other	21	20

Is there any evidence to suggest that the true proportions for type of degree held are different for each country? Use $\alpha = 0.01$.

Challenge

13.56 Marketing and Consumer Behavior In a recent *USA Today* poll, most customers shopping online indicated that shipping options are important when making buying decisions.[18] There are often several shipping options for online consumers, including overnight delivery, 3-day ground service, and free slow shipping. Suppose that, in a random sample of 500 adults in California, 135 said they prefer free slow shipping. In a random sample of 600 adults in New Jersey, 204 indicated they prefer free slow shipping.
a. Conduct a hypothesis test concerning two population proportions to determine whether there is any evidence that the proportion of adults who prefer free slow shipping is different in California and in New Jersey. Find the p value associated with this test.
b. Consider "California" and "New Jersey" as populations and "prefer" and "not preferred" as categories. Using the data given in this problem, construct a two-way frequency table and conduct a test for homogeneity of populations. Use technology to find the p value for this test.

c. What is the relationship between the value of the test statistic in part (a) and the value of the test statistic in part (b)? How are the p values related? Why do these relationships make sense?

13.57 Public Health and Nutrition The American Dental Association recently conducted a survey regarding dental hygiene habits. A random sample of elderly people was obtained, and each person was classified according to brushing and flossing frequency. The following codes and categories were used for each variable.

Code	Floss/brush frequency
1	Never
2	Once per month
3	A few times per month
4	Once per week
5	A few times per week
6	Once per day
7	More than once per day

The survey results are summarized in the following two-way frequency table. **DENTAL**

		\multicolumn{7}{c}{Floss}
		1	2	3	4	5	6	7
Brush	1	35	37	48	52	55	80	101
	2	36	38	42	47	54	75	97
	3	38	43	46	51	58	77	94
	4	33	51	42	42	51	52	86
	5	76	79	85	87	93	102	115
	6	81	41	46	78	107	103	116
	7	98	94	126	136	142	198	252

Is there any evidence to suggest that flossing frequency and brushing frequency are dependent? Conduct the appropriate hypothesis test with $\alpha = 0.01$. Find the p value associated with this test.

CHAPTER 13 SUMMARY

One-Way Frequency Table
A one-way frequency table is a method for summarizing a univariate categorical data set. The table lists each possible category and the number of times each category occurred (the observed count or frequency for each category).

Goodness-of-Fit Test
Let n_i be the number of observations falling into the ith category ($i = 1, 2, \ldots, k$), and let $n = n_1 + n_2 + \cdots + n_k$. A hypothesis test about the true category population proportions with significance level α has the form

$H_0: p_1 = p_{10}, p_2 = p_{20}, \ldots, p_k = p_{k0}$

$H_a: p_i \neq p_{i0}$ for at least one i.

TS: $X^2 = \sum_{i=1}^{k} \frac{(n_i - np_{i0})^2}{np_{i0}}$

RR: $X^2 \geq \chi^2_{\alpha, k-1}$

This test is appropriate if all expected cell counts are at least 5 ($np_{i0} \geq 5$ for all i).

Two-Way Frequency, or Contingency, Table
A two-way frequency table is a method for summarizing a bivariate categorical data set. Each *cell* contains the number of observations (observed count or frequency) in a category *combination*, or *pairing*.

Test for Homogeneity of Populations
In an $I \times J$ two-way frequency table, let n_{ij} be the observed count in the (ij) cell and let e_{ij} be the expected count in the (ij) cell. A hypothesis test for homogeneity of populations with significance level α has the form

H_0: The true category proportions are the same for all populations (homogeneity of populations).

H_a: The true category proportions are not the same for all populations.

$$\text{TS: } X^2 = \sum_{i=1}^{I} \sum_{j=1}^{J} \frac{(n_{ij} - e_{ij})^2}{e_{ij}} \quad \text{where} \quad e_{ij} = \frac{(i\text{th row total})(j\text{th column total})}{\text{grand total}} = \frac{n_{i.} \times n_{.j}}{n}$$

$$\text{RR: } X^2 \geq \chi^2_{\alpha,(I-1)(J-1)}$$

This test is appropriate if all expected cell counts are at least 5 ($e_{ij} \geq 5$ for all i and j).

Test for Independence of Two Categorical Variables

In a random sample of n individuals, suppose the values of two categorical variables are recorded. In the resulting $I \times J$ two-way frequency table, let n_{ij} be the observed count in the (ij) cell and let e_{ij} be the expected count in the (ij) cell. A hypothesis test for independence of the two categorical variables with significance level α has the form

H_0: The two variables are independent.

H_a: The two variables are dependent.

$$\text{TS: } X^2 = \sum_{i=1}^{I} \sum_{j=1}^{J} \frac{(n_{ij} - e_{ij})^2}{e_{ij}} \quad \text{where} \quad e_{ij} = \frac{(i\text{th row total})(j\text{th column total})}{\text{grand total}} = \frac{n_{i.} \times n_{.j}}{n}$$

$$\text{RR: } X^2 \geq \chi^2_{\alpha,(I-1)(J-1)}$$

This test is appropriate if all expected cell counts are at least 5 ($e_{ij} \geq 5$ for all i and j).

CHAPTER 13 EXERCISES

APPLICATIONS

13.58 Business and Management Many companies are expanding operations into Europe because of the large population and increased economic activity. However, there is less Class-A logistics space.[19] A random sample of warehouse users with plans to expand into Europe was obtained and asked to select their top logistics location in Europe. The data are given in the following table. **SPACE**

Location	Venice	Antwerp-Brussels	Rotterdam	Rhein-Ruhr	Madrid
Frequency	55	40	38	42	44

Is there any evidence to suggest that the true population proportions of preferred logistics site is different from 0.20? Use $\alpha = 0.05$.

13.59 Psychology and Human Behavior The Rescue Pet Store sells five breeds of dogs, and the owner is trying to determine whether one breed is preferred over the others. A random sample of recent dog sales was obtained, and the number of each breed purchased is given in the following table. **BREED**

Dog breed	Frequency
American bulldog	46
Collie	54
Golden retriever	32
German shepherd	30
Yorkshire terrier	46

Is there evidence to suggest that the true population proportion of sales for any breed is different from 0.20? Use $\alpha = 0.05$.

13.60 Marketing and Consumer Behavior The manager of a CVS drugstore in Madison, Wisconsin, obtained a random sample of customers who purchased adhesive bandages. The brands and frequencies are given in the following table. **BANDAGE**

Brand	Band-Aid	Curad	Nexcare	Generic
Frequency	220	215	95	510

Historical records indicate that the population proportions are Band-Aid 0.2, Curad 0.2, Nexcare 0.1, and Generic 0.5. Conduct a goodness-of-fit test to determine whether there is any evidence to suggest that the data are not consistent with the past proportions. Use $\alpha = 0.01$.

13.61 Public Policy and Political Science A Lubbock County, Texas, government official discovered some extra money in the budget that must be spent by the end of the fiscal year. A random sample of county residents was obtained, and each was asked how the money should be spent. The data and the hypothesized population proportions (from past county referendum votes) are summarized in the following one-way frequency table. **PROJECT**

Project	Frequency	p_{i0}
Road construction	103	0.3
Road resurfacing	119	0.4
Bicycle paths	40	0.1
New sidewalks	35	0.1
Park improvements	20	0.1

Is there evidence to suggest that any of the true population proportions are different from the hypothesized proportions? Use $\alpha = 0.05$.

13.62 Education and Child Development The Canadian University Survey Consortium regularly conducts research into university students' satisfaction and adjustment. As part of the survey, each first-year student is asked to select the most important reason for attending a specific university. The following table summarizes the results associated with this question from students surveyed at Carleton University and the proportions from the school's comparison group.[20] **CARLETON**

Response	Carleton frequency	Comparison group
To prepare for a specific job or career	568	0.43
To get a good job	455	0.26
To increase my knowledge in an academic field	179	0.08
To get a good general education	146	0.07
To prepare for graduate/professional school	130	0.07
To develop a broad base of skills	65	0.04
To meet parental expectations	32	0.02
Other	32	0.02
To meet new friends	16	0.01

Is there evidence to suggest that any of the Carleton true population proportions is different from the comparison group proportion? Use $\alpha = 0.01$.

13.63 Economics and Finance Many financial institutions offer customers with an Individual Retirement Account (IRA) four different plans for automatic transfer of funds from a checking or savings account to their IRA. Random samples of IRA customers from each of five different companies were obtained, and the transfer plan was recorded for each. The data are summarized in the following two-way table (Bear Sterns, BS; Commonfund, CF; Lincoln, LI; Prudential, PR; Ultimus: UL). **IRA**

		Transfer plan		
	Monthly	Quarterly	Semi-annually	Annually
BS	71	70	67	59
CF	90	51	56	61
LI	75	82	70	60
PR	69	57	78	69
UL	93	92	77	91

(Company labels the rows)

Conduct a test for homogeneity of populations with $\alpha = 0.05$. Is there any evidence to suggest that the true proportions associated with transfer plans are different for any of the populations? Justify your answer.

13.64 Psychology and Human Behavior The Ohio State Bar Association regularly surveys the legal community with regard to the economics of law practice. The resulting report includes information about attorney demographics, experience, and prevailing hourly billing rates.[21] In addition, each survey participant is asked to indicate job satisfaction and attorney category. These data are summarized in the following two-way table. **ATTORNEY**

	Attorney category		
Current satisfaction	Private practice	House counsel	Government
A great deal	109	62	101
Some	97	56	44
Very little	19	7	5

a. Is there any evidence to suggest that current satisfaction level is associated with attorney category? Use $\alpha = 0.01$.
b. Find the p value associated with the hypothesis test in part (a).
c. Which attorneys tend to be most satisfied with their job? Justify your answer.

13.65 Marketing and Consumer Behavior One of the most common home remodeling jobs involves the kitchen. In almost every house, this room is heavily used and often needs to be expanded to accommodate personal tastes. Three building-supply stores were selected, and a random sample of individuals purchasing kitchen countertops was obtained from each. The type of countertop was recorded for each person, and the data are summarized in the following two-way frequency table (Home Depot, HD; Lowe's, LO; TrueValue, TV). **COUNTERS**

	Countertop			
Supply store	Concrete	Corian	Marble	Granite
HD	52	24	37	90
LO	76	36	36	87
TV	53	43	31	78

Is there any evidence to suggest that the true proportion of each type of countertop purchased is not the same for all supply stores? Use $\alpha = 0.05$.

13.66 Demographics and Population Statistics The Scottish Household Survey is designed to gather information on the composition, characteristics, attitudes, and behavior of both Scottish households and individuals. As part of the survey, all adults are asked about the number of days they could survive on stored food supplies in an emergency. The responses are summarized in the following two-way table by tenure of household. **SURVEY**

	Household tenure		
Days	Owner occupied	Social rented	Private rented
0	18	18	8
1–2	140	89	51
3–5	473	212	95
6–9	665	171	86
10–15	298	71	19
16–25	88	18	8
≥ 26	70	12	3

Is there any evidence to suggest that the number of days they could survive is associated with household tenure? Use $\alpha = 0.01$. Interpret your results in the context of this problem.

13.67 Psychology and Human Behavior Some research suggests that music influences how much time people spend in a store. To investigate this theory, a random sample of customers at a Paramount retail store was obtained (over a long period of time), and each was classified by the amount of time spent shopping (in minutes) and the type of music played during the day. The results are given in the following table. 📊 MUSIC

	Type of music			
	Classical	Easy listening	Rock	Country
< 15	49	35	17	14
15–30	51	27	19	41
30–60	67	41	41	22
≥ 60	47	26	26	33

(Time is the row variable.)

Is there any evidence of an association between music type and time spent shopping? Conduct the appropriate hypothesis test with $\alpha = 0.01$.

13.68 Physical Sciences   Recent research suggests that the *Titanic* luxury liner broke into three sections, causing it to sink faster than was previously believed. The collision with an iceberg was a terrifying event, and some experts believe the chance of survival was associated with location aboard the vessel. The following table presents the class and survival status of passengers aboard the *Titanic*. 📊 TITANIC

	Survival status	
	Died	Survived
First	122	203
Second	167	118
Third	528	178
Crew	673	212

(Class is the row variable.)

Is there any evidence of an association between class and survival status? Conduct the appropriate hypothesis test with $\alpha = 0.05$.

EXTENDED APPLICATIONS

13.69 Marketing and Consumer Behavior   A new radio station in Boone, North Carolina, WKKY, plays a wide variety of music from 10 different genres. To assess customer preference and narrow their focus, the station obtained a random sample of listeners, who were asked to indicate their favorite type of music. The data are summarized in the following one-way frequency table. 📊 WKKY

Genre	Frequency	Genre	Frequency
Country	50	Jazz	44
Hits	62	Dance	36
Christian	62	Latin	38
Rock	68	World	42
Urban	70	Classical	35

Is there any evidence to suggest that one music genre is most preferred? Conduct the appropriate hypothesis test with $\alpha = 0.01$. Find bounds on the p value associated with this test.

13.70 Biology and Environmental Science   In a recent national litter survey, three cities were selected and similar areas were inspected. The number of discarded paper bags

found at each site was recorded along with the source. The data are summarized in the following table. 📊 LITTER

	City		
	San Francisco	Washington, D.C.	Oakland
Take-out food	35	60	82
Conv. store	40	30	78
Pharmacy	5	6	4
Grocery	60	5	5
Other	7	6	12

Is there any evidence to suggest that the proportion of discarded paper bags by source is not the same for all cities? Use $\alpha = 0.05$.

13.71 Marketing and Consumer Behavior   The Silicon Valley Bank Wine Report suggests that there is a relationship between wine sales and age group.[22] A random sample of adults purchasing wine was obtained from various states. Each person was classified by generation and retail bottle price. The summary data are given in the following two-way table. 📊 WINE

	Generation			
	Millennial	Gen X	Boomers	Matures
< $15	31	47	55	23
$15–$19	29	73	102	39
$20–$29	44	132	147	44
$30–$39	22	59	85	36
$40–$69	12	41	67	26
> $69	4	22	45	18

(Bottle price is the row variable.)

a. Is there any evidence to suggest that generation and retail bottle price of the wine purchases are dependent? Use $\alpha = 0.05$.
b. Find bounds on the p value. Use technology to confirm your answer and to find the exact p value.
c. Suppose you are marketing wine in a predominantly Millennial area. Using the table and the hypothesis test results as guides, how would you allocate your wine stock according to bottle price? Justify your answer.

LAST STEP

13.72 Is public transportation use the same for all New York boroughs?   A random sample of New York City commuters was obtained and each was classified by resident borough and the type of transportation used to get to work. The resulting frequencies are given in the following two-way table.

	Type of transportation			
	Bus	Subway	Ferry	Drive
Bronx	40	45	19	15
Brooklyn	45	52	14	8
Manhattan	30	54	17	11
Queens	47	48	15	13
Staten Island	30	43	20	14

(Borough is the row variable.)

Is there any evidence to suggest that borough residence and type of transportation are dependent? Use $\alpha = 0.05$. 📊 NEWYORK

14 Nonparametric Statistics

◀ **Looking Back**

- Recall the parametric methods discussed in previous chapters: statistical techniques based on the normality assumption.
- Remember that if any assumptions are violated, the conclusions may be invalid.

▶ **Looking Forward**

- Learn several nonparametric, or distribution-free, procedures that require very few assumptions about the underlying population(s).
- Understand the intuitive reasons for using many nonparametric test statistics.
- Compute and use ranks in several nonparametric test procedures.

Were the levels of naphthalene different in the sediment near Alabama and Florida?

The *Deepwater Horizon* oil spill in 2010 was a terrible environmental disaster in the United States. Almost 5 million barrels of oil were released into the ocean, and studies suggest that a significant amount of oil is still in the water or sediment.[1] No one can predict the long-term health effects, but in the short term the crude oil spill caused an increase in certain contaminants, especially polycyclic aromatic hydrocarbons, or PAHs. Some of these compounds are carcinogenic and may be absorbed by marine life and, therefore, become part of the food chain.

The PAH naphthalene is found in crude oil and is classified as a possible carcinogen. As part of the monitoring process, the U.S. Environmental Protection Agency (EPA) measured the amount of naphthalene in the sediment at various locations in the Gulf of Mexico. A random sample of these short-term measurements near Florida and Alabama was obtained from the EPA. The data are given in the following table[2] (in μg/kg).

| Alabama | 5.0 | 16.0 | 8.5 | 4.2 | 7.3 | 4.1 | 4.2 | 7.3 | 6.9 | 4.3 | |
| Florida | 4.2 | 14.0 | 7.9 | 8.3 | 25.0 | 8.3 | 4.1 | 8.5 | 8.3 | 8.3 | 12.0 | 4.2 |

It seems reasonable to conduct a two-sample *t* test for a difference in population means. However, it is known that naphthalene concentrations are not normally distributed. Because the normality assumption is violated, a two-sample *t* test is not valid.

The methods presented in this chapter allow for comparison of continuous distributions, with few assumptions necessary. These nonparametric procedures are handy when very little is known about the underlying distributions. A statistical test based on ranks can be used to compare the naphthalene concentrations in these two locations.

CONTENTS

Dan Anderson/EPA/Landov

14.1   The Sign Test

The normality assumption is very reasonable, because almost all distributions are normal or approximately normal.

Each of the hypothesis tests presented in the previous chapters depends on a set of assumptions. If any of the assumptions are violated, the conclusions may be invalid. Most of the statistical procedures include a normality assumption: The random sample(s) is (are) drawn from a normal distribution. Statistical techniques based on this assumption are called **parametric methods**. This chapter presents some alternative statistical techniques called **nonparametric**, or **distribution-free**, **procedures**. These techniques usually require very few assumptions about the underlying population(s).

Usually, very few assumptions are necessary for a nonparametric test to be valid. In addition, many statisticians consider the formula, or rule, for computing the test statistic in a nonparametric procedure to be more intuitive and easier to apply than a comparable parametric test. For example, the test statistic in a **sign test** is simply a count of the number of observations that are greater than the hypothesized median.

There are, however, some disadvantages to nonparametric tests. Because there are few assumptions, these procedures usually do not utilize all of the information captured in a sample. Nonparametric tests may ignore certain inherent information, for simplicity or ease of use. This means that there is a greater chance of making an error when you use a nonparametric test. Therefore, if there is ever a case in which either a parametric and a nonparametric test can be used, it is usually better to use the parametric procedure. Nonparametric tests are most useful when we cannot assume normality, or for analyzing certain non-numerical data sets.

Recall that the population median divides the distribution in half.

Suppose a random sample is obtained from a continuous (non-normal) distribution. Consider a test concerning the population median with null hypothesis $H_0: \tilde{\mu} = \tilde{\mu}_0$. If the null hypothesis is true, then approximately half of the observations should lie above $\tilde{\mu}_0$, and the other half should fall below $\tilde{\mu}_0$.

To use the *sign test*, we replace each observation above $\tilde{\mu}_0$ with a plus sign and each observation below $\tilde{\mu}_0$ with a minus sign. If H_0 is true, then the number of plus signs and the number of minus signs should be about the same. A problem arises if an observation is equal to $\tilde{\mu}_0$. Because the underlying distribution is continuous, theoretically the probability of obtaining an observation exactly equal to $\tilde{\mu}_0$ is zero. In practice, however, it is common to obtain such an observation. Any observations equal to $\tilde{\mu}_0$ are excluded from the analysis.

The binomial distribution was defined in Section 5.4.

The test statistic, X, is a count of the number of plus signs (or number of observations greater than $\tilde{\mu}_0$). If the null hypothesis is true, then the probability of a plus sign (an observation is greater than $\tilde{\mu}_0$) is 1/2. Therefore, if H_0 is true, the random variable X has a binomial distribution with number of trials equal to n, the number of observations included in the test, and $p = 1/2$: $X \sim \text{B}(n, 0.5)$. We should reject the null hypothesis for very large or very small values of X.

The Sign Test Concerning a Population Median

Suppose a random sample is obtained from a continuous distribution. A hypothesis test concerning a population median $\tilde{\mu}$ with significance level α has the form

$H_0: \tilde{\mu} = \tilde{\mu}_0$

$H_a: \tilde{\mu} > \tilde{\mu}_0, \quad \tilde{\mu} < \tilde{\mu}_0, \quad \text{or} \quad \tilde{\mu} \neq \tilde{\mu}_0$

TS: $X =$ the number of observations greater than $\tilde{\mu}_0$

RR: $X \geq c_1, \quad X \leq c_2, \quad X \geq c \quad \text{or} \quad X \leq n - c$

The critical values c_1, c_2, and c are obtained from Table I in the Appendix, Binomial Distribution Cumulative Probabilities, with parameters n and $p = 0.5$, to yield a significance level of approximately α, that is, so that $P(X \geq c_1) \leq \alpha$, $P(X \leq c_2) \leq \alpha$, and $P(X \geq c) \leq \alpha/2$.

Observations equal to $\tilde{\mu}_0$ are excluded from the analysis, and the sample size is reduced accordingly.

A CLOSER LOOK

1. If the underlying (continuous) distribution is symmetric, then $\mu = \tilde{\mu}$ and the sign test can be used to test a hypothesis about a population mean.

2. We really do not need to literally replace observations with plus or minus signs. Discard any observations equal to $\tilde{\mu}_0$, and simply count the number of observations greater than $\tilde{\mu}_0$.

3. Because the binomial distribution is discrete, we usually cannot find critical values to yield an exact level-α test. Use critical values such that the significance level is as close to α as possible but not greater than α.

Example 14.1 Unemployment Duration

DATA SET

UNEMPLOY

The duration of unemployment is often used as an indicator of the state of the U.S. economy. During 2013, a typical unemployed person found a new job after searching for approximately 36 weeks. However, near the end of the calendar year, data from the Bureau of Labor Statistics suggested that for a typical worker the duration of unemployment was increasing.[3] Suppose a random sample of 15 unemployed adults in the US was obtained in January 2014. The number of weeks each was unemployed are given in the following table.

35	40	38	37	39	43	49	39	34	41	47	35	45	38	43

Is there any evidence to suggest that the median duration of unemployment is greater than 36 weeks? Use a significance level of $\alpha = 0.10$.

Solution Trail 14.1

KEYWORDS

- Is there any evidence?
- Median duration greater than

TRANSLATION

- One-sided, right-tailed hypothesis test concerning a population median

CONCEPTS

- Sign test concerning a population median

VISION

We cannot assume anything about the shape of the underlying distribution of unemployment duration. Since the normality assumption is not justified and this is a test concerning the median, the (nonparametric) sign test is appropriate.

SOLUTION

STEP 1 The assumed median is $\tilde{\mu} = 36 \, (= \tilde{\mu}_0)$, the sample size is $n = 15$, and we will use $\alpha = 0.10$.

We are looking for any evidence that the median unemployment duration is greater than 36 weeks. Therefore, the relevant alternative hypothesis is one-sided, right-tailed.

Because the population is not assumed to be normal and this is a test concerning the median, the sign test should be used.

STEP 2 The four parts of the hypothesis test are

H_0: $\tilde{\mu} = 36$

H_a: $\tilde{\mu} > 36$

TS: $X =$ the number of observations greater than 36

RR: $X \geq c_1 = 11$

To find the critical value c_1:

a. There are no observations equal to 36, the hypothesized median. Therefore, no values are excluded from the analysis.

Table I in the Appendix, Binomial Distribution Cumulative Probabilities, $n = 15$

x	$\cdots$	p 0.50	$\cdots$
$\vdots$		$\vdots$	
8		0.6964	
9		0.8491	
10	$\cdots$	0.9408	
11		0.9824	
12		0.9963	
$\vdots$		$\vdots$	

If the null hypothesis is true, X is a binomial random variable with $n = 15$ and $p = 0.5$: $X \sim B(15, 0.5)$. We need to find a value c_1 such that $P(X \geq c_1)$ is as close to $\alpha = 0.10$ as possible without going over.

Find the smallest c_1 such that $P(X \geq c_1) \leq 0.10$.

b. Use the complement rule applied to a discrete random variable to convert this equation to cumulative probability,

$$P(X \geq c_1) = 1 - P(X < c_1) = 1 - P(X \leq c_1 - 1) \leq 0.10$$

or find the smallest c_1 such that $P(X \leq c_1 - 1) \geq 0.90$.

c. Using Table I in Appendix A, with $n = 15$ and $p = 0.5$, $c_1 - 1 = 10$. Therefore, $c_1 = 11$. The actual significance level using this critical value is

$$P(X \geq 11) = 1 - P(X < 11) = 1 - P(X \leq 10) = 1 - 0.9408 = 0.0592$$

STEP 3 Using signs, classify each observation as either greater than or less than the hypothesized median.

Observation	35	40	38	37	39	43	49	39	34	41	47	35	45	38	43
Sign	−	+	+	+	+	+	+	+	−	+	+	−	+	+	+

The value of the test statistic is the number of plus signs, or the number of observations greater than $\tilde{\mu}_0 = 36$. Therefore, $x = 12$.

STEP 4 The value of the test statistic lies in the rejection region. At the $\alpha = 0.10$ significance level, there is evidence to suggest that the population median unemployment duration is greater than 36 weeks.

STEP 5 It is unlikely that a critical value will yield the exact desired significance level, so it is often more appropriate to find a p value when using the sign test.

$$p = P(X \geq 12)$$ Definition of a p value.

$$= 1 - P(X \leq 11)$$ The complement rule; discrete random variable.

$$= 1 - 0.9824 = 0.0176$$ Table I in the Appendix, $n = 15$, $p = 0.5$.

Using the p value to draw a conclusion, because $p = 0.0176 \leq 0.10 = \alpha$, we reject the null hypothesis. There is evidence to suggest that the population median unemployment duration is greater than 36 weeks.

Figure 14.1 shows a technology solution.

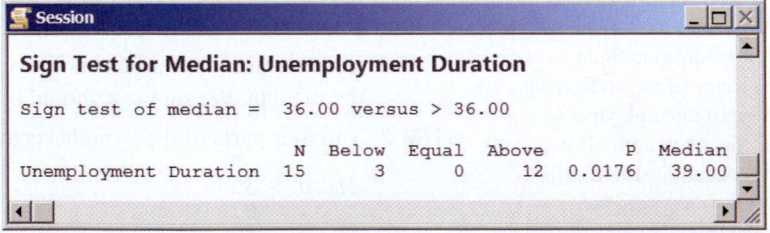

Figure 14.1 Minitab sign test output.

TRY IT NOW   GO TO EXERCISE 14.14

This is the nonparametric counterpart to a paired t test.

If the data are paired (two observations on each individual or object) and the underlying distributions are not normal, then the sign test can be used to compare population medians. Compute each pairwise difference, disregard the magnitude of the difference, and consider only the sign of the difference. Replace each positive difference with a plus

sign and each negative difference with a minus sign. Use these signs and the test procedure described above. This analysis is appropriate for comparing two population medians,
$\tilde{\mu}_1$ and $\tilde{\mu}_2$.

The Sign Test to Compare Two Population Medians

Suppose there are n independent pairs of observations such that the population of first
observations is continuous and the population of second observations is also continuous.
A hypothesis test concerning the two population medians in terms of the difference,
$\tilde{\mu}_D = \tilde{\mu}_1 - \tilde{\mu}_2$, with significance level α has the form

$H_0: \tilde{\mu}_D = \Delta_0$

$H_a: \tilde{\mu}_D > \Delta_0, \quad \tilde{\mu}_D < \Delta_0, \quad$ or $\quad \tilde{\mu}_D \neq \Delta_0$

TS: $X =$ the number of pairwise differences greater than Δ_0

RR: $X \geq c_1, \quad X \leq c_2, \quad X \geq c \quad$ or $\quad X \leq n - c$

The critical values c_1, c_2, and c are obtained from Table I in the Appendix, Binomial
Distribution Cumulative Probabilities, with parameters n and $p = 0.5$, to yield a significance level of approximately α, that is, so that $P(X \geq c_1) \leq \alpha$, $P(X \leq c_2) \leq \alpha$, and
$P(X \geq c) \leq \alpha/2$.

Differences equal to Δ_0 are excluded from the analysis, and the sample size is reduced
accordingly.

DATA SET

EGGDIET

Example 14.2 Good Reason to Eat Eggs

There have been numerous studies concerning the benefits and possible harmful effects
of eating eggs. A recent article listed 30 good reasons to eat eggs, including to protect
your eyesight.[4] A random sample of healthy male adults, ages 25–45, was obtained. The
dynamic visual acuity (DVA, logMAR scale) was used to measure each subject's ability
to detect objects while moving his head. A higher DVA score indicates better percep-
tion. Each person was put on a diet that included one egg per day. After four weeks,
each subject was tested again for dynamic visual acuity. The data are given in the following table.

Subject	1	2	3	4	5	6	7	8	9	10
Before	0.58	0.78	0.76	0.54	0.60	0.73	0.56	0.45	0.59	0.45
After	0.60	0.49	0.41	0.74	0.80	0.77	0.52	0.58	0.69	0.71

Subject	11	12	13	14	15	16	17	18	19	20
Before	0.42	0.62	0.76	0.61	0.67	0.49	0.48	0.56	0.57	0.52
After	0.58	0.70	0.73	0.43	0.52	0.70	0.47	0.44	0.64	0.68

Other research suggests that the underlying DVA score populations are not normal. Use a
sign test to compare the median DVA score before the egg diet with the median DVA
score after the egg diet. Is there any evidence to suggest that eggs help to improve eyesight, as measured by DVA? Use a significance level of 0.01.

Stockbroker/© MBI/Alamy

SOLUTION

STEP 1 The data are certainly paired: There are before and after measurements on each
individual. Typically, the before measurements are considered population 1 and
the after measurements are population 2. The underlying populations are not
normal, so a paired t test is not appropriate.

Table I in the Appendix, Binomial Distribution Cumulative Probabilities, $n = 20$

x	$\cdots$	p 0.50	$\cdots$
$\vdots$		$\vdots$	
2		0.0002	
3		0.0013	
4	$\cdots$	0.0059	
5		0.0207	
6		0.0577	
$\vdots$		$\vdots$	

What would the significance level be if $c_2 = 5$?

STEP 2 The null hypothesis is that the two population medians are equal: The egg-a-day diet has no effect on visual acuity, or

$$\tilde{\mu}_1 = \tilde{\mu}_2 \;\Rightarrow\; \tilde{\mu}_1 - \tilde{\mu}_2 = \tilde{\mu}_D = 0 \; (=\Delta_0)$$

We are searching for evidence that the egg diet improves visual acuity, so the alternative hypothesis is $\tilde{\mu}_1 < \tilde{\mu}_2 \;\Rightarrow\; \tilde{\mu}_1 - \tilde{\mu}_2 = \tilde{\mu}_D < 0$. This is a one-sided, left-tailed test.

STEP 3 The four parts of the hypothesis test are

$H_0: \tilde{\mu}_D = 0$

$H_a: \tilde{\mu}_D < 0$

TS: $X =$ the number of differences greater than 0

RR: $X \leq c_2 = 4$

To find the critical value c_2:

a. There are no differences equal to $\Delta_0 = 0$, the hypothesized difference. Therefore, no pairs are excluded from the analysis.

If the null hypothesis is true, X has a binomial distribution with $n = 20$ and $p = 0.5$: $X \sim B(20, 0.5)$. Find the largest value c_2 such that $P(X \leq c_2) \leq 0.01 = \alpha$.

b. Using Table I in the Appendix, with $n = 20$ and $p = 0.5$, the critical value is $c_2 = 4$. The actual significance level using this critical value is $P(X \leq 4) = 0.0059 \; (\leq 0.01)$.

STEP 4 Compute each pairwise difference (before − after) and determine the signs.

Subject	1	2	3	4	5	6	7	8	9	10
Before	0.58	0.78	0.76	0.54	0.60	0.73	0.56	0.45	0.59	0.45
After	0.60	0.49	0.41	0.74	0.80	0.77	0.52	0.58	0.69	0.71
Difference	−0.02	0.29	0.35	−0.20	−0.20	−0.04	0.04	−0.13	−0.10	−0.26
Sign	−	+	+	−	−	−	+	−	−	−

Subject	11	12	13	14	15	16	17	18	19	20
Before	0.42	0.62	0.76	0.61	0.67	0.49	0.48	0.56	0.57	0.52
After	0.58	0.70	0.73	0.43	0.52	0.70	0.47	0.44	0.64	0.68
Difference	−0.16	−0.08	0.03	0.18	0.15	−0.21	0.01	0.12	−0.07	−0.16
Sign	−	−	+	+	+	−	+	+	−	−

STEP 5 The value of the test statistic is the number of plus signs, or the number of differences greater than $\Delta_0 = 0$. Therefore, $x = 8$.

STEP 6 The value of the test statistic does not lie in the rejection region. At the $\alpha = 0.01$ significance level, there is no evidence to suggest that the egg diet increased the median DVA score.

STEP 7 The p value for this hypothesis test is

$p = P(X \leq 8)$ Definition of p value.

$\quad = 0.2517$ Cumulative probability; Table I in the Appendix, $n = 20$, $p = 0.5$.

Using the p value to draw a conclusion, because $p = 0.2517 > 0.01 = \alpha$, we cannot reject the null hypothesis. There is no evidence to suggest that the egg diet increased the median DVA score.

Figure 14.2 shows a technology solution.

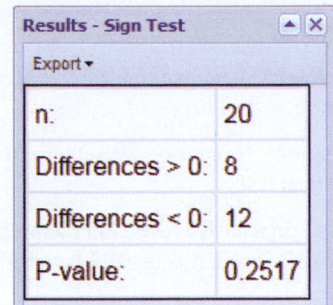

Figure 14.2 CrunchIt! two-sample sign test output.

TRY IT NOW   GO TO EXERCISE 14.20

Technology Corner

Procedure: Conduct a sign test concerning a population median.
Reconsider: Example 14.2, solution, and interpretations.

CrunchIt!

The Sign Test function is only for comparing two sample medians.

1. Enter the data from sample 1 into column Var1 and the data from sample 2 into column Var2.
2. Select Statistics; Non-parametrics; Sign Test.
3. Using the pull-down menus, select the First Variable and the Second Variable.
4. Under the Hypothesis Test tab, enter the Difference under the null hypothesis (Δ_0) and select the appropriate Alternative. Click Calculate. The results are displayed in a new window (Figure 14.2).

Minitab

Use the 1-Sample Sign function to conduct a sign test concerning a population median.

1. Enter the unemployment data into column C1.
2. Select Stat; Nonparametrics; 1-Sample Sign. Enter the Variable (column containing the data, C1), select Test median and enter the hypothesized value, 36, and select the Alternative hypothesis (greater than).
3. Click OK. The test results are displayed in a session window. Refer to Figure 14.1.

Excel

There is no built-in function to conduct a sign test. Use other functions to count the number of observations greater than the hypothesized median and to compute the p value associated with the test.

1. Enter the data into column A.
2. Count the number of observations greater than 30 using the function COUNTIF.
3. Use the function BINOM.DIST to compute the p value. See Figure 14.3.

Figure 14.3 Excel functions to conduct a sign text.

B	C
12	= COUNTIF(A1:A15,">36")
0.0176	= 1-BINOM.DIST(B1-1,15,0.5,TRUE)

SECTION 14.1 EXERCISES

Concept Check

14.1 Short Answer   A nonparametric test is also called a _____ procedure.

14.2 Short Answer   Name two advantages to using a nonparametric statistical test.

14.3 True/False   In a case where either a nonparametric or a parametric test can be used, it is usually better to use the parametric procedure.

14.4 True/False   The sign test concerning a population median is based on the number of positive values (observations greater than 0) and the number of negative values (observations less than 0).

14.5 Short Answer   Under what conditions can the sign test be used to test a hypothesis about a population mean?

14.6 True/False   The sign test is based on the binomial distribution with n trials and probability of a success $p = 0.50$.

14.7 Short Answer   Explain why we usually cannot find critical values in a sign test to yield an exact, specified significance level α.

14.8 Short Answer   In a sign test to compare two population medians, how do we treat a difference equal to Δ_0?

Practice

14.9   In each of the following problems, a data set and a null hypothesis (concerning the population median) are given. Assume that the sign test will be used, and find the value of the test statistic x. **EX14.9**

   a. $\{20, 20, 13, 20, 16, 19, 19, 11, 20, 14\}$ $H_0: \tilde{\mu} = 15$
   b. $\{66, 90, 77, 68, 70, 56, 56, 75, 57, 65, 65, 56, 66, 70, 54\}$
        $H_0: \tilde{\mu} = 70$
   c. $\{46.4, 42.6, 50.9, 49.2, 46.4, 47.6, 50.7, 49.3, 59.3,$
        $51.5, 45.3, 52.9, 54.5, 51.1, 50.7, 41.3, 57.8, 59.5,$
        $52.3, 47.2\}$ $H_0: \tilde{\mu} = 51.5$
   d. $\{-4, -7, 8, -7, 3, 9, 1, 1, 3, 10, -9, -7, -5, 8, -1,\}$
        $-9, -2, 6, 7, -6, -7, -9, 7, 9, 3\}$ $H_0: \tilde{\mu} = 0$

14.10   Use the random sample in each problem to conduct a sign test with the indicated null hypothesis, alternative hypothesis, and significance level. Find the exact p value for each test. **EX 14.10**

   a. $H_0: \tilde{\mu} = 16$, $H_a: \tilde{\mu} < 16$, $\alpha = 0.05$

13.1	17.4	13.6	18.4	11.2	15.9	14.6
13.6	13.6	14.5				

   b. $H_0: \tilde{\mu} = -25$, $H_a: \tilde{\mu} > -25$, $\alpha = 0.01$

−19	−20	−25	−28	−17	−25	−16	−11
−12	−28	−23	−11	−11	−25	−28	

c. $H_0: \tilde{\mu} = 8$, $H_a: \tilde{\mu} \neq 8$, $\alpha = 0.05$

8.90	7.68	9.41	9.47	8.98	8.41	9.51	7.40
9.66	8.17	5.32	9.06	7.67	7.87	8.16	8.77
7.01	9.22	7.41	8.57				

d. $H_0: \tilde{\mu} = 125$, $H_a: \tilde{\mu} \neq 125$, $\alpha = 0.01$

196	126	187	168	111	196	164	181	110
113	187	136	159	154	177	132	103	151
187	191	192	139	187	190	112		

14.11   Using the paired data on the website, conduct a sign test to compare population medians with $H_0: \tilde{\mu}_1 - \tilde{\mu}_2 = 0$, $H_a: \tilde{\mu}_1 - \tilde{\mu}_2 > 0$, and significance level $\alpha = 0.05$. **EX 14.11**

14.12   Using the paired data on the website, conduct a sign test to compare population medians with $H_0: \tilde{\mu}_1 - \tilde{\mu}_2 = 3$, $H_a: \tilde{\mu}_1 - \tilde{\mu}_2 \neq 3$, and significance level $\alpha = 0.01$. Find the exact p value for this test. **EX 14.12**

Applications

14.13 Medicine and Clinical Studies   Studies suggest that Parkinson's disease (PD) affects nerve endings in the heart that produce the chemical norepinephrine. Low levels of norepinephrine decrease the patient's ability to control muscle movement, hence producing the shaking symptoms associated with PD. A random sample of women with PD was obtained, and the concentration of norepinephrine (in nmol/L) was measured in each. The data are given in the following table. **PD**

1.36	1.29	1.05	1.25	1.34	1.50	1.33	1.31
1.08	1.11	1.29	1.04	1.09	1.38	1.44	

Assume that the underlying distribution is continuous. Use the sign test to determine whether there is any evidence that the median concentration of norepinephrine in women with PD is less than 1.38 nmol/L (the normal level). Use $\alpha = 0.05$.

14.14 Fuel Consumption and Cars   Disc brake pads are critical safety features on an automobile and should be replaced when the thickness has worn to approximately 2 mm. A random sample of automobiles entering a New Jersey State Inspection Station was obtained, and the brake-pad thickness on the left front tire was measured on each. Assume that the underlying distribution of brake-pad thicknesses is continuous. Use the sign test to determine whether there is any evidence that the median brake-pad thickness is less than 2 mm. Use $\alpha = 0.10$ for this test and find the exact p value. **BRAKES**

14.15 Manufacturing and Product Development   Modular homes are constructed in sections in a factory, with no delays due to weather conditions. Sections are transported to a home site on truck beds, attached together, and placed on a premade foundation. A random sample of modular-home deliveries was obtained, and the mileage from the factory to the home site was recorded for each. Assume that the underlying distribution of mileage is continuous. Use the sign test to determine whether there is any evidence that the median mileage is greater than 100. Use $\alpha = 0.05$ for this test and find the exact p value. **MODHOMES**

14.16 Manufacturing and Product Development   Grade 40s grease mohair must have a fiber diameter of approximately 24 microns. A manufacturer of mohair jackets received a large shipment of 40s-grade raw material and obtained 30 random measurements of fiber diameter. The manufacturer can only assume that the underlying distribution of fiber diameters is continuous. Is there any evidence to suggest that the median diameter is different from 24 microns? Use a significance level of $\alpha = 0.05$. **MOHAIR**

14.17 Sports and Leisure   There has been increased attention lately to sports-related head injuries. *Frontline* recently published an article about the NFL's "concussion crisis,"[5] and physicians have become more sensitive to recognizing and treating head trauma in younger children. Recent studies suggest that there may be a sex difference in memory after a concussion.[6] A random sample of female soccer players who sustained a concussion was obtained. A computer test to measure memory skills was administered, and the time (in seconds) needed to complete the test was recorded for each. Assume that the underlying distribution of time needed to complete this memory test is continuous. Is there any evidence to suggest that the median time to complete this test for women is greater than 40 seconds? Use $\alpha = 0.05$. **MEMORY**

14.18 Travel and Transportation   According to a report from TrueCar.com, the average transaction price for a new vehicle reached a record high in August 2013 of $31,252.[7] A transaction price of a vehicle is the out-the-door-price, which includes the price of the vehicle, discounts, add-ons, taxes, and license fees. A random sample of new vehicles purchased in August 2014 was obtained, and the transaction price of each was recorded. Assume that the underlying distribution of new-vehicle transaction price is continuous. **CARPRICE**
   a. Is there any evidence to suggest that the median new vehicle transaction price has increased (from ($31,252)?
   b. Why is the median, rather than the mean, new-vehicle transaction price a better measure of the center of this distribution?

14.19 Biology and Environmental Science   Chlorophyll is an important part of photosynthesis in plants and is also present in microscopic algae and other phytoplankton. A random sample of 20 locations along the western coastline of the United States was selected. Using fluorometry, the quantity of chlorophyll in the surface water was measured (in mg/m^3) in early April and in late August. The pairwise differences (April measurement − August measurement) are given in the following table. **CHLORO**

6.24	6.51	−10.96	9.52	5.39	5.88	−9.03
4.58	−8.16	6.99	4.92	−9.35	4.88	−1.99
4.96	3.90	−8.13	7.55	12.61	7.68	

The distribution of chlorophyll in surface water is assumed to be continuous but not normal. Is there any evidence to suggest that the median chlorophyll amount in surface water is different in April than in August? Use a significance level of $\alpha = 0.05$.

14.20 Travel and Transportation   Over the past several years, Times Square in New York City was made pedestrian-only, and protected bike lanes were added to five avenues in the middle of Manhattan. In addition, a new bike-share program has become extremely popular. As a result of these changes, motor vehicle traffic is actually moving more smoothly into and through Manhattan.[8] One measure of better traffic flow is the speed of a taxi ride. After carefully reviewing taxi records, a random sample of similar trips before and after the changes to Times Square were obtained. The average speed (in mph) of each trip was recorded. Use the sign test with $\alpha = 0.05$ to determine whether there is any evidence that the median speed of a taxi trip is faster after the changes to Times Square. Find the p value associated with this test. **TIMESQ**

Extended Applications

14.21 Biology and Environmental Science   Jet-engine emissions at high altitudes cause vapor trails that contribute to cloud cover and pollution, and may even affect weather patterns. An experiment was conducted to measure the concentration of volatile organic compounds (VOCs) directly behind an operating engine. A random sample of jet engines was obtained. The VOC concentration was measured (in mg/m^3) before and after a special exhaust scrubber was installed. The data are given in the following table. **JETVOC**

Before	13.0	14.9	14.0	13.4	9.8	14.9	12.0	11.2
After	11.0	13.6	12.3	9.9	10.3	10.7	9.6	9.5

Before	14.9	15.0	9.7	13.1	14.4	11.3	14.6
After	11.8	13.1	9.2	10.0	11.9	13.7	12.3

Use the sign test with $\alpha = 0.05$ to determine whether there is any evidence that the median VOC concentration is smaller when the scrubber is installed. Find the p value associated with this test.

14.22 Medicine and Clinical Studies   Calcium blockers have many effects on a person's heart rate. They are most often used to slow the heart rate in patients with atrial fibrillation. This class of medications work to decrease the heart pumping strength and, therefore, ease tension in blood vessels.[9] Suppose an experiment was conducted to determine whether calcium blockers affect heart rhythms in patients with arrhythmias. A random sample of patients was obtained, and their resting pulse rate was measured (in beats per minute). After a two-week regimen of a calcium blocker, each person's resting pulse rate was measured again. Suppose no assumptions can be made about the shape of the continuous distributions of before and after pulse rates. **CABLOCK**
   a. Conduct a sign test to determine whether there is any evidence that the median pulse rate before the calcium

blocker medication is different from the median pulse rate after the medication. Use a significance level of $\alpha = 0.01$.

  b. Do you believe the assumptions for the sign test are valid in this case? Why or why not?

14.23 Medicine and Clinical Studies The short-term survival rate following a lung transplant has improved recently. Even though patients face the risks of organ rejection and infection, approximately 78% of patients survive after the first year, and the median survival for a single-lung recipient is 4.6 years.[10] The D'Youville Hospital Center in New York City claims to have a survival rate higher than the national *average*. Suppose a random sample of single-lung transplant patients from D'Youville was

obtained. The survival (in years) after the transplant was recorded for each. Suppose no assumptions can be made about the shape of the underlying continuous distribution.   **LUNG**

  a. Conduct a sign test to determine whether there is any evidence to suggest that the median survival at D'Youville is greater than 4.6 years. Use $\alpha = 0.05$.

  b. Conduct a one-sample t test concerning a population mean to determine whether there is any evidence to suggest that the mean survival at D'Youville is greater than 4.6 years. Use $\alpha = 0.05$. How do your results compare with part (a)? Which analysis do you believe is more appropriate? Why?

14.2   The Signed-Rank Test

This test was developed in 1945 by Frank Wilcoxon and is usually called the Wilcoxon signed-rank test.

The sign test concerning a population median or to compare two population medians uses only the signs (plus or minus) of the relevant differences. The signed-rank test also utilizes the magnitude of each difference and, of course, the ranks. The test statistic does not have a common distribution, but if n is sufficiently large, a normal approximation may be used.

Suppose a random sample of size n is obtained from a continuous, symmetric distribution, and the null hypothesis is $H_0: \tilde{\mu} = \tilde{\mu}_0$.

1. Subtract $\tilde{\mu}_0$ from each observation; that is, compute the differences $x_1 - \tilde{\mu}_0$, $x_2 - \tilde{\mu}_0, \ldots, x_n - \tilde{\mu}_0$.

2. Consider the magnitude, or absolute value, of each difference; compute $|x_1 - \tilde{\mu}_0|$, $|x_2 - \tilde{\mu}_0|, \ldots, |x_n - \tilde{\mu}_0|$.

The rank of an observation is its position in the ordered list.

3. Place the absolute values in increasing order, and assign a rank to each, from smallest (rank 1) to largest (rank n).

4. Equal absolute values are assigned the mean rank of their positions in the ordered list. For example, if the fifth, sixth, and seventh absolute values were all equal, then each would be assigned the rank $(5 + 6 + 7)/3 = 6$.

5. Add the ranks associated with the positive differences.

If the null hypothesis is true, then approximately half of the observations should be above the median and approximately half below the median. Because the distribution is assumed to be symmetric, for every positive difference, there should be a corresponding negative difference of approximately the same magnitude. Therefore, the sum of the ranks associated with the positive differences should be approximately equal to the sum of the ranks associated with the negative differences. If the sum of the ranks associated with the positive differences is very large or very small, there is evidence to suggest that the population median is different from $\tilde{\mu}_0$. The dot plots in Figures 14.4 and 14.5 provide a visual interpretation of this intuitive concept.

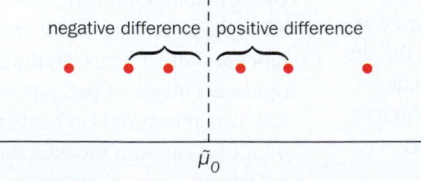

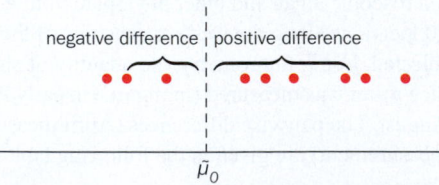

Figure 14.4 If H_0 is true, for every positive difference there should be a negative difference of about the same magnitude.

Figure 14.5 If $\tilde{\mu} > \tilde{\mu}_0$, then there should be more positive differences of large magnitude. The sum of the ranks associated with the positive difference would be large.

The Wilcoxon Signed-Rank Test

Suppose a random sample is obtained from a continuous, symmetric distribution. A hypothesis test concerning a population median $\tilde{\mu}$ with significance level α has the form

$$H_0: \tilde{\mu} = \tilde{\mu}_0$$
$$H_a: \tilde{\mu} > \tilde{\mu}_0, \qquad \tilde{\mu} < \tilde{\mu}_0, \qquad \text{or} \qquad \tilde{\mu} \neq \tilde{\mu}_0$$

Rank the absolute differences $|x_1 - \tilde{\mu}_0|, |x_2 - \tilde{\mu}_0|, \ldots, |x_n - \tilde{\mu}_0|$. Equal absolute values are assigned the mean rank for their positions.

TS: $T_+ =$ the sum of the ranks corresponding to the positive differences $x_i - \tilde{\mu}_0$

RR: $T_+ \geq c_1, \qquad T_+ \leq c_2, \qquad T_+ \geq c \qquad \text{or} \qquad T_+ \leq n(n+1) - c$

The critical values c_1, c_2, and c are obtained from Table IX in the Appendix such that $P(T_+ \geq c_1) \approx \alpha$, $P(T_+ \leq c_2) \approx \alpha$, and $P(T_+ \geq c) \approx \alpha/2$.

Differences equal to 0 ($x_i - \tilde{\mu}_0 = 0$) are excluded from the analysis, and the sample size is reduced accordingly.

The Normal Approximation: As n increases ($n \geq 20$), the statistic T_+ approaches a normal distribution with mean and variance

$$\mu_{T_+} = \frac{n(n+1)}{4} \qquad \text{and} \qquad \sigma^2_{T_+} = \frac{n(n+1)(2n+1)}{24}$$

Therefore, the random variable $Z = \dfrac{T_+ - \mu_{T_+}}{\sigma_{T_+}}$ has approximately a standard normal distribution. In this case ($n \geq 20$), Z is the test statistic and the rejection region is

RR: $Z \geq z_\alpha, \qquad Z \leq -z_\alpha, \qquad \text{or} \qquad |Z| \geq z_{\alpha/2}$

We assume the underlying population is symmetric (when using the signed-rank test), so the median is equal to the mean. Therefore, the test procedure above can be used to test a hypothesis concerning a population mean (when the underlying population is not normal but is symmetric).

DATA SET

SWEETEN

Example 14.3 Sweet River Water

Diet foods and beverages contain artificial sweeteners that are not processed by the body. Consequently, these sweeteners, for example, sucralose, cyclamate, saccharin, and acesulfame, flow into water-treatment facilities. Most of these chemicals remain unaffected by the treatment process and flow into rivers and streams. A recent study concluded that the Grand River in Ontario had the highest concentration of artificial sweeteners in the world.[11] Suppose a random sample of Grand River water was obtained from 10 different locations, and the saccharin concentration (in μg/L) was measured for each. The data are given in the following table.

5.2	7.2	3.5	4.7	5.9	3.6	2.5	5.8	4.4	3.3

Assume the underlying distribution of saccharin concentration is continuous and symmetric. Use a signed-rank test to determine whether there is any evidence that the median saccharine concentration is greater than 4 μg/L, with a significance level of 0.05.

SOLUTION

STEP 1 The distribution of saccharin concentration is assumed to be continuous and symmetric. Because this problem involves the population median, the signed-rank test is appropriate. We are looking for evidence that the population median is greater than 4; the hypothesis test is one-sided, right-tailed. The number of observations is $n = 10$, (<20); the test statistic is T_+.

STEP 2 $H_0: \tilde{\mu} = 4$

$H_a: \tilde{\mu} > 4$

TS: $T_+ = $ the sum of the ranks corresponding to the positive differences

$x_i - \tilde{\mu}_0$

RR: $T_+ \geq c_1 = 44$

We need a value for c_1 such that $P(T_+ \geq c_1) \approx 0.05$. Using Table IX in the Appendix, $P(T_+ \geq 44) = 0.0527$.

STEP 3   The following table shows the data, each difference $x_i - 4$, the absolute value of each difference $|x_i - 4|$, the rank associated with each absolute difference, and the signed rank. The positive or negative sign from the pairwise difference has been attached to the rank to create the signed rank. Note that there are no zero differences. Therefore, no observations are excluded from the analysis.

Observation	5.2	7.2	3.5	4.7	5.9	3.6	2.5	5.8	4.4	3.3
Difference	1.2	3.2	−0.5	0.7	1.9	−0.4	−1.5	1.8	0.4	−0.7
Absolute difference	1.2	3.2	0.5	0.7	1.9	0.4	1.5	1.8	0.4	0.7
Rank	6.0	10.0	3.0	4.5	9.0	1.5	7.0	8.0	1.5	4.5
Signed rank	+6.0	+10.0	−3.0	+4.5	+9.0	−1.5	−7.0	+8.0	+1.5	−4.5

To assign each rank, consider the ordered list of absolute differences and their position in the list.

Absolute difference	0.4	0.4	0.5	0.7	0.7	1.2	1.5	1.8	1.9	3.2
Position	1	2	3	4	5	6	7	8	9	10
Rank	1.5	1.5	3.0	4.5	4.5	6.0	7.0	8.0	9.0	10.0

The absolute differences in positions 1 and 2 are equal. Each is assigned the mean rank, $(1 + 2)/2 = 1.5$.

The absolute difference 0.5 is in the third position. It is assigned the rank 3.0.

The absolute differences in positions 4 and 5 are equal. Each is assigned the mean rank, $(4 + 5)/2 = 4.5$.

The absolute difference 1.2 is in the sixth position. It is assigned the rank 6.0. Similarly, 1.5, 1.8, 1.9, and 3.2 are assigned the ranks 7.0, 8.0, 9.0, and 10.0.

STEP 4   The value of the test statistic is the sum of the positive signed ranks.

$t_+ = 1.5 + 4.5 + 6.0 + 8.0 + 9.0 + 10.0 = 39.0 \ (< 44)$

STEP 5   The value of the test statistic does not lie in the rejection region. There is no evidence to suggest that the median saccharin concentration in the Grand River is greater than 4 μg/L.

Figure 14.6 shows a technology solution.

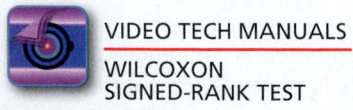

VIDEO TECH MANUALS

WILCOXON
SIGNED-RANK TEST

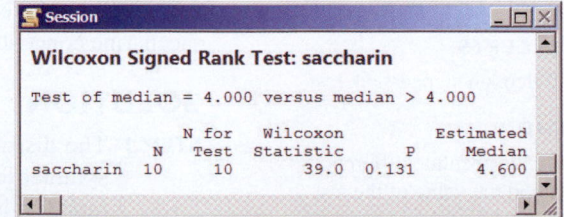

Figure 14.6 Minitab
Wilcoxon signed-rank test.

TRY IT NOW   GO TO EXERCISE 14.32

There are two kinds of *differences* here: the pairwise differences are $d_i = x_i - y_i$, and the differences used in the calculation of the test statistic are $d_i - \Delta_0$.

The signed-rank test may also be used to compare population medians when the data are paired and the underlying distributions are not normal. We assume that the distribution of the pairwise differences is continuous and symmetric, and that each pair of values is independent of all the other pairs. If the null hypothesis is $H_0: \tilde{\mu}_1 - \tilde{\mu}_2 = \tilde{\mu}_D = \Delta_0$, we apply the one-sample test procedure to the pairwise differences.

1. Subtract Δ_0 from each pairwise difference.
   That is, compute $d_i - \Delta_0$ $(i = 1, 2, \ldots, n)$.
2. Rank the absolute differences, $|d_i - \Delta_0|$ $(i = 1, 2, \ldots, n)$.
3. Find the sum of the ranks associated with the positive differences, t_+.

DATA SET

NTWORK

Solution Trail 14.4

KEYWORDS

■ Signed-rank test

■ Median greater than

TRANSLATION

■ Two-sided, right-tailed hypothesis test concerning the difference in population medians

CONCEPTS

■ Wilcoxon signed-rank test

VISION

Compute each pairwise difference and rank the absolute differences. Find the value of the test statistic, and draw the appropriate conclusion.

Example 14.4 Night-Shift Hazards

A recent study suggests that night-shift workers have a greater risk of heart attacks and strokes.[12] The body clock of those people who work evening, night, and rotating shifts becomes disrupted and may be associated with high blood pressure, high cholesterol, and diabetes. Suppose a study was conducted to compare the cholesterol level in workers on the night shift and workers on the traditional day shift in a food-processing facility. A random sample of employees on the night and day shifts were selected and paired according to sex, age, and weight. The total cholesterol level was measured (in mg/dL) in each, and the data are given in the following table.

Night shift	215	185	175	185	211	158	171	199	190	151	154	233
Day shift	203	199	168	177	176	153	173	179	187	165	145	212

Night shift	180	202	180	160	195	171	152	210	212	196	188	213
Day shift	178	190	174	167	189	156	148	200	197	204	190	224

Assume that the underlying distribution of the pairwise differences is continuous and symmetric. Use the Wilcoxon signed-rank test to determine whether there is any evidence that the median total cholesterol level is greater in night-shift workers. Use $\alpha = 0.05$.

SOLUTION

STEP 1 The underlying distributions are not assumed to be normal, but the distribution of the differences is assumed to be continuous and symmetric. The assumptions for the Wilcoxon signed-rank test are met. Note that the sign test to compare two medians could also be used, but the Wilcoxon signed-rank test is a more reliable test because it uses more of the information in the sample.

We are searching for evidence that night-shift workers have higher median total cholesterol, so the hypothesis test is one-sided, right-tailed, with $\Delta_0 = 0$.

Let population 1 be the total cholesterol level in night-shift workers and population 2 be the total cholesterol level in day-shift workers.

Because $\Delta_0 = 0$, we do not need to modify the pairwise differences. No pairwise differences are equal to zero; no pairs are excluded from the analysis.

There are $n = 24 \geq 20$ observations (pairwise differences), so the normal approximation will be used. The four parts of the hypothesis test are

$$H_0: \tilde{\mu}_1 - \tilde{\mu}_2 = \tilde{\mu}_D = \Delta_0 = 0$$
$$H_a: \tilde{\mu}_1 - \tilde{\mu}_2 = \tilde{\mu}_D > \Delta_0 = 0$$
$$\text{TS: } Z = \frac{T_+ - \mu_{T_+}}{\sigma_{T_+}}$$
$$\text{RR: } Z \geq z_\alpha = z_{0.05} = 1.6449$$

STEP 2 Compute the absolute value of each pairwise difference, and rank these numbers. Equal values are assigned the mean rank for their positions. The following table shows the original data, the pairwise differences, the absolute value of each pairwise difference, the rank associated with each pairwise difference, and the signed rank.

Night shift	Day shift	Pairwise difference	Absolute difference	Rank	Signed rank
215	203	12	12	16.5	16.5
185	199	−14	14	18.5	−18.5
175	168	7	7	9.5	9.5
185	177	8	8	11.5	11.5
211	176	35	35	24.0	24.0
158	153	5	5	6.0	6.0
171	173	−2	2	2.0	−2.0
199	179	20	20	22.0	22.0
190	187	3	3	4.0	4.0
151	165	−14	14	18.5	−18.5
154	145	9	9	13.0	13.0
233	212	21	21	23.0	23.0
180	178	2	2	2.0	2.0
202	190	12	12	16.5	16.5
180	174	6	6	7.5	7.5
160	167	−7	7	9.5	−9.5
195	189	6	6	7.5	7.5
171	156	15	15	20.5	20.5
152	148	4	4	5.0	5.0
210	200	10	10	14.0	14.0
212	197	15	15	20.5	20.5
196	204	−8	8	11.5	−11.5
188	190	−2	2	2.0	−2.0
213	224	−11	11	15.0	−15.0

STEP 3 The sum of the positive signed ranks is

$$t_+ = 2.0 + 4.0 + 5.0 + \cdots + 24.0 = 223.0$$

The mean and variance of the random variable T_+ are

$$\mu_{T_+} = \frac{n(n-1)}{4} = \frac{(24)(25)}{4} = 150.0$$

$$\sigma_{T_+}^2 = \frac{n(n+1)(2n+1)}{24} = \frac{(24)(25)(49)}{24} = 1225.0$$

The value of the test statistic is

$$z = \frac{t_+ - \mu_{T_+}}{\sigma_{T_+}} = \frac{223.0 - 150.0}{\sqrt{1225.0}} = 2.0857$$

STEP 4 The value of the test statistic lies in the rejection region; we reject the null hypothesis. At the $\alpha = 0.05$ significance level, there is evidence to suggest that the median total cholesterol level of night-shift workers is greater than the median total cholesterol level of day-shift workers.

Wilcoxon Signed Rank	
	Night shift–
	Day Shift
Test Statistic S	73.000
Prob>|S|	0.0337*
Prob>S	0.0169*
Prob<S	0.9831

Figure 14.7 JMP Wilcoxon signed-rank test.

To find the p value associated with this test, find the right-tail probability.

$$p = P(Z \geq 2.0587) \qquad \text{Definition of } p \text{ value.}$$
$$= 1 - P(Z \leq 2.0587) \qquad \text{Complement rule.}$$
$$= 1 - 0.9802 = 0.0198 \qquad \text{Cumulative probability; use Table III in the Appendix.}$$

Using the p value to draw a conclusion, because $p = 0.0198 \leq 0.05 = \alpha$, we reject the null hypothesis. There is evidence to suggest that the median total cholesterol level in night-shift workers is greater than the median total cholesterol level in day-shift workers.

Figure 14.7 shows a technology solution.

TRY IT NOW   GO TO EXERCISE 14.40

Technology Corner

Procedure: Conduct a Wilcoxon signed-rank test.
Reconsider: Example 14.3, solution, and interpretations.

CrunchIt!

Use the Wilcoxon Signed Rank function to conduct a signed-rank test concerning a population median. For paired data, use Wilcoxon Paired.

1. Enter the data into column Var1.
2. Select Statistics; Non-parametrics; Wilcoxon Signed-Rank.
3. Using the pull down menu, select the Sample column.
4. Under the Hypothesis Test tab, enter the Median under null hypothesis ($\tilde{\mu}_0$) and select the appropriate Alternative. Click Calculate. The results are displayed in a new window. See Figure 14.8.

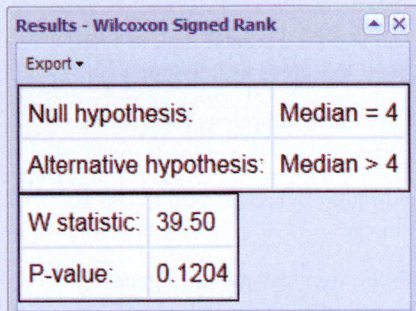

Figure 14.8 CrunchIt! Wilcoxon signed rank test.

Minitab

The built-in Wilcoxon signed-rank test is accessible in the Stat; Nonparametrics menu and may also be executed using commands in a session window. In the case of paired data, use the differences $d_i - \Delta_0$ ($i = 1, 2, \ldots, n$).

1. Enter the data into column C1.
2. Select Stat; Nonparametrics; 1-Sample Wilcoxon. Enter the Variable (column containing the data, C1), select Test median and enter the hypothesized value, 4, and select the Alternative hypothesis (greater than).
3. Click OK. The test results are displayed in a session window. Refer to Figure 14.6.

Excel

There is no built-in function to conduct a Wilcoxon signed-rank test. Use other arithmetic functions to compute the sum of the signed ranks.

1. Enter the data in column A.
2. Compute the differences, and the absolute differences using the function ABS. Use the function RANK.AVG to compute the appropriate ranks for this test. Use the IF function to compute the signed ranks.
3. Use the SUMIF function to find the value of the test statistic. See Figure 14.9.

	A	B	C	D	E
1			Absolute		Signed
2	Saccharin	Difference	difference	Rank	rank
3	5.2	1.2	1.2	6.0	6.0
4	7.2	3.2	3.2	10.0	10.0
5	3.5	-0.5	0.5	3.0	-3.0
6	4.7	0.7	0.7	4.5	4.5
7	5.9	1.9	1.9	9.0	9.0
8	3.6	-0.4	0.4	1.5	-1.5
9	2.5	-1.5	1.5	7.0	-7.0
10	5.8	1.8	1.8	8.0	8.0
11	4.4	0.4	0.4	1.5	1.5
12	3.3	-0.7	0.7	4.5	-4.5
13					
14	39.0 = SUMIF(E3:E12,">0",E3:E12)				

Figure 14.9 Excel calculations to compute the value of the test statistic in a Wilcoxon signed-rank test.

SECTION 14.2 EXERCISES

Concept Check

14.24 Fill in the Blank   The signed-rank test is based on the _____ and _____ of each difference.

14.25 True/False   The test statistic for a signed-rank test has a chi-square distribution with $n - 1$ degrees of freedom.

14.26 Short Answer   What are the assumptions for a signed-rank test?

14.27 True/False   The signed-rank test can be used to test a hypothesis about a population mean.

14.28 Short Answer   If both the sign test and the signed-rank test are appropriate, which would you use, and why?

Practice

14.29 A data set and null hypothesis are given in each of the following problems. Assume a Wilcoxon signed-rank test will be used to test H_0. Find (i) the differences, (ii) the absolute differences, and (iii) the rank associated with each absolute difference. ▥ EX 14.29

a. $H_0: \tilde{\mu} = 60$.

41	66	36	72	33	22	24	36	47
53	28	77	31	34	38	59	60	45

b. $H_0: \tilde{\mu} = 20$.

21.4	20.3	18.5	18.8	21.5	21.6	20.8	21.9
19.8	19.6	19.4	20.8	19.6	18.7	20.4	20.2
21.6	19.2	20.1	18.5	18.5	19.7	19.7	

c. $H_0: \tilde{\mu} = -5$.

-2	-8	-9	-8	-7	-2	-2	-1	-5	-4
-9	-7	-9	-7	-5	-5	-8	-8	-3	-1
-4	-6	-7	-7	-2					

d. $H_0: \tilde{\mu} = 305.4$.

296	303	271	263	288	312	260	305
250	308	315	264	254	258	274	314
279	267	312	310	309	273	269	293
264	278	255	285	272	307		

14.30 Use the random sample in each problem to conduct a Wilcoxon signed-rank test with the indicated null hypothesis, alternative hypothesis, and significance level. Find the p value associated with each test. ▥ EX14.30

a. $H_0: \tilde{\mu} = 70$, $H_a: \tilde{\mu} \neq 70$, $\alpha = 0.10$

67.1	63.5	70.1	62.5	72.2	63.7	63.7	67.7
79.1	61.6	79.7	69.5	72.2	64.5	69.3	

b. $H_0: \tilde{\mu} = 0.7,$ $H_a: \tilde{\mu} < 0.7,$ $\alpha = 0.05$

0.072	0.348	0.319	0.502	0.733	0.603	0.052
0.493	0.721	0.762	0.166	0.965	0.616	0.904
0.882	0.773	0.771	0.506	0.735		

c. $H_0: \tilde{\mu} = -45,$ $H_a: \tilde{\mu} > -45,$ $\alpha = 0.02$

-50	-35	-35	-31	-41	-32	-32
-40	-46	-35	-38	-31	-36	-32

d. $H_0: \tilde{\mu} = 450,$ $H_a: \tilde{\mu} \neq 450,$ $\alpha = 0.01$

461	424	436	485	476	457	463	409	424	450
406	402	435	457	420	410	402	438	428	450
410	418	423	466	497	492	411	426		

14.31 Using the paired data on the text website, conduct a Wilcoxon signed-rank test to compare population medians with $H_0: \tilde{\mu}_1 - \tilde{\mu}_2 = 0$, $H_a: \tilde{\mu}_1 - \tilde{\mu}_2 \neq 0$, and significance level $\alpha = 0.02$. Find the p value associated with this test. 📊 EX14.31

Applications

14.32 Medicine and Clinical Studies   A normal adult brain oxidizes approximately 120 grams of glucose per day.[13] Some researchers believe that the rate of metabolism is higher in adults whose job requires them to make many decisions daily. A random sample of basketball referees was obtained, and the brain oxidation rate of each was measured. The data are given in the following table. 📊 BRAIN

120	121	119	122	121	122	121	119	121
122	124	121	123	122	123	119	120	119

The underlying distribution of oxidation rates is assumed to be continuous and symmetric.

a. Use the Wilcoxon signed-rank test to determine whether there is any evidence that the mean oxidation rate for basketball referees is greater than 120 grams per day. Use a significance level of $\alpha = 0.05$. Find the p value for this test.

b. Why can the signed-rank test be used in this case to test a hypothesis concerning the population mean (rather than the population median)?

14.33 Medicine and Clinical Studies   A random sample of patients with a certain kidney disorder was obtained, and the renal blood-flow rate was measured (in L/min) for each. In healthy patients, the median renal blood-flow rate is known to be 3.00 L/min, and the distribution of flow rates is assumed to be continuous. Use the Wilcoxon signed-rank test to determine whether there is any evidence that the median renal blood-flow rate in patients with this kidney disorder is different from 3.00 L/min. Use $\alpha = 0.01$, and find the p value for this test. 📊 RENAL

14.34 Public Policy and Political Science   Regulations for a new office building specify that the median strength for nonmetallic, nonshrink grout should be 6000 psi when it is supporting concrete. A random sample of grout from various locations in an office building project was obtained, and the strength of each batch (in psi) was determined using the grout-cube test at 28 days. Assume the underlying distribution of grout strength is continuous and symmetric. Use the Wilcoxon signed-rank test with $\alpha = 0.01$ to determine whether there is any evidence that the median grout strength is different from 6000 psi. Find the p value for this test. 📊 GROUT

14.35 Economics and Finance   Near the end of 2013, the annual percentage rate (APR) on a typical credit card had risen to approximately 15.06%.[14] Usually the APR on student credit cards is slightly lower. To check this claim, a random sample of student credit cards was obtained and the annual percentage rate was recorded for each. Assume that the distribution of APR on student credit cards is continuous. Use the Wilcoxon signed-rank test with $\alpha = 0.01$ to determine whether there is any evidence that the median APR on student credit cards is less then 15.06. 📊 CREDIT

14.36 Sports and Leisure   During the summer of 2013, parents spent approximately \$856 per child on summer activities. This amount was greater than in 2012, and it was expected to rise again in the summer of 2014.[15] Suppose a random sample of parents with children not yet in high school was obtained. Each was asked how much they spent per child for summer 2014 activities. Assume the distribution of amount spent per child is continuous. Use the Wilcoxon signed-rank test to determine whether there is any evidence that the median amount spent per child during the summer of 2014 is greater than \$856. Use $\alpha = 0.05$. 📊 SUMMER

14.37 Economics and Finance   Two counties in Pennsylvania handle delinquent property-tax payments in different manners. The first county has very strict regulations that include interest on the unpaid balance and eventually property seizure. The second county uses a more personal touch. Someone calls or visits the property owner to discuss the overdue bill and, if necessary, a payment schedule is created. A random sample of delinquent payments in each county was obtained, and the data were paired according to the size of the original property-tax bill. The amount of time (in months) until full payment was made was recorded for each property. Assume the distribution of pairwise differences is continuous and symmetric. Use the Wilcoxon signed-rank test to determine whether there is any evidence of a difference in median collection time due to the method of handling delinquent payments. Use $\alpha = 0.10$, and find the p value associated with this test. 📊 PROPTAX

14.38 Biology and Environmental Science   The 154-day weights (in pounds) of 16 randomly selected Rambouillet lambs on a farm in Nebraska were recorded. The data are given in the following table. 📊 LAMBS

118	119	120	115	114	113	119	119
118	117	120	116	113	115	118	119

Assume the underlying distribution of weights is continuous and symmetric. Use the Wilcoxon signed-rank test to determine whether there is any evidence that the median 154-day weight is less than 118 pounds. Use a significance level of 0.05.

14.39 Public Health and Nutrition   CBC News recently reported that close to 8 million acres of farmland in China is too polluted to use for growing food. There is a high concentration of heavy metals and other chemicals in the soil, and some of these pollutants have already made it into the food chain through rice and other crops. Cadmium is a carcinogenic metal that can cause kidney damage, is absorbed by rice, and is of particular concern.[16] A random sample of farmland near Guangzhou was identified, and samples of soil were taken from each. The concentration of cadmium in each sample was measured. Assume that the distribution of the concentration of cadmium is continuous. Use the Wilcoxon signed-rank test to determine whether there is any evidence that the median cadmium concentration is greater than 1 ppm. Use $\alpha = 0.05$ and find the p value for this test. CADMIUM

Extended Applications

14.40 Travel and Transportation   A recent study was conducted to assess two methods to estimate vehicle speed on interstate highways. A random sample of vehicles on Interstate 75 along Alligator Alley in Florida was obtained. The speed of each vehicle was measured using a loop detector and an electronic toll transponder. The data (in mph) are given in the following table. HWYSPEED

Vehicle	Loop detector	Toll transponder	Vehicle	Loop detector	Toll transponder
1	66	65	2	66	66
3	70	67	4	82	80
5	74	68	6	70	67
7	63	65	8	63	61
9	80	78	10	71	68
11	82	81	12	66	65
13	72	68	14	70	68
15	65	60	16	63	62
17	66	65	18	79	79

Assume that the distribution of pairwise differences is continuous and symmetric.

   a. Use the Wilcoxon signed-rank test with $\alpha = 0.05$ to determine whether there is any evidence that the difference in median speed estimates is different from zero.

   b. Based on your conclusion in part (a), which method for estimating speed do you think the state police should use, and why?

14.41 Physical Sciences   In a study of several Chesapeake Bay tributaries, a random sample of nontidal freshwater locations on the Patuxent River was obtained. The total arsenic concentration was measured at each location (in μg/L), and the data are given in the following table. CHESBAY

0.67	0.22	0.56	0.27	0.17	0.57	0.55	0.55
0.53	0.09	0.61	0.45	0.49	0.04	0.44	0.19

Suppose the distribution of arsenic concentrations is continuous and symmetric and that the safe level of arsenic concentration is 0.30 μg/L.

   a. Conduct a sign test to determine whether there is any evidence that the median arsenic concentration in freshwater locations is greater than 0.30 μg/L. Use $\alpha = 0.05$.

   b. Conduct a signed-rank test to determine whether there is any evidence that the median arsenic concentration in freshwater locations is greater than 0.30 μg/L. Use $\alpha = 0.05$.

   c. Compare your conclusions in parts (a) and (b). Which test do you think is more accurate? Why?

14.42 Business and Management   There has been considerable discussion about dairy policy and the price of milk in the United States. A recent report compared the Federal Agriculture Reform and Risk Management Act of 2013 (DSA) and the Dairy Freedom Act (DFA) on the price of milk (in dollars per cwt). A random sample of months was selected, and the price of milk was recorded for each.[17] Assume the distribution of pairwise differences is continuous and symmetric. DAIRY

   a. Conduct a sign test to determine whether there is any evidence of a difference in the median milk prices under the two different pricing policies. Use $\alpha = 0.05$.

   b. Conduct a signed-rank test to determine whether there is any evidence of a difference in the median milk prices under the two different pricing policies. Use $\alpha = 0.05$.

   c. Compare your conclusions in parts (a) and (b). Which test do you think is more accurate? Why?

   d. If you were a dairy farmer, which pricing policy would you prefer? Why?

14.3   The Rank-Sum Test

The nonparametric Wilcoxon rank-sum test is used to compare two population medians. Suppose two independent random samples are obtained from continuous, non-normal distributions. Assume the first sample has size n_1, the second sample has size n_2, and $n_1 \leq n_2$.

Combine all of the data, for a total of $n_1 + n_2$ observations. Place these combined data in increasing order, and assign a rank to each from smallest (rank 1) to largest (rank $n_1 + n_2$). Equal values are assigned the mean rank of their positions in the ordered list (just as in the signed-rank test). Let w be the sum of the ranks associated with observations from the first sample.

Suppose, for example, that the sample sizes are equal. If $\tilde{\mu}_1 < \tilde{\mu}_2$, then the observations in the first sample will tend to be smaller than the observations in the second sample, and w will be small. If $\tilde{\mu}_1 > \tilde{\mu}_2$, then the observations in the first sample will tend to be larger than the observations in the second sample, and w will be large. There is also a normal approximation, which is valid when both n_1 and n_2 are large.

Can you figure out the smallest and the largest possible value of w?

The Wilcoxon Rank-Sum Test

Suppose two independent random samples of sizes n_1 and n_2 $(n_1 \le n_2)$ are obtained from continuous distributions. A hypothesis test concerning the two population medians with significance level α has the form

$$H_0: \tilde{\mu}_1 - \tilde{\mu}_2 = \Delta_0$$
$$H_a: \tilde{\mu}_1 - \tilde{\mu}_2 > \Delta_0, \qquad \tilde{\mu}_1 - \tilde{\mu}_2 < \Delta_0, \qquad \text{or} \qquad \tilde{\mu}_1 - \tilde{\mu}_2 \ne \Delta_0$$

Subtract Δ_0 from each observation in the first sample. Combine these differences and the observations in the second sample, and rank all of these values. Equal values are assigned the mean rank for their positions.

TS: W = the sum of the ranks corresponding to the differences from the first sample

RR: $W \ge c_1, \qquad W \le c_2, \qquad W \ge c \qquad \text{or} \qquad W \le n_1(n_1 + n_2 + 1) - c$

The critical values c_1, c_2, and c are obtained from Table X in the Appendix such that $P(W \ge c_1) \approx \alpha$, $P(W \le c_2) \approx \alpha$, and $P(W \ge c) \approx \alpha/2$.

The Normal Approximation: As n_1 and n_2 increase, the statistic W approaches a normal distribution with

$$\mu_W = \frac{n_1(n_1 + n_2 + 1)}{2} \qquad \text{and} \qquad \sigma_W^2 = \frac{n_1 n_2(n_1 + n_2 + 1)}{12}$$

Therefore, the random variable $Z = \dfrac{W - \mu_W}{\sigma_W}$ has approximately a standard normal distribution. The normal approximation is good when both n_1 and n_2 are greater than 8. In this case, Z is the test statistic and the rejection region is

RR: $Z \ge z_\alpha, \qquad Z \le -z_\alpha, \qquad \text{or} \qquad |Z| \ge z_{\alpha/2}$

A CLOSER L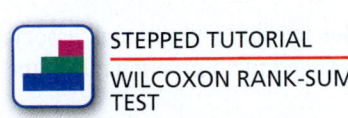OK

STEPPED TUTORIAL

WILCOXON RANK-SUM TEST

1. This general test procedure allows for any hypothesized difference between the two population medians, Δ_0. Usually, $\Delta_0 = 0$, and the null hypothesis is that the two population medians are equal.

2. If $\Delta_0 \ne 0$, we subtract this quantity from each observation in the first sample. Intuitively, if H_0 is true, this *shifts* sample 1 so that the two samples have the same median.

3. Like other nonparametric tests, this procedure is appropriate when the underlying distributions are non-normal, or when the data are ranks.

4. Suppose both underlying populations are symmetric. The Wilcoxon rank-sum test can be used to compare population means, because in each population the mean is equal to the median. In this case, one might also consider the two-sample t test, discussed in Section 10.2. However, if the underlying populations are not normal, then the t-test results are not valid.

Many software packages call this the Mann–Whitney rank-sum test.

5. Sometimes, a slightly different test statistic is used in this test procedure. The Mann–Whitney U statistic is a function of W and also approaches a normal distribution as n_1 and n_2 increase. The two statistics lead to similar conclusions.

DATA SET

ICEBREAK

Solution Trail 14.5

KEYWORDS

■ Rank-sum test

■ Difference in population medians

TRANSLATION

■ Two-sided hypothesis test concerning two population medians

CONCEPTS

■ Wilcoxon rank-sum test

VISION

Combine the samples, and rank the ordered observations. Find the value of the test statistic W and draw the appropriate conclusion.

© RIA Novosti/The Image Works

Example 14.5 Icebreakers

In December 2013 a scientific expedition ship became stuck in ice near Antarctica. At least three icebreakers attempted to reach the stranded ship but had to turn back due to weather or ice thickness. Several countries around the world maintain a fleet of icebreakers. These ships are designed with a strong hull, powerful engines, and a wide beam. Independent random samples of icebreakers from the United States and Canada were obtained, and the beam width of each was measured (in meters). The data are given in the following table.

U.S.	19.60	25.45	24.40	25.00	18.30	
Canada	24.38	17.80	19.50	24.45	19.84	19.15

Suppose the underlying distributions of icebreaker beam widths are continuous. Conduct a Wilcoxon rank-sum test to determine whether there is any evidence to suggest a difference in the population icebreaker beam widths. Use $\alpha = 0.05$.

SOLUTION

STEP 1 The samples are independent, and the populations are assumed to be continuous. We are searching for *any* difference in the medians, so $\Delta_0 = 0$.

The sample sizes are $n_1 = 5$ and $n_2 = 6$. Because n_1 and n_2 are both less than 8, the test statistic is W.

The four parts of the hypothesis test are

$H_0: \tilde{\mu}_1 - \tilde{\mu}_2 = \Delta_0 = 0$

$H_a: \tilde{\mu}_2 - \tilde{\mu}_2 \neq \Delta_0 = 0$

TS: $W =$ the sum of the ranks corresponding to the differences from the first sample

$\Delta_0 = 0$, so W is the sum of the ranks corresponding to the first (smaller) sample.

RR: $W \geq 41$ or $W \leq 5(5 + 6 + 1) - 41 = 19$

We need a value for c such that $P(W \geq c) \approx 0.05/2 = 0.025$. Using Table X in the Appendix, $P(W \geq 41) = 0.026$. Therefore, the actual significance level for this test is $2(0.026) = 0.052 \approx 0.05$.

STEP 2 The following table shows the combined samples in order and the rank associated with each value. The shaded columns correspond to observations from the first sample.

Observation	17.80	18.30	19.15	19.50	19.60	19.84	24.38	24.40	24.45	25.00	25.45
Rank	1	2	3	4	5	6	7	8	9	10	11

STEP 3 The value of the test statistic is the sum of the ranks corresponding to the observations from the first sample.

$w = 2 + 5 + 8 + 10 + 11 = 36$

The value of the test statistic does not lie in the rejection region. At the $\alpha = 0.05$ significance level, there is no evidence to suggest that the population median beam widths are different.

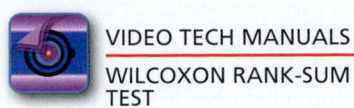

VIDEO TECH MANUALS

WILCOXON RANK-SUM TEST

Figure 14.10 shows a technology solution.

Wilcoxon / Kruskal-Wallis Tests (Rank Sums)					
			Expected		
Level	Count	Score Sum	Score	Score Mean	(Mean-Mean0)/Std0
C	6	30.000	36.000	5.00000	-1.004
US	5	36.000	30.000	7.20000	1.004
2-Sample Test, Normal Approximation					
		S	Z	Prob>|Z|	
		36	1.00416	0.3153	

Figure 14.10 JMP
Wilcoxon rank-sum test.

TRY IT NOW GO TO EXERCISE 14.55

Here is one more example that involves the normal approximation in the Wilcoxon rank-sum test.

DATA SET

OPPINDEX

Example 14.6 Economic Mobility

The Opportunity Index (OI) is a unique measure of the economic mobility in counties and states. Sixteen variables are used to derive the value of the OI, including ones that measure jobs and the local economy, education, and community health and civic life. The OI is often considered a measure of the "American dream." A higher OI suggests more social mobility and greater economic security. A random sample of counties in the Northeast and the West were obtained, and the OI was recorded for each.[18] The data are given in the following table.

Northeast	54.8	63.8	56.2	54.1	56.0	45.9	65.0
	57.6	60.6	54.8	58.2	66.4		

West	56.5	54.9	55.1	47.3	55.6	46.2	41.6
	40.0	43.0	37.3	56.5	53.3	45.7	48.3

Assume the underlying distributions of Opportunity Index in each area of the country are continuous. Use a Wilcoxon rank-sum test to determine whether there is any evidence to suggest a difference in the population median OIs. Use $\alpha = 0.01$.

Solution Trail 14.6

KEYWORDS
- Rank-sum test
- Difference in population medians
- Sample sizes: $n_1 = 12$, $n_2 = 14$

TRANSLATION
- Two-sided hypothesis test concerning two population medians
- Sample sizes greater than 8

CONCEPTS
- Wilcoxon rank-sum test
- Normal approximation

VISION
Combine the samples, and rank the ordered observations. Find the value of the test statistic Z, and draw the appropriate conclusion.

SOLUTION

STEP 1 The samples are independent, and the populations are assumed to be continuous. We are searching for *any* difference in the medians, so $\Delta_0 = 0$.

The sample sizes are $n_1 = 12$ and $n_2 = 14$. Both sample sizes are greater than eight, so the normal approximation will be used.

The four parts of the hypothesis test are

$$H_0: \tilde{\mu}_1 - \tilde{\mu}_2 = \Delta_0 = 0$$
$$H_a: \tilde{\mu}_1 - \tilde{\mu}_2 \neq \Delta_0 = 0$$
$$\text{TS: } Z = \frac{W - \mu_W}{\sigma_W}$$

RR: $|Z| \geq z_{\alpha/2} = z_{0.005} = 2.5758$

STEP 2 The following table shows the combined samples in order and the rank associated with each value. The shaded columns correspond to observations from the first sample.

Observation	37.3	40.0	41.6	43.0	45.7	45.9	46.2	47.3	48.3
Rank	1.0	2.0	3.0	4.0	5.0	6.0	7.0	8.0	9.0

Observation	53.3	54.1	54.8	54.8	54.9	55.1	55.6	56.0	56.2
Rank	10.0	11.0	12.5	12.5	14.0	15.0	16.0	17.0	18.0

Observation	56.5	56.5	57.6	58.2	60.6	63.8	65.0	66.4	
Rank	19.5	19.5	21.0	22.0	23.0	24.0	25.0	26.0	

Note: Remember that equal values are assigned the mean rank for their positions.

STEP 3 The value w is the sum of the ranks corresponding to the observations from the first sample.

$$w = 6.0 + 11.0 + 12.5 + 12.5 + 17.0 + 18.0 + 21.0 + 22.0$$
$$+ 23.0 + 24.0 + 25.0 + 26.0 = 218.0$$

The mean and variance of the random variable W are

$$\mu_W = \frac{n_1(n_1 + n_2 + 1)}{2} = \frac{12(12 + 14 + 1)}{2} = 162.0$$

$$\sigma_W^2 = \frac{n_1 n_2(n_1 + n_2 + 1)}{12} = \frac{(12)(14)(12 + 14 + 1)}{12} = 378.0$$

The value of the test statistic is

$$z = \frac{w - \mu_W}{\sigma_W} = \frac{218.0 - 162.0}{\sqrt{378.0}} = 2.88$$

The value of the test statistic lies in the rejection region. Equivalently, $p = 0.004 \leq 0.01$. The exact p value calculation and illustration is shown in Figure 14.11. At the $\alpha = 0.01$ significance level, there is evidence to suggest that the population median Opportunity Indices are different.

Figure 14.12 shows a technology solution.

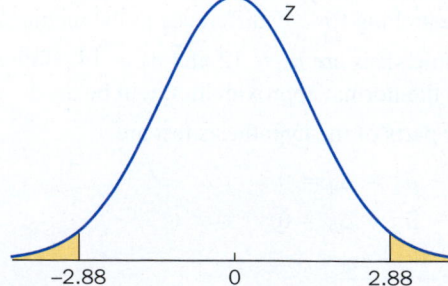

Figure 14.11 p-Value illustration:
$p = 2P(Z \geq 2.88)$
$= 0.0040 \leq 0.01 = \alpha$

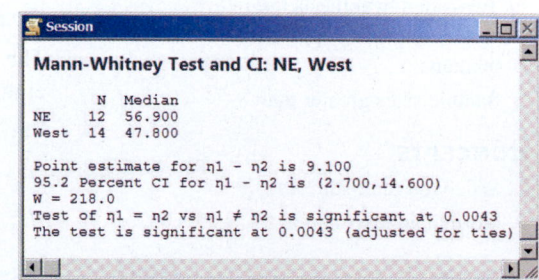

Figure 14.12 Minitab Mann-Whitney test.

TRY IT NOW   GO TO EXERCISE 14.60

Technology Corner

Procedure: Conduct a rank-sum test.
Reconsider: Example 14.6, solution, and interpretations.

CrunchIt!

Use the Mann-Whitney U function to conduct a rank-sum test.

1. Enter the NE data into column Var1 and the West data into column Var2.
2. Select Statistics; Non-parametrics; Mann-Whitney U.
3. Under the Columns tab, use the pull-down menus to select Sample 1 (Var1) and Sample 2 (Var2).
4. Under the Hypothesis Test tab enter the Location shift under the null hypothesis (0) and select the Alternative (Two-sided). Click Calculate. The results are displayed in a new window. See Figure 14.13.

Results - Mann-Whitney U

Export ▾

Null hypothesis:	Location shift = 0
Alternative hypothesis:	Location shift is not 0
W statistic:	218
P-value:	0.004296

Figure 14.13 CrunchIt! Mann-Whitney U test.

Minitab

Use the built-in function Mann-Whitney, accessible in the Stat; Nonparametrics menu or by using the appropriate commands in a session window.

1. Enter the NE data into column C1 and the West data into column C2.
2. Select Stat; Nonparametrics; Mann-Whitney. Enter the First Sample column, C1, the Second Sample column, C2, and select an Alternative hypothesis (not equal).
3. Click OK. The test results are displayed in a session window. Refer to Figure 14.12.

Excel

There is no built-in function to conduct a rank-sum test. Use other arithmetic functions to compute the ranks, the values of the test statistics, w and z, and the p value associated with this test.

1. Enter the NE data into column A and the West data into column B.
2. Use the RANK.AVG function to compute the appropriate ranks associated with each observation.
3. Sum the ranks associated with the NE observations. Find the mean and standard deviation for W and compute the value of the test statistic. Use the NORM.S.DIST function to find the p value. See Figure 14.14.

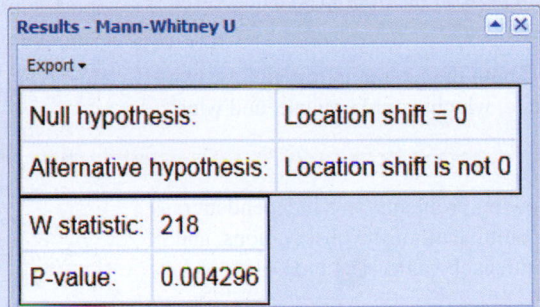

	A	B	C	D	E	F	G
1	Original Data		Ranks				Computations
2	NE	West	NE	West			
3	54.8	56.5	12.5	19.5		$w =$	218.0 = SUM(C3:C14)
4	63.8	54.9	24.0	14.0		$\mu_W =$	162.0 = (12)*(12+14+1)/2
5	56.2	55.1	18.0	15.0		$\sigma_W =$	19.44 = SQRT((12)*(14)*(12+14+1)/12)
6	54.1	47.3	11.0	8.0		$z =$	2.88 = (F3-F4)/F5
7	56.0	55.6	17.0	16.0		$p =$	0.0040 = 2*(1-NORM.S.DIST(2.88,TRUE))
8	45.9	46.2	6.0	7.0			
9	65.0	41.6	25.0	3.0			
10	57.6	40.0	21.0	2.0			
11	60.6	43.0	23.0	4.0			
12	54.8	37.3	12.5	1.0			
13	58.2	56.5	22.0	19.5			
14	66.4	53.3	26.0	10.0			
15		45.7		5.0			
16		48.3		9.0			

Figure 14.14 Excel calculations to conduct a rank-sum test.

SECTION 14.3 EXERCISES

Concept Check

14.43 True/False The sample sizes must be equal in order for the Wilcoxon rank-sum test to be valid.

14.44 Short Answer What ranks are equal values assigned in a Wilcoxon rank-sum test?

14.45 True/False The Wilcoxon rank-sum allows for any hypothesized difference between two population medians.

14.46 Fill in the Blank The Wilcoxon rank-sum test is most appropriate when the underlying distributions are

_____.

14.47 True/False The Wilcoxon rank-sum test can be used to compare two population means.

14.48 Short Answer If both the rank-sum test and the two-sample t test are appropriate, which would you use, and why?

Practice

14.49 In each of the following problems, two independent random samples are given. Combine all of the observations, and find the rank associated with each value. ⅢⅡ EX14.49

a.

Sample 1	37	21	46	29	34
Sample 2	45	42	22	41	24	39

b.

Sample 1	4.5	1.8	3.4	1.4	2.2	2.1	1.5	3.7
Sample 2	4.6	6.6	1.2	6.0	2.4	2.3	2.2	0.4
	2.5	2.7						

c. Sample 1

820	872	814	825	876	858	841	892	882
809	826	887	884	862	846	801	871	803

Sample 2

850	842	888	879	832	827	831	810	871
840	870	816	821	865	899	830	875	800
813	839	822	865	818	818			

14.50 In each of the following problems, assume a Wilcoxon rank-sum test will be used to compare population medians. Two independent random samples and the value of Δ_0 are given. Find w, the value of the test statistic. ⅢⅡ EX14.50

a. $\Delta_0 = 0$

Sample 1	90	90	87	87	81
Sample 2	83	84	80	84	87

b. $\Delta_0 = 0$

Sample 1	58	69	52	53	57	58	60	56	59
Sample 2	53	52	60	55	70	66	65	69	51
	69	66							

c. $\Delta_0 = 5$

Sample 1

36.2	33.5	32.5	33.2	32.5	34.6	39.0	39.7
32.1	39.5	38.3	30.6	32.5	38.4	38.4	

Sample 2

31.7	34.4	32.3	25.6	27.3	32.7	33.0	28.4
32.6	28.6	27.8	30.5	30.6	33.1	25.4	32.4
31.5	33.5						

14.51 In each of the following problems, assume a Wilcoxon rank-sum test will be used to compare population medians. Use the sample sizes and alternative hypothesis to find the best rejection region for the given significance level. Report the exact significance level for your rejection region.

a. $n_1 = 3$, $\quad n_2 = 5$, $\quad H_a: \tilde{\mu}_1 < \tilde{\mu}_2$, $\quad \alpha = 0.05$
b. $n_1 = 4$, $\quad n_2 = 4$, $\quad H_a: \tilde{\mu}_1 > \tilde{\mu}_2$, $\quad \alpha = 0.05$
c. $n_1 = 4$, $\quad n_2 = 10$, $\quad H_a: \tilde{\mu}_1 \neq \tilde{\mu}_2$, $\quad \alpha = 0.05$
d. $n_1 = 6$, $\quad n_2 = 8$, $\quad H_a: \tilde{\mu}_1 < \tilde{\mu}_2$, $\quad \alpha = 0.01$
e. $n_1 = 7$, $\quad n_2 = 9$, $\quad H_a: \tilde{\mu}_1 \neq \tilde{\mu}_2$, $\quad \alpha = 0.10$
f. $n_1 = 8$, $\quad n_2 = 8$, $\quad H_a: \tilde{\mu}_1 > \tilde{\mu}_2$, $\quad \alpha = 0.01$

14.52 In each of the following problems, assume a Wilcoxon rank-sum test with the normal approximation will be used to compare population medians. Find the mean, the variance, and the standard deviation for the test statistic W.

a. $n_1 = 15$, $\quad n_2 = 21$
b. $n_1 = 18$, $\quad n_2 = 18$
c. $n_1 = 11$, $\quad n_2 = 27$
d. $n_1 = 12$, $\quad n_2 = 16$
e. $n_1 = 23$, $\quad n_2 = 24$
f. $n_1 = 25$, $\quad n_2 = 30$

14.53 Two independent random samples from continuous distributions are given in the following table. Conduct a Wilcoxon rank-sum test of $H_0: \tilde{\mu}_1 = \tilde{\mu}_2$ versus $H_a: \tilde{\mu}_1 > \tilde{\mu}_2$ using a significance level of 0.01. ⅢⅡ EX14.53

Sample 1	51.8	55.6	67.6	58.7	66.2	63.1
Sample 2	57.2	51.2	58.2	56.9	57.1	63.3
	59.7	60.6				

14.54 Two independent random samples from continuous distributions are given on the text website. Conduct a Wilcoxon rank-sum test (using the normal approximation) of $H_0: \tilde{\mu}_1 = \tilde{\mu}_2$ versus $H_a: \tilde{\mu}_1 \neq \tilde{\mu}_2$ using a significance level of 0.05. ⅢⅡ EX14.54

Applications

14.55 Medicine and Clinical Studies Many complications may accompany type 2 diabetes, including heart disease, eye problems, and nerve damage. A recent study suggests that type 2 diabetes also affects bone strength.[19] Suppose a random sample of patients at the Mayo Clinic aged 50 to 80 with and without type 2 diabetes was obtained. Researchers measured the bone material strength in each person's tibia using micro indentation testing. The data are given in the following table. ▂▂ **DIABETES**

With type 2 diabetes					
1.09	1.14	1.38	1.17	1.29	1.32

Without type 2 diabetes						
1.06	1.78	1.72	1.48	1.68	1.87	1.40

Note: The indentation data are unitless, and a larger value indicates less bone strength. Use the Wilcoxon rank-sum test with $\alpha = 0.05$ to determine whether there is any evidence that patients with type 2 diabetes have less bone strength. Assume that the underlying distributions are continuous.

14.56 Technology and the Internet   Users surfing the Internet generally leave web pages very quickly, often within 10–20 seconds. Some research suggests that the first 10 seconds of a page visit are critical to a decision to read further. A random sample of users visiting similar web pages, one with advertisements and one without, was obtained. The time (in seconds) for each visit was recorded, and the data are given in the following table. ▂▂ **WEBVISIT**

With advertisements					
14.5	10.1	11.9	9.7	11.6	8.2

Without advertisements						
10.4	12.0	13.2	11.7	10.9	13.2	14.5

Use the Wilcoxon rank-sum test with $\alpha = 0.05$ to determine whether there is any evidence to suggest that the population median visit times are different. Assume the underlying distributions are continuous.

14.57 Manufacturing and Product Development   Better World Technologies claims to have produced a revolutionary new jackhammer with less vibration, reduced noise and energy input, fewer moving parts, and greater impact strength. Independent random samples of conventional jackhammers and new jackhammers were obtained. The impact strength of each, using heavy-duty concrete breakers, was measured (in ft-lb) at 75 psi. The data are given in the following table. ▂▂ **IMPACT**

Conventional jackhammer						
95	94	102	100	93	93	93

New jackhammer						
119	130	99	126	114	130	101

a. Use the Wilcoxon rank-sum test with $\alpha = 0.01$ to determine whether there is any evidence to suggest that the population median impact strength is higher for the new jackhammer than for the conventional jackhammer. Assume the underlying distributions are continuous.
b. Find the p value for this hypothesis test.

14.58 Sports and Leisure   Many hockey players use a slapshot—a short, quick, powerful swing with the stick—to try and score a goal. The speed of this shot makes it very difficult for a goalie to react and make a save. The fastest slapshot recorded was 108.8 mph in 2012 by Zdeno Chara.[20] A National Hockey League (NHL) scout believes that defensemen have faster slapshots than forwards. To test this claim, independent random samples from each group of players were obtained. Each player took a slapshot from the blue line, and the speed of the puck was recorded (in mph) using a radar gun. The data are given in the following table. ▂▂ **SLAPSHOT**

Defensemen									
94	98	97	94	94	94	91	100	98	96

Forwards									
92	88	86	94	90	85	86	95	89	85

Use the Wilcoxon rank-sum test with $\alpha = 0.01$ to determine whether there is any evidence to suggest that the population median slapshot speed of NHL defensemen is greater than the population median slapshot speed of NHL forwards. Assume the underlying distributions are continuous.

14.59 Business and Management   Kroll's South restaurant serves dairy products in two separate buffet lines. Independent random samples of the dairy products holding temperature (in °F) in each line were obtained. Use the Wilcoxon rank-sum test to determine whether there is any evidence to suggest that the population median holding temperatures are different. Use $\alpha = 0.05$, and find the p value associated with this test. ▂▂ **BUFFET**

14.60 Terrestrial Radio   Listeners have several options to traditional terrestrial radio, for example, Internet radio, music services, satellite radio, and individual playlists. Despite the loss of listeners, traditional terrestrial radio stations still run lots of commercials. A list of traditional music and talk radio stations was obtained, and a random hour during the day was selected for each. The total time for commercials during the hour (in minutes) was recorded for each. ▂▂ **RADIO**
a. Use the Wilcoxon rank-sum test to determine whether there is any evidence to suggest that the population median commercial time per hour is greater on talk radio stations than on music stations. Use $\alpha = 0.05$ and find the p value associated with this test.
b. Suppose the underlying distributions are approximately normal. Use a two-sample t test assuming equal variances to determine whether there is any evidence to suggest that the population mean commercial time per hours is greater on talk radio stations than on music stations. Use $\alpha = 0.05$ and find the p value associated with this test.
c. Which test do you think is more appropriate? Why?
d. Summarize your results—which type of station plays more commercials?

Extended Applications

14.61 Public Health and Nutrition   Independent random samples of almonds from two suppliers in the United

Kingdom were obtained, and the amount of protein (in grams) per 100 grams of edible portion was measured for each portion. SBP claims that their almonds contain more protein than any other brand. Use the Wilcoxon rank-sum test with $\alpha = 0.01$ to compare the population median amounts of protein in SBP and WTL International almonds. Do you believe the SBP claim? Why or why not? Find the p value for this test. **ALMONDS**

14.62 Fuel Consumption and Cars   New vans sold in the European Union in 2017 are targeted to have an emission rate of 175 g CO_2/km. By 2020 the manufacturers will have to reduce emissions even further.[21] A random sample of new vans sold in Germany and Denmark was obtained, and the CO_2 emission rate was measured for each. **VANS**

   a. Use the Wilcoxon signed-rank test to determine whether there is any evidence that the median CO_2 emission rate for new vans in Germany is greater than 175 g. Use $\alpha = 0.01$.
   b. Use the Wilcoxon signed-rank test to determine whether there is any evidence that the median CO_2 emission rate for new vans in Denmark is greater than 175 g. Use $\alpha = 0.01$.
   c. Use the Wilcoxon rank-sum test to compare the population median CO_2 emission rates in Germany and Denmark.
   d. Summarize your results and each country's progress toward the 2017 target emission rate.

Challenge

14.63 Biology and Environmental Science   The indoor air quality (the concentration of particles less than 2.5 microns in

diameter) in 15 bars, restaurants, and other public venues in Louisville was measured prior to the introduction of new smoking regulations. Two months after all public venues were required to be smoke-free, the air quality was measured again. The data (in $\mu g/m^3$) are given in the following table. **AIRQUAL**

Venue	1	2	3	4	5	6	7	8
Before	353	386	104	198	597	62	412	273
After	56	35	28	21	83	10	27	34

Venue	9	10	11	12	13	14	15
Before	38	156	35	87	105	101	324
After	27	31	13	26	26	18	25

Suppose the underlying before and after distributions of air quality are continuous.

   a. Conduct a sign test with $\alpha = 0.05$ to determine whether there is any evidence that the median amount of particulates decreased after the smoking regulations went into effect.
   b. Conduct a Wilcoxon signed-rank test with $\alpha = 0.05$ to determine whether there is any evidence that the median amount of particulates was greater before the smoking regulations went into effect than after.
   c. Use the Wilcoxon rank-sum test with $\alpha = 0.05$ to determine whether there is any evidence that the median amount of particulates decreased after the smoking regulations went into effect.
   d. Compare the results of these three statistical tests. Which test(s) is(are) appropriate and why?

14.4   The Kruskal-Wallis Test

The Kruskal-Wallis test is a nonparametric analysis of variance of ranks. This procedure is an alternative to the analysis of variance F test that does not require assumptions concerning normality or equal population variances. Although there is a table of critical values associated with this procedure, even for small sample sizes the test statistic has approximately a chi-square distribution.

Suppose that $k > 2$ independent random samples are obtained from continuous distributions. Assume the first sample has size n_1, the second sample has size n_2, and so on, such that the kth sample has size n_k.

Combine all of the data, for a total of $n = n_1 + n_2 + \cdots + n_k$ observations. Place these combined data in increasing order and assign a rank to each from smallest (rank 1) to largest (rank n). Equal values are assigned the mean rank for their positions in the ordered list (just as in the signed-rank test). Let r_i be the sum of the ranks associated with observations from sample i.

If all of the populations have identical distributions, then the sum of the ranks associated with each sample should be approximately equal. We expect the sum of the ranks associated with a sample from a different population to be distinct and separate from the rest. The test statistic in the Kruskal-Wallis test assesses the differences in the sums of the ranks.

The Kruskal-Wallis Test

Suppose $k > 2$ independent random samples of sizes $n_1, n_2, \ldots, n_k$ are obtained from continuous distributions. A hypothesis test concerning the general populations with significance level α has the form

H_0: The k samples are from identical populations

H_a: At least two of the populations are different

Combine all observations, and rank these values from smallest (1) to largest (n). Equal values are assigned the mean rank for their positions. Let R_i be the sum of the ranks associated with the ith sample.

$$\text{TS: } H = \left[\frac{12}{n(n+1)} \sum \frac{R_i^2}{n_i} \right] - 3(n+1)$$

Critical values for the Kruskal-Wallis test statistic are available. However, if H_0 is true and either

1. $k = 3,$ $\quad n_i \geq 6,$ $\quad (i = 1, 2, 3)$ or
2. $k > 3,$ $\quad n_i > 5,$ $\quad (i = 1, 2, \ldots, k)$

The chi-square distribution was defined in Section 8.5.

then H has an approximate chi-square distribution with $k - 1$ degrees of freedom.

RR: $H \geq \chi^2_{\alpha, k-1}$

A CLOSER LOOK

If the sample sizes are small, then the distribution of H may not be close enough to chi-square to make a reliable conclusion.

1. We reject the null hypothesis only for *large* values of the test statistic H. If the sample sizes are large enough, the rejection region is always in the right tail of the appropriate chi-square distribution.

2. If we reject the null hypothesis, there is evidence to suggest that at least two populations are different. Further analysis is needed to determine which pairs of populations are different, and how they differ; the means, medians, variances, shapes of the distributions, or other characteristics could be dissimilar.

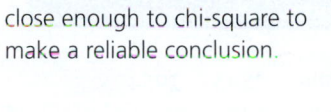
DATA SET

SHIPPING

Example 14.7 Shipping Distances

During the 2013 year-end holiday season, both UPS and FedEx acknowledged that there were many delays in delivering packages for Christmas due to increased volume and bad weather. Suppose a study was conducted to determine if package shipping distance also contributed to the delays. A random sample of packages shipped within the United States from each of three major companies was obtained, and the shipping distance (in miles) from the point of origin was recorded for each. The data are given in the following table; the sample numbers are in parentheses.

Airborne Express (1)	UPS (2)	Federal Express (3)
834	245	1617
2954	600	1538
2845	915	1298
1889	998	1580
2006	284	1968
1318	325	1526
1675	558	1002
1959	493	

Solution Trail 14.7

KEYWORDS

- Kruskal-Wallis test
- Is there any evidence?
- Populations are different

TRANSLATION

- Hypothesis test concerning the general populations

CONCEPTS

- Kruskal-Wallis test

VISION

The underlying distributions are assumed to be continuous. Use the Kruskal-Wallis test procedure; combine and rank all of the observations. Find the value of the test statistic, and draw the appropriate conclusion.

The underlying populations are assumed to be continuous. Use the Kruskal-Wallis test to determine whether there is any evidence that the package shipping distance populations are different. Use a significance level of $\alpha = 0.05$.

SOLUTION

STEP 1   There are $k = 3$ independent random samples of sizes $n_1 = 8 \geq 6$, $n_2 = 8 \geq 6$, and $n_3 = 7 \geq 6$, so the Kruskal-Wallis test statistic has an approximately chi-square distribution with $k - 1 = 3 - 1 = 2$ degrees of freedom.

STEP 2   The four parts of the hypothesis test are

H_0: The three samples are from identical populations.

H_a: At least two of the populations are different.

$$\text{TS: } H = \left[\frac{12}{n(n+1)} \sum \frac{R_i^2}{n_i} \right] - 3(n+1)$$

$$\text{RR: } H \geq \chi_{\alpha, k-1}^2 = \chi_{0.05, 2}^2 = 5.9915$$

STEP 3   There are $n = n_1 + n_2 + n_3 = 8 + 8 + 7 = 23$ total observations. The following table shows all 23 observations, sorted in ascending order, with the associated rank. Each sample is color-coded to provide a visual comparison of the ranks.

Observation	245	284	325	493	558	600	834	915
Rank	1	2	3	4	5	6	7	8

Observation	998	1002	1298	1318	1526	1538	1580	1617
Rank	9	10	11	12	13	14	15	16

Observation	1675	1889	1959	1968	2006	2845	2954
Rank	17	18	19	20	21	22	23

STEP 4   If the populations are identical, we expect the ranks to be evenly distributed among the three samples. The shaded cells suggest that there are more low ranks associated with sample 2, middle ranks associated with sample 3, and high ranks associated with sample 1; this implies that the populations are different.

All the observations, the associated ranks, and the rank sums are given in the following table.

Airborne Express (1)	Rank	UPS (2)	Rank	Federal Express (3)	Rank
834	7	245	1	1617	16
2954	23	600	6	1538	14
2845	22	915	8	1298	11
1889	18	998	9	1580	15
2006	21	284	2	1968	20
1318	12	325	3	1526	13
1675	17	558	5	1002	10
1959	19	493	4		
Rank sum	139		38		99

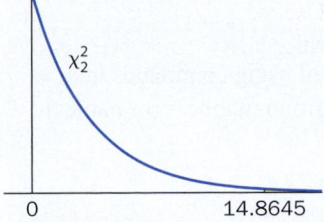

Figure 14.15 *p*-Value illustration:
$p = P(X \geq 14.8645)$
$= 0.0006 \leq 0.05 = \alpha$

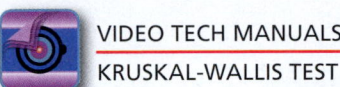

VIDEO TECH MANUALS

KRUSKAL-WALLIS TEST

STEP 5 The value of the test statistic is

$$h = \left[\frac{12}{n(n+1)} \sum \frac{R_i^2}{n_i} \right] - 3(n+1)$$

$$= \left[\frac{12}{(23)(24)} \left(\frac{139^2}{8} + \frac{38^2}{8} + \frac{99^2}{7} \right) \right] - 3(24) = 14.8645 \; (\geq 5.9915)$$

The value of the test statistic ($h = 14.8645$) lies in the rejection region. Equivalently, $p = 0.0006 \leq 0.05$. The exact p value calculation and illustration is shown in Figure 14.15. At the $\alpha = 0.05$ significance level, there is evidence to suggest that the package shipping distance populations are different.

Figures 14.16 and 14.17 show technology solutions.

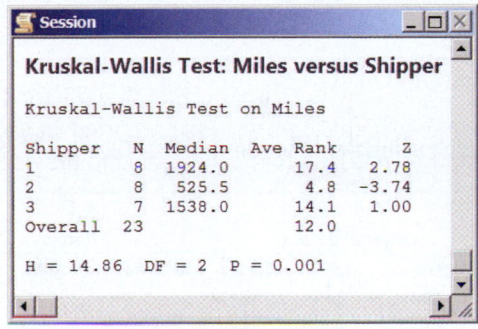

Figure 14.16 Minitab Kruskal-Wallis test results.

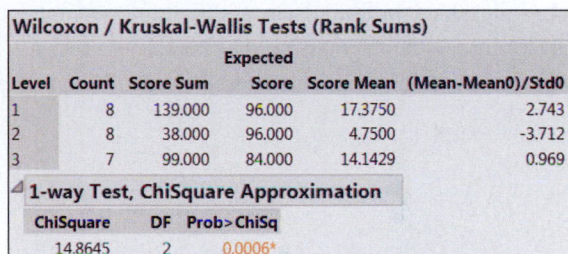

Figure 14.17 JMP Kruskal-Wallis test results.

TRY IT NOW GO TO EXERCISE 14.76

Technology Corner

Procedure: Conduct a Kruskal-Wallis test.
Reconsider: Example 14.7, solution, and interpretations.

CrunchIt!

Use Statistics; non-parametrics; Kruskal-Wallis.

1. Enter the data into columns Var1, Var2, and Var3.
2. Select Statistics; Non-parametrics; Kruskal-Wallis.
3. Check the columns containing the data. Click Calculate. The results are displayed in a new window. See Figure 14.18.

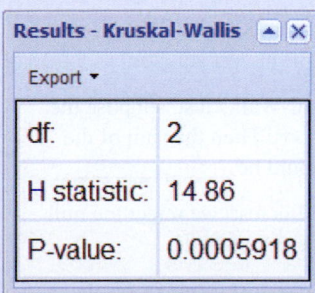

Figure 14.18 CrunchIt! Kruskal-Wallis test results.

Minitab

The Kruskal-Wallis test is accessible in the Stat; Nonparametrics menu and may also be executed using commands in a session window. The data from all samples must be in a single column, and the corresponding group numbers (or names) must be in a second column.

1. Enter the data from each sample into a separate column (C1, C2, and C3).
2. Use the command Data; Stack to place the data from all three samples into column C4 and to store the subscripts (group numbers) in column C5.
3. Select Stat; Nonparametrics; Kruskal-Wallis. Enter the Response column, C4, and the Factor column, C5.
4. Click OK. The test results are displayed in a session window. Refer to Figure 14.16.

Excel

There is no built-in function to conduct a Kruskal-Wallis test. Use other arithmetic functions to compute the ranks, the value of the test statistic h, and the p value.

1. Enter the data from each sample into a separate column (A, B, and C).
2. Compute the ranks using the function RANK.AVG.
3. Compute the value of the test statistic h. Find the p value using the function CHISQ.DIST.RT. See Figure 14.19.

	A	B	C	D	E	F
1		Original data			Ranks	
2	Airborne	UPS	FedEx	Airborne	UPS	FedEx
3	834	245	1617	7	1	16
4	2954	600	1538	23	6	14
5	2845	915	1298	22	8	11
6	1889	998	1580	18	9	15
7	2006	284	1968	21	2	20
8	1318	325	1526	12	3	13
9	1675	558	1002	17	5	10
10	1959	493		19	4	
11			R_i	139	38	99
12			R_i^2	19321	1444	9801
13			R_i^2/n_i	2415.13	180.50	1400.14
14			h	14.86		
15			p value	0.0006		

Figure 14.19 Excel calculations for a Kruskal-Wallis test.

SECTION 14.4 EXERCISES

Concept Check

14.64 Fill in the Blank The Kruskal-Wallis test is the non-parametric alternative to _____.

14.65 Short Answer For k groups in a Kruskal-Wallis test, what is the approximate distribution of the test statistic?

14.66 Fill in the Blank In a Kruskal-Wallis test, suppose the k populations have identical distributions. Then the sum of the ranks associated with each sample should be _____.

14.67 True/False In a Kruskal-Wallis test, we reject the null hypothesis only for large values of the test statistic.

14.68 True/False In a Kruskal-Wallis test, if we reject the null hypothesis, there is evidence to suggest that all k population means are different.

14.69 True/False In a Kruskal-Wallis test, the underlying population variances are assumed to be equal.

Practice

14.70 Three independent random samples are given in the following table. Assume a Kruskal-Wallis test will be used to compare populations. Find the rank sum associated with each sample. **EX14.70**

Sample 1

88	94	94	79	91	76	77	79

Sample 2

19	72	77	79	89	72	87	74	82	90

Sample 3

85	96	95	96	93	85	95	93	87	95	93	93

14.71 Five independent random samples were obtained; assume a Kruskal-Wallis test will be used to compare populations. Use the sample sizes and the rank sums in the following table to conduct this test at the $\alpha = 0.01$ level of significance.

Sample	Size	Rank sum
1	$n_1 = 12$	$r_1 = 395.5$
2	$n_2 = 14$	$r_2 = 428.5$
3	$n_3 = 12$	$r_3 = 287.0$
4	$n_4 = 16$	$r_4 = 620.0$
5	$n_5 = 20$	$r_5 = 1044.0$

14.72 Four independent random samples are given on the text website. Conduct a Kruskal-Wallis test to compare populations. Use $\alpha = 0.05$. 📊 **EX14.72**

14.73 Three independent random samples were obtained, and the data are given on the text website. Use the Kruskal-Wallis test with $\alpha = 0.05$ to test the hypothesis that all three populations are identical. 📊 **EX14.73**

Applications

14.74 Travel and Transportation A marketing manager at a rental-car agency believes that the number of miles a customer drives per week is related to the type of car rented. Independent random samples of week-long rental reservations were obtained, and the number of miles driven was recorded for each car. The summary statistics are given in the following table.

Car classification	Sample size	Rank sum
Compact car	$n_1 = 10$	$r_1 = 201.0$
Standard car	$n_2 = 10$	$r_2 = 227.0$
Luxury car	$n_3 = 12$	$r_3 = 100.0$

Use a Kruskal-Wallis test to compare these three populations. Is there any evidence to suggest that the populations are different? Use $\alpha = 0.05$, and assume the underlying populations are continuous.

14.75 Physical Sciences Raytheon is testing different propellants for the *Tomahawk* missile used by the U.S. Navy. Independent random samples of missiles were obtained, and each was totally fueled with one of four types of solid propellant. Each missile was fired at the Navy test range near Point Mugu, California, and the total distance traveled was recorded (in km). The summary statistics are given in the following table.

Solid propellant	Sample size	Rank sum
1	$n_1 = 14$	$r_1 = 525.0$
2	$n_2 = 15$	$r_2 = 355.5$
3	$n_3 = 12$	$r_3 = 373.0$
4	$n_4 = 16$	$r_4 = 399.5$

Use the Kruskal-Wallis test to determine whether there is any evidence to suggest that at least two populations of distance traveled are different. Use $\alpha = 0.05$, and assume the underlying populations are continuous.

14.76 Biology and Environmental Science The National Snow and Ice Data Center collects information on glaciers from around the world. Some of the parameters include the number of basins, the mean depth, and the primary class. A random sample of glaciers in Western Europe was obtained, and the total area (in square kilometers) of each was recorded.[22] 📊 **GLACIERS**

  a. Use the Kruskal-Wallis test with $\alpha = 0.05$ to test the hypothesis that all four glacier-area populations are identical. Assume the underlying populations are continuous.

  b. Interpret your results in part (a). Which region has on average the largest glaciers by area?

14.77 Psychology and Human Behavior Our emotional reaction to different colors affects advertising, product design, and even architecture. Three subway-tunnel walkways of identical length were painted different colors. A random sample of adults was selected in each tunnel and secretly timed (in seconds) as they walked through the tunnel. The data are given in the following table. 📊 **COLOR**

Red		Orange		Yellow		Black	
12	19	21	14	29	23	22	23
16	23	27	12	18	24	11	27
15	22	26	24	25	30	10	24
18	25	22	30	25	12	24	14

Use the Kruskal-Wallis test with $\alpha = 0.01$ to determine whether there is any evidence that the tunnel-walking-time populations are different. Assume the underlying populations are continuous.

14.78 Public Health and Nutrition Independent random samples of four types of 16-ounce steaks were obtained, and the amount of fat (in grams) in each was measured. Assume the underlying populations are continuous. Is there any evidence to suggest that the fat populations are different? Use $\alpha = 0.025$. 📊 **STEAK**

14.79 Public Policy and Political Science The U.S. Product Safety Commission issues playground specifications and guidelines to help communities build safe playgrounds. One concern is the uncompressed depth of wood chips and other loose-fill material (used as shock absorbers). Independent random samples of playgrounds in various cities were obtained, and the uncompressed depth of wood chips (in inches) was measured for each playground. The data are given in the following table. 📊 **PLAYGRND**

Atlanta			Dallas			Denver		
10.5	12.8	12.9	11.8	11.6	11.3	10.9	11.8	11.2
10.2	11.5	10.5	11.6	10.0	10.2	12.4	12.0	11.6

Assume the underlying distributions of uncompressed depths are continuous. Is there any evidence to suggest that the populations are different? Use $\alpha = 0.05$, and find bounds on the p value associated with this test.

14.80 Sports and Leisure   The New York City Marathon is one of the largest in the world, with over 50,000 finishers in 2013. The annual race takes runners through the five boroughs of New York City over 26.219 miles. A random sample of the top 100 finishers from five countries in 2013 was obtained, and the finish time for each was recorded (in minutes).[23]   **MARATHON**
a. Is there any evidence to suggest that the populations are different? Use $\alpha = 0.01$.
b. Using the ranks, which of these countries has the fastest runners? Why?

Extended Applications

14.81 Business and Management   One indication of morale in county government positions is the length of service in the current position. Independent random samples of county employees were obtained, and the length of service (in years) was recorded for each person, by job classification.   **MORALE**
a. Assume the underlying distributions are continuous. Use the Kruskal-Wallis test with $\alpha = 0.05$ to show that there

is evidence to suggest that at least two populations are different.
b. Which pair(s) of populations do you think is (are) different? Why?

14.82 Physical Sciences   In 2013 there were 1087 supernovae reported by various observation posts around the world. When possible, these exploding stars were classified by their light curves and absorption lines. A random sample of these Type I and Type II supernovae were obtained, and the magnitude of each was recorded.[24]   **SUPERNOV**
a. Assume the underlying distributions are continuous. Use the Kruskal-Wallis test with $\alpha = 0.01$ to determine if there is any evidence that the populations are different. Find the p value associated with this test.
b. Use the normal approximation in the Wilcoxon rank-sum test to determine whether there is any evidence to suggest the population median magnitudes are different. Use $\alpha = 0.01$. Find the p value associated with this test.
c. What is the relationship between the p values in parts (a) and (b)?
d. Consider the value of the test statistics found in parts (a) and (b). Compare z^2 and h. How are these values related?

14.5   The Runs Test

Some practical applications of the runs test involve the sequence of positive or negative gains of a certain stock, the strength of signals from an object in space, and monitoring stream pollution.

In all of the inference procedures presented in this text, it is very important for the sample, or samples, to be selected *randomly* from the underlying population(s). Otherwise, the results are not valid. Although it is not a test for verifying that a sample has been randomly selected, the procedure described in this section can be used to examine the *order* in which observations were drawn from a population. The **runs test** is used to assess only whether there is evidence that the sequence of observations is not random.

Suppose an usher at a Broadway play records the seating section for the next 15 patrons, O for the orchestra section and B for the balcony. If all 15 people sat in the balcony or all sat in the orchestra section, we would certainly conclude that the order of patrons entering the theater was not random. Similarly, if the first 10 sat in the balcony, and the last 5 sat in the orchestra section, we would still question whether the order was random.

To determine whether the order of observations is random, we first separate the entire sequence into smaller subsequences in which the observations are the same. Consider the sequence of theater patrons in the following table, grouped by seating section.

O O	B B B B	O	B B	O O O O	B B

The grouped subsequences of similar observations, or symbols, are called **runs**. In the sequence of observations above, there are six runs. The test for randomness is based on the total number of runs.

Definition

A **run** is a series, or subsequence, of one or more identical observations.

A CLOSER LOOK

1. The runs test is appropriate for testing whether a sequence of observations is not random. It can be used if the data can be divided into two mutually exclusive categories, for example, defective or satisfactory, working or retired, pass or fail. This test may also be used for quantitative data that can be classified into one of two categories, for example, above or below the median, or dangerous versus acceptable temperature.

Note: A run can have length 1.

2. The smallest possible number of runs in any sample is 1. This will occur if every observation in the sample falls into the same category or has the same attribute. The largest possible number of runs in a sample depends on the number of observations in each category.

3. The statistical test is based on the total number of runs. It seems reasonable that if the order of observations is random, then the total number of runs should not be very large or very small. There is a table of critical values that uses the exact distribution for the number of runs. There is also a normal approximation that can be used if the number of observations in each category is large.

The Runs Test

The runs test is a nonparametric procedure, because no assumptions are made about the underlying population.

Suppose a sample is obtained in which each observation is classified into one of two mutually exclusive categories. Assume there are m observations in one category and n observations in the other.

H_0: The sequence of observations is random.

H_a: The sequence of observations is not random.

TS: V = the total number of runs

RR: $V \geq v_1$ or $V \leq v_2$

The critical values v_1 and v_2 are obtained from Table XI in the Appendix such that

$$P(V \geq v_1) \approx \alpha/2 \quad \text{and} \quad P(V \leq v_2) \approx \alpha/2$$

The Normal Approximation: As m and n increase, the statistic V approaches a normal distribution with

$$\mu_V = \frac{2mn}{m+n} + 1 \quad \text{and} \quad \sigma_V^2 = \frac{2mn(2mn - m - n)}{(m+n)^2(m+n-1)}$$

Therefore, the random variable $Z = \dfrac{V - \mu_V}{\sigma_V}$ has approximately a standard normal distribution. The normal approximation is good when both m and n are greater than 10. In this case, Z is the test statistic and the rejection region is

RR: $|Z| \geq z_{\alpha/2}$

Example 14.8 Ways to Make a Million

DATA SET

MILLION

A recent report in *Entrepreneur*[25] listed some of the common characteristics of millionaires in the United States. For example, start with nothing and grow up in a tough neighborhood, take risks, and make sacrifices were some of the features of people who make lots of money. In addition, a large proportion of millionaires are self-employed. A sample of millionaires living in Los Angeles was obtained, and each was classified as self-employed (S) or an employee (E). The sequence of responses is given in the following table.

S	S	E	E	S	E	S	S	E	E	E	E	S

Use the runs test with $\alpha = 0.05$ to determine whether there is any evidence that the order in which the sample was selected was not random.

Solution Trail 14.8

KEYWORDS

- Any evidence?
- Order in which the sample selected was not random

TRANSLATION

- Hypothesis test to determine whether there is evidence the sequence of observations is not random

CONCEPTS

- Runs test

VISION

Use the runs test procedure: compute the total number of runs, and draw the appropriate conclusion.

For $v_1 = 11$: $P(V \geq 11) = 0.0629$, which leads to a much higher significance level.

SOLUTION

STEP 1 Each observation is classified into one of two mutually exclusive categories, self-employed (S) or an employee (E). We are looking for evidence to suggest that the sequence of observations was not random. The runs test is appropriate. There are $m = 6$ self-employed observations and $n = 8$ employed observations.

STEP 2 The four parts of the hypothesis test are

H_0: The sequence of observations is random.

H_a: The sequence of observations is not random.

TS: $V =$ the total number of runs

RR: $V \geq v_1 = 12$ or $V \leq v_2 = 4$

Find the value v_1 such that $P(V \geq v_1) \approx 0.05/2 = 0.025$. Using Table XI in the Appendix, with $m = 6$ and $n = 8$, $P(V \geq 12) = 1 - P(V \leq 11) = 1 - 0.9837 = 0.0163 \approx 0.025$. Therefore, $v_1 = 12$.

STEP 3 The following table shows the original sequence of observations separated into runs.

S	S	E	E	S	E	S	S	E	E	E	E	S

There are a total of seven runs. The value of the test statistic, $v = 7$, does not lie in the rejection region. At the $\alpha = 0.05$ significance level, there is no evidence to suggest that the order of observations was not random.

Figure 14.20 shows a technology solution.

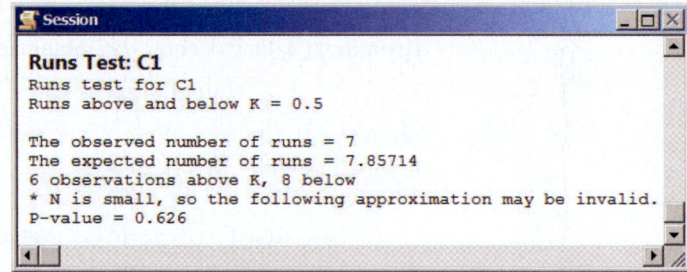

Figure 14.20 Minitab runs test.

TRY IT NOW   GO TO EXERCISE 14.96

The original data in the following example are quantitative but can be classified into two mutually exclusive categories. The number of observations in each category is large, so the normal approximation will be used in the runs test.

Example 14.9 Propane Shipping Delays

DATA SET

PROPANE

In January 2014 supplies of propane became dangerously low in Ontario and Quebec. The Canadian Propane Association issued a special news release indicating that the extreme cold was contributing to the delayed deliveries to suppliers, homes, and businesses.[26] A sample of homes that heat with propane in Ontario was selected, and the amount of propane in each tank was measured (in pounds). The data are given in the following table, in order from left to right.

22.8	95.6	106.3	20.1	23.2	76.8	18.0	61.0	151.5	29.5
24.2	95.6	18.9	68.1	118.3	130.2	21.5	69.2	24.5	38.5
102.7	23.3	72.0	39.2	83.5	22.7	134.3	97.0	17.9	33.0

Solution Trail 14.9

KEYWORDS

- Threshold value is 25 pounds
- Is there any evidence?
- Order of observations is not random
- $m = 19$, $n = 11$

TRANSLATION

- Hypothesis test to determine whether there is evidence the sequence of observations is not random with respect to the threshold value
- m and n greater than 10

CONCEPTS

- Runs test
- Normal approximation

VISION

Use the runs test procedure: compute the total number of runs, find the value of the test statistic, and draw the appropriate conclusion.

Suppose the residential threshold value for delivery of propane is 25 pounds. Is there any evidence to suggest that the order of observations is not random with respect to this threshold value? Use $\alpha = 0.05$.

SOLUTION

STEP 1 Replace each observation above the threshold value with an A and each observation below the threshold value with a B. Any observation equal to 25.0 will be excluded from the analysis and the sample size will be reduced. Each observation now falls into one of two mutually exclusive categories: above or below the delivery threshold.

STEP 2 The runs test will be used to determine whether there is evidence to suggest that the sequence of observations is not random. There are $m = 19$ observations above 25.0 and $n = 11$ observations below 25.0.

STEP 3 Because both m and n are greater than 10, the normal approximation will be used. The four parts of the hypothesis test are

H_0: The sequence of observations is random.

H_a: The sequence of observations is not random.

$$\text{TS: } Z = \frac{V - \mu_V}{\sigma_V}$$

RR: $|Z| \geq z_{\alpha/2} = z_{0.025} = 1.96$

STEP 4 Compute the number of runs using the original sequence of observations.

22.8	95.6	106.3	20.1	23.2	76.8	18.0	61.0	151.5	29.5
B	A	A	B	B	A	B	A	A	A

24.2	95.6	18.9	68.1	118.3	130.2	21.5	69.2	24.5	38.5
B	A	B	A	A	A	B	A	B	A

102.7	23.3	72.0	39.2	83.5	22.7	134.3	97.0	17.9	33.0
A	B	A	A	A	B	A	A	B	A

There are $x = 20$ runs.

STEP 5 The mean and the variance of the random variable V are

$$\mu_V = \frac{2mn}{m+n} + 1 = \frac{2(19)(11)}{19+11} + 1 = 14.9333$$

$$\sigma_V^2 = \frac{2mn(2mn - m - n)}{(m+n)^2(m+n-1)} = \frac{2(19)(11)[2(19)(11) - 19 - 11]}{(19+11)^2(19+11-1)} = 6.2139$$

The value of the test statistic is

$$z = \frac{v - \mu_V}{\sigma_V} = \frac{20 - 14.9333}{\sqrt{6.2139}} = 2.0326$$

STEP 6 Because $|z| = |2.0326| = 2.0326 \geq 1.96$, the value of the test statistic lies in the rejection region. Equivalently, $p = 0.0420 \leq 0.05$. At the $\alpha = 0.05$ significance level, there is evidence to suggest that the order of the observations is not random.

To find the p value associated with this test, find the right-tail probability and multiply by 2.

$p/2 = P(Z \geq 2.0326)$ Definition of p value.

$= 1 - P(Z \leq 2.0326)$ Complement rule.

$= 1 - 0.9790 = 0.0210$ Cumulative probability; use Table III in the Appendix.

$p = 2(0.0210) = 0.0420$ Solve for p.

The exact p value is illustrated in Figure 14.21. Figure 14.22 shows a technology solution.

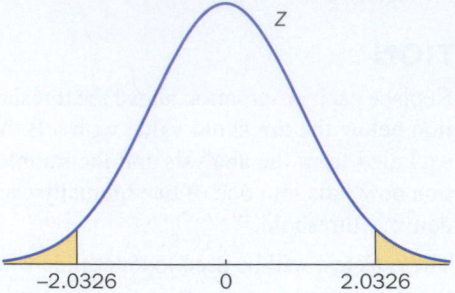

Figure 14.21 p-Value illustration:
$p = 2\, P(Z \geq 2.0326)$
$= 0.0420 \leq 0.05 = \alpha$

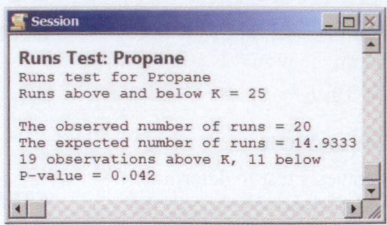

Figure 14.22 Minitab runs test.

TRY IT NOW   GO TO EXERCISE 14.103

Technology Corner

Procedure: Conduct a runs test.
Reconsider: Example 14.9, solution, and interpretations.

Minitab

The runs test is accessible in the Stat; Nonparametrics menus and may also be executed using commands in a session window. Categorical data must be converted to numerical values.

1. Enter the data into column C1.
2. Select Stat; Nonparametrics; Runs Test. Enter the Variable, the column containing the data, C1.
3. Select Above and below, and enter 25.
4. Click OK. The results are displayed in a session window. See Figure 14.22.

SECTION 14.5 EXERCISES

Concept Check

14.83 True/False   The runs test can be used to determine whether a sample has been randomly selected.

14.84 True/False   The runs test can be used if numerical data can be divided into two mutually exclusive categories.

14.85 Fill in the Blank   The smallest possible number of runs in a sample is _____.

14.86 Short Answer   What assumptions are made about the underlying distribution for a runs test to be valid?

14.87 Short Answer   When is the normal approximation in the runs test appropriate?

14.88 Short Answer   What is a run?

Practice

14.89 Find the number of runs in each sequence of observations. EX14.89
  a. A A B B B A B A B B A B
  b. G B B B B B G G B B G B G G B
  c. F F S F F S F F S S F S F F S S S F
     F F
  d. + − + − − + − − − + + − + −
     + + − − + − − +

14.90 In each of the following problems, use the values for m and n to find the critical values in a runs test with approximate significance level α. Find the exact significance level for your choice of critical values.
  a. $m = 4$, $n = 7$, $\alpha = 0.05$
  b. $m = 5$, $n = 8$, $\alpha = 0.01$

c. $m = 6$, $n = 6$, $\alpha = 0.05$
d. $m = 8$, $n = 9$, $\alpha = 0.01$

14.91 In each of the following problems, use the population median to classify each observation in the ordered sample as either above or below the median. Find the number of runs in each sequence of observations. *Note*: The order of observations is left to right, then down. **EX14.91**

a. $\tilde{\mu} = 20.0$

22.4	15.1	22.0	25.6	19.0	25.1	18.7	21.2
11.5	19.6						

b. $\tilde{\mu} = 4.00$

4.14	3.26	3.23	3.12	4.80	5.52	5.81	5.71
3.77	3.68	3.79	3.03	4.36	5.29		

c. $\tilde{\mu} = 150$

115	169	139	101	102	138	123	195	107
175	178	151	181	107	110	174	196	167

d. $\tilde{\mu} = 0.44$

0.07	0.19	0.10	0.12	0.44	0.08	0.26	0.03
0.10	0.14	0.41	0.20	0.01	0.38	0.28	0.03
0.15	0.03	0.00	0.04	0.19	0.04		

14.92 In each of the following problems, use the values for m and n to find the expected number of runs, μ_V, and the variance of the number of runs, σ_V^2.

a. $m = 10$, $n = 15$ b. $m = 15$, $n = 21$
c. $m = 2$, $n = 23$ d. $m = 26$, $n = 26$

14.93 A sample was obtained, and each observation was classified into one of two mutually exclusive categories. The sequence of observations is given in the following table. **EX14.93**

B	A	B	A	A	B	B	A	B	A
B	B	B	B	A	A	B	A	A	B

Conduct a runs test with $\alpha = 0.05$. Is there any evidence to suggest that the order of observations is not random?

14.94 A sample was obtained from a population with hypothesized median $\tilde{\mu} = 10.0$. The sequence of observations is given on the text website. **EX14.94**

a. Conduct a runs test using the normal approximation with $\alpha = 0.01$. Is there any evidence to suggest that the order of observations is not random?
b. Find the p value associated with this test.

Applications

14.95 Public Policy and Political Science During freshmen move-in day at the University of Michigan, the College Republicans staffed an information table in order to solicit new members. A sample of consecutive new members was obtained, and each was classified by sex (M or F). The ordered observations are given in the following table. **REPINFO**

M	M	M	M	M	M	M	M
F	F	M	F	M	M		

Use the runs test with $\alpha \approx 0.05$ to determine whether there is any evidence to suggest that the order of observations is not random with respect to sex.

14.96 Manufacturing and Product Development Following Hurricane Katrina, the Make It Right New Orleans charitable organization helped to rebuild homes in the Ninth Ward. Recently, homeowners have complained that the wood used in construction cannot withstand the New Orleans humidity and is rotting and growing fungus. Make It Right has blamed the company that supplied the lumber, which was guaranteed for 40 years.[27] A sample of rebuilt homes in the Ninth Ward was obtained, and each was carefully inspected for good (G) or bad (B) wood. The ordered observations are given in the following table. **MAKEITRT**

G	G	G	B	G	G	G	G	G
G	G	G	G	B	G	G	G	G

Is there any evidence to suggest that the order of observations is not random? Use the runs test with $\alpha \approx 0.05$.

14.97 Marketing and Consumer Behavior Staples sells photo paper in either glossy (G) or matte (M) finish. A sample of online customers who purchased photo paper was obtained, and each purchase was classified according to finish. The ordered observations are given in the following table. **PHOTOPPR**

M	M	G	M	G	M	G	M	G
M	G	M	M	G	M	M	G	M
G	G	G	M	M	M	G	M	M
G	G	G	M	M	M	G	G	

Is there any evidence to suggest that the order of observations is not random? Use the runs test with the normal approximation and a 0.02 level of significance.

14.98 Public Policy and Political Science A recent survey showed that approximately half of Canadians think it is acceptable, in certain cases, for the government to read private email and monitor other online activities.[28] A sample of adults in Ottawa was obtained and asked if they believe it is acceptable for the government to monitor private email. Each response was classified as yes (Y) or no (N). The sequence of responses is given in the following table. **ONLINE**

Y	Y	Y	N	Y	Y	Y	N	N	N	N	N	Y	N	N

a. Use the runs test with $\alpha \approx 0.05$ to determine whether there is any evidence that the order of observations is not random.
b. Find the p value associated with this test.

14.99 Elvis Has Left the Building More than 25 years after the death of Elvis Presley, more than 40% of Americans are still fans of "The King." Given all of the rumors, theories, and alleged sightings, 7% of Americans think Elvis is still alive.[29] A sample of visitors to Graceland was obtained and asked if they believe Elvis is still alive. There responses were classified as alive (L) or dead (D). Is there any evidence to suggest that the order of observations is not random? Use the runs test with the normal approximation and $\alpha = 0.05$. **ELVIS**

14.100 What's in Your Wallet? Perhaps due to the increase in use of debit and credit cards, only 20% of Americans always carry cash.[30] A sample of shoppers at the Quaker Bridge Mall in Lawrence, New Jersey was obtained and asked if they were carrying cash. The sequence of responses is given in the following table [cash (C) and no cash (N)]. 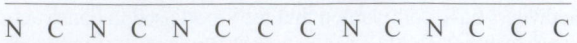 WALLET

N	C	N	C	N	C	C	C	N	C	N	C	C	C

Is there any evidence to suggest that the order of observations is not random? Use the runs test with $\alpha \approx 0.05$.

14.101 Manufacturing and Product Development
Composite decks are made of recycled wood and plastic, are low-maintenance and sturdy, and are designed to last much longer than traditional wood decks. A manufacturer of composite decks routinely checks the surface wear as part of quality control. A sample of planks was obtained, and each was measured for surface wear using a Taber tester. The ordered observations (in Taber units) are given in the following table.

204	211	188	208	190	203	203	193	204
214	176	209						

The planks are manufactured to have a median wear index of 200 units. Use the runs test with $\alpha \approx 0.02$ to determine whether there is any evidence that the order of observations is not random with respect to the median. DECKS

14.102 Education and Child Development The Kellogg School of Management at Northwestern University has no minimum GMAT score to be accepted into the MBA program. However, they reported that the *average* GMAT for the class of 2013 was 715.[31] A sample of students in this MBA program was obtained, and the GMAT score for each was recorded. Use the runs test with the normal approximation and $\alpha = 0.05$ to

determine whether there is any evidence that the order of GMAT scores is not random. GMAT

14.103 Medicine and Clinical Studies Brigham and Women's Hospital reports that approximately 20% of American males and 10% of American females will develop a kidney stone at some time in their life.[32] Some research suggests that drinking lots of sugar-sweetened soda may lead to an increased risk of developing a kidney stone. A sample of patients who were admitted to Brigham and Women's Hospital for a kidney stone was obtained. Low-dose computed tomography was used to measure the size (in ml) of each stone, and the ordered observations were recorded. Use the runs test with the normal approximation and $\alpha = 0.05$ to determine whether there is any evidence that the order of kidney stone sizes is not random with respect to the median size, 3.0 ml. Find the p value associated with this test. STONES

Challenge

14.104 Random Number Generators The purpose of this exercise is to determine whether a random number generator really produces a random sequence of observations. Using your graphing calculator or favorite statistical software:

a. Generate 100 observations from a standard normal distribution.
b. Classify each observation as either above or below the mean (and median) 0.
c. Conduct a runs test using the normal approximation to determine whether there is any evidence to suggest that the sequence of observations is not random with respect to the mean. Use $\alpha = 0.05$.

Do this 100 times. How many times did you reject the null hypothesis H_0: The sequence is random? Draw a conclusion about the sequence of observations produced by your random number generator.

14.6 Spearman's Rank Correlation

The sample correlation coefficient, r, was introduced in Chapter 12 as a measure of the strength of the linear relationship between two continuous variables. **Spearman's rank correlation coefficient** is a nonparametric alternative and is computed using ranks. Without any assumptions being made about the underlying populations, each observation is converted to a rank, and the sample correlation coefficient is computed using the ranks in place of the actual observations.

Spearman's Rank Correlation Coefficient

Suppose there are n pairs of observations $(x_1, y_1), (x_2, y_2), \ldots, (x_n, y_n)$. Rank the observations on each variable separately, from smallest to largest. Let u_i be the rank of the ith observation on the first variable and let v_i be the rank of the ith observation on the second variable. **Spearman's rank correlation coefficient**, r_S, is the sample correlation coefficient between the ranks and is computed using the equation

$$r_S = 1 - \frac{6\sum d_i^2}{n(n^2 - 1)} \tag{14.1}$$

where $d_i = u_i - v_i$.

A CLOSER LOOK

The sample correlation coefficient was defined in Section 12.2.

1. As usual, equal values within each variable are assigned the mean rank of their positions in the ordered list. Equation 14.1 is not exact when there are tied observations within either variable. In this case, one should compute r_S by finding the sample correlation coefficient between the ranks.

2. Because r_S is really a sample correlation coefficient, the value is always between -1 and $+1$. Values near -1 indicate a strong negative linear relationship, and values near $+1$ suggest a strong positive linear relationship between the ranks.

3. Remember, correlation does not imply causation; r_S is a measure of the linear association between the ranks. And a strong linear relationship between the ranks does not imply that the relationship between the original variables is also linear.

DATA SET

DRIVING

Example 14.10 Driving and the Economy

Several studies suggest that driving habits are related to certain economic indicators. For example, as the price of gas increases, it seems reasonable that people will drive fewer miles. Measures of housing starts and the stock market may also be related to driving miles. Normally, a recession and high unemployment also affect miles traveled. A random sample of months was selected, and the total miles traveled by Americans (in millions of miles) and the unemployment rate (a percentage) was recorded for each. The data are given in the following table.[33]

Month	1	2	3	4	5	6	7	8	9	10	11	12
Miles	212.74	226.30	227.70	227.90	233.66	233.28	261.62	179.54	263.06	256.37	186.83	205.98
Unemp rate	9.7	7.6	3.8	3.9	5.8	4.9	4.8	6.7	4.7	6.6	6.5	7.1

Compute Spearman's rank correlation coefficient and interpret this value.

SOLUTION

STEP 1 Let x represent the miles driven per month and let y represent the unemployment rate. For each variable, order the observations from smallest to largest and assign a rank to each value. Compute the difference between each pair of ranks.

The following table shows each observation, its associated rank, and each difference.

Month i	Miles x_i	Rank u_i	Unemp rate y_i	Rank v_i	Difference d_i
1	212.74	4	9.7	12	-8
2	226.30	5	7.6	11	-6
3	227.70	6	3.8	1	5
4	227.90	7	3.9	2	5
5	233.66	9	5.8	6	3
6	233.28	8	4.9	5	3
7	261.62	11	4.8	4	7
8	179.54	1	6.7	9	-8
9	263.06	12	4.7	3	9
10	256.37	10	6.6	8	2
11	186.83	2	6.5	7	-5
12	205.98	3	7.1	10	-7

STEP 2 There are $n = 12$ pairs and no ties within either variable. Use Equation 14.1 to compute Spearman's rank correlation coefficient.

$$r_S = 1 - \frac{6 \sum d_i^2}{n(n^2 - 1)}$$ Equation 14.1.

$$= 1 - \frac{6[(-8)^2 + (-6)^2 + \cdots + (-7)^2]}{12(12^2 - 1)}$$ Use the d_i's.

$$= 1 - \frac{2640}{1716} = -0.5385$$

STEP 3 Because $r_S = -0.5385$, there is a moderate negative linear relationship between the miles traveled ranks and the unemployment ranks. This suggests that there is a moderate negative correlation between miles traveled and unemployment rate; high unemployment rates are associated with fewer miles traveled.

Figures 14.23 and 14.24 show technology solutions.

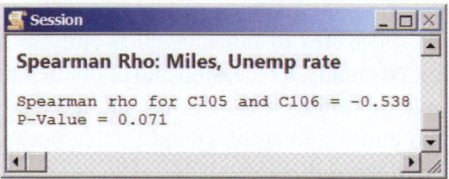

Figure 14.23 Spearman's rank correlation coefficient computed using Minitab.

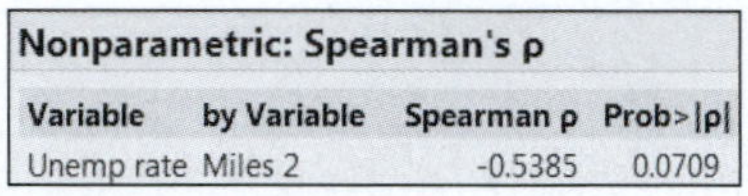

Nonparametric: Spearman's ρ			
Variable	by Variable	Spearman ρ	Prob>|ρ|
Unemp rate	Miles 2	-0.5385	0.0709

Figure 14.24 Spearman's rank correlation coefficient computed using JMP.

TRY IT NOW   GO TO EXERCISE 14.113

Technology Corner

Procedure: Compute Spearman's rank correlation coefficient.
Reconsider: Example 14.10, solution, and interpretations.

Minitab

Use the Correlation function with the appropriate Method to compute Spearman's rank correlation coefficient.

1. Enter the miles traveled data into column C1 and the unemployment rate data into column C2.
2. Select Stat; Basic Statistics; Correlation. Enter the Variables: the columns containing the data. Select Spearman rho as the Method.
3. Click OK. Spearman's rank correlation coefficient is displayed in a session window. See Figure 14.23.

Excel

There is no built-in function to compute r_S. However, rank each sample separately, and find the sample correlation coefficient between the ranks.

1. Enter the miles traveled data into column A and the unemployment rate data into column B.
2. Rank the observations in A and store the results in column C. Rank the observations in B and store the results in column D.
3. Use the function CORREL to find the sample correlation coefficient between the ranks, columns and C and D. See Figure 14.25.

	A	B	C	D
1	Miles	Rate	Miles ranks	Rate ranks
2	212.74	9.7	4	12
3	226.30	7.6	5	11
4	227.70	3.8	6	1
5	227.90	3.9	7	2
6	233.66	5.8	9	6
7	233.28	4.9	8	5
8	261.62	4.8	11	4
9	179.54	6.7	1	9
10	263.06	4.7	12	3
11	256.37	6.6	10	8
12	186.83	6.5	2	7
13	205.98	7.1	3	10
14				
15	-0.5385 = CORREL(C2:C13,D2:D13)			

Figure 14.25 Excel calculations to compute Spearman's rank correlation coefficient.

SECTION 14.6 EXERCISES

Concept Check

14.105 Short Answer What are the assumptions about the underlying populations for Spearman's rank correlation coefficient?

14.106 True/False If there are equal values with a variable, Spearman's rank correlation coefficient cannot be computed.

14.107 True/False The values of r_S are always between -1 and $+1$.

14.108 True/False A strong linear relationship between ranks implies a strong linear relationship between the original variables.

14.109 True/False If the underlying populations are normally distributed, the correlation coefficient between the original variables and Spearman's rank correlation coefficient will be the same.

Practice

14.110 Random samples of pairs of observations are given in each of the following problems. Rank the observations in each sample separately, from smallest to largest, and find each difference between the ranks, d_i. **EX14.110**

a.

Observation	1	2	3	4	5	6
Sample 1	54	17	28	69	13	49
Sample 2	113	114	139	173	145	121

b.

Observation	1	2	3	4	5	6
Sample 1	57	40	32	56	33	60
Sample 2	35	50	51	57	38	45

Observation	7	8	9
Sample 1	51	35	53
Sample 2	44	52	32

c.

Observation	1	2	3	4	5	6
Sample 1	22.5	27.0	22.8	26.5	29.9	20.8
Sample 2	27.0	30.5	21.6	27.9	33.0	22.9

Observation	7	8	9	10	11	12
Sample 1	19.3	28.3	19.5	20.3	28.2	27.9
Sample 2	24.7	22.2	24.3	26.2	25.1	29.4

Observation	13	14	15
Sample 1	27.8	31.6	25.3
Sample 2	22.6	24.8	20.3

d.

Observation	1	2	3	4	5	6
Sample 1	49.0	51.2	46.7	54.9	53.6	48.9
Sample 2	78.1	60.8	70.4	72.2	64.9	70.8

Observation	7	8	9	10	11	12
Sample 1	46.8	46.2	55.8	40.8	46.8	55.7
Sample 2	74.3	68.4	66.5	75.9	65.9	70.3

Observation	13	14	15	16	17	18
Sample 1	54.9	45.0	55.1	43.2	43.8	53.9
Sample 2	71.4	75.6	72.0	69.4	71.7	78.3

14.111 In each of the following problems, consider the data obtained in a study of the relationship between two variables, find Spearman's rank correlation coefficient, and use it to describe the relationship between the two variables. **EX14.111**

a.

x	156	268	262	206	162	166	148
y	131	258	235	189	180	262	207

b.

x	5.39	5.60	5.12	4.53	6.01	5.00
y	−15.3	−15.4	−16.5	−16.3	−14.2	−13.6

x	5.62	5.04	4.65	6.19
y	−14.9	−14.4	−15.6	−15.2

c.

x	25.51	25.80	24.10	24.19	25.52	26.16
y	24.45	26.98	25.72	27.20	28.56	25.69

x	24.79	23.95	25.14	25.09	25.35	26.64
y	27.38	25.84	24.04	25.87	25.51	26.94

x	25.63	23.84
y	26.64	26.53

d.

x	y	x	y	x	y
418	754	411	755	414	776
484	725	378	759	453	781
448	712	425	749	478	745
458	718	436	733	465	785
476	756	443	727	438	775
421	772	442	716	459	735
480	728	435	763	490	751
461	753	440	752	407	744

14.112 Consider the following data obtained in a study of the relationship between two variables. ▇ **EX14.112**

x	y	x	y
1.5	7.3	1.4	7.4
1.8	6.8	1.3	7.2
1.5	7.4	1.2	7.2
1.6	7.5	1.4	7.3
1.7	6.7	1.2	6.6

a. Rank the observations in each sample separately, and find each difference between the ranks, d_i.

b. Find the sample correlation coefficient between the ranks.

c. Find Spearman's rank correlation coefficient using the differences, d_i.

d. Explain why the values in parts (b) and (c) are different.

Applications

14.113 Travel and Transportation Dynasty Travel offers cruises to the Caribbean and often discounts unsold tickets as the sailing date approaches. These last-minute deals were studied by comparing the price of an inside stateroom (x, in dollars) with the number of days before sailing (y). A random sample of last-minute cruise deals was obtained, and the data are given in the following table.

x	998	956	763	1313	1071	1091
y	9	3	4	13	10	6

Find Spearman's rank correlation coefficient, and interpret this value. Do these data suggest that cruise prices are reduced at the last minute? Justify your answer. ▇ **CRUISES**

14.114 Biology and Environmental Science Farmers consider heavy molasses to be a high-energy food for cattle. In a recent study, the percentage of digestible energy (x) was compared with the percentage of protein (y) in heavy molasses. Random samples of various brands of molasses were obtained. The percentage of digestible energy and of protein were carefully measured. Compute Spearman's rank correlation coefficient. Use this value to describe the relationship between the percentage of digestible energy and the percentage of protein in heavy molasses. ▇ **MOLASS**

14.115 Biology and Environmental Science The bulk density of soil (x, in g/cm^3) affects plant growth, specifically, how easily roots can penetrate the soil. A group of carrot growers conducted a study to compare soil bulk density with soil texture (y), measured on a scale from 1 to 10, with 1 being very sandy soil and 10 corresponding to clay. A random sample of plots was obtained, and the data were recorded. Compute Spearman's rank correlation coefficient, and use this value to describe the relationship between these two variables. ▇ **SOIL**

14.116 Biology and Environmental Science A recent study examined the relationship between measures of egg albumen height and pH.[34] A random sample of eggs from Brown Leghorn hens was obtained, and the albumen height (x, in mm) and pH was measured for each. Compute Spearman's rank correlation coefficient. What does this value suggest about the relationship between albumen height and pH? ▇ **LEGHORN**

14.117 Nature Deficit Only a few generations ago, children played much more outside, participated in spontaneous play, communicated face to face, and learned how to compromise. Today, in fear, many parents are reluctant to let their child play outdoors alone. As a result, many children have very little contact with nature. Some research suggests that children who play regularly in natural environments tend to be healthier. A random sample of middle school children was obtained, and the number of hours per week each child played outside and the number of days of school missed due to illness was recorded. The data are given in the following table. ▇ **OUTDOOR**

Hours	3.6	6.3	5.0	7.9	8.7	4.9	6.2	10.6	6.8	8.2
Days	15	3	6	0	0	11	2	5	3	14

Compute Spearman's rank correlation coefficient. Does this value support the theory that healthier children spend more time outside? Justify your answer.

14.118 Biology and Environmental Science In January 2014 a polar vortex caused extreme weather conditions across Canada and much of the United States. Travel was treacherous, and countless flights were canceled at various airports. The severe cold at Pearson International Airport in Toronto caused at least 200 flights to be delayed, and additional police were needed to deal with irate passengers. A random sample of passengers in the airport was obtained, and the number of days delayed and the systolic blood pressure of each was recorded. ▥ **VORTEX**

  a. Compute Spearman's rank correlation coefficient.

  b. Use the value of r_S to describe the relationship between days delayed and systolic blood pressure, which is related to stress.

Extended Applications

14.119 Sports and Leisure The owner of a snow tubing resort in the Pocono Mountains believes that the total snowfall (x, in feet) during a winter is related to the number of customers (y, in thousands). A random sample of winters was obtained, and the data were recorded. ▥ **WINTER**

  a. Construct a scatter plot for these data, and describe the relationship between the two variables.

  b. Compute Spearman's rank correlation coefficient, and use this value to describe the relationship between the two variables.

  c. Explain any differences in your answers to parts (a) and (b).

14.120 Physical Sciences A certain solder joint on the undercarriage of a city bus is being tested for shear strength. A sample of solder joints was randomly selected, and the shear strength for each was measured (in N) before (x) and after (y) temperature-cycle stress. The data are given in the following table. ▥ **SOLDER**

x	71	69	66	69	62	67	69	64	73	69
y	64	56	65	62	65	64	57	63	57	63

  a. Rank the observations in each sample separately, and find each difference between the ranks, d_i.

  b. Find the sample correlation coefficient between the ranks.

  c. Find Spearman's rank correlation coefficient using the d_i's.

  d. Explain why the values in parts (b) and (c) are different, and use these results to explain the relationship between the two variables.

14.121 Too Many Selfies Advanced smartphones and easy uploads to social media sites have led to an increase in "selfies."

The dramatic increase in these self-portraits and the use of the word made "selfie" the *Oxford Dictionary* 2013 Word of the Year. Despite their popularity, a recent study suggests that those who share more self-portraits experience less intimacy with others.[35] A random sample of social media users was obtained and asked for the number of selfies they had shared within the past month. In addition, each person took the Intimacy Attitude Scale (IAS) test, which measures a person's willingness to share feelings, fears, and concerns. ▥ **SELFIES**

  a. Compute Spearman's rank correlation coefficient.

  b. Lower values on the IAS test suggest the person is less inclined to be close and personal. Explain the meaning of r_S in the context of this problem.

Challenge

14.122 Fuel Consumption and Cars Suppose there are n pairs of observations and ρ_S is the population correlation between ranks. The four parts of a hypothesis test concerning ρ_S based on a normal approximation are

H_0: $\rho_S = 0$ (no population correlation between ranks)

H_a: $\rho_S > 0$, $\rho_S < 0$,   or $\rho_S \neq 0$

TS: $Z = R_S\sqrt{n-1}$

    R_S is Spearman's rank correlation coefficient.

RR: $Z \geq z_\alpha$, $Z \leq -z_\alpha$,   or $|Z| \geq z_{\alpha/2}$

A turbocharger is a compact way to add more power to an automobile, by increasing the amount of air going into the cylinders. A study was conducted to examine the relationship between the added pressure (x, in psi) and power (y, percentage change in horsepower) provided by a certain turbocharger. A variety of automobiles was selected, and a randomly selected turbocharger was installed in each. The additional air pressure and power measurements were recorded. ▥ **TRBOCHG**

  a. Compute Spearman's rank correlation coefficient, and use this value to describe the relationship between added air pressure and change in engine power.

  b. Conduct a hypothesis test to determine whether there is any evidence that the true population correlation between ranks is greater than 0. Use $\alpha = 0.01$.

CHAPTER 14 SUMMARY

Concept	Page	Notation / Formula / Description
Sign test	682	A nonparametric test concerning a population median, based on the number of observations greater than $\tilde{\mu}_0$. It can also be used to compare two population medians, based on the number of pairwise differences greater than Δ_0.
Wilcoxon signed-rank test	691	A nonparametric test concerning a population median, based on the sum of the ranks corresponding to the positive differences $x_i - \tilde{\mu}_0$. It can also be used to compare two population medians, based on the sum of the ranks corresponding to the positive differences $d_i - \Delta_0$.
Wilcoxon rank-sum test	699	A nonparametric test concerning the difference of two population medians, based on the sum of the ranks corresponding to the differences $x_i - \Delta_0$.

Kruskal-Wallis test	707	A nonparametric test to determine whether at least two of k samples are from different populations, based on the ranks of all observations combined.
Run	712	A series, or sub-sequence, or one or more identical observations.
Runs test	713	A nonparametric test to determine whether there is evidence that a sequence of observations is not random, based on the total number of runs.
Spearman's rank correlation coefficient	718	A nonparametric alternative to the sample correlation coefficient. Each observation is converted to a rank, and the correlation coefficient is computed using the ranks in place of the actual observations.

A comparison of parametric and nonparametric procedures:

	Nonparametric procedure		Parametric procedure	
Case	Null hypothesis	Statistical test	Null hypothesis	Statistical test
One sample	$\tilde{\mu} = \tilde{\mu}_0$	Sign test	$\mu = \mu_0$	One-sample t test
	$\tilde{\mu} = \tilde{\mu}_0$	Wilcoxon signed-rank test	$\mu = \mu_0$	One-sample t test
	Sequence of observations is random	Runs test	No comparable parametric test	
Two independent samples	$\tilde{\mu}_1 - \tilde{\mu}_2 = \Delta_0$	Wilcoxon rank-sum test	$\mu_1 - \mu_2 = \Delta_0$	Two-sample t test
Paired data	$\tilde{\mu}_D = \Delta_0$	Sign test	$\mu_D = \Delta_0$	Paired t test
	$\tilde{\mu}_D = \Delta_0$	Wilcoxon signed-rank test	$\mu_D = \Delta_0$	Paired t test
		Spearman's rank correlation coefficient		Sample correlation coefficient
k Independent samples	k Samples are from identical populations	Kruskal-Wallis test	$\mu_1 = \cdots = \mu_k$	One-way ANOVA

Note: Nonparametric procedures are applicable in a very broad range of situations. However, if the underlying population is normal, then the corresponding parametric procedure is more efficient.

CHAPTER 14 EXERCISES

14 APPLICATIONS

14.123 Biology and Environmental Science Lake Powell in Colorado was created after the construction of the Glen Canyon Dam and has more than 2000 miles of shoreline. It is one of the largest man-made reservoirs in the United States and provides water to Arizona, Nevada, and California. A random sample of outflow (in cfs) per day from Lake Powell was obtained, and the data are given in the following table.[36] POWELL

13,243	11,142	12,361	13,219	13,220	11,698	10,156
10,152	8,930	9,142	10,102	10,222	8,787	10,169

To maintain an adequate water supply, the outflow from Lake Powell should be approximately 11,000 cfs per day. Assume the underlying distribution of outflow is continuous. Use the sign test with $\alpha = 0.05$ to determine whether there is any evidence that the median outflow per day is less than 11,000 cfs.

14.124 Biology and Environmental Science Farms across the country have old agricultural chemicals stored in deteriorating containment vessels that pose a threat to humans. The U.S. Department of Agriculture has started a program to clean up and properly dispose of these unusable chemicals. A sample of farms was randomly selected, and the amount of DDT still stored on each farm was recorded. The data (in kg) are given in the following table. USAD

| 3.5 | 4.3 | 7.0 | 3.0 | 6.0 | 2.1 | 2.5 | 4.9 | 5.9 | 2.4 |
| 3.5 | 8.6 | 0.8 | 1.3 | 2.9 | 2.3 | 2.1 | 1.0 | 5.5 | 1.7 |

Assume the underlying distribution of stored DDT is continuous.
   a. Use the sign test with $\alpha = 0.01$ to determine whether there is any evidence that the median amount of stored DDT on farms is greater than 2.0 kg.
   b. Find the p value associated with this test.

14.125 Economics and Finance Insurance companies recommend a personal umbrella policy for most customers, even if they already have a home and an automobile policy. Basic liability coverage may not be adequate as lawsuits become more common and jury awards escalate. A random sample of adults with umbrella policies was obtained, and the amount of coverage was recorded for each (in millions of dollars). Assume the underlying distribution of umbrella coverage amounts is continuous. Use the sign test with $\alpha = 0.05$ to determine whether there is any evidence that the median coverage amount is different from \$2.00 million. **POLICY**

14.126 The Met Several years ago the Metropolitan Opera in New York City began live-to-cinema opera transmissions in an effort to attract a greater audience. The broadcasts are now seen in 64 countries, and according to the new general manager, the Met is reaching younger people.[37] A random sample of people who attended a performance at the Metropolitan Opera was obtained and the age (in years) of each was recorded. The data are given in the following table. **OPERA**

| 52 | 54 | 67 | 55 | 68 | 63 | 56 | 62 | 52 |
| 60 | 69 | 54 | 65 | 58 | 64 | 60 | 50 | 55 |

Assume the underlying distribution of ages is continuous. Use the Wilcoxon signed-rank test with $\alpha = 0.05$ to determine whether there is any evidence that the median age of those attending a performance is less than 65 years. Find the p value for this test.

14.127 Physical Sciences The fountain beneath the Gateway Arch in St. Louis is designed to spray water 630 feet straight up, the same height as the Arch. A random sample of days was selected, and on each day the height of the spray was measured (in feet) during the fountain show. The data are given in the following table. **ARCH**

| 603 | 633 | 612 | 619 | 624 | 627 | 622 | 619 | 630 | 609 |
| 630 | 626 | 630 | 615 | 630 | 643 | 639 | 620 | 606 |

The underlying distribution is assumed to be continuous.
   a. Use the Wilcoxon signed-rank test with $\alpha = 0.01$ to determine whether there is any evidence that the median spray height is less than 630 feet.
   b. Find the p value associated with this test.
   c. Why can't this procedure be used in a test concerning the mean spray height?

14.128 Biology and Environmental Science Pelicans have a long flat bill and expandable pouch, and use a spectacular dive-bomb-type plunge into the water to capture fish. A random sample of pelicans at Tigertail Beach on Marco Island was obtained. The height of a plunge was measured (in feet) for each bird, and the data are given in the following table.

| 41 | 43 | 48 | 46 | 44 | 40 | 43 | 47 | 40 |

The underlying distribution is assumed to be continuous. Use the Wilcoxon signed-rank test with $\alpha = 0.05$ to determine whether the median plunge height is different from 42 feet. Find the p value associated with this test. **PELICAN**

14.129 Swim at your own risk Recent research suggests that exposure to disinfection by-products (DBPs) in swimming pools may increase the risk of some cancers.[38] A study was conducted to determine whether exposure to DBPs in pool water increased toxic biomarkers. A random sample of adults was obtained and the concentration of four trihalomethanes (THMs, in μ/m^3) in exhaled breath was measured before and one hour after swimming. Assume that the distribution of pairwise differences is continuous and symmetric. Use the Wilcoxon signed-rank test with $\alpha = 0.05$ to determine whether there is any evidence that the median THM concentration is greater after swimming. **DBP**

14.130 Travel and Transportation A random sample of Georgia drivers renewing their vehicle registration was obtained. Using motor vehicle records, the number of miles driven in the past year was recorded for each. The data (in thousands of miles) were classified by whether the driver had an organ donor card and are given in the following table. **DRIVERS**

Organ donor card				
13.5	13.8	14.1	13.3	14.5

No organ donor card						
13.1	12.7	12.9	13.4	14.1	12.3	13.0

Assume the underlying distributions are continuous. Is there any evidence to suggest that the median number of miles driven is different for people who carry an organ donor card and for those who do not? Use the Wilcoxon rank-sum test with $\alpha = 0.05$.

14.131 Public Health and Nutrition Irradiation is a method of food preservation and is used to reduce bacteria and microorganisms in meat, poultry, and spices. The dose of an ionizing energy source varies according to the food type and is measured in gray (Gy), the amount of radiation absorbed. In a recent study, two types of radiation sources were compared. A random sample of spices was selected, and each machine was set for the maximum allowed dose of ionizing energy. The absorbed radiation was measured for each food sample (in kGy), and the data are given in the following table. **IONIZE**

X-ray generator							
22.6	25.9	26.3	22.2	23.5	24.9	27.2	22.9

Electron accelerator							
27.5	26.3	22.1	19.5	26.7	21.6	21.2	21.4

Assume the underlying distributions are continuous.

a. Is there any evidence to suggest that there is a difference in the absorbed radiation by machine? Use the Wilcoxon rank-sum test with $\alpha = 0.10$.

b. Find the p value associated with this test.

14.132 Manufacturing and Product Development The pressure for brewing espresso is necessary to help form cream and to distinguish this drink from strong drip coffee. Random samples of espressos from each of two commercial machines were obtained, and the pressure (in atmospheres) while brewing each cup was recorded. Assume the underlying distributions are continuous. Is there any evidence to suggest that the espresso setting on each machine produces a different median pressure? Use the Wilcoxon rank-sum test with $\alpha = 0.05$, and find the p value associated with this test. ▥ **ESPRESS**

14.133 Technology and the Internet Intel and Test Devices recently began shipping a smart baby body suit called Mimo.[39] This baby onesie has an Internet connection that allows parents to monitor an infant at all times. A random sample of infants (less than 1 year old) was obtained, and each was fitted with a Mimo. The suit was used to monitor sleeping respiratory rates (breaths per minute), and each child was classified by position: on the back or on the stomach. The data are given in the following table. ▥ **ONESIE**

Respiratory rate

Back					Stomach				
56	38	55	52	54	42	55	46	51	53
47	48	38	38	45	51	44	52	35	59
38	38	44	52	46	37	44	39	54	36
43	44	45	39	44	50	50	48	43	59
49	31	51	46	39	45	39	47	54	49
					47	43	50	34	54

Is there any evidence to suggest that the respiratory rates are different for these two sleeping positions? Assume that the underlying distributions are continuous and use a Wilcoxon rank-sum test with $\alpha = 0.01$.

14.134 Sports and Leisure In the sport of speed skiing, racers ski downhill in a straight line as fast as possible. These skiers use special equipment and can reach speeds of 200 km/h. A random sample of men speed skiers was obtained from two events during 2013, and the speed was recorded for each.[40] Assume the underlying distributions are continuous. Use the Wilcoxon rank-sum test to determine if there is any evidence to suggest that the median speed is different at these two events. Use $\alpha = 0.05$. ▥ **SPDSKI**

14.135 Manufacturing and Product Development The slate in a pool table may be one piece, but because it is prone to fracturing during transporting, it is often split into three slabs. Slate slabs from three pool-table manufacturers were randomly selected. The weight of each slab (in kg) was recorded, and the summary statistics are given in the following table.

Company	Sample size	Rank sum
Brunswick	$n_1 = 12$	$r_1 = 263.0$
AMF	$n_2 = 14$	$r_2 = 185.0$
Olhausen	$n_3 = 18$	$r_3 = 542.0$

Assume the underlying weight populations are continuous. Use the Kruskal-Wallis test with $\alpha = 0.05$ to determine whether there is any evidence to suggest that at least two slate weight populations are different.

14.136 Physical Sciences Amateur radio operators (hams) communicate with one another all over the world, and many help coordinate relief efforts during a natural disaster. Unlike the 5-watt power limit on a CB radio, ham radios can have as much as 1500 watts of power. A random sample of amateur radio operators was obtained from four states. The power on each transmitter was measured (in watts), and the summary statistics are given in the following table.

State	Sample size	Rank sum
New Hampshire	$n_1 = 15$	$r_1 = 611.0$
Alabama	$n_2 = 16$	$r_2 = 605.5$
Texas	$n_3 = 18$	$r_3 = 609.0$
California	$n_4 = 20$	$r_4 = 589.5$

Assume the underlying transmitter power distributions are continuous. Use the Kruskal-Wallis test with $\alpha = 0.025$ to determine whether there is any evidence to suggest that at least two transmitter power populations are different. Find the p value associated with this test.

14.137 Steam Clean Many steam cleaners advertise a very high internal boiler temperature that produces a *dry* steam vapor and a chemical-free cleaning system. Four types of home floor steam cleaners were identified, and a random sample of each type was obtained. The temperature of the steam (in °F) was measured at the nozzle for each device. Is there any evidence to suggest that at least two of the steam cleaner nozzle temperature population distributions are different? Assume that the underlying populations are continuous, and use the Kruskal-Wallis test with $\alpha = 0.01$. ▥ **STEAMER**

14.138 Manufacturing and Product Development Sixty-five countries account for approximately 99% of the world crude steel production.[41] Four of these countries were selected, and a random sample of their crude steel production per month (in thousand tonnes) was obtained. The data are given in the following table. ▥ **STEEL**

Greece		Hungary		Slovenia		Norway	
87	81	78	37	42	50	52	49
50	102	59	56	55	52	53	46
82	109	74	84	60	49	52	20
106	31	84	88	54	60	56	54

Assume the underlying distributions of crude steel production per month are continuous. Use the Kruskal-Wallis test with

$\alpha = 0.05$ to determine whether there is any evidence to suggest that at least two production populations are different.

14.139 Twice the Comfort   What was called the tallest hotel in North America opened recently in New York City. Well, it's really two hotels, near Times Square. Floors 6–33 are a Courtyard by Marriott, and floors 37–65 are a Residence Inn by Marriott.[42] The price per room on any floor is high, due to the location near Times Square and Central Park. A sample of guests was obtained, and each was classified by hotel within the hotel, Courtyard (C) or Residence Inn (R). The ordered observations are given in the following table. **HOTELS**

R	C	C	R	C	C	R	R	C	R	C	R	C	C	C

a. Use the runs test with $\alpha \approx 0.05$ to determine whether there is any evidence to suggest that the order of observations is not random with respect to hotel (within the hotel).

b. Find the p value associated with this test.

14.140 Fuel Consumption and Cars   A sample of automobiles entering a parking garage in Monterey before 6:00 A.M. on a Monday morning was obtained, and each was classified as either foreign (F) or domestic (D). The ordered observations are given in the following table. **AUTOS**

D	F	D	F	F	D	F	F	D	F	D	F	D	D	F
D	F	D	F	F	F	F	D	F	D	F	F	D	D	F

Is there any evidence to suggest that the order of automobiles entering the parking garage is not random? Use the runs test with the normal approximation and a 0.05 level of significance.

14.141 Public Health and Nutrition   The Mayflower Health Insurance Company is conducting a survey of customers to estimate the number who have received a routine physical exam during the past year. A sample of customers was contacted by phone during the evening, and each was classified by exam (E) or no exam (N). The sequence of observations is given in the following table. **EXAM**

E	E	N	E	E	E	N	E	E	N	E	N	E	E	N
N	E	N	N	N	E	E	E	N	E	E	E	N	E	E
N	N	E	N	E	N	E	N	E	N					

Is there any evidence to suggest that the order of observations is not random? Use the runs test with the normal approximation and $\alpha = 0.01$. Find the p value associated with this test.

14.142 Public Health and Nutrition   The head of the Ontario Medical Association recently suggested that companies should stop requiring a sick note from a doctor when an employee is ill and stays home.[43] The practice of requiring a note is often demeaning and discouraging, and sick employees should stay home. A sample of Canadian companies was obtained, and each was classified as requiring a sick note (R) or not requiring any justification (N). The sequence of observations is given in the following table. **OUTSICK**

N	N	R	N	R	N	R	R	R	R	N	N	R	R	R
N	R	N	N	R	R	R	R	N	N	N	N	N	N	N
N	R	N	N	R	N	N	N	R	R	R	N	N	N	N

Is there any evidence to suggest that the order of observations is not random? Use the runs test with the normal approximation and $\alpha = 0.05$. Find the p value associated with this test.

14.143 Travel and Transportation   Interstate 95 is a major north/south highway along the eastern part of the United States. Travelers on this road have a wide variety of hotel options at most of the major exits. A random sample of hotels at the intersection of I-95 and Highway 870 in Florida was obtained. The distance (miles) from I-95 and the starting room price (dollars) for each was recorded. The data are given in the following table. **TRAVEL**

Miles	0.2	0.3	0.8	1.1	2.7	1.5	1.4	2.3
Price	67	72	163	129	125	161	144	253

Compute Spearman's rank correlation coefficient and use this value to describe the relationship between these two variables.

14.144 Public Health and Nutrition   A study was conducted to examine the relationship between the volume and the quality of health care. A random sample of patients admitted to various hospitals in Indianapolis was obtained. After being discharged, each patient was asked to complete a questionnaire regarding the quality of care, and the results were evaluated to yield a quality score (x) between 1 (bad) and 50 (good). The total number of people in the hospital when the patient was admitted (y) was also recorded. Compute Spearman's rank correlation coefficient. What does this value suggest about the relationship between the volume and the quality of health care? **CARE**

14.145 Biology and Environmental Science   A recent study investigated the possibility of using moving cars to measure the amount and intensity of rainfall, rather than traditional rain gauges.[44] A random sample of automobiles was selected, and each was equipped with an automatic wiper system. Each car was subject to 10–15 minutes of rain in a controlled laboratory setting. The wiper frequency (x, w/min) and the rain intensity (y, mm/h) were recorded for each. Compute Spearman's rank correlation coefficient. Use this value to explain the relationship between wiper frequency and rain intensity and whether you believe wiper frequency might be used to predict rainfall total. **RAIN**

14.146 Medicine and Clinical Studies   The traditional method for people to test their blood glucose level is to prick a finger with a short needle, place a drop of blood on a test strip, and then use a special measuring device. The process can be painful. A group of scientists has developed a noninvasive, painless approach to measuring glucose level using infrared laser light. A random sample of healthy adults was obtained.

The glucose level was measured in each using the traditional method and the laser light approach (both measurements in milligrams per deciliter). Assume the underlying distribution of pairwise differences is continuous and symmetric. **GLUCOSE**

a. Use the Wilcoxon signed-rank test to determine whether there is any evidence that the median glucose levels are different for the two procedures. Use $\alpha = 0.01$.

b. Using your answer in part (a), do you believe the new approach is accurate and should be recommended for people with diabetes who test their glucose level regularly? Justify your answer.

EXTENDED APPLICATIONS

14.147 Manufacturing and Product Development Most home air conditioners built before 2010 used Freon gas to provide cooling in a typical evaporation cycle. However, Freon, which contains chlorine, was phased out of new equipment by 2010 and will not be used at all by 2020. R-410A, a mixture of difluoromethane and pentafluoroethane, is now the most common refrigerant in the United States. Although an air conditioner is, theoretically, a closed system, units usually lose refrigerant and must be recharged every few years. A random sample of 10,000-BTU air conditioners was obtained, and the amount of refrigerant (in pounds) in each was carefully measured. Each unit was recharged by a technician, and the amount of refrigerant in each was measured after the service. Suppose no assumptions can be made about the shape of the continuous distributions of before and after refrigerant weights. **REFRIG**

a. Conduct a sign test to determine whether there is any evidence that the median refrigerant weight before service is less than the median refrigerant weight after service. Use a significance level of $\alpha = 0.01$.

b. Based on your results, do you believe recharging an air conditioner really results in more refrigerant in the system? Why or why not?

14.148 Manufacturing and Product Development A nail gun is a handy tool, especially if you need to install a wood floor, replace a roof, attach a Venetian blind to a metal support, or if you hit your thumb a lot with a hammer. These devices propel a nail at incredible speeds and can save time and energy. A random sample of nail guns from four companies was obtained, and the speed of the nail was measured for each gun (in feet per second, fps). Assume the underlying speed distributions are continuous. **NAILGUN**

a. Use the Kruskal-Wallis test with $\alpha = 0.05$ to determine whether there is any evidence that at least two of the nail-gun speed population distributions are different.

b. Find bounds on the p value associated with this test.

c. Suppose you would like to purchase the brand of nail gun that fires a nail at the highest speed. Which brand would you choose, and why?

14.149 Marketing and Consumer Behavior A home-building company offers a variety of styles in only two exterior finishes: vinyl siding (V) or brick (B). Immediately following an advertising campaign explaining the advantages of vinyl

siding, a sample of consecutive customers was obtained, and each was classified by the exterior finish chosen. The ordered observations are given in the following table. **SIDING**

B	V	V	V	V	V	B	V	V	V	V	V	B	V	B	B

a. Conduct a runs test with $\alpha \approx 0.05$ to determine whether there is any evidence to suggest that the order of observations is not random with respect to exterior finish.

b. Using this sample, do you believe the advertising campaign was successful? Why or why not?

14.150 Psychology and Human Behavior The manager of concessions at Fenway Park believes that the number of runs scored by the Boston Red Sox (x) is related to the number of Fenway Franks consumed by fans (y). A random sample of nine-inning games was obtained. The hot dog and run totals were recorded. **FRANKS**

a. Construct a scatter plot for these data, and describe the relationship between the two variables.

b. Rank the observations in each sample separately, and find each difference between the ranks, d_i.

c. Find the sample correlation coefficient between the ranks.

d. Find Spearman's rank correlation coefficient using the d_i's.

e. Explain why the values in parts (c) and (d) are different.

f. Suppose the manager of concessions would like to sell as many hot dogs as possible. How may runs would he or she like the Red Sox to score? Why?

14.151 Sports and Leisure In 2012 the NCAA changed a rule concerning football kickoffs. Beginning in 2012, kickoffs took place at the 35-yard line rather than the 30-yard line. There has been some concern that this rule change would affect the return yards on each kickoff. A random sample of kickoff returns from the 2011 and 2013 college football season was obtained, and the yards were recorded for each.[45] Assume the underlying distributions are continuous. **KICKOFF**

a. Use a Wilcoxon rank-sum test to determine whether there is any evidence to suggest a difference in the population median return yards. Use $\alpha = 0.05$. Explain this result in the context of the rule change.

b. Use a two-sample t test to determine whether there is any evidence to suggest a difference in the population mean return yards. Assume equal population variances and use $\alpha = 0.05$. Explain this result in the context of the rule change.

c. Which test do you believe is more appropriate? Why?

d. Suppose more return yards per kickoff leads to more exciting games. Do you believe the rule change has made the games more exciting? Justify your answer.

14.152 Boarding Pain The process of boarding a flight in any airport can be slow and agonizing. Passengers with priority boarding often crowd the gate area, and passengers with carry-on bags jostle for position in line. Because faster boarding time saves airlines money and eases anxiety, airlines are experimenting with more efficient ways to seat passengers and prepare for departure. Consider the following boarding methods.[46] **BOARD**

DIY Boarding: An electronic scan of each passenger's face removes the need for a passport check.

Positive Boarding: Passenger data is used by the airline to send messages directly to travelers concerning gates and times.

Cheese Counter Method: Passenger numbers allow one person to board at a time.

Alternate Rows: Passengers board in alternate rows, rear to front, and window to aisle.

A random sample of similar flights and airplanes using one of these four methods was selected from London's Heathrow Airport. The boarding time (in minutes) for each flight was recorded. Assume the underlying boarding time distributions are continuous.

  a. Use the Kruskal-Wallis test with $\alpha = 0.05$ to determine whether there is any evidence that at least two of the boarding time populations are different.

  b. Find the exact p value for this test.

  c. What other statistical test might be appropriate in this case? Check the assumptions and conduct this test. Compare your results with part (a).

  d. Which of these boarding methods do you think is the most efficient? Why?

LAST STEP

 14.153 Were the levels of naphthalene different in the sediment near Alabama and Florida? Following the *Deepwater Horizon* oil spill in 2012, the EPA measured the amount of naphthalene in the sediment at various locations in the Gulf of Mexico. A random sample of these short-term measurements near Florida and Alabama was obtained from the EPA. The data are given in the following table. **NAPH**

Alabama

5.0	16.0	8.5	4.2	7.3	4.1
4.2	7.3	6.9	4.3		

Florida

4.2	14.0	7.9	8.3	25.0	8.3
4.1	8.5	8.3	8.3	12.0	4.2

Use the Wilcoxon rank-sum test to determine whether there is any evidence to suggest a difference in the concentration of naphthalene in the sediment near Alabama and Florida. Use $\alpha = 0.05$.

Notes and Data Sources

CHAPTER 0

1. *Cancer Facts & Figures 2013*, American Cancer Society, Atlanta, GA, 2013, accessed January 26, 2014, http://www.cancer.org/acs/groups/content/@epidemiologysurveilance/documents/document/acspc-036845.pdf.

2. *USA Today Snapshot*, accessed January 27, 2014, http://www.usatoday.com.

3. *Age of Autism*, accessed January 28, 2014, http://www.ageofautism.com/2013/11/vaccine-injury-exposed-in-not-a-coincidence-gardasil-video-from-canary-party.html.

4. WFTX-TV, Fort Myers/Naples, FL, accessed January 28, 2014, http://www.fox4now.com/features/4inyourcorner/20-whales-found-dead-near-Marco-Island-241712821.html.

5. *Food Safety News*, accessed January 28, 2014, http://www.foodsafetynews.com/2014/01/final-report-62-sickened-in-michigan-salmonella-outbreak/#.UugI3LQo6r8.

6. *Investopedia,* accessed January 28, 2014, http://www.investopedia.com/slide-show/car-recalls.

CHAPTER 1

1. *Huffpost Healthy Living,* accessed January 3, 2013, http://www.huffingtonpost.com/rachel-lincoln-sarnoff/arsenic-in-rice_b_1911303.html.

2. *Archives of Internal Medicine*, published online October 29, 2012, doi:10.1001/2013.jamaintrnmed.46.

3. A. M. Masri and A. S. Yusuf, High incidence rate of lung cancer in oil refinery countries, *Issues*, Vol. 1, No. 1, September 2012, accessed January 3, 2013, http://www.jomb.org/index.php?m=content&c=index&a=show&catid=27&id=18.

4. Oddee, accessed on January 3, 2013, http://www.oddee.com/item_98002.aspx.

5. CalmClinic, accessed January 4, 2013, http://www.calmclinic.com/anxiety/treatment/7-foods-that-fight-anxiety.

6. The Sixth Wall Koldcast, accessed February 1, 2014, http://blog.koldcast.tv/2011/koldcast-news/10-most-watched-television-finales-of-all-time.

7. *Best Health*, accessed January 6, 2013, http://www.besthealthmag.ca/embrace-life/sleep/why-he-sleeps-better-than-you.

8. usgovernmentspending.com, accessed January 5, 2013, http://www.usgovernmentspending.com/New_York_state_spending_pie_chart.

9. Statistics Canada, accessed January 7, 2013, http://www.statcan.gc.ca/tables-tableaux/sum-som/l01/cst01/econ11b-eng.htm.

10. Shark Research Committee, accessed January 7, 2013, http://www.sharkresearchcommittee.com/statistics.htm.

11. Mail Online, accessed January 7, 2013, http://www.dailymail.co.uk/health/article-2177182/Fake-tan-cause-fertility-problems-Users-warned-lotions-harm-unborn-babies-trigger-cancer.html.

12. University of Washington Medical Center, accessed January 7, 2013, http://depts.washington.edu/uwhep/patients/hcvfaq.htm.

13. *USA Today Tech,* accessed January 7, 2013, http://usatoday30.usatoday.com/tech/products/story/2012-04-05/pew-ebook-survey/54034026/1.

14. Center for American Progress, accessed January 7, 2013, http://www.americanprogress.org/issues/labor/news/2012/07/09/11898/5-facts-about-overseas-outsourcing.

15. Autobytel, accessed January 8, 2013, http://www.autobytel.com/auto-news/features/2012-plug-in-hybrid-and-electric-car-tax-credit-information-110759.

16. MSNBC, accessed January 8, 2013, http://www.nbcnews.com/travel/travelkit/disney-world-track-visitors-wireless-wristbands-1B7874882.

17. *Men's Health News,* accessed January 12, 2013, http://news.menshealth.com/diet-soda-depression/2013/01/11.

18. *USA Today,* accessed January 13, 2013, http://www.usatoday.com/story/travel/flights/2013/01/12/analysts-dreamliner-review-could-work-to-boeings-benefit/1828913.

CHAPTER 2

1. CNN, accessed January 14, 2003, http://www.cnn.com/2012/12/06/us/florida-python-hunt/index.html.

2. Nathan's World famous Beef Hot Dogs, accessed January 16, 2013, http://216.139.227.101/interactive/nath2012/pf/page_016.pdf.

3. Forbes, accessed January 22, 2013, http://www.forbes.com/sites/brucejapsen/2012/09/06/earths-drugstore.

4. Vermont Agency of Agriculture, accessed February 3, 2013, http://vermontdairy.com/learn/number-of-farms.

5. examiner.com, accessed February 3, 2013, http://www.examiner.com/article/ga-senator-to-retire-2014.

6. Nobel Prize Facts, accessed February 3, 2013, http://www.nobelprize.org/nobel_prizes/nobelprize_facts.html.

7. Canadian Lawyer Annual Corporate Counsel Survey, accessed February 3, 2013, http://www.canadianlawyermag. com/images/stories/Surveys/2012/corpcnslsurvey% 20-%20low%20res.pdf.

8. 2012 Think Tanks and Civil Societies Program, International Relations Program, University of Pennsylvania, 1/28/13.

9. Based on information from NYU Rudin Center Survey, November 26, 2012, http://wagner.nyu.edu/blog/ rudincenter/commuting-after-hurricane-sandy-survey-results.

10. National Highway Traffic Safety Administration, accessed February 4, 2013, http://www-fars.nhtsa.dot. gov/QueryTool/QuerySection/Report.aspx.

11. Colorado Avalanche Information Center, accessed February 5, 2013, https://avalanche.state.co.us/acc/acc_stats.php.

12. 2011 Texas Hunting Incidents Analysis, Texas Parks and Wildlife, accessed February 5, 2013, http://www.tpwd. state.tx.us/publications/pwdpubs/media/pwd_rp_ k0700_1124_2011.pdf.

13. Statistics Canada, accessed February 5, 2013, http:// www.statcan.gc.ca/tables-tableaux/sum-som/l01/cst01/ legal50c-eng.htm.

14. Waterfall Database, accessed February 5, 2013, http:// www.worldwaterfalldatabase.com/tallest-waterfalls/ total-height.

15. Exit Information Guide, accessed February 5, 2013, http://www.i95exitguide.com/lodging/index.php.

16. Mayo Clinic, accessed February 5, 2013, http://www. mayoclinic.com/health/cholesterol-levels/CL00001.

17. The Engineering Toolbox, accessed February 11, 2013, http://www.engineeringtoolbox.com/light-level-rooms- d_708.html.

18. debt.org, accessed February 11, 2013, http://www.debt. org/medical/emergency-room-urgent-care-costs.

19. Port of Tacoma, accessed February 11, 2013 http://www. portoftacoma.com/Page.aspx?cid=497.

20. Great Pumpkin Commonwealth, accessed February 11, 2013, http://greatpumpkincommonwealth.com.

21. Naples-Fort Myers PDF Files Library, accessed February 11, 2013, http://www.rosnet2000.com/rx_asp/rx_ trackpdf.asp?atid=NF.

22. Science Daily, accessed February 11, 2013, http://www. sciencedaily.com/releases/2012/09/120911151835.htm.

23. Results from the 2011 Maine Sea Scallop Survey, accessed February 11, 2013, http://www.maine.gov/dmr/rm/scallops/ 2011ScallopSurveyReport.pdf.

24. Cache Metals, accessed February 12, 2013, http://www. cachemetals.com/charts/historical-silver-chart.

25. Michigan Traffic Crash Facts, accessed February 12, 2013, http://publications.michigantrafficcrashfacts. org/2011/2011MTCF_vol1.pdf.

26. U.S. NRC Strategic Plan, Fiscal Years 2008–2013.

27. Adeveloper, accessed February 17, 2013, http://www.ade- veloper.com/CPweather/Current_Vantage_Pro_Plus.htm.

28. Blue Nile, accessed February 17, 2013, http://www. bluenile.com/diamond-search?track=NavEngLoo.

29. ESPN, accessed February 17, 2013, http://espn.go.com/ nhl/statistics/player/_/stat/major-penalties/sort/penalty Minutes.

30. University of Hawaii Sea Level Center, accessed February 17, 2013, http://uhslc.soest.hawaii.edu/data/rqd.

31. Harris Interactive, accessed February 18, 2013, Ellen Dances Her Way to the Top and Is America's Favorite TV Personality.

32. Construction Noise Impact Assessment, Biological Assessment Preparation, Advanced Training Manual Version 02-2012, Washington State Department of Transportation.

33. PwC, About Paying Taxes 2013, accessed February 18, 2013, http://www.pwc.com/gx/en/ptaxes/about-paying- taxes.jhtml.

34. Energy.Gov, accessed February 18, 2013, http://energy. gov/energysaver/articles/tips-heating-and-cooling.

CHAPTER 3

1. Canadian Pacific, Key Metrics, accessed October 28, 2012, http://www.cpr.ca/en/invest-in-cp/key-metrics/ Pages/default.aspx.

2. Denali National Park and Preserve, National Park Service, accessed October 28, 2012, http://www.nps.gov/ dena/planyourvisit/upload/Weather-2011-obs.pdf.

3. D. Baker, "Why Americans Should Work Less—The Way Germans Do," *The Guardian,* July 3, 2012.

4. Wheat Production by Country in 1000 MT, Index Mundi, http://www.indexmundi.com/agriculture/?commodity= wheat accessed October 30, 2012.

5. Steam Game Stats, accessed October 30, 2012, http:// store.steampowered.com/stats.

6. CO2Now.org, accessed October 31, 2012, http://co2now. org/Current-CO2/co2-now/noaa-mauna-loa-co2-data.html.

7. Department of Marine Resources, State of Maine, accessed November 1, 2012, http://www.maine.gov/dmr/ commercialfishing/historicaldata.htm.

8. ESPN MLB, accessed November 1, 2012, http://espn. go.com/mlb/stats/pitching/_/seasontype/2/league/al/ order/false.

9. National Oceanographic Data Center, accessed November 1, 2012, http://www.nodc.noaa.gov/dsdt/cwtg/satl.htm.

10. "Cooperative Learning and Statistics Instruction," *The Journal of Statistics Education*, Vol. 5, No. 3 (1997).

11. Official London 2012 Web site, accessed November 2, 2012, http://www.london2012.com/diving/event/ women-10m/index.html.

12. "Residents Claim Area Airplane Noise Has Ramped Up," *Post City Magazine,* August 2012.

13. International Skating Union, accessed November 4, 2012, http://www.isu.org.

14. U.S. Environmental Protection Agency, accessed November 4, 2012, http://www.epa.gov/otaq/crttst.htm.

15. Infoplease, accessed November 5, 2012, http://www.infoplease.com/us/government/presidential-pardons-1789-present.html.

16. National Oceanic and Atmospheric Administration, accessed November 5, 2012, http://www.swpc.noaa.gov/ftpdir/lists/ace2/201211_ace_swepam_1h.txt.

17. U.S. Army Corps of Engineers, accessed November 5, 2012, http://www.mvn.usace.army.mil/eng/edhd/wcontrol/miss.asp.

18. National Interagency Fire Center, accessed November 6, 2012, http://www.nifc.gov/fireInfo/fireInfo_stats_YTD2012.html.

19. Nextag, accessed November 6, 2012, http://www.nextag.com/12-mp-digital-camera/products-html.

20. Bureau of Transportation, accessed November 6, 2012, http://www.bts.gov/current_topics/national_and_state_bridge_data/html/bridges_by_state.html.

21. El Dorado Weather, accessed November 6, 2012, http://www.eldoradocountyweather.com/buoy/Chesapeake%20Bay/buoy-xhtml.php.

22. CBC News, accessed November 7, 2012, http://www.cbc.ca/news/5canada/edmonton/story/2012/08/07/alberta-highways-driving-charges-long-weekend.html.

23. Guinea Pig Manual, accessed November 8, 2012, http://www.guineapigmanual.com.

24. Minnesota Department of Natural Resources, accessed November 8, 2012, http://www.dnr.state.mn.us/faq/mnfacts/state_parks.html.

25. About.com, accessed November 8, 2012, http://brooklyn.about.com/od/brooklynbridge/f/How-Long-Does-It-Take-To-Walk-Across-The-Brooklyn-Bridge.htm.

26. AirlineReporter, accessed November 13, 2012, http://www.airlinereporter.com/2012/01/how-long-does-it-take-to-build-a-boeing-777.

27. ArkansasOnline, accessed November 13, 2012, http://www.arkansasonline.com/extras/databases/2010-2011ITBS/?appSession=516336608108138.

28. TruckersReport, accessed November 13, 2013, http://www.thetruckersreport.com/facts-about-trucks.

29. Rubik's Cube World Records, accessed November 13, 2013, http://www.recordholders.org/en/list/rubik.html.

30. G. Chiva-Blanch et al., "Dealcoholized Red Wine Decreases Systolic and Diastolic Blood Pressure and Increases Plasma Nitric Oxide," *Circulation Research*, Vol. 111, pp. 959–961, 2012.

31. Run Silent Dog Sled Trips, accessed November 24, 2012, http://www.runsilent.com.

32. Flightstats, accessed November 27, 2012, http://www.flightstats.com.

33. InflationData.com.

34. US Vending, accessed November 28, 2012, http://www.usvending.com.

35. Unisys, accessed November 28, 2012, http://weather.unisys.com/hurricane/atlantic/2012.

36. Science Daily, accessed November 28, 2012, http://www.sciencedaily.com/releases/2012/09/120911151835.htm.

37. Matrix of Mnemosyne, accessed November 28, 2012, http://www.matrixbookstore.biz/trial_jury.htm.

38. Wisconsin Department of Revenue, accessed November 28, 2012, http://www.revenue.wi.gov/delqlist/nmallA.htm.

39. WSOP, accessed November 30, 2012, http://www.angelfire.com/trek/proutsy.

40. General Electric Company, accessed November 30, 2012, http://www.ge-energy.com/wind.

41. TeensHealth, accessed November 30, 2012, http://kidshealth.org/teen/food_fitness/nutrition/caffeine.html#.

42. statista, accessed November 30, 2012, http://www.statista.com/statistics/208146/number-of-subscribers-of-world-of-warcraft.

43. U.S. Geological Survey.

44. Gasland, accessed November 30, 2012, http://www.gaslandthemovie.com/whats-fracking.

45. Bloomberg BusinessWeek, accessed November 30, 2012, http://images.businessweek.com/ss/09/10/1021_americas_25_top_selling_candies/26.htm.

46. Playbill, accessed November 30, 2012, http://www.playbill.com/celebritybuzz/article/75222-Long-Runs-on-Broadway.

CHAPTER 4

1. About.com, Contests and Sweepstakes, Rare McDonald's Monopoly Game Pieces for 2012, accessed February 19, 2013, http://contests.about.com/b/2012/09/27/rare-mcdonalds-monopoly-game-pieces-for-2012.htm.

2. CBC News, accessed February 21, 2013, http://www.cbc.ca/news/canada/manitoba/story/2013/02/21/mb-police-overtime-costs-tickets-winnipeg.html.

3. Verizon Wireless, accessed February 21, 2013, http://www.verizonwireless.com/wcms/consumer/explore/choosing-a-plan.html.

4. Centers for Disease Control and Prevention, accessed February 24, 2013, http://www.cdc.gov/flu/weekly.

5. Statistics Bureau, accessed February 25, 2013, http://www.stat.go.jp/english/data/roudou/154.htm#TAB.

6. World Health Organization, Immunization Profile—Canada, accessed February 25, 2013, http://apps.who.int/immunization_monitoring/en/globalsummary/countryprofileresult.cfm?c=can.

7. USA Business Review, accessed February 25, 2013, http://www.businessreviewusa.com/business_leaders/top-ten-us-convention-centers.

8. Infodocket, accessed February 25, 2013, http://www.infodocket.com/2012/08/15/fast-facts-comscore-offers-a-look-at-the-todays-tablet-consumer.

9. APPA, accessed February 25, 2013, http://www.americanpetproducts.org/press_industrytrends.asp.

10. Discovery Fit and Health, accessed February 25, 2013, http://health.howstuffworks.com/human-body/systems/circulatory/question593.htm.

11. Based on information from the Centers for Disease Control and Prevention, accessed February 25, 2013, http://www.cdc.gov/nchs/fastats/ervisits.htm.

12. Zagat, accessed February 28, 2013, http://www.zagat.com.

13. Catalyst, accessed February 28, 2013, http://www.catalyst.org/knowledge/working-parents.

14. U.S. Census Bureau, accessed February 28, 2013, http://www.census.gov/compendia/statab/cats/population/marital_status_and_living_arrangements.html.

15. Trail Count 2012, accessed February 28, 2013, http://www.sanjoseca.gov/DocumentCenter/View/5647.

16. Summary of the 2012 Great Goliath Grouper Count, accessed February 28, 2013, "The Marine Scene," *The Southwest Florida Sea Grant Newsletter.*

17. More States Embrace Gambling to Fight Budget Woes, *USA Today,* August 29, 2012.

18. Wireless Substitution: Early Release of Estimates from the National Health Interview Survey, January-June 2012, Blumberg, S.J. and Luke, J.V., Division of Health Interview Statistics, National Center for Health Statistics, accessed March 2, 2013, http://gigaom2.files.wordpress.com/2012/12/wireless201212.pdf.

19. Autoblog, accessed March 2, 2013, http://www.autoblog.com/2012/07/31/au-survey-results-suggests-men-rely-on-gps-more-than-women.

20. U.S. Census Bureau, accessed March 2, 2013, http://www.census.gov/hhes/www/cpstables/032012/health/h01_000.htm.

21. Consumerist, accessed March 2, 2013, http://consumerist.com/2013/01/08/list-of-companies-with-worst-customer-service-scores-is-full-of-familiar-names.

22. Statistics Canada, accessed March 2, 2013, http://www.statcan.gc.ca/tables-tableaux/sum-som/l01/cst01/legal22a-eng.htm.

23. Philadelphia Injury Attorney Blog, accessed March 2, 2013, http://www.philadelphiainjuryattorneyblog.com/bb_gun_accidents.

24. Alka-Seltzer Plus, accessed March 2, 2013, http://www.alkaseltzerplus.com/asp/coldflufacts.html.

25. CNN Money, accessed March 3, 2013, http://money.cnn.com/2012/11/27/pf/cyber-monday-sales/index.html.

26. Child Trends Databank, accessed March 3, 2013, http://www.childtrendsdatabank.org/?q=node/71.

27. Centers for Disease Control and Prevention, accessed March 3, 2013, http://www.cdc.gov/vaccines/stats-surv/nis/data/tables_2011.htm.

28. Traffic Injury Research Foundation, Winter Tires: A Review of Research on Effectiveness and Use, accessed March 3, 2013, http://www.tirf.ca/publications/PDF_publications/2012_Winter_Tire_Report_7.pdf.

29. Business Pundit, accessed March 4, 2013, http://www.businesspundit.com/5-jobs-with-the-highest-fatality-rates.

30. The Garden of Eaden, accessed March 4, 2013, http://gardenofeaden.blogspot.com/2012/05/what-is-most-poisonous-snake-in-india.html.

31. The Globe and Mail, accessed March 4, 2013, http://www.theglobeandmail.com/report-on-business/careers/careers-leadership/canadians-feeling-optimistic-about-employers-hiring-investment-plans/article7367200.

32. 2008 Bay Area Earthquake Probabilities, USGS, http://earthquake.usgs.gov/regional/nca/ucerf.

33. Gallup Politics, accessed March 4, 2013, http://www.gallup.com/poll/152021/conservatives-remain-largest-ideological-group.aspx.

34. Alzheimer's Association, accessed March 4, 2013, http://www.alz.org/alzheimers_disease_facts_and_figures.asp#quickfacts.

35. Connect Your Home, accessed March 4, 2013, http://www.connectyourhome.com/news_and_articles/featured-connectyourhome-articles/introducing-the-hopper-by-dish-network-changing-the-way-americans-watch-tv.

36. PGA Tour, accessed March 4, 2013, http://www.pgatour.com/stats.html.

37. NBA, accessed March 4, 2013, http://www.nba.com/statistics/default_all_time_leaders/AllTimeLeadersFTPQuery.html?top.

38. Investopedia, accessed March 4, 2013, http://www.investopedia.com/financial-edge/1012/boomers-staying-in-debt-to-retire-in-comfort.aspx#axzz2MaJSaIO1.

39. Weird News, accessed March 4, 2013, http://www.huffingtonpost.com/2012/10/15/alien-believers-outnumber-god_n_1968259.html.

40. American Bone Marrow Donor Registry, accessed March 4, 2013, http://www.abmdr.org/faq0.aspx.

41. New Guidelines: What to Do About Unexpected Positive Tuberculin Skin Test, Curley, C., accessed March 4, 2013, http://www.ccjm.org/content/70/1/49.full.pdf.

42. Bureau of Transportation Statistics, accessed March 5, 2013, http://apps.bts.gov/xml/ontimesummarystatistics/src/ddisp/OntimeSummaryDataDisp.xml.

43. FAS Military Analysis Network, accessed March 5, 2013, http://www.fas.org/man/dod-101/sys/missile/row/aspide.htm.

44. CBS News, accessed March 5, 2013, http://www.cbsnews.com/8301-504763_162-57483789-10391704/gluten-free-diet-fad-are-celiac-disease-rates-actually-rising.

45. Yantai Best Cellar Consulting Co., accessed March 5, 2013, http://www.winechina.com/html/2013/01/201301143358.html.

46. U.S. Census Bureau, accessed March 5, 2013, http://www.census.gov/hhes/socdemo/education/data/cps/2012/tables.html.

47. Generic Seeds, accessed March 5, 2013, http://www.genericseeds.com/vegetable-garden-seed/autumn-gold-pumpkin-seeds.

48. SlideShare, accessed March 5, 2013, http://www.slideshare.net/dkjnmd/savings-bond-training-webinar-15069817.

49. About.com, Paranormal Phenomena, accessed March 5, 2013, http://paranormal.about.com/od/paranormalbasics/a/news_120214n.htm.

50. Atlantic Cities, accessed March 5, 2013, http://www.theatlanticcities.com/commute/2012/02/rise-super-commuter/1351/#.

51. NPR, Compensating Organ Donors Becomes Talk of the Nation, accessed March 5, 2013, http://www.npr.org/blogs/health/2012/05/23/153373854/compensating-organ-donors-becomes-talk-of-the-nation.

52. AC Nielsen, Nestle, Dean Foods, accessed March 6, 2013, http://www.statisticbrain.com/coffee-creamer-industry-statistics.

53. UBS, Wells Fargo, Tobacco Vapor Electronic Cigarette Association, accessed March 6, 2013, http://www.statisticbrain.com/electronic-cigarette-statistics.

54. Statistical Summary of Commercial Jet Airplane Accidents Worldwide Operations 1959-2011, accessed March 6, 2013, http://www.boeing.com/news/techissues/pdf/statsum.pdf.

55. More Men Coloring Their Hair, *Los Angeles Times*, accessed March 6, 2013, http://articles.latimes.com/2012/jan/29/image/la-ig-mens-hair-color-20120129.

56. The 2012 Statistical Abstract, The National Data Book, U.S. National Highway Safety Administration, Traffic Safety Facts, accessed March 6, 2013, http://www.census.gov/compendia/statab/cats/transportation/motor_vehicle_accidents_and_fatalities.html.

CHAPTER 5

1. Florida Department of Environmental Protection, accessed March 6, 2013, http://www.dep.state.fl.us/geology/gisdatamaps/sinkhole_database.htm.

2. Aspirin may lower deadly skin cancer risk in women, Carroll, L., MSN, accessed March 11, 2013, http://vitals.nbcnews.com/_news/2013/03/11/17240612-aspirin-may-lower-deadly-skin-cancer-risk-in-women?lite.

3. Central Blood Bank, accessed March 12, 2013, http://www.centralbloodbank.org/program-bonus-points.asp.

4. HomeInsurance.com, accessed March 12, 2013, http://homeinsurance.com/auto-insurance/faqs/what-determines-my-auto-insurance-premium.php.

5. Camden hires first members of regional police force, Laday, J., January 18, 2013, nj.com, accessed March 12, 2013, http://www.nj.com/camden/index.ssf/2013/01/camden_hires_first_members_of.html.

6. Views of Government: Key Data Points, Pew Research Center, accessed March 12, 2013, http://www.pewresearch.org/2013/03/09/views-of-government-key-data-points.

7. Relax, Twinkies likely to live on, White, M. C., NBCNews.com, Business, accessed March 12, 2013, http://www.nbcnews.com/business/relax-twinkies-likely-live-1C7121954.

8. Canadian Medical Association, accessed March 13, 2014, http://www.cma.ca/multimedia/CMA/Content_Images/Inside_cma/Statistics/09GradCountry.pdf.

9. Nearly Half of Americans Drink Soda Daily, Gallup Wellbeing, Saad, L., http://www.gallup.com/poll/156116/Nearly-Half-Americans-Drink-Soda-Daily.aspx.

10. North Carolina Education Lottery, accessed March 18, 2013, http://www.nc-educationlottery.org/instant_detail.aspx?gn=332.

11. Based on the article, How may rides do you ride a day at Disney? Budget-Travel, accessed March 18, 2013, http://www.budgettravel.com/blog/how-many-rides-do-you-ride-on-a-day-at-disney,11602.

12. Kids in Danger, accessed March 18, 2013, http://www.kidsindanger.org/product-hazards/recalls.

13. Senate Bill No 1274, Offered January 14, 2013, Virginia State Senate, accessed March 18, 2013, http://leg1.state.va.us/cgi-bin/legp504.exe?131+ful+SB1274.

14. Based on the article, Burnout in Nurses Increases Patient Infections, posted in *Medical Malpractice*, accessed March 18, 2013, http://www.kinnardclaytonandbeveridge.com/blog/2012/09/burnout-in-nurses-increases-patient-infections.shtml.

15. The Oracle of Bacon, accessed March 18, 2013, http://oracleofbacon.org/center.php.

16. About.Com Weddings, accessed March 19, 2013, http://weddings.about.com/cs/bridesandgrooms/a/numofattendants.htm.

17. How to market ice wine that can't be called ice wine, Brijbassi, A., *The Globe and Mail*, accessed March 19, 2013, http://www.theglobeandmail.com/report-on-business/small-business/sb-growth/the-challenge/how-to-market-ice-wine-that-cant-be-called-ice-wine/article9669088.

18. Pizza Industry Analysis 2013-Cost and Trends, Franchise Help, accessed March 19, 2013, http://www.franchisehelp.com/industry-reports/pizza-industry-report.

19. DailyMed, accessed March 19, 2013, http://dailymed.nlm.nih.gov/dailymed/drugInfo.cfm?id=81505#section-6.

20. Report: Childhood Poverty High in Detroit, But Teen Pregnancy Down, CBS Detroit, accessed March 19, 2013, http://detroit.cbslocal.com/2013/01/24/report-childhood-poverty-high-in-detroit-but-teen-pregnancy-down.

21. Marquette University, Tuition and Costs, accessed March 19, 2013, http://www.marquette.edu/about/studenttuition.shtml.

22. Women Moving Millions, accessed March 19, 2013, http://www.womenmovingmillions.org/how-we-do-it/facts.

23. 2013 Illinois DUI Fact Book, accessed March 20, 2013, http://www.cyberdriveillinois.com/publications/pdf_publications/dsd_a118.pdf.

24. The Blog, Real Estate Center, Texas A&M University, accessed March 20, 2013, http://blog.recenter.tamu.edu/2013/01.

25. Health Policy Snapshot, Childhood Obesity, accessed March 20, 2013, http://www.rwjf.org/content/dam/farm/reports/issue_briefs/2012/rwjf72649.

26. Travel Impact Newswire, Canadian CA's Optimism Dips in Wake of Concerns over U.S. Economy, March 8, 2013, http://www.travel-impact-newswire.com/2013/03/canadian-cas-optimism-dips-in-wake-of-concerns-over-u-s-economy/#axzz2O6Y7tZeh.

27. Theguardian, accessed March 20, 2013, http://www.guardian.co.uk/media/2013/jan/07/tom-daley-splash-itv-tv-ratings.

28. U.S. Census Bureau, Mover Rate Reaches Record Low, Census Bureau Reports, accessed March 21, 2013, http://www.census.gov/newsroom/releases/archives/mobility_of_the_population/cb11-193.html.

29. United States Parachute Association, accessed March 21, 2013, http://www.uspa.org/AboutSkydiving/SkydivingSafety/tabid/526/Default.aspx.

30. Adult ADHD (Attention Deficit/Hyperactivity Disorder), Anxiety and Depression Association of America, accessed March 21, 2013, http://www.adaa.org/understanding-anxiety/related-illnesses/other-related-conditions/adult-adhd.

31. FBI Releases 2011 Bank Crime Statistics, Federal Bureau of Investigation, accessed March 22, 2013, http://www.fbi.gov/news/pressrel/press-releases/fbi-releases-2011-bank-crime-statistics.

32. Infoworld Tech Watch, accessed March 22, 2013, http://www.infoworld.com/t/cyber-crime/malware-infects-30-percent-of-computers-in-us-199598.

33. City-Data.com, accessed March 22, 2013, http://www.city-data.com/top2/c477.html.

34. Canada's most notorious highways, CBCnews, Canada, accessed March 22, 2013, http://www.cbc.ca/news/canada/story/2013/02/01/f-highways-dangerous.html.

35. Swimming committee recommends change to false-start rule, Johnson, G., NCAA, accessed March 22, 2013, http://www.ncaa.org/wps/wcm/connect/public/NCAA/Resources/Latest+News/2011/July/Swimming+committee+recommends+change+to+false-start+rule.

36. 4 popular myths about credit unions, NBCNEWS.com, accessed March 23, 2013, http://www.nbcnews.com/business/4-popular-myths-about-credit-unions-1C8936002.

37. Rides Injure More Than 4,400 Children Per Year, Pearson, C., Huff Post Parents, accessed May 2, 2013, http://www.huffingtonpost.com/2013/05/01/rides-injury-children_n_3187571.html.

38. Earthquake Facts and Statistics, USGS, accessed March 22, 2013, http://earthquake.usgs.gov/earthquakes/eqarchives/year/eqstats.php.

39. Don't Eat Out as Often, The Simple Dollar, accessed March 22, 2013, http://www.thesimpledollar.com/2012/07/07/dont-eat-out-as-often-188365.

40. Huff Post, accessed March 22, 2013, http://www.huffingtonpost.com/2013/03/13/religion-america-decline-low-no-affiliation-report_n_2867626.html.

41. FDA proposes tightening rules for heart defibrillators, Clarke, T., nbcnews.com, accessed March 23, 2013, http://vitals.nbcnews.com/_news/2013/03/22/17416241-fda-proposes-tightening-rules-for-heart-defibrillators?lite.

42. FDIC study: outrageous overdraft fees, Bruce, L., Bankrate.com, accessed March 23, 2013, http://www.bankrate.com/finance/investing/fdic-study-outrageous-overdraft-fees-1.aspx.

43. Who's in the tower? At some regional airports soon: Nobody. Ahler, M.M., and Marsh, R., CNN Travel, accessed March 24, 2013, http://www.cnn.com/2013/03/22/travel/faa-control-tower-closures/index.html?hpt=hp_bn3.

44. International Shark Attack File, Ichthyology at the Florida Museum of Natural History, accessed March 24, 2013, http://www.flmnh.ufl.edu/fish/sharks/statistics/statsus.htm.

45. China pledges to grow broadband coverage to 70% of households, Buckley, S., engadget, http://www.engadget.com/2013/02/28/china-pledges-to-grow-broadband-coverage-to-70-of-households.

46. Cruise Ship Industry Statistics, Statistic Brain, accessed March 24, 2013, http://www.statisticbrain.com/cruise-ship-industry-statistics.

47. Canada getting rid of the penny to save costs, CNN, accessed March 24, 2013, http://www.cnn.com/2012/03/30/business/canada-penny/index.html.

48. Rotten Tomatoes by Flixter, Les Miserables, accessed March 24, 2013, http://www.rottentomatoes.com/m/1083326-les_miserables.

49. Health Buzz: Many Americans Seek Medical Diagnoses Online, McMullen, L., USNews Health, accessed March 24, 2013, http://health.usnews.com/health-news/articles/2013/01/15/35-percent-of-americans-seek-medical-diagnosis-online.

50. Knee Replacement Surgery: Does It Lead to Weight Gain? A Physical Therapist, Inc., accessed March 24, 2013, http://aphysicaltherapistinc.com/2013/01/knee-replacement-surgery-does-it-lead-to-weight-gain.

51. Toddler meals swimming in salt, CNN Health, accessed March 24, 2013, http://thechart.blogs.cnn.com/2013/03/21/meals-and-snacks-for-toddlers-heavy-in-sodium/?hpt=he_c2.

CHAPTER 6

1. Report: AT&T Wins LTE Speed Race, MetroPCS stumbles, Parker, T., Fierce Broadband Wireless, accessed March 30, 2013, http://www.fiercebroadbandwireless.com/story/report-att-wins-lte-speed-race-metropcs-stumbles/2013-02-13.

2. How Long It Took Different Groups to Vote, *The New York Times,* accessed April 1, 2013, http://www.nytimes.com/interactive/2013/02/05/us/politics/how-long-it-took-groups-to-vote.html?_r=0.

3. Drugs to Treat Insomnia, WebMD, accessed April 2, 2013, http://www.webmd.com/sleep-disorders/insomnia-medications.

4. Tourism Kingston, accessed April 2, 2013, http://www.kingstoncanada.com/en/makeaconnection/parking.asp?_mid_=3232.

5. Choose Airplane Seats, About.com, accessed April 3, 2013, http://honeymoons.about.com/od/flying/ht/choosing_airplane_seats.htm.

6. sfist, accessed April 9, 2013, http://sfist.com/2013/03/07/map_average_rent_for_1br_in_san_fra.php.

7. TheKnot.com, as reported by CNN Money, accessed April 11, 2013, http://money.cnn.com/2013/03/10/pf/wedding-cost/index.html.

8. Versa-lok, accessed April 10, 2013, http://www.versa-lok.com/products/residential-commercial/standard.

9. Science and Technology Focus, Office of Naval Research, accessed April 10, 2013, http://www.onr.navy.mil/focus/ocean/water/salinity1.htm.

10. Kashi, accessed April 11, 2013, http://www.kashi.com/products/golean_crunchy_bars_chocolate_caramel.

11. Household Products Database, U.S. Department of Health and Human Services, accessed April 10, 2013, http://hpd.nlm.nih.gov/cgi-bin/household/search?tbl=TblChemicals&queryx=111-76-2.

12. Modern Parenthood, Parker, K., and Wang, W., Pew Research Social and Demographic Trends, accessed April 14, 2013, http://www.pewsocialtrends.org/2013/03/14/modern-parenthood-roles-of-moms-and-dads-converge-as-they-balance-work-and-family.

13. Johnson Holds Off Dale Jr. for Daytona 500 Win, Bruce, K., NASCAR News and Media, accessed April 14, 2013, http://www.nascar.com/en_us/news-media/articles/2013/02/24/55th-daytona-500-finish.html.

14. U.S. Forest Service, Forest Management, accessed April 14, 2013, http://www.fs.fed.us/forestmanagement/documents/sold-harvest/documents/1905-2012_Natl_Summary_Graph.pdf.

15. Kraft Foods, accessed April 13, 2013, http://www.kraftrecipes.com/Products/ProductInfoDisplay.aspx?SiteId=1&Product=4300095369.

16. *Sports Illustrated,* Golf.com, accessed April 14, 2013, http://blogs.golf.com/presstent/2013/04/14-year-old-guan-slapped-with-slow-play-penalty-.html.

17. PGA Tour Tracks Pace of Play, GolfTalkCentral, Hoggard, R., accessed April 14, 2013, http://www.golfchannel.com/news/golftalkcentral/pga-tour-tracks-pace-of-play.

18. Utah.com, accessed April 15, 2013, http://www.utah.com/attractions/kennecott.htm.

19. Some Neighborhoods Dangerously Contaminated by Lead Fallout, *USA Today,* Young, A., and Eisler, P., accessed April 22, 2013, http://usatoday30.usatoday.com/news/nation/story/2012-04-20/smelting-lead-contamination-soil-testing/54420418/1.

20. Near Earth Object Program, National Aeronautics and Space Administration, accessed April 22, 2013, http://neo.jpl.nasa.gov/stats/wise.

21. Great Smoky Mountains, National Park Service, accessed April 22, 2013, http://www.nps.gov/grsm/naturescience/black-bears.htm.

22. Statistics Canada, Table 303-0048, accessed April 22, 2013, http://www5.statcan.gc.ca/cansim/a47.

23. The Human Memory, accessed April 24, 2013, http://www.human-memory.net/types_sensory.html.

24. Fatsecret, accessed April 26, 2013, http://www.fatsecret. com/calories-nutrition/food/noodle-soup/sodium.

25. Economic Research Service, U.S. Department of Agriculture, accessed April 24, 2013, http://www.ers.usda. gov/data-products/china-agricultural-and-economic-data/ national-and-provincial-data.aspx#.UXrBIMoVwbk.

26. 2013 Fiat 500E, *Car and Driver*, accessed April 24, 2013, http://www.caranddriver.com/news/2013-fiat-500e-photos-and-info-news.

27. Fooducate, accessed April 24, 2013, http://www.fooducate. com/app#page=product&id=2344EFDA-E10B-11DF-A102-FEFD45A4D471.

28. Average New-Car Loan a Record 65 Months in Fourth Quarter, Reuters, accessed April 24, 2013, http://www. reuters.com/article/2013/03/05/us-auto-loans-idUS-BRE9240KQ20130305.

29. Average U.S. Family Spent Nearly $3000 on Gas Last Year, AOL Autos, accessed April 24, 2013, http:// autos.aol.com/article/gas-prices-fuel-economy-income-American.

30. Washington, DC: Metrorail Trivia, tripadvisor, accessed April 24, 2013, http://www.tripadvisor.com/Travel-g28970-c56842/Washington-Dc:District-Of-Columbia: Metrorail.Trivia.html.

31. U.S. Customs and Border Protection, CBP Border Wait Times, accessed April 24, 2013, http://apps.cbp.gov/bwt.

32. Chobani Products, accessed April 24, 2013, http:// chobani.com/products/faq.

CHAPTER 6 WEB

33. Less than Half of Brazilians Favor Hosting World Cup, Poll Shows, accessed May 30, 2014, http://www. reuters.com/article/2014/04/08/us-worldcup-brazil-idUSBREA3715H20140408.

34. Radio, accessed May 31, 2014, http://radiomagonline. com/digital_radio/55_percent_cars_uk.

35. CBS DC, accessed May 31, 2014, http://washington. cbslocal.com/2014/04/22/poll-63-percent-of-americans-against-personal-drones-being-allowed-in-us-airspace.

36. *The Christian Science Monitor,* accessed May 31, 2014, http://www.csmonitor.com/USA/Education/2014/0514/ Less-than-40-percent-of-12th-graders-ready-for-college-analysis-finds.

37. Climate Change Communication, accessed May 31, 2014, http://environment.yale.edu/climate-communication/ article/american-opinion-on-climate-change-warms-up.

38. Industry Canada, accessed May 31, 2014, http://www. ic.gc.ca/eic/site/lsg-pdsv.nsf/eng/h_hn01703.html.

39. Zero Hedge, accessed May 31, 2014, http://www.zero-hedge.com/news/2014-03-24/30-survey-results-sound-false-are-actually-true.

40. Live Science, accessed May 31, 2014, http://www. livescience.com/33657-8-weird-statistics.html.

41. DunwoodyPatch, accessed May 31, 2014, http:// dunwoody.patch.com/groups/business-news/p/reports-sams-club-will-cut-2-of-its-workforce.

42. Mother Jones, accessed May 31, 2014, http://www. motherjones.com/blue-marble/2014/04/poll-science-denial-big-bang-evolution-creationism-climate-change.

CHAPTER 7

1. Foreclosure compensation checks arrive, but Anger Some Homeowners, Myers, L., and Gardella. R., accessed May 2, 2013, http://openchannel.nbcnews.com/_news/ 2013/05/02/18022071-foreclosure-compensation-checks-arrive-but-anger-some-homeowners?lite.

2. Employee Benefits, Hilmar Cheese Company, accessed May 3, 2014, http://www.hilmarcheese.com/Careers/ Employee_Benefits.

3. 2013 Feed Composition Tables, accessed May 3, 2013, http://beefmagazine.com/nutrition/nutrient-values-300-cattle-feeds.

4. American lamb, accessed May 6, 2013, http://www. americanlamb.com/lamb-101.

5. Big catfish caught on Bartlett Lake, Hungeree, accessed May 6, 2013, http://hungeree.com/animals/big-catfish-caught-on-bartlett-lake.

6. autosnout.com, accessed May 6, 2013, http://www. autosnout.com/Cars-0-60mph-List.php.

7. 3M Half Marathon, accessed May 6, 2013, http://www. wetimeraces.com/RacingSystems/Results/2013/ 3M552/3M2013.htm#results::1367880139151.

8. Jockeys, Kentucky Derby (1875-2012), Churchill Downs, accessed May 6, 2013, http://www.kentuckyderby.com/ sites/kentuckyderby.com/files/u64720/Jockey%20 Records%2C%20Kentucky%20Derby_0.pdf.

9. Toronto Rock, accessed May 6, 2013, http://www.nll. com/stats/team_instance/243876?subseason=86199.

10. Kroger's New Weapon: Infrared Cameras, Yahoo Finance, accessed May 2, 2013, http://finance.yahoo.com/news/ krogers-weapon-infrared-cameras-011800830.html.

11. Fattest countries in the world revealed: Extraordinary graphic charts the average body mass index of men and women in every country (with some surprising results), Bond, A., Mail Online, accessed May 8, 2013, http:// www.dailymail.co.uk/health/article-2301172/Fattest-countries-world-revealed-Extraordinary-graphic-charts-average-body-mass-index-men-women-country-surprising-results.html.

12. Monsoon rainfall seen average in 2013, Reuters, accessed May 8, 2013, http://in.reuters.com/article/2013/04/26/ india-monsoon-minister-idINDEE93P03S20130426.

13. Dr. Pieter Tans, NOAA/ESRL (www.esrl.noaa.gov/gmd/ccgg/trends).

14. WebMD, Information and Resources, Typhoid Fever, accessed May 8, 2013, http://www.webmd.com/a-to-z-guides/typhoid-fever.

15. topendsports, accessed May 8, 2013, http://www.topendsports.com/testing/results/vertical-jump.htm.

16. NASA, Goddard Space Flight Center, Ozone Hole Watch, accessed May 8, 2013, http://ozonewatch.gsfc.nasa.gov/meteorology/annual_data.html.

17. Blog.GasBuddy.com, accessed May 13, 2013, http://blog.gasbuddy.com/posts/Almost-1-in-4-motorists-are-shopping-for-better-auto-insurance-45-percent-switch-providers/1715-539072-1765.aspx.

18. 70 percent of Americans track their health, but most go low tech, Heusser, K.M., Gigaom, accessed May 14, 2013, http://gigaom.com/2013/01/27/70-percent-of-americans-track-health-but-most-skip-tech-and-many-just-use-their-heads.

19. The Official Microsoft Blog, accessed May 14, 2013, http://blogs.technet.com/b/microsoft_blog/archive/2013/04/17/latest-security-intelligence-report-shows-too-many-pcs-lack-antivirus-protection.aspx.

20. Inside Movies, accessed May 14, 2013, http://insidemovies.ew.com/2013/05/04/movie-trailers-show-too-much-survey.

21. ClickZ, accessed May 15, 2013, http://www.clickz.com/clickz/news/2239608/more-consumers-order-food-online-using-a-smartphone-or-tablet.

22. Medical News Today, accessed May 15, 2013, http://www.medicalnewstoday.com/articles/260022.php.

23. Yale News, accessed May 15, 2013, http://yaledailynews.com/blog/2013/02/06/yales-admission-yield-rate-ranked-9th.

24. ESPN MLB, accessed May 15, 2013, http://espn.go.com/mlb/stats/team/_/stat/fielding.

25. Huffpost TV, accessed May 15, 2013, http://www.huffingtonpost.com/2013/03/20/motorola-mobility-41-percent-dvr-content-unwatched_n_2918929.html.

26. The Ocean trash Index, accessed May 15, 2013, http://www.oceanconservancy.org/our-work/marine-debris/2012-icc-data-pdf.pdf.

27. National Geographic, accessed May 15, 2013, http://travel.nationalgeographic.com/travel/countries/canada-facts.

28. Centers for Disease Control and Prevention, accessed May 17, 2013, http://www.cdc.gov/nchs/fastats/bodymeas.htm.

29. The Tropical Rainforest Biome, accessed May 17, 2013, http://prezi.com/xki3uz46_hmb/the-tropical-rainforest-biome.

30. TheResourceSolutions.com, accessed May 17, 2013, http://theresourcesolutions.com/topics/human_body/where_is_the_smallest_bone_in_your_body.htm.

31. All About Vinyl Swimming Pool Liners, accessed May 17, 2013, http://www.poolinfo.com/Vinyl-Liners.htm.

32. hospitality.net, accessed May 19, 2013, http://www.hospitalitynet.org/news/4059582.html.

33. nbcnews, accessed May 18, 2013, http://cosmiclog.nbcnews.com/_news/2013/05/17/18326731-buggy-hordes-of-cicadas-sighted-in-virginia-but-new-york-not-yet?lite.

CHAPTER 8

1. Maryland Department of Natural Resources, accessed May 20, 2013, http://news.maryland.gov/dnr/2013/02/05/2013-midwinter-waterfowl-survey-results-are-in.

2. March Bike Pilot Evaluation and Initiatives Update, San Francisco BART, accessed May 20, 2013, http://www.bart.gov/docs/Bike_Pilot_2_Board_Presentation.pdf.

3. nbcnews.com, accessed May 20, 2013, http://vitals.nbcnews.com/_news/2013/05/20/18377639-new-sleep-pill-may-be-unsafe-at-higher-doses-fda-review-suggests?lite.

4. Manatee County Administration Dashboards, accessed May 20, 2013, http://public.mymanatee.org/dashboards/view?view=Public%20Safety&dashboard_param=65.

5. Mamma Chia, accessed May 21, 2013, http://www.mammachia.com/chia-squeeze.

6. Statistics Canada, accessed May 21, 2013, http://www.statcan.gc.ca/pub/82-625-x/2013001/article/11779-eng.htm.

7. The Geyser Observation and Study Association, accessed May 21, 2013, http://www.geyserstudy.org/geyser.aspx?pGeyserNo=CLIFF.

8. Utah Geological Survey, accessed May 22, 2013, http://geology.utah.gov/emp/energydata/coaldata.htm.

9. Huff Post Chicago, accessed May 22, 2013, http://www.huffingtonpost.com/2013/05/20/ferris-wheel-world-record_n_3306361.html.

10. Canadian Geographic, accessed May 22, 2013, http://geography.about.com/gi/o.htm?zi=1/XJ&zTi=1&sdn=geography&cdn=education&tm=95&f=10&su=p284.13.342.ip_&tt=2&bt=0&bts=0&zu=http%3A//www.canadiangeographic.ca/magazine/MA06/indepth/justthefacts.asp.

11. National Geographic, accessed May 22, 2013, http://animals.nationalgeographic.com/animals/mammals/koala.

12. Boeing, accessed May 22, 2013, http://www.newairplane.com/747/design_highlights/#/home.

13. Co.Exist, accessed May 22, 2013, http://www.fastcoexist.com/1681494/a-new-ultra-cheap-led-light-looks-and-acts-like-an-incandescent-bulb.

14. USDA National Nutrient Database for Standard Reference, accessed May 23, 2013, http://ndb.nal.usda.gov/ndb/search/list.

15. Imagine 2050, accessed May 23, 2013, http://imagine2050.newcomm.org/2013/04/01/john-morton-resigns-to-open-gourmet-snail-farm.

16. Watershed Publishing, Marketing Charts, accessed May 27, 2013, http://www.marketingcharts.com/wp/interactive/social-networking-eats-up-3-hours-per-day-for-the-average-american-user-26049.

17. Cattle.com, accessed May 31, 2013, http://www.cattle.com/markets/archive.aspx?code=NW_LS720.

18. Bike Rumor, accessed May 31, 2013, http://www.bikerumor.com/2012/08/26/2013-trek-bikes-actual-weights-for-road-mountain-bikes.

19. Sutherland, J., Liu, G., Repin, N., and Crump, T., Hospital funding policies: Average length of stay for congestive heart failure, *BCHeaPR Study Data Bulletin #15*. Vancouver: UBC Centre for Health Services and Policy Research, 2013.

20. Mirtle, J., *A Hockey Journalist's Blog*, accessed May 31, 2013, http://mirtle.blogspot.com/2013/01/2013-nhl-teams-by-weight-height-and-age.html.

21. Vehicles, accessed May 31, 2013, http://vehicles-us.blogspot.com/2013/02/longer-average-length-of-vehicle.html.

22. MSN Money, accessed May 31, 2013, http://money.msn.com/now/post.aspx?post=f3c6a0cf-33f1-43fb-9072-6e9375392905.

23. WDW Mobile App, accessed May 31, 2013.

24. Polar Bear Science, accessed May 31, 2013, http://polarbearscience.com/2013/03/10/new-chukchi-sea-polar-bear-survey-exciting-preliminary-results.

25. NBC News Travel, accessed June 23, 2013, http://www.nbcnews.com/travel/roads-less-crowded-4th-july-aaa-says-6C10393641.

26. International Business Times, accessed June 23, 2013, http://www.ibtimes.com/solar-activity-cycle-peaks-2013-more-solar-flares-expected-later-year-1302041.

27. Huffpost Healthy Living, accessed June 22, 2013, http://www.huffingtonpost.com/2013/06/11/healthy-restaurant-meals_n_3421607.html.

28. CRFA's 2013 Canadian Chef Survey, accessed June 22, 2013, http://www.crfa.ca/pdf/chefsurvey_2013_english.pdf.

29. Hyde Park Neighborhood Association, accessed June 22, 2013, http://www.austinhydepark.org/2013/04/hpna-2013-survey-results.

30. *The Wichita Eagle,* accessed June 22, 2013, http://www.kansas.com/2013/02/14/2677088/kansas-considering-restraints.html.

31. 2013 Global Management Education Graduate Survey, GMAC, accessed June 25, 2013, http://www.gmac.com/~/media/Files/gmac/Research/curriculum-insight/gmegs-2013-stats-brief.pdf.

32. Aupair World, Press kit 2012, accessed June 25, 2013.

33. Final Report, Survey of Canadians on Privacy-Related Issues, accessed June 25, 2013, http://www.priv.gc.ca/information/por-rop/2013/por_2013_01_e.pdf.

34. U.S. Department of Agriculture, Economic Research Service, U.S. milk production and related data (quarterly), accessed June 27, 2013, http://www.ers.usda.gov/data-products/dairy-data.aspx#.UcxsLJzYl8k.

35. Golden Isles of Georgia Tide Tables, accessed June 27, 2013, http://www.gacoast.com/tide1.html.

36. intellicast.com, accessed June 27, 2013, http://www.intellicast.com/Local/Observation.aspx?chart=Relative%20Humidity.

37. Bloomberg, accessed June 27, 2013, http://www.bloomberg.com/news/2012-06-07/gm-boosts-2013-volt-mileage-rating-beating-toyota-prius.html.

38. cycling news, accessed June 27, 2013, http://www.cyclingnews.com/races/tour-de-france-2012/stage-4/results.

39. Bureau of Transportation Statistics, accessed June 27, 2013, http://www.rita.dot.gov/bts/sites/rita.dot.gov.bts/files/subject_areas/airline_information/taxi_out_and_other_tarmac_times/tarmac_more_than_3_hours_april_2013.html.

40. Critical Zone Observatories, accessed June 27, 2013, http://criticalzone.org/national/data/dataset/2428.

41. The Numbers, accessed June 27, 2013, http://www.thenumbers.com/movies/records/allbudgets.php.

42. infoplease, accessed June 27, 2013, http://www.infoplease.com/encyclopedia/history/great-wall-china.html.

43. Today Health, accessed June 27, 2013, http://www.today.com/health/new-rules-make-school-junk-food-free-zone-6C10467557.

44. calorie count, accessed June 27, 2013, http://caloriecount.about.com/calories-sports-drinks-ic1444.

45. CBC British Columbia, accessed June 28, 2013, http://www.cbc.ca/bc/news/investigates.

46. Southern California Coastal Ocean Observing System, accessed June 28, 2013, http://www.sccoos.org/query/.

47. Today Entertainment, accessed June 28, 2013, http://www.today.com/entertainment/tv-binge-watching-mostly-harmless-addiction-6C10476290.

48. *Scientific American*, accessed June 28, 2013, http://www.scientificamerican.com/article.cfm?id=1000-mph-car-land-speed-record.

49. MIX 104.1, accessed June 28, 2013, http://mix1041.cbslo-cal.com/2013/05/21/do-you-hide-money-from-your-spouse.

50. Today Health, accessed June 28, 2013, http://www.today.com/health/eating-carbs-may-make-you-crave-even-more-carbs-6C10484514.

CHAPTER 9

1. *Insurance Journal*, accessed June 30, 2013, http://www.insurancejournal.com/news/national/2013/03/20/285243.htm.

2. Roy Morgan Research, accessed July 1, 2013, http://www.roymorgan.com/findings/australian-motorists-drive-average-15530km-201305090702.

3. The College Board, accessed July 1, 2013, http://professionals.collegeboard.com/testing/sat-reasoning/scores/averages.

4. Entertainment Software Association, accessed July 1, 2013, http://www.theesa.com/facts/gameplayer.asp.

5. *The Boston Globe*, accessed July 1, 2013, http://www.bostonglobe.com/sports/2013/06/08/why-baseball-games-take-long/wikaeRMGatBDGDefpbFE1H/story.html.

6. Government of Canada, Citizenship and Immigration Canada, accessed July 1, 2013, http://www.cic.gc.ca/english/information/times/canada/cit-processing.asp.

7. *Today Health*, accessed July 1, 2013, http://www.today.com/health/later-you-stay-more-you-eat-study-shows-6C10488450.

8. Institute for Energy Research, accessed July 3, 2013, http://www.instituteforenergyresearch.org/2013/06/03/americas-green-energy-problems-defective-solar-panels.

9. *University of Florida News*, accessed July 3, 2013, http://news.ufl.edu/2013/05/15/aspirin-2.

10. BuildTheBridgeNowNY.org, accessed July 3, 2013, http://buildthebridgenowny.org/faq.

11. Waikiki Roughwater Swim, accessed July 3, 2013, http://www.waikikiroughwaterswim.com/2012/2012OverallResults.html.

12. Livingstrong.com, accessed July 3, 2013, http://www.livestrong.com/article/5966-need-antioxidants-dark-chocolate.

13. National Fire Protection Association, accessed July 3, 2013, http://www.nfpa.org/codes-and-standards/document-information-pages?utm_source=feedburner&utm_medium=feed&utm_campaign=Feed%3A+nfpacodes andstandards+%28NFPA+codes+and+standards%29.

14. Canadian Tourism Human Resource Council, accessed July 3, 2013, http://cthrc.ca/en/member_area/member_news/a_panel_of_canadian_hotel_industry_it_executives_to_look_at_hotel_technology_trends.

15. drugwatch, accessed July 3, 2013, http://www.drugwatch.com/2013/02/19/hip-implants-women.

16. NOAA, accessed July 5, 2013, http://www.noaanews.noaa.gov/stories2013/20130618_deadzone.html.

17. Mayo Clinic, accessed July 5, 2013, http://www.mayoclinic.com/health/complete-blood-count/MY00476/DSECTION=results.

18. ESPN NFL, accessed July 5, 2013, http://espn.go.com/nfl/draft2013/story/_/id/8953829/2013-nfl-draft-five-year-nfl-combine-averages?utm_source=twitterfeed&utm_medium=twitter.

19. U.S. Environmental Protection Agency, accessed July 7, 2013, http://www.epa.gov/otaq/tcldata.htm.

20. International Coffee Organization Monthly Coffee Market report, May 2013, accessed July 7, 2013, http://www.ico.org/documents/cy2012-13/cmr-0513-e.pdf.

21. Net Index, accessed July 7, 2013, http://www.netindex.com/download/#.

22. Locksmith Ledger, accessed July 7, 2013, http://www.locksmithledger.com/article/10881122/locksmith-services-pricing-2013.

23. ConAgraFoods, accessed July 7, 2013, http://www.eggbeaters.com/healthy-eating-habits-about-us/compare-egg-nutrition.

24. Livingstrong.com, accessed July 7, 2013, http://www.livestrong.com/article/490608-how-much-iodine-is-in-milk.

25. NBC News Science, accessed July 8, 2013, http://www.nbcnews.com/science/antarcticas-hidden-lake-vostok-found-teem-life-6C10561955.

26. Kawasaki, accessed July 8, 2013, http://www.kawasaki.com/Products/Product-Specifications.aspx?scid=23&id=708.

27. *The New York Times*, Cheating Ourselves of Sleep, accessed July 9, 2013, http://well.blogs.nytimes.com/2013/06/17/cheating-ourselves-of-sleep.

28. gizmag, accessed July 9, 2013, http://www.gizmag.com/how-does-elon-musk-hyperloop-work/27757.

29. CNN, accessed July 10, 2013, http://www.cnn.com/2013/07/09/world/americas/toronto-train-flooding.

30. Agricorp, accessed July 10, 2013, http://www.agricorp.com/en-ca/Programs/ProductionInsurance/ForageRain-fall/Pages/RainfallDailyData.aspx?StationID=7021293&Year=2013&LANG=EN.

31. ESPN NFL, accessed July 11, 2013, http://espn.go.com/nfl/qbr/_/year/2011, 2012.

32. J2SKI, accessed July 11, 2013, http://www.j2ski.com/ski-chat-forum/posts/list/14182.page.

33. FindTheBest, accessed July 11, 2013, http://planes.findthebest.com/l/241/Airbus-A350-900.

34. Communications Satellites, accessed July 11, 2013, http://www.satellites.spacesim.org/english/function/communic/index.html.

35. Union of Concerned Scientists, accessed July 11, 2013, http://www.ucsusa.org/nuclear_weapons_and_global_security/space_weapons/technical_issues/ucs-satellite-database.html.

36. U.S. Energy Information Administration, accessed July 12, 2013, http://www.eia.gov/todayinenergy/detail.cfm?id=11831.

37. NOAA Fisheries, Office of Protected Resources, accessed July 13, 2013, http://www.nmfs.noaa.gov/pr/species/fish/bluefintuna.htm.

38. Research and Innovative Technology Administration, Bureau of Transportation Statistics, accessed July 13, 2013, http://apps.bts.gov/xml/ontimesummarystatistics/src/dstat/OntimeSummaryAirtimeData.xml.

39. Cleveland.com, Northeast Ohio, accessed July 13, 2013, http://www.cleveland.com/north-olmsted/index.ssf/2013/06/north_olmsted_mayors_court_app.html.

40. CNBC, accessed July 14, 2013, http://www.cnbc.com/id/100804842.

41. OECD, Health Policies and Data, accessed July 14, 2013, http://www.oecd.org/health/health-systems/oecdhealthdata2013-frequentlyrequesteddata.htm.

42. Rothschild, P., Grabar, S., Le Du, B., Temstet, C., Rostaqui, O., and Brezin, A. (2013), Patients subjective assessment of the duration of cataract surgery: A case series, BMJ Open 3: e002497, doi:10.1136/bmjopen-2012-002497.

43. NBC News, accessed July 14, 2013, http://www.nbcnews.com/travel/worlds-largest-building-opens-china-6C10578538.

44. Baseball Press, accessed July 15, 2013, http://www.baseballpress.com/article.php?id=1269.

45. NBC News Science, accessed July 15, 2013, http://www.nbcnews.com/science/big-lionfish-found-disturbing-depths-6C10631330.

46. SportStats, accessed July 15, 2013, http://www.sportstats.ca/displayResults.xhtml?racecode=103723.

47. National Weather Service, Climate Prediction Center, accessed July 15, 2013, http://www.cpc.ncep.noaa.gov/products/analysis_monitoring/cdus/pastdata/palmer.

48. U.S. Geological Survey Earthquake Hazards Program, accessed July 15, 2013, http://earthquake.usgs.gov/earthquakes/feed/v1.0/csv.php.

49. Diabetes Self-Management, accessed July 15, 2013, http://static.diabetesselfmanagement.com/pdfs/DSM0310_012.pdf.

50. Time News Feed, accessed July 13, 2013, http://newsfeed.time.com/2013/05/17/u-s-postal-service-releases-list-of-worst-cities-for-dog-attacks.

51. US News Health, accessed July 16, 2013, http://health.usnews.com/health-news/articles/2013/01/15/35-percent-of-americans-seek-medical-diagnosis-online.

52. *The Wall Street Journal*, accessed July 6, 2013, http://online.wsj.com/article/PR-CO-20130515-909947.html.

53. Today Money, accessed July 25, 2013, http://www.today.com/money/sometimes-boss-really-psycho-6C10732488.

54. NBC News Health, accessed July 25, 2013, http://www.nbcnews.com/health/taller-women-more-likely-get-cancer-large-study-finds-6C10746890.

55. First Read on NBC News, accessed July 25, 2013, http://firstread.nbcnews.com/_news/2013/07/24/19661695-poll-majority-more-worried-us-surveillance-goes-too-far?lite.

56. NBC News Health, accessed July 25, 2013, http://www.nbcnews.com/health/full-moon-can-mess-your-sleep-new-study-finds-6C10743979.

57. CBC News Saskatchewan, accessed July 25, 2013, http://www.cbc.ca/news/canada/saskatchewan/story/2013/07/25/business-rich-canada.html.

58. NBC News Science, accessed July 26, 2013, http://www.nbcnews.com/science/mysterious-hum-driving-people-crazy-around-world-6C10760872.

59. First Read on NBC News.com, accessed July 26, 2013, http://firstread.nbcnews.com/_news/2013/07/24/19644154-nbcwsj-poll-faith-in-dc-hits-a-low-83-percent-disapprove-of-congress.

60. CNN Tech, accessed July 26, 2013, http://www.cnn.com/2013/07/26/tech/web/impact-technology-handwriting/index.html?hpt=hp_t3.

61. Inside Higher Ed, accessed July 26, 2013, http://www.insidehighered.com/news/2012/11/12/report-shows-growth-international-enrollments-study-abroad.

62. Mingtiandi, accessed July 26, 2013, http://www.mingtiandi.com/real-estate/outbound-investment/20130707/the-chinese-are-coming-just-not-all-that-quickly.

63. CNN Money, accessed July 26, 2013, http://money.cnn.com/2013/04/26/news/economy/health-care-cost/index.html.

64. CBS New York, accessed July 26, 2013, http://newyork.cbslocal.com/2013/05/24/west-islip-students-accused-of-using-facebook-to-cheat.

65. U.S. Energy Information Administration, accessed July 28, 2013, http://www.eia.gov/analysis/studies/worldshalegas/pdf/fullreport.pdf?zscb=45967897.

66. SoLux 4100K 50W Specifications, accessed July 28, 2013, http://www.solux.net/ies_files/SoLux%204100K%2050W%20Specs.pdf.

67. Centers for Medicare & Medicaid Services, accessed July 28, 2013, http://www.cms.gov/Research-Statistics-Data-and-Systems/Statistics-Trends-and-Reports/Medicare-Provider-Charge-Data/Outpatient.html.

68. Google Public Data, Eurostat, accessed July 28, 2013, http://www.google.com/publicdata/explore?ds=ml9s8a132hlg.

69. Estes Rockets, accessed July 28, 2013, http://www.estesrockets.com/rockets/pro-series/009701-ventris-ps-rocket.

70. *The New York Times*, Health, Science, accessed July 29, 2013, http://well.blogs.nytimes.com/2013/05/30/for-new-doctors-8-minutes-per-patient/?_r=0.

71. First Read on NBC News.com, accessed July 29, 2013, http://firstread.nbcnews.com/_news/2013/07/28/19700822-ahead-of-budget-battle-more-americans-say-sequester-has-hurt?lite.

72. Genetics Home Reference, U.S. National Library of Medicine, accessed July 29, 2013, http://ghr.nlm.nih.gov/condition/parkinson-disease.

73. Association of American Railroads, accessed July 29, 2013, https://www.aar.org/safety/Documents/Freight%20Railroads%20Safely%20Moving%20Crude%20Oil.pdf.

74. Hot Air, accessed July 29, 2013, http://hotair.com/archives/2013/07/26/important-update-from-msnbc-detroit-is-fast-becoming-americas-most-libertarian-city.

75. Department of Environment and Heritage Protection, Queensland Government, accessed July 29, 2013, http://www.ehp.qld.gov.au/coastal/monitoring/waves/index.php.

76. Department of Water Resources, California Data Exchange Center, accessed July 30, 2013, http://cdec.water.ca.gov/cdecapp/resapp/resDetailOrig.action?resid=CAS.

77. Libertystone Hardscaping Systems, accessed July 30, 2013, http://www.liberty-stone.net/wall-products-stonevista-standard.html.

78. Catalyst, accessed July 30, 2013, http://www.catalyst.org/knowledge/women-ceos-fortune-1000.

CHAPTER 10

1. Marketing charts, accessed July 31, 2013, http://www.marketingcharts.com/wp/television/are-young-people-watching-less-tv-24817.

2. NBC News Business, accessed August 1, 2013, http://www.nbcnews.com/business/growing-copper-theft-epidemic-sweeping-us-6C10791941.

3. Breakdown of Deposit Savings Periods, Sam Beattie, Press Association Mediapoint, June 19, 2013.

4. Average cost of flood insurance policy in New Orleans area continues to rise, Christian Moises, New Orleans City business, March 14, 2013.

5. Ministry of Health, Surgical Wait Times, British Columbia, accessed August 1, 2013, http://www.health.gov.bc.ca/swt/faces/Wooden.jsp.

6. Top 10 Foods Highest in Iron, Health Alicious Ness.com, accessed August 2, 2013, http://www.healthaliciousness.com/articles/food-sources-of-iron.php.

7. States seek federal crackdown on mussel invaders, *USA Today*, accessed August 5, 2013, http://www.usatoday.com/story/news/nation/2013/08/05/mussel-invaders-cost-billions/2618669.

8. Forbes, accessed August 5, 2013, http://www.forbes.com/sites/tedreed/2013/05/18/will-americans-new-boarding-process-work-it-failed-at-virgin-america.

9. NHL.com, Rules, accessed August 5, 2013, http://www.nhl.com/ice/page.htm?id=26286.

10. Hardocp, accessed August 5, 2013, http://www.hardocp.com/article/2013/03/12/crysis_3_video_card_performance_iq_review/5#.Uf_7um0nb90.

11. Urban bees for hire: A thriving hive business, Reuters, accessed August 6, 2013, http://www.reuters.com/article/2013/06/19/idUS248215678920130619.

12. *Journal of Biomedical Graphics and Computing*, Volume 3, Number 3, 2013, accessed August 6, 2013, www.sciedu.ca/jbgc.

13. What's on your playlist affects how you walk, NBC News Health, accessed August 7, 2013, http://www.nbcnews.com/health/whats-your-playlist-affects-how-you-walk-6C10596898.

14. Genworth, accessed August 7, 2013, https://www.genworth.com/corporate/about-genworth/industry-expertise/cost-of-care.html.

15. CBC News, British Columbia, accessed August 7, 2013, http://www.cbc.ca/news/canada/british-columbia/story/2013/08/06/bc-coliform-false-creek.html.

16. CNN Travel, accessed August 7, 2013, http://www.cnn.com/2013/08/02/travel/20-travel-mistakes/index.html?hpt=hp_bn10.

17. *The New York Times,* Business Day, accessed August 8, 2013, http://economix.blogs.nytimes.com/2013/08/01/millennials-in-their-parents-basements/?_r=0.

18. *The New York Times,* Business Day, accessed August 8, 2013, http://www.nytimes.com/2013/07/16/business/airports-on-the-border-make-room-for-canadian-flyers.html.

19. NBC News Health, accessed August 9, 2013, http://www.nbcnews.com/health/kids-exposure-secondhand-smoke-drops-except-among-those-asthma-6C10867951.

20. Fox News, accessed August 9, 2013, http://www.foxnews.com/politics/2013/05/02/poll-shows-2-percent-voters-think-armed-revolution-might-be-needed.

21. MCB Mobile, accessed August 9, 2013, http://www.mcbmobilepayments.com/96-people-to-have-phones-39-the-internet-by-the-end-of-2013-un/#.UgUYo20nb90.

22. Public Policy Polling, accessed August 9, 2013, http://www.publicpolicypolling.com/main/2013/04/conspiracy-theory-poll-results-.html.

23. Optin Online Panel, accessed August 9, 2013, http://optinpanel.org/wp-content/media/Metro-Opt-In-17-Community-Newspapers-July.pdf.

24. Reston Association, Community Survey, Report of Results, accessed August 9, 2013, https://www.reston.org/portals/3/GENERAL/Item%20C%203%20Reston%20VA%20Report%20of%20Results%202013%20FINAL%202013-04-11.pdf.

25. Rogers, accessed August 9, 2013, http://newsroom.rogers.com/news/13-01-14/Survey_shows_Canadians_give_themselves_top_marks_for_being_tech_savvy_but_fall_short_when_put_to_the_test.aspx.

26. KevinMD.com, accessed August 11, 2013, http://www.kevinmd.com/blog/2013/07/radiation-exposure-related-rise-lawsuit-culture.html.

27. Unofficial Overall Results, China World Summit Wing Hotel Vertical Run, accessed August 12, 2013, http://www.verticalrun.cn/downloads/2013_CVR_Results_Overall_Unofficial.pdf.

28. PPIHC, accessed August 12, 2013, http://livetiming.net/ppihc.

29. Taser FOI, accessed August 13, 2013, https://docs.google.com/a/guardian.co.uk/spreadsheet/ccc?key=0At6CC4x_yBnMdE5wMnpiQ1FmNWV0QVNHWlM2WGhKTWc#gid=0.

30. TriplePundit, accessed August 13, 2013, http://www.triplepundit.com/2013/05/san-francisco-achieves-highest-recycling-rate.

31. The Week, accessed August 15, 2013, http://theweek.com/article/index/248281/its-not-your-imagination-bmw-drivers-are-the-biggest-jerks.

32. American Meteor Society, accessed August 15, 2013, http://www.amsmeteors.org/fireballs/fireball-tracking-system-analysis.

33. Predictors of indoor BTEX concentrations in Canadian residences, accessed August 15, 2013, http://www.statcan.gc.ca/pub/82-003-x/2013005/article/11793-eng.pdf.

CHAPTER 11

1. Based on information from Farm Labor, National Agricultural Statistics Service, accessed August 19, 2013, http://usda01.library.cornell.edu/usda/current/FarmLabo/FarmLabo-05-21-2013.pdf.

2. ScienceDaily, accessed August 22, 2013, http://www.sciencedaily.com/releases/2013/06/130617110931.htm.

3. National Atmospheric Deposition Program, accessed August 23, 2013, http://nadp.sws.uiuc.edu/MDN/mdndata.aspx.

4. *USA Today*, accessed August 24, 2013, http://www.usatoday.com/story/money/cars/2013/05/04/worst-traffic-cities/2127661.

5. Existing U.S. Coal Plants, accessed August 24, 2013, http://www.sourcewatch.org/index.php/Existing_U.S._Coal_Plants#SO2_pollution_and_pollution_controls.

6. Statistics Canada, accessed August 24, 2013, http://www5.statcan.gc.ca/cansim/a47.

7. USGS, Scientific Investigations Report 2013-5001, accessed August 25, 2013, http://pubs.usgs.gov/sir/2013/5001/table6.html.

8. Concur Expense IQ report 2013, accessed August 25, 2013, https://www.concur.com/sites/default/files/lp/pdfs/concur_expense_iq_report_2013_v2.pdf.

9. ESPN Home Run Tracker, accessed August 25, 2013, http://www.hittrackeronline.com/index.php?h=&p=&b=Fenway%2BPark.

10. Statistics Canada, accessed August 26, 2013, http://www.statcan.gc.ca/daily-quotidien/120510/t120510a001-eng.htm.

11. e-Power, accessed August 26, 2013, http://epower.core-mark.com/2013/04/regional-differences-influence-c-store-shopping.

12. Kaiser Health News, accessed August 27, 2013, http://www.kaiserhealthnews.org/Stories/2013/May/28/medicare-state-geographic-variation-costs.aspx.

13. World Poultry, accessed September 2, 2013, http://www.worldpoultry.net/Broilers/Nutrition/2013/3/Factors-affecting-feed-intake-of-chickens-1172230W.

14. Natural Resources Canada, accessed September 2, 2013, http://cfs.nrcan.gc.ca/pages/353.

15. Autism, accessed September 2, 2013, http://aut.sagepub.com/content/17/2/184.abstract.

16. SFGate, accessed September 4, 2013, http://blog.sfgate.com/topdown/2013/09/03/the-new-san-francisco-oakland-bay-bridge-wide-open-vistas-a-bicycle-path-and-a-6-4-billion-price-tag.

17. European Nuclear Society, accessed September 4, 2013, http://www.euronuclear.org/info/encyclopedia/n/nuclear-power-plant-world-wide.htm.

18. fox4kc.com, accessed September 4, 2013, http://fox4kc.com/2013/04/09/kan-couple-catches-record-breaking-catfish.

19. *Int.J. Environ. Res. Public Health* 2013, *10*(2), 541–555; doi:10.3390/ijerph10020541 Article Assessment of the Levels of Airborne Bacteria, Gram-Negative Bacteria, and Fungi in Hospital Lobbies Dong-Uk Park[1,*], Jeong-Kwan Yeom[2,3], Won Jae Lee[3] and Kyeong-Min Lee[1,4]

20. Based on information from Farm Labor, National Agricultural Statistics Service, accessed August 19, 2013, http://usda01.library.cornell.edu/usda/current/FarmLabo/FarmLabo-05-21-2013.pdf.

CHAPTER 12

1. Wind Turbine Noise, accessed September 7, 2013, http://ramblingsdc.net/wtnoise.html.

2. *GCM Magazine,* accessed September 7, 2013, http://www.gcsaa.org/_common/templates/GcsaaTwoColumn-Layout.aspx?id=6985&LangType=1033.

3. Mother Jones, accessed September 9, 2013, http://www. motherjones.com/environment/2013/01/lead-crime-link-gasoline.

4. EurekAlert, accessed September 13, 2013, http://www. eurekalert.org/pub_releases/2013-05/atx-sfa051313.php.

5. *Radiology*, accessed September 18, 2013, http://radiology.rsna.org/content/early/2013/06/03/radiol.13130545.abstract.

6. Climate Central, accessed October 2, 2013, http://www. climatecentral.org/news/study-finds-plant-growth-surges-as-co2-levels-rise-16094.

7. United Nations University, Working Paper Series, accessed October 2, 2013, http://www.merit.unu.edu/publications/wppdf/2013/wp2013-001.pdf.

8. *Journal of Science and Medicine in Sport*, accessed September 12, 2013, http://www.jsams.org/article/S1440-2440%2812%2900207-1/abstract.

9. Science Daily, accessed October 9, 2013, http://www. sciencedaily.com/releases/2012/10/121010102156.htm.

10. Virginia Cooperative Extension, Small Grains in 2013, accessed October 9, 2013, http://www.pubs.ext.vt.edu/CSES/CSES-62/CSES-62_pdf.pdf.

11. PubMed, accessed October 9, 2013, http://www.ncbi. nlm.nih.gov/pubmed/22077818.

12. National Data Buoy Center, National Oceanic and Atmospheric Administration, http://www.ndbc.noaa.gov.

13. Feeding America, accessed October 9, 2013, http://feedingamerica.org/press-room/press-releases/map-the-meal-gap-2013.aspx.

14. Canada.com, accessed October 11, 2013, http://o. canada.com/business/money/city-of-toronto-parking-lots-are-overflowing-costing-drivers-thousands.

15. Gao et al., Regression of NASCAR: Looking into Five Years of Jimmie Johnson, SAS Global Forum 2013, accessed October 20, 2013, http://support.sas.com/resources/papers/proceedings13/439-2013.pdf.

16. DriverAverages.com, accessed October 20, 2013, http://www.driveraverages.com/nascar_stats.

17. The Weather Underground, accessed October 20, 2013, http://www.wunderground.com/history/airport/KAPF/2013/10/20/DailyHistory.html.

18. Science Daily, accessed October 22, 2013, http://www. sciencedaily.com/releases/2013/09/130904203703.htm.

19. WRisk.com, accessed October 22, 2013, http://www. wxrisk.com/2012/11/final-winter-preview-2012-13.

20. F. He et al., Salt Intake Is Related to Soft Drink Consumption in Children and Adolescents, *Hypertension*, 51;629, 2008.

21. Owasco Watershed Lake Association, Inc.

22. USGS, accessed October 24, 2013, http://pubs.usgs.gov/ds/774.

23. European Food Safety Authority, accessed November 3, 2013, http://www.efsa.europa.eu/en/efsajournal/doc/3137.pdf.

24. *Evolution, Medicine, and Public Health,* accessed November 3, 2013, http://emph.oxfordjournals.org/content/2013/1/173.full.

25. PLOS One, accessed November 3, 2013, http://www. plosone.org/article/info%3Adoi%2F10.1371%2Fjournal. pone.0070902.

26. Newswire, accessed November 3, 2013, http://www. nielsen.com/us/en/newswire/2013/new-study-confirms-correlation-between-twitter-and-tv-ratings.html.

27. Morana RTD, accessed November 3, 2013, http://www. morana-rtd.com/e-preservationscience/2013/Luciani-13-03-2013.pdf.

28. U.S. Department of Labor, Bureau of Labor Statistics, accessed November 6, 2013, http://www.bls.gov/news. release/cewqtr.t03.htm.

29. ISRN Veterinary Science, accessed November 11, 2013, http://www.hindawi.com/isrn/veterinary.science/2013/684353.

30. MLB Team Payrolls, accessed November 12, 2013, http://www.stevetheump.com/Payrolls.htm; and Team Marketing Report, accessed November 12, 2013, http://news.cincinnati.com/assets/AB20330243.pdf.

31. Calorie Comparison of Soy Milk Brands, accessed November 13, 2013, http://www.fitsugar.com/Calorie-Comparison-Soy-Milk-Brands-24224831.

32. Natural Hazards and Earth System Sciences, accessed November 13, 2013, http://www.nat-hazards-earth-syst-sci-discuss.net/1/3449/2013/nhessd-1-3449-2013-print.pdf.

33. Financial Forecast Center, accessed November 16, 2013, http://www.forecasts.org/data/price-data.htm.

34. GCSAA, *GCM Magazine*, accessed November 16, 2013, http://www.gcsaa.org/_common/templates/GcsaaTwo-ColumnLayout.aspx?id=6985&LangType=1033.

35. Avian Conservation and Ecology, accessed November 16, 2013, http://www.ace-eco.org/vol8/iss2/art10/#collision.

36. L. Xuanyu, The Varination Tendency Analysis on Contribution Rate to Economic Growth by Production Factor Input in China, *Management Science and Engineering*, vol. 7, no. 2, pp. 16–23, accessed November 17, 2013.

37. National Weather Service Forecast Office, Honolulu, HI, accessed November 18, 2013, http://www.prh.noaa.gov/hnl/climate/PHNL_rainfall.php; and Golden Gate Weather Services, accessed November 18, 2013, http://ggweather.com/enso/oni.htm.

38. Nutrition Express, accessed November 29, 2013, http://www.nutritionexpress.com/article+index/newsletters/february+2013/showarticle.aspx?id=1817.

39. @CBS Local, accessed November 29, 2013, http://washington.cbslocal.com/2013/10/24/study-more-time-spent-online-means-less-work-sleep.

40. StateMaster, accessed November 30, 2013, http://www.statemaster.com/graph/geo_wat_are-geography-water-area; Responsive Management, accessed November 30, 2013, http://www.responsivemanagement.com/download/reports/WA_Fish_Marketing_Plan.pdf.

41. *The Annals of Occupational Hygiene*, accessed November 30, 2013, http://annhyg.oxfordjournals.org/content/48/6/547.full.

CHAPTER 12 WEB

42. Krejza, J., and Mariak, Z. Effect of Age on Cerebral Blood Flow Velocity in Patients After Aneurysmal Subarachnoid Hemorrhage, *Stroke*, vol. 33, no. 2, pp. 640–642, 2002.

43. D. G. Hall, E. J. Wenniger, and M. G. Hentz, Temperature Studies with the Asian Citrus Psyllid, *Diaphorina citri*: Cold Hardiness and Temperature Thresholds for Oviposition. *Journal of Insect Science*, vol. 11, art 83, 2011.

44. C. Moran, Stress and Emergency Work Experience: A Non-linear Relationship, *Disaster Prevention and Management*, vol. 7, issue 1, pp. 38–46, 1998.

45. Wang, S. et al., Arsenic and Fluoride Exposure in Drinking Water: Children's IQ and Growth in Shanyin County, Shanxi Province, China, *Environmental Health Perspectives*, vol. 115, no. 4, pp. 643–647, 2007.

46. Ramachandran S. et al., *Circulation*, as reported in *USA Today*, July 23, 2007.

47. Ciani E. et al., Neurochemical Correlates of Nicotine Neurotoxicity on Rat Habenulo-Interpeduncular Cholinergic Neurons, *Neuro Toxicology*, vol. 26, pp. 467–474, 2006.

48. Based on information from Study: Diet soda linked to heart risks, USA Today, 7/23/07. Couvreur et al., The Linear Relationship Between the Proportion of Fresh Grass in the Cow Diet, Milk Fatty Acid Composition, and Butter Properties, *Journal of Dairy Science*, vol. 89, pp. 1956–1969, 2006.

49. Based on a report of a study done by New Zealand's National Research Centre for Growth and Development and the University of Southhampton in Britain, *Yahoo News*, July 25, 2007.

50. Based on information in the article by S. Mishra et al, Modeling of Yarn Strength Utilization in Cotton Woven Fabrics Using Multiple Linear Regression, *Journal of Engineered Fibers and Fabrics*, vol. 9, issue 2, pp. 105–111, 2014.

51. Fisher, J. B. et al.,Wood Vessel Diameter Is Related to Elevation and Genotype in the Hawaiian Tree *Metrosideros polymorpha* (Myrtaceae), *American Journal of Botany*, vol. 94, pp. 709–715, 2007.

52. H. Riojas-Rodriguez et al., Intellectual Function in Mexican Children Living in a Mining Area and Environmentally Exposed to Manganese, *Environmental Health Perspectives*, vol. 118, pp. 1465–1470, 2010.

53. Based on information from Yanzhou Coal Mining Company Limited.

54. 2014 APTA Fact Book Appendix, accessed July 28, 2014, http://www.apta.com/resources/s.

55. Based on information from S. K. Sinnakaudan et al., Multiple Linear Regression Model for Total Bed Material Load Prediction, *Journal of Hydraulic Engineering*, May 2006.

56. D. Dumicic et al., Internet Purchases in European Union Countries: Multiple Linear Regression Approach, *International Journal of Social, Management, Economics, and Business Engineering*, vol. 8, no. 3, 834–840, 2014.

57. Based on information from J. Gidjunis, Sprawl Exceeds Reach of Hydrants, *USA Today*, August 24, 2007.

58. About 1100 Injured as Meteorite Hits Russia with Force of Atomic Bomb, Associated Press, February 15, 2013.

59. Based on information from C. Masters, Props to the New Turbos, *Time* Magazine, September 3, 2007.

60. Based on information from Y. S. Han et al., Intraocular Pressure and Influencing Systemic Health Parameters in a Korean Population, *Indian Journal of Ophthalmology*, vol. 62, issue 3, pp. 305–301, 2014.

61. Based on information from CNN Interactive, Some High Fiber Dry Cereals. Accessed August 17, 2014, http://www.cnn.com/FOOD/resources/food.for.thought/grains/cereal/compare.dry.fiber.html

62. Simulated Carrying Capacities of Fish in Norwegian Fjord, *Fisheries Oceanography*, vol. 4, no. 1, pp. 17–32, 1995.

63. Swift, P. et al., Residential Street Typology and Injury Accident Frequency, presented at the Congress for the New Urbanism, Denver, Colorado, June 1997.

CHAPTER 13

1. *Daily News*, accessed December 2, 2013, http://www.nydailynews.com/new-york/new-yorkers-havelongest-commute-times-article-1.1426047.

2. *Forbes*, accessed December 3, 2013, http://www.forbes.com/sites/hannylerner/2013/10/05/how-to-hire-the-right-employees-and-discover-the-toxic-ones.

3. Huff Post Food, accessed December 13, 2013, http://www.huffingtonpost.com/john-dick/thanksgiving-traditions_b_4297296.html.

4. American Association of Individual Investors, accessed December 15, 2013, http://www.aaii.com/sentiment-survey.

5. West Bay Yacht Club, accessed December 15, 2013, http://westbayyachtclub.org/assets/2013-wbyc-cruise-preference-survey-results.pdf.

6. Centers for Disease Control and Prevention, accessed December 15, 2013, http://www.cdc.gov/flu/weekly.

7. Statistics Canada, accessed December 15, 2013, http://www.statcan.gc.ca/pub/75-006-x/2013001/article/11775/info-eng.htm.

8. Rasmussen Reports, accessed December 16, 2013, http://www.rasmussenreports.com/public_content/lifestyle/general_lifestyle/july_2013/58_eat_at_a_restaurant_at_least_once_a_week.

9. Switchboard, accessed December 16, 2013, http://switchboard.nrdc.org/blogs/kbenfield/new_realtors_community_prefere.html.

10. The University of Maine, Miscellaneous Reports, Maine Agricultural and Forest Experiment Station, accessed December 16, 2013, http://digitalcommons.library.umaine.edu/cgi/viewcontent.cgi?article=1017&context=aes_miscreports.

11. MSN Money, accessed December 16, 2013, http://money.msn.com/investing/2013-customer-service-hall-of-shame.

12. 2013 Capital Bikeshare Member Survey Report, accessed December 16, 2013, http://capitalbikeshare.com/assets/pdf/CABI-2013SurveyReport.pdf.

13. J. G. Berry et al., Hospital Utilization and Characteristics of Patients Experiencing Recurrent Readmissions Within Children's Hospitals, *The Journal of the American Medical Association*, 305(7):682–690, 2011.

14. DraftHistory.com, accessed December 18, 2013, http://www.drafthistory.com/index.php/summary/summary.

15. Public Health Agency of Canada, Statement of Seasonal Influenza Vaccine for 2013-2014, accessed December 18, 2013, http://www.phac-aspc.gc.ca/publicat/ccdr-rmtc/13vol39/acs-dcc-4/#rec.

16. Centers for Disease Control and Prevention, Morbidity and Mortality Weekly Report, accessed December 18, 2013, http://www.cdc.gov/mmwr/preview/mmwrhtml/mm5919a2.htm.

17. City of Oakland Shot Spotter Report, October 2013, accessed December 18, 2013, http://www2.oaklandnet.com/oakca1/groups/police/documents/webcontent/oak043887.pdf.

18. USA Today Snapshot, accessed December 18, 2013, http://www.usatoday.com.

19. Europe's Most Desirable Logistics Locations, ProLogis, accessed December 19, 2013, http://www.portofantwerp.com/sites/portofantwerp/files/imce/WinningLocationsEurope_0.pdf.

20. 2013 Canadian University Survey Consortium (CUSC): First-Year Undergraduate Students, Carleton University, accessed December 18, 2013, http://oirp.carleton.ca/surveys/CUSC2013_summary.pdf.

21. The Economics of Law Practice in Ohio in 2013, Ohio State Bar Association, accessed December 18, 2013, https://www.ohiobar.org/NewsAndPublications/Documents/OSBA_EconOfLawPracticeOhio.pdf.

22. Silicon Valley Bank Wine Report, State of the Wine Industry 2013, accessed December 18, 2013, www.svb.com/wine-report-2013-pdf.

CHAPTER 14

1. How Much Damage Did the Deepwater Horizon Spill Do to the Gulf of Mexico?, *Scientific American*, accessed December 19, 2013, http://www.scientificamerican.com/article.cfm?id=how-much-damage-deepwater-horizon-gulf-mexico.

2. EPA, Coastal Water Sampling, accessed December 20, 2013, http://www.epa.gov/bpspill/water.html#mapreports.

3. YCharts, accessed December 24, 2013, http://ycharts.com/indicators/average_duration_of_unemployment.

4. Health Extremist, accessed December 24, 2013, http://www.healthextremist.com/are-eggs-good-for-you-30-reasons-to-eat-eggs.

5. League of Denial: The NFL's Concussion Crisis, accessed December 26, 2013, http://www.pbs.org/wgbh/pages/frontline/sports/league-of-denial/timeline-the-nfls-concussion-crisis.

6. NBC News Health, accessed December 26, 2013, http://www.nbcnews.com/health/women-may-have-harder-concussion-recovery-men-2D11603652.

7. *USA Today*, accessed December 26, 2013, http://www.usatoday.com/story/money/cars/2013/09/04/record-price-new-car-august/2761341.

8. StreetsBlog.org, accessed December 26, 2013, http://www.streetsblog.org/2013/09/05/after-the-addition-of-bike-lanes-and-plazas-manhattan-traffic-moves-faster.

9. American Heart Association, accessed December 26, 2013, http://www.heart.org/HEARTORG/Conditions/Arrhythmia/AboutArrhythmia/Atrial-Fibrillation-Medications_UCM_423781_Article.jsp#.

10. National Heart, Lung, and Blood Institute, accessed December 27, 2013, http://www.nhlbi.nih.gov/health/health-topics/topics/lungtxp/risks.html.

11. Yahoo News Canada, accessed December 28, 2013, http://ca.news.yahoo.com/blogs/geekquinox/ontario-grand-river-tests-highest-world-artificial-sweeteners-200300835.html.

12. Mail Online, accessed December 27, 2013, http://www.dailymail.co.uk/health/article-2179572/Night-shifts-raise-risk-heart-attacks-strokes-40-cent.html.

13. Mark's Daily Apple, accessed December 30, 2013, http://www.marksdailyapple.com/how-much-glucose-does-your-brain-really-need/#axzz2oxZxwtXj.

14. CreditCards.com, accessed on PR Newswire, December 30, 2013, http://www.prnewswire.com/news-releases/creditcardscom-weekly-credit-card-rate-report-credit-card-interest-rates-rise-to-1506-percent-232734271.html.

15. Spending & Saving Tracker, accessed December 30, 2013, http://amexspendsave.mediaroom.com/index.php?s=34135&item=19#assets_123.

16. CBC News Health, accessed December 30, 2013, http://www.cbc.ca/news/health/millions-of-acres-of-china-farmland-too-polluted-to-grow-food-1.2479102.

17. A Comparison of 2013 Dairy Policy Alternatives on Dairy Markets, accessed December 30, 2013, http://www.nmpf.org/files/AMAP2013DSADFA.pdf.

18. Opportunity Index, accessed December 31, 2013, http://opportunityindex.org/#4.00/40.00/-97.00.

19. Science Daily, accessed January 1, 2014, http://www.sciencedaily.com/releases/2013/11/131112141236.htm.

20. Sports Illustrated, Inside the NHL, accessed January 2, 2014, http://sportsillustrated.cnn.com/nhl/news/20130308/top-10-nhl-slap-shots-of-all-time.

21. European Environmental Agency, accessed January 2, 2014, http://www.eea.europa.eu/highlights/co2-emissions-from-new-vans.

22. WGMS and NSIDC, 1989, updated 2012, World Glacier Inventory. Compiled and made available by the World Glacier Monitoring Service, Zurich, Switzerland, and the National Snow and Ice Data Center, Boulder CO, U.S.A., doi:10.7265/N5/NSIDC-WGI-2012-02.

23. Flo Track, accessed January 3, 2014, http://www.flotrack.org/coverage/250963-New-York-City-Marathon-2013/article/22961-RESULTS-ING-New-York-City-Marathon-2013.

24. Supernova discovery statistics for 2013, accessed January 3, 2014, http://www.rochesterastronomy.org/sn2013/snstats.html.

25. S. Tobak, 9 Ways to Make a Million, accessed January 3, 2014, http://smallbusiness.yahoo.com/advisor/9-ways-to-make-a-million-203855561.html.

26. CBC News, Ottawa, accessed January 6, 2014, http://www.cbc.ca/news/canada/ottawa/propane-shipping-delays-hit-eastern-ontario-1.2483820.

27. Mail Online, accessed January 6, 2014, http://www.dailymail.co.uk/news/article-2531945/Brad-Pitts-Hurricane-Katrina-charity-fire-homes-rotting.html.

28. IT World Canada, accessed January 6, 2014, http://www.itworldcanada.com/article/canucks-split-on-government-reading-email-survey/84227.

29. CBS News, accessed January 6, 2014, http://www.cbsnews.com/news/the-kings-popularity-constant.

30. StarTribune, Lifestyle, accessed January 6, 2014, http://www.startribune.com/lifestyle/208529101.html.

31. Kellogg School of Management, accessed January 6, 2013, http://www.kellogg.northwestern.edu/programs/fulltimemba/faqs.aspx.

32. Science Daily, accessed January 6, 2014, http://www.sciencedaily.com/releases/2013/05/130515174407.htm.

33. P. Frase, Sociology, CUNY Graduate Center, accessed January 6, 2014, http://www.peterfrase.com/category/social-science/statistical-graphics-social-science.

34. F. G. Silversides and K. Budgell, The Relationships Among Measures of Egg Albumen Height, pH, and Whipping Volume, Poultry Science, accessed January 6, 2014, http://ps.fass.org/content/83/10/1619.long.

35. Psychology Today, accessed on January 6, 2014, http://www.psychologytoday.com/blog/our-gender-ourselves/201309/what-your-selfies-say-about-you.

36. Lake Powell Water Database, accessed January 8, 2014, http://lakepowell.water-data.com.

37. The Daily Herald, accessed January 8, 2014, http://www.thedailyherald.com/index.php?option=com_content&view=article&id=45016:a-minute-with-met-opera-chief-peter-gelb-on-live-opera-broadcasts&catid=18:entertainment&Itemid=29.

38. Environmental Health Perspectives, accessed January 8, 2014, http://www.ncbi.nlm.nih.gov/pmc/articles/PMC2974689.

39. Huff Post, accessed January 8, 2014, http://www.huffingtonpost.com/2014/01/07/mimo-baby-monitor_n_4556461.html.

40. Speedski-Info, accessed January 8, 2014, http://www.speedski-info.com/index_E.php.

41. Worldsteel Association, accessed January 9, 2014, http://www.worldsteel.org/statistics/crude-steel-production.html.

42. Yahoo News, accessed January 8, 2014, http://gma.yahoo.com/tallest-hotel-america-opens-york-city-015502341--abc-news-travel.html.

43. CBC News Toronto, accessed January 8, 2014, http://www.cbc.ca/news/canada/toronto/bosses-should-stop-asking-workers-for-sick-notes-oma-head-says-1.2489885.

44. E. Rabiei, U. Haberlandt, M. Sester, and D. Fitzner, Rainfall Estimation Using Moving Cars as Rain Gauges—Laboratory Experiments. Hydrology and Earth System Sciences, 17:4701–4712, 2013, accessed January 9, 2014.

45. College Football Statistics, accessed January 9, 2014, http://www.cfbstats.com.

46. CNN Travel, accessed January 10, 2014, http://www.cnn.com/2013/12/17/travel/four-innovative-ways-cut-boarding-planes.

Tables Appendix

Table I Binomial Distribution Cumulative Probabilities

Let X be a binomial random variable with parameters n and p: $X \sim B(n, p)$. This table contains cumulative probabilities:

$$P(X \le x) = \sum_{k=0}^{x} P(X = k) = P(X = 0) + P(X = 1) + P(X = 2) + \cdots + P(X = x).$$

$n = 5$

x	0.01	0.05	0.10	0.20	0.25	0.30	0.40	0.50	0.60	0.70	0.75	0.80	0.90	0.95	0.99
0	0.9510	0.7738	0.5905	0.3277	0.2373	0.1681	0.0778	0.0313	0.0102	0.0024	0.0010	0.0003	0.0000		
1	0.9990	0.9774	0.9185	0.7373	0.6328	0.5282	0.3370	0.1875	0.0870	0.0308	0.0156	0.0067	0.0005	0.0000	
2	1.0000	0.9988	0.9914	0.9421	0.8965	0.8369	0.6826	0.5000	0.3174	0.1631	0.1035	0.0579	0.0086	0.0012	0.0000
3		1.0000	0.9995	0.9933	0.9844	0.9692	0.9130	0.8125	0.6630	0.4718	0.3672	0.2627	0.0815	0.0226	0.0010
4			1.0000	0.9997	0.9990	0.9976	0.9898	0.9688	0.9222	0.8319	0.7627	0.6723	0.4095	0.2262	0.0490

$n = 10$

x	0.01	0.05	0.10	0.20	0.25	0.30	0.40	0.50	0.60	0.70	0.75	0.80	0.90	0.95	0.99
0	0.9044	0.5987	0.3487	0.1074	0.0563	0.0282	0.0060	0.0010	0.0001	0.0000					
1	0.9957	0.9139	0.7361	0.3758	0.2440	0.1493	0.0464	0.0107	0.0017	0.0001	0.0000	0.0000			
2	0.9999	0.9885	0.9298	0.6778	0.5256	0.3828	0.1673	0.0547	0.0123	0.0016	0.0004	0.0001	0.0000		
3	1.0000	0.9990	0.9872	0.8791	0.7759	0.6496	0.3823	0.1719	0.0548	0.0106	0.0035	0.0009	0.0000		
4		0.9999	0.9984	0.9672	0.9219	0.8497	0.6331	0.3770	0.1662	0.0473	0.0197	0.0064	0.0001	0.0000	
5		1.0000	0.9999	0.9936	0.9803	0.9527	0.8338	0.6230	0.3669	0.1503	0.0781	0.0328	0.0016	0.0001	
6			1.0000	0.9991	0.9965	0.9894	0.9452	0.8281	0.6177	0.3504	0.2241	0.1209	0.0128	0.0010	0.0000
7				0.9999	0.9996	0.9984	0.9877	0.9453	0.8327	0.6172	0.4744	0.3222	0.0702	0.0115	0.0001
8				1.0000	1.0000	0.9999	0.9983	0.9893	0.9536	0.8507	0.7560	0.6242	0.2639	0.0861	0.0043
9						1.0000	0.9999	0.9990	0.9940	0.9718	0.9437	0.8926	0.6513	0.4013	0.0956

$n = 15$

x	0.01	0.05	0.10	0.20	0.25	0.30	0.40	0.50	0.60	0.70	0.75	0.80	0.90	0.95	0.99
0	0.8601	0.4633	0.2059	0.0352	0.0134	0.0047	0.0005	0.0000							
1	0.9904	0.8290	0.5490	0.1671	0.0802	0.0353	0.0052	0.0005	0.0000						
2	0.9996	0.9638	0.8159	0.3980	0.2361	0.1268	0.0271	0.0037	0.0003	0.0000					
3	1.0000	0.9945	0.9444	0.6482	0.4613	0.2969	0.0905	0.0176	0.0019	0.0001	0.0000				
4		0.9994	0.9873	0.8358	0.6865	0.5155	0.2173	0.0592	0.0093	0.0007	0.0001	0.0000			
5		0.9999	0.9978	0.9389	0.8516	0.7216	0.4032	0.1509	0.0338	0.0037	0.0008	0.0001			
6		1.0000	0.9997	0.9819	0.9434	0.8689	0.6098	0.3036	0.0950	0.0152	0.0042	0.0008			
7			1.0000	0.9958	0.9827	0.9500	0.7869	0.5000	0.2131	0.0500	0.0173	0.0042	0.0000		
8				0.9992	0.9958	0.9848	0.9050	0.6964	0.3902	0.1311	0.0566	0.0181	0.0003	0.0000	
9				0.9999	0.9992	0.9963	0.9662	0.8491	0.5968	0.2784	0.1484	0.0611	0.0022	0.0001	
10				1.0000	0.9999	0.9993	0.9907	0.9408	0.7827	0.4845	0.3135	0.1642	0.0127	0.0006	
11					1.0000	0.9999	0.9981	0.9824	0.9095	0.7031	0.5387	0.3518	0.0556	0.0055	0.0000
12						1.0000	0.9997	0.9963	0.9729	0.8732	0.7639	0.6020	0.1841	0.0362	0.0004
13							1.0000	0.9995	0.9948	0.9647	0.9198	0.8329	0.4510	0.1710	0.0096
14								1.0000	0.9995	0.9953	0.9866	0.9648	0.7941	0.5367	0.1399

Table I Binomial Distribution Cumulative Probabilities (Continued)

n = 20

x	0.01	0.05	0.10	0.20	0.25	0.30	0.40	0.50	0.60	0.70	0.75	0.80	0.90	0.95	0.99
0	0.8179	0.3585	0.1216	0.0115	0.0032	0.0008	0.0000								
1	0.9831	0.7358	0.3917	0.0692	0.0243	0.0076	0.0005	0.0000							
2	0.9990	0.9245	0.6769	0.2061	0.0913	0.0355	0.0036	0.0002							
3	1.0000	0.9841	0.8670	0.4114	0.2252	0.1071	0.0160	0.0013	0.0000						
4		0.9974	0.9568	0.6296	0.4148	0.2375	0.0510	0.0059	0.0003						
5		0.9997	0.9887	0.8042	0.6172	0.4164	0.1256	0.0207	0.0016	0.0000					
6		1.0000	0.9976	0.9133	0.7858	0.6080	0.2500	0.0577	0.0065	0.0003	0.0000				
7			0.9996	0.9679	0.8982	0.7723	0.4159	0.1316	0.0210	0.0013	0.0002	0.0000			
8			0.9999	0.9900	0.9591	0.8867	0.5956	0.2517	0.0565	0.0051	0.0009	0.0001			
9			1.0000	0.9974	0.9861	0.9520	0.7553	0.4119	0.1275	0.0171	0.0039	0.0006			
10				0.9994	0.9961	0.9829	0.8725	0.5881	0.2447	0.0480	0.0139	0.0026	0.0000		
11				0.9999	0.9991	0.9949	0.9435	0.7483	0.4044	0.1133	0.0409	0.0100	0.0001		
12				1.0000	0.9998	0.9987	0.9790	0.8684	0.5841	0.2277	0.1018	0.0321	0.0004		
13					1.0000	0.9997	0.9935	0.9423	0.7500	0.3920	0.2142	0.0867	0.0024	0.0000	
14						1.0000	0.9984	0.9793	0.8744	0.5836	0.3828	0.1958	0.0113	0.0003	
15							0.9997	0.9941	0.9490	0.7625	0.5852	0.3704	0.0432	0.0026	
16							1.0000	0.9987	0.9840	0.8929	0.7748	0.5886	0.1330	0.0159	0.0000
17								0.9998	0.9964	0.9645	0.9087	0.7939	0.3231	0.0755	0.0010
18								1.0000	0.9995	0.9924	0.9757	0.9308	0.6083	0.2642	0.0169
19									1.0000	0.9992	0.9968	0.9885	0.8784	0.6415	0.1821

n = 25

x	0.01	0.05	0.10	0.20	0.25	0.30	0.40	0.50	0.60	0.70	0.75	0.80	0.90	0.95	0.99
0	0.7778	0.2774	0.0718	0.0038	0.0008	0.0001	0.0000								
1	0.9742	0.6424	0.2712	0.0274	0.0070	0.0016	0.0001								
2	0.9980	0.8729	0.5371	0.0982	0.0321	0.0090	0.0004	0.0000							
3	0.9999	0.9659	0.7636	0.2340	0.0962	0.0332	0.0024	0.0001							
4	1.0000	0.9928	0.9020	0.4207	0.2137	0.0905	0.0095	0.0005	0.0000						
5		0.9988	0.9666	0.6167	0.3783	0.1935	0.0294	0.0020	0.0001						
6		0.9998	0.9905	0.7800	0.5611	0.3407	0.0736	0.0073	0.0003						
7		1.0000	0.9977	0.8909	0.7265	0.5118	0.1536	0.0216	0.0012	0.0000					
8			0.9995	0.9532	0.8506	0.6769	0.2735	0.0539	0.0043	0.0001					
9			0.9999	0.9827	0.9287	0.8106	0.4246	0.1148	0.0132	0.0005	0.0000				
10			1.0000	0.9944	0.9703	0.9022	0.5858	0.2122	0.0344	0.0018	0.0002	0.0000			
11				0.9985	0.9893	0.9558	0.7323	0.3450	0.0778	0.0060	0.0009	0.0001			
12				0.9996	0.9966	0.9825	0.8462	0.5000	0.1538	0.0175	0.0034	0.0004			
13				0.9999	0.9991	0.9940	0.9222	0.6550	0.2677	0.0442	0.0107	0.0015			
14				1.0000	0.9998	0.9982	0.9656	0.7878	0.4142	0.0978	0.0297	0.0056	0.0000		
15					1.0000	0.9995	0.9868	0.8852	0.5754	0.1894	0.0713	0.0173	0.0001		
16						0.9999	0.9957	0.9461	0.7265	0.3231	0.1494	0.0468	0.0005		
17						1.0000	0.9988	0.9784	0.8464	0.4882	0.2735	0.1091	0.0023	0.0000	
18							0.9997	0.9927	0.9264	0.6593	0.4389	0.2200	0.0095	0.0002	
19							0.9999	0.9980	0.9706	0.8065	0.6217	0.3833	0.0334	0.0012	
20							1.0000	0.9995	0.9905	0.9095	0.7863	0.5793	0.0980	0.0072	0.0000
21								0.9999	0.9976	0.9668	0.9038	0.7660	0.2364	0.0341	0.0001
22								1.0000	0.9996	0.9910	0.9679	0.9018	0.4629	0.1271	0.0020
23									0.9999	0.9984	0.9930	0.9726	0.7288	0.3576	0.0258
24									1.0000	0.9999	0.9992	0.9962	0.9282	0.7226	0.2222

Table II Poisson Distribution Cumulative Probabilities

Let X be a Poisson random variable with parameter λ. This table contains cumulative probabilities:

$$P(X \le x) = \sum_{k=0}^{x} P(X = k) = P(X = 0) + P(X = 1) + P(X = 2) + \cdots + P(X = x).$$

x	λ 0.05	0.10	0.15	0.20	0.25	0.30	0.35	0.40	0.45	0.50
0	0.9512	0.9048	0.8607	0.8187	0.7788	0.7408	0.7047	0.6703	0.6376	0.6065
1	0.9988	0.9953	0.9898	0.9825	0.9735	0.9631	0.9513	0.9384	0.9246	0.9098
2	1.0000	0.9998	0.9995	0.9989	0.9978	0.9964	0.9945	0.9921	0.9891	0.9856
3		1.0000	1.0000	0.9999	0.9999	0.9997	0.9995	0.9992	0.9988	0.9982
4				1.0000	1.0000	1.0000	1.0000	0.9999	0.9999	0.9998
5								1.0000	1.0000	1.0000

x	λ 0.55	0.60	0.65	0.70	0.75	0.80	0.85	0.90	0.95	1.00
0	0.5769	0.5488	0.5220	0.4966	0.4724	0.4493	0.4274	0.4066	0.3867	0.3679
1	0.8943	0.8781	0.8614	0.8442	0.8266	0.8088	0.7907	0.7725	0.7541	0.7358
2	0.9815	0.9769	0.9717	0.9659	0.9595	0.9526	0.9451	0.9371	0.9287	0.9197
3	0.9975	0.9966	0.9956	0.9942	0.9927	0.9909	0.9889	0.9865	0.9839	0.9810
4	0.9997	0.9996	0.9994	0.9992	0.9989	0.9986	0.9982	0.9977	0.9971	0.9963
5	1.0000	1.0000	0.9999	0.9999	0.9999	0.9998	0.9997	0.9997	0.9995	0.9994
6		1.0000	1.0000	1.0000	1.0000	1.0000	1.0000	1.0000	0.9999	0.9999
7									1.0000	1.0000

x	λ 1.1	1.2	1.3	1.4	1.5	1.6	1.7	1.8	1.9	2.0
0	0.3329	0.3012	0.2725	0.2466	0.2231	0.2019	0.1827	0.1653	0.1496	0.1353
1	0.6990	0.6626	0.6268	0.5918	0.5578	0.5249	0.4932	0.4628	0.4337	0.4060
2	0.9004	0.8795	0.8571	0.8335	0.8088	0.7834	0.7572	0.7306	0.7037	0.6767
3	0.9743	0.9662	0.9569	0.9463	0.9344	0.9212	0.9068	0.8913	0.8747	0.8571
4	0.9946	0.9923	0.9893	0.9857	0.9814	0.9763	0.9704	0.9636	0.9559	0.9473
5	0.9990	0.9985	0.9978	0.9968	0.9955	0.9940	0.9920	0.9896	0.9868	0.9834
6	0.9999	0.9997	0.9996	0.9994	0.9991	0.9987	0.9981	0.9974	0.9966	0.9955
7	1.0000	1.0000	0.9999	0.9999	0.9998	0.9997	0.9996	0.9994	0.9992	0.9989
8			1.0000	1.0000	1.0000	1.0000	0.9999	0.9999	0.9998	0.9998
9						1.0000	1.0000	1.0000	1.0000	1.0000

Table II Poisson Distribution Cumulative Probabilities (Continued)

λ

x	2.1	2.2	2.3	2.4	2.5	2.6	2.7	2.8	2.9	3.0
0	0.1225	0.1108	0.1003	0.0907	0.0821	0.0743	0.0672	0.0608	0.0550	0.0498
1	0.3796	0.3546	0.3309	0.3084	0.2873	0.2674	0.2487	0.2311	0.2146	0.1991
2	0.6496	0.6227	0.5960	0.5697	0.5438	0.5184	0.4936	0.4695	0.4460	0.4232
3	0.8386	0.8194	0.7993	0.7787	0.7576	0.7360	0.7141	0.6919	0.6696	0.6472
4	0.9379	0.9275	0.9162	0.9041	0.8912	0.8774	0.8629	0.8477	0.8318	0.8153
5	0.9796	0.9751	0.9700	0.9643	0.9580	0.9510	0.9433	0.9349	0.9258	0.9161
6	0.9941	0.9925	0.9906	0.9884	0.9858	0.9828	0.9794	0.9756	0.9713	0.9665
7	0.9985	0.9980	0.9974	0.9967	0.9958	0.9947	0.9934	0.9919	0.9901	0.9881
8	0.9997	0.9995	0.9994	0.9991	0.9989	0.9985	0.9981	0.9976	0.9969	0.9962
9	0.9999	0.9999	0.9999	0.9998	0.9997	0.9996	0.9995	0.9993	0.9991	0.9989
10	1.0000	1.0000	1.0000	1.0000	0.9999	0.9999	0.9999	0.9998	0.9998	0.9997
11				1.0000	1.0000	1.0000	1.0000	1.0000	0.9999	0.9999
12									1.0000	1.0000

λ

x	3.1	3.2	3.3	3.4	3.5	3.6	3.7	3.8	3.9	4.0
0	0.0450	0.0408	0.0369	0.0334	0.0302	0.0273	0.0247	0.0224	0.0202	0.0183
1	0.1847	0.1712	0.1586	0.1468	0.1359	0.1257	0.1162	0.1074	0.0992	0.0916
2	0.4012	0.3799	0.3594	0.3397	0.3208	0.3027	0.2854	0.2689	0.2531	0.2381
3	0.6248	0.6025	0.5803	0.5584	0.5366	0.5152	0.4942	0.4735	0.4532	0.4335
4	0.7982	0.7806	0.7626	0.7442	0.7254	0.7064	0.6872	0.6678	0.6484	0.6288
5	0.9057	0.8946	0.8829	0.8705	0.8576	0.8441	0.8301	0.8156	0.8006	0.7851
6	0.9612	0.9554	0.9490	0.9421	0.9347	0.9267	0.9182	0.9091	0.8995	0.8893
7	0.9858	0.9832	0.9802	0.9769	0.9733	0.9692	0.9648	0.9599	0.9546	0.9489
8	0.9953	0.9943	0.9931	0.9917	0.9901	0.9883	0.9863	0.9840	0.9815	0.9786
9	0.9986	0.9982	0.9978	0.9973	0.9967	0.9960	0.9952	0.9942	0.9931	0.9919
10	0.9996	0.9995	0.9994	0.9992	0.9990	0.9987	0.9984	0.9981	0.9977	0.9972
11	0.9999	0.9999	0.9998	0.9998	0.9997	0.9996	0.9995	0.9994	0.9993	0.9991
12	1.0000	1.0000	1.0000	0.9999	0.9999	0.9999	0.9999	0.9998	0.9998	0.9997
13				1.0000	1.0000	1.0000	1.0000	1.0000	0.9999	0.9999
14									1.0000	1.0000

Table II Poisson Distribution Cumulative Probabilities (Continued)

x	λ 4.1	4.2	4.3	4.4	4.5	4.6	4.7	4.8	4.9	5.0
0	0.0166	0.0150	0.0136	0.0123	0.0111	0.0101	0.0091	0.0082	0.0074	0.0067
1	0.0845	0.0780	0.0719	0.0663	0.0611	0.0563	0.0518	0.0477	0.0439	0.0404
2	0.2238	0.2102	0.1974	0.1851	0.1736	0.1626	0.1523	0.1425	0.1333	0.1247
3	0.4142	0.3954	0.3772	0.3594	0.3423	0.3257	0.3097	0.2942	0.2793	0.2650
4	0.6093	0.5898	0.5704	0.5512	0.5321	0.5132	0.4946	0.4763	0.4582	0.4405
5	0.7693	0.7531	0.7367	0.7199	0.7029	0.6858	0.6684	0.6510	0.6335	0.6160
6	0.8786	0.8675	0.8558	0.8436	0.8311	0.8180	0.8046	0.7908	0.7767	0.7622
7	0.9427	0.9361	0.9290	0.9214	0.9134	0.9049	0.8960	0.8867	0.8769	0.8666
8	0.9755	0.9721	0.9683	0.9642	0.9597	0.9549	0.9497	0.9442	0.9382	0.9319
9	0.9905	0.9889	0.9871	0.9851	0.9829	0.9805	0.9778	0.9749	0.9717	0.9682
10	0.9966	0.9959	0.9952	0.9943	0.9933	0.9922	0.9910	0.9896	0.9880	0.9863
11	0.9989	0.9986	0.9983	0.9980	0.9976	0.9971	0.9966	0.9960	0.9953	0.9945
12	0.9997	0.9996	0.9995	0.9993	0.9992	0.9990	0.9988	0.9986	0.9983	0.9980
14	0.9999	0.9999	0.9998	0.9998	0.9997	0.9997	0.9996	0.9995	0.9994	0.9993
15	1.0000	1.0000	1.0000	0.9999	0.9999	0.9999	0.9999	0.9999	0.9998	0.9998
16				1.0000	1.0000	1.0000	1.0000	1.0000	0.9999	0.9999
17									1.0000	1.0000

x	λ 5.5	6.0	6.5	7.0	7.5	8.0	8.5	9.0	9.5	10.0
0	0.0041	0.0025	0.0015	0.0009	0.0006	0.0003	0.0002	0.0001	0.0001	0.0000
1	0.0266	0.0174	0.0113	0.0073	0.0047	0.0030	0.0019	0.0012	0.0008	0.0005
2	0.0884	0.0620	0.0430	0.0296	0.0203	0.0138	0.0093	0.0062	0.0042	0.0028
3	0.2017	0.1512	0.1118	0.0818	0.0591	0.0424	0.0301	0.0212	0.0149	0.0103
4	0.3575	0.2851	0.2237	0.1730	0.1321	0.0996	0.0744	0.0550	0.0403	0.0293
5	0.5289	0.4457	0.3690	0.3007	0.2414	0.1912	0.1496	0.1157	0.0885	0.0671
6	0.6860	0.6063	0.5265	0.4497	0.3782	0.3134	0.2562	0.2068	0.1649	0.1301
7	0.8095	0.7440	0.6728	0.5987	0.5246	0.4530	0.3856	0.3239	0.2687	0.2202
8	0.8944	0.8472	0.7916	0.7291	0.6620	0.5925	0.5231	0.4557	0.3918	0.3328
9	0.9462	0.9161	0.8774	0.8305	0.7764	0.7166	0.6530	0.5874	0.5218	0.4579
10	0.9747	0.9574	0.9332	0.9015	0.8622	0.8159	0.7634	0.7060	0.6453	0.5830
11	0.9890	0.9799	0.9661	0.9467	0.9208	0.8881	0.8487	0.8030	0.7520	0.6968
12	0.9955	0.9912	0.9840	0.9730	0.9573	0.9362	0.9091	0.8758	0.8364	0.7916
13	0.9983	0.9964	0.9929	0.9872	0.9784	0.9658	0.9486	0.9261	0.8981	0.8645
14	0.9994	0.9986	0.9970	0.9943	0.9897	0.9827	0.9726	0.9585	0.9400	0.9165
15	0.9998	0.9995	0.9988	0.9976	0.9954	0.9918	0.9862	0.9780	0.9665	0.9513
16	0.9999	0.9998	0.9996	0.9990	0.9980	0.9963	0.9934	0.9889	0.9823	0.9730
17	1.0000	0.9999	0.9998	0.9996	0.9992	0.9984	0.9970	0.9947	0.9911	0.9857
18		1.0000	0.9999	0.9999	0.9997	0.9993	0.9987	0.9976	0.9957	0.9928
19			1.0000	1.0000	0.9999	0.9997	0.9995	0.9989	0.9980	0.9965
20					1.0000	0.9999	0.9998	0.9996	0.9991	0.9984
21						1.0000	0.9999	0.9998	0.9996	0.9993
22							1.0000	0.9999	0.9999	0.9997
23								1.0000	0.9999	0.9999
24									1.0000	1.0000

Table III Standard Normal Distribution Cumulative Probabilities

Let Z be a standard normal random variable:
$\mu = 0$ and $\sigma = 1$.

This table contains cumulative probabilities:
$P(Z \le z)$.

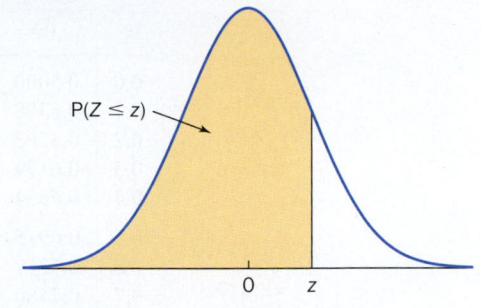

$P(Z \le z)$

z	.00	.01	.02	.03	.04	.05	.06	.07	.08	.09
−3.4	0.0003	0.0003	0.0003	0.0003	0.0003	0.0003	0.0003	0.0003	0.0003	0.0002
−3.3	0.0005	0.0005	0.0005	0.0004	0.0004	0.0004	0.0004	0.0004	0.0004	0.0003
−3.2	0.0007	0.0007	0.0006	0.0006	0.0006	0.0006	0.0006	0.0005	0.0005	0.0005
−3.1	0.0010	0.0009	0.0009	0.0009	0.0008	0.0008	0.0008	0.0008	0.0007	0.0007
−3.0	0.0013	0.0013	0.0013	0.0012	0.0012	0.0011	0.0011	0.0011	0.0010	0.0010
−2.9	0.0019	0.0018	0.0018	0.0017	0.0016	0.0016	0.0015	0.0015	0.0014	0.0014
−2.8	0.0026	0.0025	0.0024	0.0023	0.0023	0.0022	0.0021	0.0021	0.0020	0.0019
−2.7	0.0035	0.0034	0.0033	0.0032	0.0031	0.0030	0.0029	0.0028	0.0027	0.0026
−2.6	0.0047	0.0045	0.0044	0.0043	0.0041	0.0040	0.0039	0.0038	0.0037	0.0036
−2.5	0.0062	0.0060	0.0059	0.0057	0.0055	0.0054	0.0052	0.0051	0.0049	0.0048
−2.4	0.0082	0.0080	0.0078	0.0075	0.0073	0.0071	0.0069	0.0068	0.0066	0.0064
−2.3	0.0107	0.0104	0.0102	0.0099	0.0096	0.0094	0.0091	0.0089	0.0087	0.0084
−2.2	0.0139	0.0136	0.0132	0.0129	0.0125	0.0122	0.0119	0.0116	0.0113	0.0110
−2.1	0.0179	0.0174	0.0170	0.0166	0.0162	0.0158	0.0154	0.0150	0.0146	0.0143
−2.0	0.0228	0.0222	0.0217	0.0212	0.0207	0.0202	0.0197	0.0192	0.0188	0.0183
−1.9	0.0287	0.0281	0.0274	0.0268	0.0262	0.0256	0.0250	0.0244	0.0239	0.0233
−1.8	0.0359	0.0351	0.0344	0.0336	0.0329	0.0322	0.0314	0.0307	0.0301	0.0294
−1.7	0.0446	0.0436	0.0427	0.0418	0.0409	0.0401	0.0392	0.0384	0.0375	0.0367
−1.6	0.0548	0.0537	0.0526	0.0516	0.0505	0.0495	0.0485	0.0475	0.0465	0.0455
−1.5	0.0668	0.0655	0.0643	0.0630	0.0618	0.0606	0.0594	0.0582	0.0571	0.0559
−1.4	0.0808	0.0793	0.0778	0.0764	0.0749	0.0735	0.0721	0.0708	0.0694	0.0681
−1.3	0.0968	0.0951	0.0934	0.0918	0.0901	0.0885	0.0869	0.0853	0.0838	0.0823
−1.2	0.1151	0.1131	0.1112	0.1093	0.1075	0.1056	0.1038	0.1020	0.1003	0.0985
−1.1	0.1357	0.1335	0.1314	0.1292	0.1271	0.1251	0.1230	0.1210	0.1190	0.1170
−1.0	0.1587	0.1562	0.1539	0.1515	0.1492	0.1469	0.1446	0.1423	0.1401	0.1379
−0.9	0.1841	0.1814	0.1788	0.1762	0.1736	0.1711	0.1685	0.1660	0.1635	0.1611
−0.8	0.2119	0.2090	0.2061	0.2033	0.2005	0.1977	0.1949	0.1922	0.1894	0.1867
−0.7	0.2420	0.2389	0.2358	0.2327	0.2296	0.2266	0.2236	0.2206	0.2177	0.2148
−0.6	0.2743	0.2709	0.2676	0.2643	0.2611	0.2578	0.2546	0.2514	0.2483	0.2451
−0.5	0.3085	0.3050	0.3015	0.2981	0.2946	0.2912	0.2877	0.2843	0.2810	0.2776
−0.4	0.3446	0.3409	0.3372	0.3336	0.3300	0.3264	0.3228	0.3192	0.3156	0.3121
−0.3	0.3821	0.3783	0.3745	0.3707	0.3669	0.3632	0.3594	0.3557	0.3520	0.3483
−0.2	0.4207	0.4168	0.4129	0.4090	0.4052	0.4013	0.3974	0.3936	0.3897	0.3859
−0.1	0.4602	0.4562	0.4522	0.4483	0.4443	0.4404	0.4364	0.4325	0.4286	0.4247
−0.0	0.5000	0.4960	0.4920	0.4880	0.4840	0.4801	0.4761	0.4721	0.4681	0.4641

Table III Standard Normal Distribution Cumulative Probabilities (Continued)

z	.00	.01	.02	.03	.04	.05	.06	.07	.08	.09
0.0	0.5000	0.5040	0.5080	0.5120	0.5160	0.5199	0.5239	0.5279	0.5319	0.5359
0.1	0.5398	0.5438	0.5478	0.5517	0.5557	0.5596	0.5636	0.5675	0.5714	0.5753
0.2	0.5793	0.5832	0.5871	0.5910	0.5948	0.5987	0.6026	0.6064	0.6103	0.6141
0.3	0.6179	0.6217	0.6255	0.6293	0.6331	0.6368	0.6406	0.6443	0.6480	0.6517
0.4	0.6554	0.6591	0.6628	0.6664	0.6700	0.6736	0.6772	0.6808	0.6844	0.6879
0.5	0.6915	0.6950	0.6985	0.7019	0.7054	0.7088	0.7123	0.7157	0.7190	0.7224
0.6	0.7257	0.7291	0.7324	0.7357	0.7389	0.7422	0.7454	0.7486	0.7517	0.7549
0.7	0.7580	0.7611	0.7642	0.7673	0.7704	0.7734	0.7764	0.7794	0.7823	0.7852
0.8	0.7881	0.7910	0.7939	0.7967	0.7995	0.8023	0.8051	0.8078	0.8106	0.8133
0.9	0.8159	0.8186	0.8212	0.8238	0.8264	0.8289	0.8315	0.8340	0.8365	0.8389
1.0	0.8413	0.8438	0.8461	0.8485	0.8508	0.8531	0.8554	0.8577	0.8599	0.8621
1.1	0.8643	0.8665	0.8686	0.8708	0.8729	0.8749	0.8770	0.8790	0.8810	0.8830
1.2	0.8849	0.8869	0.8888	0.8907	0.8925	0.8944	0.8962	0.8980	0.8997	0.9015
1.3	0.9032	0.9049	0.9066	0.9082	0.9099	0.9115	0.9131	0.9147	0.9162	0.9177
1.4	0.9192	0.9207	0.9222	0.9236	0.9251	0.9265	0.9279	0.9292	0.9306	0.9319
1.5	0.9332	0.9345	0.9357	0.9370	0.9382	0.9394	0.9406	0.9418	0.9429	0.9441
1.6	0.9452	0.9463	0.9474	0.9484	0.9495	0.9505	0.9515	0.9525	0.9535	0.9545
1.7	0.9554	0.9564	0.9573	0.9582	0.9591	0.9599	0.9608	0.9616	0.9625	0.9633
1.8	0.9641	0.9649	0.9656	0.9664	0.9671	0.9678	0.9686	0.9693	0.9699	0.9706
1.9	0.9713	0.9719	0.9726	0.9732	0.9738	0.9744	0.9750	0.9756	0.9761	0.9767
2.0	0.9772	0.9778	0.9783	0.9788	0.9793	0.9798	0.9803	0.9808	0.9812	0.9817
2.1	0.9821	0.9826	0.9830	0.9834	0.9838	0.9842	0.9846	0.9850	0.9854	0.9857
2.2	0.9861	0.9864	0.9868	0.9871	0.9875	0.9878	0.9881	0.9884	0.9887	0.9890
2.3	0.9893	0.9896	0.9898	0.9901	0.9904	0.9906	0.9909	0.9911	0.9913	0.9916
2.4	0.9918	0.9920	0.9922	0.9925	0.9927	0.9929	0.9931	0.9932	0.9934	0.9936
2.5	0.9938	0.9940	0.9941	0.9943	0.9945	0.9946	0.9948	0.9949	0.9951	0.9952
2.6	0.9953	0.9955	0.9956	0.9957	0.9959	0.9960	0.9961	0.9962	0.9963	0.9964
2.7	0.9965	0.9966	0.9967	0.9968	0.9969	0.9970	0.9971	0.9972	0.9973	0.9974
2.8	0.9974	0.9975	0.9976	0.9977	0.9977	0.9978	0.9979	0.9979	0.9980	0.9981
2.9	0.9981	0.9982	0.9982	0.9983	0.9984	0.9984	0.9985	0.9985	0.9986	0.9986
3.0	0.9987	0.9987	0.9987	0.9988	0.9988	0.9989	0.9989	0.9989	0.9990	0.9990
3.1	0.9990	0.9991	0.9991	0.9991	0.9992	0.9992	0.9992	0.9992	0.9993	0.9993
3.2	0.9993	0.9993	0.9994	0.9994	0.9994	0.9994	0.9994	0.9995	0.9995	0.9995
3.3	0.9995	0.9995	0.9995	0.9996	0.9996	0.9996	0.9996	0.9996	0.9996	0.9997
3.4	0.9997	0.9997	0.9997	0.9997	0.9997	0.9997	0.9997	0.9997	0.9997	0.9998

Special critical values: $P(Z \geq z_\alpha) = \alpha$

α	0.10	0.05	0.025	0.01	0.005	0.001	0.0005	0.0001	
z_α	1.2816	1.6449	1.9600	2.3263	2.5758	3.0902	3.2905	3.7190	

α	0.00009	0.00008	0.00007	0.00006	0.00005	0.00004	0.00003	0.00002	0.00001
z_α	3.7455	3.7750	3.8082	3.8461	3.8906	3.9444	4.0128	4.1075	4.2649

Table IV Standardized Normal Scores

This table contains the standardized normal scores, z_i, for selected values of n.

i	n 10	20	25	30	40	50
1	−1.55	−1.87	−1.96	−2.04	−2.16	−2.24
2	−1.00	−1.40	−1.52	−1.61	−1.75	−1.85
3	−0.66	−1.13	−1.26	−1.36	−1.51	−1.62
4	−0.38	−0.92	−1.06	−1.18	−1.34	−1.46
5	−0.12	−0.74	−0.90	−1.02	−1.20	−1.33
6	0.12	−0.59	−0.76	−0.89	−1.08	−1.22
7	0.38	−0.45	−0.64	−0.78	−0.98	−1.12
8	0.66	−0.31	−0.52	−0.67	−0.88	−1.03
9	1.00	−0.19	−0.41	−0.57	−0.79	−0.95
10	1.55	−0.06	−0.30	−0.47	−0.71	−0.87
11		0.06	−0.20	−0.38	−0.63	−0.80
12		0.19	−0.10	−0.29	−0.56	−0.73
13		0.31	0.00	−0.21	−0.49	−0.67
14		0.45	0.10	−0.12	−0.42	−0.61
15		0.59	0.20	−0.04	−0.35	−0.55
16		0.74	0.30	0.04	−0.28	−0.49
17		0.92	0.41	0.12	−0.22	−0.44
18		1.13	0.52	0.21	−0.16	−0.38
19		1.40	0.64	0.29	−0.09	−0.33
20		1.87	0.76	0.38	−0.03	−0.28
21			0.90	0.47	0.03	−0.23
22			1.06	0.57	0.09	−0.18
23			1.26	0.67	0.16	−0.13
24			1.52	0.78	0.22	−0.07
25			1.96	0.89	0.28	−0.02
26				1.02	0.35	0.02
27				1.18	0.42	0.07
28				1.36	0.49	0.13
29				1.61	0.56	0.18
30				2.04	0.63	0.23
31					0.71	0.28
32					0.79	0.33
33					0.88	0.38
34					0.98	0.44
35					1.08	0.49
36					1.20	0.55
37					1.34	0.61
38					1.51	0.67
39					1.75	0.73
40					2.16	0.80
41						0.87
42						0.95
43						1.03
44						1.12
45						1.22
46						1.33
47						1.46
48						1.62
49						1.85
50						2.24

Table V Critical Values for the *t* Distribution

This table contains critical values associated with the *t* distribution, $t_{\alpha,\nu}$, defined by α and the degrees of freedom, ν.

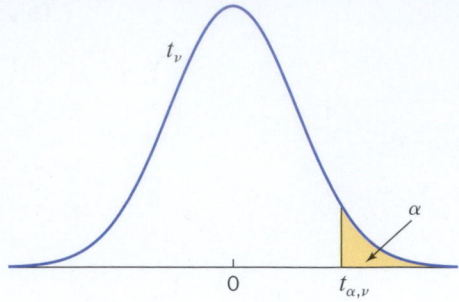

ν	0.20	0.10	0.05	0.025	0.01	0.005	0.001	0.0005	0.0001
1	1.3764	3.0777	6.3138	12.7062	31.8205	63.6567	318.3088	636.6192	3183.0988
2	1.0607	1.8856	2.9200	4.3027	6.9646	9.9248	22.3271	31.5991	70.7001
3	0.9785	1.6377	2.3534	3.1824	4.5407	5.8409	10.2145	12.9240	22.2037
4	0.9410	1.5332	2.1318	2.7764	3.7469	4.6041	7.1732	8.6103	13.0337
5	0.9195	1.4759	2.0150	2.5706	3.3649	4.0321	5.8934	6.8688	9.6776
6	0.9057	1.4398	1.9432	2.4469	3.1427	3.7074	5.2076	5.9588	8.0248
7	0.8960	1.4149	1.8946	2.3646	2.9980	3.4995	4.7853	5.4079	7.0634
8	0.8889	1.3968	1.8595	2.3060	2.8965	3.3554	4.5008	5.0413	6.4420
9	0.8834	1.3830	1.8331	2.2622	2.8214	3.2498	4.2968	4.7809	6.0101
10	0.8791	1.3722	1.8125	2.2281	2.7638	3.1693	4.1437	4.5869	5.6938
11	0.8755	1.3634	1.7959	2.2010	2.7181	3.1058	4.0247	4.4370	5.4528
12	0.8726	1.3562	1.7823	2.1788	2.6810	3.0545	3.9296	4.3178	5.2633
13	0.8702	1.3502	1.7709	2.1604	2.6503	3.0123	3.8520	4.2208	5.1106
14	0.8681	1.3450	1.7613	2.1448	2.6245	2.9768	3.7874	4.1405	4.9850
15	0.8662	1.3406	1.7531	2.1314	2.6025	2.9467	3.7328	4.0728	4.8800
16	0.8647	1.3368	1.7459	2.1199	2.5835	2.9208	3.6862	4.0150	4.7909
17	0.8633	1.3334	1.7396	2.1098	2.5669	2.8982	3.6458	3.9651	4.7144
18	0.8620	1.3304	1.7341	2.1009	2.5524	2.8784	3.6105	3.9216	4.6480
19	0.8610	1.3277	1.7291	2.0930	2.5395	2.8609	3.5794	3.8834	4.5899
20	0.8600	1.3253	1.7247	2.0860	2.5280	2.8453	3.5518	3.8495	4.5385
21	0.8591	1.3232	1.7207	2.0796	2.5176	2.8314	3.5272	3.8193	4.4929
22	0.8583	1.3212	1.7171	2.0739	2.5083	2.8188	3.5050	3.7921	4.4520
23	0.8575	1.3195	1.7139	2.0687	2.4999	2.8073	3.4850	3.7676	4.4152
24	0.8569	1.3178	1.7109	2.0639	2.4922	2.7969	3.4668	3.7454	4.3819
25	0.8562	1.3163	1.7081	2.0595	2.4851	2.7874	3.4502	3.7251	4.3517
26	0.8557	1.3150	1.7056	2.0555	2.4786	2.7787	3.4350	3.7066	4.3240
27	0.8551	1.3137	1.7033	2.0518	2.4727	2.7707	3.4210	3.6896	4.2987
28	0.8546	1.3125	1.7011	2.0484	2.4671	2.7633	3.4082	3.6739	4.2754
29	0.8542	1.3114	1.6991	2.0452	2.4620	2.7564	3.3962	3.6594	4.2539
30	0.8538	1.3104	1.6973	2.0423	2.4573	2.7500	3.3852	3.6460	4.2340
40	0.8507	1.3031	1.6839	2.0211	2.4233	2.7045	3.3069	3.5510	4.0942
50	0.8489	1.2987	1.6759	2.0086	2.4033	2.6778	3.2614	3.4960	4.0140
60	0.8477	1.2958	1.6706	2.0003	2.3901	2.6603	3.2317	3.4602	3.9621
70	0.8468	1.2938	1.6669	1.9944	2.3808	2.6479	3.2108	3.4350	3.9257
80	0.8461	1.2922	1.6641	1.9901	2.3739	2.6387	3.1953	3.4163	3.8988
90	0.8456	1.2910	1.6620	1.9867	2.3685	2.6316	3.1833	3.4019	3.8780
100	0.8452	1.2901	1.6602	1.9840	2.3642	2.6259	3.1737	3.3905	3.8616
200	0.8434	1.2858	1.6525	1.9719	2.3451	2.6006	3.1315	3.3398	3.7891
500	0.8423	1.2832	1.6479	1.9647	2.3338	2.5857	3.1066	3.3101	3.7468
∞	0.8416	1.2816	1.6449	1.9600	2.3263	2.5758	3.0902	3.2905	3.7190

Table VI Critical Values for the Chi-Square Distribution

This table contains critical values associated with the chi-square distribution, $\chi^2_{\alpha,\nu}$, defined by α and the degrees of freedom, ν.

ν	0.9999	0.9995	0.999	0.995	0.99	0.975	0.95	0.90
1	0.0^7157	0.0^6393	0.0^5157	0.0^4393	0.0002	0.0010	0.0039	0.0158
2	0.0002	0.0010	0.0020	0.0100	0.0201	0.0506	0.1026	0.2107
3	0.0052	0.0153	0.0243	0.0717	0.1148	0.2158	0.3518	0.5844
4	0.0284	0.0639	0.0908	0.2070	0.2971	0.4844	0.7107	1.0636
5	0.0822	0.1581	0.2102	0.4117	0.5543	0.8312	1.1455	1.6103
6	0.1724	0.2994	0.3811	0.6757	0.8721	1.2373	1.6354	2.2041
7	0.3000	0.4849	0.5985	0.9893	1.2390	1.6899	2.1673	2.8331
8	0.4636	0.7104	0.8571	1.3444	1.6465	2.1797	2.7326	3.4895
9	0.6608	0.9717	1.1519	1.7349	2.0879	2.7004	3.3251	4.1682
10	0.8889	1.2650	1.4787	2.1559	2.5582	3.2470	3.9403	4.8652
11	1.1453	1.5868	1.8339	2.6032	3.0535	3.8157	4.5748	5.5778
12	1.4275	1.9344	2.2142	3.0738	3.5706	4.4038	5.2260	6.3038
13	1.7333	2.3051	2.6172	3.5650	4.1069	5.0088	5.8919	7.0415
14	2.0608	2.6967	3.0407	4.0747	4.6604	5.6287	6.5706	7.7895
15	2.4082	3.1075	3.4827	4.6009	5.2293	6.2621	7.2609	8.5468
16	2.7739	3.5358	3.9416	5.1422	5.8122	6.9077	7.9616	9.3122
17	3.1567	3.9802	4.4161	5.6972	6.4078	7.5642	8.6718	10.0852
18	3.5552	4.4394	4.9048	6.2648	7.0149	8.2307	9.3905	10.8649
19	3.9683	4.9123	5.4068	6.8440	7.6327	8.9065	10.1170	11.6509
20	4.3952	5.3981	5.9210	7.4338	8.2604	9.5908	10.8508	12.4426
21	4.8348	5.8957	6.4467	8.0337	8.8972	10.2829	11.5913	13.2396
22	5.2865	6.4045	6.9830	8.6427	9.5425	10.9823	12.3380	14.0415
23	5.7494	6.9237	7.5292	9.2604	10.1957	11.6886	13.0905	14.8480
24	6.2230	7.4527	8.0849	9.8862	10.8564	12.4012	13.8484	15.6587
25	6.7066	7.9910	8.6493	10.5197	11.5240	13.1197	14.6114	16.4734
26	7.1998	8.5379	9.2221	11.1602	12.1981	13.8439	15.3792	17.2919
27	7.7019	9.0932	9.8028	11.8076	12.8785	14.5734	16.1514	18.1139
28	8.2126	9.6563	10.3909	12.4613	13.5647	15.3079	16.9279	18.9392
29	8.7315	10.2268	10.9861	13.1211	14.2565	16.0471	17.7084	19.7677
30	9.2581	10.8044	11.5880	13.7867	14.9535	16.7908	18.4927	20.5992
31	9.7921	11.3887	12.1963	14.4578	15.6555	17.5387	19.2806	21.4336
32	10.3331	11.9794	12.8107	15.1340	16.3622	18.2908	20.0719	22.2706
33	10.8810	12.5763	13.4309	15.8153	17.0735	19.0467	20.8665	23.1102
34	11.4352	13.1791	14.0567	16.5013	17.7891	19.8063	21.6643	23.9523
35	11.9957	13.7875	14.6878	17.1918	18.5089	20.5694	22.4650	24.7967
36	12.5622	14.4012	15.3241	17.8867	19.2327	21.3359	23.2686	25.6433
37	13.1343	15.0202	15.9653	18.5858	19.9602	22.1056	24.0749	26.4921
38	13.7120	15.6441	16.6112	19.2889	20.6914	22.8785	24.8839	27.3430
39	14.2950	16.2729	17.2616	19.9959	21.4262	23.6543	25.6954	28.1958
40	14.8831	16.9062	17.9164	20.7065	22.1643	24.4330	26.5093	29.0505
50	21.0093	23.4610	24.6739	27.9907	29.7067	32.3574	34.7643	37.6886
60	27.4969	30.3405	31.7383	35.5345	37.4849	40.4817	43.1880	46.4589
70	34.2607	37.4674	39.0364	43.2752	45.4417	48.7576	51.7393	55.3289
80	41.2445	44.7910	46.5199	51.1719	53.5401	57.1532	60.3915	64.2778
90	48.4087	52.2758	54.1552	59.1963	61.7541	65.6466	69.1260	73.2911
100	55.7246	59.8957	61.9179	67.3276	70.0649	74.2219	77.9295	82.3581

Table VI Critical Values for the Chi-Square Distribution (Continued)

ν	0.10	0.05	0.025	0.01	0.005	0.001	0.0005	0.0001
1	2.7055	3.8415	5.0239	6.6349	7.8794	10.8276	12.1157	15.1367
2	4.6052	5.9915	7.3778	9.2103	10.5966	13.8155	15.2018	18.4207
3	6.2514	7.8147	9.3484	11.3449	12.8382	16.2662	17.7300	21.1075
4	7.7794	9.4877	11.1433	13.2767	14.8603	18.4668	19.9974	23.5127
5	9.2364	11.0705	12.8325	15.0863	16.7496	20.5150	22.1053	25.7448
6	10.6446	12.5916	14.4494	16.8119	18.5476	22.4577	24.1028	27.8563
7	12.0170	14.0671	16.0128	18.4753	20.2777	24.3219	26.0178	29.8775
8	13.3616	15.5073	17.5345	20.0902	21.9550	26.1245	27.8680	31.8276
9	14.6837	16.9190	19.0228	21.6660	23.5894	27.8772	29.6658	33.7199
10	15.9872	18.3070	20.4832	23.2093	25.1882	29.5883	31.4198	35.5640
11	17.2750	19.6751	21.9200	24.7250	26.7568	31.2641	33.1366	37.3670
12	18.5493	21.0261	23.3367	26.2170	28.2995	32.9095	34.8213	39.1344
13	19.8119	22.3620	24.7356	27.6882	29.8195	34.5282	36.4778	40.8707
14	21.0641	23.6848	26.1189	29.1412	31.3193	36.1233	38.1094	42.5793
15	22.3071	24.9958	27.4884	30.5779	32.8013	37.6973	39.7188	44.2632
16	23.5418	26.2962	28.8454	31.9999	34.2672	39.2524	41.3081	45.9249
17	24.7690	27.5871	30.1910	33.4087	35.7185	40.7902	42.8792	47.5664
18	25.9894	28.8693	31.5264	34.8053	37.1565	42.3124	44.4338	49.1894
19	27.2036	30.1435	32.8523	36.1909	38.5823	43.8202	45.9731	50.7955
20	28.4120	31.4104	34.1696	37.5662	39.9968	45.3147	47.4985	52.3860
21	29.6151	32.6706	35.4789	38.9322	41.4011	46.7970	49.0108	53.9620
22	30.8133	33.9244	36.7807	40.2894	42.7957	48.2679	50.5111	55.5246
23	32.0069	35.1725	38.0756	41.6384	44.1813	49.7282	52.0002	57.0746
24	33.1962	36.4150	39.3641	42.9798	45.5585	51.1786	53.4788	58.6130
25	34.3816	37.6525	40.6465	44.3141	46.9279	52.6197	54.9475	60.1403
26	35.5632	38.8851	41.9232	45.6417	48.2899	54.0520	56.4069	61.6573
27	36.7412	40.1133	43.1945	46.9629	49.6449	55.4760	57.8576	63.1645
28	37.9159	41.3371	44.4608	48.2782	50.9934	56.8923	59.3000	64.6624
29	39.0875	42.5570	45.7223	49.5879	52.3356	58.3012	60.7346	66.1517
30	40.2560	43.7730	46.9792	50.8922	53.6720	59.7031	62.1619	67.6326
31	41.4217	44.9853	48.2319	52.1914	55.0027	61.0983	63.5820	69.1057
32	42.5847	46.1943	49.4804	53.4858	56.3281	62.4872	64.9955	70.5712
33	43.7452	47.3999	50.7251	54.7755	57.6484	63.8701	66.4025	72.0296
34	44.9032	48.6024	51.9660	56.0609	58.9639	65.2472	67.8035	73.4812
35	46.0588	49.8018	53.2033	57.3421	60.2748	66.6188	69.1986	74.9262
36	47.2122	50.9985	54.4373	58.6192	61.5812	67.9852	70.5881	76.3650
37	48.3634	52.1923	55.6680	59.8925	62.8833	69.3465	71.9722	77.7977
38	49.5126	53.3835	56.8955	61.1621	64.1814	70.7029	73.3512	79.2247
39	50.6598	54.5722	58.1201	62.4281	65.4756	72.0547	74.7253	80.6462
40	51.8051	55.7585	59.3417	63.6907	66.7660	73.4020	76.0946	82.0623
50	63.1671	67.5048	71.4202	76.1539	79.4900	86.6608	89.5605	95.9687
60	74.3970	79.0819	83.2977	88.3794	91.9517	99.6072	102.6948	109.5029
70	85.5270	90.5312	95.0232	100.4252	104.2149	112.3169	115.5776	122.7547
80	96.5782	101.8795	106.6286	112.3288	116.3211	124.8392	128.2613	135.7825
90	107.5650	113.1453	118.1359	124.1163	128.2989	137.2084	140.7823	148.6273
100	118.4980	124.3421	129.5612	135.8067	140.1695	149.4493	153.1670	161.3187

Table VII Critical Values for the F Distribution

This table contains values associated with the F distribution, F_{α, ν_1, ν_2}, defined by α and the degrees of freedom ν_1 and ν_2.

$\alpha = 0.05$

ν_2 \ ν_1	1	2	3	4	5	6	7	8	9	10	15	20	30	40	50	60	100
1	161.45	199.50	215.71	224.58	230.16	233.99	236.77	238.88	240.54	241.88	245.95	248.01	250.10	251.14	251.77	252.20	253.04
2	18.51	19.00	19.16	19.25	19.30	19.33	19.35	19.37	19.38	19.40	19.43	19.45	19.46	19.47	19.48	19.48	19.49
3	10.13	9.55	9.28	9.12	9.01	8.94	8.89	8.85	8.81	8.79	8.70	8.66	8.62	8.59	8.58	8.57	8.55
4	7.71	6.94	6.59	6.39	6.26	6.16	6.09	6.04	6.00	5.96	5.86	5.80	5.75	5.72	5.70	5.69	5.66
5	6.61	5.79	5.41	5.19	5.05	4.95	4.88	4.82	4.77	4.74	4.62	4.56	4.50	4.46	4.44	4.43	4.41
6	5.99	5.14	4.76	4.53	4.39	4.28	4.21	4.15	4.10	4.06	3.94	3.87	3.81	3.77	3.75	3.74	3.71
7	5.59	4.74	4.35	4.12	3.97	3.87	3.79	3.73	3.68	3.64	3.51	3.44	3.38	3.34	3.32	3.30	3.27
8	5.32	4.46	4.07	3.84	3.69	3.58	3.50	3.44	3.39	3.35	3.22	3.15	3.08	3.04	3.02	3.01	2.97
9	5.12	4.26	3.86	3.63	3.48	3.37	3.29	3.23	3.18	3.14	3.01	2.94	2.86	2.83	2.80	2.79	2.76
10	4.96	4.10	3.71	3.48	3.33	3.22	3.14	3.07	3.02	2.98	2.85	2.77	2.70	2.66	2.64	2.62	2.59
11	4.84	3.98	3.59	3.36	3.20	3.09	3.01	2.95	2.90	2.85	2.72	2.65	2.57	2.53	2.51	2.49	2.46
12	4.75	3.89	3.49	3.26	3.11	3.00	2.91	2.85	2.80	2.75	2.62	2.54	2.47	2.43	2.40	2.38	2.35
13	4.67	3.81	3.41	3.18	3.03	2.92	2.83	2.77	2.71	2.67	2.53	2.46	2.38	2.34	2.31	2.30	2.26
14	4.60	3.74	3.34	3.11	2.96	2.85	2.76	2.70	2.65	2.60	2.46	2.39	2.31	2.27	2.24	2.22	2.19
15	4.54	3.68	3.29	3.06	2.90	2.79	2.71	2.64	2.59	2.54	2.40	2.33	2.25	2.20	2.18	2.16	2.12
16	4.49	3.63	3.24	3.01	2.85	2.74	2.66	2.59	2.54	2.49	2.35	2.28	2.19	2.15	2.12	2.11	2.07
17	4.45	3.59	3.20	2.96	2.81	2.70	2.61	2.55	2.49	2.45	2.31	2.23	2.15	2.10	2.08	2.06	2.02
18	4.41	3.55	3.16	2.93	2.77	2.66	2.58	2.51	2.46	2.41	2.27	2.19	2.11	2.06	2.04	2.02	1.98
19	4.38	3.52	3.13	2.90	2.74	2.63	2.54	2.48	2.42	2.38	2.23	2.16	2.07	2.03	2.00	1.98	1.94
20	4.35	3.49	3.10	2.87	2.71	2.60	2.51	2.45	2.39	2.35	2.20	2.12	2.04	1.99	1.97	1.95	1.91
21	4.32	3.47	3.07	2.84	2.68	2.57	2.49	2.42	2.37	2.32	2.18	2.10	2.01	1.96	1.94	1.92	1.88
22	4.30	3.44	3.05	2.82	2.66	2.55	2.46	2.40	2.34	2.30	2.15	2.07	1.98	1.94	1.91	1.89	1.85
23	4.28	3.42	3.03	2.80	2.64	2.53	2.44	2.37	2.32	2.27	2.13	2.05	1.96	1.91	1.88	1.86	1.82
24	4.26	3.40	3.01	2.78	2.62	2.51	2.42	2.36	2.30	2.25	2.11	2.03	1.94	1.89	1.86	1.84	1.80
25	4.24	3.39	2.99	2.76	2.60	2.49	2.40	2.34	2.28	2.24	2.09	2.01	1.92	1.87	1.84	1.82	1.78
30	4.17	3.32	2.92	2.69	2.53	2.42	2.33	2.27	2.21	2.16	2.01	1.93	1.84	1.79	1.76	1.74	1.70
40	4.08	3.23	2.84	2.61	2.45	2.34	2.25	2.18	2.12	2.08	1.92	1.84	1.74	1.69	1.66	1.64	1.59
50	4.03	3.18	2.79	2.56	2.40	2.29	2.20	2.13	2.07	2.03	1.87	1.78	1.69	1.63	1.60	1.58	1.52
60	4.00	3.15	2.76	2.53	2.37	2.25	2.17	2.10	2.04	1.99	1.84	1.75	1.65	1.59	1.56	1.53	1.48
100	3.94	3.09	2.70	2.46	2.31	2.19	2.10	2.03	1.97	1.93	1.77	1.68	1.57	1.52	1.48	1.45	1.39

Table VII Critical Values for the *F* Distribution (Continued)

$\alpha = 0.01$

v_2	1	2	3	4	5	6	7	8	9	10	15	20	30	40	50	60	100
2	98.50	99.00	99.17	99.25	99.30	99.33	99.36	99.37	99.39	99.40	99.43	99.45	99.47	99.47	99.48	99.48	99.49
3	34.12	30.82	29.46	28.71	28.24	27.91	27.67	27.49	27.35	27.23	26.87	26.69	26.50	26.41	26.35	26.32	26.24
4	21.20	18.00	16.69	15.98	15.52	15.21	14.98	14.80	14.66	14.55	14.20	14.02	13.84	13.75	13.69	13.65	13.58
5	16.26	13.27	12.06	11.39	10.97	10.67	10.46	10.29	10.16	10.05	9.72	9.55	9.38	9.29	9.24	9.20	9.13
6	13.75	10.92	9.78	9.15	8.75	8.47	8.26	8.10	7.98	7.87	7.56	7.40	7.23	7.14	7.09	7.06	6.99
7	12.25	9.55	8.45	7.85	7.46	7.19	6.99	6.84	6.72	6.62	6.31	6.16	5.99	5.91	5.86	5.82	5.75
8	11.26	8.65	7.59	7.01	6.63	6.37	6.18	6.03	5.91	5.81	5.52	5.36	5.20	5.12	5.07	5.03	4.96
9	10.56	8.02	6.99	6.42	6.06	5.80	5.61	5.47	5.35	5.26	4.96	4.81	4.65	4.57	4.52	4.48	4.41
10	10.04	7.56	6.55	5.99	5.64	5.39	5.20	5.06	4.94	4.85	4.56	4.41	4.25	4.17	4.12	4.08	4.01
11	9.65	7.21	6.22	5.67	5.32	5.07	4.89	4.74	4.63	4.54	4.25	4.10	3.94	3.86	3.81	3.78	3.71
12	9.33	6.93	5.95	5.41	5.06	4.82	4.64	4.50	4.39	4.30	4.01	3.86	3.70	3.62	3.57	3.54	3.47
13	9.07	6.70	5.74	5.21	4.86	4.62	4.44	4.30	4.19	4.10	3.82	3.66	3.51	3.43	3.38	3.34	3.27
14	8.86	6.51	5.56	5.04	4.69	4.46	4.28	4.14	4.03	3.94	3.66	3.51	3.35	3.27	3.22	3.18	3.11
15	8.68	6.36	5.42	4.89	4.56	4.32	4.14	4.00	3.89	3.80	3.52	3.37	3.21	3.13	3.08	3.05	2.98
16	8.53	6.23	5.29	4.77	4.44	4.20	4.03	3.89	3.78	3.69	3.41	3.26	3.10	3.02	2.97	2.93	2.86
17	8.40	6.11	5.18	4.67	4.34	4.10	3.93	3.79	3.68	3.59	3.31	3.16	3.00	2.92	2.87	2.83	2.76
18	8.29	6.01	5.09	4.58	4.25	4.01	3.84	3.71	3.60	3.51	3.23	3.08	2.92	2.84	2.78	2.75	2.68
19	8.18	5.93	5.01	4.50	4.17	3.94	3.77	3.63	3.52	3.43	3.15	3.00	2.84	2.76	2.71	2.67	2.60
20	8.10	5.85	4.94	4.43	4.10	3.87	3.70	3.56	3.46	3.37	3.09	2.94	2.78	2.69	2.64	2.61	2.54
21	8.02	5.78	4.87	4.37	4.04	3.81	3.64	3.51	3.40	3.31	3.03	2.88	2.72	2.64	2.58	2.55	2.48
22	7.95	5.72	4.82	4.31	3.99	3.76	3.59	3.45	3.35	3.26	2.98	2.83	2.67	2.58	2.53	2.50	2.42
23	7.88	5.66	4.76	4.26	3.94	3.71	3.54	3.41	3.30	3.21	2.93	2.78	2.62	2.54	2.48	2.45	2.37
24	7.82	5.61	4.72	4.22	3.90	3.67	3.50	3.36	3.26	3.17	2.89	2.74	2.58	2.49	2.44	2.40	2.33
25	7.77	5.57	4.68	4.18	3.85	3.63	3.46	3.32	3.22	3.13	2.85	2.70	2.54	2.45	2.40	2.36	2.29
30	7.56	5.39	4.51	4.02	3.70	3.47	3.30	3.17	3.07	2.98	2.70	2.55	2.39	2.30	2.25	2.21	2.13
40	7.31	5.18	4.31	3.83	3.51	3.29	3.12	2.99	2.89	2.80	2.52	2.37	2.20	2.11	2.06	2.02	1.94
50	7.17	5.06	4.20	3.72	3.41	3.19	3.02	2.89	2.78	2.70	2.42	2.27	2.10	2.01	1.95	1.91	1.82
60	7.08	4.98	4.13	3.65	3.34	3.12	2.95	2.82	2.72	2.63	2.35	2.20	2.03	1.94	1.88	1.84	1.75
100	6.90	4.82	3.98	3.51	3.21	2.99	2.82	2.69	2.59	2.50	2.22	2.07	1.89	1.80	1.74	1.69	1.60

v_1

Table VII Critical Values for the F Distribution (Continued)

$\alpha = 0.01$

ν_2 \ ν_1	1	2	3	4	5	6	7	8	9	10	15	20	30	40	50	60	100
2	998.50	999.00	999.17	999.25	999.30	999.33	999.36	999.37	999.39	999.40	999.43	999.45	999.47	999.47	999.48	999.48	999.49
3	167.03	148.50	141.11	137.10	134.58	132.85	131.58	130.62	129.86	129.25	127.37	126.42	125.45	124.96	124.66	124.47	124.07
4	74.14	61.25	56.18	53.44	51.71	50.53	49.66	49.00	48.47	48.05	46.76	46.10	45.43	45.09	44.88	44.75	44.47
5	47.18	37.12	33.20	31.09	29.75	28.83	28.16	27.65	27.24	26.92	25.91	25.39	24.87	24.60	24.44	24.33	24.12
6	35.51	27.00	23.70	21.92	20.80	20.03	19.46	19.03	18.69	18.41	17.56	17.12	16.67	16.44	16.31	16.21	16.03
7	29.25	21.69	18.77	17.20	16.21	15.52	15.02	14.63	14.33	14.08	13.32	12.93	12.53	12.33	12.20	12.12	11.95
8	25.41	18.49	15.83	14.39	13.48	12.86	12.40	12.05	11.77	11.54	10.84	10.48	10.11	9.92	9.80	9.73	9.57
9	22.86	16.39	13.90	12.56	11.71	11.13	10.70	10.37	10.11	9.89	9.24	8.90	8.55	8.37	8.26	8.19	8.04
10	21.04	14.91	12.55	11.28	10.48	9.93	9.52	9.20	8.96	8.75	8.13	7.80	7.47	7.30	7.19	7.12	6.98
11	19.69	13.81	11.56	10.35	9.58	9.05	8.66	8.35	8.12	7.92	7.32	7.01	6.68	6.52	6.42	6.35	6.21
12	18.64	12.97	10.80	9.63	8.89	8.38	8.00	7.71	7.48	7.29	6.71	6.40	6.09	5.93	5.83	5.76	5.63
13	17.82	12.31	10.21	9.07	8.35	7.86	7.49	7.21	6.98	6.80	6.23	5.93	5.63	5.47	5.37	5.30	5.17
14	17.14	11.78	9.73	8.62	7.92	7.44	7.08	6.80	6.58	6.40	5.85	5.56	5.25	5.10	5.00	4.94	4.81
15	16.59	11.34	9.34	8.25	7.57	7.09	6.74	6.47	6.26	6.08	5.54	5.25	4.95	4.80	4.70	4.64	4.51
16	16.12	10.97	9.01	7.94	7.27	6.80	6.46	6.19	5.98	5.81	5.27	4.99	4.70	4.54	4.45	4.39	4.26
17	15.72	10.66	8.73	7.68	7.02	6.56	6.22	5.96	5.75	5.58	5.05	4.78	4.48	4.33	4.24	4.18	4.05
18	15.38	10.39	8.49	7.46	6.81	6.35	6.02	5.76	5.56	5.39	4.87	4.59	4.30	4.15	4.06	4.00	3.87
19	15.08	10.16	8.28	7.27	6.62	6.18	5.85	5.59	5.39	5.22	4.70	4.43	4.14	3.99	3.90	3.84	3.71
20	14.82	9.95	8.10	7.10	6.46	6.02	5.69	5.44	5.24	5.08	4.56	4.29	4.00	3.86	3.77	3.70	3.58
21	14.59	9.77	7.94	6.95	6.32	5.88	5.56	5.31	5.11	4.95	4.44	4.17	3.88	3.74	3.64	3.58	3.46
22	14.38	9.61	7.80	6.81	6.19	5.76	5.44	5.19	4.99	4.83	4.33	4.06	3.78	3.63	3.54	3.48	3.35
23	14.20	9.47	7.67	6.70	6.08	5.65	5.33	5.09	4.89	4.73	4.23	3.96	3.68	3.53	3.44	3.38	3.25
24	14.03	9.34	7.55	6.59	5.98	5.55	5.23	4.99	4.80	4.64	4.14	3.87	3.59	3.45	3.36	3.29	3.17
25	13.88	9.22	7.45	6.49	5.89	5.46	5.15	4.91	4.71	4.56	4.06	3.79	3.52	3.37	3.28	3.22	3.09
30	13.29	8.77	7.05	6.12	5.53	5.12	4.82	4.58	4.39	4.24	3.75	3.49	3.22	3.07	2.98	2.92	2.79
40	12.61	8.25	6.59	5.70	5.13	4.73	4.44	4.21	4.02	3.87	3.40	3.14	2.87	2.73	2.64	2.57	2.44
50	12.22	7.96	6.34	5.46	4.90	4.51	4.22	4.00	3.82	3.67	3.20	2.95	2.68	2.53	2.44	2.38	2.25
60	11.97	7.77	6.17	5.31	4.76	4.37	4.09	3.86	3.69	3.54	3.08	2.83	2.55	2.41	2.32	2.25	2.12
100	11.50	7.41	5.86	5.02	4.48	4.11	3.83	3.61	3.44	3.30	2.84	2.59	2.32	2.17	2.08	2.01	1.87

Table VIII Critical Values for the Studentized Range Distribution

This table contains critical values associated with the Studentized range distribution, $Q_{\alpha,k,v}$, defined by α, and the degrees of freedom k and v, where k is the number of degrees of freedom in the numerator (the number of treatment groups) and v is the number of degrees of freedom in the denominator.

$\alpha = 0.05$

v	k																		
	2	3	4	5	6	7	8	9	10	11	12	13	14	15	16	17	18	19	20
2	6.085	8.331	9.798	10.881	11.734	12.434	13.027	13.538	13.987	14.387	14.747	15.076	15.375	15.650	15.905	16.143	16.365	16.573	16.769
3	4.501	5.910	6.825	7.502	8.037	8.478	8.852	9.177	9.462	9.717	9.946	10.155	10.346	10.522	10.686	10.838	10.980	11.114	11.240
4	3.926	5.040	5.757	6.287	6.706	7.053	7.347	7.602	7.826	8.027	8.208	8.373	8.524	8.664	8.793	8.914	9.027	9.133	9.233
5	3.635	4.602	5.218	5.673	6.033	6.330	6.582	6.801	6.995	7.167	7.324	7.465	7.596	7.716	7.828	7.932	8.030	8.122	8.208
6	3.460	4.339	4.896	5.305	5.629	5.895	6.122	6.319	6.493	6.649	6.789	6.917	7.034	7.143	7.244	7.338	7.426	7.509	7.587
7	3.344	4.165	4.681	5.060	5.359	5.606	5.815	5.997	6.158	6.302	6.431	6.550	6.658	6.759	6.852	6.939	7.020	7.097	7.169
8	3.261	4.041	4.529	4.886	5.167	5.399	5.596	5.767	5.918	6.053	6.175	6.287	6.389	6.483	6.571	6.653	6.729	6.801	6.870
9	3.199	3.948	4.415	4.755	5.024	5.244	5.432	5.595	5.738	5.867	5.983	6.089	6.186	6.276	6.359	6.437	6.510	6.579	6.644
10	3.151	3.877	4.327	4.654	4.912	5.124	5.304	5.460	5.598	5.722	5.833	5.935	6.028	6.114	6.194	6.269	6.339	6.405	6.467
11	3.113	3.820	4.256	4.574	4.823	5.028	5.202	5.353	5.486	5.605	5.713	5.811	5.901	5.984	6.062	6.134	6.202	6.265	6.325
12	3.081	3.773	4.199	4.508	4.750	4.950	5.119	5.265	5.395	5.510	5.615	5.710	5.797	5.878	5.953	6.023	6.089	6.151	6.209
14	3.033	3.701	4.111	4.407	4.639	4.829	4.990	5.130	5.253	5.363	5.463	5.554	5.637	5.714	5.785	5.852	5.915	5.973	6.029
15	3.014	3.673	4.076	4.367	4.595	4.782	4.940	5.077	5.198	5.306	5.403	5.492	5.574	5.649	5.719	5.785	5.846	5.904	5.958
16	2.998	3.649	4.046	4.333	4.557	4.741	4.896	5.031	5.150	5.256	5.352	5.439	5.519	5.593	5.662	5.726	5.786	5.843	5.896
17	2.984	3.628	4.020	4.303	4.524	4.705	4.858	4.991	5.108	5.212	5.306	5.392	5.471	5.544	5.612	5.675	5.734	5.790	5.842
18	2.971	3.609	3.997	4.276	4.494	4.673	4.824	4.955	5.071	5.173	5.266	5.351	5.429	5.501	5.567	5.629	5.688	5.743	5.794
19	2.960	3.593	3.977	4.253	4.468	4.645	4.794	4.924	5.037	5.139	5.231	5.314	5.391	5.462	5.528	5.589	5.647	5.701	5.752
20	2.950	3.578	3.958	4.232	4.445	4.620	4.768	4.895	5.008	5.108	5.199	5.282	5.357	5.427	5.492	5.553	5.610	5.663	5.714
25	2.913	3.523	3.890	4.153	4.358	4.526	4.667	4.789	4.897	4.993	5.079	5.158	5.230	5.297	5.359	5.417	5.471	5.522	5.570
30	2.888	3.487	3.845	4.102	4.301	4.464	4.601	4.720	4.824	4.917	5.001	5.077	5.147	5.211	5.271	5.327	5.379	5.429	5.475
40	2.858	3.442	3.791	4.039	4.232	4.388	4.521	4.634	4.735	4.824	4.904	4.977	5.044	5.106	5.163	5.216	5.266	5.313	5.358
50	2.841	3.416	3.758	4.002	4.190	4.344	4.473	4.584	4.681	4.768	4.847	4.918	4.983	5.043	5.098	5.150	5.199	5.245	5.288
100	2.806	3.365	3.695	3.929	4.109	4.256	4.379	4.484	4.577	4.659	4.733	4.800	4.862	4.918	4.971	5.020	5.066	5.108	5.149
200	2.789	3.339	3.664	3.893	4.069	4.212	4.332	4.435	4.525	4.605	4.677	4.742	4.802	4.857	4.908	4.955	4.999	5.041	5.080
300	2.783	3.331	3.654	3.881	4.056	4.198	4.317	4.419	4.508	4.587	4.659	4.723	4.782	4.837	4.887	4.934	4.978	5.019	5.057
400	2.780	3.327	3.649	3.875	4.050	4.191	4.309	4.411	4.500	4.578	4.649	4.714	4.772	4.826	4.876	4.923	4.967	5.007	5.046
500	2.779	3.324	3.645	3.872	4.046	4.187	4.305	4.406	4.494	4.573	4.644	4.708	4.766	4.820	4.870	4.917	4.960	5.001	5.039

Table VIII Critical Values for the Studentized Range Distribution (Continued)

$\alpha = 0.01$

									k										
ν	2	3	4	5	6	7	8	9	10	11	12	13	14	15	16	17	18	19	20
2	14.035	19.019	22.293	24.717	26.628	28.199	29.528	30.677	31.687	32.585	33.395	34.129	34.802	35.421	35.995	36.529	37.028	37.496	37.937
3	8.260	10.616	12.169	13.324	14.240	14.997	15.640	16.198	16.689	17.128	17.524	17.884	18.214	18.519	18.802	19.065	19.311	19.543	19.761
4	6.511	8.118	9.173	9.958	10.582	11.099	11.539	11.925	12.264	12.566	12.840	13.089	13.318	13.530	13.726	13.909	14.081	14.242	14.394
5	5.702	6.976	7.806	8.421	8.913	9.321	9.669	9.971	10.239	10.479	10.695	10.893	11.075	11.243	11.399	11.544	11.681	11.809	11.930
6	5.243	6.331	7.033	7.556	7.974	8.318	8.611	8.869	9.097	9.300	9.485	9.653	9.808	9.951	10.084	10.208	10.325	10.434	10.538
7	4.948	5.919	6.543	7.006	7.373	7.678	7.940	8.167	8.368	8.548	8.711	8.859	8.996	9.124	9.242	9.353	9.456	9.553	9.645
8	4.745	5.635	6.204	6.625	6.960	7.238	7.475	7.681	7.864	8.028	8.177	8.312	8.437	8.552	8.659	8.760	8.854	8.942	9.026
9	4.595	5.428	5.957	6.347	6.658	6.915	7.134	7.326	7.495	7.647	7.785	7.910	8.026	8.133	8.233	8.326	8.413	8.495	8.573
10	4.482	5.270	5.769	6.136	6.428	6.669	6.875	7.055	7.214	7.356	7.485	7.603	7.712	7.813	7.906	7.994	8.076	8.153	8.226
11	4.392	5.146	5.621	5.970	6.247	6.476	6.671	6.842	6.992	7.127	7.250	7.362	7.465	7.560	7.649	7.732	7.810	7.883	7.952
12	4.320	5.046	5.502	5.836	6.101	6.321	6.507	6.670	6.814	6.943	7.060	7.167	7.265	7.356	7.441	7.520	7.594	7.664	7.731
13	4.261	4.964	5.404	5.727	5.981	6.192	6.372	6.528	6.666	6.791	6.903	7.006	7.100	7.188	7.269	7.345	7.417	7.484	7.548
14	4.210	4.895	5.322	5.634	5.881	6.085	6.258	6.409	6.543	6.664	6.772	6.871	6.962	7.047	7.125	7.199	7.268	7.333	7.394
15	4.167	4.836	5.252	5.556	5.796	5.994	6.162	6.309	6.438	6.555	6.660	6.757	6.845	6.927	7.003	7.074	7.141	7.204	7.264
16	4.131	4.786	5.192	5.488	5.722	5.915	6.079	6.222	6.348	6.461	6.564	6.658	6.743	6.824	6.897	6.967	7.032	7.093	7.151
17	4.099	4.742	5.140	5.430	5.659	5.847	6.007	6.147	6.270	6.380	6.480	6.572	6.656	6.733	6.806	6.873	6.937	6.997	7.053
18	4.071	4.703	5.094	5.379	5.603	5.787	5.944	6.081	6.201	6.309	6.407	6.496	6.579	6.655	6.725	6.791	6.854	6.912	6.967
19	4.046	4.669	5.054	5.333	5.553	5.735	5.888	6.022	6.141	6.246	6.342	6.430	6.510	6.585	6.654	6.719	6.780	6.837	6.891
20	4.024	4.639	5.018	5.293	5.509	5.687	5.839	5.970	6.086	6.190	6.285	6.370	6.449	6.523	6.591	6.654	6.714	6.770	6.823
25	3.942	4.527	4.884	5.143	5.346	5.513	5.654	5.777	5.885	5.982	6.070	6.150	6.223	6.291	6.355	6.414	6.469	6.521	6.571
30	3.889	4.454	4.799	5.048	5.242	5.401	5.536	5.653	5.756	5.848	5.932	6.008	6.078	6.142	6.202	6.258	6.311	6.360	6.407
40	3.825	4.367	4.695	4.931	5.114	5.265	5.392	5.502	5.599	5.685	5.764	5.835	5.900	5.961	6.017	6.069	6.118	6.165	6.208
50	3.787	4.316	4.634	4.863	5.040	5.185	5.308	5.414	5.507	5.590	5.665	5.734	5.796	5.854	5.908	5.958	6.005	6.050	6.092
100	3.714	4.216	4.516	4.730	4.896	5.031	5.144	5.242	5.328	5.405	5.474	5.537	5.594	5.648	5.697	5.743	5.786	5.826	5.864
200	3.714	4.216	4.516	4.730	4.896	5.031	5.144	5.242	5.328	5.405	5.474	5.537	5.594	5.648	5.697	5.743	5.786	5.826	5.864
300	3.666	4.152	4.440	4.645	4.803	4.931	5.039	5.132	5.213	5.286	5.351	5.410	5.464	5.514	5.560	5.603	5.644	5.682	5.717
400	3.661	4.144	4.431	4.634	4.791	4.919	5.026	5.118	5.199	5.271	5.335	5.394	5.448	5.498	5.543	5.586	5.626	5.664	5.699
500	3.657	4.139	4.425	4.628	4.784	4.911	5.018	5.110	5.190	5.262	5.327	5.385	5.438	5.488	5.533	5.576	5.616	5.653	5.688

Table VIII Critical Values for the Studentized Range Distribution (Continued)

α = 0.001

v	2	3	4	5	6	7	8	9	10	11	12	13	14	15	16	17	18	19	20
2	44.666	60.323	70.586	78.162	84.127	89.022	93.650	97.285	100.480	103.325	105.886	108.211	110.340	112.300	114.115	115.805	117.385	118.867	120.263
3	18.275	23.298	26.609	29.075	31.030	32.645	34.016	35.327	36.389	37.338	38.194	38.974	39.688	40.347	40.959	41.529	42.062	42.564	43.036
4	12.174	14.965	16.798	18.225	19.333	20.253	21.037	21.719	22.323	22.862	23.350	23.795	24.204	24.581	24.932	25.259	25.566	25.854	26.126
5	9.710	11.671	12.959	13.924	14.695	15.335	15.882	16.358	16.780	17.158	17.500	17.811	18.098	18.402	18.651	18.884	19.102	19.307	19.501
6	8.431	9.955	10.965	11.719	12.322	12.824	13.254	13.629	13.961	14.260	14.530	14.777	15.004	15.215	15.411	15.593	15.765	15.927	16.079
7	7.649	8.933	9.761	10.388	10.883	11.316	11.674	11.988	12.265	12.515	12.742	12.949	13.139	13.316	13.480	13.634	13.778	13.914	14.043
8	7.130	8.252	8.980	9.523	9.948	10.317	10.625	10.894	11.133	11.347	11.559	11.740	11.906	12.060	12.203	12.337	12.463	12.582	12.694
9	7.130	8.252	8.980	9.523	9.948	10.317	10.625	10.894	11.133	11.347	11.559	11.740	11.906	12.060	12.203	12.337	12.463	12.582	12.694
10	6.486	7.411	8.007	8.451	8.805	9.100	9.353	9.574	9.770	9.954	10.106	10.245	10.387	10.512	10.629	10.737	10.840	10.936	11.027
11	6.274	7.137	7.688	8.099	8.427	8.700	8.934	9.138	9.320	9.483	9.631	9.767	9.892	10.017	10.121	10.218	10.309	10.394	10.475
12	6.106	6.917	7.442	7.820	8.128	8.383	8.602	8.793	8.963	9.116	9.254	9.381	9.498	9.607	9.708	9.803	9.892	9.976	10.055
13	5.969	6.740	7.234	7.595	7.885	8.126	8.333	8.513	8.674	8.818	8.949	9.068	9.179	9.281	9.377	9.466	9.550	9.630	9.705
14	5.855	6.593	7.070	7.410	7.692	7.914	8.111	8.282	8.434	8.571	8.696	8.810	8.915	9.012	9.103	9.188	9.268	9.343	9.414
15	5.760	6.470	6.920	7.257	7.517	7.742	7.924	8.088	8.234	8.365	8.483	8.592	8.693	8.786	8.873	8.954	9.030	9.102	9.170
16	5.678	6.365	6.799	7.125	7.377	7.585	7.769	7.923	8.063	8.189	8.303	8.407	8.504	8.593	8.676	8.754	8.828	8.897	8.963
17	5.614	6.274	6.695	7.010	7.254	7.457	7.629	7.783	7.921	8.037	8.147	8.248	8.341	8.427	8.508	8.583	8.654	8.720	8.783
18	5.550	6.201	6.609	6.909	7.147	7.343	7.511	7.658	7.781	7.908	8.017	8.116	8.199	8.283	8.361	8.433	8.502	8.566	8.628
19	5.493	6.129	6.527	6.820	7.051	7.243	7.407	7.550	7.676	7.790	7.894	7.990	8.079	8.162	8.238	8.302	8.369	8.431	8.491
20	5.444	6.065	6.455	6.741	6.967	7.154	7.314	7.453	7.577	7.687	7.788	7.880	7.966	8.046	8.121	8.190	8.256	8.318	8.376
25	5.264	5.840	6.196	6.456	6.662	6.831	6.976	7.102	7.213	7.314	7.404	7.487	7.558	7.629	7.696	7.758	7.816	7.871	7.924
30	5.154	5.698	6.033	6.277	6.469	6.628	6.763	6.880	6.984	7.077	7.161	7.238	7.309	7.375	7.436	7.494	7.548	7.598	7.646
40	5.022	5.527	5.837	6.062	6.239	6.385	6.508	6.615	6.710	6.795	6.872	6.942	7.006	7.066	7.121	7.173	7.222	7.268	7.312
50	4.946	5.426	5.725	5.939	6.107	6.245	6.361	6.463	6.552	6.632	6.705	6.771	6.832	6.888	6.940	6.989	7.035	7.078	7.119
100	4.795	5.244	5.512	5.706	5.855	5.978	6.083	6.173	6.252	6.323	6.387	6.445	6.499	6.548	6.594	6.637	6.678	6.715	6.751
200	4.723	5.151	5.408	5.596	5.738	5.854	5.952	6.038	6.110	6.178	6.237	6.292	6.342	6.388	6.431	6.471	6.509	6.544	6.577
300	4.700	5.122	5.375	5.556	5.696	5.814	5.910	5.993	6.066	6.131	6.189	6.244	6.291	6.335	6.379	6.418	6.455	6.489	6.522
400	4.688	5.107	5.358	5.538	5.677	5.791	5.890	5.972	6.044	6.108	6.166	6.219	6.267	6.312	6.355	6.393	6.427	6.460	6.494
500	4.681	5.098	5.348	5.527	5.665	5.778	5.874	5.959	6.031	6.095	6.152	6.205	6.253	6.297	6.338	6.376	6.412	6.448	6.479

k

Table IX Critical Values for the Wilcoxon Signed-Rank Statistic

This table contains critical values and probabilities for the Wilcoxon signed-rank statistic T_+: n is the sample size, c_1 and c_2 are defined by $P(T_+ \leq c_1) = \alpha$, and $P(T_+ \geq c_2) = \alpha$.

n	c_1	c_2	α
1	0	1	0.5000
2	0	3	0.2500
3	0	6	0.1250
4	0	10	0.0625
	1	9	0.1250
5	0	15	0.0313
	1	14	0.0625
	2	13	0.0938
	3	12	0.1563
6	0	21	0.0156
	1	20	0.0313
	2	19	0.0469
	3	18	0.0781
	4	17	0.1094
	5	16	0.1563
7	0	28	0.0078
	1	27	0.0156
	2	26	0.0234
	3	25	0.0391
	4	24	0.0547
	5	23	0.0781
	6	22	0.1094
	7	21	0.1484
8	0	36	0.0039
	1	35	0.0078
	2	34	0.0117
	3	33	0.0195
	4	32	0.0273
	5	31	0.0391
	6	30	0.0547
	7	29	0.0742
	8	28	0.0977
	9	27	0.1250
9	0	45	0.0020
	1	44	0.0039
	2	43	0.0059
	3	42	0.0098
	4	41	0.0137
	5	40	0.0195
	6	39	0.0273
	7	38	0.0371
	8	37	0.0488
	9	36	0.0645
	10	35	0.0820
	11	34	0.1016
	12	33	0.1250

n	c_1	c_2	α
10	0	55	0.0010
	1	54	0.0020
	2	53	0.0029
	3	52	0.0049
	4	51	0.0068
	5	50	0.0098
	6	49	0.0137
	7	48	0.0186
	8	47	0.0244
	9	46	0.0322
	10	45	0.0420
	11	44	0.0527
	12	43	0.0654
	13	42	0.0801
	14	41	0.0967
	15	40	0.1162
	16	39	0.1377
11	0	66	0.0005
	1	65	0.0010
	2	64	0.0015
	3	63	0.0024
	4	62	0.0034
	5	61	0.0049
	6	60	0.0068
	7	59	0.0093
	8	58	0.0122
	9	57	0.0161
	10	56	0.0210
	11	55	0.0269
	12	54	0.0337
	13	53	0.0415
	14	52	0.0508
	15	51	0.0615
	16	50	0.0737
	17	49	0.0874
	18	48	0.1030
	19	47	0.1201
	20	46	0.1392

n	c_1	c_2	α
12	0	78	0.0002
	1	77	0.0005
	2	76	0.0007
	3	75	0.0012
	4	74	0.0017
	5	73	0.0024
	6	72	0.0034
	7	71	0.0046
	8	70	0.0061
	9	69	0.0081
	10	68	0.0105
	11	67	0.0134
	12	66	0.0171
	13	65	0.0212
	14	64	0.0261
	15	63	0.0320
	16	62	0.0386
	17	61	0.0461
	18	60	0.0549
	19	59	0.0647
	20	58	0.0757
	21	57	0.0881
	22	56	0.1018
	23	55	0.1167
	24	54	0.1331
	25	53	0.1506

n	c_1	c_2	α
13	0	91	0.0001
	1	90	0.0002
	2	89	0.0004
	3	88	0.0006
	4	87	0.0009
	5	86	0.0012
	6	85	0.0017
	7	84	0.0023
	8	83	0.0031
	9	82	0.0040
	10	81	0.0052
	11	80	0.0067
	12	79	0.0085
	13	78	0.0107
	14	77	0.0133
	15	76	0.0164
	16	75	0.0199
	17	74	0.0239
	18	73	0.0287
	19	72	0.0341
	20	71	0.0402
	21	70	0.0471
	22	69	0.0549
	23	68	0.0636
	24	67	0.0732
	25	66	0.0839
	26	65	0.0955
	27	64	0.1082
	28	63	0.1219
	29	62	0.1367
	30	61	0.1527

n	c_1	c_2	α
14	0	105	0.0001
	1	104	0.0001
	2	103	0.0002
	3	102	0.0003
	4	101	0.0004
	5	100	0.0006
	6	99	0.0009
	7	98	0.0012
	8	97	0.0015
	9	96	0.0020
	10	95	0.0026
	11	94	0.0034
	12	93	0.0043
	13	92	0.0054
	14	91	0.0067
	15	90	0.0083
	16	89	0.0101
	17	88	0.0123
	18	87	0.0148
	19	86	0.0176
	20	85	0.0209
	21	84	0.0247
	22	83	0.0290
	23	82	0.0338
	24	81	0.0392
	25	80	0.0453
	26	79	0.0520
	27	78	0.0594
	28	77	0.0676
	29	76	0.0765
	30	75	0.0863
	31	74	0.0969
	32	73	0.1083
	33	72	0.1206
	34	71	0.1338
	35	70	0.1479
	36	69	0.1629

Table IX Critical Values for the Wilcoxon Signed-Rank Statistic (Continued)

n	c_1	c_2	α	n	c_1	c_2	α	n	c_1	c_2	α	n	c_1	c_2	α
15	0	120	0.0000	16	0	136	0.0000	17	0	153	0.0000	18	0	171	0.0000
	1	119	0.0001		1	135	0.0000		1	152	0.0000		1	170	0.0000
	2	118	0.0001		2	134	0.0000		2	151	0.0000		2	169	0.0000
	3	117	0.0002		3	133	0.0001		3	150	0.0000		3	168	0.0000
	4	116	0.0002		4	132	0.0001		4	149	0.0001		4	167	0.0000
	5	115	0.0003		5	131	0.0002		5	148	0.0001		5	166	0.0000
	6	114	0.0004		6	130	0.0002		6	147	0.0001		6	165	0.0001
	7	113	0.0006		7	129	0.0003		7	146	0.0001		7	164	0.0001
	8	112	0.0008		8	128	0.0004		8	145	0.0002		8	163	0.0001
	9	111	0.0010		9	127	0.0005		9	144	0.0003		9	162	0.0001
	10	110	0.0013		10	126	0.0007		10	143	0.0003		10	161	0.0002
	11	109	0.0017		11	125	0.0008		11	142	0.0004		11	160	0.0002
	12	108	0.0021		12	124	0.0011		12	141	0.0005		12	159	0.0003
	13	107	0.0027		13	123	0.0013		13	140	0.0007		13	158	0.0003
	14	106	0.0034		14	122	0.0017		14	139	0.0008		14	157	0.0004
	15	105	0.0042		15	121	0.0021		15	138	0.0010		15	156	0.0005
	16	104	0.0051		16	120	0.0026		16	137	0.0013		16	155	0.0006
	17	103	0.0062		17	119	0.0031		17	136	0.0016		17	154	0.0008
	18	102	0.0075		18	118	0.0038		18	135	0.0019		18	153	0.0010
	19	101	0.0090		19	117	0.0046		19	134	0.0023		19	152	0.0012
	20	100	0.0108		20	116	0.0055		20	133	0.0028		20	151	0.0014
	21	99	0.0128		21	115	0.0065		21	132	0.0033		21	150	0.0017
	22	98	0.0151		22	114	0.0078		22	131	0.0040		22	149	0.0020
	23	97	0.0177		23	113	0.0091		23	130	0.0047		23	148	0.0024
	24	96	0.0206		24	112	0.0107		24	129	0.0055		24	147	0.0028
	25	95	0.0240		25	111	0.0125		25	128	0.0064		25	146	0.0033
	26	94	0.0277		26	110	0.0145		26	127	0.0075		26	145	0.0038
	27	93	0.0319		27	109	0.0168		27	126	0.0087		27	144	0.0045
	28	92	0.0365		28	108	0.0193		28	125	0.0101		28	143	0.0052
	29	91	0.0416		29	107	0.0222		29	124	0.0116		29	142	0.0060
	30	90	0.0473		30	106	0.0253		30	123	0.0133		30	141	0.0069
	31	89	0.0535		31	105	0.0288		31	122	0.0153		31	140	0.0080
	32	88	0.0603		32	104	0.0327		32	121	0.0174		32	139	0.0091
	33	87	0.0677		33	103	0.0370		33	120	0.0198		33	138	0.0104
	34	86	0.0757		34	102	0.0416		34	119	0.0224		34	137	0.0118
	35	85	0.0844		35	101	0.0467		35	118	0.0253		35	136	0.0134
	36	84	0.0938		36	100	0.0523		36	117	0.0284		36	135	0.0152
	37	83	0.1039		37	99	0.0583		37	116	0.0319		37	134	0.0171
	38	82	0.1147		38	98	0.0649		38	115	0.0357		38	133	0.0192
	39	81	0.1262		39	97	0.0719		39	114	0.0398		39	132	0.0216
	40	80	0.1384		40	96	0.0795		40	113	0.0443		40	131	0.0241
	41	79	0.1514		41	95	0.0877		41	112	0.0492		41	130	0.0269
					42	94	0.0964		42	111	0.0544		42	129	0.0300
					43	93	0.1057		43	110	0.0601		43	128	0.0333
					44	92	0.1156		44	109	0.0662		44	127	0.0368
					45	91	0.1261		45	108	0.0727		45	126	0.0407
					46	90	0.1372		46	107	0.0797		46	125	0.0449
					47	89	0.1489		47	106	0.0871		47	124	0.0494
									48	105	0.0950		48	123	0.0542
									49	104	0.1034		49	122	0.0594
									50	103	0.1123		50	121	0.0649
									51	102	0.1217		51	120	0.0708
									52	101	0.1317		52	119	0.0770
									53	100	0.1421		53	118	0.0837
									54	99	0.1530		54	117	0.0907
													55	116	0.0982
													56	115	0.1061
													57	114	0.1144
													58	113	0.1231
													59	112	0.1323
													60	111	0.1419
													61	110	0.1519

Table IX Critical Values for the Wilcoxon Signed-Rank Statistic (Continued)

n	c_1	c_2	α		n	c_1	c_2	α		n	c_1	c_2	α		n	c_1	c_2	α
19	0	190	0.0000		19	41	149	0.0145		20	0	210	0.0000		20	41	169	0.0077
	1	189	0.0000			42	148	0.0162			1	209	0.0000			42	168	0.0086
	2	188	0.0000			43	147	0.0180			2	208	0.0000			43	167	0.0096
	3	187	0.0000			44	146	0.0201			3	207	0.0000			44	166	0.0107
	4	186	0.0000			45	145	0.0223			4	206	0.0000			45	165	0.0120
	5	185	0.0000			46	144	0.0247			5	205	0.0000			46	164	0.0133
	6	184	0.0000			47	143	0.0273			6	204	0.0000			47	163	0.0148
	7	183	0.0000			48	142	0.0301			7	203	0.0000			48	162	0.0164
	8	182	0.0000			49	141	0.0331			8	202	0.0000			49	161	0.0181
	9	181	0.0001			50	140	0.0364			9	201	0.0000			50	160	0.0200
	10	180	0.0001			51	139	0.0399			10	200	0.0000			51	159	0.0220
	11	179	0.0001			52	138	0.0437			11	199	0.0001			52	158	0.0242
	12	178	0.0001			53	137	0.0478			12	198	0.0001			53	157	0.0266
	13	177	0.0002			54	136	0.0521			13	197	0.0001			54	156	0.0291
	14	176	0.0002			55	135	0.0567			14	196	0.0001			55	155	0.0319
	15	175	0.0003			56	134	0.0616			15	195	0.0001			56	154	0.0348
	16	174	0.0003			57	133	0.0668			16	194	0.0002			57	153	0.0379
	17	173	0.0004			58	132	0.0723			17	193	0.0002			58	152	0.0413
	18	172	0.0005			59	131	0.0782			18	192	0.0002			59	151	0.0448
	19	171	0.0006			60	130	0.0844			19	191	0.0003			60	150	0.0487
	20	170	0.0007			61	129	0.0909			20	190	0.0004			61	149	0.0527
	21	169	0.0008			62	128	0.0978			21	189	0.0004			62	148	0.0570
	22	168	0.0010			63	127	0.1051			22	188	0.0005			63	147	0.0615
	23	167	0.0012			64	126	0.1127			23	187	0.0006			64	146	0.0664
	24	166	0.0014			65	125	0.1206			24	186	0.0007			65	145	0.0715
	25	165	0.0017			66	124	0.1290			25	185	0.0008			66	144	0.0768
	26	164	0.0020			67	123	0.1377			26	184	0.0010			67	143	0.0825
	27	163	0.0023			68	122	0.1467			27	183	0.0012			68	142	0.0884
	28	162	0.0027			69	121	0.1562			28	182	0.0014			69	141	0.0947
	29	161	0.0031			70	120	0.1660			29	181	0.0016			70	140	0.1012
	30	160	0.0036								30	180	0.0018			71	139	0.1081
	31	159	0.0041								31	179	0.0021			72	138	0.1153
	32	158	0.0047								32	178	0.0024			73	137	0.1227
	33	157	0.0054								33	177	0.0028			74	136	0.1305
	34	156	0.0062								34	176	0.0032			75	135	0.1387
	35	155	0.0070								35	175	0.0036			76	134	0.1471
	36	154	0.0080								36	174	0.0042			77	133	0.1559
	37	153	0.0090								37	173	0.0047					
	38	152	0.0102								38	172	0.0053					
	39	151	0.0115								39	171	0.0060					
	40	150	0.0129								40	170	0.0068					

Table X Critical Values for the Wilcoxon Rank-Sum Statistic

This table contains critical values and probabilities for the Wilcoxon rank-sum statistic. W = the sum of the ranks of the m observations in the smaller sample; m and n are the sample sizes, c_1 and c_2 are defined by $P(W \leq c_1) = \alpha$, and $P(W \geq c_2) = \alpha$.

m	n	c_1	c_2	α
2	3	3	9	0.1000
2	4	3	11	0.0667
		4	10	0.1333
2	5	3	13	0.0476
		4	12	0.0952
2	6	3	15	0.0357
		4	14	0.0714
		5	13	0.1429
2	7	3	17	0.0278
		4	16	0.0556
		5	15	0.1111
2	8	3	19	0.0222
		4	18	0.0444
		5	17	0.0889
		6	16	0.1333
2	9	3	21	0.0182
		4	20	0.0364
		5	19	0.0727
		6	18	0.1091
2	10	3	23	0.0152
		4	22	0.0303
		5	21	0.0606
		6	20	0.0909
		7	19	0.1364
3	3	6	15	0.0500
		7	14	0.1000
3	4	6	18	0.0286
		7	17	0.0571
		8	16	0.1143
3	5	6	21	0.0179
		7	20	0.0357
		8	19	0.0714
		9	18	0.1250
3	6	6	24	0.0119
		7	23	0.0238
		8	22	0.0476
		9	21	0.0833
		10	20	0.1310
3	7	6	27	0.0083
		7	26	0.0167
		8	25	0.0333
		9	24	0.0583
		10	23	0.0917
		11	22	0.1333

m	n	c_1	c_2	α
3	8	6	30	0.0061
		7	29	0.0121
		8	28	0.0242
		9	27	0.0424
		10	26	0.0667
		11	25	0.0970
		12	24	0.1394
3	9	6	33	0.0045
		7	32	0.0091
		8	31	0.0182
		9	30	0.0318
		10	29	0.0500
		11	28	0.0727
		12	27	0.1045
		13	26	0.1409
3	10	6	36	0.0035
		7	35	0.0070
		8	34	0.0140
		9	33	0.0245
		10	32	0.0385
		11	31	0.0559
		12	30	0.0804
		13	29	0.1084
		14	28	0.1434
4	4	10	26	0.0143
		11	25	0.0286
		12	24	0.0571
		13	23	0.1000
4	5	10	30	0.0079
		11	29	0.0159
		12	28	0.0317
		13	27	0.0556
		14	26	0.0952
		15	25	0.1429
4	6	10	34	0.0048
		11	33	0.0095
		12	32	0.0190
		13	31	0.0333
		14	30	0.0571
		15	29	0.0857
		16	28	0.1286

m	n	c_1	c_2	α
4	7	10	38	0.0030
		11	37	0.0061
		12	36	0.0121
		13	35	0.0212
		14	34	0.0364
		15	33	0.0545
		16	32	0.0818
		17	31	0.1152
		18	30	0.1576
4	8	10	42	0.0020
		11	41	0.0040
		12	40	0.0081
		13	39	0.0141
		14	38	0.0242
		15	37	0.0364
		16	36	0.0545
		17	35	0.0768
		18	34	0.1071
		19	33	0.1414
4	9	10	46	0.0014
		11	45	0.0028
		12	44	0.0056
		13	43	0.0098
		14	42	0.0168
		15	41	0.0252
		16	40	0.0378
		17	39	0.0531
		18	38	0.0741
		19	37	0.0993
		20	36	0.1301
4	10	10	50	0.0010
		11	49	0.0020
		12	48	0.0040
		13	47	0.0070
		14	46	0.0120
		15	45	0.0180
		16	44	0.0270
		17	43	0.0380
		18	42	0.0529
		19	41	0.0709
		20	40	0.0939
		21	39	0.1199
		22	38	0.1518

m	n	c_1	c_2	α
5	5	15	40	0.0040
		16	39	0.0079
		17	38	0.0159
		18	37	0.0278
		19	36	0.0476
		20	35	0.0754
		21	34	0.1111
		22	33	0.1548
5	6	15	45	0.0022
		16	44	0.0043
		17	43	0.0087
		18	42	0.0152
		19	41	0.0260
		20	40	0.0411
		21	39	0.0628
		22	38	0.0887
		23	37	0.1234
5	7	15	50	0.0013
		16	49	0.0025
		17	48	0.0051
		18	47	0.0088
		19	46	0.0152
		20	45	0.0240
		21	44	0.0366
		22	43	0.0530
		23	42	0.0745
		24	41	0.1010
		25	40	0.1338
5	8	15	55	0.0008
		16	54	0.0016
		17	53	0.0031
		18	52	0.0054
		19	51	0.0093
		20	50	0.0148
		21	49	0.0225
		22	48	0.0326
		23	47	0.0466
		24	46	0.0637
		25	45	0.0855
		26	44	0.1111
		27	43	0.1422

Table X Critical Values for the Wilcoxon Rank-Sum Statistic (Continued)

m	n	c_1	c_2	α
5	9	15	60	0.0005
		16	59	0.0010
		17	58	0.0020
		18	57	0.0035
		19	56	0.0060
		20	55	0.0095
		21	54	0.0145
		22	53	0.0210
		23	52	0.0300
		24	51	0.0415
		25	50	0.0559
		26	49	0.0734
		27	48	0.0949
		28	47	0.1199
		29	46	0.1489
5	10	15	65	0.0003
		16	64	0.0007
		17	63	0.0013
		18	62	0.0023
		19	61	0.0040
		20	60	0.0063
		21	59	0.0097
		22	58	0.0140
		23	57	0.0200
		24	56	0.0276
		25	55	0.0376
		26	54	0.0496
		27	53	0.0646
		28	52	0.0823
		29	51	0.1032
		30	50	0.1272
		31	49	0.1548
6	6	21	57	0.0011
		22	56	0.0022
		23	55	0.0043
		24	54	0.0076
		25	53	0.0130
		26	52	0.0206
		27	51	0.0325
		28	50	0.0465
		29	49	0.0660
		30	48	0.0898
		31	47	0.1201
		32	46	0.1548

m	n	c_1	c_2	α
6	7	21	63	0.0006
		22	62	0.0012
		23	61	0.0023
		24	60	0.0041
		25	59	0.0070
		26	58	0.0111
		27	57	0.0175
		28	56	0.0256
		29	55	0.0367
		30	54	0.0507
		31	53	0.0688
		32	52	0.0903
		33	51	0.1171
		34	50	0.1474
6	8	21	69	0.0003
		22	68	0.0007
		23	67	0.0013
		24	66	0.0023
		25	65	0.0040
		26	64	0.0063
		27	63	0.0100
		28	62	0.0147
		29	61	0.0213
		30	60	0.0296
		31	59	0.0406
		32	58	0.0539
		33	57	0.0709
		34	56	0.0906
		35	55	0.1142
		36	54	0.1412
6	9	21	75	0.0002
		22	74	0.0004
		23	73	0.0008
		24	72	0.0014
		25	71	0.0024
		26	70	0.0038
		27	69	0.0060
		28	68	0.0088
		29	67	0.0128
		30	66	0.0180
		31	65	0.0248
		32	64	0.0332
		33	63	0.0440
		34	62	0.0567
		35	61	0.0723
		36	60	0.0905
		37	59	0.1119
		38	58	0.1361

m	n	c_1	c_2	α
6	10	21	81	0.0001
		22	80	0.0002
		23	79	0.0005
		24	78	0.0009
		25	77	0.0015
		26	76	0.0024
		27	75	0.0037
		28	74	0.0055
		29	73	0.0080
		30	72	0.0112
		31	71	0.0156
		32	70	0.0210
		33	69	0.0280
		34	68	0.0363
		35	67	0.0467
		36	66	0.0589
		37	65	0.0736
		38	64	0.0903
		39	63	0.1099
		40	62	0.1317
		41	61	0.1566
7	7	28	77	0.0003
		29	76	0.0006
		30	75	0.0012
		31	74	0.0020
		32	73	0.0035
		33	72	0.0055
		34	71	0.0087
		35	70	0.0131
		36	69	0.0189
		37	68	0.0265
		38	67	0.0364
		39	66	0.0487
		40	65	0.0641
		41	64	0.0825
		42	63	0.1043
		43	62	0.1297
		44	61	0.1588

m	n	c_1	c_2	α
7	8	28	84	0.0002
		29	83	0.0003
		30	82	0.0006
		31	81	0.0011
		32	80	0.0019
		33	79	0.0030
		34	78	0.0047
		35	77	0.0070
		36	76	0.0103
		37	75	0.0145
		38	74	0.0200
		39	73	0.0270
		40	72	0.0361
		41	71	0.0469
		42	70	0.0603
		43	69	0.0760
		44	68	0.0946
		45	67	0.1159
		46	66	0.1405
7	9	28	91	0.0001
		29	90	0.0002
		30	89	0.0003
		31	88	0.0006
		32	87	0.0010
		33	86	0.0017
		34	85	0.0026
		35	84	0.0039
		36	83	0.0058
		37	82	0.0082
		38	81	0.0115
		39	80	0.0156
		40	79	0.0209
		41	78	0.0274
		42	77	0.0356
		43	76	0.0454
		44	75	0.0571
		45	74	0.0708
		46	73	0.0869
		47	72	0.1052
		48	71	0.1261
		49	70	0.1496

Table X Critical Values for the Wilcoxon Rank-Sum Statistic (Continued)

m	n	c_1	c_2	α
7	10	28	98	0.0001
		29	97	0.0001
		30	96	0.0002
		31	95	0.0004
		32	94	0.0006
		33	93	0.0010
		34	92	0.0015
		35	91	0.0023
		36	90	0.0034
		37	89	0.0048
		38	88	0.0068
		39	87	0.0093
		40	86	0.0125
		41	85	0.0165
		42	84	0.0215
		43	83	0.0277
		44	82	0.0351
		45	81	0.0439
		46	80	0.0544
		47	79	0.0665
		48	78	0.0806
		49	77	0.0966
		50	76	0.1148
		51	75	0.1349
		52	74	0.1574
8	8	36	100	0.0001
		37	99	0.0002
		38	98	0.0003
		39	97	0.0005
		40	96	0.0009
		41	95	0.0015
		42	94	0.0023
		43	93	0.0035
		44	92	0.0052
		45	91	0.0074
		46	90	0.0103
		47	89	0.0141
		48	88	0.0190
		49	87	0.0249
		50	86	0.0325
		51	85	0.0415
		52	84	0.0524
		53	83	0.0652
		54	82	0.0803
		55	81	0.0974
		56	80	0.1172
		57	79	0.1393

m	n	c_1	c_2	α
8	9	36	108	0.0000
		37	107	0.0001
		38	106	0.0002
		39	105	0.0003
		40	104	0.0005
		41	103	0.0008
		42	102	0.0012
		43	101	0.0019
		44	100	0.0028
		45	99	0.0039
		46	98	0.0056
		47	97	0.0076
		48	96	0.0103
		49	95	0.0137
		50	94	0.0180
		51	93	0.0232
		52	92	0.0296
		53	91	0.0372
		54	90	0.0464
		55	89	0.0570
		56	88	0.0694
		57	87	0.0836
		58	86	0.0998
		59	85	0.1179
		60	84	0.1383
8	10	36	116	0.0000
		37	115	0.0000
		38	114	0.0001
		39	113	0.0002
		40	112	0.0003
		41	111	0.0004
		42	110	0.0007
		43	109	0.0010
		44	108	0.0015
		45	107	0.0022
		46	106	0.0031
		47	105	0.0043
		48	104	0.0058
		49	103	0.0078
		50	102	0.0103
		51	101	0.0133
		52	100	0.0171
		53	99	0.0217
		54	98	0.0273
		55	97	0.0338
		56	96	0.0416
		57	95	0.0506
		58	94	0.0610
		59	93	0.0729
		60	92	0.0864
		61	91	0.1015
		62	90	0.1185
		63	89	0.1371
		64	88	0.1577

m	n	c_1	c_2	α
9	9	45	126	0.0000
		46	125	0.0000
		47	124	0.0001
		48	123	0.0001
		49	122	0.0002
		50	121	0.0004
		51	120	0.0006
		52	119	0.0009
		53	118	0.0014
		54	117	0.0020
		55	116	0.0028
		56	115	0.0039
		57	114	0.0053
		58	113	0.0071
		59	112	0.0094
		60	111	0.0122
		61	110	0.0157
		62	109	0.0200
		63	108	0.0252
		64	107	0.0313
		65	106	0.0385
		66	105	0.0470
		67	104	0.0567
		68	103	0.0680
		69	102	0.0807
		70	101	0.0951
		71	100	0.1112
		72	99	0.1290
9	10	45	135	0.0000
		46	134	0.0000
		47	133	0.0000
		48	132	0.0001
		49	131	0.0001
		50	130	0.0002
		51	129	0.0003
		52	128	0.0005
		53	127	0.0007
		54	126	0.0011
		55	125	0.0015
		56	124	0.0021
		57	123	0.0028
		58	122	0.0038
		59	121	0.0051
		60	120	0.0066
		61	119	0.0086
		62	118	0.0110
		63	117	0.0140
		64	116	0.0175
		65	115	0.0217
		66	114	0.0267
		67	113	0.0326
		68	112	0.0394
		69	111	0.0474
		70	110	0.0564
		71	109	0.0667
		72	108	0.0782
		73	107	0.0912
		74	106	0.1055
		75	105	0.1214
		76	104	0.1388

m	n	c_1	c_2	α
10	10	55	155	0.0000
		56	154	0.0000
		57	153	0.0000
		58	152	0.0000
		59	151	0.0001
		60	150	0.0001
		61	149	0.0002
		62	148	0.0002
		63	147	0.0004
		64	146	0.0005
		65	145	0.0008
		66	144	0.0010
		67	143	0.0014
		68	142	0.0019
		69	141	0.0026
		70	140	0.0034
		71	139	0.0045
		72	138	0.0057
		73	137	0.0073
		74	136	0.0093
		75	135	0.0116
		76	134	0.0144
		77	133	0.0177
		78	132	0.0216
		79	131	0.0262
		80	130	0.0315
		81	129	0.0376
		82	128	0.0446
		83	127	0.0526
		84	126	0.0615
		85	125	0.0716
		86	124	0.0827
		87	123	0.0952
		88	122	0.1088
		89	121	0.1237
		90	120	0.1399
		91	119	0.1575

Table XI Critical Values for the Runs Test

This table contains cumulative probabilities associated with the runs test. Let m be the number of observations in one category and n the number of observations in the other category ($m \leq n$), and V be the number of runs. The values in this table are the probabilities $P(V \leq v)$ if the order of observations is random.

					v				
m	n	2	3	4	5	6	7	8	9
2	2	0.3333	0.6667	1.0000					
2	3	0.2000	0.5000	0.9000	1.0000				
2	4	0.1333	0.4000	0.8000	1.0000				
2	5	0.0952	0.3333	0.7143	1.0000				
2	6	0.0714	0.2857	0.6429	1.0000				
2	7	0.0556	0.2500	0.5833	1.0000				
2	8	0.0444	0.2222	0.5333	1.0000				
2	9	0.0364	0.2000	0.4909	1.0000				
2	10	0.0303	0.1818	0.4545	1.0000				
3	3	0.1000	0.3000	0.7000	0.9000	1.0000			
3	4	0.0571	0.2000	0.5429	0.8000	0.9714	1.0000		
3	5	0.0357	0.1429	0.4286	0.7143	0.9286	1.0000		
3	6	0.0238	0.1071	0.3452	0.6429	0.8810	1.0000		
3	7	0.0167	0.0833	0.2833	0.5833	0.8333	1.0000		
3	8	0.0121	0.0667	0.2364	0.5333	0.7879	1.0000		
3	9	0.0091	0.0545	0.2000	0.4909	0.7455	1.0000		
3	10	0.0070	0.0455	0.1713	0.4545	0.7063	1.0000		
4	4	0.0286	0.1143	0.3714	0.6286	0.8857	0.9714	1.0000	
4	5	0.0159	0.0714	0.2619	0.5000	0.7857	0.9286	0.9921	1.0000
4	6	0.0095	0.0476	0.1905	0.4048	0.6905	0.8810	0.9762	1.0000
4	7	0.0061	0.0333	0.1424	0.3333	0.6061	0.8333	0.9545	1.0000
4	8	0.0040	0.0242	0.1091	0.2788	0.5333	0.7879	0.9293	1.0000
4	9	0.0028	0.0182	0.0853	0.2364	0.4713	0.7455	0.9021	1.0000
4	10	0.0020	0.0140	0.0679	0.2028	0.4186	0.7063	0.8741	1.0000

Table XI Critical Values for the Runs Test (Continued)

m	n	2	3	4	5	6	7	8	9	10	11	12	13	14	15	16	17	18	19	20
5	5	0.0079	0.0397	0.1667	0.3571	0.6429	0.8333	0.9603	0.9921	1.0000										
5	6	0.0043	0.0238	0.1104	0.2619	0.5216	0.7381	0.9113	0.9762	0.9978	1.0000									
5	7	0.0025	0.0152	0.0758	0.1970	0.4242	0.6515	0.8535	0.9545	0.9924	1.0000									
5	8	0.0016	0.0101	0.0536	0.1515	0.3473	0.5758	0.7933	0.9293	0.9837	1.0000									
5	9	0.0010	0.0070	0.0390	0.1189	0.2867	0.5105	0.7343	0.9021	0.9720	1.0000									
5	10	0.0007	0.0050	0.0290	0.0949	0.2388	0.4545	0.6783	0.8741	0.9580	1.0000									
6	6	0.0022	0.0130	0.0671	0.1753	0.3918	0.6082	0.8247	0.9329	0.9870	0.9978	1.0000								
6	7	0.0012	0.0076	0.0425	0.1212	0.2960	0.5000	0.7331	0.8788	0.9662	0.9924	0.9994	1.0000							
6	8	0.0007	0.0047	0.0280	0.0862	0.2261	0.4126	0.6457	0.8205	0.9371	0.9837	0.9977	1.0000							
6	9	0.0004	0.0030	0.0190	0.0629	0.1748	0.3427	0.5664	0.7622	0.9021	0.9720	0.9944	1.0000							
6	10	0.0002	0.0020	0.0132	0.0470	0.1369	0.2867	0.4965	0.7063	0.8636	0.9580	0.9895	1.0000							
7	7	0.0006	0.0041	0.0251	0.0775	0.2086	0.3834	0.6166	0.7914	0.9225	0.9749	0.9959	0.9994	1.0000						
7	8	0.0003	0.0023	0.0154	0.0513	0.1492	0.2960	0.5136	0.7040	0.8671	0.9487	0.9879	0.9977	0.9998	1.0000					
7	9	0.0002	0.0014	0.0098	0.0350	0.1084	0.2308	0.4266	0.6224	0.8059	0.9161	0.9748	0.9944	0.9993	1.0000					
7	10	0.0001	0.0009	0.0064	0.0245	0.0800	0.1818	0.3546	0.5490	0.7433	0.8794	0.9571	0.9895	0.9981	1.0000					
8	8	0.0002	0.0012	0.0089	0.0317	0.1002	0.2145	0.4048	0.5952	0.7855	0.8998	0.9683	0.9911	0.9988	0.9998	1.0000				
8	9	0.0001	0.0007	0.0053	0.0203	0.0687	0.1573	0.3186	0.5000	0.7016	0.8427	0.9394	0.9797	0.9958	0.9993	1.0000	1.0000			
8	10	0.0000	0.0004	0.0033	0.0134	0.0479	0.1170	0.2514	0.4194	0.6209	0.7822	0.9031	0.9636	0.9905	0.9981	0.9998	1.0000			
9	9	0.0000	0.0004	0.0030	0.0122	0.0445	0.1090	0.2380	0.3992	0.6008	0.7620	0.8910	0.9555	0.9878	0.9970	0.9996	1.0000	1.0000		
9	10	0.0000	0.0002	0.0018	0.0076	0.0294	0.0767	0.1786	0.3186	0.5095	0.6814	0.8342	0.9233	0.9742	0.9924	0.9986	0.9998	1.0000	1.0000	
10	10	0.0000	0.0001	0.0010	0.0045	0.0185	0.0513	0.1276	0.2422	0.4141	0.5859	0.7578	0.8724	0.9487	0.9815	0.9955	0.9990	0.9999	1.0000	1.0000

v

Table XII Greek Alphabet

This table contains the Greek alphabet: the letter name, the lowercase letter, the variant of the lowercase letter where applicable, and the uppercase letter.

Name	Lowercase letter	Lowercase variant	Uppercase letter
Alpha	α		A
Beta	β		B
Gamma	γ		Γ
Delta	δ		Δ
Epsilon	ϵ	ε	E
Zeta	ζ		Z
Eta	η		H
Theta	θ	ϑ	Θ
Iota	ι		I
Kappa	κ		K
Lambda	λ		Λ
Mu	μ		M
Nu	ν		N
Xi	ξ		Ξ
Omicron	o		O
Pi	π	ϖ	Π
Rho	ρ	ϱ	R
Sigma	σ	ς	Σ
Tau	τ		T
Upsilon	υ		Y
Phi	ϕ	φ	Φ
Chi	χ		X
Psi	ψ		Ψ
Omega	ω		Ω

Answers to Odd-Numbered Exercises

CHAPTER 0

0.1 Claim, Experiment, Likelihood, Conclusion.

0.3 Keywords, Translation, Concepts, Vision.

0.5 It is possible that the rash of injuries is just lucky (or unlucky), or the injuries might be due to some other factor. Explanations will vary.

0.7 Under similar conditions, 50% of the time it rained. Answers will vary.

0.9 This is a very unusual event, probably not due to pure chance, or natural causes. It is more likely that the whales died due from something other than natural causes.

0.11 a. Pure chance, or a mechanical defect. **b.** Toyota probably believed that the runaway vehicles were not the result of pure chance. This was a very unusual event and probably due to something mechanical.

0.13 Yes. Health officials should be concerned about this outbreak. The number of cases of whooping cough was a very unusual event, probably not due to pure chance.

CHAPTER 1
Sections 1.1–1.2

1.1 True.

1.3 a. Population. **b.** Sample. **c.** Variable.

1.5 a. Descriptive. **b.** Inferential. **c.** Inferential. **d.** Inferential. **e.** Descriptive. **f.** Descriptive.

1.7 Solution Trail
Keywords: Open-heart patients; 30 patients; length of stay.
Translation: All patients; subset of all patients; characteristic of each patient.
Concepts: Population; sample, variable.
Vision: Determine the set of all objects of interest, the subset, and the attribute to be measured.
Population: All patients admitted for open-heart surgery. Sample: 30 patients selected. Variable: Length of stay.

1.9 Population: Employees at Citigroup, Inc. Sample: 35 employees selected.

1.11 Population: All families who filed claims associated with damage caused by Hurricane Sandy in 2012. Sample: 75 families selected.

1.13 a. All shark attack records. **b.** 1000 records selected. **c.** Victim group.

1.15 Population: People diagnosed with hepatitis C. Sample: 50 patients selected. Variable: Liver enzyme levels.

1.17 a. Population: All cheddar cheeses. Sample: 20 cheddar cheeses selected. **b.** Probability question: What is the probability that at least 10 of the cheddar cheeses selected are aged less than two years? Statistics question: Suppose 12 of the cheddar cheeses selected are aged less than two years. Does this suggest that the true proportion of all cheddars aged less than two years has decreased?

1.19 a. All large U.S. companies. **b.** 75 companies selected. **c.** Plan to move jobs back to the United States. **d.** Probability question: What is the probability that at most 5 companies selected plan to move jobs back to the United States? Statistics question: Suppose 10 of the companies selected plan to move jobs back to the United States. Does this suggest that the true proportion of all large companies that plan to move jobs back to the United States is greater than 0.04?

Section 1.3

1.21 False.

1.23 True.

1.25 Claim. Experiment. Likelihood. Conclusion.

1.27 a. Observational study. **b.** Sample: The students who respond to the questions. **c.** Not a random sample, only one dorm.

1.29 a. Population: All 12-ounce bottles of soda. Sample: The bottles selected. **b.** Yes, a simple random sample.

1.31 a. Observational study. **b.** Population: All Massachusetts State Police. Sample: 12 officers selected. **c.** Not a random sample, only one shift considered.

1.33 a. Obtain a list of people who have purchased this product, and assign a number to each person. Randomly select numbers from a random number table or random number generator, and ask each corresponding customer how long it took to set up the fence.

1.35 a. Assign a number to each mile-long stretch. Randomly select numbers from a random number table or random number generator. **b.** Observational study.

1.37 a. Experimental study. **b.** Variable: Which car is most comfortable? **c.** Conversation with the driver, peeking, sound of the engine, legroom.

1.39 a. Observational study. **b.** Variables: proportion of white feathers, proportion of down, proportion of other components. **c.** Randomly select stores from around the country that sell comforters. Assign a number to each store, and use a random number generator to select stores. Visit the selected stores and count the number of comforters on display. Use a random number generator to purchase one of the comforters on display.

Chapter Exercises

1.41 a. Descriptive.   **b.** Descriptive.   **c.** Inferential.
d. Inferential.

1.43 a. Descriptive.   **b.** Inferential.   **c.** Inferential.
d. Descriptive.

1.45 Population: All adults. Sample: 400 adults selected.

1.47 a. Observational study.   **b.** 1000 people called.   **c.** Not a random sample. Not everyone has a phone. Not everyone with a phone has a listed number. People may choose not to respond. People may not be home at the time of the call.

1.49 a. Experimental study.   **b.** Length of time until ice formed on the wings.   **c.** Flip a coin for each plane. If heads, assign the plane to one group. If tails, assign the plane to the other group.

1.51 a. Observational study.   **b.** Pressure on each section.
c. Not a random sample. Only the deepest areas of the tunnel were selected.

1.53 a. Observational study.   **b.** Alcohol content in each bottle.

CHAPTER 2

Section 2.1

2.1 False.

2.3 True.

2.5 a. Numerical, continuous.   **b.** Numerical, discrete.
c. Categorical.   **d.** Numerical, discrete.   **e.** Numerical, continuous.   **f.** Categorical.

2.7 a. Numerical, discrete.   **b.** Numerical, discrete.
c. Categorical.   **d.** Numerical, continuous.   **e.** Numerical, continuous.   **f.** Categorical.

2.9 a. Continuous.   **b.** Continuous.   **c.** Discrete.
d. Continuous.   **e.** Continuous.   **f.** Discrete.

2.11 a. Continuous.   **b.** Discrete.   **c.** Discrete.
d. Continuous.   **e.** Discrete.   **f.** Discrete.

2.13 a. Discrete.   **b.** Categorical.   **c.** Continuous.
d. Continuous.   **e.** Categorical.   **f.** Continuous.

Section 2.2

2.15 True.

2.17 True.

2.19

Category	Frequency	Relative requency
Comedy	7	0.1667
Drama	10	0.2381
Educational	3	0.0714
Reality	7	0.1667
Soap	10	0.2381
Sports	5	0.1190
Total	42	1.0000

2.21 a.

Issue	Frequency	Relative frequency
Salary	50	0.1250
Health insurance	100	0.2500
Retirement benefits	75	0.1875
Class size	60	0.1500
Temporary faculty	90	0.2250
Parking	25	0.0625
Total	400	1.0000

b.

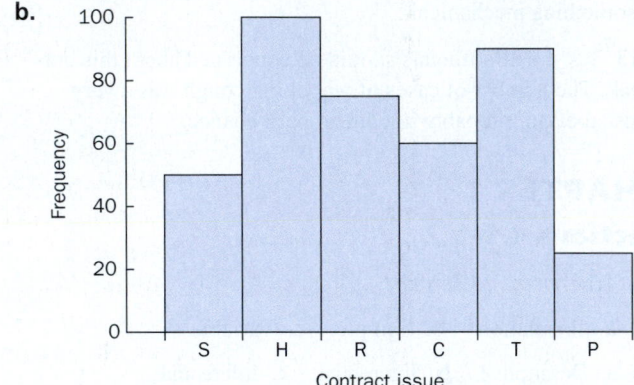

2.23 a.

Answer	Frequency	Relative frequency
VL	8	0.16
L	12	0.24
N	7	0.14
U	8	0.16
VU	15	0.30
Total	50	1.00

b.

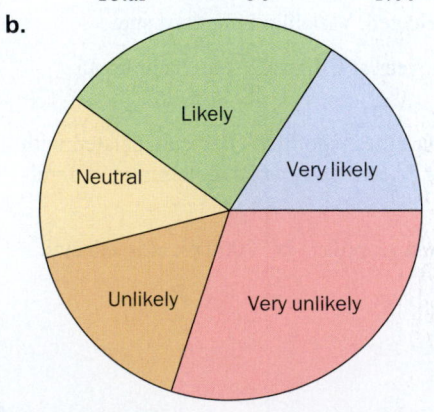

2.25 a.

Nobel Prize	Frequency	Relative frequency
Physics	194	0.2015
Chemistry	163	0.1693
Medicine	201	0.2087
Literature	209	0.2170
Peace	125	0.1298
Economic sciences	71	0.0737
Total	963	1.0000

b.

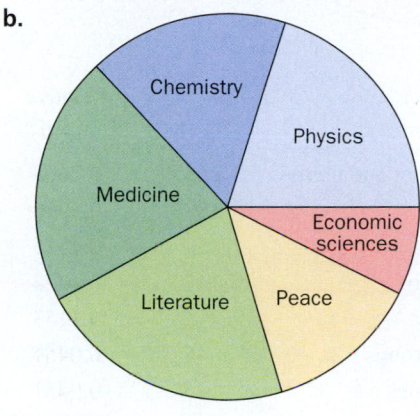

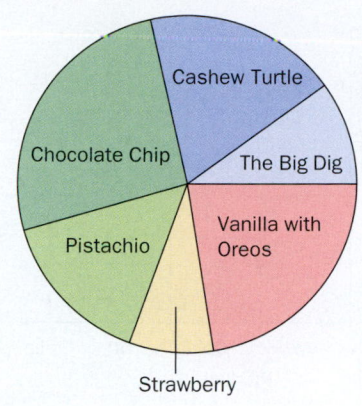

2.27 a.

Ice cream	Frequency	Relative frequency
The Big Dig	20	0.100
Cashew Turtle	37	0.185
Chocolate Chip	52	0.260
Pistachio	30	0.150
Strawberry	16	0.080
Vanilla with Oreos	45	0.225
Total	200	1.000

b.

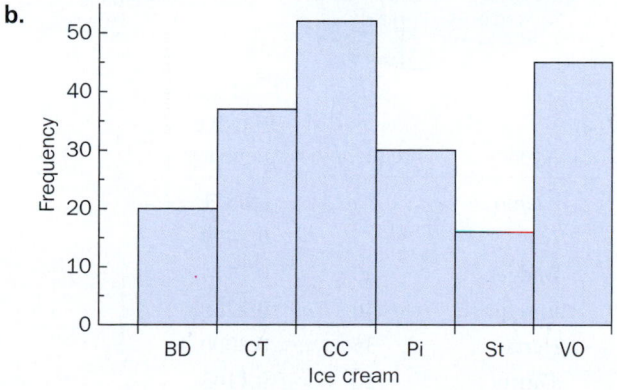

2.29 a.

Product	Frequency	Relative frequency
Alarms	75	0.2964
Training	16	0.0632
Extinguishers	13	0.0514
Pumps	6	0.0237
Sprinklers	16	0.0632
Building materials	19	0.0751
Electrical equipment	32	0.1265
Hazmat storage	22	0.0870
Security products	41	0.1621
Signaling systems	13	0.0514
Total	253	1.0000

b.

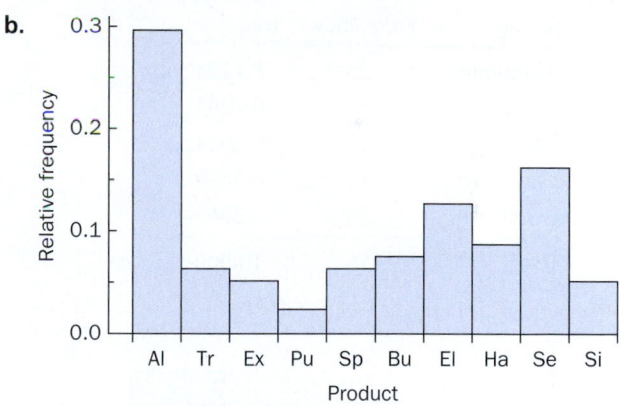

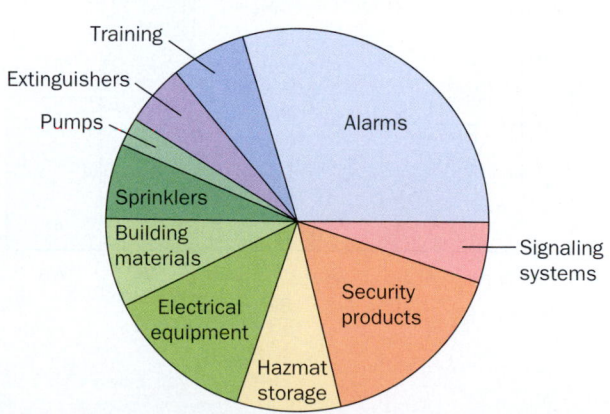

2.31

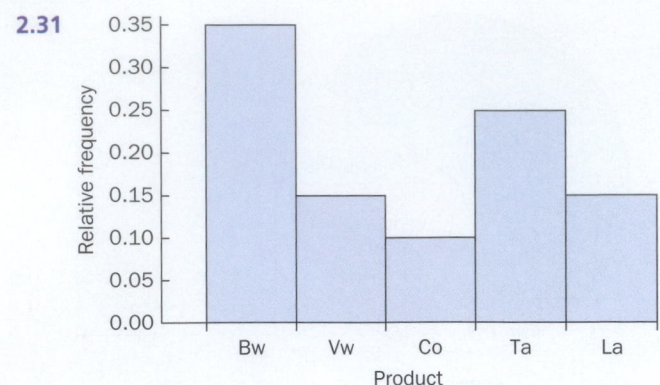

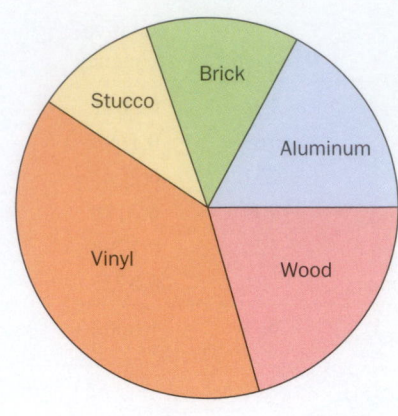

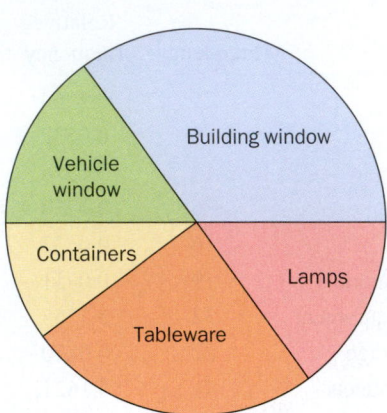

2.33 a.

Siding	Frequency	Relative frequency
Aluminum	20	0.1724
Brick	15	0.1293
Stucco	12	0.1034
Vinyl	45	0.3879
Wood	24	0.2069
Total	116	1.0000

b.

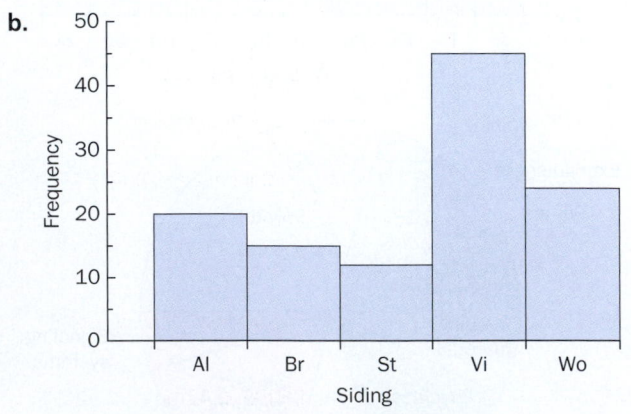

2.35 a.

Source	Frequency	Relative frequency
Official websites and alerts	265	0.2658
Social media	198	0.1986
News websites	152	0.1525
News TV/radio	147	0.1474
Friends/family	115	0.1153
Community groups	45	0.0451
Smartphone apps	45	0.0451
Other	30	0.0301
Total	997	1.0000

b.

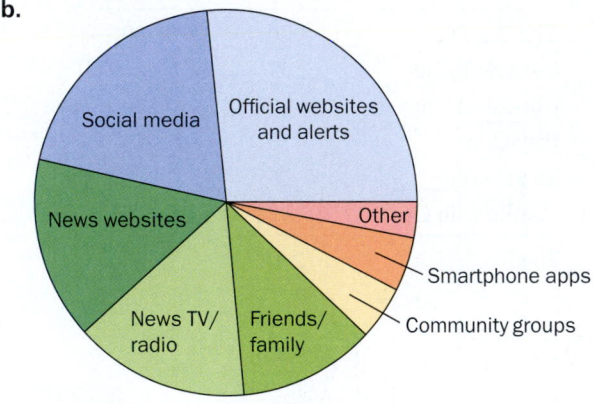

2.37 a.

Agency	Frequency	Relative frequency
Alamo	10	0.0571
Avis	25	0.1429
Budget	30	0.1714
Enterprise	40	0.2286
Hertz	35	0.2000
Thrifty	20	0.1143
Value	15	0.0857
Total	175	1.0000

b. 175 **c.** 0.5714

d.

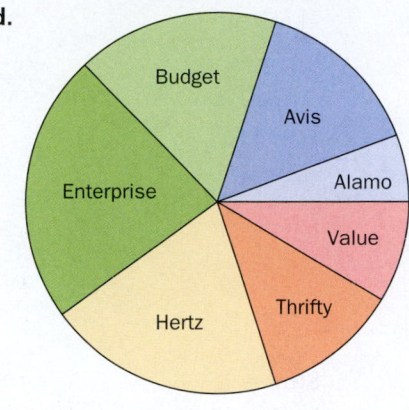

2.39 a.

Day of week	Frequency	Relative frequency
Monday	557	0.1744
Tuesday	383	0.1199
Wednesday	348	0.1090
Thursday	351	0.1099
Friday	415	0.1299
Saturday	464	0.1453
Sunday	676	0.2116
Total	3194	1.0000

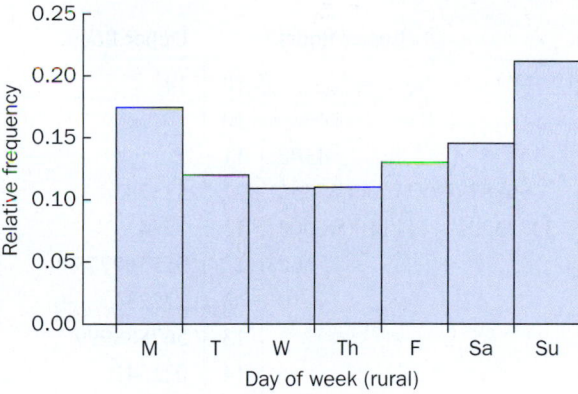

b.

Day of week	Frequency	Relative frequency
Monday	448	0.1618
Tuesday	359	0.1296
Wednesday	328	0.1185
Thursday	330	0.1192
Friday	386	0.1394
Saturday	404	0.1459
Sunday	514	0.1856
Total	2769	1.0000

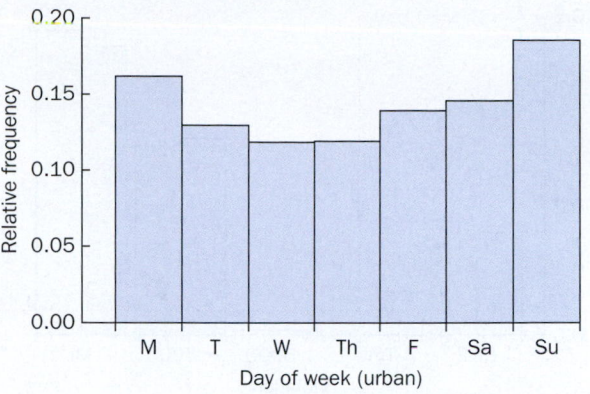

c. The shapes of the bar graphs are very similar. The highest proportion of fatal vehicle crashes occur on Sunday and Monday.

2.41 a.

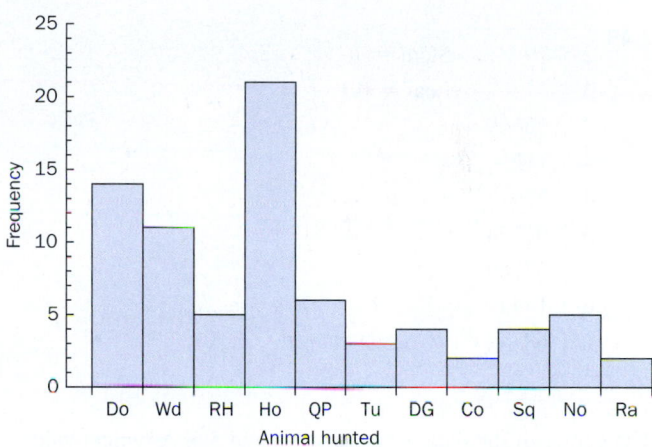

b.

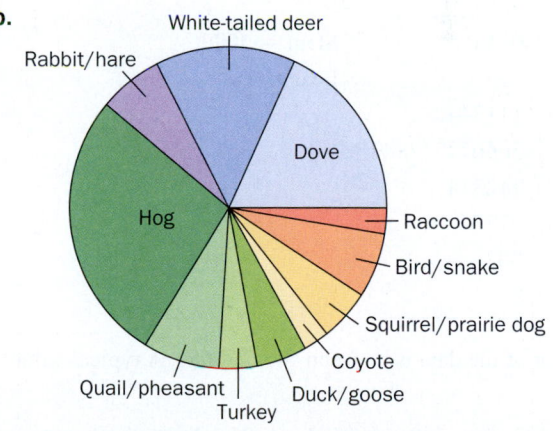

c. No. We do not know how many people are hunting each animal.

2.43 a. 0.1606, 0.0676, 0.0074, 0.3095, 0.4549
b. 0.1375, 0.0752, 0.0077, 0.2991, 0.4805

c.

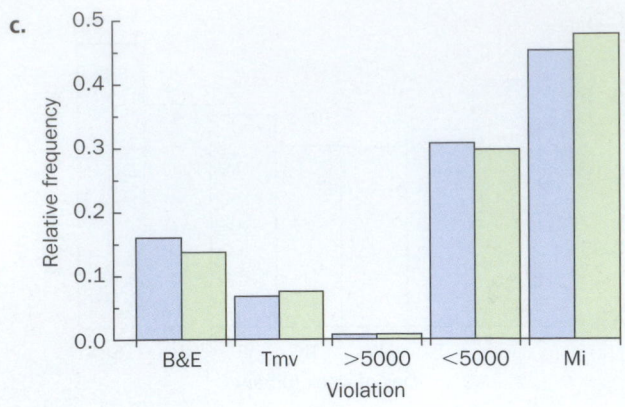

Section 2.3

2.45 False.

2.47 True.

2.49

2	79	Stem = 1
3		Leaf = 0.1
3	56669	
4	112	
4	779	
5	01124	
5	57789	
6	1444	
6	68	
7	1	

The center of the data is between 5.0 and 5.5. A typical value is 5.2.

2.51

53	0344	Stem = 100
53	799	Leaf = 1
54	111344	
54	566677777889	
55	112334	
55	67777	
56	002	
56	9	

The center of the data is between 545 and 550. A typical value is 547.

2.53 a. 543, 543, 549. **b.** 574 **c.** The data tail off slowly on the low end. **d.** There do not appear to be any outliers.

2.55 a.

0	4	Stem = 1
1	16	Leaf = 0.1
2	12779	
3	3449	
4	0111125555799	
5	699	
6	023444	
7	0123679	
8	258	
9	3	

b.

0	4	Stem = 1
1	16	Leaf = 0.1
2	2378	
3	0355	
4	001222256668	
5	00799	
6	133445	
7	1134779	
8	269	
9	3	

c. These two graphs are slightly different. However, they both suggest the same general shape, center, and variability. Typical value is 4.5.

2.57 a.

Lower floors		Upper floors
	10	14
	10	
4300	11	1
99887777777666665555	11	55578
333333222211111110000000	12	1244
665	12	55556777899
0	13	122234
	13	567888999
	14	02234
	14	5888
	15	244

Stem = 100, Leaf = 1

b. The lower-floors distribution is more compact and has, on average, smaller values. The upper-floors distribution has more variability and has, on average, larger values.

2.59 a.

1	00699	Stem = 10
2	244557	Leaf = 1
3	11245678899	
4	000011555669	
5	2259	
6	8	
7	1	

b. Typical value: 32. No outliers.

2.61 a.

2	135667	Stem = 10
3	13456777788999	Leaf = 1
4	0001122355	
5	002245566	
6		
7		
8	0	

b. Typical value: 40. One outlier: 80.

2.63 a.

14	444
14	555555555666666666777899999
15	00001222234
15	55567778999
16	001233444
16	5577789
17	23
17	55777
18	14
18	7
19	
19	
20	0

Stem = 1000
Leaf = 10

b. Typical value: 1550. One outlier: 2009.

2.65 a.

With				
	250	24	Stem = 100	
	251		Leaf = 0.1	
84	252	145		
3	253	788		
63	254	49		
8742221	255	023		
9866653110	256	149		
873	257	1248		
6553	258	23367		
4	259	3		
	260	26		
	261	2		
	262	3		

(Header: With ... Without)

b. With distribution: one peak, compact, approximately symmetric. Without distribution: unimodal, lots of variability, slightly positively skewed. It appears the humidifier does help a piano stay in tune. The "With humidifier" distribution is more compact and centered near 256.

2.67 a.

5	9	Stem = 10
6	48	Leaf = 1
7	33456668	
8	00012334467889	
9	000112467788899	
10	00011455556	
11	24666699	
12	00222	
13	34679	
14	2	

b. The shape of the distribution is approximately normal, or bell-shaped. Center: 95. Lots of variability.
c. 0.3571

Section 2.4

2.69 True.

2.71 True.

2.73 True.

2.75 a. Normal distribution.
b. Symmetric.   **c.** Skewed.

2.77

Class	Frequency	Relative frequency	Cumulative relative frequency
78–80	2	0.050	0.050
80–82	4	0.100	0.150
82–84	4	0.100	0.250
84–86	4	0.100	0.350
86–88	9	0.225	0.575
88–90	6	0.150	0.725
90–92	9	0.225	0.950
92–94	2	0.050	1.000
Total	40	1.000	

2.79

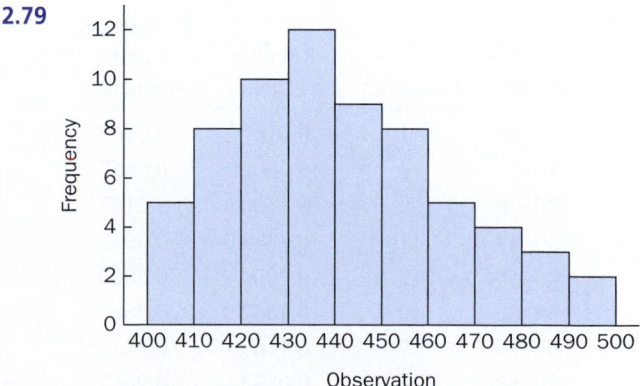

2.81

Class	Frequency	Relative frequency	Cumulative relative frequency
100–150	155	0.1938	0.1938
150–200	120	0.1500	0.3438
200–250	130	0.1625	0.5063
250–300	145	0.1813	0.6875
300–350	150	0.1875	0.8750
350–400	100	0.1250	1.0000
Total	800	1.0000	

2.83

Class	Frequency	Relative frequency	Cumulative relative frequency
0–25	150	0.150	0.150
25–50	200	0.200	0.350
50–75	175	0.175	0.525
75–100	150	0.150	0.675
100–125	125	0.125	0.800
125–150	100	0.100	0.900
150–175	75	0.075	0.975
175–200	25	0.025	1.000
Total	1000	1.000	

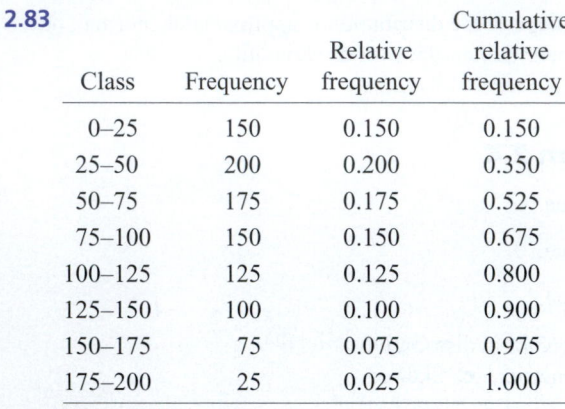

2.85 a.

Class	Frequency	Relative frequency	Cumulative relative frequency
0–10	1	0.0167	0.0167
10–20	3	0.0500	0.0667
20–30	9	0.1500	0.2167
30–40	11	0.1833	0.4000
40–50	12	0.2000	0.6000
50–60	10	0.1667	0.7667
60–70	7	0.1167	0.8833
70–80	5	0.0833	0.9667
80–90	2	0.0333	1.0000
Total	60	1.0000	

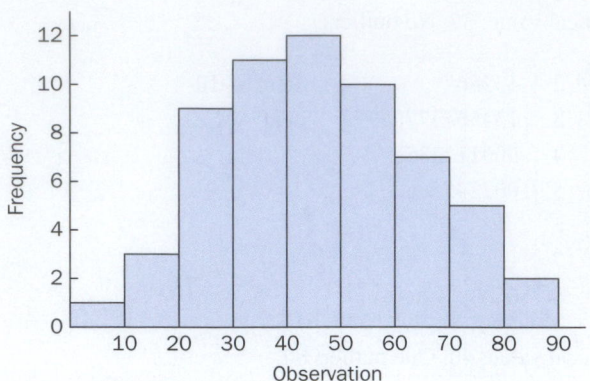

b. Unimodal, symmetric, bell-shaped, center approximately 43 and seems compact

c. $M \approx 45$

d. $Q_1 \approx 31.8$

e. $Q_3 \approx 59.0$

2.87 a.

Class	Frequency	Relative frequency	Cumulative relative frequency
0–50	5	0.1667	0.1667
50–100	9	0.3000	0.4667
100–150	5	0.1667	0.6333
150–200	3	0.1000	0.7333
200–250	2	0.0667	0.8000
250–300	1	0.0333	0.8333
300–350	4	0.1333	0.9667
350–400	0	0.0000	0.9667
400–450	0	0.0000	0.9667
450–500	0	0.0000	0.9667
500–550	1	0.0333	1.0000
Total	30	1.0000	

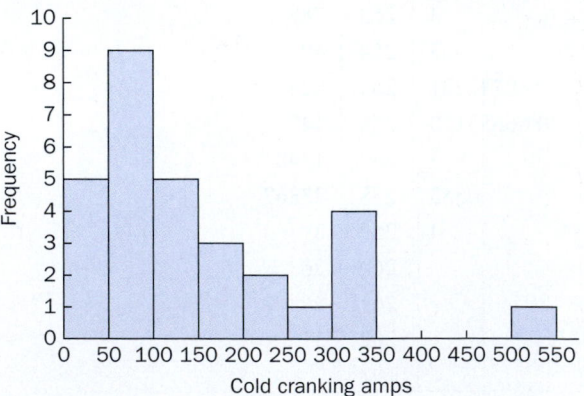

b. Positively skewed, centered near 100, lots of variability, one outlier: 514.

c. $M \approx 110$

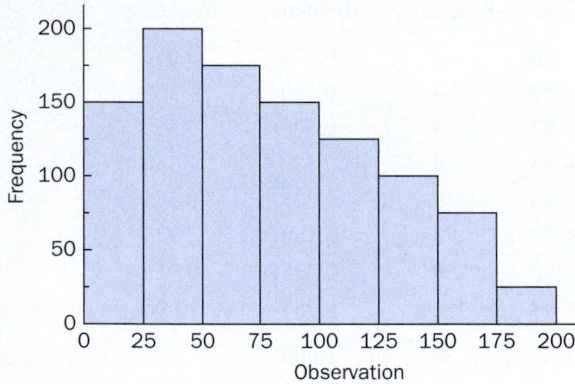

2.89 a.

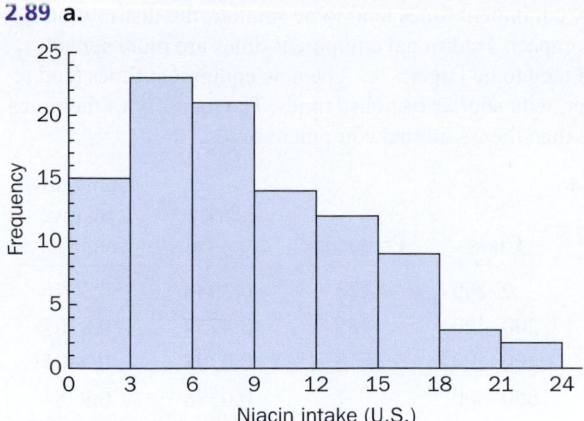

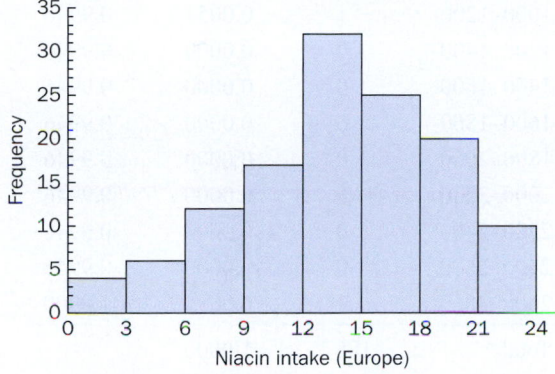

b. United States: unimodal, positively skewed. Europe: unimodal, negatively skewed. On average, it appears that Europeans have a greater daily niacin intake.

2.91 a. Solution Trail

Keywords: Histogram; shape; center; variability.

Translation: Frequency distribution; histogram, distribution.

Concepts: Histogram; description of a distribution.

Vision: Construct a frequency distribution and use this to draw a histogram and describe the shape, center, and variability of the distribution.

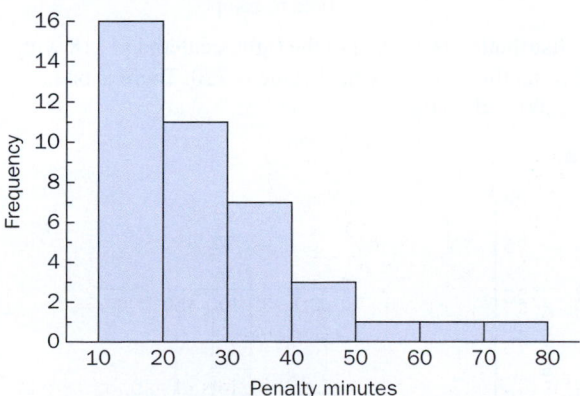

The distribution is unimodal and positively skewed. The center of the distribution is approximately 23 and there is lots of variability. **b.** $m \approx 45$. Using linear interpolation, $m \approx 46.7$

2.93 a.

Class	Frequency	Relative frequency	Cumulative relative frequency
100–105	10	0.050	0.050
105–110	75	0.375	0.425
110–115	40	0.200	0.625
115–120	25	0.125	0.750
120–125	20	0.100	0.850
125–130	15	0.075	0.925
130–135	10	0.050	0.975
135–140	5	0.025	1.000
Total	200	1.000	

b.

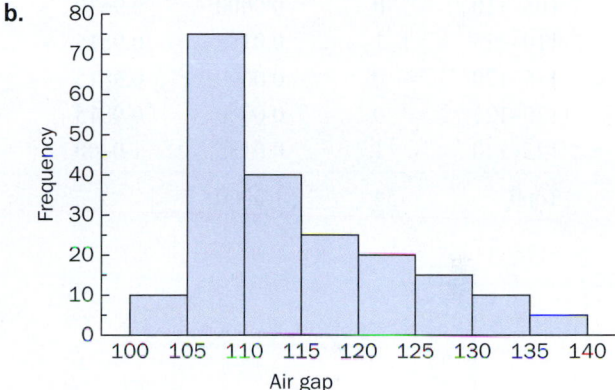

c. 0.425

2.95 a.

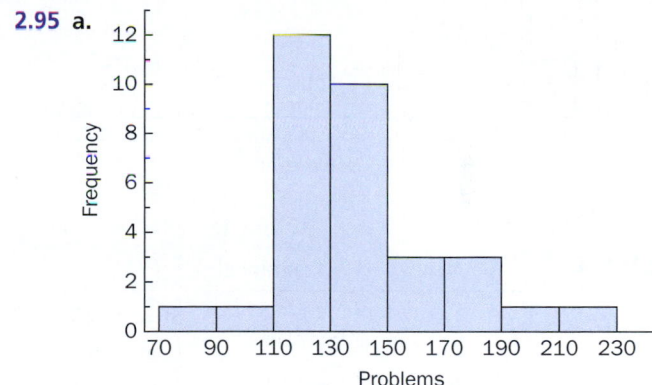

b. The distribution is centered around 133, is skewed slightly to the right, and has a lot of variability. **c.** $Q_1 \approx 120.5$, $Q_3 \approx 149.75$. **d.** There should be 16 values between Q_1 and Q_3. There are 16 values between Q_1 and Q_3.

Chapter Exercises

2.97 a.

1	5789	Stem = 0.10
2	01134	Leaf = 0.01
2	5567889999	
3	0011223344	
3	55566679	
4	01234	
4	89	
5	2	

b. Approximately bell-shaped, center around 0.32, little variability.

2.99 a.

Class	Frequency	Relative frequency	Cumulative relative frequency
70–75	4	0.0741	0.0741
75–80	12	0.2222	0.2963
80–85	21	0.3889	0.6852
85–90	7	0.1296	0.8148
90–95	4	0.0741	0.8889
95–100	3	0.0556	0.9444
100–105	1	0.0185	0.9630
105–110	0	0.0000	0.9630
110–115	1	0.0185	0.9815
115–120	0	0.0000	0.9815
120–125	0	0.0000	0.9815
125–130	1	0.0185	1.0000
Total	54	1.0000	

b.

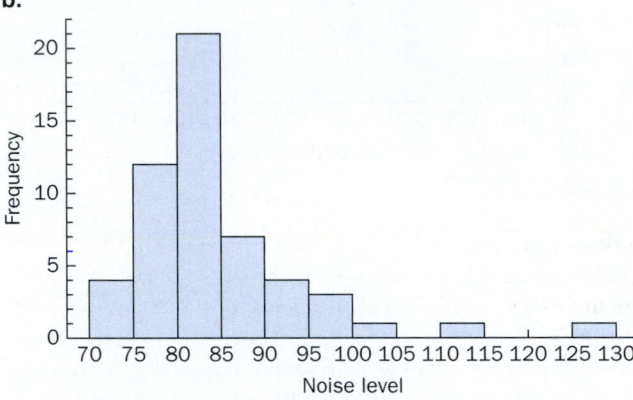

c. 0.2963   **d.** 0.1852

2.101 a.

New		Traditional	
	0	89	Stem = 10
	1	5	Leaf = 0.1
87660	2	6	
765443200	3	34	
99774444322	4	02568	
110	5	579	
43	6	122888	
	7	33567	
	8	033589	
	9	178	
	10	034	
	11	3	
	12	033	
	13	03	
	14	16	

b. New equipment times tend to be smaller; the distribution is more compact. Traditional equipment times are more spread out and tend to be larger.   **c.** The new equipment times tend to be better, with shorter response times. The majority of the times are less than the traditional equipment times.

2.103 a.

Class	Frequency	Relative frequency	Cumulative relative frequency
0–200	73	0.3946	0.3946
200–400	88	0.4757	0.8703
400–600	13	0.0703	0.9405
600–800	7	0.0378	0.9784
800–1000	2	0.0108	0.9892
1000–1200	1	0.0054	0.9946
1200–1400	0	0.0000	0.9946
1400–1600	0	0.0000	0.9946
1600–1800	0	0.0000	0.9946
1800–2000	0	0.0000	0.9946
2000–2200	0	0.0000	0.9946
2200–2400	0	0.0000	0.9946
2400–2600	0	0.0000	0.9946
2600–2800	1	0.0054	1.0000
Total	185	1.0000	

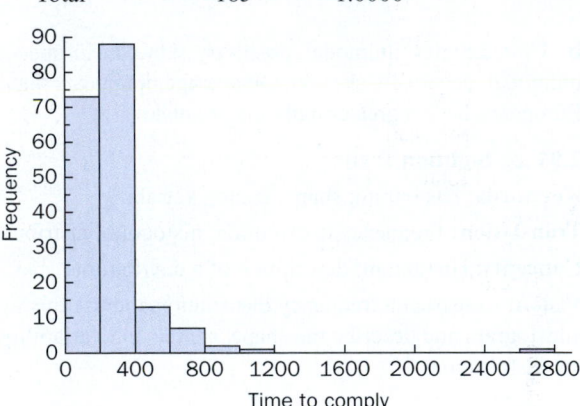

b. The distribution is skewed to the right, centered at 218 with a lot of variability.   **c.** A typical value is 220. There is one outlier: 2600.   **d.** 0.027

2.105 a.

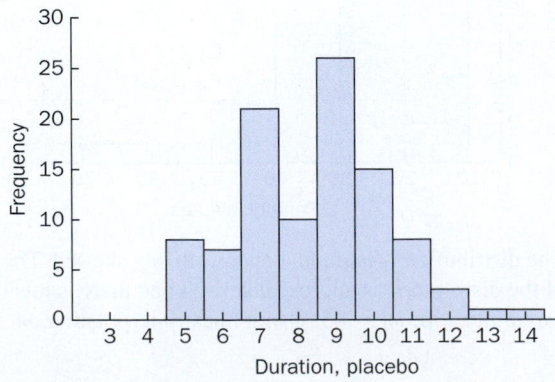

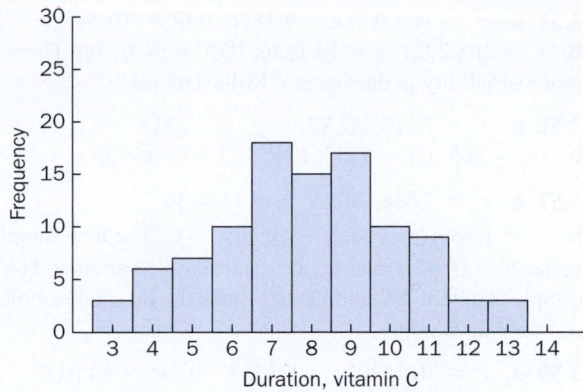

b. Both graphs appear to be centered at about the same duration. Both appear to be symmetric and bell-shaped. The placebo durations are slightly more compact than the vitamin C durations.
c. There is no graphical evidence to suggest that vitamin C reduced the duration.

2.107 a.

Class	Frequency	Relative frequency	Cumulative relative frequency
100–110	1	0.02	0.02
110–120	2	0.04	0.06
120–130	9	0.18	0.24
130–140	7	0.14	0.38
140–150	9	0.18	0.56
150–160	9	0.18	0.74
160–170	11	0.22	0.96
170–180	1	0.02	0.98
180–190	1	0.02	1.00
Total	50	1.00	

b.

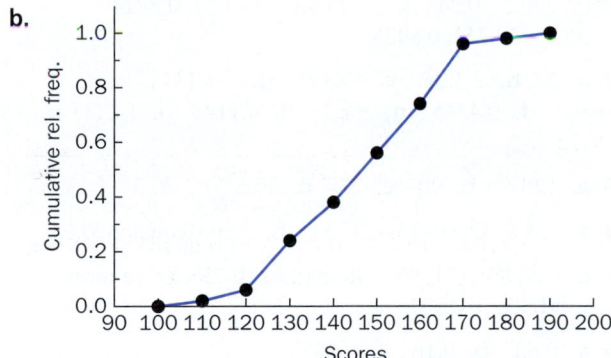

2.109 a.

Class	Frequency	Relative frequency	Cumulative relative frequency
2–4	7	0.14	0.14
4–6	15	0.30	0.44
6–8	11	0.22	0.66
8–10	7	0.14	0.80
10–12	5	0.10	0.90
12–14	4	0.08	0.98
14–16	1	0.02	1.00
Total	50	1.00	

Stem	Leaf	
2	8	Stem = 1
3	357789	Leaf = 0.1
4	01236667	
5	2222889	
6	013447	
7	01455	
8	1333	
9	345	
10	579	
11	11	
12	02	
13	26	
14	2	

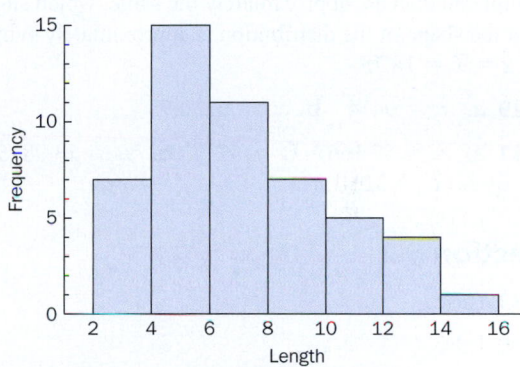

b. The stem-and-leaf plot and the histogram suggest that the data are positively skewed. Most of the observations are in the lower tail of the distribution; the upper tail has few observations. The center of the distribution (the middle number) is approximately 6.0. Most of the observations are between 2 and 7, but the distribution is spread out between 2.8 and 14.2. (The largest python captured was, in fact, 14.3 feet, and is not part of the data set.) The proportion of observations less than 14 is 0.98. There are no real outliers, observations very far away from the rest.

CHAPTER 3

Section 3.1

3.1 a. Center, variability   **b.** Centered or clustered

3.3 a. 82   **b.** 3474   **c.** 32   **d.** 2779   **e.** 164   **f.** 164

3.5 a. 105.7   **b.** 13.1852   **c.** 6.9583   **d.** 0.1232
e. −2.4933   **f.** 17.7432

3.7 a. 6.6667, 7   **b.** 6.6364, 9   **c.** 10.6889, 7.7
d. −107.69, −109.1

3.9 a. Skewed left.   **b.** Symmetric.   **c.** Skewed left.
d. Skewed left.

3.11 a. 6   **b.** 0   **c.** No mode.

3.13 a. 68.5238   **b.** 67.0   **c.** Slightly skewed right.

3.15 a. 813.556, 804.5   **b.** 924.667, 804.5. The mean is greater, pulled in the direction of the new, higher value. The median stays the same.

3.17 a. $\bar{x} = 619.5, \tilde{x} = 620.0$ **b.** 619.1667
c. Approximately symmetric.

3.19 a. 1.9225, 1.4738   **b.** The median is a better measure of central tendency. There is one outlier pulling the mean in its direction.

3.21 a. 81.9, 81.0   **b.** 81.75   **c.** 84

3.23 a. 8.1979, 8.5   **b.** 8.5   **c.** 26.2333. The new mean is 3.2 times the original mean: (3.2)(8.1979).

3.25 a. 1634.58   **b.** 19,614.9. The new mean is 12 times the original mean: (12)(1634.58).

3.27 a. 20.0444, 18.75   **b.** The sample mean and the sample median are approximately the same, which suggests that the shape of the distribution is approximately symmetric.
c. $\bar{x} = \tilde{x} = 18.75$

3.29 a. $x_5 = 9414$ **b.** $x_5 = 6250.75$

3.31 a. $\bar{x}_F = 57.1667, \tilde{x}_F = 57.5$ **b.** $\bar{x}_C = 13.9815$
c. $\bar{x}_C = (\bar{x}_F - 32)/1.8$

Section 3.2

3.33 False.

3.35 False.

3.37 True.

3.39 a. $s^2 = 323.7757, s = 17.9938$
b. $s^2 = 479.7322, s = 21.9028$
c. $s^2 = 31.3735, s = 5.6012$
d. $s^2 = 2.4892, s = 1.5777$

3.41 a. $Q_1 = 20, Q_3 = 35, IQR = 15$
b. $Q_1 = 2.3, Q_3 = 7.75, IQR = 5.45$
c. $Q_1 = -21, Q_3 = -13, IQR = 8$
d. $Q_1 = 44.1, Q_3 = 59.2, IQR = 15.1$

3.43 a. Increases.   **b.** Increases.   **c.** Does not affect.
d. Does not affect.

3.45 a. $s = 279.0969$ **b.** $IQR = 529$ **c.** Probably IQR because there are several observations that are very large and some that are very small.

3.47 a. $s^2 = 4.2958, s = 2.0726, IQR = 2.00$
b. $s^2 = 12.2632, s = 3.5019, IQR = 5.50$
c. The More than two hours data set has more variability.

3.49 a. $Q_1 = 170, Q_3 = 1187, IQR = 1017$
b. $s^2 = 471,674.1238, s = 686.7854$ **c.** $IQR = 1086.5$,
$s^2 = 991,485.9333$ **d.** IQR is approximately the same. Adding one more outlier to the data set does not change the middle 50% much. However, s^2 is sensitive to outliers, and is much larger in the expanded data set.

3.51 a. East: CV = 3.3932, CQV = 2.6149; West:
CV = 16.5767, CQV = 13.7452   **b.** The West-side development data have more variability.

3.53 a. $s^2 = 54.5763, s = 7.3876, IQR = 10.5$
b. $s^2 = 365.2921, s = 19.1126, IQR = 26.0$ **c.** There is more variability in the General Mills data set.

3.55 a. $s^2 = 3,618,447.73, s = 1902.2218$.
b. $Q_1 = 265, Q_3 = 1832, IQR = 1567$ **c.** $\sum(x_i - \bar{x}) = 0$

3.57 a. $s^2 = 2,086,130.87, s = 1444.34$
b. $s^2 = 1,689,766.008, s = 1299.91$ **c.** The new sample variance is $(0.9)^2$ times the original sample variance. The new sample standard deviation is 0.9 times the original sample standard deviation.

3.59 a. $s^2 = 70.1449, s = 8.3753$ **b.** $s^2 = 14.0181$,
$s = 3.7441$ **c.** The new sample variance is $(0.44704)^2$ times the original sample variance. The new sample standard deviation is 0.44704 times the original sample standard deviation.

3.61 a. $s^2 = 3175.5667, s = 56.3522$ **b.** $s^2 = 3175.5667$,
$s = 56.3522$ **c.** Same.   **d.** $s_y^2 = s_x^2$ and $s_y = s_x$

3.63 $s_y^2 = a^2 s_x^2, s_y = |a| s_x$

3.65 a. $\hat{p} = 0.9063$ **b.** $s^2 = 0.0877 = \frac{32}{31}(\hat{p} - \hat{p}^2)$
c. $\sigma^2 = 0.0850 = \hat{p} - \hat{p}^2$.

Section 3.3

3.67 True.

3.69 True.

3.71 False.

3.73 True.

3.75 a. (40.0, 60.0), 0.75   **b.** (320.5, 383.5), 0.8889
c. (11.4, 22.6), 0.6094   **d.** (18.2125, 54.7875), 0.6735
e. (95.5, 220.5), 0.84   **f.** (−55.35, −54.65), 0.8724
g. (−56.35, 59.75), 0.8025

3.77 a. 3   **b.** −1.25   **c.** 0.8333   **d.** −1.1111
e. 1.6563   **f.** 0.4545   **g.** −2.2   **h.** 4.1143   **i.** 1.1111
j. 5.8125

3.79 a. 120.5   **b.** 90   **c.** 22   **d.** 30.5   **e.** 20.5   **f.** 3525

3.81 a. (18.8, 32.4), (15.4, 35.8)   **b.** Approximately 0.68

3.83 a. (77, 89), (71, 95)   **b.** At least 0.75   **c.** At most 0.1111   **d.** 0.95, 0.003

3.85 a. 0.68   **b.** 0.16   **c.** 0.8385

3.87 a. Of students nationwide, 7th–9th grade Carlisle High School students scored the same or better than 53% in mathematics and 69% in science on the ITBS.   **b.** The median.
c. The seventh-grader did better than 99% of those who took the exam.

3.89 a. Solution Trail

Keywords: Is there any reason to believe the general manager's claim is false?

Translation: Draw a conclusion.

Concepts: Inference procedure; z-score.

Vision: Compute the z-score for the waiting time and determine if it is reasonable, between −3 and +3.

Claim: $\mu = 11$ ($\sigma = 2.5$, distribution approximately normal)

Experiment: $x = 13$

Likelihood: $z = (13 - 11)/2.5 = 0.80$

Conclusion: This is a reasonable z-score. There is no evidence to suggest the manager's claim is false.   **b.** Claim: $\mu = 11$ ($\sigma = 2.5$, distribution approximately normal)

Experiment: $x = 20$

Likelihood: $z = (20 - 11)/2.5 = 3.6$

Conclusion: This is a very unusual observation. There is evidence to suggest the manager's claim is false.

3.91 a. $\bar{x} = 486.73$, $s = 37.42$ **b.** 0.60, 1.00, 1.00
c. These proportions are not close to the empirical rule proportions, which suggests that the shape of the distribution is not normal.

3.93 a. $\bar{x} = 120.3$, $s = 9.9672$ **b.** -0.7324, 0.3712, 2.0768, -0.5318, -0.5318, 0.8729, -0.7324, 0.8729, -0.8327, -0.8327 **c.** $\bar{z} = 0$, $s_z = 1.0$
d. Predictions: $\bar{z} = 0$, $s_z = 1.0$.

Proof:

$$\sum_{i=1}^{n} z_i = \sum_{i=1}^{n} \frac{(x_i - \bar{x})}{s} = \frac{1}{s}\left(\sum_{i=1}^{n} x_i - \sum_{i=1}^{n} \bar{x}\right)$$

$$= \frac{1}{s}\left(\sum_{i=1}^{n} x_i - n\bar{x}\right)$$

$$= \frac{1}{s}\left(\sum_{i=1}^{n} x_i - n\frac{1}{n}\sum_{i=1}^{n} x_i\right) = \frac{1}{s} \cdot 0 = 0$$

Therefore, $\bar{z} = \frac{1}{n}(0) = 0$

$$s_z^2 = \frac{1}{n-1}\sum(z_i = \bar{z})^2 = \frac{1}{n-1}\sum z_i^2$$

$$= \frac{1}{n-1}\sum\left(\frac{x_i - \bar{x}}{s}\right)^2$$

$$= \frac{1}{s^2}\left[\frac{1}{n-1}\sum(x_i - \bar{x})^2\right]$$

$$= \frac{1}{s^2}s^2 = 1 \Rightarrow s_z = 1$$

3.95 The interval two standard deviations from the mean is $(48, 96)$. A good minimum guaranteed life is 48 months.

Section 3.4

3.97 False.

3.99 False.

3.101 a.

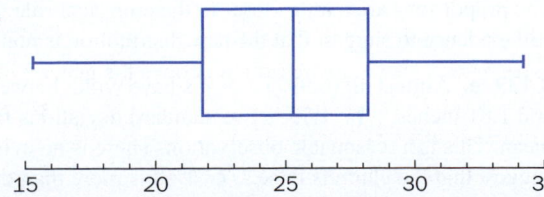

b.

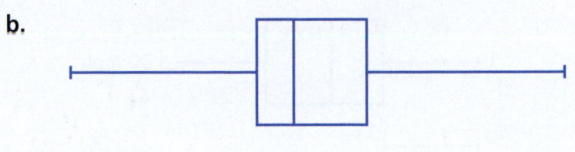

c.

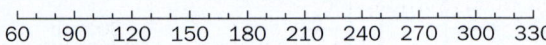

d.

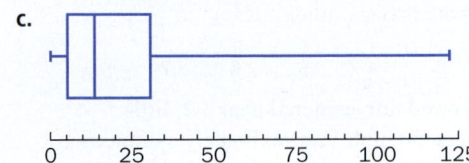

3.103 a. Neither.   **b.** Mild outlier.   **c.** Neither.
d. Extreme outlier.   **e.** Mild outlier.   **f.** Neither.

3.105

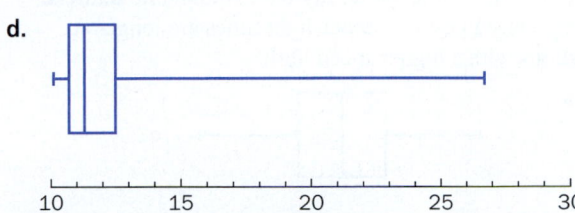

Centered near 48.5, positively skewed, two mild outliers.

3.107 Approximately symmetric, centered near 1560, two mild outliers.

3.109 Males: Slightly skewed right, centered near 29, lots of variability, 1 mild outlier. Females: Slightly skewed right, centered near 29, little variability, 1 mild outlier. Both centered near 29, both slightly skewed right. Female data are more compact.

3.111 a.

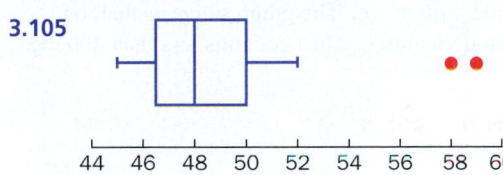

Positively skewed, centered near 34, compact except for the two extreme outliers.

b.

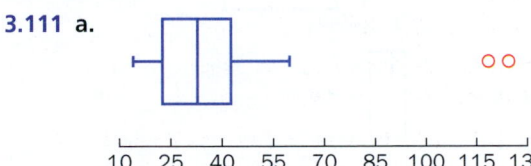

The modified box plot is more descriptive. The standard box plot hides information in the right tail of the distribution.

3.113

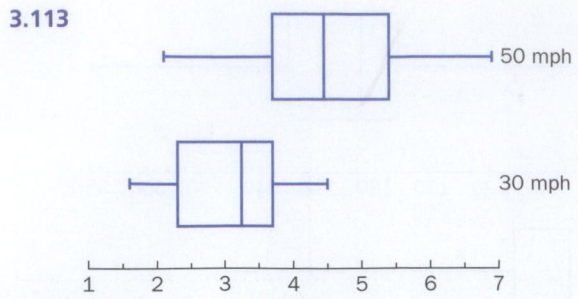

30 mph: Slightly skewed left, centered near 3.2, little variability, no outliers. 50 mph: Approximately symmetric, centered near 4.5, more variability than 30 mph, no outliers. The box plots suggest the amber-light times are longer at intersections with a higher speed limit.

3.115 a.

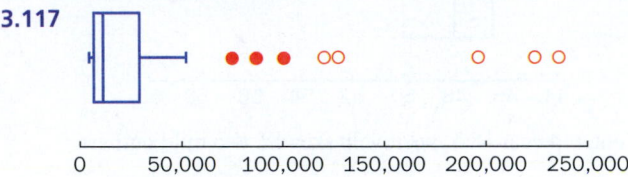

b. Approximately symmetric, centered near 380, lots of variability, 1 mild outlier.   **c.** The graph suggests that, on average, a 400-mg vitamin C tablet contains less than 400 mg.

3.117

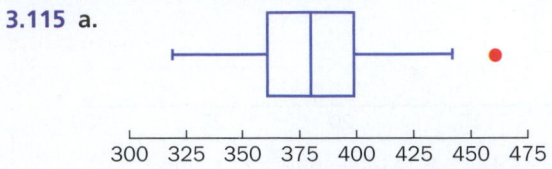

Centered near 13,000, lots of variability, skewed to the right, 3 mild outliers, 5 extreme outliers.

Chapter Exercises

3.119 a.

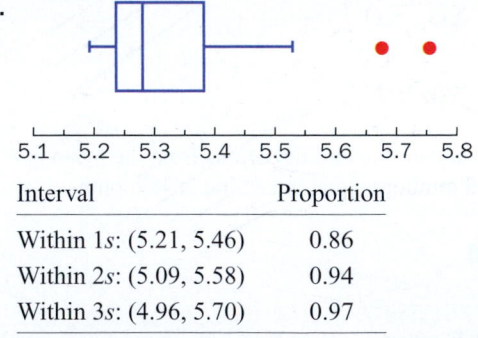

b. Miami: centered near 24.5, little variability, approximately symmetric, 2 mild outliers. Denver: centered near 27, lots of variability, approximately symmetric, no outliers.   **c.** Both distributions approximately symmetric. Denver has more variability and the values are, on average, larger.

3.121 a. $z = 0.5714$. This suggests that 1.54 is a reasonable observation. This is not an unusual generating capacity.
b. $z = -2.8571$. This z-score indicates that 1.3 is almost three standard deviations from the mean. Therefore, this is an unusual generating capacity.

3.123 a. $\bar{x} = 3.0003$, $\tilde{x} = 2.995$ **b.** The mean is approximately equal to the median, which suggests that the distribution

is approximately symmetric.   **c.** $\bar{x}_{\text{tr}(0.10)} = 3.0046$. A trimmed mean is not necessary. The distribution is approximately symmetric, and there are no extreme outliers.

3.125 a. $\bar{x} = 8.8731$, $s^2 = 8.0668$, $s = 2.8402$ **b.** 0.73, 0.96, 1.00. These proportions are close to the empirical rule proportions, so there is no evidence to suggest that the distribution is non-normal.   **c.** 11.5

3.127 a. $\bar{x} = 458.2083$, $s^2 = 2231.4764$, $s = 47.2385$
b.

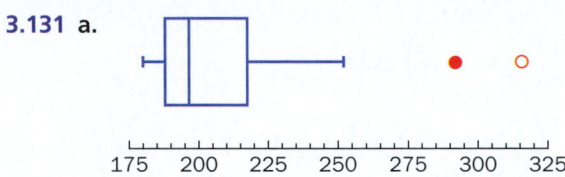

Three mild outliers: 359, 551, 566   **c.** $p_9 = 409$. 400 is in the 9th percentile.   **d.** At least 0.75 of the observations lie in the interval (363.73, 552.69). At least 0.89 of the observations lie in the interval (316.49, 599.92).

3.129 a. $\bar{x} = 1.656$, $\tilde{x} = 1.45$, $s^2 = 0.9841$, $s = 0.9920$.
b. $p_{40} = 1.3$, $p_{80} = 2.2$.   **c.** 4.8 is over three standard deviations from the mean. This is an unlikely magnitude.

3.131 a.

b.

Interval	Proportion
Within 1s: (179.932, 237.679)	0.92
Within 2s: (151.058, 266.553)	0.94
Within 3s: (122.184, 295.427)	0.97

These proportions are far enough away from those in the empirical rule to suggest that the data are non-normal.
c.

Interval	Proportion
Within 1s: (5.21, 5.46)	0.86
Within 2s: (5.09, 5.58)	0.94
Within 3s: (4.96, 5.70)	0.97

The box plot suggests that the distribution is positively skewed. The proportions are a little closer to the empirical rule. There is still evidence to suggest that the new distribution is non-normal.

3.133 a. Almost all (0.997) 2 $\times$ 4's have width between 1.69 and 1.81 inches.   **b.** 1.79 is two standard deviations from the mean. This is a reasonable observation. There is no evidence to suggest that the claim is false.   **c.** 1.68 is more than three standard deviations from the mean. This is a very unusual observation. There is evidence to suggest that the claim is false.

3.135 a. $\bar{x} = 1777.1136$, $\tilde{x} = 1288$ **b.** $s_x^2 = 1,853,111.58$, $s_x = 1361.29$

Interval	Proportion
Within $1s$: (415.82, 3138.40)	0.88
Within $2s$: (-945.47, 4499.69)	0.93
Within $3s$: (-2306.76, 5860.98)	0.95

These proportions are far enough away from those in the empirical rule to suggest that the data are non-normal.

c. $Q_1 = 940$, $Q_3 = 1835.5$, IQR $= 895.5$.

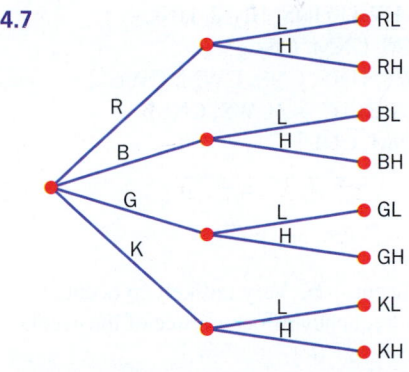

750 1750 2750 3750 4750 5750 6750 7750

The distribution is centered near 1300, has lots of variability, is positively skewed, with several mild and extreme outliers. This description agrees with parts (a) and (b).

d. *Phantom of the Opera*: 10,807 performances as of January 14, 2014. This value should increase the values of the sample mean, median, variance, and standard deviation.

$\bar{y} = 1845.01 > \bar{x}$, $\tilde{y} = 1295 > \tilde{x}$, $s_y^2 = 2,452,146.9 > s_x^2$, $s_y = 1565.93 > s_x$

CHAPTER 4

Section 4.1

4.1 False.

4.3 True.

4.5 a. Not. **b.** Or. **c.** And.

4.7

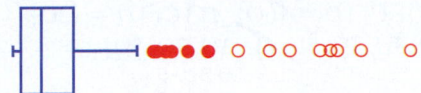

S = {RL, RH, BL, BH, GL, GH, KL, KH}

4.9 52

4.11 a. $B \cap C = \{1, 3\}$ **b.** $B \cap D = \{5, 7, 9\}$
c. $A \cap B = \{\}$ **d.** $A \cap C = \{0, 2, 4\}$
e. $(B \cap C)' = \{0, 2, 4, 5, 6, 7, 8, 9\}$
f. $B' \cup C' = \{0, 2, 4, 5, 6, 7, 8, 9\}$

4.13 a. $A \cup B \cup D = \{a, b, c, d, e, f, g, h, j, k\}$
b. $B \cup C \cup D = \{a, b, c, d, e, f, g, h, i, j, k\}$
c. $B \cap C \cap D = \{\}$ **d.** $A \cap B \cap C = \{c\}$

4.15

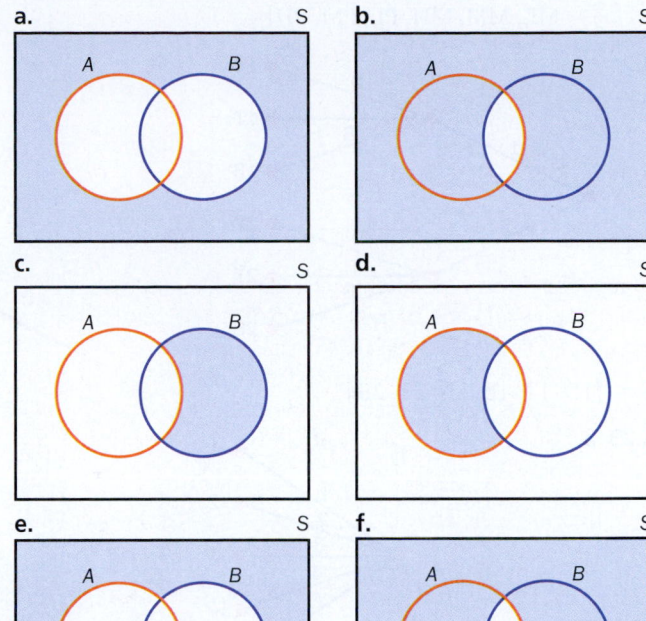

4.17 a. $A = \{YNN, NYN, NNY\}$,
$B = \{YNN, NYN, NNY\}$,
$C = \{YNN, NYN, NNY, YYN, YNY, NYY, YYY\}$,
$D = \{YYY, YYN, YNY, NYY\}$
b. $A \cup D = \{NNY, NYN, NYY, YNN, YNY, YYN, YYY\}$
c. $D' = \{NNN, NNY, NYN, YNN\}$
d. $B \cap C = \{NNY, NYN, YNN\}$
e. $D = \{YYY, YYN, YNY, NYY\}$

4.19 a.

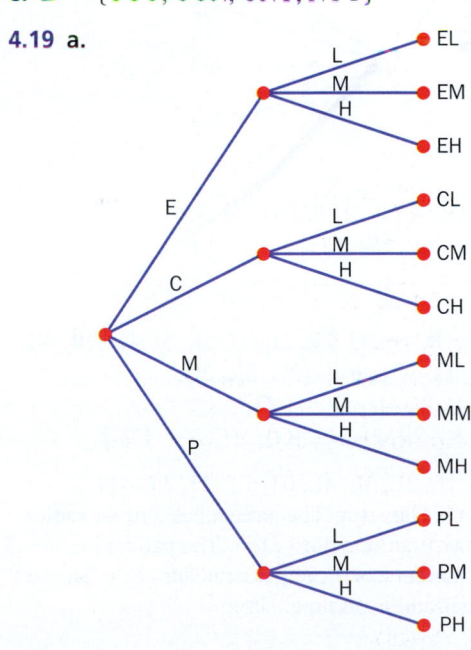

b. $S = \{$EL, EM, EH, CL, CM, CH,
ML, MM, MH, PL, PM, PH$\}$

4.21

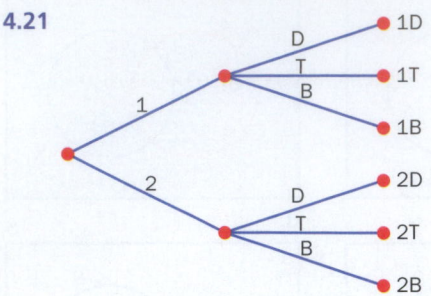

$S = \{$1D, 1T, 1B, 2D, 2T, 2B$\}$

4.23 a.

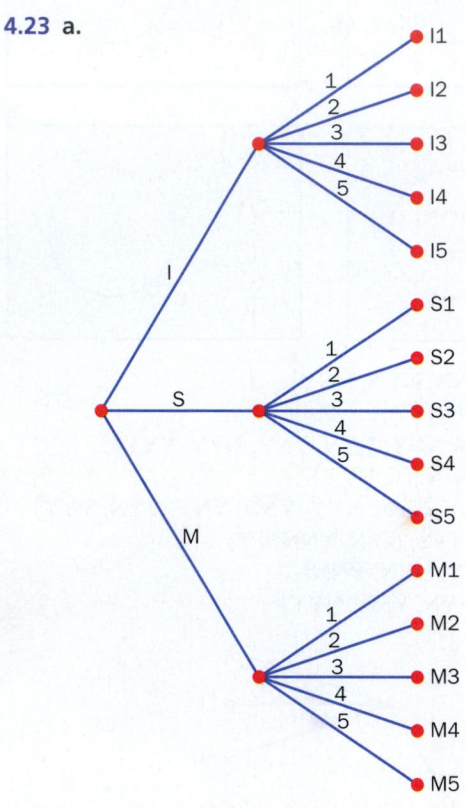

b. $S = \{$I1, I2, I3, I4, I5, S1, S2, S3, S4, S5,
M1, M2, M3, M4, M5$\}$

4.25 66

4.27 a. $S = \{$1G, 1R, 1I, 2G, 2R, 2I, 3G, 3R, 3I, 4G, 4R, 4I$\}$
b. $A = \{$1G, 2G, 3G, 4G$\}$, $B = \{$2G, 2R, 2I$\}$,
$C = \{$3G, 3R, 3I, 1I, 2I, 4I$\}$, $D = \{$4G$\}$
c. $A \cup B = \{$1G, 2G, 2R, 2I, 3G, 4G$\}$, $A \cap B = \{$2G$\}$

4.29 a. $S = \{$0L, 1L, 2L, 3L, 4L, 0T, 1T, 2T, 3T, 4T$\}$
b. $A = $ the patient is late. $B = $ The patient has 3 or 4 cavities.
$C = $ The patient has 1 or 3 cavities. $D = $ The patient has 0
cavities. $E = $ The patient has 0 cavities or is late. $F = $ The
patient has 4 cavities and is on time.

4.31 a. $S = \{$MCY, MCD, MCR, MNY, MND, MNR, FCY,
FCD, FCR, FNY, FND, FNR$\}$ **b.** $A = $ The customer is male.
$B = $ The customer orders fresh-cut fries and is a senior.
$C = $ The customer is young.
$D = $ The customer did not order fresh-cut fries.

4.33 a. There are infinitely many outcomes in this experiment.   **b.** Some of the possible outcomes in this experiment
are H, BH, BBH, BBBH, BBBBH.

4.35 a. $S = \{$1U, 2U, 3U, 4U, 5U, 6U, 1O, 2O, 3O,
4O, 5O, 6O$\}$ **b.** $B' = \{$4U, 5U, 6U, 4O, 5O, 6O$\}$
$A \cup B = \{$1U, 2U, 3U, 1O, 2O, 3O, 4O, 5O, 6O$\}$
$A \cap B = \{$1O, 2O, 3O$\}$ $C \cap D = \{$6O$\}$ $A \cap C \cap D = \{$2O$\}$
$(A \cap D)' = \{$1U, 2U, 3U, 4U, 5U, 6U, 1O, 3O, 5O$\}$

4.37 a.

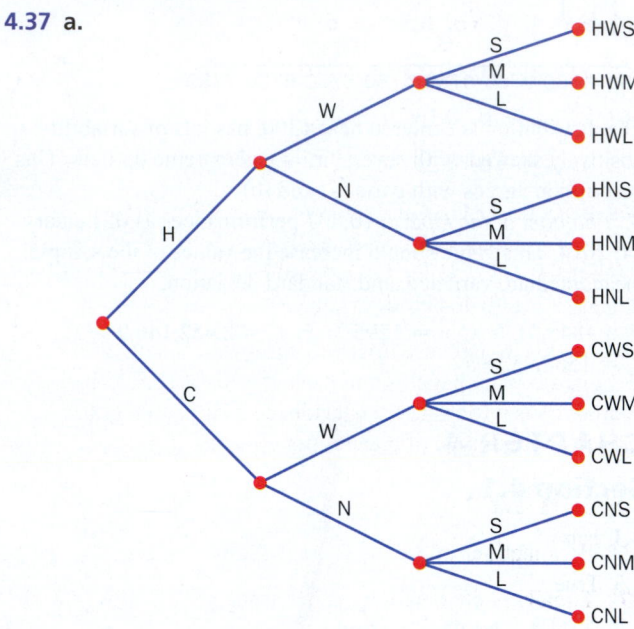

b. $S = \{$HWS, HWM, HWL, HNS, HNM, HNL,
CWS, CWM, CWL, CNS, CNM, CNL$\}$
c. $A \cup B = \{$HWS, CWS, HNS, CNS, CWM, CWL,
CNM, CNL$\}$ $B \cup C = S$ $B \cap C = \{$CWS, CNS$\}$
$C' = \{$SWM, CWL, CNM, CNL$\}$

Section 4.2

4.39 a. Very likely to occur.   **b.** Very unlikely to occur.
c. The limiting relative frequency of occurrence of the event.

4.41 False.

4.43 $P(A \cap B)$

4.45 a. 0.5   **b.** 0.3333   **c.** 0.3333   **d.** 0.5

4.47 a. 0.85   **b.** 0.15   **c.** 0.4   **d.** 0.7

4.49 a. 0.532   **b.** 0.468   **c.** 0.594   **d.** 0.771

4.51

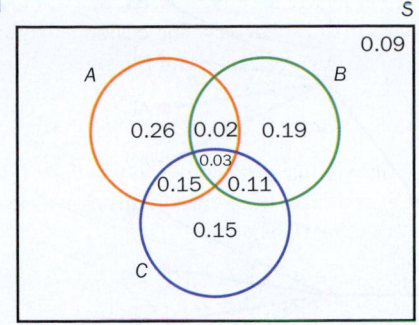

4.53 a. 0.20   **b.** 0.40   **c.** 0.60   **d.** 0.10

4.55 a. 10   **b.** 0.7   **c.** 0.6   **d.** 0.6364, 0.6364

4.57 a. 0.4305   **b.** 0.5434   **c.** 0.9784

4.59 a. 1000   **b.** 0.01   **c.** 0.002   **d.** 0.001

4.61 a. 0.84   **b.** 0.16   **c.** 0.40, 0.14

4.63 a. 0.29   **b.** 0.25   **c.** 0.24   **d.** 0.95

4.65 a. G_1G_2, G_1G_3, G_1G_4, G_1G_5, G_1G_6, G_1B_1, G_1B_2, G_2G_3, G_2G_4, G_2G_5, G_2G_6, G_2B_1, G_2B_2, G_3G_4, G_3G_5, G_3G_6, G_3B_1, G_3B_2, G_4G_5, G_4G_6, G_4B_1, G_4B_2, G_5G_6, G_5B_1, G_5B_2, G_6B_1, G_6B_2, B_1B_2
b. 0.5357   **c.** 0.4643   **d.** 0.0357

4.67 0.000976. Because this probability is so small, all 10 people buying a plain bagel is a rare event. This suggests that the assumption is wrong. There is evidence to suggest that the demand for each type of bagel is not equal.

Section 4.3

4.69 Permutation.

4.71 Equally likely outcomes experiment.

4.73 a. 1,680   **b.** 1,663,200   **c.** 11,880   **d.** 3,628,800
e. 10   **f.** 1   **g.** 72   **h.** 380   **i.** 9,900

4.75 362,880

4.77 390,700,800

4.79 6840

4.81 a. 64,000   **b.** 0.0156   **c.** 59,280, 0.0121

4.83 a. 216   **b.** 144   **c.** 36

4.85 a. 0.3626   **b.** 0.0088   **c.** 0.6374

4.87 a. 19,958,400   **b.** 0.0152   **c.** 0.0909

4.89 a. 455   **b.** 0.0220   **c.** 0.2637   **d.** 0.80

4.91 a. 40,320   **b.** 0.125

4.93 a. 2550   **b.** 0.84   **c.** 0.4706

4.95 a. 20   **b.** P(two girls selected) = 0.10. Because this probability is so small, there is evidence to suggest that the process was not random.

4.97 a. 3,628,800   **b.** 0.0222   **c.** 0.2   **d.** 0.0079

4.99 a. 1326   **b.** 0.0045   **c.** 0.0588   **d.** 0.2353

4.101 a. 4   **b.** 8   **c.** 16   **d.** 2^n

4.103 $(n - 1)!$

4.105 a. 1,345,860,629,046,814,650   **b.** 0.00001103
c. 0.0000017862

Section 4.4

4.107 False.

4.109 a. True.   **b.** True.   **c.** False.

4.111 a. Unconditional.   **b.** Conditional.   **c.** Unconditional.
d. Unconditional.   **e.** Conditional.

4.113 a. Valid.   **b.** Rows: 0.17, 0.22, 0.21, 0.21, 0.19.
Columns: 0.74, 0.26.   **c.** 0.12, 0.07, 0.02   **d.** 0.1923,
0.9048, 0   **e.** 0.21 = 0.17 + 0.04

4.115 a. 0.466, 0.299   **b.** 0.135, 0.215   **c.** 0.3258,
0.4893, 0.4491   **d.** 0.534 = 0.145 + 0.174 + 0.215

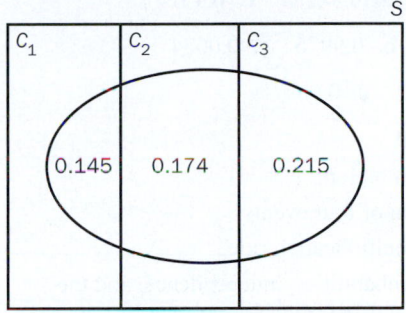

4.117 a. 0.13, 0.36, 0.62. These events are not mutually exclusive and exhaustive.   **b.** 0, 0.15   **c.** 0.2419, 0.4167
d. 0.2031, 0.7126, 0   **e.** 0.3056, 0.2778, 0.4167. Given B has occurred, either 3, 4, or 5 must occur.

4.119 a. 0.0784   **b.** 0.0588   **c.** 0.2353   **d.** 0.25

4.121 a. 0.50   **b.** 0.3333   **c.** 0.6667

4.123 a. 0.4439   **b.** 0.1629   **c.** 0.5324

4.125 a. 0.3441   **b.** 0.2672   **c.** 0.1062   **d.** 0.2471
e. 0.3779

4.127 a. 0.5385   **b.** 0.4615   **c.** 0.3

4.129 a. 0.1166   **b.** 0.2930   **c.** 0.7070

4.131 a. 0.144   **b.** 0.4493   **c.** 0.7843
d. P(North | (Good ∪ Fair)) = 0.1814
P(Midwest | (Good ∪ Fair)) = 0.3733
P(South | (Good ∪ Fair)) = 0.4453
The customer is most likely from the South.

4.133 a.

		Arrival mode			
		Bus	Car	Walk	
Lunch	Carries	625	466	142	1233
	Buys	345	122	500	967
		970	588	642	2200

b. 0.2118   **c.** 0.3557   **d.** 0.7003   **e.** Walk.

4.135 a. 0.0535   **b.** 0.5429   **c.** 0.5699   **d.** 0.2810
e. 0.4563

Section 4.5

4.137 $P(A) \cdot P(B)$

4.139 True.

4.141 a. Dependent.   **b.** Dependent.   **c.** Independent.
d. Dependent.

4.143 a. 0.085, 0.66, 0.165   **b.** 0.0527, 0.38, 0.323
c. Not enough information to determine independence or
dependence.

4.145 a. 0.1, 0.18, 0.12   **b.** Not enough information to
determine independence or dependence.
c. 0.75. No, $P(B|A) + P(C|A) + P(D|A) = 1$

4.147 a. $P(A') = 0.65, P(C|A) = 0.18$,
$P(B|A') = 0.36$ **b.** 0.0630, 0.234   **c.** 0.451

4.149 a. 0.00000144   **b.** 0.9976   **c.** 0.0024

4.151 a. 0.04   **b.** 0.32   **c.** 0.96

4.153 a. Solution Trail
Keywords: All four fault regions.
Translation: Intersection of four events.
Concepts: Probability multiplication rule.
Vision: Use the given probabilities, independence, and the
probability multiplication rule for the intersection of four
events.
0.000117   **b.** 0.4970   **c.** 0.5030   **d.** 0.8177

4.155 a. 0.0002856   **b.** 0.3424   **c.** 0.0847

4.157 a. 0.2646   **b.** 0.1554   **c.** 0.3402

4.159 a. 0.7531   **b.** 0.0057   **c.** 0.2412

4.161 a. 0.008   **b.** 0.512   **c.** 0.384

4.163 a. 0.24   **b.** 0.695   **c.** 0.0072

4.165 a. 0.0475   **b.** 0.0665   **c.** 0.7143

4.167 a. 0.992   **b.** 0.008   **c.** 6

4.169 a. $P(L) = 0.366, P(T'|D) = 0.774, P(T|L) = 0.156$,
$P(B'|D \cap T) = 0.545, P(B'|D \cap T') = 0.622$,
$P(B'|L \cap T) = 0.105, P(B'|L \cap T') = 0.005$
b. 0.0652   **c.** 0.3909   **d.** 0.5803

4.171 $P(A|R) = 0.2830$ $P(B|R) = 0.3396$ $P(C|R) = 0.3774$.
The salesperson most likely stayed at Hotel C.

Chapter Exercises

4.173 a. 1140   **b.** 0.0447   **c.** 0.4035

4.175 a.

b. $S = \{$AW, AL, AT, AO, NW, NL, NT, NO, SW,
SL, ST, SO$\}$
c. $E = \{$SW, SL, ST, SO$\}, F = \{$AW, NW, SW$\}$,
$G = \{$AO, NW, NL, NT, NO, SO$\}, H = \{$AL$\}$
d. $E \cup F = \{$AW, NW, SL, SO, ST, SW$\}$,
$F \cap G = \{$NW$\}, H' = \{$AO, AT, AW, NL, NO, NT,
NW, SL, SO, ST, SW$\}$
e. $E \cup H' = \{$AO, AT, AW,
NL, NO, NT, NW, SL, SO, ST, SW$\}$,
$E \cup F \cup G' = \{$AL, AT, AW, NW, SL, SO, ST, SW$\}$,
$F \cup G' = \{$AL, AT, AW, NW, SL, ST, SW$\}$

4.177 a. 0.3   **b.** 0.81   **c.** 0.5263

4.179 a. 0.95   **b.** 0.05   **c.** 0.25   **d.** 0.2143
e. 0.3025

4.181 a. 0.8   **b.** 0.042   **c.** 0.168

4.183 a. 0.7225   **b.** 0.0225   **c.** 0.0811

4.185 a. 0.6   **b.** 0.248   **c.** 0.2686

4.187 a. 0.0187   **b.** 0.1575   **c.** 0.3701

4.189 a. 0.0001464   **b.** 0.3102
c. Claim: $P(C_i) = 0.11$
Experiment: None of the 4 colors her hair.
Likelihood: P(0 color her hair) = 0.6274
Conclusion: This probability is large, so there is no evidence to
suggest that the study's claim is false.

4.191 a. 0.0643   **b.** 0.1106   **c.** 0.1501   **d.** 0.0114

4.193 a. A simulation produced the following awards:
1, 2, 1, 5, 3, 1, 1, 1, 1, 1, 1. One person won 5 or more free
nights.

b.

n	Rel. freq.	n	Rel. freq.	n	Rel. freq.
50	0.0600	100	0.0900	150	0.0800
200	0.0950	250	0.0800	300	0.0567
350	0.0629	400	0.0775	450	0.0844
500	0.0600	550	0.0618	600	0.0733
650	0.0492	700	0.0586	750	0.0640
800	0.0713	850	0.0541	900	0.0633
950	0.0642	1000	0.0540	1050	0.0590
1100	0.0755	1150	0.0635	1200	0.0692
1250	0.0640	1300	0.0677	1350	0.0681
1400	0.0671	1450	0.0517	1500	0.0680
1550	0.0516	1600	0.0613	1650	0.0709
1700	0.0729	1750	0.0674	1800	0.0661
1850	0.0595	1900	0.0658	1950	0.0585
2000	0.0635				

c.

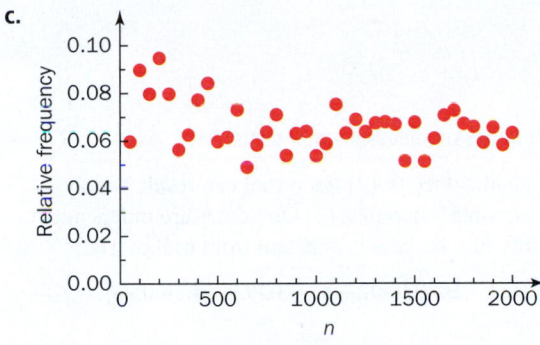

d. An estimate of the probability of winning 5 or more free nights is 0.063. **e.** 0.0625

CHAPTER 5
Section 5.1

5.1 True.

5.3 False.

5.5 Sample space, real numbers.

5.7 Measurement.

5.9 a. Discrete. **b.** Continuous. **c.** Continuous.
d. Discrete. **e.** Discrete. **f.** Continuous. **g.** Discrete.
h. Discrete.

5.11 a. Discrete. **b.** Continuous. **c.** Continuous.
d. Discrete. **e.** Continuous. **f.** Discrete.

5.13 a. $S = \{MM, MW, MB, MG, WM, WW, WB,$
$WG, BM, BW, BB, BG, GM, GW, GB, GG\}$
b. 0, 1, 2. Discrete. X can assume only a finite number of values.

5.15 a. Discrete. **b.** Continuous. **c.** Discrete.
d. Continuous.

5.17 Continuous. Measuring acceleration.

5.19 a. Continuous. **b.** Continuous. **c.** Discrete.
d. Continuous. **e.** Discrete. **f.** Discrete **g.** Discrete.
h. Continuous.

Section 5.2

5.21 True.

5.23 All the possible values of X, the probability associated with each value.

5.25 a. 0.07 **b.** 0.62, 0.42 **c.** 0.7 **d.** 0.38

5.27 a. 0.65, 0.35 **b.** 0.6 **c.** 0.4615

d.

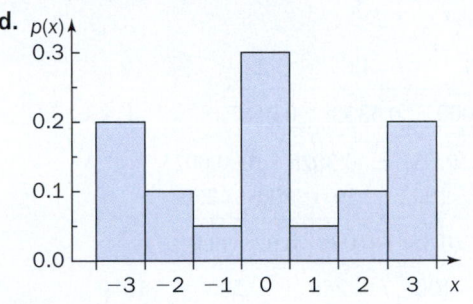

5.29

x	1	2	3	4
p(x)	0.01	0.08	0.27	0.64

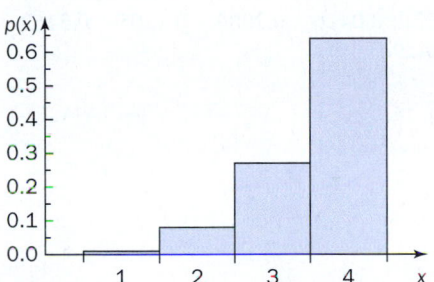

5.31 a.

x	1	2	3	4	5	6
p(x)	0.0179	0.0536	0.1071	0.1786	0.2679	0.3750

$0 \le p(x) \le 1$ and $\sum p(x) = 1$.

b. 0.1786 **c.** 0.9286 **d.** 0.2857

e.

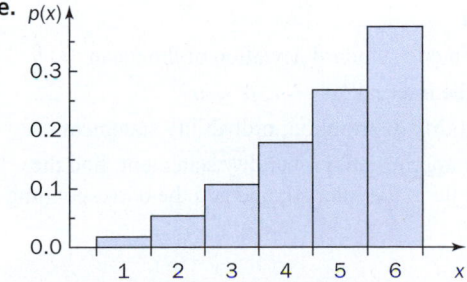

5.33 a. 0.35 **b.** 0.95 **c.** 0.04 **d.** 0.01 **e.** 0.5775

5.35 a.
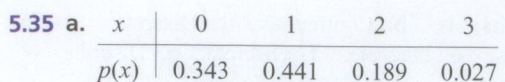

x	0	1	2	3
$p(x)$	0.343	0.441	0.189	0.027

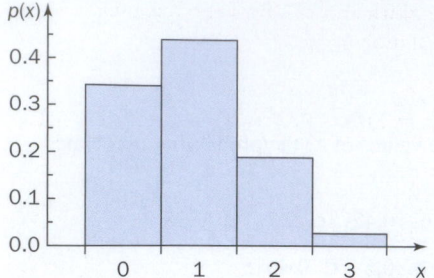

b. 0.027   **c.** 0.657

5.37

x	0	1	2
$p(x)$	0.4000	0.5333	0.0667

5.39 a. 0.1   **b.** 0.65   **c.** 0.3025   **d.** 0.0023
e.

y	50	100	150	200	250
$p(y)$	0.55	0.35	0.07	0.02	0.01

5.41 a.

m	100	250	500	1000
$p(m)$	0.0667	0.1333	0.4667	0.3333

b. 0.1111

5.43 a.

y	0	1	2	3	4
$p(y)$	0.0027	0.0416	0.2089	0.4305	0.3162

b. 0.9973   **c.** 0.5576

Section 5.3

5.45 True.

5.47 True.

5.49 $\sum\limits_{\text{all } x} [f(x) \cdot p(x)]$

5.51 $\mu = 7.20, \sigma^2 = 8.96, \sigma = 2.9933$

5.53 a. $\mu = 0, \sigma^2 = 270, \sigma = 16.4317$
b. 1   **c.** 0.55, 0.45

5.55 a. $\mu = 7.35, \sigma^2 = 24.6275, \sigma = 4.9626$
b. 0.4   **c.** 0.95

5.57 a. $\mu = 9.04$ **b.** $\sigma^2 = 3.0184, \sigma = 1.7374$
c. Solution Trail
Keywords: Within one standard deviation of the mean.
Translation: In the interval $(\mu - \sigma, \mu + \sigma)$.
Concepts: Probability distribution; probability statement.
Vision: Write the appropriate probability statement, find the value(s) of X that lie in the interval, and add the corresponding probabilities.
0.62   **d.** 0.82

5.59 a. $\mu = 15.55, \sigma^2 = 40.4475, \sigma = 6.3598$
b. 0.82   **c.** 0.92   **d.** 386

5.61 a. $\mu = 7.998, \sigma^2 = 6.706, \sigma = 2.5896$ **b.** 0.6
c. 0.000144

5.63 a. $\mu = 2.9797, \sigma^2 = 0.4183, \sigma = 0.6468$ **b.** 0.8002
c. 10.9594, 1.6732, 1.2935

5.65 a. $p(x) \geq 0$ for all x and $\sum\limits_{\text{all } x} p(x) = 1$.   **b.** $\mu = 0,$
$\sigma^2 = 5.2632, \sigma = 2.2942$ **c.** 0.0028

5.67 a. $\mu = 0.6, \sigma^2 = 0.24, \sigma = 0.4899$ **b.** $\mu = 0.7,$
$\sigma^2 = 0.21, \sigma = 0.4583$ **c.** $\mu = 0.8, \sigma^2 = 0.16, \sigma = 0.4000$
d. $\mu = p, \sigma = p(1 - p), \sigma = \sqrt{p(1 - p)}$ **e.** $p = q = 0.5$

5.69 $E[(X - \mu)^2]$
$= \sum\limits_{\text{all } x} (x - \mu)^2 p(x) = \sum\limits_{\text{all } x} (x^2 + 2x\mu + \mu^2) p(x)$
$= \sum\limits_{\text{all } x} x^2 p(x) - 2\mu \sum\limits_{\text{all } x} x p(x) + \mu^2 \sum\limits_{\text{all } x} p(x)$
$= E(X^2) - 2\mu\mu + \mu^2 = E(X^2) - \mu^2$

Section 5.4

5.71 False.

5.73 True.

5.75 True.

5.77 The number of successes in n trials.

5.79 (1) n identical trials. (2) Each trial can result in only one of two possible outcomes. (3) Outcomes are independent. (4) Probability of a success is constant from trial to trial.

5.81 a. 0.2361   **b.** 0.0802   **c.** 0.0393   **d.** 0.0566
e. 0.7638

5.83 a. ≈ 1 **b.** 0.9995   **c.** 0.5118   **d.** 0.8047

5.85 a. $\mu = 12, \sigma^2 = 7.2, \sigma = 2.6833$
b. (9.3167, 14.6833), (6.6334, 17.3666), (3.9501, 20.0499)
c. 0.0009   **d.** 0.0172

5.87 a. 0.0432   **b.** 0.9999   **c.** 18   **d.** 0.1230

5.89 a. 24   **b.** 0.9744   **c.** 0.0095

5.91 a. $\mu = 45, \sigma^2 = 4.5, \sigma = 2.1213$ **b.** 0.9421
c. 0.8304   **d.** Claim: $p = 0.9$ $\Rightarrow$ $X \sim B(50, 0.9)$
Experiment: $x = 41$
Likelihood: $P(X \leq 41) = 0.0579$
Conclusion: There is no evidence to suggest the claim is false.

5.93 a. 0.1722   **b.** 0.2348
c. Claim: $p = 0.17$ $\Rightarrow$ $X \sim B(35, 0.17)$
Experiment: $x = 3$
Likelihood: $P(X \leq 3) = 0.1315$
Conclusion: There is no evidence to suggest the claim is false.

5.95 a. 0.2599   **b.** 0.4705   **c.** 0.5205

5.97 a. 0.1297   **b.** 0.3438
c. Claim: $p = 0.11$ $\Rightarrow$ $X \sim B(50, 0.11)$
Experiment: $x = 12$
Likelihood: $P(X \geq 12) = 0.0069$ (≤ 0.05)
Conclusion: There is evidence to suggest the claim is false.

5.99 a. 0.3269 **b.** $\mu = 4.72$, $P(X < \mu) = 0.4729$
c. 0.0097

5.101 a. 0.2880, 0.0640 **b.** 0.9744 **c.** 0.9898 **d.** $n \geq 8$

5.103 a. 0.0014 **b.** 0.9659 **c.** 0.9064

5.105 a. 0.1933 **b.** 0.9435 **c.** 0.1121 **d.** 0.00005268

Section 5.5

5.107 True.

5.109 True.

5.111 In a hypergeometric experiment, the probability of a success changes on each trial.

5.113 a. 0.25 **b.** 0.4290 **c.** 0.0563

5.115 a. 0.4679 **b.** 0.1125 **c.** 0.3606 **d.** 0.9597

5.117 a. 4, 5, 6, 7, 8 **b.** $\mu = 6$, $\sigma^2 = 0.8$, $\sigma = 0.8944$
c. 0.2462 **d.** 0.0385

5.119 a. Solution Trail

Keywords: Mean number of robberies per day; 4.32.

Translation: Fixed time; $\lambda = 4.32$.

Concepts: Poisson probability distribution.

Vision: The mean of the Poisson distribution is given. Write a probability statement and convert to cumulative probability if necessary.

0.1241 **b.** 0.0325 **c.** 0.00017689

5.121 a. 0.1743 **b.** 0.7199 **c.** 0.0358 **d.** 0.2957

5.123 a. 0.0047 **b.** 0.3134 **c.** 0.3012

5.125 a. 0.0821 **b.** 0.0042 **c.** 0.9580

5.127 a. Solution Trail

Keywords: 15 lobstermen; 5 fined; 4 selected at random.

Translation: $N = 15$; $M = 5$ successes; 10 failures; sample size $n = 4$.

Concepts: Hypergeometric probability distribution.

Vision: Sampling without replacement, and the population is finite. Write a probability statement involving a hypergeometric random variable and use the appropriate formula.

0.3297 **b.** 0.0037 **c.** 0.8462

5.129 a. 0.3679 **b.** 0.9810
c. Claim: $\lambda = 1$ $\Rightarrow$ X is a Poisson RV.

Experiment: $x = 6$

Likelihood: $P(X \geq 6) = 0.0005942$

Conclusion: There is evidence to suggest the claim is false.

5.131 a. 0.1839 **b.** 0.9963
c. Claim: $\lambda = 1$ $\Rightarrow$ X is a Poisson RV.

Experiment: $x = 3$

Likelihood: $P(X \geq 3) = 0.0803$

Conclusion: There is no evidence to suggest the claim is false.

5.133 a. 0.0091 **b.** 0.1954 **c.** 0.0959

5.135 a. 0.4431 **b.** 0.9819 **c.** 0.4290, 0.9782 **d.** 0.4096, 0.9728 **e.** In a hypergeometric experiment, the probability of a success changes on each trial. As N increases with M/N constant, the hypergeometric probabilities approach the corresponding binomial probabilities.

5.137

a	$P(X = a)$	$P(Y = 1)/a$	$P(X \leq a)$	$1 - P(Y = 0)$
1	0.4	0.4	0.4	0.4
2	0.24	0.24	0.64	0.64
3	0.144	0.144	0.784	0.784
4	0.0864	0.0864	0.8704	0.8704
5	0.0518	0.0518	0.9222	0.9222
6	0.0311	0.0311	0.9533	0.9533
7	0.0187	0.0187	0.9720	0.9720
8	0.0112	0.0112	0.9832	0.9832
9	0.0067	0.0067	0.9899	0.9899
10	0.0040	0.0040	0.9940	0.9940

$P(X = a) = P(Y = 1)/a$; $P(X \leq a) = 1 - P(Y = 0)$.

$$\frac{P(Y = 1)}{a} = \frac{\binom{a}{1}(0.4)^1(0.6)^{a-1}}{a} = \frac{a(0.4)(0.6)^{a-1}}{a}$$

$$= (0.4)(0.6)^{a-1} = P(X = a)$$

$$1 - P(Y = 0) = 1 - \binom{a}{0}(0.4)^0(0.6)^a = 1 - (1)(1)(0.6)^a$$

$$= 1 - (0.6)^a = P(X \leq a)$$

Chapter Exercises

5.139 a. $\mu = 20$, $\sigma^2 = 4$, $\sigma = 2$ **b.** 0.8909
c. Claim: $p = 0.8$ $\Rightarrow$ $X \sim B(25, 0.8)$

Experiment: $x = 21$

Likelihood: $P(X \geq 21) = 0.4207$

Conclusion: There is no evidence to suggest the supervisor's claim is false.

5.141 a. 0.0573 **b.** 12.5 **c.** 0.1887

5.143 a. 0.1291 **b.** 0.8687 **c.** 0.0139

5.145 a. 0.0774 **b.** 0.8041 **c.** 0.0238 **d.** 0.6476

5.147 a. 0.2205 **b.** 0.0567 **c.** 1.878×10^{-12}

5.149 a. 0.0451 **b.** 0.4826
c. Claim: $p = 0.81$ $\Rightarrow$ $X \sim B(30, 0.81)$

Experiment: $x = 18$

Likelihood: $P(X \leq 18) = 0.0062$

Conclusion: There is evidence to suggest the claim (81%) is false.

5.151 a. 0.0498 **b.** 0.6472
c. Claim: $\lambda = 3$ $\Rightarrow$ X is a Poisson RV.

Experiment: $x = 6$

Likelihood: $P(X \geq 6) = 0.0839$

Conclusion: There is no evidence to suggest the claim is false. There is no evidence to suggest the number of Caf-Pows per show has changed.

5.153 a. 0.1254   **b.** 0.2677, 0.6550
c. 0.0604, 0.1538, 0.5000

5.155 a. 0.1042   **b.** $w = 9$
c. Claim: $p = 0.30 \Rightarrow X \sim B(40, 0.30)$
Experiment: $x = 16$
Likelihood: $P(X \geq 16) = 0.1151$
Conclusion: There is no evidence to suggest the claim is false. There is no evidence to suggest that the proportion of patients with knee joint replacement surgery who experience a weight gain has changed.

5.157 a. Answers will vary.

Complete shipments	Relative frequency	$p(y)$
0	0.0000	0.0000
1	0.0000	0.0007
2	0.0060	0.0052
3	0.1600	0.0234
4	0.0790	0.0701
5	0.1300	0.1471
6	0.2020	0.2207
7	0.2640	0.2365
8	0.1850	0.1774
9	0.0830	0.0887
10	0.0320	0.0266
11	0.0030	0.0036

b.

Complete shipments	Relative frequency	$p(y)$
0	0.0000	0.0000
1	0.0000	0.0000
2	0.0000	0.0001
3	0.0000	0.0008
4	0.0050	0.0040
5	0.0140	0.0142
6	0.0390	0.0392
7	0.0960	0.0840
8	0.1210	0.1417
9	0.1980	0.1889
10	0.1910	0.1983
11	0.1590	0.1623
12	0.1050	0.1014
13	0.0580	0.0468
14	0.0120	0.0150
15	0.0020	0.0030
16	0.0000	0.0003

c. 0.1612, 0.1538, 0.2735, 15

CHAPTER 6

Section 6.1

6.1 False.

6.3 True.

6.5 For a continuous random variable, probability is the area under the probability density function. Suppose X is a continuous random variable with probability density function $f(x)$. $P(a \leq X \leq b)$ is the area under the graph of $y = f(x)$, above the x axis, and between the lines $x = a$ and $x = b$.

6.7 a.

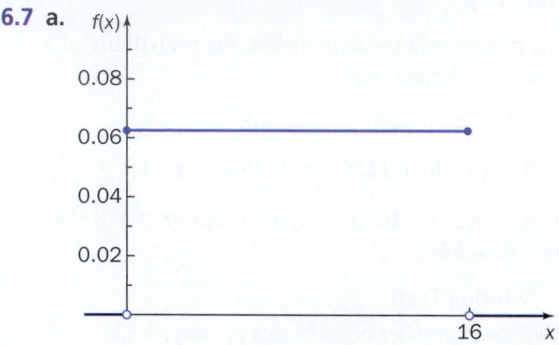

b. $\mu = 8, \sigma^2 = 21.3333, \sigma = 4.6188$ **c.** 0.75
d. 0.625   **e.** 0.4375

6.9 a. $\mu = 75, \sigma^2 = 208.3333, \sigma = 14.4338$ **b.** 0.5774
c. 0   **d.** $c = 60$

6.11 a. 0.0625   **b.** 0.4375   **c.** 0   **d.** 0.3125   **e.** 0.4444
f. 2.8284. The distribution is not symmetric.

6.13 a. Solution Trail
Keywords: Uniform distribution.
Translation: Uniform random variable; $a = 5.085$ and $b = 5.155$.
Concepts: Uniform probability distribution; probability is area.
Vision: Write the appropriate probability statement and find the corresponding area under the density curve.
0.2143   **b.** 0.3   **c.** 0.0571

6.15 a. 0.5   **b.** 0.3333   **c.** 37.5   **d.** 0.1667

6.17 a. $f(x) \geq 0$ and total area is
$(10)(0.05) + \frac{1}{2}(20)(0.05) = 1$
b. 0.25   **c.** 0.125   **d.** 15.8579
e. 0.1884, 0.2976, 0.0314

6.19 a. $f(x) \geq 0$ and total area is 1.   **b.** 0.75   **c.** 0.75
d. 0.25

6.21 a. $f(x) \geq 0$ and total area is $\frac{1}{2}(5)(0.4) = 1$
b. 0.75   **c.** 0.20   **d.** 3.8820   **e.** 0.04
f. 0.0340

6.23 a. 0.3333   **b.** $\mu = 2, 0$ **c.** 3.75
d. 0.00026, 0

6.25 a.

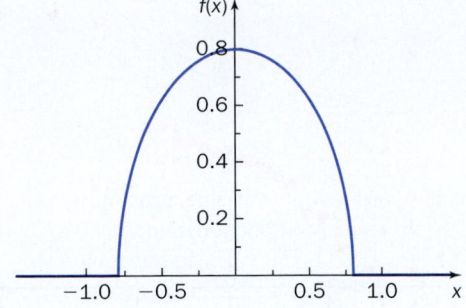

b. 0.8183

Section 6.2

6.27 True.

6.29 Standardization.

6.31

a.

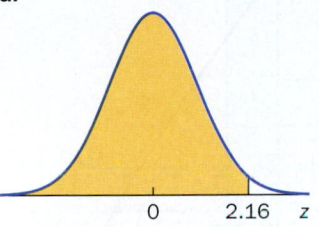

$P(Z \le 2.16) = 0.9846$

b.

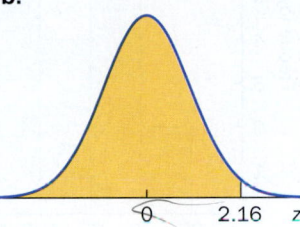

$P(Z < 2.16) = 0.9846$

c.

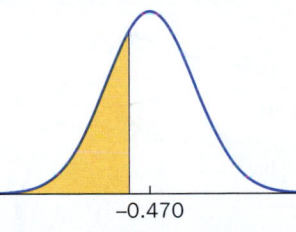

$P(Z \le -0.47) = 0.3192$

d.

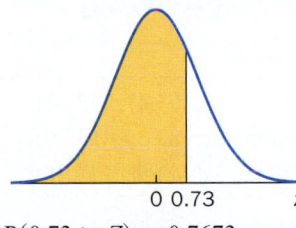

$P(0.73 > Z) = 0.7673$

e.

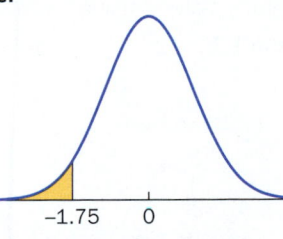

$P(-1.75 \ge Z) = 0.0401$

f.

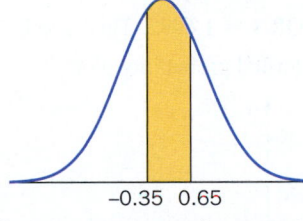

$P(-0.35 \le Z \le 0.65) = 0.3790$

g.

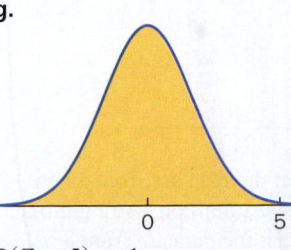

$P(Z < 5) \approx 1$

h.

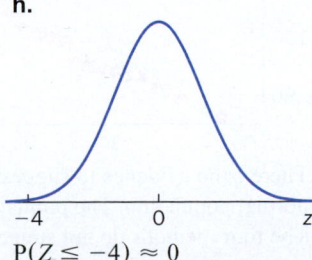

$P(Z \le -4) \approx 0$

i.

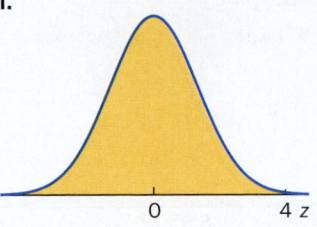

$P(Z \le 4) \approx 1$

6.33 a. 0.6827 **b.** 0.9544 **c.** 0.9974. These are the empirical rule probabilities.

6.35

a.   **b.**

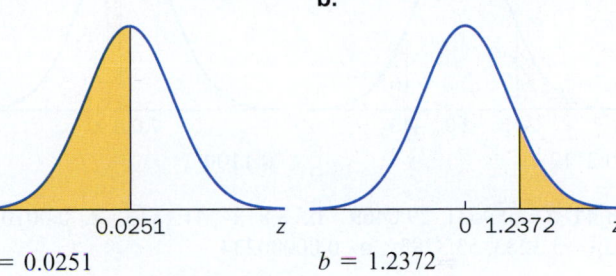

$b = 0.0251$ $b = 1.2372$

c.   **d.**

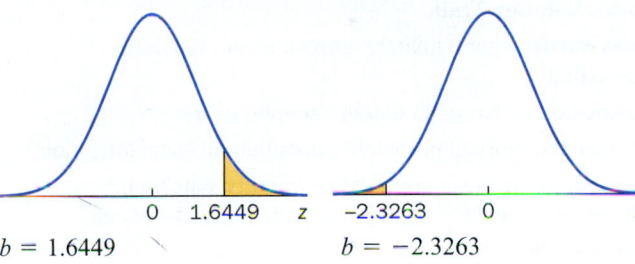

$b = 1.6449$ $b = -2.3263$

e.   **f.**

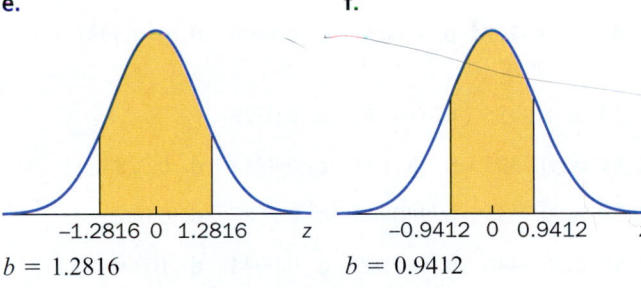

$b = 1.2816$ $b = 0.9412$

6.37 a. $-0.6745, 0.6745$ **b.** $-2.6980, 2.6980$ **c.** 0.0070
d. $-4.7215, 4.7215$ **e.** 0.00000234

6.39

a.   **b.**

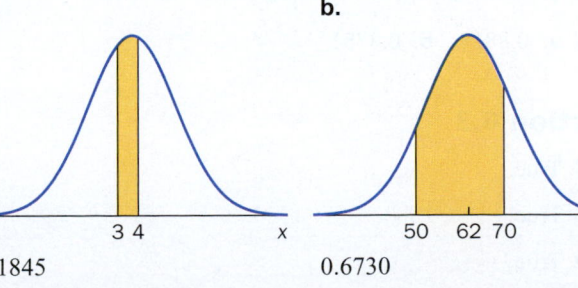

0.1845   0.6730

c.

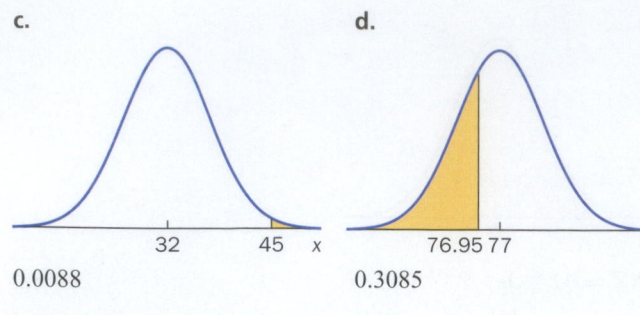

0.0088

d.

0.3085

e.

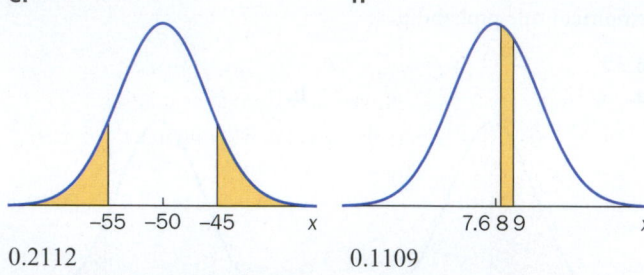

0.2112

f.

0.1109

6.41 a. 20.9531, 29.0469   **b.** 8.8124, 41.1876   **c.** 0.0070
d. −3.3283, 53.3283   **e.** 0.00000234

6.43 a. 0.0431   **b.** 0.80   **c.** 0.0112

6.45 Solution Trail

Keywords: Approximately normal; mean; standard deviation.

Translation: Normal random variable; μ; σ.

Concepts: Normal probability distribution; standardization.

Vision: Write the appropriate probability statement, standardize, and use cumulative probability associated with Z if necessary.

0.1556   **b.** 0.3377   **c.** 0.0173

6.47 a. 0.0272   **b.** 0.0020   **c.** 0.4999   **d.** 34.6493, 35.3507, quartiles

6.49 a. 0.3085   **b.** 0.4796   **c.** 0.0228

6.51 a. 0.7335   **b.** 0.7107   **c.** 0.9233   **d.** 0.2294

6.53 a. 0.7333   **b.** 0.0548   **c.** 0.2587   **d.** 0.0525

6.55 a. 0.3446   **b.** 0.2195   **c.** 0.0044   **d.** 103.6

6.57 a. 0.7273   **b.** 0.0013   **c.** 0.0118   **d.** 0.3664

6.59 a. 2.3   **b.** 0.9851   **c.** 0.0024

6.61 a. 89.4   **b.** 0.0039   **c.** 83.7, 86.3   **d.** 83.25

6.63 a. 0.8818   **b.** 0.3781

Section 6.3

6.65 True.

6.67 True.

6.69 True.

6.71

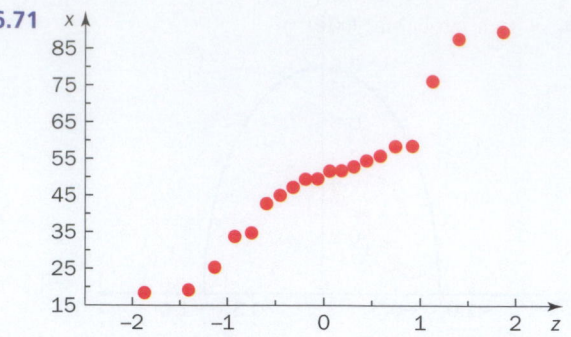

The curved graph and the possible outliers suggest the data are from a nonnormal population.

6.73 Frequency histogram:

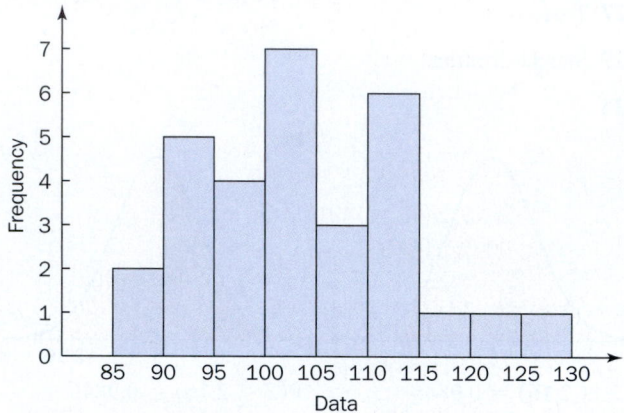

Backward empirical rule:
$\bar{x} = 103.41$, $s = 9.94$

Interval	Proportion
Within $1s$: (93.47, 113.36)	0.70
Within $2s$: (83.53, 123.30)	0.97
Within $3s$: (73.58, 133.25)	1.00

The actual proportions are not far enough away from those given by the empirical rule to suggest that the data are from a non-normal population.

$IQR/s = 1.6492$. This ratio is close to 1.3.

Normal probability plot:

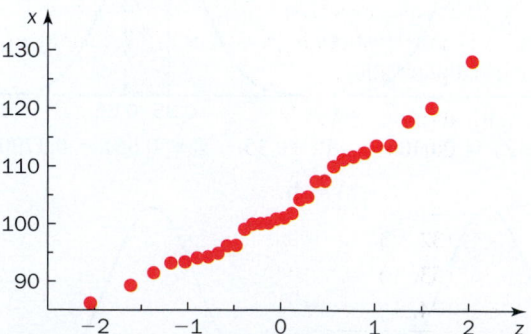

There is no evidence to suggest that the data are from a non-normal population. The points appear to fall on a straight line. The four methods do not suggest that the data are from a non-normal population.

6.75 a. Normal probability plot:

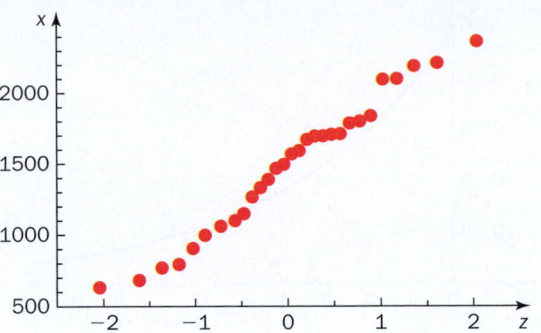

There is a slight curve in the plot near the bottom left. However, there is probably not enough evidence to suggest non-normality.

b. Backward empirical rule:
$\bar{x} = 1480$, $s = 482.75$

Interval	Proportion
Within 1s: (997.25, 1962.75)	0.67
Within 2s: (514.50, 2445.50)	1.00
Within 3s: (31.75, 2928.25)	1.00

The actual proportions are not far enough away from those given by the empirical rule to suggest that the data are from a non-normal population.

6.77 Frequency histogram:

Retail price

There are only 10 observations. The histogram is not very revealing. There is no overwhelming evidence that the data are from a non-normal population.

Backward empirical rule:
$\bar{x} = 111.925$, $s = 20.7842$

Interval	Proportion
Within 1s: (91.14, 132.71)	0.60
Within 2s: (70.36, 153.49)	1.00
Within 3s: (49.57, 174.28)	1.00

There is no overwhelming evidence that the data are from a non-normal distribution.

IQR/s = 1.429. The ratio is pretty close to 1.3.

Normal probability plot:

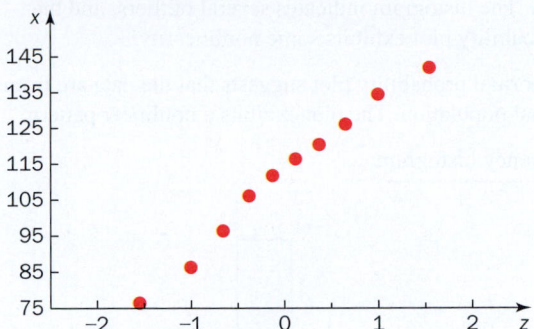

The points appear to fall along a straight line. There is no evidence to suggest that the data are from a non-normal population.

The four methods indicate that there is no evidence to suggest the data are from a non-normal population.

6.79 Frequency histogram:

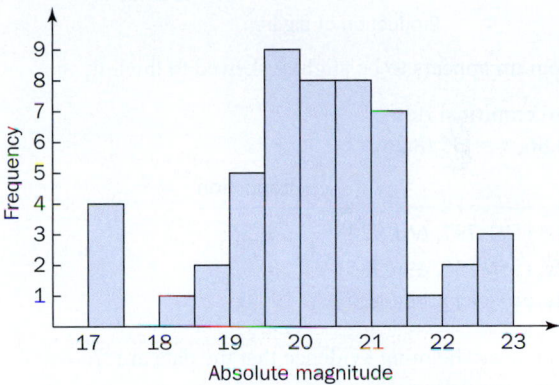

Absolute magnitude

The histogram suggests that the data are from a non-normal population.

Backward empirical rule:
$\bar{x} = 20.156$, $s = 1.3408$

Interval	Proportion
Within 1s: (18.82, 21.50)	0.74
Within 2s: (17.47, 22.84)	0.92
Within 3s: (16.13, 24.18)	1.00

There is no overwhelming evidence that the data are from a non-normal distribution.

IQR/s = 1.1188. The ratio is pretty close to 1.3.

Normal probability plot:

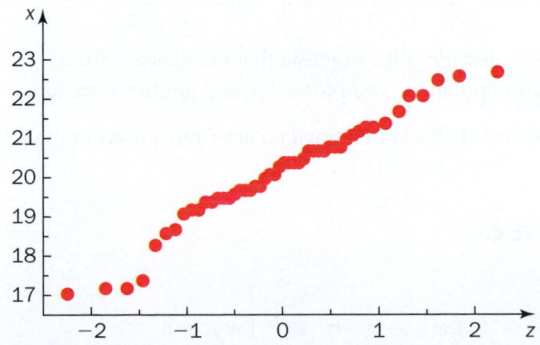

There is some evidence that these data are from a non-normal distribution. The histogram indicates several outliers, and the normal probability plot exhibits some nonlinearity.

6.81 The normal probability plot suggests that the data are from a non-normal population. The plot exhibits a nonlinear pattern.

6.83 Frequency histogram:

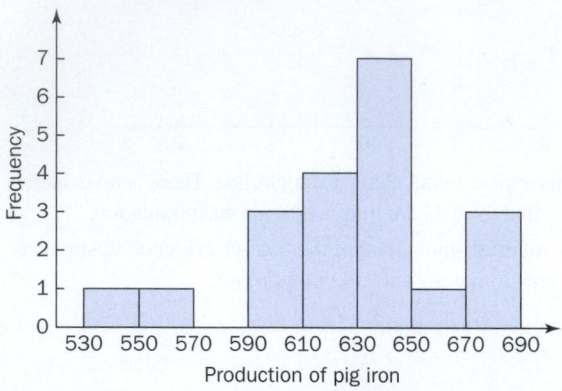

Production of pig iron

The histogram appears to be slightly skewed to the left.

Backward empirical rule:
$\bar{x} = 626.86$, $s = 35.0626$

Interval	Proportion
Within $1s$: (591.797, 661.923)	0.75
Within $2s$: (556.735, 696.985)	0.95
Within $3s$: (521.672, 732.048)	1.00

There is no overwhelming evidence that the data are from a non-normal population.

$IQR/s = 0.8813$. This ratio suggests that the data are from a non-normal population.

Normal probability plot:

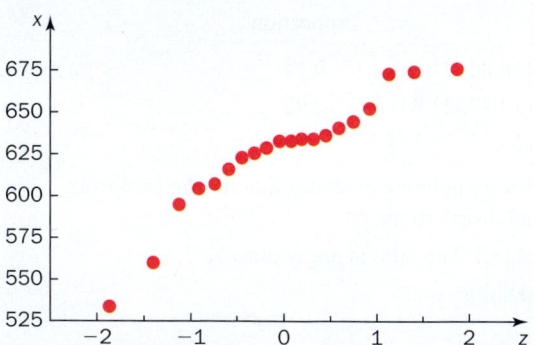

The normal probability plot suggests that the data are from a non-normal population. The plot exhibits a nonlinear pattern.

The four methods suggest that the data are from a non-normal population.

Section 6.4

6.85 True.

6.87 a. $1 - e^{-\lambda x}$ for $x \geq 0$ **b.** $e^{-\lambda x}$ for $x \geq 0$

6.89 a.

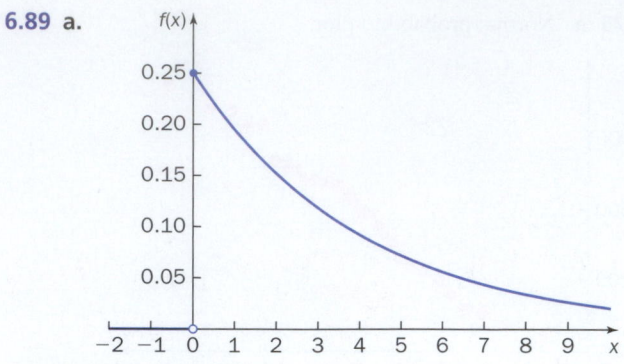

b. 0.0099 **c.** 0.9851

6.91 0.08

6.93 a. 0.00016236 **b.** 0.2728 **c.** 0.2469 **d.** 0.0780

6.95 a. Solution Trail

Keywords: Exponential distribution; $\lambda = 0.01$; within five minutes.

Translation: Exponential probability distribution with parameter 0.01; less than or equal to five minutes.

Concepts: Exponential probability distribution.

Vision: Write a cumulative probability statement and use the appropriate formula.

0.0488 **b.** 0.0861 **c.** 0.7408

6.97 a. 0.3935 **b.** 0.7165 **c.** 0.0360

6.99 a. 15 **b.** 0.9502 **c.** 20.79 **d.** 0.7364

6.101 a. 0.7788 **b.** 0.8825, 0.9200, 0.9394 **c.** 0.9984

6.103 a. 0.4512 **b.** 231.0491 **c.** 998.5774

Chapter Exercises

6.105 a. 2.5 **b.** 0.1813 **c.** 0.1353 **d.** 9.7801

6.107 a. 0.0401 **b.** 0.2417 **c.** 0.9796 **d.** 0.000149

6.109 Frequency histogram:

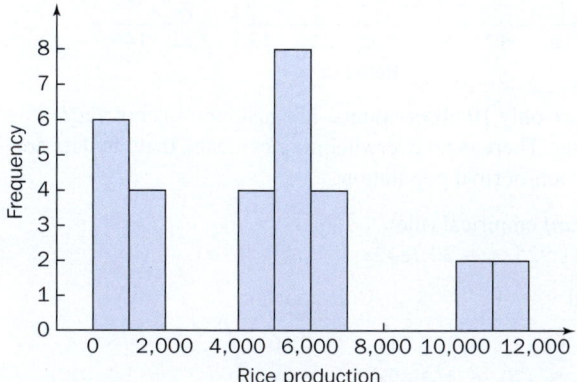

Rice production

Backward empirical rule:
$\bar{x} = 4620.07$, $s = 3229.74$

Interval	Proportion
Within $1s$: (1390.3, 7849.8)	0.60
Within $2s$: (−1839.4, 11079.5)	0.97
Within $3s$: (−5069.1, 14309.3)	1.00

IQR/s = 1.576.

Normal probability plot:

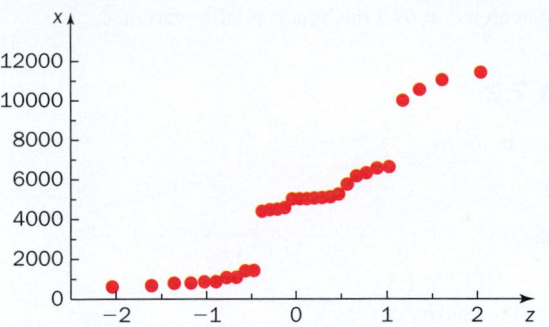

The four methods suggest that the data are from a non-normal population.

6.111 a. 0.3467   **b.** 0.0912   **c.** 0.0023   **d.** 0.0294

6.113 Frequency histogram:

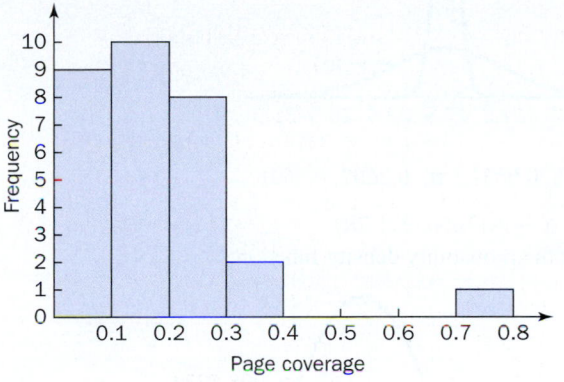

Backward empirical rule:
$\bar{x}$ = 0.1730, s = 0.1289

Interval	Proportion
Within 1s: (0.0441, 0.3019)	0.87
Within 2s: (−0.0847, 0.4307)	0.97
Within 3s: (−0.2136, 0.5596)	0.97

IQR/s = 1.2415

Normal probability plot:

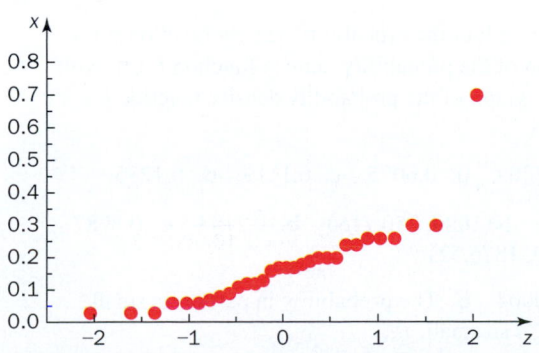

There is some evidence that the data are from a non-normal population. The histogram and the normal probability plot indicate an outlier, and the backward empirical rule proportions are inconsistent.

6.115 a. 0.0985   **b.** 0.6437   **c.** 0.0049

6.117 a. 0.0295   **b.** 0.6630   **c.** (47.36, 82.64)   **d.** 0.0234

6.119 a. 0.5   **b.** 0.0923   **c.** 0.0007

6.121 a.

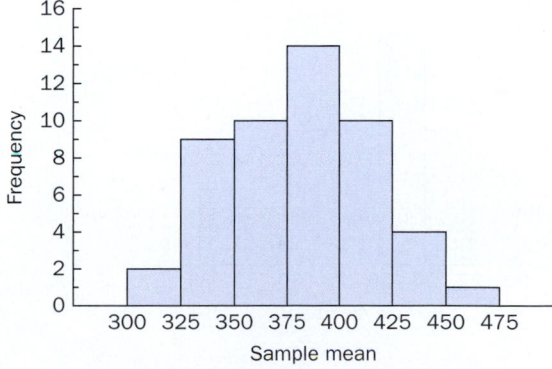

b. 0.25   **c.** 0.36   **d.** 0.32   **e.** 0.25

6.123 a. 0.75   **b.** 0.5, 1.0, 1.5, 2.0   **c.** 0.0026   **d.** 0.25

6.125 a. 0.1056   **b.** 0.9392, 0.0608   **c.** 0.6376   **d.** 0.0184

CHAPTER 7

Section 7.1

7.1 Population. Sample.

7.3 True.

7.5 Obtain many values of the statistic and construct a histogram.

7.7 $\dbinom{N}{n}$

7.9 a. Statistic.   **b.** Parameter.   **c.** Statistic.   **d.** Parameter.   **e.** Statistic.

7.11 a. Statistic.   **b.** Parameter.   **c.** Parameter.   **d.** Parameter.   **e.** Statistic.

7.13 a. Random samples will vary.   **b.** Frequency histogram:

c. Approximately normal. Approximate mean: 380
d. μ = 379.7, almost the same.

7.15 a. Distribution of $\tilde{X}$:

$\tilde{x}$	0.0	0.5	1.0	1.5	2.0
$p(\tilde{x})$	0.2500	0.2500	0.1625	0.1500	0.1100

$\tilde{x}$	2.5	3.0	3.5	4.0
$p(\tilde{x})$	0.0450	0.0200	0.0100	0.0025

b. $\mu_{\tilde{X}} = 0.95$, $\sigma_{\tilde{X}}^2 = 0.7238$, $\sigma_{\tilde{X}} = 0.8508$

7.17 a. Distribution of $\overline{X}$:

$\overline{x}$	6.90	6.95	7.05	8.25	8.35
$p(\overline{x})$	0.1	0.1	0.1	0.1	0.2

$\overline{x}$	8.40	8.45	8.50	9.80
$p(\overline{x})$	0.1	0.1	0.1	0.1

b. Distribution of the total:

t	13.8	13.9	14.1	16.5	16.7
$p(t)$	0.1	0.1	0.1	0.1	0.2

t	16.8	16.9	17.0	19.6
$p(t)$	0.1	0.1	0.1	0.1

7.19 a. 0.7875

b. Distribution of S^2:

s^2	0.0	0.5	2.0	4.5
$p(s^2)$	0.365	0.405	0.180	0.050

c. 0.7875, same.

7.21 a. Distribution of the maximum, M:

m	21	22	26
$p(m)$	0.1667	0.3333	0.5000

b. Distribution of the total, T:

t	42	43	47	48
$p(t)$	0.1667	0.3333	0.3333	0.1667

7.23 a. $\mu = 25.5$ **b.** Random samples will vary.
c. Frequency histogram:

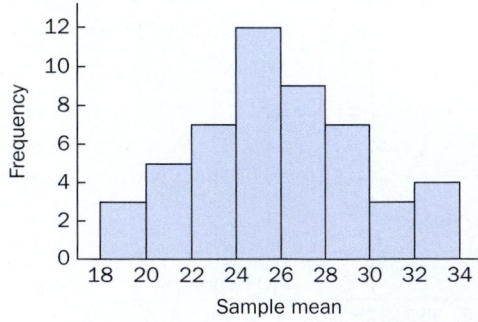

The histogram is centered near the population mean, 25.5.

7.25 a. S_2 is better. It is centered at θ and has smaller variance. **b.** S_2 is better. Both estimators are centered at θ,

but S_2 has smaller variance. **c.** S_2 is better. Both estimators are centered at θ, but S_2 has smaller variance. **d.** This is a difficult decision. I'll take S_2. Even though the distribution of S_2 is not centered at θ, it has much smaller variance.

Section 7.2

7.27 a. μ **b.** σ^2/n

7.29 True.

7.31 False.

7.33 a. $\overline{X} \sim N(17.5, 1.5)$
b. Graph of the density curves:

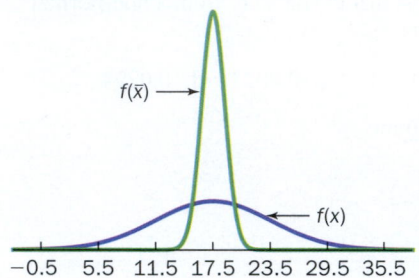

c. 0.2798, 0.0021 **d.** 0.2602, 0.8691

7.35 a. $X \overset{\bullet}{\sim} N(1000, 277.78)$
Graph of the probability density function:

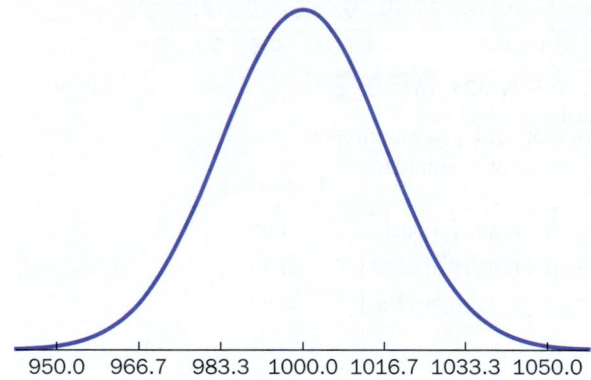

b. 0.9332 **c.** 0.9641 **d.** 0.6826 **e.** (967.33, 1032.67)

7.37 Blue: graph of the probability density function for X. Green: graph of the probability density function for $\overline{X}$ with $n = 5$. Red: graph of the probability density function for $\overline{X}$ with $n = 15$.

7.39 a. 0.2209 **b.** 0.0075 **c.** 0.1119 **d.** 0.1235

7.41 a. $\overline{X} \sim N(1000, 250^2/15)$ **b.** 0.2193 **c.** 0.8787
d. (1623.49, 1876.52)

7.43 a. 0.0004 **b.** The probability in part (a) is small because $\sigma_{\overline{X}}$ is so small.
c. Claim: $\mu = 12 \implies X \overset{\bullet}{\sim} N(12, 0.3^2/100)$
Experiment: $\overline{x} = 11.9$
Likelihood: $P(\overline{X} \le 11.9) = 0.0004$

Conclusion: There is no evidence to suggest that the claim is false, that the mean weight of bags of potato chips is less than 12 ounces.

7.45 a. 0.0142   **b.** 0.0031   **c.** (87.5, 90.5)

7.47 a. 0.9431   **b.** 0.0569   **c.** 0.2635

7.49 a. 0.0432   **b.** 0.3318
c. Claim: $\mu = 28$ $\Rightarrow$ $\overline{X} \overset{\bullet}{\sim} N(28, 7^2/50)$
Experiment: $\bar{x} = 29.75$
Likelihood: $P(\overline{X} \geq 29.75) = 0.0385$
Conclusion: There is evidence to suggest that the claim is false, that the mean vertical leap has increased.

7.51 a. $\overline{X}_5 \sim N(3.7, 0.2)$, $\overline{X}_{20} \sim N(3.7, 0.05)$
b. Blue curve: X; red curve: $\overline{X}_5$; green curve: $\overline{X}_{20}$.

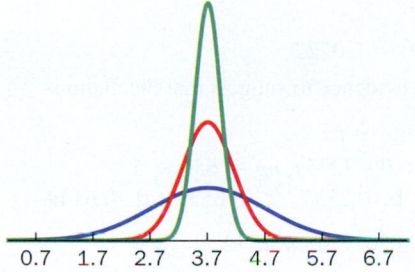

c. 0.2420, 0.0588, 0.0009   **d.** 0.0797, 0.1769, 0.3453

7.53 a. $T \overset{\bullet}{\sim} N(480, 22.5)$ **b.** 0.00001241   **c.** 0.9650
d. 0.1030

7.55 a. $T \overset{\bullet}{\sim} N(325, 16)$ **b.** 0.3085   **c.** 0.1055   **d.** 329.15

7.57 a. Table of the population mean and the probability of accepting the entire shipment:

μ	Prob	μ	Prob	μ	Prob
1.86	0.0001	1.88	0.0016	1.90	0.0176
1.92	0.1031	1.94	0.3368	1.96	0.6632
1.98	0.8953	2.00	0.9648	2.02	0.8953
2.04	0.6632	2.06	0.3368	2.08	0.1031
2.10	0.0176	2.12	0.0016	2.14	0.0001

b. Graph of the OC curve:

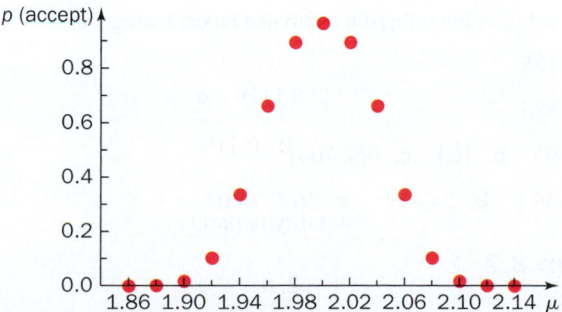

7.59 a. $\mu = 30.85$ **b.** Answers will vary.

c. Histogram of the sample means:

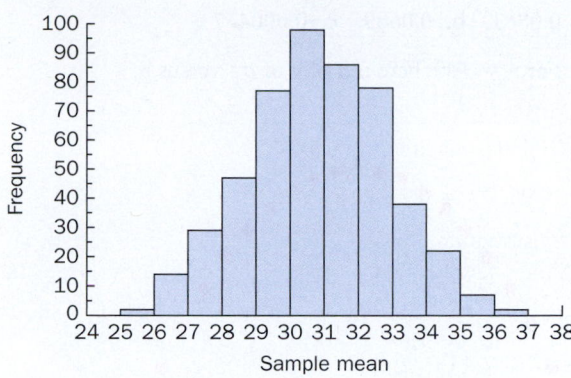

The shape of the histogram is approximately normal, centered near the population mean 30.85.

Section 7.3

7.61 False.

7.63 True.

7.65 a. $np = 25 \geq 5$, $n(1 - p) = 75 \geq 5$,
$\widehat{P} \overset{\bullet}{\sim} N(0.25, 0.0019)$ **b.** $np = 135 \geq 5$,
$n(1 - p) = 15 \geq 5$, $\widehat{P} \overset{\bullet}{\sim} N(0.90, 0.0006)$
c. $np = 75 \geq 5$, $n(1 - p) = 25 \geq 5$,
$\widehat{P} \overset{\bullet}{\sim} N(0.75, 0.0019)$ **d.** $np = 850 \geq 5$,
$n(1 - p) = 150 \geq 5$, $\widehat{P} \overset{\bullet}{\sim} N(0.85, 0.0001275)$
e. $np = 30 \geq 5$, $n(1 - p) = 4970 \geq 5$,
$\widehat{P} \overset{\bullet}{\sim} N(0.006, 0.000001928)$

7.67 a. 0.8145   **b.** 0.9101   **c.** 0.3237   **d.** 0.9747

7.69 a. 0.2275   **b.** 0.2853   **c.** 0.0353
d. $Q_1 = 0.2408$, $Q_3 = 0.2592$

7.71 a. $\widehat{P} \overset{\bullet}{\sim} N(0.24, 0.000912)$ **b.** 0.0927
c. 0.0489   **d.** 0.1697

7.73 a. $\widehat{P} \overset{\bullet}{\sim} N(0.46, 0.003312)$ **b.** 0.1486
c. 0.5690   **d.** (0.3118, 0.6082)

7.75 a. 0.0881   **b.** 0.9634
c. Claim: $p = 0.125$ $\Rightarrow$ $\widehat{P} \overset{\bullet}{\sim} N(0.125, 0.0007292)$
Experiment: $\hat{p} = 0.06$
Likelihood: $P(\widehat{P} \leq 0.06) = 0.0080$
Conclusion: There is evidence to suggest that the claim is false, that the proportion of children in this age group with allergies is less than 0.125.

7.77 a. 0.0745   **b.** 0.2799
c. Claim: $p = 0.10$ $\Rightarrow$ $\widehat{P} \overset{\bullet}{\sim} N(0.10, 0.0003)$
Experiment: $\hat{p} = 0.16$
Likelihood: $P(\widehat{P} \geq 0.16) = 0.000266$
Conclusion: There is evidence to suggest that the claim is false, that the funding rate has increased.

7.79 a. 0.0468   **b.** 0.1366   **c.** 0.4655

7.81 a. 0.0066 **b.** 0.8781

7.83 a. 0.0863 **b.** 0.0649 **c.** 0.000477

7.85 a. For $n = 100$, here is a plot of $\sigma_{\hat{P}}^2$ versus p:

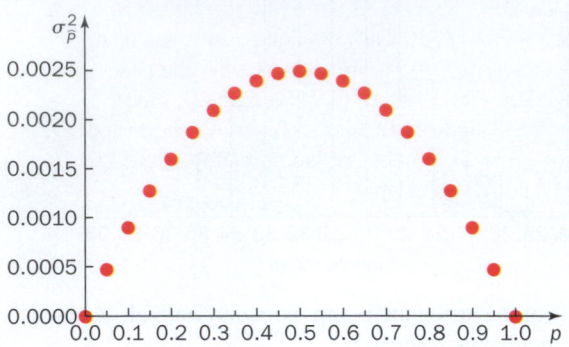

The graph suggests that the variance is a maximum when $p = 0.50$.

Chapter Exercises

7.87 Claim: $\mu = 0.5 \Rightarrow \overline{X} \stackrel{\bullet}{\sim} N(0.5, 0.0008)$

Experiment: $\bar{x} = 0.6$

Likelihood: $P(\overline{X} \geq 0.6) = 0.0002$

Conclusion: There is evidence to suggest that the claim is false, that the mean coefficient of static friction is greater than 0.5.

7.89 a. Distribution of M:

m	1	2	3	4	5	6
$p(m)$	0.0004	0.1020	0.0740	0.3420	0.3280	0.1536

b. 4.356, 1.3013, 1.1407

7.91 a. 0.0455 **b.** 0.3010

c. Claim: $\mu = 7.5 \Rightarrow \overline{X} \stackrel{\bullet}{\sim} N(7.5, 0.0875)$

Experiment: $\bar{x} = 8.1$

Likelihood: $P(\overline{X} \geq 8.1) = 0.0213$

Conclusion: There is evidence to suggest that the claim is false, that the mean oxygen produced is greater than 7.5 ml/h.

d. 0.2151, 0.1534

Claim: $\mu = 7.5 \Rightarrow \overline{X} \stackrel{\bullet}{\sim} N(7.5, 0.4018)$

Experiment: $\bar{x} = 8.1$

Likelihood: $P(\overline{X} \geq 8.1) = 0.1719$

Conclusion: There is no evidence to suggest that the claim is false.

7.93 a. 0.0095 **b.** 0.2075 **c.** 135.25

7.95 a. Statistic **b.** Parameter **c.** Statistic **d.** Parameter **e.** Statistic **f.** Statistic

7.97 a. f_1: underlying distribution; f_2: distribution of the sample mean. **b.** f_1: underlying distribution; f_2: distribution of the sample mean. **c.** f_1: underlying distribution; f_2: distribution of the sample mean. **d.** f_1: underlying distribution; f_2: distribution of the sample mean.

7.99 a. $\hat{P} \stackrel{\bullet}{\sim} N(0.12, 0.000422)$

Graph of the probability density function:

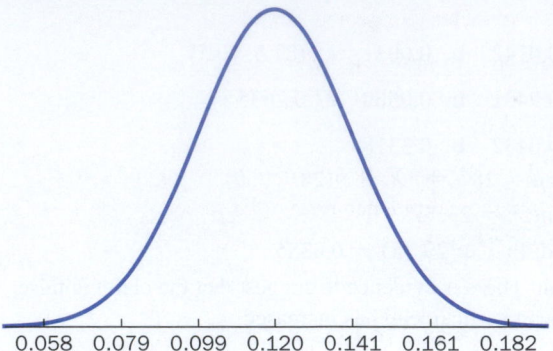

b. 0.0722 **c.** 0.0906

d. Claim: $p = 0.12 \Rightarrow \hat{P} \stackrel{\bullet}{\sim} N(0.12, 0.000422)$

Experiment: $\hat{p} = 0.09$

Likelihood: $P(\hat{P} \leq 0.09) = 0.0722$

Conclusion: There is no evidence to suggest that the claim is false.

7.101 a. $\hat{P} \stackrel{\bullet}{\sim} N(0.65, 0.0002275)$, $np = 650 \geq 5$, $n(1 - p) = 350 \geq 5$. **b.** 0.2537 **c.** 0.6539 **d.** 0.6149

7.103 a. 0.5480 **b.** 0.0201 **c.** $n \geq 72$

7.105 a. $\overline{X} \stackrel{\bullet}{\sim} N(90, 6.084)$ **b.** ≈ 0 **c.** ≈ 0 **d.** $n \geq 74$

7.107 a. $\overline{X} \stackrel{\bullet}{\sim} N(800, 125^2/40)$

b. Claim: $\mu = 800 \Rightarrow \overline{X} \stackrel{\bullet}{\sim} N(800, 125^2/40)$

Experiment: $\bar{x} = 755$.

Likelihood: $P(\overline{X} \leq 755) = 0.0114$

Conclusion: There is evidence to suggest that the true mean check amount is less than \$800.

CHAPTER 8

Section 8.1

8.1 True.

8.3 Unbiased, small variance.

8.5 Minimum variance unbiased estimator.

8.7 $\hat{\theta}_2$: unbiased and small variance.

8.9 The value of the unbiased estimator is, on average, θ.

8.11 0.8125

8.13 0.1886

8.15 a. 95 **b.** 104 **c.** (95, 104)

8.17 a. 30.5 **b.** 24.4615 **c.** 26.5, 34.0

Section 8.2

8.19 False.

8.21 True.

8.23 Wider.

8.25 The probability that the confidence interval encloses the population parameter in repeated samplings.

8.27 a. (13.507, 17.693)   **b.** (6232.2, 6411.8)
c. (−51.06, −40.50)   **d.** (0.0763, 0.0827)
e. (36.287, 39.073)

8.29 a. 9.7   **b.** 95% CI: (8.55, 10.85); 99.9% CI: (8.40, 11.0). For a higher confidence level (all else being equal), the CI has to be larger.

8.31 (4.637, 5.343)

8.33 a. (9.3713, 9.6851)   **b.** (9.3220, 9.7344)
c. For a larger confidence level, the critical value is also larger. This change makes the confidence interval in part (b) wider.

8.35 a. (92.8006, 111.1994)
b. Solution Trail

Keywords: Is there any evidence?

Translation: Inference procedure.

Concepts: $100(1 - \alpha)$% CI for a population mean when σ is known.

Vision: If the CI in part (a) includes 95, then there is no evidence to suggest that μ is more than 95. Use the four-step inference procedure.

Claim: $\mu = 95$

Experiment: $\bar{x} = 102$

Likelihood: A 99% confidence interval or interval of likely values for μ: (92.8006, 111.1994).

Conclusion: Because this CI includes 95, there is no evidence to suggest that μ is greater than 95.

8.37 a. (19.422, 22.078)
b. Yes. The CI does not include 23.

8.39 a. (44,204.59, 50,795.41)
b. 22   **c.** 196

8.41 a. (0.1248, 0.1372)
b. It's close, but $1/8 = 0.125$ is captured by the CI in part (a). Therefore, there is no evidence to suggest that the true mean is greater than 0.125. The town should not embark on the safety program.

8.43 a. (794.44, 801.56)
b. No. The CI in part (a) includes 800.
c. (796.48, 799.52). Now, there is evidence to suggest that the true mean brightness is less than 800. The CI does not include 800.

8.45 a. (122,495.89, 127,904.11)
b. (144,999.95, 166,800.05)
c. Yes, The CIs do not overlap.

8.47 a. Football, (61.1087, 70.4313); basketball, (49.4270, 58.3730); hockey, (64.8986, 72.0014).

b. There is evidence to suggest that the mean coping skills level is different for football and basketball players. The CIs do not overlap.
c. Football, 191; basketball, 151; hockey: 101.

8.49 a. Cashew, (5.0591, 5.2809); filbert, (4.0737, 4.4063); pecan, (2.3367, 2.8633). Cashews and pecans: yes. The CIs do not overlap. Filberts and pecans: yes. The CIs do not overlap.   **b.** Cashew, (4.9852, 5.3548); filbert, (3.9628, 4.5172); pecan, (2.1611, 3.0389), Cashews and pecans: yes. The CIs do not overlap. Filberts and pecans: yes. The CIs do not overlap.

8.51 a. Upper bound: $\bar{x} + z_\alpha \dfrac{\sigma}{\sqrt{n}}$

b. Lower bound: $\bar{x} - z_\alpha \dfrac{\sigma}{\sqrt{n}}$ **c.** $\mu < 2.4467$

8.53 This is the *best* CI because it is the narrowest $100(1 - \alpha)$% CI for μ.

Section 8.3

8.55 True.

8.57 False.

8.59 α, ν (the degrees of freedom)

8.61 a. 1.4759   **b.** 0.8569   **c.** 2.8609   **d.** 2.3646
e. 2.9467   **f.** 5.2076   **g.** 3.7676   **h.** 22.2037   **i.** 1.7959

8.63 a. 1.0931   **b.** 1.5286   **c.** 3.1534   **d.** 2.4377
e. 2.6981   **f.** 1.9921   **g.** 2.1150   **h.** 2.8652   **i.** 1.6794

8.65 a. (0.1908, 0.2772)   **b.** (217.5618, 301.6382)
c. (19.005, 26.695)   **d.** (367.9725, 393.8275)
e. (72.4005, 103.7995)

8.67 a. (0.2345, 0.2963)   **b.** No. The CI in part (a) includes 0.25.   **c.** Frequency histogram:

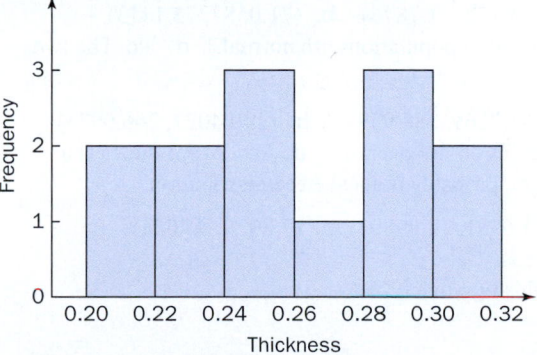

Backward empirical rule:
$\bar{x} = 0.2654$, $s = 0.0384$

Interval	Proportion
Within 1s: (0.2271, 0.3038)	0.64
Within 2s: (0.1887, 0.3421)	1.00
Within 3s: (0.1503, 0.3805)	1.00

$IQR/s = 1.615$

Normal probability plot:

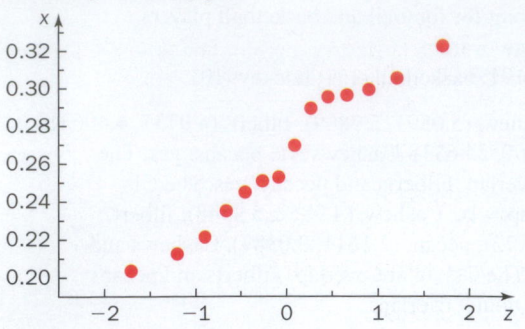

There is evidence to suggest that the data are from a non-normal population.

8.69 a. (262.3195, 270.0971)   **b.** Yes. The CI does not include 275.4.

8.71 a. (15.3816, 18.0184)
b. Solution Trail

Keywords: Is there any evidence?

Translation: Inference procedure.

Concepts: A $100(1 - \alpha)$% CI for a population mean when σ is unknown.

Vision: If 20 lies in (or is greater than) the CI from part (a), then there is no evidence to suggest that the true mean length is over 20 miles.

There is no evidence to suggest that the true mean length is over 20 miles. The CI in part (a) suggests that the true mean length is under 20 miles.

8.73 a. (29.2575, 29.7908)   **b.** Yes. The CI does not contain 30.

8.75 a. (7.8303, 11.9697)   **b.** Yes. The CI is below, and does not include, 12.

8.77 a. (59.1498, 84.5168)   **b.** (38.6665, 58.1668)
c. Yes. The CIs do not overlap.

8.79 a. (68.6122, 71.7878)   **b.** (71.0587, 73.1413)
c. The underlying populations are normal.   **d.** No. The two CIs overlap.

8.81 a. (281.7269, 288.9397)   **b.** (249.4027, 266.0973)
c. Yes. The CIs do not overlap.   **d.** Yes. Atrial flutter is a measurement, probably not a skewed distribution.

8.83 a. (11.6981, 14.2989)   **b.** (7.8028, 9.7044)
c. Rural areas:
Frequency histogram:

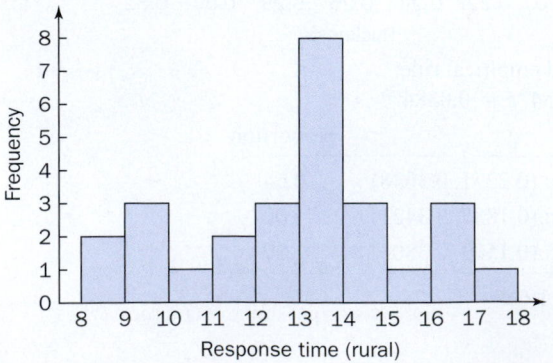

The shape of the distribution is approximately normal.

Backward empirical rule:
$\bar{x} = 12.9985$, $s = 2.4317$

Interval	Proportion
Within $1s$: (10.5668, 15.4302)	0.67
Within $2s$: (8.1351, 17.8619)	1.00
Within $3s$: (5.7035, 20.2936)	1.00

The actual proportions are fairly close to those given by the empirical rule.

IQR/s = 1.2131
This ratio is fairly close to 1.3.

Normal probability plot:

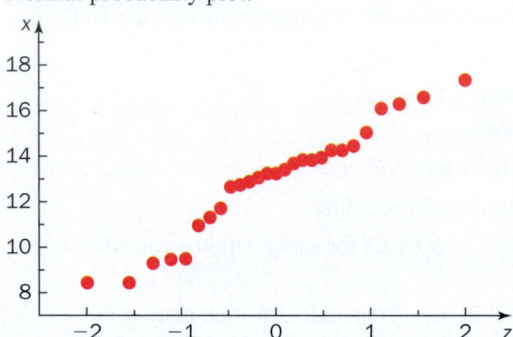

There is some evidence of a curve in this plot near the bottom left.

There is insufficient evidence to suggest that the data are from a non-normal population.

City areas:
Frequency histogram:

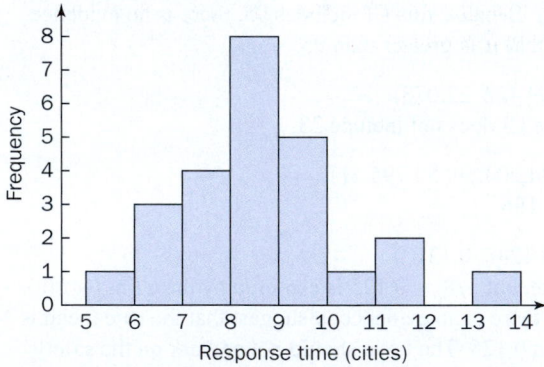

The shape of this distribution is approximately normal.

Backward empirical rule:
$\bar{x} = 8.7536$, $s = 1.6997$

Interval	Proportion
Within $1s$: (7.0539, 10.4533)	0.72
Within $2s$: (5.3543, 12.1529)	0.96
Within $3s$: (3.6546, 13.8526)	1.00

The actual proportions are close to those given by the empirical rule.

IQR/s = 1.1531
This ratio is fairly close to 1.3.

Normal probability plot:

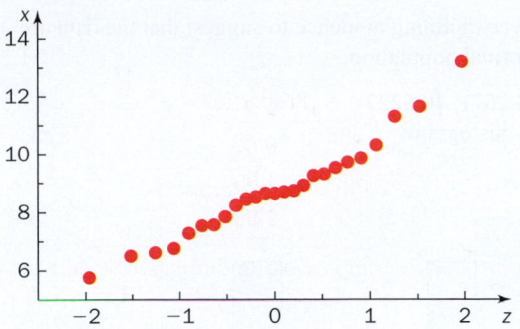

The points lie along a fairly straight line.

There is insufficient evidence to suggest that the data are from a non-normal population.

d. Yes. The CIs do not overlap.

8.85 a. Men, (36.8592, 40.9408); Women, (34.1608, 37.0392). **b.** No. The CIs overlap. **c.** (240.1594, 274.8406) is an interval in which we are 99% confident the true mean distance traveled lies.

8.87 a. Upper bound: $\bar{x} + t_{\alpha, n-1} \dfrac{\sigma}{\sqrt{n}}$

Lower bound: $\bar{x} - t_{\alpha, n-1} \dfrac{\sigma}{\sqrt{n}}$ **b.** $\mu > 19.5949$

Section 8.4

8.89 A large sample CI for a population proportion is an interval in which we are fairly certain the true population proportion lies. If the hypothesized value of p does not lie in the CI, there is evidence to suggest that the true population proportion is different from this value.

8.91 False.

8.93

	$n\hat{p}$	$n(1 - \hat{p})$	Approx. normal Yes	No
a.	85	20	X	
b.	1645	105	X	
c.	220	5	X	
d.	3	180		X
e.	350	27	X	
f.	478	2		X

8.95 a. (0.7187, 0.7798) **b.** (0.8543, 0.9005) **c.** (0.3761, 0.5886) **d.** (0.8223, 0.9090) **e.** (0.0449, 0.1212)

8.97 a. 461 **b.** 97 **c.** 338,244 **d.** 271 **e.** 68

8.99 a. Increases. **b.** Increases. **c.** Decreases. **d.** Decreases.

8.101 a. Solution Trail
Keywords: 90% CI for the true proportion; 575 homes; 235 had gas water heaters.
Translation: 90% CI for p; $n = 575$; $x = 235$.

Concepts: A large sample $100(1 - \alpha)\%$ CI for a population proportion.
Vision: Check the nonskewness criterion, find the appropriate critical value, and use Equation 8.13.
(0.3750, 0.4424) **b.** 1025

8.103 a. (0.0524, 0.1876) **b.** 542

8.105 a. (0.5458, 0.6485) **b.** (0.2574, 0.3540) **c.** (0.2330, 0.3270)

8.107 a. (0.2575, 0.3914) **b.** 1153

8.109 a. (0.2901, 0.3509)
b. Solution Trail
Keywords: Is there any evidence?
Translation: Inference procedure.
Concepts: A large sample $100(1 - \alpha)\%$ CI for a population proportion.
Vision: If 0.30 lies in the CI from part (a), there is no evidence to suggest that the population proportion has changed.

There is no evidence to suggest that the true proportion has changed, because 0.30 is included in the CI (just barely).

8.111 Northeast, (0.7313, 0.8687); Midwest, (0.7510, 0.8722); South Central, (0.7084, 0.8332); South Atlantic, (0.7850, 0.8758); West, (0.7801, 0.8811). **b.** Northeast; sample size is the smallest.

8.113 a. MBA, (0.5377, 0.6931); MAcc, (0.7013, 0.8145); MFin, (0.3315, 0.4750). **b.** Yes. The CI for MFin does not overlap either of the other CIs.

8.115 a. (0.0645, 0.1177) **b.** (0.0746, 0.1270) **c.** No. The CIs overlap.

8.117 a. Treatment, (0.1005, 0.1619); Placebo, (0.0506, 0.1442). **b.** No. The CIs overlap. **c.** Treatment, (0.0398, 0.0936); Placebo, (0.0145, 0.1024). **d.** No. The CIs overlap.

8.119 a. Upper bound: $\hat{p} + z_\alpha \sqrt{\dfrac{\hat{p}(1 - \hat{p})}{n}}$

Lower bound: $\hat{p} - z_\alpha \sqrt{\dfrac{\hat{p}(1 - \hat{p})}{n}}$

b. $p < 0.7061$

8.121 Plot of n versus $\hat{p}$:

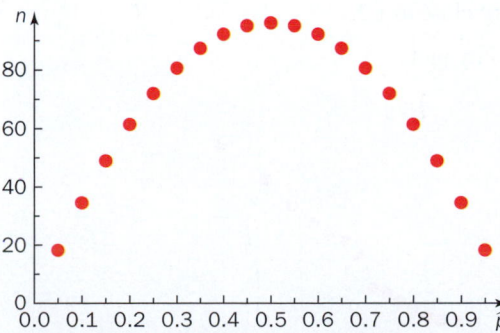

The points appear to lie along a parabola. n is largest when $\hat{p} = 0.50$.

Section 8.5

8.123 False.

8.125 False.

8.127 a. 9.2364 **b.** 61.0983 **c.** 26.2962 **d.** 35.4789
e. 3.0535 **f.** 7.2609 **g.** 11.6886 **h.** 1.7349

8.129 a. 10.2829, 35.4789 **b.** 17.8867, 61.5812
c. 2.5582, 23.2093 **d.** 18.4927, 43.7730
e. 0.4844, 11.1433 **f.** 14.4012, 70.5881

8.131 a. (1.347, 11.2313), (1.1588, 3.3513)
b. (31.2940, 197.4147), (5.5941, 14.0504)
c. (31.9769, 183.4121), (5.6548, 13.5430)
d. (3.0994, 26.5425), (1.7605, 5.1519)
e. (18.5739, 64.0850), (4.3097, 8.0053)
f. (5.8969, 36.9707), (2.4284, 6.0804)

8.133 a. (2.5912, 8.2250) **b.** (1.6097, 2.8679)

8.135 a. (1.7229, 8.9829)
b. Frequency histogram:

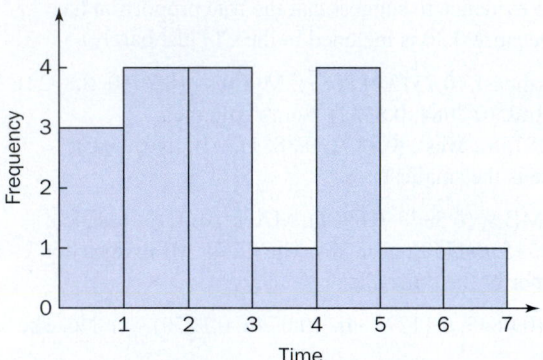

The shape of the distribution does not appear to be normal.

Backward empirical rule:
$\bar{x} = 2.8294$, $s = 1.8401$

Interval	Proportion
Within $1s$: (0.9894, 4.6695)	0.67
Within $2s$: (−0.8507, 6.5096)	0.94
Within $3s$: (−2.6908, 8.3497)	1.00

The actual proportions are close to those given by the empirical rule.
$IQR/s = 1.3152$
This ratio is very close to 1.3.

Normal probability plot:

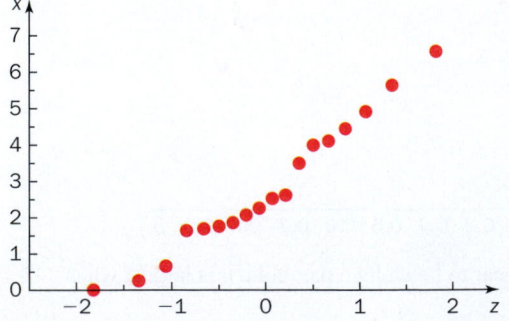

There is a slight curve to the points in this scatter plot.

There is no overwhelming evidence to suggest that the data are from a non-normal population.

8.137 a. (14.2571, 40.6222)
b. Frequency histogram:

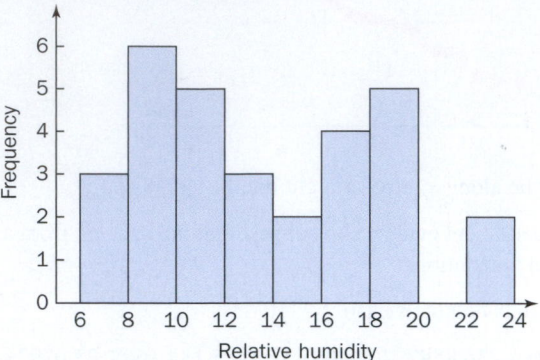

The shape of the distribution does not appear to be normal.

Backward empirical rule:
$\bar{x} = 13.0667$, $s = 4.7411$

Interval	Proportion
Within $1s$: (8.3256, 17.8078)	0.60
Within $2s$: (3.5844, 22.5489)	0.97
Within $3s$: (−1.1567, 27.2900)	1.00

The actual proportions are close to those given by the empirical rule.
$IQR/s = 1.6874$
This ratio is not close to 1.3.

Normal probability plot:

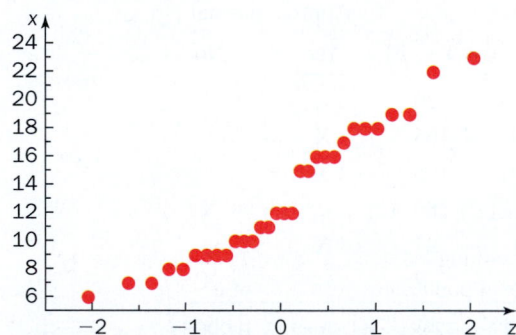

There is a slight curve to the points in this scatter plot.

There is evidence to suggest that the data are from a non-normal population.

8.139 a. (4.1702, 23.9560) **b.** (18.1418, 171.9068)
c. No. The CIs overlap.

8.141 a. (0.5611, 2.2381) **b.** (15.5896, 71.8413)
c. Veteran. The CI for the variance for yardage gained is much narrower and covers smaller values.

8.143 a. (1589.5501, 7536.9105)   **b.** (39.8692, 86.8154)
c. Frequency histogram:

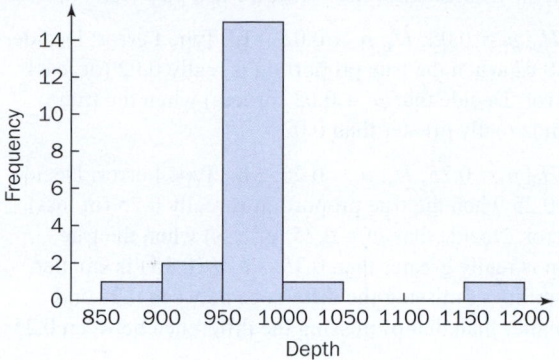

The shape of the distribution does not appear to be normal.

Backward empirical rule:
$\bar{x} = 977.8$, $s = 55.025$

Interval	Proportion
Within 1s: (922.7750, 1032, 8250)	0.90
Within 2s: (867.7501, 1087.8499)	0.95
Within 3s: (812.7251, 1142.8749)	0.95

The actual proportions are not close to those given by the empirical rule.
IQR/s = 0.4180
This ratio is not close to 1.3.

Normal probability plot:

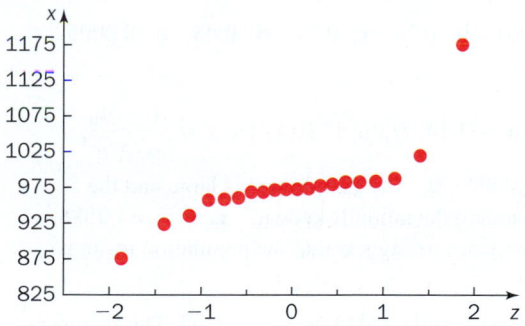

There is a distinct curve to the points in this scatter plot.

There is overwhelming evidence to suggest that the data are from a non-normal population.

8.145 a. $(1.648 \times 10^{16}, 7.6471 \times 10^{16})$
b. (0.3240, 1.5033)   **c.** No. The CI includes 1.

8.147 a. (6.5503, 23.3373)   **b.** No. The CI includes 12.

8.149 a. (0.9514, 2.7108)   **b.** (0.5268, 1.8176)
c. No. The CIs overlap.

8.151 a. (0.0238, 0.2783), (94.1776, 627.5906)
b. (0.0400, 0.2152), (122.6953, 515.8717)   **c.** Column water vapor: no. The CIs overlap. IB: no. The CIs overlap.

8.153 a. (26.9090, 365.8732)   **b.** (629.0011, 3384.2599)
c. Yes. The CIs do not overlap.

8.155 a. CI for σ^2:

$$\left(\frac{s^2(n-1)}{z_{\alpha/2}\sqrt{2(n-1)} + (n-1)} < \sigma^2 < \frac{s^2(n-1)}{-z_{\alpha/2}\sqrt{2(n-1)} + (n-1)} \right)$$

b. i. (11.3763, 25.1102)   **ii.** (11.6739, 26.7267)
iii. The interval based on the normal distribution is wider because this is based on an approximate distribution.

Chapter Exercises

8.157 a. (2.0964, 3.7636)   **b.** 98

8.159 a. (86.5634, 89.1638)   **b.** (4.3620, 22.4793)
c. Yes. The CI for the population mean does not include 90.225.

8.161 a. (117.4437, 122.8063)   **b.** No. The CI includes 120.

8.163 a. (0.5960, 0.6442)   **b.** 3908

8.165 a. (395.77, 424.73)   **b.** (749.86, 2380.24)
c. (27.38, 48.79)

8.167 a. (70.7980, 80.0020)   **b.** No. The CI includes 78.4.

8.169 a. (0.0392, 0.0888)   **b.** (0.0568, 0.1057)
c. No. The CIs overlap.

8.171 a. (0.4387, 0.4941)   **b.** (0.3108, 0.3590)
c. Yes. The CIs do not overlap.

8.173 a. (49.1185, 55.4815)   **b.** (49.1981, 55.4019) The CI based on the Z distribution is slightly wider.
c. (84.0416, 168.9512)

8.175 a. (0.00095, 0.00112)   **b.** (0.00290, 0.00312)
c. Yes. The CIs do not overlap.

8.177 a. (8.1572, 13.3628)   **b.** Because the entire confidence interval is less than 35, there is evidence to suggest that the true mean concentration of ultrafine particles during a 24-hour period in Kansas City is less than the EPA standard.

CHAPTER 9

Section 9.1

9.1 True.

9.3 False.

9.5 There is evidence to suggest that the alternative hypothesis is true. There is no evidence to suggest that the alternative hypothesis is true.

9.7 a. Valid, null hypothesis.   **b.** Invalid.
c. Valid, alternative hypothesis.   **d.** Invalid.   **e.** Invalid.
f. Valid, alternative hypothesis.   **g.** Valid, alternative hypothesis.   **h.** Valid, null hypothesis.

9.9 a. Valid.   **b.** Valid.   **c.** Invalid. The null hypothesis should be stated so that p (a parameter) equals a single value.
d. Invalid. The null and alternative hypotheses are always about a parameter, not a statistic.

9.11 a. Not permissible. We never *accept* a null hypothesis. We fail to reject the null hypothesis.   **b.** Permissible.
c. Not permissible. We conduct a hypothesis test to try and find evidence in favor of the alternative hypothesis.
d. Permissible.   **e.** Permissible.   **f.** Permissible.

9.13 $H_0: p = 0.11$, $H_a: p > 0.11$

9.15 $H_0: \mu = 4.25$, $H_a: \mu < 4.25$

9.17 $H_0: \sigma^2 = 32$, $H_a: \sigma^2 < 32$

9.19 $H_0: \mu = 178$ (minutes), $H_a: \mu < 178$

9.21 $H_0: p = 0.65$, $H_a: p > 0.65$

9.23 $H_0: \mu = 2000$, $H_a: \mu < 2000$

9.25 $H_0: \mu = 525$, $H_a: \mu < 525$

9.27 $H_0: \mu = 25$, $H_a: \mu < 25$

9.29 $H_0: \mu = 1925$, $H_a: \mu < 1925$

9.31 $H_0: \mu = 2666$, $H_a: \mu > 2666$

Section 9.2

9.33 False.

9.35 True.

9.37 a. Type I error.   **b.** Correct decision.   **c.** Type II error.
d. Type I error.

9.39 a. Type I error.   **b.** Type II error.   **c.** Correct decision.
d. Correct decision.

9.41 There is always a chance of making a mistake in any hypothesis test because we never examine the entire population, only a sample.

9.43 a. $H_0: \mu = 140{,}000$, $H_a: \mu > 140{,}000$.   **b.** Type I error: Decide that the mean is greater than 140,000 when the true mean is 140,000 (or less). Type II error: Decide that the mean is 140,000 (or less) when the true mean is greater than 140,000.
c. Drivers are more angry. Tolls do not have to be raised.
d. Transportation officials are more angry. They really need additional tolls to fund planned repairs.

9.45 a. $H_0: \mu = 0.65$, $H_a: \mu > 0.65$.   **b.** Type I error: Decide that $\mu > 0.65$ when the true mean is really 0.65 (or less). Type II error: Decide that $\mu = 0.65$ (or less) when the true mean is really greater than 0.65.   **c.** Type II error. If the mean current velocity is greater than 0.65, it is unsafe for swimmers.
d. Type I error. The race would be canceled, but the mean current is really safe.

9.47 a. Type I error: Decide that $\mu > 0.4$ when the true mean is really 0.4 (or less). Type II error: Decide that $\mu = 0.4$ (or less) when the true mean is really greater than 0.4.   **b.** Type II error. Chocolate is really increasing the level of antioxidants, but there is no evidence.

9.49 a. $H_0: \mu = 1200$, $H_a: \mu < 1200$.   **b.** Type I error: Decide that $\mu < 1200$ when the true mean is really 1200 (or greater). That is, decide that the helmets do not meet the standard when they really do. Type II error: Decide that $\mu = 1200$ (or greater) when the true mean is really less than 1200. That is, decide that the helmets meet the standard when they really do not.

9.51 a. $H_0: p = 0.02$, $H_a: p > 0.02$.   **b.** Type I error: Decide that $p > 0.02$ when the true proportion is really 0.02 (or less). Type II error: Decide that $p = 0.02$ (or less) when the true proportion is really greater than 0.02.

9.53 a. $H_0: p = 0.25$, $H_a: p > 0.25$.   **b.** Type I error: Decide that $p > 0.25$ when the true proportion is really 0.25 (or less). Type II error: Decide that $p = 0.25$ (or less) when the true proportion is really greater than 0.25.   **c.** $\beta(0.35)$ is smaller. The probability of missing the difference between 0.25 and 0.35 is smaller than that of missing the difference between 0.25 and 0.27.

9.55 a. $H_0: \mu = 6400$, $H_a: \mu > 6400$.   **b.** $\alpha = 0.1$. This allows a larger error on the side of safety.

Section 9.3

9.57 False.

9.59 False.

9.61 True.

9.63 a. $Z = \dfrac{\overline{X} - 45.6}{15/\sqrt{16}}$ **b. i.** $Z \geq 2.3263$ **ii.** $Z \geq 1.96$
iii. $Z \geq 1.6449$ **iv.** $Z \geq 1.2816$ **v.** $Z \geq 2.5758$
vi. $Z \geq 3.2905$

9.65 a. 0.05   **b.** 0.005   **c.** 0.02   **d.** 0.01   **d.** 0.001
e. 0.0001

9.67 a. 0.0001   **b.** 0.20   **c.** 0.01   **d.** 0.05   **e.** 0.0005
f. 0.002

9.69 a. $H_0: \mu = 3.14$; $H_a: \mu < 3.14$; TS: $Z = \dfrac{\overline{X} - \mu_0}{\sigma/\sqrt{n}}$;
RR: $Z \leq -3.0902$ **b.** The sample size is large and the population standard deviation is known.   **c.** $z = -1.2588$. There is no evidence to suggest that the population mean is less than 3.14.

9.71 a. The test is right-tailed with $\alpha = 0.05$. The rejection region should be RR: $Z \geq 1.6449$.   **b.** The numerator in the test statistic is reversed. It should be $\overline{X} - \mu_0$.   **c.** We never accept the null hypothesis, and we cannot prove the null hypothesis is true. We simply fail to find evidence in favor of the alternative hypothesis.   **d.** The null hypothesis is always stated with an equals sign.   **e.** The probabilities of type I and type II errors are inversely related. However, this does not mean they sum to 1.

9.73 Solution Trail

Keywords: Is there any evidence; decreased; $\sigma = 65$.

Translation: Conduct a one-sided, left-tailed test about a population mean.

Concepts: Hypothesis test concerning a population mean when σ is known.

Vision: Use the template for a one-sided, left-tailed test about μ. The underlying population distribution is unknown, but n is large and σ is known. Determine the appropriate alternative hypothesis and the corresponding rejection region, find the value of the test statistic, and draw a conclusion.

$H_0: \mu = 335$; $H_a: \mu < 335$; TS: $Z = \dfrac{\overline{X} - \mu_0}{\sigma/\sqrt{n}}$;

RR: $Z \leq -1.6449$

$z = -0.9098$. There is no evidence to suggest that the mean CO_2 emissions has decreased.

9.75 $H_0: \mu = 12.4$; $H_a: \mu < 12.4$; TS: $Z = \dfrac{\overline{X} - \mu_0}{\sigma/\sqrt{n}}$;

RR: $Z \leq -1.96$

$z = -1.0722$. There is no evidence to suggest that the mean water table is less than 12.4 feet.

9.77 Solution Trail

Keywords: Is there any evidence; greater than; population standard deviation 8.6.

Translation: Conduct a one-sided, right-tailed test about a population mean; $\sigma = 8.6$.

Concepts: Hypothesis test concerning a population mean when σ is known.

Vision: Use the template for a one-sided, right-tailed test about μ. The underlying population distribution is unknown, but n is large and σ is known. Determine the appropriate alternative hypothesis and the corresponding rejection region, find the value of the test statistic, and draw a conclusion.

$H_0: \mu = 126.96$; $H_a: \mu > 126.96$; TS: $Z = \dfrac{\overline{X} - \mu_0}{\sigma/\sqrt{n}}$;

RR: $Z \geq 1.6449$

$z = 2.2578 \geq 1.6449$. There is evidence to suggest that the mean composite indicator is greater than 126.96.

9.79 $H_0: \mu = 21$; $H_a: \mu > 21$; TS: $Z = \dfrac{\overline{X} - \mu_0}{\sigma/\sqrt{n}}$;

RR: $Z \geq 2.5758$

$z = 2.9332 \geq 2.5758$. There is evidence to suggest the mean thickness is greater than 21.

9.81 $H_0: \mu = 14.0$; $H_a: \mu > 14.0$; TS: $Z = \dfrac{\overline{X} - \mu_0}{\sigma/\sqrt{n}}$;

RR: $Z \geq 3.0902$

$z = 0.0354$. There is no evidence to suggest that the mean response time has increased.

9.83 $H_0: \mu = 68$; $H_a: \mu > 68$; TS: $Z = \dfrac{\overline{X} - \mu_0}{\sigma/\sqrt{n}}$;

RR: $Z \geq 1.6449$

$z = 1.4310$. There is no evidence to suggest that the mean locksmith service charge is greater than \$68.

9.85 $H_0: \mu = 5$; $H_a: \mu < 5$; TS: $Z = \dfrac{\overline{X} - \mu_0}{\sigma/\sqrt{n}}$;

RR: $Z \leq -2.0538$

$z = 0.3275 -2.0538$. There is no evidence to suggest that the mean amount of protein is less than 5 grams.

9.87 $H_0: \mu = 88$; $H_a: \mu < 88$; TS: $Z = \dfrac{\overline{X} - \mu_0}{\sigma/\sqrt{n}}$;

RR: $Z \leq -2.3263$

$z = -2.61 \leq -2.3263$. There is evidence to suggest that the mean iodine concentration is less than 88.

9.89 a. $H_0: \mu = 0.23$; $H_a: \mu < 0.23$; TS: $Z = \dfrac{\overline{X} - \mu_0}{\sigma/\sqrt{n}}$;

RR: $Z \leq -2.3263$

$z = -4.8189 \leq -2.3263$. There is evidence to suggest that the mean level of HC emissions is less than 0.23. **b.** A type I error is more important to the company because in this case they will be building the biodiesel fuel plant when the biodiesel fuel does not really decrease the mean level of HC emissions. Therefore, the company will want to use a very small significance level.

9.91 a. $H_0: \mu = 15.5$; $H_a: \mu < 15.5$; TS: $Z = \dfrac{\overline{X} - \mu_0}{\sigma/\sqrt{n}}$;

RR: $Z \leq -2.3263$

$z = -3.0407 \leq -2.3263$. There is evidence to suggest that the mean weight of pretzel packages is less than 15.5 ounces. **b.** The sample size is very large and σ is small.

9.93 a. $H_0: \mu = 714$; $H_a: \mu \neq 714$; TS: $Z = \dfrac{\overline{X} - \mu_0}{\sigma/\sqrt{n}}$;

RR: $|Z| \geq 1.96$

$z = -1.5943$. There is no evidence to suggest that the mean monthly water usage is different from 714 cubic feet. **b.** The standard deviation is very large.

9.95 a. $H_0: \mu = 12$; $H_a: \mu > 12$; TS: $Z = \dfrac{\overline{X} - \mu_0}{\sigma/\sqrt{n}}$;

RR: $Z \geq 2.3263$

$z = 1.4599$. There is no evidence to suggest that the mean moisture content is greater than 12%. **b.** 0.9120

9.97 a. $H_0: \mu = 496$; $H_a; \mu > 496$; TS: $Z = \dfrac{\overline{X} - \mu_0}{\sigma/\sqrt{n}}$;

RR: $Z > 2.3263$

$z = 0.2977$. There is no evidence to suggest that the true mean weight of passengers and gear is greater than 496 pounds. **b.** 0.1073

Section 9.4

9.99 False.

9.101 True.

9.103 Reject the null hypothesis.

9.105 a. 0.0307 **b.** 0.0054 **c.** 0.1151 **d.** 0.2843 **e.** 0.00005 **f.** 0.8729

9.107 a. 0.0767 **b.** 0.1527 **c.** 0.0099 **d.** 0.7114 **e.** 0.0003 **f.** 0.3953

9.109 a. 0.0059. Reject. **b.** 0.0823. Do not reject. **c.** 0.0113. Do not reject. **d.** 0.5675. Do not reject. **e.** 0.1031. Do not reject. **f.** 0.0031. Do not reject.

9.111 $H_0: \mu = 10$; $H_a: \mu > 10$; TS: $Z = \dfrac{\bar{X} - \mu_0}{\sigma/\sqrt{n}}$

$z = 2.6904$, $p = 0.0036 \leq 0.05$. There is evidence to suggest that the mean is greater than 10.

9.113 $H_0: \mu = 1$; $H_a: \mu \neq 1$; TS: $Z = \dfrac{\bar{X} - \mu_0}{\sigma/\sqrt{n}}$

$z = 1.7678$, $p = 0.0771 > 0.01$. There is no evidence to suggest that the mean is different from 1.

9.115 $H_0: \mu = 5.5$; $H_a: \mu > 5.5$; TS: $Z = \dfrac{\bar{X} - \mu_0}{\sigma/\sqrt{n}}$

$z = 2.2692$, $p = 0.0116 \leq 0.05$. There is evidence to suggest that the mean is greater than 5.5.

9.117 $H_0: \mu = 8000$; $H_a: \mu > 8000$; TS: $Z = \dfrac{\bar{X} - \mu_0}{\sigma/\sqrt{n}}$

$z = 1.4265$, $p = 0.0769 > 0.05$. There is no evidence to suggest that the mean is greater than 8000.

9.119 $H_0: \mu = 80$; $H_a: \mu < 80$; TS: $Z = \dfrac{\bar{X} - \mu_0}{\sigma/\sqrt{n}}$

$z = -2.4007$, $p = 0.0082 \leq 0.01$. There is evidence to suggest that the mean is less than 80, that the manufacturer's claim is false.

9.121 **a.** $H_0: \mu = 1600$; $H_a: \mu < 1600$; TS: $Z = \dfrac{\bar{X} - \mu_0}{\sigma/\sqrt{n}}$

$z = -0.9565$, $p = 0.1694 > 0.01$. There is no evidence to suggest that the mean is less than 1600.
b. p-Value illustration:

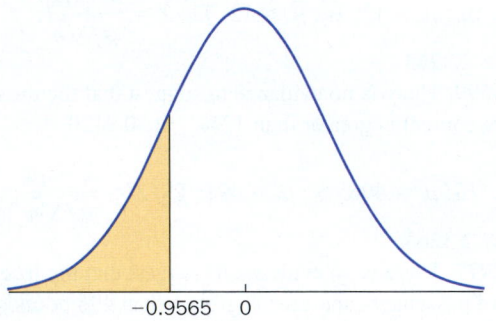

$-0.9565 \quad 0$

9.123 **a.** $H_0: \mu = 35,800$; $H_a: \mu > 35,800$; TS: $Z = \dfrac{\bar{X} - \mu_0}{\sigma/\sqrt{n}}$

$z = 1.0029$, $p = 0.1580 > 0.05$. There is no evidence to suggest that the mean is greater than 35,800.
b. p-Value illustration:

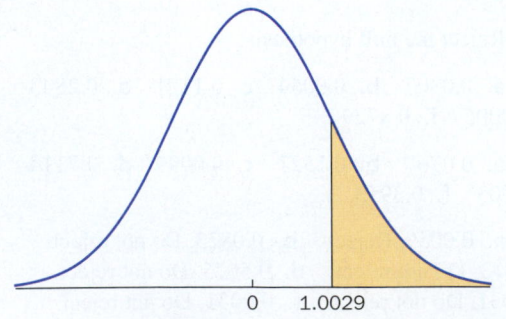

$0 \quad 1.0029$

Section 9.5

9.125 True.

9.127 False.

9.129 **a.** $T = \dfrac{\bar{X} - \mu_0}{S/\sqrt{n}}$ **b.** **(i)** $T \geq 3.3649$
ii. $T \geq 2.0739$ **iii.** $T \geq 1.7459$ **iv.** $T \geq 1.3125$
v. $T \geq 4.2968$ **vi.** $T \geq 6.4420$

9.131 **a.** $T = \dfrac{\bar{X} - \mu_0}{S/\sqrt{n}}$ **b.** **i.** $|T| \geq 3.1058$
ii. $|T| \geq 1.3304$ **iii.** $|T| \geq 2.0595$ **iv.** $|T| \geq 1.7033$
v. $|T| \geq 5.9588$ **vi.** $|T| \geq 2.3961$

9.133 **a.** 0.10 **b.** 0.001 **c.** 0.005 **d.** 0.01

9.135 **a.** $0.01 \leq p \leq 0.025$ **b.** $0.005 \leq p \leq 0.01$
c. $p < 0.0001$ **d.** $0.05 \leq p \leq 0.10$

9.137 **a.** $0.05 \leq p \leq 0.10$ **b.** $0.001 \leq p \leq 0.005$
c. $p < 0.0001$ **d.** $0.10 \leq p \leq 0.20$

9.139 **a.** $H_0: \mu = 57.71$; $H_a: \mu > 57.71$;
TS: $T = \dfrac{\bar{X} - \mu_0}{S/\sqrt{n}}$; RR: $T \geq 2.7638$ **b.** $\bar{x} = 59.3082$,
$s = 1.6037$, $t = 3.3053 \geq 2.7638$. There is evidence to suggest that the mean is greater than 57.71.
c. 0.0040

9.141 **a.** Should be S, not σ in the denominator.
b. Should be $\sqrt{25}$, not $\sqrt{24}$. **c.** Should be a two-sided rejection region. **d.** $0.01 \leq p \leq 0.025$

9.143 $H_0: \mu = 303$; $H_a: \mu > 303$; TS: $T = \dfrac{\bar{X} - \mu_0}{S/\sqrt{n}}$;

RR: $T \geq 2.1448$
$t = 2.2465 \geq 2.1448$; $p = 0.0207 \leq 0.025$. There is evidence to suggest that the mean airborne time is greater than 303.

9.145 **Solution Trail**

Keywords: Is there any evidence; increased; $s = 15.59$; normal random sample.

Translation: Conduct a one-sided, right-tailed test about μ.

Concepts: Hypothesis test concerning a population mean when σ is unknown.

Vision: Use the template for a one-sided, right-tailed t test about μ. The underlying population is assumed to be normal, and σ is unknown. Determine the appropriate alternative hypothesis and the corresponding rejection region, find the value of the test statistic, and draw a conclusion.

$H_0: \mu = 64.98$; $H_a: \mu > 64.98$; TS: $T = \dfrac{\bar{X} - \mu_0}{S/\sqrt{n}}$;

RR: $T \geq 2.6503$
$t = 1.3128 < 2.6503$; $p = 0.1060 > 0.01$. There is no evidence to suggest that the mean fine has increased.

9.147 H_0: $\mu = 23.1$; H_a: $\mu > 23.1$; TS: $T = \dfrac{\bar{X} - \mu_0}{S/\sqrt{n}}$;

RR: $T \geq 2.8965$

$t = 2.1429 < 2.8965$; $p = 0.0322 > 0.01$. There is no evidence to suggest that the mean hotel room occupation rate has increased. If the test is significant, one cannot conclude the ad campaign caused the increase.

9.149 H_0: $\mu = 3.2$; H_a: $\mu \neq 3.2$; TS: $T = \dfrac{\bar{X} - \mu_0}{S/\sqrt{n}}$;

RR: $|T| \geq 2.0860$

$t = -0.2770$; $|t| < 2.0860$; $p = 0.7846 > 0.05$. There is no evidence to suggest that the mean physician density is different from 3.2.

9.151 a. H_0: $\mu = 9$; H_a: $\mu > 9$; TS: $T = \dfrac{\bar{X} - \mu_0}{S/\sqrt{n}}$;

RR: $T \geq 1.8595$

$t = 0.8517 < 1.8595$; $p = 0.2096 > 0.05$. There is no evidence to suggest that the mean floor area is greater than 9 million square feet.
b. $p > 0.20$

9.153 a. H_0: $\mu = 1381$; H_a: $\mu \neq 1381$;

TS: $T = \dfrac{\bar{X} - \mu_0}{S/\sqrt{n}}$; RR: $|T| \geq 2.1199$

$t = 2.4969$; $|t| \geq 2.1199$; $p = 0.0238 \leq 0.05$. There is evidence to suggest that the true mean loss as a result of a residential break-in has changed from \$1381.
b. $0.02 \leq p \leq 0.05$
c. *p*-Value illustration:

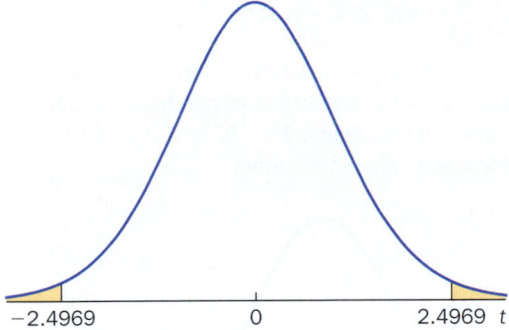

$-2.4969 \qquad 0 \qquad 2.4969 \; t$

9.155 a. H_0: $\mu = 159{,}350$; H_a: $\mu > 159{,}350$;

TS: $T = \dfrac{\bar{X} - \mu_0}{S/\sqrt{n}}$; RR: $T \geq 2.7638$

$t = 1.4855 < 2.7638$; $p = 0.0841 > 0.01$. There is no evidence to suggest that the mean methane output per well per day has increased.
b. $0.05 \leq p \leq 0.10$

9.157 a. H_0: $\mu = 13.5$; H_a: $\mu > 13.5$; TS: $T = \dfrac{\bar{X} - \mu_0}{S/\sqrt{n}}$;

RR: $T \geq 2.4999$

$t = 3.0384 \geq 2.4999$; $p = 0.0029 \leq 0.01$. There is evidence to suggest that the mean length of lionfish in this area is greater than 13.5 inches.   **b.** $0.001 \leq p \leq 0.005$

9.159 a. H_0: $\mu = 4.75$; H_a: $\mu < 4.75$; TS: $T = \dfrac{\bar{X} - \mu_0}{S/\sqrt{n}}$;

RR: $T \leq -2.4121$

$t = -2.4416 \leq -2.4121$; $p = 0.0093 \leq 0.01$. There is evidence to suggest that the population mean width of little-necks has decreased.   **b.** This test is statistically significant even though the sample mean (4.66) is so close to the assumed population mean (4.75), because the sample size is fairly large ($n = 46$) and the sample standard deviation is quite small ($s = 0.25$) relative to the sample mean.
c. $0.005 \leq p \leq 0.01$

9.161 a. H_0: $\mu = 1.0$; H_a: $\mu > 1.0$; TS: $T = \dfrac{\bar{X} - \mu_0}{S/\sqrt{n}}$;

RR: $T \geq 1.7033$

$t = 2.0264 \geq 1.7033$; $p = 0.0264 \leq 0.05$. There is evidence to suggest that the true mean magnitude is greater than 1.0.
b. $0.025 \leq p \leq 0.05$

9.163 a. H_0: $\mu = 65$; H_a: $\mu > 65$; TS: $T = \dfrac{\bar{X} - \mu_0}{S/\sqrt{n}}$;

RR: $T \geq 2.6503$

$t = 0.4795 < 2.6503$; $p = 0.3198 > 0.01$. There is no evidence to suggest that the mean amount of aspartame is greater than 65 mg.   **b.** $p \geq 0.20$

9.165 a. H_0: $\lambda = 4$; H_a: $\lambda < 4$; TS $X = $ the number of people locked out of their rooms; RR: $X \leq 0$ ($\alpha \approx 0.0183$).
b. $x = 2$. There is no evidence to suggest that the mean number of people locked out of their rooms per day is less than 4.
c. $P(X \leq 2) = 0.2381$

Section 9.6

9.167 True.

9.169 False.

9.171 The hypothesis test is not valid.

9.173

	RR	Value of the TS	Conclusion
a.	$Z \geq 1.6449$	0.3033	Do not reject.
b.	$Z \geq 1.2816$	1.4882	Reject.
c.	$Z \geq 2.3263$	2.3327	Reject.
d.	$Z \geq 1.9600$	0.6872	Do not reject.
e.	$Z \geq 2.3263$	0.3307	Do not reject.

9.175

	RR	Value of the TS	Conclusion
a.	$|Z| \geq 2.2414$	-2.0142	Do not reject.
b.	$|Z| \geq 2.3263$	2.3451	Reject.
c.	$|Z| \geq 1.9600$	-2.0294	Reject.
d.	$|Z| \geq 2.8070$	-1.4025	Do not reject.
e.	$|Z| \geq 2.5758$	2.6455	Reject.

9.177

	Value of the TS	p Value	Conclusion
a.	−2.0285	0.0213	Reject.
b.	0.0574	0.5229	Do not reject.
c.	−0.7611	0.2233	Do not reject.
d.	−2.2166	0.0133	Reject.
e.	−2.6187	0.0044	Reject.

9.179 a. 500, 16, 0.02.
b. $np_0 = 10 \geq 5$, $n(1 − p_0) = 490 \geq 5$.
The large-sample test is appropriate.

c. $H_0: p = 0.02$; $H_a: p > 0.02$; TS: $Z = \dfrac{\hat{P} − p_0}{\sqrt{\frac{p_0(1 − p_0)}{n}}}$;

RR: $Z \geq 1.6449$
$z = 1.9166 \geq 1.6449$. There is evidence to suggest that the population proportion is greater than 0.02. **d.** 0.0276

9.181 a. 225, 189, 0.90.
b. $np_0 = 202.5 \geq 5$, $n(1 − p_0) = 22.5 \geq 5$. The large-sample test is appropriate.

c. $H_0: p = 0.90$; $H_a: p \neq 0.90$; TS: $Z = \dfrac{\hat{P} − p_0}{\sqrt{\frac{p_0(1 − p_0)}{n}}}$;

RR: $|Z| \geq 1.96$
$z = −3.0 \leq −1.96$. There is evidence to suggest that the population proportion is different from 0.90. **d.** 0.0027

9.183 Solution Trail

Keywords: Is there any evidence; greater than; 1250 women; 502 developed cancer; 38%.

Translation: Conduct a one-sided, right-tailed test about p; $n = 1250$; $x = 502$; $p_0 = 0.38$.

Concepts: Large-sample hypothesis test concerning a population proportion.

Vision: Check the nonskewness criterion. Use the template for a one-sided, right-tailed test about p. Use $\alpha = 0.01$ to find the critical value, compute the value of the test statistic, and draw a conclusion.

$H_0: p = 0.38$; $H_a: p > 0.38$; TS: $Z = \dfrac{\hat{P} − p_0}{\sqrt{\frac{p_0(1 − p_0)}{n}}}$;

RR: $Z \geq 2.3263$
$z = 1.5733 < 2.3263$; $p = 0.0578 > 0.01$.
There is no evidence to suggest that the proportion of taller women who develop cancer is greater than 0.38.

9.185 $H_0: p = 0.50$; $H_a: p > 0.50$; TS: $Z = \dfrac{\hat{P} − p_0}{\sqrt{\frac{p_0(1 − p_0)}{n}}}$;

RR: $Z \geq 1.6449$
$z = 2.9212 \geq 1.6449$; $p = 0.0017 \leq 0.05$.
There is evidence to suggest that more than half of all people have trouble sleeping during a full moon.

9.187 $H_0: p = 0.40$; $H_a: p < 0.40$; TS: $Z = \dfrac{\hat{P} − p_0}{\sqrt{\frac{p_0(1 − p_0)}{n}}}$;

RR: $Z \leq −2.3263$
$z = −2.4244 \leq −2.3263$; $p = 0.0077 \leq 0.01$.
There is evidence to suggest that fewer than 40% of all residents are satisfied with the local government. This politician should enter the race for mayor.

9.189 $H_0: p = 0.10$; $H_a: p < 0.10$; TS: $Z = \dfrac{\hat{P} − p_0}{\sqrt{\frac{p_0(1 − p_0)}{n}}}$;

RR: $Z \leq −1.6449$
$z = −0.3333 > −1.6449$; $p = 0.3694 > 0.05$.
There is no evidence to suggest that the proportion of rent-controlled apartments is less than 0.10.

9.191 $H_0: p = 1/3$; $H_a: p \neq 1/3$; TS: $Z = \dfrac{\hat{P} − p_0}{\sqrt{\frac{p_0(1 − p_0)}{n}}}$;

RR: $|Z| \geq 1.96$
$z = 1.3528$; $|z| < 1.96$; $p = 0.1761 > 0.05$.
There is no evidence to suggest that the proportion of Americans who do not use handwriting is different from 1/3.

9.193 a. $H_0: p = 0.95$; $H_a: p < 0.95$;

TS: $Z = \dfrac{\hat{P} − p_0}{\sqrt{\frac{p_0(1 − p_0)}{n}}}$; RR: $Z \leq −1.6449$

$z = −2.2711 \leq −1.6449$; $p = 0.0116 \leq 0.05$.
There is evidence to suggest that the proportion of all tractor batteries that last at least three years is less than 0.95.
b. $p = 0.0116$. **c.** There is evidence to suggest that less than 95% of all batteries last at least three years, so the company should not implement the new warranty.

9.195 a. $H_0: p = 0.23$; $H_a: p < 0.23$;

TS: $Z = \dfrac{\hat{P} − p_0}{\sqrt{\frac{p_0(1 − p_0)}{n}}}$; RR: $Z \leq −1.6449$

$z = −2.1443 \leq −1.6449$; $p = 0.0160 \leq 0.05$. There is evidence to suggest that the proportion of purchases of U.S. homes by Canadians is less than 0.23. **b.** $p = 0.0160$.
c. Critical value and p-value illustration:

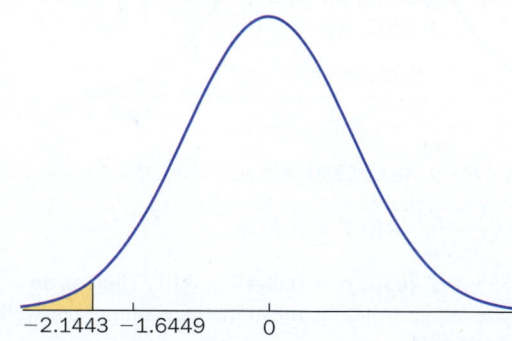

−2.1443 −1.6449 0

9.197 a. $H_0: p = 0.51$; $H_a: p < 0.51$
TS: $X =$ the number of students who admit to cheating out of the 25.
RR: $X \leq 7$.
b. $x = 9$. There is no evidence to suggest that the true proportion of Long Island students who admit to cheating is less than 0.51.

c. Assuming $X \sim B(25, 0.51)$:
$p = P(X \leq 9) = 0.0964$

Section 9.7

9.199 False.

9.201 False.

9.203 a. $X^2 = \dfrac{(n-1)S^2}{\sigma_0^2}$

b. i. $X^2 \geq 19.6751$ **ii.** $X^2 \geq 31.5264$
iii. $X^2 \geq 40.2894$ **iv.** $X^2 \geq 37.9159$
v. $X^2 \geq 20.5150$ **vi.** $X^2 \geq 42.5793$

9.205 a. $X^2 = \dfrac{(n-1)S^2}{\sigma_0^2}$

b.

Rejection region		
i. $X^2 \leq 23.6543$	or	$X^2 \geq 58.1201$
ii. $X^2 \leq 13.7867$	or	$X^2 \geq 53.6720$
iii. $X^2 \leq 5.8957$	or	$X^2 \geq 49.0108$
iv. $X^2 \leq 5.2293$	or	$X^2 \geq 30.5779$
v. $X^2 \leq 10.3909$	or	$X^2 \geq 56.8923$
vi. $X^2 \leq 1.0636$	or	$X^2 \geq 7.7794$

9.207 a. 0.0005 **b.** 0.01 **c.** 0.025 **d.** 0.005

9.209 a. $0.01 \leq p \leq 0.025$ **b.** $0.05 \leq p \leq 0.10$
c. $0.001 \leq p \leq 0.005$ **d.** $p \leq 0.0001$

9.211 a. $0.02 \leq p \leq 0.05$ **b.** $0.0002 \leq p \leq 0.001$
c. $0.05 \leq p \leq 0.10$ **d.** $0.001 \leq p \leq 0.002$

9.213 a. $H_0: \sigma^2 = 36.8$; $H_a: \sigma^2 \neq 36.8$;
TS: $X^2 = \dfrac{(n-1)S^2}{\sigma_0^2}$; RR: $X^2 \leq 6.2621$ or $X^2 \geq 27.4884$
b. $s^2 = 105.8633$, $\chi^2 = 43.1508 \geq 27.4884$. There is evidence to suggest that the population variance is different from 36.8. **c.** $0.0002 \leq p \leq 0.01$

9.215 $H_0: \sigma^2 = 0.25$; $H_a: \sigma^2 > 0.25$;
TS: $X^2 = \dfrac{(n-1)S^2}{\sigma_0^2}$; RR: $X^2 \geq 48.6024$
$\chi^2 = 42.3455 < 48.6024$; $p = 0.1541 > 0.05$.
There is no evidence to suggest that the population variance is greater than 0.25.

9.217 $H_0: \sigma^2 = 324$; $H_a: \sigma^2 > 324$; TS: $X^2 = \dfrac{(n-1)S^2}{\sigma_0^2}$;
RR: $X^2 \geq 24.7250$
$\chi^2 = 15.7814 < 24.7250$; $p = 0.1494 > 0.01$.
There is no evidence to suggest that the variance in stress is more than 324. There is no evidence to refute the manufacturer's claim.

9.219 Solution Trial

Keywords: Is there any evidence; variance; greater than

Translation: Conduct a one-sided, right tailed test about σ^2.

Concepts: Hypothesis test concerning a population variance.

Vision: The underlying population is assumed to be normal. Use the template for a one-sided, right-tailed test about σ^2. Use $\alpha = 0.01$ to find the critical value. Compute the value of the test statistic, and draw a conclusion.

$H_0: \sigma^2 = 62.5$; $H_a: \sigma^2 > 62.5$; TS: $X^2 = \dfrac{(n-1)S^2}{\sigma_0^2}$;
RR: $X^2 \geq 21.6660$
$\chi^2 = 10.0944 < 21.6660$; $p = 0.3429 > 0.01$.
There is no evidence to suggest that the variance is greater than 62.5.

9.221 $H_0: \sigma^2 = 0.3^2$; $H_a: \sigma^2 > 0.3^2$; TS: $X^2 = \dfrac{(n-1)S^2}{\sigma_0^2}$;
RR: $X^2 \geq 37.6525$
$\chi^2 = 32.4900 < 37.6525$; $p = 0.1443 > 0.05$.
There is no evidence to suggest that the standard deviation is greater than 0.3.

9.223 $H_0: \sigma^2 = 22.5$; $H_a: \sigma^2 < 22.5$;
TS: $X^2 = \dfrac{(n-1)S^2}{\sigma_0^2}$; RR: $X^2 \leq 23.2686$
$\chi^2 = 24.9600 > 23.2686$; $p = 0.0832 > 0.05$.
There is no evidence to suggest that the variance in bull-riding time has decreased.

9.225 a. $H_0: \sigma^2 = 0.36$; $H_a: \sigma^2 > 0.36$;
TS: $X^2 = \dfrac{(n-1)S^2}{\sigma_0^2}$; RR: $X^2 \geq 30.1435$
$\chi^2 = 22.1667 < 30.1435$; $p = 0.2760 > 0.05$.
There is no evidence to suggest that the population variance is greater than 0.36. **b.** $p > 0.10$

9.227 a. $H_0: \sigma^2 = 230$; $H_a: \sigma^2 > 230$;
TS: $X^2 = \dfrac{(n-1)S^2}{\sigma_0^2}$; RR: $X^2 \geq 38.8851$
$\chi^2 = 21.9352 < 38.8851$; $p = 0.6922 > 0.05$.
There is no evidence to suggest that the variance is less than 230. **b.** $p > 0.10$

9.229 a. $H_0: \sigma^2 = 2.75^2$; $H_a: \sigma^2 > 2.75^2$;
TS: $X^2 = \dfrac{(n-1)S^2}{\sigma_0^2}$; RR: $X^2 \geq 32.6706$
$\chi^2 = 40.3239 \geq 32.6706$; $p = 0.0068 \leq 0.05$.
There is evidence to suggest that the population variance in wingspan is greater than 2.75^2 mm. **b.** $0.005 \leq p \leq 0.01$

9.231 a. $H_0: \sigma^2 = 49$; $H_a: \sigma^2 \neq 49$;
TS: $X^2 = \dfrac{(n-1)S^2}{\sigma_0^2}$; RR: $X^2 \leq 9.5908$ or $X^2 \geq 34.1696$.
b. $S^2 \leq 23.4975$ or $S^2 \geq 83.7155$.
c. $23.4975 < 56 < 83.7155$; $p = 0.5917 > 0.05$. There is no evidence to suggest that the population variance in thickness is different from 49.
d. $s^2 = 15.6 \leq 23.4975$; $p = 0.0034 \leq 0.05$. There is evidence to suggest that the population variance in thickness is different from 49.

Chapter Exercises

9.233 a. $H_0: \mu = 1.6$; $H_a: \mu \neq 1.6$; TS: $Z = \dfrac{\bar{x} - \mu_0}{\sigma/\sqrt{n}}$;

RR: $|Z| \geq 2.5758$
$z = 1.7265$. $|z| < 2.5758$.
There is no evidence to suggest that the population mean is different from 1.6; there is no evidence to suggest that the machine is malfunctioning.   **b.** 0.0843

9.235 $H_0: \mu = 1.0$; $H_a: \mu \neq 1.0$; TS: $Z = \dfrac{\bar{x} - \mu_0}{\sigma/\sqrt{n}}$;

RR: $|Z| \geq 1.96$
$z = 1.3435$; $|z| < 1.96$.
There is no evidence to suggest that the mean weight is different from 1.   **b.** 0.1791

9.237 $H_0: \mu = 13$; $H_a: \mu \neq 13$; TS: $T = \dfrac{\bar{X} - \mu_0}{\sigma/\sqrt{n}}$;

RR: $|T| \geq 2.2622$
a. $t = -0.3623$; $|t| < 2.2622$; $p = 0.7255 > 0.05$. There is no evidence to suggest that the mean diameter is different from 13. The process should not be stopped.
b. $t = 2.8109$; $|t| \geq 2.2622$; $p = 0.0203 \leq 0.05$. There is evidence to suggest that the mean diameter is different from 13. The process should be stopped.
c. I would prefer a large significance level. This would mean a smaller type II error, failing to realize the mean diameter is different from 13, and less of a chance of costly engine damage.

9.239 a. 0.20, 1500, 0.23
b. $(1500)(0.20) = 300 \geq 5$; $(1500)(0.80) = 1200 \geq 5$

c. $H_0: p = 0.20$; $H_a: p > 0.20$; TS: $Z = \dfrac{\hat{P} - p_0}{\sqrt{\frac{p_0(1 - p_0)}{n}}}$;

RR: $Z \geq 2.3263$
$z = 2.9047 \geq 2.3263$; $p = 0.0018 \leq 0.01$.
There is evidence to suggest that the population proportion of companies that require this information is greater than 0.20.

9.241 a. $H_0: p = 0.22$; $H_a: p > 0.22$;

TS: $Z = \dfrac{\hat{P} - p_0}{\sqrt{\frac{p_0(1 - p_0)}{n}}}$; RR: $|Z| \geq 1.96$

$z = -1.8098$; $|z| < 1.96$. There is no evidence to suggest that the proportion of Americans affected by the sequester has changed.   **b.** 0.0703

9.243 a. $H_0: p = 0.74$; $H_a: p < 0.74$;

TS: $Z = \dfrac{\hat{P} - p_0}{\sqrt{\frac{p_0(1 - p_0)}{n}}}$; RR: $Z \leq -1.6449$

$z = -2.0996 \leq -1.6449$. There is evidence to suggest that the proportion of crude-oil spills that involve less than five gallons has decreased.   **b.** Fewer spills involve less than five gallons. Therefore, more spills are larger.   **c.** $p = 0.0179$

9.245 $H_0: \sigma^2 = 0.50$; $H_a: \sigma^2 < 0.50$;

TS: $X^2 = \dfrac{(n - 1)S^2}{\sigma_0^2}$; RR: $X^2 \leq 12.4426$

$\chi^2 = 15.6 > 12.4426$; $p = 0.2589$.
There is no evidence to suggest that the population variance is less than 0.50.

9.247 a. $H_0: \mu = 58$; $H_a: \mu < 58$; TS: $T = \dfrac{\bar{X} - \mu_0}{S/\sqrt{n}}$;

RR: $T \leq -1.7109$
$t = -2.1191 \leq -1.7109$. There is evidence to suggest that the mean police response time has decreased.   **b.** $0.01 \leq p \leq 0.025$

9.249 a. $H_0: \mu = 1$; $H_a: \mu > 1$; TS: $T = \dfrac{\bar{X} - \mu_0}{S/\sqrt{n}}$;

RR: $T \geq 2.6810$
$t = 3.5555 \geq 2.6810$. There is evidence to suggest that the mean significant wave height is greater than 1 meter.
b. $0.001 \leq p \leq 0.005$

9.251 a. $H_0: \mu = 4{,}500{,}000$; $H_a: \mu > 4{,}500{,}000$;

TS: $T = \dfrac{\bar{X} - \mu_0}{S/\sqrt{n}}$; RR: $T \geq 2.3060$

$t = 1.0895 < 2.3060$. There is no evidence to suggest that the mean amount of oil stored is above the safety level.
b. $0.10 \leq p \leq 0.20$

9.253 a. $H_0: p = 0.92$; $H_a: p < 0.92$;

TS: $Z = \dfrac{\hat{P} - p_0}{\sqrt{\frac{p_0(1 - p_0)}{n}}}$; RR: $Z \leq -2.3263$

$z = -1.2511 > -2.3263$. There is no evidence to suggest that the true proportion of assisted-living patients who are satisfied is less than 0.92.   **b.** $p = 0.1055$.
c. Critical value and p-value illustration:

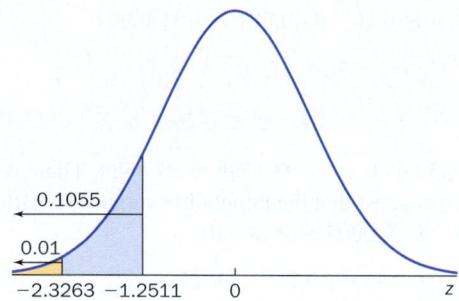

9.255 a. $H_0: \sigma^2 = 4$; $H_a: \sigma^2 > 4$; TS: $X^2 = \dfrac{(n - 1)S^2}{\sigma_0^2}$;

RR: $X^2 \geq 26.2962$
$\chi^2 = 31.00 \geq 26.2962$. There is evidence to suggest that the opulation variance is greater than 4 pounds.
c. $0.01 \leq p \leq 0.025$

9.257 $H_0: p = 0.24$; $H_a: p > 0.24$; TS: $Z = \dfrac{\hat{P} - p_0}{\sqrt{\frac{p_0(1 - p_0)}{n}}}$;

RR: $Z \geq 1.6449$
$z = 1.3421 < 1.6449$; $p = 0.0898 > 0.05$.
There is no evidence to suggest that more than 24% of all Liberty Mutual policyholders view padding as acceptable. The insurance company will not raise automobile rates in this case.

CHAPTER 10

Section 10.1

10.1 True.

10.3 True.

10.5 False.

10.7 a. $\mu_1 - \mu_2 = 0$ **b.** $\mu_1 - \mu_2 < 0$ **c.** $\mu_1 - \mu_1 \neq 7$
d. $\mu_1 - \mu_2 > -4$ **e.** $\mu_1 - \mu_2 \neq 0$ **f.** $\mu_1 - \mu_2 = 10$

10.9 a. $H_0: \mu_1 - \mu_2 = 0$; $H_a: \mu_1 - \mu_2 > 0$;
TS: $Z = \dfrac{(\overline{X}_1 - \overline{X}_2) - 0}{\sqrt{\frac{\sigma_1^2}{n_1} + \frac{\sigma_2^2}{n_2}}}$; RR: $Z \geq 1.6449$
b. $z = 2.0951 \geq 1.6449$. There is evidence to suggest that population mean 1 is greater than population mean 2.
c. 0.0181

10.11 a. $H_0: \mu_1 - \mu_2 = 0$; $H_a: \mu_1 - \mu_2 \neq 0$;
TS: $Z = \dfrac{(\overline{X}_1 - \overline{X}_2) - 0}{\sqrt{\frac{\sigma_1^2}{n_1} + \frac{\sigma_2^2}{n_2}}}$; RR: $|Z| \geq 3.2905$
b. $z = -1.9379$; $|z| < 3.2905$. There is no evidence to suggest that population mean 1 is different from population mean 2.
c. No. Both sample sizes are large.

10.13 a. $\overline{X}_1 - \overline{X}_2$ is normal with mean 15, variance 2.4286, and standard deviation 1.5584.
b. Probability distribution:

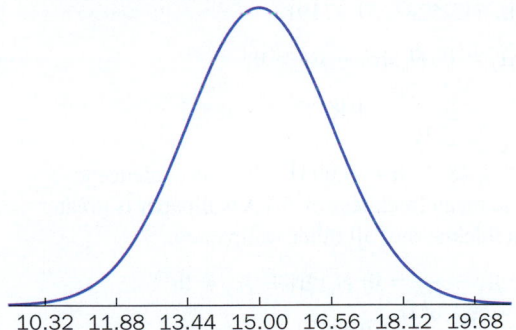

10.32 11.88 13.44 15.00 16.56 18.12 19.68

c. 0.0997 **d.** 0.2063 **e.** 0.2605

10.15 Solution Trail

Keywords: Is there any evidence; different from; known variances; independent random samples; sample sizes.

Translation: Conduct a two-sided test to compare μ_1 and μ_2.

Concepts: Hypothesis test concerning two population means when variances are known.

Vision: Use the template for this hypothesis test. The samples are random and independent, the sample sizes are large, and the population variances are known. Use a two-sided alternative hypothesis and the corresponding rejection region, find the value of the test statistic, and draw a conclusion.

$H_0: \mu_1 - \mu_2 = 0$; $H_a: \mu_1 - \mu_2 \neq 0$; TS: $Z = \dfrac{(\overline{X}_1 - \overline{X}_2) - 0}{\sqrt{\frac{\sigma_1^2}{n_1} + \frac{\sigma_2^2}{n_2}}}$;

RR: $|Z| \geq 2.5758$
$z = -3.0814$; $|z| \geq 2.5758$; $p = 0.0021$.

There is evidence to suggest that the mean Nordstrom gift-certificate value is different from the mean Macy's gift-certificate value.

10.17 a. $(-48.8911, 64.6911)$ **b.** Because 0 is included in the CI, there is no evidence to suggest that the mean dollar amount of copper stolen differs for the two states.

10.19 a. $H_0: \mu_1 - \mu_2 = 5$; $H_a: \mu_1 - \mu_2 < 5$;
TS: $Z = \dfrac{(\overline{X}_1 - \overline{X}_2) - 0}{\sqrt{\frac{\sigma_1^2}{n_1} + \frac{\sigma_2^2}{n_2}}}$; RR: $Z \leq -2.3263$
$z = -0.7546 > -2.3263$. There is no evidence to refute the claim; there is no evidence to suggest that the difference in mean weights is less than 5 pounds. **b.** 0.2252
c. The normality assumption is not necessary, because both sample sizes are large.

10.21 a. $(-7.1844, -3.4756)$ **b.** There is evidence to suggest that the mean number of revolutions per minute for the Altura fan is greater than the mean number of revolutions per minute for the Hampton fan; 0 is not included in the CI, and the CI is less than 0.

10.23 a. $H_0: \mu_1 - \mu_2 = 0$; $H_a: \mu_1 - \mu_2 \neq 0$;
TS: $Z = \dfrac{(\overline{X}_1 - \overline{X}_2) - 0}{\sqrt{\frac{\sigma_1^2}{n_1} + \frac{\sigma_2^2}{n_2}}}$; RR: $|Z| \geq 1.96$
$z = 1.9666 \geq 1.96$. There is evidence to suggest that the population mean airspeeds are different, just barely. **b.** 0.0492

10.25 a. $H_0: \mu_1 - \mu_2 = 0$; $H_a: \mu_1 - \mu_2 \neq 0$;
TS: $Z = \dfrac{(\overline{X}_1 - \overline{X}_2) - 0}{\sqrt{\frac{\sigma_1^2}{n_1} + \frac{\sigma_2^2}{n_2}}}$; RR: $|Z| \geq 2.5758$
$z = -1.2536$; $|z| < 2.5758$; $p = 0.2100$. There is no evidence to suggest that the population mean magnesium level in each serving of baked beans and potatoes is different.
b. $z = -1.8214$; $p = 0.0685$. There is still no evidence to refute the claim. **c.** $n_1 = n_2 = 76$

10.27 a. $H_0: \mu_1 - \mu_2 = 0$; $H_a: \mu_1 - \mu_2 \neq 0$;
TS: $Z = \dfrac{(\overline{X}_1 - \overline{X}_2) - 0}{\sqrt{\frac{\sigma_1^2}{n_1} + \frac{\sigma_2^2}{n_2}}}$; RR: $|Z| \geq 1.96$
$z = -1.1404$; $|z| < 1.96$. There is no evidence to suggest that the mean times are different. **b.** 0.2541
c. $n_1 = n_2 = 205$

10.29 a. $n_1 = n_2 = n = \dfrac{(z_{\alpha/2})^2(\sigma_1^2 + \sigma_2^2)}{B^2}$
b. $n_1 = n_2 = 39$ **c.** $B = 4.9776 \leq 5$

Section 10.2

10.31 $\sigma^2\left(\dfrac{1}{n_1} + \dfrac{1}{n_2}\right)$

10.33 If the assumptions are not entirely true, the hypothesis test is still very reliable.

10.35 True.

10.37 a. H_a: $\mu_1 - \mu_2 = 0$; H_a: $\mu_1 - \mu_2 < 0$;

TS: $T = \dfrac{(\overline{X}_1 - \overline{X}_2) - 0}{\sqrt{S_p^2(\frac{1}{n_1} + \frac{1}{n_2})}}$

RR: $T \le -1.7011$

$t = -0.1181 > -1.7011$. There is no evidence to suggest that μ_1 is less than μ_2. **b.** $p > 0.20$

10.39 a. H_a: $\mu_1 - \mu_2 = 0$; H_a: $\mu_1 - \mu_2 \ne 0$;

TS: $T = \dfrac{(\overline{X}_1 - \overline{X}_2) - 0}{\sqrt{S_p^2(\frac{1}{n_1} + \frac{1}{n_2})}}$ RR: $|T| \ge 2.0484$

$t = 3.0096 \ge 2.0484$. There is evidence to suggest that the two population means are different.
b. $0.002 \le p \le 0.01$

10.41 a. 24 **b.** 15 **c.** 33 **d.** 62

10.43 a. RR: $|T| \ge 2.0796$, $t = 1.8502$. There is no evidence to suggest that the two population means are different.
b. RR: $|T'| \ge 2.1009$, $t' = 2.6994 \ge 2.1009$. There is evidence to suggest that the two population means are different.
c. The sample standard deviations suggest unequal variances. Therefore, the test in part (b), seems more appropriate.

10.45 a. H_a: $\mu_1 - \mu_2 = 0$; H_a: $\mu_1 - \mu_2 \ne 0$;

TS: $T = \dfrac{(\overline{X}_1 - \overline{X}_2) - 0}{\sqrt{S_p^2(\frac{1}{n_1} + \frac{1}{n_2})}}$ RR: $|T| \ge 2.0687$

$t = -0.9209$; $|t| < 2.0687$. There is no evidence to suggest a difference in the population mean deviations from perfect flatness for these two rotor brands. **b.** $0.20 \le p \le 0.40$

10.47 Solution Trail

Keywords: Is there any evidence; larger; underlying distributions normal.

Translation: Conduct a one-sided test to compare μ_1 and μ_2; variances are unknown and assumed equal.

Concepts: Hypothesis test concerning two population means when variances are unknown but equal.

Vision: Use a one-sided alternative hypothesis and the corresponding rejection region, find the pooled estimate of the common variance, compute the value of the test statistic, and draw a conclusion.
H_a: $\mu_1 - \mu_2 = 0$; H_a: $\mu_1 - \mu_2 > 0$;

TS: $T = \dfrac{(\overline{X}_1 - \overline{X}_2) - 0}{\sqrt{S_p^2(\frac{1}{n_1} + \frac{1}{n_2})}}$ RR: $T \ge 1.6955$

$t = 1.8126 \ge 1.6955$; $p = 0.0398 \le 0.05$. There is evidence to suggest that the population mean size of quagga mussels is larger in Lake Texoma than in Lake Mead.

10.49 a. H_a: $\mu_1 - \mu_2 = 0$; H_a: $\mu_1 - \mu_2 \ne 0$;

TS: $T = \dfrac{(\overline{X}_1 - \overline{X}_2) - 0}{\sqrt{S_p^2(\frac{1}{n_1} + \frac{1}{n_2})}}$ RR: $|T| \ge 2.6259$

$t = -3.0863$; $|t| \ge 2.25259$; $p = 0.0026 \le 0.01$. There is evidence to suggest that the population mean sagittal diameter of women's biceps tendons is different from that of men's biceps tendons. **b.** $(-0.4928, -0.1072)$

10.51 H_a: $\mu_1 - \mu_2 = 0$; H_a: $\mu_1 - \mu_2 < 0$;

TS: $T = \dfrac{(\overline{X}_1 - \overline{X}_2) - 0}{\sqrt{S_p^2(\frac{1}{n_1} + \frac{1}{n_2})}}$ RR: $T \le -1.6794$

$t = -1.4916 > -1.6794$; $p = 0.0714 > 0.05$. There is no evidence to suggest that the population mean boarding time for American is less than the population mean boarding time for US Airways.

10.53 a. Solution Trail

Keywords: Is there any evidence; larger mean; underlying populations are normal.

Translation: Conduct a one-sided, right-tailed test to compare μ_1 and μ_2; variances are unknown and assumed unequal.

Concepts: Hypothesis test concerning two population means when variances are unknown and unequal.

Vision: Use the template for this hypothesis test. The samples are random and independent, the underlying populations are normal, and the population variances are unknown and unequal. Use the one-sided, right-tailed alternative and the corresponding rejection region. Find the value of the test statistic and draw a conclusion.

$\mu_1 - \mu_2 = 0$; H_a: $\mu_1 - \mu_2 > 0$; TS: $T' = \dfrac{(\overline{X}_1 - \overline{X}_2) - 0}{\sqrt{\frac{S_1^2}{n_1} + \frac{S_2^2}{n_2}}}$

RR: $T' \ge 2.4922$
$t' = 3.5202 \ge 2.4922$; $p = 0.0009$. There is evidence to suggest that the population mean amount of coating for Fero is greater than the population mean amount of coating for Cintex. **b.** $(13.5281, 51.8719)$

10.55 $\mu_1 - \mu_2 = 0$; H_a: $\mu_1 - \mu_2 > 0$;

TS: $T' = \dfrac{(\overline{X}_1 - \overline{X}_2) - 0}{\sqrt{\frac{S_1^2}{n_1} + \frac{S_2^2}{n_2}}}$ RR: $T' \ge 2.4851$

$t' = 2.9891 \ge 2.4851$; $p = 0.0031$. There is evidence to suggest that the mean thickness of Aries wallpaper is greater than the mean thickness of all other wallpapers.

10.57 a. H_a: $\mu_1 - \mu_2 = 0$; H_a: $\mu_1 - \mu_2 \ne 0$;

TS: $T = \dfrac{(\overline{X}_1 - \overline{X}_2) - 0}{\sqrt{S_p^2(\frac{1}{n_1} + \frac{1}{n_2})}}$ RR: $|T| \ge 2.0345$

$t = -3.2025$; $|t| \ge 2.0345$. There is evidence to suggest that the mean honey harvest per hive is different in the two states.
b. $0.002 \le p \le 0.01$; $p = 0.0030$

10.59 a. H_a: $\mu_1 - \mu_2 = 0$; H_a: $\mu_1 - \mu_2 \ne 0$;

TS: $T = \dfrac{(\overline{X}_1 - \overline{X}_2) - 0}{\sqrt{S_p^2(\frac{1}{n_1} + \frac{1}{n_2})}}$ RR: $|T| \ge 2.8982$

$t = -3.2218$; $|t| \ge 2.8982$; $p = 0.005$. There is evidence to suggest that the population mean shading coefficients are different. **b.** $(-0.2659, -0.0141)$ **c.** Because 0 is not included in the CI, there is evidence to suggest that the population mean shading coefficients are different. This agrees with the conclusion in part (a).

10.61 a. The assumption of equal variances seems reasonable. The sample variances are close.
b. H_a: $\mu_1 - \mu_2 = 0$; H_a: $\mu_1 - \mu_2 \ne 0$;

TS: $T = \dfrac{(\overline{X}_1 - \overline{X}_2) - 0}{\sqrt{S_p^2\left(\frac{1}{n_1} + \frac{1}{n_2}\right)}}$ RR: $|T| \geq 2.6846$

$t = 0.9943$; $|t| < 2.6846$; $p = 0.3252$. There is no evidence to suggest that the mean burn times are different.   **c.** $p \geq 0.20$

10.63 a. H_a: $\mu_1 - \mu_2 = 0$; H_a: $\mu_1 - \mu_2 < 0$;
TS: $T = \dfrac{(\overline{X}_1 - \overline{X}_2) - 0}{\sqrt{S_p^2\left(\frac{1}{n_1} + \frac{1}{n_2}\right)}}$ RR: $T \leq -2.5280$

$t = -3.2199 \leq -2.5280$; $p = 0.0021$. There is evidence to suggest that the population mean lifetime of the new fuel rod is greater than the population mean lifetime of the old fuel rod.
b. $(-8.7402, -0.5398)$. This CI supports part (a); 0 is not included in the CI.

10.65 You should reject the null hypothesis approximately 5 out of every 100 trials. This is true for any value of σ_2. However, as σ_2 gets farther from σ_1, you should reject the null hypothesis more often. Robust means that our results are still 95% accurate, even when the underlying assumptions are not met.

Section 10.3

10.67 True.

10.69 False.

10.71 The correct analysis is based on a distribution with greater variability and is more conservative.

10.73 a. Paired; patients (arms).   **b.** Paired; lathes.
c. Independent; 20-year-old males, 70-year-old males.
d. Paired; style of modular home.   **e.** Independent; frequent flyers on United, frequent flyers on Delta.

10.75 H_0: $\mu_D = 0$; H_a: $\mu_D > 0$; TS: $T = \dfrac{\overline{D} - \Delta_0}{S_D/\sqrt{n}}$;
RR: $T \geq 1.7459$
$t = 1.9270 \geq 1.7459$; $p = 0.0360 \leq 0.05$. There is evidence to suggest that population mean 1 is greater than population mean 2.

10.77 a. H_0: $\mu_D = 0$; H_a: $\mu_D \neq 0$; TS: $T = \dfrac{\overline{D} - \Delta_0}{S_D/\sqrt{n}}$;
RR: $|T| \geq 2.8609$
$t = 3.1458$; $|t| \geq 2.8609$; $p = 0.0053$. There is evidence to suggest that population mean 1 is different from population mean 2.   **b.** $0.002 \leq p \leq 0.01$

10.79 a. Handgun.   **b.** H_0: $\mu_D = 0$; H_a: $\mu_D < 0$;
TS: $T = \dfrac{\overline{D} - \Delta_0}{S_D/\sqrt{n}}$; RR: $T \leq -3.3649$
$t = -0.4966 > -3.3649$; $p = 0.3203 > 0.01$. There is no evidence to suggest that the population mean muzzle velocity of a clean gun is greater than the population mean muzzle velocity of a dirty gun.   **c.** $p > 0.20$

10.81 Solution Trail

Keywords: Concentration measured again; differences; is there any evidence; improves air quality.

Translation: Before and after measurements on the same businesses; conduct a one-sided, right-tailed test to compare the before and after concentrations.

Concepts: Hypothesis test concerning two population means when data are paired.

Vision: The data are paired and the differences are given. Use a one-sided alternative hypothesis and the corresponding rejection region, find the value of the test statistic, and draw a conclusion.
H_0: $\mu_D = 0$; H_a: $\mu_D > 0$; TS: $T = \dfrac{\overline{D} - \Delta_0}{S_D/\sqrt{n}}$;
RR: $T \geq 1.6991$
$t = 1.5491 < 1.6991$; $p = 0.0661 > 0.05$. There is no evidence to suggest that the population mean concentration of particulate matter before filtration is greater than the population mean concentration of particulate matter after filtration.

10.83 H_0: $\mu_D = 0$; H_a: $\mu_D > 0$; TS: $T = \dfrac{\overline{D} - \Delta_0}{S_D/\sqrt{n}}$;
RR: $T \geq 2.1604$
$t = 0.3132 < 2.1604$; $p = 0.3795 > 0.025$. There is no evidence to suggest that the population mean ammonia-ion concentration before treatment is greater than the population mean ammonia-ion concentration after treatment.

10.85 a. Adult.   **b.** H_0: $\mu_D = 0$; H_a: $\mu_D > 0$;
TS: $T = \dfrac{\overline{D} - \Delta_0}{S_D/\sqrt{n}}$; RR: $T \geq 2.5669$
$t = 2.8425 \geq 2.5669$; $p = 0.0056 \leq 0.01$. There is evidence to suggest that the population mean steps per minute listening to hard rock is greater than the population mean steps per minute listening to ballads.

10.87 H_0: $\mu_D = 0$; H_a: $\mu_D > 0$; TS: $T = \dfrac{\overline{D} - \Delta_0}{S_D/\sqrt{n}}$;
RR: $T \geq 3.3962$
$t = 5.4503 \geq 3.3962$;
$p = 0.00000363 \leq 0.001$. There is evidence to suggest that the population mean porosity before treatment is greater than the population mean porosity after treatment.

10.89 a. H_0: $\mu_D = 0$; H_a: $\mu_D < 0$; TS: $T = \dfrac{\overline{D} - \Delta_0}{S_D/\sqrt{n}}$;
RR: $T \leq -3.4210$
$t = -3.4924 \leq -3.4210$; $p = 0.0008 \leq 0.001$. There is evidence to suggest that the electric-grill population mean fat content is less than the frying-pan population mean fat content.   **b.** $0.0005 \leq p \leq 0.001$

10.91 a. H_0: $\mu_D = 0$; H_a: $\mu_D > 0$; TS: $T = \dfrac{\overline{D} - \Delta_0}{S_D/\sqrt{n}}$;
RR: $T \geq 2.0150$
$t = 2.4360 \geq 2.0150$; $p = 0.0295$. There is evidence to suggest that the population mean cloud point before the additive is greater than the population mean cloud point after the additive.
b. $\mu_1 - \mu_2 = 0$; H_a: $\mu_1 - \mu_2 > 0$; TS: $T' = \dfrac{(\overline{X}_1 - \overline{X}_2) - 0}{\sqrt{\frac{S_1^2}{n_1} + \frac{S_2^2}{n_2}}}$

RR: $T' \geq 1.8595$

$t' = 1.5211 < 1.8595$; $p = 0.0834$. There is no evidence to suggest that the population mean cloud point before the additive is greater than the population mean cloud point after the additive.   **c.** The conclusions are different. The test statistics have the same numerator but different denominators.

10.93 a. $H_0: \mu_D = 0$; $H_a: \mu_D \neq 0$; TS: $T = \dfrac{\overline{D} - \Delta_0}{S_D/\sqrt{n}}$;

RR: $|T| \geq 2.0301$

$t = -1.4236$; $|t| < 2.0301$; $p = 0.1634$. There is no evidence to suggest that the mean price of a ticket on this route differs whether purchased 21 or 49 days in advance.
b. $0.10 \leq p \leq 0.20$ **c.** $(-22.7777, 3.9999)$. This CI supports the conclusion in part (a). The interval includes 0.

Section 10.4

10.95 True.

10.97 True.

10.99 True.

10.101

	Mean	Variance	Standard deviation	Probability
a.	-0.020	0.000579	0.0241	0.0034
b.	0.040	0.001751	0.0418	0.0280
c.	0.010	0.001557	0.0395	0.7804
d.	-0.070	0.000858	0.0293	0.8472
e.	-0.120	0.000275	0.0166	0.9999
f.	0.041	0.000914	0.0302	0.9527

10.103 a. $z = 1.1137$, $p = 0.1327$. Do not reject H_0.
b. $z = -1.8938$, $p = 0.0291$. Do not reject H_0.
c. $z = -2.2705$, $p = 0.0232$. Reject H_0.
d. $z = 0.5012$, $p = 0.6163$. Do not reject H_0.

10.105 a. $(-0.0972, 0.0390)$ **b.** $(-0.0710, -0.0052)$
c. $(-0.1376, 0.1289)$ **d.** $(0.0168, 0.0598)$

10.107 Solution Trail

Keywords: Is there any evidence; different from; proportion.

Translation: Conduct a two-sided hypothesis test about $p_1 - p_2$; $\Delta_0 = 0$.

Concepts: Large-sample hypothesis test concerning two population proportions when $\Delta_0 = 0$.

Vision: Check the large-sample assumptions. Use the template for a two-sided, right-tailed test concerning $p_1 - p_2$ when $\Delta_0 = 0$. Use $\alpha = 0.05$ to find the critical value, compute the value of the test statistic, and draw a conclusion.
$H_0: p_1 - p_2 = 0$; $H_a: p_1 - p_2 \neq 0$;

TS: $Z = \dfrac{\widehat{P}_1 - \widehat{P}_2}{\sqrt{\widehat{P}_c(1 - \widehat{P}_c)(\frac{1}{n_1} + \frac{1}{n_2})}}$;   RR: $|Z| \geq 1.96$

$z = -0.4264$; $|z| < 1.96$; $p = 0.6698 > 0.05$. There is no evidence to suggest that the proportion of voters in the West

who believe an armed revolution might be necessary is different from the proportion of voters in the East.

10.109 a. $H_0: p_1 - p_2 = 0$; $H_a: p_1 - p_2 > 0$;

TS: $Z = \dfrac{\widehat{P}_1 - \widehat{P}_2}{\sqrt{\widehat{P}_c(1 - \widehat{P}_c)(\frac{1}{n_1} + \frac{1}{n_2})}}$; RR: $Z \geq 3.0902$

$z = 3.2086 \geq 3.0902$. There is evidence to suggest that the population proportion of 18–29-year-olds who believe movies are getting better is greater than the population proportion of 30–49-year-olds who believe movies are getting better.
b. 0.0007

10.111 a. $n_1\widehat{p}_1 = 530 \geq 5$, $n_1(1 - \widehat{p}_1) = 525 \geq 5$,
$n_2\widehat{p}_2 = 825 \geq 5$, $n_2(1 - \widehat{p}_2) = 838 \geq 5$
b. $H_0: p_1 - p_2 = 0$; $H_a: p_1 - p_2 > 0$;

TS: $Z = \dfrac{\widehat{P}_1 - \widehat{P}_2}{\sqrt{\widehat{P}_c(1 - \widehat{P}_c)(\frac{1}{n_1} + \frac{1}{n_2})}}$; RR: $Z \geq 2.3263$

$z = 0.3190 < 2.3263$. There is no evidence to suggest that the population proportion of carpoolers crossing the George Washington Bridge is greater than the population proportion of carpoolers using the Lincoln Tunnel.   **c.** 0.3749

10.113 a. $\widehat{p}_1 = 0.2784$, $\widehat{p}_2 = 0.2321$
b. $H_0: p_1 - p_2 = 0$; $H_a: p_1 - p_2 \neq 0$;

TS: $Z = \dfrac{\widehat{P}_1 - \widehat{P}_2}{\sqrt{\widehat{P}_c(1 - \widehat{P}_c)(\frac{1}{n_1} + \frac{1}{n_2})}}$; RR: $|Z| \geq 2.5758$

$z = 1.1773$; $|z| < 2.5758$; $p = 0.2391 > 0.01$. There is no evidence to suggest that the population proportion of people who experience relief from the antihistamine is different from the population proportion of people who experience relief from butterbur extract.

10.115 a. $\widehat{p}_1 = 0.0755$, $\widehat{p}_2 = 0.0992$
b. $n_1\widehat{p}_1 = 8 \geq 5$, $n_1(1 - \widehat{p}_1) = 98 \geq 5$,
$n_2\widehat{p}_2 = 12 \geq 5$, $n_2(1 - \widehat{p}_2) = 109 \geq 5$
c. $H_0: p_1 - p_2 = 0$; $H_a: p_1 - p_2 \neq 0$;

TS: $Z = \dfrac{\widehat{P}_1 - \widehat{P}_2}{\sqrt{\widehat{P}_c(1 - \widehat{P}_c)(\frac{1}{n_1} + \frac{1}{n_2})}}$; RR: $|Z| \geq 1.96$

$z = -0.6286$; $|z| < 1.96$; $p = 0.5296 > 0.05$. There is no evidence to suggest that the population proportion of defective lenses is different for Process A than for Process B.

10.117 $H_0: p_1 - p_2 = 0$; $H_a: p_1 - p_2 \neq 0$;

TS: $Z = \dfrac{\widehat{P}_1 - \widehat{P}_2}{\sqrt{\widehat{P}_c(1 - \widehat{P}_c)(\frac{1}{n_1} + \frac{1}{n_2})}}$; RR: $|Z| \geq 1.96$

$z = -1.0974$; $|z| < 1.96$; $p = 0.2725 > 0.05$. There is no evidence to suggest that the true proportion of residents who feel safe after dark is different in the two districts.

10.119 a. $H_0: p_1 - p_2 = 0$; $H_a: p_1 - p_2 < 0$;

TS: $Z = \dfrac{\widehat{P}_1 - \widehat{P}_2}{\sqrt{\widehat{P}_c(1 - \widehat{P}_c)(\frac{1}{n_1} + \frac{1}{n_2})}}$; RR: $Z \leq -1.6449$

$z = -3.0677 \leq -1.6449$; $p = 0.0011 \leq 0.05$. There is evidence to suggest that the population proportion of 18- to 29-year-olds who obtain news every day from nightly news shows is less than the population proportion of 30- to

49-year-olds who obtain news every day from nightly network news shows.

b. $H_0: p_1 - p_2 = 0$; $H_a: p_1 - p_2 < 0$;

$$\text{TS: } Z = \frac{\hat{P}_1 - \hat{P}_2}{\sqrt{\hat{P}_c(1 - \hat{P}_c)\left(\frac{1}{n_1} + \frac{1}{n_2}\right)}}; \text{ RR: } Z \leq -2.3263$$

$z = -3.7319 \leq -2.3263$; $p = 0.0001 \leq 0.01$. There is evidence to suggest the population proportion of 18- to 29-year-olds who obtain news every day from cable news shows is less than the population proportion of 30- to 49-year-olds who obtain news every day from cable news shows.

c. $H_0: p_1 - p_2 = 0$; $H_a: p_1 - p_2 \neq 0$;

$$\text{TS: } Z = \frac{\hat{P}_1 - \hat{P}_2}{\sqrt{\hat{P}_c(1 - \hat{P}_c)\left(\frac{1}{n_1} + \frac{1}{n_2}\right)}}; \text{ RR: } |Z| \geq 2.8070$$

$z = -2.0501$; $|z| < 2.8070$; $p = 0.0404 > 0.005$. There is no evidence to suggest that the population proportion of 18- to 29-year-olds who obtain news every day from the Internet is different from the population proportion of 30- to 49-year-olds who obtain news every day from the Internet.

10.121 $H_0: p_1 - p_2 = 0.10$; $H_a: p_1 - p_2 > 0.10$;

$$\text{TS: } Z = \frac{(\hat{P}_1 - \hat{P}_2) - \Delta_0}{\sqrt{\frac{\hat{P}_1(1 - \hat{P}_1)}{n_1} + \frac{\hat{P}_2(1 - \hat{P}_2)}{n_2}}} \text{ RR: } Z \geq 2.3263$$

$z = 0.8038 < 2.3263$; $p = 0.2108 > 0.01$. There is no evidence to suggest that the population proportion of homeowners planning a landscaping project is more than 0.10 greater than the population proportion of condominium owners planning a landscaping project.

Section 10.5

10.123 False.

10.125 False.

10.127 The value of S_1^2 is, on average, σ_1^2. The value of S_2^2 is, on average, σ_2^2.

10.129 a. 2.20 **b.** 3.15 **c.** 3.58 **d.** 4.99 **e.** 0.23 **f.** 0.12 **g.** 0.34 **h.** 0.25

10.131 a. $H_0: \sigma_1^2 = \sigma_2^2$; $H_a: \sigma_1^2 > \sigma_2^2$;
TS: $F = S_1^2/S_2^2$; RR: $F \geq 1.94$
b. $f = 2.5448 \geq 1.94$. There is evidence to suggest that population variance 1 is greater than population variance 2.
c. $0.01 \leq p \leq 0.05$. p-Value illustration:

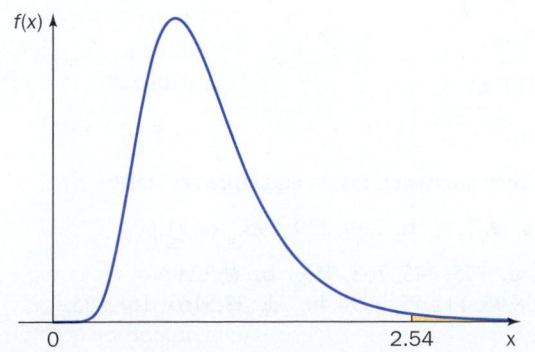

10.133 a. $H_0: \sigma_1^2 = \sigma_2^2$; $H_a: \sigma_1^2 \neq \sigma_2^2$;
TS: $F = S_1^2/S_2^2$; RR: $F \leq 0.17$ or $F \geq 4.54$
b. $f = 4.8259 \geq 4.54$; $p = 0.0074$. There is evidence to suggest that population variance 1 is different from population variance 2. **c.** $0.002 \leq p \leq 0.02$

10.135 a. (0.3254, 3.5608) **b.** (0.4712, 5.8451)
c. (0.2703, 2.3458) **d.** (0.2843, 2.5079)

10.137 $H_0: \sigma_1^2 = \sigma_2^2$; $H_a: \sigma_1^2 > \sigma_2^2$; TS: $F = S_1^2/S_2^2$;
RR: $F \geq 2.39$
$f = 4.6944 \geq 2.39$; $p = 0.0019 \leq 0.05$. There is evidence to suggest that the population variance in the aerosol light absorption coefficient is greater in Africa than in South America.

10.139 $H_0: \sigma_1^2 = \sigma_2^2$; $H_a: \sigma_1^2 \neq \sigma_2^2$;
TS: $F = S_1^2/S_2^2$; RR: $F \leq 0.43$ or $F \geq 2.41$
$f = 0.4960$; $0.43 < f < 2.41$; $p = 0.1025 > 0.05$. There is no evidence to suggest that the population variance in checked-baggage weight is different for these two airlines.

10.141 a. $H_0: \sigma_1^2 = \sigma_2^2$; $H_a: \sigma_1^2 > \sigma_2^2$;
TS: $F = S_1^2/S_2^2$; RR: $F \geq 2.60$
$f = 2.8681 \geq 2.60$; $p = 0.0054 \leq 0.01$. There is evidence to suggest that the population variance in winning times for an ordinary race is greater than the population variance in winning times for a stakes race. **b.** (1.1014, 7.4689)

10.143 a. $H_0: \sigma_1^2 = \sigma_2^2$; $H_a: \sigma_1^2 \neq \sigma_2^2$;
TS: $F = S_1^2/S_2^2$; RR: $F \leq 0.40$ or $F \geq 2.53$
$f = 0.5784$; $0.40 < f < 2.53$. There is no evidence to suggest a difference in the population variance of times for men and women who finished this vertical run.
b. $p \geq 0.05$; $p \approx 0.2418$

10.145 $H_0: \sigma_1^2 = \sigma_2^2$; $H_a: \sigma_1^2 < \sigma_2^2$; TS: $F = S_1^2/S_2^2$;
RR: $F \leq 0.42$
$f = 0.3237 \leq 0.42$; $p = 0.0014 \leq 0.01$. There is evidence to suggest that the variability in item weight at Jefferson is greater that at New York Avenue.

10.147 $H_0: \sigma_1^2 = \sigma_2^2$; $H_a: \sigma_1^2 \neq \sigma_2^2$;
TS: $F = S_1^2/S_2^2$; RR: $F \leq 0.30$ or $F \geq 3.83$
$f = 1.1104$; $0.30 < f < 3.83$; $p = 0.8976 > 0.05$. There is no evidence to suggest a difference in the variability of the span of suspension bridges in China and the United States.

10.149 a. $H_0: \sigma_1^2 = \sigma_2^2$; $H_a: \sigma_1^2 \neq \sigma_2^2$;
TS: $F = S_1^2/S_2^2$; RR: $F \leq 0.31$ or $F \geq 3.53$
$f = 2.7128$; $0.31 < f < 3.53$. There is no evidence to suggest that the population variance in saccharin amount for the two teas is different. **b.** $0.02 \leq p \leq 0.10$; $p \approx 0.0621$

10.151 a. $H_0: \sigma_1^2 = \sigma_2^2$; $H_a: \sigma_1^2 \neq \sigma_2^2$;

$$\text{TS: } Z = \frac{(S_1^2/S_2^2) - [(n_2 - 1)/(n_2 - 3)]}{\sqrt{\frac{2(n_2 - 1)^2(n_1 + n_2 - 4)}{(n_1 - 1)(n_2 - 3)^2(n_2 - 5)}}}; \text{ RR: } |Z| \geq 1.96$$

$z = 1.7059$; $|z| < 1.96$; $p = 0.0880 > 0.05$. There is no evidence to suggest that the two population variances are different. **b.** $H_0: \sigma_1^2 = \sigma_2^2$; $H_a: \sigma_1^2 \neq \sigma_2^2$;
TS: $F = S_1^2/S_2^2$; RR: $F \leq 0.48$ or $F \geq 2.07$

$f = 1.7763$; $0.48 < f < 2.07$; $p = 0.1212 > 0.05$. There is no evidence to suggest that the two population variances are different.

Chapter Exercises

10.153 $H_0: p_1 - p_2 = 0$; $H_a: p_1 - p_2 > 0$;

TS: $Z = \dfrac{\hat{P}_1 - \hat{P}_2}{\sqrt{\hat{P}_c(1 - \hat{P}_c)\left(\frac{1}{n_1} + \frac{1}{n_2}\right)}}$; RR: $Z \geq 2.3263$

$z = 3.5139 \geq 2.3263$; $p = 0.0002 \leq 0.01$. There is evidence to suggest that the true proportion of chest hits is greater in Lancashire than in West Mercia.

10.155 $H_0: \mu_1 - \mu_2 = 0$; $H_a: \mu_1 - \mu_2 \neq 0$;

TS: $T = \dfrac{(\overline{X}_1 - \overline{X}_2) - 0}{\sqrt{S_p^2\left(\frac{1}{n_1} + \frac{1}{n_2}\right)}}$ RR: $|T| \geq 2.7045$

$t = 1.3083$; $|t| < 2.7045$; $p = 0.1982 > 0.01$. There is no evidence to suggest that the mean amount of sap from trees in New York is different from the mean amount of sap from trees in Vermont.

10.157 a. Archer. **b.** $H_0: \mu_D = 0$; $H_a: \mu_D < 0$;

TS: $T = \dfrac{\overline{D} - \Delta_0}{S_D/\sqrt{n}}$; RR: $T \leq -1.7959$

$t = -1.1829 > -1.7959$; $p = 0.1309 > 0.05$. There is no evidence to suggest that the population mean speed of a carbon arrow is less than the population mean speed of an aluminum arrow. **c.** $0.10 \leq p \leq 0.20$.

10.159 $H_a: \mu_1 - \mu_2 = 0$; $H_a: \mu_1 - \mu_2 > 0$;

TS: $T = \dfrac{(\overline{X}_1 - \overline{X}_2) - 0}{\sqrt{S_1^2\left(\frac{1}{n_1} + \frac{1}{n_2}\right)}}$ RR: $T \geq 2.4286$

$t = 2.1140 < 2.4286$; $p = 0.0206 > 0.01$. There is no evidence to suggest that the mean pier length in California is greater than the mean pier length in Florida.

10.161 $H_0: \mu_D = 0$; $H_a: \mu_D \neq 0$; TS: $T = \dfrac{\overline{D} - \Delta_0}{S_D/\sqrt{n}}$;

RR: $|T| \geq 2.0930$

$t = 0.4957$; $|t| < 2.0930$; $p = 0.6258 > 0.05$. There is no evidence to suggest that the population mean moisture content of bulk grain measured by chemical reaction and by distillation is different.

10.163 a. $\hat{p}_1 = 0.1143$, $\hat{p}_2 = 0.0952$.
$n_1\hat{p}_1 = 16 \geq 5$, $n_1(1 - \hat{p}_1) = 124 \geq 5$,
$n_2\hat{p}_2 = 12 \geq 5$, $n_2(1 - \hat{p}_2) = 114 \geq 5$
b. $H_0: p_1 - p_2 = 0$; $H_a: p_1 - p_2 \neq 0$;

TS: $Z = \dfrac{\hat{P}_1 - \hat{P}_2}{\sqrt{\hat{P}_c(1 - \hat{P}_c)\left(\frac{1}{n_1} + \frac{1}{n_2}\right)}}$; RR: $|Z| \geq 2.5758$

$z = 0.5054$; $|z| < 2.5758$; $p = 0.6133 > 0.01$. There is no evidence to suggest that the population proportion of stations in noncompliance with the law is different near Los Angeles and near San Francisco.

10.165 a. $H_0: \sigma_1^2 = \sigma_2^2$; $H_a: \sigma_1^2 \neq \sigma_2^2$;
TS: $F = S_1^2/S_2^2$; RR: $F \leq 0.29$ or $F \geq 3.78$

$f = 0.1038 \leq 0.29$; $p = 0.0001 \leq 0.02$. There is evidence to suggest that the two population variances are different.
b. $H_0: \mu_1 - \mu_2 = 0$; $H_a: \mu_1 - \mu_2 < 0$;

TS: $T' = \dfrac{(\overline{X}_1 - \overline{X}_2) - 0}{\sqrt{\left(\frac{S_1^2}{n_1} + \frac{S_2^2}{n_2}\right)}}$; RR: $T' \leq -1.7531$

$t' = -1.8697 \leq -1.7531$; $p = 0.0406 \leq 0.05$. There is evidence to suggest that the population mean time to complete Form 1040 for the lower income level is less than the population mean time to complete Form 1040 for the higher income level. $0.025 \leq p \leq 0.05$.

10.167 a. $\hat{p}_1 = 0.3793$, $\hat{p}_2 = 0.2876$.
$n_1\hat{p}_1 = 132 \geq 5$, $n_1(1 - \hat{p}_1) = 216 \geq 5$,
$n_2\hat{p}_2 = 65 \geq 5$, $n_2(1 - \hat{p}_2) = 161 \geq 5$
b. $(0.0137, 0.1697)$ **c.** There is evidence to suggest that the population proportion of online investors is different for these two portfolio classifications. 0 is not in the CI.

10.169 For each test:
$H_a: \mu_1 - \mu_2 = 0$; $H_a: \mu_1 - \mu_2 \neq 0$; TS: $T = \dfrac{(\overline{X}_1 - \overline{X}_2) - 0}{\sqrt{S_p^2\left(\frac{1}{n_1} + \frac{1}{n_2}\right)}}$
Single detached versus Double/duplex:
RR: $|T| \geq 2.7633$. $t = -1.5545$; $|t| < 2.7633$;
$p = 0.1313 > 0.01$. There is no evidence to suggest that the population mean toluene levels are different.
Single detached versus Apartment:
RR: $|T| \geq 2.7564$. $t = -3.2994$; $|t| \geq 2.7564$;
$p = 0.0026 \leq 0.01$. There is evidence to suggest that the population mean toluene levels are different.
Double/duplex versus Apartment:
RR: $|T| \geq 2.7564$. $t = -2.1737$; $|t| < 2.7564$;
$p = 0.0380 > 0.01$. There is no evidence to suggest that the population mean toluene levels are different.

10.171 $H_0: p_1 - p_2 = 0$; $H_a: p_1 - p_2 < 0$;

TS: $Z = \dfrac{\hat{P}_1 - \hat{P}_2}{\sqrt{\hat{P}_c(1 - \hat{P}_c)\left(\frac{1}{n_1} + \frac{1}{n_2}\right)}}$; RR: $Z \geq -1.6449$

$z = -1.7661 \leq -1.6449$; $p = 0.0387 \leq 0.05$. There is evidence to suggest that the population proportion of obese Denver residents is less than the population proportion of obese New Orleans residents. This suggests that the adults living in Denver are thinner.

CHAPTER 11

Section 11.1

11.1 True.

11.3 False.

11.5 σ^2

11.7 They are easier, faster, and more accurate.

11.9 a. 9, 7, 6 **b.** 260, 229, 195 **c.** 21,602

11.11 a. 775, 745, 768, 753 **b.** 465,193
c. 2808.95, 112.55, 2696.40 **d.** 37.5167, 168.525
e. 0.2226

11.13
ANOVA summary table:

Source of variation	Sum of squares	Degrees of freedom	Mean square	F	p-Value
Factor	584.1	4	145.0250	0.57	0.6857
Error	12,062.1	47	256.6404		
Total	12,646.2	51			

11.15 a. $H_0: \mu_1 = \mu_2 = \mu_3 = \mu_4$;
$H_a: \mu_i \neq \mu_j$ for some $i \neq j$;
TS: $F = $ MSA/MSE; RR: $F \geq 3.10$

b.
ANOVA summary table:

Source of variation	Sum of squares	Degrees of freedom	Mean square	F	p-Value
Factor	67,000.33	3	22,333.44	12.39	0.0001
Error	36,039.67	20	1,801.98		
Total	103,040.00	23			

c. $f = 12.39 \geq 3.10$; $p = 0.0001 \leq 0.05$. There is evidence to suggest that at least two of the population means are different.

11.17 $H_0: \mu_1 = \mu_2 = \mu_3 = \mu_4 = \mu_5$;
$H_a: \mu_i \neq \mu_j$ for some $i \neq j$;
TS: $F = $ MSA/MSE; RR: $F \geq 2.48$
$f = 7.35 \geq 2.48$; $p = 0.00004 \leq 0.01$. There is evidence to suggest that at least two population mean times are different.

11.19 $H_0: \mu_1 = \mu_2 = \mu_3$;
$H_a: \mu_i \neq \mu_j$ for some $i \neq j$;
TS: $F = $ MSA/MSE; RR: $F \geq 5.19$
$f = 6.35 \geq 5.19$; $p = 0.0041 \leq 0.01$. There is evidence to suggest that at least two of the population mean weights are different.

11.21 $H_0: \mu_1 = \mu_2 = \mu_3 = \mu_4$;
$H_a: \mu_i \neq \mu_j$ for some $i \neq j$;
TS: $F = $ MSA/MSE; RR: $F \geq 4.09$
$f = 6.58 \geq 4.09$; $p = 0.0006 \leq 0.01$. There is evidence to suggest that at least two of the population mean sulfur dioxide emission rates are different.

11.23 $H_0: \mu_1 = \mu_2 = \mu_3 = \mu_4 = \mu_5$;
$H_a: \mu_i \neq \mu_j$ for some $i \neq j$;
TS: $F = $ MSA/MSE; RR: $F \geq 2.69$
$f = 2.91 \geq 2.69$; $p = 0.0380 \leq 0.05$. There is evidence to suggest that at least two of the population mean daily production values are different.

11.25 $H_0: \mu_1 = \mu_2 = \mu_3 = \mu_4$;
$H_a: \mu_i \neq \mu_j$ for some $i \neq j$;
TS: $F = $ MSA/MSE; RR: $F \geq 3.10$
$f = 0.81 < 3.10$; $p = 0.5050 > 0.05$. There is no evidence to suggest that at least two population mean thaw depths are different.

11.27 $H_0: \mu_1 = \mu_2 = \mu_3 = \mu_4 = \mu_5$;
$H_a: \mu_i \neq \mu_j$ for some $i \neq j$;
TS: $F = $ MSA/MSE; RR: $F \geq 2.58$
$f = 6.01 \geq 2.58$; $p = 0.0006 \leq 0.05$. There is evidence to suggest that at least two population mean flash points are different.

11.29 a. $H_0: \mu_1 = \mu_2 = \mu_3 = \mu_4$;
$H_a: \mu_i \neq \mu_j$ for some $i \neq j$;
TS: $F = $ MSA/MSE; RR: $F \geq 3.01$
$f = 3.40 \geq 3.01$; $p = 0.0339 \leq 0.05$. There is evidence to suggest that at least two of the population mean pressures are different. **b.** Recommend Holder, because this broom has the highest mean pressure.

11.31 a. $H_0: \mu_1 = \mu_2 = \mu_3 = \mu_4 = \mu_5$;
$H_a: \mu_i \neq \mu_j$ for some $i \neq j$;
TS: $F = $ MSA/MSE; RR: $F \geq 5.36$
$f = 21.58 \geq 5.36$; $p = 0.00000000008 \leq 0.001$. There is evidence to suggest that at least two population mean weights are different.
b. Buy at Weis, because these bags have the largest sample mean weight.

11.33 a. $H_a: \mu_1 - \mu_2 = 0$; $H_a: \mu_1 - \mu_2 \neq 0$;
TS: $T = \dfrac{(\overline{X}_1 - \overline{X}_2) - 0}{\sqrt{S_p^2 \left(\frac{1}{n_1} + \frac{1}{n_2} \right)}}$; RR: $|T| \geq 2.0739$
$t = -2.3462$; $|t| > 2.0739$; $p = 0.0284 \leq 0.05$. There is evidence to suggest that the population mean widths are different.
b. $H_0: \mu_1 = \mu_2$;
$H_a: \mu_i \neq \mu_j$ for some $i \neq j$;
TS: $F = $ MSA/MSE; RR: $F \geq 4.30$
$f = 5.50 \geq 4.30$; $p = 0.0284 \leq 0.05$. There is evidence to suggest that the population mean widths are different.
c. $t^2 = f$. The p values are the same. This makes sense, because these two procedures are testing the exact same hypothesis.

Section 11.2

11.35 True.

11.37 False.

11.39 $\dfrac{k(k-1)}{2}$

11.41 There is no evidence to suggest any difference among population means.

11.43 a. 3.609 **b.** 3.791 **c.** 4.634 **d.** 4.863 **e.** 6.469

11.45 a.

$\overline{x}_{1.}$	$\overline{x}_{2.}$	$\overline{x}_{3.}$	$\overline{x}_{4.}$
-33.44	-14.83	0.48	4.30

b.

$\overline{x}_{3.}$	$\overline{x}_{2.}$	$\overline{x}_{4.}$	$\overline{x}_{1.}$
1.30	1.41	1.50	1.62

c.

$\bar{x}_{4\cdot}$	$\bar{x}_{2\cdot}$	$\bar{x}_{3\cdot}$	$\bar{x}_{1\cdot}$	$\bar{x}_{5\cdot}$
51.92	54.21	60.80	64.35	64.85

11.47

Difference	Bonferroni CI	Significantly different
$\mu_1 - \mu_2$	(−7.62, 3.64)	No
$\mu_1 - \mu_3$	(−2.54, 9.72)	No
$\mu_1 - \mu_4$	(−11.80, −0.54)	Yes
$\mu_2 - \mu_3$	(−0.55, 10.71)	No
$\mu_2 - \mu_4$	(−9.81, 1.45)	No
$\mu_3 - \mu_4$	(−14.89, −3.63)	Yes

11.49

Difference	Bonferroni CI	Significantly different
$\mu_1 - \mu_2$	(−573.84, 3.28)	No
$\mu_1 - \mu_3$	(−586.20, −24.88)	Yes
$\mu_2 - \mu_3$	(−285.67, 245.15)	No

11.51 a. $0.001 \le p \le 0.01$. There is evidence to suggest that at least two population means are different.
b.

Difference	Bonferroni CI	Significantly different
$\mu_1 - \mu_2$	(0.44, 7.35)	Yes
$\mu_1 - \mu_3$	(−3.56, 3.36)	No
$\mu_1 - \mu_4$	(−1.86, 5.06)	No
$\mu_2 - \mu_3$	(−7.45, −0.54)	Yes
$\mu_2 - \mu_4$	(−5.76, 1.16)	No
$\mu_3 - \mu_4$	(−1.76, 5.16)	No

11.53 a. $p < 0.0001$
b.

Difference	Bonferroni CI	Significantly different
$\mu_1 - \mu_2$	(0.87, 1.25)	Yes
$\mu_1 - \mu_3$	(0.56, 0.95)	Yes
$\mu_1 - \mu_4$	(0.20, 0.59)	Yes
$\mu_2 - \mu_3$	(−0.49, −0.12)	Yes
$\mu_2 - \mu_4$	(−0.85, −0.48)	Yes
$\mu_3 - \mu_4$	(−0.55, 0.17)	Yes

$\bar{x}_{2\cdot}$	$\bar{x}_{3\cdot}$	$\bar{x}_{4\cdot}$	$\bar{x}_{1\cdot}$
0.2567	0.5601	0.9206	1.3164

11.55 a. $H_0: \mu_1 = \mu_2 = \mu_3 = \mu_4$;
$H_a: \mu_i \ne \mu_j$ for some $i \ne j$;
TS: $F = MSA/MSE$; RR: $F \ge 3.10$
$f = 10.03 \ge 3.10$; $p = 0.0003 \le 0.05$. There is evidence to suggest that at least two population means are different.

b.

Difference	Bonferroni CI	Significantly different
$\mu_1 - \mu_2$	(−15.05, 0.31)	No
$\mu_1 - \mu_3$	(−6.53, 8.83)	No
$\mu_1 - \mu_4$	(−18.76, −3.40)	Yes
$\mu_2 - \mu_3$	(0.84, 16.20)	Yes
$\mu_2 - \mu_4$	(−11.40, 3.96)	No
$\mu_3 - \mu_4$	(−19.91, −4.55)	Yes

c.

$\bar{x}_{3\cdot}$	$\bar{x}_{1\cdot}$	$\bar{x}_{2\cdot}$	$\bar{x}_{4\cdot}$
65.55	66.70	74.07	77.78

11.57 a. $H_0: \mu_1 = \mu_2 = \mu_3$;
$H_a: \mu_i \ne \mu_j$ for some $i \ne j$;
TS: $F = MSA/MSE$; RR: $F \ge 3.32$
$f = 30.85 \ge 3.32$; $p = 0.00000005 \le 0.05$. There is evidence to suggest that at least two population means are different.

b.

Difference	Tukey CI	Significantly different
$\mu_1 - \mu_2$	(32.84, 63.58)	Yes
$\mu_1 - \mu_3$	(17.74, 49.10)	Yes
$\mu_2 - \mu_3$	(−29.77, 0.19)	No

c.

$\bar{x}_{2\cdot}$	$\bar{x}_{3\cdot}$	$\bar{x}_{1\cdot}$
67.80	82.59	116.01

11.59 a. $H_0: \mu_1 = \mu_2 = \mu_3 = \mu_4 = \mu_5$;
$H_a: \mu_i \ne \mu_j$ for some $i \ne j$;
TS: $F = MSA/MSE$; RR: $F \ge 4.43$
$f = 12.82 \ge 4.43$; $p = 0.00002 \le 0.01$. There is evidence to suggest that at least two population means are different.

b.

Difference	Bonferroni CI	Significantly different
$\mu_1 - \mu_2$	(−4.66, 0.54)	No
$\mu_1 - \mu_3$	(−3.36, 1.84)	No
$\mu_1 - \mu_4$	(−4.36, 0.84)	No
$\mu_1 - \mu_5$	(−8.04, −2.84)	Yes
$\mu_2 - \mu_3$	(−1.30, 3.90)	No
$\mu_2 - \mu_4$	(−2.30, 2.90)	No
$\mu_2 - \mu_5$	(−5.98, −0.78)	Yes
$\mu_3 - \mu_4$	(−3.60, 1.60)	No
$\mu_3 - \mu_5$	(−7.28, −2.08)	Yes
$\mu_4 - \mu_5$	(−6.28, −1.08)	Yes

c.

$\bar{x}_{1\cdot}$	$\bar{x}_{3\cdot}$	$\bar{x}_{4\cdot}$	$\bar{x}_{2\cdot}$	$\bar{x}_{5\cdot}$
14.28	15.04	16.04	16.34	19.72

11.61 $H_0: \mu_1 = \mu_2 = \mu_3 = \mu_4$;
$H_a: \mu_i \ne \mu_j$ for some $i \ne j$;

TS: $F = $ MSA/MSE; RR: $F \geq 4.26$
$f = 19.51 \geq 4.26$; $p = 0.00000003 \leq 0.01$. There is evidence to suggest that at least two population means are different.

b.

Difference	Tukey CI	Significantly different
$\mu_1 - \mu_2$	(0.28, 1.45)	Yes
$\mu_1 - \mu_3$	(−0.29, 0.88)	No
$\mu_1 - \mu_4$	(0.64, 1.81)	Yes
$\mu_2 - \mu_3$	(−1.16, −0.01)	No
$\mu_2 - \mu_4$	(−0.23, 0.94)	No
$\mu_3 - \mu_4$	(0.35, 1.52)	Yes

c. The inconsistent results are between groups 2 and 3, and groups 2 and 4. We would expect evidence to suggest these pairs of population means are different. Note that the Tukey CI for $\mu_2 - \mu_3$ just barely includes 0.

11.63 H_0: $\mu_1 = \mu_2 = \mu_3 = \mu_4$;
H_a: $\mu_i \neq \mu_j$ for some $i \neq j$;
TS: $F = $ MSA/MSE; RR: $F \geq 2.76$
$f = 6.31 \geq 2.76$; $p = 0.0012 \leq 0.05$. There is evidence to suggest that at least two of the population mean attenuation values are different.

b.

Difference	Bonferroni CI	Significantly different
$\mu_1 - \mu_2$	(−1.623, −0.384)	Yes
$\mu_1 - \mu_3$	(−0.768, 0.471)	No
$\mu_1 - \mu_4$	(−1.061, 0.178)	No
$\mu_1 - \mu_5$	(−1.333, −0.094)	Yes

Polystyrene and PVC are significantly different from the control.

Section 11.3

11.65 False.

11.67 a. $(a-1)(b-1)$, $ab(n-1)$ **b.** $a-1$, $ab(n-1)$ **c.** $b-1$, $ab(n-1)$

11.69 a. 2, 3, 3 **b.** $t_{11.} = 21$, $t_{12.} = 31$, $t_{13.} = 37$, $t_{21.} = 28$, $t_{22.} = 32$, $t_{23.} = 48$ **c.** $t_{1..} = 89$, $t_{2..} = 108$, $t_{.1.} = 49$, $t_{.2.} = 63$, $t_{.3.} = 85$

11.71 a. SST = 481.880, SSA = 160.820, SSB = 76.827, SS(AB) = 45.480, SSE = 198.753 **b.** MSA = 53.6067, MSB = 38.4133, MS(AB) = 7.5800, MSE = 8.2814 **c.** $f_A = 6.47$, $f_B = 4.64$, and $f_{AB} = 0.92$ **d.** $f_{AB} = 0.92 < 2.51$; $p = 0.5010 > 0.05$. There is no evidence of interaction.
$f_A = 6.47 \geq 3.01$; $p = 0.0023 \leq 0.05$. There is evidence of an effect due to factor A.
$f_B = 4.64 \geq 3.40$; $p = 0.0198 \leq 0.05$. There is evidence of an effect due to factor B.

11.73

Source of variation	Sum of squares	Degrees of freedom	Mean square	F	p-Value
Factor A	162.64	4	40.66	2.53	0.0476
Factor B	156.54	2	78.27	4.86	0.0103
Interaction	144.23	8	18.03	1.12	0.3596
Error	1206.87	75	16.09		
Total	1670.28	89			

$f_{AB} = 1.12 < 2.06$; $p = 0.3596 > 0.05$; there is no evidence of interaction. $f_A = 2.53 \geq 2.49$; $p = 0.0476 \leq 0.05$; there is evidence of an effect due to factor A. $f_B = 4.86 \geq 3.12$; $p = 0.0103 \leq 0.05$; there is evidence of an effect due to factor B.

11.75 a.

Source of variation	Sum of squares	Degrees of freedom	Mean square	F	p-Value
Type	90.14	2	45.07	3.02	0.0580
Restaurant	266.49	3	88.83	5.96	0.0015
Interaction	56.59	6	9.43	0.63	0.7034
Error	715.60	48	14.91		
Total	1128.82	59			

b. 60 **c.** $f_{AB} = 0.63 < 2.29$; $p = 0.7034 > 0.05$; there is no evidence of interaction. The other two hypothesis tests can be conducted as usual. **d.** $f_A = 3.02 < 3.19$; $p = 0.0580 > 0.05$; there is no evidence of an effect due to type. $f_B = 5.96 \geq 2.80$; $p = 0.0015 \leq 0.05$; there is evidence of an effect due to restaurant.

11.77 a.

Source of variation	Sum of squares	Degrees of freedom	Mean square	F	p-Value
Flock size	93.84	1	93.84	4.85	0.0375
Temperature	29.73	3	9.91	0.51	0.6779
Interaction	6.06	3	2.02	0.10	0.9567
Error	464.55	24	19.36		
Total	594.19	31			

b. $f_{AB} = 0.10 < 3.01$; $p = 0.9567 > 0.05$; there is no evidence of interaction. $f_A = 4.85 > 4.26$; $p = 0.0375 \leq 0.05$; there is evidence of an effect due to flock size. $f_B = 0.51 < 3.01$; $p = 0.6779 > 0.05$; there is no evidence of an effect due to temperature.

11.79

Source of variation	Sum of squares	Degrees of freedom	Mean square	F	p-Value
Office	30.25	2	15.13	0.40	0.6761
Type	216.00	1	216.00	5.66	0.0286
Interaction	2.25	2	1.13	0.03	0.9705
Error	687.50	18	38.19		
Total	936.00	23			

$f_{AB} = 0.03 < 3.89; p = 0.9710 > 0.05$; there is no evidence of interaction. $f_A = 0.40 < 3.89; p = 0.6787 > 0.05$; there is no evidence of an effect due to office. $f_B = 5.66 \geq 4.75; p = 0.0287 \leq 0.05$; there is evidence of an effect due to type of development.

11.81 a.

Source of variation	Sum of squares	Degrees of freedom	Mean square	F	p-Value
Cover type	1.54	3	0.51	0.87	0.4770
State	3.12	3	1.04	1.76	0.1953
Interaction	7.11	9	0.79	1.34	0.2919
Error	9.45	16	0.59		
Total	21.22	31			

$f_{AB} = 1.34 < 3.78; p = 0.2934 > 0.01$; there is no evidence of interaction.
b. $f_A = 0.87 < 5.29; p = 0.4774 > 0.01$; there is no evidence of an effect due to cover type. $f_B = 1.76 < 5.29; p = 0.1954 > 0.01$; there is no evidence of an effect due to state.

11.83 $f_{AB} = 1.68 < 2.36; p = 0.1542 > 0.05$; there is no evidence of interaction. $f_A = 7.23 > 3.26$; $p = 0.0023 \leq 0.05$; there is evidence of an effect due to region. $f_B = 19.71 \geq 2.87; p < 0.0001$; there is evidence of an effect due to road marking quality.

11.85 a.

Source of variation	Sum of squares	Degrees of freedom	Mean square	F	p-Value
Gender	261.33	1	261.33	10.52	0.0024
Altitude	241.17	3	80.39	3.24	0.0321
Interaction	106.17	3	35.39	1.43	0.2497
Error	993.33	40	24.83		
Total	1602.00	47			

$f_{AB} = 1.43 < 2.84; p = 0.2497 > 0.05$; there is no evidence of interaction.
b. $f_A = 10.52 \geq 4.08; p = 0.0024 \leq 0.05$; there is evidence of an effect due to gender. $f_B = 3.24 \geq 2.84$; $p = 0.0321 \leq 0.05$; there is evidence of an effect due to altitude.

11.87 a.

Source of variation	Sum of squares	Degrees of freedom	Mean square	F	p-Value
ADHD	72.45	1	72.45	57.75	< 0.0001
County	38.72	17	2.28	1.82	0.0290
Interaction	15.70	17	0.92	0.74	0.7627
Error	225.82	180	1.25		
Total	352.69	215			

$f_{AB} = 0.74 < 1.68; p = 0.7363 > 0.05$; there is no evidence of interaction.
b. $f_A = 57.75 \geq 3.89; p < 0.0001$; there is evidence of an effect due to ADHD diagnosis. $f_B = 1.82 > 1.68$; $p = 0.0290 \leq 0.05$; there is evidence of an effect due to county.
c. Diagnosis of ADHD is associated with an increase in age at diagnosis for ASD. The mean age of all those not diagnosed with ADHD was 3.72, and with ADHD, 4.30.

Chapter Exercises

11.89

Source of variation	Sum of squares	Degrees of freedom	Mean square	F	p-Value
Factor	106,568	3	35,522.67	4.67	0.0124
Error	151,997	20	7,599.85		
Total	258,565	23			

$f = 4.67 \geq 3.10; p = 0.0124 \leq 0.05$; there is evidence to suggest that at least two of the population mean wattages of spotlights are different.

11.91

Difference	Bonferroni CI	Significantly different
$\mu_1 - \mu_2$	(−4.38, 8.58)	No
$\mu_1 - \mu_3$	(−4.08, 8.88)	No
$\mu_1 - \mu_4$	(1.62, 14.58)	Yes
$\mu_2 - \mu_3$	(−6.18, 6.78)	No
$\mu_2 - \mu_4$	(−0.48, 12.48)	No
$\mu_3 - \mu_4$	(−0.78, 12.18)	No

$\bar{x}_{4.}$	$\bar{x}_{3.}$	$\bar{x}_{2.}$	$\bar{x}_{1.}$
98.2	104.3	104.6	106.7

11.93 a. $H_0: \mu_1 = \mu_2 = \mu_3 = \mu_4$;
$H_a: \mu_i \neq \mu_j$ for some $i \neq j$;
TS: $F = MSA/MSE$; RR: $F \geq 3.24$
$f = 6.23 \geq 3.24; p = 0.0052 \leq 0.05$. There is evidence to suggest that at least two of the population mean step heights are different.

b.

Difference	Bonferroni CI	Significantly different
$\mu_1 - \mu_2$	(0.0139, 0.2892)	Yes
$\mu_1 - \mu_3$	(−0.0346, 0.2407)	No
$\mu_1 - \mu_4$	(−0.0276, 0.2477)	No
$\mu_2 - \mu_3$	(−0.1862, 0.0892)	No
$\mu_2 - \mu_4$	(−0.1792, 0.0962)	No
$\mu_3 - \mu_4$	(−0.1307, 0.1447)	No

$\bar{x}_{2.}$	$\bar{x}_{4.}$	$\bar{x}_{3.}$	$\bar{x}_{1.}$
0.2299	0.2714	0.2784	0.3815

11.95

Source of variation	Sum of squares	Degrees of freedom	Mean square	F	p-Value
Lanes	735.06	3	245.02	16.78	< 0.0001
Diff Elev	113.56	3	37.85	2.59	0.0635
Interaction	325.81	9	36.20	2.48	0.0206
Error	701.00	48	14.60		
Total	1875.44	63			

$f_{AB} = 2.48 \geq 2.08$; $p = 0.0206 \leq 0.05$; there is evidence of interaction.

$f_A = 16.78 \geq 2.80$; $p < 0.0001$; there is probably an effect due to lanes.

$f_B = 2.59 < 2.80$; $p = 0.0635 > 0.05$; there is interaction present, so the effect due to differential elevation is inconclusive.

11.97 a. $H_0: \mu_1 = \mu_2 = \mu_3 = \mu_4 = \mu_5$;
$H_a: \mu_i \neq \mu_j$ for some $i \neq j$;
TS: $F = MSA/MSE$; RR: $F \geq 3.79$
$f = 4.07 \geq 3.79$; $p = 0.0069 \leq 0.05$. There is evidence to suggest that at least two of the population mean catfish weights are different.

b.

Difference	Tukey CI	Significantly different
$\mu_1 - \mu_2$	(−33.32, 2.63)	No
$\mu_1 - \mu_3$	(−33.24, 3.46)	No
$\mu_1 - \mu_4$	(−36.41, −1.09)	Yes
$\mu_1 - \mu_5$	(−38.76, 1.28)	No
$\mu_2 - \mu_3$	(−16.45, 17.36)	No
$\mu_2 - \mu_4$	(−19.55, 12.74)	No
$\mu_2 - \mu_5$	(−22.10, 15.30)	No
$\mu_3 - \mu_4$	(−20.42, 12.70)	No
$\mu_3 - \mu_5$	(−22.91, 15.21)	No
$\mu_4 - \mu_5$	(−18.39, 18.40)	No

$\bar{x}_{1.}$	$\bar{x}_{3.}$	$\bar{x}_{2.}$	$\bar{x}_{5.}$	$\bar{x}_{4.}$
20.70	35.59	36.05	39.44	39.45

Recommend Campground. On average, the largest catfish are caught at this location.

11.99 $f_{AB} = 1.78 < 2.64$; $p = 0.0850 > 0.01$; there is no evidence of interaction. $f_A = 7.59 \geq 4.04$; $p = 0.0002 \leq 0.01$; there is evidence of an effect due to island. $f_B = 1.92 < 4.04$; $p = 0.1330 > 0.01$; there is no evidence of an effect due to season.

11.101 a. $H_0: \mu_1 = \mu_2 = \mu_3 = \mu_4 = \mu_5 = \mu_6$;
$H_a: \mu_i \neq \mu_j$ for some $i \neq j$;
TS: $F = MSA/MSE$; RR: $F \geq 3.18$
$f = 5.18 \geq 3.18$; $p = 0.0003 \leq 0.01$. There is evidence to suggest that at least two population mean glutathione levels are different.

b.

Difference	Tukey CI	Significantly different
$\mu_1 - \mu_2$	(1.81, 50.89)	Yes
$\mu_1 - \mu_3$	(−15.89, 33.19)	No
$\mu_1 - \mu_4$	(−3.39, 45.69)	No
$\mu_1 - \mu_5$	(14.36, 63.44)	Yes
$\mu_1 - \mu_6$	(−3.09, 45.99)	No
$\mu_2 - \mu_3$	(−42.24, 6.84)	No
$\mu_2 - \mu_4$	(−29.74, 19.34)	No
$\mu_2 - \mu_5$	(−11.99, 37.09)	No
$\mu_2 - \mu_6$	(−29.44, 19.64)	No
$\mu_3 - \mu_4$	(−12.04, 37.04)	No
$\mu_3 - \mu_5$	(5.71, 54.79)	Yes
$\mu_3 - \mu_6$	(−11.74, 37.34)	No
$\mu_4 - \mu_5$	(−6.79, 42.29)	No
$\mu_4 - \mu_6$	(−24.24, 24.84)	No
$\mu_5 - \mu_6$	(−41.99, 7.09)	No

$\bar{x}_{5.}$	$\bar{x}_{2.}$	$\bar{x}_{6.}$	$\bar{x}_{4.}$	$\bar{x}_{3.}$	$\bar{x}_{1.}$
510.70	523.25	528.15	528.45	540.95	549.60

c.

Difference	Bonferroni CI	Significantly different
$\mu_1 - \mu_2$	(4.18, 48.52)	Yes
$\mu_1 - \mu_3$	(−13.52, 30.82)	No
$\mu_1 - \mu_4$	(−1.02, 43.32)	No
$\mu_1 - \mu_5$	(16.73, 61.07)	Yes
$\mu_1 - \mu_6$	(−0.72, 43.62)	No

There is no consistent evidence that pain relievers lower glutathione levels.

11.103 A one-way ANOVA will be used with $\alpha = 0.05$ to determine whether there is any difference among the regions in the population mean number of hours worked.

Source of variation	Sum of squares	Degrees of freedom	Mean square	F	p-Value
Factor	291.166	3	97.055	5.55	0.0021
Error	978.536	56	17.474		
Total	1269.702	59			

$H_0: \mu_1 = \mu_2 = \mu_3 = \mu_4$;
$H_a: \mu_i \neq \mu_j$ for some $i \neq j$;
TS: $F = MSA/MSE$; RR: $F \geq F_{0.05,3,56} = 2.77$
$f = 97.055/17.474 = 5.55 \geq 2.77$
$p = P(F \geq 5.55) = 0.0021 \leq 0.05$

There is evidence to suggest that at least two population mean number of hours worked are different.

Consider the Tukey 95% confidence intervals.

Difference	Tukey CI	Significantly different
$\mu_1 - \mu_2$	$(-4.94, \ \ 3.15)$	No
$\mu_1 - \mu_3$	$(-9.35, -1.26)$	Yes
$\mu_1 - \mu_4$	$(-8.20, -0.11)$	Yes
$\mu_2 - \mu_3$	$(-8.46, -0.37)$	Yes
$\mu_2 - \mu_4$	$(-7.30, \ \ 0.78)$	No
$\mu_3 - \mu_4$	$(-2.89, \ \ 5.20)$	No

Graph of results:

	$\bar{x}_{1.}$	$\bar{x}_{2.}$	$\bar{x}_{4.}$	$\bar{x}_{3.}$
Sample mean	38.6	39.5	42.8	43.9

There is evidence to suggest that the population mean number of hours worked is different in the Northeast II and California regions, Northeast II and Mountain I regions, and the Cornbelt I and Mountain I regions.

CHAPTER 12
Section 12.1

12.1 Deterministic and random.

12.3 Homogeneity of variance.

12.5 The mean value of Y for $x = x^*$; an observed value of Y for $x = x^*$.

12.7 SST = SSR + SSE

12.9 False.

12.11 a. Appropriate, slope negative. **b.** Not appropriate, relationship is not linear. **c.** Appropriate, slope zero. **d.** Not appropriate, no linear relationship.

12.13 a. 615 **b.** 3.6; the coefficient on the independent variable. **c.** 0.1217

12.15 a. $y = 41.7004 - 14.8744x$ **b.** -568.15

12.17 a. Scatter plot and regression line:

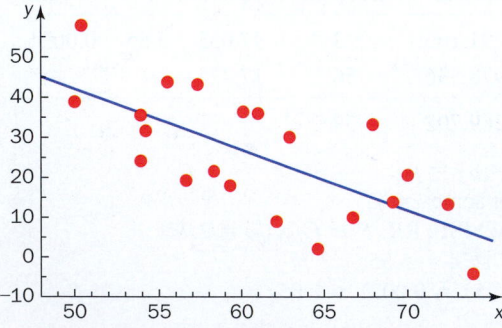

A simple linear regression model seems reasonable. The points appear to fall near a straight line. **b.** $y = 117.9145 - 1.5169x$
c. 23.8657

d. ANOVA summary table:

Source of variation	Sum of squares	Degrees of freedom	Mean square	F
Regression	2290.75	1	2290.75	18.18
Error	2393.78	19	125.99	
Total	4684.53	20		

12.19 a. 2.45 **b.** 0.90 **c.** 0.3446

12.21 a. $y = 3.7167 - 1.7016x$ **b.** 1.5897

12.23 a. 6.181 **b.** 14.866 **c.** 0.2044 **d.** -0.188. The expected growth in employment is -0.188%. That is, employment is expected to decrease.

12.25 a. Scatter plot and regression line:

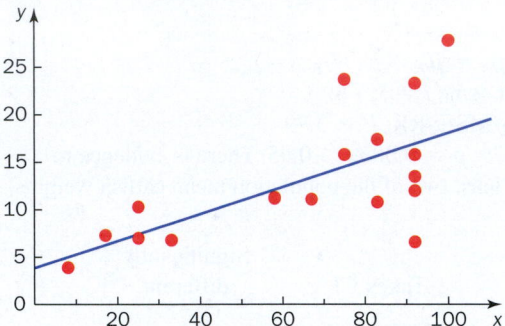

b. $y = 3.8619 + 0.1423x$ **c.** 14.5358

12.27 a. $y = 24.4297 - 58.9778x$ **b.** 6.7364 **c.** 0.1599

12.29 a. Scatter plot:

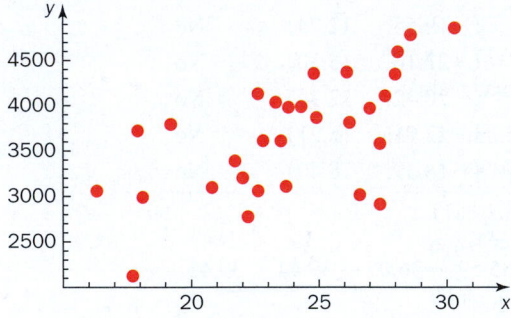

b. $y = 1004.68 + 111.9421x$ **c.** 3803.2335 **d.** Yes, the points fall along a fairly straight line and the value of the F statistic is high (suggesting a small p value).

12.31 a. $y = 0.4629 + 2.8059x$
b. ANOVA summary table:

Source of variation	Sum of squares	Degrees of freedom	Mean square	F
Regression	2.5478	1	2.5478	9.87
Error	2.5822	10	0.2582	
Total	5.1300	11		

c. $r^2 = 0.497$. Approximately 50% of the variation in the data is explained by the regression model. **d.** 1.26

12.33 a. Scatter plot of y versus x_1:

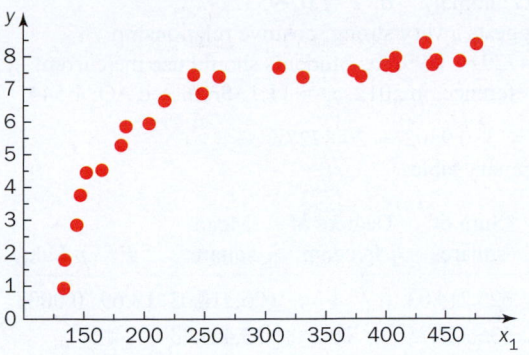

As x_1 increases, the value of y tends to level off. The relationship appears to be logarithmic.
b. Values of x_2: 1.3863, 3.9512, 4.7185, 4.7791, 2.8904, 3.1355, 5.6240, 5.7366, 5.8493, 2.7081, 4.0254, 4.4773, 5.2040, 5.4972, 1.6094, 5.312, 5.7462, 3.5835, 4.3175, 4.8903, 5.3083, 5.5215, 5.6021, 5.7170, 5.8081
c. Scatter plot of y versus x_2:

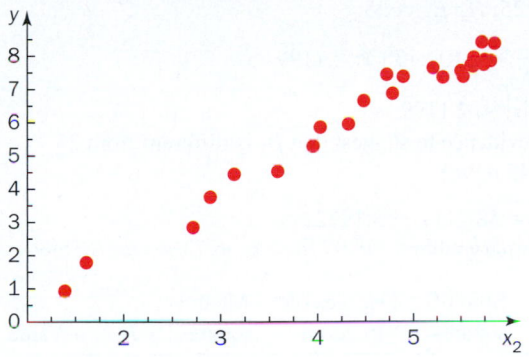

d. $y = -0.8059 + 1.5603x_2$
ANOVA summary table:

Source of variation	Sum of squares	Degrees of freedom	Mean square	F	p-Value
Regression	103.1611	1	103.1611	741.04	< 0.0001
Error	3.2019	23	0.1392		
Total	106.3629	24			

Section 12.2

12.35 Testing the hypothesis $H_0: \beta_1 = 0$.

12.37 False.

12.39 Between -1 and $+1$.

12.41 a. ANOVA summary table:

Source of variation	Sum of squares	Degrees of freedom	Mean square	F	p-Value
Regression	11691.9	1	11691.90	2.58	0.1219
Error	104372.1	23	4537.92		
Total	116064.0	24			

b. H_0: There is no significant linear relationship.
H_a: There is a significant linear relationship.
TS: $F = MSR/MSE$; RR: $F \geq 4.28$
$f = 2.58 < 4.28$; $p = 0.1219 > 0.05$.
There is no evidence of a significant linear relationship.
c. 0.1007   **d.** 0.3174

12.43 a. $H_0: \beta_1 = 0$; $H_a: \beta_1 \neq 0$; TS: $T = B_1/S_{B_1}$;
RR: $|T| \geq 2.0796$
$t = 2.2980 \geq 2.0796$; $p = 0.0319 \leq 0.05$.
There is evidence to suggest that $\beta_1 \neq 0$, the regression line is significant.   **b.** (0.4209, 8.4361)
c. Yes. The CI does not include 0.

12.45 a. H_0: There is no significant linear relationship.
H_a: There is a significant linear relationship.
TS: $F = MSR/MSE$; RR: $F \geq 4.60$
$f = 4.14 < 4.60$. There is no evidence of a significant linear relationship. $p > 0.05, p = 0.0614$
b. $H_0: \beta_1 = 0$; $H_a: \beta_1 \neq 0$; TS: $T = B_1/S_{B_1}$;
RR: $|T| \geq 2.1448$
$t = 2.0337$; $|t| < 2.1448$. There is no evidence to suggest that β_1 is different from 0. $0.05 \leq p \leq 0.10$. 0.0614   **c.** $t^2 = f$
d. Same. These two hypothesis tests are testing the same null hypothesis.

12.47 a. $y = 68.0071 - 8.7993x$

Source of variation	Sum of squares	Degrees of freedom	Mean square	F	p-Value
Regression	7,462.54	1	7,462.54	10.35	0.0324
Error	2,884.77	4	721.19		
Total	10,347.31	5			

b. H_0: There is no significant linear relationship.
H_a: There is a significant linear relationship.
TS: $F = MSR/MSE$; RR: $F \geq 7.71$
$f = 10.35 \geq 7.71$; $p = 0.0324 \leq 0.05$.
There is evidence of a significant linear relationship.
c. 0.7212
d. -0.8492. Negative relationship. b_1 is negative.
e. $r^2 = 0.7212$

12.49 a. $y = 2.9430 + 0.6925x$

Source of variation	Sum of squares	Degrees of freedom	Mean square	F	p-Value
Regression	0.1062	1	0.1062	0.40	0.5403
Error	3.4909	13	0.2685		
Total	3.5971	14			

b. H_0: There is no significant linear relationship.
H_a: There is a significant linear relationship.
TS: $F = MSR/MSE$; RR: $F \geq 9.07$
$f = 0.40 < 9.07$; $p = 0.5403 > 0.01$.
There is no evidence of a significant linear relationship.
c. $(-6.7570, 12.6430)$. No. The CI includes 0.

12.51 a. $r = -0.6434$ **b.** There is a negative linear relationship between mean annual temperature and depth of permafrost layer. As the mean annual temperature increases, the depth of the permafrost layer decreases.

12.53 a. Scatter plot of y versus x:

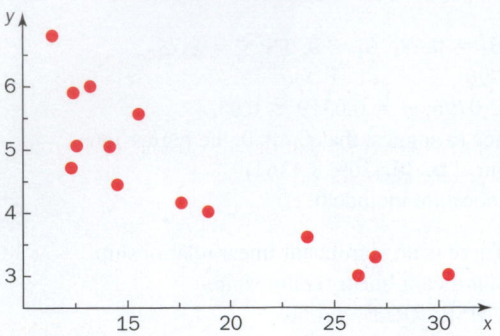

There seems to be a negative linear relationship. As average finish increases, total winnings decreases.   **b.** $r = -0.8847$. This is consistent with part (a). r is negative, and close to -1.
c. $y = 7.4920 - 0.1605x$
ANOVA summary table:

Source of variation	Sum of squares	Degrees of freedom	Mean square	F	p-Value
Regression	15.5939	1	15.5939	46.84	< 0.0001
Error	4.3282	13	0.3329		
Total	19.9221	14			

d. 9.3

12.55 a. Scatter plot of y versus x:

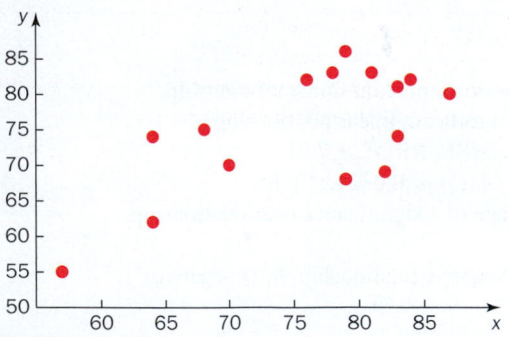

b. $r = 0.6863$ **c.** Positive.   **d.** $y = 23.5751 + 0.6787x$. This supports the results in part (c). The sign of $\hat{\beta}_1$ is positive.

12.57 a. Scatter plot of y versus x:

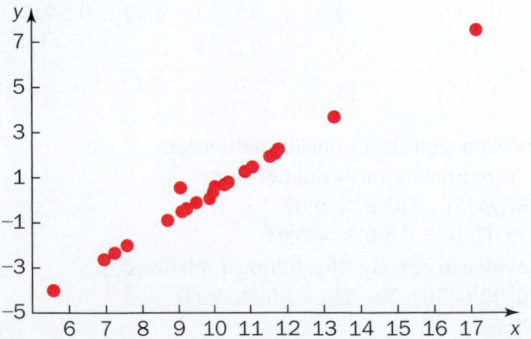

There is a strong positive relationship between Eurasian snow cover and AO anomaly.   **b.** $r = 0.9953$.
This also suggests a very strong, positive relationship.
c. $y = -9.4729 + 0.9925x$. Students should use the current winter. For reference, in 2012, $x = 11.1$. Predicted AO: 1.544

12.59 a. $y = 330.9302 + 26.8727x$
ANOVA summary table:

Source of variation	Sum of squares	Degrees of freedom	Mean square	F	p-Value
Regression	625,214.03	1	625,214.03	18.69	0.0005
Error	535,323.75	16	33,457.73		
Total	1,160,537.78	17			

b. H_0: There is no significant linear relationship.
H_a: There is a significant linear relationship.
TS: $F = $ MSR/MSE; RR: $F \geq 4.49$
$f = 18.69 \geq 4.49$; there is evidence of a significant linear relationship $p < 0.001$.

c. $H_0: \beta_1 = 25$; $H_a: \beta_1 \neq 25$;
TS: $T = \dfrac{B_1 - 25}{S_{B_1}}$; RR: $|T| \geq 2.1199$
$t = 0.3012; |t| < 2.1199$.
There is no evidence to suggest that β_1 is different from 25.
d. (8.7156, 45.0298)

12.61 a. $y = 58.2111 - 3.1972x$
ANOVA summary table:

Source of variation	Sum of squares	Degrees of freedom	Mean square	F	p-Value
Regression	1,840.00	1	1,840.00	1.55	0.2339
Error	16,642.94	14	1,188.78		
Total	18,482.94	15			

b. H_0: There is no significant linear relationship.
H_a: There is a significant linear relationship.
TS: $F = $ MSR/MSE; RR: $F \geq 8.86$
$f = 1.55 < 8.86; p = 0.2339 > 0.01$.
There is no evidence of a significant linear relationship.
c. Scatter plot of y versus x:

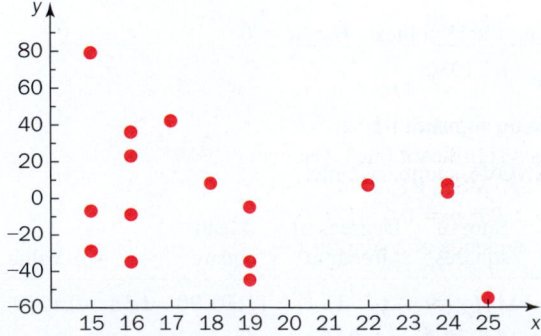

$r = -0.3155$ **d.** No. There is no evidence of a significant linear relationship.

12.63 a. Scatter plot of y versus x:

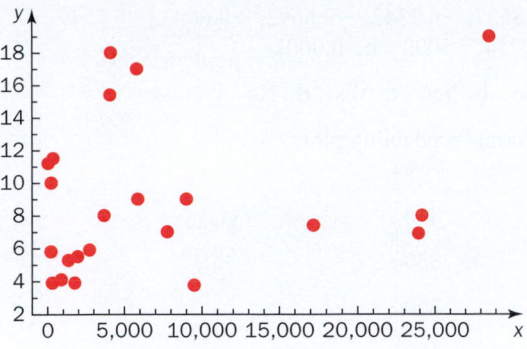

There does not appear to be any relationship between x and y. There is very slight evidence that as x increases, y increases.

b. $r = 0.2438$. Yes, $|r|$ is close to 0.   **c.** No. The scatter plot and r suggest a weak positive linear relationship at best.

d. $y = 7.9620 + 0.0001x$.

H_0: There is no significant linear relationship.
H_a: There is a significant linear relationship.
TS: $F = $ MSR/MSE; RR: $F \geq 4.35$
$f = 1.2642 < 4.35; p = 0.2742 > 0.05$.
There is no evidence of a significant linear relationship. This is consistent with part (c).

Section 12.3

12.65 The mean value of Y for $x = x^*$; an observed value of Y for $x = x^*$.

12.67 False.

12.69 $x^* = \bar{x}$.

12.71 a. $(-100.3043, -94.0097)$, 6.2946
b. $(-103.2959, -94.4725)$, 8.8233   **c.** 31.9 is farther from the mean, $\bar{x} = 30.891$, than 31.5.

12.73 a. $y = -0.9215 + 1.2552x$
ANOVA summary table:

Source of variation	Sum of squares	Degrees of freedom	Mean square	F	p-Value
Regression	29.3171	1	29.3271	4.65	0.0745
Error	37.8679	6	6.3113		
Total	67.1950	7			

H_0: There is no significant linear relationship.
H_a: There is a significant linear relationship.
TS: $F = $ MSR/MSE; RR: $F \geq 5.99$ ($\alpha = 0.05$)
$f = 4.65 < 5.99; p = 0.0745 > 0.05$.
There is no evidence of a significant linear relationship.
b. 2.5122
c. $H_0: y^* = 4; H_a: y^* > 4$;

$$\text{TS: } T = \frac{(B_0 + B_1 x^*) - y_0^*}{S\sqrt{(1/n) + [(x^* - \bar{x})^2/S_{xx}]}}; \text{ RR: } T \geq 1.9432$$

$t - 2.7188 \geq 1.9432; p = 0.0173 \leq 0.05$.
There is evidence to suggest that the mean value of Y for $x = 6$ is greater than 4.
d. $(2.2405, 17.2543)$

12.75 a. 22.8726   **b.** $(18.5067, 27.2385)$
c. $H_0: y^* = 30; H_a: y^* > 30$;

$$\text{TS: } T = \frac{(B_0 + B_1 x^*) - y_0^*}{S\sqrt{(1/n) + [(x^* - \bar{x})^2/S_{xx}]}}; \text{ RR: } T \geq 2.6245$$

$t = -1.2783 < 2.6245$.
There is no evidence to suggest that the mean value of Y for $x = 0.55$ is greater than 30.

12.77 a. ANOVA summary table:

Source of variation	Sum of squares	Degrees of freedom	Mean square	F	p-Value
Regression	87.1673	1	87.1673	5.72	0.0286
Error	258.8706	17	15.2277		
Total	346.0379	18			

H_0: There is no significant linear relationship.
H_a: There is a significant linear relationship.
TS: $F = $ MSR/MSE; RR: $F \geq 4.45$ ($\alpha = 0.05$)
$f = 5.72 \geq 4.45; p = 0.0286 \leq 0.05$.
There is evidence of a significant linear relationship.
Because $\hat{\beta}_1 > 0$, there is a positive linear relationship between tire weight and tire diameter. As the tire diameter increases, so does the tire weight.
b. 24.3260   **c.** $(11.6212, 35.0575)$ **d.** $H_0: \beta_0 = 0$;

$$H_a: \beta_0 \neq 0; \text{ TS: } T = \frac{B_0}{S_{B_0}}; \text{ RR: } |T| \geq 2.8982$$

$t = -1.4469; |t| < 2.8982; p = 0.1661 > 0.01$.
There is no evidence to suggest that β_0 is different from 0, that the true regression line does not pass through the origin. This result makes sense because a tire with 0 diameter should have weight 0.

12.79 a. $y = 505.2149 + 1.0141x$
b. 606.6292

c. $H_0: y^* = 625; H_a: y^* > 625$;

$$\text{TS: } T = \frac{(B_0 + B_1 x^*) - y_0^*}{S\sqrt{(1/n) + [(x^* - \bar{x})^2/S_{xx}]}}; \text{ RR: } T \geq 1.7459$$

($\alpha = 0.05$)
$t = 0.1871 < 1.7396; p = 0.4270 > 0.05$.
There is no evidence to suggest that the mean value of Y for $x = 120$ is greater than 625.
d. 93.4633

12.81 a. $y = 7.1834 - 0.6207x$. Because $\hat{\beta}_1 < 0$, as parasite stress increases, AD burden decreases.
b. H_0: There is no significant linear relationship.
H_a: There is a significant linear relationship.
TS: $F = $ MSR/MSE; RR: $F \geq 4.41$ ($\alpha = 0.05$)
$f = 6.61 \geq 4.41; p = 0.0192 \leq 0.05$.
There is evidence of a significant linear relationship.
c. $(2.9980, 4.4172)$

12.83 a. Scatter plot of y versus x:

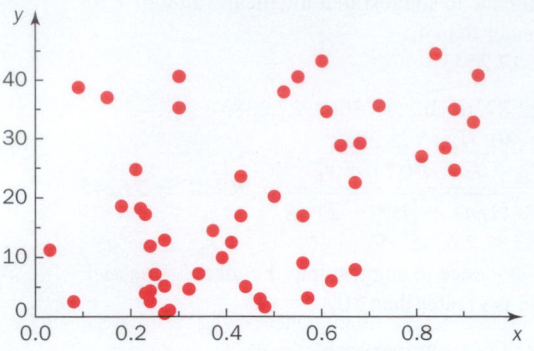

There appears to be a weak positive linear relationship. This supports the study; there is a relationship between linguistic diversity and annual road fatalities.   **b.** $y = 7.7428 + 25.3696x$ $\hat{\beta}_1 > 0$, which supports the conclusion that there is a weak positive linear relationship.   **c.** (12.1524, 20.2295)

12.85 a. ANOVA summary table:

Source of variation	Sum of squares	Degrees of freedom	Mean square	F	p-Value
Regression	853.50	1	853.50	8.99	0.0103
Error	1234.23	13	94.94		
Total	2087.73	14			

H_0: There is no significant linear relationship.
H_a: There is a significant linear relationship.
TS: $F = $ MSR/MSE; RR: $F \geq 4.67$ ($\alpha = 0.05$)
$f = 8.99 \geq 4.67$; $p = 0.0103 \leq 0.05$.
There is evidence of a significant linear relationship. As CPI increases, so does ESI.   **b.** 56.8902   **c.** (40.1554, 87.2630). No. The PI does not include 90.

12.87 a. Scatter plot of y versus x:

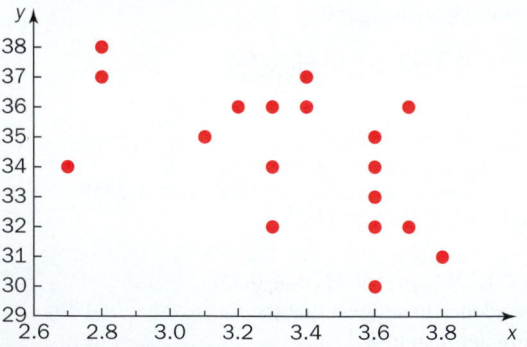

There is a weak negative linear relationship.
b. $y = 46.91 - 3.7x$ **c.** (27.5775, 38.8625)
d. (29.2183, 40.1817)   **e.** 3.7 is farther from the mean than 3.3.

Section 12.4

12.89 $\hat{e}_i = y_i - \hat{y}_i$

12.91 The points will fall along an approximate straight line.

12.93 The variance is not constant.

12.95 a. 8.8557, -6.3342, -9.6692, -8.6693, -6.7549, 4.3457, 10.7254, 7.5009   **b.** 0.0001

12.97 a. No   **b.** Yes   **c.** Yes   **d.** No

12.99 a. Normal probability plot:

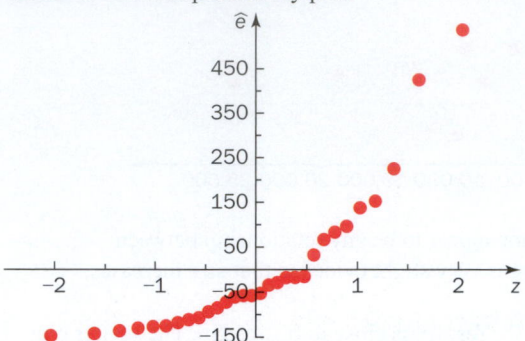

b. There is quite a bit of evidence to suggest that the random error terms are not normal. The plot is curved (not linear) and has several outliers.

12.101 a. 29.4156, -23.1316, -13.0912, -3.76574, -10.2607, 4.23426, -10.7988, -23.6600, -34.7145, -1.91207, 8.00498, -13.6152, 3.62451, -10.4535, 61.2238, -2.54092, 21.9411, 19.4999
b. Scatter plot of residuals versus x:

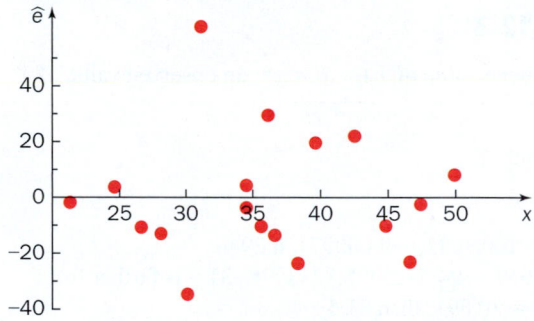

There is some evidence of a violation in the regression-model assumptions. The variance in the error terms does not appear to be constant.

12.103 a. Normal probability plot:

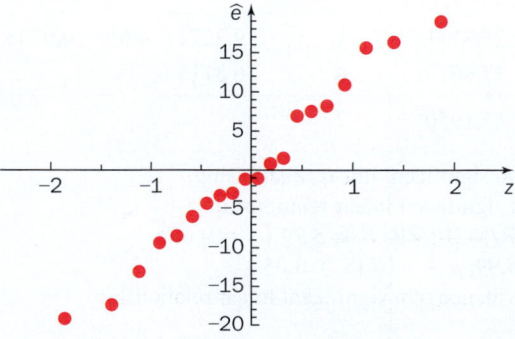

b. There is little evidence to suggest that the error terms are not normal.

12.105 a. $15.2847, 17.5013, -16.9531, -9.3904, 15.6096, -17.6644, 3.5458, -2.3565, -2.9870, -2.5900. \sum \hat{e}_i = 0$

b. Normal probability plot:

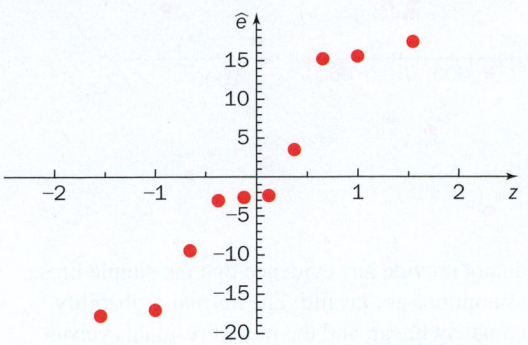

There is evidence to suggest that the random error terms are not normal. There is a nonlinear pattern in the graph.

12.107 a. $3.7425, -1.3811, 3.5112, -6.3979, 2.3609, 5.9687, -5.3974, 3.0674, -2.0416, 0.4215, 3.5257, -2.8021, 2.5082, 2.7378, -6.4356, -2.3193, -1.6682, -3.3047, 0.7335, 3.1704. \sum \hat{e}_i = 0$

b. Scatter plot of residuals versus x:

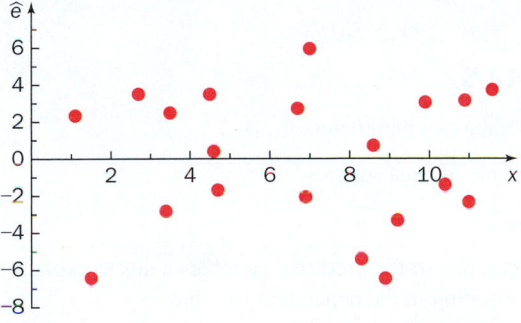

There is no evidence of a violation in the simple linear regression assumptions. The points appear to be randomly scattered and there is no discernible pattern.

12.109 a. $y = 0.0535 + 0.0065x$. Residuals: $0.0092, -0.0214, 0.0097, 0.0002, -0.0089, -0.0006, -0.0044, 0.0133, 0.0071, -0.0042$

b. Normal probability plot:

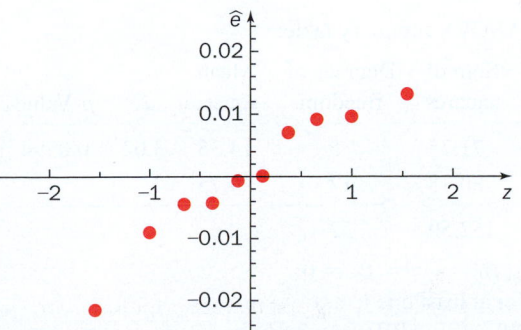

Scatter plot of residuals versus x:

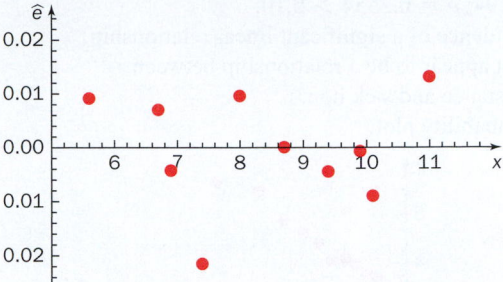

c. There is no overwhelming evidence to suggest that the simple linear regression assumptions are invalid.

12.111 a. $y = -8.4546 + 3.3981x$
Residuals: $6.3086, 5.3283, -4.8369, -0.0698, -0.7688, 4.2894, -6.1573, 3.3281, 7.3477, -2.5456, -6.4194, -4.2058, 1.2506, 7.9688, -7.0991, 0.0465, -4.6233, 4.3086, 1.6873, -5.1379$

b. Normal probability plot:

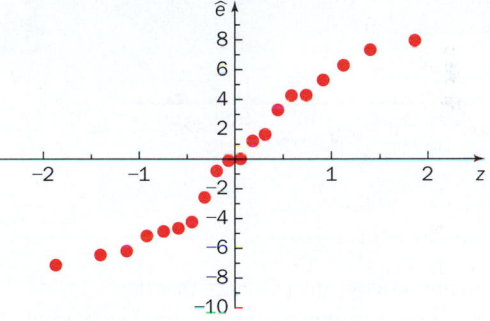

Scatter plot of residuals versus x:

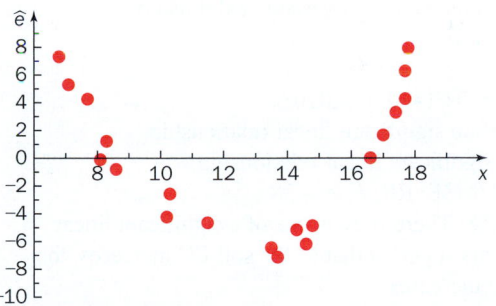

c. There is evidence to suggest that the simple linear regression assumptions are invalid. The normal probability plot is nonlinear. The plot of residuals versus the predictor variable has a distinct nonlinear pattern. To improve the regression model, add a quadratic term.

12.113 a. $y = 29.8441 - 0.0902x$
Residuals: $-2.3173, 0.4801, 4.3291, 5.6679, -5.7685, -2.4002, -2.9638, 4.3217, -9.5126, 9.9386, 2.8631, -10.8588, 10.6753, -7.3247, 3.9681, 6.5095, 3.2315, 7.6071, -4.9564, -9.7612, -1.4978, -1.7004, 1.9460, -2.8662, 0.3898$

b. H_0: There is no significant linear relationship.
H_a: There is a significant linear relationship.

TS: $F = \text{MSR/MSE}$; RR: $F \geq 2.94$
$f = 1.36 < 2.94$; $p = 0.2554 > 0.10$.
There is no evidence of a significant linear relationship.
There does not appear to be a relationship between
commuting distance and sick hours.
c. Normal probability plot:

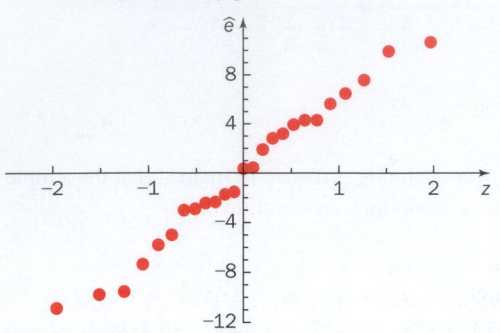

Scatter plot of residuals versus x:

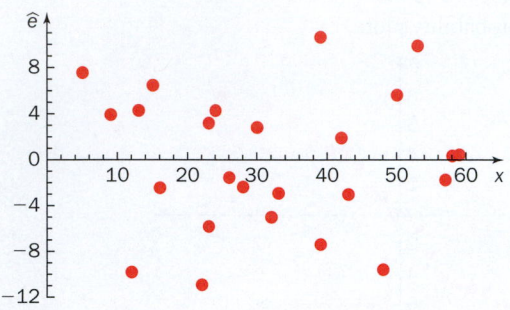

d. The graphs do not provide any evidence that the
simple linear regression assumptions are invalid. The normal
probability plot is approximately linear, and the plot of
residuals versus the predictor variable exhibits no
discernible pattern.

12.115 a. $y = 94.1430 + 0.0768x$
b. H_0: There is no significant linear relationship.
H_a: There is a significant linear relationship.
TS: $F = \text{MSR/MSE}$; RR: $F \geq 4.38$
$f = 7.69 \geq 4.38$. There is evidence of a significant linear
relationship. This suggests that as the soil EC increases, the
corn yield also increases.
$0.01 \leq p \leq 0.05$. $p = 0.0121$.

c. Normal probability plot:

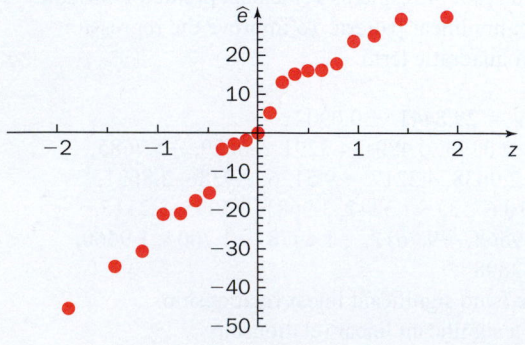

Scatter plot of residuals versus x:

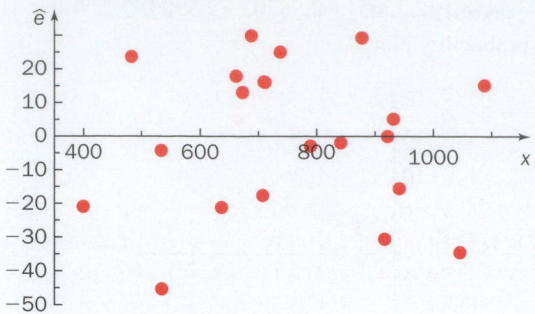

The graphs do not provide any evidence that the simple linear
regression assumptions are invalid. The normal probability
plot is approximately linear, and the plot of residuals versus
the predictor variable exhibits no discernible pattern.

12.117

$$\sum_{i=1}^{n}(y_i - \hat{y}_i) = \sum_{i=1}^{n}\left[y_i - (\hat{\beta}_0 + \hat{\beta}_1 x_i)\right]$$

$$= \sum_{i=1}^{n}\left[y_i - (\bar{y} - \hat{\beta}_1\bar{x} + \hat{\beta}_1 x_i)\right]$$

$$= \sum_{i=1}^{n}\left[y_i - \bar{y} - \hat{\beta}_1\sum_{i=1}^{n}(x_i - \bar{x})\right]$$

$$= n\bar{y} - n\bar{y} - \hat{\beta}_1(n\bar{x} - n\bar{x}) = 0$$

Section 12.5

12.119 The unknown parameters $\beta_0, \beta_1, \ldots, \beta_k$.

12.121 Principle of least squares.

12.123 False.

12.125 At least one of the predictor variables helps to explain
some of the variation in the dependent variable.

12.127 This assumes that only one test is conducted.

12.129 The mean value of Y for $x = x^*$ and an estimate of an
observed value of Y for $x = x^*$.

12.131 a. 482.4 **b.** As x_1 increases, y increases.
c. -5.3 **d.** 0.1680

12.133 a. $y = 12.7786 + 1.9638x_1 + 7.4479x_2$
b. 168.2178

12.135 a. ANOVA summary table:

Source of variation	Sum of squares	Degrees of freedom	Mean square	F	p-Value
Regression	71.75	5	14.35	3.02	0.0394
Error	80.75	17	4.75		
Total	152.50	22			

b. 5 **c.** H_0: $\beta_1 = \cdots = \beta_5 = 0$;
H_a: $\beta_i \neq 0$ for at least one i;
TS: $F = \text{MSR/MSE}$; RR: $F \geq 2.81$
$f = 3.02 \geq 2.81$. There is evidence to suggest that at least one
of the regression coefficients is different from 0.

$0.01 \leq p \leq 0.05$. **d.** $r^2 = 0.4705$. Approximately 47% of the variation in y is explained by this regression model.

12.137
a. $y = -46.2192 - 4.2153x_1 + 16.4785x_2 + 1.1186x_3$
b. $H_0: \beta_1 = \beta_2 = \beta_3 = 0$;
$H_a: \beta_i \neq 0$ for at least one i;
TS: $F = MSR/MSE$; RR: $F \geq 4.76$
$f = 32.80 \geq 4.76$; $p = 0.0004 \leq 0.05$. There is evidence to suggest that at least one of the regression coefficients is different from 0. The overall regression is significant.
d. $\beta_1: t = -5.8458$, $p = 0.0011$. $\beta_2: t = 3.9533$, $p = 0.0075$. $\beta_3: t = 0.6198$, $p = 0.5582$. x_1 and x_2 are significant predictors.
e. $(-117.7319, 25.2936)$. There is no evidence to suggest that the constant regression coefficient is different from 0. The CI includes 0.

12.139
a. $y = 7.5139 - 14.9684x_1 + 2.9118x_2 - 0.9704x_3$
b. Normal probability plot:

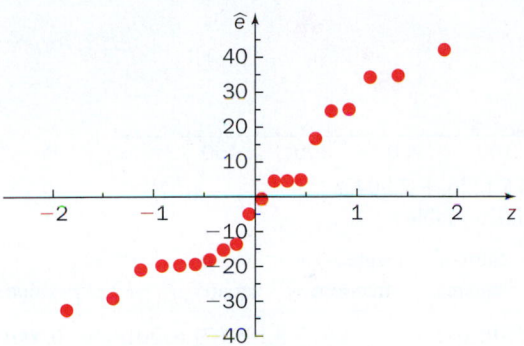

The graph suggests that the random error terms are not normal. The pattern is nonlinear.
b. Scatter plot of residuals versus x_1:

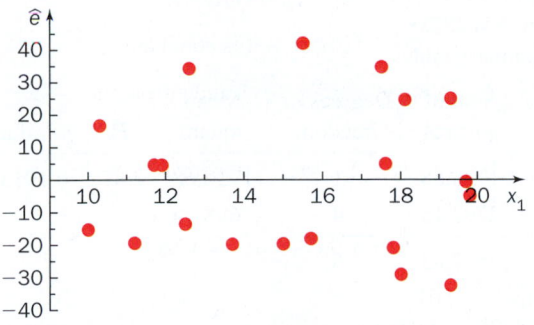

Scatter plot of residuals versus x_2:

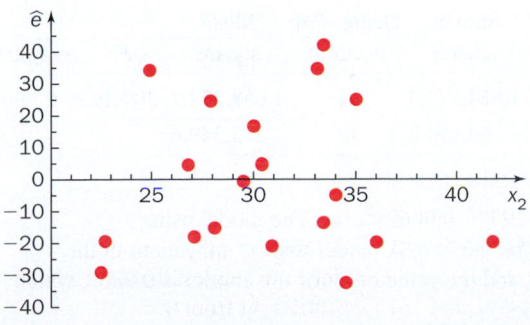

Scatter plot of residuals versus x_3:

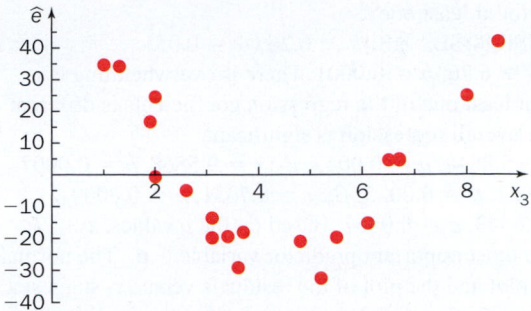

This graph suggests a violation in the regression assumptions. Include a quadratic term, x_3^2, to improve the model.

12.141
a. $y = -0.0565 - 0.0007x_1 + 0.0002x_2 + 0.000002x_3$
$H_0: \beta_1 = \beta_2 = \beta_3 = 0$;
$H_a: \beta_i \neq 0$ for at least one i;
TS: $F = MSR/MSE$; RR: $F \geq 3.29$
$f = 3.24 < 3.28$; $p = 0.0519$. It's really close, but at the $\alpha = 0.05$ level, there is no evidence to suggest that at least one of the regression coefficients is different from 0. The overall regression is not significant.
b. 0.0622 **c.** The t tests suggest that the price of gold is the most important predictor.

12.143 a. ANOVA summary table:

Source of variation	Sum of squares	Degrees of freedom	Mean square	F	p-Value
Regression	799.95	3	266.65	29.95	< 0.0001
Error	160.23	18	8.90		
Total	960.18	21			

$H_0: \beta_1 = \beta_2 = \beta_3 = 0$;
$H_a: \beta_i \neq 0$ for at least one i;
TS: $F = MSR/MSE$; RR: $F \geq 3.16$ ($\alpha = 0.05$)
$f = 29.95 \geq 3.16$. There is evidence to suggest that at least one of the regression coefficients is different from 0. The overall regression is significant.
b. As temperature increases, the bonding strength decreases. As contact area increases, the bonding strength increases. As wood density increases, the bonding strength decreases.
c. $\beta_1: t = -2.382$, $0.02 \leq p \leq 0.05$.
$\beta_2: t = 8.473$, $p < 0.0002$. $\beta_3: t = -1.0736$, $0.20 \leq p \leq 0.40$. Temperature and contact area are the most important (significant) predictor variables.

12.145 a.
$y = 0.0510 - 0.0028x_1 + 0.3247x_2 + 0.2777x_3$
$\quad + 0.2188x_4 - 0.1029x_5$
b. ANOVA summary table:

Source of variation	Sum of squares	Degrees of freedom	Mean square	F	p-Value
Regression	6.4709	5	1.2942	172.18	< 0.0001
Error	0.0301	4	0.0075		
Total	6.5010	9			

$H_0: \beta_1 = \cdots = \beta_5 = 0$;
$H_a: \beta_i \neq 0$ for at least one i;
TS: $F = MSR/MSE$; RR: $F \geq 6.26$ ($\alpha = 0.05$)
$f = 172.18 \geq 6.26$; $p < 0.0001$. There is overwhelming evidence that at least one of the regression coefficients is different from 0. The overall regression is significant.
c. $\beta_1: t = -5.8848$, $p = 0.0042$. $\beta_2: t = 9.5568$, $p = 0.0007$. $\beta_3: t = 6.7921$, $p = 0.0025$. $\beta_4: t = 8.7331$, $p = 0.0009$. $\beta_5: t = -2.6543$, $p = 0.0567$. Based on the p values, x_2 is the single most important predictor variable. **d.** The normal probability plot and the plot of the residuals versus x_5 suggest a slight violation in the model assumptions. However, with only 10 observations, it is difficult to be certain.

12.147 a. $y = 2447.7015 + 51.2511x_1 - 10.8445x_2$
b. $H_0: \beta_1 = \beta_2 = 0$;
$H_a: \beta_i \neq 0$ for at least one i;
TS: $F = MSR/MSE$; RR: $F \geq 6.93$
$f = 8.19 \geq 6.93$; $p = 0.0057 \leq 0.01$. There is evidence that at least one of the regression coefficients is different from 0. The overall regression is significant.
c. As the percentage of sea ice extent increases, the size of the penguin colony population also increases. As the number of stormy days increases, the size of the penguin colony population decreases.
d. $r^2 = 0.5771$. Approximately 58% of the variation in the data is explained by the regression model. **e.** 2708.7832

12.149 a. ANOVA summary table:

Source of variation	Sum of squares	Degrees of freedom	Mean square	F	p-Value
Regression	342.970	5	68.594	5.79	0.0004
Error	462.340	39	11.855		
Total	805.310	44			

$H_0: \beta_1 = \cdots = \beta_5 = 0$;
$H_a: \beta_i \neq 0$ for at least one i;
TS: $F = MSR/MSE$; RR: $F \geq 2.46$ ($\alpha = 0.05$)
$f = 5.79 \geq 2.48$; $p = 0.0004 \leq 0.05$. There is evidence that at least one of the regression coefficients is different from 0. $r^2 = 0.4259$. The overall model is significant. However, only approximately 43% of the variation in the data is explained by the regression model.
b. $\beta_1: t = 2.91$, $p = 0.0059$. $\beta_2: t = -1.76$, $p = 0.0862$. $\beta_3: t = 0.11$, $p = 0.9130$. $\beta_4: t = -4.04$, $p = 0.0002$. $\beta_5: t = -0.83$, $p = 0.4116$. The p values suggest that the most important predictor is x_4, total hardness. Using the signs of the estimated regressions coefficients, as total dissolved solids increases, chloride in groundwater increases; as iron increases, chloride in groundwater decreases; as nitrate increases, chloride in groundwater increases; as total hardness increases, chloride in groundwater decreases; as depth of the sample increases, chloride in groundwater decreases.

12.151 a.
$y = 9.3465 - 0.0618x_1 - 0.0812x_2 + 0.9117x_2 - 0.0837x_4$
b. $H_0: \beta_1 = \cdots = \beta_4 = 0$;

$H_a: \beta_i \neq 0$ for at least one i;
TS: $F = MSR/MSE$; RR: $F \geq 3.06$
$f = 3.70 \geq 3.06$; $p = 0.0275 \leq 0.05$. There is evidence that at least one of the regression coefficients is different from 0. $r^2 = 0.4965$.
c. $\beta_1: t = -0.9881$, $p = 0.3388$. $\beta_2: t = -1.6693$, $p = 0.1158$. $\beta_3: t = 3.3806$, $p = 0.0041$. $\beta_4: t = -1.2554$, $p = 0.2286$. Only x_3 is a significant predictor variable.
d. $H_0: \beta_3 = 0.93$; $H_a: \beta_1 < 0.93$;
TS: $T = (B_3 - 0.93)/S_{B_3}$; RR: $T \leq -1.7531$.
$t = -0.0679 > -1.7531$; $p = 0.4733 > 0.05$. There is no evidence to suggest that $\beta_3 < 0.93$. **e.** (1.0539, 1.7159)

12.153 a. Scatter plot of y versus x:

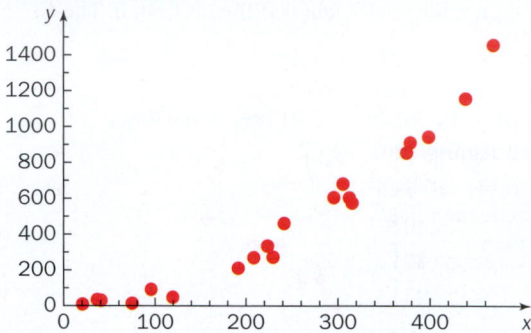

b. $y = -215.1226 + 2.9043x$
ANOVA summary table:

Source of variation	Sum of squares	Degrees of freedom	Mean square	F	p-Value
Regression	3,105,671.6	1	3,105,671.6	201.69	<0.0001
Error	277,162.9	18	15,397.9		
Total	3,382,834.5	19			

c. Small depths (≤ 150):
$y = 7.4516 + 0.4828x$
ANOVA summary table:

Source of variation	Sum of squares	Degrees of freedom	Mean square	F	p-Value
Regression	1720.68	1	1720.68	2.47	0.1915
Error	2792.15	4	698.04		
Total	4512.83	5			

Large depths (> 150):
$y = -598.5726 + 4.0383x$
ANOVA summary table:

Source of variation	Sum of squares	Degrees of freedom	Mean square	F	p-Value
Regression	1,654,572.7	1	1,654,572.7	309.29	< 0.0001
Error	64,194.7	12	5,349.6		
Total	1,718,767.4	13			

d. $y = 7.3195 + 0.0062x^2$ **e.** The model using x^2 seems to be the best. A model with x^2 appears to fit the scatter plot, and the value of r^2 for this model is 0.9869, which is very high.

Chapter Exercises

12.155 a. $y = 2.8408 + 2.1231x$ **b.** 24.0716 **c.** 7.0870

12.157 a. $y = 302.3637 - 0.0629x$
ANOVA summary table:

Source of variation	Sum of squares	Degrees of freedom	Mean square	F	p-Value
Regression	44,580.81	1	44,580.81	21.15	0.0001
Error	48,475.33	23	2,107.62		
Total	93,056.14	24			

b. H_0: There is no significant linear relationship.
H_a: There is a significant linear relationship.
TS: $F = MSR/MSE$; RR: $F \geq 7.88$
$f = 21.15 \geq 7.88$; $p = 0.0001 \leq 0.01$.
There is evidence of a significant linear relationship.
c. 0.4791
d. The overall regression is significant and explains approximately 48% of the variation in the model. Omega-3 is a good predictor of beta amyloids blood level.

12.159 a. Scatter plot of y versus x:

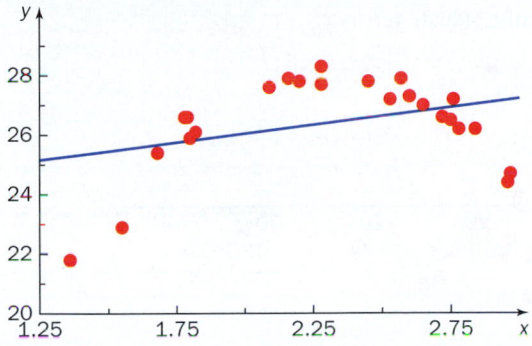

The relationship between x and y appears to be quadratic.

b. $y = 23.6853 + 1.1767x$
H_0: There is no significant linear relationship.
H_a: There is a significant linear relationship.
TS: $F = MSR/MSE$; RR: $F \geq 4.30$
$f = 3.08 < 4.30$; $p = 0.0932 > 0.05$. There is no evidence of a significant linear relationship.
d. Scatter plot of residuals versus x:

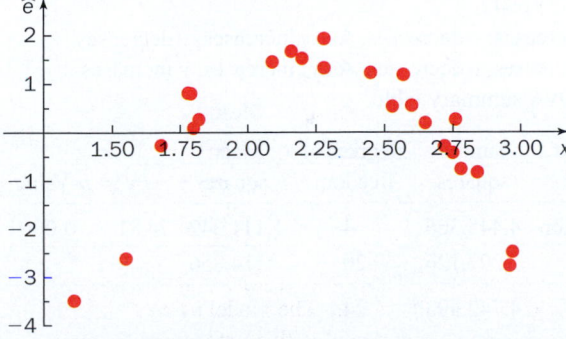

e. The graphs in part (a) and (c) suggest the model could be improved by adding a quadratic term, x^2.

12.161 a. $y = 8.3870 + 0.6024x$
b. ANOVA summary table:

Source of variation	Sum of squares	Degrees of freedom	Mean square	F	p-Value
Regression	67.39	1	67.39	2.29	0.1370
Error	1414.44	48	29.47		
Total	1481.83	49			

c. H_0: $\beta_1 = 0$; H_a: $\beta_1 \neq 0$;
TS: $T = B_1/S_{B_1}$; RR: $|T| \geq 2.0106$.
$t = 1.5123$; $|t| < 2.0106$; $p = 0.1370 > 0.05$.
There is no evidence to suggest that $\beta_1 \neq 0$, the regression line is not significant.
d. No. There is no significant relationship between rating and price.

12.163 a. $y = 4.3207 + 1.2584x$
ANOVA summary table:

Source of variation	Sum of squares	Degrees of freedom	Mean square	F	p-Value
Regression	104.4777	1	104.4777	27.25	< 0.0001
Error	107.3632	28	3.8344		
Total	211.8409	29			

b. The later that shoppers start shopping, the more money they spend. Retailers should entice shoppers to start shopping later in the day.
c. (2.7212, 10, 9539)
d. H_0: $y^* = 7.25$; H_a: $y^* > 7.25$;

$$\text{TS: } T = \frac{(B_0 + B_1 x^*) - y_0^*}{S\sqrt{(1/n) + [(x^* - \bar{x})^2/S_{xx}]}}; \text{ RR: } T \geq 1.7011$$

$t = 2.3553 \geq 1.7011$; $p = 0.0129 \leq 0.05$. There is evidence to suggest that the true mean spent is greater than 7.25 ($725).

12.165 a. Scatter plot of y (cicada density) versus x (mole density):

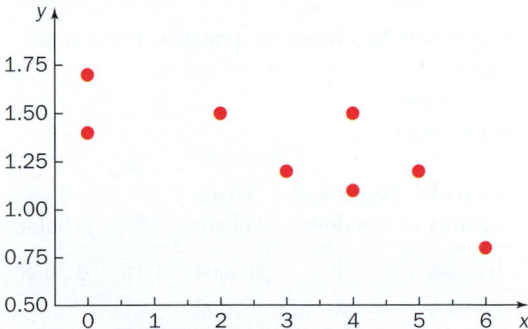

There appears to be a negative linear relationship.
b. $y = 1.6096 - 0.0924x$
H_0: $\beta_1 = 0$; H_a: $\beta_1 \neq 0$;
TS: $T = B_1/S_{B_1}$; RR: $|T| \geq 2.3646$.
$t = -2.6441$; $|t| \geq 2.3646$; $p = 0.0332 \leq 0.05$. There is evidence to suggest that $\beta_1 \neq 0$, the regression line is significant.

c. Normal probability plot:

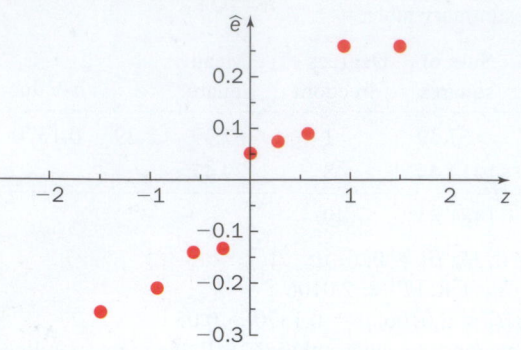

Scatter plot of residuals versus x:

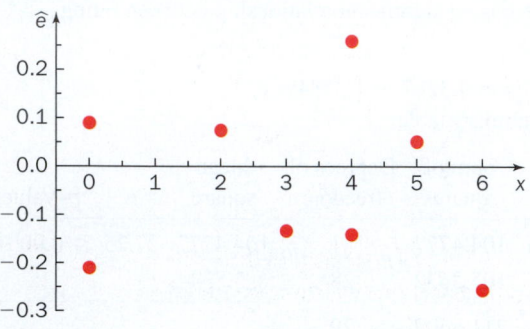

The normal probability plot suggests a violation in the normality assumption.

12.167 a. Scatter plot of y versus x:

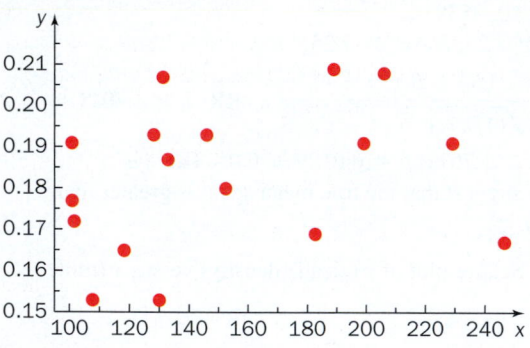

There does not appear to be a linear relationship. The scatter plot appears random.

b. $y = 0.1673 + 0.0001x$
ANOVA summary table:

Source of variation	Sum of squares	Degrees of freedom	Mean square	F	p-Value
Regression	0.0004	1	0.0004	1.15	0.3008
Error	0.0047	15	0.0003		
Total	0.0050	16			

c. The F test is not significant ($p = 0.3008$). There is no evidence to suggest a significant linear relationship.

12.169 a. Estimated regression line:
$y = 14.8391 - 0.2436x$

ANOVA summary table:

Source of variation	Sum of squares	Degrees of freedom	Mean square	F	p-Value
Regression	228.53	1	228.53	7.32	0.0095
Error	1466.91	47	31.21		
Total	1695.44	48			

There is evidence to suggest a significant linear relationship ($p = 0.0095$).
Normal probability plot:

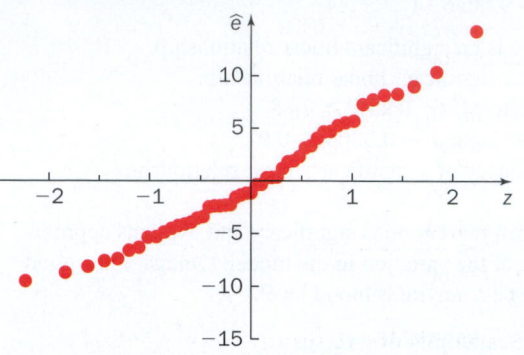

Scatter plot of residuals versus x:

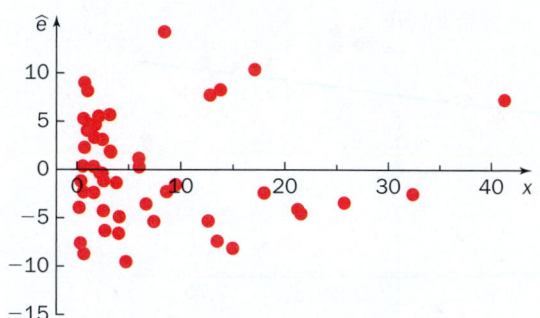

There appears to be a violation in the assumption of constant variance. The sign of $\hat{\beta}_1$ does not seem appropriate. One would think that as the percentage of freshwater increases, the fishing participation rate would also increase, and that the sign of $\hat{\beta}_1$ would be positive.

12.171 a. Estimated regression line:
$y = 7274.5117 - 971.4403x_1 - 69.2220x_2 - 64.1724x_3 + 32.1604x_4$
As x_1 increases, y decreases. As x_2 increases, y decreases. As x_3 increases, y decreases. As x_4 increases, y increases.
b. ANOVA summary table:

Source of variation	Sum of squares	Degrees of freedom	Mean square	F	p-Value
Regression	4,445,366	4	1,111,342	74.81	< 0.0001
Error	297,126	20	14,856		
Total	4,742,493	24			

$H_0: \beta_1 = \cdots = \beta_4 = 0$;
$H_a: \beta_i \neq 0$ for at least one i;

TS: $F = $ MSR/MSE;   RR: $F \geq 2.87$ ($\alpha = 0.05$).
$f = 74.81 \geq 2.87; p < 0.05.$
There is evidence to suggest that at least one of the
regression coefficients is different from 0. The overall
regression is significant. $p = 0.0000000000096665.$
c. $\beta_1: t = -1.2730, p = 0.2176. \ \beta_2: t = -0.7246,$
$p = 0.4771. \ \beta_3: t = -12.4987, p < 0.0001.$
$\beta_4: t = 12.4091, p < 0.0001.$
The variables x_3 and x_4 are significant predictors.
d. The plots suggest there is a possible outlier and there is
some evidence of nonconstant variance.

12.173 a. $y = 90.2677 - 0.1633x$
ANOVA summary table:

Source of variation	Sum of squares	Degrees of freedom	Mean square	F	p-Value
Regression	5771.86	1	5771.86	27.91	0.0007
Error	1654.14	8	206.77		
Total	7426.00	9			

b. $r^2 = 0.777.$ Approximately 78% of the variation in noise
level is explained by this regression model.
c. 33.1205, $(-3.45, 69.69)$
d. Normal probability plot:

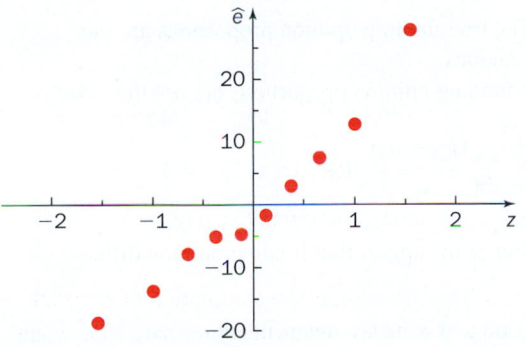

Scatter plot of residuals versus x:

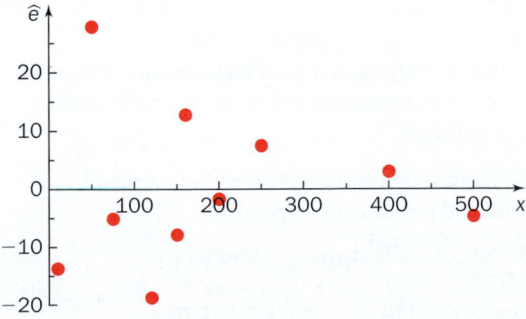

There is some evidence that the variance in the random error
terms decreases as distance increases. However, for $n = 10$
observations, the pattern is not distinct enough to be evidence
of a violation in the simple linear regression assumptions.

CHAPTER 13

Section 13.1

13.1 False.

13.3 Small.

13.5 At least 5.

13.7 150, 140, 310, 230, 170

13.9 a. $H_0: p_1 = 0.175, p_2 = 0.171, p_3 = 0.162, p_4 = 0.225,$
$p_5 = 0.202, p_6 = 0.065;$
$H_a: p_i \neq p_{i0}$ for at least one i;
TS: $X^2 = \sum_{i=1}^{6} (n_i - e_i)^2/e_i$; RR: $X^2 \geq 11.0705.$
b. $X^2 = 2.7994 < 11.0705.$ There is no evidence to suggest
that any of the population proportions differs from its
hypothesized value.   **c.** $p > 0.10, (p = 0.7309)$

13.11 $H_0: p_1 = 0.25, p_2 = 0.15, p_3 = 0.10, p_4 = 0.50;$
$H_a: p_i \neq p_{i0}$ for at least one i;
TS: $X^2 = \sum_{i=1}^{4} (n_i - e_i)^2/e_i$; RR: $X^2 \geq 7.8147.$
$\chi^2 = 11.6 \geq 7.8147; p = 0.0089 \leq 0.05.$ There is evidence to
suggest that at least one of the population proportions differs
from its hypothesized value.

13.13 $H_0: p_1 = 0.390, p_2 = 0.305, p_3 = 0.305;$
$H_a: p_i \neq p_{i0}$ for at least one i;
TS: $X^2 = \sum_{i=1}^{3} (n_i - e_i)^2/e_i$; RR: $X^2 \geq 9.2103.$
$\chi^2 = 14.6318 \geq 9.2103; p = 0.0007 \leq 0.01.$ There is evidence
to suggest that at least one of the percentages during this week
is different from the long-term percentages.

13.15 $H_0: p_1 = 0.10, p_2 = 0.20, p_3 = 0.25, p_4 = 0.05,$
$p_5 = 0.40; H_a: p_i \neq p_{i0}$ for at least one i;
TS: $X^2 = \sum_{i=1}^{5} (n_i - e_i)^2/e_i$; RR: $X^2 \geq 9.4877.$
$\chi^2 = 10.5927 \geq 9.4877; p = 0.0315 \leq 0.05.$ There is evidence
to suggest that at least one of the population proportions differs
from its hypothesized value.

13.17 $H_0: p_1 = 0.14, p_2 = 0.06, p_3 = 0.37, p_4 = 0.30,$
$p_5 = 0.13; H_a: p_i \neq p_{i0}$ for at least one i;
TS: $X^2 = \sum_{i=1}^{5} (n_i - e_i)^2/e_i$; RR: $X^2 \geq 9.4877.$
$\chi^2 = 6.0884 < 9.4877; p = 0.1926 > 0.05.$ There is no
evidence to suggest that any of the population proportions
differs from its hypothesized value.

13.19 $H_0: p_1 = 0.09, p_2 = 0.18, p_3 = 0.07, p_4 = 0.07,$
$p_5 = 0.09, p_6 = 0.03, p_7 = 0.24, p_8 = 0.15, p_9 = 0.03,$
$p_{10} = 0.05; H_a: p_i \neq p_{i0}$ for at least one i;
TS: $X^2 = \sum_{i=1}^{10} (n_i - e_i)^2/e_i$; RR: $X^2 \geq 21.6660.$
$\chi^2 = 41.6934 \geq 21.6660; p < 0.0001.$ There is evidence to
suggest that at least one of the population proportions differs
from its hypothesized value.

13.21 H_0: $p_1 = 0.40$, $p_2 = 0.30$, $p_3 = 0.15$, $p_4 = 0.15$; H_a: $p_i \neq p_{i0}$ for at least one i;
TS: $X^2 = \sum_{i=1}^{4} (n_i - e_i)^2/e_i$; RR: $X^2 \geq 11.3449$.
$\chi^2 = 4.9622 < 11.3449$; $p = 0.1746 > 0.01$. There is no evidence to suggest that any of the population proportions differs from its hypothesized value.

13.23 a. H_0: $p_1 = 0.52$, $p_2 = 0.24$, $p_3 = 0.14$, $p_4 = 0.06$, $p_5 = 0.04$; H_a: $p_i \neq p_{i0}$ for at least one i;
TS: $X^2 = \sum_{i=1}^{5} (n_i - e_i)^2/e_i$; RR: $X^2 \geq 9.4877$.
$\chi^2 = 26.6083 \geq 9.4877$. There is evidence to suggest that at least one of the population proportions differs from its hypothesized value. **b.** $p = 0.000024$ **c.** The two housing types contributing most to the chi-square statistic are single-family detached home, large yard, and apartment or condominium. It could be that retirees in Florida like both of these housing types.

13.25 a. 0.13, 0.17, 0.34, 0.36.
b. H_0: $p_1 = 0.13$, $p_2 = 0.17$, $p_3 = 0.34$, $p_4 = 0.36$; H_a: $p_i \neq p_{i0}$ for at least one i;
TS: $X^2 = \sum_{i=1}^{4} (n_i - e_i)^2/e_i$; RR: $X^2 \geq 11.3449$.
$\chi^2 = 5.6952 < 11.3449$; $p = 0.1274 > 0.01$. There is no evidence to suggest that any of the population proportions changed in 2014.

13.27 H_0: $p_1 = 0.78$, $p_2 = 0.11$, $p_3 = 0.04$, $p_4 = 0.02$, $p_5 = 0.01$, $p_6 = 0.02$, $p_7 = 0.02$; H_a: $p_i \neq p_{i0}$ for at least one i;
TS: $X^2 = \sum_{i=1}^{7} (n_i - e_i)^2/e_i$; RR: $X^2 \geq 12.5916$.
$\chi^2 = 1.0594 < 12.5916$; $p = 0.9833 > 0.05$. There is no evidence to suggest that any of the historical proportions have changed.

13.29 a. 0.0228, 0.1359, 0.3413, 0.3413, 0.1359, 0.0228
b. H_0: $p_1 = 0.0228$, $p_2 = 0.1359$, $p_3 = 0.3413$, $p_4 = 0.3413$, $p_5 = 0.1359$, $p_6 = 0.0228$; H_a: $p_i \neq p_{i0}$ for at least one i;
TS: $X^2 = \sum_{i=1}^{6} (n_i - e_i)^2/e_i$; RR: $X^2 \geq 11.0705$.
$\chi^2 = 3.88 < 110,705$; $p = 0.5668 > 0.05$.
There is no evidence to suggest that any of the population proportions differs from its hypothesized value. There is no evidence to suggest that the observed weights do not fit the hypothesized distribution.

Section 13.2

13.31 Side-by-side bar chart, stacked bar chart.

13.33 False.

13.35 True.

13.37 a. 12.5916   **b.** 15.0863   **c.** 14.4494   **d.** 26.1245

13.39

		Category			
	1	2	3	4	
Population 1	18	14	18	15	65
Population 2	25	21	16	12	74
Population 3	32	33	26	28	119
	75	68	60	55	258

13.41 H_0: Variable 1 and variable 2 are independent. H_a: Variable 1 and variable 2 are dependent.
TS: $X^2 = \sum_{i=1}^{4} \sum_{j=1}^{4} \frac{(n_{ij} - e_{ij})}{e_{ij}}$; RR: $X^2 \geq 16.9190$
$\chi^2 = 16.8226 < 16.9190$; $p = 0.0516 > 0.05$.
There is just barely no evidence to suggest that the two categorical variables are dependent.

13.43 H_0: The true product proportions are the same for all stores.
H_a: The true product proportions are not the same for all stores.
TS: $X^2 = \sum_{i=1}^{3} \sum_{j=1}^{4} \frac{(n_{ij} - e_{ij})}{e_{ij}}$; RR: $X^2 \geq 16.8119$
$\chi^2 = 9.2674 < 16.8119$; $p = 0.1591 > 0.01$.
There is no evidence to suggest that the true proportion of each favorite differs by grocery store.

13.45 H_0: The true funding opinion proportions are the same for all regions.
H_a: The true funding opinion proportions are not the same for all regions.
TS: $X^2 = \sum_{i=1}^{4} \sum_{j=1}^{4} \frac{(n_{ij} - e_{ij})}{e_{ij}}$; RR: $X^2 \geq 23.5894$
$\chi^2 = 17.8692 < 23.5894$; $p = 0.0367 \leq 0.05$.
There is evidence to suggest that funding opinion differs by region.

13.47 H_0: Food and wine are independent.
H_a: Food and wine are dependent.
TS: $X^2 = \sum_{i=1}^{2} \sum_{j=1}^{2} \frac{(n_{ij} - e_{ij})}{e_{ij}}$; RR: $X^2 \geq 7.8794$
$\chi^2 = 11.279 \geq 7.8794$; $p = 0.0008 \leq 0.05$.
There is evidence to suggest that food and wine are dependent. This suggests that diners are still following the traditional food-and-wine pairings.

13.49 a. H_0: Draft round and position are independent.
H_a: Draft round and position are dependent.
TS: $X^2 = \sum_{i=1}^{4} \sum_{j=1}^{3} \frac{(n_{ij} - e_{ij})}{e_{ij}}$; RR: $X^2 \geq 16.8119$
$\chi^2 = 23.4839 \geq 16.8119$; $p = 0.0006 \leq 0.01$.
There is evidence to suggest that draft round and position are dependent.
b. Fewer than expected special-teams players are drafted in rounds 1 and 2.

13.51 H_0: Age group and influenza type are independent. H_a: Age group and influenza type are dependent.

TS: $X^2 = \sum_{i=1}^{5} \sum_{j=1}^{4} \frac{(n_{ij} - e_{ij})}{e_{ij}}$; RR: $X^2 \geq 26.2170$

$\chi^2 = 2505.6007 \geq 26.2170$; $p < 0.0001$.
There is overwhelming evidence to suggest that age group and influenza type are dependent.

13.53 a. H_0: Day and time of day are independent. H_a: Day and time of day are dependent.

TS: $X^2 = \sum_{i=1}^{7} \sum_{j=1}^{3} \frac{(n_{ij} - e_{ij})}{e_{ij}}$; RR: $X^2 \geq 21.0261$

$\chi^2 = 30.7707 \geq 21.0261$; $p = 0.0021 \leq 0.05$.
There is evidence to suggest that day of the week and time of day are dependent.
b. The assumptions are not met. One cell has an expected count less than 5.
c. Monday, daytime.

13.55 H_0: The true proportions for type of degree are the same for each country. H_a: The true proportions for type of degree are different for each country.

TS: $X^2 = \sum_{i=1}^{2} \sum_{j=1}^{7} \frac{(n_{ij} - e_{ij})}{e_{ij}}$; RR: $X^2 \geq 22.4577$

$\chi^2 = 24.4007 \geq 22.4577$; $p = 0.0004 \leq 0.001$.
There is evidence to suggest that the true proportions for type of degree are different for each country.

13.57 H_0: Flossing frequency and brushing frequency are independent. H_a: Flossing frequency and brushing frequency are dependent.

TS: $X^2 = \sum_{i=1}^{7} \sum_{j=1}^{7} \frac{(n_{ij} - e_{ij})}{e_{ij}}$; RR: $X^2 \geq 58.6192$

$\chi^2 = 62.4388 \geq 58.6192$; $p = 0.0041 \leq 0.01$.
There is evidence to suggest that flossing frequency and brushing frequency are dependent.

Chapter Exercises

13.59 H_0: $p_i = 0.20$; H_a: $p_i \neq p_{i0}$ for at least one i;

TS: $X^2 = \sum_{i=1}^{5} (n_i - e_i)^2/e_i$; RR: $X^2 \geq 9.4877$.

$\chi^2 = 10.0769 \geq 9.4877$; $p = 0.0392 \leq 0.05$.

There is evidence to suggest that at least one of the population proportion of sales of dog breed differs from 0.20.

13.61 H_0: $p_1 = 0.3$, $p_2 = 0.4$, $p_3 = 0.1$, $p_4 = 0.1$, $p_5 = 0.1$ H_a: $p_i \neq p_{i0}$ for at least one i;

TS: $X^2 = \sum_{i=1}^{5} (n_i - e_i)^2/e_i$; RR: $X^2 \geq 9.4877$.

$\chi^2 = 7.9711 < 9.4877$; $p = 0.0926 > 0.05$.
There is no evidence to suggest that any of the true population proportions differs from its hypothesized value.

13.63 H_0: The true population proportions associated with transfer plans are the same for all companies. H_a: The true population proportions associated with transfer plans are not the same for all companies.

TS: $X^2 = \sum_{i=1}^{5} \sum_{j=1}^{4} \frac{(n_{ij} - e_{ij})}{e_{ij}}$; RR: $X^2 \geq 21.0261$

$\chi^2 = 18.3478 < 21.0261$; $p = 0.1055 > 0.05$.
There is no evidence to suggest that the true proportions associated with transfer plans are different for any of the populations.

13.65 H_0: The true proportions of each countertop type is the same for all supply stores. H_a: The true proportions of each countertop type is not the same for all supply stores.

TS: $X^2 = \sum_{i=1}^{3} \sum_{j=1}^{4} \frac{(n_{ij} - e_{ij})}{e_{ij}}$; RR: $X^2 \geq 12.5916$

$\chi^2 = 10.2014 < 12.5916$; $p = 0.1164 > 0.05$.
There is no evidence to suggest that the true proportion of each type of countertop purchased differs for supply stores.

13.67 H_0: Type of music and time spent shopping are independent. H_a: Type of music and time spent shopping are dependent.

TS: $X^2 = \sum_{i=1}^{4} \sum_{j=1}^{4} \frac{(n_{ij} - e_{ij})}{e_{ij}}$; RR: $X^2 \geq 21.6660$

$\chi^2 = 26.6081 \geq 21.6660$; $p = 0.0016 \leq 0.01$.
There is evidence to suggest an association between music type and time spent shopping.

13.69 H_0: $p_i = 0.10$; H_a: $p_i \neq p_{i0}$ for at least one i;

TS: $X^2 = \sum_{i=1}^{10} (n_i - e_i)^2/e_i$; RR: $X^2 \geq 27.8772$.

$\chi^2 = 32.9803 \geq 27.8772$; $p = 0.0001 \leq 0.01$.
There is evidence to suggest that at least one of the population proportions of music genre is different from 0.10, that at least one music genre is preferred more than the others.

13.71 a. H_0: Generation and bottle price are independent. H_a: Generation and bottle price are dependent.

TS: $X^2 = \sum_{i=1}^{6} \sum_{j=1}^{4} \frac{(n_{ij} - e_{ij})}{e_{ij}}$; RR: $X^2 \geq 24.9958$

$\chi^2 = 28.7552 \geq 24.9958$.
There is evidence to suggest that generation is associated with bottle price.
b. $0.01 \leq p \leq 0.025$. $p = 0.0173$.
c. Allocate the wine stock toward the middle-priced wines.

CHAPTER 14

Section 14.1

14.1 Distribution-free.

14.3 True.

14.5 The underlying distribution is continuous and symmetric.

14.7 Because the binomial distribution is discrete.

14.9 a. 7 **b.** 3 **c.** 6 **d.** 13

14.11 $x = 15$, $p = 0.0038 \leq 0.05$. There is evidence to suggest that $\tilde{\mu}_1 > \tilde{\mu}_2$.

14.13 $H_0: \tilde{\mu} = 1.38$; $H_a: \tilde{\mu} < 1.38$;
TS: $X =$ the number of observations greater than 1.38.
$x = 2$, $p = 0.0065 \leq 0.05$.
There is evidence to suggest that the population median concentration of norephinephrine in women with PD is less than 1.38 nmol/L.

14.15 $H_0: \tilde{\mu} = 100$; $H_a: \tilde{\mu} > 100$;
TS: $X =$ the number of observations greater than 100.
$x = 17$, $p = 0.0320 \leq 0.05$.
There is evidence to suggest that the median mileage is greater than 100.

14.17 $H_0: \tilde{\mu} = 40$; $H_a: \tilde{\mu} > 40$;
TS: $X =$ the number of observations greater than 40.
$x = 11$, $p = 0.4119 > 0.05$.
There is no evidence to suggest that the median time to complete this test is greater than 40 seconds.

14.19 $H_0: \tilde{\mu}_1 - \tilde{\mu}_2 = 0$; $H_a: \tilde{\mu}_1 - \tilde{\mu}_2 \neq 0$;
TS: $X =$ the number of pairwise difference greater than 0.
$x = 14$, $p = 0.1153 > 0.05$.
There is no evidence to suggest that the median chlorophyll amount in surface water is different in April and August.

14.21 $H_0: \tilde{\mu}_1 - \tilde{\mu}_2 = 0$; $H_a: \tilde{\mu}_1 - \tilde{\mu}_2 > 0$;
TS: $X =$ the number of pairwise differences greater than 0.
$x = 13$, $p = 0.0037 \leq 0.05$.
There is evidence to suggest that the median VOC concentration is smaller when the scrubber is installed.

14.23 a. $H_0: \tilde{\mu} = 4.6$; $H_a: \tilde{\mu} > 4.6$;
TS: $X =$ the number of observations greater than 4.6.
$x = 15$, $p = 0.0669 > 0.05$.
There is no evidence to suggest that the median survival is greater than 4.6 years.

b. $H_0: \mu = 4.6$; $H_a: \mu > 4.6$; TS: $T = \dfrac{\overline{X} - \mu_0}{S/\sqrt{n}}$;
RR: $T \geq 1.7109$
$t = 2.6824 \geq 1.7109$; $p = 0.0065 \leq 0.05$.
There is evidence to suggest that the median survival is greater than 4.6 years. This result is contrary to part (a). The analysis in part (a) seems more reasonable. It seems unlikely that the survival-time distribution is normally distributed.

Section 14.2

14.25 False.

14.27 True.

14.29 a.

Difference	Absolute difference	Rank
−19	19	9.0
−7	7	4.0
6	6	3.0
−32	32	16.0
−24	24	11.5
17	17	8.0
12	12	5.0
−29	29	15.0
−27	27	14.0
−26	26	13.0
−38	38	18.0
−22	22	10.0
−36	36	17.0
−1	1	2.0
−24	24	11.5
0	0	1.0
−13	13	6.0
−15	15	7.0

b.

Difference	Absolute difference	Rank
1.4	1.4	16.0
−0.2	0.2	2.5
1.6	1.6	21.5
0.3	0.3	5.0
−0.4	0.4	8.0
−0.8	0.8	12.0
−1.5	1.5	18.5
−0.6	0.6	10.0
0.1	0.1	1.0
−1.2	1.2	14.0
0.8	0.8	12.0
−1.5	1.5	18.5
1.5	1.5	18.5
−0.4	0.4	8.0
−1.5	1.5	18.5
1.6	1.6	21.5
−1.3	1.3	15.0
−0.3	0.3	5.0
0.8	0.8	12.0
0.4	0.4	8.0
−0.3	0.3	5.0
1.9	1.9	23.0
0.2	0.2	2.5

c.

Difference	Absolute difference	Rank
3.0	3.0	16.5
−4.0	4.0	23.0
1.0	1.0	5.0
−3.0	3.0	16.5
−2.0	2.0	9.5
−1.0	1.0	5.0
−4.0	4.0	23.0
−4.0	4.0	23.0
−2.0	2.0	9.5
−3.0	3.0	16.5
−2.0	2.0	9.5
−2.0	2.0	9.5
−2.0	2.0	9.5
0.0	0.0	2.0
3.0	3.0	16.5
3.0	3.0	16.5
0.0	0.0	2.0
3.0	3.0	16.5
−3.0	3.0	16.5
4.0	4.0	23.0
−3.0	3.0	16.5
0.0	0.0	2.0
2.0	2.0	9.5
1.0	1.0	5.0
4.0	4.0	23.0

d.

Difference	Absolute difference	Rank
−9.4	9.4	10.0
−55.4	55.4	30.0
−26.4	26.4	15.0
−41.4	41.4	23.5
−2.4	2.4	3.0
2.6	2.6	4.0
−38.4	38.4	22.0
−27.4	27.4	16.0
−34.4	34.4	20.0
9.6	9.6	11.0
6.6	6.6	7.5
−50.4	50.4	28.0
−42.4	42.4	25.0
41.4	41.4	23.5
4.6	4.6	6.0
−20.4	20.4	14.0
−17.4	17.4	13.0
−51.4	51.4	29.0
3.6	3.6	5.0
−33.4	33.4	19.0
6.6	6.6	7.5
−47.4	47.4	27.0
−32.4	32.4	18.0
1.6	1.6	2.0
−45.4	45.4	26.0
−31.4	31.4	17.0
−36.4	36.4	21.0
−0.4	0.4	1.0
8.6	8.6	9.0
−12.4	12.4	12.0

14.31 $H_0: \tilde{\mu}_1 - \tilde{\mu}_2 = 0$; $H_a: \tilde{\mu}_1 - \tilde{\mu}_2 \neq 0$;
TS: $T_+ =$ the sum of the ranks corresponding to the positive differences $d_i - 0$;
RR: $T_+ \leq 43$ or $T_+ \geq 167$
$t_+ = 150.5$; $p = 0.0896 > 0.02$. There is no evidence to suggest that $\tilde{\mu}_1$ is different from $\tilde{\mu}_2$.

14.33 $H_0: \tilde{\mu} = 3$; $H_a: \tilde{\mu} \neq 3$;
TS: $T_+ =$ the sum of the ranks corresponding to the positive differences $x_i - 3$;
RR: $T_+ \leq 37$ or $T_+ \geq 173$
$t_+ = 30.5 \leq 37$; $p = 0.006 \leq 0.01$. There is evidence to suggest that the median renal blood-flow rate is different from 3.

14.35 $H_0: \tilde{\mu} = 15.06$; $H_a: \tilde{\mu} < 15.06$;
TS: $Z = \dfrac{T_+ - \mu_{T_+}}{\sigma_{T_+}}$; RR: $Z \leq -2.3263$
$z = -2.9335 \leq -2.3263$; $p = 0.0017 \leq 0.01$.
There is evidence to suggest that the median APR is less than 15.06.

14.37 $H_0: \tilde{\mu}_1 - \tilde{\mu}_2 = 0$; $H_a: \tilde{\mu}_1 - \tilde{\mu}_2 \neq 0$;
TS: $Z = \dfrac{T_+ - \mu_{T_+}}{\sigma_{T_+}}$; RR: $|Z| \geq 1.6449$
$z = -0.9256$; $|z| < 1.6449$; $p = 0.3547 > 0.10$.
There is no evidence to suggest a difference in median collection times.

14.39 $H_0: \tilde{\mu} = 1$; $H_a: \tilde{\mu} \neq 1$;
TS: $Z = \dfrac{T_+ - \mu_{T_+}}{\sigma_{T_+}}$; RR: $Z \geq 1.6449$
$z = 1.9088 \geq 1.6499$; $p = 0.0281 \leq 0.05$. There is evidence to suggest that the median cadmium concentration is greater than 1.

14.41 a. $H_0: \tilde{\mu} = 0.30$; $H_a: \tilde{\mu} > 0.30$;
TS: $X =$ the number of observations greater than 0.30. $x = 6$;
$p = 0.2272 > 0.05$. There is no evidence to suggest that the median arsenic concentration is greater than 0.30.
b. $H_0: \tilde{\mu} = 0.30$; $H_a: \tilde{\mu} > 0.30$;
TS: $T_+ =$ the sum of the ranks corresponding to the positive differences $x_i - 0.30$;
RR: $T_+ \geq 101$; $t_+ = 105.5 \geq 101$; $p = 0.0253 \leq 0.05$. There is evidence to suggest that the median arsenic concentration is greater than 0.30.
c. The conclusions in parts (a) and (b) are not the same. The test in part (b) is probably more accurate because it uses more of the information in the sample than the test in part (a).

Section 14.3

14.43 False.

14.45 True.

14.47 True, if the populations are symmetric.

14.49 a.

Sample 1		Sample 2	
Obs	Rank	Obs	Rank
37	6	45	10
21	1	42	9
46	11	22	2
29	4	41	8
34	5	24	3
		39	7

b.

Sample 1		Sample 2	
Obs	Rank	Obs	Rank
4.5	15.0	4.6	16.0
1.8	5.0	6.6	18.0
3.4	13.0	1.2	2.0
1.4	3.0	6.0	17.0
2.2	7.5	2.4	10.0
2.1	6.0	2.3	9.0
1.5	4.0	2.2	7.5
3.7	14.0	0.4	1.0
		2.5	11.0
		2.7	12.0

c.

Sample 1		Sample 2	
Obs	Rank	Obs	Rank
820	11.0	850	25.0
809	4.0	840	21.0
872	33.0	813	6.0
826	15.0	842	23.0
814	7.0	870	30.0
887	39.0	839	20.0
825	14.0	888	40.0
884	38.0	816	8.0
876	35.0	822	13.0
862	27.0	879	36.0
858	26.0	821	12.0
846	24.0	865	28.5
841	22.0	832	19.0
801	2.0	865	28.5
892	41.0	818	9.5
871	31.5	827	16.0
882	37.0	899	42.0
803	3.0	818	9.5
		831	18.0
		830	17.0
		810	5.0
		875	34.0
		871	31.5
		800	1.0

14.51 a. $W \leq 7$, $\alpha = 0.0357$
b. $W \geq 24$, $\alpha = 0.0571$
c. $W \leq 16$ or $W \geq 44$, $\alpha = 0.0540$
d. $W \leq 27$, $\alpha = 0.0100$
e. $W \leq 44$ or $W \geq 75$, $\alpha = 0.1142$
f. $W \geq 90$, $\alpha = 0.0103$

14.53 $H_0: \tilde{\mu}_1 - \tilde{\mu}_2 = 0$; $H_a: \tilde{\mu}_1 - \tilde{\mu}_2 > 0$;
TS: $W = $ the sum of the ranks corresponding to the smaller sample; RR: $W \geq 63$
$w = 51 < 63$; $p > 0.1412$. There is no evidence to suggest that $\tilde{\mu}_1 > \tilde{\mu}_2$.

14.55 $H_0: \tilde{\mu}_1 - \tilde{\mu}_2 = 0$; $H_a: \tilde{\mu}_1 - \tilde{\mu}_2 < 0$;
TS: $W = $ the sum of the ranks corresponding to the smaller sample; RR: $W \leq 30$
$w = 27 \leq 30$; $p = P(W \leq 27) = 0.0175 \leq 0.05$. There is evidence to suggest that the median visit times are different.

14.57 a. $H_0: \tilde{\mu}_1 - \tilde{\mu}_2 = 0$; $H_a: \tilde{\mu}_1 - \tilde{\mu}_2 < 0$;
TS: $W = $ the sum of the ranks corresponding to the smaller sample; RR: $W \leq 35$
$w = 31 \leq 35$. There is evidence to suggest that the median impact strength is higher for the new jackhammer. **b.** 0.0030

14.59 $H_0: \tilde{\mu}_1 - \tilde{\mu}_2 = 0$; $H_a: \tilde{\mu}_1 - \tilde{\mu}_2 \neq 0$;
TS: $Z = \dfrac{W - \mu_W}{\sigma_W}$; RR: $|Z| \geq 1.96$

$z = 1.8981$; $|z| < 1.96$; $p = 0.0577 > 0.05$. There is no evidence to suggest that the population median holding temperatures are different.

14.61 $H_0: \tilde{\mu}_1 - \tilde{\mu}_2 = 0$; $H_a: \tilde{\mu}_1 - \tilde{\mu}_2 < 0$;
TS: $Z = \dfrac{W - \mu_W}{\sigma_W}$; RR: $Z \leq -2.3263$

$z = -3.9975 \leq -2.3263$; $p < 0.0001$. There is evidence to suggest that the population median amount of protein in SBP is greater than the population median amount of protein in WTL.

14.63 a. $H_0: \tilde{\mu}_1 - \tilde{\mu}_2 = 0$; $H_a: \tilde{\mu}_1 - \tilde{\mu}_2 > 0$;
TS: $X = $ the number of observations greater than 0;
RR: $X \geq 11$ ($\alpha = 0.0592$)
$x = 15 \geq 11$. There is evidence to suggest that the median amount of particulates after the smoking regulations is less than the median amount before the regulations.
b. $H_0: \tilde{\mu}_1 - \tilde{\mu}_2 = 0$; $H_a: \tilde{\mu}_1 - \tilde{\mu}_2 > 0$;
TS: $T_+ = $ the sum of the ranks corresponding to the positive differences $d_i - 0$; RR: $T_+ \geq 89$ ($\alpha = 0.0535$)
$t_+ = 120 \geq 89$. There is evidence to suggest that the median amount of particulates after the smoking regulations is less than the median amount before the regulations.
c. $H_0: \tilde{\mu}_1 - \tilde{\mu}_2 = 0$; $H_a: \tilde{\mu}_1 - \tilde{\mu}_2 > 0$;
TS: $Z = \dfrac{W - \mu_W}{\sigma_W}$; RR: $Z \geq 1.6449$

$z = 4.4382 \geq 1.6449$; $p < 0.0001$. There is evidence to suggest that the median amount of particulates after the smoking regulations is less than the median amount before the regulations.
d. All three tests lead to the same conclusion. The rank-sum test, however, should not be used, because the samples are dependent.

Section 14.4

14.65 χ^2_{k-1}

14.67 True.

14.69 False.

14.71 H_0: The five samples are from identical populations; H_a: At least two of the populations are different;

TS: $H = \left[\dfrac{12}{n(n+1)}\sum\dfrac{r_i^2}{n_i}\right] - 3(n+1)$; RR: $H \geq 13.2767$;

$h = 16.1591 \geq 13.2767$; $p = 0.0028 \leq 0.01$. There is evidence to suggest that at least two of the populations are different.

14.73 H_0: The three samples are from identical populations; H_a: At least two of the populations are different;

TS: $H = \left[\dfrac{12}{n(n+1)}\sum\dfrac{r_i^2}{n_i}\right] - 3(n+1)$; RR: $H \geq 5.9915$

$h = 6.5184 \geq 5.9915$; $p = 0.0384 \leq 0.05$. There is evidence to suggest that at least two of the populations are different.

14.75 H_0: The four samples are from identical populations; H_a: At least two of the populations are different;

TS: $H = \left[\dfrac{12}{n(n+1)}\sum\dfrac{r_i^2}{n_i}\right] - 3(n+1)$; RR: $H \geq 7.8147$

$h = 6.3338 < 7.8147$; $p = 0.0965 > 0.05$. There is no evidence to suggest that the populations are different.

14.77 H_0: The four samples are from identical populations; H_a: At least two of the populations are different;

TS: $H = \left[\dfrac{12}{n(n+1)}\sum\dfrac{r_i^2}{n_i}\right] - 3(n+1)$; RR: $H \geq 11.3449$

$h = 3.6953 < 11.3449$; $p = 0.2963 > 0.01$. There is no evidence to suggest that the populations are different.

14.79 H_0: The three samples are from identical populations; H_a: At least two of the populations are different;

TS: $H = \left[\dfrac{12}{n(n+1)}\sum\dfrac{r_i^2}{n_i}\right] - 3(n+1)$; RR: $H \geq 5.9915$

$h = 1.3713 < 5.9915$; $p = 0.5038 > 0.05$. There is no evidence to suggest that the uncompressed depth populations are different. $p > 0.10$.

14.81 H_0: The three samples are from identical populations; H_a: At least two of the populations are different;

TS: $H = \left[\dfrac{12}{n(n+1)}\sum\dfrac{r_i^2}{n_i}\right] - 3(n+1)$; RR: $H \geq 5.9915$

$h = 22.5890 \geq 5.9915$; $p < 0.0001$. There is evidence to suggest that at least two of the length-of-service populations are different. **b.** Public safety and communications. Support services and communications. The corresponding rank sums are very far apart.

Section 14.5

14.83 False.

14.85 1

14.87 When both m and n are greater than 10.

14.89 a. 8 **b.** 8 **c.** 11 **d.** 15

14.91 a. 8 **b.** 5 **c.** 8 **d.** 1

14.93 H_0: The sequence of observations is random; H_a: The sequence of observations is not random. TS: $V =$ the number of runs. RR: $V \leq 6$ or $V \geq 16$ $v = 13$. There is no evidence to suggest that the order of observations is not random.

14.95 H_0: The sequence of observations is random; H_a: The sequence of observations is not random. TS: $V =$ the number of runs. RR: $V \leq 3$ or $V \geq 7$ ($\alpha = 0.3682$) $v = 5$. There is no evidence to suggest that the order of observations is not random.

14.97 H_0: The sequence of observations is random; H_a: The sequence of observations is not random.

TS: $z = \dfrac{V - \mu_V}{\sigma_V}$; RR: $|Z| \geq 2.3263$

$z = 1.2546$; $|z| < 2.3263$; $p = 0.2096 > 0.02$. There is no evidence to suggest that the order of observations is not random.

14.99 H_0: The sequence of observations is random; H_a: The sequence of observations is not random.

TS: $Z = \dfrac{V - \mu_V}{\sigma_V}$; RR: $|Z| \geq 1.96$

$z = 1.1051$; $|z| < 1.96$; $p = 0.2691 > 0.05$. There is no evidence to suggest that the order of observations is not random.

14.101 H_0: The sequence of observations is random; H_a: The sequence of observations is not random. TS: $V =$ the number of runs. RR: $V \leq 2$ or $V \geq 9$ ($\alpha = 0.0711$) $v = 9$. There is evidence to suggest the order of observations is not random at the $\alpha = 0.0711$ level, but not for $\alpha = 0.02$.

14.103 H_0: The sequence of observations is random; H_a: The sequence of observations is not random.

TS: $z = \dfrac{V - \mu_V}{\sigma_V}$; RR: $|Z| \geq 1.96$

$z = -2.0370$; $|z| \geq 1.96$. There is evidence to suggest that the order of observations is not random with respect to the median. $p = 0.0416$.

Section 14.6

14.105 No assumptions.

14.107 True.

14.109 False.

14.111 a. 0.5357; moderate positive relationship. **b.** 0.3333; weak positive relationship. **c.** 0.0637; no definitive relationship. **d.** 0.2922; weak negative relationship.

14.113 0.7714. There is a positive relationship between x and y. As the price of a stateroom increases, so does the number of days before sailing. Therefore, this suggests that cruise prices are reduced at the last minute.

14.115 -0.4429. There is a weak negative relationship between bulk density of soil and soil texture. As the bulk density of soil increases, the texture of the soil tends to decrease.

14.117 -0.4634. This weak negative relationship suggests that the more time a child plays outdoors, the fewer number of school days are missed.

14.119 a. Scatter plot of y versus x:

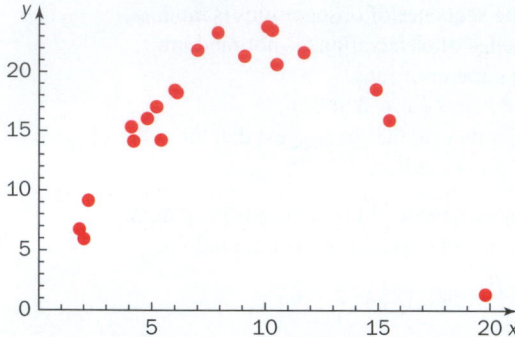

The scatter plot suggests a quadratic relationship between total snowfall and number of customers.   **b.** 0.4842. This value suggests a weak to moderate positive linear relationship between the variables.   **c.** Spearman's rank correlation coefficient can only test for a linear relationship between two variables. If there is some other type of relationship between the variables, as in this case, the value of R_S can be misleading.

14.121 a. -0.3634.   **b.** $r_S < 0$, which suggests that as the number of selfies increases, the IAS score decreases. Therefore, people who share more selfies tend to be less inclined to be close and personal.

Chapter Exercises

14.123 $H_0: \tilde{\mu} = 11,000;\ H_a: \tilde{\mu} < 11,000;$
TS: the number of observations greater than 11,000.
RR: $X \le 3\ (\alpha = 0.0287)$
$x = 6 > 3;\ p = 0.3953 > 0.05$. There is no evidence to suggest that the median outflow is less than 11,000 cfs.

14.125 $H_0: \tilde{\mu} = 2;\ H_a: \tilde{\mu} \ne 2;$
TS: the number of observations greater than 2.
RR: $X \le 5$ or $X \ge 15\ (\alpha = 0.0414)$
$x = 15 \ge 15;\ p = 0.0414 \le 0.05$. There is evidence to suggest that the median coverage amount is different from 2.

14.127 a. $H_0: \tilde{\mu} = 630;\ H_a: \tilde{\mu} < 630;$
TS: $T_+ =$ the sum of the ranks corresponding to the positive differences $x_i - 630$. RR: $T_+ \le 20\ (\alpha = 0.0108)$
$t_+ = 17.5 \le 20$. There is evidence to suggest that the median spray height is less than 630.   **b.** 0.0062   **c.** The distribution is not assumed to be symmetric.

14.129 $H_0: \tilde{\mu}_1 - \tilde{\mu}_2 = 0;\ H_a: \tilde{\mu}_1 - \tilde{\mu}_2 < 0;$
TS: The sum of the ranks corresponding to the positive differences $d_i - 0$. RR: $T_+ \le 47\ (\alpha = 0.0494)$
$t_+ = 25 \le 47;\ p = 0.0033 \le 0.05$. There is evidence to suggest that the median THM concentration is greater after swimming.

14.131 a. $H_0: \tilde{\mu}_1 - \tilde{\mu}_2 = 0;\ H_a: \tilde{\mu}_1 - \tilde{\mu}_2 \ne 0;$
TS: $W =$ the sum of the ranks corresponding to the first sample.
RR: $W \le 52$ or $W \ge 84\ (\alpha = 0.1048)$
$w = 78.5$. There is no evidence to suggest that there is a difference in the absorbed radiation by machine.   **b.** 0.2786

14.133 $H_0: \tilde{\mu}_1 - \tilde{\mu}_2 = 0;\ H_a: \tilde{\mu}_1 - \tilde{\mu}_2 \ne 0;$
TS: $Z = \dfrac{W - \mu_W}{\sigma_W}$; RR: $|Z| \ge 2.5758$
$z = -1.1156;\ |z| < 2.5758;\ p = 0.2646 > 0.01$. There is no evidence to suggest that the median respiratory rates are different.

14.135 H_0: The three samples are from identical populations; H_a: At least two of the populations are different;
TS: $H = \left[\dfrac{12}{n(n+1)} \Sigma \dfrac{r_i^2}{n_i} \right] - 3(n+1)$; RR: $H \ge 5.9915$
$h = 13.6603 \ge 5.9915;\ p = 0.0011 \le 0.05$. There is evidence to suggest that at least two slate weight populations are different.

14.137 H_0: The four samples are from identical populations; H_a: At least two of the populations are different;
TS: $H = \left[\dfrac{12}{n(n+1)} \Sigma \dfrac{r_i^2}{n_i} \right] - 3(n+1)$; RR: $H \ge 11.3449$
$h = 14.2557 \ge 11.3449;\ p = 0.0026 \le 0.01$. There is evidence to suggest that at least two nozzle temperature populations are different.

14.139 a. H_0: The sequence of observations is random. H_a: The sequence of observations is not random.
TS: $V =$ the number of runs; RR: $v \le 4$ or $V \ge 12$ $(\alpha = 0.0560)$
$v = 10$. There is no evidence to suggest that the order of observations is not random.   **b.** $p = 0.4756$

14.141 H_0: The sequence of observations is random. H_a: The sequence of observations is not random.
TS: $Z = \dfrac{V - \mu_V}{\sigma_V}$; RR: $|Z| \ge 2.5758$
$z = 1.7872;\ |z| < 2.5758;\ p = 0.0739$. There is no evidence to suggest that the order of observations is not random.

14.143 0.7785. This value suggests a moderate positive linear relationship. As the number of miles from I-95 increases, so does the starting room price of a hotel room.

14.145 0.8660. This value suggests a strong positive linear relationship. Wiper frequency might be used to predict rainfall total.

14.147 a. $H_0: \tilde{\mu}_1 - \tilde{\mu}_2 = 0;\ H_a: \tilde{\mu}_1 - \tilde{\mu}_2 < 0;$
TS: $X =$ the number of pairwise differences greater than 0.
RR: $X \le 5\ (\alpha = 0.0207)$
$x = 15$. There is no evidence to suggest that the median refrigerant weight before service is less than the median refrigerant weight after service.   **b.** Because we cannot reject the null

hypothesis, there is no evidence to suggest that recharging an air conditioner increases the amount of refrigerant in the system.

14.149 **a.** H_0: The sequence of observations is random. H_a: The sequence of observations is not random. TS: $V = $ the number of runs; RR: $V \leq 4$ or $V \geq 11$ ($\alpha = 0.0710$) $v = 7$. There is no evidence to suggest that the order of observations is not random with respect to exterior finish. **b.** Using the runs test, we cannot tell if the advertising campaign was successful. And, we don't know the historical proportion of home builders who use vinyl. Therefore, we cannot tell whether this proportion has increased.

14.151 **a.** H_0: $\tilde{\mu}_1 - \tilde{\mu}_2 = 0$; H_a: $\tilde{\mu}_1 - \tilde{\mu}_2 \neq 0$;
TS: $Z = \dfrac{W - \mu_W}{\sigma_W}$; RR: $|Z| \geq 1.96$
$z = 1.7118$; $|z| < 1.96$; $p = 0.0869 > 0.05$. There is no evidence to suggest that the population median return yards are different.

b. H_a: $\mu_1 - \mu_2 = 0$, H_a: $\mu_1 - \mu_2 \neq 0$;
TS: $T = \dfrac{(\bar{X}_1 - \bar{X}_2) - 0}{\sqrt{S_p^2 \left(\dfrac{1}{n_1} + \dfrac{1}{n_2} \right)}}$; RR: $|T| \geq 1.9824$
$t = 2.3391$; $|t| \geq 1.9824$; $p = 0.0212 \leq 0.05$. There is evidence to suggest that the population mean return yards are different. **c.** The Wilcoxon rank-sum test is more appropriate, because the underlying distributions are probably not normal. **d.** No. Both the mean and the median return yards were larger in 2011.

14.153 H_0: $\tilde{\mu}_1 - \tilde{\mu}_2 = 0$; H_a: $\tilde{\mu}_1 - \tilde{\mu}_2 \neq 0$;
TS: $Z = \dfrac{W - \mu_W}{\sigma_W}$; RR: $|Z| \geq 1.96$
$z = -1.3847$; $|z| < 1.96$; $p = 0.1661 > 0.05$. There is no evidence to suggest that the population median naphthalene concentrations are different.

Index

Table V Critical Values for the *t* Distribution

This table contains critical values associated with the *t* distribution, $t_{\alpha,\nu}$, defined by α and the degrees of freedom, ν.

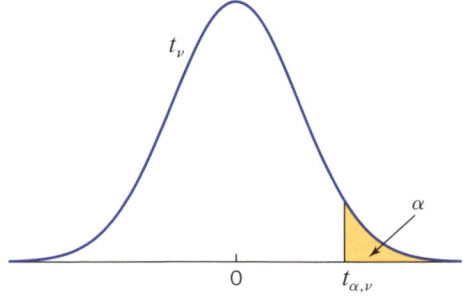

ν	0.20	0.10	0.05	0.025	0.01	0.005	0.001	0.0005	0.0001
1	1.3764	3.0777	6.3138	12.7062	31.8205	63.6567	318.3088	636.6192	3183.0988
2	1.0607	1.8856	2.9200	4.3027	6.9646	9.9248	22.3271	31.5991	70.7001
3	0.9785	1.6377	2.3534	3.1824	4.5407	5.8409	10.2145	12.9240	22.2037
4	0.9410	1.5332	2.1318	2.7764	3.7469	4.6041	7.1732	8.6103	13.0337
5	0.9195	1.4759	2.0150	2.5706	3.3649	4.0321	5.8934	6.8688	9.6776
6	0.9057	1.4398	1.9432	2.4469	3.1427	3.7074	5.2076	5.9588	8.0248
7	0.8960	1.4149	1.8946	2.3646	2.9980	3.4995	4.7853	5.4079	7.0634
8	0.8889	1.3968	1.8595	2.3060	2.8965	3.3554	4.5008	5.0413	6.4420
9	0.8834	1.3830	1.8331	2.2622	2.8214	3.2498	4.2968	4.7809	6.0101
10	0.8791	1.3722	1.8125	2.2281	2.7638	3.1693	4.1437	4.5869	5.6938
11	0.8755	1.3634	1.7959	2.2010	2.7181	3.1058	4.0247	4.4370	5.4528
12	0.8726	1.3562	1.7823	2.1788	2.6810	3.0545	3.9296	4.3178	5.2633
13	0.8702	1.3502	1.7709	2.1604	2.6503	3.0123	3.8520	4.2208	5.1106
14	0.8681	1.3450	1.7613	2.1448	2.6245	2.9768	3.7874	4.1405	4.9850
15	0.8662	1.3406	1.7531	2.1314	2.6025	2.9467	3.7328	4.0728	4.8800
16	0.8647	1.3368	1.7459	2.1199	2.5835	2.9208	3.6862	4.0150	4.7909
17	0.8633	1.3334	1.7396	2.1098	2.5669	2.8982	3.6458	3.9651	4.7144
18	0.8620	1.3304	1.7341	2.1009	2.5524	2.8784	3.6105	3.9216	4.6480
19	0.8610	1.3277	1.7291	2.0930	2.5395	2.8609	3.5794	3.8834	4.5899
20	0.8600	1.3253	1.7247	2.0860	2.5280	2.8453	3.5518	3.8495	4.5385
21	0.8591	1.3232	1.7207	2.0796	2.5176	2.8314	3.5272	3.8193	4.4929
22	0.8583	1.3212	1.7171	2.0739	2.5083	2.8188	3.5050	3.7921	4.4520
23	0.8575	1.3195	1.7139	2.0687	2.4999	2.8073	3.4850	3.7676	4.4152
24	0.8569	1.3178	1.7109	2.0639	2.4922	2.7969	3.4668	3.7454	4.3819
25	0.8562	1.3163	1.7081	2.0595	2.4851	2.7874	3.4502	3.7251	4.3517
26	0.8557	1.3150	1.7056	2.0555	2.4786	2.7787	3.4350	3.7066	4.3240
27	0.8551	1.3137	1.7033	2.0518	2.4727	2.7707	3.4210	3.6896	4.2987
28	0.8546	1.3125	1.7011	2.0484	2.4671	2.7633	3.4082	3.6739	4.2754
29	0.8542	1.3114	1.6991	2.0452	2.4620	2.7564	3.3962	3.6594	4.2539
30	0.8538	1.3104	1.6973	2.0423	2.4573	2.7500	3.3852	3.6460	4.2340
40	0.8507	1.3031	1.6839	2.0211	2.4233	2.7045	3.3069	3.5510	4.0942
50	0.8489	1.2987	1.6759	2.0086	2.4033	2.6778	3.2614	3.4960	4.0140
60	0.8477	1.2958	1.6706	2.0003	2.3901	2.6603	3.2317	3.4602	3.9621
70	0.8468	1.2938	1.6669	1.9944	2.3808	2.6479	3.2108	3.4350	3.9257
80	0.8461	1.2922	1.6641	1.9901	2.3739	2.6387	3.1953	3.4163	3.8988
90	0.8456	1.2910	1.6620	1.9867	2.3685	2.6316	3.1833	3.4019	3.8780
100	0.8452	1.2901	1.6602	1.9840	2.3642	2.6259	3.1737	3.3905	3.8616
200	0.8434	1.2858	1.6525	1.9719	2.3451	2.6006	3.1315	3.3398	3.7891
500	0.8423	1.2832	1.6479	1.9647	2.3338	2.5857	3.1066	3.3101	3.7468
∞	0.8416	1.2816	1.6449	1.9600	2.3263	2.5758	3.0902	3.2905	3.7190

Table III Standard Normal Distribution Cumulative Probabilities

Let Z be a standard normal random variable:
$\mu = 0$ and $\sigma = 1$.

This table contains cumulative probabilities:
$P(Z \le z)$.

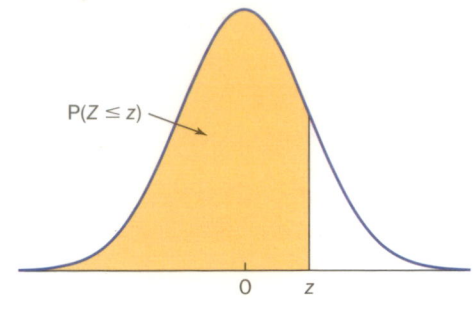

$P(Z \le z)$

z	.00	.01	.02	.03	.04	.05	.06	.07	.08	.09
−3.4	0.0003	0.0003	0.0003	0.0003	0.0003	0.0003	0.0003	0.0003	0.0003	0.0002
−3.3	0.0005	0.0005	0.0005	0.0004	0.0004	0.0004	0.0004	0.0004	0.0004	0.0003
−3.2	0.0007	0.0007	0.0006	0.0006	0.0006	0.0006	0.0006	0.0005	0.0005	0.0005
−3.1	0.0010	0.0009	0.0009	0.0009	0.0008	0.0008	0.0008	0.0008	0.0007	0.0007
−3.0	0.0013	0.0013	0.0013	0.0012	0.0012	0.0011	0.0011	0.0011	0.0010	0.0010
−2.9	0.0019	0.0018	0.0018	0.0017	0.0016	0.0016	0.0015	0.0015	0.0014	0.0014
−2.8	0.0026	0.0025	0.0024	0.0023	0.0023	0.0022	0.0021	0.0021	0.0020	0.0019
−2.7	0.0035	0.0034	0.0033	0.0032	0.0031	0.0030	0.0029	0.0028	0.0027	0.0026
−2.6	0.0047	0.0045	0.0044	0.0043	0.0041	0.0040	0.0039	0.0038	0.0037	0.0036
−2.5	0.0062	0.0060	0.0059	0.0057	0.0055	0.0054	0.0052	0.0051	0.0049	0.0048
−2.4	0.0082	0.0080	0.0078	0.0075	0.0073	0.0071	0.0069	0.0068	0.0066	0.0064
−2.3	0.0107	0.0104	0.0102	0.0099	0.0096	0.0094	0.0091	0.0089	0.0087	0.0084
−2.2	0.0139	0.0136	0.0132	0.0129	0.0125	0.0122	0.0119	0.0116	0.0113	0.0110
−2.1	0.0179	0.0174	0.0170	0.0166	0.0162	0.0158	0.0154	0.0150	0.0146	0.0143
−2.0	0.0228	0.0222	0.0217	0.0212	0.0207	0.0202	0.0197	0.0192	0.0188	0.0183
−1.9	0.0287	0.0281	0.0274	0.0268	0.0262	0.0256	0.0250	0.0244	0.0239	0.0233
−1.8	0.0359	0.0351	0.0344	0.0336	0.0329	0.0322	0.0314	0.0307	0.0301	0.0294
−1.7	0.0446	0.0436	0.0427	0.0418	0.0409	0.0401	0.0392	0.0384	0.0375	0.0367
−1.6	0.0548	0.0537	0.0526	0.0516	0.0505	0.0495	0.0485	0.0475	0.0465	0.0455
−1.5	0.0668	0.0655	0.0643	0.0630	0.0618	0.0606	0.0594	0.0582	0.0571	0.0559
−1.4	0.0808	0.0793	0.0778	0.0764	0.0749	0.0735	0.0721	0.0708	0.0694	0.0681
−1.3	0.0968	0.0951	0.0934	0.0918	0.0901	0.0885	0.0869	0.0853	0.0838	0.0823
−1.2	0.1151	0.1131	0.1112	0.1093	0.1075	0.1056	0.1038	0.1020	0.1003	0.0985
−1.1	0.1357	0.1335	0.1314	0.1292	0.1271	0.1251	0.1230	0.1210	0.1190	0.1170
−1.0	0.1587	0.1562	0.1539	0.1515	0.1492	0.1469	0.1446	0.1423	0.1401	0.1379
−0.9	0.1841	0.1814	0.1788	0.1762	0.1736	0.1711	0.1685	0.1660	0.1635	0.1611
−0.8	0.2119	0.2090	0.2061	0.2033	0.2005	0.1977	0.1949	0.1922	0.1894	0.1867
−0.7	0.2420	0.2389	0.2358	0.2327	0.2296	0.2266	0.2236	0.2206	0.2177	0.2148
−0.6	0.2743	0.2709	0.2676	0.2643	0.2611	0.2578	0.2546	0.2514	0.2483	0.2451
−0.5	0.3085	0.3050	0.3015	0.2981	0.2946	0.2912	0.2877	0.2843	0.2810	0.2776
−0.4	0.3446	0.3409	0.3372	0.3336	0.3300	0.3264	0.3228	0.3192	0.3156	0.3121
−0.3	0.3821	0.3783	0.3745	0.3707	0.3669	0.3632	0.3594	0.3557	0.3520	0.3483
−0.2	0.4207	0.4168	0.4129	0.4090	0.4052	0.4013	0.3974	0.3936	0.3897	0.3859
−0.1	0.4602	0.4562	0.4522	0.4483	0.4443	0.4404	0.4364	0.4325	0.4286	0.4247
−0.0	0.5000	0.4960	0.4920	0.4880	0.4840	0.4801	0.4761	0.4721	0.4681	0.4641